Preparation for College

MATHEMATICS

D. FRANKLIN WRIGHT

Editor: Susan Niese

Co-Editor: Nina Waldron

Vice President, Development: Marcel Prevuznak

Production Editor: Kim Cumbie

Editorial Assistant: Joseph Miller

Layout Design: Tracy Carr, Nancy Derby, Rachel A. I. Link, Jennifer Moran, Tee Jay Zajac

Layout Production: E. Jeevan Kumar, D. Kanthi, U. Nagesh, B. Syamprasad

Copy Editors: Jessica Ballance, Danielle C. Bess, Joshua Falter, Margaret Gibbs, Taylor Hamrick, Mary Katherine Huffman, Rebecca Hughes, Justin Lamothe, Sojwal Pohekar, Eric Powers, William J. Radjewski, Jake Stauch, Joseph Tracy, Claudia Vance, Colin Williams, Barry Wright, III

Answer Key Editors: Mary Katherine Huffman, Justin Lamothe, Sojwal Pohekar, Eric Powers, William J. Radjewski, Claudia Vance, Colin Williams

Art: Jennifer Guerette, Rachel A. I. Link, Jennifer Moran, Ayvin Samonte, Tee Jay Zajac

Cover Design: Tee Jay Zajac

Photograph Credits: BigStockPhoto.com, iStockPhoto.com, and Digital Vision

A division of Quant Systems, Inc.

546 Long Point Road, Mt. Pleasant, SC 29464

Library of Congress Control Number: 2011929303

Printed in the United States of America

ISBN:
Student Textbook: 978-1-941552-44-5

Table of Contents

Preface

Purpose and Style

The purpose of Preparation for College Mathematics is to provide students with a review of basic arithmetic, an introduction to algebra, and a learning tool that will help them:

1. review basic arithmetic skills,
2. develop reasoning and problem-solving skills,
3. become familiar with algebraic notation,
4. understand the connections between arithmetic and algebra,
5. develop basic algebra skills,
6. provide a smooth transition from arithmetic through prealgebra to algebra, and
7. achieve satisfaction in learning so that they will be encouraged to continue their education in mathematics.

The writing style gives carefully worded, thorough explanations that are direct, easy to understand, and mathematically accurate. The use of color, boldface, subheadings, and shaded boxes helps students understand and reference important topics.

Each topic is developed in a straightforward step-by-step manner. Each section contains many detailed examples to lead students successfully through the exercises and help them develop an understanding of the related concepts. Practice Problems with answers are provided in almost every section to allow students to "warm up" and to provide instructors with immediate classroom feedback.

Algebra skills and topics from geometry are integrated within the discussions and problems. In particular, Chapters 1-3 introduce basic arithmetic skills while providing students with an introduction to solving equations. Chapter 4 provides an in-depth study of geometrical concepts (perimeter, area, volume, and so on). From Chapter 5 on, the text concentrates on developing useful algebraic skills and concepts.

Students are encouraged to use calculators when appropriate and explicit directions and diagrams are provided as they relate to a scientific calculator, as well as to a TI-84 Plus graphing calculator.

The NCTM and AMATYC curriculum standards have been taken into consideration in the development of the topics throughout the text.

Special Features

In each chapter:

- Mathematics @ Work! presents a brief discussion related to a concept developed in the coming chapter. Sometimes these sections are challenging and may be better understood after the student completes a portion, or all, of the chapter.

- Learning Objectives are listed at the beginning of each section and are used to highlight headers in the section to emphasize the topic being discussed.

- Chapter Test Exercises appear in each chapter and are presented in a testing format.

- Cumulative Review exercises appear in each chapter beyond Chapter 1 to provide continuous, cumulative review.

In the exercise sets:

- Writing and Thinking About Mathematics exercises encourage students to express their ideas, interpretations, and understanding in writing.

- Collaborative Learning Exercises are designed to be done in interactive groups.

Features

Chapter Openers:

Each chapter begins with a chapter table of contents and an engaging preview of an application of the chapter's material.

Math @ Work:

A practical application of the chapter's material designed to pique student interest and give an idea of why the topics to be studied are useful.

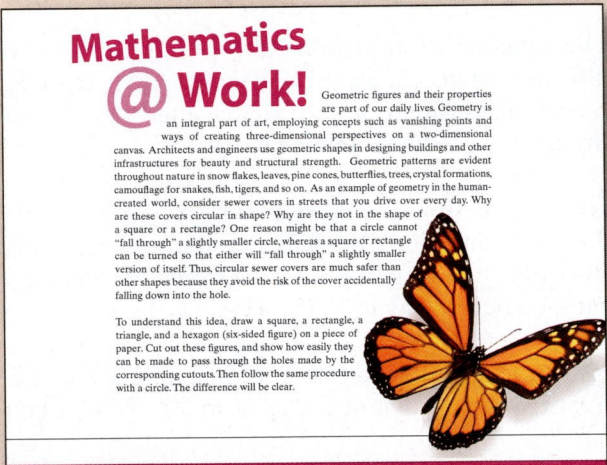

Objectives:

The objectives provide students with a clear and concise list of the main concepts and methods taught in each section, enabling students to focus their time and effort on the most important topics. Objectives have corresponding buttons located in the section text where the topic is introduced for ease of reference.

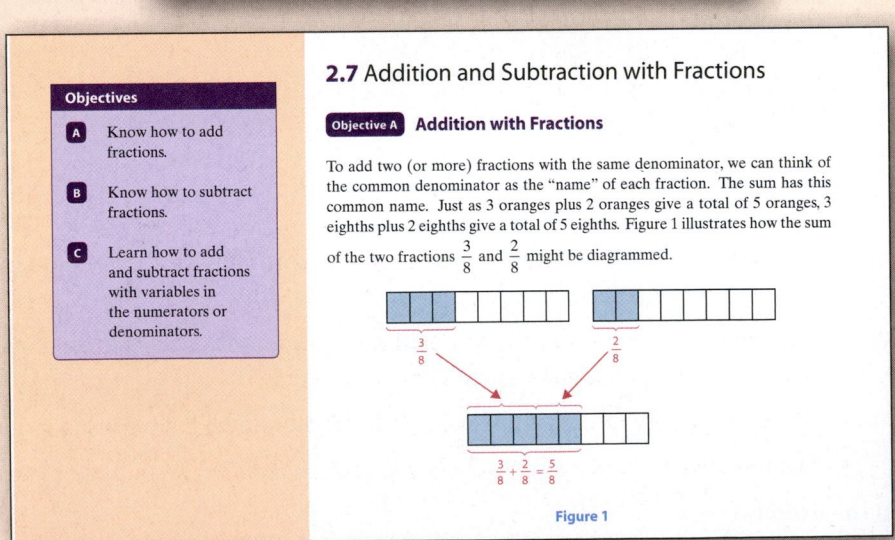

As the table indicates, the number being divided is called the **dividend**; the number dividing the dividend is called the **divisor**, and the result of division is called the **quotient**.

Division does not always involve factors (or exact divisors), and we need a more general idea and method for performing division. The more general method is a process known as the **division algorithm*** (or the method of **long division**) and is illustrated in Example 7. The **remainder** must be less than the divisor.

Note: If the remainder is 0, then both the divisor and quotient are **factors** (or **divisors**) of the dividend. See Example 8.

Example 7

Division with Whole Numbers

Find the quotient and remainder for $9385 \div 45$.

Solution

Step 1:
$$45\overline{)9385}$$
with 2 above, 90 below, remainder 3

By trial, divide 40 into 90 or 4 into 9.

Write the result, 2, in the hundreds position of the quotient.

Multiply this result by the divisor and subtract.

Step 2:
$$45\overline{)9385}$$
with 20 above, 90, 38, 0, 385

45 will not divide into 38. So, write 0 in the tens column and multiply $0 \cdot 45 = 0$.

Do not forget to write the 0 in the quotient.

Step 3:
$$45\overline{)9385}$$
quotient 208 ← quotient, dividend, 90, 38, 0, 385, 360, 25 ← remainder
divisor

Choose a trial divisor of 8 for the ones digit. Since $8 \cdot 45 = 360$ and 360 is smaller than 385, 8 is the desired number.

Check: Multiply the quotient by the divisor and then add the remainder. This sum should equal the dividend.

$$
\begin{array}{r}
208 \leftarrow \text{quotient} \\
\times\ 45 \leftarrow \text{divisor} \\
\hline
1040 \\
832 \\
\hline
9360
\end{array}
\qquad
\begin{array}{r}
9360 \\
+\ 25 \leftarrow \text{remainder} \\
\hline
9385 \leftarrow \text{dividend}
\end{array}
$$

Now work margin exercise 7.

*An algorithm is a process or pattern of steps to be followed in solving a problem or, sometimes, only a certain part of a problem.

7. Find the quotient and remainder: $7982 \div 31$.

Examples:

Examples are denoted with titled headers indicating the problem-solving skill being presented. Each section contains many carefully explained examples with appropriate tables, diagrams, and graphs. Examples are presented in an easy to understand, step-by-step fashion and annotated with notes for additional clarification.

notes

Notes boxes throughout the text point out important information that will help deepen student understanding of the topics. Often these will be helpful hints about subtle details in the definitions that many students do not notice upon first glance.

notes

1 is **neither** a prime nor a composite number. $1 = 1 \cdot 1$, and 1 is the only factor of 1. 1 does not have **exactly two different** factors, and it does not have more than two different factors.

To Find the Prime Factorization of a Composite Number

1. Factor the composite number into any two factors.

2. Factor each factor that is not prime.

3. Continue this process until all factors are prime.

The **prime factorization** of a number is a factorization of that number using only prime factors.

Definition Boxes:

Straightfoward definitions are presented in highly-visible boxes for easy reference.

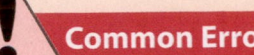

Common Error

DO NOT multiply the base and the exponent.

DO multiply the base by itself.

$10^2 = 10 \cdot 2$ **WRONG**

$6^3 = 6 \cdot 3$ **WRONG**

$10^2 = 10 \cdot 10$ **CORRECT**

$6^3 = 6 \cdot 6 \cdot 6$ **CORRECT**

Practice Problems

Solve the following equations.

1. $2\sqrt{x+4} = x+1$

2. $\sqrt{3x+1} + 1 = \sqrt{x}$

3. $\sqrt[3]{2x-9} + 4 = 3$

4. $\sqrt{2x-5} = -1$

Practice Problem Answers

1. $x = 5$ **2.** no solution   **3.** $x = 4$ **4.** no solution

Example 10

Subtraction with a Calculator

Use your calculator to find the following difference.

$$\begin{array}{r} 678{,}025 \\ -\ \underline{483{,}975} \end{array}$$

Solution

To subtract numbers on a calculator, you will need to find the subtraction key, **−**, and the equal sign key, **=**.

To subtract the values, press the keys

6 **7** **8** **0** **2** **5** **−** **4** **8** **3** **9** **7** **5** .

Then press **=** .

The display will show **194050** as the answer.

Now work margin exercises 9 and 10.

We need to be careful about relying on calculators. One frequent problem with using a calculator is that numbers might be entered incorrectly. Thus, the use of estimation, even just a mental estimation, can be very helpful in verifying answers.

Exercises 2.6

Solve the following. See Example 1.

1. What is the reciprocal of $\frac{13}{25}$?

2. To divide by any nonzero number, multiply by its _____.

3. The quotient of $0 \div \frac{6}{7}$ is _____.

4. The quotient of $\frac{6}{7} \div 0$ is _____.

5. Show that the phrases "ten divided by two" and "ten divided by one-half" do not have the same meaning.

6. Show that the phrases "twelve divided by three" and "twelve divided by one-third" do not have the same meaning.

7. Explain in your own words why division by 0 is undefined.

8. Give three examples that illustrate that division is not a commutative operation.

Find the following quotients. Reduce to lowest terms whenever possible. See Examples 3 through 7.

9. $\frac{5}{8} \div \frac{3}{5}$
10. $\frac{1}{3} \div \frac{1}{5}$
11. $\frac{2}{11} \div \frac{1}{7}$
12. $\frac{2}{7} \div \frac{1}{2}$

13. $\frac{3}{14} \div \frac{3}{14}$
14. $\frac{5}{8} \div \frac{5}{8}$
15. $\frac{3}{4} \div \frac{4}{3}$
16. $\frac{9}{10} \div \frac{10}{9}$

17. $\frac{15}{20} \div (-3)$
18. $\frac{14}{20} \div (-7)$
19. $\frac{25}{40} \div 10$
20. $\frac{36}{80} \div 9$

21. $\frac{7}{8} \div 0$
22. $0 \div \frac{5}{6}$
23. $0 \div \frac{1}{2}$
24. $\frac{15}{64} \div 0$

25. $\frac{-16}{35} \div \frac{2}{7}$
26. $\frac{-15}{27} \div \frac{5}{9}$
27. $\frac{-15}{24} \div \frac{-25}{18}$
28. $\frac{-36}{25} \div \frac{-24}{20}$

29. $\frac{34b}{21a} \div \frac{17b}{14}$
30. $\frac{16x}{20y} \div \frac{18x}{10y}$
31. $\frac{20x}{21y} \div \frac{10y}{14x}$
32. $\frac{19a}{24b} \div \frac{5a}{8b}$

Writing & Thinking

89. Explain in your own words, why the domains of the two composite functions $f(g(x))$ and $g(f(x))$ might not be the same. Give an example of two functions that illustrate this possibility.

90. Explain briefly why a function must be one-to-one to have an inverse.

Collaborative Learning

70. Do you know how to find the gas mileage (miles per gallon) that your car is using? If you are not sure, proceed as follows and compare your mileage with other students in the class. (Your car might need some work if the mileage is not consistent or it is much worse than other similar sized cars.)

   Step 1: Fill up your gas tank and write down the mileage indicated on the odometer.

   Step 2: Drive the car for a few days.

   Step 3: Fill up your gas tank again and write down the number of gallons needed to fill the tank and the new mileage indicated on the odometer.

   Step 4: Find the number of miles that you drove by subtracting the new and old numbers indicated on the odometer.

   Step 5: Divide the number of miles driven by the number of gallons needed to fill the tank. This number is your gas mileage (miles per gallon).

Exercises:

Each section includes a variety of exercises to give the students much-needed practice applying and reinforcing the skills they learned in the section. The exercises progress from relatively easy problems to more difficult problems.

Writing and Thinking:

In this feature, students are given an opportunity to independently explore and expand on concepts presented in the chapter. These questions will foster a better understanding of the concepts learned within each section.

Collaborative Learning:

In this feature, students are encouraged to work with others to further explore and apply concepts learned in the chapter. These questions will help students realize that they see many mathematical concepts in the world around them every day.

Index of Key Terms and Ideas:

Each chapter contains an index highlighting the main concepts within the chapter. This summary gives complete definitions and concise steps to solve particular types of problems. Page numbers are also given for easy reference.

Answer Key:

Located in the back of the book, the answer key provides all answers to the margin exercises, odd answers to all section exercises, and all answers to exercises in the Chapter Tests and Cumulative Reviews. This allows students to check their work to ensure that they are accurately applying the methods and skills they have learned.

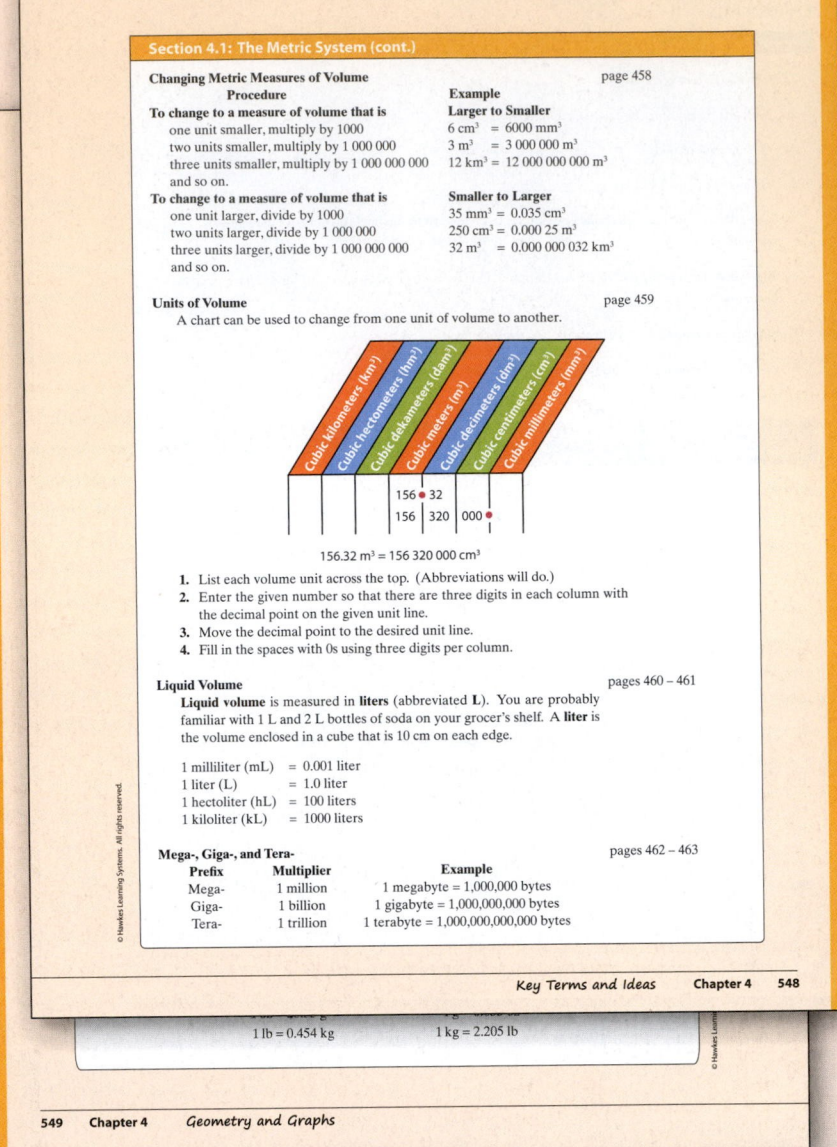

Section 4.1: The Metric System (cont.)

Changing Metric Measures of Volume
Procedure — page 458

To change to a measure of volume that is
one unit smaller, multiply by 1000
two units smaller, multiply by 1 000 000
three units smaller, multiply by 1 000 000 000
and so on.

Example
Larger to Smaller
$6 \text{ cm}^3 = 6000 \text{ mm}^3$
$3 \text{ m}^3 = 3\,000\,000 \text{ mm}^3$
$12 \text{ km}^3 = 12\,000\,000\,000 \text{ m}^3$

To change to a measure of volume that is
one unit larger, divide by 1000
two units larger, divide by 1 000 000
three units larger, divide by 1 000 000 000
and so on.

Smaller to Larger
$35 \text{ mm}^3 = 0.035 \text{ cm}^3$
$250 \text{ cm}^3 = 0.000\,25 \text{ m}^3$
$32 \text{ m}^3 = 0.000\,000\,032 \text{ km}^3$

Units of Volume — page 459
A chart can be used to change from one unit of volume to another.

Cubic kilometers (km³) Cubic hectometers (hm³) Cubic dekameters (dam³) Cubic meters (m³) Cubic decimeters (dm³) Cubic centimeters (cm³) Cubic millimeters (mm³)

156 • 32
156 320 000 •

$156.32 \text{ m}^3 = 156\,320\,000 \text{ cm}^3$

1. List each volume unit across the top. (Abbreviations will do.)
2. Enter the given number so that there are three digits in each column with the decimal point on the given unit line.
3. Move the decimal point to the desired unit line.
4. Fill in the spaces with 0s using three digits per column.

Liquid Volume — pages 460 – 461
Liquid volume is measured in **liters** (abbreviated **L**). You are probably familiar with 1 L and 2 L bottles of soda on your grocer's shelf. A **liter** is the volume enclosed in a cube that is 10 cm on each edge.

1 milliliter (mL) = 0.001 liter
1 liter (L) = 1.0 liter
1 hectoliter (hL) = 100 liters
1 kiloliter (kL) = 1000 liters

Mega-, Giga-, and Tera- — pages 462 – 463

Prefix	Multiplier	Example
Mega-	1 million	1 megabyte = 1,000,000 bytes
Giga-	1 billion	1 gigabyte = 1,000,000,000 bytes
Tera-	1 trillion	1 terabyte = 1,000,000,000,000 bytes

Key Terms and Ideas Chapter 4 548

1 lb = 0.454 kg   1 kg = 2.205 lb

Chapter 11: Rational Expressions

Section 11.1: Rational Expressions

Margin Exercises 1. a. $x \neq \dfrac{1}{5}$ **b.** $x \neq 3, 4$ **c.** no restrictions. **2. a.** $\dfrac{2}{5}, (x \neq 3)$ **b.** $\dfrac{x^2 + 5x + 25}{x + 5}, (x \neq -5, 5)$ **c.** $-1, (x \neq 5)$

3. a. $\dfrac{x^4}{6y^6}, (x \neq 0, y \neq 0)$ **b.** $\dfrac{x+3}{x^2}, (x \neq 0, 3)$ **c.** $\dfrac{x-1}{x(x+1)}$ or $\dfrac{x-1}{x^2+x}, (x \neq -1, 0, 1)$ **d.** $\dfrac{x^2 - x - 30}{3x + 9}, (x \neq -3, 2, 5)$

4. a. $\dfrac{1}{4x^2y}$ **b.** $\dfrac{-x(x^2 - xy + y^2)}{y^2}$ or $\dfrac{-x^3 + x^2y - xy^2}{y^2}$ **c.** $\dfrac{(x-3)(x-5)}{(3x+1)(3x+1)}$ or $\dfrac{x^2 - 8x + 15}{9x^2 + 6x + 1}$

Exercises 1. $\dfrac{3x}{4y}; x \neq 0, y \neq 0$ **3.** $\dfrac{2x^3}{3y^3}; x \neq 0, y \neq 0$ **5.** $\dfrac{1}{x-3}; x \neq 0, 3$ **7.** $7; x \neq 2$ **9.** $-\dfrac{3}{4}; x \neq 3$ **11.** $\dfrac{2x}{y}; x \neq -\dfrac{2}{3}, y \neq 0$

13. $\dfrac{x}{x-1}; x \neq -6, 1$ **15.** $\dfrac{x^2 - 3x + 9}{x - 3}; x \neq -3, 3$ **17.** $\dfrac{x-3}{y-2}; y \neq -2, 2$ **19.** $\dfrac{x^2 + 2x + 4}{y + 5}; x \neq 2, y \neq -5$ **21.** $\dfrac{ab}{6y}$

23. $\dfrac{8x^2y^3}{15}$ **25.** $\dfrac{x+3}{x}$ **27.** $\dfrac{x-1}{x+1}$ **29.** $-\dfrac{1}{x-8}$ **31.** $\dfrac{x-2}{x}$ **33.** $\dfrac{4x+20}{x(x+1)}$ **35.** $\dfrac{x}{(x+3)(x-1)}$ **37.** $-\dfrac{x+4}{x(x+1)}$

39. $\dfrac{x+2y}{(x-3y)(x-2y)}$ **41.** $\dfrac{x-1}{x(2x-1)}$ **43.** $\dfrac{1}{x+1}$ **45.** $\dfrac{x+2}{x-2}$ **47.** $\dfrac{1}{3xy^6}$ **49.** $\dfrac{6y^7}{x^4}$ **51.** $\dfrac{x}{12}$ **53.** $\dfrac{6x+18}{x^2}$ **55.** $\dfrac{6}{5x}$

57. $\dfrac{3x+1}{x+1}$ **59.** $\dfrac{x-2}{2x-1}$ **61.** $-\dfrac{x+4}{x(2x-1)}$ **63.** $\dfrac{x+1}{x-1}$ **65.** $\dfrac{6x^3 - x^2 + 1}{x^2(4x-3)(x-1)}$ **67.** $\dfrac{x^2 + 4x + 4}{x^2(2x-5)}$ **69.** $\dfrac{x^2 - 6x + 5}{(x-7)(x-2)(x+7)}$

Chapter 1: Test

Solve the following problems.

1. Write the number 3,075,122 in words.

2. Write the number fifteen million, three hundred twenty thousand, five hundred eighty-four in standard notation.

Round as indicated.

3. 675 (to the nearest ten)

4. 13,620 (to the nearest thousand)

Find the correct value of n and state the property of addition or multiplication that is illustrated.

5. $16 + 52 = 52 + n$

6. $18(20 \cdot 3) = (18 \cdot 20) \ n$

7. $74 + n = 74$

8. $6 + (2 + 4) = (6 + 2) + n$

Estimate the answer, then find the answer. Do all of your work with pencil and paper. Do not use a calculator.

9. $\begin{array}{r} 9587 \\ 345 \\ + \quad 2075 \\ \hline \end{array}$

10. $\begin{array}{r} 45,872 \\ 589,651 \\ + \quad 235,003 \\ \hline \end{array}$

11. $\begin{array}{r} 7046 \\ - \ 4562 \\ \hline \end{array}$

12. $\begin{array}{r} 80,000 \\ - 28,830 \\ \hline \end{array}$

13. $\begin{array}{r} 47 \\ \times 25 \\ \hline \end{array}$

14. $\begin{array}{r} 2593 \\ \times \quad 86 \\ \hline \end{array}$

Cumulative Review: Chapters 1 - 2

Answer the following questions.

1. Given the expression 5^3 :
   a. name the base,
   b. name the exponent, and
   c. find the value of the expression.

2. Match each expression with the best estimate of its value.
   _____ a. $175 + 92 + 96 + 125$ A. 200
   _____ b. $8465 \div 41$ B. 500
   _____ c. $32 \cdot 48$ C. 1000
   _____ d. $5484 - 4380$ D. 1500

3. Multiply mentally: $70(\ 9000 \ ) = $ _____.

4. Round 176,200 to the nearest ten thousand.

5. Evaluate by using the rules for order of operations:
   $$5^2 + (12 \cdot 5 \div 2 \cdot 3) - 60 \cdot 4$$

6. a. List all the prime numbers less than 25.
   b. List all the squares of these prime numbers.

7. Find the LCM of 45, 75, and 105.

8. Find the GCD for $\{120, 150\}$

9. The value of $0 \div 19$ is _____, whereas the value of $19 \div 0$ is _____.

10. Find the mean of 45, 36, 54, and 41.

Perform the indicated operations. Reduce all fractions to lowest terms.

11. $\begin{array}{r} 8376 \\ 3749 \\ + \ 2150 \\ \hline \end{array}$

12. $\begin{array}{r} 1563 \\ - \ 975 \\ \hline \end{array}$

13. $\begin{array}{r} 751 \\ \times \ 38 \\ \hline \end{array}$

14. $14\overline{)2856}$

15. $-\dfrac{3}{20} \div 6$

16. $\dfrac{5}{18}\left(\dfrac{3}{10}\right)\left(\dfrac{4}{21}\right)$

17. $-\dfrac{5}{6} - \left(-\dfrac{7}{8}\right)$

18. $\dfrac{7}{20} - \dfrac{3}{5}$

19. $\begin{array}{r} 38\dfrac{4}{5} \\ + \ 29\dfrac{7}{12} \\ \hline \end{array}$

20. $\begin{array}{r} -13\dfrac{3}{20} \\ + \ 6\dfrac{5}{8} \\ \hline \end{array}$

21. $3\dfrac{2}{3} \div \left(-2\dfrac{5}{11}\right)$

22. $-4\dfrac{3}{8} + \left(-2\dfrac{1}{2}\right)$

Simplify each of the following expressions.

23. $\dfrac{1\dfrac{3}{10} + 2\dfrac{4}{5}}{2\dfrac{3}{5} - 1\dfrac{1}{3}}$

24. $1\dfrac{1}{5} - 11\dfrac{7}{10} + 6\dfrac{2}{3} \div 2\dfrac{1}{2}$

25. $\dfrac{x}{5} \cdot \dfrac{1}{3} - \dfrac{2}{5} \div 3$

Content

Chapter 1, Whole Numbers, Integers, and Introduction to Solving Equations, reviews the fundamental operations of addition, subtraction, multiplication, and division with whole numbers. Estimating results is used to develop better understanding of whole number concepts, and applications help to reinforce the need for these ideas and skills in problems such as finding averages and making purchases. Exponents are introduced and the rules for order of operations are used to evaluate expressions with more than one operation. This chapter also develops the ideas of number lines and absolute value. This early introduction of integers allows positive and negative numbers to be used throughout the text in evaluations, solving equations, and various applications.

Chapter 2, Fractions, Mixed Numbers, and Proportions, discusses the concepts of divisibility and factors and their relationship to finding prime factorizations, which are then used to develop the skills needed for finding the least common multiple of a set of numbers or algebraic expressions. These ideas form the basis for the work with fractions and mixed numbers. Emphasis is placed on using prime numbers to reduce fractions and to find the least common denominator (LCD) when adding or subtracting fractions. Area is introduced, and the rules for order of operations are applied to evaluate expressions with fractions and mixed numbers. The last two sections deal with solving equations involving fractions and proportions.

Chapter 3, Decimal Numbers, Percents, and Square Roots, reviews reading and writing decimal numbers as well as operating with decimal numbers. Rounding decimals is used to help in estimating results in addition, subtraction, multiplication, and division. Section 3.4 introduces the statistical ideas of mean, median, mode, and range. The last several sections then introduce the concept and applications of percent, including discount, sales tax, commission, profit, and simple and compound interest. Section 3.9 finishes the chapter with an introduction to square roots. The end of the chapter allows students to become more familiar with the power and convenience of their calculators, as this tool is seen as necessary for evaluating square roots of numbers that are not perfect squares, as well as for calculating compound interest.

Chapter 4, Geometry and Graphs, begins with an introduction to the metric and U.S. Customary systems of measurement. Then geometric concepts are introduced through angles and angle measurement. Topics include perimeter, area, volume, and surface area. Triangles are shown to be classified in terms of angle measure and lengths of sides. Similar triangles and congruent triangles are discussed. Square roots are presented and shown to be related to right triangles through the Pythagorean Theorem. Finally, section 4.9 introduces the concept of reading graphs.

Chapter 5, Algebraic Expressions, Linear Equations, and Applications, shows how arithmetic concepts, through the use of variables and signed numbers, can be generalized with algebraic expressions. Algebraic expressions are simplified by combining like terms and evaluated by using the rules for order of operations. As a lead-in to interpreting and understanding word problems, a section involving translating English phrases and algebraic expressions is included. The chapter then goes on to develop the techniques for solving linear (or first-degree) equations in a step-by-step manner over three sections. Techniques include combining like terms and use of the distributive property. Finally, the chapter explores common formulas and how to work with them. Applications relate to number problems, consecutive integers, percent, distance, interest, and mean.

Chapter 6, Graphing Linear Equations, allows for the introduction of graphing in two dimensions and the ideas and notation related to functions. Included are complete discussions on the three basic forms for equations of lines in a plane: the standard form, the slope-intercept form, and the point-slope form. Slope is discussed for parallel and perpendicular lines and treated as a rate of change. Functions are introduced and the vertical line test is used to tell whether or not a graph represents a function. Use of a TI-84 graphing calculator is an integral part of this introduction to functions as well as part of graphing linear inequalities in the last section.

Chapter 7, Exponents, Polynomials, and Factoring Techniques, studies the properties of exponents in depth. The remainder of the chapter is concerned with definitions and operations related to polynomials. Included are the FOIL method of multiplication with two binomials, special products of binomials, and the division algorithm. This chapter also discusses methods of factoring polynomials, including finding common monomial factors, factoring by grouping, factoring trinomials by grouping and by trial-and-error, and factoring special products. A special section provides students with tips on determining which type of factoring to use and provides extra exercises to allow them to practice related skills. The topic of solving quadratic equations is introduced, and quadratic equations are solved by factoring only. Applications with quadratic equations are included and involve topics such as the use of function notation to represent area, the Pythagorean Theorem, and consecutive integers.

Chapter 8, Systems of Linear Equations, covers solving systems of two equations in two variables and systems of three equations in three variables. The basic methods of graphing, substitution, and addition are included along with matrices and Gaussian elimination. Double subscript notation is used with matrices for an easy transition to the use of matrices in solving systems of equations with the TI-84 Plus calculator. Applications involve mixture, interest, work, algebra, and geometry. The last section discusses half-planes and graphing systems of linear inequalities, again including the use of a graphing calculator.

Chapter 9, Roots and Radicals, discusses simplifying radicals along with the use of calculators to find approximate values of expressions with radicals. Arithmetic with radicals includes addition, subtraction, multiplication, and rationalizing denominators. Methods for solving equations with radicals are developed. The rules for using rational exponents (fractional exponents) and simplifying expressions with rational exponents are discussed. A section on functions with radicals shows how to analyze the domain and range of radical functions and how to graph these functions by using a graphing calculator.

Chapter 10, Exponential and Logarithmic Functions, begins with a section on the algebra of functions and leads to the development of the composition of functions and methods for finding the inverses of one-to-one functions. This introduction lays the groundwork for understanding the relationship between exponential functions and logarithmic functions. While the properties of real exponents and logarithms are presented completely, most numerical calculations are performed with the aid of a calculator.

Chapter 11, Rational Expressions, provides still more practice with factoring and shows how to use factoring to operate with rational expressions. Included are the topics of multiplication, division, addition, and subtraction with rational expressions, simplifying complex fractions, and solving equations containing rational expressions. Applications are related to work, distance-rate-time, and variation.

Chapter 12, Complex Numbers and Quadratic Equations, introduces complex numbers along with the basic operations of addition, subtraction, multiplication, and division. These skills are then used to further explore solving quadratic equations. A review of solving quadratic equations by factoring is presented, followed by an introduction of the methods of using the square root property and completing the square. The quadratic formula is developed by completing the square and students are encouraged to use the most efficient method for solving any particular quadratic equation. Applications are related to the Pythagorean theorem, projectiles, geometry, and cost.

Practice and Review

There are more than 9500 margin exercises, completion examples, practice problems, section exercises, and review exercises overall. Section exercises are carefully chosen and graded, proceeding from easy exercises to more difficult ones. Each chapter includes a Chapter Test and a Cumulative Review (beginning with Chapter 2). Also, each chapter contains an Index of Key Terms and Ideas.

Many sections have special exercises entitled "Writing and Thinking about Mathematics" and "Collaborative Learning Exercises." These exercises are an important part of the text and provide a chance for each student to improve communication skills, to develop a deeper understanding of general concepts, and to communicate his or her ideas to the instructor. Written responses can be a great help to the instructor in identifying just what students do and do not understand. Many of these questions are designed for the student to investigate ideas other than those presented in the text with responses that are to be based on each student's own experiences and perceptions.

Answers to the odd-numbered exercises, all margin exercises, all Chapter Test questions, and all Cumulative Review questions are provided in the back of the book.

Acknowledgements

I would like to thank Editor Susan Niese. Also, thanks to Editor Nina Waldron, Production Editor Kim Cumbie, and Vice President of Development Marcel Prevuznak for their invaluable assistance in the development and production of this text.

Thank you to the mathematics faculty at Kirkwood Community College for their collaboration and guidance.

Many thanks go to the following reviewers who offered constructive and critical comments:

Elaine Arrington at University of Montana, Darcee Bex at South Louisiana Community College, Cindy Bond at Butler Community College, Randall Dorman at Cochise College, Valeree Falduto at Lynn University, Debra Gupton at New River Community College, Kimberly Haughee at University of Montana, Billie Havens at Walla Walla Community College, Rebecca Heiskell at Mountain View College, Bobbie Jo Hill at Coastal Bend College, Sandee House at Georgia Perimeter College, Marjorie Hunter at Butler Community College, Joanne Kendall at North Harris Montgomery Community College, Joanne Koratich at Muskegon Community College, Jim Martin at Cochise College, Val Mohanakumar at Hillsborough Community College, Dr. Carol Okigbo at Minnesota State University, Gabriel Perrow at Eastern Maine Community College, Amy Rexrode at Navarro College, Harriete Roadman at New River Community College, Connie Rost at South Louisiana Community College, Jim Sheff at Spoon River College, Dr. Melanie Smith at Bishop State Community College, Nan Strebeck at Navarro College, Emily Whaley at Georgia Perimeter College.

Finally, special thanks go to Dr. James Hawkes for his support in this first edition and his willingness to commit so many resources to guarantee a top-quality product for students and teachers.

D. Franklin Wright

To the Student

The goal of this text and of your instructor is for you to succeed in your study of mathematics. Certainly, you should make this your goal as well. What follows is a brief discussion about developing good work habits and using the features of this text to your best advantage. For you to achieve the greatest return on your investment of time and energy you should practice the following three rules of learning:

1. Reserve a block of time for study every day.
2. Study what you don't know.
3. Don't be afraid to make mistakes.

How to Use This Book

The following seven-step guide will not only make using this book a more worthwhile and efficient task, but it will help you benefit more from classroom lectures or the assistance that you receive in a math lab.

1. Try to look over the assigned section(s) before attending class or lab.

2. Read examples carefully.

3. Work margin exercises when asked to do so throughout the lessons.

4. Work the section exercises faithfully as they are assigned.

5. Use the Writing and Thinking About Mathematics questions as an opportunity to explore the way you think about mathematics.

6. Use the Chapter Tests to practice for the tests that are actually given in class or lab.

7. Study the Cumulative Reviews to help you retain the skills that you have acquired in studying earlier chapters.

How to Prepare for an Exam

Gaining Skill and Confidence

The stress that many students feel while trying to succeed in mathematics is what you have probably heard called "math anxiety." It is a real-life phenomenon, and many students experience such a high level of anxiety during mathematics exams in particular that they simply cannot perform to the best of their abilities. It is possible to overcome this stress simply by building your confidence in your ability to do mathematics and by minimizing your fears of making mistakes.

No matter how much it may seem that in mathematics you must either be right or wrong, with no middle ground, you should realize that you can be learning just as much from the times that you make mistakes as you can from the times that your work is correct. Success will come. Don't think that making mistakes at first means that you'll never be any good at mathematics. Learning mathematics requires lots of practice. Most importantly, it requires a true confidence in yourself and in the fact that with practice and persistence the mistakes will become fewer, the successes will become greater, and you will be able to say, "I can do this."

Showing What You Know

If you have attended class or lab regularly, taken good notes, read your textbook, kept up with homework exercises, and asked for help when it was needed, then you have already made significant progress in preparing for an exam and conquering any anxiety. Here are a few other suggestions to maximize your preparedness and minimize your stress.

1. Give yourself enough time to review. You will generally have several days advance notice before an exam. Set aside a block of time each day with the goal of reviewing a manageable portion of the material that the test will cover. Don't cram!

2. Work many problems to refresh your memory and sharpen you skills. Go back and redo selected exercises from all of your homework assignments.

3. Reread the text and your notes, and use the Chapter Index of Key Terms and Ideas to recap major ideas and test yourself by going back over problems.

4. Be sure that you are well-rested so that you can be alert and focused during the exam.

5. Don't study up to the last minute. Give yourself some time to wind down before the exam. This will help you to organize your thoughts and feel more calm as the test begins.

6. As you take the test, realize that its purpose is not to trick you, but to give you and your instructor an accurate idea of what you have learned. Good study habits, a positive attitude, and confidence in your own ability will be reflected in your performance on any exam.

7. Finally, you should realize that your responsibility does not end with taking the exam. When your instructor returns your corrected exam, you should review your instructor's comments and any mistakes that you might have made. Take the opportunity to learn from this important feedback about what you have accomplished, where you could work harder, and how you can best prepare for future exams.

HAWKES LEARNING:
Preparation for College Mathematics

Hawkes Learning specializes in interactive courseware with a unique mastery-based approach to student learning. The courseware is designed to help you develop a solid foundation of skills and has been proven to increase your overall success. Within each homework lesson you will find three learning modes: Learn, Practice, and Certify.

Learn: Learn is a multimedia presentation that includes the information you need to successfully answer each question in your assignment. Each lesson includes definitions, rules, properties, and examples, along with instructional videos.

Practice: Practice gives you unlimited opportunities to practice the types of problems you will receive in Certify. In Practice, you have access to learning aids through the Interactive Tutor. Step-By-Step breaks a problem down into smaller steps; Solution offers guided solutions to every problem; and Explain Error gives targeted feedback specific to your mistake.

Certify: This is the credit component of your homework! You will answer your problem set by using your knowledge and the foundation you built in Learn and Practice. You will have the opportunity to try again with no penalty if you do not demonstrate Mastery in your initial attempt(s). Pay close attention to any due dates or benchmarks assigned by your instructor.

Video: View instructional videos anytime, anywhere at HawkesTV.com.

Feel free to contact support for questions or technical help.

SUPPORT:

Support: support.hawkeslearning.com

Chat: chat.hawkeslearning.com

E-mail: support@hawkeslearning.com

Phone: 843.571.2825

Mathematics @ Work!

Everyone has heard someone say, "I'm just no good at math." Chances are, you may have even said it yourself. What many people do not realize is that math is a language. And, just as with any other foreign language, math requires practice and perseverance. Thus to succeed in this course, you will need to be diligent and remember that you can often learn more from your mistakes than if you get the correct answer on the first try. Once you succeed, you will have acquired a set of skills you can make use of for the rest of your life. No matter what job you seek after graduation, numbers will play some role in that position. For example, you may need to keep track of the number of sales you make, the measurements of the building you construct, the quantity of medicine you measure out for a sick child, or the amounts you charge your clients for services.

As a student striving to succeed, you will want to keep track of your progress in this course. One way to do this is to find the average of your test scores towards the end of the semester. For example, say you received the following scores on the first four tests.

78   85   94   83

What is your current test average for the class?

For more problems like these, see Section 1.7.

Whole Numbers, Integers, and Introduction to Solving Equations

1.1 Introduction to Whole Numbers

General Remarks to the Student

This text is written for your enjoyment! That's right, you are going to have a good time studying mathematics because you are going to be successful. A positive attitude will be your greatest asset. Algebra is just arithmetic with letters, and this text will reinforce your basic understanding of arithmetic concepts with whole numbers, fractions, decimals, and percents and show how these same concepts can be applied to algebra. Then you will be ready and primed to be successful in future courses in mathematics. Hopefully, you will also see how important and useful mathematics is in your daily life. The main goals are to teach you how to think and how to solve problems. All **you** have to do to attain these goals is:

1. **Read the book.**
2. **Study every day.**
3. **Do not be afraid to make mistakes.**
4. **Ask lots of questions.**

Good luck! (**Hint:** If you follow these four procedures you won't need luck.)

Objective A Reading and Writing Whole Numbers

The **whole numbers** are the number 0 and the **natural numbers** (also called the **counting numbers**). We use $\mathbb{N}$ to represent the set of natural numbers and $\mathbb{W}$ to represent the set of whole numbers.

> **Whole Numbers**
>
> The **whole numbers** are the natural (or counting) numbers and the number 0.
>
> $$\text{Natural numbers} = \mathbb{N} = \{\,1, 2, 3, 4, 5, 6, 7, 8, 9, 10, 11, \dots\,\}$$
> $$\text{Whole numbers} = \mathbb{W} = \{\,0, 1, 2, 3, 4, 5, 6, 7, 8, 9, 10, 11, \dots\,\}$$
>
> Note that 0 is a whole number but not a natural number.

The three dots (called an **ellipsis**) in the definition indicate that the pattern continues without end.

To write a whole number in **standard notation** (or **standard form**) we use a **place value system** that depends on three things:

1. the ten digits: 0, 1, 2, 3, 4, 5, 6, 7, 8, 9,
2. the placement of each digit, and
3. the value of each place.

Figure 1 shows the value of the first ten places in the **decimal system** we use. (The decimal system will be discussed in more detail in a later chapter.) Every three places constitutes a **period** and the digits in each period are separated with commas. A number with 4 or fewer digits need not have any commas.

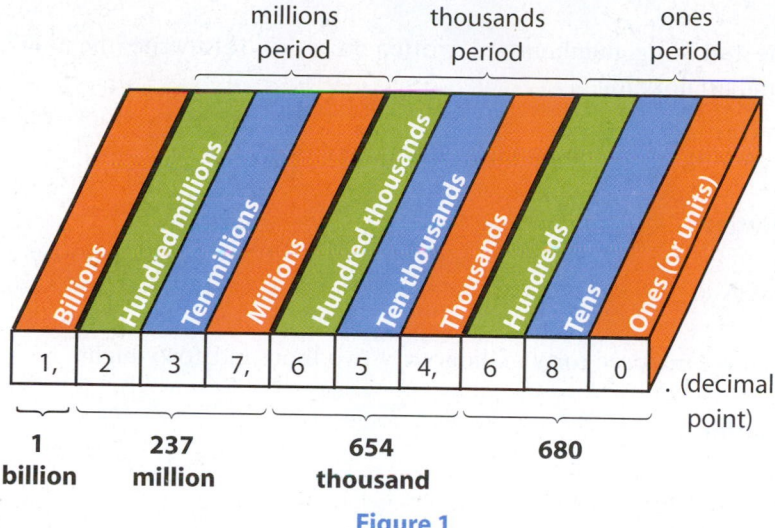

Figure 1

To read (or write) a number in this system, start from the left and use the name of the number (three digits or less) in each period. For example,

7,839,076,532 is read as:

seven billion, eight hundred thirty-nine million, seventy-six thousand, five hundred thirty-two. (Note that a hyphen is used when writing two-digit numbers larger than twenty.)

Note that the word **and** does not appear as part of reading (or writing) any whole number. The word **and** indicates the decimal point. You may put a decimal point to the right of the digits in a whole number if you choose, but it is not necessary unless digits are written to the right of the decimal point. (These ideas will be discussed in a later chapter.)

Example 1

Reading and Writing Whole Numbers

The following numbers are written in standard notation. Write them in words.

a. 25,380

Solution

twenty-five thousand, three hundred eighty

b. 3,400,562

Solution

three million, four hundred thousand, five hundred sixty-two

Now work margin exercise 1.

1. Write each of the following numbers in words.

a. 32,450,090

b. 5784

2. Write each of the following numbers in standard notation.

a. six thousand, forty-one

b. nine billion, four hundred eighty-three thousand

Example 2

Writing Whole Numbers in Standard Notation

The following numbers are written in words. Rewrite them in standard notation.

a. twenty-seven thousand, three hundred thirty-six

Solution

27,336

b. three hundred forty million, sixty-two thousand, forty-eight

Solution

340,062,048

Note: 0s must be used to fill out a three-digit period.

Now work margin exercise 2.

Objective B ## Rounding Whole Numbers

To **round** a given number means to find another number close to the given number. The desired place of accuracy must be stated. For example, if you were asked to round 762, you might say 760 or 800. Either answer could be correct, depending on whether the accuracy was to be the nearest ten or the nearest hundred. Rounded numbers are quite common and acceptable in many situations.

For example, a credit card application might ask for your approximate income: 0 - $10,000, $10,001 - $20,000, $20,001 - $30,000, $30,001 - $40,000, and so on, but not your exact income. Even the IRS allows rounding to the nearest dollar when calculating income tax. In effect, rounded numbers are simply approximations of the given numbers.

In Example 3, we use **number lines** as visual aids in understanding the rounding process. On number lines, the whole numbers are used to label equally-spaced points, usually to the right of the point labeled 0. (We will see in a later chapter that negative numbers are used to label points to the left of 0.)

Example 3

Rounding Whole Numbers

a. Round 762 to the nearest ten.

Solution

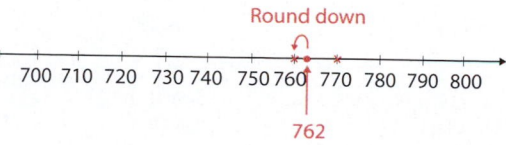

We see that 762 is closer to 760 than to 770. Thus 762 rounds to 760 (to the nearest ten).

b. Round 762 to the nearest hundred.

Solution

Also, 762 is closer to 800 than to 700. Thus 762 rounds to 800 (to the nearest hundred).

Now work margin exercise **3.**

3. Use a number line to round each number as indicated.

    a. 583 to the nearest ten

    b. 269 to the nearest hundred

    c. 9732 to the nearest thousand

    d. 9732 to the nearest hundred

Using number lines as aids to understanding is fine, but for practical purposes, the following rule is more useful.

Rounding Rule for Whole Numbers

1. Look at the single digit just to the right of the digit that is in the place of desired accuracy.
2. **If this digit is 5 or greater**, make the digit in the desired place of accuracy one larger and replace all digits to the right with zeros. All digits to the left remain unchanged unless a 9 is made one larger; then the next digit to the left is increased by 1.
3. **If this digit is less than 5**, leave the digit that is in the place of desired accuracy as it is, and replace all digits to the right with zeros. All digits to the left remain unchanged.

4. Round each number as indicated.

a. 7350 to the nearest hundred

b. 29,736 to the nearest thousand

c. 137,800 to the nearest hundred thousand

d. 28,379,200 to the nearest million

Example 4

Rounding Whole Numbers

a. Round 6849 to the nearest hundred.

Solution

6849   6849   6800

Place of desired accuracy.   Look at one digit to the right; 4 is less than 5.   Leave 8 and fill in zeros.

So, 6849 rounds to 6800 (to the nearest hundred).

b. Round 3500 to the nearest thousand.

Solution

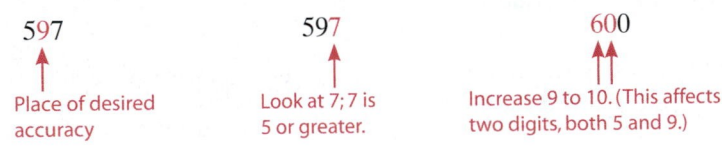

3500   3500   4000

Place of desired accuracy   Look at 5; 5 is 5 or greater.   Increase 3 to 4 (one larger) and fill in zeros.

So, 3500 rounds to 4000 (to the nearest thousand).

c. Round 597 to the nearest ten.

Solution

597   597   600

Place of desired accuracy   Look at 7; 7 is 5 or greater.   Increase 9 to 10. (This affects two digits, both 5 and 9.)

So, 597 rounds to 600 (to the nearest ten).

Now work margin exercise 4.

Exercises 1.1

Write each of the following numbers in words. See Example 1.

1. 893

2. 525

3. 7831

4. 9061

5. 8,957,000

6. 2,745,000

7. 6,006,006

8. 3,003,003

9. 25,000,000

10. 163,042,000

11. 67,940,300

12. 810,895

13. 305,444

14. 7,830,000,452

15. 19,462,541,300

Write each of the following numbers in standard notation. See Example 2.

16. seventy-five

17. eighty-four

18. seven hundred thirty-six

19. six hundred six

20. thirty-seven thousand, one hundred ten

21. one hundred fourteen thousand, two hundred twenty-five

22. one hundred thirteen million, seventy-five thousand, four hundred sixty-seven

23. two hundred million, sixteen thousand, fifty-three

24. three billion, one hundred thirty-two million, seven hundred fifty-nine

25. fourteen billion, sixty-three million, one hundred seventy-two thousand, eighty-one

26. thirty-seven million

27. fourteen thousand, fourteen

28. seventy-two billion, one hundred three thousand

29. fifteen million, fifteen thousand, fifteen

30. two billion, two hundred million, twenty thousand, two

Round as indicated. Draw a number line, graph the given number on the line with a heavy dot and mark the rounded number with an ×. See Examples 3 and 4.

31. 78 (nearest ten)

32. 94 (nearest ten)

33. 655 (nearest ten)

34. 382 (nearest ten)

35. 479 (nearest hundred)

36. 258 (nearest hundred)

37. 1957 (nearest hundred)

38. 2005 (nearest hundred)

39. 4399 (nearest thousand)

40. 11,388 (nearest thousand)

Round each number to the nearest ten.

41. 783   **42.** 51   **43.** 625   **44.** 7603

Round each number to the nearest hundred.

45. 296   **46.** 876   **47.** 981   **48.** 24,784

Round each number to the nearest thousand.

49. 7912   **50.** 5500   **51.** 7499   **52.** 47,800

Round each number to the nearest ten thousand.

53. 78,429   **54.** 135,000   **55.** 1,073,200   **56.** 5,276,000

Round each number to the nearest hundred thousand.

57. 119,200   **58.** 184,900   **59.** 3,625,000   **60.** 1,312,500

Complete the following problems.

61. Venezuela: The Republic of Venezuela covers an area of 352,143 square miles, or about 912,050 square kilometers. Round each of these numbers to the nearest ten thousand.

62. Earth to Sun: The distance from the earth to the sun is about 149,730,000 kilometers. Round this number to the nearest million.

63. **Wyoming:** According to the 2010 U.S. Census Bureau data, Wyoming was the least populous state in the United States with 563,626 people. Vermont was the second least populous state with 625,741 people. Round each number to the nearest thousand.

64. **Largest Cities:** The two largest cities in the world are Tokyo and Mexico City. According to the Japanese Bureau of Citizens and Cultural Affairs, Tokyo has 13,161,751 people in 2010. The Mexican National Institute of Statistics reported Mexico City's population as 8,851,080 in 2010. Round each number to the nearest million.

1.2 Addition and Subtraction with Whole Numbers

A Be able to add with whole numbers and understand the terms addend and sum.

B Learn the properties of additon.

C Understand the concept of perimeter.

D Be able to subtract with whole numbers and understand the meaning of the term difference.

E Learn how to estimate sums and differences with rounded numbers.

F Learn how to add and subtract with a calculator.

Objective A **Addition with Whole Numbers**

The operation of addition with whole numbers is indicated either by writing the numbers horizontally, separated by (+) signs, or by writing the numbers vertically in columns with instructions to add. For example, the following two notations represent the same addition problem.

$$17 + 4 + 132 \qquad \text{or} \qquad \begin{array}{r} 17 \\ 4 \\ + 132 \\ \hline \end{array}$$

The numbers being added are called **addends**, and the result of the addition is called the **sum**.

$$\text{Add:} \quad \begin{array}{r} 17 \\ 4 \\ + 132 \\ \hline 153 \end{array} \quad \begin{array}{l} \text{addend} \\ \text{addend} \\ \text{addend} \\ \text{sum} \end{array}$$

Be sure to keep the digits aligned (in column form) so that you will be adding units to units, tens to tens, and so on. The speed and accuracy with which you add depends on your having memorized the basic addition facts of adding single-digit numbers.

The sum of several numbers can be found by writing the numbers vertically so that the place values are lined up in columns. Note the following:

1. If the sum of the digits in one column is more than 9,
    a. write the ones digit in that column, and
    b. carry the other digit as a number to be added to the next column to the left.

2. Look for combinations of digits that total 10 (See Example 1).

Example 1

Addition with Whole Numbers

To add the following numbers, we note the combinations that total 10 to find the sums quickly.

$$\begin{array}{r} 2 \\ 217 \\ 389 \\ 634 \\ + 536 \\ \hline 6 \end{array}$$

add $7 + 9 + 4 + 6 = 7 + 9 + 10$
$= 26$

Carry the 2 from the 26.

$$\begin{array}{r} 12 \\ 217_{10} \\ 389 \\ 634 \\ +\ 536 \\ \hline 76 \end{array}$$

add $2 + 1 + 8 + 3 + 3 = 10 + 1 + 3 + 3$
$= 17$

└ Carry the 1 from the 17.

$$\begin{array}{r} 12 \\ 217 \\ 10 \to 389 \\ 634 \\ +\ 536 \\ \hline 1776 \end{array}$$

add $1 + 2 + 3 + 6 + 5 = 10 + 2 + 5$
$= 17$

Now work margin exercise **1.**

Objective B — Properties of Addition

There are several properties of the operation of addition. To state the general form of each property, we introduce the notation of a **variable**. As we will see, variables not only allow us to state general properties, they also enable us to set up equations to help solve many types of applications.

> ## Variable
>
> A **variable** is a symbol (generally a letter of the alphabet) that is used to represent an unknown number or any one of several numbers.

In the following statements of the properties of addition, the set of whole numbers is the **replacement set** for each variable. (The replacement set for a variable is the set of all possible values for that variable.)

> ## Commutative Property of Addition
>
> For any whole numbers a and b, $\boldsymbol{a + b = b + a}$.
>
> For example, $33 + 14 = 14 + 33$.
>
> (The **order** of the numbers in addition can be reversed.)

> ## Associative Property of Addition
>
> For any whole numbers a, b and c, $\boldsymbol{(a + b) + c = a + (b + c)}$.
>
> For example, $(6 + 12) + 5 = 6 + (12 + 5)$.
>
> (The **grouping** of the numbers can be changed.)

1. Find the sum.

$$\begin{array}{r} 463 \\ 905 \\ 344 \\ +\ 437 \\ \hline \end{array}$$

Additive Identity Property

For any whole number a, $a + 0 = a$.

For example, $8 + 0 = 8$.

(The sum of a number and 0 is that same number.)

The number 0 is called the **additive identity**.

Example 2

Properties of Addition

Each of the properties of addition is illustrated.

a. $40 + 3 = 3 + 40$ commutative property of addition

As a check, we see that $40 + 3 = 43$ and $3 + 40 = 43$.

b. $2 + (5 + 9) = (2 + 5) + 9$ associative property of addition

As a check, we see that $2 + (14) = (7) + 9 = 16$.

c. $86 + 0 = 86$ additive identity property

2. Find each sum and tell which property of addition is illustrated.

a. $(9+1) + 7 = 9 + (1+7)$

b. $25 + 4 = 4 + 25$

c. $39 + 0$

3. Find the value of each variable that will make each statement true and name the property used.

a. $36 + y = 12 + 36$

b. $17 + n = 17$

c. $(x+8)+4=10+(8+4)$

Completion Example 3

Properties of Addition

Use your knowledge of the properties of addition to find the value of the variable that will make each statement true. State the name of the property used.

Solutions

		Value	Property
a. $14 + n = 14$	$n =$ _____	_____
b. $3 + x = 5 + 3$	$x =$ _____	_____
c. $(1 + y) + 2 = 1 + (7 + 2)$	$y =$ _____	_____

Now work margin exercises 2 and 3.

Completion Example Answers

3. **a.** $n = 0$; additive identity
   b. $x = 5$; commutative property of addition
   c. $y = 7$; associative property of addition

Objective C **The Concept of Perimeter**

The **perimeter** of a geometric figure is the distance around the figure. The perimeter is found by adding the lengths of the sides of the figure, and is measured in linear units such as inches, feet, yards, miles, centimeters, or meters. The perimeter of the triangle (a geometric figure with three sides) shown in Figure 1 is found by adding the measures of the three sides. Be sure to label the answers for the perimeter of any geometric figure with the correct units of measurement.

4. Find the perimeter of the rectangle.

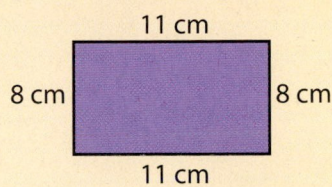

Example 4

Finding Perimeter

To find the perimeter, find the sum of the lengths of the sides.

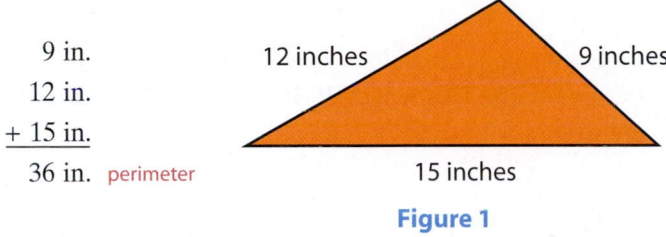

$$\begin{array}{r} 9 \text{ in.} \\ 12 \text{ in.} \\ + 15 \text{ in.} \\ \hline 36 \text{ in.} \quad \text{perimeter} \end{array}$$

Figure 1

Now work margin exercise 4.

Objective D **Subtraction with Whole Numbers**

Suppose we know that the sum of two numbers is 20 and one of the numbers is 17. What is the other number? We can represent the problem in the following format.

$$17 \quad + \quad \underline{} \quad = \quad 20$$

addend   missing addend   sum

This kind of "addition" problem is called **subtraction** and can be written as follows.

$$20 \quad - \quad 17 \quad = \quad \underline{}$$

sum   minus   addend   missing addend (or difference)

or

$$\begin{array}{r} 20 \\ - 17 \\ \hline 3 \end{array}$$ ← minuend
← subtrahend
← difference or missing addend

In other words, subtraction is reverse addition. To subtract, we must know how to add. The missing addend is called the **difference** between the sum (now called the **minuend**) and one addend (now called the **subtrahend**). In this case, we know that the difference is 3, since

$$17 + 3 = 20.$$

we write

$$20 - 17 = 3.$$

However, finding a difference such as $742 - 259$, where the numbers are larger, is more difficult because we do not know readily what to add to 259 to get 742. To find such a difference, we use the technique for subtraction developed in Example 5 that uses place values and borrowing.

5. Find the difference.

$$\begin{array}{r} 342 \\ -\ 187 \\ \hline \end{array}$$

Example 5

Subtraction with Whole Numbers

Find the difference.

$$\begin{array}{r} 742 \\ -\ 259 \\ \hline \end{array}$$

Solution

Step 1: Since 2 is smaller than 9, borrow 10 from 40 and add this 10 to 2 to get 12. This leaves 30 in the tens place; cross out 4 and write 3.

10 borrowed from 40

$$\begin{array}{r} 7\ \overset{3}{\cancel{4}}\ {}^{1}2 \\ -\ \ \ 2\ 5\ \ 9 \\ \hline \end{array}$$

Step 2: Since 3 is smaller than 5, borrow 100 from 700. This leaves 600, so cross out 7 and write 6.

$$\begin{array}{r} \overset{6}{\cancel{7}}\ \overset{{}^{1}3}{\cancel{4}}\ {}^{1}2 \\ -\ \ \ 2\ 5\ \ 9 \\ \hline \end{array}$$

Step 3: Now subtract.

$$\begin{array}{r} \overset{6}{\cancel{7}}\ \overset{{}^{1}3}{\cancel{4}}\ {}^{1}2 \\ -\ \ \ 2\ 5\ \ 9 \\ \hline 4\ 8\ \ 3 \end{array}$$

Check: Add the difference to the subtrahend. The sum should be the minuend.

$$\begin{array}{r} 483 \\ +259 \\ \hline 742 \end{array}$$

Now work margin exercise 5.

Example 6

Buying a Car

In pricing a new car, Jason found that he would have to pay a base price of $15,200 plus $1025 in taxes and $575 for license fees. If the bank loaned him $10,640, how much cash would Jason need to buy the car?

Solution

The solution involves both addition and subtraction. First, we add Jason's expenses and then we subtract the amount of the bank loan.

Expenses		**Cash Needed**	
$15,200	base price	$ 16,800	total expenses
1025	taxes	− 10,640	bank loan
+ 575	license fees	$ 6 160	cash
$16,800	total expenses		

Jason would need $6160 in cash to buy the car.

Now work margin exercise 6.

6. In pricing a new motorcycle, Alex found that he would have to pay a base price of $10,470 plus $630 in taxes and $70 for the license fee. If the bank loaned him $7530, how much cash would Alex need to buy the motorcycle?

Objective E   Estimating Sums and Differences with Whole Numbers

One use for rounded numbers is to **estimate** an answer (or to find an **approximate** answer) before any calculations are made with the given numbers. You may find that you do this mentally in everyday calculations. In this manner, major errors can be spotted at a glance (possibly a wrong key is pushed on a calculator). Usually, when an error is suspected, we simply repeat the calculations and find the error.

To estimate an answer means to use rounded numbers in a calculation to form an idea of what the size of the actual answer should be. In some situations an estimated answer may be sufficient. For example, a shopper may simply estimate the total cost of purchases to be sure that he or she has enough cash to cover the cost.

To Estimate an Answer

1. Round each number to the place of the **leftmost digit**.
2. Perform the indicated operation with these rounded numbers.

notes

Special Note About Estimating Answers

There are several methods for estimating answers. Methods that give reasonable estimates for addition and subtraction may not be as appropriate for multiplication and division. You may need to use your judgment, particularly about the desired accuracy. Discuss different ideas with your instructor and fellow students. You and your instructor may choose to follow a completely different set of rules for estimating answers than those discussed here. Estimating techniques are "flexible," and we will make a slight adjustment to our leftmost digit rule with division. In any case, remember that an estimate is just that, an estimate.

7. Estimate the sum and then find the sum.

```
  156
   83
+  75
```

Example 7

Estimating a Sum

Estimate the sum; then find the sum.

```
   68
  925
+ 487
```

Solution

Note that in this example numbers are rounded to different places because they are of different sizes. That is, the leftmost digit is not in the same place for all numbers. Another method might be to round each number to the tens place (or hundreds place). This other method might give a more accurate estimate. However, the method chosen here is relatively fast and easy to follow and will be used in the discussions throughout the text.

a. Estimate the sum by first rounding each number to the place of the leftmost digit and then adding. In actual practice, many of these steps can be done mentally.

68	$\longrightarrow$	70	rounded value of 68
925	$\longrightarrow$	900	rounded value of 925
+ 487	$\longrightarrow$	+ 500	rounded value of 487
		1470	estimated sum

b. Now we find the sum, knowing that the answer should be close to 1470.

```
  1 2
   68
  925
+ 487
 ----
 1480
```
This sum is very close to 1470.

Now work margin exercise 7.

Example 8

Estimating a Difference

Estimate the difference, then find the difference.

$$
\begin{array}{r}
2783 \\
- 975 \\
\hline
\end{array}
$$

Solution

a. Round each number to the place of the leftmost digit and subtract using these rounded numbers.

2783	→	3000	rounded value of 2783
− 975	→	− 1000	rounded value of 975
		2000	estimated difference

b. Now we find the difference, knowing that the difference should be close to 2000.

$$
\begin{array}{r}
{}^{1}2\,{}^{1}7\,{}^{7}\!\!\not{8}\,{}^{1}3 \\
- 9\,7\,5 \\
\hline
1\,8\,0\,8
\end{array}
$$

The difference is close to 2000.

Now work margin exercise 8.

Caution

The use of rounded numbers to approximate answers demands some understanding of numbers in general and some judgment as to how "close" an estimate can be to be acceptable. In particular, when all numbers are rounded up or all numbers are rounded down, the estimate might not be "close enough" to detect a large error. Still, the process is worthwhile and can give "quick" and useful estimates in many cases.

8. Estimate the difference and then find the difference.

$$
\begin{array}{r}
2685 \\
- 847 \\
\hline
\end{array}
$$

notes

■ **A Special Note on Calculators**

■ The use of calculators is discussed throughout this text. However, you should be aware that a **calculator cannot take the place of**
■ **understanding mathematical concepts**. A calculator is a useful aid for speed and accuracy, but it is limited to doing only what you command
■ it to do. A calculator does not think!

Operations with whole numbers (addition and subtraction in this section) can be accomplished accurately with hand-held calculators, as long as none of the numbers involved (including answers) has more digits than the screen on the calculator can display (eight to ten digits on most calculators). Calculators are particularly helpful when there are several numbers in a problem and/or the numbers are very large. There are many scientific calculators and graphing calculators on the market and most of these can be used for the problems in this course. However, in this text, we have chosen to use a scientific calculator in the examples and illustrations in earlier chapters, and then a TI-84 plus graphing calculator in later chapters for more advanced operations.

The sequence of steps used here may not be exactly the same for your calculator. In that case, you may need to visit with your instructor and study the manual that comes with your calculator. Example 9 illustrates addition and Example 10 illustrates subtraction with a scientific calculator.

Example 9

Addition with a Calculator

Find the following sum with the use of a calculator.

$$\begin{array}{r} 899 \\ 743 \\ 625 \\ +592 \\ \hline \end{array}$$

Solution

To add numbers on a calculator, you will need to find the addition key, $+$, and the equal sign key, $=$.

To add the values, press the keys

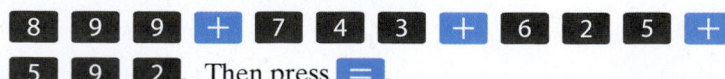

$8\ 9\ 9\ +\ 7\ 4\ 3\ +\ 6\ 2\ 5\ +$
$5\ 9\ 2$. Then press $=$.

The display will show **2859** as the answer.

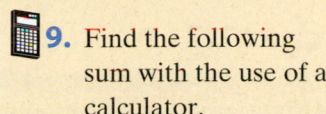

Example 10

Subtraction with a Calculator

Use your calculator to find the following difference.

$$678{,}025$$
$$-\ 483{,}975$$

Solution

To subtract numbers on a calculator, you will need to find the subtraction key, [−], and the equal sign key, [=].

To subtract the values, press the keys

Then press [=].

The display will show **194050** as the answer.

Now work margin exercises 9 and 10.

We need to be careful about relying on calculators. One frequent problem with using a calculator is that numbers might be entered incorrectly. Thus, the use of estimation, even just a mental estimation, can be very helpful in verifying answers.

9. Find the following sum with the use of a calculator.

$$842$$
$$578$$
$$789$$
$$+\ 154$$

10. Use your calculator to find the following difference.

$$58{,}240$$
$$-\ 48{,}563$$

Exercises 1.2

1. The number _____ is called the additive identity.

2. Give two examples of the commutative property of addition.

3. Give two examples of the associative property of addition.

Name the property of addition that is illustrated in each exercise. See Example 2.

4. $20 + 0 = 20$

5. $8 + 74 = 74 + 8$

6. $13 + 42 = 42 + 13$

7. $(3 + 7) + 11 = 3 + (7 + 11)$

8. $15 + (2 + 16) = (15 + 2) + 16$

Find the following sums and differences. Do not use a calculator. See Examples 1 and 5.

9.
```
   76
   18
 + 56
```

10.
```
  164
  235
+ 394
```

11.
```
  875
  756
  206
+ 290
```

12.
```
  1452
  1468
   237
+  702
```

13.
```
  700
− 104
```

14.
```
  5070
− 4375
```

15.
```
  7,045,213
− 2,743,521
```

16.
```
   521
   895
  1456
+ 5035
```

17.
```
  530
  800
  465
  324
+  90
```

18.
```
  15,643
   3 125
   1 416
   3 836
+ 15,070
```

19.
```
  58,853
 180,099
+ 62,894
```

20.
```
  92,003
  70,022
 163,987
+ 95,634
```

21.
```
  830,900
−  74,985
```

22.
```
  6,000,000
− 1,475,475
```

23. $395 + 1065 + 72 + 3144$

24. $186 + 92 + 1077 + 85 + 17$

25. $732,082 - 416,083$

26. $17,055 - 13,264$

First estimate the answers by using rounded numbers; then find the actual sum or difference. Do not use a calculator. See Examples 7 and 8.

27.
$$\begin{array}{r} 85 \\ 64 \\ +75 \\ \hline \end{array}$$

28.
$$\begin{array}{r} 98 \\ 56 \\ +61 \\ \hline \end{array}$$

29.
$$\begin{array}{r} 851 \\ 763 \\ 572 \\ +95 \\ \hline \end{array}$$

30.
$$\begin{array}{r} 6521 \\ 5742 \\ 3215 \\ +1020 \\ \hline \end{array}$$

31.
$$\begin{array}{r} 8652 \\ -1076 \\ \hline \end{array}$$

32.
$$\begin{array}{r} 63,504 \\ -52,200 \\ \hline \end{array}$$

33.
$$\begin{array}{r} 10,435 \\ -8748 \\ \hline \end{array}$$

34.
$$\begin{array}{r} 74,503 \\ -33,086 \\ \hline \end{array}$$

35. Mileage Driven: Mr. Swanson kept the mileage records indicated in the table shown here. How many miles did he drive during the 6 months?

Month	Mileage
January	456
February	398
March	340
April	459
May	504
June	485

36. College Expenses: During 4 years of college, June estimated her total yearly expenses for tuition, books, fees, and housing as $13,540, $15,200, $14,357, and $15,430 for each year. What were her total estimated expenses for 4 years of schooling?

37. Checking Account: If you had $980 in your checking account and you wrote a check for $358 and made a deposit of $225, what would be the balance in the account?

38. Car Pricing: In pricing a four-door car, Pat found she would have to pay a base price of $30,500 plus $2135 in taxes and $1006 for license fees. For a two-door sedan car of the same make, she would pay a base price of $25,000 plus $1750 in taxes and $825 for license fees. Including all expenses, how much cheaper was the two door model?

39. Sylvia's TV: The cost of repairing Sylvia's TV set would be $350 for parts (including tax) plus $105 for labor. To buy a new set, she would pay $670 plus $40 in sales tax, and a friend of hers would pay her $90 for her old set. How much more would Sylvia have to pay to buy a new set than to have her old set repaired?

> **Remember that the perimeter of a geometric figure can be found by adding the lengths of the sides of the figure. Find the perimeter of each of the following geometric figures with the given dimensions. See Example 4.**

40. a square (all four sides have the same length)

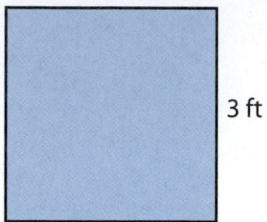

3 ft

41. a rectangle (opposite sides have the same length)

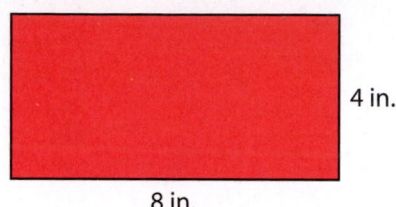

4 in.

8 in.

42. a triangle (a three sided figure)

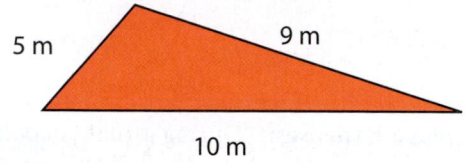

5 m   9 m

10 m

43. a triangle (a three sided figure)

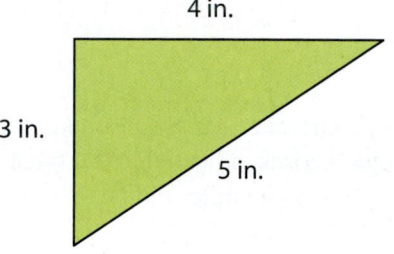

4 in.

3 in.

5 in.

44. a regular hexagon (all six sides have the same length)

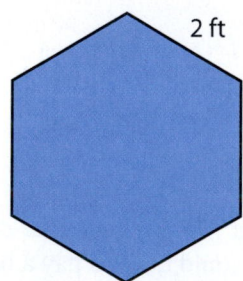

2 ft

45. a trapezoid (two sides are parallel)

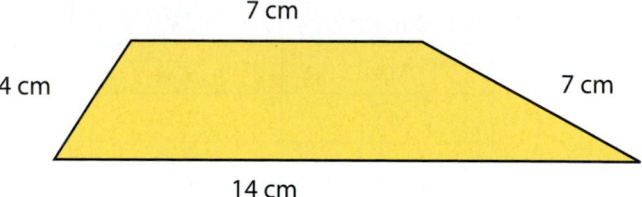

7 cm

4 cm   7 cm

14 cm

46. Find the perimeter of a parking lot that is in the shape of a rectangle 50 yards wide and 75 yards long.

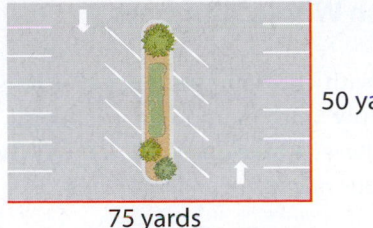

50 yards

75 yards

47. A window is in the shape of a triangle placed on top of a square. The length of each of two equal sides of the triangle is 24 inches and the third side is 36 inches long. The length of each side of the square is 36 inches long. Find the perimeter of the window.

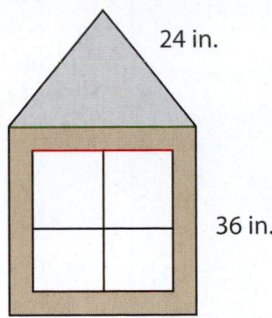

24 in.

36 in.

48. Draw what you think is the shape of a **regular octagon**. Where have you seen this shape?

🖩 **Use your calculator to perform the indicated operations. See Examples 9 and 10.**

49. $635 + 984 + 235$

50. $98,765 + 25,436 + 205$

51. $750,438 - 72,895$

52. $1,095,005 - 567,890$

53. $13,090,872 - 10,145,288$

54. $1,890,147,833 + 5,970,285,020$

55. $195 + 460 + 578 + 1052 + 380 + 71$

Objectives

A Be able to multiply and understand the following terms: factor and product.

B Learn the properties of multiplication.

C Understand the concept of area.

D Be able to divide and understand the following terms: divisor, dividend, quotient, and remainder.

E Learn how to estimate products and quotients with rounded numbers.

F Learn how to multiply and divide with a calculator.

Objective A   **Multiplication with Whole Numbers**

The process of repeated addition with the same number can be shortened considerably by learning to multiply and memorizing the basic facts of multiplication with single-digit numbers. For example, if you buy five 6-packs of soda, to determine the total number of cans of soda purchased, you can add five 6's:

$$6 + 6 + 6 + 6 + 6 = 30.$$

Or you can multiply 5 times 6 and get the same result:

$$5 \cdot 6 = 30.$$

The raised dot indicates multiplication.

In either case, you know that you have 30 cans of soda.

Multiplication is generally much easier than repeated addition, particularly if the numbers are large. The result of multiplication is called a **product**, and the two numbers being multiplied are called **factors** of the product. In the following example, the repeated addend (175) and the number of times it is used (4) are both **factors** of the **product** 700.

$$175 + 175 + 175 + 175 = 4 \cdot 175 = 700$$

factor factor   product

Several notations can be used to indicate multiplication. In this text, to avoid possible confusion between the letter x (used as a variable) and the $\times$ sign, we will use the raised dot and parentheses much of the time.

Symbols for Multiplication

Symbol		Example
$\cdot$	raised dot	$4 \cdot 175$
$(\ )$	numbers inside or next to parentheses	$4(175)$ or $(4)175$ or $(4)(175)$
$\times$	cross sign	4×175 or $\begin{array}{r} 175 \\ \times\ 4 \\ \hline \end{array}$

As with addition, the operation of multiplication has several properties. Multiplication is **commutative** and **associative**, and the number **1** is the **multiplicative identity**. Also, multiplication by 0 always gives a product of 0, and this fact is called the **zero-factor law**.

Commutative Property of Multiplication

For any two whole numbers *a* and *b*, $a \cdot b = b \cdot a$.

For example, $3 \cdot 4 = 4 \cdot 3$.

(The **order** of multiplication can be reversed.)

Associative Property of Multiplication

For any whole numbers *a*, *b*, and *c*, $(a \cdot b) \cdot c = a \cdot (b \cdot c)$.

For example, $(7 \cdot 2) \cdot 5 = 7 \cdot (2 \cdot 5)$.

(The **grouping** of the numbers can be changed.)

Multiplicative Identity Property

For any whole number *a*, $a \cdot 1 = a$.

For example, $6 \cdot 1 = 6$.

(The product of any number and 1 is that same number.)

The number 1 is called the **multiplicative identity**.

Zero Factor Law

For any whole number *a*, $a \cdot 0 = 0$.

For example, $63 \cdot 0 = 0$.

(The product of a number and 0 is always 0.)

1. Find the product and identify which property of multiplication is being illustrated.

   a. $8 \times 12 = 12 \times 8$

   b. $0 \cdot 57$

   c. $35 \cdot 1$

   d. $7 \cdot (3 \cdot 2) = (7 \cdot 3) \cdot 2$

2. Find the value of the variable that will make each statement true and name the property of multiplication that is used.

   a. $25 \cdot 1 = x$

   b. $14 \cdot 5 = 5 \cdot n$

   c. $(9 \cdot 2) \cdot y = 9 \cdot (2 \cdot 1)$

   d. $t \cdot 17 = 0$

Example 1

Properties of Multiplication

Each of the properties of multiplication is illustrated.

a. $5 \times 6 = 6 \times 5$ commutative property of multiplication

As a check, we see that $5 \times 6 = 30$ and $6 \times 5 = 30$.

b. $2 \cdot (5 \cdot 9) = (2 \cdot 5) \cdot 9$ associative property of multiplication

As a check, we see that $2 \cdot (45) = 90$ and $(10) \cdot 9 = 90$.

c. $8 \cdot 1 = 8$ multiplicative identity property

d. $196 \cdot 0 = 0$ zero-factor law

Completion Example 2

Properties of Multiplication

Use your knowledge of the properties of multiplication to find the value of the variable that will make each statement true. State the name of the property used.

	Value	Property
a. $24 \cdot n = 24$	$n =$ _____	_____
b. $3 \cdot x = 5 \cdot 3$	$x =$ _____	_____
c. $(7 \cdot y) \cdot 2 = 7 \cdot (5 \cdot 2)$	$y =$ _____	_____
d. $82 \cdot t = 0$	$t =$ _____	_____

Now work margin exercises 1 and 2.

Completion Example Answers

2. **a.** $n = 1$; multiplicative identity property
   b. $x = 5$; commutative property of multiplication
   c. $y = 5$; associative property of multiplication
   d. $t = 0$; zero-factor law

Example 3

Properties of Multiplication

The following products illustrate how the commutative and associative properties of multiplication are related to the method of multiplication of numbers that end with 0's.

a. $7 \cdot 90 = 7(9 \cdot 10)$
   $= (7 \cdot 9)10$
   $= 63 \cdot 10$
   $= 630$

b. $6 \cdot 800 = 6(8 \cdot 100)$
   $= (6 \cdot 8) \cdot 100$
   $= 48 \cdot 100$
   $= 4800$

c. $50 \cdot 700 = (5 \cdot 10)(7 \cdot 100)$
   $= (5 \cdot 7)(10 \cdot 100)$
   $= 35 \cdot 1000$
   $= 35,000$

d. $200 \cdot 4000 = (2 \cdot 100)(4 \cdot 1000)$
   $= (2 \cdot 4)(100 \cdot 1000)$
   $= 8 \cdot 100,000$
   $= 800,000$

Now work margin exercise 3.

To understand the technique for multiplying two whole numbers, we use expanded notation, the method of multiplication by powers of 10, and the following property, called the **distributive property of multiplication over addition** (or sometimes simply the **distributive property**).

Distributive Property of Multiplication Over Addition

If a, b, and c are whole numbers, then

$$a(b+c) = a \cdot b + a \cdot c.$$

3. Find the following products.

a. $7 \cdot 9000$

b. $500 \cdot 300$

c. $80 \cdot 2000$

d. $900 \cdot 5000$

For example, to multiply $3(50+2)$, we can add first and then we can multiply.

$$3(50+2) = 3(52) = 156$$

But we can also multiply first and then add, in the following manner.

$$3(50+2) = 3 \cdot 50 + 3 \cdot 2 \qquad \text{This step is called the distributive property.}$$
$$= 150 + 6 \qquad\qquad \text{150 and 6 are called partial products.}$$
$$= 156$$

Or, vertically,

$$52 \;=\; 50 + 2$$
$$\underline{\times\ \ 3} \;=\; \underline{\times \qquad 3}$$
$$150 + 6 \;=\; 156 \qquad \text{Use the distributive property to multiply } 3 \cdot 2 \text{ and } 3 \cdot 50.$$

partial products   product

Multiplication is explained step by step in Example 4 and then shown in the standard form in Example 5.

Example 4

Multiplication with Whole Numbers

The steps in multiplication are shown in finding the product $37 \cdot 27$.

Step 1:

$$\begin{array}{r} 4 \\ 37 \\ \times\ 27 \\ \hline 9 \end{array}$$

4 carried from 49.

Multiply: $7 \cdot 7 = 49$.

Step 2:

$$\begin{array}{r} 4 \\ 37 \\ \times\ 27 \\ \hline 259 \end{array}$$

Now multiply: $7 \cdot 3 = 21$.
And add the 4: $4 + 21 = 25$.

Step 3:

$$\begin{array}{r} 1 \\ 37 \\ \times\ 27 \\ \hline 259 \\ 4 \end{array}$$

1 carried from the 14.

Next, multiply: $2 \cdot 7 = 14$.

Write the 4 in the tens column because you are actually multiplying $20 \cdot 7 = 140$. Generally, the 0 is not written.

Step 4:

$$\overset{1}{37}$$
$$\times\ 27$$
$$\overline{259}$$
$$74$$

Now multiply: 2 · 3 = 6.
Then add 1: 6 + 1 = 7.

Step 5:

$$37$$
$$\times\ 27$$
$$\overline{259}$$
$$74$$
$$\overline{999}$$

Add to find the final product.

Example 5

Multiplication with Whole Numbers

The standard form of multiplication is used here to find the product 93 · 46.

$$\overset{1\ 1}{93}$$
$$\times\ 46$$
$$\overline{558}$$ ⬅   6 · 3 = 18. Write 8, carry 1. 6 · 9 = 54. Add 1: 54 + 1 = 55.
$$\underline{372}$$ ⬅   4 · 3 = 12. Write 2, carry 1. 4 · 9 = 36. Add 1: 36 + 1 = 37.
$$4278$$ ⬅   product

Now work margin exercises 4 and 5.

Objective C   **The Concept of Area**

Area is the measure of the interior, or enclosed region, of a plane surface and is measured in **square units**. The concept of area is illustrated in Figure 1 in terms of square inches.

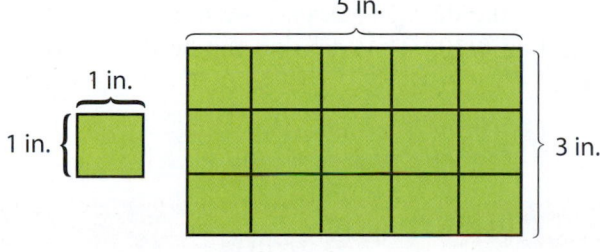

Figure 1

The area is 15 square inches.

---

4. Use the step-by-step method shown in Example 4 to find the product.

$$26$$
$$\times\ 34$$

5. Use the standard form of multiplication, as shown in Example 5, to find the product 15 · 32.

Some of the units of area in the metric system are square meters, square decimeters, square centimeters, and square millimeters. In the U.S. customary system, some of the units of area are square feet, square inches, and square yards.

6. Find the area of an elementary school playground that is in the shape of a rectangle 52 yards wide and 73 yards long.

Example 6

Finding Area

The area of a rectangle (measured in square units) is found by multiplying its length by its width. Find the area of a rectangular plot of land with dimensions as shown here.

186 ft

92 ft

Solution

To find the area, we multiply $186 \cdot 92$.

$$
\begin{array}{r}
186 \\
\times\ \ 92 \\
\hline
372 \\
1674\ \ \\
\hline
17,112\ \ \text{square feet}
\end{array}
$$

Now work margin exercise 6.

Objective D **Division with Whole Numbers**

We know that $8 \cdot 10 = 80$ and that 8 and 10 are **factors** of 80. They are also called **divisors** of 80. In division, we want to find how many times one number is contained in another. How many 8's are in 80? There are 10 eights in 80, and we say that 80 **divided** by 8 is 10 (or $80 \div 8 = 10$). In this case we know the results of the division because we are already familiar with the related multiplication, $8 \cdot 10$. Thus division can be thought of as the reverse of multiplication.

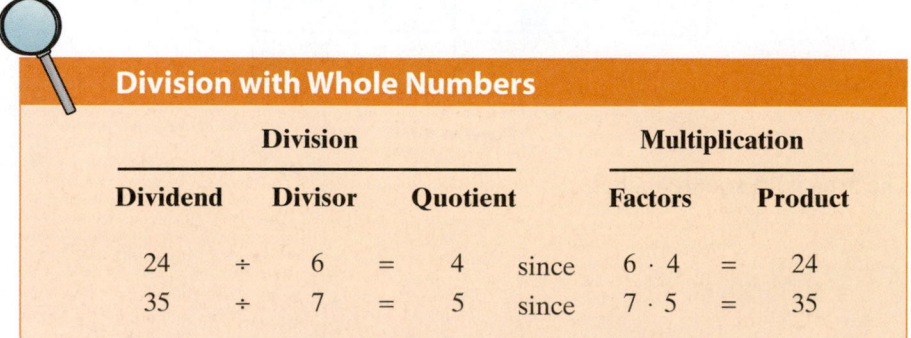

Division with Whole Numbers					
Division				**Multiplication**	
Dividend	Divisor	Quotient		Factors	Product
24	÷ 6	= 4	since	6 · 4	= 24
35	÷ 7	= 5	since	7 · 5	= 35

As the table indicates, the number being divided is called the **dividend**; the number dividing the dividend is called the **divisor**, and the result of division is called the **quotient**.

Division does not always involve factors (or exact divisors), and we need a more general idea and method for performing division. The more general method is a process known as the **division algorithm*** (or the method of **long division**) and is illustrated in Example 7. The **remainder** must be less than the divisor.

Note: If the remainder is 0, then both the divisor and quotient are **factors** (or **divisors**) of the dividend. See Example 8.

7. Find the quotient and remainder: $7982 \div 31$.

Example 7

Division with Whole Numbers

Find the quotient and remainder for $9385 \div 45$.

Solution

Step 1:

$$
\begin{array}{r}
2 \\
45\overline{)9385} \\
90 \\
\hline
3
\end{array}
$$

By trial, divide 40 into 90 or 4 into 9.

Write the result, 2, in the hundreds position of the quotient.

Multiply this result by the divisor and subtract.

Step 2:

$$
\begin{array}{r}
20 \\
45\overline{)9385} \\
90 \\
\hline
38 \\
0 \\
\hline
385
\end{array}
$$

45 will not divide into 38. So, write 0 in the tens column and multiply $0 \cdot 45 = 0$.

Do not forget to write the 0 in the quotient.

Step 3:

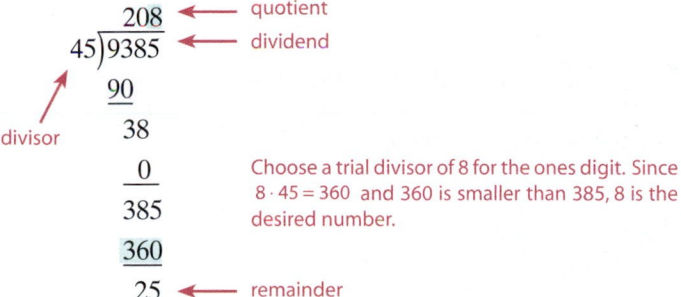

quotient ← 208

dividend ← 9385

divisor

$$
\begin{array}{r}
208 \\
45\overline{)9385} \\
90 \\
\hline
38 \\
0 \\
\hline
385 \\
360 \\
\hline
25
\end{array}
$$

remainder ← 25

Choose a trial divisor of 8 for the ones digit. Since $8 \cdot 45 = 360$ and 360 is smaller than 385, 8 is the desired number.

Check: Multiply the quotient by the divisor and then add the remainder. This sum should equal the dividend.

$$
\begin{array}{r}
208 \\
\times 45 \\
\hline
1040 \\
832 \\
\hline
9360
\end{array}
\qquad
\begin{array}{r}
9360 \\
+ 25 \\
\hline
9385
\end{array}
$$

208 ← quotient

× 45 ← divisor

+ 25 ← remainder

9385 ← dividend

Now work margin exercise 7.

*An algorithm is a process or pattern of steps to be followed in solving a problem or, sometimes, only a certain part of a problem.

If the remainder is 0 in a division problem, then both the divisor and the quotient are **factors** (or **divisors**) of the dividend. We say that both factors **divide exactly** into the dividend. As a check, note that the product of the divisor and quotient will be the dividend.

8. Use long division to show that both 26 and 68 are factors (or divisors) of 1768.

Example 8

Factors

Show that 19 and 43 are **factors** (or **divisors**) of 817.

Solution

In order to show that 19 and 43 are factors of 817, we divide. Both $817 \div 19$ and $817 \div 43$ will give a remainder of 0 if indeed these numbers are factors of 817. We only need to try one of them to find out if the remainder is 0.

$$
\begin{array}{r}
43 \\
19{\overline{\smash{)}817}} \\
\underline{76} \\
57 \\
\underline{57} \\
0
\end{array}
$$

Check:

$$
\begin{array}{r}
43 \\
\times\ 19 \\
\hline
387 \\
\underline{43} \\
817
\end{array}
$$

Thus 19 and 43 are indeed factors of 817.

Now work margin exercise 8.

Example 9

Plumbing

A plumber purchased 32 pipe fittings. What was the price of one fitting if the bill was $512 before taxes?

Solution

We need to know how many times 32 is contained in 512.

$$\begin{array}{r} 16 \\ 32\overline{)512} \\ \underline{32} \\ 192 \\ \underline{192} \\ 0 \end{array}$$

The price for one fitting was $16.

Now work margin exercise 9.

9. A painter bought 26 gallons of paint for $390. What was the price per gallon of paint before taxes?

For completeness, we close this section with two rules about division involving 0. These rules will be discussed here and again in a later chapter.

> ### Division with 0
>
> **Case 1:** If a is any nonzero whole number, then $0 \div a = 0$. (In fraction form, $\dfrac{0}{a} = 0$.)
>
> **Case 2:** If a is any whole number, then $a \div 0$ is **undefined**. (In fraction form, $\dfrac{a}{0}$ is undefined.)

The following discussion of these two cases is based on the fact that division is defined in terms of multiplication. For example,

$$14 \div 7 = 2 \quad \text{because} \quad 14 = 2 \cdot 7.$$

Since fractions can be used to indicate division, we can also write:

$$\frac{14}{7} = 2 \quad \text{because} \quad 14 = 2 \cdot 7.$$

(**Note:** Fractions are discussed in detail in a later chapter.)

Reasoning in the same manner, we can divide 0 by a nonzero number:

$$0 \div 6 = 0 \quad \text{because} \quad 0 = 6 \cdot 0.$$

Again, in fraction form:

$$\frac{0}{6} = 0 \quad \text{because} \quad 0 = 6 \cdot 0.$$

However, if $6 \div 0 = x$, then $6 = 0 \cdot x$. But this is not possible because $0 \cdot x = 0$ for any number x. Thus, $6 \div 0$ is **undefined**. That is, there is no number that can be multiplied by 0 to give a product of 6.

In fraction form: $\dfrac{6}{0}$ is **undefined**.

For the problem $0 \div 0 = x$, we have $0 = 0 \cdot x$, which is a true statement for any value of x. Since there is not a unique value for x, we agree that $0 \div 0$ is **undefined** also.

In fraction form: $\dfrac{0}{0}$ is **undefined.**

That is, **division by 0 is not possible**.

Estimating Products and Quotients

Products and quotients can be estimated, just as sums and differences can be estimated. We will use the "rule" of rounding to the left most digit of each number before performing an operation as stated in Section 1.2. Example 10 illustrates the estimating procedure with multiplication.

Example 10

Estimating a Product

Find the product $63 \cdot 49$, but first estimate the product.

Solution

a. Estimate the product.

63	⟶	60	rounded value of 63
× 49	⟶	× 50	rounded value of 49
		3000	estimated product

b. Notice that, in rounding, 63 was rounded down and 49 was rounded up. Thus, we can expect the actual product to be reasonably close to 3000.

$$
\begin{array}{r}
63 \\
\times\ 49 \\
\hline
567 \\
252 \\
\hline
3087
\end{array}
$$

The actual product is close to 3000.

Adjustment to the Procedure of Estimating Answers

As was stated earlier in Section 1.2, the "rule" of using numbers rounded to the left most digit is flexible. There are times when adjusting to using two digits instead of one gives simpler calculations and more accurate estimates. This adjustment is generally more applicable with division than with addition, subtraction, or multiplication. In general, estimating answers involves basic understanding of numbers and intuitive judgment, and there is no one best way to estimate answers.

Example 11 illustrates how the adjustment to the rule makes sense with division.

Example 11

Estimating a Quotient

a. Estimate the quotient $148,062 \div 26$.

Solution

$$148,062 \div 26 \longrightarrow 150,000 \div 30$$

In this case, we rounded using the two left most digits for 148,062 because 150,000 can be divided evenly by 30.

```
      5 000   ←— approximate quotient
30)150,000
   150
     00
     00
     ___
      00
      00
      ___
       00
       00
       ___
        0    ←— remainder
```

b. Find the quotient $148,062 \div 26$.

Solution

The quotient should be near 5000.

```
      5 694   ←— actual quotient
26)148,062
   130
    18 0
    15 6
    ____
     2 46
     2 34
     ____
      122
      104
      ___
       18   ←— remainder
```

Now work margin exercises 10 and 11.

10. Estimate the following product, then find the product.

$41 \cdot 72$

11. Estimate the following quotient, then find the quotient.

$39,024 \div 48$

12. Estimate the following quotient, then find the quotient. What is the difference between the estimate and the actual quotient?

$$16\overline{)344}$$

Completion Example 12

Completing Division

Find the quotient and remainder.

$$
\begin{array}{r}
32 \\
21\overline{)6857} \\
63 \\
\hline
55 \\
\underline{} \\
\underline{} \\
\underline{} \\
\ \text{remainder}
\end{array}
$$

Now estimate the quotient by mentally dividing rounded numbers: _____.

Is your estimate close to the actual quotient? _____

What is the difference? _____

Now work margin exercise **12.**

Objective F **Multiplication and Division with Calculators**

Note that the key for multiplication on the calculator is marked with the $\times$ sign but a star (*) appears on the display. Similarly, for division, the key for division is marked with the $\div$ sign but a slash (/) appears on the display.

Completion Example Answers

12.

$$
\begin{array}{r}
326 \\
21\overline{)6857} \\
63 \\
\hline
55 \\
\underline{42} \\
137 \\
\underline{126} \\
11\ \ \text{remainder}
\end{array}
\qquad
\begin{array}{r}
350\ \ \text{estimate} \\
20\overline{)7000}
\end{array}
$$

Yes, the estimate is close.
The difference is $350 - 326 = 24$.

Example 13

 **Multiplication with a Calculator**

Use your calculator to find the product: $\begin{array}{r} 572 \\ \times\, 635 \end{array}$.

Solution

To multiply numbers on a calculator, you will need to find the multiplication key, $\boxed{\times}$, and the equal sign key, $\boxed{=}$.

To multiply the values, press the keys

 .
Then press $\boxed{=}$.

The display will read **363220**.

Example 14

 Division with a Calculator

Use your calculator to find the quotient: $3695 \div 48$.

Solution

To divide the numbers on a calculator, you will need to find the division key, $\boxed{\div}$, and the equal sign key, $\boxed{=}$.

To divide the values, press the keys

 .
Then press $\boxed{=}$.

The display will read **76.97916667**.

If the quotient is not a whole number (as is the case here), the calculator gives rounded decimal number answers. (These decimal numbers will be discussed in Chapter 5.) Thus calculators do have some limitations in terms of whole numbers. The whole number part of the display (76) is the correct whole number part of the quotient, but we do not know the remainder R.
Thus $3695 \div 48 = 76$ with an undetermined remainder.

Now work margin exercises 13 and 14.

Can the whole number remainder R be found when dividing with a calculator that does not have special keys for fractions and remainders? The answer is "Yes," and we can proceed as the following discussion indicates.

To check long division with whole numbers, we multiply the divisor and quotient and then add the remainder. This product should be the dividend. Now consider the problem of finding the whole number remainder R when the division gives a decimal number as in Example 14.

 13. Use your calculator to find the product.

$\begin{array}{r} 371 \\ \times\, 582 \end{array}$

14. Use your calculator to find the quotient.

$4387 \div 29$

If we multiply the whole number in the quotient (76) times the divisor (48), we get the following.

$$
\begin{array}{rl}
48 & \text{divisor} \\
\times\ 76 & \text{whole number part of quotient} \\
\hline
288 & \\
336 & \\
\hline
3648 & \text{product}
\end{array}
$$

This product is not equal to the dividend (3695) because we have not added on the remainder, since we do not know what it is. However, thinking in reverse, we can find the remainder by subtracting this product from the dividend. Thus the remainder is found as follows.

$$
\begin{array}{rl}
3695 & \text{dividend} \\
-\ 3648 & \text{product} \\
\hline
47 & \text{whole number remainder}
\end{array}
$$

Example 15

 Finding the Quotient and Remainder

Use a calculator to divide $1857 \div 17$, and then find the quotient and remainder as whole numbers.

Solution

With a calculator, $1857 \div 17 = 109.2352941$. Multiply the whole number part of the quotient (109) by the divisor (17):

$$109 \cdot 17 = 1853.$$

To find the remainder, subtract this product from the dividend:

$$1857 - 1853 = 4.$$

Thus the quotient is 109 and the remainder is 4.

Example 16

Acres of Land

One acre of land is equal to 43,560 square feet. How many acres are in a piece of land in the shape of a rectangle 8712 feet by 2000 feet?

Solution

Step 1: Find the area in square feet.

$8712 \cdot 2000 = 17{,}424{,}000$ sq ft

Step 2: Find the number of acres by division.

$17{,}424{,}000 \div 43{,}560 = 400$ acres

Thus the land has a measure of 400 acres.

Now work margin exercises 15 and 16.

15. Use a calculator to find the quotient and remainder as whole numbers.

$247{,}512 \div 25$

16. How many acres are in a piece of land in the shape of a rectangle 3267 feet by 4000 feet?

Exercises 1.3

1. Factor

Find the value of each variable and state the name of the property of multiplication that is illustrated. See Examples 1 and 2.

2. $(3 \cdot 2) \cdot 7 = 3 \cdot (x \cdot 7)$

3. $(5 \cdot 10) \cdot y = 5 \cdot (10 \cdot 7)$

4. $25 \cdot a = 0$

5. $34 \cdot n = 34$

6. $0 \cdot 51 = n$

7. $8 \cdot 10 = 10 \cdot x$

Use the distributive property to find each of the following products. Show each step.

8. $5(2+6)$

9. $4(7+9)$

10. $2(3+5)$

11. $7(3+4)$

12. $6(4+8)$

Find the following products. See Examples 4 and 5.

13.
$$\begin{array}{r} 28 \\ \times\ 5 \\ \hline \end{array}$$

14.
$$\begin{array}{r} 72 \\ \times\ 7 \\ \hline \end{array}$$

15.
$$\begin{array}{r} 93 \\ \times\ 41 \\ \hline \end{array}$$

16.
$$\begin{array}{r} 67 \\ \times\ 52 \\ \hline \end{array}$$

17.
$$\begin{array}{r} 320 \\ \times\ 102 \\ \hline \end{array}$$

18.
$$\begin{array}{r} 450 \\ \times\ 203 \\ \hline \end{array}$$

Find the following quotients mentally.

19. $30 \div 5$

20. $48 \div 6$

21. $24 \div 3$

22. $63 \div 9$

23. $6 \div 6$

24. $7 \div 7$

25. $28 \div 7$

26. $72 \div 8$

27. $44 \div 4$

28. $0 \div 3$

29. $80 \div 10$

30. $11 \div 0$

31. $6\overline{)42}$

32. $5\overline{)20}$

33. $0\overline{)23}$

34. $13\overline{)0}$

35. $16\overline{)129}$

36. $20\overline{)305}$

37. $18\overline{)207}$

38. $68\overline{)207}$

39. $102\overline{)918}$

40. $213\overline{)4560}$

41. Show that 35 and 45 are both factors of 1575 by using long division.

42. Show that 22 and 32 are both factors of 704 by using long division.

43. Show that 56 and 39 are both factors of 2184 by using long division.

44. Show that 125 and 157 are both factors of 19,625 by using long division.

45. Round each number and estimate the product by performing the calculations mentally.

    a. $19 \cdot 19$

    b. $6(82)$

    c. $11(31)$

    d. $(9)(77)$

46. Round each number and estimate the product by performing the calculations mentally.

    a. $57 \cdot 500$

    b. $57(50)$

    c. $57 \cdot 5$

    d. $(57)(5000)$

47. Round the divisor and dividend as necessary and estimate the quotient by dividing mentally.

    a. $8\overline{)810}$

    b. $5\overline{)32}$

    c. $33\overline{)12,000}$

    d. $18\overline{)3800}$

48. Round the dividend and estimate the quotient by dividing mentally.

    a. $3\overline{)860}$

    b. $3\overline{)86,000}$

    c. $3\overline{)86}$

    d. $3\overline{)8600}$

49.   66
         × 4

50.   53
         × 6

51.   98
         × 12

52.   39
         × 15

53.   81
         × 36

54.   126
         × 102

55.   420
         × 204

56.   508
         × 119

Use your judgment to estimate each quotient; then find the quotient and the remainder. Understand that, depending on how the divisor and dividend are rounded, different estimates are possible. See Example 11.

57. 18)216

58. 13)260

59. 49)993

60. 50)3065

61. 96)3056

62. 103)4768

63. 54)83,276

64. 82)98,762

In each of the following word problems you must decide what operation (or operations) to perform. Do not be afraid of making a mistake. Problems like these help you learn to think mathematically. If an answer does not make sense to you, try another approach.

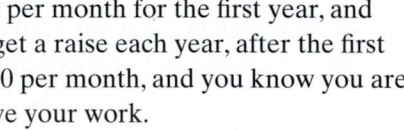

65. **New Job:** On your new job, your salary is to be $3200 per month for the first year, and you are to get a raise each year, after the first year, of $500 per month, and you know you are going to love your work.

   a. What will you earn during your first year?

   b. What total amount will you earn during the first two years on this job?

   c. During the first five years on the job?

66. **Rent Increase:** You rent an apartment with three bedrooms for $1250 per month, and you know that the rent will increase each year after the first year by $50 per month.

   a. What will you pay in rent for the first year?

   b. What total amount will you pay in rent during the first 3-year period?

   c. During the first 5-year period?

67. **Pool Size:** A rectangular pool measures 36 feet long by 18 feet wide. Find the area of the pool in square feet.

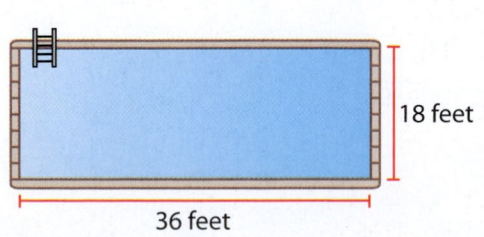

18 feet

36 feet

68. Land Size: If you know that the area of a rectangular plot of land is 39,000 square feet and one side measures 312 feet, what is the length of the other side of the lot? (Remember that area of a rectangle is found by multiplying length times width.)

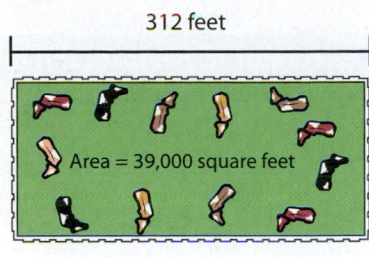

312 feet

Area = 39,000 square feet

69. Office Typists: Three typists in one office can type a total of 30 pages of data in 2 hours. How many pages can they type in a workweek of 40 hours?

70. Jelly Jars: A case of grape jelly contains 18 jars and each jar contains 14 ounces of jelly.
a. How many jars of jelly are there in 15 cases?
b. How many 8 ounce servings are contained in one case?

Solve the following problems as indicated.

71. A **triangle** is a plane figure with three sides. The **height** is the distance from the tip (called a **vertex**) to the opposite side (called a **base**). The area of a triangle can be found by multiplying the length of an altitude by the length of the base and dividing by 2. Find the area of the triangle with a base of 10 feet and an altitude of 6 feet.

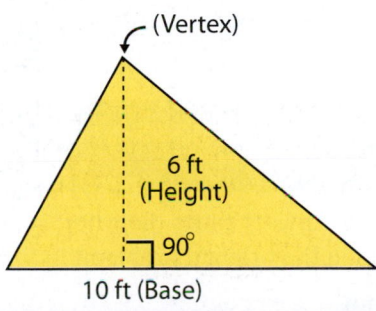

(Vertex)

6 ft (Height)

90°

10 ft (Base)

72. Refer to Exercise 71 for the method of finding the area of a triangle. In a **right triangle** (one in which two sides are perpendicular to each other), the two perpendicular sides can be treated as the base and height. Find the area of a right triangle in which the two perpendicular sides have lengths of 12 inches and 5 inches. (The longest side in a right triangle is called the **hypotenuse**.)

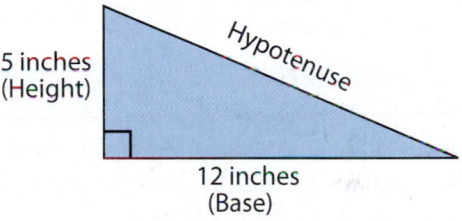

5 inches (Height)

Hypotenuse

12 inches (Base)

Solve the following word problems.

73. Wall Painting: A painting is mounted in a rectangular frame (16 inches by 24 inches) and hung on a wall. How many square inches of wall space will the framed painting cover?

74. Land Acres: Large amounts of land area are measured in **acres**, not usually in square feet. (Metric units of land area are **ares** and **hectares**.) One acre is equal to 43,560 square feet or 4840 square yards. See Example 16.

a. Find the number of acres in a piece of land in the shape of a rectangle 4356 feet by 1000 feet.

b. Find the number of acres in a piece of land in the shape of a right triangle with perpendicular sides of length 500 yards by 968 yards. (See Exercises 71 and 72.)

75. $7054(934)$ **76.** $2031 \cdot 745$ **77.** $28{,}080 \div 36$

78. $16{,}300 \div 25$ **79.** $2748 \cdot 350$ **80.** $5072 \cdot 3487$

81. $101{,}650 \div 107$ **82.** $5{,}430{,}642 \div 651$

83. Income: If your income for 1 year was $30,576, what was your monthly income?

84. Textbooks: A school bought 3045 new textbooks for a total cost of $79,170. What was the price of each book?

85. Building: A rectangular lot for a house measures 210 feet long by 175 feet wide. Find the area of the lot in square feet. (The area of a rectangle is found by multiplying length times width.)

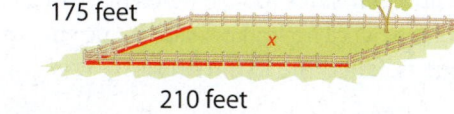

175 feet

210 feet

86. Driving: If you drive your car to work 205 days a year and the distance from your house to work is 13 miles, how many miles each year do you drive your car to and from work?

Remember: To find the whole number remainder, first multiply the divisor times the whole number part of the quotient, and then subtract this product from the dividend in the original problem.

87. $3568 \div 15$ **88.** $9563 \div 6$

89. $25{,}483 \div 35$ **90.** $36{,}245 \div 19$

91. $15{,}767 \div 21$ **92.** $93{,}442 \div 71$

93. $1{,}835{,}350 \div 133$ **94.** $2{,}795{,}632 \div 275$

1.4 Problem Solving with Whole Numbers

Objective A **Basic Strategy for Solving Word Problems**

In this section, the applications are varied and sometimes involve combinations of the four operations of addition, subtraction, multiplication, and division. You must read each problem carefully and decide, using your previous experiences, just what operations to perform. Learn to enjoy the challenge of word problems and be aware of the following precepts:

1. **At least try something.**
2. **If you do nothing, you learn nothing.**
3. **Even by making errors in technique or judgment, you are learning what does not work.**
4. **Do not be embarrassed by mistakes.**

The problems discussed in this section come under the following headings: number problems, consumer items, checking, geometry, and reading graphs. The steps in the basic strategy listed here will help give an organized approach regardless of the type of problem.

Basic Strategy for Solving Word Problems

1. Read each problem carefully until you understand the problem and know what is being asked. Discard any irrelevant information.
2. Draw any type of figure or diagram that might be helpful, and decide what operations are needed.
3. Perform these operations to find the unknown value(s).
4. Mentally check to see if your answer is reasonable and see if you can think of another more efficient or more interesting way to do the same problem.

Number problems usually contain key words or phrases that tell what operation or operations are to be performed with given numbers. Learn to look for these key words.

Key Words that Indicate Operations

Addition	Subtraction	Multiplication	Division
add	subtract (from)	multiply	divide
sum	difference	product	quotient
plus	minus	times	ratio
more than	less than	twice	
increased by	decreased by		

Example 1

Number Problems

Find the sum of the product of 78 and 32 and the difference between 7600 and 2200.

Solution

As an aid, we rewrite the problem and emphasize the key words by printing them in boldface. (You might want to do the same by underlining the key words on your paper.)

Find the **sum** of the **product** of 78 and 32 and the **difference** between 7600 and 2200.

After reading the problem carefully, note that the sum cannot be found until both the product and the difference are found.

Product	**Difference**

$$
\begin{array}{r}
78 \\
\times\ 32 \\
\hline
156 \\
234 \\
\hline
2496
\end{array}
\qquad
\begin{array}{r}
7600 \\
-\ 2200 \\
\hline
5400
\end{array}
$$

Now we find the sum.

Sum

$$
\begin{array}{r}
2496 \\
+\ 5400 \\
\hline
7896
\end{array}
$$

Thus the requested sum is 7896.

(As a quick mental check, use rounded numbers: $80 \cdot 30 = 2400$ and $8000 - 2000 = 6000$ and the sum $2400 + 6000 = 8400$ seems reasonably close to the result 7896.)

(**Note:** the word **and** does not indicate any operation. **And** is used twice in this problem as a grammatical connector.)

Example 2

Number Problems

If the quotient of 1265 and 5 is decreased by the sum of 120 and 37, what is the difference?

Solution

Again, we rewrite the problem and emphasize the key words by printing them in boldface.

If the **quotient** of 1265 and 5 is **decreased by** the **sum** of 120 and 37, what is the **difference**?

To find the difference, first find the quotient and sum:

Quotient	Sum	Difference
$\begin{array}{r} 253 \\ 5\overline{)1265} \\ \underline{10} \\ 26 \\ \underline{25} \\ 15 \\ \underline{15} \\ 0 \end{array}$	$\begin{array}{r} 120 \\ +\ 37 \\ \hline 157 \end{array}$	$\begin{array}{r} 253 \\ -157 \\ \hline 96 \end{array}$

The difference between the quotient and the sum is 96.

Now work margin exercises 1 and 2.

Example 3

Financing a Car

Mr. Fujimoto bought a used car for $15,000 for his son. The salesperson included $1050 for taxes and $480 for license fees. If he made a down payment of $3500 and financed the rest through his credit union, how much did he finance?

Solution

First, find the total cost by adding the expenses. Then subtract the down payment. (Note that the words "add" and "subtract" do not appear in the statement of the problem. Only real-life experience tells us that these are the operations to be performed.)

1. Find the difference between the sum of 279 and 150 and the product of 42 and 9.

2. If the quotient of 978 and 3 is increased by the difference between 693 and 500, what is the sum?

3. Mrs. Spencer bought a new car for her daughter for $27,550. The salesperson added $1600 for taxes and $475 for license fees. If she made a down payment of $4500 and financed the rest through her bank, how much did she finance?

4. In December, Kurt opened a checking account and deposited $2000 into it. He also made another deposit of $200. During that month, he wrote checks for $57, $120, and $525, and used his debit card for purchases totaling $630. What was his balance at the end of the month?

Add Expenses		Subtract Down Payment	
$15,000		$16,530	total expenses
1 050		− 3 500	down payment
+ 480		$13,030	to be financed
$16,530	total expenses		

Mr. Fujimoto financed $13,030.

Example 4

Balance of a Checking Account

In June, Ms. Maxwell opened a checking account and deposited $5280. During the month, she made another deposit of $800 and wrote checks for $135, $450, $775, and $72. What was the balance in her account at the end of the month?

Solution

To find the balance, we find the difference between the sum of the deposit amounts and the sum of the check amounts.

$ 5280
+ 800
$ 6080   sum of deposits

$ 135
450
775
+ 72
$ 1432   sum of checks

$ 6080
− 1432
$ 4648   balance

The balance in the account was $4648 at the end of the month.

(Mental check: Using rounded numbers, we find that a balance of around $4500 is close.)

Ms. Maxwell's Check Register for Month of June				BALANCE
DATE	TRANSACTION DESCRIPTION	SUBTRACTIONS	ADDITIONS	$0.00
6/4	Deposit		5280 00	5280 00 / 5280 00
6/10	Deposit		800 00	800 00 / 6080 00
6/16	check– car payment	135 00		135 00 / 5945 00
6/17	check– Tracy's Department Store	450 00		450 00 / 5495 00
6/22	check– credit card payment	775 00		775 00 / 4720 00
6/30	check– Big Fresh Supermarket	72 00		72 00 / 4648 00

Now work margin exercises 3 and 4.

Example 5

Area of a Picture

A rectangular picture is mounted in a rectangular frame with a border (called a mat). (A rectangle is a four-sided figure with opposite sides equal and all four angles equal 90° each.) If the picture is 14 inches by 20 inches, the frame is 18 inches by 24 inches, and the width of the frame is 1 inch, what is the area of the mat?

Solution

In this case, a figure is very helpful. Also, remember that the area of a rectangle is found by multiplying length and width.

The area of the mat will be the difference between the areas of the two inner rectangles. Since the frame is one inch thick on each edge, we must subtract 2 from the length and width of the frame to get the dimensions of the mat. Therefore, the mat is 16 inches by 22 inches.

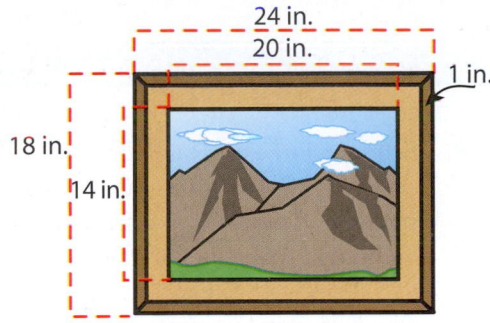

Larger Area	Smaller Area	Area of Mat
16	14	3⁵2
× 22	× 20	−2 8 0
32	280 square inches	7 2 square inches
32		
352 square inches		

The area of the mat is 72 square inches.

Now work margin exercise 5.

---

5. A square picture is mounted in a square frame with a border. (A square is a four-sided figure with all sides equal and all four angles equal 90° each.) If the picture is 6 inches by 6 inches and the frame is 8 inches by 8 inches, what is the area of the border (or mat).

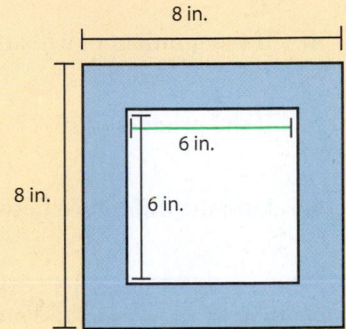

Exercises 1.4

1. Find the **sum** of the three numbers 845, 960, and 455. Then **subtract** 690. What is the **quotient** if the **difference** is **divided by** 2?

2. The **difference** between 9000 and 1856 is **added to** 635. If this **sum** is **multiplied by** 400, what is the **product**?

3. Find the **sum** of the **product** of 23 and 47 and the **product** of 220 and 3.

4. If the **quotient** of 670 and 5 is **decreased by** 120, what is the **difference**?

5. If the **product** of 23, 17, and 5 is **increased by** the **quotient** of 1400 and 4, what is the **sum**?

6. Find the **difference** between the **product** of 30 and 62 and the **sum** of 532 and 1037.

7. **TV Payments:** To purchase a new 27-in. television set with remote control and stereo that sells for $1300 including tax, Mr. Daley paid $200 down and the remainder in five equal monthly payments. What were his monthly payments?

8. **School Clothes:** Ralph decided to go shopping for school clothes before college started in the fall. How much did he spend if he bought four pairs of pants for $75 a pair, five shirts for $45 each, two pairs of socks for $6 a pair, and two pairs of shoes for $98 a pair?

9. **Dining Room Set:** To purchase a new dining room set for $1500, Mrs. Thomas has to pay $120 in sales tax. The dining room set consisted of a table and six chairs. If she made a deposit of $324, how much did she still owe?

10. **Statistics Class Expenses:** For a class in statistics, Anthony bought a new graphing calculator for $86, special graph paper for $8, flash drive for $10, a textbook for $105, and a workbook for $37. What did he spend for this class? The flash drive was writable and able to hold 1 GB of memory.

11. **Gym Class Expenses:** For her physical education class, Paula bought a pair of running shoes for $84, two pairs of socks for $5 a pair, one pair of shorts for $26, and two shirts for $15 each. What were her expenses for this class if the class was to meet at 8:00 A.M. each morning?

12. **Sports Car Cost:** Lynn decided to buy a new sports car with a six-cylinder engine. She could buy a blue standard model for $21,500 plus $690 in taxes and $750 in fees, or she could buy a red custom model for $22,300 plus $798 in taxes and $880 in fees. If the manufacturer was giving rebates of $1200 on the red model and $900 on the blue model, which car was cheaper for her to buy? How much cheaper?

13. New Account Balance: If you opened a checking account with $975 and then wrote checks for $25, $45, $122, $8, and $237, what would be your balance?

14. End of Month Balance: On July 1, Mel was 24 years old and had a balance of $840 in his checking account. During the month, he made deposits of $490, $43, $320, and $49. He wrote checks for $487, $75, $82, and $146. What was the balance in his checking account at the end of the month?

15. Fifth Month Balance: During a 5-month period, Jeff, who goes skydiving once a month, made deposits of $3520, $4560, $1595, $3650, and $2894. He wrote checks totaling $9732. If his beginning balance was $1230, what was the balance in his account at the end of the 5-month period?

16. Weekly Balance: Your friend is 5 feet tall and weighs 103 pounds. She has a checking balance of $4325 at the beginning of the week and writes checks for $55, $37, $560, $278, and $531. What is her balance at the end of the week?

Geometry

17. A triangle has three sides: a base of 6 feet, a height of 8 feet, and a third side of 10 feet. This triangle is called a *right triangle* because one angle is 90°.

   a. Find the perimeter of (distance around) the triangle.

   b. Find the area of the triangle in square feet.

   (To find the area, multiply base times height and then divide by 2.)

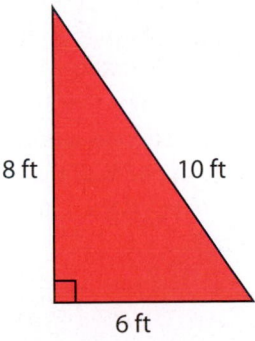

18. a. Find the perimeter of (distance around) a rectangle that has a width of 18 meters and a length of 25 meters.

   b. Also find its area (multiply length times width).

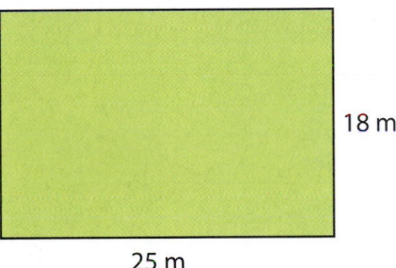

19. A regular hexagon is a six-sided figure with all six sides equal and all six angles equal. Find the perimeter of a regular hexagon with one side of 29 centimeters.

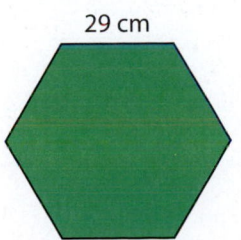

29 cm

20. An isosceles triangle (two sides equal) is placed on one side of a square as shown in the figure. If each of the two equal sides of the triangle is 14 inches long and the square is 18 inches on each side, what is the perimeter of the figure?

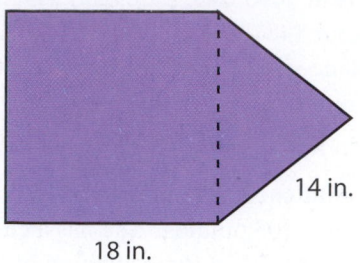

14 in.

18 in.

21. A square that is 10 inches on a side is placed inside a rectangle that has a width of 20 inches and a length of 24 inches. What is the area of the region inside the rectangle that surrounds the square?
(Find the area of the shaded region in the figure.)

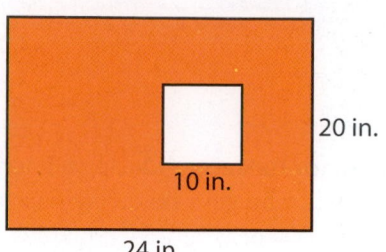

20 in.

10 in.

24 in.

22. A flag is in the shape of a rectangle that is 4 feet by 6 feet with a right triangle in one corner. The right triangle is 1 foot high and 2 feet long.

a. If the triangle is purple and the rest of the flag is blue and white, what is the area of the flag that is purple?

b. What is the area of the flag that is blue and white?

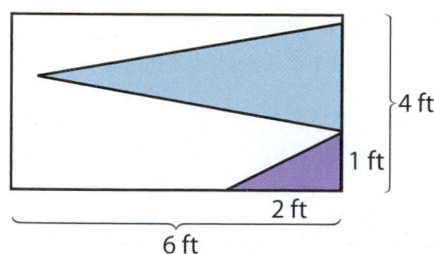

4 ft

1 ft

2 ft

6 ft

23. A pennant in the shape of a right triangle is 10 inches by 24 inches by 26 inches. A yellow square that is 3 inches on a side is inside the triangle and contains the team logo. The remainder of the triangle is orange. What is the area of the orange colored part of the pennant?

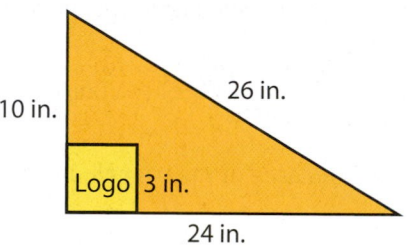

10 in.

26 in.

Logo 3 in.

24 in.

24. Three walls that are 8 feet by 15 feet are to be in a room painted by a light tan color. There is a 3 feet by 4 feet window space cut in one of the walls that will not be painted. What is the total area of the three walls that is to be painted?

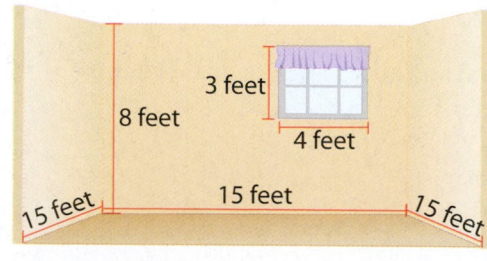

3 feet

8 feet

4 feet

15 feet   15 feet   15 feet

25. Four walls measuring 8 feet by 10 feet are to be wallpapered. There is a door measuring 3 feet by 7 feet in one wall that will not be wallpapered. What is the total area that is to be wallpapered?

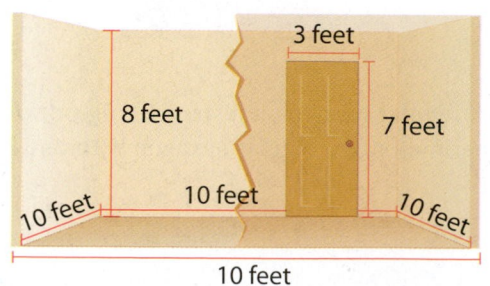

3 feet

8 feet

7 feet

10 feet   10 feet   10 feet

10 feet

26. A triangle with sides of 5 inches, 5 inches, and 6 inches is given. If the height is 4 inches (as shown), find:

    a. the perimeter and

    b. the area of the triangle (See Exercise **17b.**)

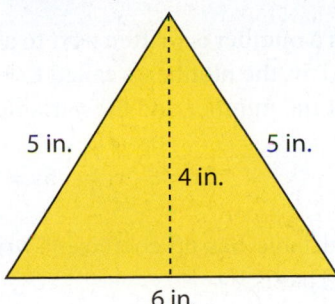

5 in.   5 in.

4 in.

6 in.

Writing & Thinking

27. Four steps: In your own words (without looking at the text), write down the four steps listed as Basic Strategy for Solving Word Problems. Then compare your list with that in the text. Were you close? How well did you read the text?

28. Use your calculator to find each of the following results. Explain, in your own words, what these results indicate.

    a. $17 \div 0$

    b. $0 \div 30$

    c. $0 \div 21$

    d. $23 \div 0$

29. Determine whether the following statement is true or false: $\dfrac{0}{0} = 1$. Briefly describe your reasoning.

30. Account Balance: Do you have a checking account? If you do, explain briefly how you balance your account at the end of each month when you receive the statement from your bank. If you do not have a checking account, then talk with a friend who does and explain how he or she balances his or her account each month.

1.5 Solving Equations with Whole Numbers

Objectives

A Become familiar with the terms equation, solution, constant, and coefficient.

B Learn how to solve equations by using the addition principle, subtraction principle, and the division principle.

Objective A   Notation and Terminology

When a number is written next to a variable (see definition on page 12), such as $3x$ and $5y$, the number is called the **coefficient** of the variable, and the meaning is that the number and the variable are to be multiplied. Thus,

$$3x = 3 \cdot x, \qquad 5y = 5 \cdot y, \qquad \text{and} \qquad 7a = 7 \cdot a.$$

If a variable has no coefficient written, then the coefficient is understood to be 1. That is,

$$x = 1x, \qquad y = 1y, \qquad \text{and} \qquad z = 1z.$$

A single number, such as 10 or 36, is called a **constant**.

This type of algebraic notation and terminology is a basic part of understanding how to set up and solve equations. In fact, understanding how to set up and solve equations is one of the most important skills in mathematics, and we will be developing techniques for solving equations throughout this text.

An **equation** is a statement that two expressions are equal. For example,

$$12 = 5 + 7, \qquad 18 = 9 \cdot 2, \qquad 14 - 3 = 11, \qquad \text{and} \qquad x + 6 = 10$$

are all equations. Equations involving variables, such as

$$x - 8 = 30 \qquad \text{and} \qquad 2x = 18$$

are useful in solving word problems and applications involving unknown quantities.

If an equation has a variable, then numbers may be substituted for the variable, and any number that gives a true statement is called a **solution** to the equation.

1. a. Show that 8 is a solution to the equation $9 + x = 17$.

b. Show that 11 is not a solution to the equation $9 + x = 17$.

Example 1

Solutions to an Equation

a. Show that 4 is a solution to the equation $x + 6 = 10$.

Solution

Substituting 4 for x gives $4 + 6 = 10$ which is true.
Thus 4 is a solution to the equation.

b. Show that 6 is **not** a solution to the equation $x + 6 = 10$.

Solution

Substituting 6 for x gives the statement $6 + 6 = 10$ which is false.
Thus 6 is not a solution to the equation.

Now work margin exercise 1.

As was stated in Section 1.2, a **replacement set** of a variable is the set of all possible values for that variable. In many applications the replacement set for a variable is clear from the context of the problem. For example, if a problem asks how many people attended a meeting, then the replacement set for any related variable will be the set of whole numbers. That is, there is no sense to allow a negative number of people or a fraction of a person at a meeting. In Example 2, the replacement set is limited to a specific set of numbers for explanation purposes.

Equation, Solution, Solution Set

An **equation** is a statement that two expressions are equal.

A **solution** of an equation is a number that gives a true statement when substituted for the variable.

A **solution set** of an equation is the set of all solutions of the equation.

(**Note:** In this chapter, each equation will have only one number in its solution set. In later chapters, you will see equations that have more than one solution.)

Example 2

Replacement Set

Given the equation $x + 5 = 9$, determine which number from the replacement set $\{3, 4, 5, 6\}$ is the solution of the equation.

Solution

Each number is substituted for x in the equation until a true statement is found. If no true statement is found, then the equation has no solution in this replacement set.

Substitute 3 for x: $3 + 5 = 9$. This statement is false. Continue to substitute values for x.

Substitute 4 for x: $4 + 5 = 9$. This statement is true. Therefore, 4 is the solution of the equation.

There is no need to substitute 5 or 6 because the solution has been found.

Now work margin exercise 2.

2. Given that $x + 13 = 22$, determine which number from the replacement set $\{6, 7, 8, 9\}$ is the solution of the equation.

Generally, the replacement set of a variable is a large set, such as the set of all whole numbers, and substituting numbers one at a time for the variable would be ridiculously time consuming. In this section, and in many sections throughout the text, we will discuss methods for **finding solutions of equations** (or **solving equations**). We begin with equations such as

$$x + 5 = 8, \qquad x - 13 = 7, \qquad \text{and} \qquad 4x = 20,$$

or, more generally, equations of the forms

$$x + b = c, \qquad x - b = c, \qquad \text{and} \qquad ax = c$$

where x is a variable and a, b, and c represent constants.

Objective B **Basic Principles for Solving Equations**

The following principles can be used to find the solution to an equation.

Basic Principles for Solving Equations

In the three basic principles stated here, A, B, and C represent numbers (constants) or expressions that may involve variables.

1. The **addition principle:** The equations $A = B$ and $A + C = B + C$ have the same solutions.

2. The **subtraction principle:** The equations $A = B$ and $A - C = B - C$ have the same solutions.

3. The **division principle:** The equations $A = B$ and $\dfrac{A}{C} = \dfrac{B}{C}$ (where $C \neq 0$) have the same solutions.

Essentially, these three principles say that if we perform the same operation on both sides of an equation, the resulting equation will have the same solution as the original equation. Equations with the same solution sets are said to be **equivalent**.

---

notes

■ Although we do not discuss fractions in detail until Chapter 2, we will
■ need the concept of division in fraction form when solving equations
■ and will assume that you are familiar with the fact that a number
   divided by itself is 1. For example,

$$3 \div 3 = \frac{3}{3} = 1 \qquad \text{and} \qquad 17 \div 17 = \frac{17}{17} = 1.$$

■ In this manner, we will write expressions such as the following:

$$\frac{5x}{5} = \frac{5}{5}x = 1 \cdot x = x.$$

As illustrated in the following examples, the variable may appear on the right side of an equation as well as on the left. Also, as these examples illustrate, **the objective in solving an equation is to isolate the variable on one side of the equation. That is, we want the variable on one side of the equation by itself with a coefficient of 1.**

A balance scale gives an excellent visual aid in solving equations. The idea is that, to keep the scales in balance, whatever is done to one side of the scales must be done to the other side. This is, of course, what is done in solving equations.

a.    The seesaw is balanced.

   The seesaw remains balanced if 2 is subtracted from each side.

   The seesaw is balanced but x is not isolated.

b.    The seesaw is balanced.

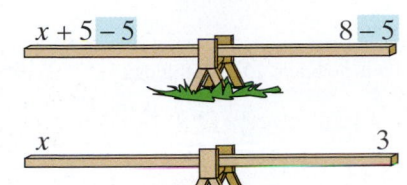   The seesaw remains balanced if 5 is subtracted from each side.

The seesaw is balanced and x is isolated. The solution is $x = 3$.

Example 3

Solving Equations

a. Solve the equation $x + 5 = 14$.

Solution

$$x + 5 = 14 \qquad \text{Write the equation.}$$
$$x + 5 - 5 = 14 - 5 \qquad \text{Using the subtraction principle, subtract 5 from both sides.}$$
$$x + 0 = 9 \qquad \text{Simplify both sides.}$$
$$x = 9 \qquad \text{solution}$$

b. Solve the equation $17 = x + 6$.

Solution

$$17 = x + 6 \qquad \text{Write the equation.}$$
$$17 - 6 = x + 6 - 6 \qquad \text{Using the subtraction principle, subtract 6 from both sides.}$$
$$11 = x + 0 \qquad \text{Simplify both sides.}$$
$$11 = x \qquad \text{solution}$$

Note that, in each step of the solution process, equations are written below each other **with the equal signs aligned**. This format is generally followed in solving all types of equations throughout all of mathematics.

Example 4

Solving Equations

a. Solve the equation $y - 2 = 12$.

Solution

$y - 2 = 12$	Write the equation.
$y - 2 + 2 = 12 + 2$	Using the addition principle, add 2 to both sides.
$y + 0 = 14$	Simplify both sides. Note that $-2 + 2 = 0$.
$y = 14$	solution

b. Solve the equation $25 = y - 20$.

Solution

$25 = y - 20$	Write the equation.
$25 + 20 = y - 20 + 20$	Using the addition principle, add 20 to both sides.
$45 = y + 0$	Simplify both sides.
$45 = y$	solution

Example 5

Solving Equations

a. Solve the equation $3n = 24$.

Solution

$3n = 24$	Write the equation.
$\dfrac{3n}{3} = \dfrac{24}{3}$	Using the division principle, divide both sides by the coefficient, 3. Note that in solving equations, the fraction form of division is used.
$1 \cdot n = 8$	Simplify by performing the division on both sides.
$n = 8$	solution

b. Solve the equation $38 = 19y$.

Solution

$38 = 19y$	Write the equation.
$\dfrac{38}{19} = \dfrac{19y}{19}$	Using the division principle, divide both sides by the coefficient, 19.
$2 = 1 \cdot y$	Simplify by performing the division on both sides.
$2 = y$	solution

c. Solve the equation $2x + 3 + 5 = 20$.

Solution

$2x + 3 + 5 = 20$	Write the equation.
$2x + 8 = 20$	Simplify.
$2x + 8 - 8 = 20 - 8$	Using the subtraction principle, subtract 8 from both sides.
$2x + 0 = 12$	Simplify.
$2x = 12$	Simplify.
$\dfrac{2x}{2} = \dfrac{12}{2}$	Using the division principle, divide both sides by the coefficient 2.
$1 \cdot x = 6$	Simplify.
$x = 6$	solution

Now work margin exercises 3 through 5.

3. a. $x + 7 = 19$

  b. $37 = x + 21$

4. a. $y - 53 = 29$

  b. $40 = y - 7$

5. a. $9y = 36$

  b. $2x + 10 - 8 = 14$

  c. $17 + 29 - 7 = 3n$

Exercises 1.5

In each exercise, an equation and a replacement set of numbers for the variable are given. Substitute each number into the equation until you find the number that is the solution of the equation. See Example 2.

1. $x - 4 = 12$ $\{10, 12, 14, 16, 18\}$

2. $x - 3 = 5$ $\{4, 5, 6, 7, 8, 9\}$

3. $25 = y + 7$ $\{18, 19, 20, 21, 22\}$

4. $13 = y + 12$ $\{0, 1, 2, 3, 4, 5\}$

5. $7x = 105$ $\{0, 5, 10, 15, 20\}$

6. $72 = 8n$ $\{6, 7, 8, 9, 10\}$

7. $13n = 39$ $\{0, 1, 2, 3, 4, 5\}$

8. $14x = 56$ $\{0, 1, 2, 3, 4, 5\}$

Use either the addition principle, the subtraction principle, or the division principle to solve the following equations. Show each step, and keep the equal signs aligned vertically in the format used in the text. See Examples 3 through 5.

9. $x + 12 = 16$

10. $x + 35 = 65$

11. $27 = x + 14$

12. $32 = x + 10$

13. $y + 9 = 9$

14. $y + 18 = 18$

15. $y - 9 = 9$

16. $y - 18 = 18$

17. $22 = n + 12$

18. $44 = n + 15$

19. $75 = n - 50$

20. $100 = n - 50$

21. $8x = 32$

22. $9x = 72$

23. $13y = 52$

24. $15y = 75$

25. $84 = 21n$

26. $99 = 11n$

27. $x + 14 = 20$

28. $15 = x - 3$

29. $42 = y - 5$

30. $73 = y + 4$

In each of the following exercises, perform any indicated arithmetic operations before using the principles to solve the equations.

31. $x + 14 - 1 = 17 + 3$

32. $x + 20 + 3 = 6 + 40$

33. $7x = 25 + 10$

34. $6x = 14 + 16$

35. $3x = 17 + 19$

36. $2x = 72 - 28$

37. $21 + 30 = 17y$

38. $16 + 32 = 16y$

39. $31 + 15 - 10 = 4x$

40. $17 + 20 - 3 = 17y$

41. $3n = 43 - 40 + 6$

42. $2n = 26 + 3 - 1$

43. $23 - 23 = 5x$

44. $6 - 6 = 3x$

45. $5 + 5 - 5 - 5 = 4x$

46. $5 + 5 + 5 = 3x$

47. $6 + 6 + 6 = 3x$

48. $4y = 7 + 7 + 7 + 7$

49. $5n = 2 + 2 + 2 + 2 + 2$

50. $8x = 17 - 17$

1.6 Exponents and Order of Operations

Objective A **Terminology of Exponents**

We know that repeated addition of the same number is shortened by using multiplication. For example,

$$3 + 3 + 3 + 3 + 3 = 5 \cdot 3 = 15 \quad \text{and} \quad 16 + 16 + 16 + 16 = 4 \cdot 16 = 64.$$

factors product   factors product

The result is called the **product**, and the numbers that are multiplied are called **factors** of the product.

In a similar manner, repeated multiplication by the same number can be shortened by using **exponents**. For example, if 3 is used as a factor 5 times, we can write

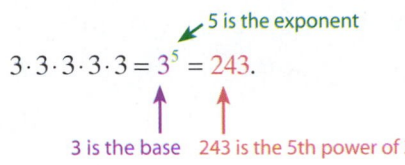

5 is the exponent

$$3 \cdot 3 \cdot 3 \cdot 3 \cdot 3 = 3^5 = 243.$$

3 is the base   243 is the 5th power of 3

In the equation $3^5 = 243$, the number 243 is the **power**, 3 is the **base**, and 5 is the **exponent**. (Exponents are written slightly to the right and above the base.) The expression 3^5 is an **exponential expression** and is read "3 to the fifth power." We can also say that "243 is the fifth power of 3."

Example 1

Exponents

In order to illustrate exponential notation, below we show several products written with repeated multiplication and the equivalent exponential expressions.

With Repeated Multiplication	**With Exponents**
a. $7 \cdot 7 = 49$	$7^2 = 49$
b. $3 \cdot 3 = 9$	$3^2 = 9$
c. $2 \cdot 2 \cdot 2 = 8$	$2^3 = 8$
d. $10 \cdot 10 \cdot 10 \cdot 10 = 10,000$	$10^4 = 10,000$

Now work margin exercise 1.

Objectives

A Understand the terms base, exponent, and power.

B Know how to evaluate expressions with exponents, including 1 and 0 as exponents.

C Learn how to use a calculator to evaluate exponential expressions.

D Know the rules for order of operations and be able to apply these rules when evaluating numerical expressions.

1. Rewrite the following expressions using exponents and find the products.

a. $100 \cdot 100 \cdot 100$

b. $2 \cdot 2 \cdot 2 \cdot 2$

c. $1 \cdot 1 \cdot 1$

d. $8 \cdot 8$

Exponent and Base

A whole number n is an **exponent** if it is used to tell how many times another whole number a is used as a factor. The repeated factor a is called the **base** of the exponent. Symbolically,

$$\underbrace{a \cdot a \cdot a \cdot \ldots \cdot a \cdot a}_{n \text{ factors}} = a^n.$$

! Common Error

DO NOT multiply the base and the exponent.	**DO** multiply the base by itself.
$10^2 = 10 \cdot 2$ **WRONG**	$10^2 = 10 \cdot 10$ **CORRECT**
$6^3 = 6 \cdot 3$ **WRONG**	$6^3 = 6 \cdot 6 \cdot 6$ **CORRECT**

In expressions with exponent 2, the base is said to be **squared**. In expressions with exponent 3, the base is said to be **cubed**.

2. Write out the following expressions.

a. $6^3 = 216$

b. $9^2 = 81$

Example 2

Reading Exponents

a. $8^2 = 64$ is read "eight squared is equal to sixty-four."

b. $5^3 = 125$ is read "five cubed is equal to one hundred twenty-five."

Expressions with exponents other than 2 or 3 are read as the base "to the _____ power." For example,

$2^5 = 32$ is read "two to the fifth power is equal to thirty–two."

Now work margin exercise 2.

notes

- There is usually some confusion about the use of the word "power." Since 2^5 is read "two to the fifth power," it is natural to think of 5 as the power. This is not true. A power is not an exponent. A power is the product indicated by a base raised to an exponent. Thus, for the equation $2^5 = 32$, think of the phrase "two to the fifth power" in its entirety. The corresponding power is the product, 32.

To improve your speed in factoring and in working with fractions and simplifying expressions, you should memorize all the squares of the whole numbers from 1 to 20. These numbers are called **perfect squares**. The following table lists these perfect squares.

Number (n)	1	2	3	4	5	6	7	8	9	10
Square (n^2)	1	4	9	16	25	36	49	64	81	100

Number (n)	11	12	13	14	15	16	17	18	19	20
Square (n^2)	121	144	169	196	225	256	289	324	361	400

Objective B   **One and Zero as Exponents**

Note that if there is no exponent written, then the exponent is understood to be 1. That is, any number or algebraic expression raised to the first power is equal to itself. Thus

$$8 = 8^1, \quad 72 = 72^1, \quad x = x^1, \quad \text{and} \quad 6y = 6y^1.$$

The Exponent 1

For any whole number a, $a = a^1$.

The use of 0 as an exponent is a special case and, as we will see later, is needed to simplify and evaluate algebraic expressions. To understand the meaning of 0 as an exponent, study the following patterns of numbers. These patterns lead to the meaning of 2^0, 3^0, and 5^0 and the general expression of the form a^0.

$$
\begin{array}{lll}
2^4 = 16 & 3^4 = 81 & 5^4 = 625 \\
2^3 = 8 & 3^3 = 27 & 5^3 = 125 \\
2^2 = 4 & 3^2 = 9 & 5^2 = 25 \\
2^1 = 2 & 3^1 = 3 & 5^1 = 5 \\
2^0 = ? & 3^0 = ? & 5^0 = ?
\end{array}
$$

Notice that, in the column of powers of 2, each number can be found by dividing the previous number by 2. Therefore, for 2^0, we might guess that $2^0 = \dfrac{2}{2} = 1.$ Similarly, $3^0 = \dfrac{3}{3} = 1$ and $5^0 = \dfrac{5}{5} = 1.$ Thus we are led to the following statement about 0 as an exponent.

The Exponent 0

If a is any nonzero whole number, then $a^0 = 1$.

(**Note:** The expression 0^0 is undefined.)

The calculator has keys marked as 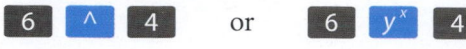. These keys can be used to find powers. The x^2 key will find squares and the caret key, $\wedge$, or power key, y^x, can be used to find squares and the values of other exponential expressions. The caret key and power key function the same way, and either can be used as shown below. In this text, we will use the caret key in most directions.

To find 6^4, press the keys

The display will appear as **1296**.

To find 27^2,

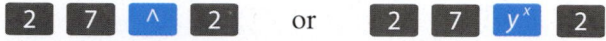

The display will appear as **729**.

Alternatively, we could have entered:

Example 3

Evaluating Exponents with a Calculator

Use a calculator to find the value of each of the following exponential expressions.

a. 9^5

Solution

Step 1: Enter 9 . (This is the base.)
Step 2: Press the caret key $\wedge$.
Step 3: Enter 5 . (This is the exponent.)
Step 4: Press the = key.

The display should read **59049**.

b. 41^4

Solution

Step 1: Enter 4 1 .
Step 2: Press the caret key $\wedge$.
Step 3: Enter 4 .
Step 4: Press the = key.

The display should read **2825761**.

c. 250^2

Solution

Step 1: Enter 2 5 0 .
Step 2: Press the x^2 key.
Step 3: Press the = key.

The display should read **62500**.

Now work margin exercise 3.

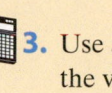**3.** Use a calculator to find the value of each of the following exponential expressions.

a. 7^5

b. 56^3

c. 37^2

Objective D **Rules for Order of Operations**

Mathematicians have agreed on a set of rules for the order of operations in simplifying (or evaluating) any numerical expression involving addition, subtraction, multiplication, division, and exponents. These rules are used in all branches of mathematics and computer science to ensure that there is only one correct answer regardless of how complicated an expression might be. For example, the expression $5 \cdot 2^3 - 14 \div 2$ might be evaluated as follows.

$$5 \cdot 2^3 - 14 \div 2 = 5 \cdot 8 - 14 \div 2$$
$$= 40 - 14 \div 2$$
$$= 40 - 7$$
$$= 33 \qquad \textbf{CORRECT}$$

However, if we were simply to proceed from left to right as in the following steps, we would get an entirely different answer.

$$5 \cdot 2^3 - 14 \div 2 = 10^3 - 14 \div 2$$
$$= 1000 - 14 \div 2$$
$$= 986 \div 2$$
$$= 493 \qquad \textbf{INCORRECT}$$

By the standard set of rules for order of operations, we can conclude that **the first answer (33) is the accepted correct answer** and that **the second answer (493) is incorrect**.

notes

- **Special Note about the Difference between Evaluating Expressions and Solving Equations**

- Some students seem to have difficulty distinguishing between evaluating expressions and solving equations. In evaluating expressions, we are simply trying to find the numerical value of an expression. In solving equations, we are trying to find the value of some unknown number represented by a variable in an equation. These two concepts are not the same. In solving equations, we use the addition principle, the subtraction principle, and the division principle. In evaluating expressions, we use the rules for order of operations. These rules are programmed into nearly every calculator.

Rules for Order of Operations

1. First, simplify within grouping symbols, such as parentheses ( ), brackets [ ], braces { }, radical signs $\sqrt{\ }$, absolute value bars | |, or fraction bars. Start with the innermost grouping.
2. Second, evaluate any numbers or expressions raised to exponents.
3. Third, moving from **left to right**, perform any multiplications or divisions in the order in which they appear.
4. Fourth, moving from **left to right**, perform any additions or subtractions in the order in which they appear.

These rules are very explicit and should be studied carefully. Note that in Rule 3, neither multiplication nor division has priority over the other. Whichever of these operations occurs first, moving **left to right**, is done first. In Rule 4, addition and subtraction are handled in the same way. Unless they occur within grouping symbols, **addition and subtraction are the last operations to be performed**.

For example, consider the relatively simple expression $2 + 5 \cdot 6$. If we make the mistake of adding before multiplying, we get

$$2 + 5 \cdot 6 = 7 \cdot 6 = 42. \qquad \textbf{INCORRECT}$$

The correct procedure yields

$$2 + 5 \cdot 6 = 2 + 30 = 32. \qquad \textbf{CORRECT}$$

A well-known mnemonic device for remembering the rules for order of operations is the following.

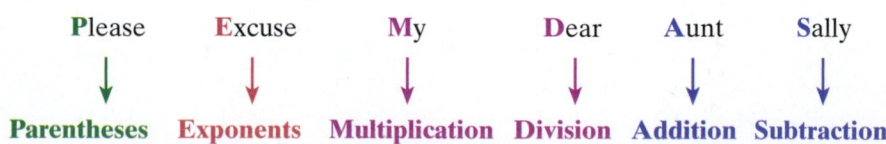

Please	**E**xcuse	**M**y	**D**ear	**A**unt	**S**ally
↓	↓	↓	↓	↓	↓
Parentheses	**Exponents**	**Multiplication**	**Division**	**Addition**	**Subtraction**

notes

- Even though the mnemonic **PEMDAS** is helpful, remember that multiplication and division are performed as they appear in the expression, left to right. Also, addition and subtraction are performed as they appear in the expression, left to right.

The following examples show how to apply these rules. In some cases, more than one step can be performed at the same time. This is possible when parts are separated by + or − signs or are within separate symbols of inclusion, as illustrated in Examples 4 – 6. **Work through each example step by step**, **and rewrite the examples on a separate sheet of paper.**

Example 4

Evaluating an Expression

Evaluate the expression $24 \div 8 + 4 \cdot 2 - 6$.

Solution

$24 \div 8 + 4 \cdot 2 - 6$ Divide before multiplying in this case. Remember to move left to right.

$= 3 + 4 \cdot 2 - 6$ Multiply before adding or subtracting.

$= 3 + 8 - 6$ Add before subtracting in this case. Remember to move left to right.

$= 11 - 6$ Subtract.

$= 5$

Example 5

Evaluating an Expression

Evaluate the expression $5 - 18 \div 9 \cdot 2 + 4(7)$.

Solution

$5 - 18 \div 9 \cdot 2 + 4(7)$ Divide.

$= 5 - 2 \cdot 2 + 4(7)$ Multiply.

$= 5 - 4 + 4(7)$ Multiply.

$= 5 - 4 + 28$ Subtract.

$= 1 + 28$ Add.

$= 29$

Example 6

Evaluating an Expression

Evaluate the expression $30 \div 10 \cdot 2^3 + 3(6-2)$.

Solution

$$30 \div 10 \cdot 2^3 \ + \ 3(6-2) \quad \text{Operate within parentheses.}$$

$$= \ 30 \div 10 \cdot 2^3 \ + \ 3(4) \quad \text{Find the power.}$$

$$= \ 30 \div 10 \cdot 8 \ + \ 3(4) \quad \text{Divide.}$$

$$= \ 3 \cdot 8 \ + \ 3(4) \quad \text{Multiply in each part separated by +.}$$

$$= \ 24 \ + \ 12 \quad \text{Add.}$$

$$= \ 36$$

Evaluate the expressions.

4. $3 \cdot 2^3 - 10 \div 2$

5. $19 + 5(3 - 1)$

6. $2(5^2 - 1) - 4 + 3 \cdot 2^3$

7. $45 + 18(2) \div 3 - 17$

Completion Example 7

Evaluating an Expression

Evaluate the expression $(14 - 10)[(5 + 3^2) \div 2 + 5]$.

Solution

$$(14 - 10)[(5 + 3^2) \div 2 + 5] \quad \text{Operate within parentheses and find the power.}$$

$$= (\underline{\quad})[(5 + \underline{\quad}) \div 2 + 5] \quad \text{Operate within parentheses.}$$

$$= (\underline{\quad})[(\underline{\quad}) \div 2 + 5] \quad \text{Divide within brackets.}$$

$$= (\underline{\quad})[\underline{\quad} + 5] \quad \text{Add within brackets.}$$

$$= (\underline{\quad})[\underline{\quad}] \quad \text{Multiply.}$$

$$= \underline{\quad}$$

Now work margin exercises 4 through 7.

Completion Example Answers

7. $= (4)[(5 + 9) \div 2 + 5]$

$\quad = (4)[(14) \div 2 + 5]$

$\quad = (4)[7 + 5]$

$\quad = (4)[12]$

$\quad = 48$

Completion Example 8

Evaluating an Expression

Evaluate the expression $9(2^2 - 1) - 17 + 6 \cdot 2^2$.

Solution

$9(2^2 - 1) - 17 + 6 \cdot 2^2$

$= 9(\underline{} - 1) - 17 + 6 \cdot \underline{}$

$= 9(\underline{}) - 17 + 6 \cdot \underline{}$

$= \underline{} - 17 + \underline{}$

$= \underline{} + \underline{}$

$= \underline{}$

Now work margin exercise 8.

8. Evaluate the expression.

$6^2 \div 4 + 15 \cdot 2$

Some calculators are programmed to follow the rules for order of operations. To test if your calculator is programmed in this manner, let's evaluate $7 + 4 \cdot 2$. Press the keys [7] [+] [4] [×] [2], and then [=]. If the answer on your calculator screen shows 15, then you do have a built in order of operations. This means that most of the time, you can type in the problem as it is written with the calculator performing the order of operations in a proper manner. If the calculator gave you an answer of 22, then you will need to be careful when entering problems.

If your calculator has parentheses buttons, [(] and [)], you can evaluate the expression $((6 + 2) + (8 + 1)) \div 9$, by pressing the keys

 [(] [6] [+] [2] [)] [+] [(] [8] [+] [1] [)] [÷] [9].

Then press [=].

The display will read **9**.

Completion Example Answers

8. $= 9(4 - 1) - 17 + 6 \cdot 4$

$= 9(3) - 17 + 6 \cdot 4$

$= 27 - 17 + 24$

$= 10 + 24$

$= 34$

Exercises 1.6

For each of the following exponential expressions: a. name the base , b. name the exponent, and c. find the value of the exponential expression.

1. 2^3
a. _____
b. _____
c. _____

2. 5^3
a. _____
b. _____
c. _____

3. 4^2
a. _____
b. _____
c. _____

4. 6^2
a. _____
b. _____
c. _____

5. 9^2
a. _____
b. _____
c. _____

6. 7^3
a. _____
b. _____
c. _____

7. 11^2
a. _____
b. _____
c. _____

8. 2^6
a. _____
b. _____
c. _____

9. 3^5
a. _____
b. _____
c. _____

10. 2^4
a. _____
b. _____
c. _____

11. 19^0
a. _____
b. _____
c. _____

12. 22^0
a. _____
b. _____
c. _____

13. 1^6
a. _____
b. _____
c. _____

14. 1^{57}
a. _____
b. _____
c. _____

15. 24^1
a. _____
b. _____
c. _____

16. 13^1
a. _____
b. _____
c. _____

17. 20^3
a. _____
b. _____
c. _____

18. 15^2
a. _____
b. _____
c. _____

19. 30^2
a. _____
b. _____
c. _____

20. 40^2
a. _____
b. _____
c. _____

Find a base and exponent form for each of the following numbers without using the exponent 1. See Example 1.

21. 16

22. 4

23. 25

24. 32

25. 49

26. 121

27. 36

28. 100

29. 1000

30. 81

31. 8

32. 125

33. 216

34. 343

35. 243

36. 10,000

37. 100,000

38. 625

39. 27

40. 64

Rewrite the following products by using exponents. See Example 1.

41. $6 \cdot 6 \cdot 6 \cdot 6 \cdot 6$

42. $7 \cdot 7 \cdot 7 \cdot 7$

43. $11 \cdot 11 \cdot 11$

44. $13 \cdot 13 \cdot 13$

45. $2 \cdot 2 \cdot 2 \cdot 3 \cdot 3$

46. $2 \cdot 2 \cdot 5 \cdot 5 \cdot 5$

47. $2 \cdot 3 \cdot 3 \cdot 11 \cdot 11$

48. $5 \cdot 5 \cdot 5 \cdot 7 \cdot 7$

49. $3 \cdot 3 \cdot 3 \cdot 7 \cdot 7 \cdot 7$

50. $2 \cdot 2 \cdot 2 \cdot 2 \cdot 11 \cdot 11 \cdot 13 \cdot 13$

Find the following perfect squares. Write as many of them as you can from memory.

51. 8^2

52. 3^2

53. 11^2

54. 14^2

55. 20^2

56. 15^2

57. 13^2

58. 18^2

59. 30^2

60. 50^2

Use your calculator to find the value of each of the following exponential expressions. See Example 3.

61. 52^2

62. 35^2

63. 25^4

64. 32^4

65. 125^2

66. 47^3

67. 5^7

68. 2^{10}

Use the rules for order of operations to find the value of each of the following expressions. See Examples 4 through 8.

69. $6 + 5 \cdot 3$

70. $18 + 2 \cdot 5$

71. $20 - 4 \div 4$

72. $6 - 15 \div 3$

73. $32 - 14 + 10$

74. $25 - 10 + 11$

75. $18 \div 2 - 1 - 3 \cdot 2$

76. $6 \cdot 3 \div 2 - 5 + 13$

77. $2 + 3 \cdot 7 - 10 \div 2$

78. $14 \cdot 2 \div 7 \div 2 + 10$

79. $(2 + 3 \cdot 4) \div 7 - 2$

80. $(2 + 3) \cdot 4 \div 5 - 4$

81. $14(2 + 3) - 65 - 5$

82. $13(10 - 7) - 20 - 19$

83. $2 \cdot 5^2 - 8 \div 2$

84. $16 \div 2^4 + 9 \div 3^2$

85. $(2^3 + 2) \div 5 + 7^2 \div 7$

86. $18 - 9(3^2 - 2^3)$

87. $(4 + 3)^2 + (2 + 3)^2$

88. $(2 + 1)^2 + (4 + 1)$

89. $8 \div 2 \cdot 4 - 16 \div 4 \cdot 2 + 3 \cdot 2^2$

90. $50 \div 2 \cdot 5 - 5^3 \div 5 + 5$

91. $(10 + 1)[(5 - 2)^2 + 3(4 - 3)]$

92. $(12-2)[4(6-3)+(4-3)^2]$

93. $100+2[3(4^2-6)+2^3]$

94. $75+3[2(3+6)^2-10^2]$

95. $16+3[17+2^3 \div 2^2-4]$

96. $10^3-2[(13+3) \div 2^4+18 \div 3^2]$

97. $12^3-1\left[\left(3^2+5\right)^2\right]$

98. $10^4-1\left[\left(10+5\right)^2-5^2\right]$

99. $30 \div 2-11+2(5-1)^3$

100. $2(15-6+4) \div 13 \cdot 2+1^0$

Writing & Thinking

101. **Evaluate Expression:** Use your calculator to evaluate the expression 0^0. What was the result? State, in your own words, the meaning of the result.

102. **a.** What mathematics did you use today? (Yes, you did use some mathematics. I know you did).

   b. What mathematics did you use this week?

   c. What mathematical thinking did you use to solve some problem (not in the text) today?

1.7 Introduction to Polynomials

Objective A   **Definition of a Polynomial**

Algebraic expressions that are numbers, powers of variables, or products of numbers and powers of variables are called **terms**. That is, an expression that involves only multiplication and/or division with variables and constants is called a **term**. Remember that a number written next to a variable indicates multiplication, and the number is called the **coefficient** of the variable.

algebraic terms in one variable:

$$5x, \qquad 25x^3, \qquad 6y^4, \qquad \text{and} \qquad 48n^2$$

algebraic terms in two variables:

$$3xy, \qquad 4ab, \qquad 6xy^5, \qquad \text{and} \qquad 10x^2y^2$$

A term that consists of only a number, such as 3 or 15, is called a **constant** or a **constant term**. In this text, we will discuss terms in only one variable and constants.

Monomial

A **monomial in x** is a term of the form

$$kx^n \qquad \text{where } k \text{ and } n \text{ are whole numbers.}$$

n is called the **degree** of the monomial, and k is called the **coefficient**.

Now consider the fact that we can write constants, such as 7 and 63, in the form

$$7 = 7 \cdot 1 = 7x^0 \qquad \text{and} \qquad 63 = 63 \cdot 1 = 63x^0.$$

Thus we say that a nonzero constant is a **monomial of 0 degree**. (Note that a monomial may be in a variable other than x.) Since we can write $0 = 0x^2 = 0x^5 = 0x^{63}$, we say that **0 is a monomial of no degree**.

In algebra, we are particularly interested in the study of expressions such as

$$5x + 7, \qquad 3x^2 - 2x + 4, \qquad \text{and} \qquad 6a^4 + 8a^3 - 10a$$

that involve sums and/or differences of terms. These algebraic expressions are called **polynomials**.

Polynomial

A **polynomial** is a monomial or the sum or difference of monomials.

The **degree of a polynomial** is the largest of the degrees of its terms.

Objectives

A   Learn the meaning of the term polynomial.

B   Know how to evaluate a polynomial for a given value of the variable by following the rules for order of operations.

Generally, for easy reading, a polynomial is written so that the degrees of its terms either decrease from left to right or increase from left to right. If the degrees decrease, we say that the terms are written in **descending order**. If the degrees increase, we say that the terms are written in **ascending order**. For example,

$3x^4 + 5x^2 - 8x + 34$ is a fourth-degree polynomial written in descending order.

$15 - 3y + 4y^2 + y^3$ is a third-degree polynomial written in ascending order.

For consistency, and because of the style used in operating with polynomials (see Chapter 8), the polynomials in this text will be written in descending order. Note that this is merely a preference, and a polynomial written in any order is acceptable and may be a correct answer to a problem.

notes

- As stated earlier concerning terms, we will limit our discussions in this text to polynomials in only one variable. In future courses in mathematics, you may study polynomials in more than one variable.

Some forms of polynomials are used so frequently that they have been given special names, as indicated in the following box.

Classification of Polynomials

Description	Name	Example
Polynomial with one term	Monomial	$5x^3$
Polynomial with two terms	Binomial	$7x - 23$
Polynomial with three terms	Trinomial	$a^2 + 5a + 6$

Example 1

Degree of Polynomials

Name the degree and type of each of the following polynomials.

a. $5x - 10$

Solution

$5x - 10$ is a first-degree binomial.

b. $3x^4 + 5x^3 - 4$

Solution

$3x^4 + 5x^3 - 4$ is a fourth-degree trinomial.

c. $17y^{20}$

Solution

$17y^{20}$ is a twentieth-degree monomial.

Now work margin exercise 1.

1. Name the degree and type of each of the following polynomials.

a. $8x$

b. $13 - 6x^{18}$

c. $7x^6 - 4x^3 + 2$

Objective B Evaluating Polynomials

To evaluate a polynomial for a given value of the variable:

1. substitute that value for the variable wherever it occurs in the polynomial and,
2. follow the rules for order of operations.

Example 2

Evaluating a Polynomial

Evaluate the polynomial $4a^2 + 5a - 12$ for $a = 3$.

Solution

Substitute 3 for a wherever a occurs, and follow the rules for order of operations.

$$
\begin{aligned}
4a^2 + 5a - 12 \quad &= 4 \cdot 3^2 + 5 \cdot 3 - 12 \\
&= 4 \cdot 9 + 5 \cdot 3 - 12 \\
&= 36 + 15 - 12 \\
&= 39
\end{aligned}
$$

Example 3

Evaluating a Polynomial

Evaluate the polynomial $3x^3 + x^2 - 4x + 5$ for $x = 2$.

Solution

$$
\begin{aligned}
3x^3 + x^2 - 4x + 5 &= 3 \cdot 2^3 + 2^2 - 4 \cdot 2 + 5 \\
&= 3 \cdot 8 + 4 - 4 \cdot 2 + 5 \\
&= 24 + 4 - 8 + 5 \\
&= 28 - 8 + 5 \\
&= 20 + 5 \\
&= 25
\end{aligned}
$$

Evaluate the polynomials.

2. $x^3 + 7x - 20$ for $x = 4$

3. $n^2 + 3n + 11$ for $n = 7$

4. $5y^2 - y + 42$ for $y = 2$

Example 4

Evaluating a Polynomial

Evaluate the polynomial $x^2 + 5x - 36$ for $x = 4$.

Solution

$$
\begin{aligned}
x^2 + 5x - 36 &= 4^2 + 5(4) - 36 \\
&= 16 + 5(4) - 36 \\
&= 16 + 20 - 36 \\
&= 0
\end{aligned}
$$

Note that a raised dot or parentheses can be used to indicate multiplication.

Now work margin exercises 2 through 4.

Exercises 1.7

Name the degree and type of each of the following polynomials.

1. $x^2 + 5x + 6$

2. $5x^2 + 8x - 3$

3. $x^2 + 2x$

4. $3x^2 + 7x$

5. $13x^{12}$

6. $20x^5$

7. $y^4 + 2y^2 + 1$

8. $9y^3 + 8y - 10$

9. $a^5 + 5a^3 - 16$

10. $2a^4 + 5a^3 - 3a^2$

Evaluate each of the following polynomials for $x = 2$. See Examples 2 through 4.

11. $x^2 + 5x + 6$

12. $5x^2 + 8x - 3$

13. $x^2 + 2x$

14. $3x^2 + 7x$

15. $13x^{12}$

16. $20x^5$

Evaluate each of the following polynomials for $y = 3$.

17. $y^2 + 5y - 15$

18. $3y^2 + 6y - 11$

19. $y^3 + 3y^2 + 9y$

20. $y^2 + 7y - 12$

21. $y^3 - 9y$

22. $y^4 - 2y^2 + 6y - 2$

Evaluate each of the following polynomials for $a = 1$.

23. $2a^3 - a^2 + 6a - 4$

24. $2a^4 - a^3 + 17a - 3$

25. $3a^3 + 3a^2 - 3a - 3$

26. $5a^2 + 5a - 10$

27. $a^3 + a^2 + a + 2$

28. $a^4 + 6a^3 - 4a^2 + 7a - 1$

Evaluate each of the following polynomials for $c = 5$.

29. $c^3 + 3c^2 + 9c$

30. $c^2 + 7c - 12$

31. $4c^2 - 8c + 14$

32. $3c^2 + 15c + 150$

33. $4c^3 + 9c^2 - 4c + 1$

34. $c^3 + 8c - 20$

35. Evaluate the polynomial $x^2 + 5x + 4$ for each value of x in the following table.

x	Value of $x^2 + 5x + 4$
0	
1	
2	
3	
4	
5	

36. Evaluate the polynomial $x^3 + 7x + 10$ for each value of x in the following table.

x	Value of $x^3 + 7x + 10$
0	
1	
2	
3	
4	
5	

37. Evaluate the polynomial $2y^2 + 3y + 8$ for each value of y in the following table.

y	Value of $2y^2 + 3y + 8$
0	
1	
2	
3	
4	
5	

38. Evaluate the polynomial $3a^3 + 5a^2 - a$ for each value of a in the following table.

a	Value of $3a^3 + 5a^2 - a$
0	
1	
2	
3	
4	
5	

39. $3(3 + 4) - 2(1 + 5)$

40. $(14 + 6) \div 2 + 3 \cdot 4$

41. $16 - 4 + 2(5 + 8)$

42. $14 + 6 \div 2 + 3 \cdot 4$

43. $(3 \cdot 4 - 2 \cdot 5)(2 \cdot 6 + 3)$

44. $(3^2 - 3 \cdot 2)^2 - 5 + 1$

45. $(3^2 + 4^2)^3 - 10^2$

46. $(2 \cdot 8 - 3)^2(3 + 4 \cdot 2)^3$

47. $(5 + 3 \cdot 5)(15 \div 3 \cdot 5)$

48. $3 \cdot 5^2 + (3 \cdot 5)^2$

49. Use your calculator to evaluate each polynomial for the given value of the variable.

   a. $16x^4 + x^2 - 7x + 133$ for $x = 12$

   b. $4y^3 - 4y^2 + 15y - 100$ for $y = 10$

   c. $5x^4 - x^3 + 14x^2 - 10$ for $x = 8$

   d. $13a^3 + 5a^2 - 10a - 25$ for $a = 5$

Writing & Thinking

50. **Most Interesting:** Of all the topics and ideas discussed so far in Chapter 1, which one(s) did you find the most interesting? Explain briefly using complete sentences in paragraph form.

51. **Most Difficult:** Of all the topics and ideas discussed so far in Chapter 1, which one(s) did you find the most difficult to learn? Explain briefly using complete sentences in paragraph form.

52. **Life Problem:** Discuss how you used mathematics to solve some problem in your life today. (You know you did!)

1.8 Introduction to Integers

notes

▪ **Special Note from the Author about Calculators**

▪ All of the operations with integers can be done with or without a calculator. However, at this beginning stage, it is recommended you use a calculator as little as possible to develop a thorough understanding of operating with negative numbers.

Objectives

A. Be able to graph a set of integers on a number line.

B. Understand and be able to read inequality symbols such as < and >.

C. Know the meaning of the absolute value of an integer.

D. Be able to graph an inequality on a number line.

Objective A **Number Lines**

The concepts of positive and negative numbers occur frequently in our daily lives:

	Negative	Zero	Positive
Temperatures are recorded as:	below zero	zero	above zero
The stock market will show:	a loss	no change	a gain
Altitude can be measured as:	below sea level	sea level	above sea level
Businesses will report:	losses	no gain	profits

In the remainder of this chapter, we will develop techniques for understanding and operating with positive and negative numbers. We begin with the graphs of numbers on **number lines**. We generally use horizontal lines for number lines. For example, choose some point on a horizontal line and label it with the number 0 (Figure 1).

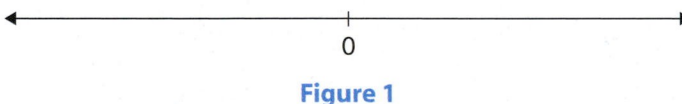

Figure 1

Now choose another point on the line to the right of 0 and label it with the number 1 (Figure 2).

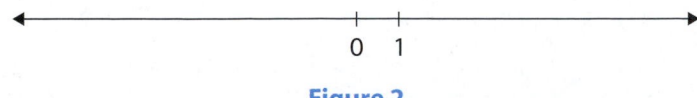

Figure 2

Points corresponding to all whole numbers are now determined and are all to the right of 0. That is, the point for 2 is the same distance from 1 as 1 is from 0, and so on (Figure 3).

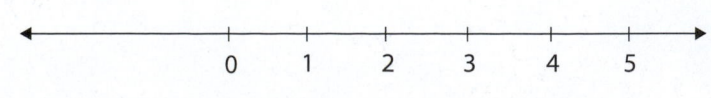

Figure 3

The graph of a number is the point on a number line that corresponds to that number, and the number is called the **coordinate** of the point. The terms "number" and "point" are used interchangeably when dealing with number lines. Thus, we might refer to the "point" 6. The graphs of numbers are indicated by marking the corresponding points with large dots .

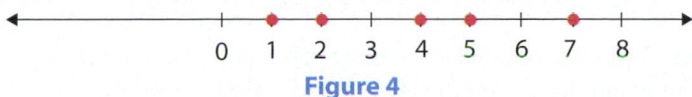

Figure 4

Figure 4 shows the graph of the set of numbers $S = \{ 1, 2, 4, 5, 7 \}$

(**Note:** Even though other numbers are marked, only those with a large dot are considered to be "graphed.")

The point one unit to the left of 0 is the opposite of 1. It is called "negative 1" and is symbolized as –1. Similarly, the opposite of 2 is called "negative 2" and is symbolized as –2, and so on (Figure 5).

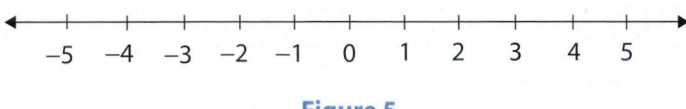

Figure 5

(**Note: Additive inverse** is the technical term for the opposite of a number. We will discuss this in more detail in a later section).

Integers

The set of **integers** is the set of whole numbers and their opposites (or additive inverses).

$$\text{Integers} = \mathbb{Z} = \{ \ldots, -3, -2, -1, 0, 1, 2, 3, \ldots \}$$

The counting numbers (all whole numbers except 0) are called **positive integers** and may be written as +1, +2, +3, and so on; the opposites of the counting numbers are called **negative integers**. 0 is neither positive nor negative (Figure 6).

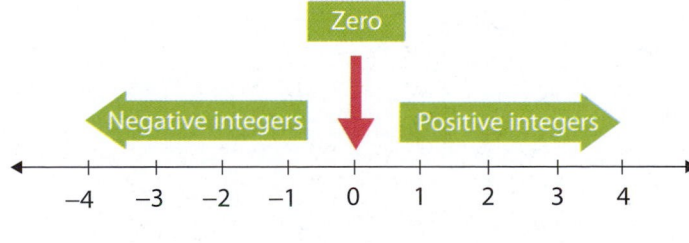

Figure 6

Note the following facts about integers.

1. The opposite of a positive integer is a negative integer.
For example, $-(+1) = -1$ and $-(+6) = -6$.

↑   ↑
opposite of +1   opposite of +6

2. The opposite of a negative integer is a positive integer.
For example, $-(-2) = +2 = 2$ and $-(-8) = +8 = 8$.

↑   ↑
opposite of −2   opposite of −8

3. The opposite of 0 is 0.
(That is, $-0 = +0 = 0$. This shows that -0 should be thought of as the opposite of 0 and not as "negative 0." Remember the number 0 is neither positive nor negative, and 0 is its own additive inverse.)

---

notes

There are many types of numbers other than integers that can be graphed on number lines. We will study these types of numbers in later chapters. Therefore, you should be aware that the set of integers does not include all of the positive numbers or all of the negative numbers.

---

1. Find the opposite of the following integers.

a. −10

b. −8

c. +17

Example 1

Opposites

Find the opposite of the following integers.

a. −5

Solution

$-(-5) = 5$

b. −11

Solution

$-(-11) = 11$

c. +14

Solution

$-(+14) = -14$

*Now work margin exercise **1.***

Example 2

Graphing Integers

Graph the set of integers $B = \{-3, -1, 0, 1, 3\}$.

Solution

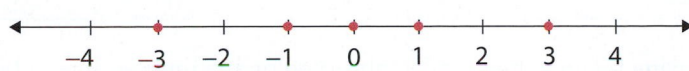

A set of numbers is infinite if it is so large that it cannot be counted. The set of whole numbers and the set of integers are both infinite. To graph an infinite set of integers, three dots are marked above the number line to indicate that the pattern shown is to continue without end. Example 3 illustrates this technique.

Example 3

Graphing Integers

Graph the set of integers $C = \{2, 4, 6, 8, \dots\}$.

Solution

The three dots above the number line indicate that the pattern in the graph continues without end.

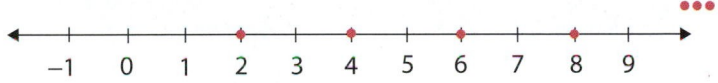

Now work margin exercises 2 and 3.

Objective B **Inequality Symbols**

On a horizontal number line, **smaller numbers are always to the left of larger numbers**. Each number is smaller than any number to its right and larger than any number to its left. We use the following inequality symbols to indicate the order of numbers on the number line.

Symbols for Order

$<$ is read "is less than" $\leq$ is read "is less than or equal to"

$>$ is read "is greater than" $\geq$ is read "is greater than or equal to"

2. Graph the set of integers $\{-2, -1, 0\}$ on a number line.

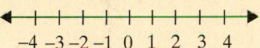

3. Graph the set of integers $\{\dots, -7, -5, -3, -1\}$ on a number line.

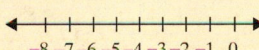

	Using <		Using >	
a.	$3 < 5$	3 is less than 5	$5 > 3$	5 is greater than 3
b.	$-2 < 0$	-2 is less than 0	$0 > -2$	0 is greater than -2
c.	$-4 < -1$	-4 is less than -1	$-1 > -4$	-1 is greater than -4

The following relationships can be observed on the number line in Figure 7.

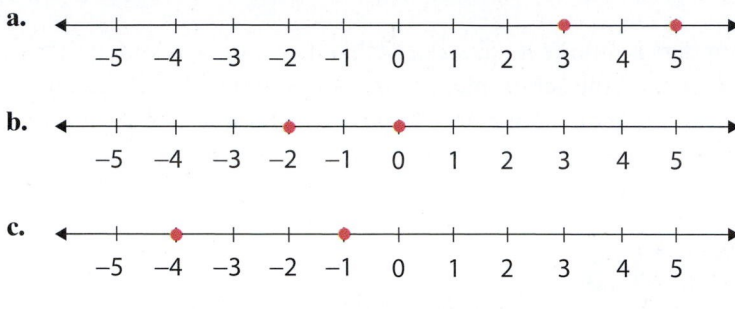

Figure 7

One useful idea implied by the previous discussion is that the symbols < and > can be read either from right to left or from left to right. For example, we might read

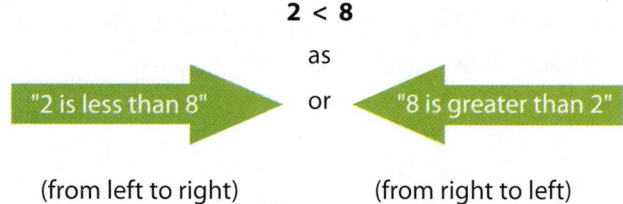

Remember that, from left to right, ≥ is read "greater than or equal to," and ≤ is read "less than or equal to." Thus, the symbols ≥ and ≤ allow for both equality and inequality. That is, if > **or** = is true, then ≥ is true.

For example,

$6 \geq -13$ and $6 \geq 6$ are both true statements.

$6 = -13$ is false, but $6 > -13$ is true, which means $6 \geq -13$ is true and

$6 > 6$ is false, but $6 = 6$ is true, which means $6 \geq 6$ is true.

Example 4

Correcting False Statements

Determine whether each of the following statements is true or false. Rewrite any false statement so that it is true. There may be more than one change that will make a false statement correct.

a. $4 \leq 12$

Solution

$4 \leq 12$ is true since 4 is less than 12.

b. $4 \leq 4$

Solution

$4 \leq 4$ is true since 4 is equal to 4.

c. $4 < 0$

Solution

$4 < 0$ is false. We can change the inequality to read $4 > 0$ or $0 < 4$.

d. $-7 \geq 0$

Solution

$-7 \geq 0$ is false. We can write $-7 \leq 0$ or $0 \geq -7$.

Now work margin exercise 4.

4. Determine whether each of the following statements is true or false. Rewrite any false statement so that it is true. There may be more than one change that will make a false statement correct.

a. $6 \geq 7$

b. $0 \geq 0$

c. $-8 < 0$

d. $2 \leq -2$

Objective C **Absolute Value**

Another concept closely related to **signed numbers** (positive and negative numbers and zero) is that of **absolute value**, symbolized with two vertical bars, | |. (**Note:** We will see later that the definition given here for the absolute value of integers is valid for any type of number on a number line.) We know, from working with number lines, that any integer and its opposite lie the same number of units from zero. For example +5 and −5 are both 5 units from 0 (Figure 8). The + and − signs indicate direction, and the 5 indicates distance. Thus $|-5| = |+5| = 5$.

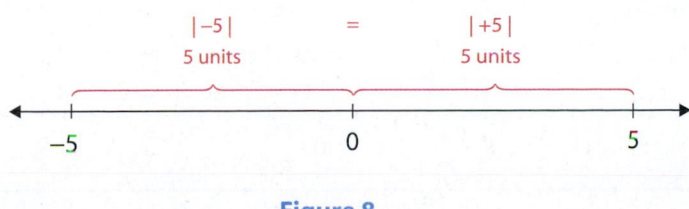

Figure 8

Absolute Value

The **absolute value** of an integer is its distance from 0. Symbolically, for any integer a,

$$\begin{cases} \text{If } a \text{ is a positive integer or } 0, \; |a| = a. \\ \text{If } a \text{ is a negative integer, } |a| = -a. \end{cases}$$

The absolute value of an integer is never negative.

notes

- When a represents a negative number, the symbol $-a$ represents a positive number. That is, the opposite of a negative number is a positive number. For example,

$$\text{If } a = -5, \quad \text{then} \quad -a = -(-5) = +5.$$

- Similarly,

$$\text{If } x = -3, \quad \text{then} \quad -x = -(-3) = +3.$$

$$\text{If } y = -8, \quad \text{then} \quad -y = -(-8) = +8.$$

- For these examples, we have

$$-a = +5, \quad -x = +3 \quad \text{and} \quad -y = +8.$$

Remember to read $-x$ as "the opposite of x" and not "negative x," because $-x$ may not be a negative number. In summary,

1. If x represents a positive number, then $-x$ represents a negative number.
2. If x represents a negative number, then $-x$ represents a positive number.

Example 5

Absolute Value

$$|-2| = -(-2) = 2$$

Solution

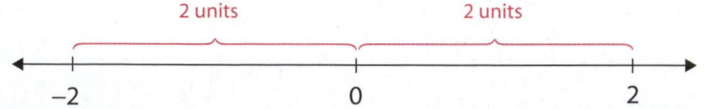

Example 6

Absolute Value

$|0| = -0 = +0 = 0$

Solution

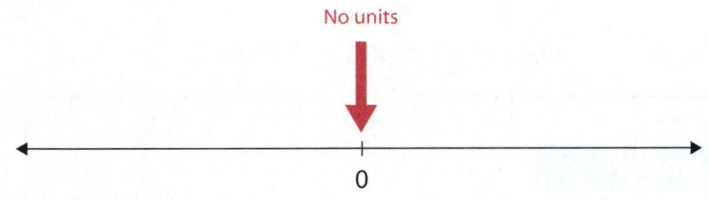

No units

0

Example 7

Absolute Value

True or false: $|-9| \leq 9$

Solution

True, since $|-9| = 9$ and $9 \leq 9$. (Remember that since $9 = 9$ the statement $9 \leq 9$ is true.)

Now work margin exercises 5 through 7.

Find the absolute value:

5. $|-1|$

6. $|+4|$

7. True or false:
$|-20| \leq 20$

Example 8

Absolute Value

If $|x| = 7$, what are the possible values for x?

Solution

Since $|-7| = 7$ and $|7| = 7$, then $x = -7$ or $x = 7$.

8. If $|z| = 3$, what are the possible values of z?

9. If $|y| = -19$, what are the possible values of y?

Example 9

Absolute Value

If $|a| = -2$, what are the possible values for a?

Solution

There are no values of a for which $|a| = -2$. The absolute value of every nonzero number is positive.

Now work margin exercises 8 and 9.

Objective D ## Graphing Inequalities

Inequalities, such as $x \geq 3$ and $y < 1$, where the replacement set for x and y is the set of integers, have an infinite number of solutions. Any integer greater than or equal to 3 is one of the solutions for $x \geq 3$ and any integer less than 1 is one of the solutions for $y < 1$. The solution sets can be represented as

$$\{\,3, 4, 5, 6, \ldots\,\} \quad \text{and} \quad \{\,\ldots, -4, -3, -2, -1, 0\,\}, \text{respectively.}$$

The graphs of the solutions to such inequalities are indicated by marking large dots over a few of the numbers in the solution sets and then putting three dots above the number line to indicate the pattern is to continue without end (see Figure 9).

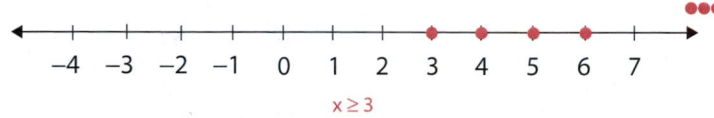

$x \geq 3$

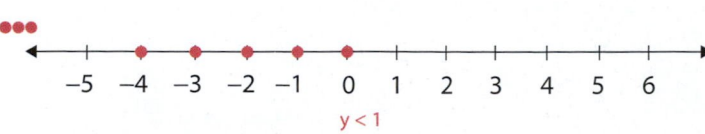

$y < 1$

Figure 9

The solution sets and the corresponding graphs for inequalities that have absolute values can be finite or infinite, depending on the nature of the inequality. Example 10 and Completion Example 11 illustrate both situations. Look these examples over carefully to fully understand the thinking involved with absolute value inequalities.

Example 10

Finding the Possible Integer Values

If $|x| \geq 4$, what are the possible integer values for x? Graph these integers on a number line.

Solution

There are an infinite number of integers 4 or more units from 0, both negative and positive. These integers are $\{\ldots, -7, -6, -5, -4, 4, 5, 6, 7, \ldots\}$.

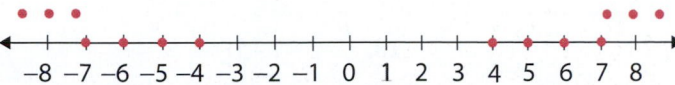

Completion Example 11

Finding the Possible Integer Values

If $|x| < 4$, what are the possible integer values for x? Graph these integers on a number line.

Solution

The integers that are less than 4 units from 0 have absolute values less than 4. These integers are _____.

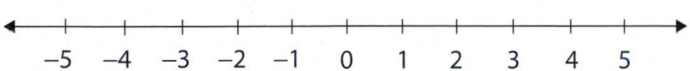

Now work margin exercises 10 and 11.

10. If $|x| \geq 1$, what are the possible integer values for x? Graph these integers on a number line.

11. If $|x| < 3$, what are the possible integer values for x? Graph these integers on a number line.

Completion Example Answers

11. These integers are –3, –2, –1, 0, 1, 2, 3.

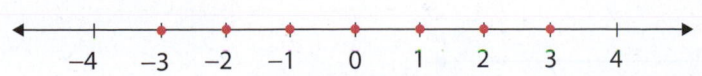

Exercises 1.8

1. 3 _____ 0

2. 7 _____ −1

3. 10 _____ −10

4. −3 _____ −2

5. −6 _____ 0

6. −2 _____ −1

7. −20 _____ −19

8. −67 _____ −50

9. $|-4|$ _____ 4

10. $|7|$ _____ −7

11. $|-8|$ _____ −8

12. −15 _____ $|-15|$

Determine whether each statement is true or false. If the statement is false, change the inequality or equality symbol so that the statement is true. (There may be more than one change that will make a false statement correct.) See Example 4.

13. $0 = -0$

14. $-9 < -10$

15. $-2 < 0$

16. $8 > |-8|$

17. $|2| = -2$

18. $-5 = |-5|$

19. $-3 < |-3|$

20. $-5 < 5$

21. $|-3| < 3$

22. $16 > |-16|$

23. $-4 = |-4|$

24. $7 = |-7|$

Graph each of the following sets of numbers on a number line. See Examples 2 and 3.

25. $\{1, 3, 4, 6\}$

26. $\{-3, -2, 0, 1, 3\}$

27. $\{-2, -1, 0, 2, 4\}$

28. $\{-5, -4, -3, -2, 0, 1\}$

29. All whole numbers less than 4

30. All integers less than 4

31. All integers greater than or equal to −2

32. All negative integers greater than −4

33. All whole numbers less than 0

34. All integers less than or equal to 0

35. $|x| = 5$ **36.** $|y| = 8$ **37.** $|a| = 2$ **38.** $|a| = 0$

39. $|y| = -6$ **40.** $|a| = -1$ **41.** $|x| = 23$ **42.** $|x| = 105$

43. $a > 3$

<--->

44. $x \leq 2$

<--->

45. $|x| > 5$

<--->

46. $|a| > 6$

<--->

47. $|x| \leq 2$

<--->

48. $|y| \geq 5$

<--->

49. $|a|$ is (never, sometimes, always) equal to a.

50. $|x|$ is (never, sometimes, always) equal to $-x$.

51. $|y|$ is (never, sometimes, always) equal to a positive integer.

52. $|x|$ is (never, sometimes, always) greater than x.

Writing & Thinking

53. Explain, in your own words, how an expression such as $-y$ might represent a positive number.

54. Explain, in your own words, the meaning of absolute value.

55. Explain, in your own words, why

  a. there are no values for x where $|x| < -5$.

  b. the solution set for $|x| > -5$ is all integers.

   Show your explanations to a friend to see if he or she can understand your reasoning.

1.9 Addition with Integers

Objective A An Intuitive Approach

We have discussed the operations of addition, subtraction, multiplication, and division with whole numbers. In the next three sections, we will discuss the basic rules and techniques for these same four operations with integers.

As an intuitive approach to addition with integers, consider an open field with a straight line marked with integers (much like a football field is marked every 10 yards). Imagine that a ball player stands at 0 and throws a ball to the point marked 4 and then stands at 4 and throws the ball 3 more units in the positive direction (to the right). Where will the ball land? (See Figure 1.)

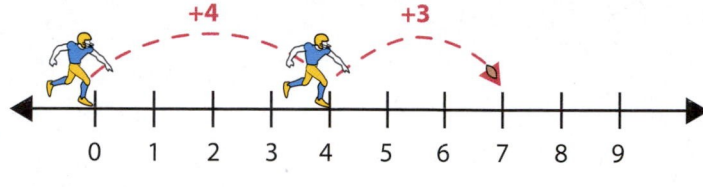

Figure 1

The ball will land at the point marked +7. We have, essentially, added the two positive integers +4 and +3.

$$(+4) + (+3) = +7 \qquad \text{or} \qquad 4 + 3 = 7$$

Now, if the same player **stands at 0** and throws the ball the same distances in the opposite direction (to the left), where will the ball land on the second throw? The ball will land at −7. We have just added two negative numbers, −4 and −3. (See Figure 2.)

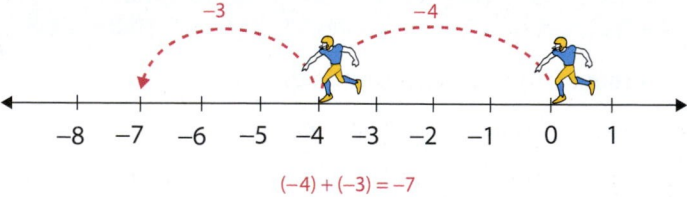

$$(-4) + (-3) = -7$$

Figure 2

Sums involving both positive and negative integers are illustrated in Figures 3 and 4.

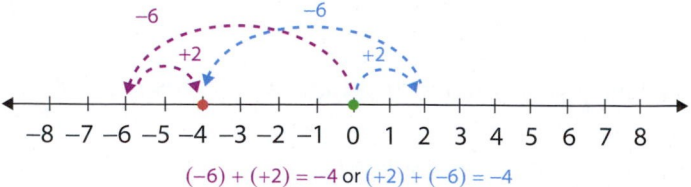

$$(-6) + (+2) = -4 \text{ or } (+2) + (-6) = -4$$

Figure 3

Objectives

A Understand intuitively how to add integers.

B Know how to add integers.

C Learn how to add integers in a vertical format.

D Be able to determine whether or not a particular integer is a solution to an equation.

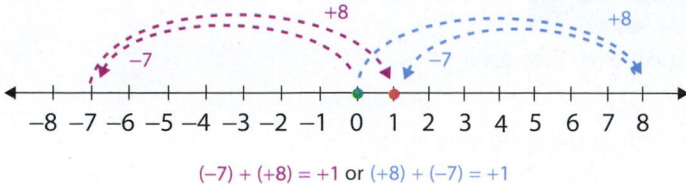

$$(-7) + (+8) = +1 \text{ or } (+8) + (-7) = +1$$

Figure 4

Although we are concerned here mainly with learning the techniques of adding integers, we note that addition with integers is **commutative** (as illustrated in Figures 3 and 4) and **associative**.

To help in understanding the rules for addition, we make the following suggestions for reading expressions with the + and − signs

+ used as the sign of a number is read "positive"
+ used as an operation is read "plus"
− used as the sign of a number is read "negative" and "opposite"
− used as an operation is read "minus" (See Section 1.10)

In summary:

1. The sum of two positive integers is positive.

$$(+5) \quad + \quad (+4) \quad = \quad +9$$

positive plus positive   positive

2. The sum of two negative integers is negative.

$$(-2) \quad + \quad (-3) \quad = \quad -5$$

negative plus negative   negative

3. The sum of a positive integer and a negative integer may be negative or positive (or zero), depending on which number is farther from 0.

$$(+3) \quad + \quad (-7) \quad = \quad -4$$

positive plus negative   negative

$$(+8) \quad + \quad (-2) \quad = \quad +6$$

positive plus negative   positive

1. Find each of the following sums.

a. $(+3) + (+5)$

b. $(+11) + (-2)$

c. $(-4) + (-6)$

Example 1

Addition with Integers

Find each of the following sums.

a. $(-15) + (-5)$

Solution

$(-15) + (-5) = -20$

b. $(+10) + (-2)$

Solution

$(+10) + (-2) = +8$

c. $(+4) + (+11)$

Solution

$(+4) + (+11) = +15$

d. $(-12) + (+5)$

Solution

$(-12) + (+5) = -7$

e. $(-90) + (+90)$

Solutions

$(-90) + (+90) = 0$

Now work margin exercise **1.**

Objective B **Rules for Addition with Integers**

Saying that one number is farther from 0 than another number is the same as saying that the first number has a larger absolute value. With this basic idea, the rules for adding integers can be written out formally in terms of absolute value.

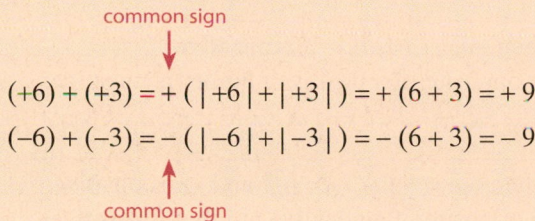

Rules for Addition with Integers

1. To add two integers with like signs, add their absolute values and use the common sign.

common sign

$$(+6) + (+3) = + (\, |+6| + |+3| \,) = + (6 + 3) = +9$$

$$(-6) + (-3) = - (\, |-6| + |-3| \,) = - (6 + 3) = -9$$

common sign

2. To add two integers with unlike signs, subtract their absolute values (the smaller from the larger) and use the sign of the integer with the larger absolute value.

because $|-15| > |+10|$

$$(-15) + (+10) = - (\, |-15| - |+10| \,) = - (15 - 10) = -5$$

$$(+15) + (-10) = + (\, |+15| - |-10| \,) = + (15 - 10) = +5$$

because $|+15| > |-10|$

If two numbers are the same distance from 0 in opposite directions, then each number is the **opposite** (or **additive inverse**) of the other, and their sum is 0. Addition with any integer and its opposite (or additive inverse) always yields the sum of 0. The idea of "opposite" rather than "negative" is very important for understanding both addition and subtraction (see Section 1.10) with integers. The following definition and discussion clarify this idea.

Additive Inverse

The **opposite** of an integer is called its **additive inverse**.

The sum of any integer and its additive inverse is 0.

Symbolically, for any integer a,

$$a + (-a) = 0.$$

As an example, $(+20) + (-20) = + (\, |+20| - |-20| \,) = + (20 - 20) = 0.$

notes

- The symbol −*a* should be read as "the opposite of *a*." Since *a* is a variable, −*a* might be positive, negative, or 0. For example:

  If *a* = 10, then −*a* = −10 and −*a* is **negative**.
  That is, the opposite of a positive number is negative.

- However,

  if *a* = −7, then −*a* = −(−7) = +7 and −*a* is **positive**.
  That is, the opposite of a negative number is positive.

2. Find the additive inverse (opposite) of each integer.

a. 9

b. −4

c. −28

Example 2

Finding the Additive Inverse

Find the additive inverse (opposite) of each integer.

a. 5

Solution

The additive inverse of 5 is − 5, and 5 + (− 5) = 0.

b. − 2

Solution

The additive inverse of − 2 is + 2, and (− 2) + (+ 2) = 0.

c. − 15

Solution

The additive inverse of − 15 is + 15, and (− 15) + (+ 15) = 0.

Now work margin exercise 2.

Objective C Addition in a Vertical Format

Equations in algebra are almost always written horizontally, so addition (and subtraction) with integers is done much of the time in the horizontal format. However, there are situations (such as in long division) where sums (and differences) are written in a vertical format with one number directly under another, as illustrated in Example 3.

notes

■ The positive sign (+) may be omitted when writing positive numbers, but the negative sign (−) must always be written for negative numbers. Thus, if there is no sign in front of an integer, the integer is understood to be positive.

Example 3

Addition with Integers

One technique for adding several integers is to mentally add the positive and negative integers separately and then add these results. (We are, in effect, using the commutative and associative properties of addition.)

Find the sums.

a. -20
          7
          -30

Solution

$\begin{array}{r} -20 \\ 7 \\ -30 \\ \hline -43 \end{array}$ or $\begin{array}{r} -50 \\ 7 \\ \hline -43 \end{array}$

b. 42
          -10
          -3

Solution

$\begin{array}{r} 42 \\ -10 \\ -3 \\ \hline 29 \end{array}$ or $\begin{array}{r} 42 \\ -13 \\ \hline 29 \end{array}$

3. Find the sums.

a. -10
        5
    $\underline{-40}$

b. 57
       -5
    $\underline{-1}$

c. -10
        -5
        -1
     $\underline{-2}$

c. -6
        -7
        -8
     $\underline{-9}$

Solution

  -6
  -7
  -8
$\underline{-9}$
-30

Now work margin exercise 3.

Objective D **Determining Integer Solutions to Equations**

We can use our knowledge of addition with integers to determine whether a particular integer is a solution to an equation.

Example 4

Determining the Solution

Determine whether or not the given integer is a solution to the given equation by substituting for the variable and adding.

a. $x + 8 = -2;\ x = -10$

Solution

$$x + 8 = -2$$
$$-10 + 8 \overset{?}{=} -2$$
$$-2 = -2$$

-10 **is** a solution.

b. $x + (-5) = -6;\ x = 1$

Solution

$$x + (-5) = -6$$
$$+1 + (-5) \overset{?}{=} -6$$
$$-4 \neq -6$$

$+1$ **is not** a solution.

c. $17 + y = 0$; $y = -17$

Solution

$$17 + y = 0$$
$$17 + (-17) \overset{?}{=} 0$$
$$0 = 0$$

-17 **is** a solution.

Now work margin exercise 4.

4. Determine whether $x = -5$ is a solution to each equation.

a. $x + 14 = 9$

b. $x + 5 = 0$

c. $x + (-25) = 9$

Exercises 1.9

Name the additive inverse (opposite) of each integer. See Example 2.

1. 15

2. 28

3. −40

4. −32

5. −9

6. 11

Evaluate each of the indicated sums. See Example 1.

7. $8 + (-10)$

8. $19 + (-22)$

9. $(-3) + (-5)$

10. $(-4) + (-7)$

11. $(-1) + (-16)$

12. $(-6) + (-2)$

13. $(-9) + (+9)$

14. $(+43) + (-43)$

15. $(+73) + (-73)$

16. $-10 + (+10)$

17. $-11 + (-5) + (-2)$

18. $(-3) + (-6) + (-4)$

19. $-25 + (-30) + (+10)$

20. $+36 + (-12) + (-1)$

21. $(-35) + (+18) + (+17)$

22. $12 + (+14) + (-16)$

23. $-33 + (+13) + (-12)$

24. $-36 + (-2) + (+40)$

25. $9 + (-4) + (-5) + (+3)$

26. $12 + (-5) + (-10) + (+8)$

27. $-15 + (-3) + (+6) + (-7)$

28. $-26 + (+3) + (-15) + (+25)$

29. $-110 + (-25) + (+125)$

30. $-120 + (-30) + (-5)$

Find each of the indicated sums in a vertical format. See Example 3.

31.
$$\begin{array}{r} -32 \\ 8 \\ \hline \end{array}$$

32.
$$\begin{array}{r} -53 \\ 19 \\ \hline \end{array}$$

33.
$$\begin{array}{r} 37 \\ -6 \\ \hline \end{array}$$

34.
$$\begin{array}{r} -80 \\ -8 \\ \hline \end{array}$$

35.
$$\begin{array}{r} 102 \\ -21 \\ -5 \\ \hline \end{array}$$

36.
$$\begin{array}{r} 130 \\ -45 \\ -32 \\ \hline \end{array}$$

37.
$$\begin{array}{r} -210 \\ -200 \\ 100 \\ \hline \end{array}$$

38.
$$\begin{array}{r} -108 \\ -105 \\ -330 \\ \hline \end{array}$$

39.
$$\begin{array}{r} 35 \\ 2 \\ -5 \\ -5 \\ \hline \end{array}$$

40.
$$\begin{array}{r} -56 \\ -3 \\ -1 \\ 3 \\ \hline \end{array}$$

41.
$$\begin{array}{r} 28 \\ 9 \\ -9 \\ -28 \\ \hline \end{array}$$

42.
$$\begin{array}{r} 76 \\ 12 \\ -12 \\ -76 \\ \hline \end{array}$$

43. Name two numbers that are

a. six units from 0 on a number line.

b. five units from 10 on a number line.

c. nine units from 3 on a number line.

44. Name two numbers that are

a. three units from 7 on a number line.

b. eight units from 5 on a number line.

c. three units from −2 on a number line.

Find each of the indicated sums. Be sure to find the absolute values first.

45. $13 + |-5|$

46. $|-2| + (-5)$

47. $|-10| + |-4|$

48. $|-7| + (+7)$

49. $|-18| + |+17|$

50. $|-14| + |-6|$

Determine whether or not the given integer is a solution to the equation by substituting the given value for the variable and adding. See Example 4.

51. $x + 5 = 7;\ x = -2$

52. $x + 6 = 9;\ x = -3$

53. $x + (-3) = -10;\ x = -7$

54. $-5 + x = -13;\ x = -8$

55. $y + (24) = 12;\ y = -12$

56. $y + (35) = -2;\ y = -37$

57. $z + (-18) = 0;\ z = 18$

58. $z + (27) = 0;\ z = -27$

59. $a + (-3) = -10;\ a = 7$

60. $a + (-19) = -29;\ a = 10$

Choose the response that correctly completes each statement. In each problem, give two examples that illustrate your reasoning.

61. If x and y are integers, then $x + y$ is (never, sometimes, always) equal to 0.

62. If x and y are integers, then $x + y$ is (never, sometimes, always) negative.

63. If x and y are integers, then $x + y$ is (never, sometimes, always) positive.

64. If x is a positive integer and y is a negative integer, then $x + y$ is (never, sometimes, always) equal to 0.

65. If x and y are both positive integers, then $x + y$ is (never, sometimes, always) equal to 0.

66. If x and y are both negative integers, then $x + y$ is (never, sometimes, always) equal to 0.

67. If x is a negative integer, then $-x$ is (never, sometimes, always) negative.

68. If x is a positive integer, then $-x$ is (never, sometimes, always) negative.

69. Describe in your own words the conditions under which the sum of two integers will be 0.

70. Explain how the sum of the absolute values of two integers might be 0. (Is this possible?)

Use a calculator to find the value of each expression.

71. $6790 + (-5635) + (-4560)$

72. $-8950 + (-3457) + (-3266)$

73. $-10{,}890 + (-5435) + (25{,}000) + (-11{,}250)$

74. $29{,}842 + (-5854) + (-12{,}450) + (-13{,}200)$

75. $72{,}456 + (-83{,}000) + 63{,}450 + (-76{,}000)$

76. $(4783 + 5487 + (-734)) + (7125 + (-8460))$

77. $(750 + 320 + (-400)) + (325 + (-500))$

78. $((-500) + (-300) + 400) + ((-75) + (-20))$

1.10 Subtraction with Integers

Subtraction with whole numbers is defined in terms of addition. For example, the difference $20 - 14$ is equal to 6 because $14 + 6 = 20$. If we are restricted to whole numbers, then the difference $14 - 20$ cannot be found because there is no whole number that we can add to 20 and get 14. Now that we are familiar with integers and addition with integers, we can reason that

$$14 - 20 = -6 \text{ because } 14 + (-20) = -6.$$

That is, we can now define subtraction in terms of addition with integers so that a larger number may be subtracted from a smaller number. The definition of subtraction with integers involves the concept of additive inverses (opposites) as discussed in Section 1.9. We rewrite the definition here for convenience and easy reference.

Additive Inverse

The **opposite** of an integer is called its **additive inverse**. The sum of an integer and its additive inverse is 0. Symbolically, for any integer a,

$$a + (-a) = 0.$$

Recall that the symbol $-a$ should be read as "the opposite of a." Since a is a variable, $-a$ might be positive, negative, or 0.

Subtraction with integers is defined in terms of addition. The distinction is that now we can subtract larger numbers from smaller numbers, and we can subtract negative and positive integers from each other. In effect, we can now study differences that are negative. For example,

$5 - 8 = -3$	because	$5 + (-8) = -3$
$10 - 12 = -2$	because	$10 + (-12) = -2$
$10 - (-6) = 16$	because	$10 + (+6) = 16$
$4 - (-1) = 5$	because	$4 + (+1) = 5$
$-5 - (-8) = 3$	because	$-5 + (+8) = 3.$

Difference

For any integers a and b, the **difference** between a and b is defined as follows:

$$a - b = a + (-b)$$

↑ minus   ↑ opposite

This expression is read, "a minus b is equal to a plus the opposite of b."

Objectives

A   Know how to subtract integers.

B   Use the relationship between addition and subtraction of integers to simplify notation.

C   Learn how to subtract integers in a vertical format.

D   Be able to determine whether or not a particular integer is a solution to an equation.

Since a and b are variables that may themselves be positive, negative, or 0, the definition automatically includes other forms as follows:

$$a - (-b) = a + (+b) = a + b \quad \text{and} \quad -a - b = -a + (-b).$$

The following examples illustrate how to apply the definition of subtraction. Remember, to subtract an integer you add its opposite.

Example 1

Subtraction with Integers

Find the following differences.

a. $(-1) - (-5)$
b. $(-2) - (-7)$
c. $(-10) - (+3)$
d. $(+4) - (+5)$
e. $(-6) - (-6)$

Solutions

a. $(-1) - (-5) = (-1) + (+5) = 4$
b. $(-2) - (-7) = (-2) + (+7) = 5$
c. $(-10) - (+3) = (-10) + (-3) = -13$
d. $(+4) - (+5) = (+4) + (-5) = -1$
e. $(-6) - (-6) = (-6) + (+6) = 0$

Objective B Simplifying Notation

The interrelationship between addition and subtraction with integers allows us to eliminate the use of so many parentheses. The following two interpretations of the same problem indicate that both interpretations are correct and lead to the same answer.

$$5 - 18 \quad = \quad 5 \quad - \quad (+18) \quad = \quad -13$$

5   minus   positive 18   negative 13

$$5 - 18 \quad = \quad 5 \quad + \quad (-18) \quad = \quad -13$$

5   plus   negative 18   negative 13

Thus the expression $5 - 18$ can be thought of as subtraction or addition. In either case, the answer is the same. Understanding that an expression such as $5 - 18$ (with no parentheses) can be thought of as addition or subtraction takes some practice, but it is quite important because the notation is commonly used in all mathematics textbooks. Study the following examples carefully.

Example 2

Subtraction with Integers

a. $5 - 18 = -13$ think: $(+5) + (-18)$

b. $-5 - 18 = -23$ think: $(-5) + (-18)$

c. $-16 - 3 = -19$ think: $(-16) + (-3)$

d. $25 - 20 = 5$ think: $(+25) + (-20)$

e. $32 - 33 = -1$ think: $(+32) + (-33)$

Now work margin exercises 1 and 2.

In Section 1.9, we discussed the fact that addition with integers is both commutative and associative. That is, for integers a, b, and c,

$$a + b = b + a \qquad \text{commutative property of addition}$$

and $\qquad a + (b + c) = (a + b) + c. \qquad$ associative property of addition

However, as the following examples illustrate, subtraction is **not commutative** and **not associative**.

$$13 - 5 = 8 \qquad \text{and} \qquad 5 - 13 = -8$$

So, $13 - 5 \neq 5 - 13$, and in general,

$$a - b \neq b - a. \qquad (\textbf{Note: } \neq \text{ is read "is not equal to."})$$

Similarly,

$$8 - (3 - 1) = 8 - (2) = 6 \qquad \text{and} \qquad (8 - 3) - 1 = 5 - 1 = 4$$

So, $8 - (3 - 1) \neq (8 - 3) - 1$, and in general,

$$a - (b - c) \neq (a - b) - c.$$

Thus the order that we write the numbers in subtraction and the use of parentheses can be critical. Be careful when you write any expression involving subtraction.

Objective C   Subtraction in a Vertical Format

As with addition, integers can be written vertically in subtraction. One number is written underneath the other, and the sign of the integer being subtracted is changed. That is, we add the opposite of the integer that is subtracted.

Find the following differences.

1. a. $(-4) - (-4)$

  b. $(-3) - (+8)$

  c. $(-3) - (-8)$

2. a. $2 - 3$

  b. $-6 - 12$

  c. $-10 - 2$

  d. $30 - 36$

Subtract the bottom number from the top number.

3.
$$\begin{array}{r} -\,27 \\ -(+\,19) \\ \hline \end{array}$$

4.
$$\begin{array}{r} -\,20 \\ -(+87) \\ \hline \end{array}$$

5.
$$\begin{array}{r} -56 \\ -(-56) \\ \hline \end{array}$$

Example 3

Subtraction with Integers

To subtract, add the opposite of the bottom number.

$$\begin{array}{r} -\,45 \\ -(+\,32) \\ \hline \end{array} \qquad \begin{array}{r} -45 \\ +(-32) \\ \hline -77 \end{array}$$ sign is changed

difference

Example 4

Subtraction with Integers

To subtract, add the opposite of the bottom number.

$$\begin{array}{r} -56 \\ -\left(+10\right) \\ \hline \end{array} \qquad \begin{array}{r} -56 \\ +\left(-10\right) \\ \hline -66 \end{array}$$ sign is changed

difference

Example 5

Subtraction with Integers

To subtract, add the opposite of the bottom number.

$$\begin{array}{r} -\,20 \\ -(-\,95) \\ \hline \end{array} \qquad \begin{array}{r} -\,20 \\ +\left(+\,95\right) \\ \hline +\,75 \end{array}$$ sign is changed

difference

Now work margin exercises 3 through 5.

Objective D   **Determining the Solutions to Equations**

Now both addition and subtraction can be used to help determine whether or not a particular integer is a solution to an equation. (This skill will prove useful in evaluating formulas and solving equations.)

Example 6

Determining the Solution

Determine whether or not the given integer is a solution to the given equation by substituting for the variable and then evaluating.

a. $x - (-6) = -10$; $x = -14$

Solution

$$x - (-6) = -10$$
$$-14 - (-6) \stackrel{?}{=} -10$$
$$-14 + (+6) \stackrel{?}{=} -10$$
$$-8 \neq -10$$

-14 is **not** a solution.

b. $7 - y = -1$; $y = 8$

Solution

$$7 - y = -11$$

$$7 - 8 \stackrel{?}{=} -1$$
$$-1 = -1$$

8 **is** a solution.

c. $a - 12 = -2$; $a = -10$

Solution

$$a - 12 = -2$$
$$-10 - 12 \stackrel{?}{=} -2$$
$$-22 \neq -2$$

-10 **is not** a solution.

Now work margin exercise **6.**

6. Determine whether or not the given integer is a solution to the given equation by substituting for the variable and then evaluating.

a. $x - (-2) = 5$; $x = 3$

b. $2 - y = -13$; $y = 15$

c. $a - 6 = -3$; $a = 3$

Exercises 1.10

Perform the indicated operations. See Example 3.

1. **a.** $6 - 14$
    b. $-6 - 14$

2. **a.** $17 - 5$
    b. $-17 - 5$

3. **a.** $-10 - 19$
    b. $10 - 19$

4. **a.** $-25 - 3$
    b. $25 - 3$

5. **a.** $12 - 20$
    b. $-12 - 20$

6. **a.** $-15 - 18$
    b. $15 - 18$

Perform the indicated operations.

7. $9 - 3$

8. $10 - 7$

9. $-5 - (-10)$

10. $-8 - (-4)$

11. $-7 + (-13)$

12. $(-2) + (-15)$

13. $-14 - 20$

14. $-17 - 30$

15. $0 - (-4)$

16. $0 - (-9)$

17. $0 - 4$

18. $0 - 9$

19. $-2 - 3 + 11$

20. $-5 - 4 + 6$

21. $-6 - 5 - 4$

22. $-3 - (-3) + (-2)$

23. $-7 + (-7) + (-8)$

24. $-1 + 12 - 10$

25. $-8 - 11$

26. $-5 - 11$

27. $13 - 20$

28. $17 - 22$

29. $-6 - 7 - 8$

30. $-40 + 15$

31. $-16 + 15 - 1$

32. $25 - 13 - 12$

33. $18 - 11 - 7$

34. $-7 - (-10) + 3$

35. $-10 - (-5) + 5$

36. $15 + (-4) + (-5) - (-3)$

37. $-17 + (-14) - (-3)$

38. $-80 - 70 - 30 + 100$

39. $-18 - 17 - 9 + (-1)$

40. $-21 - 5 - 8 + (-2)$

41. $-4 - (-3) + (-5) + 7 - 2$

42. $-6 - (-8) + 10 - 3$

43. $26 - 13 - 17 + 5 - 20$

44. $35 - 22 - 8 + 7 - 30$

Subtract the bottom number from the top number. See Examples 3, 4, and 5.

45. $\begin{array}{r} 37 \\ -(45) \\ \hline \end{array}$

46. $\begin{array}{r} 20 \\ -(36) \\ \hline \end{array}$

47. $\begin{array}{r} -21 \\ -(-10) \\ \hline \end{array}$

48. $\begin{array}{r} -46 \\ -(-25) \\ \hline \end{array}$

49. $\begin{array}{r} -16 \\ -(-16) \\ \hline \end{array}$

50. $\begin{array}{r} -24 \\ -(-24) \\ \hline \end{array}$

51. $\begin{array}{r} 105 \\ -(-22) \\ \hline \end{array}$

52. $\begin{array}{r} 202 \\ -(-35) \\ \hline \end{array}$

Find the differences indicated in each table.

53.

Value of a	Value of b	Difference between a and b
a	b	$a - b$
10	8	_____
10	10	_____
10	12	_____
10	14	_____

54.

Value of a	Value of b	Difference between a and b
a	b	$a - b$
12	-8	_____
12	-10	_____
12	-12	_____
12	-14	_____

55.

Value of a	Value of b	Difference between a and b
a	b	$a - b$
-10	8	_____
-10	10	_____
-10	12	_____
-10	14	_____

56.

Value of a	Value of b	Difference between a and b
a	b	$a - b$
-12	-6	_____
-12	-7	_____
-12	-8	_____
-12	-9	_____

Insert the symbol in the blank that will make each statement true: <, >, or =.

57. $-7 + (-2)$ _____ $-4 - 6$

58. $-6 + (-3)$ _____ $-2 - 2$

59. $0 - 8$ _____ $0 - (-8)$

60. $0 - 12$ _____ $0 - (-2)$

61. $9 - (-3)$ _____ $9 - 3$

62. $4 - (-1)$ _____ $4 - 1$

63. $-6 - 6$ _____ $0 - 12$

64. $-10 - 10$ _____ $0 - 20$

Determine whether or not the given number is a solution to the equation. Substitute the number for the variable and perform the indicated operations mentally. See Example 6.

65. $x - 7 = -9$; $x = -2$

66. $x - 4 = -10$; $x = -6$

67. $a + 5 = -10$; $a = -15$

68. $a - 3 = -12$; $a = -9$

69. $x + 13 = 13$; $x = 0$

70. $x + 21 = 21$; $x = 0$

71. $3 - y = 6$; $y = 3$

72. $5 - y = 10$; $y = 5$

73. $22 + x = -1$; $x = -23$

74. $30 + x = -2$; $x = -32$

Solve the following word problems.

75. Temperature: Beginning with a temperature of 8° above zero, the temperature was measured hourly for 4 hours. It rose 3°, dropped 7°, dropped 2°, and rose 1°. What was the final temperature recorded?

76. Diet Plan: George and his wife went on a diet plan for 5 weeks. During these 5 weeks, George lost 5 pounds, gained 2 pounds, lost 4 pounds, lost 6 pounds, and gained 3 pounds. What was his total loss or gain for these 5 weeks? If he weighed 225 pounds when he started the diet, what did he weigh at the end of the 5-week period?

77. NASDAQ: In a 5-day week, the NASDAQ stock market posted a gain of 145 points, a loss of 100 points, a loss of 82 points, a gain of 50 points, and a gain of 25 points. If the NASDAQ started the week at 4200 points, what was the market at the end of the week?

78. Dow Jones: In a 5-day week, the Dow Jones stock market mean showed a gain of 32 points, a gain of 10 points, a loss of 20 points, a loss of 2 points, and a loss of 25 points. What was the net change in the stock market for the week? If the Dow started the week at 10,530 points, what was the mean at the end of the week?

79. Football Game: In 10 running plays in a football game, the halfback gained 2 yards, gained 12 yards, lost 5 yards, lost 3 yards, gained 22 yards, gained 3 yards, gained 7 yards, lost 2 yards, gained 4 yards, and gained 45 yards. What was his net yardage for the game? This was a good game for him but not his best.

Use a calculator to find the value of each expression.

80. **a.** $14{,}655 - 15{,}287$

  b. $14{,}655 - (-15{,}287)$

81. **a.** $22{,}456 - 35{,}000$

  b. $22{,}456 - (-35{,}000)$

82. **a.** $-203{,}450 - 16{,}500 + 45{,}600$

  b. $-203{,}450 - (-16{,}500) + (-45{,}600)$

83. **a.** $500{,}000 - 1{,}043{,}500 - 250{,}000 + 175{,}500$

  b. $500{,}000 - (-1{,}043{,}500) - (-250{,}000) + (-175{,}500)$

84. **a.** $345,300 - 42,670 - 356,020 - 250,000 + 321,000$

b. $345,300 - (-42,670) - (-356,020) - (-250,000) + (-321,000)$

85. **a.** What should be added to -7 to get a sum of -12?

b. What should be added to -7 to get a sum of 12?

86. **a.** What should be subtracted from -10 to get a difference of 20?

b. What should be subtracted from -10 to get a difference of -20?

87. From the sum of -3 and -5, subtract the sum of -20 and 13. What is the difference?

88. Find the difference between the sum of 6 and -10 and the sum of -15 and -4.

Writing & Thinking

89. Explain why the expression $-x$ might represent a positive number, a negative number, or even 0.

90. Under what conditions can the difference of two negative numbers be a positive number?

A Know the rules for multiplication with integers and be able to multiply with integers.

B Know the rules for division with integers and be able to divide with integers.

C Be able to apply the rules for order of operations to expressions with integers.

1.11 Multiplication, Division, and Order of Operations with Integers

Objective A **Multiplication with Integers**

We know that multiplication with whole numbers is a shorthand form of repeated addition. Multiplication with integers can be thought of in the same manner. For example,

$$6 + 6 + 6 + 6 + 6 = 5 \cdot 6 = 30$$

and

$$(-6) + (-6) + (-6) + (-6) + (-6) = 5(-6) = -30$$

and

$$(-3) + (-3) + (-3) + (-3) = 4(-3) = -12.$$

As we have just seen, repeated addition with a negative integer results in a product of a positive integer and a negative integer. Therefore, because the sum of negative integers is negative, we can reason that the **product of a positive integer and a negative integer is negative**.

Example 1

Multiplication with Integers

a. $5(-4) = -20$

b. $3(-15) = -45$

c. $7(-11) = -77$

d. $100(-2) = -200$

e. $-1(9) = -9$

Now work margin exercise 1.

1. Find each product.

a. $5(-3)$

b. $-8(4)$

c. $-1(3)$

The product of two negative integers can be explained in terms of opposites and the fact that, for any integer a,

$$-a = -1(a).$$

That is, the opposite of a can be treated as -1 times a. Consider, for example, the product $(-2)(-5)$. We can proceed as follows:

$$(-2)(-5) = -1(2)(-5) \qquad \text{\color{red}{$-2 = -1(2)$}}$$
$$= -1[2(-5)] \qquad \text{\color{red}{associative property of multiplication}}$$
$$= -1[-10] \qquad \text{\color{red}{multiply}}$$
$$= -[-10] \qquad \text{\color{red}{-1 times a number is its opposite}}$$
$$= +10 \qquad \text{\color{red}{opposite of -10}}$$

Although one example does not prove a rule, this process is such that we can use it as a general procedure and come to the following correct conclusion:

The product of two negative numbers is positive.

Example 2

Multiplication with Integers

a. $(-6)(-4) = +24$

b. $-7(-9) = +63$

c. $-8(-10) = +80$

d. $0(-15) = 0$ and $0(+15) = 0$ and $0(0) = 0$

As Example **2d.** illustrates, **the product of 0 and any integer is 0**.

The rules for multiplication with integers can be summarized as follows.

Rules for Multiplication with Integers

If a and b are positive integers, then

1. The product of two positive integers is positive. $a \cdot b = ab$

2. The product of two negative integers is positive. $(-a)(-b) = ab$

3. The product of a positive integer and a negative integer is negative.
   $a(-b) = -ab$ and $(-a)(b) = -ab$

4. The product of 0 and any integer is 0.
   $a \cdot 0 = 0$ and $(-a) \cdot 0 = 0$

The **commutative** and **associative** properties for multiplication are valid for multiplication with integers, just as they are with whole numbers. Thus, to find the product of more than two integers, we can multiply any two and continue to multiply until all the integers have been multiplied.

Find each product.

2. a. $-6(-4)$

  b. $-19(0)$

  c. $-10(-2)$

3. a. $3(-20)(-1)(-1)$

  b. $-9(7)(-2)$

  c. $(-5)^3$

Example 3

Multiplication with Integers

a. $(-5)(-3)(10) = \left[(-5)(-3)\right](10)$
$$= \left[+15\right](10)$$
$$= +150$$

b. $(-2)(-2)(-2)(3) = +4(-2)(3)$
$$= -8(3)$$
$$= -24$$

c. $(-3)(5)(-6)(-1)(-1) = -15(-6)(-1)(-1)$
$$= (+90)(-1)(-1)$$
$$= (-90)(-1) \qquad \text{or} = (+90)(+1)$$
$$= +90$$

d. $(-4)^3 = (-4)(-4)(-4)$
$$= (+16)(-4)$$
$$= -64$$

e. $(-3)^2 = (-3)(-3)$
$$= +9$$

Now work margin exercises 2 and 3.

Objective B **Division with Integers**

For the purposes of this discussion, we will indicate division such as $a \div b$ in the fraction form, $\dfrac{a}{b}$. (Operations with fractions will be discussed in detail in the next chapter.) Thus a division problem such as $56 \div 8$ can be indicated in the fraction form as $\dfrac{56}{8}$.

Since division is defined in terms of multiplication, we have

$$\frac{56}{8} = 7 \quad \text{because} \quad 56 = 8 \cdot 7.$$

Another Meaning of $\dfrac{a}{b} = x$

For integers a, b, and x (where $b \neq 0$),

$$\frac{a}{b} = x \quad \text{means that} \quad a = b \cdot x.$$

notes

In this section, we are emphasizing the rules for signs when multiplying and dividing with integers. With this emphasis in mind, the problems are set up in such a way that the quotients are integers. However, these rules are valid for many types of numbers, and we will use them with fractions and decimals in later chapters.

Example 4

Division with Integers

a. $\dfrac{+35}{+5} = +7$ because $+35 = (+5)(+7)$.

b. $\dfrac{-35}{-5} = +7$ because $-35 = (-5)(+7)$.

c. $\dfrac{+35}{-5} = -7$ because $+35 = (-5)(-7)$.

d. $\dfrac{-35}{+5} = -7$ because $-35 = (+5)(-7)$.

The rules for division with integers can be stated as follows.

Rules for Division with Integers

If a and b are positive integers, then

1. The quotient of two positive integers is positive.

$$\frac{a}{b} = +\frac{a}{b}$$

2. The quotient of two negative integers is positive.

$$\frac{-a}{-b} = +\frac{a}{b}$$

3. The quotient of a positive integer and a negative integer is negative.

$$\frac{-a}{b} = -\frac{a}{b} \quad \text{and} \quad \frac{a}{-b} = -\frac{a}{b}$$

In summary:

when the signs are alike, the quotient is positive.

when the signs are not alike, the quotient is negative.

In Section 1.3, we stated that division by 0 is undefined. We explain it again here in terms of multiplication with fraction notation.

Division by 0 Is Not Defined

Case 1: Suppose that $a \neq 0$ and $\dfrac{a}{0} = x$. Then, by the meaning of division, $a = 0 \cdot x$. But this is not possible because $0 \cdot x = 0$ for any value of x, and we stated that $a \neq 0$.

Case 2: Suppose that $\dfrac{0}{0} = x$. Then, by the meaning of division, $0 = 0 \cdot x$ is true for all values of x. But we must have a unique answer for division. Therefore, we conclude that, in every case, division by 0 is not defined.

The following mnemonic device may help you remember the rules for division by 0:

$\dfrac{0}{K}$ is "**OK**". 0 **can** be in the **numerator**.

$\dfrac{K}{0}$ is a "**KO**" (knockout). 0 **cannot** be in the **denominator**.

Find each quotient.

4. a. $\dfrac{-30}{+10}$

b. $\dfrac{+40}{-10}$

c. $\dfrac{32}{8}$

5. a. $\dfrac{0}{5}$

b. $\dfrac{26}{0}$

c. $\dfrac{1}{0}$

Example 5

Division Involving 0

a. $\dfrac{0}{-2} = 0$ because $0 = -2(0)$.

b. $\dfrac{7}{0}$ is undefined. If $\dfrac{7}{0} = x$, then $7 = 0 \cdot x$. But this is not possible because $0 \cdot x = 0$ for any value of x.

c. $\dfrac{-32}{0}$ is undefined. There is no number whose product with 0 is -32.

d. $\dfrac{0}{0}$ is undefined. Suppose you think that $\dfrac{0}{0} = 1$ because $0 = 0 \cdot 1$. This is certainly true. However, someone else might reason in an equally valid way that $\dfrac{0}{0} = 92$. Since there can be no unique value, we conclude that $\dfrac{0}{0}$ is undefined.

Now work margin exercises 4 and 5.

Objective C **Rules for Order of Operations**

The rules for order of operations were discussed in section 1.6 and are restated here for convenience and easy reference.

> ### Rules for Order of Operations
>
> 1. First, simplify within grouping symbols, such as parentheses ( ), brackets [ ], braces { }, radicals signs $\sqrt{\ }$, absolute value bars | |, or fraction bars. Start with the innermost grouping.
> 2. Second, evaluate any numbers or expressions raised to exponents.
> 3. Third, moving from **left to right**, perform any multiplications or divisions in the order in which they appear.
> 4. Fourth, moving from **left to right**, perform any additions or subtractions in the order in which they appear.

Now that we know how to operate with integers, we can apply these rules in evaluating expressions that involve negative numbers as well as positive numbers.

Observe carefully the use of parentheses around negative numbers in Examples 6 and 7. Two operation signs are not allowed together. For example, do not write $+-$ or $\div \cdot$ or $+\div$ without using numbers and/or parentheses with the operations.

Example 6

Order of Operations with Integers

Evaluate the expression $27 \div (-9) \cdot 2 - 5 + 4(-5)$.

Solution

$27 \div (-9) \cdot 2 - 5 + 4(-5)$ Divide.

$= -3 \cdot 2 - 5 + 4(-5)$ Multiply.

$= -6 - 5 + 4(-5)$ Multiply.

$= -6 - 5 - 20$ Add. (Remember we are adding negative numbers.)

$= -11 - 20$ Add.

$= -31$

Find the value of the expression by using the rules for order of operations.

6. $15 \div 5 + 10 \cdot 2$

7. $4 \div 2^2 + 3 \cdot 2^2$

Example 7

Order of Operations with Integers

Evaluate the expression $9 - 11\left[\left(3 - 5^2\right) \div 2 + 5\right]$.

Solution

$$9 - 11\left[\left(3 - 5^2\right) \div 2 + 5\right] \quad \text{Evaluate the power.}$$
$$= 9 - 11\left[\left(3 - 25\right) \div 2 + 5\right] \quad \text{Operate within parentheses.}$$
$$= 9 - 11\left[\left(-22\right) \div 2 + 5\right] \quad \text{Divide within brackets.}$$
$$= 9 - 11\left[-11 + 5\right] \quad \text{Add within brackets.}$$
$$= 9 - 11\left[-6\right] \quad \text{Multiply, in this case, } -11 \text{ times } -6.$$
$$= 9 + 66 \quad \text{Add.}$$
$$= 75$$

We can also write

$$= 9 - 11\left[-6\right]$$
$$= 9 - (11)\left[-6\right]$$
$$= 9 - (-66)$$
$$= 9 + (+66)$$
$$= 75$$

The result is the same.

Now work margin exercises 6 and 7.

Completion Example 8

Order of Operations with Integers

Evaluate $9\left(-1+2^2\right)-7-6\cdot2^2$.

Solution

$$9\left(-1+2^2\right)-7-6\cdot2^2$$
$$=9\left(-1+\underline{}\right)-7-6\cdot\underline{}$$
$$=9\left(\underline{}\right)-7-6\cdot\underline{}$$
$$=\underline{}-7-6\cdot\underline{}$$
$$=\underline{}-7-\underline{}$$
$$=\underline{}-\underline{}$$
$$=\underline{}$$

Now work margin exercise 8.

8. Evaluate

$$2\left(-1+3^2\right)-4-3\cdot2^3.$$

Completion Example Answers

8. $9\left(-1+2^2\right)-7-6\cdot2^2$

$=9(-1+4)-7-6\cdot4$

$=9(3)-7-6\cdot4$

$=27-7-6\cdot4$

$=27-7-24$

$=20-24$

$=-4$

Exercises 1.11

Find the following products. See Examples 1, 2, and 3.

1. $4(-6)$

2. $-6(4)$

3. $-8(3)$

4. $-2(-7)$

5. $14(2)$

6. $21(3)$

7. $-2(9)$

8. $(-6)(-3)$

9. $-10(5)$

10. $9(-4)$

11. $0(-5)$

12. $0(-1)$

13. $(-2)(-1)(-7)$

14. $(-5)(3)(-4)$

15. $(-3)(7)(-5)$

16. $(-2)(-5)(-3)$

17. $(-2)(-2)(-8)$

18. $(-3)(-3)(-3)$

19. $(-5)(0)(-6)$

20. $(-9)(0)(-8)$

21. $(-1)(-1)(-1)$

22. $(-2)(-2)(-2)(-2)$

23. $(-5)(-2)(5)(-1)$

24. $(-2)(-3)(-11)(-5)$

25. $(-10)(-2)(-3)(-1)$

26. $(-1)(-2)(-6)(-3)(-1)(-5)(-4)(0)$

27. $(-2)(-3)(-4)(-1)(-4)(-8)(-2)(0)$

28.
   a. $(-1)^7$
   b. $(-1)^{10}$
   c. $(-1)^0$
   d. $(-1)^{31}$

29.
   a. $(-1)^6$
   b. $(-1)^{11}$
   c. $(-1)^{14}$
   d. $(-1)^0$

30. Can you make a general statement relating the sign of a product and the number of negative factors?

Find the following quotients. See Examples 4 and 5.

31. $\dfrac{-12}{3}$

32. $\dfrac{-18}{6}$

33. $\dfrac{14}{-7}$

34. $\dfrac{-28}{7}$

35. $\dfrac{-30}{5}$

36. $\dfrac{50}{-5}$

37. $\dfrac{-20}{5}$

38. $\dfrac{-35}{5}$

39. $\dfrac{39+3}{-3}$

40. $\dfrac{16+8}{-8}$

41. $\dfrac{18+9}{-9}$

42. $\dfrac{9+27}{-9}$

43. $\dfrac{-36-9}{-9}$

44. $\dfrac{-13-13}{-13}$

45. $\dfrac{-28-4}{-4}$

46. $\dfrac{-13-26}{-13}$

47. $\dfrac{0}{2}$

48. $\dfrac{0}{8}$

49. $\dfrac{35}{0}$

50. $\dfrac{56}{0}$

51. If the product of –15 and –6 is added to the product of –32 and 5, what is the sum?

52. If the quotient of –56 and 8 is subtracted from the product of –12 and –3, what is the difference?

53. **a.** What number should be added to –5 to get a sum of –22?

  b. What number should be added to +5 to get a sum of –22?

54. **a.** What number should be added to –39 to get a sum of –13?

  b. What number should be added to +39 to get a sum of –13?

Find the value of each expression by following the rules for order of operations. See Examples 6, 7, and 8.

55. $15 \div (-5) \cdot 2 - 8$

56. $20 \cdot 2 \div (-10) + 4(-5)$

57. $3^3 \div (-9) \cdot 4 + 5(-2)$

58. $4^2 \div (-8)(-2) + 6(-15 + 3 \cdot 5)$

59. $5^2 \div (-5)(-3) - 2(4 \cdot 3)$

60. $-23 + 16(-3) - 10 + 2(-5)^2$

61. $-27 \div (-3) - 14 - 4(-2)^3$

62. $-36 \div (-2)^2 + 15 - 2(16 - 17)$

63. $-25 \div (-5)^2 + 35 - 3(24 - 5^2)^2$

64. $7 \cdot 2^3 + 3^2 - 4^2 - 4(6 - 2 \cdot 3)$

65. $13 - 2\big[(19 - 14) \div 5 + 3(-4) - 5\big]$

66. $2 - 5\big[(-10) \div (-5) \cdot 2 - 35\big]$

67. $6 - 20\big[(-15) \cdot 2 + 4(-5) - 1\big]$

68. $8 - 9\big[(-39) \div (-13) + 14 \cdot 2 - 7 - (-5)\big]$

69. $(7 - 10)\big[49 \div (-7) + 20 \cdot 3 - 4 \cdot 15 - (-10)\big]$

70. $(9 - 11)\big[(-10)^2 \cdot 2 + 6(-5)^2 - 10^3\big]$

Use a calculator to find the value of each expression.

71. $(-72)^2 - 35(16)$

72. $(15)^3 - 60 + 13(-5)$

73. $(15 - 20)(13 - 14)^2 + 50(-2)$

74. $(32 - 50)^2 (15 - 22)^2 (17 - 20)$

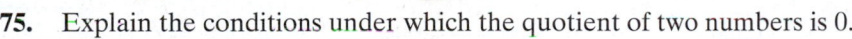

Writing & Thinking

75. Explain the conditions under which the quotient of two numbers is 0.

76. Explain, in your own words, why division by 0 is not a valid arithmetic operation.

1.12 Applications: Change in Value and Mean

Objectives

A Understand how the change between two integer values can be found by subtracting the beginning value from the end value.

B Be able to find the mean of multiple integers.

Objective A **Change in Value**

Subtraction with integers can be used to find the change in value between two readings of measures such as temperatures, distances, and altitudes.

To calculate the change between two integer values, including direction (negative for down, positive for up), use the following rule: **first find the end value, then subtract the beginning value.**

$$(\text{change in value}) = (\text{end value}) - (\text{beginning value})$$

Example 1

Temperature Changes

On a cold day at a ski resort, the temperature dropped from a high of 25°F at 1 P.M. to a low of −10°F at 2 A.M. What was the change in temperature?

Solution

$$\underset{\text{end temperature}}{-10°} \quad \underset{-}{-} \quad \underset{\text{beginning temperature}}{(25°)} \quad \underset{=}{=} \quad \underset{\text{change}}{-35°}$$

The temperature dropped 35°, so the change was −35°F.

Now work margin exercise 1.

1. On a cold day at a ski resort, the temperature dropped from a high of 21°F at 1:30 P.M. to a low of −17°F at 2:00 A.M. What was the change in temperature?

Example 2

Altitude of a Rocket

A rocket was fired from a silo 1000 feet below ground level. If the rocket attained a height of 15,000 feet, what was its change in altitude?

Solution

The end value was +15,000 feet.

The beginning value was −1000 feet since the rocket was below ground level.

$$\begin{aligned} \text{Change} &= 15,000 - (-1000) \\ &= 15,000 + 1000 \\ &= 16,000 \text{ feet} \end{aligned}$$

The change in altitude was 16,000 feet.

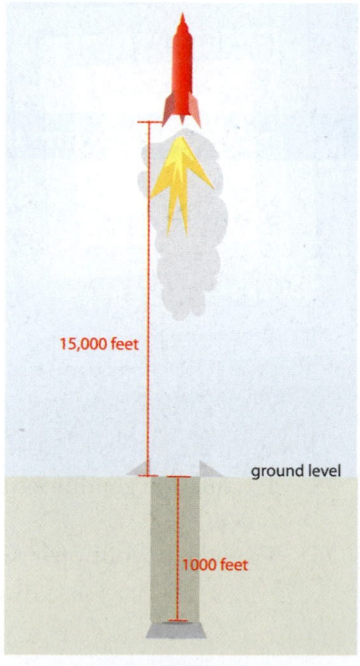

15,000 feet

ground level

1000 feet

Now work margin exercise 2.

2. A rocket was fired from a silo 2500 feet below ground level. If the rocket attained a height of 18,000 feet, what was its change in altitude?

Mean

A topic closely related to addition and division with integers is the **mean** (also called the **arithmetic average** or just **average**) of a set of integers. Your grade in this course may be based on the mean of your exam scores. Newspapers and magazines report mean income, mean life expectancy, mean sales, mean attendance at sporting events, and so on. The mean of a set of integers is one representation of the concept of the "middle" of the set. It may or may not actually be one of the numbers in the set.

Mean

The **mean** of a set of numbers is the value found by adding the numbers in the set, then dividing this sum by the number of numbers in the set.

In this section, we will use only sets of integers for which the mean is an integer. However, in later chapters, we will remove this restriction and find that a mean can be calculated for types of numbers other than integers.

Example 3

Mean Temperature

Find the mean noonday temperature for the 5 days on which the temperatures at noon were 10°F, –2°F, 5°F, –7°F, and –11°F.

Solution

Find the sum of the temperatures and divide the sum by 5.

$$\frac{-5}{5} = -1°F \quad \text{mean}$$

10
–2
5
–7
–11
―――
–5 sum

The mean noonday temperature was –1°F (or 1°F below 0).

Now work margin exercise 3.

3. Find the mean noonday temperature for the 5 days on which the temperatures at noon were 70°F, 71°F, 73°F, 76°F, and 80°F.

4. On an art history exam, two students scored 96, five scored 88, one scored 72, and six scored 81. What is the mean score for the class on this exam?

Example 4

Test Scores

On an art history exam, two students scored 95, five scored 86, one scored 78, and six scored 75. What is the mean score for the class on this exam?

Solution

$$
\begin{array}{cccc}
95 & 86 & 78 & 75 \\
\times\, 2 & \times\, 5 & \times\, 1 & \times\, 6 \\
\hline
190 & 430 & 78 & 450
\end{array}
$$

We multiplied rather than wrote down all 14 scores. Thus the sum of the four products represents the sum of 14 exam scores. This sum is divided by 14 to find the mean.

$$
\begin{array}{r}
190 \\
430 \\
78 \\
+\,450 \\
\hline
1148
\end{array}
$$

$$
\begin{array}{r}
82 \quad \text{mean score} \\
14\overline{)1148} \\
\underline{112} \\
28 \\
\underline{28} \\
0
\end{array}
$$

The class mean is 82 points.

Now work margin exercise 4.

Example 5

Hybrid Car Sales

The table below shows the number of hybrid cars sold in the U.S. from the year 2003 to 2010. Find the mean number of car sales per year over the 8-year period.

Number of Hybrid Cars Sold Per Year	
2003	47,500
2004	88,000
2005	212,000
2006	254,500
2007	352,200
2008	313,500
2009	292,500
2010	274,200

Source: www.hybridcars.com

a. What year had the highest number of hybrid car sales?

Solution

From the table, we can see that 2007 had the highest number of hybrid car sales with 352,200 cars sold.

b. What year had the lowest number of hybrid car sales?

Solution

From the table, we can see that 2003 had the lowest number of hybrid car sales with 47,500 cars sold.

c. Find the mean number of car sales per year over the 8-year period.

Solution

In order to find the mean number of cars sold per year over the 8 years, find the sum of yearly sales and divide by 8.

$$
\begin{array}{r}
47{,}500 \\
88{,}000 \\
212{,}000 \\
254{,}500 \\
352{,}200 \\
313{,}500 \\
292{,}500 \\
+\ 274{,}200 \\
\hline
1{,}834{,}400
\end{array}
\qquad
\begin{array}{r}
229{,}300 \quad \text{mean sales} \\
8\overline{)1{,}834{,}400} \\
\underline{16} \\
23 \\
\underline{16} \\
74 \\
\underline{72} \\
2\ 4 \\
\underline{2\ 4} \\
0
\end{array}
$$

The mean hybrid sales per year over the 8-year period is 229,300 cars.

Now work margin exercise 5.

5. Using the data from example 5, Find the difference between the highest and lowest number of hybrid car sales.

Exercises 1.12

1. **Lisa's Age:** Lisa was a very nice woman with 14 grandchildren. She was born in 1888 and died on her birthday in 1960. How old was she when she died?

2. **Bank Deposit:** Your bank statement indicates that you are overdrawn on your checking account by $253. How much must you deposit to bring the checking balance up to $400? Your separate savings account balance is $862.

3. **Plane Altitude:** A pilot flew a plane from an altitude of 20,000 feet to an altitude of 1500 feet. What was the change in altitude? The plane was flying over the Mojave desert at 500 miles per hour.

4. **Mountain Temperature:** At noon the temperature at the top of a mountain was 30°F. By midnight the temperature was –6°F. What was the change in temperature from noon to midnight? Eight inches of snow fell in that same 12-hour period.

5. **Submarine Missile:** A test missile was fired from a submarine from 300 feet below sea level. If the missile reached a height of 16,000 feet before exploding, what was the change in altitude of the missile?

6. **Bake Sale:** The members of the Math Club sold cake at a bake sale and made a profit of $250. If their expenses totaled $80, what was the total amount of the sales? The chocolate cake sold out in 2 hours.

7. **Dow Jones Index:** At the end of a 5-day week the Dow Jones stock index was at 10,495 points. If the Dow started the week at 10,600 points, what was the change in the index for that week?

8. **Stock Price Change:** On Monday the stock of a computer company sold at a market price of $56 per share. By Friday the stock had dropped to $48 per share. Find the change in the price of the stock. The president of the company earns a salary of $210,000 annually.

Find the mean of each of the following sets of integers.

9. –10, –20, 14, 34, –18

10. 56, –64, –38, 58, –12

11. –6, –8, –7, –4, –4, –5, –6, –8

12. –25, 30, –15, –6, –26, 18

Solve the following word problems.

13. **Mean Price Per Share:** Alicia bought shares of two companies on the stock market. She paid $9000 for 90 shares in one company and $6600 for 110 shares in another company. What was the mean price per share for the 200 shares?

14. **Bench Press Weight:** In a weight lifting program two men bench pressed 300 pounds, three men bench pressed 350 pounds, and 5 men bench pressed 400 pounds. What was the mean number of pounds that these men bench pressed?

15. Class Score: On an English exam two students scored 95, six students scored 90, three students scored 80, and one student scored 50. What was the mean score for the class?

16. Speech Score: In a speech class the students graded each other on a particular assignment. (Generally students are harder on each other and themselves than the instructor is. Why, do you think, that this is the case?) On this speech, three students scored 60, three scored 70, five scored 80, five scored 82, and four scored 85. What was the mean score on this speech?

17. Ski Resort Temperatures: The temperature reading for 30 days at a ski resort were recorded as follows: (All temperatures were measured in degrees Fahrenheit.)

28	24	22	10	−2	−5	−2	12	10	15
−6	5	20	13	−2	−6	−15	−18	−10	8
−1	7	20	21	32	30	22	12	3	−7

What was the mean temperature recorded for the 30 days?

18. Exam Score: On a multiple choice exam with 25 questions, the instructor scored the exam as follows: 4 points for each correct answer, −1 point for each blank and −3 points for each wrong answer.

a. What was the maximum score possible on the exam?

b. What was the score of a student who had 18 correct answers, 3 blanks, and 4 wrong answers?

19. Golf: In the 2010 Masters Tournament, American Phil Mickleson won by shooting scores of 67, 71, 67, and 67 in the four rounds of play.

a. What was his mean score in each round?

b. Par was 72 on that course. How many strokes under par was his mean score?

20. Golf: In the same tournament mentioned in Exercise 19, fellow American Anthony Kim finished in third place by shooting 68, 70, 73, and 65 in the four rounds of play.

a. What was his mean score in each round?

b. How many strokes under par was his mean score?

21. Box Office Income: The approximate box office gross income (to the nearest thousand dollars) of the 5 top-grossing movies of 2010 are listed below. Find the mean gross income of these 5 movies.
Source: boxofficemojo.com

Movie	U.S. Gross ($)
1. Toy Story 3	415,005,000
2. Alice in Wonderland	334,191,000
3. Iron Man 2	312,433,000
4. The Twilight Saga: Eclipse	300,532,000
5. Harry Potter and the Deathly Hallows Part 1	295,001,000

22. **Top Global Brands:** The top 5 most valuable brands as of 2010 are listed below. Find the mean value of 5 brands. **Source:** interbrand.com

Brand Name	Brand Value (in millions)
1. Coca Cola	$70,452
2. IBM	$64,727
3. Microsoft	$60,895
4. Google	$42,808
5. General Electric	$33,578

23. **Outside an Airplane:** Suppose that the temperature outside an airplane decreases 2°C for every 500 feet the plane increases in altitude. Find the change in temperature (as a signed number) outside a plane that increases its altitude from 10,000 feet to 15,500 feet.

24. **Outside a Hot Air Balloon:** Suppose that the temperature outside a hot air balloon decreases 3°F for every 500 feet the balloon increases in altitude. Find the change in temperature (as a signed number) outside a balloon that decreases its altitude from 8000 feet to 6500 feet.

Mentally estimate the value of each expression by using rounded numbers. Then use a calculator to find the value of each expression. (Estimates may vary.)

25. $[270 + 257 + 300 - 507] \div 4$

26. $[(-650) + (-860) + (-530)] \div 3$

27. $[(-53,404) + (25,803) + (-10,005) + (-35,002)] \div 4$

28. $[(-62,496) + (-72,400) + (53,000) + (-16,000)] \div 4$

29. $[-21,540 - 10,200 + 36,300 - 25,000 + 12,400] \div 10$

30. $[317,164 - 632,000 - 427,000 + 250,500] \div 8$

Writing & Thinking

31. The formula for converting a temperature, C, in degrees celcius to an equivalent temperature, F, in degrees Fahrenheit is $F = \dfrac{9C}{5} + 32$. Use your calculator and this formula to answer the following questions.

a. Which is warmer: –10°C or +10°F?

b. Which is colder: –20°C or –20°F?

c. Convert –40°C into an equivalent Fahrenheit temperature.

1.13 Introduction to Like Terms and Polynomials

Objective A **Combining Like Terms**

Objectives

A Recognize like terms and unlike terms.

B Be able to evaluate a polynomial when an integer is substituted for the variable.

C Understand when to place parentheses around negative numbers.

A single number is called a **constant**. Any constant, variable, or the product of constants and variables is called a **term**. Examples of terms are

$$17, \quad -2, \quad 3x, \quad 5xy, \quad -3x^2, \quad \text{and} \quad 14a^2b^3.$$

A number written next to a variable (as in $9x$) or a variable written next to another variable (as in xy) indicates multiplication. In the term $4x^2$, the constant 4 is called the **numerical coefficient** of x^2 (or simply the **coefficient** of x^2).

Like terms are terms that contain the same variables raised to the same powers. Whatever power a variable is raised to in one term, it is raised to the same power in other like terms. Constants are like terms.

Like Terms	
$-6, 12, 132$	are like terms because each term is a constant.
$-2a, 15a,$ and $3a$	are like terms because each term contains the same variable a, raised to the same power, 1.
$5xy^2$ and $3xy^2$	are like terms because each term contains the same two variables, x and y, where x is first-degree in both terms and y is second-degree in both terms.

Unlike Terms	
$7x$ and $8x^2$	are unlike terms (**not** like terms) because the variable x is not of the same power in both terms.
$6ab^2$ and $2a^2b$	are **not** like terms because the variables are not of the same power in both terms.

If no number is written next to a variable, then the coefficient is understood to be 1. For example,

$$x = 1 \cdot x, \quad a^3 = 1 \cdot a^3, \quad \text{and} \quad xy = 1 \cdot xy.$$

If a negative sign ($-$) is written next to a variable, then the coefficient is understood to be -1. For example,

$$-y = -1 \cdot y, \quad -x^2 = -1 \cdot x^2, \quad \text{and} \quad -xy = -1 \cdot xy.$$

1. From the following, pick out the like terms:
$5, 10x, -1, -5x, 4x^2z, -3x, 2x^2z,$ and 0.

Example 1

Like Terms

From the following list of terms, pick out the like terms.

$$6, \quad 2x, \quad -10, \quad -x, \quad 3x^2z, \quad -5x, \quad 4x^2z, \quad \text{and} \quad 0.$$

Solutions

$6, -10,$ and 0 are like terms. All are constants.

$2x, -x,$ and $-5x$ are like terms.

$3x^2z$ and $4x^2z$ are like terms.

Now work margin exercise **1.**

As we have discussed in a previous section, expressions such as

$$9 + x, \quad 3y^3 - 4y, \quad \text{and} \quad 2x^2 + 8x - 10$$

are not terms. These algebraic sums of terms are called **polynomials**, and our objective here is to learn to simplify polynomials that contain like terms. This process is called **combining like terms**. For example, by combining like terms, we can write

$$8x + 3x = 11x \quad \text{and} \quad 4n - n + 2n = 5n.$$

To understand how to combine like terms, we need the **distributive property** as it applies to integers (see Section 1.3). (**Note:** The distributive property is a property of addition and multiplication and not of any particular type of number. As we will see later, the distributive property applies to all types of numbers, including integers, fractions, and decimals.)

Distributive Property of Multiplication over Addition

For integers a, b, and c, $a(b + c) = ab + ac$.

Some examples of multiplication, using the distributive property, are:

$$
\begin{array}{rcl cr}
4(x + 3) & = & 4 \cdot x + 4 \cdot 3 & = & 4x + 12 \\
9(a - 2) & = & 9 \cdot a + 9(-2) & = & 9a - 18 \\
-8(x + 6) & = & -8 \cdot x + (-8) \cdot 6 & = & -8x - 48 \\
-8(y - 1) & = & -8 \cdot y + (-8) \cdot (-1) & = & -8y + 8
\end{array}
$$

Now consider the following analysis that leads to the method for combining like terms:

By using the commutative property of multiplication, along with the distributive property, we can write

$$a(b + c) = ab + ac = ba + ca = (b + c)a$$

Thus, by using the distributive property in a "reverse" sense, we have combined the like terms $8x$ and $3x$.

$$8x + 3x = (8 + 3)x = 11x.$$

Similarly, we can write,

$$4n - n + 2n = (4 - 1 + 2)n = 5n \quad \text{(Note that −1 is the coefficient of −n.)}$$

and $17y - 14y = (17 - 14)y = 3y.$

Example 2

Simplifying Polynomials

Simplify each of the following polynomials by combining like terms whenever possible.

a. $8x + 10x + 1$

Solution

$$\begin{aligned} 8x + 10x + 1 &= (8 + 10)x + 1 \\ &= 18x + 1 \end{aligned}$$

Use the distributive property with $8x + 10x$. Note that the constant 1 is not combined with $18x$ because they are not like terms.

b. $5x^2 - x^2 + 6x^2 + 2x - 7x$

Solution

$$\begin{aligned} 5x^2 - x^2 + 6x^2 + 2x - 7x &= (5 - 1 + 6)x^2 + (2 - 7)x \\ &= 10x^2 - 5x \end{aligned}$$

c. $2(x + 5) + 3(x - 1)$

Solution

$$\begin{aligned} 2(x + 5) + 3(x - 1) &= 2x + 10 + 3x - 3 \\ &= (2 + 3)x + 10 - 3 \\ &= 5x + 7 \end{aligned}$$

Use the distributive property.

Use the commutative property of addition.

d. $-3(x + 2) - 4 (x - 2)$

Solution

$$\begin{aligned} -3(x + 2) - 4(x - 2) &= -3x - 6 - 4x + 8 \\ &= (-3 - 4)x - 6 + 8 \\ &= -7x + 2 \end{aligned}$$

Use the distributive property.

2. Simplify each polynomial by combining like terms whenever possible.

a. $7x + 9x$

b. $6x - 7a - a$

c. $3x^2 + x^2 - 5b + 2b$

d. $3x^2 - 5x^2 + x^2$

e. $9y^3 + 4y^2 - 3y - 1$

Solution

$9y^3 + 4y^2 - 3y - 1$ This expression is already simplified, since it has no like terms to combine.

Now work margin exercise **2.**

After some practice, you should be able to combine like terms by adding the corresponding coefficients mentally. The use of the distributive property is necessary in beginning development and serves to lay a foundation for simplifying more complicated expressions in later courses.

Objective B **Evaluating Polynomials and Other Expressions**

In Section 1.7, we evaluated polynomials for given whole number values of the variables. Now, since we know how to operate with integers, we can evaluate polynomials and other algebraic expressions for given integer values of the variables. Also, we can make the process of evaluation easier by first combining like terms, whenever possible.

To Evaluate a Polynomial or Other Algebraic Expression

1. Combine like terms, if possible.
2. Substitute the values given for any variables.
3. Follow the rules for order of operations.

(**Note:** Terms separated by + and – signs may be evaluated at the same time. Then the value of the expression can be found by adding and subtracting from left to right.)

Example 3

Simplifying and Evaluating Polynomials

First, simplify the polynomial $x^3 - 5x^2 + 2x^2 + 3x - 4x - 15 + 3$ then evaluate it for $x = -2$.

Solution

First, simplify the polynomial by combining like terms.

$$x^3 - 5x^2 + 2x^2 + 3x - 4x - 15 + 3$$
$$= x^3 + (-5 + 2)x^2 + (3 - 4)x - 15 + 3$$
$$= x^3 - 3x^2 - x - 12$$

Now substitute -2 for x (**using parentheses around -2 to be sure the signs are correct**) and evaluate this simplified polynomial by following the rules for order of operations. (**Note:** terms separated by $+$ and $-$ signs may be evaluated at the same time.)

$$
\begin{aligned}
x^3 - 3x^2 - x - 12 &= (-2)^3 - 3(-2)^2 - (-2) - 12 \\
&= -8 - 3(4) - (-2) - 12 \\
&= -8 - 12 + 2 - 12 \\
&= -30
\end{aligned}
$$

Example 4

Evaluating Polynomials

Evaluate the expression $5ab + ab + 8a - a$ for $a = 2$ and $b = -3$.

Solution

Combine like terms.

$$
\begin{aligned}
5ab + ab + 8a - a &= (5 + 1)ab + (8 - 1)a \\
&= 6ab + 7a
\end{aligned}
$$

Substituting 2 for a and -3 for b in this simplified expression and following the rules for order of operations gives

$$6ab + 7a = 6(2)(-3) + 7(2) = -36 + 14 = -22.$$

First, simplify the following algebraic expressions and then evaluate the expressions for $x = -2$ and $y = 3$.

3. $3xy - x^2 + 5xy$

4. $7x + 5xy - 10x + 4x$

5. $4x^2 - 3x + 2x^2 + 4x$

Completion Example Answer

5.
$$5x^2 - 2x^2 + x + 3x^2 - 2x = (5 - 2 + 3)x^2 + (1 - 2)x$$
$$= 6x^2 - 1x$$

$$6x^2 - x = 6(-1)^2 - 1(-1)$$
$$= 6(+1) + 1$$
$$= 6 + 1$$
$$= 7$$

Consider evaluating the expressions $4x^2 + x^2$ and $6x^2 - 7x^2$ for $x = -2$. In both cases, we combine like terms first:

$$4x^2 + x^2 = 5x^2$$

and $6x^2 - 7x^2 = -x^2.$ Note that for $-7x^2$ the coefficient is -7.

Now evaluate for $x = -2$ by writing -2 in parentheses as follows.

$$5x^2 = 5(-2)^2 = 5(4) = 20$$ Note that for $5x^2$ the coefficient is 5.

and $-x^2 = -1(-2)^2 = -1(4) = -4$ Note that for $-x^2$ the coefficient is -1.

These examples illustrate the importance of using parentheses around negative numbers so that fewer errors will be made with signs. Equally important is the fact that in expressions such as $-x$ and $-x^2$ the coefficient in each case is understood to be -1. That is,

$$-x = -1 \cdot x \qquad \text{and} \qquad -x^2 = -1 \cdot x^2.$$

As just illustrated, we can evaluate both expressions for $x = -3$ by using parentheses:

$$-x = -1 \cdot x = -1 \cdot (-3) = +3$$

and $-x^2 = -1 \cdot x^2 = -1 \cdot (-3)^2 = -1 \cdot 9 = -9.$

Notice that, in the second case, -3 is squared first, then multiplied by the coefficient -1. The square is on x, not $-x$. In effect, the exponent 2 takes precedence over the coefficient -1. This is consistent with the rules for order of operations. However, there is usually some confusion when numbers are involved in place of variables. Study the differences in the following evaluations.

$-5^2 = -1 \cdot 5^2 = -1 \cdot 25 = -25$ and, with parentheses, $(-5)^2 = (-5)(-5) = +25$

$-6^2 = -1 \cdot 6^2 = -1 \cdot 36 = -36$ and, with parentheses, $(-6)^2 = (-6)(-6) = +36$

With parentheses around -5 and -6, the results are entirely different.

These examples are shown to emphasize the importance of using parentheses around negative numbers when evaluating expressions. Without the parentheses, an evaluation can be dramatically changed and lead to wrong answers. We state the following results for the exponent 2. (The results are the same for all positive even exponents.)

In general, except for $x = 0$,

1. $-x^2$ is negative. $\left(-5^2 = -25\right)$

2. $(-x)^2$ is positive. $\left((-5)^2 = 25\right)$

3. $-x^2 \neq (-x)^2.$

Exercises 1.13

Use the distributive property to multiply each of the following expressions.

1. $6(x+3)$

2. $4(x-2)$

3. $-7(x+5)$

4. $-9(y-4)$

5. $-11(n+3)$

6. $2(n-6)$

Simplify each of the following expressions by combining like terms whenever possible. See Example 2.

7. $2x+3x$

8. $9x-4x$

9. $y+y$

10. $-x-x$

11. $-n-n$

12. $9n-9n$

13. $5x-7x+2x$

14. $-40a+20a+20a$

15. $3y-5y-y$

16. $12x+6-x-2$

17. $8x-14+x-10$

18. $6y+3-y+8$

19. $13x^2+2x^2+2x+3x-1$

20. $x^2-3x^2+5x-3+1$

21. $3x^2-4x+2+x^2-5$

22. $-4a+5a^2+4a+5a^2+6$

23. $3(x-1)+2(x+1)$

24. $3(y+4)+5(y-2)$

25. $-2(x-1)+6(x-5)$

26. $5ab+7a-4ab+2ab$

27. $2xy+4x-3xy+7x-y$

28. $5x^2y-3xy^2+2$

29. $5x^2y-3x^2y+2$

30. $7ab^2+8ab^2-2ab+5ab$

31. $7ab^2+8a^2b-2ab+5ab$

32. $2(x^2-x+3)+5(x^2+2x-3)$

33. $2(x+a-1)+3(x-a-1)$

34. $8xy^2+2xy^2-7xy+xy+2x$

35. $3xyz+2xyz-7xyz+14-20$

First simplify each polynomial, and then evaluate the polynomial for $x=-2$ and $y=-1$. See Examples 3 and 5.

36. $2x^2-3x^2+5x-8+1$

37. $5x^2-4x+2-x^2+3$

38. $y^2+2y^2+2y-3y$

39. $y^2+y^2-8y+2y-5$

40. x^3-2x^2+x-3x

41. $x^3-3x^3+x+x^2-2x$

42. $y^3+3y^3+5y-4y^2+1$

43. $7y^3+4y^2+6+y^2-12$

44. $2(x^2-3x-5)+3(x^2+5x-4)$

45. $5(y^2-4y+3)-2(y^2-2y+10)$

46. $a^2 - a + a^2 - a$

47. $a^3 - 2a^3 - 3a + a - 7$

48. $5ab - 7a + 4ab + 2b$

49. $2ab + 4b - 3a + ab - b$

50. $2(a + b - 1) + 3(a - b + 2)$

51. $5(a - b + 1) + 4(a + b - 5)$

52. $4ac + 16ac - 4c + 2c$

53. $4abc + 2abc - 6ab + ab - bc + 3bc$

54. $-9abc + 3abc - abc + 2ab - bc + 3bc + 15$

55. $16bc^2 - 15bc^2 + 3ab^2 + ab^2 + a^2b^2c^2$

56. $14(a + 7) - 15(b + 6) + 2(c - 3)$

57. $12(a - 3) + 8(b - 2) - 3(c + 4)$

58. $20(a + b + c) - 10(a + b + c)$

59. $16(a - b + c) + 16(-a + b - c)$

60. $7(a + 2b - 3c) - 7(-a - 2b + 3c)$

61. Find the value of each of the following expressions.

    a. -4^2 **b.** $(-4)^2$ **c.** -4^3 **d.** $(-4)^3$

62. Find the value of each of the following expressions.

    a. -5^3 **b.** $(-5)^3$ **c.** -5^4 **d.** $(-5)^4$

63. Find the value of each of the following expressions.

    a. $(-6)^3$ **b.** $(-1)^7$ **c.** -7^3 **d.** -2^5

64. Find the value of each of the following expressions for $x = 2$ and $y = -3$.

    a. $-x^2$ **b.** $-y^2$ **c.** $(-x)^2$ **d.** $(-y)^2$

65. Find the value of each of the following expressions for $x = 2$ and $y = -3$.

    a. $-x$ **b.** $-y$ **c.** $-5x$ **d.** $-7y$

1.14 Solving Equations with Integers

Objectives

A Be able to solve equations with integer coefficients and integer solutions.

B Revisit the basic principles for solving equations.

Objective A **Negative Constants, Negative Coefficients, and Negative Solutions**

In Section 1.5, we discussed solving equations of the forms

$$x + b = c, \qquad x - b = c, \qquad \text{and} \qquad ax = c,$$

where x is a variable and a, b, and c represent constants. Each of these equations is called a **first-degree equation in x** because the exponent of the variable is 1 in each case. The equations in Section 1.5 were set up so that the coefficients, constants, and solutions to the equations were whole numbers. Now that we have discussed integers and operations with integers, we can discuss equations that have negative constants, negative coefficients, and negative solutions.

For convenience and easy reference, the following definitions are repeated from Section 1.5.

> ## Equation, Solution, Solution Set
>
> An **equation** is a statement that two expressions are equal.
>
> A **solution** to an equation is a number that gives a true statement when substituted for the variable.
>
> A **solution set** of an equation is the set of all solutions of the equation.

1. Solve the following problems.

a. Show that –10 is a solution to the equation $x - 3 = -13$.

b. Show that 2 is not a solution to the equation $x - 3 = -13$.

Example 1

Solutions to an Equation

a. Show that –5 **is** a solution to the equation $x - 6 = -11$.

$$x - 6 = -11$$
$$\overset{?}{-5 - 6 = -11}$$
$$-11 = -11$$

Therefore, –5 **is** a solution.

b. Show that 6 **is not** a solution to the equation $x - 6 = -11$.

$$x - 6 = -11$$
$$\overset{?}{6 - 6 = -11}$$
$$0 \neq -11$$

Therefore, 6 **is not** a solution.

*Now work margin exercise **1**.*

Restating the Basic Principles for Solving Equations

Another adjustment that can be made, since we now have an understanding of integers, is that the addition and subtraction principles discussed in Section 1.5 can be stated as one principle, the addition principle. Because subtraction can be treated as the algebraic sum of numbers and expressions, we know that

$$A - C = A + (-C).$$

That is, subtraction of a number can be thought of as addition of the opposite of that number.

The following principles can be used to find the solution to first-degree equations of the forms

$$x + b = c, \qquad x - b = c, \qquad \text{and} \qquad ax = c.$$

> ### Basic Principles for Solving Equations
>
> In the three basic principles stated here, A, B, and C represent numbers (constants) or expressions that may involve variables.
>
> 1. The **addition principle**:   The equations $A = B$
>   and $A + C = B + C$
>   have the same solutions.
>
> 2. The **subtraction principle**:   The equations $A = B$
>   and $A - C = B - C$
>   have the same solutions.
>
> 3. The **division principle**:   The equations $A = B$
>   and $\dfrac{A}{C} = \dfrac{B}{C}$ (where $C \neq 0$)
>   have the same solutions.
>
> Essentially, these three principles say that if we perform the same operation on both sides of an equation, the resulting equation will have the same solution as the original equation. Equations with the same solution sets are said to be **equivalent**.

Note: In the principles just stated, the letters A, B, and C can represent negative numbers as well as positive numbers.

Remember: The objective in solving an equation is to isolate the variable with coefficient 1 on one side of the equation, left side or right side.

Example 2

Solving Equations

a. Solve the equation $x + 5 = -14$.

Solution

$$x + 5 = -14 \qquad \text{Write the equation.}$$
$$x + 5 - 5 = -14 - 5 \qquad \text{Using the addition principle, add } -5 \text{ to both sides.}$$
$$x + 0 = -19 \qquad \text{Simplify both sides.}$$
$$x = -19 \qquad \text{solution}$$

b. Solve the equation $11 = x - 6$.

Solution

$$11 = x - 6 \qquad \text{Write the equation.}$$
$$11 + 6 = x - 6 + 6 \qquad \text{Using the addition principle, add 6 to both sides.}$$
$$17 = x + 0 \qquad \text{Simplify both sides.}$$
$$17 = x \qquad \text{solution}$$

Note that in both equations in Example 2, a constant was added to both sides of the equation. In part **a.**, a negative number was added, and in part **b.**, a positive number was added.

As in Section 1.5, we will need the concept of division in fraction form when solving equations and will assume that you are familiar with the fact that a number divided by itself is 1. This fact includes negative numbers. For example,

$$(-3) \div (-3) = \frac{(-3)}{(-3)} = 1 \quad \text{and} \quad (-8) \div (-8) = \frac{(-8)}{(-8)} = 1.$$

In this manner, we will write expressions such as the following:

$$\frac{-5x}{-5} = \frac{-5}{-5}x = 1 \cdot x.$$

In Example 3 the division principle is illustrated with positive and negative constants and positive and negative coefficients.

Example 3

Solving Equations

a. Solve the equation $4n = -24$.

Solution

$4n = -24$	Write the equation.
$\dfrac{4n}{4} = \dfrac{-24}{4}$	Using the division principle, divide both sides by the coefficient 4. Note that in solving equations, the fraction form of division is used.
$1 \cdot n = -6$	Simplify by performing the division on both sides.
$n = -6$	solution

b. Solve the equation $42 = -21n$.

Solution

$42 = -21n$	Write the equation.
$\dfrac{42}{-21} = \dfrac{-21n}{-21}$	Using the division principle, divide both sides by the coefficient -21.
$-2 = 1 \cdot n$	Simplify by performing the division on both sides.
$-2 = n$	solution

In Example 4, we illustrate solving equations in which we combine like terms before applying the principles. (**Note:** In later chapters we discuss solving equations involving fractions and decimal numbers and the use of more steps in finding the solutions. In every case, the basic principles and techniques are the same as those used here.)

Solve each of the following equations.

2. a. $x + 9 = 2$

b. $23 = x - 5$

3. a. $8n = 64$

b. $38 = -2n$

4. a. $n + 14 - 10 = 7 - 2$

b. $-28 = 8y - y$

Example 4

Solving Equations

a. Solve the equation $y - 10 + 11 = 7 - 8$.

Solution

$y - 10 + 11 = 7 - 8$	Write the equation.
$y + 1 = -1$	Combine like terms.
$y + 1 - 1 = -1 - 1$	Using the addition principle, add -1 to both sides.
$y + 0 = -2$	Simplify both sides.
$y = -2$	solution

b. Solve the equation $-15 = 4y - y$.

Solution

$-15 = 4y - y$	Write the equation.
$-15 = 3y$	Combine like terms.
$\dfrac{-15}{3} = \dfrac{3y}{3}$	Using the division principle, divide both sides by 3.
$-5 = 1 \cdot y$	Simplify both sides.
$-5 = y$	solution

Now work margin exercises 2 through 4.

Exercises 1.14

In each exercise an equation and a replacement set of numbers for the variable are given. Substitute each number into the equation until you find the number that is the solution of the equation.

1. $x + 4 = -8$ $\{-10, -12, -14, -16, -18\}$

2. $x + 3 = -5$ $\{-4, -5, -6, -7, -8, -9\}$

3. $-32 = y - 7$ $\{-20, -21, -22, -23, -24, -25\}$

4. $-13 = y - 12$ $\{0, -1, -2, -3, -4, -5\}$

5. $7x = -105$ $\{0, -5, -10, -15, -20\}$

6. $-72 = 8n$ $\{-6, -7, -8, -9, -10\}$

7. $-13n = 39$ $\{0, -1, -2, -3, -4, -5\}$

8. $-14x = 56$ $\{0, -1, -2, -3, -4, -5\}$

Use either the addition principle or the division principle to solve the following equations. Show each step and keep the equal signs aligned vertically as in the format used in the text. Combine like terms whenever necessary. See Examples 2 through 4.

9. $x - 2 = -16$

10. $x - 35 = -45$

11. $-27 = x + 10$

12. $-42 = x + 32$

13. $y + 9 = -9$

14. $y + 18 = -18$

15. $y + 9 = 0$

16. $y - 18 = 0$

17. $12 = n + 12$

18. $15 = n + 15$

19. $-75 = n + 30$

20. $-100 = n + 20$

21. $-8x = -32$

22. $-9x = -72$

23. $-13y = 65$

24. $-15y = 30$

25. $-84 = 4n$

26. $-88 = 11n$

27. $x + 1 = -20$

28. $-12 = x - 3$

29. $-83 = y + 13$

30. $3x - 2x + 6 = -20$

31. $4x - 3x + 3 = 16 + 30$

32. $7x + 3x = 30 + 10$

33. $6x + 2x = 14 + 26$

34. $3x - x = -27 - 19$

35. $2x + 3x = -62 - 28$

36. $-21 + 30 = -3y$

37. $-16 - 32 = -6y$

38. $-35 - 15 - 10 = 4x - 14x$

39. $-17 + 20 - 3 = 5x$

40. $5x + 2x = -32 + 12 + 20$

41. $x - 4x = -25 + 3 + 22$

42. $3y + 10y - y = -13 + 11 + 2$

43. $5y + y - 2y = 14 + 6 - 20$

44. $6y + y - 3y = -20 - 4$

45. $4x - 10x = -35 + 1 - 2$

46. $6x - 5x + 3 = -21 - 3$

47. $10n - 3n - 6n + 1 = -41$

48. $-5n - 3n + 2 = 36 - 2$

49. $-2n - 6n - n - 10 = -11 + 100$

Writing & Thinking

50. Explain, in your own words, why combining the addition principle and the subtraction principle as stated in Section 1.5 into one principle (the addition principle as stated in the section) is mathematically correct.

51. Name one topic discussed in Chapter 1 that you found particularly interesting. Explain why you think it is important. (Write only a short paragraph, and be sure to use complete sentences.)

52. Explain briefly, in your own words, how the concepts of perimeter and area are different.

Section 1.1: Introduction to Whole Numbers

Whole Numbers   pages 3 – 5

The whole numbers are the natural (or counting) numbers and the number 0.

    Natural numbers: $\mathbb{N}$ =   { 1, 2, 3, 4, 5, 6, 7, 8, 9, 10, 11, . . . }

    Whole numbers: $\mathbb{W}$ =   { 0, 1, 2, 3, 4, 5, 6, 7, 8, 9, 10, 11, . . . }

Rounding Whole Numbers   pages 5 – 6

To **round** a given number means to find another number close to the given number.

Rounding Rule for Whole Numbers   pages 6 – 7

1. Look at the single digit just to the right of the digit that is in the place of desired accuracy.

2. If this digit is 5 or greater, make the digit in the desired place of accuracy one larger, and replace all digits to the right with zeros. All digits to the left remain unchanged unless a 9 is made one larger; then the next digit to the left is increased by 1.

3. If this digit is less than 5, leave the digit that is in the place of desired accuracy as it is, and replace all digits to the right with zeros. All digits to the left remain unchanged.

Section 1.2: Addition and Subtraction with Whole Numbers

Addition with Whole Numbers   pages 11 – 12

Addends are the numbers being added.
The **sum** is the result of the addition.

Variable   page 12

A **variable** is a symbol (generally a letter of the alphabet) that is used to represent an unknown number or any one of several numbers.

Properties of Addition   pages 12 – 13

    Commutative Property: For any whole numbers a and b, $a + b = b + a$.

    Associative Property: For any whole numbers a, b, and c, $(a + b) + c = a + (b + c)$.

    Additive Identity Property: For any whole number a, $a + 0 = a$.

Perimeter   page 14

The **perimeter** of a geometric figure is the distance around the figure.

Subtraction   pages 14 – 16

Subtraction is reverse addition. The difference is the result of subtracting one addend from the sum. The **subtrahend** is the number being subtracted and the **minuend** is the number from which the subtrahend is subtracted.

Section 1.2: Addition and Subtraction with Whole Numbers (cont.)

To Estimate an Answer:
page 16 – 18
1. Round each number to the place of the leftmost digit.
2. Perform the indicated operation with these rounded numbers.

Section 1.3: Multiplication and Division with Whole Numbers

Multiplication with Whole Numbers
page 25
Multiplication is repeated addition.

Properties of Multiplication
pages 26 – 28
Commutative Property: For any whole numbers a and b, $a \cdot b = b \cdot a$.
Associative Property: For any whole numbers a, b, and c, $(a \cdot b) \cdot c = a \cdot (b \cdot c)$.
Multiplicative Identity Property: For any whole number a, $a \cdot 1 = a$.
Zero-Factor Law: For any whole number a, $a \cdot 0 = 0$.
Distributive Property of Multiplication Over Addition: If a, b, and c, are whole numbers, then $a(b + c) = a \cdot b + a \cdot c$.

Area
pages 30 – 31
Area is the measure of the interior, or enclosed region, of a plane surface and is measured in square units.

Division With Whole Numbers
pages 31 – 32
Division can be thought of as the reverse of multiplication.
The number being divided is called the **dividend**.
The number dividing the dividend is called the **divisor**.
The result of division is called the **quotient**.

Division with 0
page 34
Case 1: If a is any nonzero whole number, then $0 \div a = 0$.
Case 2: If a is any nonzero whole number, then $a \div 0$ is **undefined**.

Section 1.4: Problem Solving with Whole Numbers

Basic Strategy for Solving Word Problems
page 46
1. Read each problem carefully until you understand the problem and know what is being asked. Discard any irrelevant information.
2. Draw any type of figure or diagram that might be helpful and decide what operations are needed.
3. Perform the required operations.
4. Mentally check to see if your answer is reasonable and see if you can think of another more efficient or more interesting way to do the same problem.

Section 1.4: Problem Solving with Whole Numbers (cont.)

Key Words that Indicate Operations   page 46

    Addition: add, sum, plus, more than, increased by

    Subtraction: subtract (from), difference, minus, less than, decreased by

    Multiplication: multiply, product, times, twice

    Division: divide, quotient, ratio

Section 1.5: Solving Equations with Whole Numbers

Equations, Solution, and Solution Set   page 55 – 56

    An **equation** is a statement that two expressions are equal.

    A **solution** of an equation is a number that gives a true statement when substituted for the variable.

    A **solution set** of an equation is the set of all solutions of the equation.

Basic Principles for Solving Equations   pages 57 – 60

    1. **The Addition Principle:** The equations $A = B$ and $A + C = B + C$ have the same solutions.

    2. **The Subtraction Principle:** The equations $A = B$ and $A - C = B - C$ have the same solutions.

    3. **The Division Principle:** The equations $A = B$ and $\dfrac{A}{C} = \dfrac{B}{C}$ (where $C \neq 0$) have the same solutions.

    Equations with the same solution sets are said to be **equivalent**.

Section 1.6: Exponents and Order of Operations

Exponents   page 62

    A whole number n is an **exponent** if it is used to tell how many times another whole number a is used as a factor.

    The repeated factor, a, is called the **base** of the exponent.

Squared and cubed   page 63

    In expressions with exponent 2, the base is said to be **squared**. In expressions with exponent 3, the base is said to be **cubed**.

The Exponent 1   page 64

    For any whole number a, $a = a^{1}$.

The Exponent 0   page 64

    If a is any nonzero whole number, then $a^{0} = 1$.

    (**Note:** The expression 0^{0} is undefined.)

Rules for Order of Operations   pages 66 – 70

1. First, simplify within grouping symbols, such as parentheses ( ), brackets [ ], braces { }, radical signs $\sqrt{}$, absolute value bars $|\ |$, or fraction bars. Start with the innermost grouping.
2. Second, evaluate any numbers or expressions raised to exponents.
3. Third, moving from **left to right**, perform any multiplications or divisions in the order in which they appear.
4. Fourth, moving from **left to right**, perform any additions or subtractions in the order in which they appear.

Section 1.7: Introduction to Polynomials

Polynomials   page 74

A **monomial in x** is a term of the form kx^n, where k and n are whole numbers; n is called the degree of the monomial, and k is called the coefficient.

A **polynomial** is a monomial or the sum or difference of monomials.

The **degree of a polynomial** is the largest of the degrees of its terms.

Classification of Polynomials   page 75

A **monomial** is a polynomial with one term.
A **binomial** is a polynomial with two terms.
A **trinomial** is a polynomial with three terms.

Evaluating a Polynomial   page 76

To evaluate a polynomial for a given value of the variable
1. substitute that value for the variable wherever it occurs in the polynomial and
2. follow the rules for order of operations.

Section 1.8: Introduction to Integers

Integers {..., –3, –2, –1, 0, 1, 2, 3, ...}   pages 82 – 84

The set of integers is the set of whole numbers and their opposites (or additive inverses).

Positive and Negative Integers   pages 82 – 83

The counting numbers are called **positive integers** and may be written +1, +2, +3, and so on; the opposites of the counting numbers are called **negative integers**. 0 is neither positive nor negative.

Inequality Symbols   pages 84 – 86

Symbols for order:
< is read "is less than"   ≤ is read "is less than or equal to"
> is read "is greater than"   ≥ is read "is greater than or equal to"

Section 1.8: Introduction to Integers (cont.)

Absolute Value   pages 86 – 88

The **absolute value** of an integer is its distance from 0. Symbolically, for

any integer $a,$ $\begin{cases} \text{If } a \text{ is a positive integer or } 0, |a| = a. \\ \text{If } a \text{ is a negative integer, } |a| = -a. \end{cases}$

The absolute value of an integer is never negative.

Section 1.9: Addition with Integers

Rules for Addition with Integers   pages 95 -96

1. To add two integers with like signs, add their absolute values and use the common sign.
2. To add two integers with unlike signs, subtract their absolute values (the smaller from the larger) and use the sign of the integer with the larger absolute value.

Additive Inverse   pages 96 – 97

The **opposite** of an integer is called its **additive inverse**. The sum of any integer and its additive inverse is 0. Symbolically for any integer, $a,$

$$a + (-a) = 0.$$

Section 1.10: Subtraction with Integers

Difference   page 104 – 105

For any integers a and b, the **difference** between a and b is defined as follows.

$$a - b = a + (-b)$$

Section 1.11: Multiplication, Division, and Order of Operations with Integers

Rules for Multiplication with Integers   page 114– 115

If a and b are positive integers, then

1. The product of two positive integers is positive: $a \cdot b = ab$.
2. The product of two negative integers is positive: $(-a)(-b) = ab$.
3. The product of a positive integer and a negative integer is negative: $a(-b) = -ab$ and $(-a)(b) = -ab$.
4. The product of 0 and any integer is 0: $a \cdot 0 = 0$ and $(-a) \cdot 0 = 0$.

Rules for Division with Integers   page 116

If a and b are positive integers, then

1. The quotient of two positive integers is positive: $\dfrac{a}{b} = +\dfrac{a}{b}$.

2. The quotient of two negative integers is positive: $\dfrac{-a}{-b} = +\dfrac{a}{b}$.

3. The quotient of a positive integer and a negative integer is negative:

    $\dfrac{-a}{b} = -\dfrac{a}{b}$ and $\dfrac{a}{-b} = -\dfrac{a}{b}$.

Division by 0 is Not Defined   page 117

Case 1: Suppose that $a \neq 0$ and $\dfrac{a}{0} = x$. Then, by the meaning of division, $a = 0 \cdot x$. But this is not possible because $0 \cdot x = 0$ for any value of x, and we stated that $a \neq 0$.

Case 2: Suppose that $\dfrac{0}{0} = x$. Then, by the meaning of division, $0 = 0 \cdot x$ is true for all values of x. But we must have a unique answer for division. Therefore we conclude that, in every case, division by 0 is not defined.

Section 1.12: Applications: Change in Value and Mean

Change in Value   page 123

(change in value) = (end value) − (beginning value)

Mean   page 124

The **mean** of a set of numbers is the value found by adding the numbers in the set, then dividing this sum by the number of numbers in the set.

Section 1.13: Introduction to Like Terms and Polynomials

Constants and Terms   page 130

    A single number is called a **constant**. Any constant or variable or the product of constants and variables is called a **term**.

Coefficient   page 130

    In the term $-4x^2$, the constant -4 is called the **numerical coefficient** of x^2.

Like Terms   page 130

    Like Terms are terms that contain the same variables (if any) raised to the same powers.

Distributive Property of Multiplication over Addition   pages 131 – 133

    For integers $a, b,$ and $c,$ $a(b+c) = ab + ac$.

To Evaluate a Polynomial or Other Algebraic Expression: pages 133 – 134

    1. Combine like terms, if possible.
    2. Substitute the values given for any variables.
    3. Follow the rules for order of operations.
    Note: Terms separated by + and – signs may be evaluated at the same time. Then the value of the expression can be found by adding and subtracting from left to right.

Section 1.14: Solving Equations with Integers

Objective in Solving Equations   pages 140 – 141

    The objective in solving an equation is to isolate the variable with coefficicint 1 on one side of the equation, left side or right side.

Chapter 1: Test

Solve the following problems.

1. Write the number 3,075,122 in words.

2. Write the number fifteen million, three hundred twenty thousand, five hundred eighty-four in standard notation.

Round as indicated.

3. 675 (to the nearest ten)

4. 13,620 (to the nearest thousand)

Find the correct value of *n* and state the property of addition or multiplication that is illustrated.

5. $16 + 52 = 52 + n$

6. $18(20 \cdot 3) = (18 \cdot 20) \cdot n$

7. $74 + n = 74$

8. $6 + (2 + 4) = (6 + 2) + n$

Estimate the answer, then find the answer. Do all of your work with pencil and paper. Do not use a calculator.

9.
$$\begin{array}{r} 9587 \\ 345 \\ + \ 2075 \\ \hline \end{array}$$

10.
$$\begin{array}{r} 45{,}872 \\ 589{,}651 \\ + \ 235{,}003 \\ \hline \end{array}$$

11.
$$\begin{array}{r} 7046 \\ - \ 4562 \\ \hline \end{array}$$

12.
$$\begin{array}{r} 80{,}000 \\ - \ 28{,}830 \\ \hline \end{array}$$

13.
$$\begin{array}{r} 47 \\ \times \ 25 \\ \hline \end{array}$$

14.
$$\begin{array}{r} 2593 \\ \times \ \ \ 86 \\ \hline \end{array}$$

15. $463\overline{)78{,}950}$

16. If the quotient of 56,000 and 7 is subtracted from the product of 22 and 3000, what is the difference?

17. A rectangular swimming pool is surrounded by a concrete border. The swimming pool is 50 meters long and 25 meters wide, and the concrete border is 3 meters wide all around the pool.

    a. Find the perimeter of the swimming pool.

    b. Find the perimeter of the rectangle formed by the outside of the border around the pool.

    c. Find the area of the concrete border.

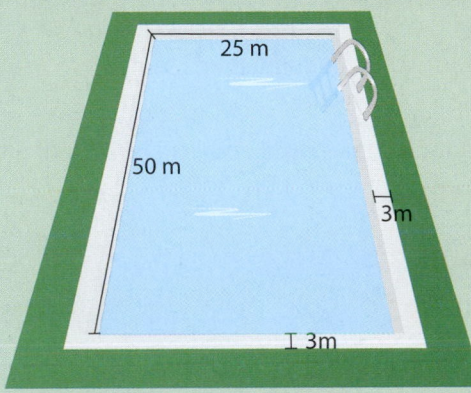

18. Mr. Powers is going to buy a new silver SUV for $43,000. He will pay taxes of $2500 and license fees of $1580. He plans to make a down payment of $8000 and finance the remainder at his bank. How much will he finance if he makes a salary of $85,000?

Solve the following equations.

19. $x - 10 = 32$ **20.** $x + 25 = 60$ **21.** $17 + 34 = 17y$

22. For the exponential expression 4^3 name:

    a. the base

    b. the exponent

    c. the value of the expression

23. Use your calculator to find the value of each of the following exponential expressions.

    a. 8^5 **b.** 19^3 **c.** 142^0

Use the rules for order of operations to evaluate each of the following expressions.

24. $20 + 6 \div 3 \cdot 2$ **25.** $14 \div 7 \cdot 2 + 14 \div (7 \cdot 2)$

26. $15 + 9 \div 3 - 10$ **27.** $18 \div 2 + 5 \cdot 2^3 - 6 \div 3$

28. $12 \div 3 \cdot 2 - 8 \div 2^3 + 5$ **29.** $6 + 3(7^2 - 5) - 11^2$

Name the degree and type of each of the following polynomials.

30. $3y^2 + 15$ **31.** $5y^4 - 2y^2 + 4$

32. $5x^2 + 6x - 8$

33. $2x^5 + 3x^4 - 4x^3 + 3x^2 - 5x - 1$

34. State whether each of the following numbers is a perfect square, and explain your reasoning.

    a. 100   **b.** 80   **c.** 225

35. Explain, in your own words, why $7 \div 0$ is undefined.

36. Find the area of the orange shaded region. (**Remember:** The area of a triangle is found by multiplying the base times the height and dividing by two.)

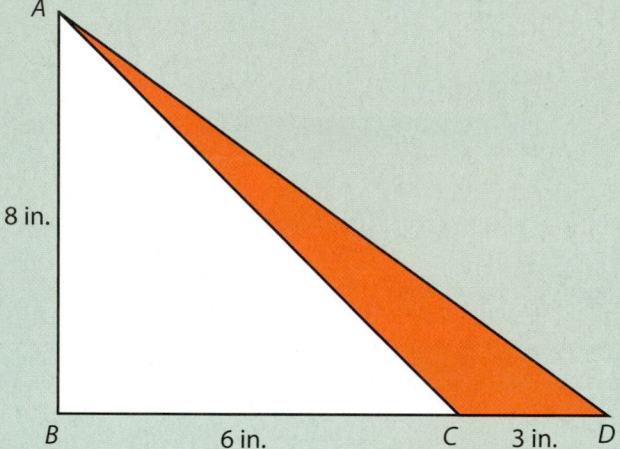

37. What two numbers are 8 units from 5 on a number line? Draw a number line and graph the numbers.

38. What two numbers are 13 units from −10 on a number line? Draw a number line and graph the numbers.

39. Graph the following set of integers on a number line: $\{ -2, -1, 0, 3, 4 \}$

40. What are the possible integer values of n if $|n| < 3$?

41. Graph the set of whole numbers less than 4.

42. $-15 + (-2) + (14)$

43. $-120 - 75 + 30$

44. $45 + 25 - 70 - 40$

45. $-35 + 20 - 17 + 30$

46. $(-3)(-5)(-4)(7)$

47. $(-8)(-3)(-7)(-56)(0)$

48. $(5)(-2)(-4)(12)$

49. $\dfrac{-32}{4}$

50. $\dfrac{0}{1}$

51. $\dfrac{-48}{-12}$

52. $\dfrac{-100-5}{7}$

Find the mean of each of the following sets of numbers.

53. $60, -120, -35, -42, -38$

54. $56, 92, 84, -60$

55. A jet pilot flew her plane, with a wingspan of 70 feet, from an altitude of 32,000 feet to an altitude of 5000 feet. What was the change in altitude?

56. If the quotient of -51 and 17 is subtracted from the product of -21 and $+3$, what is the difference?

57. In one football game the quarterback threw passes that gained 10 yards, gained 22 yards, lost 5 yards, lost 2 yards, gained 15 yards, gained 13 yards, and lost 3 yards. If 3 passes were incomplete, what was the mean gain (or loss) for the 10 pass plays?

58. A highway patrol car clocked 25 cars at the following speeds.

65	60	55	54	63	62	68	70	72	68
64	60	55	70	75	69	58	57	68	65
65	60	65	50	82					

(The last one got a ticket.) Use a calculator to find the mean speed (in miles per hour) of the 25 cars.

59. The graph shows the low temperature in degrees Fahrenheit for 7 consecutive days at a ski resort. Find the mean low temperature for these 7 days.

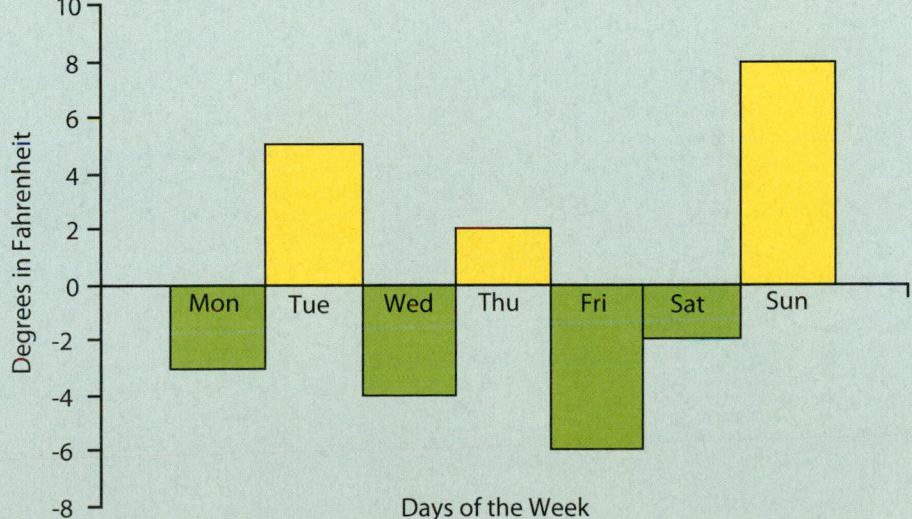

60. The heights (rounded to the nearest 10 units) of the 5 highest mountains in the world are listed in the following table. Find the mean heights of these mountains in both feet and meters.
(**Source:** Chinese National Geography.)

Mountain/Location	First Ascent	Feet	Meters
1. Everest/Nepal/China	May 29, 1953	29,030	8850
2. K2 (Chogori)/Pakistan/China	July 31, 1954	28,240	8610
3. Kangchenjunga/Nepal/India	May 25, 1955	28,160	8590
4. Lhotse/Nepal/China	May 18, 1956	27,930	8520
5. Makalu 1/Nepal/China	May 15, 1955	27,760	8460

Follow the rules for order of operations and evaluate each of the following expressions.

61. $(-19+10)\left[8^2 \div 2 \cdot 3 - 16(6)\right]$

62. $60 - 2\left[35 + 6(-3) - 10^2 + 3 \cdot 5^2\right]$

Combine like terms (if possible) in each of the following polynomials.

63. $7x + 3x - 5x$

64. $8x^2 + 9x - 10x^2 + 15 - x$

65. $y^2 - 3y^3 + 2y^2 + 5y - 9y$

66. $5y^2 + 3y - 2y - 6 - y$

Evaluate each of the following expressions for $x = -3$ and $y = 2$.

67. $14xy + 5x^2 - 6y$

68. $7x^2y + 8xy^2 - 11xy - 2x - 3y$

Solve each of the following equations.

69. $x - 28 = -30$

70. $-12y = -36$

71. $15x - 14x + 14 = -14$

72. $4y - 6y = 42 - 44$

Mathematics @ Work!

Politics and the political process affect everyone in some way. In local, state or national elections, registered voters make decisions about who will represent them and make choices about various ballot measures. In major issues at the state and national levels, pollsters use mathematics (in particular, statistics and statistical methods) to indicate attitudes and to predict, within certain percentages, how the electorate will vote. When there is an important election in your area, read the papers and magazines and listen to the television reports for mathematically related statements predicting the outcome. Consider the following situation.

There are 8000 registered voters in Brownsville, and $\frac{3}{8}$ of these voters live in neighborhoods on the north-side of town. A survey indicates that $\frac{4}{5}$ of these north-side voters are in favor of a bond measure for constructing a new recreation facility that would largely benefit their neighborhoods. Also, $\frac{7}{10}$ of the registered voters from all other parts of town are in favor of the measure. We might then want to know the following numbers.

    a. How many north-side voters favor the bond measure?

    b. How many voters in the town favor the bond measure?

For more problems like these, see Section 2.2, Exercise 99.

Fractions, Mixed Numbers, and Proportions

2.1 Tests for Divisibility

Objectives

A Know the rules for testing divisibility by 2, 3, 5, 6, 9, and 10.

B Be able to apply the concept of divisibility to products of whole numbers.

Objective A **Tests For Divisibility (2, 3, 5, 6, 9, 10)**

In our work with factoring and fractions, we will need to be able to divide quickly and easily by small numbers. Since we will be looking for factors, we will want to know if a number is **exactly divisible** (remainder 0) by some number **before** actually dividing. There are simple tests that can be performed mentally to determine whether a number is divisible by 2, 3, 5, 6, 9, or 10 **without actually dividing**.

For example, can you tell (without dividing) if 585 is divisible by 2? By 3? Note that we are not trying to find the quotient, only to determine whether 2 or 3 is a factor of 585. The answer is that 585 **is not** divisible by 2 and **is** divisible by 3.

$$
\begin{array}{r}
292 \\
2\overline{)585} \\
\underline{4} \\
18 \\
\underline{18} \\
05 \\
\underline{4} \\
\mathbf{1}
\end{array}
\quad \text{remainder}
\qquad
\begin{array}{r}
195 \\
3\overline{)585} \\
\underline{3} \\
28 \\
\underline{27} \\
15 \\
\underline{15} \\
\mathbf{0}
\end{array}
\quad \text{remainder}
$$

Thus the number 2 **is not a factor** of 585. However, $3 \cdot 195 = 585$, and 3 and 195 **are factors** of 585.

Tests for Divisibility of Integers by 2, 3, 5, 6, 9, and 10
For 2: If the last digit (units digit) of an integer is 0, 2, 4, 6, or 8, then the integer is divisible by 2.
For 3: If the sum of the digits of an integer is divisible by 3, then the integer is divisible by 3.
For 5: If the last digit of an integer is 0 or 5, then the integer is divisible by 5.
For 6: If the integer is divisible by both 2 and 3, then it is divisible by 6.
For 9: If the sum of the digits of an integer is divisible by 9, then the integer is divisible by 9.
For 10: If the last digit of an integer is 0, then the integer is divisible by 10.

There are other quick tests for divisibility by other numbers such as 4, 7, and 8. (See Exercise 41 for divisibility by 4.)

Even and Odd Integers

Even integers are divisible by 2. (If an integer is divided by 2 and the remainder is 0, then the integer is even.)

Odd integers are not divisible by 2. (If an integer is divided by 2 and the remainder is 1, then the integer is odd.)

Note: Every integer is either even or odd.

The **even** integers are

$$\ldots, -10, -8, -6, -4, -2, 0, 2, 4, 6, 8, 10, \ldots$$

The **odd** integers are

$$\ldots, -11, -9, -7, -5, -3, -1, 1, 3, 5, 7, 9, 11, \ldots$$

If the units digit of an integer is one of the even digits (0, 2, 4, 6, 8), then the integer is divisible by 2, and therefore, it is an even integer.

Example 1

Testing Divisibility

a. 1386 is divisible by 2 since the units digit is 6, an even digit. Thus 1386 is an even integer.

b. 7701 is divisible by 3 because $7 + 7 + 0 + 1 = 15$, and 15 is divisible by 3.

c. 23,365 is divisible by 5 because the units digit is 5.

d. 9036 is divisible by 6 since it is divisible by both 2 and 3. (The sum of the digits is $9 + 0 + 3 + 6 = 18$, and 18 is divisible by 3. The units digit is 6 so 9036 is divisible by 2.)

e. 9567 is divisible by 9 because $9 + 5 + 6 + 7 = 27$ and 27 is divisible by 9.

f. 253,430 is divisible by 10 because the units digit is 0.

Now work margin exercise 1.

1. Test the divisibility of the following examples.

   a. Does 6 divide 8034? Explain why or why not.

   b. Does 3 divide 206? Explain why or why not.

   c. Is 4065 divisible by 3? By 9?

   d. Is 12,375 divisible by 5? By 10?

> ## Note about Terminology
>
> The following sentences are simply different ways of saying the same thing.
>
> 1. 1386 is **divisible by** 2.
> 2. 2 is a **factor** of 1386.
> 3. 2 **divides** 1386.
> 4. 2 **divides into** 1386.
> 5. 2 is a **divisor** of 1386.

In examples 2 and 3, all six tests for divisibility are used to determine which of the numbers 2, 3, 5, 6, 9, and 10 will divide into each number.

Example 2

Testing for Divisibility of 3430

The number 3430 is

a. divisible by 2 (units digit is 0, an even digit);

b. not divisible by 3 ($3 + 4 + 3 + 0 = 10$ and 10 is not divisible by 3);

c. divisible by 5 (units digit is 0);

d. not divisible by 6 (to be divisible by 6, it must be divisible by both 2 and 3, but 3430 is not divisible by 3);

e. not divisible by 9 ($3 + 4 + 3 + 0 = 10$ and 10 is not divisible by 9);

f. divisible by 10 (units digit is 0).

Example 3

Testing for Divisibility of −5718

The number −5718 is

a. divisible by 2 (units digit is 8, an even digit);

b. divisible by 3 ($5 + 7 + 1 + 8 = 21$ and 21 is divisible by 3);

c. not divisible by 5 (units digit is not 0 or 5);

d. divisible by 6 (divisible by both 2 and 3);

e. not divisible by 9 ($5 + 7 + 1 + 8 = 21$ and 21 is not divisible by 9);

f. not divisible by 10 (units digit is not 0).

Completion Example 4

Testing for Divisibility of −375

The number −375 is

a. divisible by 5 because _____.

b. divisible by 3 because _____.

c. not divisible by 6 because _____.

Completion Example 5

Testing for Divisibility of 612

The number 612 is

a. divisible by 6 because _____.

b. divisible by 9 because _____.

c. not divisible by 10 because _____.

Now work margin exercises 2 through 5.

Objective B ## Divisibility of Products

Now consider a number that is written as a product of several factors. For example,

$$3 \cdot 4 \cdot 5 \cdot 10 \cdot 12 = 7200.$$

To better understand how products and factors are related, consider the problem of finding the quotient 7200 divided by 150.

$$7200 \div 150 = ?$$

Determine which of the numbers 2, 3, 5, 6, 9, and 10 divides into each of the following numbers.

2. 1742

3. 8020

4. 33,031

5. 2400

Completion Example Answers

    4. a. the units digit is 5.
      b. the sum of the digits is 15, and 15 is divisible is 3.
      c. −375 is not even.
    5. a. 612 is divisible by both 2 and 3.
      b. the sum of the digits is 9, and 9 is divisible by 9.
      c. the units digit is not 0.

By using the commutative and associative properties of multiplication (i.e. rearranging and grouping factors), we can see that

$$3 \cdot 5 \cdot 10 = 150 \qquad \text{and} \qquad 7200 = (3 \cdot 5 \cdot 10) \cdot (4 \cdot 12) = 150 \cdot 48.$$

Thus

$$7200 \div 150 = 48.$$

In general, for any product of two or more integers, each integer is a factor of the total product. Also, the product of any combination of factors is a factor of the total product. This means that every one of the following products is a factor of 7200.

$3 \cdot 4 = 12$	$3 \cdot 5 = 15$	$3 \cdot 10 = 30$
$3 \cdot 12 = 36$	$4 \cdot 5 = 20$	$4 \cdot 10 = 40$
$4 \cdot 12 = 48$	$5 \cdot 10 = 50$	$5 \cdot 12 = 60$
$10 \cdot 12 = 120$	$3 \cdot 4 \cdot 5 = 60$	$3 \cdot 4 \cdot 10 = 120$
$3 \cdot 4 \cdot 12 = 144$	$3 \cdot 5 \cdot 10 = 150$	and so on

Proceeding in this manner, taking all combinations of the given factors two at a time, three at a time, and four at a time, we will find 25 combinations. Not all of these combinations lead to **different** factors. For example, the list so far shows both $5 \cdot 12 = 60$ and $3 \cdot 4 \cdot 5 = 60$. In fact, by using techniques involving prime factors (prime numbers are discussed in a future section), we can show that there are 54 different factors of 7200, including 1 and 7200 itself.

Thus, by grouping the factors 3, 4, 5, 10, and 12 in various ways and looking at the remaining factors, we can find the quotient if 7200 is divided by any particular product of a group of the given factors. We have shown how this works in the case of $7200 \div 150 = 48$. For $7200 \div 144$, we can write

$$7200 = 3 \cdot 4 \cdot 5 \cdot 10 \cdot 12 = (3 \cdot 4 \cdot 12) \cdot (5 \cdot 10) = 144 \cdot 50,$$

which shows that both 144 and 50 are factors of 7200 and $7200 \div 144 = 50$. (Also, of course, $7200 \div 50 = 144$.)

Example 6

Divisibility of a Product

Does 54 divide the product $4 \cdot 5 \cdot 9 \cdot 3 \cdot 7$? If so, how many times?

Solution

Since $54 = 9 \cdot 6 = 9 \cdot 3 \cdot 2$, we factor 4 as $2 \cdot 2$ and rearrange the factors to find 54.

$$\begin{aligned} 4 \cdot 5 \cdot 9 \cdot 3 \cdot 7 &= 2 \cdot 2 \cdot 5 \cdot 9 \cdot 3 \cdot 7 \\ &= (9 \cdot 3 \cdot 2) \cdot (2 \cdot 5 \cdot 7) \\ &= 54 \cdot 70 \end{aligned}$$

Thus 54 does divide the product, and it divides the product 70 times. Notice that we do not even need to know the value of the product, just the factors.

Example 7

Divisibility of a Product

Does 26 divide the product $4 \cdot 5 \cdot 9 \cdot 3 \cdot 7$? If so, how many times?

Solution

26 does **not** divide $4 \cdot 5 \cdot 9 \cdot 3 \cdot 7$ because $26 = 2 \cdot 13$ and 13 is not a factor of the product.

Now work margin exercises 6 and 7.

6. Show that 39 divides the product $3 \cdot 5 \cdot 8 \cdot 13$ and find the quotient without actually dividing.

7. Explain why 35 does not divide the product $2 \cdot 3 \cdot 5 \cdot 13$.

Exercises 2.1

1. 82

2. 92

3. 81

4. 210

5. −441

6. −344

7. 544

8. 370

9. 575

10. 675

11. 711

12. −801

13. −640

14. 780

15. 1253

16. 1163

17. −402

18. −702

19. 7890

20. 6790

21. 777

22. 888

23. 45,000

24. 35,000

25. 7156

26. 8145

27. −6948

28. −8140

29. −9244

30. −7920

31. 33,324

32. 23,587

33. −13,477

34. −15,036

35. 722,348

36. 857,000

37. 635,000

38. 437,200

39. 117,649

40. 131,074

41. A test for divisibility by 4:

   If the number formed by the last two digits of a number is divisible by 4, then the number is divisible by 4. For example, 5144 is divisible by 4 because 44 is divisible by 4. (Mentally dividing gives $44 \div 4 = 11$. Using long division, we would find that $5144 \div 4 = 1286$.)

   Use the test for divisibility by 4 (just discussed) to determine which, if any, of the following numbers is divisible by 4.

   a. 3384   b. 9672   c. 386,162   d. 4954   e. 78,364

42. A digit is missing in each of the following numbers. Find a value for the missing digit so that the resulting number will be

   a. divisible by 3. (There may be several correct answers or no correct answer.)

   i. 32 ☐ 5   ii. 54 ☐ 3   iii. 7 ☐ ,488   iv. 7 ☐ ,457

   b. divisible by 9. (There may be several correct answers or no correct answer.)

   i. 32 ☐ 5   ii. 54 ☐ 3   iii. 19, ☐ 25   iv. 33,6 ☐ 2

43. A digit is missing in each of the following numbers. Find a value for the missing digit so that the resulting number will be

a. divisible by 9. (There may be several correct answers or no correct answer.)

i. 34 ☐ 4   **ii.** 53,12 ☐   **iii.** 4 ☐ ,475   **iv.** 73 ☐ ,432

b. divisible by 6. (There may be several correct answers or no correct answer.)

i. 34 ☐ 4   **ii.** 53,12 ☐   **iii.** 4 ☐ ,475   **iv.** 73 ☐ ,432

44. Discuss the following statement:

An integer may be divisible by 5 and not divisible by 10. Give two examples to illustrate your point.

45. If an integer is divisible by both 3 and 9, will it necessarily be divisible by 27? Explain your reasoning, and give two examples to support this reasoning.

46. If an integer is divisible by both 2 and 4, will it necessarily be divisible by 8? Explain your reasoning and give two examples to support this reasoning.

Determine whether each of the given numbers divides (is a factor of) the given product. If so, tell how many times it divides the product. Find each product, and make a written statement concerning the divisibility of the product by the given number. See Examples 6 and 7.

47. 6; $2 \cdot 3 \cdot 3 \cdot 7$ **48.** 14; $2 \cdot 3 \cdot 3 \cdot 7$

49. 10; $2 \cdot 3 \cdot 5 \cdot 9$ **50.** 25; $2 \cdot 3 \cdot 5 \cdot 7$

51. 25; $3 \cdot 5 \cdot 7 \cdot 10$ **52.** 35; $2 \cdot 2 \cdot 3 \cdot 3 \cdot 7$

53. 45; $2 \cdot 3 \cdot 3 \cdot 7 \cdot 15$ **54.** 36; $2 \cdot 3 \cdot 4 \cdot 5 \cdot 13$

55. 40; $2 \cdot 3 \cdot 5 \cdot 7 \cdot 12$ **56.** 45; $3 \cdot 5 \cdot 7 \cdot 10 \cdot 21$

57. **a.** If a whole number is divisible by both 3 and 5, then it will be divisible by 15. Give two examples.

  b. However, a number might be divisible by 3 and not by 5. Give two examples.

  c. Also, a number might be divisible by 5 and not 3. Give two examples.

58. **a.** If a number is divisible by 9, then it will be divisible by 3. Explain in your own words why this statement is true.

  b. However, a number might be divisible by 3 and not by 9. Give two examples that illustrate this statement.

Collaborative Learning

59. In groups of three to four students, use a calculator to evaluate 20^{10} and 10^{20}. Discuss what you think is the meaning of the notation on the display.

  (**Note:** The notation is a form of a notation called **scientific notation** and is discussed in detail in Chapter 5. Different calculators may use slightly different forms.)

2.2 Prime Numbers

Objective A   **Prime Numbers and Composite Numbers**

Objectives

A   Understand the concepts of prime and composite numbers.

B   Use the Sieve of Eratosthenes to identify prime numbers.

C   Be able to determine whether a number is prime or composite.

The positive integers are also called the **counting numbers** (or **natural numbers**). Every counting number, except the number 1, has at least two distinct factors. The following list illustrates this idea. Note that, in this list, every number has **at least two** factors, but 7, 13, and 19 have **exactly two** factors.

Counting Numbers	Factors
7	1, 7
12	1, 2, 3, 4, 6, 12
13	1, 13
19	1, 19
20	1, 2, 4, 5, 10, 20
25	1, 5, 25
86	1, 2, 43, 86

Our work with fractions will be based on the use and understanding of counting numbers that have exactly two different factors. Such numbers (for example, 7, 13, and 19 in the list above) are called **prime numbers**. (**Note:** In later discussions involving negative integers and negative fractions, we will treat a negative integer as the product of −1 and a counting number.)

Prime Numbers

A **prime number** is a counting number greater than 1 that has exactly two different factors (or divisors), itself and 1.

Composite Numbers

A **composite number** is a counting number with more than two different factors (or divisors).

Thus, in the previous list, 7, 13, and 19 are prime numbers and 12, 20, 25, and 86 are composite numbers.

notes

■   1 is **neither** a prime nor a composite number. $1 = 1 \cdot 1$, and 1 is the only factor of 1. 1 does not have **exactly two different** factors, and it does
■   not have more than two different factors.

1. Explain why 19, 31, and 41 are prime.

Example 1

Prime Numbers

Some prime numbers:

2: 2 has exactly two different factors, 1 and 2.
3: 3 has exactly two different factors, 1 and 3.
11: 11 has exactly two different factors, 1 and 11.
37: 37 has exactly two different factors, 1 and 37.

Now work margin exercise **1.**

---

2. Explain why 14, 42, and 12 are composite.

Example 2

Composite Numbers

Some composite numbers:

15: 1, 3, 5, and 15 are all factors of 15.
39: 1, 3, 13, and 39 are all factors of 39.
49: 1, 7, and 49 are all factors of 49.
50: 1, 2, 5, 10, 25, and 50 are all factors of 50.

Now work margin exercise **2.**

Objective B **The Sieve of Eratosthenes**

One method used to find prime numbers involves the concept of **multiples**. To find the multiples of a counting number, multiply each of the counting numbers by that number. The multiples of 2, 3, 5, and 7 are listed here.

Counting Numbers:	1,	2,	3,	4,	5,	6,	7,	8,	...
Multiples of 2:	2,	4,	6,	8,	10,	12,	14,	16,	...
Multiples of 3:	3,	6,	9,	12,	15,	18,	21,	24,	...
Multiples of 5:	5,	10,	15,	20,	25,	30,	35,	40,	...
Multiples of 7:	7,	14,	21,	28,	35,	42,	49,	56,	...

All of the multiples of a number have that number as a factor. Therefore, none of the multiples of a number, except possibly that number itself, can be prime. A process called the **Sieve of Eratosthenes** involves eliminating multiples to find prime numbers as the following steps describe. The description here illustrates finding the prime numbers less than 50.

Step 1: List the counting numbers from 1 to 50 as shown. 1 is neither prime nor composite, so cross out 1. Then, circle the next listed number that is not crossed out, and cross out all of its listed multiples. In this case, circle 2 and cross out all the multiples of 2.

1	②	3	4	5	6	7	8	9	10
11	12	13	14	15	16	17	18	19	20
21	22	23	24	25	26	27	28	29	30
31	32	33	34	35	36	37	38	39	40
41	42	43	44	45	46	47	48	49	50

Step 2: Circle 3 and cross out all the multiples of 3. (Some of these, such as 6 and 12, will already have been crossed out.)

1	②	③	4	5	6	7	8	9	10
11	12	13	14	15	16	17	18	19	20
21	22	23	24	25	26	27	28	29	30
31	32	33	34	35	36	37	38	39	40
41	42	43	44	45	46	47	48	49	50

Step 3: Circle 5 and cross out the multiples of 5. Then circle 7 and cross out the multiples of 7. Continue in this manner and the prime numbers will be circled and composite numbers crossed out (up to 50).

1	②	③	4	⑤	6	⑦	8	9	10
⑪	12	⑬	14	15	16	⑰	18	⑲	20
21	22	㉓	24	25	26	27	28	㉙	30
㉛	32	33	34	35	36	㊲	38	39	40
㊶	42	㊸	44	45	46	㊼	48	49	50

This last table shows that the prime numbers less than 50 are

2, 3, 5, 7, 11, 13, 17, 19, 23, 29, 31, 37, 41, 43, and **47**.

Two Important Facts about Prime Numbers

1. **Even Numbers:** 2 is the only even prime number.
2. **Odd Numbers:** All other prime numbers are odd numbers. But, not all odd numbers are prime.

Objective C Determining Prime Numbers

An important mathematical fact is that **there is no known pattern or formula for determining prime numbers**. In fact, only recently have computers been used to show that very large numbers previously thought to be prime are actually composite. The previous discussion has helped to develop an understanding of the nature of prime numbers and showed some prime and composite numbers in list form. However, writing a list every time to determine whether a number is prime or not would be quite time consuming and unnecessary. The following procedure of dividing by prime numbers can be used to determine whether or not relatively small numbers (say, less than 1000) are prime. If a prime number smaller than the given number is found to be a factor (or divisor), then the given number is composite. (**Note:** You still should memorize the prime numbers less than 50 for convenience and ease with working with fractions.)

Divide the number by progressively larger **prime numbers** $(2, 3, 5, 7, 11,$ and so forth) until:

1. You find a remainder of 0 (meaning that the prime number is a factor and the given number is composite); or
2. You find a quotient smaller than the prime divisor (meaning that the given number has no smaller prime factors and is therefore prime itself). (See example 5.)

notes

Reasoning that if a composite number was a factor, then one of its prime factors would have been found to be a factor in an earlier division, we divide only by prime numbers – that is, there is no need to divide by a composite number.

3. Is 999 a prime number?

Example 3

Determining Prime Numbers

Is 305 a prime number?

Solution

Since the units digit is 5, the number 305 is divisible by 5 (by the divisibility test in Section 2.1) and is not prime. 305 is a composite number.

Now work margin exercise 3.

Example 4

Determining Prime Numbers

Is 247 prime?

Solution

Mentally, note that the tests for divisibility by the prime numbers 2, 3, and 5 fail.

Divide by 7:
$$\begin{array}{r} 35 \\ 7\overline{)247} \\ \underline{21} \\ 37 \\ \underline{35} \\ 2 \end{array}$$

Divide by 11:

$$\begin{array}{r} 22 \\ 11{\overline{)247}} \\ \underline{22} \\ 27 \\ \underline{22} \\ 5 \end{array}$$

Divide by 13:

$$\begin{array}{r} 19 \\ 13{\overline{)247}} \\ \underline{13} \\ 117 \\ \underline{117} \\ 0 \end{array}$$

247 is composite and not prime. In fact, $247 = 13 \cdot 19$ and 13 and 19 are factors of 247.

Now work margin exercise 4.

4. Is 289 a prime number?

Example 5

Determining Prime Numbers

Is 109 prime?

Solution

Mentally, note that the tests for divisibility by the prime numbers 2, 3, and 5 fail.

Divide by 7:

$$\begin{array}{r} 15 \\ 7{\overline{)109}} \\ \underline{7} \\ 39 \\ \underline{35} \\ 4 \end{array}$$

Divide by 11:

$$\begin{array}{r} 9 \\ 11{\overline{)109}} \\ \underline{99} \\ 10 \end{array}$$ ← The quotient 9 is less than the divisor 11.

There is no point in dividing by larger numbers such as 13 or 17 because the quotient would only get smaller, and if any of these larger numbers were a divisor, the smaller quotient would have been found to be a divisor earlier in the procedure. Therefore, 109 is a prime number.

Now work margin exercise 5.

5. Determine whether 221 is prime or composite. Explain each step.

6. Determine whether 239 is prime or composite. Explain each step.

Completion Example 6

Determining Prime Numbers

Is 199 prime or composite?

Solution

Tests for 2, 3, and 5 all fail.

Divide by 7: $7\overline{)199}$ Divide by 11: $11\overline{)199}$

Divide by 13: $13\overline{)199}$ Divide by _____ : _____ $\overline{)199}$

199 is a _____ number.

Now work margin exercise **6.**

7. Find two factors of 270 whose sum is 59.

Example 7

Applications

One interesting application of factors of counting numbers (very useful in beginning algebra) involves finding two factors whose sum is some specified number. For example, find two factors for 75 such that their product is 75 and their sum is 20.

Solution

The factors of 75 are 1, 3, 5, 15, 25, and 75, and the pairs whose products are 75 are

$$1 \cdot 75 = 75 \qquad 3 \cdot 25 = 75 \qquad 5 \cdot 15 = 75.$$

Thus the numbers we are looking for are 5 and 15 because

$$5 \cdot 15 = 75 \qquad \text{and} \qquad 5 + 15 = 20.$$

Now work margin exercise **7.**

Completion Example Answers

6. Divide by 7:
$$\begin{array}{r} 28 \\ 7\overline{)199} \\ \underline{14} \\ 59 \\ \underline{56} \\ 3 \end{array}$$

Divide by 11:
$$\begin{array}{r} 18 \\ 11\overline{)199} \\ \underline{11} \\ 89 \\ \underline{88} \\ 1 \end{array}$$

Divide by 13:
$$\begin{array}{r} 15 \\ 13\overline{)199} \\ \underline{13} \\ 69 \\ \underline{65} \\ 4 \end{array}$$

Divide by 17:
$$\begin{array}{r} 11 \\ 17\overline{)199} \\ \underline{17} \\ 29 \\ \underline{17} \\ 12 \end{array}$$

199 is a prime number.

Exercises 2.2

List the first five multiples for each of the following numbers.

1. 3   **2.** 6   **3.** 8   **4.** 10   **5.** 11

6. 15   **7.** 20   **8.** 30   **9.** 41   **10.** 61

First, list all the prime numbers less than 50, and then decide whether each of the following numbers is prime or composite by dividing by prime numbers. If the number is composite, find at least two pairs of factors whose product is that number. See Examples 3 through 6.

11. 23   **12.** 41   **13.** 47   **14.** 67

15. 55   **16.** 65   **17.** 97   **18.** 59

19. 205   **20.** 502   **21.** 719   **22.** 517

23. 943   **24.** 1073   **25.** 551   **26.** 961

Two numbers are given. Find two factors of the first number such that their product is the first number and their sum is the second number. For example, for the numbers 45, 18, you would find two factors of 45 whose sum is 18. The factors are 3 and 15, since $3 \cdot 15 = 45$ and $3 + 15 = 18$. See Example 7.

27. 24, 11   **28.** 12, 7   **29.** 16, 10   **30.** 24, 14

31. 20, 12   **32.** 20, 9   **33.** 36, 13   **34.** 36, 15

35. 51, 20   **36.** 57, 22   **37.** 25, 10   **38.** 16, 8

39. 72, 27   **40.** 52, 17   **41.** 55, 16   **42.** 66, 17

43. 63, 24   **44.** 65, 18   **45.** 75, 28   **46.** 60, 17

47. Find two integers whose product is 60 and whose sum is –16.

48. Find two integers whose product is 75 and whose sum is –20.

49. Find two integers whose product is –20 and whose sum is –8.

50. Find two integers whose product is –30 and whose sum is –7.

51. Find two integers whose product is –50 and whose sum is 5.

Writing & Thinking

52. Discuss the following two statements:

   a. All prime numbers are odd.

   b. All positive odd numbers are prime.

53. **a.** Explain why 1 is not prime and not composite.

   b. Explain why 0 is not prime and not composite.

54. Numbers of the form $2^N - 1$, where N is a prime number, are sometimes prime. These prime numbers are called "Mersenne primes" (after Marin Mersenne, 1588 – 1648). Show that for $N = 2, 3, 5, 7$, and 13 the numbers $2^N - 1$ are prime, but for $N = 11$ the number $2^N - 1$ is composite. (**Historical Note:** In 1978, with the use of a computer, two students at California State University at Hayward proved that the number $2^{21,701} - 1$ is prime.)

2.3 Prime Factorization

Finding a Prime Factorization

In all our work with fractions (including multiplication, division, addition, and subtraction), we will need to be able to factor whole numbers so that all of the factors are prime numbers. This is called finding the **prime factorization** of a whole number. For example, to find the prime factorization of 63, we start with any two factors of 63.

$$63 = 9 \cdot 7 \qquad \text{Although 7 is prime, 9 is not prime.}$$

Factoring 9 gives

$$63 = 3 \cdot 3 \cdot 7. \qquad \text{Now all of the factors are prime.}$$

This last product, $3 \cdot 3 \cdot 7$, is the **prime factorization** of 63.

Note: Because multiplication is a commutative operation, the factors may be written in any order. Thus we might write

$$63 = 21 \cdot 3$$
$$= 3 \cdot 7 \cdot 3.$$

The prime factorization is the same even though the factors are in a different order. For consistency, we will generally write the factors in order, from smallest to largest, as in the first case ($3 \cdot 3 \cdot 7$).

notes

For the purposes of prime factorization, a negative integer will be treated as a product of –1 and a positive integer. This means that only prime factorizations of positive integers need to be discussed at this time.

Regardless of the method used, you will always arrive at the same factorization for any given composite number. That is, **there is only one prime factorization for any composite number**. This fact is called the **fundamental theorem of arithmetic**.

The Fundamental Theorem of Arithmetic

Every composite number has exactly one prime factorization.

Objectives

A Be able to find the prime factorization of a composite number.

B Be able to list all factors of a composite number.

To Find the Prime Factorization of a Composite Number

1. Factor the composite number into any two factors.
2. Factor each factor that is not prime.
3. Continue this process until all factors are prime.

The **prime factorization** of a number is a factorization of that number using only prime factors.

Many times, the beginning factors needed to start the process for finding a prime factorization can be found by using the tests for divisibility by 2, 3, 5, 6, 9, and 10 discussed in Section 2.1. This was one purpose for developing these tests, and you should review them or write them down for easy reference.

Example 1

Prime Factorization

Find the prime factorization of 90.

Solution

$$90 = \quad 9 \quad \cdot \quad 10$$

$$= 3 \cdot 3 \cdot 2 \cdot 5$$

Since the units digit is 0, we know that 10 is a factor.

9 and 10 can both be factored so that each factor is a prime number. This is the prime factorization.

OR

$$90 = 3 \cdot 30$$

$$= 3 \cdot 10 \cdot 3$$

$$= 3 \cdot 2 \cdot 5 \cdot 3$$

3 is prime, but 30 is not.

10 is not prime.

All factors are prime.

Note: The final prime factorization was the same in both factor trees even though the first pair of factors was different.

Since multiplication is commutative, the order of the factors is not important. What is important is that all the factors are prime. Writing the factors in order, we can write $90 = 2 \cdot 3 \cdot 3 \cdot 5$ or, with exponents, $90 = 2 \cdot 3^2 \cdot 5$.

Example 2

Prime Factorization

Find the prime factorizations of each number.

a. 65

Solution

$$65 \ = \ 5 \ \cdot \ 13$$

5 is a factor because the units digit is 5. Since both 5 and 13 are prime, $5 \cdot 13$ is the prime factorization.

b. 72

Solution

$$
\begin{aligned}
72 \ &= \ 8 \ \cdot \ 9 \\
&= \ 2 \ \cdot \ 4 \ \cdot \ 3 \ \cdot \ 3 \\
&= \ 2 \ \cdot \ 2 \ \cdot \ 2 \ \cdot \ 3 \ \cdot \ 3 \\
&= \ 2^3 \cdot 3^2
\end{aligned}
$$

9 is a factor because the sum of the digits is 9.

using exponents

c. 294

Solution

$$
\begin{aligned}
294 \ &= \ 2 \ \cdot \ 147 \\
&= \ 2 \ \cdot \ 3 \ \cdot \ 49 \\
&= \ 2 \ \cdot \ 3 \ \cdot \ 7 \ \cdot \ 7 \\
&= \ 2 \cdot 3 \cdot 7^2
\end{aligned}
$$

2 is a factor because the units digit is even.

3 is a factor of 147 because the sum of the digits is divisible by 3.

using exponents

If we begin with the product $294 = 6 \cdot 49$, we see that the prime factorization is the same.

$$
\begin{aligned}
294 \ &= \ 6 \ \cdot \ 49 \\
&= \ 2 \ \cdot \ 3 \ \cdot \ 7 \ \cdot \ 7 \\
&= \ 2 \cdot 3 \cdot 7^2
\end{aligned}
$$

Completion Example 3

Prime Factorization

Find the prime factorization of 60.

Solution

$60 = \quad 6 \qquad \cdot$

$= \quad 2 \quad \cdot \quad 3 \quad \cdot \quad \underline{} \cdot \underline{}$

$= \underline{} \qquad$ using exponents

Completion Example 4

Prime Factorization

Find the prime factorization of 308.

Solution

$308 = \quad 4 \qquad \cdot$

$= \quad \underline{} \cdot \underline{} \quad \cdot \quad \underline{} \cdot \underline{}$

$= \underline{} \qquad$ using exponents

Now work margin exercises 1 through 4.

Find the prime factorization of the following numbers.

1. 74

2. 52

3. 460

4. 616

Objective B | ## Factors of a Composite Number

Once a prime factorization of a number is known, all the factors (or divisors) of that number can be found. For a number to be a factor of a composite number, it must be either 1, the number itself, one of the prime factors, or the product of two or more of the prime factors. (See the discussion of divisibility of products in section 2.1 for a similar analysis.)

Factors of Composite Numbers

The only factors (or divisors) of a composite number are:

1. 1 and the number itself,
2. each prime factor, and
3. products formed by all combinations of the prime factors (including repeated factors).

Completion Example Answers

3. $60 = 6 \cdot 10$

$= 2 \cdot 3 \cdot 2 \cdot 5$

$= 2^2 \cdot 3 \cdot 5$ using exponents

4. $308 = 4 \cdot 77$

$= 2 \cdot 2 \cdot 7 \cdot 11$

$= 2^2 \cdot 7 \cdot 11$ using exponents

Example 5

Factorization

Find all the factors of 60.

Solution

Since $60 = 2 \cdot 2 \cdot 3 \cdot 5$, the factors are
a. 1 and 60 (1 and the number itself),

b. 2, 3, and 5 (each prime factor),

c. $2 \cdot 2 = 4$, $2 \cdot 3 = 6$, $2 \cdot 5 = 10$, $3 \cdot 5 = 15$,
   $2 \cdot 2 \cdot 3 = 12$, $2 \cdot 2 \cdot 5 = 20$,   and $2 \cdot 3 \cdot 5 = 30$.

The twelve factors of 60 are
1, 60, 2, 3, 5, 4, 6, 10, 12, 15, 20, and 30.
These are the only factors of 60.

Completion Example 6

Factorization

Find all the factors of 154.

Solution

The prime factorization of 154 is $2 \cdot 7 \cdot 11$.
a. Two factors are 1 and _____.

b. Three prime factors are _____, _____, and _____.

c. Other factors are
   _____ · _____ = _____   _____ · _____ = _____

   _____ · _____ = _____.

The factors of 154 are _____.

Now work margin exercises 5 and 6.

Find all the factors of the following numbers.

5. 63

6. 54

Completion Example Answers

6. a. Two factors are 1 and 154.
b. Three prime factors are 2, 7, and 11.
c. Other factors are
   $2 \cdot 7 = 14$ $2 \cdot 11 = 22$ $7 \cdot 11 = 77$.
The factors of 154 are 1, 154, 2, 7, 11, 14, 22, and 77.

Exercises 2.3

In your own words, define the following terms.

1. prime number

2. composite number

Find the prime factorization for each of the following numbers. Use the tests for divisibility for 2, 3, 5, 6, 9, and 10, whenever they help, to find beginning factors. See Examples 1 through 4.

3. 20	**4.** 44	**5.** 32	**6.** 45
7. 50	**8.** 36	**9.** 70	**10.** 80
11. 51	**12.** 162	**13.** 62	**14.** 125
15. 99	**16.** 94	**17.** 37	**18.** 43
19. 120	**20.** 225	**21.** 196	**22.** 289
23. 361	**24.** 400	**25.** 65	**26.** 91
27. 1000	**28.** 10,000	**29.** 100,000	**30.** 1,000,000
31. 600	**32.** 700	**33.** 107	**34.** 211
35. 309	**36.** 505	**37.** 165	**38.** 231
39. 675	**40.** 135	**41.** 216	**42.** 1125
43. 1692	**44.** 2200	**45.** 676	**46.** 2717

47. 42

48. 24

49. 300

50. 700

51. 66

52. 96

53. 78

54. 130

55. 150

56. 175

57. 90

58. 275

59. 1001

60. 715

61. 585

62. 2310

Collaborative Learning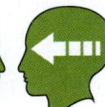

In groups of three to four students, discuss the theorem in Exercise 63 and answer the related questions. Discuss your answers in class.

63. In higher level mathematics, number theorists have proven the following theorem:

Write the prime factorization of a number in exponential form. Add 1 to each exponent. The product of these sums is the number of factors of the original number.

For example, $60 = 2^2 \cdot 3 \cdot 5 = 2^2 \cdot 3^1 \cdot 5^1$. Adding 1 to each exponent and forming the product gives

$$(2+1)(1+1)(1+1) = 3 \cdot 2 \cdot 2 = 12$$

and there are twelve factors of 60.

Use this theorem to find the number of factors of each of the following numbers. Find as many of these factors as you can.

a. 500   **b.** 660   **c.** 450   **d.** 148,225

2.4 Least Common Multiple (LCM)

Objectives

A Be able to use prime factorizations to find the least common multiple of a set of counting numbers.

B Be able to use prime factorizations to find the least common multiple of a set of algebraic terms.

C Recognize the application of the LCM concept in a word problem.

D Be able to use prime factorizations to find the greatest common divisor of a set of counting numbers.

Objective A Finding the LCM of a Set of Counting (or Natural) Numbers

The ideas discussed in this section are used throughout our development of fractions and mixed numbers. Study these ideas and the related techniques carefully because they will make your work with fractions much easier and more understandable.

Recall from a previous section that the multiples of an integer are the products of that integer with the counting numbers. Our discussion here is based entirely on multiples of positive integers (or counting numbers). Thus the first multiple of any counting number is the number itself, and all other multiples are larger than that number. We are interested in finding common multiples, and more particularly, the **least common multiple** for a set of counting numbers. For example, consider the lists of multiples of 8 and 12 shown here.

Counting Numbers: 1,   2,   3,   4,   5,   6,   7,   8,   9,   10, 11, ...

Multiples of 8:   8,   16, ⟨24⟩, 32,   40,   ⟨48⟩,   56,   64,   ⟨72⟩,   80, 88, ...

Multiples of 12:   12, ⟨24⟩, 36, ⟨48⟩,   60,   ⟨72⟩,   84,   ⟨96⟩, 108, ⟨120⟩, 132, ...

The common multiples of 8 and 12 are 24, 48, 72, 96, 120, ... **The least common multiple (LCM) is 24.**

Listing all the multiples, as we just did for 8 and 12, and then choosing the least common multiple (LCM) is not very efficient. The following technique involving prime factorizations is generally much easier to use.

To Find the LCM of a Set of Counting Numbers

1. Find the prime factorization of each number.
2. Identify the prime factors that appear in any one of the prime factorizations.
3. Form the product of these primes using each prime the most number of times it appears in any one of the prime factorizations.

Your skill with this method depends on your ability to find prime factorizations quickly and accurately. With practice and understanding, this method will prove efficient and effective. STAY WITH IT!

notes

There are other methods for finding the LCM, maybe even easier to use at first. However, just finding the LCM is not our only purpose, and the method outlined here allows for a solid understanding of using the LCM when working with fractions.

Example 1

Finding the Least Common Multiple

Find the least common multiple (LCM) of 8, 10, and 30.

Solution

a. prime factorizations:

$8 = 2 \cdot 2 \cdot 2$ three 2's
$10 = 2 \cdot 5$ one 2, one 5
$30 = 2 \cdot 3 \cdot 5$ one 2, one 3, one 5

b. Prime factors that are present are 2, 3, and 5.

c. The most of each prime factor in any one factorization:

Three 2's   (in 8)
One 3   (in 30)
One 5   (in 10 and in 30)

$$LCM = 2 \cdot 2 \cdot 2 \cdot 3 \cdot 5$$
$$= 2^3 \cdot 3 \cdot 5 = 120$$

120 is the LCM and therefore, the smallest number divisible by 8, 10, and 30.

Example 2

Finding the Least Common Multiple

Find the LCM of 27, 30, 35, and 42.

Solution

a. prime factorizations:

$27 = 3 \cdot 3 \cdot 3$ three 3's
$30 = 2 \cdot 3 \cdot 5$ one 2, one 3, one 5
$35 = 5 \cdot 7$ one 5, one 7
$42 = 2 \cdot 3 \cdot 7$ one 2, one 3, one 7

b. Prime factors present are 2, 3, 5, and 7.

c. The most of each prime factor in any one factorization is shown here.

one 2	(in 30 and in 42)
three 3's	(in 27)
one 5	(in 30 and in 35)
one 7	(in 35 and in 42)

$$LCM = 2 \cdot 3 \cdot 3 \cdot 3 \cdot 5 \cdot 7$$
$$= 2 \cdot 3^3 \cdot 5 \cdot 7$$
$$= 1890$$

1890 is the smallest number divisible by all four of the numbers 27, 30, 35, and 42.

Find the LCM for the following set of counting numbers.

1. 140, 168

2. 14, 18, 6, 21

3. 30, 40, 70

Completion Example 3

Finding the Least Common Multiple

Find the LCM for 15, 24, and 36.

Solution

a. prime factorizations:

15 = _____
24 = _____
36 = _____

b. The only prime factors are _____, _____, and _____.

c. The most of each prime factor in any one factorization:

_____   (in 24)
_____   (in 36)
_____   (in 15)

LCM = _____ = _____ (using exponents)

= _____

Now work margin exercises **1 through 3.**

Completion Example Answers

3. a. prime factorizations:
15 = 3 · 5
24 = 2 · 2 · 2 · 3
36 = 2 · 2 · 3 · 3
b. The only prime factors are 2, 3, and 5.
c. The most of each prime factor in any one factorization:
three 2's   (in 24)
two 3's   (in 36)
one 5   (in 15)
LCM = 2 · 2 · 2 · 3 · 3 · 5 = $2^3 \cdot 3^2 \cdot 5 = 360$

In Example 2, we found that 1890 is the LCM for 27, 30, 35, and 42. This means that 1890 is a multiple of each of these numbers, and each is a factor of 1890. To find out how many times each number will divide into 1890, we could divide by using long division. However, by looking at the prime factorization of 1890 (which we know from Example 2) and the prime factorization of each number, we find the quotients without actually dividing. We can group the factors as follows:

$$1890 = \underbrace{(3 \cdot 3 \cdot 3)}_{27} \cdot \underbrace{(2 \cdot 5 \cdot 7)}_{70} = \underbrace{(2 \cdot 3 \cdot 5)}_{30} \cdot \underbrace{(3 \cdot 3 \cdot 7)}_{63}$$

$$= \underbrace{(5 \cdot 7)}_{35} \cdot \underbrace{(2 \cdot 3 \cdot 3 \cdot 3)}_{54} = \underbrace{(2 \cdot 3 \cdot 7)}_{42} \cdot \underbrace{(3 \cdot 3 \cdot 5)}_{45}$$

So,

27 divides into 1890 70 times,
30 divides into 1890 63 times,
35 divides into 1890 54 times, and
42 divides into 1890 45 times.

Completion Example 4

Finding the Least Common Multiple

a. Find the LCM for 18, 36, and 66, and
b. Tell how many times each number divides into the LCM.

Solutions

a.
$$18 = \underline{\hspace{2cm}}$$
$$36 = \underline{\hspace{2cm}} \Big\} \text{LCM} = \underline{\hspace{2cm}}$$
$$66 = \underline{\hspace{2cm}}$$
$$= \underline{\hspace{2cm}} = 396$$

b.
$$396 = \underline{\hspace{2cm}} = (2 \cdot 3 \cdot 3) \cdot \underline{\hspace{2cm}} = 18 \cdot \underline{\hspace{1cm}}$$
$$396 = \underline{\hspace{2cm}} = (2 \cdot 2 \cdot 3 \cdot 3) \cdot \underline{\hspace{2cm}} = 36 \cdot \underline{\hspace{1cm}}$$
$$396 = \underline{\hspace{2cm}} = (2 \cdot 3 \cdot 11) \cdot \underline{\hspace{2cm}} = 66 \cdot \underline{\hspace{1cm}}$$

Now work margin exercise 4.

4. Find the LCM for 20, 35, and 70; then tell how many times each number divides into the LCM.

Completion Example Answers

4.a.
$$18 = 2 \cdot 3 \cdot 3$$
$$36 = 2 \cdot 2 \cdot 3 \cdot 3 \Big\} \text{LCM} = 2 \cdot 2 \cdot 3 \cdot 3 \cdot 11$$
$$66 = 2 \cdot 3 \cdot 11$$
$$= 2^2 \cdot 3^2 \cdot 11 = 396$$

b. $396 = 2 \cdot 2 \cdot 3 \cdot 3 \cdot 11 = (2 \cdot 3 \cdot 3) \cdot (2 \cdot 11) = 18 \cdot 22$
$396 = 2 \cdot 2 \cdot 3 \cdot 3 \cdot 11 = (2 \cdot 2 \cdot 3 \cdot 3) \cdot (11) = 36 \cdot 11$
$396 = 2 \cdot 2 \cdot 3 \cdot 3 \cdot 11 = (2 \cdot 3 \cdot 11) \cdot (2 \cdot 3) = 66 \cdot 6$

Algebraic expressions that are numbers, powers of variables, or products of numbers and powers of variables are called **terms**. Note that a number written next to a variable or two variables written next to each other indicate multiplication. For example,

$$3, 4ab, 25a^3, 6xy^2, 48x^3y^4$$

are all algebraic terms. In each case where multiplication is indicated, the number is called the **numerical coefficient** of the term. A term that consists of only a number, such as 3, is called a **constant** or a **constant term**.

Another approach to finding the LCM, particularly useful when algebraic terms are involved, is to write each prime factorization in exponential form and proceed as follows.

Step 1: Find the prime factorization of each term in the set and write it in exponential form, including variables.

Step 2: Find the largest power of each prime factor present in any of the prime factorizations. Remember that, if no exponent is written, the exponent is understood to be 1.

Step 3: The LCM is the product of these powers.

Example 5

Finding the LCM of Algebraic Terms

Find the LCM of the terms $6a$, a^2b, $4a^2$, and $18b^3$.

Solution

Write each prime factorization in exponential form (including variables) and multiply the largest powers of each prime factor, including variables.

$$6a \quad = 2 \cdot 3 \cdot a$$
$$a^2b \ = a^2 \cdot b$$
$$4a^2 \ = 2^2 \cdot a^2$$
$$18b^3 = 2 \cdot 3^2 \cdot b^3$$

$$LCM = 2^2 \cdot 3^2 \cdot a^2 \cdot b^3 = 36a^2b^3$$

Example 6

Finding the LCM of Algebraic Terms

Find the LCM of the terms $16x, 25xy^3, 30x^2y$.

Solution

Write each prime factorization in exponential form (including variables) and multiply the largest powers of each prime factor, including variables.

$$
\left.
\begin{array}{l}
16x \quad = 2^4 \cdot x \\
25xy^3 = 5^2 \cdot x \cdot y^3 \\
30x^2y = 2 \cdot 3 \cdot 5 \cdot x^2 \cdot y
\end{array}
\right\}
\quad \text{LCM} = 2^4 \cdot 3 \cdot 5^2 \cdot x^2 \cdot y^3 = 1200x^2y^3
$$

Completion Example 7

Finding the LCM of Algebraic Terms

Find the LCM of the terms $8xy, 10x^2$, and $20y$.

Solution

$$
\left.
\begin{array}{l}
8xy \quad = \underline{\hspace{2cm}} \\
10x^2 = \underline{\hspace{2cm}} \\
20y \quad = \underline{\hspace{2cm}}
\end{array}
\right\}
\quad \text{LCM} = \underline{\hspace{2cm}}
$$

$$= \underline{\hspace{2cm}}$$

Now work margin exercises 5 through 7.

Objective C **An Application**

Many events occur at regular intervals of time. Weather satellites may orbit the earth once every 10 hours or once every 12 hours. Delivery trucks arrive once a day or once a week at department stores. Traffic lights change once every 3 minutes or once every 2 minutes. The periodic frequency with which such events occur can be explained in terms of the least common multiple, as illustrated in Example 8.

Find the LCM for each set of algebraic terms.

5. $14xy, 10y^2, 25x$

6. $15ab^2, ab^3, 20a^3$

7. $5a, 6ab, 14b^2$

Completion Example Answers

7.
$$
\left.
\begin{array}{l}
8xy \quad = 2^3 \cdot x \cdot y \\
10x^2 = 2 \cdot 5 \cdot x^2 \\
20y \quad = 2^2 \cdot 5 \cdot y
\end{array}
\right\}
\quad \text{LCM} = 2^3 \cdot 5 \cdot x^2 \cdot y
$$

$$= 40x^2y$$

8. A walker and two joggers begin using the same track at the same time. Their lap times are 6, 3, and 5 minutes, respectively.

a. In how many minutes will they be together at the starting place?

b. How many laps will each person have completed at this time?

Example 8

Weather Satellites

Suppose three weather satellites – **A**, **B**, and **C** – are orbiting the earth at different times. Satellite **A** takes 24 hours, **B** takes 18 hours, and **C** takes 12 hours. If they are directly above each other now, as shown in part **a.** of the figure below, in how many hours will they again be directly above each other in the position shown in part **a.** ? How many orbits will each satellite have made in that time?

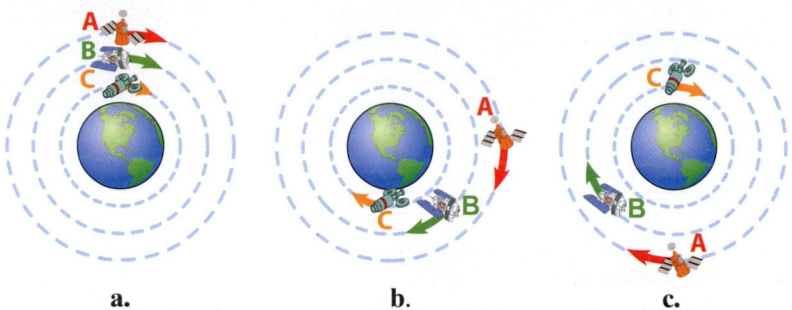

| **a.** | **b.** | **c.** |
| Beginning Positions | Positions after 6 hours | Positions after 12 hours |

Solution

Study the diagrams shown above. When the three satellites are again in the initial position shown, each will have made some number of complete orbits. Since **A** takes 24 hours to make one complete orbit, the solution must be a multiple of 24. Similarly, the solution must be a multiple of 18 and a multiple of 12 to allow for complete orbits of satellites **B** and **C**.

The solution is the LCM of 24, 18, and 12.

$$\left.\begin{array}{l} 24 = 2^3 \cdot 3 \\ 18 = 2 \cdot 3^2 \\ 12 = 2^2 \cdot 3 \end{array}\right\} \quad \text{LCM} = 2^3 \cdot 3^2 = 72$$

Thus the satellites will align again at the position shown in part **a.** in 72 hours (or 3 days). Note that:

Satellite A will have made 3 orbits: $24 \cdot \mathbf{3} = 72$

Satellite B will have made 4 orbits: $18 \cdot \mathbf{4} = 72$

Satellite C will have made 6 orbits: $12 \cdot \mathbf{6} = 72.$

Now work margin exercise 8.

Objective D **Greatest Common Divisor (GCD)**

Consider the two numbers 12 and 18. Is there a number (or numbers) that will divide into **both** 12 and 18? To help answer this question, the divisors for 12 and 18 are listed on the next page.

Set of divisors for 12: {1, 2, 3, 4, 6, 12}
Set of divisors for 18: {1, 2, 3, 6, 9, 18}

The **common divisors** for 12 and 18 are 1, 2, 3, and 6. The **greatest common divisor (GCD)** for 12 and 18 is 6. That is, of all the common divisors of 12 and 18, 6 is the largest divisor.

Example 9

Finding All Divisors and the Greatest Common Divisor

List the divisors of each number in the set {36, 24, 48} and find the greatest common divisor (GCD).

Solution

Set of divisors for 36: {**1, 2, 3, 4, 6,** 9, **12,** 18, 36}
Set of divisors for 24: {**1, 2, 3, 4, 6,** 8, **12,** 24}
Set of divisors for 48: {**1, 2, 3, 4, 6,** 8, **12,** 16, 24, 48}

The common divisors are **1, 2, 3, 4, 6,** and **12. GCD = 12**.

Now work margin exercise 9.

9. List the divisors of each number in the set {45, 60, 90} and find the greatest common divisor (GCD).

The Greatest Common Divisor

The **greatest common divisor** (GCD)* of a set of natural numbers is the largest natural number that will divide into all the numbers in the set.

As Example 9 illustrates, listing the divisors of each number before finding the GCD can be tedious and difficult. **The use of prime factorizations leads to a simple technique for finding the GCD.**

Technique for Finding the GCD of a Set of Counting Numbers

1. Find the prime factorization of each number.
2. Find the prime factors common to all factorizations.
3. Form the product of these primes, using each prime the number of times it is common to all factorizations.
4. This product is the GCD. If there are no primes common to all factorizations, the GCD is 1.

*The greatest common divisor is, of course, the greatest common factor, and the GCD could be called the **greatest common factor**, and be abbreviated GCF.

Example 10

Finding the GCD

Find the GCD for $\{36, 24, 48\}$.

Solution

$$\left.\begin{array}{l} 36 = \mathbf{2} \cdot \mathbf{2} \cdot 3 \cdot \mathbf{3} \\ 24 = \mathbf{2} \cdot \mathbf{2} \cdot 2 \cdot \mathbf{3} \\ 48 = \mathbf{2} \cdot \mathbf{2} \cdot 2 \cdot 2 \cdot \mathbf{3} \end{array}\right\} \quad \text{GCD} = 2 \cdot 2 \cdot 3 = 12$$

In all the prime factorizations, the factor 2 appears twice and the factor 3 appears once. The GCD is 12.

Example 11

Finding the GCD

Find the GCD for $\{360, 75, 30\}$.

Solution

$$\left.\begin{array}{l} 360 = 36 \cdot 10 = 4 \cdot 9 \cdot 2 \cdot 5 = 2 \cdot 2 \cdot 2 \cdot \mathbf{3} \cdot 3 \cdot \mathbf{5} \\ 75 = 3 \cdot 25 = \mathbf{3} \cdot \mathbf{5} \cdot 5 \\ 30 = 6 \cdot 5 = 2 \cdot \mathbf{3} \cdot \mathbf{5} \end{array}\right\} \quad \text{GCD} = 3 \cdot 5 = 15$$

Each of the factors 3 and 5 appears only once in all the prime factorizations. The GCD is 15.

Example 12

Finding the GCD

Find the GCD for $\{168, 420, 504\}$.

Solution

$$\left.\begin{array}{l} 168 = 8 \cdot 21 = \mathbf{2} \cdot \mathbf{2} \cdot 2 \cdot \mathbf{3} \cdot \mathbf{7} \\ 420 = 10 \cdot 42 = 2 \cdot 5 \cdot 6 \cdot 7 \\ \quad\quad = \mathbf{2} \cdot \mathbf{2} \cdot \mathbf{3} \cdot 5 \cdot \mathbf{7} \\ 504 = 4 \cdot 126 \cdot = 2 \cdot 2 \cdot 6 \cdot 21 \\ \quad\quad = \mathbf{2} \cdot \mathbf{2} \cdot 2 \cdot \mathbf{3} \cdot 3 \cdot \mathbf{7} \end{array}\right\} \quad \text{GCD} = 2 \cdot 2 \cdot 3 \cdot 7 = 84$$

In **all** prime factorizations, 2 appears twice, 3 once, and 7 once. The GCD is 84.

If the GCD of two numbers is 1 (that is, they have no common prime factors), then the two numbers are said to be **relatively prime**. The numbers themselves may be prime or they may be composite.

Example 13

Relatively Prime Numbers

Find the GCD for $\{15, 8\}$.

Solution

$$15 = 3 \cdot 5$$
$$8 = 2 \cdot 2 \cdot 2$$
GCD = 1   8 and 15 are relatively prime.

Example 14

Relatively Prime Numbers

Find the GCD for $\{20, 21\}$.

Solution

$$20 = 2 \cdot 2 \cdot 5$$
$$21 = 3 \cdot 7$$
GCD = 1   20 and 21 are relatively prime.

Now work margin exercises **10 through 14.**

Find the GCD for the following sets of numbers.

10. $\{26, 52, 104\}$

11. $\{49, 343, 147\}$

12. $\{60, 90, 210\}$

13. $\{26, 19\}$

14. $\{33, 343\}$

Exercises 2.4

1. List the first twelve multiples of each number.

   a. 5　　　　　　　　　　b. 6　　　　　　　　　　c. 15

2. From the lists you made in Exercise 1, find the least common multiple for the following pairs of numbers.

   a. 5 and 6　　　　　　　b. 5 and 15　　　　　　c. 6 and 15

Find the LCM of each of the following sets of counting numbers. See Examples 1 through 3.

3. 3, 5, 7　　　　　　4. 2, 7, 11　　　　　　5. 6, 10

6. 9, 12　　　　　　　7. 2, 3, 11　　　　　　8. 3, 5, 13

9. 4, 14, 35　　　　　10. 10, 12, 20　　　　　11. 50, 75

12. 30, 70　　　　　　13. 20, 90　　　　　　14. 50, 80

15. 28, 98　　　　　　16. 10, 15, 35　　　　　17. 6, 12, 15

18. 34, 51, 54　　　　19. 22, 44, 121　　　　20. 15, 45, 90

21. 14, 28, 56　　　　22. 20, 50, 100　　　　23. 30, 60, 120

24. 35, 40, 72　　　　25. 144, 169, 196　　　26. 225, 256, 324

27. 81, 256, 361

In Exercises 28 – 37,
   a. find the LCM of each set of numbers, and
   b. state the number of times each number divides into the LCM. See Example 4.

28. 10, 15, 25　　　　29. 6, 24, 30　　　　　30. 10, 18, 90

31. 12, 18, 27　　　　32. 20, 28, 45　　　　　33. 99, 121, 143

34. 125, 135, 225, 250　　　　　35. 40, 56, 160, 196

36. $35, 49, 63, 126$

37. $45, 56, 98, 99$

Find the LCM of each of the following sets of algebraic terms. See Examples 5 through 7.

38. $25xy, 40xyz$

39. $40xyz, 75xy^2$

40. $20a^2b, 50ab^3$

41. $16abc, 28a^3b$

42. $10x, 15x^2, 20xy$

43. $14a^2, 10ab^2, 15b^3$

44. $20xyz, 25xy^2, 35x^2z$

45. $12ab^2, 28abc, 21bc^2$

46. $16mn, 24m^2, 40mnp$

47. $30m^2n, 60mnp^2, 90np$

48. $13x^2y^3, 39xy, 52xy^2z$

49. $27xy^2z, 36xyz^2, 54x^2yz$

50. $45xyz, 125x^3, 150y^2$

51. $33ab^3, 66b^2, 121$

52. $15x, 25x^2, 30x^3, 40x^4$

53. $10y, 20y^2, 30y^3, 40y^4$

54. $6a^5, 18a^3, 27a, 45$

55. $12c^4, 95c^2, 228$

56. $99xy^3, 143x^3, 363$

57. $18abc^2, 27ax, 30ax^2y, 34a^2bc$

58. $25axy, 35x^5y, 55a^2by^2, 65a^3x^2$

Solve each of the following word problems.

59. Security: Three security guards meet at the front gate for coffee before they walk around inspecting buildings at a manufacturing plant. The guards take 15, 20, and 30 minutes, respectively, for the inspection trip.

   a. If they start at the same time, in how many minutes will they meet again at the front gate for coffee?

   b. How many trips will each guard have made?

60. Aeronautics: Two astronauts miss connections at their first rendezvous in space.

   a. If one astronaut circles the earth every 15 hours and the other every 18 hours, in how many hours will they rendezvous again at the same place?

   b. How many orbits will each astronaut have made before the second rendezvous?

61. Trucking: Three truck drivers have dinner together whenever all three are at the routing station at the same time. The first driver's route takes 6 days, the second driver's route takes 8 days, and the third driver's route takes 12 days.

   a. How frequently do the three drivers have dinner together?

   b. How frequently do the first two drivers meet?

62. Lawn Care: Three neighbors mow their lawns at different intervals during the summer months. The first one mows every 5 days, the second every 7 days, and the third every 10 days.

   a. How frequently do they mow their lawns on the same day?

   b. How many times does each neighbor mow in between the times when they all mow together?

63. **Shipping:** Four ships leave the port on the same day. They take 12, 15, 18, and 30 days, respectively, to sail their routes and reload cargo. How frequently do these four ships leave this port on the same day?

64. **Sales Travel:** Four women work for the same book company selling textbooks. They leave the home office on the same day and take 8 days, 12 days, 14 days, and 15 days, respectively, to visit schools in their own sales regions.

  a. In how many days will they all meet again at the home office?

  b. How many sales trips will each have made in this time?

Find the GCD for each of the following sets of numbers. See Examples 10 through 14.

65. {12, 8} **66.** {16, 28} **67.** {85, 51} **68.** {20, 75} **69.** {20, 30}

70. {42, 48} **71.** {15, 21} **72.** {27, 18} **73.** {18, 24} **74.** {77, 66}

75. {182, 184} **76.** {110, 66} **77.** {8, 16, 64} **78.** {121, 44} **79.** {28, 52, 56}

80. {98, 147} **81.** {60, 24, 96} **82.** {33, 55, 77} **83.** {25, 50, 75} **84.** {30, 78, 60}

85. {17, 15, 21} **86.** {520, 220} **87.** {14, 55} **88.** {210, 231, 84} **89.** {140, 245, 420}

State whether each set of numbers is relatively prime using the GCD. If it is not relatively prime, state the GCD.

90. {35, 24} **91.** {11, 23} **92.** {14, 36} **93.** {72, 35} **94.** {42, 77}

95. {8, 15} **96.** {20, 21} **97.** {16, 22} **98.** {16, 51} **99.** {10, 27}

Writing & Thinking

100. Explain, in your own words, why each number in a set divides the LCM of that set of numbers.

101. Explain, in your own words, why the LCM of a set of numbers is greater than or equal to each number in the set.

2.5 Introduction to Fractions

Objective A **Rational Numbers**

Fractions that can be simplified so that the numerator (top number) and the denominator (bottom number) are integers, with the denominator not 0, have the technical name **rational numbers**. (**Note:** There are other types of numbers that can be written in the form of fractions in which the numerator and denominator may be numbers other than integers. These numbers are called **irrational numbers** and will be studied in later chapters.)

Previously we have discussed the fact that **division by 0 is undefined**. This fact is indicated in the following definition of a rational number by saying that the denominator cannot be 0. We say that the denominator b is **nonzero** or that $b \neq 0$ (b is not equal to 0).

> ### Rational Numbers
>
> A **rational number** is a number that can be written in the form $\dfrac{a}{b}$, where a and b are integers and $b \neq 0$.
>
> $$\dfrac{a}{b} \quad \begin{array}{l} \leftarrow \quad \text{numerator} \\ \leftarrow \quad \text{denominator} \end{array}$$

The following diagram illustrates the fact that whole numbers and integers are also rational numbers. However, there are rational numbers that are not integers or whole numbers.

Examples of rational numbers are 5, 0, $\dfrac{1}{2}$, $\dfrac{3}{4}$, $\dfrac{-9}{10}$, and $\dfrac{17}{-3}$. Note that in fraction form, $5 = \dfrac{5}{1}$ and $0 = \dfrac{0}{1}$.

In fact, we see that every integer can be written in fraction form with a denominator of 1. We can write:

$$0 = \dfrac{0}{1}, 1 = \dfrac{1}{1}, 2 = \dfrac{2}{1}, 3 = \dfrac{3}{1}, 4 = \dfrac{4}{1}, \text{ and so on.}$$

$$-1 = \dfrac{-1}{1}, -2 = \dfrac{-2}{1}, -3 = \dfrac{-3}{1}, -4 = \dfrac{-4}{1}, \text{ and so on.}$$

That is, as illustrated in the following diagram, **every integer is also a rational number**.

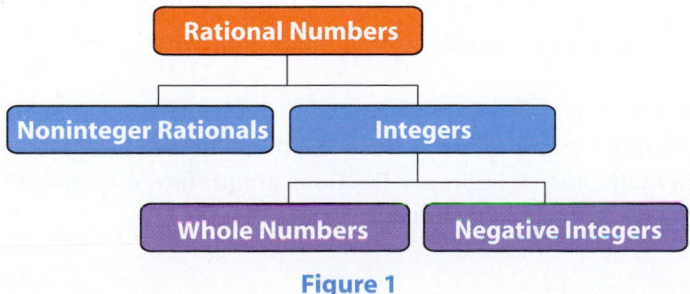

Figure 1

Objectives

A Recognize that the term rational number is the technical term for fraction.

B Learn how to multiply fractions.

C Determine what to multiply by in order to build a fraction to higher terms.

D Determine how to reduce a fraction to lowest terms.

E Be able to multiply fractions and reduce at the same time.

Unless otherwise stated, we will use the terms **fraction** and **rational number** to mean the same thing. Fractions can be used to indicate:

1. equal parts of a whole or
2. division: the numerator is divided by the denominator.

1. Translate the following into fractional form:

a. 5 of 9 equal parts

b. 10 of 11 equal parts

c. 54 divided by 6

Example 1

Understanding Fractions

a. $\frac{1}{2}$ can mean 1 of 2 equal parts.

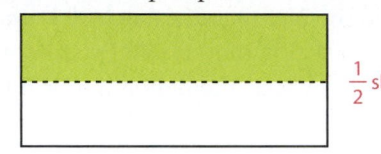

$\frac{1}{2}$ shaded

b. $\frac{3}{4}$ can mean 3 of 4 equal parts.

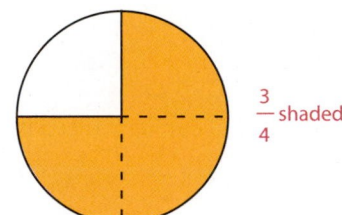

$\frac{3}{4}$ shaded

c. Previously, we used the fraction form of a number to indicate division. For example,

$$\frac{-35}{7} = -5 \text{ and } \frac{39}{13} = 3.$$

Now work margin exercise **1.**

Proper and Improper Fractions

A **proper fraction** is a fraction in which the numerator is less than the denominator.

An **improper fraction** is a fraction in which the numerator is greater than or equal to the denominator.

examples of proper fractions: $\frac{2}{3}$, $\frac{29}{60}$, $\frac{7}{8}$, and $\frac{1}{2}$.

examples of improper fractions: $\frac{5}{3}$, $\frac{14}{14}$, $\frac{89}{22}$, and $\frac{250}{100}$.

Unfortunately the term "improper fraction" is somewhat misleading because there is nothing "improper" about such fractions. **In fact, in algebra and other courses in mathematics, improper fractions are preferred to mixed numbers, and we will use them throughout this text.**

Rule for the Placement of Negative Signs

If a and b are integers and $b \neq 0$, then

$$-\frac{a}{b} = \frac{-a}{b} = \frac{a}{-b}.$$

Example 2

Placement of Negative Signs in Fractions

a. $\quad -\dfrac{1}{3} = \dfrac{-1}{3} = \dfrac{1}{-3}$

b. $\quad -\dfrac{36}{9} = \dfrac{-36}{9} = \dfrac{36}{-9} = -4$

Now work margin exercise 2.

2. Write two alternate forms of $-\dfrac{9}{80}$.

Objective B ## Multiplying Rational Numbers (or Fractions)

Now we state the rule for multiplying fractions and discuss the use of the word " of " to indicate multiplication by fractions. Remember that any integer can be written in fraction form with a denominator of 1.

To Multiply Fractions

1. Multiply the numerators.
2. Multiply the denominators.

$$\frac{a}{b} \cdot \frac{c}{d} = \frac{a \cdot c}{b \cdot d}$$

Finding the product of two fractions can be thought of as finding one fractional part **of** another fraction. For example,

when we multiply $\dfrac{1}{2}$ by $\dfrac{1}{4}$ we are finding $\dfrac{1}{2}$ **of** $\dfrac{1}{4}$.

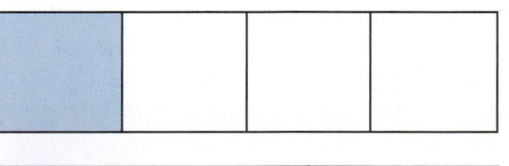

$\dfrac{1}{4}$

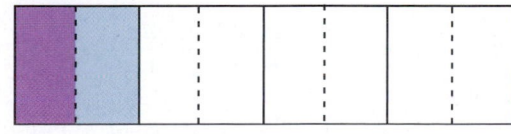

$\dfrac{1}{2}$ of $\dfrac{1}{4} = \dfrac{1}{8}$

Thus we see from the diagram that $\dfrac{1}{2}$ of $\dfrac{1}{4}$ is $\dfrac{1}{8}$, and from the definition,

$$\frac{1}{2} \cdot \frac{1}{4} = \frac{1 \cdot 1}{2 \cdot 4} = \frac{1}{8}.$$

Example 3

Multiplying Rational Numbers

Find the product of $\frac{1}{3}$ and $\frac{2}{7}$.

Solution

$$\frac{1}{3} \cdot \frac{2}{7} = \frac{1 \cdot 2}{3 \cdot 7} = \frac{2}{21}$$

Example 4

Multiplying Rational Numbers

Find $\frac{2}{5}$ of $\frac{7}{3}$.

Solution

$$\frac{2}{5} \cdot \frac{7}{3} = \frac{14}{15}$$

3. Find the product of

$\frac{2}{3}$ and $\frac{8}{11}$

4. Find $\frac{3}{4}$ of $\frac{5}{8}$

5. Find each product.

a. $\frac{1}{7} \cdot \frac{3}{5} \cdot 4$

b. $\frac{0}{4} \cdot \frac{11}{16}$

c. $\frac{4}{5} \cdot 7$

Example 5

Multiplying Rational Numbers

a. $\dfrac{4}{13} \cdot 3 = \dfrac{4}{13} \cdot \dfrac{3}{1} = \dfrac{4 \cdot 3}{13 \cdot 1} = \dfrac{12}{13}$

b. $\dfrac{9}{8} \cdot 0 = \dfrac{9}{8} \cdot \dfrac{0}{1} = \dfrac{9 \cdot 0}{8 \cdot 1} = \dfrac{0}{8} = 0$

c. $4 \cdot \dfrac{5}{3} \cdot \dfrac{2}{7} = \dfrac{4}{1} \cdot \dfrac{5}{3} \cdot \dfrac{2}{7} = \dfrac{40}{21}$

*Now work margin exercises **3** through **5**.*

Both the commutative property and the associative property of multiplication apply to rational numbers (or fractions).

Commutative Property of Multiplication
If $\dfrac{a}{b}$ and $\dfrac{c}{d}$ are rational numbers, then $\dfrac{a}{b} \cdot \dfrac{c}{d} = \dfrac{c}{d} \cdot \dfrac{a}{b}$.

Example 6

The Commutative Property of Multiplication

$$\frac{3}{5}\cdot\frac{2}{7}=\frac{2}{7}\cdot\frac{3}{5}$$ illustrates the commutative property of multiplication

$$\frac{3}{5}\cdot\frac{2}{7}=\frac{6}{35}\quad\text{and}\quad\frac{2}{7}\cdot\frac{3}{5}=\frac{6}{35}$$

Associative Property of Multiplication

If $\dfrac{a}{b}$, $\dfrac{c}{d}$ and $\dfrac{e}{f}$ are rational numbers, then

$$\frac{a}{b}\cdot\frac{c}{d}\cdot\frac{e}{f}=\frac{a}{b}\cdot\left(\frac{c}{d}\cdot\frac{e}{f}\right)=\left(\frac{a}{b}\cdot\frac{c}{d}\right)\cdot\frac{e}{f}.$$

Example 7

The Associative Property of Multiplication

$$\left(\frac{1}{2}\cdot\frac{3}{2}\right)\cdot\frac{3}{11}=\frac{1}{2}\cdot\left(\frac{3}{2}\cdot\frac{3}{11}\right)$$ illustrates the associative property of multiplication

$$\left(\frac{1}{2}\cdot\frac{3}{2}\right)\cdot\frac{3}{11}=\frac{3}{4}\cdot\frac{3}{11}=\frac{9}{44}\quad\text{and}\quad\frac{1}{2}\cdot\left(\frac{3}{2}\cdot\frac{3}{11}\right)=\frac{1}{2}\cdot\frac{9}{22}=\frac{9}{44}$$

Now work margin exercises 6 and 7.

Objective C   **Raising Fractions to Higher Terms**

We know that 1 is the multiplicative identity for integers; that is, $a\cdot 1=a$ for any integer a. The number 1 is also the multiplicative identity for rational numbers, since

$$\frac{a}{b}\cdot\mathbf{1}=\frac{a}{b}\cdot\frac{\mathbf{1}}{\mathbf{1}}=\frac{a\cdot 1}{b\cdot 1}=\frac{a}{b}.$$

Multiplicative Identity

1. For any rational number $\dfrac{a}{b}$, $\dfrac{a}{b}\times\mathbf{1}=\dfrac{a}{b}$

2. $\dfrac{a}{b}=\dfrac{a}{b}\cdot\mathbf{1}=\dfrac{a}{b}\cdot\dfrac{k}{k}=\dfrac{a\cdot k}{b\cdot k}$ where $k\neq 0$ $\left(1=\dfrac{k}{k}\right)$.

Find each product and state the property of multiplication that is illustrated.

6. $\dfrac{3}{4}\cdot 9=9\cdot\dfrac{3}{4}$

7. $\dfrac{1}{3}\left(\dfrac{1}{2}\cdot\dfrac{1}{4}\right)=\left(\dfrac{1}{3}\cdot\dfrac{1}{2}\right)\cdot\dfrac{1}{4}$

In effect, the value of a fraction is unchanged if both the numerator and the denominator are multiplied by the same nonzero integer k, which is the same as multiplying by 1. This fact allows us to perform the following two operations with fractions:

1. **raise a fraction to higher terms** (find an equal fraction with a larger denominator) and

2. **reduce a fraction to lower terms** (find an equal fraction with a smaller denominator)

Example 8

Raising Fractions to Higher Terms

Raise $\dfrac{3}{4}$ to higher terms as indicated: $\dfrac{3}{4} = \dfrac{?}{24}$.

Solution

We know that $4 \cdot 6 = 24$, so we use $1 = \dfrac{k}{k} = \dfrac{6}{6}$.

$$\frac{3}{4} = \frac{3}{4} \cdot 1 = \frac{3}{4} \cdot \frac{?}{?} = \frac{3}{4} \cdot \frac{6}{6} = \frac{18}{24}.$$

---

8. Raise $\dfrac{5}{6}$ to higher terms as indicated: $\dfrac{5}{6} = \dfrac{?}{36}$.

9. Raise $\dfrac{3}{7}$ to higher terms as indicated: $\dfrac{3}{7} = \dfrac{?}{42x}$.

Example 9

Raising Fractions to Higher Terms

Raise $\dfrac{9}{10}$ to higher terms as indicated: $\dfrac{9}{10} = \dfrac{?}{30x}$.

Solution

We know that $10 \cdot 3x = 30x$, so we use $k = 3x$.

$$\frac{9}{10} = \frac{9}{10} \cdot 1 = \frac{9}{10} \cdot \frac{?}{?} = \frac{9}{10} \cdot \frac{3x}{3x} = \frac{27x}{30x}.$$

Now work margin exercises 8 and 9.

Objective D **Reducing Fractions to Lowest Terms**

Changing a fraction to lower terms is the same as **reducing** the fraction. **A fraction is reduced to lowest terms if the numerator and the denominator have no common factor other than +1 or −1.**

To Reduce a Fraction to Lowest Terms

1. Factor the numerator and denominator into prime factors.

2. Use the fact that $\dfrac{k}{k} = 1$ and "divide out" all common factors.

   Note: Reduced fractions might be improper fractions.

Example 10

Reducing Fractions to Lowest Terms

Reduce $\dfrac{-16}{-28}$ to lowest terms.

Solution

If both numerator and denominator have a factor of -1, then we "divide out" -1 and treat the fraction as positive, since $\dfrac{-1}{-1} = 1$.

$$\frac{-16}{-28} = \frac{-1 \cdot 2 \cdot 2 \cdot 2 \cdot 2}{-1 \cdot 2 \cdot 2 \cdot 7} = \frac{-1}{-1} \cdot \frac{2}{2} \cdot \frac{2}{2} \cdot \frac{4}{7}$$

$$= 1 \cdot 1 \cdot 1 \cdot \frac{4}{7} = \frac{4}{7}$$

Example 11

Reducing Fractions to Lowest Terms

Reduce $\dfrac{60a^2}{72ab}$ to lowest terms.

Solution

$$\frac{60a^2}{72ab} = \frac{6 \cdot 10 \cdot a \cdot a}{8 \cdot 9 \cdot a \cdot b} = \frac{2 \cdot 3 \cdot 2 \cdot 5 \cdot a \cdot a}{2 \cdot 2 \cdot 2 \cdot 3 \cdot 3 \cdot a \cdot b}$$

$$= \frac{2}{2} \cdot \frac{2}{2} \cdot \frac{3}{3} \cdot \frac{a}{a} \cdot \frac{5 \cdot a}{2 \cdot 3 \cdot b}$$

$$= 1 \cdot 1 \cdot 1 \cdot 1 \cdot \frac{5a}{6b} = \frac{5a}{6b}$$

As a shortcut, we do not generally write the number 1 when reducing the form $\frac{k}{k}$, as we did in Examples 10 and 11. We do use cancel marks to indicate dividing out common factors in numerators and denominators. **But remember that these numbers do not simply disappear. Their quotient is understood to be 1 even if the 1 is not written.**

Example 12

Reducing Fractions to Lowest Terms

Reduce $\frac{45}{36}$ to lowest terms.

Solution

a. Using prime factors, we have

$$\frac{45}{36} = \frac{\cancel{3} \cdot \cancel{3} \cdot 5}{2 \cdot 2 \cdot \cancel{3} \cdot \cancel{3}} = \frac{5}{4}.$$ Note that the answer is a reduced improper fraction.

b. Larger common factors can be divided out, but we must be sure that we have the largest common factor.

$$\frac{45}{36} = \frac{5 \cdot \cancel{9}}{4 \cdot \cancel{9}} = \frac{5}{4}$$

Completion Example 13

Reducing Fractions to Lowest Terms

Reduce $\frac{52}{65}$ to lowest terms.

Solution

$$\frac{52}{65} = \frac{2 \cdot 2 \cdot \rule{1cm}{0.4pt}}{5 \cdot \rule{0.6cm}{0.4pt}} = \rule{2cm}{0.4pt}$$

***Now work margin exercises* 10 through 13.**

Objective E   **Multiplying and Reducing Fractions at the Same Time**

Now we can multiply fractions and reduce all in one step by using prime factors (or other common factors). If you have any difficulty understanding how to multiply and reduce, use prime factors. By using prime factors, you can be sure that you have not missed a common factor and that your answer is reduced to lowest terms.

Completion Example Answers

13. $\frac{52}{65} = \frac{2 \cdot 2 \cdot \cancel{13}}{5 \cdot \cancel{13}} = \frac{4}{5}$

Reduce each fraction to lowest terms.

10. $\frac{45}{75}$

11. $\frac{66x^2}{44xy}$

12. $\frac{-56}{120}$

13. $\frac{35}{46}$

Examples 14, 15, and 16 illustrate how to multiply and reduce at the same time by factoring the numerators and the denominators. Note that **if all the factors in the numerator or denominator divide out, then 1 must be used as a factor**. (See Examples 15 and 16.)

Example 14

Multiplying and Reducing Fractions

Multiply and reduce to lowest terms.

$$-\frac{18}{35} \cdot \frac{21}{12} = -\frac{2 \cdot \cancel{3} \cdot 3 \cdot 3 \cdot \cancel{7}}{5 \cdot \cancel{7} \cdot \cancel{2} \cdot 2 \cdot \cancel{3}} = -\frac{9}{10}$$

Example 15

Multiplying and Reducing Fractions

Multiply and reduce to lowest terms.

$$\frac{17}{50} \cdot \frac{25}{34} \cdot 8 = \frac{17 \cdot 25 \cdot 8}{50 \cdot 34 \cdot 1}$$

$$= \frac{\cancel{17} \cdot \cancel{25} \cdot \cancel{2} \cdot \cancel{2} \cdot 2}{\cancel{2} \cdot \cancel{25} \cdot \cancel{2} \cdot \cancel{17} \cdot 1}$$

In this example, 25 is a common factor that is not a prime number.

$$= \frac{2}{1} = 2$$

Example 16

Multiplying and Reducing Fractions

Multiply and reduce to lowest terms.

$$\frac{3x}{35} \cdot \frac{15y}{9xy^3} \cdot \frac{7}{2y} = \frac{3 \cdot x \cdot 3 \cdot 5 \cdot y \cdot 7}{5 \cdot 7 \cdot 3 \cdot 3 \cdot x \cdot y \cdot y \cdot y \cdot 2 \cdot y}$$

$$= \frac{\cancel{3} \cdot \cancel{x} \cdot \cancel{3} \cdot \cancel{5} \cdot \cancel{y} \cdot \cancel{7}}{\cancel{5} \cdot \cancel{7} \cdot \cancel{3} \cdot \cancel{3} \cdot \cancel{x} \cdot \cancel{y} \cdot y \cdot y \cdot 2 \cdot y} = \frac{1}{2y^3}$$

Multiply and reduce to lowest terms.

14. $\dfrac{15}{63} \cdot \dfrac{9}{5}$

15. $\dfrac{14x^2}{20} \cdot \dfrac{25y}{28} \cdot \dfrac{4}{15x}$

16. $8 \cdot \dfrac{19}{100} \cdot \dfrac{24}{38} \cdot \dfrac{5}{18}$

17. $\dfrac{33a^2}{28} \cdot \dfrac{6b}{88ab^2} \cdot \dfrac{5}{2}$

Completion Example 17

Multiplying and Reducing Fractions

Multiply and reduce to lowest terms.

$$\frac{55a^2}{26} \cdot \frac{8b}{44ab^2} \cdot \frac{91}{35} = \frac{55 \cdot a \cdot a \cdot 8 \cdot b \cdot 91}{26 \cdot 44 \cdot a \cdot b \cdot b \cdot 35}$$

$$= \underline{\hspace{3cm}}$$

$$= \underline{\hspace{3cm}}$$

Now work margin exercises 14 through 17.

Many students are familiar with another method of multiplying and reducing at the same time which is to divide numerators and denominators by common factors whether they are prime or not. If these factors are easily determined, then this method is probably faster. But common factors are sometimes missed with this method, whereas they are not missed with the prime factorization method. In either case, be careful and organized. The problems from Examples 14 and 15 have been worked again in Examples 18 and 19 using the division method.

Example 18

The Division Method

$$-\frac{\overset{3}{\cancel{18}}}{35} \cdot \frac{\overset{3}{\cancel{21}}}{\underset{2}{\cancel{12}}} = -\frac{9}{10}$$

6 divides into both 18 and 12.
7 divides into both 21 and 35.

Example 19

The Division Method

Multiply and reduce to lowest terms.

18. $-\dfrac{18}{9} \cdot \dfrac{27}{6}$

19. $\dfrac{13}{20} \cdot \dfrac{65}{39} \cdot 15$

$$\frac{17}{50} \cdot \frac{25}{34} \cdot 8 = \frac{\overset{1}{\cancel{17}}}{\underset{2}{\cancel{50}}} \cdot \frac{\overset{1}{\cancel{25}}}{\underset{2}{\cancel{34}}} \cdot \frac{\overset{\overset{2}{\cancel{4}}}{\cancel{8}}}{1} = \frac{2}{1} = 2$$

17 divides into both 17 and 34.
25 divides into both 25 and 50.
2 divides into both 8 and 2.
2 divides into both 4 and 2.

Now work margin exercises 18 and 19.

Completion Example Answers

17. $\dfrac{55a^2}{26} \cdot \dfrac{8b}{44ab^2} \cdot \dfrac{91}{35} = \dfrac{55 \cdot a \cdot a \cdot 8 \cdot b \cdot 91}{26 \cdot 44 \cdot a \cdot b \cdot b \cdot 35}$

$$= \frac{\cancel{5} \cdot \cancel{11} \cdot a \cdot \cancel{a} \cdot \cancel{2} \cdot \cancel{2} \cdot \cancel{2} \cdot \cancel{b} \cdot \cancel{7} \cdot \cancel{13}}{\cancel{2} \cdot \cancel{13} \cdot \cancel{2} \cdot \cancel{2} \cdot \cancel{11} \cdot \cancel{a} \cdot b \cdot \cancel{b} \cdot \cancel{5} \cdot \cancel{7}}$$

$$= \frac{a}{b}$$

Example 20

Buying Computer Disks

If you had $25 and you spend $15 to buy computer disks, what fraction of your money did you spend for computer disks? What fraction do you still have?

Solution

a. The fraction you spent is $\dfrac{15}{25} = \dfrac{3 \cdot 5}{5 \cdot 5} = \dfrac{3}{5}$.

b. Since you still have $10, the fraction you still have is $\dfrac{10}{25} = \dfrac{2 \cdot 5}{5 \cdot 5} = \dfrac{2}{5}$.

Now work margin exercise 20.

20. If you had $42 and spent $30 to buy books, what fraction of your money did you spend on books? What fraction do you still have?

Exercises 2.5

Answer the following questions:

a. What fraction of each figure is shaded?

b. What fraction of each figure is not shaded? See Example 1.

1.

2.

3.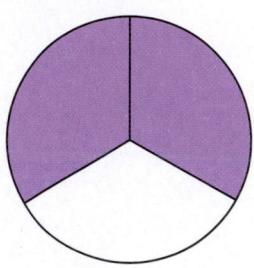

4.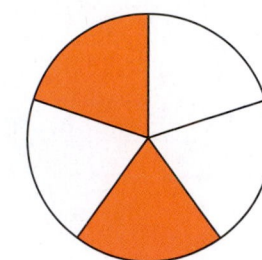

5.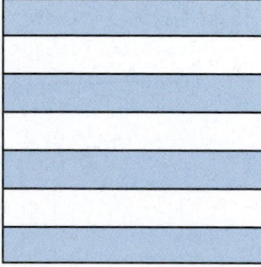

6.

7.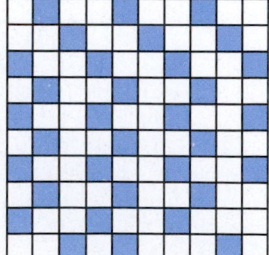

8.

9. What is the value, if any, of each of the following expressions?

a. $\dfrac{0}{6} \cdot \dfrac{5}{7}$
b. $\dfrac{3}{10} \cdot \dfrac{0}{2}$
c. $\dfrac{5}{0}$
d. $\dfrac{16}{0}$

10. What is value, if any, of each of the following expressions?

a. $\dfrac{32}{0} \cdot \dfrac{6}{0}$
b. $\dfrac{1}{5} \cdot \dfrac{0}{3}$
c. $\dfrac{0}{8} \cdot \dfrac{0}{9}$
d. $\dfrac{1}{0} \cdot \dfrac{0}{1}$

11. Find $\dfrac{1}{5}$ of $\dfrac{3}{5}$.
12. Find $\dfrac{1}{4}$ of $\dfrac{1}{4}$.
13. Find $\dfrac{2}{3}$ of $\dfrac{2}{3}$.

14. Find $\dfrac{1}{4}$ of $\dfrac{5}{7}$.
15. Find $\dfrac{6}{7}$ of $-\dfrac{3}{5}$.
16. Find $\dfrac{1}{3}$ of $-\dfrac{1}{4}$.

17. Find $\dfrac{1}{2}$ of $\dfrac{1}{2}$.
18. Find $\dfrac{1}{9}$ of $\dfrac{2}{3}$.

Find the following products. See Example 5.

19. $\dfrac{3}{4} \cdot \dfrac{3}{4}$
20. $\dfrac{5}{6} \cdot \dfrac{5}{6}$
21. $\dfrac{1}{9} \cdot \dfrac{4}{3}$
22. $\dfrac{4}{5} \cdot \dfrac{3}{7}$

23. $\dfrac{0}{5} \cdot \dfrac{2}{3}$
24. $\dfrac{3}{8} \cdot \dfrac{0}{2}$
25. $-4 \cdot \dfrac{3}{5}$
26. $-7 \cdot \dfrac{5}{6}$

27. $\dfrac{1}{3} \cdot \dfrac{1}{10} \cdot \dfrac{1}{5}$
28. $\dfrac{4}{7} \cdot \dfrac{2}{5} \cdot \dfrac{6}{13}$
29. $\dfrac{5}{8} \cdot \dfrac{5}{8} \cdot \dfrac{5}{8}$
30. $\dfrac{3}{10} \cdot \dfrac{9}{10} \cdot \dfrac{7}{10}$

Tell which property of multiplication is illustrated. See Examples 6 and 7.

31. $\dfrac{a}{b} \cdot \dfrac{11}{14} = \dfrac{11}{14} \cdot \dfrac{a}{b}$

32. $\dfrac{3}{8} \cdot \left(\dfrac{7}{5} \cdot \dfrac{1}{2} \right) = \left(\dfrac{3}{8} \cdot \dfrac{7}{5} \right) \cdot \dfrac{1}{2}$

33. $\dfrac{7}{5} \cdot \left(\dfrac{13}{11} \cdot \dfrac{6}{5} \right) = \left(\dfrac{7}{5} \cdot \dfrac{13}{11} \right) \cdot \dfrac{6}{5}$

34. $7 \cdot \dfrac{2}{3} = \dfrac{2}{3} \cdot 7$

35. $\dfrac{3}{4} = \dfrac{3}{4} \cdot \dfrac{?}{?} = \dfrac{?}{12}$

36. $\dfrac{2}{3} = \dfrac{2}{3} \cdot \dfrac{?}{?} = \dfrac{?}{12}$

37. $\dfrac{6}{7} = \dfrac{6}{7} \cdot \dfrac{?}{?} = \dfrac{?}{14}$

38. $\dfrac{5a}{8} = \dfrac{5a}{8} \cdot \dfrac{?}{?} = \dfrac{?}{16}$

39. $\dfrac{3n}{-8} = \dfrac{3n}{-8} \cdot \dfrac{?}{?} = \dfrac{?}{40}$

40. $\dfrac{-3x}{16y} = \dfrac{-3x}{16y} \cdot \dfrac{?}{?} = \dfrac{?}{80y}$

41. $\dfrac{-5x}{13} = \dfrac{-5x}{13} \cdot \dfrac{?}{?} = \dfrac{?}{39y}$

42. $\dfrac{-3a}{5b} = \dfrac{-3a}{5b} \cdot \dfrac{?}{?} = \dfrac{?}{25b^2}$

43. $\dfrac{4a^2}{17b} = \dfrac{4a^2}{17b} \cdot \dfrac{?}{?} = \dfrac{?}{34ab}$

44. $\dfrac{-3x}{1} = \dfrac{-3x}{1} \cdot \dfrac{?}{?} = \dfrac{?}{5x}$

45. $\dfrac{-9x^2}{10y^2} = \dfrac{-9x^2}{10y^2} \cdot \dfrac{?}{?} = \dfrac{?}{100y^3}$

46. $\dfrac{-7xy^2}{-10} = \dfrac{-7xy^2}{-10} \cdot \dfrac{?}{?} = \dfrac{?}{100xy}$

47. $\dfrac{24}{30}$

48. $\dfrac{14}{36}$

49. $\dfrac{22}{65}$

50. $\dfrac{60x}{75x}$

51. $\dfrac{-26y}{39y}$

52. $\dfrac{-7n}{28n}$

53. $\dfrac{-27}{56x^2}$

54. $\dfrac{-12}{35y}$

55. $\dfrac{34x}{-51x^2}$

56. $\dfrac{-30x}{45x^2}$

57. $\dfrac{51x^2}{6x}$

58. $\dfrac{6y^2}{-51y}$

59. $\dfrac{-54a^2}{-9ab}$

60. $\dfrac{-24a^2b}{-100a}$

61. $\dfrac{66xy^2}{84xy}$

62. $\dfrac{-14xyz}{-63xz}$

63. $\dfrac{-7}{8} \cdot \dfrac{4}{21}$

64. $\dfrac{-23}{36} \cdot \dfrac{20}{46}$

65. $9 \cdot \dfrac{7}{24}$

66. $8 \cdot \dfrac{5}{12}$

67. $\dfrac{-9a}{10b} \cdot \dfrac{35a}{40} \cdot \dfrac{65b}{15a}$

68. $\dfrac{-42x}{52xy} \cdot \dfrac{-27}{22x} \cdot \dfrac{33}{9}$

69. $\dfrac{75x^2}{8y^2} \cdot \dfrac{-16xy}{36} \cdot 9 \cdot \dfrac{-7y}{25x}$

70. $\dfrac{17n}{8m} \cdot \dfrac{-5mn}{42n} \cdot \dfrac{18}{51} \cdot 4m^2$

71. $\dfrac{-3}{4} \cdot \dfrac{7}{2} \cdot \dfrac{22}{54} \cdot 18$

72. $\dfrac{-69}{15} \cdot \dfrac{30}{8} \cdot \dfrac{-14}{46} \cdot 9$

73. If you have $10 and you spend $7 for a sandwich and drink, what fraction of your money have you spent on food? What fraction do you still have?

74. In a class of 30 students, 6 received a grade of A. What fraction of the class did not receive an A?

75. A glass is 8 inches tall. If the glass is $\frac{3}{4}$ full of water, what is the height of the water in the glass?

76. Suppose that a ball is dropped from a height of 40 feet and that each time the ball bounces it bounces back to $\frac{1}{2}$ the height it dropped. How high will the ball bounce on its third bounce?

40 ft

1/2 of previous bounce

77. If you go on a bicycle trip of 75 miles in the mountains and $\frac{1}{5}$ of the trip is downhill, what fraction of the trip is not downhill? How many miles are not downhill?

78. A study showed that $\frac{3}{10}$ of the students in an elementary school were left-handed. If the school had an enrollment of 600 students, how many were left-handed?

Writing & Thinking

79. One-third of the hexagon is shaded. Copy the hexagon and shade one-third in as many different ways that you can. How many ways do you think that this can be done? Discuss your solutions in class.

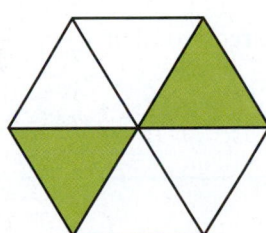

Objectives

A Recognize and find the reciprocal of a number or algebraic expression.

B Know that to divide you must multiply by the reciprocal of the divisor.

2.6 Division with Fractions

Objective A **Reciprocals**

If the product of two fractions is 1, then the fractions are called **reciprocals** of each other. For example,

$\frac{3}{4}$ and $\frac{4}{3}$ are reciprocals because $\frac{3}{4} \cdot \frac{4}{3} = \frac{12}{12} = 1$, and

$-\frac{9}{2}$ and $-\frac{2}{9}$ are reciprocals because $\left(-\frac{9}{2}\right) \cdot \left(-\frac{2}{9}\right) = +\frac{18}{18} = 1$.

Reciprocal of an Algebraic Expression

The **reciprocal** of $\frac{a}{b}$ is $\frac{b}{a}$ $(a \neq 0$ and $b \neq 0)$.

The product of a nonzero number and its reciprocal is always 1.

$$\frac{a}{b} \cdot \frac{b}{a} = 1$$

notes

- **The number 0 has no reciprocal.**

- That is, $\frac{0}{1}$ has no reciprocal because $\frac{1}{0}$ is undefined.

Example 1

Finding the Reciprocal of a Fraction

The reciprocal of $\frac{15}{7}$ is $\frac{7}{15}$.

$$\frac{15}{7} \cdot \frac{7}{15} = \frac{15 \cdot 7}{7 \cdot 15} = 1$$

Example 2

Finding the Reciprocal of an Algebraic Expression

The reciprocal of $6x$ is $\dfrac{1}{6x}$.

$$6x \cdot \frac{1}{6x} = \frac{6x}{1} \cdot \frac{1}{6x} = 1$$

*Now work margin exercises **1** and **2**.*

Objective B   **Division with Fractions**

To understand how to divide with fractions, we write the division in fraction form, where the numerator and denominator are themselves fractions. For example,

$$\frac{2}{3} \div \frac{5}{7} = \frac{\dfrac{2}{3}}{\dfrac{5}{7}}.$$

Now we multiply the numerator and the denominator (both fractions) by the **reciprocal of the denominator**. This is the same as multiplying by 1 and does not change the value of the expression. The reciprocal of $\dfrac{5}{7}$ is $\dfrac{7}{5}$, so we multiply both the numerator and the denominator by $\dfrac{7}{5}$.

$$\frac{2}{3} \div \frac{5}{7} = \frac{\dfrac{2}{3} \cdot \dfrac{7}{5}}{\dfrac{5}{7} \cdot \dfrac{7}{5}} = \frac{\dfrac{2}{3} \cdot \dfrac{7}{5}}{1} = \frac{2}{3} \cdot \frac{7}{5}$$

divisor is now 1

Thus a division problem has been changed into a multiplication problem:

$$\frac{2}{3} \div \frac{5}{7} = \frac{2}{3} \cdot \frac{7}{5} = \frac{14}{15}.$$

Division with Fractions

To divide by any nonzero number, multiply by its reciprocal. In general,

$$\frac{a}{b} \div \frac{c}{d} = \frac{a}{b} \cdot \frac{d}{c} \text{ where } b, c, d \neq 0.$$

Note: $\dfrac{a}{b} \div \dfrac{c}{d} \neq \dfrac{b}{a} \cdot \dfrac{c}{d}$

The reciprocal used is always the divisor.

Find the reciprocal of each number or expression.

1. $\dfrac{11}{18}$

2. $\dfrac{2}{7x}$

Example 3

Dividing Fractions

Divide: $\dfrac{3}{4} \div \dfrac{2}{3}$.

Solution

The reciprocal of $\dfrac{2}{3}$ is $\dfrac{3}{2}$, so we multiply by $\dfrac{3}{2}$.

$$\frac{3}{4} \div \frac{2}{3} = \frac{3}{4} \cdot \frac{3}{2} = \frac{9}{8}$$

Divide.

3. $\dfrac{4}{9} \div \dfrac{1}{2}$

4. $\dfrac{9}{14} \div 5$

Example 4

Dividing Fractions

Divide: $\dfrac{3}{4} \div 4$.

Solution

The reciprocal of 4 is $\dfrac{1}{4}$, so we multiply by $\dfrac{1}{4}$.

$$\frac{3}{4} \div 4 = \frac{3}{4} \cdot \frac{1}{4} = \frac{3}{16}$$

Now work margin exercises 3 and 4.

As with any multiplication problem, we reduce whenever possible by factoring numerators and denominators.

Example 5

Dividing and Reducing Fractions

Divide and reduce to lowest terms: $\dfrac{16}{27} \div \dfrac{8}{9}$.

Solution

The reciprocal of $\dfrac{8}{9}$ is $\dfrac{9}{8}$. Reduce by factoring.

$$\frac{16}{27} \div \frac{8}{9} = \frac{16}{27} \cdot \frac{9}{8} = \frac{\cancel{8} \cdot 2 \cdot \cancel{9}}{3 \cdot \cancel{9} \cdot \cancel{8}} = \frac{2}{3}$$

Example 6

Dividing and Reducing Fractions

Divide and reduce to lowest terms: $\dfrac{9a}{10} \div \dfrac{9a}{10}$.

Solution

We expect the result to be 1 because we are dividing a number by itself.

$$\frac{9a}{10} \div \frac{9a}{10} = \frac{9a}{10} \cdot \frac{10}{9a} = \frac{90a}{90a} = 1$$

Completion Example 7

Dividing and Reducing Fractions

Divide and reduce to lowest terms: $\dfrac{-4x}{13} \div \dfrac{20}{39x}$.

Solution

$$\frac{-4x}{13} \div \frac{20}{39x} = \frac{-4x}{13} \cdot \underline{\hspace{1.5cm}} = \underline{\hspace{2cm}} = \underline{\hspace{1.5cm}}$$

Now work margin exercises 5 through 7.

If the product of two numbers is known and one of the numbers is also known, then the other number can be found by **dividing the product by the known number**. For example, with whole numbers, suppose the product of two numbers is 24 and one of the numbers is 8. What is the other number? Since $24 \div 8 = 3$, the other number is 3.

Divide as indicated and reduce to lowest terms.

5. $\dfrac{1}{5} \div \dfrac{1}{5}$

6. $\dfrac{5}{13y} \div \dfrac{25}{52y}$

7. $\dfrac{5x}{18} \div \dfrac{19}{9x}$

Completion Example Answers

7. $\dfrac{-4x}{13} \div \dfrac{20}{39x} = \dfrac{-4x}{13} \cdot \dfrac{39x}{20} = -\dfrac{\cancel{4} \cdot x \cdot 3 \cdot \cancel{13} \cdot x}{\cancel{13} \cdot \cancel{4} \cdot 5} = -\dfrac{3x^2}{5}$

Example 8

Dividing the Product by the Known Number

The result of multiplying two numbers is $\dfrac{14}{15}$. If one of the numbers is $\dfrac{2}{5}$, what is the other number?

Solution

Divide the product $\left(\text{which is } \dfrac{14}{15}\right)$ by the number $\dfrac{2}{5}$.

$$\frac{14}{15} \div \frac{2}{5} = \frac{14}{15} \cdot \frac{5}{2} = \frac{7 \cdot \cancel{2} \cdot \cancel{5}}{3 \cdot \cancel{5} \cdot \cancel{2}} = \frac{7}{3}$$

The other number is $\dfrac{7}{3}$.

Check by multiplying: $\dfrac{2}{5} \cdot \dfrac{7}{3} = \dfrac{14}{15}$

8. The result of multiplying two numbers is $\dfrac{9}{10}$. If one of the numbers is $\dfrac{8}{25}$, what is the other number?

9. If the product of $\dfrac{2}{3}$ with another number is $\dfrac{1}{9}$, what is the other number?

Example 9

Dividing the Product by the Known Number

If the product of $\dfrac{1}{4}$ and another number is $-\dfrac{11}{18}$, what is the other number?

Solution

Divide the product by the given number.

$$-\frac{11}{18} \div \frac{1}{4} = \frac{-11}{18} \cdot \frac{4}{1} = \frac{-11 \cdot \cancel{2} \cdot 2}{\cancel{2} \cdot 3 \cdot 3} = -\frac{22}{9}$$

The other number is $-\dfrac{22}{9}$.

Note that we could have anticipated that this number would be negative because the product $\left(-\dfrac{11}{18}\right)$ is negative and the given number $\left(\dfrac{1}{4}\right)$ is positive.

Now work margin exercises 8 and 9.

Example 10

A box contains 30 pieces of candy. This is $\frac{3}{5}$ of the maximum amount of this candy the box can hold.

a. Is the maximum amount of candy the box can hold more or less than 30 pieces?

Solution

The maximum number of pieces of candy is more than 30.

b. If you want to multiply $\frac{3}{5}$ times 30, would the product be more or less than 30?

Solution

Less than 30.

c. What is the maximum number of pieces of candy the box can hold?

Solution

To find the maximum number of pieces, divide:

$$30 \div \frac{3}{5} = \frac{\overset{10}{\cancel{30}}}{1} \cdot \frac{5}{\underset{1}{\cancel{3}}} = 50$$

The maximum number of pieces the box will hold is 50.

Now work margin exercise 10.

10. A truck is towing 3000 pounds. This is $\frac{2}{3}$ of the truck's towing capacity.

a. Can the truck tow 4000 pounds?

b. If you were to multiply $\frac{2}{3}$ times 3000, would the product be more or less than 3000?

c. What is the maximum towing capacity of the truck?

Exercises 2.6

1. What is the reciprocal of $\frac{13}{25}$?

2. To divide by any nonzero number, multiply by its _____.

3. The quotient of $0 \div \frac{6}{7}$ is _____.

4. The quotient of $\frac{6}{7} \div 0$ is _____.

5. Show that the phrases "ten divided by two" and "ten divided by one-half" do not have the same meaning.

6. Show that the phrases "twelve divided by three" and "twelve divided by one-third" do not have the same meaning.

7. Explain in your own words why division by 0 is undefined.

8. Give three examples that illustrate that division is not a commutative operation.

Find the following quotients. Reduce to lowest terms whenever possible. See Examples 3 through 7.

9. $\frac{5}{8} \div \frac{3}{5}$

10. $\frac{1}{3} \div \frac{1}{5}$

11. $\frac{2}{11} \div \frac{1}{7}$

12. $\frac{2}{7} \div \frac{1}{2}$

13. $\frac{3}{14} \div \frac{3}{14}$

14. $\frac{5}{8} \div \frac{5}{8}$

15. $\frac{3}{4} \div \frac{4}{3}$

16. $\frac{9}{10} \div \frac{10}{9}$

17. $\frac{15}{20} \div (-3)$

18. $\frac{14}{20} \div (-7)$

19. $\frac{25}{40} \div 10$

20. $\frac{36}{80} \div 9$

21. $\frac{7}{8} \div 0$

22. $0 \div \frac{5}{6}$

23. $0 \div \frac{1}{2}$

24. $\frac{15}{64} \div 0$

25. $\frac{-16}{35} \div \frac{2}{7}$

26. $\frac{-15}{27} \div \frac{5}{9}$

27. $\frac{-15}{24} \div \frac{-25}{18}$

28. $\frac{-36}{25} \div \frac{-24}{20}$

29. $\frac{34b}{21a} \div \frac{17b}{14}$

30. $\frac{16x}{20y} \div \frac{18x}{10y}$

31. $\frac{20x}{21y} \div \frac{10y}{14x}$

32. $\frac{19a}{24b} \div \frac{5a}{8b}$

33. $14x \div \dfrac{1}{7x}$ **34.** $25y \div \dfrac{1}{5y}$ **35.** $-30a \div \dfrac{1}{10}$ **36.** $-50b \div \dfrac{1}{2}$

37. $\dfrac{29a}{50} \div \dfrac{31a}{10}$ **38.** $\dfrac{92}{7a} \div \dfrac{46a}{11}$ **39.** $\dfrac{-33x^2}{32} \div \dfrac{11x}{4}$ **40.** $\dfrac{-26x^2}{35} \div \dfrac{39x}{40}$

Solve each of the following word problems. See Examples 8 through 10.

41. The product of $\dfrac{5}{6}$ with another number is $\dfrac{2}{5}$.

     a. Which number is the product?

     b. What is the other number?

42. The result of multiplying two numbers is 150.

     a. Is 150 a product or a quotient?

     b. If one of the numbers is $\dfrac{15}{7}$, what is the other one?

43. The product of two numbers is 210.

     a. If one of the numbers is $\dfrac{2}{3}$, do you expect the other number to be larger than 210 or smaller than 210?

     b. What is the other number?

44. An airplane is carrying 180 passengers. This is $\dfrac{9}{10}$ of the capacity of the airplane.

     a. Is the capacity of the airplane more or less than 180?

     b. If you were to multiply 180 times $\dfrac{9}{10}$ would the product be more or less than 180?

     c. What is the capacity of the airplane?

45. A bus is carrying 60 passengers. This is $\dfrac{4}{5}$ of the capacity of the bus.

     a. Is the capacity of the bus more or less than 60?

     b. If you were to multiply 60 times $\dfrac{4}{5}$, would the product be more or less than 60?

     c. What is the capacity of the bus?

46. What is the quotient if $\dfrac{1}{4}$ of 36 is divided by $\dfrac{2}{3}$ of $\dfrac{5}{8}$?

47. What is the quotient if $\dfrac{3}{8}$ of 64 is divided by $\dfrac{3}{5}$ of $\dfrac{15}{16}$?

48. The continent of Africa covers approximately 11,707,000 square miles. This is $\dfrac{1}{5}$ of the land area of the world. What is the approximate total land area of the world?

49. The student senate has 75 members, and $\frac{7}{15}$ of these are women. A change in the senate constitution is being considered, and at the present time (before debating has begun), a survey shows that $\frac{3}{5}$ of the women and $\frac{4}{5}$ of the men are in favor of this change.

   a. How many women are on the student senate?

   b. How many women on the senate are in favor of the change?

   c. If the change requires a $\frac{2}{3}$ majority vote in favor to pass, would the constitutional change pass if the vote were taken today?

   d. By how many votes would the change pass or fail?

50. The tennis club has 250 members, and they are considering putting in a new tennis court. The cost of the new court is going to involve an assessment of $200 for each member. Of the seven-tenths of the members who live near the club, $\frac{3}{5}$ of them are in favor of the assessment. However, $\frac{2}{3}$ of the members who live some distance away are not in favor of the assessment.

   a. If a vote were taken today, would more than one-half of the members vote for or against the new court?

   b. By how many votes would the question pass or fail if more than one-half of the members must vote in favor for the question to pass?

51. There are 4000 registered voters in Roseville, and $\frac{3}{8}$ of these voters are registered Democrats. A survey indicates that $\frac{2}{3}$ of the registered Democrats are in favor of Measure A and $\frac{3}{5}$ of the other registered voters are in favor of this measure.

   a. How many of the voters are registered Democrats?

   b. How many of the voters are not registered Democrats?

   c. How many of the registered Democrats favor Measure A?

   d. How many of the registered voters favor Measure A?

52. There are 3000 students at Mountain High School, and $\frac{1}{4}$ of these students are seniors. If $\frac{3}{5}$ of the seniors are in favor of the school forming a debating team and $\frac{7}{10}$ of the remaining students (not seniors) are also in favor of forming a debating team, how many students do not favor this idea?

53. A manufacturing plant is currently producing 6000 steel rods per week. Because of difficulties getting materials, this number is only $\frac{3}{4}$ of the plant's potential production.

    a. Is the potential production number more or less than 6000 rods?

    b. If you were to multiply $\frac{3}{4}$ times 6000, would the product be more or less than 6000?

    c. What is the plant's potential production?

54. A grove of orange trees was struck by an off-season frost and the result was a relatively poor harvest. This year's crop was 10,000 tons of oranges which is about $\frac{4}{5}$ of the usual crop.

    a. Is the usual crop more or less than 10,000 tons of oranges?

    b. If you were to multiply 10,000 times $\frac{4}{5}$, would the product be more or less than 10,000?

    c. About how many tons of oranges are usually harvested?

55. Due to environmental considerations homeowners in a particularly dry area have been asked to use less water than usual. One home is currently using 630 gallons per day. This is $\frac{7}{10}$ of the usual amount of water used in this home.

    a. Is the usual amount of water used more or less than 630 gallons?

    b. If you were to multiply $\frac{7}{10}$ times 630, would the product be more or less than 630?

    c. What is the usual amount of water used in this home?

2.7 Addition and Subtraction with Fractions

Objective A   Addition with Fractions

To add two (or more) fractions with the same denominator, we can think of the common denominator as the "name" of each fraction. The sum has this common name. Just as 3 oranges plus 2 oranges give a total of 5 oranges, 3 eighths plus 2 eighths give a total of 5 eighths. Figure 1 illustrates how the sum of the two fractions $\frac{3}{8}$ and $\frac{2}{8}$ might be diagrammed.

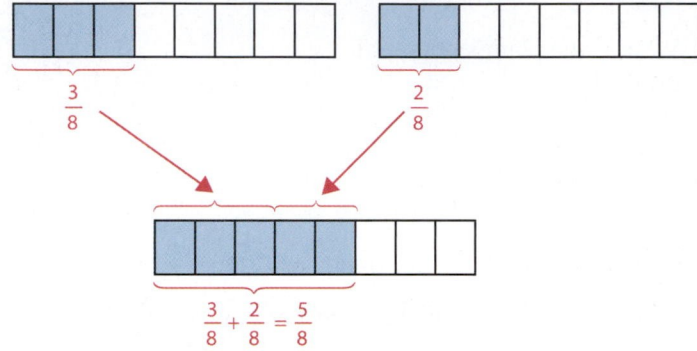

Figure 1

To Add Two (or More) Fractions with the Same Denominator

1. Add the numerators.
2. Keep the common denominator. $\dfrac{a}{b} + \dfrac{c}{b} = \dfrac{a+c}{b}$
3. Reduce, if possible.

Example 1

Adding Two Fractions with the Same Denominator

$$\frac{1}{5} + \frac{2}{5} = \frac{1+2}{5} = \frac{3}{5}$$

Example 2

Adding Three Fractions with the Same Denominator

$$\frac{1}{11} + \frac{2}{11} + \frac{3}{11} = \frac{1+2+3}{11} = \frac{6}{11}$$

Now work margin exercise 1 and 2.

Find each sum.

1. $\dfrac{1}{7} + \dfrac{3}{7}$

2. $\dfrac{1}{9} + \dfrac{4}{9} + \dfrac{3}{9}$

We know that fractions to be added will not always have the same denominator. Figure 2 illustrates how the sum of the two fractions $\frac{1}{3}$ and $\frac{2}{5}$ might be diagrammed.

Figure 2

To find the result of adding these two fractions, a common unit (that is, a common denominator) is needed. By dividing each third into five parts and each fifth into three parts, we find that the common unit is fifteenths. Now the two fractions can be added as shown in Figure 3.

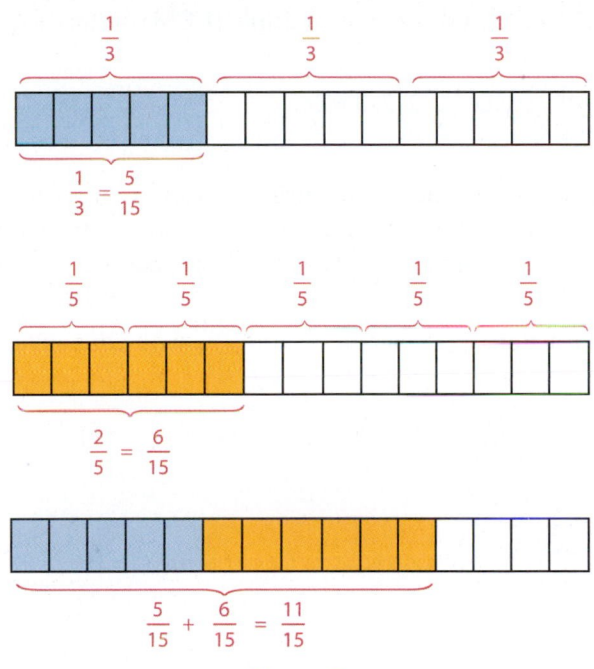

Figure 3

Of course, drawing diagrams every time two or more fractions are to be added would be difficult and time consuming. A better way is to find the least common denominator of all the fractions, change each fraction to an equivalent fraction with this common denominator, and then add. In Figure 3, the least common denominator was the product $3 \cdot 5 = 15$. In general, **the least common denominator (LCD) is the least common multiple (LCM) of the denominators**. (See Section 2.4.)

To Add Fractions with Different Denominators

1. Find the least common denominator (LCD).
2. Change each fraction into an equivalent fraction with that denominator.
3. Add the new fractions.
4. Reduce, if possible.

Example 3

Adding Two Fractions with Different Denominators

Find the sum: $\dfrac{3}{8} + \dfrac{13}{12}$.

Solution

a. Find the LCD. Remember that the least common denominator (LCD) is the least common multiple (LCM) of the denominators.

$$\left.\begin{array}{l} 8 = 2 \cdot 2 \cdot 2 \\ 12 = 2 \cdot 2 \cdot 3 \end{array}\right\} \text{LCD} = 2 \cdot 2 \cdot 2 \cdot 3 = 24$$

Note: You might not need to use prime factorizations to find the LCD. If the denominators are numbers that are familiar to you, then you might be able to find the LCD simply by inspection.

b. Find fractions equivalent to $\dfrac{3}{8}$ and $\dfrac{13}{12}$ with denominator 24.

$$\frac{3}{8} = \frac{3}{8} \cdot \frac{3}{3} = \frac{9}{24}$$

$$\frac{13}{12} = \frac{13}{12} \cdot \frac{2}{2} = \frac{26}{24}$$

c. Add

$$\frac{3}{8} + \frac{13}{12} = \frac{9}{24} + \frac{26}{24} = \frac{9 + 26}{24} = \frac{35}{24}$$

d. The fraction $\dfrac{35}{24}$ is in lowest terms because 35 and 24 have only +1 and −1 as common factors.

Example 4

Adding Two Fractions with Different Denominators

Find the sum: $\dfrac{7}{45} + \dfrac{7}{36}$.

Solution

a. Find the LCD.

$$\left.\begin{array}{l} 45 = 3 \cdot 3 \cdot 5 \\ 36 = 2 \cdot 2 \cdot 3 \cdot 3 \end{array}\right\} \begin{array}{l} LCD = 2 \cdot 2 \cdot 3 \cdot 3 \cdot 5 = 180 \\ \quad = (3 \cdot 3 \cdot 5)(2 \cdot 2) = 45 \cdot 4 \\ \quad = (2 \cdot 2 \cdot 3 \cdot 3)(5) = 36 \cdot 5 \end{array}$$

b. Steps **b.**, **c.**, and **d.** from Example 3 can be written together in one step.

$$\frac{7}{45} + \frac{7}{36} = \frac{7}{45} \cdot \frac{4}{4} + \frac{7}{36} \cdot \frac{5}{5}$$

$$= \frac{28}{180} + \frac{35}{180} = \frac{63}{180}$$

$$= \frac{\cancel{3} \cdot \cancel{3} \cdot 7}{2 \cdot 2 \cdot \cancel{3} \cdot \cancel{3} \cdot 5} = \frac{7}{20}$$

Note that, in adding fractions, we also may choose to write them vertically. The process is the same.

$$\frac{7}{45} = \frac{7}{45} \cdot \frac{4}{4} = \frac{28}{180}$$

$$+\frac{7}{36} = \frac{7}{36} \cdot \frac{5}{5} = \frac{35}{180}$$

$$\frac{63}{180} = \frac{\cancel{3} \cdot \cancel{3} \cdot 7}{2 \cdot 2 \cdot \cancel{3} \cdot \cancel{3} \cdot 5} = \frac{7}{20}$$

Now work margin exercises 3 and 4.

Example 5

Adding Three Fractions with Different Denominators

Find the sum: $6 + \dfrac{9}{10} + \dfrac{3}{1000}$.

Solution

a. LCD = 1000. All the denominators are powers of 10, and 1000 is the largest. Write 6 as $\dfrac{6}{1}$.

Find each sum.

3. $\dfrac{1}{6} + \dfrac{3}{10}$

4. $\dfrac{1}{4} + \dfrac{3}{8} + \dfrac{7}{10}$

5. Find the sum.

$$5 + \frac{3}{10} + \frac{7}{100}$$

b. $6 + \dfrac{9}{10} + \dfrac{3}{1000} = \dfrac{6}{1} \cdot \dfrac{1000}{1000} + \dfrac{9}{10} \cdot \dfrac{100}{100} + \dfrac{3}{1000}$

$$= \dfrac{6000}{1000} + \dfrac{900}{1000} + \dfrac{3}{1000}$$

$$= \dfrac{6903}{1000}$$

Note: The only prime factors of 1000 are 2's and 5's. And since neither 2 nor 5 is a factor of 6903, the answer will not reduce.

Now work margin exercise 5.

Common Error

The following common error must be avoided.

Find the sum $\dfrac{3}{2} + \dfrac{1}{6}$.

Wrong Solution

You **cannot** cancel across the + sign.

$$\dfrac{\overset{1}{\cancel{3}}}{2} + \dfrac{1}{\underset{2}{\cancel{6}}} = \dfrac{1}{2} + \dfrac{1}{2} = 1$$

Correct Solution

Use LCD = 6.

$$\dfrac{3}{2} + \dfrac{1}{6} = \dfrac{3}{2} \cdot \dfrac{3}{3} + \dfrac{1}{6} = \dfrac{9}{6} + \dfrac{1}{6} = \dfrac{10}{6}$$

NOW reduce.

$$\dfrac{10}{6} = \dfrac{5 \cdot \cancel{2}}{3 \cdot \cancel{2}} = \dfrac{5}{3}$$

2 is a factor in both the numerator and the denominator.

Both the commutative and associative properties of addition apply to fractions.

Commutative Property of Addition

If $\dfrac{a}{b}$ and $\dfrac{c}{d}$ are fractions, then $\dfrac{a}{b} + \dfrac{c}{d} = \dfrac{c}{d} + \dfrac{a}{b}$.

Associative Property of Addition

If $\dfrac{a}{b}$, $\dfrac{c}{d}$ and $\dfrac{e}{f}$ are fractions, then

$$\frac{a}{b}+\frac{c}{d}+\frac{e}{f}=\frac{a}{b}+\left(\frac{c}{d}+\frac{e}{f}\right)=\left(\frac{a}{b}+\frac{c}{d}\right)+\frac{e}{f}.$$

Objective B **Subtraction with Fractions**

Finding the difference between two fractions with a common denominator is similar to finding the sum. The numerators are simply subtracted instead of added. Just as with addition, the common denominator "names" each fraction.

Figure 4 shows how the difference between $\dfrac{4}{5}$ and $\dfrac{1}{5}$ might be diagrammed.

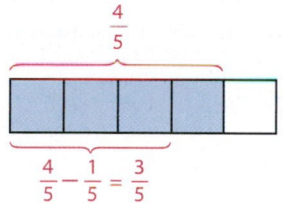

Figure 4

To Subtract Two Fractions with the Same Denominator

1. Subtract the numerators.
2. Keep the common denominator. $\dfrac{a}{b}-\dfrac{c}{b}=\dfrac{a-c}{b}$
3. Reduce, if possible.

Example 6

Subtracting Two Fractions with the Same Denominator

Find the difference: $\dfrac{9}{10}-\dfrac{7}{10}$.

Solution

$$\frac{9}{10}-\frac{7}{10}=\frac{9-7}{10}=\frac{2}{10}=\frac{\cancel{2}\cdot 1}{\cancel{2}\cdot 5}=\frac{1}{5}$$

The difference is reduced just as any fraction is reduced.

Find each difference.

6. $\dfrac{8}{15} - \dfrac{6}{15}$

7. $\dfrac{8}{13} - \dfrac{14}{13}$

Example 7

Subtracting Two Fractions with the Same Denominator

Find the difference: $\dfrac{6}{11} - \dfrac{19}{11}$.

Solution

$$\frac{6}{11} - \frac{19}{11} = \frac{6-19}{11} = \frac{-13}{11} = -\frac{13}{11}$$

The difference is negative because a larger number is subtracted from a smaller number.

Now work margin exercises 6 and 7.

To Subtract Fractions with Different Denominators

1. Find the least common denominator (LCD).
2. Change each fraction to an equivalent fraction with that denominator.
3. Subtract the new fractions.
4. Reduce, if possible.

Example 8

Subtracting Two Fractions with Different Denominators

Find the difference: $\dfrac{24}{55} - \dfrac{4}{33}$.

Solution

a. Find the LCD.

$$\left.\begin{array}{l} 55 = 5 \cdot 11 \\ 33 = 3 \cdot 11 \end{array}\right\} \ \text{LCD} = 3 \cdot 5 \cdot 11 = 165$$

b. Find equal fractions with denominator 165, subtract, and reduce, if possible.

$$\frac{24}{55} - \frac{4}{33} = \frac{24}{55} \cdot \frac{3}{3} - \frac{4}{33} \cdot \frac{5}{5} = \frac{72-20}{165}$$

$$= \frac{52}{165} = \frac{2 \cdot 2 \cdot 13}{3 \cdot 5 \cdot 11} = \frac{52}{165}$$

The answer $\dfrac{52}{165}$ does not reduce because there are no common prime factors in the numerator and denominator.

Example 9

Subtracting Two Fractions with Different Denominators

Subtract: $1 - \dfrac{13}{15}$

Solution

a. Since 1 can be written $\dfrac{15}{15}$, the LCD is 15.

b. $1 - \dfrac{13}{15} = \dfrac{15}{15} - \dfrac{13}{15} = \dfrac{15 - 13}{15} = \dfrac{2}{15}$

When two or more negative numbers are added, the expression is often called an **algebraic sum**. This is a way to help avoid confusion with subtraction. Remember, that subtraction can always be described in terms of addition.

Completion Example 10

Finding the Algebraic Difference

Find the algebraic difference as indicated: $-\dfrac{7}{12} - \dfrac{9}{20}$.

Solution

a. Find the LCD.

$$\left. \begin{array}{l} 12 = \underline{\hspace{2cm}} \\ 20 = \underline{\hspace{2cm}} \end{array} \right\} \text{LCD} = \underline{\hspace{1.5cm}} = \underline{\hspace{1.5cm}}$$

b. $-\dfrac{7}{12} - \dfrac{9}{20} = -\dfrac{7}{12} \cdot \underline{\hspace{1cm}} - \dfrac{9}{20} \cdot \underline{\hspace{1cm}}$

$$= \underline{\hspace{1cm}} - \underline{\hspace{1cm}}$$

$$= \underline{\hspace{1cm}} = \underline{\hspace{1cm}}$$

$$= \underline{\hspace{1cm}} = \underline{\hspace{1cm}}$$

Now work margin exercises 8 through 10.

Find each difference.

8. $\dfrac{19}{38} - \dfrac{21}{19}$

9. $1 - \dfrac{9}{20}$

10. $-\dfrac{5}{12} - \dfrac{8}{18}$

Completion Example Answers

10. a. Find the LCD.

$$\left. \begin{array}{l} 12 = 2 \cdot 2 \cdot 3 \\ 20 = 2 \cdot 2 \cdot 5 \end{array} \right\} \text{LCD} = 2 \cdot 2 \cdot 3 \cdot 5 = 60$$

b. $-\dfrac{7}{12} - \dfrac{9}{20} = -\dfrac{7}{12} \cdot \dfrac{5}{5} - \dfrac{9}{20} \cdot \dfrac{3}{3}$

$$= -\dfrac{35}{60} - \dfrac{27}{60}$$

$$= \dfrac{-35 - 27}{60} = \dfrac{-62}{60}$$

$$= -\dfrac{2 \cdot 31}{2 \cdot 30} = -\dfrac{31}{30}$$

Addition and Subtraction with Fractions   Section 2.7   230

Now we will consider adding and subtracting fractions that contain variables in the numerators and/or the denominators. There are two basic rules that we must remember at all times.

1. Addition or subtraction can be accomplished only if the fractions have a common denominator.
2. A fraction can be reduced only if the numerator and denominator have a common factor.

Example 11

Adding Fractions Containing Variables

Find the sum: $\dfrac{1}{a} + \dfrac{2}{a} + \dfrac{3}{a}$.

Solution

The fractions all have the same denominator, so we simply add the numerators and keep the common denominator.

$$\frac{1}{a} + \frac{2}{a} + \frac{3}{a} = \frac{1+2+3}{a} = \frac{6}{a}$$

Find the indicated algebraic sums.

11. $\dfrac{5}{c} + \dfrac{6}{c} + \dfrac{3}{c}$

12. $\dfrac{2}{x} + \dfrac{3}{5}$

Example 12

Adding Fractions Containing Variables

Find the sum: $\dfrac{3}{x} + \dfrac{7}{8}$.

Solution

The two fractions have different denominators. The LCD is the product of the two denominators: $8x$.

$$\frac{3}{x} + \frac{7}{8} = \frac{3}{x} \cdot \frac{8}{8} + \frac{7}{8} \cdot \frac{x}{x} = \frac{24}{8x} + \frac{7x}{8x} = \frac{24+7x}{8x}$$

Now work margin exercises 11 and 12.

Do not try to reduce the answer in Example 12.

In the fraction $\dfrac{24+7x}{8x}$, neither 8 nor x is a **factor** of the numerator.

We will see in a future chapter, how to factor and reduce algebraic fractions when there is a common factor in the numerator and denominator. For example, by using the distributive property we can find **factors** and reduce each of the following fractions:

a. $\dfrac{24+8x}{8x} = \dfrac{8(3+x)}{8x} = \dfrac{\cancel{8}(3+x)}{\cancel{8}x} = \dfrac{3+x}{x}$

b. $\dfrac{3y-15}{6y} = \dfrac{3(y-5)}{3\cdot 2y} = \dfrac{\cancel{3}(y-5)}{\cancel{3}\cdot 2y} = \dfrac{y-5}{2y}$

Example 13

Subtracting Fractions Containing Variables

Find the difference: $\dfrac{y}{6} - \dfrac{5}{18}$

Solution

The LCD is 18.

$$\frac{y}{6} - \frac{5}{18} = \frac{y}{6}\cdot\frac{3}{3} - \frac{5}{18} = \frac{3y-5}{18}$$

Notice that this fraction cannot be reduced because there are no common factors in the numerator and denominator.

Example 14

Subtracting Fractions Containing Variables

Simplify the following expression: $\dfrac{a}{6} - \dfrac{1}{3} - \dfrac{1}{12}$

Solution

$$\frac{a}{6} - \frac{1}{3} - \frac{1}{12} = \frac{a}{6}\cdot\frac{2}{2} - \frac{1}{3}\cdot\frac{4}{4} - \frac{1}{12} = \frac{2a}{12} - \frac{4}{12} - \frac{1}{12} = \frac{2a-4-1}{12} = \frac{2a-5}{12}$$

Notice that this fraction cannot be reduced because there are no common factors in the numerator and denominator.

Now work margin exercises 13 and 14.

Find the indicated algebraic sums or differences.

13. $y - \dfrac{3}{4} - \dfrac{1}{2}$

14. $\dfrac{n}{2} + \dfrac{5}{6} + \dfrac{1}{3}$

15. Subtract: $x - \dfrac{5}{6}$.

16. Fred's total income for the year was \$30,000 and he spent $\dfrac{1}{3}$ of his income on traveling and $\dfrac{1}{10}$ of his income on entertainment. What total amount did he spend on these two items?

Completion Example 15

Subtracting Fractions Containing Variables

Subtract: $a - \dfrac{2}{7}$

Solution

The LCD is _____ .

$$a - \frac{2}{7} = \frac{a}{1} \cdot \underline{\quad} - \frac{2}{7} = \underline{\quad}$$

Now work margin exercise 15.

Example 16

Keith's Budget

If Keith's total income for the year was \$36,000 and he spent $\dfrac{1}{4}$ of his income on rent and $\dfrac{1}{12}$ of his income on his car, what total amount did he spend on these two items?

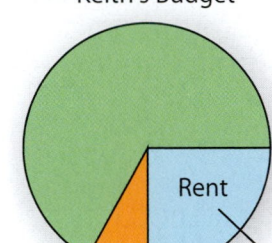

Keith's Budget

Rent $\dfrac{1}{4}$

Car $\dfrac{1}{12}$

Solution

We can add the two fractions, and then multiply the sum by \$36,000. (Or, we can multiply each fraction by \$36,000, and then add the results. We will get the same answer either way.) The LCD is 12.

$$\frac{1}{4} + \frac{1}{12} = \frac{1}{4} \cdot \frac{3}{3} + \frac{1}{12} = \frac{3}{12} + \frac{1}{12} = \frac{4}{12} = \frac{\cancel{4} \cdot 1}{\cancel{4} \cdot 3} = \frac{1}{3}$$

Now multiply $\dfrac{1}{3}$ times \$36,000.

$$\frac{1}{\cancel{3}_1} \cdot \overset{12,000}{\cancel{36,000}} = 12,000$$

Keith spent a total of \$12,000 on rent and his car.

Now work margin exercise 16.

Completion Example Answers

15. The LCD is 7.

$$a - \frac{2}{7} = \frac{a}{1} \cdot \frac{7}{7} - \frac{2}{7} = \frac{7a - 2}{7}$$

Exercises 2.7

Find the indicated products and sums. Reduce if possible. See Example 1.

1. a. $\dfrac{1}{3} \cdot \dfrac{1}{3}$ b. $\dfrac{1}{3} + \dfrac{1}{3}$ 2. a. $\dfrac{1}{5} \cdot \dfrac{1}{5}$ b. $\dfrac{1}{5} + \dfrac{1}{5}$

3. a. $\dfrac{2}{7} \cdot \dfrac{2}{7}$ b. $\dfrac{2}{7} + \dfrac{2}{7}$ 4. a. $\dfrac{3}{10} \cdot \dfrac{3}{10}$ b. $\dfrac{3}{10} + \dfrac{3}{10}$

5. a. $\dfrac{5}{4} \cdot \dfrac{5}{4}$ b. $\dfrac{5}{4} + \dfrac{5}{4}$ 6. a. $\dfrac{7}{100} \cdot \dfrac{7}{100}$ b. $\dfrac{7}{100} + \dfrac{7}{100}$

7. a. $\dfrac{8}{9} \cdot \dfrac{8}{9}$ b. $\dfrac{8}{9} + \dfrac{8}{9}$ 8. a. $\dfrac{9}{12} \cdot \dfrac{9}{12}$ b. $\dfrac{9}{12} + \dfrac{9}{12}$

9. a. $\dfrac{6}{10} \cdot \dfrac{6}{10}$ b. $\dfrac{6}{10} + \dfrac{6}{10}$ 10. a. $\dfrac{5}{8} \cdot \dfrac{5}{8}$ b. $\dfrac{5}{8} + \dfrac{5}{8}$

Find the indicated sums and differences, and reduce all answers to lowest terms. (Note: In some cases you may want to reduce fractions before adding or subtracting. In this way, you may be working with a smaller common denominator.) See Examples 1 through 10.

11. $\dfrac{3}{14} + \dfrac{3}{14}$ 12. $\dfrac{1}{10} + \dfrac{3}{10}$ 13. $\dfrac{4}{6} + \dfrac{4}{6}$

14. $\dfrac{7}{15} + \dfrac{2}{15}$ 15. $\dfrac{14}{25} - \dfrac{6}{25}$ 16. $\dfrac{7}{16} - \dfrac{5}{16}$

17. $\dfrac{1}{15} - \dfrac{4}{15}$ 18. $\dfrac{1}{12} - \dfrac{7}{12}$ 19. $\dfrac{3}{8} - \dfrac{5}{16}$

20. $\dfrac{2}{5} + \dfrac{3}{10}$ 21. $\dfrac{2}{7} + \dfrac{4}{21} + \dfrac{1}{3}$ 22. $\dfrac{2}{39} + \dfrac{1}{3} + \dfrac{4}{13}$

23. $\dfrac{5}{6} - \dfrac{1}{2}$ 24. $\dfrac{2}{3} - \dfrac{1}{4}$ 25. $-\dfrac{5}{14} - \dfrac{5}{7}$

26. $-\dfrac{1}{3} - \dfrac{2}{15}$ 27. $\dfrac{20}{35} - \dfrac{24}{42}$ 28. $\dfrac{14}{45} - \dfrac{12}{30}$

29. $\dfrac{7}{10} - \dfrac{78}{100}$ 30. $\dfrac{29}{100} - \dfrac{3}{10}$ 31. $\dfrac{2}{3} + \dfrac{3}{4} - \dfrac{5}{6}$

32. $\dfrac{1}{5} + \dfrac{1}{10} - \dfrac{1}{4}$ 33. $\dfrac{1}{15} - \dfrac{2}{27} + \dfrac{1}{45}$ 34. $\dfrac{72}{105} - \dfrac{2}{45} + \dfrac{15}{21}$

35. $\dfrac{5}{6} - \dfrac{50}{60} + \dfrac{1}{3}$ 36. $\dfrac{15}{45} - \dfrac{9}{27} + \dfrac{1}{5}$ 37. $-\dfrac{1}{2} - \dfrac{3}{4} - \dfrac{1}{100}$

38. $-\dfrac{1}{4} - \dfrac{1}{8} - \dfrac{7}{100}$ 39. $\dfrac{1}{10} - \left(-\dfrac{3}{10}\right) - \dfrac{4}{15}$ 40. $\dfrac{1}{8} - \left(-\dfrac{5}{12}\right) - \dfrac{3}{4}$

Change 1 into the form $\dfrac{k}{k}$ so that the fractions will have a common denominator, then find the difference. See Example 9.

41. $1 - \dfrac{13}{16}$ **42.** $1 - \dfrac{3}{7}$ **43.** $1 - \dfrac{5}{8}$ **44.** $1 - \dfrac{2}{3}$

Add the fractions in the vertical format. See Example 4.

45.
$$\dfrac{3}{5}$$
$$\dfrac{7}{15}$$
$$+\ \dfrac{5}{6}$$

46.
$$\dfrac{4}{27}$$
$$\dfrac{5}{18}$$
$$+\ \dfrac{1}{9}$$

47.
$$\dfrac{7}{24}$$
$$\dfrac{7}{16}$$
$$+\ \dfrac{7}{12}$$

48.
$$\dfrac{5}{9}$$
$$\dfrac{2}{3}$$
$$+\ \dfrac{4}{15}$$

Find the indicated algebraic sums or differences. See Examples 11 and 12.

49. $\dfrac{x}{3} + \dfrac{1}{5}$ **50.** $\dfrac{x}{2} + \dfrac{3}{5}$ **51.** $\dfrac{y}{6} - \dfrac{1}{3}$

52. $\dfrac{y}{2} - \dfrac{2}{7}$ **53.** $\dfrac{1}{x} + \dfrac{4}{3}$ **54.** $\dfrac{3}{8} + \dfrac{5}{x}$

Find the indicated algebraic sums or differences. See Example 13 through 15.

55. $a - \dfrac{7}{8}$ **56.** $a - \dfrac{5}{16}$ **57.** $\dfrac{a}{10} + \dfrac{3}{100}$

58. $\dfrac{9}{10} + \dfrac{a}{100}$ **59.** $\dfrac{x}{5} + \dfrac{1}{4} + \dfrac{2}{5}$ **60.** $\dfrac{x}{5} + \dfrac{7}{2} + \dfrac{5}{3}$

61. $\dfrac{3}{5} + \dfrac{2}{x} + \dfrac{5}{x}$ **62.** $\dfrac{6}{7} + \dfrac{1}{x} + \dfrac{7}{x}$ **63.** $\dfrac{2}{3} - \dfrac{2}{n} - \dfrac{1}{n}$

64. $\dfrac{n}{6} - \dfrac{1}{8} - \dfrac{2}{3}$ **65.** $\dfrac{n}{5} - \dfrac{1}{5} - \dfrac{3}{5}$ **66.** $\dfrac{x}{15} - \dfrac{1}{3} - \dfrac{2}{15}$

67. $\dfrac{7}{x} + \dfrac{3}{5} + \dfrac{1}{5}$ **68.** $\dfrac{1}{a} + \dfrac{5}{8} + \dfrac{1}{12}$

69. Find the product of $-\dfrac{9}{10}$ and $\dfrac{2}{3}$. Decrease the product by $-\dfrac{5}{3}$. What is the difference?

70. Find the quotient of $-\dfrac{3}{4}$ and $\dfrac{15}{16}$. Add $-\dfrac{3}{10}$ to the quotient. What is the sum?

71. Sam's income is \$3300 a month and he plans to budget $\frac{1}{3}$ of his income for rent and $\frac{1}{10}$ of his income for food.

   a. What fraction of his income does he plan to spend on these two items?

   b. What amount of money does he plan to spend each month on these two items?

Sam's Budget

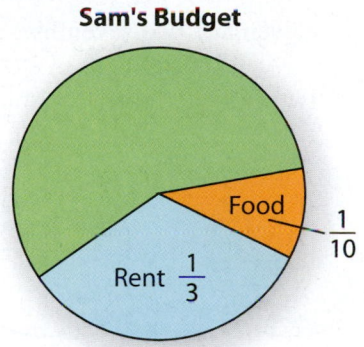

72. There were over 292 million cell phone users in the United States in 2010. Of those mobile phone users, $\frac{321}{1000}$ use Verizon Wireless as their service provider, $\frac{326}{1000}$ use AT&T, $\frac{171}{1000}$ use Sprint, and $\frac{115}{1000}$ use T-Mobile. What fraction of Americans use another service provider?

2010 Mobile Phone Service Providers

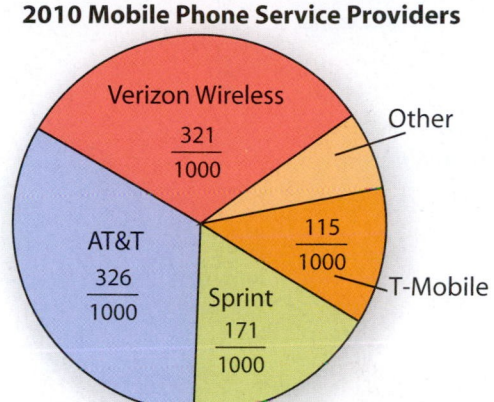

73. A watch has a rectangular-shaped display screen that is $\frac{3}{4}$ inch by $\frac{1}{2}$ inch. The display screen has a border of silver that is $\frac{1}{10}$ inch thick. What are the dimensions (length and width) of the face of the watch (including the silver border)?

74. Four postal letters weigh $\frac{1}{2}$ ounce, $\frac{1}{5}$ ounce, $\frac{3}{10}$ ounce, and $\frac{9}{10}$ ounce. What is the total weight of the letters?

Writing & Thinking

75. Pick one problem in this section that gave you some difficulty. Explain briefly why you had difficulty and why you think that you can better solve problems of this type in the future.

2.8 Introduction to Mixed Numbers

Objectives

A Understand that a mixed number is the sum of a whole number and a proper fraction.

B Be able to change mixed numbers into improper fractions.

C Be able to change improper fractions into mixed numbers.

Objective A **Mixed Number Notation**

A **mixed number** is the **sum** of a whole number and a proper fraction. By convention, we usually write the whole number and the fraction side by side without the plus sign. For example,

$$6 + \frac{3}{5} = 6\frac{3}{5}$$ Read "six and three-fifths".

$$11 + \frac{2}{7} = 11\frac{2}{7}$$ Read "eleven and two-sevenths".

$$2 + \frac{1}{4} = 2\frac{1}{4} = \frac{9}{4}$$ Read "two and one-fourth" or "nine-fourths."

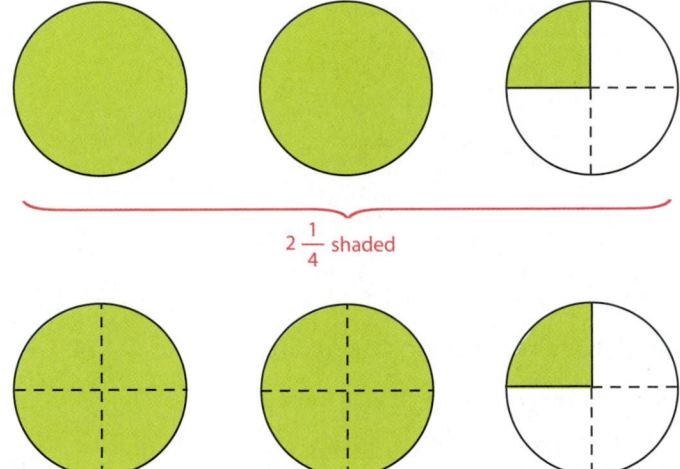

$2\frac{1}{4}$ shaded

$2\frac{1}{4} = \frac{9}{4}$ shaded

Remember that a mixed number indicates addition and not multiplication.

Mixed Number

$$6\frac{3}{5} = 6 + \frac{3}{5}$$ not the same as

Multiplication

$$6\left(\frac{3}{5}\right) = \frac{6}{1} \cdot \frac{3}{5} = \frac{18}{5}$$

Objective B **Changing Mixed Numbers to Fraction Form**

Generally, people are familiar with mixed numbers and use them frequently. For example, "A carpenter sawed a board into two pieces, each $6\frac{3}{4}$ feet long" or "I talked with my brother for $1\frac{1}{2}$ hours today." However, while mixed numbers are commonly used in many situations, they are not convenient for the operations of multiplication and division. These operations are more easily accomplished by first changing each mixed number to the form of an improper fraction.

Changing a Mixed Number to an Improper Fraction

To change a mixed number to an improper fraction, add the whole number and the fraction using a common denominator. Remember, the whole number can be written with 1 as a denominator. For example,

$$2 + \frac{3}{4} = \frac{2}{1} + \frac{3}{4} = \frac{2}{1} \cdot \frac{4}{4} + \frac{3}{4} = \frac{11}{4}.$$

Example 1

Changing Mixed Numbers to Improper Fractions

$$6\frac{3}{5} = 6 + \frac{3}{5} = \frac{6}{1} \cdot \frac{5}{5} + \frac{3}{5} = \frac{30}{5} + \frac{3}{5} = \frac{30+3}{5} = \frac{33}{5}$$

Example 2

Changing Mixed Numbers to Improper Fractions

$$11\frac{2}{7} = 11 + \frac{2}{7} = \frac{11}{1} \cdot \frac{7}{7} + \frac{2}{7} = \frac{77}{7} + \frac{2}{7} = \frac{77+2}{7} = \frac{79}{7}$$

Now work margin exercise 1 and 2.

Express each mixed number as an improper fraction.

1. $4\frac{5}{8}$

2. $9\frac{3}{4}$

There is a pattern to changing mixed numbers into improper fractions that leads to a familiar shortcut. Since the denominator of the whole number is always 1, the LCD is always the denominator of the fraction part. Therefore, in Example 1, the LCD was 5, and we multiplied the whole number 6 by 5. Similarly, in Example 2, we multiplied the whole number 11 by 7. After each multiplication, we added the numerator of the fraction part of the mixed number and used the common denominator. This process is summarized as the following shortcut.

Shortcut for Changing Mixed Numbers to Fraction Form

1. Multiply the whole number by the denominator of the fraction part.
2. Add the numerator of the fraction part to this product.
3. Write this sum over the denominator of the fraction.

Example 3

The Shortcut Method

Use the shortcut method to change $6\frac{3}{5}$ to an improper fraction.

Solution

Step 1: Multiply the whole number by the denominator: $6 \cdot 5 = 30$.

Step 2: Add the numerator: $30 + 3 = 33$.

Step 3: Write this sum over the denominator: $6\frac{3}{5} = \frac{33}{5}$. This is the same answer as in Example 1.

Example 4

The Shortcut Method

Change $8\frac{9}{10}$ to an improper fraction.

Solution

Multiply $8 \cdot 10 = 80$ and add 9: $80 + 9 = 89$.

Write 89 over 10 as follows: $8\frac{9}{10} = \frac{89}{10}$.

$$8 \overset{80+9}{\underset{8 \cdot 10}{\frac{9}{10}}} = \frac{8 \cdot 10 + 9}{10} = \frac{89}{10}$$

Completion Example 5

Changing Mixed Numbers to Improper Fractions

Change $11\frac{2}{3}$ to an improper fraction.

Solution

Multiply $11 \cdot 3 = $ _____ and add 2: _____ + _____ = _____ .

Write this sum, _____, over the denominator _____. Therefore,

$11\frac{2}{3} = $ _____ .

Now work margin exercises 3 through 5.

Use the shortcut method to change the mixed number to an improper fraction.

3. $10\frac{4}{5}$

4. $7\frac{2}{9}$

5. $3\frac{5}{32}$

Completion Example Answers

5. Multiply $11 \cdot 3 = 33$ and add 2: $33 + 2 = 35$
Write this sum, 35, over the denominator 3. Therefore, $11\frac{2}{3} = \frac{35}{3}$.

Objective C **Changing Improper Fractions to Mixed Numbers**

To reverse the process (that is, to change an improper fraction to a mixed number), we use the fact that a fraction can indicate division.

Changing an Improper Fraction to a Mixed Number

1. Divide the numerator by the denominator to find the whole number part of the mixed number.
2. Write the remainder over the denominator as the fraction part of the mixed number.

Example 6

Changing Improper Fractions to Mixed Numbers

Change $\dfrac{67}{5}$ to a mixed number.

Solution

Divide 67 by 5.

$$
\begin{array}{r}
13 \\
5\overline{)67} \\
5 \\
\hline
17 \\
15 \\
\hline
2
\end{array}
$$

whole number part

remainder

$$\frac{67}{5} = 13 + \frac{2}{5} = 13\frac{2}{5}$$

Example 7

Changing Improper Fractions to Mixed Numbers

Change $\dfrac{85}{2}$ to a mixed number.

Solution

Divide 85 by 2.

$$
\begin{array}{r}
42 \\
2\overline{)85} \\
8 \\
\hline
05 \\
4 \\
\hline
1
\end{array}
$$

whole number part

remainder

$$\frac{85}{2} = 42 + \frac{1}{2} = 42\frac{1}{2}$$

Now work margin exercises 6 and 7.

Change each improper fraction to a mixed number.

6. $\dfrac{20}{11}$

7. $\dfrac{149}{6}$

Changing an improper fraction to a mixed number is not the same as reducing it. **Reducing** involves finding common factors in the numerator and denominator. **Changing to a mixed number** involves division of the numerator by the denominator. Common factors are not involved. In any case, the fraction part of a mixed number should be in reduced form. To ensure this, we can follow either of the following two procedures.

1. Reduce the improper fraction first, then change this fraction to a mixed number.

2. Change the improper fraction to a mixed number first, then reduce the fraction part.

Each procedure is illustrated in Example 8.

Use both procedures shown in Example 8 to change each improper fraction to a mixed number, and show that the same answer is received through both procedures.

8. $\dfrac{46}{10}$

Example 8

Reducing Improper Fractions and Changing to Mixed Number

Change $\dfrac{34}{12}$ to a mixed number with the fraction part reduced.

Solution

a. Reduce first.

$$\frac{34}{12} = \frac{\cancel{2} \cdot 17}{\cancel{2} \cdot 6} = \frac{17}{6}.$$

Then change to a mixed number.

$$6\overline{)17} \quad \frac{2}{}$$
$$\underline{12}$$
$$5$$

$$\frac{17}{6} = 2\frac{5}{6}.$$

b. Change to a mixed number.

$$12\overline{)34} \quad \frac{2}{}$$
$$\underline{24}$$
$$10$$

$$\frac{34}{12} = 2\frac{10}{12}.$$

Then reduce.

$$2\frac{10}{12} = 2\frac{\cancel{2} \cdot 5}{\cancel{2} \cdot 6} = 2\frac{5}{6}.$$

Both procedures give the same result, $2\dfrac{5}{6}$.

Now work margin exercise 8.

Exercises 2.8

Write each mixed number as the sum of a whole number and a proper fraction.

1. $5\frac{1}{2}$ 2. $3\frac{7}{8}$ 3. $10\frac{11}{15}$ 4. $16\frac{2}{3}$

5. $7\frac{1}{6}$ 6. $9\frac{5}{12}$ 7. $32\frac{1}{16}$ 8. $50\frac{3}{4}$

Write each fraction in the form of an improper fraction reduced to lowest terms.

9. $\frac{28}{10}$ 10. $\frac{42}{35}$ 11. $\frac{26}{12}$ 12. $\frac{75}{60}$

13. $\frac{112}{63}$ 14. $\frac{51}{34}$ 15. $\frac{85}{30}$

Change each mixed number to fraction form and reduce if possible. See Examples 1 through 5.

16. $4\frac{3}{4}$ 17. $3\frac{5}{8}$ 18. $1\frac{2}{15}$ 19. $5\frac{3}{5}$

20. $2\frac{1}{4}$ 21. $15\frac{1}{3}$ 22. $10\frac{2}{3}$ 23. $12\frac{1}{2}$

24. $7\frac{1}{2}$ 25. $31\frac{1}{5}$ 26. $9\frac{1}{3}$ 27. $7\frac{2}{5}$

28. $3\frac{1}{10}$ 29. $5\frac{3}{10}$ 30. $6\frac{3}{20}$ 31. $19\frac{1}{2}$

32. $4\frac{257}{1000}$ 33. $3\frac{931}{1000}$ 34. $2\frac{63}{100}$ 35. $7\frac{7}{100}$

Change each fraction to mixed number form with the fraction part reduced to lowest terms. See Examples 6 through 8.

36. $\frac{28}{12}$ 37. $\frac{45}{30}$ 38. $\frac{75}{60}$ 39. $\frac{105}{14}$

40. $\frac{96}{80}$ 41. $\frac{80}{64}$ 42. $\frac{48}{32}$ 43. $\frac{87}{51}$

44. $\frac{51}{34}$ 45. $\frac{70}{14}$ 46. $\frac{65}{13}$ 47. $\frac{125}{100}$

48. **Stock Market:** In 5 days, the price of stock in Microsoft rose $\frac{1}{4}$, fell $\frac{5}{8}$, rose $\frac{3}{4}$, rose $\frac{1}{2}$, and rose $\frac{3}{8}$ of a dollar. What was the net gain in the price of this stock over these 5 days?

49. **Lumber:** Three pieces of lumber are stacked, one on top of the other. If two pieces are each $\frac{3}{4}$ inch thick and the third piece is $\frac{1}{2}$ inch thick, how many inches high is the stack?

50. **Biology:** A tree in the Grand Teton National Park in Wyoming grew $\frac{1}{2}$ foot, $\frac{2}{3}$ foot, $\frac{3}{4}$ foot, and $\frac{5}{8}$ foot in 4 consecutive years. How many feet did the tree grow during these 4 years?

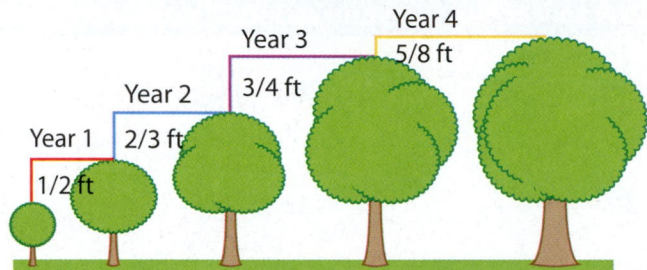

Use your calculator to help in changing the following mixed numbers to fraction form.

51. $18\frac{53}{97}$

52. $16\frac{31}{45}$

53. $20\frac{15}{37}$

54. $32\frac{26}{41}$

55. $19\frac{25}{47}$

56. $72\frac{27}{35}$

57. $89\frac{171}{300}$

58. $206\frac{128}{501}$

59. $602\frac{321}{412}$

2.9 Multiplication and Division with Mixed Numbers

Objectives

A   Multiply with mixed numbers.

B   Find the area of a rectangle and the area of a triangle.

C   Find the volume of a three-dimensional rectangular solid.

D   Divide with mixed numbers.

Objective A   **Multiplication with Mixed Numbers**

The fact that a mixed number is the sum of a whole number and a fraction does not help in understanding multiplication (or division) with mixed numbers. The simplest way to multiply mixed numbers is to change each mixed number to an improper fraction and then multiply. Thus multiplication with mixed numbers is the same as multiplication with fractions. We use prime factorizations and reduce as we multiply.

To Multiply Mixed Numbers

1. Change each number to fraction form.
2. Factor the numerator and denominator of each improper fraction, and then reduce and multiply.
3. Change the answer to a mixed number or leave it in fraction form. (The choice sometimes depends on what use is to be made of the answer.)

Example 1

Multiplying Mixed Numbers

Find the product: $\left(1\frac{1}{2}\right)\left(2\frac{1}{5}\right)$.

Solution

Change each mixed number to fraction form, then multiply the fractions.

$$\left(1\frac{1}{2}\right)\left(2\frac{1}{5}\right) = \left(\frac{3}{2}\right)\left(\frac{11}{5}\right) = \frac{3\cdot11}{2\cdot5} = \frac{33}{10} \quad \text{or} \quad 3\frac{3}{10}$$

Now work margin exercise 1.

Example 2

Multiplying and Reducing Mixed Numbers

Multiply and reduce to lowest terms: $4\frac{2}{3}\cdot1\frac{1}{7}\cdot2\frac{1}{16}$.

Solution

$$4\frac{2}{3}\cdot1\frac{1}{7}\cdot2\frac{1}{16} = \frac{14}{3}\cdot\frac{8}{7}\cdot\frac{33}{16} = \frac{\cancel{2}\cdot\cancel{7}\cdot\cancel{8}\cdot\cancel{3}\cdot11}{\cancel{3}\cdot\cancel{7}\cdot\cancel{8}\cdot\cancel{2}} = \frac{11}{1} = 11$$

1. Find the product.

$$\left(3\frac{2}{3}\right)\left(1\frac{1}{2}\right)$$

2. Multiply and reduce to lowest terms.

$$3\frac{3}{4} \cdot 1\frac{1}{6} \cdot 4\frac{2}{5}$$

3. Find the product of

$$-\frac{7}{30}, 4\frac{2}{3}, \text{ and } 2\frac{1}{7}.$$

Example 3

Multiplying and Reducing Mixed Numbers

Find the product of $-\frac{5}{33}$, $3\frac{1}{7}$, and $6\frac{1}{8}$.

Solution

$$-\frac{5}{33} \cdot 3\frac{1}{7} \cdot 6\frac{1}{8} = \frac{-5}{33} \cdot \frac{22}{7} \cdot \frac{49}{8} = \frac{-5 \cdot \cancel{2} \cdot \cancel{11} \cdot \cancel{7} \cdot 7}{3 \cdot \cancel{11} \cdot \cancel{7} \cdot \cancel{2} \cdot 4}$$

$$= \frac{-35}{12} = -2\frac{11}{12}$$

Now work margin exercises 2 and 3.

Large mixed numbers can be multiplied in the same way as smaller mixed numbers. The products will be large numbers, and a calculator may be helpful in changing the mixed numbers to improper fractions.

Completion Example 4

Multiplying and Reducing Mixed Numbers

Find the product and write it as a mixed number with the fraction part in lowest terms: $2\frac{1}{6} \cdot 1\frac{3}{5} \cdot 1\frac{1}{34}$.

Solution

$$2\frac{1}{6} \cdot 1\frac{3}{5} \cdot 1\frac{1}{34} = \frac{13}{6} \cdot \frac{\rule{1cm}{0.4pt}}{\rule{1cm}{0.4pt}} \cdot \frac{\rule{1cm}{0.4pt}}{\rule{1cm}{0.4pt}}$$

$$= \frac{13 \cdot \rule{2cm}{0.4pt}}{2 \cdot 3 \cdot \rule{2cm}{0.4pt}}$$

$$= \frac{\rule{0.8cm}{0.4pt}}{\rule{0.8cm}{0.4pt}} = \frac{\rule{0.8cm}{0.4pt}}{\rule{0.8cm}{0.4pt}}$$

Completion Example Answers

4. $2\frac{1}{6} \cdot 1\frac{3}{5} \cdot 1\frac{1}{34} = \frac{13}{6} \cdot \frac{8}{5} \cdot \frac{35}{34}$

$$= \frac{13 \cdot \cancel{2} \cdot \cancel{2} \cdot 2 \cdot \cancel{5} \cdot 7}{\cancel{2} \cdot 3 \cdot \cancel{5} \cdot \cancel{2} \cdot 17}$$

$$= \frac{182}{51} = 3\frac{29}{51}$$

Completion Example 5

Multiplying Mixed Numbers

Multiply: $20\frac{3}{4} \cdot 19\frac{1}{5}$.

Solution

$$20\frac{3}{4} \cdot 19\frac{1}{5} = \frac{83}{4} \cdot \underline{\quad}$$

$$= \underline{\quad} = \underline{\quad} = \underline{\quad}$$

Now work margin exercises 4 and 5.

Earlier, we discussed the idea that finding a fractional part of a number indicates multiplication. The key word is **of**, and we will find that this same idea is related to decimals and percents in later chapters. For example, we are familiar with expressions such as "take off 40% **of** the list price," and "three-tenths **of** our net monthly income goes to taxes." This concept is emphasized again here with mixed numbers.

> ## notes
>
> ■ To find a fraction **of** a number means to **multiply** the number by the fraction.
>
> ■

Example 6

Fractional Parts of Mixed Numbers

Find $\frac{3}{5}$ of 40.

Solution

$$\frac{3}{5} \cdot 40 = \frac{3}{\cancel{5}_1} \cdot \frac{\cancel{40}^{8}}{1} = 24$$

Completion Example Answers

5. $20\frac{3}{4} \cdot 19\frac{1}{5} = \frac{83}{4} \cdot \frac{96}{5}$

$$= \frac{7968}{20} = 398\frac{8}{20} = 398\frac{2}{5}$$

Find each product and write the product in mixed number form with the fraction part reduced to lowest terms.

4. $\left(2\frac{1}{4}\right)\left(7\frac{1}{3}\right)\left(\frac{4}{55}\right)$

5. $\left(3\frac{1}{2}\right)\left(4\frac{4}{7}\right)$

6. Find $\frac{4}{7}$ of 21.

7. Find $\frac{1}{3}$ of $-7\frac{1}{2}$.

8. Find the area of a rectangle with sides of length $10\frac{7}{12}$ inches and $8\frac{2}{3}$ inches.

Example 7

Fractional Parts of Mixed Numbers

Find $\frac{2}{3}$ of $5\frac{1}{4}$.

Solution

$$\frac{2}{3} \cdot 5\frac{1}{4} = \frac{2}{3} \cdot \frac{21}{4} = \frac{2 \cdot 3 \cdot 7}{3 \cdot 2 \cdot 2} = \frac{7}{2} \text{ or } 3\frac{1}{2}$$

Now work margin exercise 6 and 7.

Objective B **Area**

Area is the measure of the interior of a closed plane surface. For a rectangle, the area is the product of the length times the width. (In the form of a formula, $A = lw$.) For a triangle, the area is one-half the product of the base times its height. (In the form of a formula, $A = \frac{1}{2}bh$.) These concepts and formulas are true regardless of the type of number used for the lengths. Thus we can use these same techniques for finding the areas of rectangles and triangles whose dimensions are mixed numbers. **Remember that area is measured in square units.**

Example 8

Area of a Rectangle

Find the area of the rectangle with sides of length $5\frac{1}{2}$ feet and $3\frac{3}{4}$ feet.

Solution

The area is the product of the length times the width and we can use the formula $A = lw$.

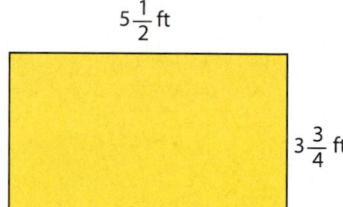

$5\frac{1}{2}$ ft

$3\frac{3}{4}$ ft

$$A = 5\frac{1}{2} \cdot 3\frac{3}{4} = \frac{11}{2} \cdot \frac{15}{4}$$
$$= \frac{165}{8} = 20\frac{5}{8} \text{ square feet (or ft}^2\text{)}.$$

Now work margin exercise 8.

Example 9

Area of a Triangle

Find the area of the triangle with base 6 in. and height $5\frac{1}{4}$ in.

9. Find the area of a triangle with base $6\frac{3}{5}$ inches and height 4 inches.

Solution

The area can be found by using the formula $A = \frac{1}{2}bh$.

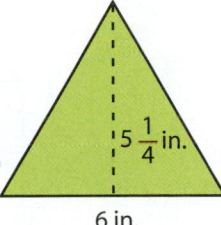
$5\frac{1}{4}$ in.

6 in.

$$A = \frac{1}{2} \cdot 6 \cdot 5\frac{1}{4} = \frac{1}{\cancel{2}_1} \cdot \frac{\cancel{6}^3}{1} \cdot \frac{21}{4} = \frac{63}{4} = 15\frac{3}{4}$$

The area of the triangle is $15\frac{3}{4}$ square inches (or in.²).

Now work margin exercise 9.

Objective C **Volume**

Volume is the measure of space enclosed by a three-dimensional figure and is measured in **cubic units.** The concept of volume is illustrated in Figure 1 in terms of cubic inches.

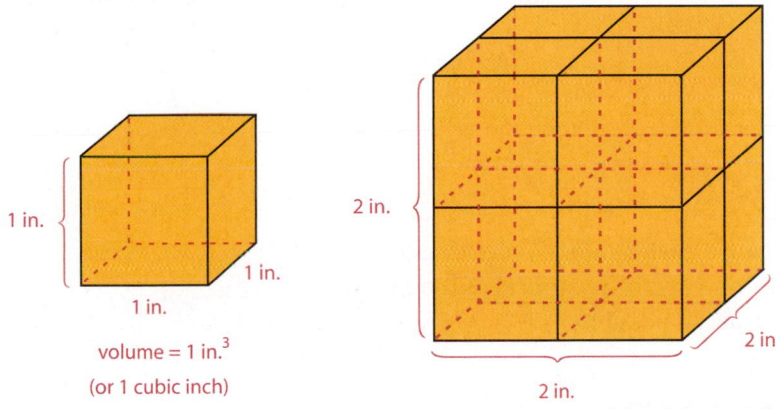

1 in.
1 in.
1 in.
volume = 1 in.³
(or 1 cubic inch)

2 in.
2 in.
2 in.
volume = length · width · height = 2 · 2 · 2 = 8 in.³
There are a total of 8 cubes that are each 1 in.³ for a total of 8 in.³
Figure 1

In the metric system, some of the units of volume are cubic meters (m³), cubic decimeters (dm³), cubic centimeters (cm³), and cubic millimeters (mm³). In the U.S. customary system, some of the units of volume are cubic feet (ft³), cubic inches (in.³), and cubic yards (yd³).

In this section, we will discuss only the volume of a rectangular solid. The volume of such a solid is the product of its length times its width times its height. (In the form of a formula, $V = lwh$.)

10. Find the volume of a rectangular solid with dimensions of $9\frac{3}{4}$ inches by $8\frac{1}{2}$ inches by 14 inches.

Example 10

Volume of a Rectangular Solid

Find the volume of a rectangular solid with dimensions 8 inches by 4 inches by $12\frac{1}{2}$ inches.

Solution

$V = lwh$

$V = 8 \cdot 4 \cdot 12\frac{1}{2}$

$ = \frac{8}{1} \cdot \frac{4}{1} \cdot \frac{25}{2}$

$ = 400 \text{ in.}^3$

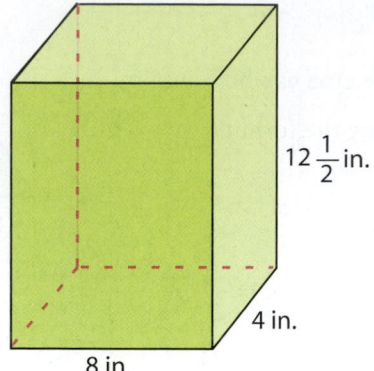

$12\frac{1}{2}$ in.

4 in.

8 in.

Now work margin exercise 10.

Objective D **Division with Mixed Numbers**

Division with mixed numbers is the same as division with fractions, as discussed in earlier. Simply change each mixed number to an improper fraction before dividing. Recall that **to divide by any nonzero number, multiply by its reciprocal**. That is, for $c \neq 0$ and $d \neq 0$, the **reciprocal** of $\frac{c}{d}$ is $\frac{d}{c}$, and

$$\frac{a}{b} \div \frac{c}{d} = \frac{a}{b} \cdot \frac{d}{c}.$$

To Divide with Mixed Numbers

1. Change each number to fraction form.
2. Multiply by the reciprocal of the divisor.
3. Reduce, if possible.

Example 11

Dividing Mixed Numbers

Divide: $6 \div 7\frac{7}{8}$.

Solution

First, change the mixed number $7\frac{7}{8}$ to the improper fraction $\frac{63}{8}$ and then multiply by its reciprocal.

$$6 \div 7\frac{7}{8} = \frac{6}{1} \div \frac{63}{8} = \frac{6}{1} \cdot \frac{8}{63} = \frac{2 \cdot \cancel{3} \cdot 8}{1 \cdot \cancel{3} \cdot 3 \cdot 7} = \frac{16}{21}$$

Completion Example 12

Dividing Mixed Numbers

Find the quotient: $3\dfrac{1}{15} \div 1\dfrac{1}{5}$.

Solution

$$3\frac{1}{15} \div 1\frac{1}{5} = \frac{46}{15} \div \underline{\hspace{1cm}} = \frac{46}{15} \cdot \underline{\hspace{1cm}}$$

$$= \frac{2 \cdot 23 \cdot \underline{\hspace{1cm}}}{3 \cdot 5 \cdot \underline{\hspace{1cm}}} = \underline{\hspace{1cm}} = \underline{\hspace{1cm}}$$

Now work margin exercises 11 and 12.

Example 13

Dividing the Product by the Known Number

The product of two numbers is $-5\dfrac{1}{8}$. One of the numbers is $\dfrac{3}{4}$. What is the other number?

Solution

Reasoning that $\dfrac{3}{4}$ is to be multiplied by some unknown number, we can write

$$\frac{3}{4} \cdot ? = -5\frac{1}{8}.$$

To find the missing number, **divide** $-5\dfrac{1}{8}$ by $\dfrac{3}{4}$. Do **not** multiply.

$$-5\frac{1}{8} \div \frac{3}{4} = \frac{-41}{8} \cdot \frac{4}{3} = \frac{-41 \cdot \cancel{4}}{2 \cdot \cancel{4} \cdot 3} = \frac{-41}{6} = -6\frac{5}{6}$$

11. Divide: $24 \div 2\dfrac{2}{3}$.

12. Find the quotient:

$$4\frac{2}{3} \div 2\frac{1}{9}.$$

Completion Example Answers

12. $3\dfrac{1}{15} \div 1\dfrac{1}{5} = \dfrac{46}{15} \div \dfrac{6}{5} = \dfrac{46}{15} \cdot \dfrac{5}{6}$

$$= \frac{\cancel{2} \cdot 23 \cdot \cancel{5}}{3 \cdot \cancel{5} \cdot \cancel{2} \cdot 3} = \frac{23}{9} = 2\frac{5}{9}$$

13. The product of two numbers is $9\dfrac{5}{6}$. One of the numbers is $-1\dfrac{2}{3}$. What is the other number?

The other number is $-6\dfrac{5}{6}$.

OR, reasoning again that $\dfrac{3}{4}$ is to be multiplied by some unknown number, we can write an equation and solve the equation. If x is the unknown number, then

$$\frac{3}{4}\cdot x = -5\frac{1}{8}$$

$$\frac{4}{3}\cdot\frac{3}{4}\cdot x = \frac{4}{3}\left(-\frac{41}{8}\right)$$

Multiply both sides by $\dfrac{4}{3}$, the reciprocal of the coefficient.

$$1\cdot x = \frac{\overset{1}{\cancel{4}}}{3}\left(-\frac{41}{\underset{2}{\cancel{8}}}\right) = -\frac{41}{6}$$

Note that this is the same as dividing by $\dfrac{3}{4}$.

$$x = -\frac{41}{6}. \qquad \left(\text{or } x = -6\frac{5}{6}\right)$$

Check:

$$\frac{3}{4}\left(-6\frac{5}{6}\right) = \frac{3}{4}\left(\frac{-41}{6}\right) = \frac{-41}{8} = -5\frac{1}{8}$$

Now work margin exercise 13.

Exercises 2.9

Find the indicated products and write your answers in mixed number form. See Examples 1 through 5.

1. $\left(2\frac{4}{5}\right)\left(1\frac{1}{7}\right)$

2. $\left(2\frac{1}{3}\right)\left(3\frac{1}{2}\right)$

3. $5\frac{1}{3}\left(2\frac{1}{2}\right)$

4. $2\frac{1}{4}\left(3\frac{1}{9}\right)$

5. $\left(9\frac{1}{3}\right)3\frac{3}{4}$

6. $\left(1\frac{3}{5}\right)1\frac{1}{4}$

7. $\left(-11\frac{1}{4}\right)\left(-2\frac{2}{15}\right)$

8. $\left(-6\frac{2}{3}\right)\left(-5\frac{1}{7}\right)$

9. $4\frac{3}{8}\cdot2\frac{4}{5}$

10. $12\frac{1}{2}\cdot3\frac{1}{3}$

11. $9\frac{3}{5}\cdot3\frac{3}{7}$

12. $6\frac{3}{8}\cdot2\frac{2}{17}$

13. $5\frac{1}{4}\cdot7\frac{2}{3}$

14. $\left(4\frac{3}{4}\right)\left(2\frac{1}{5}\right)\left(1\frac{1}{7}\right)$

15. $\left(-6\frac{3}{16}\right)\left(2\frac{1}{11}\right)\left(5\frac{3}{5}\right)$

16. $-7\frac{1}{3}\cdot5\frac{1}{4}\cdot6\frac{2}{7}$

17. $\left(-2\frac{5}{8}\right)\left(-3\frac{2}{5}\right)\left(-1\frac{3}{4}\right)$

18. $\left(-2\frac{1}{16}\right)\left(-4\frac{1}{3}\right)\left(-1\frac{3}{11}\right)$

19. $1\frac{3}{32}\cdot1\frac{1}{7}\cdot1\frac{1}{25}$

20. $1\frac{5}{16}\cdot1\frac{1}{3}\cdot1\frac{1}{5}$

 Use a calculator as an aid in finding the following products. Write the answers as mixed numbers. See Example 4.

21. $24\frac{1}{5}\cdot35\frac{1}{6}$

22. $72\frac{3}{5}\cdot25\frac{1}{6}$

23. $42\frac{5}{6}\left(-30\frac{1}{7}\right)$

24. $75\frac{1}{3}\cdot40\frac{1}{25}$

25. $\left(-36\frac{3}{4}\right)\left(-17\frac{5}{12}\right)$

26. $25\frac{1}{10}\cdot31\frac{2}{3}$

27. Find $\frac{2}{3}$ of 90.

28. Find $\frac{1}{4}$ of 60.

29. Find $\frac{3}{5}$ of 100.

30. Find $\frac{5}{6}$ of 240.

31. Find $\frac{1}{2}$ of $3\frac{3}{4}$.

32. Find $\frac{9}{10}$ of $\frac{15}{21}$.

Find the indicated quotients, and write your answers in mixed number form. See Examples 12 and 13.

33. $3\frac{1}{2}\div\frac{7}{8}$

34. $3\frac{1}{10}\div2\frac{1}{2}$

35. $6\frac{3}{5}\div2\frac{1}{10}$

36. $\left(-2\frac{1}{7}\right)\div2\frac{1}{17}$

37. $-6\frac{3}{4}\div4\frac{1}{2}$

38. $7\frac{1}{5}\div(-3)$

39. $7\dfrac{1}{3} \div (-4)$ **40.** $7\dfrac{1}{3} \div \left(-\dfrac{1}{4}\right)$ **41.** $\left(-1\dfrac{1}{32}\right) \div 3\dfrac{2}{3}$

42. $\left(-6\dfrac{3}{11}\right) \div \left(-\dfrac{3}{4}\right)$ **43.** $-10\dfrac{2}{7} \div \left(-4\dfrac{1}{2}\right)$ **44.** $2\dfrac{2}{49} \div 3\dfrac{3}{14}$

45. Telephones: A telephone pole is 64 feet long and $\dfrac{5}{16}$ of the pole must be underground.

  a. What fraction of the pole is above ground?

  b. Is more of the pole above ground or underground?

  c. How many feet of the pole must be underground?

  d. How many feet are above ground?

(**Hint:** When dealing with fractions of a whole item, think of this item as corresponding to the whole number 1.)

46. Driving: If you drive your car to work $6\dfrac{3}{10}$ miles one way 5 days a week, how many miles do you drive each week going to and from work?

47. Reading: A man can read $\dfrac{1}{5}$ of a book in 3 hours.

  a. What fraction of the book can he read in 6 hours?

  b. If the book contains 450 pages, how many pages can he read in 6 hours?

  c. How long will he take to read the entire book?

48. Squares: The perimeter of a square can be found by multiplying the length of one side by 4. The area can be found by squaring the length of one side. Find

  a. the perimeter and

  b. the area of a square if the length of one side is $8\dfrac{2}{3}$ inches.

49. Rectangles: Find

  a. the perimeter and

  b. the area of a rectangle that has sides of length $5\dfrac{7}{8}$ meters and $4\dfrac{1}{2}$ meters.

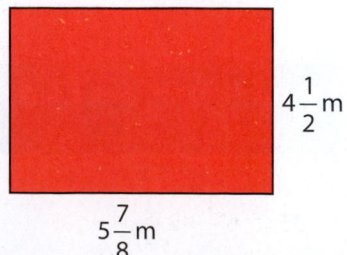

$4\dfrac{1}{2}$ m

$5\dfrac{7}{8}$ m

50. Triangles: Find the area of the triangle in the figure shown here.

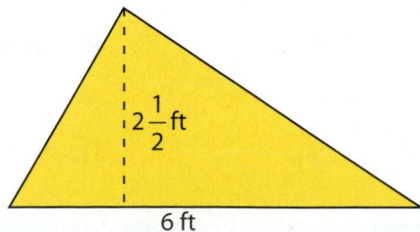

$2\dfrac{1}{2}$ ft

6 ft

51. Triangles: A right triangle is a triangle with one right angle (measures 90°). The two sides that form the right angle are called legs, and they are perpendicular to each other. The longest side is called the hypotenuse. The legs can be treated as the base and height of the triangle. Find the area of the right triangle shown here.

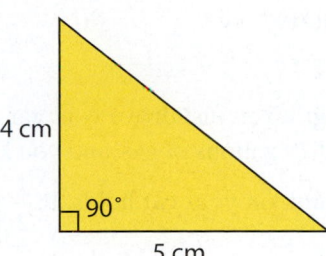

4 cm

90°

5 cm

52. Rectangular Solids: Find the volume of the rectangular solid (a box) that is $6\frac{1}{2}$ inches long, $5\frac{3}{4}$ inches wide, and $10\frac{2}{5}$ inches high.

53. Groceries: A cereal box is $19\frac{3}{10}$ centimeters long by $5\frac{3}{5}$ centimeters wide and $26\frac{9}{10}$ centimeters high. What is the volume of the box in cubic centimeters?

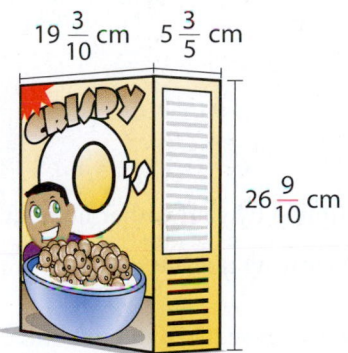

$19\frac{3}{10}$ cm $5\frac{3}{5}$ cm

$26\frac{9}{10}$ cm

54. Car Sales: You are looking at a used car, and the salesman tells you that the price has been lowered to $\frac{4}{5}$ of its original price.

a. Is the current price more or less than the original price?

b. What was the original price if the sale price is $2400?

55. Shopping: The sale price of a coat is $240. This is $\frac{3}{4}$ of the original price.

a. Is the sale price more or less than the original price?

b. What was the original price?

56. Airplanes: An airplane that flies a 430-mile route between two cities is carrying 180 passengers. This is $\frac{2}{3}$ of its capacity.

a. Is the capacity more or less than 180?

b. What is the capacity of the airplane?

57. **Buses:** A bus makes several trips each day from local cities to the beach. On one such trip the bus has 40 passengers. This is $\frac{2}{3}$ of the capacity of the bus.

a. Is the capacity more or less than 40?

b. What is the capacity of the bus?

58. **Shopping:** A clothing store is having a sale on men's suits. The sale rack contains 130 suits, which constitutes $\frac{2}{3}$ of the store's inventory of men's suits.

a. Is the inventory more or less than 130 suits?

b. What is the number of suits in the inventory?

59. **Traveling:** You are planning a trip of 506 miles (round trip), and you know that your car means $25\frac{3}{10}$ miles per gallon of gas and that the gas tank on your car holds $15\frac{1}{2}$ gallons of gas.

a. How many gallons of gas will you use on this trip?

b. If the gas you buy costs $\$2\frac{7}{8}$ per gallon, how much should you plan to spend on this trip for gas?

60. **Traveling:** You just drove your car 450 miles and used 20 gallons of gas, and you know that the gas tank on your car holds $16\frac{1}{2}$ gallons of gas.

a. What is the most number of miles you can drive on one tank of gas?

b. If the gas you buy costs $\$3\frac{1}{10}$ per gallon, what would you pay to fill one-half of your tank?

Writing & Thinking

61. Suppose that a fraction between 0 and 1, such as $\frac{1}{2}$ or $\frac{2}{3}$, is multiplied by some other number. Give brief discussions and several examples in answering each of the following questions.

a. If the other number is a positive fraction, will this product always be smaller than the other number?

b. If the other number is a positive whole number, will this product always be smaller than the other number?

c. If the other number is a negative number (integer or fraction), will this product ever be smaller than the other number?

2.10 Addition and Subtraction with Mixed Numbers

Objective A **Addition with Mixed Numbers**

Since a mixed number represents the sum of a whole number and a fraction, two or more mixed numbers can be added by adding the fraction parts and the whole numbers separately.

To Add Mixed Numbers

1. Add the fraction parts.
2. Add the whole numbers.
3. Write the answer as a mixed number with the fraction part less than 1.

Example 1

Adding Mixed Numbers with the Same Denominator

Find the sum: $5\dfrac{2}{9}+8\dfrac{5}{9}$.

Solution

We can write each number as a sum and then use the commutative and associative properties of addition to treat the whole numbers and fraction parts separately.

$$5\frac{2}{9}+8\frac{5}{9}=5+\frac{2}{9}+8+\frac{5}{9}$$

$$=(5+8)+\left(\frac{2}{9}+\frac{5}{9}\right)$$

$$=13+\frac{7}{9}=13\frac{7}{9}$$

Or, vertically,

$$\begin{array}{r} 5\dfrac{2}{9} \\ +\ 8\dfrac{5}{9} \\ \hline 13\dfrac{7}{9} \end{array}$$

Objectives

A Be able to add mixed numbers.

B Be able to subtract mixed numbers.

C Know how to add and subtract with positive and negative mixed numbers.

D Use the method of changing mixed numbers to improper fractions to add and subtract mixed numbers.

Example 2

Adding Mixed Numbers with Different Denominators

Add: $35\frac{1}{6} + 22\frac{7}{18}$.

Solution

In this case, the fractions do not have the same denominator. The LCD is 18.

$$35\frac{1}{6} + 22\frac{7}{18} = 35 + \frac{1}{6} + 22 + \frac{7}{18}$$

$$= (35 + 22) + \left(\frac{1}{6} \cdot \frac{3}{3} + \frac{7}{18}\right)$$

$$= 57 + \left(\frac{3}{18} + \frac{7}{18}\right)$$

$$= 57\frac{10}{18} = 57\frac{5}{9} \qquad \text{Reduce the fraction part.}$$

1. Find the sum: $8\frac{1}{5} + 3\frac{3}{5}$

2. Add: $5\frac{1}{2} + 3\frac{1}{5}$

3. Find the sum:

$12\frac{3}{4} + 16\frac{2}{3}$

Example 3

Adding Mixed Numbers with Different Denominators

Find the sum: $6\frac{2}{3} + 13\frac{4}{5}$.

Solution

The LCD is 15.

$$
\begin{array}{rcccl}
6\frac{2}{3} & = & 6\frac{2}{3} \cdot \frac{5}{5} & = & 6\frac{10}{15} \\
+\ 13\frac{4}{5} & = & 13\frac{4}{5} \cdot \frac{3}{3} & = & 13\frac{12}{15} \\
\hline
& & & & 19\frac{22}{15} = 19 + 1\frac{7}{15} = 20\frac{7}{15}
\end{array}
$$

fraction is greater than one   changed to a mixed number

Now work margin exercises 1 through 3.

Example 4

Adding Mixed Numbers and Whole Numbers

Add: $5\frac{3}{4} + 9\frac{3}{10} + 2$.

Solution

The LCD is 20. Since the whole number 2 has no fraction part, we must be careful to align the whole numbers if we write the numbers vertically.

$$5\frac{3}{4} = 5\frac{3}{4} \cdot \frac{5}{5} = 5\frac{15}{20}$$

$$9\frac{3}{10} = 9\frac{3}{10} \cdot \frac{2}{2} = 9\frac{6}{20}$$

$$\underline{+\ 2} = \underline{2} = \underline{2}$$

$$16\frac{21}{20} = 16 + 1\frac{1}{20} = 17\frac{1}{20}$$

fraction is greater than one   changed to a mixed number

Now work margin exercise 4.

4. Add.　　8

$$+\ 7\frac{3}{11}$$

Objective B　**Subtraction with Mixed Numbers**

Subtraction with mixed numbers also involves working with the fraction parts and whole numbers separately.

To Subtract Mixed Numbers

1. Subtract the fraction parts.
2. Subtract the whole numbers.

Example 5

Subtracting Mixed Numbers with the Same Denominator

Find the difference: $10\frac{3}{7} - 6\frac{2}{7}$.

Solution

$$10\frac{3}{7} - 6\frac{2}{7} = (10 - 6) + \left(\frac{3}{7} - \frac{2}{7}\right)$$

$$= 4 + \frac{1}{7} = 4\frac{1}{7}$$

Or, vertically,

$$10\frac{3}{7}$$
$$-\ 6\frac{2}{7}$$
$$\overline{4\frac{1}{7}}$$

Now work margin exercise 5.

5. Find the difference.

$$6\frac{3}{5} - 2\frac{2}{5}$$

6. Subtract. $11\dfrac{1}{2} - 4\dfrac{2}{7}$

Example 6

Subtracting Mixed Numbers with Different Denominators

Subtract: $13\dfrac{4}{5} - 7\dfrac{1}{3}$.

Solution

The LCD for the fraction parts is 15.

$$
\begin{aligned}
13\dfrac{4}{5} &= 13\dfrac{4}{5}\cdot\dfrac{3}{3} = 13\dfrac{12}{15}\\
-\;7\dfrac{1}{3} &= -7\dfrac{1}{3}\cdot\dfrac{5}{5} = -7\dfrac{5}{15}\\
&\hphantom{= -7\dfrac{1}{3}\cdot\dfrac{5}{5} = } 6\dfrac{7}{15}
\end{aligned}
$$

Now work margin exercise 6.

Sometimes the fraction part of the number being subtracted is larger than the fraction part of the first number. By "borrowing" the whole number 1 from the whole number part, we can rewrite the first number as a whole number plus an improper fraction. Then the subtraction of the fraction parts can proceed as before.

If the Fraction Part Being Subtracted is Larger than the First Fraction

1. "Borrow" the whole number 1 from the first whole number.
2. Add this 1 to the first fraction. (This will always result in an improper fraction that is larger than the fraction being subtracted.)
3. Now subtract.

Example 7

Subtracting Mixed Numbers by Borrowing

Find the difference: $4\dfrac{2}{9} - 1\dfrac{5}{9}$.

Solution

$$
\begin{aligned}
&4\dfrac{2}{9}\\
-\;&1\dfrac{5}{9}
\end{aligned}
$$

$\dfrac{5}{9}$ is larger than $\dfrac{2}{9}$, so "borrow" 1 from 4.

Rewrite.

$4\dfrac{2}{9}$ as $3 + 1 + \dfrac{2}{9} = 3 + 1\dfrac{2}{9} = 3\dfrac{11}{9}$

$$4\frac{2}{9} = 3+1\frac{2}{9} = 3\frac{11}{9}$$
$$-1\frac{5}{9} = -1\frac{5}{9} = -1\frac{5}{9}$$
$$2\frac{6}{9} = 2\frac{2}{3}$$

Example 8

Subtracting Mixed Numbers from Whole Numbers

Subtract: $10 - 7\frac{5}{8}$.

Solution

The fraction part of the whole number 10 is understood to be 0. Borrow $1 = \frac{8}{8}$ from 10.

$$10 = 9\frac{8}{8}$$
$$-7\frac{5}{8} = -7\frac{5}{8}$$
$$2\frac{3}{8}$$

Example 9

Subtracting Mixed Numbers by Borrowing

Find the difference: $76\frac{5}{12} - 29\frac{13}{20}$.

Solution

First, find the LCD and then borrow 1 if necessary.

$$\left.\begin{array}{l} 12 = 2\cdot2\cdot3 \\ 20 = 2\cdot2\cdot5 \end{array}\right\} \text{LCD} = 2\cdot2\cdot3\cdot5 = 60$$

$$76\frac{5}{12} = 76\frac{5}{12}\cdot\frac{5}{5} = 76\frac{25}{60} = 75\frac{85}{60} \qquad \text{Borrow } 1 = \frac{60}{60}.$$
$$-29\frac{13}{20} = -29\frac{13}{20}\cdot\frac{3}{3} = -29\frac{39}{60} = -29\frac{39}{60}$$
$$46\frac{46}{60} = 46\frac{23}{30}$$

Find each difference.

7. $10\dfrac{1}{8} - 3\dfrac{5}{8}$

8. $8 - 4\dfrac{7}{9}$

9. $\begin{array}{r} 21\dfrac{3}{4} \\ -17\dfrac{7}{8} \\ \hline \end{array}$

10. $8\dfrac{9}{10} - 2\dfrac{1}{5}$

Completion Example 10

Subtracting Mixed Numbers by Borrowing

Find the difference: $12\dfrac{3}{4} - 7\dfrac{9}{10}$.

Solution

$$12\dfrac{3}{4} = 12\ \underline{\quad} = 11\ \underline{\quad}$$

$$-7\dfrac{9}{10} = -7\ \underline{\quad} = -7\ \underline{\quad}$$

$$\underline{\quad}\ \text{difference}$$

Now work margin exercises 7 through 10.

Objective C **Positive and Negative Mixed Numbers**

The rules of addition and subtraction of positive and negative mixed numbers follow the same rules of the addition and subtraction of integers. We must be particularly careful when adding numbers with unlike signs and a negative number that has a larger absolute value than the positive number. In such a case, to find the difference between the absolute values, we must deal with the fraction parts just as in subtraction. That is, the fraction part of the mixed number being subtracted must be smaller than the fraction part of the other mixed number. This may involve "borrowing 1." The following examples illustrate two possibilities.

Example 11

Finding the Algebraic Sum of Mixed Numbers

Find the algebraic sum: $-10\dfrac{1}{2} - 11\dfrac{2}{3}$.

Solution

Both numbers have the same sign. Add the absolute values of the numbers, and use the common sign. In this example, the answer will be negative.

Completion Example Answers

10. $12\dfrac{3}{4} = 12\dfrac{15}{20} = 11\dfrac{35}{20}$

$-7\dfrac{9}{10} = -7\dfrac{18}{20} = -7\dfrac{18}{20}$

$\rule{3cm}{0.4pt}$

$4\dfrac{17}{20}$

$$10\frac{1}{2} \;=\; 10\frac{1}{2}\cdot\frac{3}{3} \;=\; 10\frac{3}{6}$$

$$+\;11\frac{2}{3} \;=\; 11\frac{2}{3}\cdot\frac{2}{2} \;=\; 11\frac{4}{6}$$

$$21\frac{7}{6} = 22\frac{1}{6}$$

Thus $-10\frac{1}{2}-11\frac{2}{3}=-22\frac{1}{6}$.

Example 12

Finding the Algebraic Sum of Mixed Numbers

Find the algebraic sum: $-15\frac{1}{4}+6\frac{2}{5}$.

Solution

The numbers have opposite signs. Find the difference between the absolute values of the numbers, and use the sign of the number with the larger absolute value. In this example, the answer will be negative.

$$15\frac{1}{4} \;=\; 15\frac{1}{4}\cdot\frac{5}{5} \;=\; 15\frac{5}{20} \;=\; 14\frac{25}{20}$$

$$-6\frac{2}{5} \;=\; -6\frac{2}{5}\cdot\frac{4}{4} \;=\; -6\frac{8}{20} \;=\; -6\frac{8}{20}$$

$$8\frac{17}{20}$$

Thus

$$-15\frac{1}{4}+6\frac{2}{5}=-8\frac{17}{20}.$$

Now work margin exercises 11 and 12.

Find each algebraic sum.

11. $-1\frac{1}{5}-7\frac{6}{11}$

12. $-42\frac{7}{12}+33\frac{8}{9}$

Objective D **Optional Approach to Adding and Subtracting Mixed Numbers**

An optional approach to adding and subtracting mixed numbers is to simply do all the work with improper fractions as we did earlier with multiplication and division. By using this method, we are never concerned with the size of any of the numbers or whether they are positive or negative. You may find this approach much easier. In any case, addition and subtraction can be accomplished only if the denominators are the same. So you must still be concerned about least common denominators. For comparison purposes, Examples 13 and 14 are the same problems as in Examples 1 and 3.

Example 13

Adding Mixed Numbers Using Improper Fractions

Find the sum: $5\dfrac{2}{9}+8\dfrac{5}{9}$.

Solution

First, change each number into its corresponding improper fraction, then add.

$$5\frac{2}{9}+8\frac{5}{9}=\frac{47}{9}+\frac{77}{9}=\frac{124}{9}=13\frac{7}{9}$$

Find the sum using improper fractions.

13. $6\dfrac{5}{12}+7\dfrac{3}{8}$

14. $3\dfrac{7}{12}+8\dfrac{2}{3}$

Example 14

Adding Mixed Numbers Using Improper Fractions

Find the sum: $6\dfrac{2}{3}+13\dfrac{4}{5}$.

Solution

First, find the common denominator. Change each number into its corresponding improper fraction, change each fraction to an equivalent fraction with the common denominator, and add these fractions.

The LCD is 15.

$$6\frac{2}{3}+13\frac{4}{5}=\frac{20}{3}+\frac{69}{5}=\frac{20}{3}\cdot\frac{5}{5}+\frac{69}{5}\cdot\frac{3}{3}=\frac{100}{15}+\frac{207}{15}=\frac{307}{15}=20\frac{7}{15}$$

Now work margin exercises **13 and 14.**

15. Subtract using improper fractions.

$$8\frac{4}{5}-11\frac{2}{13}$$

Example 15

Subtracting Mixed Numbers Using Improper Fractions

Subtract: $13\dfrac{1}{6}-17\dfrac{1}{4}$.

Solution

The LCD is 12.

$$13\frac{1}{6}-17\frac{1}{4}=\frac{79}{6}-\frac{69}{4}=\frac{79}{6}\cdot\frac{2}{2}-\frac{69}{4}\cdot\frac{3}{3}=\frac{158}{12}-\frac{207}{12}=-\frac{49}{12}=-4\frac{1}{12}$$

Now work margin exercise **15.**

Exercises 2.10

Find each sum. See Examples 1 through 4.

1. $\quad 5\dfrac{1}{3}$
$\quad +\ 4\dfrac{2}{3}$

2. $\quad 7\dfrac{1}{2}$
$\quad +\ 6\dfrac{1}{2}$

3. $\quad 9$
$\quad +\ 2\dfrac{3}{10}$

4. $\quad 18$
$\quad +\ 1\dfrac{4}{5}$

5. $\quad 10\dfrac{3}{4}$
$\quad +\ \dfrac{1}{6}$

6. $\quad 3\dfrac{1}{4}$
$\quad +\ 5\dfrac{3}{8}$

7. $\quad 13\dfrac{2}{7}$
$\quad +\ 4\dfrac{1}{28}$

8. $\quad 11\dfrac{3}{10}$
$\quad +\ 16\dfrac{5}{6}$

9. $\quad 12\dfrac{3}{4}$
$\quad +\ 8\dfrac{5}{16}$

10. $\quad 13\dfrac{7}{10}$
$\quad +\ 2\dfrac{1}{6}$

11. $\quad 6\dfrac{7}{8}$
$\quad +\ 6\dfrac{7}{8}$

12. $\quad 7\dfrac{3}{8}$
$\quad +\ 5\dfrac{7}{12}$

13. $\dfrac{3}{8}+2\dfrac{1}{12}+3\dfrac{1}{6}$

14. $3\dfrac{5}{7}+1\dfrac{1}{14}+4\dfrac{3}{5}$

15. $13\dfrac{1}{20}+8\dfrac{1}{15}+9\dfrac{1}{2}$

16. $15\dfrac{1}{40}+5\dfrac{7}{24}+6\dfrac{1}{8}$

Find each difference. See Examples 5 through 10.

17. $\quad 12\dfrac{2}{5}$
$\quad -\ 4$

18. $\quad 8\dfrac{9}{10}$
$\quad -\ 3$

19. $\quad 15$
$\quad -\ 6\dfrac{5}{6}$

20. $\quad 11$
$\quad -\ 1\dfrac{2}{7}$

21. $\quad 8\dfrac{2}{3}$
$\quad -\ 6\dfrac{2}{3}$

22. $\quad 20\dfrac{3}{4}$
$\quad -\ 16\dfrac{3}{4}$

23. $\quad 6\dfrac{9}{10}$
$\quad -\ 3\dfrac{3}{5}$

24. $\quad 10\dfrac{5}{16}$
$\quad -\ 7\dfrac{1}{4}$

25. $\quad 21\dfrac{3}{4}$
$\quad -\ 14\dfrac{7}{12}$

26. $\quad 17\dfrac{3}{7}$
$\quad -\ 4\dfrac{15}{28}$

27. $\quad 18\dfrac{3}{7}$
$\quad -\ 15\dfrac{2}{3}$

28. $\quad 9\dfrac{3}{10}$
$\quad -\ 8\dfrac{1}{2}$

29. $26\dfrac{2}{3}$
$-5\dfrac{7}{8}$
___

30. $71\dfrac{5}{12}$
$-55\dfrac{7}{16}$
___

31. $187\dfrac{3}{20}$
$-133\dfrac{7}{15}$
___

32. $17-6\dfrac{2}{3}$

33. $20-4\dfrac{3}{7}$

34. $1-\dfrac{3}{4}$

35. $1-\dfrac{5}{16}$

36. $2-\dfrac{7}{8}$

37. $2-\dfrac{3}{5}$

38. $6-\dfrac{1}{2}$

39. $8-\dfrac{3}{4}$

40. $10-\dfrac{9}{10}$

Find the following algebraic sums and differences. See Examples 11 and 12.

41. $-6\dfrac{1}{2}-5\dfrac{3}{4}$

42. $-2\dfrac{1}{4}-3\dfrac{1}{5}$

43. $-3-2\dfrac{3}{8}$

44. $-7-4\dfrac{2}{5}$

45. $-12\dfrac{2}{3}-\left(-5\dfrac{7}{8}\right)$

46. $-6\dfrac{1}{2}-\left(-10\dfrac{3}{4}\right)$

47. $-2\dfrac{1}{6}-\left(-15\dfrac{3}{5}\right)$

48. $-7\dfrac{3}{8}-\left(-4\dfrac{3}{16}\right)$

Find the following algebraic sums and differences. See Examples 11 and 12.

49. $-17\dfrac{2}{3}$
$+14\dfrac{2}{15}$
___

50. $-9\dfrac{1}{9}$
$+2\dfrac{5}{18}$
___

51. $-30\dfrac{5}{6}$
$-20\dfrac{2}{15}$
___

52. $-16\dfrac{3}{4}$
$-11\dfrac{5}{6}$
___

53. $-12\dfrac{1}{2}$
$-17\dfrac{2}{5}$
$-15\dfrac{1}{4}$
___

54. $-6\dfrac{1}{3}$
$-8\dfrac{3}{7}$
$-4\dfrac{1}{9}$
___

Solve each of the following word problems.

55. **Carpentry:**   A board is 16 feet long. If two pieces are cut from the board, one $6\dfrac{3}{4}$ feet long and the other $3\dfrac{1}{2}$ feet long, what is the length of the remaining piece of the original board?

56. **Swimming Pools:**   A swimming pool contains $495\dfrac{2}{3}$ gallons of water. If $35\dfrac{4}{5}$ gallons evaporate, how many gallons of water are left in the pool?

57. Triangles: A triangle has sides of $3\frac{1}{8}$ inches, $2\frac{2}{3}$ inches, and $4\frac{3}{4}$ inches. What is the perimeter of (distance around) the triangle in inches?

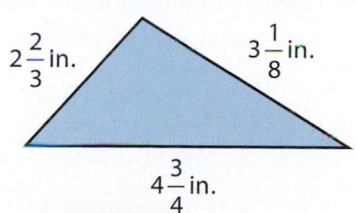

58. Quadrilaterals: A quadrilateral has sides of $5\frac{1}{2}$ meters, $3\frac{7}{10}$ meters, $4\frac{9}{10}$ meters, and $6\frac{3}{5}$ meters. What is the perimeter of (distance around) the quadrilateral in meters?

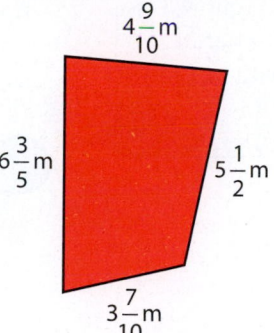

59. Pentagons: A pentagon (five-sided figure) has sides of $3\frac{3}{8}$ centimeters, $5\frac{1}{2}$ centimeters, $6\frac{1}{4}$ centimeters, $9\frac{1}{10}$ centimeters, and $4\frac{7}{8}$ centimeters. What is the perimeter (distance around) in centimeters of the pentagon?

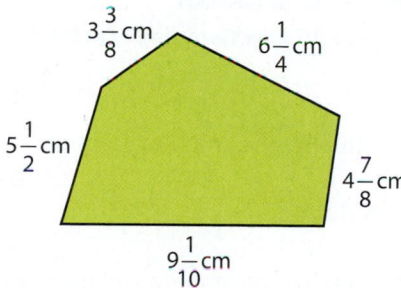

60. Concert Tours: The 5 top-grossing concert tours of 2010 were as follows. **Source:** The Los Angeles Times

Band/Artist	Concert Tour Revenue (in millions of dollars)
1. Bon Jovi	$201\frac{1}{10}$
2. AC/DC	177
3. U2	$160\frac{9}{10}$
4. Lady Gaga	$133\frac{3}{5}$
5. Metallica	110

a. What total amount did these five concert tours earn?

b. How much more did AC/DC earn than Lady Gaga?

c. How much more did Bon Jovi earn than U2?

d. What was the mean earnings of these five tours?

2.11 Complex Fractions and Order of Operations

Objectives

A   Know how to simplify complex fractions.

B   Know how to follow the rules for order of operations with mixed numbers.

Objective A   **Simplifying Complex Fractions**

A **complex fraction** is a fraction in which the numerator or denominator or both contain one or more fractions or mixed numbers. To simplify a complex fraction, we treat the fraction bar as a symbol of inclusion, such as parentheses, similar for both the numerator and denominator. The procedure is outlined as follows.

To Simplify a Complex Fraction

1. Simplify the numerator so that it is a single fraction, possibly an improper fraction.
2. Simplify the denominator so that it also is a single fraction, possibly an improper fraction.
3. Divide the numerator by the denominator, and reduce if possible.

Example 1

Simplifying Complex Fractions

Simplify the complex fraction $\dfrac{\frac{2}{3}+\frac{1}{6}}{1-\frac{1}{3}}$.

Solution

Simplify the numerator and denominator separately, and then divide.

$$\frac{2}{3}+\frac{1}{6}=\frac{4}{6}+\frac{1}{6}=\frac{5}{6} \quad \text{numerator}$$

$$1-\frac{1}{3}=\frac{3}{3}-\frac{1}{3}=\frac{2}{3} \quad \text{denominator}$$

So

$$\frac{\frac{2}{3}+\frac{1}{6}}{1-\frac{1}{3}} \;=\; \frac{\frac{5}{6}}{\frac{2}{3}} \;=\; \frac{5}{6}\div\frac{2}{3} \;=\; \frac{5}{\overset{2}{\cancel{6}}}\cdot\frac{\overset{1}{\cancel{3}}}{2} \;=\; \frac{5}{4} \;=\; 1\frac{1}{4}.$$

Special Note About the Fraction Bar

The complex fraction in Example 1 could have been written as:

$$\frac{\dfrac{2}{3}+\dfrac{1}{6}}{1-\dfrac{1}{3}}=\left(\frac{2}{3}+\frac{1}{6}\right)\div\left(1-\frac{1}{3}\right).$$

Thus, the fraction bar in a complex fraction serves the same purpose as two sets of parentheses, one surrounding the numerator and the other surrounding the denominator.

Example 2

Simplifying Complex Fractions

Simplify the complex fraction $\dfrac{3\dfrac{4}{5}}{\dfrac{3}{4}+\dfrac{1}{5}}$.

Solution

$$3\frac{4}{5}=\frac{19}{5}$$ Change the mixed number to an improper fraction.

$$\frac{3}{4}+\frac{1}{5}=\frac{15}{20}+\frac{4}{20}=\frac{19}{20}$$ Add the fractions in the denominator.

So

$$\frac{3\dfrac{4}{5}}{\dfrac{3}{4}+\dfrac{1}{5}}=\frac{\dfrac{19}{5}}{\dfrac{19}{20}}=\frac{19}{5}\cdot\frac{20}{19}=\frac{19\cdot4\cdot\cancel{5}}{\cancel{5}\cdot\cancel{19}}=\frac{4}{1}=4.$$

Example 3

Simplifying Complex Fractions

Simplify the complex fraction $\dfrac{\dfrac{2}{3}}{-5}$.

Solution

In this case, division is the only operation to be performed. Note that -5 is an integer and can be written as $\dfrac{-5}{1}$.

$$\frac{\dfrac{2}{3}}{-5}=\frac{\dfrac{2}{3}}{\dfrac{-5}{1}}=\frac{2}{3}\cdot\frac{1}{-5}=-\frac{2}{15}$$

Now work margin exercises 1 through 3.

Simplify the following complex fractions.

1. $\dfrac{\dfrac{3}{4}+\dfrac{1}{2}}{2-\dfrac{7}{8}}$

2. $\dfrac{2\dfrac{1}{3}}{\dfrac{1}{4}+\dfrac{1}{3}}$

3. $\dfrac{\dfrac{3}{10}}{-3}$

Objective B Order of Operations

With complex fractions, we know to simplify the numerator and denominator separately. That is, we treat the fraction bar as a grouping symbol. More generally, expressions that involve more than one operation are evaluated by using the rules for order of operations. These rules were discussed and used in earlier sections and are listed again here for easy reference.

Rules for Order of Operations

1. First, simplify within grouping symbols, such as parentheses $(\ )$, brackets $[\ ]$, braces $\{\ \}$, radicals signs $\sqrt{}$, absolute value bars $|\ |$, or fraction bars. Start with the innermost grouping.
2. Second, evaluate any numbers or expressions raised to exponents.
3. Third, moving from **left to right**, perform any multiplications or divisions in the order in which they appear.
4. Fourth, moving from **left to right**, perform any additions or subtractions in the order in which they appear.

Example 4

Using the Order of Operations

Use the rules for order of operations to simplify the following expression:

$$\frac{3}{5} \cdot \frac{1}{6} + \frac{1}{4} \div \left(2\frac{1}{2}\right)^2 .$$

Solution

$$\frac{3}{5} \cdot \frac{1}{6} + \frac{1}{4} \div \left(2\frac{1}{2}\right)^2 = \frac{3}{5} \cdot \frac{1}{6} + \frac{1}{4} \div \left(\frac{5}{2}\right)^2$$

$$= \frac{3}{5} \cdot \frac{1}{6} + \frac{1}{4} \div \left(\frac{25}{4}\right)$$

$$= \frac{\cancel{3}}{5} \cdot \frac{1}{\cancel{6}_2} + \frac{1}{\cancel{4}} \cdot \frac{\cancel{4}}{25}$$

$$= \frac{1}{10} + \frac{1}{25}$$

$$= \frac{5}{50} + \frac{2}{50} = \frac{7}{50}$$

Example 5

Using the Order of Operations

Simplify the expression $x + \dfrac{1}{3} \cdot \dfrac{2}{5}$ in the form of a single fraction, using the rules for order of operations.

Solution

First, multiply the two fractions, and then add x in the form $\dfrac{x}{1}$. The LCD is 15.

$$x + \frac{1}{3} \cdot \frac{2}{5} = x + \frac{2}{15} = \frac{x}{1} \cdot \frac{15}{15} + \frac{2}{15} = \frac{15x}{15} + \frac{2}{15} = \frac{15x+2}{15}$$

Now work margin exercises 4 and 5.

Another topic related to order of operations is that of **mean**. As discussed previously, the mean of a set of numbers can be found by adding the numbers in the set, then dividing this sum by the number of numbers in the set. Now we can find the mean of mixed numbers as well as integers.

Example 6

Finding the Mean

Find the mean of $1\dfrac{1}{2}$, $\dfrac{5}{8}$, and $2\dfrac{1}{4}$.

Solution

Finding this mean is the same as evaluating the expression:

$$\left(1\frac{1}{2} + \frac{5}{8} + 2\frac{1}{4}\right) \div 3$$

Find the sum first, and then divide by 3.

$$1\frac{1}{2} = 1\frac{4}{8}$$
$$\frac{5}{8} = \frac{5}{8}$$
$$+2\frac{1}{4} = 2\frac{2}{8}$$
$$3\frac{11}{8} = 4\frac{3}{8}$$

$$4\frac{3}{8} \div 3 = \frac{35}{8} \cdot \frac{1}{3} = \frac{35}{24} = 1\frac{11}{24}$$

Alternatively, using improper fractions, we can write

$$\left(1\frac{1}{2} + \frac{5}{8} + 2\frac{1}{4}\right) \div 3 = \left(\frac{3}{2} + \frac{5}{8} + \frac{9}{4}\right) \div \frac{3}{1} = \left(\frac{3}{2} \cdot \frac{4}{4} + \frac{5}{8} + \frac{9}{4} \cdot \frac{2}{2}\right) \cdot \frac{1}{3}$$

$$= \left(\frac{12}{8} + \frac{5}{8} + \frac{18}{8}\right) \cdot \frac{1}{3} = \frac{35}{8} \cdot \frac{1}{3} = \frac{35}{24} = 1\frac{11}{24}.$$

Now work margin exercise 6.

Use the rules for order of operations to simplify the following expressions.

4. $\dfrac{5}{6} + 2^3 - \dfrac{2}{3} \div 3\dfrac{4}{5}$

5. $x + \dfrac{1}{7} \div \dfrac{3}{14}$

6. Find the mean of the following mixed numbers.

$$5\frac{7}{9}, 2\frac{1}{4}, 4\frac{5}{6}$$

Exercises 2.11

Solve the following.

1. Complete the following sentence.

   A complex fraction is

2. Rewrite each of the following complex fractions as a division problem with two sets of parentheses.

   a. $\dfrac{\dfrac{3}{4}+\dfrac{1}{6}}{2+\dfrac{2}{3}}$

   b. $\dfrac{5\dfrac{1}{2}}{\dfrac{1}{7}-\dfrac{x}{10}}$

   c. $\dfrac{\dfrac{5}{8}-\dfrac{5}{6}}{-4}$

Simplify each of the following complex fractions. See Examples 1 through 3.

3. $\dfrac{\dfrac{3}{4}}{7}$

4. $\dfrac{\dfrac{5}{8}}{3}$

5. $\dfrac{\dfrac{4}{3x}}{\dfrac{8}{9x}}$

6. $\dfrac{\dfrac{a}{6}}{\dfrac{2a}{3}}$

7. $\dfrac{-2\dfrac{1}{3}}{-1\dfrac{2}{5}}$

8. $\dfrac{-7\dfrac{3}{4}}{-5\dfrac{7}{11}}$

9. $\dfrac{\dfrac{2}{3}+\dfrac{1}{5}}{4\dfrac{1}{2}}$

10. $\dfrac{3-\dfrac{1}{2}}{1-\dfrac{1}{3}}$

11. $\dfrac{\dfrac{5}{8}-\dfrac{1}{2}}{\dfrac{1}{8}-\dfrac{3}{16}}$

12. $\dfrac{\dfrac{5}{12}+\dfrac{1}{15}}{6}$

13. $\dfrac{3\dfrac{1}{5}-1\dfrac{1}{10}}{3\dfrac{1}{2}-1\dfrac{3}{10}}$

14. $\dfrac{6\dfrac{1}{10}-3\dfrac{1}{15}}{2\dfrac{1}{5}+1\dfrac{1}{2}}$

15. $\dfrac{1\dfrac{7}{12}+2\dfrac{1}{3}}{\dfrac{1}{5}-\dfrac{2}{15}}$

16. $\dfrac{-12x}{\dfrac{1}{2}+\dfrac{1}{10}}$

17. $\dfrac{-15y}{\dfrac{1}{4}-\dfrac{7}{8}}$

18. $\dfrac{\dfrac{1}{2}-1\dfrac{3}{4}}{\dfrac{2}{3}+\dfrac{3}{4}}$

19. $\dfrac{\dfrac{7}{8}-\dfrac{19}{20}}{\dfrac{1}{10}-\dfrac{11}{20}}$

20. $\dfrac{3\dfrac{1}{4}}{4\dfrac{1}{5}+1\dfrac{1}{2}}$

Use the rules for order of operations to simplify each of the following expressions. See Examples 4 through 6.

21. $\dfrac{3}{5}\cdot\dfrac{1}{9}+\dfrac{1}{5}\div 2^2$

22. $\dfrac{1}{2}\div\dfrac{1}{4}-\dfrac{2}{3}\cdot 18+5$

23. $\dfrac{2}{3}\div\dfrac{7}{12}-\dfrac{2}{7}+\left(\dfrac{1}{2}\right)^2$

24. $3\dfrac{1}{2}\cdot 5\dfrac{1}{3}+\dfrac{5}{12}\div\dfrac{15}{16}$

25. $\dfrac{5}{8}\div\dfrac{1}{10}+\left(\dfrac{1}{3}\right)^2\cdot\dfrac{3}{5}$

26. $\dfrac{5}{9}-\dfrac{1}{3}\cdot\dfrac{2}{3}+6\dfrac{1}{10}$

27. $2\dfrac{1}{2} \cdot 3\dfrac{1}{5} \div \dfrac{3}{4} + \dfrac{7}{10}$

28. $\dfrac{4}{9} + \dfrac{1}{6} \div \dfrac{1}{2} \cdot \dfrac{11}{12} - 7\dfrac{1}{2}$

29. $-5\dfrac{1}{7} \div (2+1)^3$

30. $\left(\dfrac{1}{3} - 2\right) \div \left(1 - \dfrac{1}{3}\right)^2$

31. $x - \dfrac{1}{5} - \dfrac{2}{3}$

32. $x + \dfrac{3}{4} + 2\dfrac{1}{2}$

33. $y + \dfrac{1}{7} + \dfrac{1}{6}$

34. $y - \dfrac{4}{5} - 3\dfrac{1}{3}$

35. $\dfrac{2a}{5} + \dfrac{3}{5} \cdot \dfrac{3}{7}$

36. $\dfrac{a}{3} \cdot \dfrac{1}{2} - \dfrac{2}{3} \div 1\dfrac{1}{3}$

37. $\dfrac{1}{x} \cdot \dfrac{3}{7} - \dfrac{1}{7} \div \dfrac{1}{2}$

38. $\dfrac{2}{x} \cdot \dfrac{3}{4} - \dfrac{5}{6} \cdot 15$

39. Find the mean of $2\dfrac{1}{2}, 3\dfrac{2}{5}, 5\dfrac{3}{4}$, and $6\dfrac{1}{10}$.

40. Find the mean of $-7\dfrac{3}{5}, -4\dfrac{3}{5}, -8\dfrac{1}{3}$, and $-3\dfrac{1}{18}$.

41. If $\dfrac{4}{5}$ of 80 is divided by $\dfrac{3}{4}$ of 90, what is the quotient?

42. If $\dfrac{2}{3}$ of $2\dfrac{7}{10}$ is added to $\dfrac{1}{2}$ of $5\dfrac{1}{4}$, what is the sum?

43. If the square of $\dfrac{1}{4}$ is subtracted from the square of $\dfrac{3}{10}$, what is the difference?

44. If the square of $\dfrac{1}{3}$ is added to the square of $-\dfrac{3}{4}$, what is the sum?

45. The sum of $\dfrac{4}{5}$ and $\dfrac{2}{15}$ is to be divided by the difference between $2\dfrac{1}{4}$ and $\dfrac{7}{8}$.

  a. Write this expression in the form of a complex fraction.

  b. Write an equivalent expression in a form using two pairs of parentheses and a division sign. Do not evaluate either expression.

46. The sum of $-4\dfrac{1}{2}$ and $-1\dfrac{3}{4}$ is to be divided by the sum of $-5\dfrac{1}{10}$ and $6\dfrac{1}{2}$.

  a. Write this expression in the form of a complex fraction.

  b. Write an equivalent expression in a form using two pairs of parentheses and a division sign. Do not evaluate either expression.

47. Consider any number between 0 and 1. If you square this number, will the result be larger or smaller than the original number? Is this always the case? Explain your answer.

48. Consider any number between −1 and 0. If you square this number, will the result be larger or smaller than the original number? Is this always the case? Explain your answer.

2.12 Solving Equations with Fractions

Objective A **Solving First-Degree Equations Containing Fractions**

In Section 1.5, we discussed solving first-degree equations with whole number constants and coefficients. In Section 1.14, we again discussed solving first-degree equations but included integers as possible constants and coefficients. In these two sections we applied either the addition principle or the division principle, but we did not need to apply both principles to solve any one equation. Equation-solving skills are a very important part of mathematics, and we continue to build these skills in this section by developing techniques for solving equations in which

1. fractions (positive and negative) are possible constants and coefficients, and
2. fractions (as well as mixed numbers) are possible solutions.

notes

Solving equations will be discussed further in future sections. In each of these sections, the equations will involve ideas related to the topics in the corresponding chapter. In this way, solving equations will become more interesting and more applications can be discussed.

Recall that if an equation contains a variable, then a **solution** to the equation is a number that gives a true statement when substituted for the variable, and **solving the equation means finding all the solutions of the equation**. For example, if we substitute $x = -3$ in the equation $5x + 14 = -1$ we get $5(-3) + 14 = -1$, which is a **true statement**.

Thus -3 is the solution to the equation $5x + 14 = -1$.

First-Degree Equation

A **first-degree equation in x** (or **linear equation in x**) is any equation that can be written in the form

$$ax + b = c$$

where a, b, and c are constants and $a \neq 0$.

(**Note:** A variable other than x may be used.)

Every First-Degree Equation has Exactly One Solution

A fundamental fact of algebra, stated here without proof, is that **every first-degree equation has exactly one solution**. Therefore, if we find any one solution to a first-degree equation, then that is the only solution.

To find the solution of a first-degree equation, we apply the principles below.

Principles Used in Solving a First-Degree Equation

In the two basic principles stated here, A and B represent algebraic expressions or constants. C represents a constant, and C is not 0 in the multiplication principle.

1. The **Addition Principle**:

   The equations
   $A = B$, and
   $A + C = B + C$
   have the same solutions.

2. The **Multiplication Principle**:

   The equations
   $A = B$,
   $$\frac{A}{C} = \frac{B}{C},$$
   $$\frac{1}{C} \cdot A = \frac{1}{C} \cdot B \text{ (where } C \neq 0)$$
   have the same solutions.

Essentially, these two principles say that if we perform the same operation to both sides of an equation, the resulting equation will have the same solution as the original equation.

These principles were stated in an earlier section as the **addition principle** and the **division principle**. The **multiplication principle** is the same as the **division principle**. As the Multiplication Principle shows, division by a nonzero number C is the same as multiplication by its reciprocal, $\frac{1}{C}$. For example, to solve $5x = 35$, use either method as shown by the following diagram.

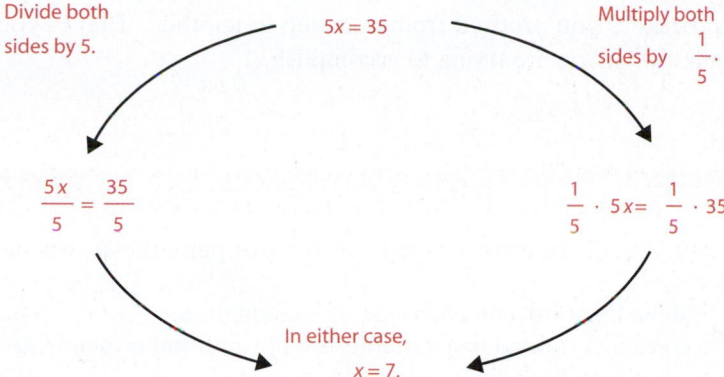

Divide both sides by 5.

$5x = 35$

Multiply both sides by $\frac{1}{5}$.

$$\frac{5x}{5} = \frac{35}{5}$$

$$\frac{1}{5} \cdot 5x = \frac{1}{5} \cdot 35$$

In either case, $x = 7$.

The process is further illustrated in the following examples. Study them carefully.

Example 1

Solving Equations Using the Addition Principle

Solve the equation $x + 17 = -12$.

Solution

$$x + 17 = -12 \qquad \text{Write the equation.}$$
$$x + 17 - 17 = -12 - 17 \qquad \text{Using the addition principle, add } -17 \text{ to both sides.}$$
$$x + 0 = -29$$
$$x = -29 \qquad \text{Simplify}$$

Example 2

Solving Equations Using the Multiplication Principle

Solve the equation $8y = 248$.

Solution

We can divide both sides by 8, as we did previously. However, here we show the same results by multiplying both sides by $\frac{1}{8}$.

$$8y = 248 \qquad \text{Write the equation.}$$
$$\frac{1}{\cancel{8}} \cdot \overset{1}{\cancel{8}} \, y = \frac{1}{\cancel{8}} \cdot \overset{31}{\cancel{248}} \qquad \text{Multiply both sides by } \frac{1}{8}.$$
$$1 \cdot y = 31 \qquad \text{Simplify}$$
$$y = 31$$

Now work margin exercises 1 and 2.

Solve each of the following equations.

1. $x + 14 = -6$

2. $5x = 210$

More difficult problems may involve several steps. Keep in mind the following general process as you proceed from one step to another. That is, you must keep in mind what you are trying to accomplish.

To Solve Equations

1. Apply the distributive property to remove parentheses whenever necessary.
2. Combine like terms on each side of the equation.
3. If a constant is added to a variable, use the addition principle to add its opposite to both sides of the equation.
4. If a variable has a constant coefficient other than 1, use the multiplication (or division) principle to divide both sides by that coefficient (that is, in effect, multiply both sides by the reciprocal of that coefficient).
5. Remember that the object is to isolate the variable on one side of the equation with a coefficient of 1.

Checking can be done by substituting the solution found into the original equation to see if the resulting statement is true. If it is not true, review your procedures to look for any possible errors.

Example 3

Solving Equations

Solve the equation $5x - 9 = 61$.

Solution

$$5x - 9 = 61 \qquad \text{Write the equation.}$$
$$5x - 9 + \mathbf{9} = 61 + \mathbf{9} \qquad \text{Add 9 to both sides.}$$
$$5x + 0 = 70 \qquad \text{simplify}$$
$$5x = 70 \qquad \text{simplify}$$
$$\frac{\cancel{5}x}{\cancel{5}} = \frac{70}{5} \qquad \text{Divide both sides by 5.}$$
$$x = 14 \qquad \text{solution}$$

Check:

$$5x - 9 = 61$$
$$5(14) - 9 \overset{?}{=} 61$$
$$70 - 9 \overset{?}{=} 61$$
$$61 = 61$$

Example 4

Solving Equations

Solve the equation $3(x - 4) + x = -20$.

Solution

$3(x - 4) + x = -20$	Write the equation.
$3x - 12 + x = -20$	Apply the distributive property.
$4x - 12 = -20$	Combine like terms.
$4x - 12 + 12 = -20 + 12$	Add 12 to both sides.
$4x = -8$	simplify
$\dfrac{\cancel{4}x}{\cancel{4}} = \dfrac{-8}{4}$	Divide both sides by 4.
$x = -2$	simplify

Check:

$$3(x - 4) + x = -20$$
$$3(-2 - 4) + (-2) \overset{?}{=} -20$$
$$3(-6) - 2 \overset{?}{=} -20$$
$$-18 - 2 \overset{?}{=} -20$$
$$-20 = -20$$

Now work margin exercises 3 and 4.

Solve each of the following equations.

3. $4x + 22 = 18$

4. $2(x - 5) + 3x = 20$

Objective B ## Equations with Fractions

notes

Special Note on Fractional Coefficients

An expression such as $\dfrac{1}{5}x$ can be thought of as a product.

$$\frac{1}{5}x = \frac{1}{5} \cdot \frac{x}{1} = \frac{1 \cdot x}{5 \cdot 1} = \frac{x}{5}$$

Thus $\dfrac{1}{5}x$ and $\dfrac{x}{5}$ have the same meaning, and $\dfrac{1}{9}x$ and $\dfrac{x}{9}$ have the same meaning.

Similarly, $\dfrac{2}{3}x = \dfrac{2x}{3}$ and $\dfrac{7}{8}x = \dfrac{7x}{8}$.

5. Solve each of the following equations by multiplying both sides of the equation by the reciprocal of the coefficient.

a. $\dfrac{-5}{8}y = 1$

b. $-4 = \dfrac{8}{7}x$

Example 5

Solving Equations with Fractional Coefficients

Solve the following equations by multiplying both sides of the equation by the reciprocal of the coefficient.

a. $\dfrac{2}{3}x = 14$

Solution

$$\frac{2}{3}x = 14$$

$$\frac{3}{2} \cdot \frac{2}{3}x = \frac{3}{2} \cdot 14 \qquad \textcolor{red}{\text{Multiply each side by } \frac{3}{2}.}$$

$$1 \cdot x = 21$$

$$x = 21$$

b. $-\dfrac{7n}{10} = 35$

Solution

$$-\frac{7n}{10} = 35$$

$$\left(-\frac{10}{7}\right)\left(-\frac{7n}{10}\right) = \left(-\frac{10}{7}\right) \cdot 35 \qquad \textcolor{red}{\text{Multiply each side by } -\frac{10}{7}.}$$

$$\left(-\frac{10}{7}\right)\left(-\frac{7}{10}\right) \cdot n = \left(-\frac{10}{7}\right) \cdot \frac{35}{1}$$

$$1 \cdot n = -50$$

$$n = -50$$

Now work margin exercise 5.

More generally, equations with fractions can be solved by first multiplying each term in the equation by the LCM (least common multiple) of all the denominators. **The object is to find an equivalent equation with integer constants and coefficients that will be easier to solve.** That is, we generally find working with integers easier than working with fractions.

Example 6

Solving Equations with Fractional Coefficients

Solve the equation $\dfrac{5}{8}x - \dfrac{1}{4}x = \dfrac{3}{10}$.

Solution

For the denominators 8, 4, and 10, the LCM is 40. Multiply both sides of the equation by 40, apply the distributive property, and reduce to get all integer coefficients and constants.

$$\frac{5}{8}x - \frac{1}{4}x = \frac{3}{10}$$

$$40\left(\frac{5}{8}x - \frac{1}{4}x\right) = 40\left(\frac{3}{10}\right) \quad \text{\textcolor{red}{Multiply both sides by 40.}}$$

$$40\left(\frac{5}{8}x\right) - 40\left(\frac{1}{4}x\right) = 40\left(\frac{3}{10}\right) \quad \text{\textcolor{red}{Apply the distributive property.}}$$

$$25x - 10x = 12$$

$$15x = 12$$

$$\frac{\cancel{15}x}{\cancel{15}} = \frac{12}{15}$$

$$x = \frac{4}{5}$$

Example 7

Solving Equations with Fractional Coefficients

Solve the equation $\dfrac{x}{2} + \dfrac{3x}{4} - \dfrac{5}{3} = 0$.

Solution

$$\frac{x}{2} + \frac{3x}{4} - \frac{5}{3} = 0$$

$$12\left(\frac{x}{2} + \frac{3x}{4} - \frac{5}{3}\right) = 12(0) \quad \text{\textcolor{red}{Multiply both sides by 12, the LCM of 2, 4, and 3.}}$$

$$12\left(\frac{x}{2}\right) + 12\left(\frac{3x}{4}\right) - 12\left(\frac{5}{3}\right) = 12(0) \quad \text{\textcolor{red}{Apply the distributive property.}}$$

$$6x + 9x - 20 = 0$$

$$15x - 20 = 0$$

$$15x - 20 + 20 = 0 + 20$$

$$15x = 20$$

$$\frac{\cancel{15}x}{\cancel{15}} = \frac{20}{15}$$

$$x = \frac{4}{3} \quad \text{or} \quad x = 1\frac{1}{3}$$

Now work margin exercises 6 and 7.

Note that the solution can be in the form of an improper fraction or a mixed number. **Either form is correct and is acceptable mathematically.** In general, improper fractions are preferred in algebra and the mixed number form is more appropriate if the solution indicates a measurement.

Solve the following equations.

6. $\dfrac{3}{4}x - \dfrac{1}{3}x = \dfrac{3}{8}$

7. $\dfrac{x}{5} + \dfrac{2x}{3} - \dfrac{1}{2} = 0$

8. Use the solution steps in Example 8 to solve.

$4(x+3)-2x = 44$

Completion Example 8

The Solution Process

Explain each step in the solution process shown here.

Equation	Explanation
$3(n-5)-n = 1$	Write the equation.
$3n-15-n = 1$	_____
$2n-15 = 1$	_____
$2n-15+15 = 1+15$	_____
$2n = 16$	_____
$n = 8$	_____

Now work margin exercise 8.

Completion Example Answers

	Equation	Explanation
8.	$3(n-5)-n = 1$	Write the equation.
	$3n-15-n = 1$	Apply the distributive property.
	$2n-15 = 1$	Combine like terms.
	$2n-15+15 = 1+15$	Add 15 to both sides.
	$2n = 16$	Simplify.
	$n = 8$	Divide both sides by 2 and simplify.

Exercises 2.12

Give an explanation (or a reason) for each step in the solution process. See Example 8.

1. $3x - x = -10$

   $2x = -10$

   $\dfrac{2x}{2} = \dfrac{-10}{2}$

   $x = -5$

   _____

   _____

   _____

   _____

2. $4x - 12 = -10$

   $4x - 12 + 12 = -10 + 12$

   $4x = 2$

   $\dfrac{4x}{4} = \dfrac{2}{4}$

   $x = \dfrac{1}{2}$

   _____

   _____

   _____

   _____

   _____

3. $\dfrac{1}{2}x + \dfrac{1}{5} = 3$

   $10\left(\dfrac{1}{2}x\right) + 10\left(\dfrac{1}{5}\right) = 10(3)$

   $5x + 2 = 30$

   $5x + 2 - 2 = 30 - 2$

   $5x = 28$

   $\dfrac{5x}{5} = \dfrac{28}{5}$

   $x = \dfrac{28}{5} \left(\text{or } x = 5\dfrac{3}{5}\right)$

   _____

   _____

   _____

   _____

   _____

   _____

   _____

4. $\dfrac{y}{3} - \dfrac{2}{3} = 7$

   $3\left(\dfrac{y}{3}\right) - 3\left(\dfrac{2}{3}\right) = 3(7)$

   $y - 2 = 21$

   $y - 2 + 2 = 21 + 2$

   $y = 23$

   _____

   _____

   _____

   _____

   _____

5. $x + 13 = 25$

6. $y - 20 = 30$

7. $y - 35 = 20$

8. $x + 10 = -22$

9. $2x + 3 = 13$

10. $3x - 1 = 20$

11. $5n - 2 = 33$

12. $7n - 5 = 23$

13. $5x - 2x = -33$

14. $3x - x = -12$

15. $6x - 2x = -24$

16. $3y - 2y = -5$

17. $5y - 2 - 4y = -6$

18. $7x + 14 - 10x = 5$

19. $5(n - 2) = -24$

20. $2(n + 1) = -3$

21. $4(y - 1) = -6$

22. $3(x + 2) = -10$

23. $16x + 23x - 5 = 8$

24. $71y - 62y + 3 = -33$

25. $2(x + 9) + 10 = 3$

26. $4(x - 3) - 4 = 0$

27. $x - 5 - 4x = -20$

28. $x - 6 - 6x = 24$

29. $5x + 2(6 - x) = 10$

30. $3y + 2(y + 1) = 4$

31. $\dfrac{2}{3}y - 5 = 21$

32. $\dfrac{3}{4}x + 2 = 17$

33. $\dfrac{1}{5}x + 10 = -32$

34. $\dfrac{1}{2}y - 3 = -23$

35. $\dfrac{1}{2}x - \dfrac{2}{3} = 11$

36. $\dfrac{2}{3}y - \dfrac{1}{5} = -2$

37. $\dfrac{y}{4} + \dfrac{2}{3} = -\dfrac{1}{6}$

38. $\dfrac{x}{8} + \dfrac{1}{6} = -\dfrac{1}{10}$

39. $\dfrac{5}{8}y - \dfrac{1}{4} = \dfrac{1}{3}$

40. $\dfrac{3}{5}y - 4 = \dfrac{1}{5}$

41. $\dfrac{x}{7} + \dfrac{1}{3} = \dfrac{1}{21}$

42. $\dfrac{n}{3} - 6 = \dfrac{2}{3}$

43. $\dfrac{n}{5} - \dfrac{1}{5} = \dfrac{1}{5}$

44. $\dfrac{x}{15} - \dfrac{2}{15} = -\dfrac{2}{15}$

45. $\dfrac{3}{4} = \dfrac{1}{5}x - \dfrac{5}{8}$

46. $\dfrac{1}{2} = \dfrac{1}{3}x + \dfrac{4}{15}$

47. $\dfrac{7}{8} = \dfrac{3}{4}x - \dfrac{5}{8}$

48. $\dfrac{1}{10} = \dfrac{4}{5}x + \dfrac{3}{10}$

49. $-\dfrac{5}{6} = \dfrac{2}{3}n + \dfrac{1}{6}$

50. $-\dfrac{2}{7} = \dfrac{1}{5}x - \dfrac{3}{5}$

51. $\dfrac{3x}{5} + \dfrac{2x}{5} = -\dfrac{1}{10}$

52. $\dfrac{5}{8}x - \dfrac{3}{4}x = -\dfrac{1}{10}$

53. $\dfrac{5n}{6} - \dfrac{n}{12} = -\dfrac{2}{3}$

54. $\dfrac{y}{7} + \dfrac{y}{28} = \dfrac{3}{4}$

55. $\dfrac{5}{6}y - \dfrac{7}{8}y = \dfrac{5}{12}$

Use this method to solve each of the following equations. See Example 5.

Sometimes an equation is simplified so that the variable has a fractional coefficient and all constants are on the other side of the equation. In such cases, the equation can be solved in one step by multiplying both sides by the reciprocal of the coefficient. This will give 1 as the coefficient of the variable. For example,

$$\frac{2}{3}x = 14$$

$$\frac{3}{2} \cdot \frac{2}{3}x = \frac{3}{2} \cdot 14$$

$$1 \cdot x = 21$$

$$x = 21$$

56. $\dfrac{3}{4}x = 15$

57. $\dfrac{5}{8}x = 40$

58. $\dfrac{7}{10}y = -28$

59. $\dfrac{2}{3}y = -30$

60. $\dfrac{4}{5}x = -\dfrac{2}{3}$

61. $\dfrac{5}{6}x = -\dfrac{5}{8}$

62. $-\dfrac{1}{2}n = \dfrac{3}{4}$

63. $-\dfrac{2}{5}n = \dfrac{1}{3}$

64. $\dfrac{7}{8}x = -\dfrac{1}{2}$

65. $-\dfrac{2}{9}x = -18$

2.13 Ratios and Proportions

Objective A **Understanding Ratios**

We know two meanings for fractions:

1. to indicate a part of a whole

$$\frac{7}{8} \quad \text{means} \quad \frac{7 \text{ indicated parts/pieces}}{8 \text{ total parts/pieces}}$$

2. to indicate division.

$$\frac{3}{8} \quad \text{means} \quad 3 \div 8 \quad \text{or} \quad 8\overline{)3.000}$$

$$
\begin{array}{r}
0.375 \\
8\overline{)3.000} \\
\underline{2\,4} \\
60 \\
\underline{56} \\
40 \\
\underline{40} \\
0
\end{array}
$$

A third use of fractions is to compare two quantities. Such a comparison is called a **ratio**. For example,

$$\frac{3}{4} \quad \text{might mean} \quad \frac{3 \text{ dollars}}{4 \text{ dollars}} \quad \text{or} \quad \frac{3 \text{ hours}}{4 \text{ hours}}$$

> ### Ratio
>
> A **ratio** is a comparison of two quantities by division.
> The ratio of a to b can be written as
>
> $$\frac{a}{b} \quad \text{or} \quad a:b \quad \text{or} \quad a \text{ to } b.$$

Ratios have the following characteristics.

1. Ratios can be reduced, just as fractions can be reduced.

2. Whenever the units of the numbers in a ratio are the same, then the ratio has no units. We say the ratio is an **abstract number**.

3. When the numbers in a ratio have different units, then the numbers must be labeled to clarify what is being compared. Such a ratio is called a **rate**.

For example, the ratio of 55 miles : 1 hour $\left(\text{or } \dfrac{55 \text{ miles}}{1 \text{ hour}}\right)$ is a rate of 55 miles per hour (or 55 mph).

Example 1

Ratios

Compare the quantities 30 students and 40 chairs as a ratio.

Solution

Since the units (students and chairs) are not the same, the units must be written in the ratio. We can write

a. $\dfrac{30 \text{ students}}{40 \text{ chairs}}$

b. 30 students : 40 chairs

c. 30 students to 40 chairs

Furthermore, the same ratio can be simplified by reducing. Since $\dfrac{30}{40} = \dfrac{3}{4}$, we can write the ratio as $\dfrac{3 \text{ students}}{4 \text{ chairs}}$.

The reduced ratio can also be written as

3 students : 4 chairs   or   3 students to 4 chairs.

Example 2

Ratio

Write the comparison of 2 feet to 3 yards as a ratio.

Solution

a. We can write the ratio as $\dfrac{2 \text{ feet}}{3 \text{ yards}}$.

b. Another procedure is to change to common units. Because 1 yd = 3 ft, we have 3 yd = 9 ft and we can write the ratio as an abstract number:

$$\frac{2 \text{ feet}}{3 \text{ yards}} = \frac{2 \text{ feet}}{9 \text{ feet}} = \frac{2}{9} \quad \text{or} \quad 2 : 9 \quad \text{or} \quad 2 \text{ to } 9.$$

Now work margin exercises 1 and 2.

1. Compare 12 apples and 15 oranges as a ratio.

2. Write as a ratio reduced to lowest terms:

3 quarters to 1 dollar

Example 3

Batting Averages

During baseball season, major league players' batting averages are published in the newspapers. Suppose a player has a batting average of 0.250. What does this indicate?

Solution

A batting average is a ratio (or rate) of hits to times at bat. Thus, a batting average of 0.250 means

$$0.250 = \frac{250}{1000} = \frac{250 \text{ hits}}{1000 \text{ times at bat}}.$$

Reducing gives

$$0.250 = \frac{250}{1000} = \frac{250 \cdot 1}{250 \cdot 4} = \frac{1}{4} = \frac{1 \text{ hit}}{4 \text{ times at bat}}.$$

This means that we can expect this player to hit successfully once for every 4 times he comes to bat.

Example 4

Ratios

What is the reduced ratio of 300 centimeters (cm) to 2 meters (m)? (In the metric system, there are 100 centimeters in 1 meter.)

Solution

Since 1 m = 100 cm, we have 2 m = 200 cm. Thus the ratio is

$$\frac{300 \text{ cm}}{2 \text{ m}} = \frac{300 \text{ cm}}{200 \text{ cm}} = \frac{3}{2}.$$

We can also write the ratio as 3 : 2 or 3 to 2.

Now work margin exercises 3 and 4.

Objective B Price per Unit

When you buy a new battery for your car, you usually buy just one battery. However, when if you buy flashlight batteries, you probably buy two or four. So, in cases such as these, the price for one unit (one battery) is clearly marked or understood. However, when you buy groceries, the same item may be packaged in two (or more) different sizes. Since you want to get the most for your money, you want the better (or best) buy. The following box explains how to calculate the **price per unit** (or **unit price**).

Write as a ratio reduced to lowest terms.

3. Inventory shows 500 washers and 4000 bolts. What is the ratio of washers to bolts?

4. 36 inches to 5 feet.

To Find the Price per Unit

1. Set up a ratio (usually in fraction form) of price to units.
2. Divide the price by the number of units.

notes

In the past, many consumers did not understand the concept of price per unit or know how to determine such a number. Now, most states have a law that grocery stores must display the price per unit for certain goods they sell so that consumers can be fully informed.

In the following examples, we show the division process as you would write it on paper. However, you may choose to use a calculator to perform these operations. In either case, your first step should be to write each ratio so you can clearly see what you are dividing and that all ratios are comparing the same types of units.

Notice that the comparisons being made in the examples and exercises are with different amounts of the same brand and quality of goods. Any comparison of a relatively expensive brand of high quality with a cheaper brand of lower quality would be meaningless.

Example 5

Price per Unit

A 16-ounce jar of grape jam is priced at $3.99 and a 9.5-ounce jar of the same jam is $2.69. Which is the better buy?

Solution

We write two ratios of price to units, divide, and compare the results to decide which buy is better. In this problem, we convert dollars to cents ($3.99 = 399¢ and $2.69 = 269¢) so that the results will be ratios of cents per ounce.

a.

$$
\begin{array}{r}
24.93 \\
16\overline{)399.00} \\
\underline{32} \\
79 \\
\underline{64} \\
150 \\
\underline{144} \\
60 \\
\underline{48}
\end{array}
$$

or 24.9¢ per ounce

b. $\dfrac{269¢}{9.5\text{ oz}}$ →

$$
\begin{array}{r}
28.31 \\
9.5.\overline{)269.0.00} \\
\underline{190} \\
790 \\
\underline{760} \\
300 \\
\underline{285} \\
150 \\
\underline{95}
\end{array}
$$

or 28.3¢ per ounce
for the 9.5-ounce jar

Thus, the larger jar (16 ounces for $3.99) is the better buy because the price per ounce is less.

Example 6

Price per Unit

Pancake syrup comes in three different sized bottles:

36 fluid ounces for $5.29
24 fluid ounces for $4.69
12 fluid ounces for $3.99.

Find the price per fluid ounce for each size of bottle and tell which is the best buy.

Solution

After each ratio is set up, the numerator is changed from dollars and cents to just cents so that the division will yield cents per ounce.

a.

$$
\dfrac{\$5.29}{36\text{ oz}} = \dfrac{529¢}{36\text{ oz}} \quad\longrightarrow\quad
\begin{array}{r}
14.69 \\
36\overline{)529.00} \\
\underline{36} \\
169 \\
\underline{144} \\
250 \\
\underline{216} \\
340 \\
\underline{324} \\
16
\end{array}
$$

So the largest bottle comes on at 14.7¢ per ounce, or 14.7¢/oz. **Note:** "14.7¢/oz." is read "14.7¢ per ounce".

b.

$$\frac{\$4.69}{24 \text{ oz}} = \frac{469¢}{24 \text{ oz}} \longrightarrow$$

$$
\begin{array}{r}
19.54 \\
24\overline{)469.00} \\
\underline{24} \\
229 \\
\underline{216} \\
130 \\
\underline{120} \\
100 \\
\underline{96}
\end{array}
$$

The medium-sized bottle costs 19.54¢/oz.

c.

$$\frac{\$3.99}{12 \text{ oz}} = \frac{399¢}{12 \text{ oz}} \longrightarrow$$

$$
\begin{array}{r}
33.25 \\
12\overline{)399.00} \\
\underline{36} \\
39 \\
\underline{36} \\
30 \\
\underline{24} \\
60 \\
\underline{60}
\end{array}
\quad \text{or } 33.3¢/\text{oz}
$$

The small bottle comes in at 33.3¢/oz.
The largest container (36 fluid ounces) is the best buy.

Now work margin exercises 5 and 6.

In each of the examples, the larger (or largest) amount of units was the better (or best) buy. In general, larger amounts are less expensive because the manufacturer wants you to buy more of the product. However, the larger amount is not always the better buy because of other considerations such as packaging or the consumer's individual needs. For example, people who do not use much pancake syrup may want to buy a smaller bottle. Even though they pay more per unit, it's more economical in the long run not to have to throw any away.

Special Comment on the Term *per*

The student should be aware that the term **per** can be interpreted to mean **divided by**. For example,

cents **per** ounce	means	cents **divided by** ounces
dollars **per** pound	means	dollars **divided by** pounds
miles **per** hour	means	miles **divided by** hours
miles **per** gallon	means	miles **divided by** gallons

5. Which is a better buy: a 2-liter bottle of soda for $1.09 or a 3-liter bottle for $1.49?

6. Which is a better buy: a 12.5-ounce box of cereal for $2.50 or a 16-ounce box for $3.00? What is the price per ounce (to the nearest tenth of a cent) for each?

Ratios and Proportions　　Section 2.13　　**290**

Understanding Proportions

Consider the equation $\frac{3}{6} = \frac{4}{8}$. This statement (or equation) says that two ratios are equal. Such an equation is called a **proportion**. As we will see, proportions may be true or false.

Proportions

A **proportion** is a statement that two ratios are equal. In symbols,

$$\frac{a}{b} = \frac{c}{d} \text{ is a proportion.}$$

A proportion has four **terms**.

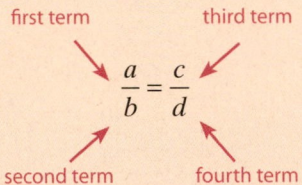

The first and fourth terms (a and d) are called the **extremes**.
The second and third terms (b and c) are called the **means**.
The product of the extremes and the product of the means together are called **cross products**.

To help in remembering which terms are the extremes and which terms are the means, think of a general proportion written with colons, as shown here.

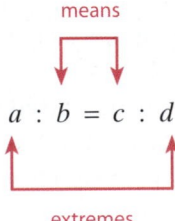

With this form, you can see that a and d are the two end terms; thus the name **extremes** might seem more reasonable.

Example 7

Finding Means and Extremes

In the proportion $\frac{8.4}{4.2} = \frac{10.2}{5.1}$, tell which numbers are the extremes and which are the means.

Solution

8.4 and 5.1 are the extremes.

4.2 and 10.2 are the means.

Completion Example 8

Finding Means and Extremes

In the proportion $\dfrac{2\frac{1}{2}}{10} = \dfrac{3\frac{1}{4}}{13}$, tell which numbers are the extremes and which are the means.

Solution

_____ and _____ are the extremes.

_____ and _____ are the means.

Now work margin exercises 7 and 8.

Identify the means and extremes in the following proportions.

7. $\dfrac{1.3}{1.5} = \dfrac{1.82}{2.1}$

8. $\dfrac{5\frac{1}{2}}{2} = \dfrac{8\frac{1}{4}}{3}$

Objective D **Identifying True Proportions**

A proportion is an equation. However, some proportions are false. In a false proportion, the two sides of the equation do not have equal values. In order to say that a proportion is true, we must verify that both sides of the equation are in fact equal. We use cross products to do this.

Identifying True Proportions

In a true proportion, the cross products are equal. In symbols,

$$\frac{a}{b} = \frac{c}{d} \quad \text{if and only if} \quad a \cdot d = b \cdot c,$$

where $b \neq 0$ and $d \neq 0$.

Note that in a proportion the terms can be any of the types of numbers that we have studied: whole numbers, integers, fractions, or mixed numbers. Thus, all the related techniques for multiplying these types of numbers should be reviewed at this time.

Completion Example Answers

8. $2\frac{1}{2}$ and 13 are the extremes.

   10 and $3\frac{1}{4}$ are the means.

Example 9

Identifying True Proportions

Determine whether the proportion $\dfrac{9}{13} = \dfrac{4.5}{6.5}$ is true or false.

Solution

$$\begin{array}{r} 6.5 \\ \times\ 9 \\ \hline 58.5 \end{array}$$ ← product of extremes

$$\begin{array}{r} 4.5 \\ \times\ 13 \\ \hline 13\,5 \\ 45 \\ \hline 58.5 \end{array}$$ ← product of means

Since $9(6.5) = 13(4.5)$, the proportion is true.

Example 10

Identifying True Proportions

Is the proportion $\dfrac{5}{8} = \dfrac{7}{10}$ true or false?

Solution

$5 \cdot 10 = 50 \quad \text{and} \quad 8 \cdot 7 = 56$

Since $50 \neq 56$, the proportion is false.

Example 11

Identifying True Proportions

Is the proportion $\dfrac{\frac{3}{5}}{\frac{3}{4}} = \dfrac{12}{15}$ true or false?

Solution

The extremes are $\dfrac{3}{5}$ and 15, and their product is

$$\frac{3}{5} \cdot 15 = \frac{3}{5} \cdot \frac{15}{1} = 9.$$

The means are $\dfrac{3}{4}$ and 12, and their product is

$$\frac{3}{4} \cdot 12 = \frac{3}{4} \cdot \frac{12}{1} = 9.$$

Since the cross products are equal, the proportion is true.

Completion Example 12

Identifying True Proportions

Is the proportion $\dfrac{1}{4} : \dfrac{2}{3} = 9 : 24$ true or false?

Solution

a. The extremes are _____ and _____ .
The means are _____ and _____ .

b. The product of the extremes is _____ .
The product of the means is _____ .

c. The proportion is _____ because the cross products are _____ .

Now work margin exercises 9 through 12.

Determine whether the following proportions are true or false.

9. $\dfrac{3.2}{5} = \dfrac{4}{6.25}$

10. $\dfrac{2}{6} = \dfrac{3}{9}$

11. $\dfrac{\frac{4}{5}}{\frac{1}{3}} = \dfrac{4}{3}$

12. $4 : 18 = 10 : 45$

Objective E **Finding the Unknown Term in a Proportion**

Proportions can be used to solve certain types of word problems. In these problems, a proportion is set up in which one of the terms in the proportion is not known and the solution to the problem is the value of this unknown term. In this section, we will use the fact that in a true proportion, the cross products are equal as a method for finding the unknown term in a proportion.

Variable

A **variable** is a symbol (generally a letter of the alphabet) that is used to represent an unknown number or any one of several numbers. The set of possible values for a variable is called its **replacement set**.

Completion Example Answers

12. a. The extremes are $\dfrac{1}{4}$ and 24.

The means are $\dfrac{2}{3}$ and 9.

b. The product of the extremes is $\dfrac{1}{4} \cdot 24 = 6$.

The product of the means is $\dfrac{2}{3} \cdot 9 = 6$.

c. The proportion is true because the cross products are equal.

To Find the Unknown Term in a Proportion

1. In a proportion, the unknown term is represented with a variable (some letter such as x, y, w, A, B, etc.).
2. Write an equation that sets the cross products equal.
3. Divide both sides of the equation by the number multiplying the variable. (This number is called the **coefficient** of the variable.)

The resulting equation will have a coefficient of 1 for the variable and will give the missing value for the unknown term in the proportion.

notes

- As you work through each problem, be sure to write each new equation below the previous equation in the same format shown in the examples. (Arithmetic that cannot be done mentally should be performed to one side, and the results written in the next equation.)
- This format carries over into solving all types of equations at all levels of mathematics. Also, when a number is written next to a variable, such as in $3x$ or $4y$, the meaning is to multiply the number by the value of the variable. That is, $3x = 3 \cdot x$ and $4y = 4 \cdot y$. The number 3 is the coefficient of x and 4 is the coefficient of y.

notes

- **About the Form of Answers**

- In general, if the numbers in the problem are in fraction form, then the answer will be in fraction form. If the numbers are in decimal form, then the answer will be in decimal form.

Example 13

Finding the Unknown Term in a Proportion

Find the value of x if $\dfrac{3}{6} = \dfrac{5}{x}$.

Solution

In this case, you may be able to see that the correct value of x is 10 since

$\dfrac{3}{6}$ reduces to $\dfrac{1}{2}$ and $\dfrac{1}{2} = \dfrac{5}{10}$.

However, not all proportions involve such simple ratios, and the following general method of solving for the unknown is important.

$\dfrac{3}{6} = \dfrac{5}{x}$ Write the proportion.

$3 \cdot x = 6 \cdot 5$ Set the cross products equal.

$\dfrac{3 \cdot x}{3} = \dfrac{30}{3}$ Divide both sides by 3, the coefficient of the variable.

$\dfrac{\cancel{3} \cdot x}{\cancel{3}} = \dfrac{30}{3}$ Reduce both sides to find the solution.

$x = 10$

Example 14

Finding the Unknown Term in a Proportion

Find the value of y if $\dfrac{6}{16} = \dfrac{y}{24}$.

Solution

Note that the variable may appear on the right side of the equation as well as on the left side of the equation. In either case, **we divide both sides of the equation by the coefficient of the variable**.

$\dfrac{6}{16} = \dfrac{y}{24}$ Write the proportion.

$6 \cdot 24 = 16 \cdot y$ Set the cross products equal.

$\dfrac{6 \cdot 24}{16} = \dfrac{16 \cdot y}{16}$ Divide both sides by 16, the coefficient of y.

$\dfrac{6 \cdot 24}{16} = \dfrac{\cancel{16} \cdot y}{\cancel{16}}$ Reduce both sides to find the value of y.

$9 = y$

Alternative Solution

Reduce the fraction $\dfrac{6}{16}$ before solving the proportion.

$$\frac{6}{16} = \frac{y}{24}$$
Write the proportion.

$$\frac{3}{8} = \frac{y}{24}$$
Reduce the fraction: $\dfrac{6}{16} = \dfrac{3}{8}$.

$$3 \cdot 24 = 8 \cdot y$$
Proceed to solve as before.

$$\frac{3 \cdot 24}{8} = \frac{\cancel{8} \cdot y}{\cancel{8}}$$

$$9 = y$$

Example 15

Finding the Unknown Term in a Proportion

Find w if $\dfrac{w}{7} = \dfrac{20}{\frac{2}{3}}$.

Solution

$$\frac{w}{7} = \frac{20}{\frac{2}{3}}$$
Write the proportion.

$$\frac{2}{3} \cdot w = 7 \cdot 20$$
Set the cross products equal.

$$\frac{\frac{\cancel{2}}{\cancel{3}} \cdot w}{\frac{\cancel{2}}{\cancel{3}}} = \frac{7 \cdot 20}{\frac{2}{3}}$$
Divide each side by the coefficient $\dfrac{2}{3}$.

$$w = \frac{7}{1} \cdot \frac{20}{1} \cdot \frac{3}{2} = 210$$
Simplify. Remember, to divide by a fraction, multiply by its reciprocal.

Completion Example 16

Finding the Unknown Term in a Proportion

Find A if $\dfrac{A}{3} = \dfrac{7.5}{6}$.

Solution

$$\frac{A}{3} = \frac{7.5}{6}$$

$$6 \cdot A = \underline{\hspace{2cm}}$$

$$\frac{\cancel{6} \cdot A}{\cancel{6}} = \underline{\hspace{2cm}}$$

$$A = \underline{\hspace{2cm}}$$

Now work margin exercises 13 through 16.

Solve each proportion for the unknown term:

13. $\dfrac{12}{x} = \dfrac{9}{15}$

14. $\dfrac{3}{10} = \dfrac{R}{100}$

15. $\dfrac{\frac{1}{2}}{6} = \dfrac{5}{x}$

16. $\dfrac{x}{2.3} = \dfrac{3.2}{9.2}$

As illustrated in Examples 17 and 18, when the coefficient of the variable is a fraction, we can multiply both sides of the equation by the reciprocal of this coefficient. This is the same as dividing by the coefficient, but fewer steps are involved.

Example 17

Finding the Unknown Term in a Proportion

Find x if $\dfrac{x}{1\frac{1}{2}} = \dfrac{1\frac{2}{3}}{3\frac{1}{3}}$.

Solution

$$\frac{x}{1\frac{1}{2}} = \frac{1\frac{2}{3}}{3\frac{1}{3}}$$ Write the proportion.

$$\frac{10}{3} \cdot x = \frac{\cancel{3}}{2} \cdot \frac{5}{\cancel{3}}$$ Write each mixed number as an improper fraction. and simplify by multipling it by the reciprocal of the denominator on each side.

$$\frac{10}{3} \cdot x = \frac{5}{2}$$

$$\frac{\cancel{3}}{\cancel{10}} \cdot \frac{\cancel{10}}{\cancel{3}} \cdot x = \frac{3}{10} \cdot \frac{5}{2}$$ Multiply each side by $\frac{3}{10}$, the reciprocal of $\frac{10}{3}$.

$$x = \frac{3}{\underset{2}{\cancel{10}}} \cdot \frac{\cancel{5}}{2} = \frac{3}{4}$$

Completion Example Answers

16. a. $\dfrac{A}{3} = \dfrac{7.5}{6}$

b. $6 \cdot A = 3 \cdot 7.5$

c. $\dfrac{\cancel{6} \cdot A}{\cancel{6}} = \dfrac{22.5}{6}$

d. $A = 3.75$

Solve each proportion for the
unknown term.

17. $\dfrac{z}{6} = \dfrac{4\frac{1}{2}}{7\frac{1}{2}}$

18. $\dfrac{\frac{8}{3}}{5} = \dfrac{\frac{1}{5}}{z}$

Finding the Unknown Term in a Proportion

Find y if $\dfrac{2\frac{1}{2}}{6} = \dfrac{3}{y}$.

Solution

$$\dfrac{2\frac{1}{2}}{6} = \dfrac{3}{y}$$

$$\dfrac{5}{2} \cdot y = \underline{\hspace{3cm}} \qquad \left(2\frac{1}{2} = \dfrac{5}{2} \right)$$

$$\dfrac{\cancel{2}}{\cancel{5}} \cdot \dfrac{\cancel{5}}{\cancel{2}} \cdot y = \dfrac{2}{5} \cdot \underline{\hspace{3cm}}$$

$$y = \underline{\hspace{2cm}}$$

Now work margin exercises 17 and 18.

Completion Example Answers

18. a. $\dfrac{2\frac{1}{2}}{6} = \dfrac{3}{y}$

b. $\dfrac{5}{2} \cdot y = 3 \cdot 6$

c. $\dfrac{\cancel{2}}{\cancel{5}} \cdot \dfrac{\cancel{5}}{\cancel{2}} \cdot y = \dfrac{2}{5} \cdot 18$

d. $y = \dfrac{36}{5} \left(\text{or } 7\dfrac{1}{5} \right)$

Practice Problems

In the following proportions, identify the extremes and the means.

1. $\dfrac{2}{3} = \dfrac{6}{9}$

2. $\dfrac{80}{100} = \dfrac{4}{5}$

Solve the following proportions.

3. $\dfrac{3}{5} = \dfrac{R}{100}$

4. $\dfrac{2\frac{1}{2}}{8} = \dfrac{25}{x}$

Write the following comparisons as ratios reduced to lowest terms. **Note:** There is more than one form of the correct answer.

5. 2 teachers to 24 students

6. 90 wins to 72 losses

Find the price per unit for both **a.** and **b.** and determine which is the better deal.

7. a. A 6-pack of cola for $2.99   **b.** A 12-pack of cola for $5.49

Practice Problem Answers

1. The extremes are 2 and 9; the means are 3 and 6.
2. The extremes are 80 and 5; the means are 100 and 4.
3. $R = 60$ **4.** $x = 80$
5. 1 teacher : 12 students   **6.** 5 wins : 4 losses
7. a. 49.8¢/can **b.** 45.8¢/can; **b.** is the better deal.

Exercises 2.13

Write the following comparisons as ratios reduced to lowest terms. Use common units in the numerator and denominator whenever possible. See Examples 1 through 4.

1. 1 dime to 4 nickels

2. 5 nickels to 3 quarters

3. 5 dollars to 5 quarters

4. 6 dollars to 50 dimes

5. 250 miles to 5 hours

6. 270 miles to 4.5 hours

7. 18 inches to 2 feet

8. 36 inches to 2 feet

9. 8 days to 1 week

10. 21 days to 4 weeks

11. $200 in profit to $500 invested

12. $200 in profit to $1000 invested

13. **Blood Types:** About 28 out of every 100 African-Americans have type-A blood. Express this fact as a ratio in lowest terms.

14. **Nutrition:** A serving of four home-baked chocolate chip cookies weighs 40 grams and contains 12 grams of fat. What is the ratio, in lowest terms, of fat grams to total grams?

15. **Standardized Testing:** In recent years, 18 out of every 100 students taking the SAT (Scholastic Aptitude Test) have scored 600 or above on the mathematics portion of the test. Write the ratio, in lowest terms, of the number of scores 600 or above to the number of scores below 600.

16. **Weather:** In a recent year, Albany, NY, reported a total of 60 clear days, the rest being cloudy or partly cloudy. For a 365-day year, write the ratio, in lowest terms, of clear days to cloudy or partly cloudy days.

Find the unit price (to the nearest tenth of a cent) of each of the following items and tell which is the better (or best) buy. See Examples 5 and 6.

17. sugar
    4 lb at $2.79
    10 lb at $5.89

18. sliced bologna
    8 oz at $1.79
    12 oz at $1.99

19. coffee beans
    1.75 oz at $1.99
    12 oz at $7.99

20. coffee
    11.5 oz at $3.99
    39 oz at $8.99

21. cottage cheese
    16 oz at $2.69
    32 oz at $4.89

22. large trash bags
    18 at $4.99
    28 at $6.59

23. honey
12 oz at $3.49
24 oz at $7.99
40 oz at $10.99

24. apple sauce
16 oz at $1.69
23 oz at $2.09
48 oz at $3.19

25. aluminum foil
200 sq ft at $7.39
75 sq ft at $3.69
50 sq ft at $3.19
25 sq ft at $1.49

26. mayonnaise
8 oz at $1.59
16 oz at $2.69
32 oz at $3.89
64 oz at $6.79

27. In the proportion $\dfrac{7}{8} = \dfrac{476}{544}$,

    a. the extremes are _____ and _____ .

    b. the means are _____ and _____ .

28. In the proportion $\dfrac{x}{y} = \dfrac{w}{z}$,

    a. the extremes are _____ and _____ .

    b. the means are _____ and _____ .

    c. the two terms _____ and _____ cannot be 0.

Determine whether each proportion is true or false by comparing the cross products. See Examples 3 through 6.

29. $\dfrac{5}{6} = \dfrac{10}{12}$

30. $\dfrac{2}{7} = \dfrac{5}{17}$

31. $\dfrac{7}{21} = \dfrac{4}{12}$

32. $\dfrac{6}{15} = \dfrac{2}{5}$

33. $\dfrac{5}{8} = \dfrac{12}{17}$

34. $\dfrac{12}{15} = \dfrac{20}{25}$

35. $\dfrac{5}{3} = \dfrac{15}{9}$

36. $\dfrac{6}{8} = \dfrac{15}{20}$

37. $\dfrac{2}{5} = \dfrac{4}{10}$

38. $\dfrac{3}{5} = \dfrac{6}{10}$

39. $\dfrac{125}{1000} = \dfrac{1}{8}$

40. $\dfrac{3}{8} = \dfrac{375}{1000}$

41. $\dfrac{8\frac{1}{2}}{2\frac{1}{3}} = \dfrac{4\frac{1}{4}}{1\frac{1}{6}}$

42. $\dfrac{6\frac{1}{5}}{1\frac{1}{7}} = \dfrac{3\frac{1}{10}}{\frac{8}{14}}$

43. $\dfrac{7}{16} = \dfrac{3\frac{1}{2}}{8}$

44. $\dfrac{10}{17} = \dfrac{5}{8\frac{1}{2}}$

45. $3 : 6 = 5 : 10$

46. $1\frac{1}{4} : 1\frac{1}{2} = \frac{1}{4} : \frac{1}{2}$

47. $6 : 1.56 = 2 : 0.52$

48. $3.75 : 3 = 7.5 : 6$

Solve for the variable in each of the following proportions. See Examples 1 through 6.

49. $\dfrac{3}{6} = \dfrac{6}{x}$

50. $\dfrac{7}{21} = \dfrac{y}{6}$

51. $\dfrac{5}{7} = \dfrac{x}{28}$

52. $\dfrac{4}{10} = \dfrac{5}{x}$

53. $\dfrac{8}{B} = \dfrac{6}{30}$ **54.** $\dfrac{7}{B} = \dfrac{5}{15}$ **55.** $\dfrac{1}{2} = \dfrac{x}{100}$ **56.** $\dfrac{3}{4} = \dfrac{x}{100}$

57. $\dfrac{A}{3} = \dfrac{7}{2}$ **58.** $\dfrac{x}{100} = \dfrac{1}{20}$ **59.** $\dfrac{3}{5} = \dfrac{60}{D}$ **60.** $\dfrac{3}{16} = \dfrac{9}{x}$

61. $\dfrac{\frac{1}{2}}{x} = \dfrac{5}{10}$ **62.** $\dfrac{\frac{2}{3}}{3} = \dfrac{y}{127}$ **63.** $\dfrac{\frac{1}{3}}{x} = \dfrac{5}{9}$ **64.** $\dfrac{\frac{3}{4}}{7} = \dfrac{3}{z}$

65. $\dfrac{\frac{1}{8}}{6} = \dfrac{\frac{1}{2}}{w}$ **66.** $\dfrac{\frac{1}{6}}{5} = \dfrac{5}{w}$ **67.** $\dfrac{\frac{1}{4}}{} = \dfrac{1\frac{1}{2}}{y}$ **68.** $\dfrac{\frac{1}{5}}{} = \dfrac{x}{2\frac{1}{2}}$

69. $\dfrac{5}{x} = \dfrac{2\frac{1}{4}}{27}$ **70.** $\dfrac{x}{3} = \dfrac{16}{3\frac{1}{5}}$ **71.** $\dfrac{3.5}{2.6} = \dfrac{10.5}{B}$ **72.** $\dfrac{4.1}{3.2} = \dfrac{x}{6.4}$

73. $\dfrac{7.8}{1.3} = \dfrac{x}{0.26}$ **74.** $\dfrac{7.2}{y} = \dfrac{4.8}{14.4}$ **75.** $\dfrac{A}{42} = \dfrac{65}{100}$ **76.** $\dfrac{A}{595} = \dfrac{6}{100}$

> **Now that you are familiar with proportions and the techniques for finding the unknown term in a proportion, check your general understanding by choosing the answer (using mental calculations only) that seems the most reasonable to you in each of the following exercises.**

77. Given the proportion $\dfrac{x}{100} = \dfrac{1}{4}$, which of the following values seems the most reasonable for x?

    a. 10   **b.** 25   **c.** 50   **d.** 75

78. Given the proportion $\dfrac{x}{200} = \dfrac{1}{10}$, which of the following values seems the most reasonable for x?

    a. 10   **b.** 20   **c.** 30   **d.** 40

79. Given the proportion $\dfrac{3}{5} = \dfrac{60}{D}$, which of the following values seems the most reasonable for D?

    a. 50   **b.** 80   **c.** 100   **d.** 150

80. Given the proportion $\dfrac{4}{10} = \dfrac{20}{x}$, which of the following values seems the most reasonable for x?

    a. 10   **b.** 30   **c.** 40   **d.** 50

81. Given the proportion $\dfrac{1.5}{3} = \dfrac{x}{6}$, which of the following values seems the most reasonable for x?

    a. 1.5   **b.** 2.5   **c.** 3.0   **d.** 4.5

82. Given the proportion $\dfrac{2\frac{1}{3}}{x} = \dfrac{4\frac{2}{3}}{10}$, which of the following values seems the most reasonable for x?

    a. $4\frac{2}{3}$ **b.** 5   **c.** 20   **d.** 40

Section 2.1: Tests for Divisibility

Tests for Divisibility of Integers by 2, 3, 5, 6, 9, and 10   page 161

For 2: If the last digit (units digit) of an integer is 0, 2, 4, 6, or 8, then the integer is divisible by 2.

For 3: If the sum of the digits of an integer is divisible by 3, then the integer is divisible by 3.

For 5: If the last digit of an integer is 0 or 5, then the integer is divisible by 5.

For 6: If the integer is divisible by both 2 and 3, then it is divisible by 6.

For 9: If the sum of the digits of an integer is divisible by 9, then the integer is divisible by 9.

For 10: If the last digit of an integer is 0, then the integer is divisible by 10.

Even and Odd Integers   page 162

Even integers are divisible by 2. (If an integer is divided by 2 and the remainder is 0, then the integer is even.)

Odd integers are not divisible by 2. (If an integer is divided by 2 and the remainder is 1, then the integer is odd.)

(**Note:** Every integer is either even or odd.)

Section 2.2: Prime Numbers

Prime Number   pages 170 – 171

A **prime number** is a whole number greater than 1 that has exactly two different factors (or divisors) – itself and 1.

Composite Number   pages 170 – 171

A **composite number** is a counting number with more than two different factors (or divisors).

Note: 1 is **neither** a prime nor a composite number. $1 = 1 \cdot 1$, and 1 is the only factor of 1. 1 does not have **exactly two different** factors, and it does not have more than two different factors.

Multiples   pages 171 – 172

To find multiples of a counting number, multiply each of the counting numbers by that number.

Two Important Facts about Prime Numbers   pages 173 – 175

1. **Even Numbers:** 2 is the only even prime number.
2. **Odd Numbers:** All other prime numbers are odd numbers. But, not all odd numbers are prime.

Section 2.2: Prime Numbers (cont.)

To Determine Whether a Number is Prime   page 175

Divide the number by progressively larger **prime numbers** (2, 3, 5, 7, 11, and so forth) until:

1. You find a remainder of 0 (meaning that the prime number is a factor and the given number is composite); or

2. You find a quotient smaller than the prime divisor (meaning that the given number has no smaller prime factors and is therefore prime itself).

Note: Reasoning that if a composite number were a factor, then one of its prime factors would have been found to be a factor in an earlier division, we divide only by prime numbers – that is, there is no need to divide by a composite number.

Section 2.3: Prime Factorization

The Fundamental Theorem of Arithmetic   page 178

Every composite number has exactly one prime factorization.

To Find the Prime Factorization of a Composite Number   pages 179 – 181

1. Factor the composite number into any two factors.

2. Factor each factor that is not prime.

3. Continue this process until all factors are prime.

The **prime factorization** is the product of all the prime factors.

Factors of Composite Numbers   pages 181 – 182

The only factors (or divisors) of a composite number are:

1. 1 and the number itself,

2. each prime factor, and

3. products formed by all combinations of the prime factors (including repeated factors).

Section 2.4: Least Common Multiple (LCM)

To Find the LCM of a Set of Counting Numbers or Algebraic Terms   page 185 – 187

1. Find the prime factorization of each term in the set and write it in exponential form, including variables.

2. Find the largest power of each prime factor present in all of the prime factorizations.

3. The LCM is the product of these powers.

Technique for Finding the GCD of a Set of Counting Numbers   pages 189 – 190

1. Find the prime factorization of each number.

2. Find the prime factors common to all factorizations.

3. Form the product of these primes, using each prime the number of times it is common to all factorizations.

4. This product is the GCD. If there are no primes common to all factorizations, the GCD is 1.

Greatest Common Divisor   pages 191 – 193

The **greatest common divisor** (GCD) of a set of natural numbers is the largest natural number that will divide into all the numbers in the set.

Relatively Prime   pages 193 –194

If the GCD of two numbers is 1 (that is, they have no common prime factors), then the two numbers are said to be **relatively prime**.

Section 2.5: Introduction to Fractions

Rational Number   pages 198 – 199

A **rational number** is a number that can be written in the form $\dfrac{a}{b}$ where a and b are integers and $b \neq 0$.

Proper and Improper Fractions   page 199

A **proper fraction** is a fraction in which the numerator is less than the denominator.

An **improper fraction** is a fraction in which the numerator is greater than the denominator.

Rule for the Placement of Negative Signs   page 200

If a and b are integers and $b \neq 0$, then $-\dfrac{a}{b} = \dfrac{-a}{b} = \dfrac{a}{-b}$.

To Multiply Fractions   pages 200 – 201

1. Multiply the numerators. $\dfrac{a}{b} \cdot \dfrac{c}{d} = \dfrac{a \cdot c}{b \cdot d}$
2. Multiply the denominators.

Commutative Property of Multiplication   pages 201 – 202

If $\dfrac{a}{b}$ and $\dfrac{c}{d}$ are rational numbers, then $\dfrac{a}{b} \cdot \dfrac{c}{d} = \dfrac{c}{d} \cdot \dfrac{a}{b}$.

Associative Property of Multiplication   page 202

If $\dfrac{a}{b}, \dfrac{c}{d}$ and $\dfrac{e}{f}$ are rational numbers, then

$$\dfrac{a}{b} \cdot \dfrac{c}{d} \cdot \dfrac{e}{f} = \dfrac{a}{b} \cdot \left(\dfrac{c}{d} \cdot \dfrac{e}{f} \right) = \left(\dfrac{a}{b} \cdot \dfrac{c}{d} \right) \cdot \dfrac{e}{f}.$$

Multiplicative Identity   pages 202 – 203

1. For any rational number $\dfrac{a}{b}, \dfrac{a}{b} \cdot 1 = \dfrac{a}{b}$.

2. $\dfrac{a}{b} = \dfrac{a}{b} \cdot 1 = \dfrac{a}{b} \cdot \dfrac{k}{k} = \dfrac{a \cdot k}{b \cdot k}$ where $k \neq 0, \left(1 = \dfrac{k}{k} \right)$.

To Reduce a Fraction to Lowest Terms   pages 203 – 205

1. Factor the numerator and denominator into prime factors.

2. Use the fact that $\dfrac{k}{k} = 1$ and "divide out" all common factors.

Note: Reduced fractions might be improper fractions.

Section 2.6: Division with Fractions

Reciprocal of an Algebraic Expression pages 213–214

The **reciprocal of** $\dfrac{a}{b}$ is $\dfrac{b}{a}$ $(a \neq 0 \text{ and } b \neq 0)$.

The product of a nonzero number and its reciprocal is always 1.

$$\frac{a}{b} \cdot \frac{b}{a} = 1$$

Note: The number 0 has no reciprocal. That is $\dfrac{0}{1}$ has no reciprocal because $\dfrac{1}{0}$ is undefined.

Division with Fractions pages 214 – 216

To divide by any nonzero number, multiply by its reciprocal. In general,

$$\frac{a}{b} \div \frac{c}{d} = \frac{a}{b} \cdot \frac{d}{c}$$

where $b, c, d \neq 0$.

Section 2.7: Addition and Subtraction with Fractions

To Add Two (or More) Fractions with the Same Denominator page 223
1. Add the numerators.
2. Keep the common denominator. $\dfrac{a}{b} + \dfrac{c}{b} = \dfrac{a+c}{b}$
3. Reduce, if possible.

To Add Fractions with Different Denominators pages 225 – 227
1. Find the least common denominator (LCD).
2. Change each fraction into an equivalent fraction with that denominator.
3. Add the new fractions.
4. Reduce, if possible.

Commutative Property of Addition page 227

If $\dfrac{a}{b}$ and $\dfrac{c}{d}$ are fractions, then $\dfrac{a}{b} + \dfrac{c}{d} = \dfrac{c}{d} + \dfrac{a}{b}$.

Associative Property of Addition page 228

If $\dfrac{a}{b}$, $\dfrac{c}{d}$ and $\dfrac{e}{f}$ are fractions, then $\dfrac{a}{b} + \left(\dfrac{c}{d} + \dfrac{e}{f} \right) = \left(\dfrac{a}{b} + \dfrac{c}{d} \right) + \dfrac{e}{f}$.

To Subtract Fractions with the Same Denominator pages 228 – 229
1. Subtract the numerators.
2. Keep the common denominator. $\dfrac{a}{b} - \dfrac{c}{b} = \dfrac{a-c}{b}$
3. Reduce, if possible.

To Subtract Fractions with Different Denominators pages 229 – 230
1. Find the least common denominator (LCD).
2. Change each fraction into an equivalent fraction with that denominator.
3. Subtract the new fractions.
4. Reduce, if possible.

Section 2.8: Introduction to Mixed Numbers

Mixed Numbers   page 237

A **mixed number** is the sum of a whole number and a proper fraction.

Shortcut for Changing Mixed Numbers to Fraction Form   pages 238 – 239

1. Multiply the whole number by the denominator of the fraction part.
2. Add the numerator of the fraction part to this product.
3. Write this sum over the denominator of the fraction.

Changing an Improper Fraction to a Mixed Number   pages 240 – 241

1. Divide the numerator by the denominator to find the whole number part of the mixed number.
2. Write the remainder over the denominator as the fraction part of the mixed number.

Section 2.9: Multiplication and Division with Mixed Numbers

To Multiply Mixed Numbers   pages 244 – 246

1. Change each number to fraction form.
2. Factor the numerator and denominator of each improper fraction, and then reduce and multiply.
3. Change the answer to a mixed number or leave it in fraction form. (The choice sometimes depends on what use is to be made of the answer.)

Note: To find a fraction **of** a number means to **multiply** the number by the fraction.

Area and Volume   page 247 – 249

Area is the measure of the interior of a closed plane surface. For a rectangle, the area is the product of the length times the width. (In the form of a formula, $A = lw$.) For a triangle, the area is one-half the product of the base times its height. (In the form of a formula, $A = \frac{1}{2}bh$.)

Volume is the measure of space enclosed by a three-dimensional figure and is measured in **cubic units.** The volume of such a solid is the product of its length times its width times its height. (In the form of a formula, $V = lwh$.)

To Divide with Mixed Numbers   pages 249 – 250

1. Change each number to fraction form.
2. Multiply by the reciprocal of the divisor.
3. Reduce, if possible.

Section 2.10: Addition and Subtraction with Mixed Numbers

To Add Mixed Numbers   pages 256 – 258
1. Add the fraction parts.
2. Add the whole numbers.
3. Write the answer as a mixed number with the fraction part less than 1.

To Subtract Mixed Numbers   pages 258 – 259
1. Subtract the fraction parts.
2. Subtract the whole numbers.

If the Fraction Part Being Subtracted is Larger Than the First Fraction   pages 259 – 261
1. "Borrow" the whole number 1 from the first whole number.
2. Add this 1 to the first fraction. (This will always result in an improper fraction that is larger than the fraction being subtracted.)
3. Now subtract.

Section 2.11: Complex Fractions and Order of Operations

To Simplify a Complex Fraction: page 267
1. Simplify the numerator so that it is a single fraction, possibly an improper fraction.
2. Simplify the denominator so that it also is a single fraction, possibly an improper fraction.
3. Divide the numerator by the denominator, and reduce if possible.

Special Note about the Fraction Bar   pages 268
   The fraction bar in a complex fraction serves the same purpose as two sets of parentheses, one surrounding the numerator and the other surrounding the denominator.

Rules for Order of Operations   pages 269 – 270
1. First, simplify within grouping symbols, such as parentheses $(\ )$, brackets $[\ ]$, braces $\{\ \}$, radicals signs $\sqrt{\ }$, absolute value bars $|\ |$, or fraction bars. Start with the innermost grouping.
2. Second, evaluate any numbers or expressions raised to exponents.
3. Third, moving from left to right, perform any multiplications or divisions in the order in which they appear.
4. Fourth, moving from left to right, perform any additions or subtractions in the order in which they appear.

First-Degree Equation in x   page 274

A **first-degree equation in x** (or **linear equation in x**) is any equation that can be written in the form $ax + b = c$, where a, b, and c are constants and $a \neq 0$.

(**Note:** A variable other than x may be used.)

A fundamental fact of algebra is that **every first-degree equation has exactly one solution**. Therefore, if we find any one solution to a first-degree equation, then that is the only solution.

Principles Used in Solving a First-Degree Equation   pages 275 – 276

In the two basic principles stated here, A and B represent algebraic expressions or constants. C represents a constant, and C is not 0 in the multiplication principle.

1. The **addition principle:** The equations $A = B$ and $A + C = B + C$ have the same solutions.
2. The **multiplication principle:** The equations $A = B$ and

$$\frac{A}{C} = \frac{B}{C} \text{ or } \frac{1}{C} \cdot A = \frac{1}{C} \cdot B \text{ (where } C \neq 0 \text{) have the same solutions.}$$

Essentially, these two principles say that if we perform the same operation to both sides of an equation, the resulting equation will have the same solution as the original equation.

Understand How to Solve Equations   pages 277 – 278

1. Apply the distributive property to remove parentheses whenever necessary.
2. Combine like terms on each side of the equation.
3. If a constant is added to a variable, use the addition principle to add its opposite to both sides of the equation.
4. If a variable has a constant coefficient other than 1, use the multiplication (or division) principle to divide both sides by that coefficient (that is, in effect, multiply both sides by the reciprocal of that coefficient).
5. Remember that the objective is to isolate the variable on one side of the equation with a coefficient of 1.

Ratio   pages 285 – 287

A **ratio** is a comparison of two quantities by division. The ratio of a to b can be written as

$$\frac{a}{b} \qquad \text{or} \qquad a : b \qquad \text{or} \qquad a \text{ to } b.$$

Price per Unit

To calculate the **price per unit** (or unit price):

1. Set up a ratio (usually in fraction form) of price to units.
2. Divide the price by the number of units.

Proportions   page 291

A **proportion** is a statement that two ratios are equal. In symbols, $\frac{a}{b} = \frac{c}{d}$
is a proportion.

A proportion has four terms:

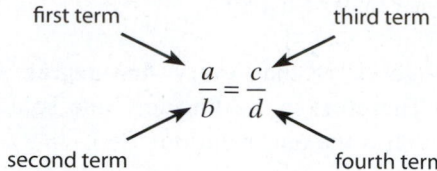

The first and fourth terms (a and d) are called the **extremes.**

The second and third terms (b and c) are called the **means.**

The product of the extremes and the product of the means together are
called the **cross products**.

Identifying True Proportions   pages 292 – 294

In a true proportion, the cross products are equal.

To Find the Unknown Term in a Proportion: pages 294 – 298

1. In the proportion, the unknown term is represented with a variable (some
   letter such as x, y, w, A, B, etc.).
2. Write an equation that sets the cross products equal.
3. Divide both sides of the equation by the number multiplying the variable.
   (This number is called the **coefficient** of the variable.)

The resulting equation will have a coefficient of 1 for the variable and will
give the missing value for the unknown term in the proportion.

Chapter 2: Test

Using the tests for divisibility, tell which of the numbers 2, 3, 5, 6, 9, and 10 (if any) will divide the following numbers.

1. 612

2. 190

3. 1169

Answer the following questions.

4. List the multiples of 11 that are less than 100.

5. Find all the even prime numbers that are less than 10,000.

6. Identify all the prime numbers in the following list: 1, 2, 4, 6, 9, 11, 13, 15, 37, 40, 42

7. **a.** True or False: 25 divides the product $5 \cdot 4 \cdot 3 \cdot 2 \cdot 1$. Explain your answer in terms of factors.

b. True or False: 27 divides the product $(2 \cdot 5 \cdot 6 \cdot 7 \cdot 9)$. Explain your answer in terms of factors.

Find the prime factorization of each number.

8. 80

9. 180

10. 225

11. **a.** Find the prime factorization of 90.

b. Find all the factors of 90.

Find the GCD of each of the following sets of numbers.

12. $\{48, 56\}$

13. $\{30, 75\}$

14. $\{12, 42, 90\}$

Find the LCM of each of the following sets of numbers or algebraic terms.

15. $15, 24, 35$

16. $30xy, 40x^3, 45y^2$

17. $8a^4, 14a^2, 21a$

Answer the following questions.

18. Two long-distance runners practice on the same track and start at the same point. One takes 54 seconds and the other takes 63 seconds to go around the track once.

a. If each continues at the same pace, in how many seconds will they meet again at the starting point?

b. How many laps will each have run at that time?

c. Briefly discuss the idea of whether or not they will meet at some other point on the track as they are running.

19. A rational number can be written as a fraction in which the numerator and denominator are _____ and the denominator is not _____.

20. Give two examples using rational numbers that illustrate the commutative property of multiplication.

21. The value of $\dfrac{0}{-16}$ is _____, and value of $\dfrac{-16}{0}$ is _____.

Reduce to lowest terms.

22. $\dfrac{108}{90}$

23. $\dfrac{48x}{-216x^2}$

24. $\dfrac{75a^3b}{60ab^2}$

Perform the indicated operations. Reduce all fractions to lowest terms.

25. $\dfrac{21}{16} \cdot \dfrac{13}{15} \cdot \dfrac{5}{7}$

26. $\dfrac{-25xy^2}{35xy} \cdot \dfrac{21x^2}{-30y^2}$

27. $45x \div \dfrac{1}{9x}$

28. $\dfrac{-36a^2}{50} \div \dfrac{63ab}{30}$

29. $\dfrac{11}{20a} + \dfrac{5}{20a}$

30. $\dfrac{2}{9} - \dfrac{7}{12}$

31. $-\dfrac{5}{6} + \dfrac{1}{2} + \dfrac{3}{4}$

32. $\dfrac{1}{6} - \left(-\dfrac{7}{8}\right)$

Answer the following questions.

33. A computer's hard disk contains 50 gigabytes of information. This amount is $\dfrac{2}{3}$ of the capacity of the disk. What is the capacity of the disk?

34. Find the reciprocal of $2\dfrac{1}{3}$.

Change each improper fraction to a mixed number with the fraction part reduced to lowest terms.

35. $\dfrac{54}{8}$

36. $\dfrac{342}{100}$

Change each mixed number to the form of an improper fraction.

37. $-5\dfrac{1}{100}$

38. $2\dfrac{12}{13}$

Answer the following questions.

39. Find $\dfrac{3}{10}$ of $7\dfrac{2}{9}$.

40. Find the mean of $3\dfrac{1}{3}$, $4\dfrac{1}{4}$ and $5\dfrac{1}{5}$.

Simplify each of the following expressions.

41.
$$6\dfrac{5}{14}$$
$$+\ \dfrac{19}{21}$$

42.
$$23\dfrac{1}{10}$$
$$-\ 8\dfrac{7}{15}$$

43.
$$-3\dfrac{5}{6}$$
$$-4\dfrac{1}{2}$$
$$-7\dfrac{2}{3}$$

44. $2\dfrac{2}{7} \cdot 5\dfrac{3}{5} + \dfrac{3}{5} \div \left(-\dfrac{1}{2}\right)$

45. $\dfrac{x}{3} \cdot \dfrac{1}{3} - \dfrac{2}{9} \div 2$

46. $\dfrac{1\dfrac{1}{9} + \dfrac{5}{18}}{-1\dfrac{1}{2} - 2\dfrac{2}{3}}$

Solve the following equations.

47. $3(x-1)+2x=12$

48. $-\dfrac{7}{4}n+\dfrac{2}{5}=\dfrac{1}{6}$

49. $\dfrac{y}{5}+\dfrac{2y}{9}-\dfrac{14}{15}=0$

50. $-4=\dfrac{2}{3}x+\dfrac{1}{4}x+\dfrac{1}{8}$

Solve the following word problems.

51. An artist is going to make a rectangular-shaped pencil drawing for a customer. The drawing is to be $3\dfrac{1}{4}$ inches wide and $4\dfrac{1}{2}$ inches long, and there is to be a mat around the drawing that is $1\dfrac{1}{2}$ inches wide.

   a. Find the perimeter of the matted drawing.

   b. Find the area of the matted drawing. (Sometimes an artist will charge for the size of a work as much as for the actual work itself.)

52. The result of multiplying two numbers is $14\dfrac{5}{8}$. One of the numbers is $\dfrac{3}{4}$.

   a. Do you expect the other number to be smaller or larger than $14\dfrac{5}{8}$?

   b. What is the other number?

53. Find the volume of a tissue box that measures $10\dfrac{9}{10}$ centimeters by $11\dfrac{1}{10}$ centimeters by $13\dfrac{5}{10}$ centimeters.

54. The ratio 7 to 6 can be expressed as the fraction _____.

55. In the proportion $\dfrac{3}{4}=\dfrac{75}{100}$, 3 and 100 are called the _____.

56. Is the proportion $\dfrac{4}{6}=\dfrac{9}{14}$ true or false? Give a reason for your answer.

Write the following comparisons as ratios reduced to lowest terms. Use common units whenever possible.

57. 3 weeks to 35 days

58. 6 nickels to 3 quarters

59. 220 miles to 4 hours

60. Find the unit prices (to the nearest cent) and tell which is the better buy for a pair of socks: 5 pairs of socks for $12.50 or 3 pairs of socks for $8.25.

Solve the following proportions.

61. $\dfrac{9}{17} = \dfrac{x}{51}$

62. $\dfrac{3}{5} = \dfrac{10}{y}$

63. $\dfrac{50}{x} = \dfrac{0.5}{0.75}$

64. $\dfrac{2.25}{y} = \dfrac{1.5}{13}$

65. $\dfrac{\frac{3}{8}}{x} = \dfrac{9}{20}$

66. $\dfrac{\frac{1}{3}}{x} = \dfrac{5}{\frac{1}{6}}$

Cumulative Review: Chapters 1 - 2

Answer the following questions.

1. Given the expression 5^3 :
   a. name the base,
   b. name the exponent, and
   c. find the value of the expression.

2. Match each expression with the best estimate of its value.
   _____ a. $175 + 92 + 96 + 125$ A. 200
   _____ b. $8465 \div 41$ B. 500
   _____ c. $32 \cdot 48$ C. 1000
   _____ d. $5484 - 4380$ D. 1500

3. Multiply mentally: $70(\ 9000 \ ) = $ _____.

4. Round 176,200 to the nearest ten thousand.

5. Evaluate by using the rules for order of operations:
$$5^2 + (12 \cdot 5 \div 2 \cdot 3) - 60 \cdot 4$$

6. a. List all the prime numbers less than 25.

   b. List all the squares of these prime numbers.

7. Find the LCM of 45, 75, and 105.

8. Find the GCD for $\{120, 150\}$

9. The value of $0 \div 19$ is _____, whereas the value of $19 \div 0$ is _____.

10. Find the mean of 45, 36, 54, and 41.

Perform the indicated operations. Reduce all fractions to lowest terms.

11. $\begin{array}{r} 8376 \\ 3749 \\ + \ 2150 \\ \hline \end{array}$

12. $\begin{array}{r} 1563 \\ - \ 975 \\ \hline \end{array}$

13. $\begin{array}{r} 751 \\ \times \ 38 \\ \hline \end{array}$

14. $14 \overline{)2856}$

15. $-\dfrac{3}{20} \div 6$

16. $\dfrac{5}{18}\left(\dfrac{3}{10}\right)\left(\dfrac{4}{21}\right)$

17. $-\dfrac{5}{6} - \left(-\dfrac{7}{8}\right)$

18. $\dfrac{7}{20} - \dfrac{3}{5}$

19. $\begin{array}{r} 38\dfrac{4}{5} \\ + \ 29\dfrac{7}{12} \\ \hline \end{array}$

20. $\begin{array}{r} -13\dfrac{3}{20} \\ + \ 6\dfrac{5}{8} \\ \hline \end{array}$

21. $3\dfrac{2}{3} \div \left(-2\dfrac{5}{11}\right)$

22. $-4\dfrac{3}{8} \div \left(-2\dfrac{1}{2}\right)$

Simplify each of the following expressions.

23. $\dfrac{1\dfrac{3}{10} + 2\dfrac{4}{5}}{2\dfrac{3}{5} - 1\dfrac{1}{3}}$

24. $1\dfrac{1}{5} - 11\dfrac{7}{10} + 6\dfrac{2}{3} \div 2\dfrac{1}{2}$

25. $\dfrac{x}{5} \cdot \dfrac{1}{3} - \dfrac{2}{5} \div 3$

Solve each of the following equations.

26. $x + 17 = -10$

27. $5y - 14 = 21$

28. $\dfrac{2}{3}x + \dfrac{5}{6} = \dfrac{1}{2}$

29. $\dfrac{n}{12} - \dfrac{1}{3} = \dfrac{3}{4}$

30. $2(x - 14) = -12$

31. $-3(y + 16) = -17$

32. $20x + 3 = -15$

33. $6n - 2(n + 5) = 46$

34. $\dfrac{2}{5}x + \dfrac{1}{10} = \dfrac{1}{2}$

35. $\dfrac{y}{4} + \dfrac{2y}{8} = -\dfrac{1}{16}$

36. $\dfrac{1}{2}x + \dfrac{1}{6}x = \dfrac{2}{3}$

37. $\dfrac{5}{7}x + \dfrac{1}{14}x = -\dfrac{3}{28}$

Solve the following word problems.

38. **Shopping:** The discount price of a new television set is $652. This price is $\dfrac{4}{5}$ of the original price

  a. Is the original price more or less than $652?

  b. What was the original price?

39. **Driving:** You know that the gas tank on your pickup truck holds 24 gallons, and the gas gauge reads that the tank is $\dfrac{3}{8}$ full.

  a. How many gallons will be needed to fill the tank?

  b. Do you have enough cash to fill the tank if you have $50 with you and gas costs $3\dfrac{3}{10}$ per gallon?

40. Find the perimeter of a triangle with sides of length $6\dfrac{5}{8}$ meters, $3\dfrac{3}{4}$ meters, and $5\dfrac{7}{10}$ meters.

41. For a rectangular-shaped swimming pool that is $35\dfrac{1}{2}$ feet long and $22\dfrac{3}{4}$ feet wide, find:

  a. the perimeter and

  b. the area of the pool.

Solve the following proportions.

42. $\dfrac{5}{7} = \dfrac{x}{3\dfrac{1}{2}}$

43. $\dfrac{0.5}{1.6} = \dfrac{0.1}{A}$

Mathematics @ Work!

Many people are familiar with decimal numbers because of sports. Batting averages in baseball and save percentages in hockey are calculated to the nearest thousandth. Quarterback ratings in football and average points per game in basketball are calculated to the nearest tenth. Decimal numbers are also used to record many of the world's sport speed records. For example, the men's swimming record for the 100 meter freestyle is 46.91 seconds by Cesar Cielo of Brazil and the women's swimming record is 52.07 seconds by Germany's Britta Steffen. Decimal numbers are necessary to solve problems like the one below.

In 1996, Lance Armstrong was diagnosed with testicular cancer that had spread to his brain and lungs. Against all odds, in 1999 he won his first Tour de France bicycling race by 7 minutes 37 seconds. In 2005 he won the Tour for a record seventh consecutive time. The race covered 2242 miles in all kinds of weather conditions and over mountain passes, switchback roads and winding paths. His time was 86 hours 15 minutes 2 seconds. This time was 4 minutes 40 seconds better than the second place finisher, Ivan Basso of Italy. How would you find the average speed of these two cyclists?

See Exercises 68 and 69 in Section 3.3.

Decimal Numbers, Percents, and Square Roots

3.1 Reading, Writing, and Rounding Decimal Numbers

Objectives

A Learn to read and write decimal numbers.

B Learn how to round decimal numbers to indicated places of accuracy.

Objective A Reading and Writing Decimal Numbers

Technically, what we write to represent numbers are symbols or notations called **numerals**. **Numbers** are abstract ideas represented by numerals. There are many notations for numbers. For example, the Romans used V to represent five and the Alexandrian Greek system used capital epsilon, E, to represent five. In the Hindu-Arabic system that we use today, the symbol 5 represents five. We will not emphasize the distinction between numbers and numerals (or symbols), but you should remember the fact that numbers are abstract ideas, and we use symbols to represent these ideas so that we can communicate in a meaningful manner.

The common **decimal notation** uses a **place value system** and a **decimal point**, with whole numbers written to the left of the decimal point and fractions written to the right of the decimal point. We will say that numbers represented by decimal notation are **decimal numbers**. The values of several places in this decimal system are shown in Figure 1 below.

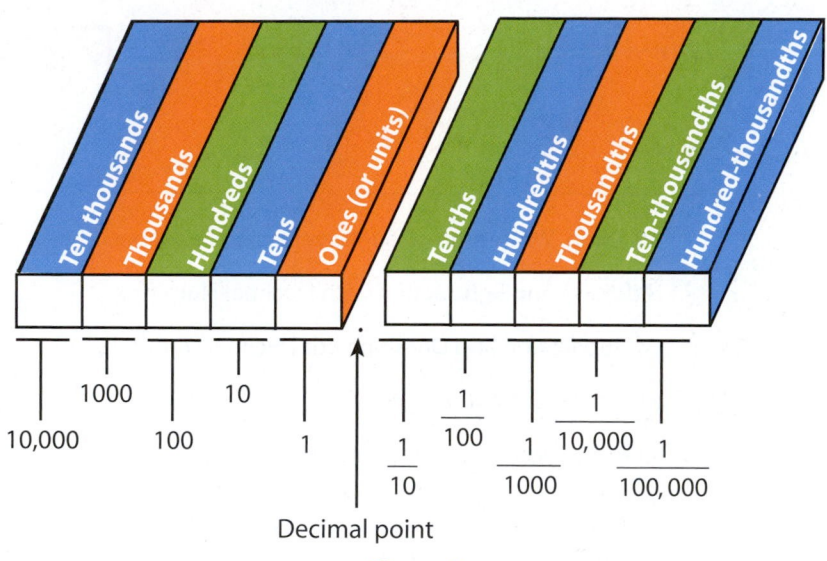

Figure 1

There are three classifications of decimal numbers.

1. **finite (or terminating) decimal numbers,**
2. **infinite repeating decimal numbers, and**
3. **infinite nonrepeating decimal numbers.**

In arithmetic, we are used to dealing with finite (or terminating) decimal numbers, and these are the numbers we will discuss and operate with in Sections 3.1 through 3.3. In Section 3.5, we will show how fractions are related to infinite repeating decimals.

In reading a fraction such as $\dfrac{147}{1000}$, we read the numerator as a whole number ("one hundred forty- seven") and then attach the name of the denominator ("thousand**ths**"). Note that **ths** (or **th**) is used to indicate the fraction. This same procedure is followed with numbers written in decimal notation.

$$\frac{147}{1000} = 0.147 \qquad \text{Read "one hundred forty-seven thousandths."}$$

$$2\frac{36}{100} = 2.36 \qquad \text{Read "two and thirty-six hundredths."}$$

If there is no whole number part, as in the first example just shown, then 0 is commonly written to the left of the decimal point. Writing the 0 is not always necessary. However, writing the 0 does sometimes avoid confusion with periods at the end of sentences when decimal numbers are written in sentences. In general, decimal numbers are read and written according to the following convention.

To Read or Write a Decimal Number

1. Read or write the whole number.
2. Read or write **and** in place of the decimal point.
3. Read or write the fraction part as a whole number with the name of the place of the last digit on the right.

Example 1

Reading and Writing Decimal Numbers

Write $72\dfrac{8}{10}$ in decimal notation and in words.

Solution

72.8   in decimal notation

seventy-two and eight tenths   in words

And indicates the decimal point; the digit 8 is in the tenths position.

Write the following mixed number in decimal notation and in words.

1. $52\dfrac{42}{100}$

2. $12\dfrac{3}{1000}$

Example 2

Reading and Writing Decimal Numbers

Write $9\dfrac{63}{1000}$ in decimal notation and in words.

Solution

One 0 must be inserted as a placeholder.

9.063   in decimal notation

nine and sixty-three thousandths   in words

And indicates the decimal point; the digit 3 is in the thousandths position.

Now work margin exercises 1 and 2.

notes

1. The **ths** (or **th**) at the end of a word indicates a fraction part (a part to the right of the decimal point).

   seven hundred = 700
   seven hundred**ths** = 0.07

2. The hyphen (-) indicates one word.

   three hundred thousand = 300,000
   three hundred-thousand**ths** = 0.00003

Example 3

Reading and Writing Decimal Numbers

Write fifteen hundredths in decimal notation.

Solution

0.15

Note that the digit 5 is in the hundredths position.

Example 4

Reading and Writing Decimal Numbers

Write four hundred and two thousandths in decimal notation.

Solution

Two 0's are inserted as placeholders.

400.002   The digit 2 is in the thousandths position.

Example 5

Reading and Writing Decimal Numbers

Write four hundred two thousandths in decimal notation.

Solution

0.402

Note carefully how the use of **and** in the phrase in Example 4 gives it a completely different meaning from the phrase in this example.

Now work margin exercises 3 through 5.

Write the following words in decimal numbers.

3. ten and four thousandths

4. six hundred and five tenths

5. seven hundred and seven thousandths

Objective B Rounding Decimal Numbers

Measuring devices, such as rulers, meter sticks, speedometers, micrometers, and surveying transits, give only approximate measurements. Whether the units are large (such as miles and kilometers) or small (such as inches and centimeters), there are always smaller, more accurate units (such as eighths of an inch and millimeters) that could be used to indicate a measurement. We are constantly dealing with approximate (or rounded) numbers in our daily lives. If a recipe calls for 1.5 cups of flour and the cook puts in 1.53 cups (or 1.47 cups), the result will still be reasonably tasty. In fact, the measures of all ingredients will have been approximations.

There are several rules for rounding decimal numbers. The IRS, for example, allows rounding to the nearest dollar on income tax forms. A technique sometimes used in cases when many numbers are involved is to round to the nearest even digit at some particular place value. The rule chosen in a particular situation depends on the use of the numbers and whether there might be some sort of penalty for an error. In this text, we will use the following rules for rounding.

1. Look at the single digit just to the right of the place of desired accuracy.
2. If this digit is 5 or greater, make the digit in the desired place of accuracy one larger and replace all digits to the right with zeros. All digits to the left remain unchanged unless a 9 is made one larger, and then the next digit to the left is increased by 1.
3. If this digit is less than 5, leave the digit in the desired place of accuracy as it is and replace all digits to the right with zeros. All digits to the left remain unchanged.
4. Trailing 0's **to the right of the place of accuracy** must be dropped so that the place of accuracy is clearly understood. If a rounded number has a 0 in the desired place of accuracy, then that 0 remains.

Example 6

Rounding Decimal Numbers

Round 18.649 to the nearest tenth.

Solution

The next digit to the right is 4.

18.649

6 is in the tenths position.

Since 4 is less than 5, leave the 6 and replace 4 and 9 with 0's.

18.649 rounds to **18.6** to the nearest tenth. Note that the trailing 0's in 18.600 are dropped to indicate the position of accuracy.

Example 7

Rounding Decimal Numbers

Round 5.83971 to the nearest thousandth.

Solution

The next digit to the right is 7.

5.83971

9 is in the thousandths position.

Since 7 is greater than 5, make 9 one larger and replace 7 and 1 with 0's. (Making 9 one larger gives 10, which affects the digit 3, too.)

5.83971 rounds to **5.840** to the nearest thousandth, and two trailing 0's are dropped.

Completion Example 8

Rounding Decimal Numbers

Round 2.00643 to the nearest ten-thousandth.

Solution

The digit in the ten-thousandths position is _____.

The next digit to the right is _____.

Since ____ is less than 5, leave ____ as it is and replace ____ with a 0.

2.00643 rounds to _____ to the nearest _____.

Completion Example 9

Rounding Decimal Numbers

Round 9653 to the nearest hundred.

Solution

The decimal point is understood to be to the right of _____.

The digit in the hundreds position is _____.

The next digit to the right is _____.

Since _____ is equal to 5, change the _____ to _____ and replace _____ and _____ with 0's.

So, 9653 rounds to _____ to the nearest hundred.

Now work margin exercises 6 through 9.

6. Round 8.637 to the nearest tenth.

7. Round 5.042 to the nearest hundredth.

8. Round 0.01792 to the nearest thousandth.

9. Round 239.53 to the nearest ten.

Completion Example Answers

8. 4; 3; Since 3 is less than 5, leave 4 as it is and replace 3 with a 0.; 2.00643 rounds to 2.0064 to the nearest ten-thousandth.

9. 3; 6; 5; Since 5 is equal to 5, change the 6 to a 7 and replace 5 and 3 with 0's.; So 9653 rounds to 9700 to the nearest hundred.

notes

- The 0's must **not** be dropped in a whole number. Every 0 to the right of the desired place of accuracy to the right of the decimal point **must** be dropped.

Exercises 3.1

Write the following mixed numbers in decimal notation. See Examples 1 and 2.

1. $6\dfrac{5}{10}$

2. $82\dfrac{3}{100}$

3. $19\dfrac{75}{1000}$

4. $100\dfrac{25}{100}$

5. $62\dfrac{547}{1000}$

Write the following decimal number in mixed number form. Do not reduce the fractional part.

6. 2.57

7. 13.02

8. 38.004

9. 200.6

10. 50.001

Write the following words in decimal notation. See Examples 3 through 5.

11. four tenths

12. fifteen thousandths

13. twenty-three hundredths

14. five and twenty-eight hundredths

15. five and twenty-eight thousandths

16. seventy-three and three hundred forty-one thousandths

17. six hundred and sixty-six hundredths

18. six hundred and sixty-six thousandths

19. three thousand four hundred ninety-five and three hundred forty-two thousandths

20. seven thousand five hundred and eighty-three ten-thousandths

Write the following decimal numbers in words.

21. 0.9

22. 0.53

23. 6.05

24. 6.004

25. 50.007

26. 19.102

27. 800.009

28. 0.809

29. 5000.005

30. 25.4538

Fill in the blanks to correctly complete each statement. See Examples 8 and 9.

31. Round 34.78 to the nearest tenth.

  a. The digit in the tenths position is _____.

  b. The next digit to the right is _____.

  c. Since _____ is greater than 5, change _____ to _____ and replace _____ with 0.

  d. So 34.78 rounds to _____ to the nearest tenth.

32. Round 3.00652 to the nearest ten-thousandth.

  a. The digit in the ten-thousandths position is _____.

  b. The next digit to the right is _____.

  c. Since _____ is less than 5, leave _____ as it is and replace _____ with 0.

  d. So 3.00652 rounds to _____ to the nearest _____ .

Round each of the following decimal numbers to the nearest tenth.

33. 89.016   **34.** 8.555   **35.** 18.123

36. 0.076   **37.** 14.332   **38.** 46.444

Round each of the following decimal numbers to the nearest hundredth.

39. 0.385   **40.** 0.296   **41.** 7.997

42. 13.1345   **43.** 0.0764   **44.** 6.0035

Round each of the following decimal numbers to the nearest thousandth.

45. 0.0572   **46.** 0.6338   **47.** 0.00191

48. 20.76963   **49.** 32.4578   **50.** 1.66666

Round each of the following decimal numbers to the nearest whole number (or nearest unit).

51. 479.32   **52.** 7.8   **53.** 163.5

54. 701.41   **55.** 300.3   **56.** 29.999

Round each of the following decimal numbers to the nearest hundred.

57. 5163   **58.** 6475   **59.** 76,523.2

60. 435.7   **61.** 453.7   **62.** 1572.36

Round each of the following decimal numbers to the nearest thousand.

63. 62,375   **64.** 75,445   **65.** 103,499

66. 4,500,766   **67.** 7,305,438   **68.** 573,333.15

69. 0.0007582 (nearest hundred-thousandth)   **70.** 78,419 (nearest ten thousand)

In the following exercises, write the decimal numbers that are not whole numbers in words.

71. **Units of Length:** One yard is equal to 36 inches. One yard is also approximately equal to 0.914 meter. One meter is approximately equal to 1.09 yards. One meter is also approximately equal to 39.37 inches. (Thus a meter is longer than a yard by about 3.37 inches.)

72. **Units of Length:** One foot is equal to 12 inches. One foot is also equal to 30.48 centimeters. One square foot is approximately 0.093 square meter.

73. Water Weight: One quart of water weighs approximately 2.0825 pounds.

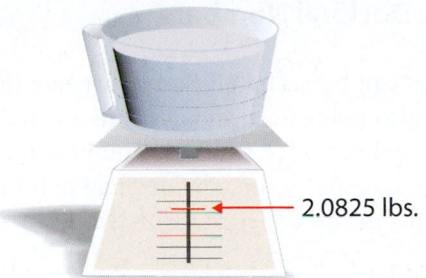

2.0825 lbs.

74. States: The largest state in the United States is Alaska, which covers approximately 656.4 thousand square miles. The second largest state is Texas, with approximately 268.6 thousand square miles. Alaska is more than 10 times the size of Wisconsin (twenty-third in size), with about 65.5 thousand square miles.

656.4 thousand sq mi

268.6 thousand sq mi

65.5 thousand sq mi

75. Pi: The number π is approximately equal to 3.14159.

76. Euler's Number: The number *e* (used in higher-level mathematics) is approximately equal to 2.71828.

77. Aging: An interesting fact about aging is that the longer you live, the longer you can expect to live. A white male of age 40 can expect to live 35.8 more years; of age 50, can expect to live 26.9 more years; of age 60 can expect to live 18.9 more years; of age 70 can expect to live 12.3 more years; and of age 80 can expect to live 7.2 more years. (This same phenomenon is true of men and women of all races.)

78. The Sun: The mean distance from the Sun to Earth is about 92.9 million miles and from the Sun to Venus is about 67.24 million miles. One period of revolution of the Earth about the Sun takes 365.2 days, and one period of revolution of Venus about the Sun takes 224.7 days.

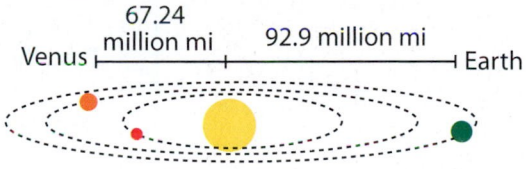

67.24 million mi   92.9 million mi

Venus   Earth

79. Unicycle: The tallest unicycle ever ridden was 114.8 feet tall, and was ridden by Sem Abrahams (with a safety wire suspended from an overhead crane) for a distance of 28 feet in Pontiac, Michigan on January 29, 2004.
Source: http://semcycle.biz/record/

80. World Records: 9.58 seconds for 100 meters (by Usain Bolt, Jamaica, 2009); 19.19 seconds for 200 meters (by Usain Bolt, Jamaica, 2009); 43.18 seconds for 400 meters (by Michael Johnson, USA, 1999). **Source:** Wikipedia.com

3.2 Addition and Subtraction with Decimal Numbers

Objectives

A Know how to add with decimal numbers.

B Know how to subtract with decimal numbers.

C Learn how to add and subtract with positive and negative decimal numbers.

D Be able to estimate sums and differences by using rounded decimal numbers.

Objective A   **Addition with Decimal Numbers**

Addition with decimal numbers can be accomplished by writing the decimal numbers one under the other and keeping the decimal points aligned vertically. In this way, the whole numbers will be added to whole numbers, tenths added to tenths, hundredths to hundredths, and so on. The decimal point in the sum is in line with the decimal points in the addends. Thus, we have the format shown here.

$$\begin{array}{r} 2.357 \\ +\ 6.14 \\ \hline 8.497 \end{array}$$

In the number 6.14, 0's may be written to the right of the last digit in the fraction part to help keep the digits in the correct line. This will not change the value of any number or the sum.

> ### To Add Decimal Numbers
> 1. Write the addends in a vertical column.
> 2. Keep the decimal points aligned vertically.
> 3. Keep digits with the same position value aligned. (Zeros may be filled in as aids.)
> 4. Add the numbers, just as with whole numbers, keeping the decimal point in the sum aligned with the other decimal points.

1. Find the sum:

$$17 + 8.61 + 5.004 + 29.19.$$

Example 1

Adding Decimal Numbers

Find the sum: 17 + 4.88 + 50.033 + 0.6.

Solution

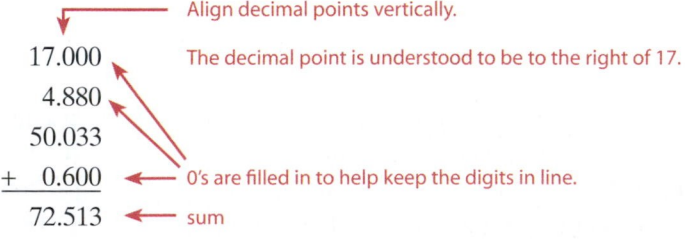

Align decimal points vertically.

The decimal point is understood to be to the right of 17.

0's are filled in to help keep the digits in line.

sum

Now work margin exercise **1.**

Example 2

Combining Like Terms

Simplify by combining like terms: $8.3x + 9.42x + 25.07x$.

Solution

The method for combining like terms is to use the distributive property with decimal number coefficients just as with integer coefficients.

$$8.3x + 9.42x + 25.07x = (8.3 + 9.42 + 25.07)x$$
$$= 42.79x$$

$$\begin{array}{r} 8.30 \\ 9.42 \\ +\ 25.07 \\ \hline 42.79 \end{array}$$

Now work margin exercise 2.

2. Combine like terms.

$$45.2x + 2.08x + 3.5x$$

Objective B ## Subtraction with Decimal Numbers

To Subtract Decimal Numbers

1. Write the numbers in a vertical column.
2. Keep the decimal points aligned vertically.
3. Keep digits with the same position value aligned. (Zeros may be filled in as aids.)
4. Subtract, just as with whole numbers, keeping the decimal point in the difference aligned with the other decimal points.

Example 3

Subtracting Decimal Numbers

Find the difference: $21.715 - 14.823$.

Solution

$$\begin{array}{r} 21.715 \\ -\ 14.823 \\ \hline 6.892 \end{array}$$ difference

Example 4

Subtracting Decimal Numbers

At the bookstore, Mrs. Gonzalez bought a text for $55, art supplies for $32.50, and computer supplies for $29.25. If tax was $9.34, how much change did she receive from a gift certificate worth $150?

3. Find the difference:

$500 - 342.1$.

4. Mr. Jones had $1000. He wrote checks for $280.25, $640.32, $54.19, and $16.56. How much does he have left?

Solution

a. Find the total of her expenses including tax.

$$\begin{array}{r} \$55.00 \\ 32.50 \\ 29.25 \\ +\ \ 9.34 \\ \hline \$126.09 \end{array} \text{ total}$$

b. Subtract the answer in part **a.** from $150.

$$\begin{array}{r} \$150.00 \\ -\ \ 126.09 \\ \hline \$23.91 \end{array}$$

She received $23.91 in change.

Now work margin exercises 3 and 4.

Objective C **Positive and Negative Decimal Numbers**

Decimal numbers can be positive and negative, just as integers and other rational numbers can be positive and negative. Some positive and negative decimal numbers are illustrated on the number line in Figure 1.

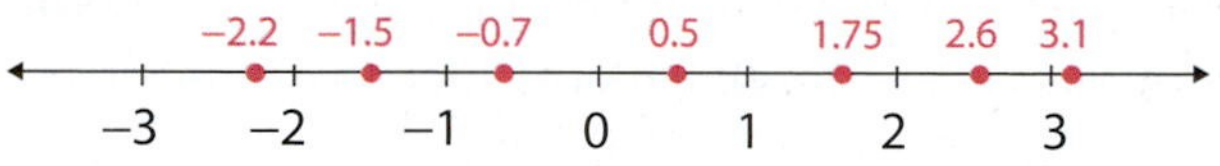

Figure 1

The rules for operating with positive and negative decimal numbers are the same as those for operating with integers (Chapter 1) and fractions and mixed numbers (Chapters 2). Examples 5 and 6 illustrate these ideas.

5. Find the difference:

$-310.5 - (-275.32)$.

Example 5

Subtracting Signed Decimal Numbers

Find the difference: $-200.36 - (-45.87)$.

Solution

$-200.36 - (-45.87) = -200.36 + 45.87$

$$\begin{array}{r} -200.36 \\ +\ \ 45.87 \\ \hline -154.49 \end{array}$$

Now work margin exercise 5.

Example 6

Solving Equations involving Decimal Numbers

Solve the equation: $x - 25.67 = 6.25 - 7.3$.

Solution

$$x - 25.67 = 6.25 - 7.3$$
$$x - 25.67 = -1.05$$
$$x - 25.67 + 25.67 = -1.05 + 25.67$$
$$x = 24.62$$

Now work margin exercise 6.

6. Solve the equation:

$x + 37.6 = 4.8 - 7.25$.

Objective D   ## Estimating Sums and Differences

By rounding each number to the place of the **last nonzero digit on the left** and adding (or subtracting) these rounded numbers, we can estimate (or approximate) the answer before the actual calculations are done. (See Section 1.2 for estimating answers with whole numbers.) This technique of estimating is especially helpful when working with decimal numbers, where the incorrect placement of a decimal point can change an answer dramatically.

Example 7

Estimating the Sum of Decimal Numbers

First estimate the sum $84 + 3.53 + 62.71$; then find the sum.

Solution

a. Estimate by adding rounded numbers.
(Each number is rounded to the place of the leftmost nonzero digit in that number.)

Actual value		Rounded value
84.00	$\longrightarrow$	80
3.53	$\longrightarrow$	4
62.71	$\longrightarrow$	+ 60
		144 estimate

b. Find the actual sum.

$$
\begin{array}{r}
84.00 \\
3.53 \\
+ 62.71 \\
\hline
150.24
\end{array}
$$

Now work margin exercise 7.

7. First, estimate the sum $5.63 + 16.8 + (-35.47)$; then find the sum.

In Example 7, the estimated sum and the actual sum are reasonably close (the difference is about 6). This leads us to have confidence in the answer.

If, for example, 84 had been written as 0.84 (instead of 84.00), then addition would have given a sum as follows.

decimal in the wrong place ⟶

$$
\begin{array}{r}
0.84 \\
3.53 \\
+\ 62.71 \\
\hline
67.08
\end{array}
$$

Here, the difference between the estimate of 144 and the wrong sum of 67.08 is over 70 (a relatively large amount), and an error should be suspected.

8. Estimate the difference 1240.68 − 38.09; then find the difference.

Completion Example 8

Estimating the Difference of Decimal Numbers

Estimate the difference 132.418 − 17.526; then find the difference.

Solution

a. Estimate:

$$
\begin{array}{r}
100 \\
-\ 20 \\
\hline
\end{array}
$$
estimate

b. Actual difference:

$$
\begin{array}{r}
132.418 \\
-\ 17.526 \\
\hline
\end{array}
$$
actual difference

Now work margin exercise 8.

Completion Example 9

Subtracting Decimal Numbers

Samantha bought a pair of shoes for $47.50, a jacket for $25.60 (on sale), and a pair of slacks for $39.75. (Mentally estimate both answers before you actually calculate them.)

Completion Example Answers

8. a. Estimate:

$$
\begin{array}{r}
100 \\
-\ 20 \\
\hline
80
\end{array}
$$
estimate

b. Actual difference:

$$
\begin{array}{r}
132.418 \\
-\ 17.526 \\
\hline
114.892
\end{array}
$$
actual difference

a. How much did she spend? (Tax was included in the prices.)

Solution

$$
\begin{array}{r}
\$47.50 \\
25.60 \\
+\ \ 39.75 \\
\hline
\end{array}
$$ expenses

b. What was her change from $120?

Solution

$$
\begin{array}{r}
\$120.00 \\
-\ \ \rule{2cm}{0.4pt} \\
\hline
\end{array}
$$ expenses

\rule{3cm}{0.4pt} change

Did you estimate her expenses as $120 and her change as $0?

Now work margin exercise **9.**

Example 10

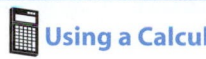

Using a Calculator

Use a calculator to perform the following operations.

a. 4852.621 + 5443.1   **b.** 84.3613 − 46.98

Solution

a. To add the values given, press the keys

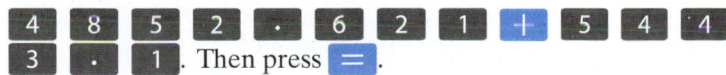

. Then press [=] .

The display will read **10295.721**.

b. To subtract the values given, press the keys

. Then press [=] .

The display will read **37.3813**.

Now work margin exercise **10.**

Completion Example Answers

9. a.
$$
\begin{array}{r}
\$47.50 \\
25.60 \\
+\ 39.75 \\
\hline
\$112.85 \ \text{expenses}
\end{array}
$$

b.
$$
\begin{array}{r}
\$120.00 \\
-112.85 \quad \text{expenses} \\
\hline
\$7.15 \quad \text{change}
\end{array}
$$

9. Fred bought cheese, crackers, and a soft drink for $2.25, $3.49, and $0.35, respectively. How much was his total? What was his change from $10? (Mentally estimate both answers before you actually calculate them.)

10. Use a calculator to perform the following operations.

a. 6743.87 − 2179.391

b. 13.7859 + 174.22

Exercises 3.2

Find each of the indicated sums. Estimate your answers, before performing the actual calculations. Check to see that your sums are close to the estimated values. See Example 7.

1. $0.7 + 0.3 + 2.3$

2. $6 + 5.1 + 0.8$

3. $0.69 + 4.91 + 0.05$

4. $3.577 + 16.563 + 25.01$

5. $4.0085 + 0.054 + 0.7 + 0.03$

6. $43.655 + 9.33 + 12 + 30.1$

7.
$$\begin{array}{r} 47.3 \\ 42.08 \\ +\ 28.005 \\ \hline \end{array}$$

8.
$$\begin{array}{r} 1.007 \\ 20.332 \\ +\ 4.992 \\ \hline \end{array}$$

9.
$$\begin{array}{r} 107.39 \\ 64.904 \\ 335.056 \\ +\ 210.8 \\ \hline \end{array}$$

10.
$$\begin{array}{r} 5.0015 \\ 2.334 \\ 0.3075 \\ +\ 3.8771 \\ \hline \end{array}$$

Find each of the indicated differences. First estimate the difference mentally. See Example 8.

11. $5.3 - 3.75$

12. $17.82 - 8.9$

13. $29.6 - 13.71$

14. $1.0054 - 0.03$

15. $78.015 - 23.069$

16. $45.002 - 43.008$

17.
$$\begin{array}{r} 22.426 \\ -\ 17.538 \\ \hline \end{array}$$

18.
$$\begin{array}{r} 4.8 \\ -\ 0.0023 \\ \hline \end{array}$$

19.
$$\begin{array}{r} 31.007 \\ -\ 0.543 \\ \hline \end{array}$$

20.
$$\begin{array}{r} 40 \\ -\ 6.425 \\ \hline \end{array}$$

Simplify each expression by combining like terms. See Example 2.

21. $8.3x + x - 22.7x$

22. $9.54x - x - 12.82x$

23. $85.7y - 22.3y - 17.9y$

24. $77.5y - 34.1y - 56.3y$

25. $0.3x - x$

26. $-0.07x + x$

27. $13.4t - 22.7t + 15.6t$

28. $167.1s - 200s + 45.3s$

29. $2.1x + 8.2x - y - 3.1y$

30. $-8.4x - 3.7x + 2y - 0.1y$

Solve each of the following equations. See Example 6.

31. $x + 34.8 = 7.5 - 80$

32. $x - 27.9 = 18.3 - 20$

33. $16.5 + y = 30 - 15.2$

34. $20.1 + y = 16.8 + 50$

35. $z - 8.65 = 22.3 - 12.56$

36. $z - 7.64 = 15.23 - 45.6$

37. $w + 67.5 = 88.54 - 17.2$

38. $w - 47.35 = 9.6 - 143$

39. New Car: Martin wants to buy a new car for $18,000. He has talked to the loan officer at his credit union and knows that they will loan him $12,600. He must also pay $350 for a license fee and $1200 for taxes. What amount of cash will he need to buy the car?

40. Haircut: Terri wants to have a haircut and a manicure. She knows that the haircut will cost $38.00 and the manicure will cost $15.50. If she plans to tip the hair stylist $8.00, how much change should she receive from $70?

41. Architect: An architect's scale drawing shows a rectangle that measures 2.25 inches on one side and 3.75 inches on another side. What is the perimeter (distance around) of the rectangle in the drawing?

42. Grain: In 2010, U.S. farmers produced 180.268 million bushels of barley, 12.447 million bushels of corn, and 2208.391 million bushels of wheat. (**Source:** US Department of Agriculture, National Agricultural Statistics Survey)

  a. What was the total U.S. production of these three grains in 2010?

  b. How many more bushels of wheat were produced than barley?

43. Livestock: In 2010, U.S. farmers owned the following amounts of livestock: 35.685 million calves, 112.989 million hogs and sows, and 3.6 million sheep. (**Source:** US Department of Agriculture, National Agricultural Statistics Survey)

  a. What was the total amount of these livestock in 2010?

  b. How many more calves than sheep were there?

44. Orbit: The eccentricity of a planet's orbit is the measure of how much the orbit varies from a perfectly circular pattern. Earth's orbit has an eccentricity of 0.017, and Pluto's orbit has an eccentricity of 0.254. How much greater is Pluto's eccentricity of orbit than Earth's?

45. Rain: Albany, New York, receives a mean rainfall of 35.74 inches and 65.5 inches of snow. Charleston, South Carolina, receives a mean rainfall of 51.59 inches and 0.6 inches of snow. What is the difference between the mean rainfalls in Charleston and Albany? What is the difference between the mean snow falls in Charleston and Albany?

46. Check Book: Suppose your checking account shows a balance of $382.35 at the beginning of the month. During the month, you make deposits of $580.00, $300.00, $182.50, and $45.00, and you write checks for $85.35, $210.50, $43.75, and $650. Find the end-of-the-month balance in your account.

| Albany, NY | |
| Charleston, SC | |

💧 = 10 inches of rain
❄ = 10 inches of snow

		YOUR CHECKBOOK REGISTER				
Check No.	Date	Transaction Description	Payment (−)	(√)	Deposit (+)	Balance
		Balance brought forward				$382.35
	11-4	Deposit			580.00	580.00
326	11-4	Electric Bill	85.35			85.35
327	11-6	Wally's Computer Shop	210.50			210.50
	11-7	Deposit			300.00	300.00
328	11-20	Groceries	43.75			43.75
	11-22	Deposit			182.50	182.50
	11-30	Deposit			45.00	45.00
329	12-1	BJ's Auto Repair	650.00			650.00
		True Balance:				

47. If the sum of 33.72 and 19.63 is subtracted from the sum of 92.61 and 35.7, what is the difference?

48. What is the sum if −17.43 is added to the difference between 16.3 and 20.1?

 Use a calculator to find the value of each expression.

49. $6.5678 + 7.9213 − 4.5682$

50. $67.341 − 56.329 + 43.892$

51. $107.65 − 35.202 + 17.5004$

52. $230 + 34.2 − 300.5 + 32.15$

53. $0.98455 − 0.356218 + 0.0326$

54. $0.003407 + 0.076331 − 1$

55. $892.783 + 861.459 − 347.442$

56. $788.383 − 806.525 − 100.2$

57. $346.672 − (35.783 − 200.1)$

58. $5482.1 − (845.92 − 906.53)$

59. $(28.74 − 32.569) + (45.93 − 76.451)$

60. $(67.349 + 93.65) − (72.914 + 123.5)$

3.3 Multiplication and Division with Decimal Numbers

Objectives

A Know how to multiply decimal numbers and place the decimal point correctly in the product.

B Learn how to multiply decimal numbers mentally by powers of 10.

C Know how to divide decimal numbers and place the decimal point correctly in the product.

D Learn how to divide decimal numbers mentally by powers of 10.

E Be able to estimate products and quotients with rounded decimal numbers.

F Understand how to solve word problems involving decimal numbers.

Objective A Multiplication with Decimal Numbers

When decimal numbers are added or subtracted, the decimal points are aligned so that digits in the tenths position are added (or subtracted) to digits in the tenths position, digits in the hundredths position are added (or subtracted) to digits in the hundredths position, and so on. In this way fractions with a common denominator are added (or subtracted). However, when fractions are multiplied, there is no need to have a common denominator. For example, in adding $\frac{1}{2} + \frac{3}{8}$, we need the common denominator 8:

$$\frac{1}{2} + \frac{3}{8} = \frac{1}{2} \cdot \frac{\mathbf{4}}{\mathbf{4}} + \frac{3}{8} = \frac{4}{8} + \frac{3}{8} = \frac{7}{8}.$$ But, in multiplying, there is no need for a

common denominator: $\frac{1}{2} \cdot \frac{3}{8} = \frac{3}{16}.$

Thus, since there is no concern about common denominators when multiplying fractions, the same is true for multiplying decimals. **Therefore, there is no need to keep the decimal points in line for multiplication with decimals**.

Decimal numbers are multiplied in the same manner as whole numbers, but with the added concern of the correct placement of the decimal point in the product. Two examples are shown here, in both fraction form and decimal form, to illustrate how the decimal point is to be placed in the product.

Products in Fraction Form

$$\frac{4}{10} \cdot \frac{6}{100} = \frac{24}{1000}$$

$$\frac{5}{1000} \cdot \frac{7}{100} = \frac{35}{100,000}$$

Products in Decimal Form

$$\begin{array}{r} .4 \quad \leftarrow \text{1 place} \\ \times \quad .06 \quad \leftarrow \text{2 places} \\ \hline .024 \end{array}$$ total of 3 places (thousandths)

$$\begin{array}{r} .005 \quad \leftarrow \text{3 places} \\ \times \quad .07 \quad \leftarrow \text{2 places} \\ \hline .00035 \end{array}$$ total of 5 places (hundred-thousandths)

The following rule states how to multiply positive decimal numbers and place the decimal point in the product. (**For negative decimal numbers, the rule is the same except for the determination of the sign**.)

To Multiply Positive Decimal Numbers

1. Multiply the two numbers as if they were whole numbers.
2. Count the total number of places to the right of the decimal points in both numbers being multiplied.
3. Place the decimal point in the product so that the number of places to the right is the same as that found in step 2.

1. Multiply:

(5.716)(52.01).

2. Find the product:

(−2.3)(0.02).

Example 1

Multiplying Positive Decimal Numbers

Multiply: 2.435×4.1.

Solution

$$
\begin{array}{r}
2.435 \quad \longleftarrow \text{3 places} \\
\times \quad 4.1 \quad \longleftarrow \text{1 place} \\
\hline
2435 \\
97400 \\
\hline
9.9835 \quad \longleftarrow \text{4 places in the product}
\end{array}
$$

total of 4 places (ten-thousandths)

Now work margin exercise **1.**

Example 2

Multiplying Signed Decimal Numbers

Find the product: $(-0.126)(0.003)$.

Solution

Since the signs are not alike, the product will be negative.

$$
\begin{array}{r}
-0.126 \quad \longleftarrow \text{3 places} \\
0.003 \quad \longleftarrow \text{3 places} \\
\hline
-0.000378 \quad \longleftarrow \text{6 places in the product}
\end{array}
$$

total of 6 places

Note that three 0's had to be inserted between the 3 and the decimal point in the product to get a total of 6 decimal places.

Now work margin exercise **2.**

Objective B **Multiplication by Powers of 10**

The following general guidelines can be used to multiply decimal numbers (including whole numbers) by powers of 10.

To Multiply a Decimal Number by a Power of 10

1. Move the decimal point to the right.
2. Move it the same number of places as the number of 0's in the power of 10.

Multiplication by **10** moves the decimal point **one** place **to the right**;

Multiplication by **100** moves the decimal point **two** places **to the right**;

Multiplication by **1000** moves the decimal point **three** places **to the right**; and so on.

Example 3

Multiplying by Powers of 10

The following products illustrate multiplication by powers of 10.

a. $10(9.35) = 93.5$ Move decimal point 1 place to the right.

b. $100(9.35) = 935$ Move decimal point 2 places to the right.

c. $100(163) = 16,300$ Move decimal point 2 places to the right.

d. $1000(0.8723) = 872.3$ Move decimal point 3 places to the right.

e. $10^2(87.5) = 8750$ Exponent tells how many places to move the decimal point. Move decimal point 2 places.

f. $10^3(4.86591) = 4865.91$ Move decimal point 3 places to the right.

Now work margin exercise 3.

3. Find each product by performing the operation mentally.

a. $10(8.36)$

b. $100(-0.9735)$

c. $10^3(14.82)$

Objective C ## Division with Decimal Numbers

The process of division with decimal numbers is, in effect, the same as division with whole numbers with the added concern of where to place the decimal point in the quotient. This is reasonable because whole numbers are decimal numbers, and we would not expect a great change in a process as important as division. Division with whole numbers gives a quotient and possibly a remainder.

$$
\begin{array}{r}
23. \\
40\overline{)950.} \\
80 \\
\hline
150 \\
120 \\
\hline
30
\end{array}
$$

divisor → 40 dividend ← 950. quotient ← 23. remainder ← 30

By adding 0's onto the dividend, we can continue to divide and get a decimal quotient other than a whole number.

$$
\begin{array}{r}
23.75 \\
40\overline{)950.00} \\
\underline{80} \\
150 \\
\underline{120} \\
30\,0 \\
\underline{28\,0} \\
2\,00 \\
\underline{2\,00} \\
0
\end{array}
$$

quotient is a decimal number ⟵
0's added on ⟵
divisor ⟶

If the divisor is a decimal number rather than a whole number, multiply both the divisor and dividend by a power of 10 so that the new divisor is a whole number. For example, we can write

$$6.2\overline{)63.86} \text{ as } \frac{63.86}{6.2} \cdot \frac{10}{10} = \frac{638.6}{62}.$$

This means that

$$6.2\overline{)63.86} \text{ is the same as } 62\overline{)638.6}.$$

Similarly, we can write

$$1.23\overline{)4.6125} \text{ as } \frac{4.6125}{1.23} \cdot \frac{100}{100} = \frac{461.25}{123} \text{ or } 123\overline{)461.25}.$$

To Divide Decimal Numbers

1. Move the decimal point in the divisor to the right so that the divisor is a whole number.
2. Move the decimal point in the dividend the same number of places to the right.
3. Place the decimal point in the quotient directly above the new decimal point in the dividend.
4. Divide just as with whole numbers.

notes

1. In moving the decimal point, you are multiplying by a power of 10.

2. Be sure to place the decimal point in the quotient **before actually dividing.**

Example 4

Dividing Decimal Numbers

Find the quotient: $63.86 \div 6.2$.

Solution

a. Write down the numbers.

$$6.2\overline{)63.86}$$

b. Move both decimal points one place to the right so that the divisor becomes a whole number. Then place the decimal point in the quotient.

$$6.2.\overline{)63.8.6}$$ ← Decimal point in quotient

c. Proceed to divide as with whole numbers.

$$
\begin{array}{r}
10.3 \\
62.\overline{)638.6} \\
\underline{62} \\
18 \\
\underline{0} \\
186 \\
\underline{186} \\
0
\end{array}
$$

Example 5

Dividing Decimal Numbers

Find the quotient: $4.6125 \div 1.23$.

Solution

a. Write down the numbers.

$$1.23\overline{)4.6125}$$

Find the following quotients.

4. $936 \div 0.5$

5. $8.208 \div 7.2$

b. Move both decimal points two places to the right so that the divisor becomes a whole number. Then place the decimal point in the quotient.

← decimal point in quotient

$$1.23.\overline{)4.61.25}$$

c. Proceed to divide as with whole numbers.

$$
\begin{array}{r}
3.75 \\
123.\overline{)461.25} \\
369 \\
\hline
92\,2 \\
86\,1 \\
\hline
6\,15 \\
6\,15 \\
\hline
0
\end{array}
$$

Now work margin exercises 4 and 5.

If the remainder is eventually 0, then the quotient is a **terminating decimal number**. We will see in Section 3.4 that if the remainder is never 0, then the quotient is an **infinite repeating decimal number**. That is, the quotient will be a repeating pattern of digits. To avoid an infinite number of steps, which can never be done anyway, we generally agree to some place of accuracy for the quotient before the division is performed. If the remainder is not 0 by the time this place of accuracy is reached in the quotient, then we divide one more place and round the quotient.

When the Remainder is Not Zero

1. Decide first how many decimal places are to be in the quotient.
2. Divide until the quotient is one digit to the right of the place of desired accuracy.
3. Using this last digit, round the quotient to the desired place of accuracy.

Example 6

Dividing Decimal Numbers

Find the quotient $82.3 \div 2.9$ to the nearest tenth.

Solution

Divide until the quotient is in hundredths (one place more than tenths); then round to tenths.

```
                    hundredths
                    read approximately
          28.37   ≈ 28.4
   2.9.)82.3.00      ←  Add 0's as needed.
        58
        24 3
        23 2
           110
            8 7
           2 30
           2 03
             27
```

$82.3 \div 2.9 \approx 28.4$ accurate to the nearest tenth

Example 7

Dividing Decimal Numbers

Find the quotient $1.935 \div 3.3$ to the nearest hundredth.

Solution

```
                  thousandths
          0.586
   3.3.)1.9.350
        16 5
        2 85
        2 64
          210
          198
           12
```

$1.935 \div 3.3 \approx 0.59$ accurate to the nearest hundredth

Now work margin exercises 6 and 7.

6. Find the quotient $83.541 \div 5.6$ to the nearest tenth.

7. Find the quotient $34.66 \div 7$ to the nearest hundredth.

Division By Powers of 10

Earlier, we found that multiplication by powers of 10 can be accomplished by moving the decimal point to the right. Division by powers of 10 can be accomplished by moving the decimal point to the left.

To Divide a Decimal Number by a Power of 10

1. Move the decimal point **to the left**.
2. Move it the same number of places as the number of 0's in the power of 10.

Division by **10** moves the decimal point **one** place **to the left**;

Division by **100** moves the decimal point **two** places **to the left**;

Division by **1000** moves the decimal point **three** places **to the left**; and so on.

Two general guidelines will help you to understand working with powers of 10.

1. Multiplication by a power of 10 will make a number larger, so move the decimal point to the right.

2. Division by a power of 10 will make a number smaller, so move the decimal point to the left.

Example 8

Dividing by Powers of 10

The following quotients illustrate division by powers of 10.

a. $5.23 \div 100 = \dfrac{5.23}{100} = 0.0523$ Move decimal point 2 places to the left.

b. $817 \div 10 = \dfrac{817}{10} = 81.7$ Move decimal point 1 place to the left.

c. $495.6 \div 10^3 = 0.4956$ Move decimal point 3 places to the left. The exponent tells how many places to move the decimal point.

d. $286.5 \div 10^2 = 2.865$ Move decimal point 2 places to the left.

Now work margin exercise 8.

Estimating Products and Quotients

Estimating with multiplication and division can be used to help in placing the decimal point in the product and in the quotient to verify the reasonableness of the answer. The technique is to **round all numbers to the place of the leftmost nonzero digit and then operate with these rounded numbers.**

Example 9

Estimating Products of Decimal Numbers

For the product $(0.358)(6.2)$:

a. Estimate the product.

Solution

Estimate by multiplying rounded numbers.

$$
\begin{array}{r}
0.4 \\
\times\ \ 6 \\
\hline
2.4
\end{array}
$$

0.358 rounded
6.2 rounded
$\longleftarrow$ estimate

b. Find the product.

Solution

Find the actual product.

$$
\begin{array}{r}
0.358 \\
\times\ \ \ 6.2 \\
\hline
716 \\
2148\ \ \\
\hline
2.2196
\end{array}
$$

$\longleftarrow$ actual product

The estimated product 2.4 helps place the decimal point correctly in the product 2.2196. Thus an answer of 0.22196 or 22.196 would indicate an error in the placement of the decimal point, since the answer should be near 2.4 (or between 2 and 3).

Now work margin exercise 9.

Example 10

Estimating Quotients of Decimal Numbers

a. Estimate the quotient $6.1 \div 0.312$.

Solution

Using $6.1 \approx 6$ and $0.312 \approx 0.3$, estimate the quotient.

$$
0.3.\overline{)6.0.}\ \ \ \ \ \ \ \ \overset{20.}{}
$$

$\longleftarrow$ estimate

9. For the product $(0.42)(7.25)$:

a. Estimate the product.

b. Find the product.

10. First estimate the quotient $8.2 \div 1.92$, then find the quotient to the nearest tenth.

b. Find the quotient to the nearest tenth.

Solution

Find the quotient to the nearest tenth.

$$
\begin{array}{r}
19.55 \\
0.312.\overline{)6.100.00} \\
312 \\
\overline{2\,980} \\
2\,808 \\
\overline{1720} \\
1560 \\
\overline{1600} \\
1560 \\
\overline{40}
\end{array}
$$

$6.1 \div 0.312 \approx 19.6$

Now work margin exercise 10.

Objective F **Applications**

Some word problems may involve several operations with decimal numbers. The words do not usually say directly to add, subtract, multiply, or divide. Experience and reasoning abilities are needed to decide which operations to perform with the given numbers. Example 11 illustrates a problem that involves several steps and how estimating can provide a check for a reasonable answer.

Example 11

Buying a Car

You can buy a car for $8500 cash, or you can make a down payment of $1700 and then pay $616.67 each month for 12 months. How much can you save by paying cash?

Solution

a. Find the amount paid in monthly payments by multiplying the amount of each payment by 12. In this case, judgement dictates that we do not want to lose two full monthly payments in our estimate, so we use 12 and do not round to 10.

	Estimate		Actual Amount	

Estimate	Actual Amount
$ 600	$ 616.67
× 12	× 12
1200	123334
600	61667
$7200	$7400.04 paid in monthly payments

b. Find the total amount paid by adding the down payment to the answer in part **a.**

Estimate		Actual Amount	
$ 2000	down payment	$1700.00	down payment
+ 7200	monthly payments	+ 7400.04	monthly payments
$ 9200	total paid	$9100.04	total paid

c. Find the savings by subtracting $8500 from the answer in part **b.**

Estimate	Actual Amount
$ 9200.00	$ 9100.04
− $ 8500.00	− $ 8500.00
$ 700.00	$ 600.04 savings by paying cash

The $700 estimate is very close to the actual savings of $600.04.

Now work margin exercise 11.

We know from a previous section that the mean of a set of numbers can be found by adding the numbers, and then dividing the sum by the number of addends. Another meaning of the term mean is in the sense of "a mean speed of 52 miles per hour" or "the mean number of miles per gallon of gas." This kind of mean can be found by division. If we know the total amount of a quantity (distance, dollars, gallons of gas, etc.) and a number of units (time, items bought, miles, etc.), then we can find the mean amount per unit by dividing the total amount by the number of units.

Think of "per" as indicating division. Thus miles per hour means miles divided by hours.

Example 12

Fuel Efficiency

The gas tank of a car holds 18 gallons of gasoline. Approximately how many miles per gallon does the car mean if it will go 470 miles on one tank of gas?

11. You can buy a lawnmower for $450 cash, or you can make a down payment of $50 and then pay $25 each month for two years. How much can you save by paying cash?

12. A batter had 299 hits in 114 games. Estimate the batter's mean number of hits per game.

Solution

Miles per gallon means miles divided by gallons. Since the question calls only for an approximate answer, rounded values can be used.

$18 \approx 20$ gal and $470 \approx 500$ miles

$$
\begin{array}{r}
25 \quad \text{miles per gallon} \\
20\overline{)500} \\
40 \\
\hline
100 \\
100 \\
\hline
0
\end{array}
$$

The car gets a mean of about 25 miles per gallon.

Now work margin exercise 12.

If a mean amount per unit is known, then a corresponding total amount can be found by multiplying. For example, if you jog a mean of 5 miles per hour, then the distance you jog can be found by multiplying your mean speed by the time you spend jogging.

Example 13

Jogging

If you jog at a mean speed of 4.8 miles per hour, how far will you jog in 3.2 hours?

Solution

Multiply the mean speed by the number of hours.

$$
\begin{array}{r}
4.8 \quad \text{miles per hour} \\
\times \ 3.2 \quad \text{hours} \\
\hline
9\,6 \\
14\,4 \\
\hline
1\,5.3\,6 \quad \text{miles}
\end{array}
$$

You will jog 15.36 miles in 3.2 hours.

Now work margin exercise 13.

13. If your boat has a mean speed of 15.3 miles per hour, how far will you travel in 1.8 hours?

In some cases, a business will advertise the total price of its items, including sales tax. You will see a phrase such as "tax included." If, for example, the sales tax is figured at 0.06 times the actual price, you can find the price you are paying for the item by dividing the total price by 1.06. (The number 1.06 represents the actual price plus 0.06 times the actual price.) If you buy gas for your car, the price at the gas pump includes all types of taxes (state, federal, local, etc.). If these taxes are figured at, say 0.45 times the price of a gallon of gas, then the actual price of the gas sold to you can be found by dividing the price at the pump by the number 1.45.

Example 14

Price of Gasoline

Suppose that the price of a gallon of gas is stated as $3.15 at the pump and the station owner tells you that the taxes you are paying are figured at 0.45 times the price he is actually charging for a gallon of gas. What price (to the nearest penny) is he actually charging for a gallon of gas?

Solution

To find the actual price of a gallon of gas, before taxes, divide the total price by 1.45.

$$
\begin{array}{r}
2.172 \\
1.45\overline{)\,3.15.000} \\
\underline{2\ 90} \\
25\ 0 \\
\underline{14\ 5} \\
10\ 50 \\
\underline{10\ 15} \\
350 \\
\underline{290} \\
60
\end{array}
$$

So, the actual cost of the gas is about $2.17 per gallon before taxes.

Now work margin exercise 14.

Example 15

 Using a Calculator

Use a calculator to perform the following operations and write the answer accurate to the nearest ten-thousandth.

a. $4.631(0.235)(2.3)$ **b.** $19.36 \div 5.4$

Solution

a. To multiply the values given, press the keys

. Then press ⟨=⟩.

The display will read **2.5031**.

b. To divide the values given, press the keys

. Then press ⟨=⟩.

The display will read **3.5852**.

Now work margin exercise 15.

---

14. If a hot dog costs $2.25 (before taxes) at a little league hockey game and taxes were 0.06 times the actual price, how much (to the nearest cent) does the hot dog cost?

15. Use a calculator to perform the following operations and write the answer accurate to the nearest ten-thousandth.

a. $7.5(4.138)(0.732)$

b. $32.89 \div 6.4$

Exercises 3.3

Estimate each of the following products mentally by using rounded values.

1.

    a. $(0.92)(0.81)$ **b.** $(33.6)(0.11)$ **c.** $(0.64)(9.71)$

    d. $(0.22)(0.26)$ **e.** $(4.7)(1.1)$

2.

    a. $(1.62)(0.03)$ **b.** $1.62(0.003)$ **c.** $1.62(0.03)$

    d. $1.62(3)$

Estimate each of the following quotients by using rounded values.

3.

    a. $3.1\overline{)6.36}$ **b.** $0.1\overline{)211.5}$ **c.** $3.6\overline{)282.4}$

    d. $18.2\overline{)132.9}$ **e.** $3.1\overline{)0.0636}$

4.

    a. $28.34 \div 0.003$ **b.** $28.34 \div 0.03$ **c.** $28.34 \div 0.3$

    d. $28.34 \div 3$

Find each of the indicated products. See Examples 1 and 2.

5. $(0.5)(0.7)$ **6.** $(0.2)(0.8)$ **7.** $6(1.8)$

8. $9(3.5)$ **9.** $(0.02)(0.02)$ **10.** $(0.03)(0.03)$

11. $5.6(-0.02)$ **12.** $4.8(-0.25)$ **13.** $8(-0.125)$

14. $4(-0.125)$ **15.** $4.29(0.001)$ **16.** $5.48(0.02)$

17.
$$\begin{array}{r} 5.09 \\ \times\ 0.4 \\ \hline \end{array}$$

18.
$$\begin{array}{r} 0.137 \\ \times\ 0.09 \\ \hline \end{array}$$

19.
$$\begin{array}{r} 0.0312 \\ \times\ 0.065 \\ \hline \end{array}$$

20.
$$\begin{array}{r} 84.105 \\ \times\ 0.111 \\ \hline \end{array}$$

21.
$$\begin{array}{r} 16.002 \\ \times\ 0.102 \\ \hline \end{array}$$

22.
$$\begin{array}{r} 93.1 \\ \times\ 0.57 \\ \hline \end{array}$$

Find each of the indicated quotients. See Examples 4 and 5.

23. $6.48 \div 2$ **24.** $2.73 \div 3$ **25.** $3.95 \div (-5)$

26. $28 \div 5.6$ **27.** $(-0.054) \div 9$ **28.** $-80.24 \div 0.04$

29. $9.4\overline{)6.429}$

30. $0.37\overline{)5.682}$

31. $1.64\overline{)35}$

32. $2.7\overline{)5.483}$

33. $13\overline{)65.582}$

34. $3.381\overline{)6}$

35. $31\overline{)71}$

36. $1.62\overline{)0.0116}$

37. $0.03\overline{)6.275}$

38. $10(0.619)$

39. $100(3.76)$

40. $100(0.455)$

41. $10^3(0.95)$

42. $10^3(0.005)$

43. $10^5(7.3)$

44. $98.5 \div 100$

45. $26.483 \div 1000$

46. $\dfrac{169}{10}$

47. $\dfrac{1.78}{1000}$

48. $\dfrac{3.25}{100}$

49. Find
   a. the perimeter and
   b. the area of a square with sides 13.4 inches long.

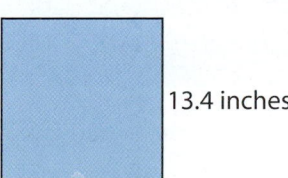

13.4 inches

50. **Television:**
   a. If the sale price of a new flat screen TV is $683 and sales tax is figured at 0.08 times the price, approximately what total amount is paid for the television set?
   b. What is the exact amount paid for the television set?

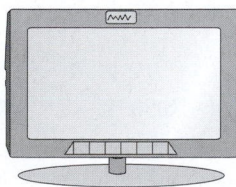

51. **Used Car:** To buy a used car, you can pay $2045.50 cash or put $300 down and make 18 monthly payments of $114.20. How much do you save by paying cash?

52. **Gas Mileage**
   a. If a car averages 25.3 miles per gallon, about how far will it go on 19 gallons of gas?
   b. Exactly how far will the car go on 19 gallons of gas?

53. Gas Mileage:

   a. If a car travels 330 miles on 15 gallons of gas, approximately how many miles does it travel per gallon?

   b. Exactly how many miles per gallon does it average?

54. Purchasing a Car: If the total price of a car was $33,075 including tax at 0.05 times the list price, you can find the list price by dividing the total price by 1.05. What was the list price?
(**Note**: 1.05 represents the list price plus 0.05 times the list price.)

55. Chill: If the total price of a refrigerator was $874.50 including tax at 0.06 times the list price, you can find the list price by dividing the total price by 1.06. What was the list price? (**Note**: 1.06 represents the list price plus 0.06 times the list price.)

56. Mortgages: Suppose that the total interest that will be paid on a 30-year mortgage for a home loan of $150,000 is going to be $480,000. What will be the payment each month if the payments are to pay off both the loan and the interest?

57. Fly: About how long will it take an airplane to fly from Los Angeles to New York if the distance is approximately 3000 miles and the airplane averages 465 miles per hour?

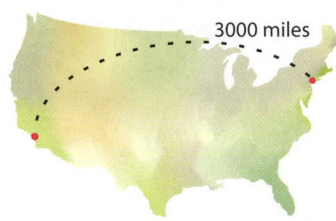
3000 miles

58. Bicycling:

   a. If a bicyclist rode 150.6 miles in 11.3 hours, about how fast did she ride in miles per hour?

   b. What was her average speed in miles per hour (to the nearest tenth)?

59. The Bears: Walter Payton played football for the Chicago Bears for 13 years. In those years he carried the ball 3838 times for a total of 16,726 yards. What was his average yardage per carry (to the nearest tenth)?

Use a TI-84 Plus calculator to find the value of each of the following expressions (accurate to the nearest ten-thousandth).

60. $3.521(0.124)(-1.2)$

61. $67.3(42.44)(-2.7)$

62. $34.6(-0.02)(-3.577)$

63. $14.5(-72.16)(-10.563)$

64. $5.7 \div 0.214$

65. $3.457 \div (-0.12)$

66. $\dfrac{67.152}{3.54}$

67. $\dfrac{-4580}{19.64}$

68. Alberto Contador won the 2010 Tour de France by traveling about 2262.66 miles in 91 hours, 58 minutes, 48 seconds. Andy Schleck finished only 39 seconds behind him for second place. Third place went to Denis Menchov, who finished 2 minutes, 1 second behind Contador. In order to find the average speed of these riders, you need to first change each of their times into the decimal form of an hour and then divide each time into 2262.66 to find the average speed in miles per hour.

(Remember that there are 60 seconds in a minute and 60 minutes in an hour.)

a. What was the mean speed of Alberto Contador?

b. What was the mean speed of Andy Schleck?

c. What was the mean speed of Denis Menchov?

69. Tour de France: In 2005, Lance Armstrong won the Tour de France for a record seventh consecutive times. To win, he traveled a total distance of 2242 miles in 86 hours 15 minutes 2 seconds. The second place finisher, Ivan Basso of Italy, was 4 minutes 40 seconds behind.

a. What was Lance Armstrong's mean speed?

b. What was Ivan Basso's mean speed?

Collaborative Learning

70. Do you know how to find the gas mileage (miles per gallon) that your car is using? If you are not sure, proceed as follows and compare your mileage with other students in the class. (Your car might need some work if the mileage is not consistent or it is much worse than other similar sized cars.)

Step 1: Fill up your gas tank and write down the mileage indicated on the odometer.

Step 2: Drive the car for a few days.

Step 3: Fill up your gas tank again and write down the number of gallons needed to fill the tank and the new mileage indicated on the odometer.

Step 4: Find the number of miles that you drove by subtracting the new and old numbers indicated on the odometer.

Step 5: Divide the number of miles driven by the number of gallons needed to fill the tank. This number is your gas mileage (miles per gallon).

3.4 Measures of Center

Objectives

A Understand the meaning of the statistical terms mean, median, mode, and range.

B Learn how to find the mean, median, mode, and range of a set of data items.

Objective A **Statistical Terms**

The October 18, 2005 Los Angeles Times reported that the **median price** of a home in Southern California was $475,000. By counties, the median price of homes were as follows.

San Bernardino: $352,000	Los Angeles: $494,000
Orange County: $610,000	Riverside County: $391,000
Ventura County: $604,000	San Diego County: $498,000

So, if you are moving to Southern California or already live there, these are very interesting numbers; but what do they mean? (Do you know the median price of homes in your area?) Just what is a median, and is it different from a mean? Both the mean and the median are statistics that we are going to study in this section.

Statistics is the study of how to gather, organize, analyze, and interpret information. A **statistic** is a particular measure or characteristic of a part (a **sample**) from a larger collection of items called the **population** of interest. The population can be a collection of people, animals, objects, or numbers related to information of interest.

In this section we will study only four numerical statistics that are easily found or calculated: **mean, median, mode**, and **range**. The mean, median, and mode are measures that describe the "average" or "middle" of a set of data. The range is a measure that describes how "spread out" the data are. Other measures that you might read about in newspapers and magazines and will certainly study in a course in statistics are

standard deviation and **variance** (both measures of how "spread out" data are),

z-score (a measure that compares numbers that are expressed in different units; for example, your test score on a mathematics exam and your performance in a physical exercise), and

correlation coefficient (a measure of how two different types of data might be related; for example, the relationship between a person's income and the number of years of schooling they have).

You will need a semester or two of algebra in order to be able to study and understand these and other statistics, so keep working hard. The following terms and their definitions are necessary for understanding the ideas and problems in this section. They are listed here for easy reference.

Terms Used in the Study of Statistics

Data: Value(s) measuring some information of interest
             (We will consider only numerical data.)

Statistic: A single number describing some characteristic of the data

Mean: The arithmetic average of the data
             (Find the sum of all the data and divide by the number of data
             items.)

Median: The middle of the data after the data have been arranged in
             order
             (The median may or may not be one of the data items.)

Mode: The data item(s) that appears most frequently
             (A set of data may have more than one mode.)

Range: The difference between the largest and smallest data items

---

Objective B   ## Finding the Mean, Median, Mode, and Range of a Set of Data

Two sets of data, group A and group B, are shown here and used in the discussion and calculations in Examples 1 through 3.

Group A: Body Temperature (in Fahrenheit degrees) of 8 People							
96.4°	98.6°	98.7°	99.8°	99.2°	101.2°	98.6°	97.1°

Group B: The Time (in minutes) of 11 Movies					
100 min	90 min	113 min	110 min	88 min	90 min
155 min	88 min	105 min	93 min	90 min	

1. Find the mean movie time for the 11 movies in group B, rounded to nearest minute.

Group B	
100 min	90 min
113 min	110 min
88 min	90 min
155 min	88 min
105 min	93 min
90 min	

Example 1

Finding the Mean

Find the mean body temperature for the 8 people in group A.

Solution

The mean is the average of the data. Therefore, we add the 8 temperatures and divide the sum by 8. (You can, of course, perform these calculations with a calculator.)

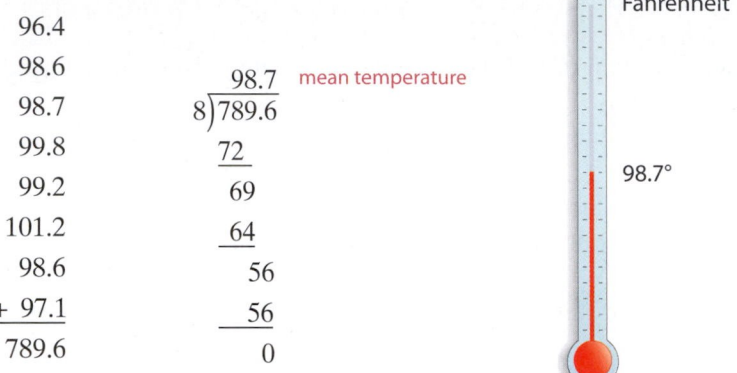

```
   96.4
   98.6              98.7   mean temperature
   98.7          8)789.6
   99.8            72
   99.2            69
  101.2            64
   98.6            56
 + 97.1            56
  789.6             0
```

The mean temperature is 98.7°F.

Now work margin exercise **1.**

The **median** is another way to measure the "middle" of a set. In a set of **ranked data** (data arranged in order, smallest to largest or largest to smallest), the median is the middle value. As we will see, the determination of this value depends on whether there is an odd number of data items or an even number of data items.

To Find the Median

1. Rank the data. (Arrange the data in order, either from smallest to largest or largest to smallest.)

2. The median can be found by counting from the top down (or from the bottom up) to the position $\frac{n+1}{2}$ where n represents the number of data items.
   a. If there are an **odd** number of items, the median is the middle item.
   b. If there are an **even** number of items, the median is the value found by calculating the mean of the two middle items.
   (**Note:** This value may or may not be in the data.)

Example 2

Finding the Median

Find the median temperature for the 8 people in group A and the median time for the movies in group B.

Solution

First, we rank both sets of data in order from smallest to largest.

Group A **(Temperatures)**	**Group B** **(Movie Times)**
1. 96.4	1. 88
2. 97.1	2. 88
3. 98.6	3. 90
4. 98.6 ← The median is between 98.6 and 98.7.	4. 90
5. 98.7	5. 90
6. 99.2	6. 93 ← The median is 93.
7. 99.8	7. 100
8. 101.2	8. 105
	9. 110
	10. 113
	11. 155

Group A has 8 items. With the formula for the position $\frac{n+1}{2} = \frac{8+1}{2} = \frac{9}{2} = 4.5,$ the median is the average of the items in the fourth and fifth positions.

$$\text{median temperature} = \frac{98.6 + 98.7}{2} = \frac{197.3}{2} = 98.65°F$$

(Note that 8 is an **even** number and the median is the mean of the two middle items.)

Group B has 11 items. With the formula for the position $\frac{n+1}{2} = \frac{11+1}{2} = \frac{12}{2} = 6,$ the median is the item in the sixth position.

$$\text{median movie time} = 93 \text{ minutes}$$

(Note that 11 is an **odd** number and the median is the middle item.)

(**Comment**: Note that in group A, the median is not one of the data items; while in group B, the median is one of the data items.)

Now work margin exercise 2.

2. 12 students took a make-up exam for a history class, and their scores are listed below. Find the median score of the 12 students.

86	77
79	67
92	84
98	91
93	88
95	77

Once the data have been ranked (as was done in Example 2), the mode (if there is one) and the range are easily determined. Remember that the mode is the item that occurs most frequently, and the range is the difference between the largest and smallest items in each set of data.

3. Refer to the test scores in margin exercise 2. Find the mode and the range of the scores.

Example 3

Finding the Mode and Range

For both group A and group B, find

a. the mode, and
b. the range.

Solution

a. From the ranked data in Example 2, we can see the most frequent item in each group.

For group A, the mode is 98.6°. (98.6° occurs twice and no other item occurs more than once.)

For group B, the mode is 90 minutes. (90 min. occurs three times and no other item occurs more than twice.)

b. Again referring to the ranked data in Example 2, we can calculate each range as follows.

range = (largest value) − (smallest value)

group A range = $101.2° − 96.4° = 4.8°$

group B range = $155 − 88 = 67$ minutes

Now work margin exercise 3.

Completion Example 4

Computing a Goal Test Score

Suppose that your grade in this class is based on 5 exam scores: 4 sectional exams and 1 comprehensive final exam, with each exam scored on a basis of 100 points. On the first 4 exams, you have scores of 85, 78, 82, and 70. What is the lowest score you could get on the final exam and still earn a grade of B in the course? (Assume that to get a B your mean score must be between 80 and 89.)

Solution

Solve the problem by first finding the total number of points needed to obtain a mean of 80 on 5 exams. Then calculate the number of points accumulated on the first four exams. Finally, subtract the number of points accumulated from the total needed. This will give the number of points needed on the final exam.

a. Find the total number of points needed for a B. Since the 5 exams are to have a mean score of 80 (or more), then the total number of points must be at least the following profuct.

$$80$$
$$\underline{\times\ 5}$$
total points (or more) needed for a B

b. Calculate the number of points you have accumulated on the first 4 exams.

$$85$$
$$78$$
$$82$$
$$\underline{+\ 70}$$
points accumulated

c. To obtain a mean of 80 (or more) on your 5 exams, you need the following score on the final exam.

$$400$$ total Points needed
$$\underline{-}$$ points accumulated
$$\underline{}$$ points needed on the final exam

Thus you need at least a score of _____ on your final exam to earn a grade of B for the course. (**Note**: We will see an algebraic approach to solving problems of this type in a later section.)

Now work margin exercise 2.

> # notes
>
> ■ Of the four statistics mentioned in this section, the mean and median are most commonly used. Many people who use statistics professionally
> ■ believe that the mean (or average) is relied on too much in reporting central tendencies for data such as income, housing costs, and taxes,
> ■ because a few very high or very low items can distort the picture of a central tendency. For example, the median of 93 minutes for the movies
> ■ in group B is probably more representative than the mean, which is 102 minutes. This is so because the one high time of 155 minutes
> ■ raises the mean considerably, whereas the median is not affected by this one extreme outer value. When you read an article in a magazine
> ■ or newspaper that reports means or medians, you should now have a better understanding of the implications of these statistical measures..
> ■

Completion Example Answers
a. 400 total points (or more) needed for a B
b. 315 points accumulated
c. 315 points accumulated; 85 points needed on the final exam
Thus you need at least a score of 85 on your final exam to earn a grade of B for the course.

4. Suppose that your grade in this class is based on 4 exam scores: 3 sectional exams and 1 comprehensive final exam, with each exam scored on a basis of 100 points. On the first 3 exams you have scores of 93, 95, and 75. What is the lowest score you could get on the final exam and still earn a grade of A in the course? (Assume that to get an A, you must average 90 or above.)

Exercises 3.4

Find the a. mean, b. median, c. mode (if any), and d. range of the given data.

1. **Test Scores:** Dr. Wright recorded the following nine test scores for students in his statistics course.

   95   82   85   71   65   85   62   77   98

2. **Presidents:** The ages of the first five U.S. presidents on the date of their inaugurations were as follows. (The presidents were Washington, Adams, Jefferson, Madison, and Monroe.)

   57   61   57   57   58

3. **Income:** Family incomes in a survey of eight students are as follows.

   $35,000 $28,000 $42,000 $71,000

   $63,000 $36,000 $51,000 $63,000

4. **College Tuition:** Resident tuition charged by 10 colleges in North Carolina in 2010 are listed in the chart below. Round to the nearest tenth, if necessary. **Source:** collegeboard.com

   $4088 $5175 $5076 $3639 $3756

   $6529 $4479 $5922 $5138 $5124

5. **College Tuition:** Nonresident tuition charged by 10 colleges in North Carolina in 2010 are as follows. **Source:** collegeboard.com

   $16,487 $13,234 $17,831 $13,276 $14,220

   $19,064 $15,052 $24,736 $16,185 $14,721

6. **College Students:** The following list contains the ages of 20 students surveyed in a college chemistry class.

   18   23   23   23   22   18   21   20   18   20

   19   20   21   19   23   36   35   26   17   24

7. **Travel:** The distances from Chicago to selected cities are shown below.

Boston:	980 miles
Cleveland:	345 miles
Dallas:	930 miles
Denver:	1050 miles
Detroit:	280 miles
Indianapolis:	190 miles
Los Angeles:	2110 miles
San Francisco:	2210 miles
New Orleans:	950 miles
Miami:	1390 miles
Seattle:	2050 miles

8. Air Travel: Passengers (to the nearest thousand in 2009) in the world's 10 busiest airports:

Atlanta, Hartsfield: 88,032,000   Paris, Charles de Gaulle: 57,907,000

London, Heathrow: 66,038,000   Los Angeles, LAX: 56,521,000

Beijing, PEK: 65,372,000   Dallas/Ft Worth, DFW: 56,030,000

Chicago, O'Hare: 64,158,000   Frankfurt–Main: 50,933,000

Tokyo, Haneda: 61,904,000   Denver, DEN: 50,167,000

Source: Airports Council International

9. The volume (in thousands of cubic meters) of the world's 5 largest dams are as follows. (**Source:** infoplease)

Three Gorges (China): 39,300,000   Syncrude Tailings, Canada: 540,000

Chapeton (Argentina): 296,200   Pati, Argentina: 238,200

New Cornelia Tailings (United States): 209,500

a. Mean = _____   **b.** Median = _____

c. Mode = _____   **d.** Range = _____

10. The number of volumes (to the nearest thousand) in top college libraries in the United States is as follows. **Source:** www.insidecollege.com

Harvard: 15,827,000   Yale: 12,369,000
U of Illinois: 10,525,000   UC Berkeley: 10,094,000
U of Texas: 9,022,000   U of Michigan: 8,273,000
UCLA: 8,157,000   Columbia: 9,455,000

a. Mean = _____   **b.** Median = _____

c. Mode = _____   **d.** Range = _____

11. The winning margin for each of the first 40 Super Bowls:

25	19	9	16	3	21	7	17	10	4	18	17	4	12
17	5	10	29	22	36	19	32	4	45	1	13	35	17
23	10	14	7	15	7	27	3	27	3	3	11		

a. Mean = _____   **b.** Median = _____

c. Mode = _____   **d.** Range = _____

12. The Medal of Honor is the nation's highest military award for uncommon valor by men and women in battle. The medals awarded by war are as follows. **Source:** Congressional Medal of Honor Society

Civil War: 1520   Indian Wars (1861-1898): 428
Korean Expedition (1871): 15   Spanish-American War: 109
Philippines/Samoa (1899-1913): 91   Boxer Rebellion (1900): 59
Dominican Republic (1904): 3   Nicaragua (1911): 2
Mexico (Veracruz) (1914): 55   Haiti (1915): 6
Miscellaneous (1861-1920): 166   World War I: 124
Haitian Action (1919-20): 2   Korean War: 131
World War II: 433   Somalia: 2
Vietnam War: 238

a. Mean = _____   **b.** Median = _____

c. Mode = _____   **d.** Range = _____

13. Exams: Suppose that you are to take four hourly exams and a final exam in your chemistry class. Each exam has a maximum of 100 points, and you must average between 75 and 82 points to receive a passing grade of C. If you have scores of 83, 65, 70, and 78 on the hourly exams, what is the minimum score you can make on the final exam and receive a grade of C? (First explain your strategy in solving this problem. Then solve the problem.)

14. Final Exam: Suppose that the instructor in the class in Exercise 13 has informed the class that the lowest of your hourly exam scores will be replaced by your score on the final exam provided that your final exam score is higher. (That is, the final exam score may be counted twice.) Now, what is the minimum score that you can make on the final exam and still receive a grade of C? (First, explain your strategy in solving this problem. Then solve the problem.)

15. Farms: The number of farms in the U. S. is decreasing. The following graph shows the number of farms (in millions) for each decade since 1940. As you can tell from the graph the number of farms seems to be leveling off somewhat. [Possible point of discussion in class: Does this mean that there are fewer acres being farmed? Or do you think that it means that the farms are simply larger (being owned by large corporations rather than individual farmers)?]

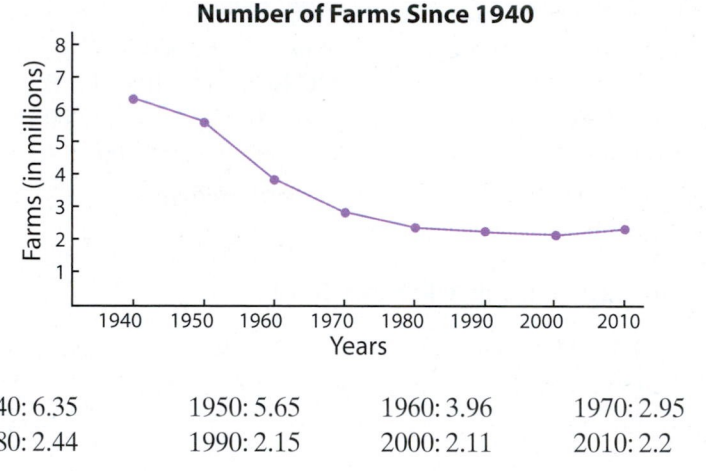

Number of Farms Since 1940

| 1940: 6.35 | 1950: 5.65 | 1960: 3.96 | 1970: 2.95 |
| 1980: 2.44 | 1990: 2.15 | 2000: 2.11 | 2010: 2.2 |

a. Find the mean number of farms from 1940 to 2010. _____

b. Find the mean number of farms from 1980 to 2000. _____

16. Banking: Financial institutions sometimes fail. Most are insured by the FDIC (Federal Deposit Insurance Corporation). In the years from 2001 to 2010 the following numbers of financial institutions have failed.

Year	2001	2002	2003	2004	2005	2006	2007	2008	2009	2010
Number Failed	4	11	3	4	0	0	3	25	140	157

Find the following statistics for closures over these ten years:

a. Mean = _____

b. Median = _____

c. Mode = _____

d. Range = _____

17. Military: The armed forces of countries are measured in several ways: active troops, reserve troops, tanks, navy (carriers, cruisers, frigates, destroyers, submarines), and combat aircraft. For 2009, the following countries had active troops in the following numbers:

Afghanistan: 93,800	Egypt: 468,500	China: 2,285,000
France: 352,771	U.S.: 1,580,255	Chile: 60,560

a. What was the mean number of active troops for these countries? Round your answer to the nearest tenth. _____

b. What was the median number of active troops for these countries? _____

18. Government: Each state pays its governor a salary. As of 2010, the governors/ salaries from ten states were as follows:

New York: $179,000	Michigan: $177,000	California: $206,500
New Jersey: $175,000	Virginia: $166,000	Vermont: $142,542
Illinois: $177,500	Connecticut: $150,000	Washington: $166,891
Maryland: $150,000		

Find the following statistics related to these ten salaries:

a. Mean = _____ **b.** Median = _____

c. Mode = _____ **d.** Range = _____

Writing & Thinking

19. Your grade point average (GPA) is a form of a **weighted average**. That is, 4 units of A counts more than 4 units of B. The most common weight for grades is A, 4 points; B, 3 points; C, 2 points; D, 1 point; and F, 0 points. To find a GPA,

1. Multiply the points for each grade by the number of units for the course.

2. Find the sum of these products.

3. Divide this sum by the total number of units taken.

Find the GPA (to the nearest tenth) for each of the following situations.

a. 3 units of A in astronomy, 4 units of B in geometry, 3 units of C in sociology, and 4 units of A in biology

b. 5 units of B in history, 4 units of C in calculus, 3 units of A in computer science, and 4 units of D in geology

c. Your own GPA for the last semester (or your anticipated GPA for this semester).

Collaborative Learning

20. With the class separated into teams of 2 to 4 students, each team is to go on campus and survey 50 students and ask each student how many minutes it takes him or her to drive to school from home.

a. Each team is to find the mean, median, mode, and range for the 50 responses.

b. Each team is to bring all the data to class, and the class is to pool the information and find the mean, median, mode, and range for the pooled data.

c. The class is to discuss the results of the individual teams and the pooled data and what use such information might have for the administration of the college.

3.5 Decimal Numbers, Fractions, and Scientific Notation

Objective A   Decimals and Fractions

In this section, we will discuss the fact that certain decimal numbers (terminating and infinite repeating decimals), fractions, and mixed numbers are simply different forms of the same type of number – namely, **rational numbers**. We also will show how operations can be performed with various combinations of these numbers, review absolute value, and explore the concept of scientific notation.

From Section 3.1, we know that decimal numbers can be written in fraction form with denominators that are powers of 10. For example,

$$0.75 = \frac{75}{100} \text{ and } 0.016 = \frac{16}{1000}.$$

In each case, the denominator is the value of the position of the rightmost digit. This leads to the following method for changing a decimal number to fraction form.

Changing from Decimal Form to Fraction Form

A decimal number with digits to the right of the decimal point can be written in fraction form by writing a fraction with:

1. A **numerator** that consists of the whole number formed by all the digits of the decimal number, and
2. A **denominator** that is the power of 10 that corresponds to the rightmost digit. (For example, a denominator of 100 corresponds to hundredths.).

In the following examples, each decimal number is changed to fraction form and then reduced by using the factoring techniques discussed in Chapter 2 for reducing fractions.

Example 1

Changing a Decimal Number to a Fraction

a. $0.75 = \frac{75}{100} = \frac{3 \cdot \cancel{5} \cdot \cancel{5}}{2 \cdot 2 \cdot \cancel{5} \cdot \cancel{5}} = \frac{3}{4}$

   ↑ hundredths

b. $0.36 = \frac{36}{100} = \frac{\cancel{4} \cdot 9}{\cancel{4} \cdot 25} = \frac{9}{25}$

   ↑ hundredths

c. $0.085 = \frac{85}{1000} = \frac{\cancel{5} \cdot 17}{\cancel{5} \cdot 200} = \frac{17}{200}$

   ↑ thousandths

Objectives

A   Be able to change decimal numbers to fractions and fractions to decimal numbers.

B   Understand how to work with decimal numbers and fractions in the same problem.

C   Know the definition of absolute value as it pertains to decimal numbers.

D   Know how to read and write decimal numbers in scientific notation.

1. Change each decimal number to a fraction reduced to lowest terms.

a. 0.35

b. 0.125

c. 2.4

2. Change $\frac{1}{5}$ to a decimal number.

---

d. $3.8 = \dfrac{38}{10} = \dfrac{\cancel{2} \cdot 19}{\cancel{2} \cdot 5} = \dfrac{19}{5}$

tenths

We can also write 3.8 as a mixed number.

$$3.8 = 3\frac{8}{10} = 3\frac{4}{5}$$

*Now work margin exercise **1**.*

Changing from Fraction Form to Decimal Form

A fraction can be written in decimal form by dividing the numerator by the denominator.

1. If the remainder is 0, the decimal number is said to be **terminating**.

2. If the remainder is not 0, the decimal number is said to be **nonterminating**.

The following examples illustrate fractions that convert to terminating decimals.

Example 2

Changing a Fraction to a Decimal Number

Change $\frac{5}{8}$ to a decimal.

Solution

$$\begin{array}{r} 0.625 \\ 8\overline{)5.000} \\ \underline{4\,8} \\ 20 \\ \underline{16} \\ 40 \\ \underline{40} \\ 0 \end{array}$$

$$\frac{5}{8} = 0.625$$

*Now work margin exercise **2**.*

 Using a Calculator to Change Fractions to Decimal Numbers
To change a fraction to a decimal number you, you will need to find the division key, $\boxed{\div}$, and the equal sign key, $\boxed{=}$.

To change the fraction given in example 2, press the keys
$\boxed{5}$ $\boxed{\div}$ $\boxed{8}$. Then press $\boxed{=}$.

The display will read **0.625**.

Example 3

Changing a Fraction to a Decimal Number

Change $-\dfrac{4}{5}$ to a decimal number.

Solution

$-\dfrac{4}{5}$ also can be written $\dfrac{-4}{5}$ or $\dfrac{4}{-5}$. In any case, the result of the division will be negative. For convenience, we simply divide 4 by 5 and then write the negative sign in the quotient.

$$5)\overline{4.0} \qquad -\dfrac{4}{5} = -0.8$$
$$\begin{array}{r} 0.8 \\ \underline{4\,0} \\ 0 \end{array}$$

Thus, $-\dfrac{4}{5} = -0.8$ which is a terminating decimal number.

Now work margin exercise 3.

Nonterminating decimal numbers can be **repeating** or **nonrepeating**. A nonterminating (or infinite) repeating decimal number has a repeating pattern to its digits. Every fraction with an integer numerator and a nonzero integer denominator (called a **rational number**) is either a terminating decimal number or a repeating decimal number. Nonterminating (or infinite), nonrepeating decimal numbers are called **irrational numbers** and are discussed in later sections.

The following examples illustrate nonterminating, repeating decimal numbers.

Example 4

Changing a Fraction to a Decimal Number

Change $\dfrac{1}{3}$ to a decimal number.

Solution

$$3)\overline{1.000}$$
$$\begin{array}{r} 0.333 \\ \underline{9} \\ 10 \\ \underline{9} \\ 10 \\ \underline{9} \\ 1 \end{array}$$

The 3 will repeat without end.

Continuing to divide will give a remainder of 1 each time.

We write

$\dfrac{1}{3} = 0.333...$ The ellipsis (...) means "and so on" or that the digits continue without stopping.

Now work margin exercise 4.

3. Change $-\dfrac{9}{16}$ to a decimal number.

4. Change $\dfrac{2}{9}$ to a decimal number.

5. Change $\dfrac{1}{7}$ to a decimal number.

© Hawkes Learning Systems. All rights reserved.

Example 5

Changing a Fraction to a Decimal Number

Change $\dfrac{3}{7}$ to a decimal number.

Solution

$$
\begin{array}{r}
0.428571 \\
7\overline{)3.000000} \\
\underline{2\,8} \\
20 \\
\underline{14} \\
60 \\
\underline{56} \\
40 \\
\underline{35} \\
50 \\
\underline{49} \\
10 \\
\underline{7} \\
3
\end{array}
$$

The six digits will repeat in the same pattern without end.

The remainders will repeat in sequence: 2, 6, 4, 5, 1, 3, and so on. Therefore, the digits in the quotient will also repeat.

We can write

$$\frac{3}{7} = 0.428571428571428571...$$

Now work margin exercise 5.

Another way to write repeating decimal numbers is to write a **bar** over the repeating pattern of digits. Thus

$$\frac{1}{3} = 0.\overline{3} \ \text{ and } \ \frac{3}{7} = 0.\overline{428571}.$$

Objective B **Operations with Both Decimal Numbers and Fractions**

To evaluate expressions that contain both decimal numbers and fractions, we can change all fractions to decimal form or change all decimal numbers to fraction form. In cases where the decimal form of a fraction involves more than four decimal places or is a nonterminating decimal number, we will agree to round the decimal number to thousandths.

Example 6

Operating with Both Decimal Numbers and Fractions

Find the sum $10\dfrac{1}{2} + 7.64 + 3\dfrac{4}{5}$

a. in decimal form, and   **b.** in fraction form.

Solution

a. Change each fraction to decimal form and then add. In this example, each fraction has a terminating decimal form.

$$10\frac{1}{2} = 10.50$$
$$7.64 = 7.64$$
$$+\quad 3\frac{4}{5} = 3.80$$
$$\overline{\qquad\qquad 21.94}$$

b. Change the decimal number to a fraction; then find a common denominator and add.

$$10\frac{1}{2} = 10\frac{1}{2} = 10\frac{1}{2} = 10\frac{25}{50}$$
$$7.64 = 7\frac{64}{100} = 7\frac{16}{25} = 7\frac{32}{50}$$
$$+\quad 3\frac{4}{5} = 3\frac{4}{5} = 3\frac{4}{5} = 3\frac{40}{50}$$
$$\overline{\qquad\qquad\qquad 20\frac{97}{50} = 21\frac{47}{50}}$$

The answers for parts **a.** and **b.** must be equal, and they are:

$$21\frac{47}{50} = 21\frac{94}{100} = 21.94.$$

6. Find the sum $5\frac{1}{2} + 7.46$ in decimal form and in fraction form.

7. Find the quotient $7\frac{1}{8} \div (-19)$ in decimal form.

Example 7

Operating with Both Decimal Numbers and Fractions

Find the quotient $4\frac{3}{8} \div (-10)$ in decimal form.

Solution

Since $4\frac{3}{8} = 4.375$, we can divide mentally as

$$4\frac{3}{8} \div (-10) = \frac{4.375}{-10} = -0.4375.$$

Now work margin exercises 6 and 7.

8. Find the difference $\dfrac{7}{8} - \dfrac{4}{5}$ in fraction form and in decimal form.

Example 8

Operating with Both Decimal Numbers and Fractions

Find the difference $\dfrac{3}{7} - \dfrac{1}{3}$

a. in fraction form, and   **b.** in decimal form.

Solution

a. Both numbers are already in fraction form and the LCD = 21. We can proceed as follows.

$$\frac{3}{7} - \frac{1}{3} = \frac{3}{7} \cdot \frac{3}{3} - \frac{1}{3} \cdot \frac{7}{7} = \frac{9}{21} - \frac{7}{21} = \frac{2}{21}$$

b. As we have seen earlier in this section, both numbers are nonterminating decimal numbers. Therefore, to subtract in decimal form, we agree beforehand to round each decimal number to a certain place and accept an approximate answer. In this example we write each decimal number accurate to thousandths and subtract.

$\dfrac{3}{7} = 0.429$ (rounded to thousandths) and $\dfrac{1}{3} = 0.333$ (rounded to thousandths)

Subtraction gives the following difference.

$$\begin{array}{r} 0.429 \\ -\,0.333 \\ \hline 0.096 \end{array}$$

Note that 0.096 is an approximation of $\dfrac{2}{21} = 0.\overline{095238}$.

Now work margin exercise 8.

Example 9

Operating with Both Decimal Numbers and Fractions

Evaluate the expression $x^2 + 3x - 2y$ for $x = \dfrac{3}{4}$ and $y = 4.2$.

Solution

Changing $\dfrac{3}{4}$ to decimal form and substituting $x = \dfrac{3}{4} = 0.75$ and $y = 4.2$ we have

$$x^2 + 3x - 2y = (0.75)^2 + 3(0.75) - 2(4.2)$$
$$= 0.5625 + 2.25 - 8.4$$
$$= -5.5875.$$

Changing 4.2 to fraction form and substituting $x = \dfrac{3}{4}$ and $y = 4.2 = \dfrac{42}{10} = \dfrac{21}{5}$, we have the following.

$$x^2 + 3x - 2y = \left(\frac{3}{4}\right)^2 + 3\left(\frac{3}{4}\right) - 2\left(\frac{21}{5}\right)$$

$$= \frac{9}{16} + \frac{9}{4} - \frac{42}{5}$$

$$= \frac{9}{16} \cdot \frac{5}{5} + \frac{9}{4} \cdot \frac{20}{20} - \frac{42}{5} \cdot \frac{16}{16}$$

$$= \frac{45}{80} + \frac{180}{80} - \frac{672}{80}$$

$$= -\frac{447}{80} = -5\frac{47}{80}$$

Because the decimal numbers in this example were all terminating decimal numbers, we get exactly the same answer whether we use the fraction form or the decimal form of each number. In this case,

$$-5\frac{47}{80} = -5.5875.$$

If rounded decimal numbers are used, then the answers will not be exactly the same (only close).

Now work margin exercise 9.

Objective C **Absolute Value**

The definition of absolute value was given in Section 1.8 in terms of integers. This definition is also valid for all decimal numbers.

Absolute Value

The **absolute value** of a decimal number is its distance from 0. The absolute value of a decimal number is nonnegative. We can express the definition symbolically, for any decimal number a, as follows.

$$\begin{cases} \text{If } a \text{ is positive or 0, then } |a| = a. \\ \text{If } a \text{ is negative, then } |a| = -a. \end{cases}$$

10. Find the absolute value of each of the following expressions.

a. $\left|\dfrac{8}{9}\right|$

b. $|-0.007|$

Example 10

Absolute Value

a. $|-5.32| = 5.32$

b. $|16.7| = 16.7$

c. $\left|-\dfrac{3}{4}\right| = \dfrac{3}{4}$

Now work margin exercise 10.

Decimal Numbers, Fractions, and Scientific Notation **Section 3.5** 374

Objective D ## Scientific Notation

A basic application of integer exponents occurs in scientific disciplines, such as astronomy and biology, when very large and very small numbers are involved. For example, the distance from the earth to the sun is approximately 93,000,000 miles, and the approximate radius of a carbon atom is 0.0000000077 centimeters.

In **scientific notation** (an option in all scientific and graphing calculators), **decimal numbers are written as the product of a number greater than or equal to 1 and less than 10, and an integer power of 10**. In scientific notation there is just one digit to the left of the decimal point. For example,

$$250,000 = 2.5 \times 10^5 \quad \text{and} \quad 0.000000345 = 3.45 \times 10^{-7}.$$

The exponent tells how many places the decimal point is to be moved and in what direction. If the exponent is positive, the decimal point is moved to the right.

$$5.6 \times 10^4 = 5.6000.\qquad \text{4 places right}$$

A negative exponent indicates that the decimal point should move to the left.

$$4.9 \times 10^{-3} = 0.004.9 \qquad \text{3 places left}$$

Scientific Notation

If N is a decimal number, then in **scientific notation**

$$N = a \times 10^n \quad \text{where } 1 \le a < 10 \text{ and } n \text{ is an integer.}$$

Example 11

Decimal Numbers in Scientific Notation

Write the following decimal numbers in scientific notation.

a. 8,720,000

Solution

$8{,}720{,}000 = 8.72 \times 10^{6}$ 8.72 is between 1 and 10

To check, move the decimal point 6 places to the right and get the original number.

$8.72 \times 10^{6} = 8.720000. = 8{,}720{,}000$
$$\phantom{8.72 \times 10^{6} = 8.}\underset{1\ 2\ 3\ 4\ 5\ 6}{\curvearrowright}$$

b. 0.000000376

Solution

$0.000000376 = 3.76 \times 10^{-7}$ 3.76 is between 1 and 10

To check, move the decimal point 7 places to the left and get the original number.

$3.76 \times 10^{-7} = 0.0000003.76 = 0.000000376$
$$\phantom{3.76 \times 10^{-7} = 0.}\underset{7\ 6\ 5\ 4\ 3\ 2\ 1}{\curvearrowleft}$$

Now work margin exercise **11.**

11. a. 63,900,000

   b. 0.00000245

Exercises 3.5

1. 0.7

2. 0.6

3. 0.5

4. 0.48

5. 0.016

6. 0.125

7. 8.35

8. 7.2

Change each decimal number to fraction form (or mixed number form), and reduce it if possible. See Example 1.

9. 0.125

10. 1.25

11. 1.8

12. 0.375

13. −2.75

14. −3.45

15. 0.33

16. 0.029

Change each fraction to decimal form. If the decimal number is nonterminating, write it using the bar notation over the repeating pattern of digits. See Examples 2 through 5.

17. $\dfrac{2}{3}$

18. $\dfrac{5}{16}$

19. $\dfrac{-1}{7}$

20. $\dfrac{-5}{9}$

21. $\dfrac{11}{16}$

22. $\dfrac{9}{16}$

23. $\dfrac{7}{11}$

24. $\dfrac{5}{18}$

Change each fraction to decimal form rounded to the nearest thousandth. See Examples 2 through 5.

25. $\dfrac{5}{24}$

26. $\dfrac{1}{32}$

27. $\dfrac{11}{12}$

28. $\dfrac{13}{16}$

29. $\dfrac{40}{7}$

30. $\dfrac{20}{9}$

31. $\dfrac{25}{-11}$

32. $\dfrac{17}{-23}$

33. $\dfrac{1}{7} + 0.355 + \dfrac{2}{3}$

34. $\dfrac{1}{3} + \dfrac{3}{10} + 3.452$

35. $\dfrac{7}{9} + \dfrac{3}{5} + 0.418$

36. $\dfrac{1}{2} + 8\dfrac{1}{6} + 2\dfrac{11}{50}$

37. $25.03 + 35 + 6\dfrac{3}{10} + 4\dfrac{89}{100}$

38. $\dfrac{3}{100} + 8\dfrac{3}{5} + 1\dfrac{3}{4}$

39. $1 - 0.125 - 3.75$

40. $3 - 0.78 - 1.25$

41. $0.375 - 1$

42. $\dfrac{5}{8} - 1$

43. $\left(\dfrac{3}{10}\right)^2 (0.63)$

44. $\left(2\dfrac{1}{10}\right)^2 (1.5)^2$

45. $\left(-1\dfrac{3}{8}\right)(2.1)(3.6)$

46. $\left(2\dfrac{1}{4}\right)\left(-3\dfrac{1}{2}\right)(4.1)$

47. $72.186 \div \dfrac{3}{5}$

48. $917 \div \dfrac{1}{4}$

49. $\left|-3\dfrac{3}{4}\right| - |21.3|$

50. $\left|22\dfrac{4}{5}\right| + |-5.8|$

51. $\left|-5\dfrac{1}{2}\right| + \left|3\dfrac{1}{2}\right| - |-10.7|$

52. $|-4.72| \div \dfrac{8}{9}$

Evaluate each of the following expressions for $x = 1.5 = \dfrac{3}{2}$ and $y = \dfrac{2}{3}$. (Do not use $\dfrac{2}{3}$ in decimal form because you will get only an approximate answer depending on the number of decimal places you use.) Leave the answers in fraction or mixed number form.

53. $x^2 + 3x - 4$

54. $y^2 - 5y + 6$

55. $2x^2 - x - 3$

56. $3y^2 - 5y + 2$

57. $xy^3 - xy$

58. $x^2y^2 + 2xy - 5$

59. $x^3 + 3x^2 + 3x + 1$

60. $x^2 + 6xy + 9y^2$

Write each of the following numbers in scientific notation.

61. 180,000

62. 76,200

63. 0.000124

64. 0.000000912

65. 890

66. 0.4321

67. A light-year (the distance traveled by light in 1 year) is 5,880,000,000,000 miles. Write this number in scientific notation.

68. Light travels approximately 1.86×10^5 miles per second.

    a. Write this number in decimal notation.

    b. Write the equivalent number of miles per minute in scientific notation.

    c. Write the equivalent number of miles per hour in scientific notation.

69. Worms: Nematode sea worms are the most numerous of all sea and land animals, with an estimated population of 4×10^{25}. Write this number in decimal notation.

70. Waste: In 2009, the average American generated 4.34 pounds of municipal solid waste each day. The U.S. population in 2009 was about 306,000,000. Using these facts, find the number of pounds of municipal solid waste produced each day in the United States in 2009. Write your answer in both decimal notation and scientific notation.

71. Water: In 2007, Seattle's per capita water usage was 97 gallons per day. According to the US Census, the population estimate for Seattle in 2007 was 594,210. Using these facts, find the number of gallons of water used each day in Seattle in 2007. Write your answer in both decimal notation and scientific notation.

72. Debt: In 2011, the national debt of the United States was about $14,313,126,000,000. The population of the U.S. at this time was about 311,197,000. What would the average American have needed to contribute at that time to pay off the debt? Write your answer in both decimal notation and in scientific notation.

73. Energy: In 2008, the state of Texas was number one in total consumed energy in the U.S. with 1.15522×10^{16} Btu (British thermal units) of energy, while the state of Vermont was last in energy consumption with 1.544×10^{14} Btu. The U.S. consumed a total of 9.93821×10^{16} Btu. Write these three numbers in decimal notation.

74. About how many minutes does it take for light to reach Earth from the Sun if the distance is about 93,000,000 miles and light travels 186,000 miles per second?

75. Give five examples of values for x and y for which

    a. $|x + y| < |x| + |y|$ and for which

    b. $|x + y| = |x| + |y|$

    c. and **d.** Make a general statement that you believe to be true about the values of x and y in statements **a.** and **b.**

76. If someone made a statement that "$|x + y| > |x| + |y|$ is never possible for any value of x and y," would you believe this statement? Explain why or why not.

77. With **a.**, **b.**, and **c.** as examples, explain in your own words how you can tell quickly when one decimal number is larger (or smaller) than another decimal number.

    a. The decimal number 2.765274 is larger than the decimal number 2.763895.

    b. The decimal number 17.345678 is larger than the decimal number 17.345578.

    c. The decimal number 0.346973 is larger than the decimal number 0.346972.

Objectives

A Understand that percent mean hundredths.

B Know how to compare investments by using percent of profit.

C Be able to change percentages to decimal numbers and decimal numbers to percentages.

D Be able to change fractions to percentages and percentages to fractions.

3.6 Basics of Percent

Objective A **Percent Means Hundredths**

The word **percent** comes from the Latin *per centum*, meaning "per hundred." So, **percent means hundredths**, or **percent is the ratio of a number to 100**. The symbol % is called the **percent symbol** (or **percent sign**). As we shall see, this symbol can be treated as equivalent to the fraction $\frac{1}{100}$.

For example,

$$\frac{25}{100} = 25\left(\frac{1}{100}\right) = 25\% \quad \text{and} \quad \frac{70}{100} = 70\left(\frac{1}{100}\right) = 70\%.$$

In Figure 1, the large square is partitioned into 100 small squares, and each small square represents 1%, or $1 \cdot \frac{\mathbf{1}}{\mathbf{100}}$, of the large square. Thus the shaded portion of the large square is

$$\frac{40}{100} = 40 \cdot \frac{1}{100} = 40\%.$$

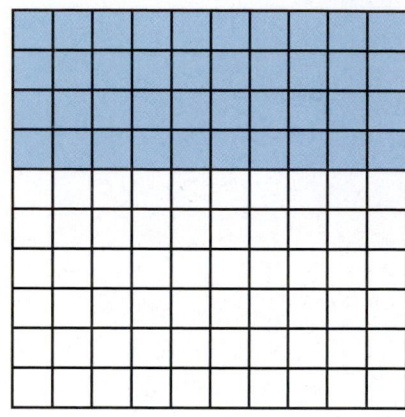

Figure 1

If a fraction has a denominator of 100, then (with no change in the placement of a decimal point or fraction in the numerator) the numerator can be read as a percent by dropping the denominator, 100, and adding on the % symbol.

Example 1

Calculating a Fraction to a Percentage

Each fraction is changed to a percentage.

a. $\frac{30}{100} = 30 \cdot \frac{1}{100} = 30\%$ Remember that percent means hundredths.

b. $\frac{45}{100} = 45\%$ The % symbol indicates hundreds or $\frac{1}{100}$.

c. $\frac{7.3}{100} = 7.3\%$ Note that the decimal point is not moved.

d. $\dfrac{20\frac{1}{2}}{100} = 20\frac{1}{2}\% \quad \text{or} \quad 20.5\%$ Note that the fraction $\frac{1}{2}$ is part of the answer.

e. $\dfrac{250}{100} = 250\%$ If the numerator is larger than 100, then the number is larger than 1 and it is more than 100%.

f. $\dfrac{100}{100} = 1 = 100\%$ All of something is 100% of that thing.

Now work margin exercise 1.

Objective B ## Percent of Profit

Percent of profit is the ratio of money made to money invested. Generally, two investments do not involve the same amount of money, and therefore, the comparative success of each investment cannot be based on the amount of profit. In comparing investments, the investment with the greater percent of profit is considered the better investment.

The use of percentages gives an effective method of comparison because each ratio has the same denominator (100).

Example 2

Percent of profit

Calculate the percent of profit for both **a.** and **b.**, and tell which is the better investment.

a. $200 made as profit by investing $500

b. $270 made as profit by investing $900

Solutions

In each case, find the ratio of dollars of profit to dollars invested and reduce the ratio so that it has a denominator of 100. Do not reduce to lowest terms.

a. $\dfrac{\$200 \text{ profit}}{\$500 \text{ invested}} = \dfrac{\cancel{5} \cdot 40}{\cancel{5} \cdot 100} = \dfrac{40}{100} = 40\%$

b. $\dfrac{\$270 \text{ profit}}{\$900 \text{ invested}} = \dfrac{\cancel{9} \cdot 30}{\cancel{9} \cdot 100} = \dfrac{30}{100} = 30\%$

Investment **a.** is better than investment **b.** because 40% is larger than 30%. Obviously, $270 profit is more than $200 profit, but the money risked ($900) is also greater.

Now work margin exercise 2.

1. Change each fraction to a percentage.

a. $\dfrac{89}{100}$

b. $\dfrac{4.2}{100}$

c. $\dfrac{35\frac{3}{4}}{100}$

2. Calculate percent of profit for both **a.** and **b.** and tell which is the better investment.

a. $630 made as a profit by investing $1800

b. $280 made as a profit by investing $700

3. Tell which is the better investment.

a. $300 made as a profit by investing $6000

b. $550 made as a profit by investing $10,000

Completion Example 3

Percent of Profit

Which of the following is the better investment?

a. An investment of $1000 that makes a profit of $150.

Solution

$$\frac{\$\underline{\hspace{1cm}}\ \text{profit}}{\$1000\ \text{invested}} = \frac{\underline{\hspace{1cm}}}{\underline{\hspace{1cm}} \cdot 100} = \underline{\hspace{1cm}}\%$$

b. An investment of $800 that makes a profit of $128.

Solution

$$\frac{\$\underline{\hspace{1cm}}\ \text{profit}}{\$800\ \text{invested}} = \frac{\underline{\hspace{1cm}}}{\underline{\hspace{1cm}} \cdot 100} = \underline{\hspace{1cm}}\%$$

Thus investment _____ is better because _____% is larger than _____%

Now work margin exercise 3.

Objective C **Decimal Numbers and Percents**

One way to change a decimal number to a percent is to first change the decimal to fraction form with denominator 100 and then change the fraction to percent form.

For example:

Decimal Form		Fraction Form		Percent Form
0.33	=	$\dfrac{33}{100} = 33 \cdot \dfrac{1}{100}$	=	33%
0.74	=	$\dfrac{74}{100} = 74 \cdot \dfrac{1}{100}$	=	74%
0.2	=	$\dfrac{2}{10} = \dfrac{20}{100} = 20 \cdot \dfrac{1}{100}$	=	20%
0.625	=	$\dfrac{62.5}{100} = 62.5 \cdot \dfrac{1}{100}$	=	62.5%

By noting the way that the decimal point is moved in these four examples, we can make the change directly (and more efficiently) by using the following rule regardless of how many digits there are in the decimal number.

Completion Example Answers

3. **a.** $\dfrac{\$150\ \text{profit}}{\$1000\ \text{invested}} = \dfrac{\cancel{10} \cdot 15}{\cancel{10} \cdot 100} = 15\%$

b. $\dfrac{\$128\ \text{profit}}{\$800\ \text{invested}} = \dfrac{\cancel{8} \cdot 16}{\cancel{8} \cdot 100} = 16\%$

Thus investment **b.** is better because 16% is larger than 15%.

To Change a Decimal to a Percentage

Step 1: Move the decimal point two places to the right.
Step 2: Add the % symbol.

These two steps have the effect of multiplying by 100 and then dividing by 100. Thus the number is not changed. Just the form is changed.

Example 4

Changing a Decimal Number to a Percentage

Change each decimal number to an equivalent percent.

a. 0.254

Solution

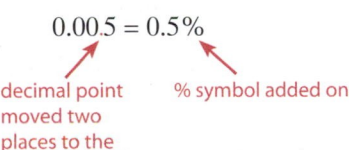

$$0.25.4 = 25.4\%$$

decimal point moved two places to the right   % symbol added on

b. 0.005

Solution

$$0.00.5 = 0.5\%$$ Note that this is less than 1%.

decimal point moved two places to the right   % symbol added on

c. 1.5

Solution

$$1.50. = 150\%$$ Note that this is more than 100%.

decimal point moved two places to the right   % symbol added on

d. 0.2

Solution

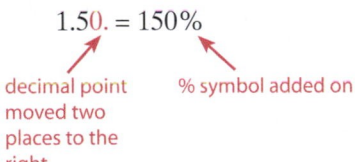

$$0.20. = 20\%$$

decimal point moved two places to the right   % symbol added on

Now work margin exercise 4.

4. Change each decimal number to an equivalent percentage.

a. 0.35

b. 2.17

c. 0.871

d. 0.1

To change percents to decimals, we reverse the procedure for changing decimals to percentages. For example,

$$38\% = 38\left(\frac{1}{100}\right) = \frac{38}{100} = 0.38.$$

As indicated in the following statement, the same result can be found by noting the placement of the decimal point.

To Change a Percentage to a Decimal Number

Step 1: Move the decimal point two places to the left.
Step 2: Delete the % symbol.

5. Change each percent to a decimal number.

a. 21.3%

b. 0.6%

c. 175%

d. 85%

Example 5

Changing a Percentage to a Decimal Number

Change each percent to an equivalent decimal.

a. 38%

Solution

$$38.\% \quad = \quad 0.38 \quad \longleftarrow \text{ \% symbol deleted}$$

↑ understood decimal point

↑ decimal point moved two places left

b. 16.2%

Solution

$$16.2\% = 0.162$$

c. 100%

Solution

$$100\% = 1.00 = 1$$

d. 0.25%

Solution

$$0.25\% = 0.0025 \qquad \text{The percent is less than 1\%, and the decimal is less than 0.01.}$$

Now work margin exercise 5.

The following relationships between decimal numbers and percentages provide helpful guidelines when changing from one form to the other.

A Decimal Number that is

1. less than 0.01 is less than 1%.
2. between 0.01 and 0.10 is between 1% and 10%.
3. between 0.10 and 1.00 is between 10% and 100%.
4. more than 1.00 is more than 100%.

Objective D **Fractions and Percentages**

As we discussed earlier, if a fraction has a denominator of 100, the fraction can be changed to a percentage by writing the numerator and adding on the % symbol. If the denominator is one of the factors of 100,

$$1, 2, 4, 5, 10, 20, 25, \text{ or } 50, \qquad \text{(factors of 100)}$$

we can easily write it in an equivalent form with denominator 100 and then change it to a percent. For example,

$$\frac{3}{4} = \frac{3}{4} \cdot \frac{25}{25} = \frac{75}{100} = 75\% \qquad \text{(Note that } 4 \cdot 25 = 100.\text{)}$$

$$\frac{2}{5} = \frac{2}{5} \cdot \frac{20}{20} = \frac{40}{100} = 40\% \qquad \text{(Note that } 5 \cdot 20 = 100.\text{)}$$

$$\frac{13}{50} = \frac{13}{50} \cdot \frac{2}{2} = \frac{26}{100} = 26\% \qquad \text{(Note that } 50 \cdot 2 = 100.\text{).}$$

However, most fractions do not have denominators that are factors of 100. A more general approach (easily applied with calculators) is to change the fraction to decimal form by dividing the numerator by the denominator, and then change the decimal number to a percentage.

To Change a Fraction to a Percentage

Step 1: Change the fraction to a decimal number.
(Divide the numerator by the denominator.)
Step 2: Change the decimal number to a percentage.

Example 6

Changing a Fraction to a Percentage

Change $\dfrac{5}{8}$ to a percentage.

Solution

First divide 5 by 8 to get the decimal number form. (This can be done with a calculator.)

$$\begin{array}{r} 0.625 \\ 8\overline{)5.000} \\ \underline{48} \\ 20 \\ \underline{16} \\ 40 \\ \underline{40} \\ 0 \end{array}$$

Change 0.625 to a percent.

$$\frac{5}{8} = 0.625 = 62.5\%$$

Example 7

Changing a Fraction to a Percentage

Change $\dfrac{11}{20}$ to a percentage.

Solution

Divide.

$$\begin{array}{r} 0.55 \\ 20\overline{)11.00} \\ \underline{10\ 0} \\ 1\ 00 \\ \underline{1\ 00} \\ \end{array}$$

Thus, $\dfrac{11}{20} = 0.55 = 55\%$.

Or, we can note that 20 is a factor of 100 and write

$$\frac{11}{20} = \frac{11}{20} \cdot \frac{5}{5} = \frac{55}{100} = 55\%.$$

Completion Example 8

Changing a Fraction to a Percentage

Change $3\frac{3}{5}$ to a percent.

Solution

Since $3\frac{3}{5}$ is larger than 1, the percent will be more than _____%.

$$3\frac{3}{5} = \frac{18}{5}$$

Divide.

$$5\overline{)18.0}$$

$$\underline{}$$
$$30$$

$$\underline{}$$
$$0$$

Now, $3\frac{3}{5} = \frac{18}{5} = $ _____ $= $ _____ %.

Completion Example Answers

8. Since $3\frac{3}{5}$ is larger than 1, the percent will be more than 100%.

$$\begin{array}{r} 3.6 \\ 5\overline{)18.0} \\ \underline{15} \\ 30 \\ \underline{30} \\ \end{array}$$

So $3\frac{3}{5} = \frac{18}{5} = 3.6 = 360\%$

Change the following fractions to the equivalent percent.

6. $\dfrac{3}{5}$

7. $\dfrac{27}{50}$

8. $2\dfrac{4}{5}$

9. $\dfrac{4}{9}$

Example 9

Changing a Fraction to a Percentage

Change $\dfrac{1}{3}$ to a percent.

Solution

With a calculator, $\dfrac{1}{3} = 0.333333333$. Rounding gives

$\dfrac{1}{3} = 0.333 = 33.3\%$.   This answer is not exact.

To be exact, we can divide and leave the answer with a fraction.

$$\begin{array}{r} 0.33\frac{1}{3} \\ 3\overline{)1.00} \\ \underline{9} \\ 10 \\ \underline{9} \\ 1 \end{array}$$
$\dfrac{1}{3} = 0.33\dfrac{1}{3} = 33\dfrac{1}{3}\%$ or $33.\overline{3}\%$

$33\dfrac{1}{3}\%$ is exact and 33.3% is rounded.

Both answers are acceptable. However, you should remember that 33.3% is a rounded answer.

Now work margin exercises 6 through 9.

Example 10

College Graduates

In the U. S. in 1900, there were about 27,000 college graduates (received bachelor's degrees), of which about 22,000 were men. In 2003, there were about 1,300,000 college graduates, of which about 550,000 were men.

a. Find the percentage of college graduates that were men in each of those years, which is the number of men graduates divided by total number of graduates in each year.

Solution

Divide with a calculator.

For 1900, $\dfrac{22,000}{27,000} \approx 0.8148 = 81.48\%$ of college graduates were men.

For 2003, $\dfrac{550,000}{1,300,000} \approx 0.4231 = 42.31\%$ of college graduates were men.

b. What was the percentage growth of women college graduates from 1900 to 2003?

Solution

First, find the number of women college graduates in each of the given years.
Then find the growth in women college graduates.
Then find the percent growth by dividing the growth by the original number.

10. Use Example 10 to answer the following questions.

a. Find the percentage of college graduates that were women in 1900 and 2003.

b. What was the total percent increase of graduates from 1900 to 2003?

In 1900		**In 2003**	
27,000	total graduates	1,300,000	total graduates
− 22,000	men graduates	− 550,000	men graduates
5 000	women graduates	750,000	woman graduates

Growth		**Percent Growth**
750,000	women graduates in 2003	
− 5,000	women graduates in 1900	$\dfrac{745,000}{5000} = 149 = 14,900\%$
745,000	growth in women graduates	

Now work margin exercise **10.**

To Change a Percentage to a Fraction or Mixed Number

Step 1: Write the percent as a fraction with denominator 100 and delete the % symbol.
Step 2: Reduce the fraction.

Example 11

Changing a Percentage to a Fraction

Change each percentage to an equivalent fraction in reduced form.

a. 28%

Solution

$$28\% = \frac{28}{100} = \frac{\overset{1}{\cancel{4}} \cdot 7}{\underset{1}{\cancel{4}} \cdot 25} = \frac{7}{25}$$

b. $9\frac{1}{4}\%$

Solution

$$9\frac{1}{4}\% = \frac{9\frac{1}{4}}{100} = \frac{\frac{37}{4}}{100} = \frac{37}{4} \cdot \frac{1}{100} = \frac{37}{400}$$

c. 735%

Solution

$$735\% = \frac{735}{100} = \frac{\overset{1}{\cancel{5}} \cdot 147}{\underset{1}{\cancel{5}} \cdot 20} = \frac{147}{20} \text{ or } 7\frac{7}{20}$$

Common Error

The fractions $\frac{1}{4}$ and $\frac{1}{2}$ are often confused with the percents $\frac{1}{4}\%$ and $\frac{1}{2}\%$. The differences can be clarified using decimal numbers.

Percent	Decimal	Fraction
$\frac{1}{4}\%$ (or 0.25%)	0.0025	$\frac{25}{10,000} = \frac{1}{400}$
$\frac{1}{2}\%$ (or 0.5%)	0.005	$\frac{5}{1000} = \frac{1}{200}$
25%	0.25	$\frac{25}{100} = \frac{1}{4}$
50%	0.50	$\frac{50}{100} = \frac{1}{2}$

Thus

$$\frac{1}{4} = 0.25 \qquad \text{and} \qquad \frac{1}{4}\% = 0.0025$$

$0.25 \neq 0.0025$.

Similarly,

$$\frac{1}{2} = 0.50 \qquad \text{and} \qquad \frac{1}{2}\% = 0.005$$

$0.50 \neq 0.005$.

You can think of $\frac{1}{4}$ as being one-fourth of a dollar (a quarter) and $\frac{1}{4}\%$ as being one-fourth of a penny. $\frac{1}{2}$ can be thought of as one-half of a dollar and $\frac{1}{2}\%$ as one-half of a penny.

Completion Example 12

Changing a Percentage to a Fraction

Change $37\dfrac{1}{2}\%$ to an equivalent fraction in reduced form.

Solution

$$37\dfrac{1}{2}\% = \dfrac{}{100} = \dfrac{}{100}$$

$$= \underline{} \cdot \dfrac{1}{100} = \underline{} = \underline{}$$

Now work margin exercises 11 and 12.

Some percents are so common that their decimal and fraction equivalents should be memorized. Their fractional values are particularly easy to work with, and many times calculations involving these fractions can be done mentally.

Common Percentage – Decimal – Fraction Equivalents

$$1\% = 0.01 = \dfrac{1}{100}$$

$$25\% = 0.25 = \dfrac{1}{4}$$

$$50\% = 0.50 = \dfrac{1}{2}$$

$$75\% = 0.75 = \dfrac{3}{4}$$

$$100\% = 1.00 = 1$$

$$33\dfrac{1}{3}\% = 0.33\dfrac{1}{3} = \dfrac{1}{3}$$

$$66\dfrac{2}{3}\% = 0.66\dfrac{2}{3} = \dfrac{2}{3}$$

$$12\dfrac{1}{2}\% = 0.125 = \dfrac{1}{8}$$

$$37\dfrac{1}{2}\% = 0.375 = \dfrac{3}{8}$$

$$62\dfrac{1}{2}\% = 0.625 = \dfrac{5}{8}$$

$$87\dfrac{1}{2}\% = 0.875 = \dfrac{7}{8}$$

---

11. Change each percentage to an equivalent fraction or mixed number with all fractions reduced.

a. 32%

b. 335%

c. 90.5%

12. Change $63\dfrac{2}{5}\%$ to an equivalent fraction in reduced form.

Completion Example Answers

12. $37\dfrac{1}{2}\% = \dfrac{37\dfrac{1}{2}}{100} = \dfrac{\dfrac{75}{2}}{100} = \dfrac{75}{2} \cdot \dfrac{1}{100} = \dfrac{3 \cdot \cancel{25}}{2 \cdot 4 \cdot \cancel{25}} = \dfrac{3}{8}$

Exercises 3.6

What percentage of each square is shaded?

1.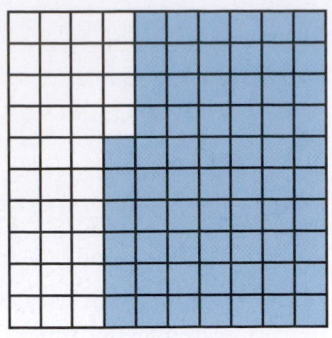

2.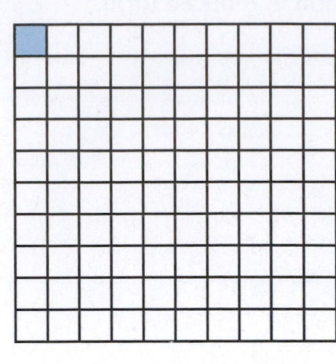

3.

4.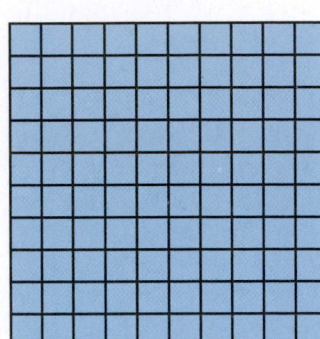

5.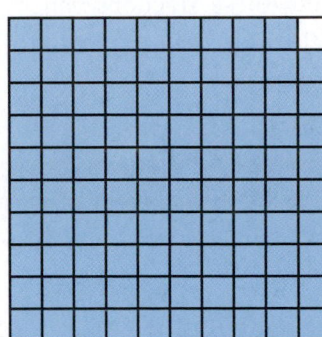

6.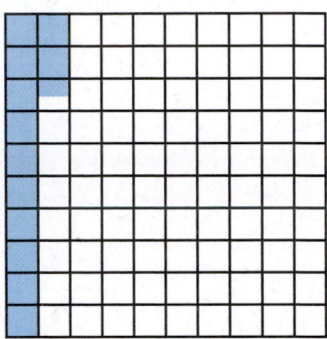

Change the following fractions to percentages.

7. $\dfrac{30}{100}$

8. $\dfrac{90}{100}$

9. $\dfrac{48}{100}$

10. $\dfrac{125}{100}$

11. $\dfrac{16.3}{100}$

12. $\dfrac{0.5}{100}$

13. $\dfrac{24\frac{1}{2}}{100}$

14. $\dfrac{17\frac{3}{4}}{100}$

Write the ratio of profit to investment as hundredths and compare the percentages. In part c., tell which investment is better, a. or b.

15. a. A profit of $38 on a $200 investment.

    b. A profit of $51 on a $300 investment.

    c. Which investment is better?

16. a. A profit of $70 on a $700 investment.

    b. A profit of $100 on a $1000 investment.

    c. Which investment is better?

17. **a.** A profit of $150 on a $3000 investment.

   b. A profit of $160 on a $4000 investment.

   c. Which investment is better?

18. **a.** A profit of $300 on a $2000 investment.

   b. A profit of $480 on a $4000 investment.

   c. Which investment is better?

Change the following decimal numbers to percentages.

19. 0.03	**20.** 0.3	**21.** 3.0	**22.** 0.52
23. 0.055	**24.** 0.004	**25.** 1.75	**26.** 2.3
27. 1.08	**28.** 2	**29.** 0.36	**30.** 0.5

Change the following percentages to decimal numbers.

31. 2%	**32.** 8%	**33.** 22%	**34.** 15%
35. 25%	**36.** 150%	**37.** 10.1%	**38.** 12.5%
39. $6\frac{1}{2}\%$	**40.** $15\frac{1}{4}\%$	**41.** 80%	**42.** $219\frac{3}{4}\%$

Change the following fractions and mixed numbers to percentages.

43. $\frac{1}{20}$	**44.** $\frac{7}{10}$	**45.** $\frac{24}{25}$	**46.** $\frac{1}{9}$
47. $\frac{1}{7}$	**48.** $\frac{5}{6}$	**49.** $1\frac{3}{8}$	**50.** $2\frac{1}{15}$

Change the following percentages to fractions or mixed numbers in reduced form.

51. 5%	**52.** 50%	**53.** $12\frac{1}{2}\%$	**54.** $16\frac{2}{3}\%$
55. 120%	**56.** 140%	**57.** 0.2%	**58.** 0.75%

Fraction Form	Decimal Number Form	Percent Form
59. $\dfrac{7}{8}$	**b.** _____	**c.** _____
60. $\dfrac{19}{20}$	**b.** _____	**c.** _____
61. a.	0.06	**c.** _____
62. a.	1.8	**c.** _____
63. a.	**b.** _____	24%
64. a.	**b.** _____	$66\dfrac{2}{3}\%$

Solve the following problems.

65. The interest rate on a loan is 15%. Change 15% to a decimal number.

66. The sales commission for the clerk in a retail store is figured at $8\dfrac{1}{2}\%$. Change $8\dfrac{1}{2}\%$ to a decimal.

67. To calculate what your maximum monthly house payment should be, a banker multiplied your monthly income by 0.32. Change 0.32 to a percentage.

68. Suppose that the state motor vehicle licensing fee is figured by multiplying the cost of your car by 0.009. Change 0.009 to a percentage.

69. A department store offers a 35% discount during a special sale on dresses. Change 35% to a fraction reduced to lowest terms.

70. The discount you earn by paying cash is found by multiplying the amount of your purchase by 0.025. Change 0.025 to a percentage.

71. **California:** In the 2000 census, California was ranked the most populous state with about 33,872,000 people. Wyoming was the least populous state with about 494,000 people, and the entire U.S. population was about 281,422,000. In 2010 census, California was again the most populous state with about 37,254,000 people. Wyoming was the least populous state with about 564,000 people, and the entire U.S. population was about 312,471,000.

a. By what percentage did the population of California grow in the ten years from one census to the other?

b. By what percentage did the population of Wyoming grow?

c. By what percentage did the U.S. population grow?

d. What percentage of the entire U.S. population lived in California in 2010?

72. In the United States as a whole, males are outnumbered by a ratio of 100 to 95.8. What percentage of the U. S. population is female?

73. A credit card company charges 20.71% interest per year for credit. Write this interest rate in decimal number form.

74. Three pairs of fractions are given. In each case, tell which fraction would be easier to change to a percentage mentally, and explain your reasoning.

a. $\dfrac{3}{20}$ or $\dfrac{1}{8}$

b. $\dfrac{1}{10}$ or $\dfrac{1}{9}$

c. $\dfrac{5}{12}$ or $\dfrac{4}{5}$

Writing & Thinking

75. Use your calculator to find the decimal number form for each of the two fractions $\dfrac{3}{11}$ and $\dfrac{2}{3}$. Does your calculator round at the place of the last digit on the display? Explain how you can tell.

(**Note:** Some calculators will round and others will "truncate" (simply cut off) the remaining digits. Check with your instructor to find out which calculators used in class round and which calculators truncate.)

3.7 Applications of Percent

The Problem-Solving Process

Objectives

A Become familiar with the problem-solving process.

B Learn the meaning of the terms discount and sales tax and be able to calculate these values.

C Understand the concept of commission.

D Understand the concepts of profit and percent of profit.

George Pòlya (1887 – 1985), a famous professor at Stanford University, studied the process of discovery learning. Among his many accomplishments, he developed the following four-step process as an approach to problem solving.

1. Understand the problem.
2. Devise a plan.
3. Carry out the plan.
4. Look back over the results.

For a complete discussion of these ideas, see *How To Solve It* by Pòlya (Princeton University Press, 1945, 2nd edition, 1957).

There are a variety of types of applications discussed throughout this text and subsequent courses in mathematics, and you will find these four steps helpful as guidelines for understanding and solving all of them. Applying the necessary skills to solve exercises, such as adding fractions or solving equations, is not the same as accumulating the knowledge to solve problems. **Problem solving can involve careful reading, reflection, and some original or independent thought**.

Basic Steps for Solving Word Problems

1. Understand the problem. For example,
   a. Read the problem.
   b. Understand all the words.
   c. If it helps, restate the problem in your own words.
   d. Be sure that there is enough information.

2. Devise a plan using, for example, one or all of the following:
   a. Guess, estimate, or make a list of possibilities.
   b. Draw a picture or diagram.
   c. Use a variable and form an equation.

3. Carry out the plan. For example,
   a. Try all the possibilities you have listed.
   b. Study your picture or diagram for insight into the solution.
   c. Solve any equation that you may have set up.

4. Look back over the results. For example,
   a. Can you see an easier way to solve the problem?
   b. Does your solution actually work? Does it make sense in terms of the wording of the problems? Is it reasonable?
   c. If there is an equation, check your answer in the equation.

In this section, we will be concerned with applications that involve percent and the types of percent problems that were discussed in Section 3.6. The use of a calculator is recommended, although you should keep in mind that a calculator is a tool to enhance, and not replace, the necessary skills and abilities related to problem solving. Also, your personal experience, knowledge, and general understanding of problem-solving situations will determine the level of difficulty of some problems. The following examples illustrate some basic strategies for solving applications with percent.

Objective B **Discount and Sales Tax**

To attract customers or to sell goods that have been in stock for some time, retailers and manufacturers offer a **discount**, a reduction in the **original selling price**. The new, reduced price is called the **sale price**, and the discount is the difference between the original price and the sale price. The **rate of discount** (or **percent of discount**) is a percent of the original price.

Discounts are sometimes advertised as a percent "off" the selling price. For example,

"The White Sale is now 20% off the original price."

or

"Computers are on sale at a discount of 15% off the list price."

Sales tax is a tax charged on goods sold by retailers, and it is assessed by states and cities for income to operate various services. The **rate of the sales tax** varies from state to state (or even city to city in some cases). In fact, some states do not have a sales tax at all.

Terms Related to Discount and Sales Tax	
Discount:	reduction in original selling price
Sale price:	original selling price minus the discount
Rate of discount:	percent of original price to be discounted
Sales tax:	tax based on actual selling price
Rate of sales tax:	percent of actual selling price

As you study the examples and work through the exercises, be sure to **label each amount of money as to what it represents (such as original price, discount, sale price, sales tax, profit, and so forth)**. This will help in organizing the results and help in determining what operations to perform.

1. A new front-load washing machine that regularly sells for $880 is on sale at a 30% discount.

   a. What is the amount of the discount?

   b. What is the sale price?

Example 1

Calculating Discount

A new refrigerator that regularly sells for $1200 is on sale at a 20% discount. What is the amount of the discount? What is the sale price?

Solution

Step 1: Read the problem carefully. Do you understand all the words?

Step 2: Make a plan. The plan here is

    a. to find the amount of the discount, then

    b. subtract this amount from the original price to find the sale price.

Step 3: Carry out the plan as shown below by multiplying and then subtracting.

Step 4: Check to see that the answer makes sense.

    (For example, "Does the sale price seem to be 20% less than the original price?")

a. Find the discount: 20% of 1200 is _____.

$$0.20 \cdot 1200 = A$$
$$240 = A$$

1200	original price
0.20	rate of discount
240.00	discount

The discount is $240.

b. Find the sale price. The problem does not say specifically how to find the sale price. We know from experience, however, the meaning of the word "discount" and that a discount is subtracted from the original price.

$1200.00	original price
− 240.00	discount
960.00	sale price

The sale price is $960.00.

Now work margin exercise 1.

Example 2

Calculating Final Cost

If sales tax is calculated at $7\frac{1}{4}\%$, what is the final cost of the refrigerator in Example 1?

Solution

Step 1: Read the problem carefully.

Step 2: Make a plan. The plan is to find the tax on the answer from Example 1 and then add the tax to the sale price to find the final cost.

Step 3: Carry out the plan as shown below by multiplying and then adding.

Step 4: Check to see that the answer is reasonable. (For example, "Is the total of the sale price plus the tax still less than the original price?" This would certainly indicate an error.)

Find the amount of the sales tax $\left(7\frac{1}{4}\% = 7.25\% = 0.0725 \right)$

$7\frac{1}{4}\%$ of \$960 is _____.

$$0.0725 \cdot 960 = A$$
$$69.60 = A$$

$$\begin{array}{r} 960 \\ 0.0725 \\ \hline 4800 \\ 1\,920 \\ 67\,20 \\ \hline 69.6000 \end{array}$$

The sales tax is \$69.60.

From experience, we know to add tax to the sale price to find the final cost.

$$\begin{array}{r} \$\ 960.00 \\ +\quad 69.60 \\ \hline \$1029.60 \end{array}$$

The final cost of the refrigerator is \$1029.60.

Now work margin exercise 2.

3. Leather bean bag chairs are on sale at the local department store for $48.80. If this was a discount of 20%, what was the original price?

Example 3

Calculating Original Price

Large fluffy towels were on sale at a discount of 30%. If the sale price was $8.40, what was the original price?

Solution

Step 1: Read the problem carefully. Read it two or three times until you understand the nature of the problem and all the terms. This problem involves critical thinking before any calculation.

Step 2: Make a plan. First realize that we are **not** trying to find the discount. We already know the sale price. We need to realize that the sale price is 70% of the original price. (100% − 30% = 70%)

Therefore, the plan is to set up a percent problem and solve the related equation.

(**Note:** Do **not** take 30% of $8.40. The sale price of $8.40 is **not** 30% of the original price.)

Step 3: Carry out the plan as shown below by setting up the equation $0.70\,B = 8.40$ and then solving the equation.

Step 4: Check to see that the answer makes sense.

(For example, "Is the original price more than $8.40?")

70% of _____ is $8.40

$$0.70 \cdot B = 8.40$$
$$\frac{0.70 \cdot B}{0.70} = \frac{8.40}{0.70}$$
$$B = \$12.00$$

$$\begin{array}{r} 12. \\ 0.70.)\overline{8.40.} \\ \underline{70} \\ 140 \\ \underline{140} \\ 0 \end{array}$$

The original price of the towels was $12.00 each.

Now work margin exercise 3.

Objective C **Commissions**

A **commission** is a fee paid to an agent or salesperson for a service. Commissions are usually a percent of a negotiated contract or a percent of sales. In some cases, salespeople earn a straight commission on what they sell. In other cases, as illustrated in Example 4, the salesperson earns a base salary plus a commission on sales above a certain level.

Example 4

Calculating Commission

Susan earns a salary of $1100 a month plus a commission of 8% on whatever she sells after she has sold $8500 in merchandise. Her co-worker sold $15,000 in merchandise that month. What did Susan earn the month she sold $22,500 in merchandise?

Solution

Step 1: Read the problem carefully. Do you understand all the terms?

Step 2: Make a plan. Find the commission and add it to the salary. Remember, the commission is made only on the amount over $8500.

Step 3: Carry out the plan as shown below by finding the base for the commission. Multiply this base by 8% and add the result to $1100.

Step 4: Make sure that the answer is reasonable.
(For example, "Is the income about right for someone who makes a salary of $1100 a month?" Is the income over $1100?)

First, find the base for her commission. Since the commission is based on what she sells over $8500, we subtract $8500 from her sales.

$$\begin{array}{r} \$22,500 \\ -\ \ 8,500 \\ \hline \$14,000 \end{array}$$

Now find the amount of the commission by finding 8% of the base.

$A = 8\%$ of $\$14,000$

$A = 0.08 \cdot 14,000$

$A = \$1120$

Now add the amount of the commission to her salary to find her income for the month.

$$\begin{array}{r} \$1100.00 \\ +\ 1120.00 \\ \hline \$2220.00 \end{array}$$

She earned $2220 for the month.

Now work margin exercise 4.

You should read each problem carefully, and be aware that sometimes numerical data given in a problem is not related to the question posed by the problem. In Example 4, $15,000 sold by her co-worker is simply extra information. The intention is to help you develop better reasoning skills.

4. A salesperson earns a salary of $1500 a month plus a 10% commission on whatever he or she sells after he or she has sold $7700 in merchandise. What will the salesperson earn if he or she sold $11,000 of merchandise in one month?

Manufacturers and retailers are concerned with the profit on each item produced or sold. In this sense, the **profit on an item** is the difference between the selling price to the customer and the cost to the company. For example, suppose that a department store can buy a certain light fixture from a manufacturer for $40 (the store's cost) and sell the fixture for $50 (the selling price). The profit for the store is the difference between the selling price and the cost of the fixture. That is,

$$\text{profit} = \$50 - \$40 = \$10.$$

The **percent of profit is a ratio**. And, as the following discussion indicates, this ratio can be based either on cost or on selling price.

Terms Related to Profit

Profit: The difference between selling price and cost.
profit = selling price − cost

Percent of Profit: There are two types; both are ratios with **profit in the numerator**.

1. Percent of profit **based on cost** is the ratio of profit to cost:
    $$\frac{\text{Profit}}{\text{Cost}} = \% \text{ of profit based on cost.} \quad \text{(Cost is in the denominator.)}$$

2. Percent of profit **based on selling price** is the ratio of profit to selling price:
    $$\frac{\text{Profit}}{\text{Selling price}} = \% \text{ of profit based on selling price.}$$
    (Selling price is in the denominator.)

The fact that a percent of profit is a ratio can be seen by looking at the formula for percent problems: $R \cdot B = A$. By dividing both sides of this equation by B, we get the following result.

$$R \cdot B = A$$

$$\frac{R \cdot \cancel{B}}{\cancel{B}} = \frac{A}{B}$$

$$R = \frac{A}{B}$$

Thus the rate (or percent) is a ratio. Remember this: $R = \dfrac{A}{B}$.

Example 5

Calculating Profit and Percent of Profit

A retail store markets calculators that cost the store $45 each and are sold to customers for $60 each.

a. What is the profit on each calculator?

Solution

First find the profit.

$$
\begin{array}{r}
\$60.00 \\
-\ 45.00 \\
\hline
\$15.00
\end{array}
$$

The profit is $15 per calculator.

For **b.** and **c.**, use a ratio to find each percent of profit, and change the fraction to a percent.

b. What is the percent of profit based on cost?

Solution

For profit based on cost, remember that cost is in the denominator.

$$\frac{\$15 \text{ profit}}{\$45 \text{ cost}} = \frac{1}{3} = 33\frac{1}{3}\%$$

c. What is the percent of profit based on selling price?

Solution

For profit based on selling price, remember that selling price is in the denominator.

$$\frac{\$15 \text{ profit}}{\$60 \text{ selling price}} = \frac{1}{4} = 25\%$$

Now work margin exercise 5.

5. A music store sells certain CDs for $15 each when the CDs actually cost the store $6.

a. What is the profit on each CD?

b. What is the percent of profit based on cost?

c. What is the percent of profit based on selling price?

notes

Percent of profit **based on cost** is always higher than percent of profit **based on selling price** because the selling price is larger than the cost. The business community reports whichever percent serves its purpose better. Your responsibility as an investor or consumer is to know which percent is reported and what it means to you.

Remember to write the steps of a problem and the results in a neat, organized form even if you perform the actual calculations with a calculator.

Exercises 3.7

The following problems may involve several calculations, and calculators are recommended. Follow Pòlya's four-step process for problem solving.

Step 1: Make sure that you read the problem until you understand it. Write down the known information.

Step 2: Make a plan for solving the problem.

Step 3: Then follow through with your plan.

Step 4: Go over your answer to see if it makes sense to you. In some cases you may think of a better plan (or another plan) after you have solved the problem. This type of thinking may prove useful in solving future problems.

1. **Commission:** A realtor works on 6% commission. What is his commission on a house he sold for $125,000?

2. **Commission:** A car saleswoman earns a commission of 7% on each car she sells. How much did she earn on the sale of a car for $12,500?

3. **Salary:** A sales clerk receives a monthly salary of $500 plus a commission of 6% on all sales over $3500. What did the clerk earn the month she sold $8000 in merchandise?

4. **Bonus:** A computer programmer was told that he would be given a bonus of 5% of any money his programs could save the company. How much would he have to save the company to earn a bonus of $500?

5. **Tax:** The property taxes on a house were $1050. What was the tax rate if the house was valued at $70,000?

6. **Tax:** If sales tax is figured at 7.25%, how much tax will be added to the total purchase price of three textbooks, priced at $25.00, $35.00, and $52.00?

7. **Basketball:** In one season a basketball player missed 15% of her free throws. How many free throws did she make if she attempted 160 free throws?

8. **Discount:** An auto dealer paid $8730 for a large order of special parts. This was not the original price. The amount paid reflects a 3% discount off the original price because the dealer paid cash. What was the original price of the parts?

9. **Test:** A student missed 6 problems on a mathematics test and received a grade of 85%. If all the problems were of equal value, how many problems were on the test?

10. **Football:** A kicker on a professional football team made 45 of 48 field goal attempts.

    a. What percent of his attempts did he make?

    b. What percent did he miss? Would you keep this player on your team or trade for a new kicker?

11. **Sale:** At a department store linen sale, sheets were originally marked $22.50 and pillowcases were originally marked $7.50. What is the sale price of

    a. sheets and

    b. pillowcases if each item is discounted 25% from the original marked price? Round your answer to the nearest cent.

12. **Loan:** You want to purchase a new home for $122,000. The bank will loan you 80% of the purchase price. How much will the bank loan you? This amount is called your mortgage and you will pay it off over several years with interest. For example, a 30-year loan will probably cost you a total of more than 3 times the original loan amount.

13. Selling Your House: Suppose you sell your three bedroom home for $100,000 and you owe the balance of the mortgage of $55,000 to the bank. You pay a real estate agent a fee of 6% of the selling price and other fees and taxes that total $1500. How much cash do you have after the sale? (You may have to pay income taxes later.)

14. Sale: The Golf Pro Shop had a set of 10 golf clubs that were marked on sale for $860. This was a discount of 20% off the original selling price.

  a. What was the original selling price?

  b. If the clubs cost the Golf Pro Shop $602, what was its profit?

  c. What was the shop's percent of profit based on cost?

  d. What was the percent of profit based on the sale price?

15. Profit: The cost of a 20-inch television set to a store owner was $450, and she sold the set for $630.

  a. What was her profit?

  b. What was her percent of profit based on cost?

  c. What was her percent of profit based on selling price?

16. Selling Price: A car dealer bought a five year old used car for $2500. He marked up the price so that he would make a profit of 25% based on his cost.

  a. What was the selling price?

  b. If the customer paid 8% of the selling price in taxes and fees, what was the customer's total cost for the car?

17. Income: In one year Mr. Hill, who is 35 years old, earned $32,000. He spent $9600 on rent, $10,240 on food, and $3840 on taxes. What percent of his income did he spend on each of those items?

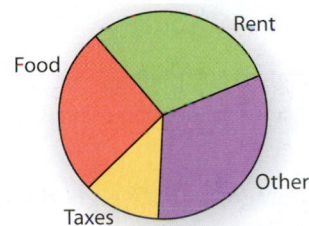

18. Discount: To get more subscribers, a book club offered three books with an original total selling price of $61.75 for a special price of $12.35.

  a. What was the amount of the discount?

  b. Based on the original selling price of these books, what was the rate of discount on these three books? Do you think this would be a reasonable thing for the book club to do? Why?

19. Books: The author of a book was told that she would have to cut the number of pages by 14% for the book to sell at a popular price and still make a profit.

  a. If these cuts were made, what percent of the original number of pages was in the final version?

  b. If the finished book contained 258 pages, how many pages were in the original form?

  c. How many pages were cut?

20. Dresses: A department store received a shipment of dresses together with the bill for $2856.50. Some of the dresses were not as ordered, however, and were returned at once. The value of the returned dresses was $340.15. The terms of the billing provided the store with a 2% discount if it paid cash within 2 weeks. What did the store pay for the dresses it kept if it paid cash within 2 weeks?

21. a. A new computer has two DVD disk drives and a 1000 gigabyte hard disk drive with 1.5 gigabytes of RAM. If the sales tax on the computer was $153 and tax was figured at 6% of the selling price, what was the selling price?

b. What did the customer pay for the computer?

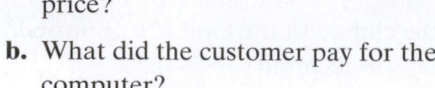 **22. New Car:** The total discount on a new car was $1499.40. This included a rebate from the manufacturer of $1000.

a. What was the original price of the car if the total discount was 7% of the original price?

b. What would a customer pay for the car if taxes were 5% of the final selling price and license fees were $642.60?

23. Discount: The discount on a new coat was $175. This was a discount of 20%.

a. What was the original selling price of the coat?

b. What was the sale price?

c. What would a customer pay for the coat if a 7.25% sales tax was added to the sale price?

24. Discount: Blank CDs were on sale for $27.00 (90 CDs come in one package). This price was a discount of 10% off the original price.

a. Was the original price more than $27 or less than $27?

b. What was the original price?

25. Discount: In the roofing business, shingles are sold by the "square," which is enough material to cover a 10 ft by 10 ft square (or 100 square feet). A roofing supplier has a closeout on shingles at a 40% discount.

a. If the original price was $260 per square, what is the sale price per square?

b. How much would a roofer pay for 35 squares?

Writing & Thinking

26. One shoe salesman worked on a straight 9% commission. His friend worked on a salary of $400 plus a 5% commission.

a. How much did each salesman make during the month in which each sold $7500 worth of shoes?

b. What percent more did the salesman who made the most make? Explain why there is more than one answer to part **b.**

27. A man weighed 200 pounds. He lost 20 pounds in 3 months. Then he gained back 20 pounds 2 months later.

a. What percent of his weight did he lose in the first 3 months?

b. What percent of his weight did he gain back? The loss and the gain are the same, but the two percents are different. Explain why.

Collaborative Learning

28. With the class separated into teams of two to four students, each team is to analyze the following problem and decide how to answer the related questions. Then each team leader is to present the team's answers and related ideas to the class for general discussion.

Jerry works in a bookstore and gets a salary of $500 per month plus a commission of 3% on whatever he sells over $2000. Wilma works in the same store, but she has decided to work on a straight 8% commission.

a. At what amount of sales will Jerry and Wilma make the same amount of money?

b. Up to that point, who would be making more?

c. After that point, who would be making more? Explain briefly. (If you were offered a job at this bookstore, which method of payment would you choose?)

3.8 Simple and Compound Interest

Objective A   **Understanding Simple Interest (*I = Prt*)**

The following terms and ideas are related to interest.

Interest:	money paid for the use of money
Principal:	money that is invested or borrowed
Rate:	percent of interest (stated as an annual interest rate)

Regardless of whether you are a borrower, lender, or investor, the calculations for finding interest are the same. Although interest rates can vary from year to year (or even daily), the concept of interest is the same throughout the world.

A **note** is a loan for a period of 1 year or less, and the interest earned (or paid) is called **simple interest**. A note involves only one payment at the end of the term of the note and includes both principal and interest. If the borrower wants the money for more than 1 year, terms can be arranged in which the borrower pays the interest at the end of 1 year and then renegotiates a new note (or extension of the old note) for some new period of time. Eventually, the borrower must pay back the original loan amount plus any interest earned.

The following formula is used to calculate simple interest.

Formula for Calculating Simple Interest

$I = Prt$ where

I = Interest (earned or paid)

P = Principal (the amount invested or borrowed)

r = Rate of interest (stated as an annual or yearly rate)

t = Time (in years or fraction of a year)

Although the rate of interest is generally given in the form of a percentage, the calculations in the formula are made by changing the percentage into decimal form or fraction form.

Example 1

Calculating Simple Interest

You want to borrow $2000 at 12% interest for 3 years. How much interest would you pay?

Solution

$P = \$2000$

$r = 12\% = 0.12$ (in decimal form) or $\dfrac{12}{100}$ (in fraction form)

$t = 3$ years

Now, use the formula $I = Prt$, with

$$I = 2000 \cdot 0.12 \cdot 3 = 2000 \cdot 0.36 = \$720.00.$$

You would pay $720 in interest if you borrowed $2000 for 3 years at 12%.

Example 2

Calculating Simple Interest

Carmen loaned $500 to a friend for 6 months at an interest rate of 8%. How much will her friend pay her at the end of the 6 months?

Solution

$$P = \$500 \qquad r = 8\% \qquad t = 6 \text{ months} = \frac{6}{12} \text{ year} = \frac{1}{2} \text{ year}$$

The interest is found by using the formula $I = Prt$.

$$I = 500 \cdot \overset{0.04}{\cancel{0.08}} \cdot \frac{1}{\cancel{2}} = 500 \cdot 0.04 = \$20.00$$

The interest is $20 and the total amount to be paid at the end of 6 months is

principal + interest = $500 + $20 = $520.

Example 3

Calculating Simple Interest

What principal would you need to invest to earn $450 in interest in 6 months if the rate of interest was 9%?

Solution

In this problem we know the interest, rate, and time, and want to find the principal P.

$$I = \$450 \qquad r = 9\% = 0.09 \qquad t = 6 \text{ months } = \frac{1}{2} \text{ year}$$

Substitute into the formula $I = Prt$ and solve for P.

$$450 = P \cdot 0.09 \cdot \frac{1}{2}$$

$$2 \cdot 450 = P \cdot 0.09 \cdot \frac{1}{\cancel{2}} \cdot \cancel{2} \qquad \text{Multiply both sides by 2.}$$

$$900 = P \cdot 0.09$$

$$\frac{900}{0.09} = \frac{P \cdot \cancel{0.09}}{\cancel{0.09}} \qquad \text{Divide both sides by 0.09.}$$

$$10,000 = P$$

You would need a principal amount of $10,000 invested at 9% to make $450 in 6 months.

1. You want to borrow $3500 at 15% interest for 1 year. How much interest would you pay?

2. Stacey loaned her aunt $2000 at 10% interest. How much interest did she earn if her aunt paid her back after 9 months?

3. What principal would you need to invest to earn $170 in interest in 5 months if the rate of interest was 6%?

4. What interest rate would you be paying if you borrowed $3000 for 1 year and paid $480 in interest?

Completion Example 4

Calculating Simple Interest

Find the rate of interest that would be paid if $50 interest was to be earned on $2000 in 3 months.

Solution

The unknown quantity is _____ .

$I = \$$ _____ .
$P = \$$ _____ .
$t = 3$ months = _____ year.

Substituting into the formula $I = Prt$ gives

$$50 = \underline{\quad} \cdot r \cdot \frac{1}{4}$$

$$50 = \underline{\quad} \cdot r$$

$$\frac{50}{\underline{\quad}} = \frac{\underline{\quad} \cdot r}{\underline{\quad}}$$

$$\underline{\quad} = r$$

The interest rate would be _____%.

Now work margin exercises 1 through 4.

Objective B Using the Formula $I = Prt$ Repeatedly To Calculate Compound Interest

Interest paid on interest earned is called **compound interest**. To calculate compound interest, we can calculate the simple interest for each period of time that interest is compounded using a **new principal for each calculation**. This new principal is **the previous principal plus the earned interest**. The calculations can be performed in a step-by-step manner, as indicated in the following outline:

Completion Example Answers

4.   The unknown quantity is r.

$I = \$50$ $\qquad$ $P = \$2000$ $\qquad$ $t = 3$ months $= \frac{1}{4}$ year

Substituting into the formula $I = Prt$ gives

$$50 = 2000 \cdot r \cdot \frac{1}{4}$$

$$50 = 500 \cdot r \qquad \text{Simplify the right-hand side.}$$

$$\frac{50}{500} = \frac{500 \cdot r}{500} \qquad \text{Divide both sides by 500, the coefficient of } r.$$

$$0.10 = r$$

The interest rate would be 10%.

1. Using the formula $I = Prt$, calculate the simple interest where $t = \dfrac{1}{n}$ and n is the number of periods per year for compounding. For example:

For compounding annually, $n = 1$ and $t = \dfrac{1}{1} = 1$

For compounding semiannually, $n = 2$ and $t = \dfrac{1}{2}$

For compounding quarterly, $n = 4$ and $t = \dfrac{1}{4}$

For compounding monthly, $n = 12$ and $t = \dfrac{1}{12}$

2. Add this interest to the principal to create a new value for the principal.

3. Repeat steps 1 and 2 however many times the interest is to be compounded.

In Example 5 compound interest is calculated in the step–by–step manner just outlined. This process is, to say the least, somewhat laborious and time-consuming. However, it does serve to develop a basic understanding of the concept of compound interest. After this example, formulas are used for calculating compound interest, inflation, depreciation, and current value.

Example 5

Calculating Compound Interest

If an account is compounded monthly ($n = 12$) at 6%, how much interest will a principal of $5500 earn in 3 months?

Solution

Since the compounding is monthly, use $t = \dfrac{1}{n} = \dfrac{1}{12}$ in the formula $I = Prt$.

And, since the period is three months, calculate the interest three times. Calculate each time with a new principal.

a. First period: the principal is $P = \$5500$.

$$I = 5500 \cdot 0.06 \cdot \frac{1}{12} = 330 \cdot \frac{1}{12} = \$27.50 \qquad \text{interest for first period}$$

b. Second period: the new principal is $P = \$5500 + \$27.50 = \$5527.50$.

$$I = 5527.50 \cdot 0.06 \cdot \frac{1}{12} = 331.65 \cdot \frac{1}{12} \approx \$27.64 \qquad \text{interest for second period}$$

5. If an account is compounded quarterly at 8%, how much interest will be earned in 9 months on an investment of $4000? What will be the balance in the account?

c. Third period: the new principal is $P = \$5527.50 + \$27.64 = \$5555.14$.

$$I = 5555.14 \cdot 0.06 \cdot \frac{1}{12} \approx 333.31 \cdot \frac{1}{12} \approx \$27.78 \quad \text{interest for third period}$$

The new principal is $P = \$5610.55 + \$27.78 = \$5582.92$.

To find the total interest earned in three months, find the sum of the interest for each period, or subtract the original principal from the final principal as follows.

$\$\ 27.50$ interest		$\$\ 5582.92$ final principal
27.64 interest	OR	-5500.00 original principal
$+\quad 27.78$ interest		$\$\ 82.92$ total interest earned
$\$\ 82.92$ total interest earned		

Now work margin exercise 5.

Calculating Compound Interest with the Formula

The steps outlined in Example 5 illustrate how to adjust the principal for each new time period of the compounding process and how the interest increases for each new time period. The interest is greater each time period because the new adjusted principal (the old principal plus the interest over the previous time period) is greater. In this process, we have calculated the actual interest for each time period. This process is simplified by using the compound interest formula.

The following compound interest formula does not calculate the actual interest for each time period. In fact, this formula does not calculate the interest directly. This formula does calculate the total accumulated **amount** (also called the **future value** of the principal).

Compound Interest Formula

When interest is compounded, the total **amount** A accumulated (including principal and interest) is given by the formula

$$A = P\left(1 + \frac{r}{n}\right)^{nt} \quad \text{where}$$

P = the principal
r = the annual interest rate
t = the length of time in years
n = the number of compounding periods in 1 year

Examples 6 and 8 illustrate, step-by-step, how to use a calculator in working with the formula. Example 7 illustrates how to find the total interest earned.

Example 6

 Calculating Compound Interest

Max invested $4500 at 9% to be compounded monthly. What will be the amount in his account in 5 years?

Solution

$P = \$4500,\quad r = 9\% = 0.09,\quad n = 12$ times per year,$\quad t = 5$ years

Substituting into the formula gives the following.

$$A = 4500\left(1 + \frac{0.09}{12}\right)^{12(5)}$$

$$= 4500(1 + 0.0075)^{60}$$

$$= 4500(1.0075)^{60}$$

$$= \$7045.56 \quad (\text{rounded to nearest cent})$$

A scientific calculator automatically follows the rules for order of operations. Therefore, you can enter all of the numbers and parentheses just as you see them placed in the formula. You can enter the exponent at 60 or enter it as the product of 12 times 5.

The product must be in parentheses as (12 * 5).

Step 1: Enter the numbers and parentheses as follows.

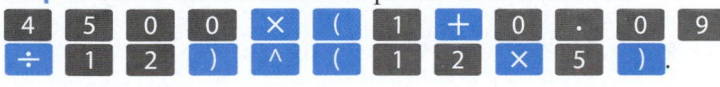

Step 2: Press ⬛. The display should read 7045.564621.

Whether we do the calculations step-by-step or let the calculator do all the steps internally, we find that the amount in Max's account in 5 years will be $7045.56.

Now work margin exercise 6.

Total Interest Earned

To find the total interest earned on an investment that has earned interest by compounding, subtract the initial principal from the accumulated amount.

$$I = A - P$$

6. Dana invested $3000 in a special account at her bank at 4.5% interest to be compounded monthly. What will be the amount in her account after 4 years?

7. How much interest did Dana earn in the investment described in margin exercise 6?

Example 7

Calculating Total Interest Earned

How much interest did Max earn in the investment described in Example 6?

Solution

$$I \quad = A - P$$

$$= 7045.56 - 4500.00$$

$$= 2545.56$$

Max earned $2545.56 in interest.

Now work margin exercise 7.

Completion Example 8

  Using the Compound Interest Formula

a. Use the compound interest formula to find the value of $6000 invested for 4 years if it is compounded quarterly at 12%.

Solution

With a scientific calculator, follow the steps outlined in the first part of Example 6 with

$$P = \$6000, \quad r = 12\% = 0.12, \quad n = 4, \quad t = 4.$$

Substituting into the formula gives

$$A = 6000\left(1 + \underline{}\right)^{4(\underline{})}$$

$$= 6000\left(1 + \underline{}\right)^{\overline{}}$$

$$= 6000\left(1.\underline{}\right)^{\overline{}}$$

$$\approx 6000\left(\underline{}\right)$$

$$\approx \underline{}$$

The value (or amount) will be $\underline{}$.

Completion Example Answers

a. Substituting into the formula, gives

$$A = 6000\left(1 + \frac{0.12}{4}\right)^{4 \cdot 4}$$

$$= 6000\left(1 + 0.03\right)^{16}$$

$$= 6000\left(1.03\right)^{16}$$

$$\approx 6000\left(1.604706439\right)$$

$$\approx 9628.238$$

The value (or amount) will be $9628.24.

b. Find the amount of interest earned.

Solution

The interest earned will be

$I = \$\underline{\hspace{2cm}} - \$6000.00 = \$\underline{\hspace{2cm}}.$

Now work margin exercise 8.

8. a. Use the compound interest formula to find the value of \$10,000 invested for 5 years if it is compounded every 6 months at 9% interest.

b. Find the amount of interest earned.

Objective D **Inflation and Depreciation**

Inflation (also known as the **cost-of-living index**) is a measure of your relative purchasing power. For example, if inflation is at 6% annually, then you will need to increase your income by 6% each year to be able to afford to buy the same items (or to live in the same lifestyle) that you did the year before. Many worker bargaining committees tie their salary requests to the government's cost-of-living index each year.

Inflation can be treated in the same manner as interest compounded annually (once a year). That is, we can use the compound interest formula with $n = 1$. For example, consider what you will be paying for a \$3.00 loaf of bread in 20 years if annual inflation is at 5%.

$$A = 3.00\left(1 + \frac{0.05}{1}\right)^{1 \cdot 20} = 3.00(1 + 0.05)^{20} = 7.96 \quad \text{(rounded to the nearest cent)}$$

Note that this is not a super-improved loaf of bread; it is just the same bread you are eating today. In fact, if you get raises in pay of 3% each year you may not be able to afford a whole loaf. Consider the \$3.00 inflated at 3% annually for 20 years.

$$A = 3.00(1 + 0.03)^{20} = 5.42$$

Thus you could afford to pay \$5.42 but not \$7.96 for the bread. Example 9 illustrates this idea by considering the annual growth in a salary based on an inflation rate that is assumed to remain steady at 6% each year.

Formula for Inflation

The formula for the accumulated amount A due to **inflation** is the same as the formula for compound interest with $n = 1$.

$$A = P(1 + r)^t$$

Completion Example Answers

b. The interest will be
$I = \$9628.24 - \$6000.00 = \$3629.24.$

9. Suppose that your income is $1800 per month or $21,600 per year and that you will receive a cost-of-living raise each year. If inflation is steady at 4%, find your income in 15 years.

Example 9

 Calculating Inflation

Suppose that your income is $2400 per month (or $28,800 per year) and that you will receive a cost-of-living raise each year. If inflation is steady at 6% each year, find your income in:

a. 5 years **b.** 10 years **c.** 20 years.

Solution

We answer this question by using the formula $A = 28800(1+0.06)^t$ and let $t = 5$, $t = 10$, and $t = 20$ and round to the nearest cent.

for $t = 5$: $\qquad A = 28800(1+0.06)^5 \approx 38{,}540.90$

for $t = 10$: $\qquad A = 28800(1+0.06)^{10} \approx 51{,}576.41$

for $t = 20$: $\qquad A = 28800(1+0.06)^{20} \approx 92{,}365.50$

Thus, if you stay on the same job (or a similar job) for 20 years and inflation is at 6%, you will need an income of $92,365.50 to have the same buying power you have today if your annual income is $28,800.

[Try this based on different cost-of-living indexes and with different base salaries. You will probably be very surprised. This will help you to understand the effect of compound interest.]

Now work margin exercise 9.

Depreciation is the decrease in value of an item. Depreciation is used to determine the value of property and machinery (also called **capital goods**) for income tax purposes, and if we know the original value and the rate of depreciation of an item, we can calculate its current market value. For example, if a car was purchased for $15,000 and it depreciates 20% each year, then its value each year is 80% of its value the previous year.

value after 1 year: $15,000(0.80) = $12,000

first year

value after 2 years: $\left[\$15{,}000(0.80)\right](0.80) = \$12{,}000(0.80) = \$9600$

or

$$\$15{,}000(0.80)^2 = 15{,}000(0.64) = \$9600.$$

Formula for Depreciation

The **current value**, *V*, of an item due to **depreciation** can be determined from the formula

$$V = P(1-r)^t,$$

where

P = the original value
r = the annual rate of depreciation (in decimal form)
t = the time in years

Example 10

Calculating Depreciation

Suppose that a certain make of automobile depreciates 15% each year. Find the current market value of one of these automobiles if it is 5 years old and its original cost was $40,000.

Solution

$P = \$40,000, \quad r = 15\% = 0.15, \quad t = 5$ years

Using the formula for depreciation, we have the following.

$$V = P(1-r)^t = 40000(1-0.15)^5$$

Putting this in the calculator gives us 17748.2125.

The current market value of the automobile is $17,748.21.

Now work margin exercise 10.

10. Suppose that a boat depreciates at a rate of 18% each year. Find the current market value of a boat if it is 7 years old and its original cost was $13,000.

Exercises 3.8

Answer the following word problems.

1. How much simple interest would be paid on a loan of $1000 at 15% for 3 months? (**Note:** 3 months = $\frac{3}{12}$ year = $\frac{1}{4}$ year.)

2. Paul loaned his uncle $1500 at an interest rate of 10% for 9 months. How much simple interest did he earn? (**Note:** 9 months = $\frac{9}{12}$ year = $\frac{3}{4}$ year.)

3. How much simple interest is earned on a loan of $5000 at 11% for a period of 6 months? (**Note:** 6 months = $\frac{6}{12}$ year = $\frac{1}{2}$ year.)

4. If you borrow $750 for 2 years at 18%, how much simple interest will you pay?

5. Find the simple interest paid on a savings account at $1800 for 4 months at 8%.

6. A savings account of $3200 is left for 1 year and draws interest at a rate of 7%.
   a. How much interest is earned?
   b. What is the balance in the account at the end of the 1-year period?

7. For how many months must you leave $1000 in a savings account at 5.5% to earn $11.00 in interest? (**Hint:** Find the value of t in the formula $I = Prt$.)

8. What is the rate of interest charged if a 3-month loan of $2500 is paid off with $2562.50? (**Hint:** The payoff is principal plus interest. So first find the interest earned.)

9. What is the rate of interest charged if a 9-month loan of $3000 is paid off with $3225.00? (**Hint:** The payoff is principal plus interest. So first find the interest earned.)

10. A savings account of $25,000 is drawing interest at 8%.
    a. How much interest will be earned in 6 months?
    b. How long will it take for the account to earn $1500 in interest?

11. You buy an oven on sale with the price reduced from $500 to $450. The store has allowed you to keep the oven for 1 year without making a payment. However, they are charging you interest on what you owe at a rate of 18%.
    a. How much will you end up paying for the oven if you pay the total amount owed in 3 months?
    b. How much did you save by buying the oven on sale?
    c. For what reason did the store allow you to keep the oven for 1 year before paying?

12. A new computer is on sale for $2200, and no payments need to be made for 6 months. The original price was $2500. The terms are that the buyer will pay the total amount owed at the end of 6 months, and the interest rate is 21%.
    a. How much will the buyer end up paying for the computer?
    b. How much will the buyer save by paying cash on the purchase date instead of waiting 6 months to pay what is owed?
    c. If the buyer waits 6 months to pay, how much will he or she save from the original price by buying the computer on sale?

13. Determine the missing item in each row and complete the table.

Principal	Rate	Time	Interest
$ 400	16%	3 months	**a.**$_____
b.$_____	15%	4 months	$ 5
$ 560	12%	**c.**_____	$ 5.60
$ 2700	**d.**_____	1 year	$ 25.50

14. Determine the missing item in each row and complete the table.

Principal	Rate	Time	Interest
$ 1000	$10\frac{1}{2}$%	2 months	**a.**$_____
$ 800	$13\frac{1}{2}$%	**b.**_____	$ 18.00
$ 2000	**c.**_____	9 months	$ 172.50
d.$_____	7.5%	1 year	$ 85.00

Show the calculations for each period of compounding in the step-by-step manner illustrated in Example 5.

15. a. If a bank compounds interest quarterly at 8% on a certificate of deposit, what will your $7500 deposit be worth in 6 months?

b. In 1 year?

16. a. If an account is compounded quarterly at a rate of 6%, how much interest will be earned on $5000 in 1 year?

b. What will be the total amount in the account?

17. You borrowed $5000 and agreed to make equal payments of $1000 each plus interest over the next 5 years. Interest is at a rate of 8% based only on what you owe.

a. How much interest will you pay?

b. How much interest would you have paid if you did not make the annual payments and paid only the $5000 plus interest compounded annually at the end of 5 years?

18. An amount of $9000 is deposited in a savings account and interest is compounded monthly at 10%. What will be the balance of the account in 6 months?

19. a. How much interest will be earned on a savings account of $3000 in 2 years if interest is compounded annually at 7.25%?

b. If interest is compounded semiannually?

20. If interest is compounded semiannually at 10%, what will be the value of $15,000 in $1\frac{1}{2}$ years?

 Use the following formulas whenever they apply.

$$A = P\left(1+\frac{r}{n}\right)^{nt}, \quad I = A - P, \quad A = P(1+r)^t, \quad \text{and} \quad V = P(1-r)^t$$

21. a. What will be the value of a $20,000 savings account at the end of 3 years if interest is calculated at 10% compounded annually?

b. Suppose the interest is compounded semiannually. What is the value? Is the value the same?

c. If not, explain why not in your own words.

22. a. Calculate the interest earned in one year on $10,000 compounded monthly at 14%.

b. What is the difference between this and simple interest at 14% for 1 year?

23. Suppose that $50,000 is invested in a certificate of deposit for 5 years and the interest rate is 8%. What will be the interest earned if it is compounded monthly?

24. The value of your house today is $125,000. If inflation is at 5%, what will be its value, to the nearest thousand dollars, in 30 years?

25. The current prices of four items are given below. What will be their prices in 10 years if inflation is at 6%?
   a. An SUV: $41,000

   b. A television set: $1850

   c. A textbook: $95.00

   d. A cup of coffee: $1.25

26. If a new pickup truck is valued at $18,000, what will be its value in 3 years if it depreciates 22% each year?

27. Suppose that an apartment complex is purchased for $1,500,000. For property tax purposes, the land is considered to be 30% of the value of the property. For income tax purposes, the owners are allowed to depreciate the value of the buildings (capital goods) by 5% per year. What will be the value of the apartment complex (buildings and land) in 10 years? (**Note:** This will not be the market value, but it will form the basis for capital gains taxes when the property is sold.)

28. a. Calculate the interest you would pay on a 1-year loan of $7500 at 18% if the interest were compounded every 3 months and you made no monthly payments.

b. If you made payments of $1000 (against principal and interest) every 3 months and paid the balance plus interest owed at the end of the year, how much interest would you pay?

Find the amount (*A*) and the interest earned (*I*) for the given information.

	Compounding Period	Principle (*P*)	Annual Rate	Time	Amount (*A*)	Interest Earned (*I*) $I = A - P$
29.	Quarterly	$1000	10%	5 yr	a. _____	b. _____
30.	Monthly	$1000	10%	5 yr	a. _____	b. _____
31.	Semiannually	$1000	10%	5 yr	a. _____	b. _____
32.	Monthly	$5000	7.5%	10 yr	a. _____	b. _____
33.	Semiannually	$25,000	8%	20 yr	a. _____	b. _____
34.	Yearly	$25,000	12%	20 yr	a. _____	b. _____

Solve the following inflation and depreciation problems. See Examples 9 and 10.

35. Rent: Kevin currently spends $1300 per month on rent and utilities. If the current annual inflation rate is 3%, how much should he plan to spend on rent and utilities each month 2 years from now?

36. Groceries: In 2011, the average price for a gallon of milk was $3.00, and the average price for a loaf of bread was $2.50. If the inflation rate is 6% per year, how much will these items cost in 2016?

37. Income: Sam receives a cost-of-living raise each year. If inflation was steady at 5% annually, and his current income is $56,800, how much money did he make 4 years ago?

38. New Car: Brenda's car is valued at $24,000 today. She plans to sell her current car and buy a new car in 5 years. If the car depreciates at 12% per year, how much will money will her car be worth when she is ready to upgrade?

39. Used Boat: Stephen has a fishing boat that he bought 3 years ago. He has decided to sell it, and the boat is valued at $8,500. If the yearly rate of depreciation for his boat is 13.2%, how much did he originally pay for the boat?

40. Truck: Stan bought a new truck last year for $29,900. This year, he decided that he wants to trade it in for a smaller car. He can resell the truck for $26,500. What was the rate of depreciation for the year?

41. Use your calculator and choose the values of *t* (in years) to use in the formula for compound interest until you find how many years of weekly compounding at 8% are needed for an investment of $5000 to **approximately** double in value. Write down the values you chose for *t*, why you chose those particular values, and the corresponding accumulated values of money. Explain why you agree or disagree with the idea that $10,000 would double in a shorter time. (**Note:** This is a type of trial-and-error exercise. There is no best way to do the problem. However, with some practice, you should develop an understanding of compound interest so that you can do this type of problem in a more efficient manner each time.)

t	$\left(1+\dfrac{0.08}{52}\right)^{52t}$	A

42. **a.** What will be the value of $10,000 compounded weekly at 10% for 3 years?

b. Use your calculator and choose values for *t* to use in the formula until you find approximately how many years of weekly compounding are needed for the value to accumulate to $20,000.

t	$\left(1+\dfrac{0.10}{52}\right)^{52t}$	A

Collaborative Learning

With the class separated into teams of two to four students, each team is to analyze the following problem related to compound interest. Each team leader is to discuss the results found by the team and how the team arrived at these results. A general classroom discussion should follow with the class coming to an understanding of the concepts of **present value** and **future value**.

43. Suppose that you would like to set aside some money today for your child's college education. Your child is 3 years old and will be going to college (you hope) at the age of 18. What amount should you invest today (called the **present value**) at 8% compounded quarterly to accumulate $40,000 (called the **future value**) for your child's education?

3.9 Square Roots

Objective A   **Square Roots and Real Numbers**

Objectives

A   Understand the concepts of square roots and real numbers.

B   Learn how to use a calculator to find square roots.

A number is **squared** when it is multiplied by itself. For example,

$$5^2 = 5 \cdot 5 = 25$$

$$(-20)^2 = (-20)(-20) = 400$$

$$\text{and } (3.2)^2 = (3.2)(3.2) = 10.24.$$

If an integer is squared, the result is called a **perfect square**. Thus 25 and 400 are both perfect squares and 10.24 is not a perfect square. Table 1 shows the perfect squares found by squaring the positive integers from 1 to 20.

Squares of Positive Integers from 1 to 20 (Perfect Squares)				
$1^2 = 1$	$5^2 = 25$	$9^2 = 81$	$13^2 = 169$	$17^2 = 289$
$2^2 = 4$	$6^2 = 36$	$10^2 = 100$	$14^2 = 196$	$18^2 = 324$
$3^2 = 9$	$7^2 = 49$	$11^2 = 121$	$15^2 = 225$	$19^2 = 361$
$4^2 = 16$	$8^2 = 64$	$12^2 = 144$	$16^2 = 256$	$20^2 = 400$

Table 1

Now we want to reverse the process of squaring. That is, given a number, we want to find a number that when squared will result in the given number. This is called **finding the square root** of the given number. For example, find the **square root** of 49. What number squared will result in 49?

Since

$$7^2 = 49, \text{ the number 7 is called the } \textbf{square root} \text{ of 49,}$$

and since

$$10^2 = 100, \text{ the number 10 is called the } \textbf{square root} \text{ of 100.}$$

We write

$$\sqrt{49} = 7 \quad \text{and} \quad \sqrt{100} = 10.$$

Similarly,

$$\sqrt{25} = 5 \quad \text{since} \quad 5^2 = 25,$$
$$\sqrt{121} = 11 \quad \text{since} \quad 11^2 = 121, \text{ and}$$
$$\sqrt{1} = 1 \quad \text{since} \quad 1^2 = 1.$$

The symbol $\sqrt{}$ is called a **radical sign**.

The number under the radical sign is called the **radicand**.

The complete expression, such as $\sqrt{49}$, is called a **radical**.

Each perfect square has two square roots, one positive and one negative. For example, since $(-6)^2 = 36$ and $6^2 = 36$, both -6 and 6 are square roots of 36. However, to distinguish between the two square roots, the positive square root, 6, is called the **principal square root**. And, unless otherwise stated, the term **square root** is understood to mean only the positive (or principal) square root. Thus

$$\sqrt{36} = 6 \qquad \text{and} \qquad -\sqrt{36} = -6.$$

As a special case, for 0, we have

$$\sqrt{0} = -\sqrt{0} = 0.$$

Table 2 contains the square roots of the perfect square whole numbers from 1 to 400. (Note that the values shown in Table 2 are found by reversing the process illustrated in Table 1.)

Squares Roots of Perfect Squares from 1 to 400				
$\sqrt{1} = 1$	$\sqrt{25} = 5$	$\sqrt{81} = 9$	$\sqrt{169} = 13$	$\sqrt{289} = 17$
$\sqrt{4} = 2$	$\sqrt{36} = 6$	$\sqrt{100} = 10$	$\sqrt{196} = 14$	$\sqrt{324} = 18$
$\sqrt{9} = 3$	$\sqrt{49} = 7$	$\sqrt{121} = 11$	$\sqrt{225} = 15$	$\sqrt{361} = 19$
$\sqrt{16} = 4$	$\sqrt{64} = 8$	$\sqrt{144} = 12$	$\sqrt{256} = 16$	$\sqrt{400} = 20$

Table 2

1. Refer to Table 2 to find the following square roots.

a. $\sqrt{36}$

b. $\sqrt{324}$

c. $\sqrt{196}$

Example 1

Finding Square Roots

Refer to Table 2 to find the following square roots.

a. $\sqrt{100}$

Solution

$\sqrt{100} = 10$

b. $\sqrt{169}$

Solution

$\sqrt{169} = 13$

c. $-\sqrt{121}$

Solution

$-\sqrt{121} = -11$

d. $-\sqrt{361}$

Solution

$-\sqrt{361} = -19$

Now work margin exercise 1.

Many square roots, such as $\sqrt{2}$, $\sqrt{3}$, and $\sqrt{24}$ (square roots of numbers other than perfect squares), are **irrational numbers**, and their decimal forms are **infinite nonrepeating decimals**. In fact, computers have been used to calculate π accurate to over 1 billion decimal places!

Real numbers are numbers that are either rational or irrational. **That is, every rational number and every irrational number is a real number.** The relationships among the various types of numbers are shown in the diagram in Figure 3.

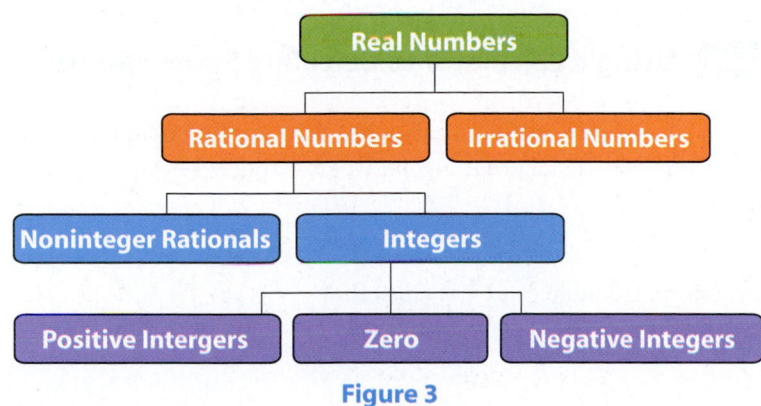

Figure 3

The definition of square root can be stated in terms of real numbers.

Square Root

For any real number a and any nonnegative real number b, a is a **square root** of b if $a^2 = b$. If a is positive, then we write $a = \sqrt{b}$.

Thus $\left(\sqrt{b}\right)^2 = b$.

Example 2

Evaluating Expressions Containing Square Roots

Find the value of each of the following expressions.

a. $\left(\sqrt{81}\right)^2$

Solution

$\left(\sqrt{81}\right)^2 = \left(9\right)^2 = 81$

b. $\left(\sqrt{25}\right)^2$

Solution

$\left(\sqrt{25}\right)^2 = \left(5\right)^2 = 25$

2. Find the value of each expression.

a. $\left(\sqrt{64}\right)^2$

b. $\left(\sqrt{10}\right)^2$

c. $\left(\sqrt{3}\right)^2$

Solution

Even though we do not know the exact decimal value of $\sqrt{3}$, its square must be 3. Thus $\left(\sqrt{3}\right)^2 = 3$.

Now work margin exercise 2.

Objective B Using a Calculator to Calculate Square Roots

The values of most square roots can be only approximated with decimal numbers. Consider the fact that 5 is between 4 and 9. Using inequality signs we can write

$$4 < 5 < 9.$$

Thus it seems reasonable (and it is true) that

$$\sqrt{4} < \sqrt{5} < \sqrt{9},$$

and we have

$$2 < \sqrt{5} < 3.$$

Therefore, even though we may not know the exact value of $\sqrt{5}$, we do know that it is between 2 and 3.

Decimal number approximations to $\sqrt{5}$ are shown here to illustrate more accurate estimates.

$$(2.2)^2 = 4.84 < 5$$
$$(2.23)^2 = 4.9729 < 5$$
$$(2.236)^2 = 4.999696 < 5$$
$$(2.237)^2 = 5.004169 > 5$$

Thus we can now say that $\sqrt{5}$ is between 2.236 and 2.237. Or, symbolically,

$$2.236 < \sqrt{5} < 2.237.$$

Example 3

 Evaluating Square Roots Using a Calculator

Find $\sqrt{2}$ rounded to the nearest thousandth by using a calculator.

Solution

To find the square root of a number, you will need to find the square root key, $\boxed{\sqrt{}}$, and the equal sign key, $\boxed{=}$.

Therefore, press $\boxed{\sqrt{}}$ followed by $\boxed{2}$. Then press $\boxed{=}$.

The display will read 1.414213562.

Rounding to the nearest thousandth, we will have **1.414**.

Example 4

 Evaluating Square Roots Using a Calculator

Find $\sqrt{18}$ rounded to the nearest thousandth by using a calculator.

Solution

Following the steps from Example 2, press $\boxed{\sqrt{}}$ followed by $\boxed{1}$ $\boxed{8}$. Then press $\boxed{=}$.

The display will read 4.242640687.

Rounding to the nearest thousandth, we will have **4.243**.

Now work margin exercises 3 and 4.

Use a calculator to find the following square roots accurate to the nearest thousandth.

3. $\sqrt{5}$

4. $\sqrt{7.5}$

Exercises 3.9

State whether or not each number is a perfect square.

1. 121   **2.** 144   **3.** 48   **4.** 16

5. 400   **6.** 256   **7.** 225   **8.** 45

9. 40   **10.** 196

Find the value of each of the following expressions. See Example 2.

11. $\left(\sqrt{7}\right)^2$ **12.** $\left(\sqrt{8}\right)^2$ **13.** $\left(\sqrt{36}\right)^2$ **14.** $\left(\sqrt{100}\right)^2$

15. $\left(\sqrt{16}\right)^2$ **16.** $\left(\sqrt{49}\right)^2$ **17.** $\left(\sqrt{21}\right)^2$ **18.** $\left(\sqrt{39}\right)^2$

19. $\left(\sqrt{206}\right)^2$ **20.** $\left(\sqrt{352}\right)^2$

21. **a.** Do you think that $\sqrt{39}$ is more than 6 or less than 6? Explain your reasoning.

     b. Do you think that $\sqrt{39}$ is more than 7 or less than 7? Explain your reasoning.

22. **a.** Do you think that $\sqrt{18}$ is more than 4 or less than 4? Explain your reasoning.

     b. Do you think that $\sqrt{18}$ is more than 5 or less than 5? Explain your reasoning.

23. Show that $(1.4142)^2 < 2$ and $(1.4143)^2 > 2$. Tell what these two facts indicate about $\sqrt{2}$.

24. Show that $(3.1622)^2 < 10$ and $(3.1623)^2 > 10$. Tell what these two facts indicate about $\sqrt{10}$.

For part a., use your understanding of square roots to estimate the value of each square root. Then use your calculator to find the value of each square root accurate to the nearest ten-thousandth.

25. **a.** The nearest integers to $\sqrt{23}$ are _____ and _____ .
     b. Find the value of $\sqrt{23}$.

26. **a.** The nearest integers to $\sqrt{28}$ are _____ and _____ .
     b. Find the value of $\sqrt{28}$.

27. a. $\sqrt{13}$ is between the two integers _____ and _____ .

    b. Find the value of $\sqrt{13}$.

28. a. $\sqrt{11}$ is between the two integers _____ and _____ .

    b. Find the value of $\sqrt{11}$.

29. a. The nearest integers to $\sqrt{72}$ are _____ and _____ .

    b. The value of $\sqrt{72}$ is _____ .

30. a. The nearest integers to $\sqrt{50}$ are _____ and _____ .

    b. The value of $\sqrt{50}$ is _____ .

31. a. The nearest integers to $\sqrt{60}$ are _____ and _____ .

    b. The value of $\sqrt{60}$ is _____ .

32. a. $\sqrt{80}$ is between the integers _____ and _____ .

    b. The value of $\sqrt{80}$ is _____ .

33. a. $\sqrt{95}$ is between the integers _____ and _____ .

    b. The value of $\sqrt{95}$ is _____ .

 Use your calculator to find the value of each of the following expressions accurate to 4 decimal places, or the nearest ten-thousandth. See Example 3 and 4.

34. $1+\sqrt{2}$

35. $1-\sqrt{2}$

36. $2-\sqrt{5}$

37. $3+2\sqrt{3}$

38. $4-3\sqrt{2}$

39. $1+2\sqrt{6}$

40. $\sqrt{2}+\sqrt{3}$

41. $\sqrt{25}-\sqrt{36}$

42. $\sqrt{16}+\sqrt{4}$

Writing & Thinking

43. Cubes of integers are called perfect cubes. For example, because $1^3 = 1$, $2^3 = 8$, and $9^3 = 729$, the numbers 1, 8, 27, and 343 are perfect cubes. Find all the perfect cubes from 1 to 1000.

44. Cube roots are symbolized by $\sqrt[3]{}$. For example because $5^3 = 125$ we have $\sqrt[3]{125} = 5$. Find the following cube roots.

    a. $\sqrt[3]{8}$ **b.** $\sqrt[3]{27}$

    c. $\sqrt[3]{64}$ **d.** $\sqrt[3]{343}$

Section 3.1: Reading, Writing, and Rounding Decimal Numbers

Decimal Numbers   page 321

A number that is represented by decimal notation is a **decimal number**.

3 Classifications of Decimal Numbers   page 321
1. finite (or terminating) decimal numbers
2. infinite repeating decimal numbers
3. infinite nonrepeating decimal numbers

To Read or Write a Decimal Number   page 322
1. Read or write the whole number.
2. Read or write **and** in place of the decimal point.
3. Read or write the fraction part as a whole number with the name of the place of the last digit on the right.

Rule for Rounding Decimal Numbers   pages 325 – 326
1. Look at the single digit just to the right of the place of desired accuracy.
2. If this digit is 5 or greater, make the digit in the desired place of accuracy one larger and replace all digits to the right with zeros. All digits to the left remain unchanged unless a 9 is made one larger, and then the next digit to the left is increased by 1.
3. If this digit is less than 5, leave the digit in the desired place of accuracy as it is and replace all digits to the right with zeros. All digits to the left remain unchanged.
4. Trailing 0's to the right of the place of accuracy must be dropped so that the place of accuracy is clearly understood. If a rounded number has a 9 in the desired place of accuracy, then that 0 remains.

Section 3.2: Addition and Subtraction with Decimal Numbers

To Add Decimal Numbers: page 331
1. Write the addends in a vertical column.
2. Keep the decimal points aligned vertically.
3. Keep digits with the same position value aligned. (Zeros may be filled in as aids.)
4. Add the numbers, keeping the decimal point in the sum aligned with the other decimal points.

To Subtract Decimal Numbers: pages 332 – 333
1. Write the numbers in a vertical column.
2. Keep the decimal points aligned vertically.
3. Keep digits with the same position value aligned. (Zeros may be filled in as aids.)
4. Subtract, keeping the decimal point in the difference aligned with the other decimal points.

To Multiply Positive Decimal Numbers: pages 340 – 341
1. Multiply the two numbers as if they were whole numbers.
2. Count the total number of places to the right of the decimal points in both numbers being multiplied.
3. Place the decimal point in the product so that the number of places to the right is the same as that found in step 2.

To Multiply a Decimal Number by a Power of 10:   pages 341 – 342
1. Move the decimal point to the right.
2. Move it the same number of places as the number of 0's in the power of 10.

Multiplication by **10** moves the decimal point **one** place **to the right**;
Multiplication by **100** moves the decimal point **two** places **to the right**;
Multiplication by **1000** moves the decimal point **three** places **to the right**; and so on.

To Divide Decimal Numbers: page 342 – 344
1. Move the decimal point in the divisor to the right so that the divisor is a whole number.
2. Move the decimal point in the dividend the same number of places to the right.
3. Place the decimal point in the quotient directly above the new decimal point in the dividend.
4. Divide just as with whole numbers.

When the Remainder is Not Zero   pages 345 – 346
1. Decide how many decimal places are to be in the quotient.
2. Divide until the quotient is one digit past the place of desired accuracy.
3. Using this last digit, round the quotient to the desired place of accuracy.

To Divide a Decimal Number by a Power of 10:   page 347
1. Move the decimal point **to the left**.
2. Move it the same number of places as the number of 0's in the power of 10.

Division by **10** moves the decimal point **one** place **to the left**.
Division by **100** moves the decimal point **two** places **to the left**.
Division by **1000** moves the decimal point **three** places **to the left**; and so on.

Section 3.4: Measures of Center

Statistic   page 357

A **statistic** is a particular measure or characteristic of a part of a **sample** from a larger collection of items called the **population**.

Terms Used in the Study of Statistics   page 356

Data: Value(s) measuring some information of interest.

Statistic: A single number describing some characteristic of the data.

Mean: The arithmetic average of the data.

Median: The middle of the data after the data have been arranged in order.

Mode: The single data item that appears most frequently.

Range: The difference between the largest and smallest data items.

To Find the Median: page 359 – 360

1. Rank the data. (Arrange the data in order, either from smallest to largest or largest to smallest.)
2. The median can be found by counting from the top down (or from the bottom up) to the position $\frac{n+1}{2}$ where n represents the number of data items.
   a. If there are an **odd** number of items, the median is the middle item.
   b. If there are an **even** number of items, the median is the value found by averaging the two middle items. (**Note:** This value may or may not be in the data.)

Section 3.5: Decimal Numbers, Fractions, and Scientific Notation

Changing from Decimal Form to Fraction Form: pages 368 – 369

A decimal number with digits to the right of the decimal point can be written in fraction form by writing a fraction with

1. A **numerator** that consists of the whole number formed by all the digits of the decimal number and
2. A **denominator** that is the power of the 10 that names the rightmost digit.

Changing from Fraction Form to Decimal Form: page 369

A fraction can be written in decimal form by dividing the numerator by the denominator.

1. If the remainder is 0, the decimal is said to be **terminating**.
2. If the remainder is not 0, the decimal is said to be **nonterminating**.

Absolute Value   page 374

The **absolute value** of a decimal number is its distance from 0. The absolute value of a decimal number is nonnegative. We can express the definition symbolically, for any decimal number a, as follows:

$$\begin{cases} \text{If } a \text{ is positive or } 0, \text{ then } |a| = a. \\ \text{If } a \text{ is negative, then } |a| = -a. \end{cases}$$

Scientific Notation   pages 375 – 376

In **scientific notation**, decimal numbers are written as the product of a number between 1 and 10 and an integer power of 10.

Section 3.6: Basics of Percent

Percent Means Hundredths   pages 381 – 382

Percent means hundredths, or the ratio of a number to 100.

Percent of Profit   page 382

Percent of profit is the ratio of money made to money invested.

To Change a Decimal Number to a Percent   pages 383 – 384

1. Move the decimal point two places to the right.
2. Add the % symbol.

To Change a Percent to a Decimal Number   page 385

1. Move the decimal point two places to the left.
2. Delete the % symbol.

A Decimal Number That Is   page 386

1. less than 0.01 is less than 1%.
2. between 0.01 and 0.10 is between 1% and 10%.
3. between 0.10 and 1.00 is between 10% and 100%.
4. more than 1.00 is more than 100%.

To Change a Fraction to a Percent   page 386 – 387

1. Change the fraction to a decimal number. (Divide the numerator by the denominator.)
2. Change the decimal number to a percent.

To Change a Percent to a Fraction or Mixed Number   pages 390 – 391

1. Write the percent as a fraction with denominator 100 and delete the % symbol.
2. Reduce the fraction.

Section 3.7: Applications of Percent

Terms Related to Discount and Sales Tax   pages 398 – 399
    Discount: reduction in original selling price.
    Sale price: original selling price minus the discount.
    Rate of discount: percent of original price to be discounted.
    Sales tax: tax based on actual selling price.
    Rate of sales tax: percent of actual selling price.

Commission   pages 401 – 402
    A commission is a fee paid to an agent or salesperson for a service.

Terms Related to Profit   pages 403 – 404
    Profit: the difference between selling price and cost.
    Percent of Profit: there are two types; both are ratios with profit in the numerator.
    1. Percent of profit **based on cost** is the ratio of profit to cost.

$$\frac{\text{Profit}}{\text{Cost}} = \% \text{ of profit based on cost. (Cost is in the denominator.)}$$

    2. Percent of profit **based on selling price** is the ratio of profit to selling price.

$$\frac{\text{Profit}}{\text{Selling price}} = \% \text{ of profit based on selling price.}$$

Section 3.8: Simple and Compound Interest

Formula for Calculating Simple Interest   pages 409 – 410
    $I = Prt$
    where
    I = **Interest** (earned or paid)
    P = **Principal**
    r = **Rate** of interest (states an annual or yearly rate)
    t = **Time** (in years or fraction of a year)

Compound Interest Formula   pages 413 – 414
    When interest is compounded, the total **amount** A accumulated (including principal and interest) is given by the formula

$$A = P\left(1 + \frac{r}{n}\right)^{nt}$$

    where
    P = the principal
    r = the annual interest rate
    t = the length of time in years
    n = the number of compounding periods in 1 year

Total Interest Earned   page 414 – 415

To find the total interest earned on an investment that has earned interest by compounding, subtract the initial principal from the accumulated amount.

$$I = A - P$$

Formula for Inflation   page 416 – 417

The formula for the accumulated amount A due to **inflation** is the same as the formula for compound interest with $n = 1$.

$$A = P(1+r)^t$$

Formula for Depreciation   page 418

The **current value** V of an item due to **depreciation** can be determined from the following formula.

$$V = P(1-r)^t$$

where

P = the original value

r = the annual rate of depreciation (in decimal form)

t = the time in years

Section 3.9: Square Roots

Square Root   page 424

For any real number a and any nonnegative real number b, a is a **square root** of b if $a^2 = b$. If a is positive, then we write $a = \sqrt{b}$. Thus $\left(\sqrt{b}\right)^2 = b$.

Chapter 3: Test

Solve the following problems as indicated.

1. Write 32.064

   a. in words and

   b. as a mixed number with the fraction part reduced to lowest terms.

2. Write two and thirty-two thousandths in decimal notation.

Round each number as indicated.

		Nearest Tenth	Nearest Hundredth	Nearest Thousandth	Nearest Ten
3.	216.7049	**a.** _____	**b.** _____	**c.** _____	**d.** _____
4.	73.01485	**a.** _____	**b.** _____	**c.** _____	**d.** _____

Perform the indicated operation.

5. $85.9 + 36.963$

6. $946.75 - 1073.24$

7. $|82.1| + |-200| - |-76.83|$

8. $13\frac{2}{5} + 10 + 27.316$

9. $(1.93)(-2.75)$

10. $(0.1)(0.02)(0.03)$

11. $\dfrac{82.17}{10^3}$

12. $1\frac{1}{1000} - \frac{3}{4} - 0.0328$

13. $1000(82.17)$

14. $\begin{array}{r} 18.41 \\ \times\ 0.587 \\ \hline \end{array}$

Find each quotient to the nearest hundredth.

15. $0.31\overline{)8}$

16. $5.6\overline{)92.35}$

Solve the following problems as indicated.

17. Ten cars had the following miles-per-gallon ratings:

    26, 21, 20, 34, 20, 30, 25, 25, 25, 28

    Find the following statistics for these ten data items.

    a. Mean = _____

    b. Median = _____

    c. Mode = _____

    d. Range = _____

18. **a.** Write 67,000,000 in scientific notation.

    b. Write 8.3×10^{-5} in decimal notation.

19. Use $x = -2.1$ and $y = \frac{3}{4}$ to evaluate each of the following expressions.

a. $2x^2 - 3x - 1$

b. $y^2 - 5y + 1.2$

Perform the indicated operation.

20. Find the value of each expression.

a. $\left(\sqrt{121}\right)^2$

b. $\left(\sqrt{5}\right)^2$

21. Use a calculator to find the value of each expression accurate to the nearest ten-thousandth.

a. $\sqrt{300}$

b. $1 + 2\sqrt{3}$

22. MPG: Your car gets 20.5 miles per gallon of gas and the tank holds 17.5 gallons of gas.

a. How far can your car travel (to the nearest mile) on a full tank of gas?

b. How many gallons of gas (to the nearest gallon) should you plan to use on a trip of 500 miles?

Change each number to a percent.

23. $\frac{102}{100}$

24. 0.0725

25. 0.005

26. $\frac{3}{20}$

27. $\frac{3}{8}$

Change each percent to a decimal number.

28. 24%

29. 6.8%

30. $5\frac{1}{2}\%$

Change each percent to a fraction or mixed number. Reduce to lowest terms.

31. 160%

32. 9.6%

33. $62\frac{1}{2}\%$

Solve the following word problems.

34. Real Estate: A real estate company charges a commission of 6% for the sale of a home and pays the agent 60% of this fee. How much will the agent earn on the sale of a home for $150,000?

Did you read the problem at least twice? Do you understand all of the terms used?

What is your plan to solve the problem?

Carry out this plan. Show all of your work.

Does the answer seem reasonable to you?

35. New Sofa: A customer received a discount of 3% on the purchase of a new sofa because she paid cash. She paid $2425 in cash.

    a. Was the original price more or less than $2425?

    b. The amount $2425 is what percent of the original price?

    c. What was the original price?

    d. How much did she save by paying cash?

36. Used Car: Leona bought a used car for $1500 and replaced parts and made repairs that cost her another $500. Then she sold the car for $2600.

    a. What was her percent of profit based on cost?

    b. What was her percent of profit based on selling price?

37. Interest Rate: A note of $2000 is given for 9 months and will be paid off with a total of $2135. What is the rate of interest?

38. Interest Rate: An amount of $10,000 is deposited in a savings account, and interest is compounded monthly at 6.5%.

    a. What will be the balance in the account in 5 years?

    b. How much interest will have been earned?

39. Inflation: The current prices of three items are given. What will be their prices in 15 years if inflation is at 5%?

    a. a shirt: $35.00   **b.** a book: $50.00   **c.** a house: $100,000

Cumulative Review: Chapters 1 - 3

Solve the following as indicated.

1. The number 1 is called the multiplicative _____ because for any real number a, $a \cdot 1 =$ _____.

2. Explain why subtraction is not a commutative operation.

3. **a.** Find the product of the two numbers 2.78 and 3.59, then round the product to the nearest tenth.

   b. Round the two numbers 2.78 and 3.59 to the nearest tenth. Find the product of the rounded numbers and round this product to the nearest tenth.

   c. The two answers in parts **a.** and **b.** are different. Briefly explain why and what this tells you about calculating with rounded numbers.

4. **a.** First, estimate the quotient $945.7 \div 10.4$; then

   b. Find the quotient accurate to the nearest hundredth.

Evaluate each of the following expressions.

5. $3^2 - 5^2$

6. $28 + 5(2^3 + 25) \div 11 \cdot 2$

7. $24 + \left[16 - \left(6^2 \cdot 1 - 20\right)\right]$

8. $36 \div 4 - 9 \cdot 2^2$

9. Find $\dfrac{3}{4}$ of 76.

Solve the following as indicated.

10. Evaluate the polynomial $3x^2 + 15$ for $x = -2$.

11. Find the prime factorization of each number.
    a. 65   **b.** 144   **c.** 270

12. List the multiples of 6 that are less than 100.

13. Find the LCM for each set of terms.
    a. 28, 70, 80   **b.** $24a^2$, $15ab^3$, $20a^2b^2$

14. **a.** Does 10 divide the product $7 \cdot 6 \cdot 5 \cdot 3$?

    b. If so, what is the quotient?

15. Solve the proportion $\dfrac{A}{100} = \dfrac{0.125}{8}$.

16. Change each decimal number to fraction form or mixed number form, and reduce if possible.

    a. 0.015   **b.** 6.32   **c.** −1.05

Evaluate (or simplify) each of the following expressions. Reduce all fractions.

17. $\dfrac{27}{35} \cdot \dfrac{2}{3} \cdot \dfrac{7}{18}$

18. $\dfrac{x}{3} \div \dfrac{2}{3} + \dfrac{1}{4}$

19. $3.6 - 7\dfrac{1}{2} + |-2.6| - \left| 5\dfrac{1}{4} \right|$

20. $\left(\dfrac{1}{2} \right)^2 \left(\dfrac{16}{15} \right) + \dfrac{6}{5} \div 2$

Solve the following equations.

21. $3(a + 5) - 8a = 4$

22. $\dfrac{1}{2}x + \dfrac{3}{5} = -\dfrac{1}{10}$

Solve the following.

23. **a.** Write the number 865,000 in scientific notation.

    b. Write the number 3.41×10^{-3} in decimal notation.

24. Find **a.** the perimeter and **b.** the area of a rectangular playground that is 100 yards long and 65 yards wide.

25. **a.** Do you think that $\sqrt{29}$ is more than or less than 5? Explain your reasoning.

    b. Use your calculator to find the value of $\sqrt{29}$ accurate to the nearest thousandth.

26. **Stop Sign:** The pole for a stop sign is 12 feet long.

    a. If $\dfrac{5}{16}$ of the length of the pole is to be below the ground, what fraction of the length of the pole is to be above ground?

    b. How many feet are to be above ground?

27. **Computer Sale:** An advertised sale on computers declares $\dfrac{1}{3}$ off the original price. The advertised price is $1500.

    a. Was the original price less than or more than $1500?

    b. What was the original price?

28. **Savings Interest:** What will be the interest earned in one year on a savings account of $2300 that pays simple interest at $6\dfrac{1}{4}\%$?

29. **Earned Interest:** If a principal of $10,000 is invested for 2 years and interest is compounded quarterly at 8%, how much interest is earned?

30. Simple Interest: Determine the missing item in each row if the interest is simple interest.

Principal	Rate	Time	Interest
$500	12%	6 months	a. _____
$200	18%	b. _____	$ 18
$1000	c. _____	9 months	$ 150

31. Men's Suits: At its annual close-out sale a clothing store had men's suits that had originally sold for $350 marked down to $250.

    a. What was the rate of discount based on the original selling price?

    b. If each suit cost the store $200, what was the percent of profit based on cost?

    c. What was the percent of profit based on the sale price?

32. Car Wash: Marty can wash the car in $\frac{3}{4}$ of an hour, and Larry can wash the car in 1.2 hours. How many minutes faster is Marty at washing the car?

33. Table Lamps: Find the average (to the nearest cent) of the following set of prices for table lamps: $25.50; $32.75; $56.00; $30.00; $16.50; $72.75

34. Electric Usage: Find the average usage per day (to the nearest hundredth) for each billing period of electric usage given the following data for the same period over three years:

For November:	This year	Last year	2 years ago
Kilowatt-hour (kWh) used	373.00	455.00	279.00
Number of days	29	28	28
Average usage per day	_____	_____	_____
For December:	**This year**	**Last year**	**2 years ago**
Kilowatt-hour (kWh) used	364.00	321.00	554.00
Number of days	31	30	33
Average usage per day	_____	_____	_____

(How does this data compare to the electric usage at your house?)

35. Used Car: Your friend plans to buy a used car for $16,000. He can either save up and pay cash or make a down payment of $1600 and 48 monthly payments of $420. What would you recommend that he should do? Why?

36. Water Usage: The water usage at your home for 6 billing periods is shown in the following graph. On the graph, each unit is equivalent to 748 gallons of water.

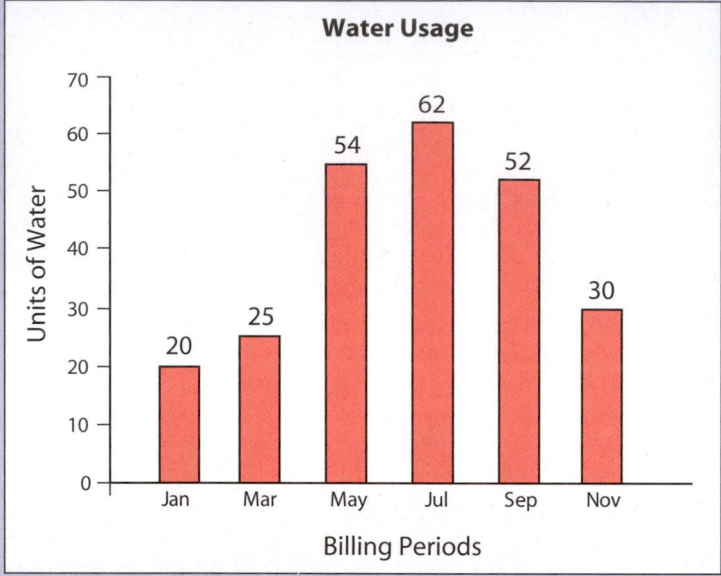

Find the following statistics (in gallons) for the six billing periods.

a. Mean = _____

b. Median = _____

c. Mode = _____

d. Range = _____

37. Rectangular Solid: Find **a.** the volume and **b.** the surface area of the rectangular solid with dimensions as shown.

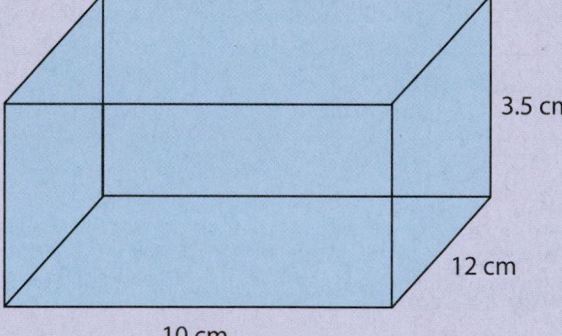

Mathematics @ Work!

Geometric figures and their properties are part of our daily lives. Geometry is an integral part of art, employing concepts such as vanishing points and ways of creating three-dimensional perspectives on a two-dimensional canvas. Architects and engineers use geometric shapes in designing buildings and other infrastructures for beauty and structural strength. Geometric patterns are evident throughout nature in snow flakes, leaves, pine cones, butterflies, trees, crystal formations, camouflage for snakes, fish, tigers, and so on. As an example of geometry in the human-created world, consider sewer covers in streets that you drive over every day. Why are these covers circular in shape? Why are they not in the shape of a square or a rectangle? One reason might be that a circle cannot "fall through" a slightly smaller circle, whereas a square or rectangle can be turned so that either will "fall through" a slightly smaller version of itself. Thus, circular sewer covers are much safer than other shapes because they avoid the risk of the cover accidentally falling down into the hole.

To understand this idea, draw a square, a rectangle, a triangle, and a hexagon (six-sided figure) on a piece of paper. Cut out these figures, and show how easily they can be made to pass through the holes made by the corresponding cutouts. Then follow the same procedure with a circle. The difference will be clear.

Geometry and Graphs

A Recognize measures of length in the metric system, and be able to change between metric measures of length.

B Recognize measures of area in the metric system, and be able to change between metric measures of area.

C Know the metric units of measurement for mass.

D Know the metric units of measurement for volume.

E Know the metric units of measurement for liquid volume.

F Become familiar with the metric prefixes mega-, giga-, and tera-.

4.1 The Metric System

Objective A **Metric Units of Length**

The **metric system** of measurement is used by about 90% of the people in the world. The United States is the only major industrialized country still committed to the U.S. customary system (formerly called the English system). Even in the United States, the metric system is used in many fields of study and business, such as medicine, science, the computer industry, and the military. Industries involved in international trade must be familiar with the metric system.

The **meter** is the basic unit of length in the metric system. Smaller and larger units are named by putting a prefix in front of the basic unit – for example, **centi**meter and **kilo**meter. The prefixes we will use are shown in boldface print in Table 1. Other prefixes that indicate extremely small units are micro-, nano-, pico-, femto- and atto-. Prefixes that indicate extremely large units are mega-, giga-, and tera-. You are probably familiar with computer terms such as megabytes and gigabytes.

Metric Measures of Length		
1 **milli**meter (mm)	= 0.001 meter	1 m = 1000 mm
1 **centi**meter (cm)	= 0.01 meter	1 m = 100 cm
1 **deci**meter (dm)	= 0.1 meter	1 m = 10 dm
1 meter (m)	= 1.0 meter (the basic unit)	
1 **deca**meter (dam)	= 10 meters	
1 **hecto**meter (hm)	= 100 meters	
1 **kilo**meter (km)	= 1000 meters	

Table 1

As indicated in Table 1, the metric units of length are related to each other by powers of 10. That is, simply multiply by 10 to get the equivalent measure expressed in the next lower (or smaller) unit. Thus you will have **more** of a **smaller** unit. For example,

$$1 \text{ m} = 10 \text{ dm} = 100 \text{ cm} = 1000 \text{ mm}.$$

Conversely, divide by 10 to get the equivalent measure expressed in the next higher (or larger) unit. Thus you will have **less** of a **larger** unit. For example,

$$1 \text{ mm} = 0.1 \text{ cm} = 0.01 \text{ dm} = 0.001 \text{ m}.$$

Changing Metric Measures of Length

Procedure	Example
To change to a measure of length that is: one unit smaller, multiply by 10	**Larger to Smaller** 6 cm = 60 mm
two units smaller, multiply by 100	3 m = 300 cm
three units smaller, multiply by 1000 and so on.	12 m = 12,000 mm
To change to a measure of length that is: one unit larger, divide by 10	**Smaller to Larger** 35 mm = 3.5 cm
two units larger, divide by 100	250 cm = 2.5 m
three units larger, divide by 1000 and so on.	32 m = 0.032 km

Example 1

Converting Metric Units of Length

The following examples illustrate how to change from larger to smaller units of length in the metric system.

a. $5.6 \text{ m} = 100(5.6) \text{ cm} = 560 \text{ cm}$

b. $5.6 \text{ m} = 1000(5.6) \text{ mm} = 5600 \text{ mm}$

c. $23.5 \text{ cm} = 10(23.5) \text{ mm} = 235 \text{ mm}$

d. $1.42 \text{ km} = 1000(1.42) \text{ m} = 1420 \text{ m}$

Example 2

Converting Metric Units of Length

The following examples illustrate how to change from smaller to larger units of length in the metric system.

a. $375 \text{ cm} = \left(\dfrac{375}{100}\right) \text{m} = 3.75 \text{ m}$

b. $375 \text{ mm} = \left(\dfrac{375}{1000}\right) \text{m} = 0.375 \text{ m}$

c. $1055 \text{ m} = \left(\dfrac{1055}{1000}\right) \text{km} = 1.055 \text{ km}$

Now work margin exercises 1 and 2.

Change the following units as indicated.

1. a. 35 m = _____ cm

  b. 6.4 cm = _____ mm

  c. 0.123 m = _____ dm

  d. 6 m = _____ mm

2. a. 5.9 m = _____ km

  b. 320 mm = _____ m

  c. 6500 cm = _____ m

  d. 70 m = _____ dam

Another technique for changing units in the metric system can be illustrated with the concept of a number line with the metric prefixes below in order.

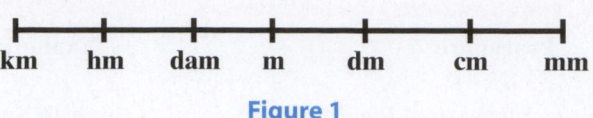

Figure 1

Simply move the decimal point in the direction of change.

Example 3

Converting Metric Units of Length

Change 56 cm to the equivalent measure in meters.

Solution

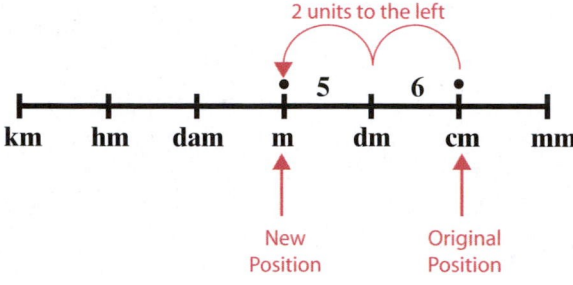

Thus, 56 cm = 0.56 m.

3. Change 65 decimeters to the equivalent measure in decameters.

4. Change 43 meters to an equivalent measure in centimeters.

Example 4

Converting Metric Units of Length

Change 13.5 m to an equivalent measure in millimeters.

Solution

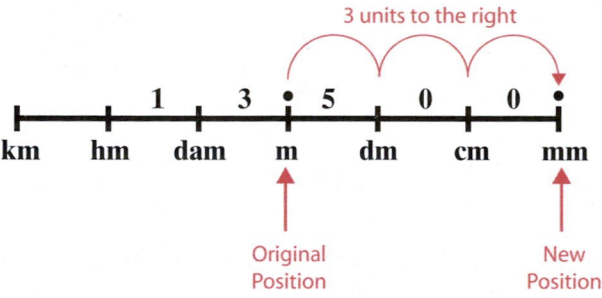

Thus, 13.5 m = 13 500 mm.

Now work margin exercises 3 and 4.

notes

In the metric system,

1. A 0 is written to the left of the decimal point if there is no whole number part (0.287 m).

2. No commas are used in writing numbers. If a number has more than four digits (left or right of the decimal point), the digits are grouped in threes from the decimal point with a space between the groups (25 000 m or 0.000 34 m).

Objective B   **Metric Units of Area**

Recall from earlier discussion that **area** is a measure of the interior, or enclosure, of a surface. For example, the two rectangles in Figure 2 have different areas because they have different amounts of interior space. That is, different amounts of space are enclosed by the sides of the figures.

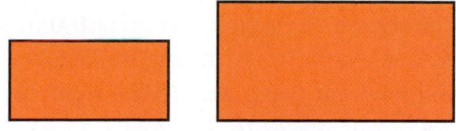

These two rectangles have different areas.
Figure 2

Also recall that area is measured in **square units**. In metric units, a square that is 1 centimeter long on each side is said to have an area of 1 square centimeter (or **1 cm²**). Figure 3 illustrates the area concept with a rectangle of area of 28 cm².

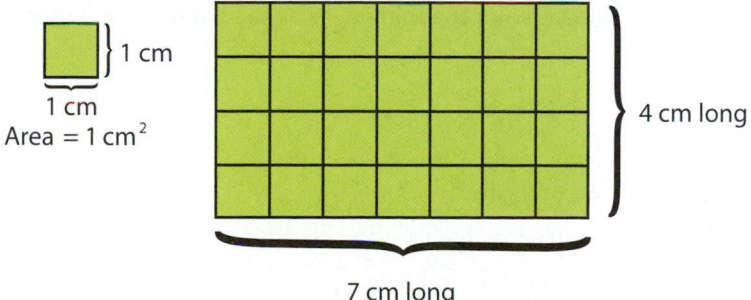

1 cm
1 cm
Area = 1 cm²

4 cm long

7 cm long

There are 28 squares that are each 1 cm² in the large rectangle.
Figure 3

Table 2 shows metric area measures useful for relatively small areas. For example, the area of a tennis court might be measured in square meters, while the area of this page of paper might be measured in square centimeters. Other measures, listed in Table 3, are used for measuring land.

Metric Measures of Small Areas
$1 \text{ cm}^2 = 100 \text{ mm}^2$
$1 \text{ dm}^2 = 100 \text{ cm}^2 = 10\,000 \text{ mm}^2$
$1 \text{ m}^2 = 100 \text{ dm}^2 = 10\,000 \text{ cm}^2 = 1\,000\,000 \text{ mm}^2$

Table 2

notes

- Units are treated much in the same way as the numbers they describe. That is, when finding area, the units in Figure 3 operate as follows.
- $$\text{cm} \times \text{cm} = \text{cm}^2 = \text{square centimeters}$$

Note that each unit of area in the metric system is 100 times the next smaller unit of area – not just 10 times, as it is with length. Also, remember to multiply by 100, move the decimal point two places **to the right**; to multiply by 10,000, move the decimal point four places **to the right**, and so on. Remember that to divide by 100, move the decimal point two places **to the left**; to divide by 10,000, move the decimal point four places **to the left**; and so on.

Changing Metric Measures of Area

Procedure	Example
To change to a measure of area that is:	**Larger to Smaller**
one unit smaller, multiply by 100	$8.2 \text{ cm}^2 = 820 \text{ mm}^2$
two units smaller, multiply by 10,000	$3.56 \text{ m}^2 = 35\,600 \text{ cm}^2$
three units smaller, multiply by 1,000,000 and so on.	$43 \text{ km}^2 = 43\,000\,000 \text{ m}^2$
To change to a measure of area that is:	**Smaller to Larger**
one unit larger, divide by 100	$75 \text{ mm}^2 = 0.75 \text{ cm}^2$
two units larger, divide by 10,000	$52 \text{ cm}^2 = 0.0052 \text{ m}^2$
three units larger, divide by 1,000,000 and so on.	$65 \text{ m}^2 = 0.000\,065 \text{ km}^2$

Example 5

Equivalent Metric Measures of Area

a. A square 1 centimeter on each side encloses 100 square millimeters: $1 \text{ cm}^2 = 100 \text{ mm}^2$

Area $= 1 \text{ cm}^2$ Area $= 1 \text{ cm}^2 = 100 \text{ mm}^2$

b. A square 1 decimeter (10 cm) on a side encloses 10 000 square millimeters: $1 \text{ dm}^2 = 100 \text{ cm}^2 = 10\,000 \text{ mm}^2$

1 dm = 10 cm = 100 mm

1 dm = 10 cm = 100 mm

$1 \text{ dm}^2 = 100 \text{ cm}^2 = 10\,000 \text{ mm}^2$

Example 6

Converting Metric Measures of Area

The following examples illustrate how to change from larger to smaller units of area in the metric system.

a. $5 \text{ cm}^2 = 5\left(1 \text{ cm}^2\right) = 5\left(100 \text{ mm}^2\right) = 500 \text{ mm}^2$

Note: mm^2 is one unit smaller than cm^2.

b. $3 \text{ dm}^2 = 3\left(1 \text{ dm}^2\right) = 3\left(10\,000 \text{ mm}^2\right) = 30\,000 \text{ mm}^2$

Note: mm^2 is two units smaller than dm^2.

c. $9.8 \text{ km}^2 = 9.8\left(1 \text{ km}^2\right) = 9.8\left(1\,000\,000 \text{ m}^2\right) = 9\,800\,000 \text{ m}^2$

Note: m^2 is three units smaller than km^2.

Change the units as indicated.

5. $1 \text{ m}^2 = \underline{\hspace{1.5cm}} \text{ dm}^2$

6. a. $23 \text{ cm}^2 = \underline{\hspace{1.5cm}} \text{ mm}^2$

   b. $86 \text{ hm}^2 = \underline{\hspace{1.5cm}} \text{ dam}^2$

   c. $0.06 \text{ dam}^2 = \underline{\hspace{1cm}} \text{ m}^2$

7. a. $5200 \text{ mm}^2 = \underline{\hspace{1cm}} \text{ cm}^2$

   b. $360 \text{ dm}^2 = \underline{\hspace{1.5cm}} \text{ m}^2$

   c. $4200 \text{ dam}^2 = \underline{\hspace{1cm}} \text{ km}^2$

8. How many hectares are in $49\,500 \text{ km}^2$?

Example 7

Converting Metric Measures of Area

The following examples illustrate how to change from smaller to larger units of area in the metric system.

a. $1.4 \text{ m}^2 = \dfrac{1.4}{100} \text{ dam}^2 = 0.014 \text{ dam}^2$

   Note: dam^2 is one unit larger than m^2.

b. $3500 \text{ mm}^2 = \dfrac{3500}{100} \text{ cm}^2 = 35 \text{ cm}^2$

   Note: cm^2 is one unit larger than mm^2.

c. $3500 \text{ mm}^2 = \dfrac{3500}{1\,000\,000} \text{ m}^2 = 0.0035 \text{ m}^2$

   Note: m^2 is three units larger than mm^2.

Now work margin exercises 5 through 7.

A square with each side 10 meters long encloses an area of 1 **are (a)**. A **hectare (ha)** is 100 ares. The are and hectare are used to measure land area in the metric system.

Metric Measurements of Land Area
$1 \text{ a} = 100 \text{ m}^2$
$1 \text{ ha} = 100 \text{ a} = 10\,000 \text{ m}^2$

Table 3

Example 8

Converting Metric Measurements of Land Area

How many ares are in 1 km^2? (**Note:** For comparison, 1 km is about 0.6 mile, so 1 km^2 is about $0.6 \text{ mi} \times 0.6 \text{ mi} = 0.36 \text{ mi}^2$.)

Solution

Remember that 1 km = 1000 m, so

$$1 \text{ km}^2 = 1 \text{ km} \times 1 \text{ km} = 1000 \text{ m} \times 1000 \text{ m}$$
$$= 1\,000\,000 \text{ m}^2$$
$$= 10\,000 \text{ a}$$

Now work margin exercise 8.

Objective C **Mass (Weight)**

Mass is the amount of material in an object. Regardless of where the object is in space, its mass remains the same. (See Figure 4.) **Weight** is the force of the Earth's gravitational pull on an object. The farther an object is from Earth, the less the gravitational pull of the Earth. Thus astronauts experience weightlessness in space, but their mass is unchanged.

The two objects have the same mass
and balance on an equal arm balance,
regardless of their location in space.

Figure 4

Because most of us do not stray far from the Earth's surface, weight and mass will be used interchangeably in this text. Thus a **mass** of 20 kilograms will be said to **weigh** 20 kilograms.

The basic unit of mass in the metric system is the **kilogram***, about 2.2 pounds. In some fields, such as medicine, the **gram** (about the mass of a paper clip) is more convenient as a basic unit than the kilogram.

Large masses, such as loaded trucks and railroad cars, are measured by the **metric ton** (1000 kilograms, or about 2200 pounds). (See Tables 4 and 5.)

Measures of Mass	
1 **milli**gram (mg)	= 0.001 gram
1 **centi**gram (cg)	= 0.01 gram
1 **deci**gram (dg)	= 0.1 gram
1 gram (g)	= 1.0 gram
1 **deka**gram (dag)	= 10 grams
1 **hecto**gram (hg)	= 100 grams
1 **kilo**gram (kg)	= 1000 grams
1 metric ton (t)	= 1000 kilograms

Table 4

Equivalent Measures of Mass	
1000 mg = 1 g	0.001 g = 1 mg
1000 g = 1 kg	0.001 kg = 1 g
1000 kg = 1 t	0.001 t = 1 kg
1 t = 1000 kg = 1,000,000 g = 1,000,000,000 mg	

Table 5

---

* Technically, a kilogram is the mass of a certain cylinder of platinum-iridium alloy kept by the International Bureau of Weights and Measures in Paris.

Originally, the basic unit was a gram, defined to be the mass of 1 cm³ of distilled water at 4° Celsius. This mass is still considered accurate for many purposes, so that

1 cm³ of water has a mass of 1 g.
1 dm³ of water has a mass of 1 kg.
1 m³ of water has a mass of 1000 kg, or 1 metric ton.

Units of Mass

A chart can be used to change from one unit of mass to another.

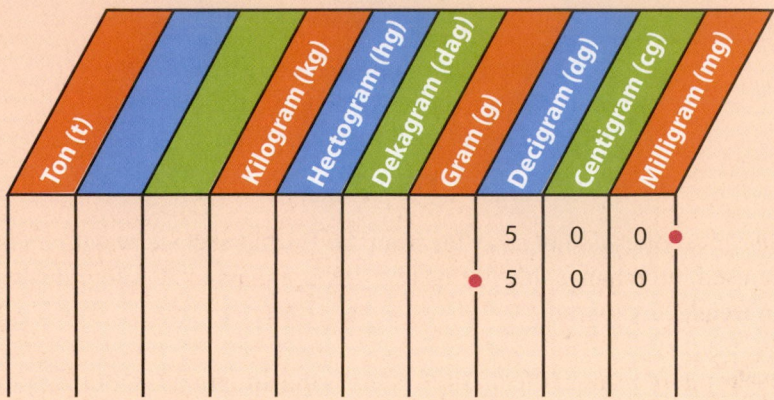

500 mg = 0.5 g

1. List each unit across the top.
2. Enter the given number so that there is one digit in each column with the decimal point on the given unit line.
3. Move the decimal point to the desired unit line.
4. Fill in the spaces with 0s using one digit per column.

The use of the chart following Example 9 shows how the following equivalent measures can be found.

Example 9

Equivalent Measures of Mass

a. 23 mg   =   0.023 g

b. 6 g   =   6000 mg

c. 49 kg   =   49 000 g

d. 5 t   =   5000 kg

e. 70 kg   =   0.07 t

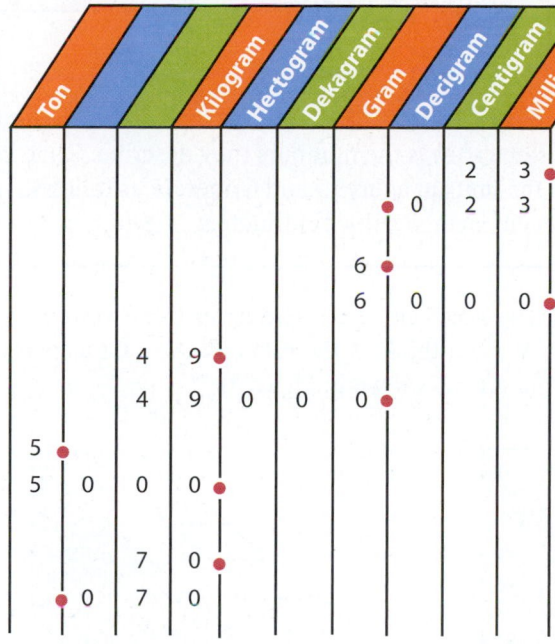

a. Move the decimal point to g and fill in 0s. 23 mg = 0.023 g.

b. Move the decimal point to mg and fill in 0s. 6 g = 6000 mg.

c. Move the decimal point to g and fill in 0s. 49 kg = 49 000 g.

d. Move the decimal point to kg and fill in 0s. 5 t = 5000 kg.

e. Move the decimal point to t and fill in 0s. 70 kg = 0.070 t.

Make your own chart on a piece of paper and see if you agree with the results in Example 10.

Example 10

Equivalent Measures of Mass

a. 60 mg = 0.06 g

b. 135 mg = 0.135 g

c. 5700 kg = 5.7 t

d. 100 g = 0.1 kg

e. 78 kg = 78 000 g

Now work margin exercises 9 and 10.

Objective D **Volume**

Volume is a measure of the space enclosed by a three-dimensional figure and is measured in **cubic units**. The volume or space contained within a cube that is 1 centimeter on each edge is **one cubic centimeter**, or 1 cm³, as shown in Figure 5. A cubic centimeter is about the size of a sugar cube.

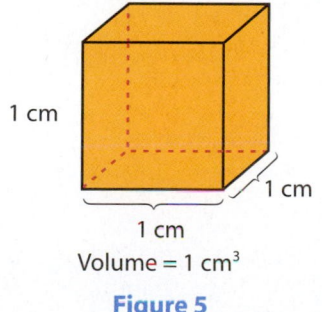

1 cm
1 cm
1 cm
Volume = 1 cm³

Figure 5

9. a. Change 121 kilograms to grams.

b. Change 3500 kilograms to tons.

c. Change 4 576 000 grams to tons.

d. Change 6700 milligrams to grams.

10. a. Change 43 kilograms to grams.

b. Change 250 kilograms to tons.

c. Change 23 grams to milligrams.

A rectangular solid that has edges of 3 cm, 2 cm and 5 cm has a volume of
$3 \text{ cm} \times 2 \text{ cm} \times 5 \text{ cm} = 30 \text{ cm}^3$. We can think of the rectangular solid as being
three layers of ten cubic centimeters, as shown in Figure 6.

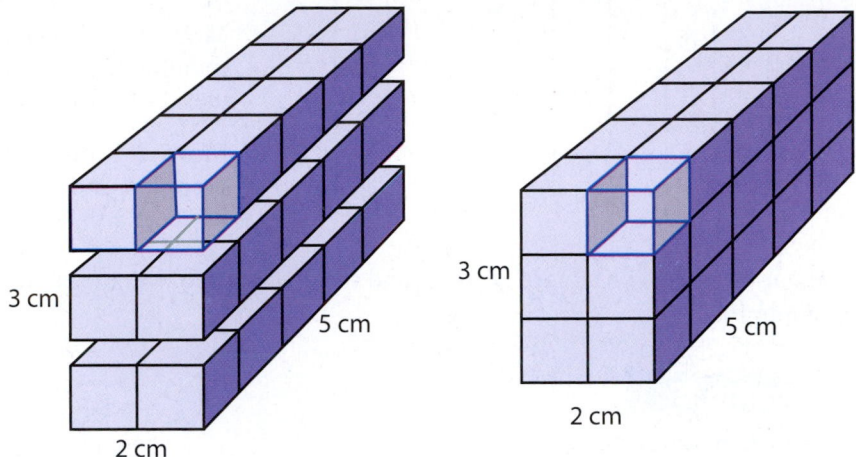

Figure 6

Metric Measures of Volume
$1 \text{ cm}^3 = 1000 \text{ mm}^3$
$1 \text{ dm}^3 = 1000 \text{ cm}^3 = 1\,000\,000 \text{ mm}^3$
$1 \text{ m}^3 = 1000 \text{ dm}^3 = 1\,000\,000 \text{ cm}^3 = 1\,000\,000\,000 \text{ mm}^3$

Table 6

Changing Metric Measures of Volume

Procedure	Example

To change to a measure of volume that is:
one unit smaller, multiply by 1000

Larger to Smaller
$6 \text{ cm}^3 = 6000 \text{ mm}^3$

two units smaller, multiply by 1,000,000

$3 \text{ m}^3 = 3\,000\,000 \text{ cm}^3$

three units smaller, multiply by 1,000,000,000
and so on.

$12 \text{ m}^3 = 12\,000\,000\,000 \text{ mm}^3$

To change to a measure of volume that is:
one unit larger, divide by 1000,

Smaller to Larger
$35 \text{ mm}^3 = 0.035 \text{ cm}^3$

two units larger, divide by 1,000,000,

$250 \text{ cm}^3 = 0.000\,25 \text{ m}^3$

three units larger, divide by 1,000,000,000
and so on.

$32 \text{ m}^3 = 0.000\,000\,032 \text{ km}^3$

Example 11

Converting Metric Measures of Volume

a. $15 \text{ cm}^3 = 15\left(1 \text{ cm}^3\right) = 15\left(1000 \text{ mm}^3\right) = 15\,000 \text{ mm}^3$

b. $4.1 \text{ dm}^3 = 4.1\left(1 \text{ dm}^3\right) = 4.1\left(1000 \text{ cm}^3\right) = 4100 \text{ cm}^3$

c. $22.6 \text{ m}^3 = 22.6\left(1 \text{ m}^3\right) = (22.6)\left(1\,000\,000 \text{ cm}^3\right) = 22\,600\,000 \text{ cm}^3$

d. $4 \text{ m}^3 = 4\left(1 \text{ m}^3\right) = 4\left(1000 \text{ dm}^3\right) = 4000 \text{ dm}^3 = 4000\left(1000 \text{ cm}^3\right)$
$= 4\,000\,000 \text{ cm}^3$

Example 12

Converting Metric Measures of Volume

The following examples illustrate how to change from smaller to larger units of volume in the metric system.

a. $8 \text{ cm}^3 = \dfrac{8}{1000} \text{ dm}^3 = 0.008 \text{ dm}^3$

b. $0.8 \text{ mm}^3 = \dfrac{0.8}{1000} \text{ cm}^3 = 0.0008 \text{ cm}^3$

Now work margin exercises 11 and 12.

Again, we can use a chart; however, this time **there must be three digits in each column.**

11. Change the units as indicated.

a. $9 \text{ cm}^3 = $ _____ mm^3

b. $7.5 \text{ m}^3 = $ _____ dm^3

c. $0.25 \text{ dm}^3 = $ _____ mm^3

d. $0.035 \text{ hm}^3 = $ _____ m^3

12. Change the units as indicated.

a. $0.54 \text{ mm}^3 = $ _____ cm^3

b. $5982 \text{ cm}^3 = $ _____ m^3

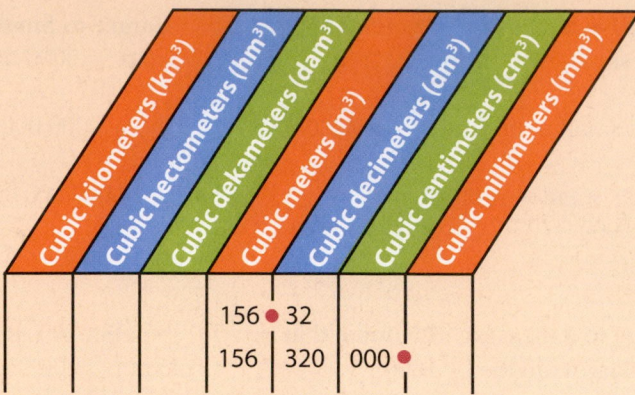

Units of Volume

A chart can be used to change from one unit of volume to another.

Cubic kilometers (km³) · Cubic hectometers (hm³) · Cubic dekameters (dam³) · Cubic meters (m³) · Cubic decimeters (dm³) · Cubic centimeters (cm³) · Cubic millimeters (mm³)

| 156 • 32 | | |
| 156 | 320 | 000 • |

156.32 m³ = 156 320 000 cm³

1. List each volume unit across the top. (Abbreviations will do.)
2. Enter the given number so that there are **three** digits in each column with the decimal point on the given unit line.
3. Move the decimal point to the desired unit line.
4. Fill in the spaces with 0s using three digits per column.

Example 13

Equivalent Measures of Volume

a. 15 cm³   =   15 000 mm³

b. 4.1 dm³   =   4100 cm³

c. 8 dm³   =   0.008 m³

d. 22.6 m³   =   22 600 000 cm³

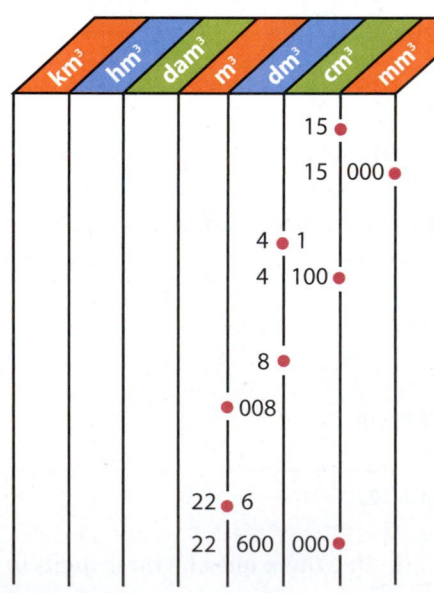

a. Move decimal point to mm³ and fill in 0s.
15 cm³ = 15 000 mm³

b. Move decimal point to cm³ and fill in 0s.
4.1 dm³ = 4100 cm³

c. Move decimal point to m³ and fill in 0s.
8 dm³ = 0.008 m³

d. Move decimal point to cm³ and fill in 0s.
22.6 m³ = 22 600 000 cm³

Make your own chart on a piece of paper and see if you agree with the results in Example 14.

Example 14

Equivalent Measures of Volume

a. $3.7 \text{ dm}^3 = 3700 \text{ cm}^3$

b. $0.8 \text{ m}^3 = 0.0008 \text{ dam}^3$

c. $4 \text{ m}^3 = 4000 \text{ dm}^3 = 4\,000\,000 \text{ cm}^3 = 4\,000\,000\,000 \text{ mm}^3$

Now work margin exercises 13 and 14.

Of particular interest is the volume of a cube that is 1 decimeter along each edge. The volume of such a cube (see Figure 7) is **1 cubic decimeter** (or **1 dm^3**). In terms of cubic centimeters, this same cube has a volume

$$10 \text{ cm} \times 10 \text{ cm} \times 10 \text{ cm} = 1000 \text{ cm}^3.$$

$$\text{Thus, } 1 \text{ dm}^3 = 1000 \text{ cm}^3.$$

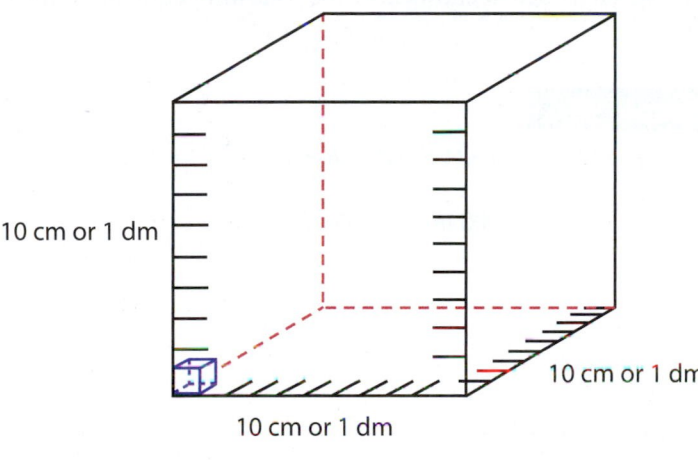

10 cm or 1 dm

10 cm or 1 dm

10 cm or 1 dm

Figure 7

Objective E **Liquid Volume**

The **liquid volume** contained in a cube that is 10 cm along each edge (as illustrated in Figure 7) is called a **liter (L)**. That is,

$$1 \text{ L} = 1000 \text{ cm}^3 \quad \text{and} \quad 1 \text{ mL} = 1 \text{ cm}^3.$$

The basic measurement of liquid volume in the metric system is called a **liter** (abbreviated L). You are probably familiar with 1 L and 2 L bottles of soda on your grocer's shelf. A **liter** is the volume enclosed in a cube that is 10 cm on each edge. So 1 liter is equal to

$$10 \text{ cm} \times 10 \text{ cm} \times 10 \text{ cm} = 1000 \text{ cm}^3 \quad \text{or} \quad 1 \text{ liter} = 1000 \text{ cm}^3.$$

That is, the box shown in Figure 7 would hold 1 liter of liquid.

The prefixes kilo-, hecto-, deka-, deci-, centi-, and milli- all indicate the same part of a liter as they do of the meter. **One digit per column** will be helpful for changing units. The centiliter (cL), deciliter(dL), and dekaliter (daL) are not commonly used and are not included in the tables or exercises.

13. a. Change 750 cubic millimeters to cubic centimeters.

b. Change 19 cubic decimeters to cubic centimeters.

c. Change 1.6 cubic meters to cubic centimeters.

d. Change 63.7 cubic meters to cubic decimeters.

14. a. Change 6.3 cubic centimeters to cubic meters.

b. Change 192 cubic decimeters to cubic centimeters.

Measures of Liquid Volume

1 **milli**liter (mL)	=	0.001 liter
1 liter (L)	=	1.0 liter
1 **hecto**liter (hL)	=	100 liters
1 **kilo**liter (kL)	=	1000 liters

Table 7

Equivalent Measures of Volume

1000 mL = 1 L		1 mL = 1 cm³
1000 L = 1 kL		1 L = 1 dm³
10 hL = 1 kL		1 kL = 1 m³

Table 8

The use of the chart below shows how the following equivalent measures can be found.

Example 15

Equivalent Measures of Liquid Volume

a. 6 L   =   6000 mL = 0.06 hL

b. 500 mL   =   0.5 L

c. 3 kL   =   3000 L

d. 72 hL   =   7.2 kL

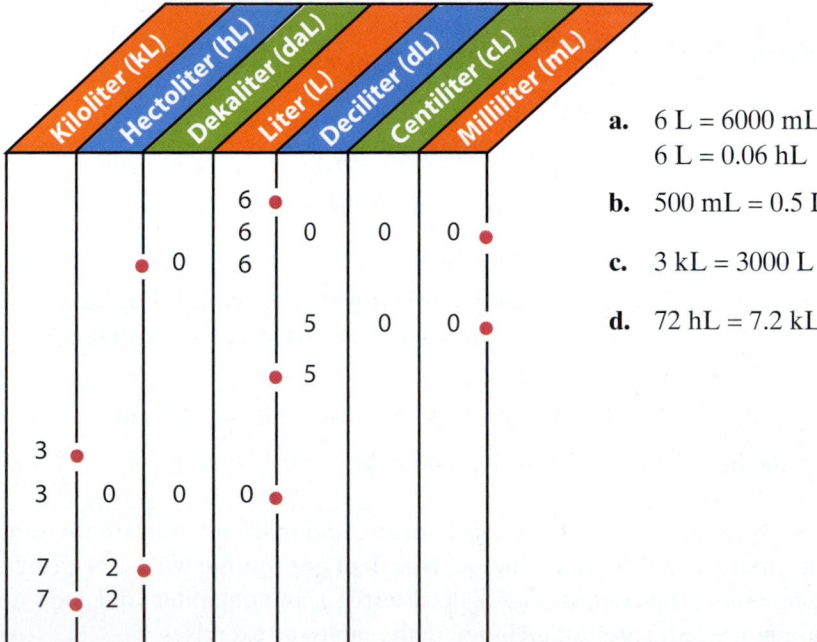

a. 6 L = 6000 mL
   6 L = 0.06 hL

b. 500 mL = 0.5 L

c. 3 kL = 3000 L

d. 72 hL = 7.2 kL

There is an interesting "crossover" relationship between liquid volume measures and cubic volume measures. Since, 1 L = 1000 mL and 1L = 1000 cm³ we have **1 mL = 1 cm³**. Also, 1 kL = 1000 L = 1 000 000 cm³ and 1 000 000 cm³ = 1 m³. This gives **1 kL = 1000 L = 1 m³**.

Use Table 7 to confirm the following conversions.

Example 16

Equivalent Measures of Liquid Volume

a. 6000 mL = 6 L

b. 3.2 L = 3200 mL

c. 60 hL = 6 kL

d. 637 mL = 0.637 L

e. 70 mL = 70 cm³

f. 3.8 kL = 3.8 m³

Now work margin exercises 15 and 16.

Objective F **Mega-, Giga-, and Tera-**

You may be familiar with the the metric prefixes mega-, giga-, and tera- because they often appear in today's technology world. For example, you may be familiar with megapixels on your camera, gigahertz for computing speed, or terabytes of computer memory. However, you may not understand what these prefixes tell you about those amounts.

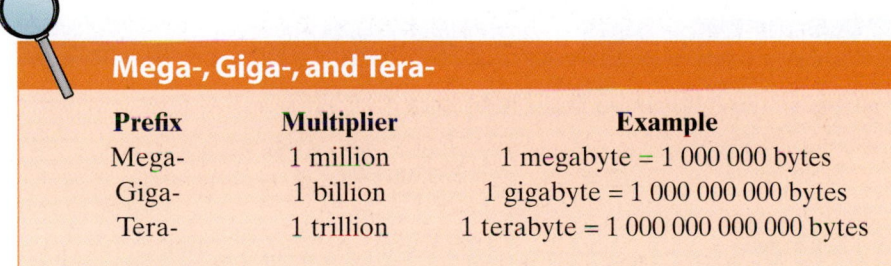

Mega-, Giga-, and Tera-		
Prefix	**Multiplier**	**Example**
Mega-	1 million	1 megabyte = 1 000 000 bytes
Giga-	1 billion	1 gigabyte = 1 000 000 000 bytes
Tera-	1 trillion	1 terabyte = 1 000 000 000 000 bytes

Notice that each prefix has an associated multiplier. When converting from a larger unit to a smaller unit, multiply by this number. Similarly, when converting from a smaller unit to a larger unit, divide by this multiplier.

15. a. Change 9.37 liters to hectoliters.

b. Change 353 milliliters to liters.

c. Change 12 kiloliters to liters.

16. a. Change 1952 milliliters to liters.

b. Change 124 milliliters to cubic centimeters.

c. Change 19.75 kiloliters to cubic meters.

17. a. How many hertz are in 4.3 gigahertz?

b. How many megatons are in 2 125 000 metric tons?

Example 17

The Prefixes Mega-, Giga-, and Tera-

a. How many pixels are in 12 megapixels?

Solution

Because the prefix mega- is a multiplier of 1 million and a megapixel is a larger unit than a pixel, we must multiply 12 by 1 000 000 in order to obtain an equivalent measurement in pixels.

$$12 \cdot 1\,000\,000 = 12\,000\,000$$

Thus, there are 12 000 000 pixels in 12 megapixels.

b. How many terabytes are in 17 458 000 000 000 bytes?

Solution

Because the prefix tera- is a multiplier of 1 trillion and a byte is a smaller unit than a terabyte, we must divide 17 458 000 000 000 by 1 000 000 000 000 in order to obtain an equivalent measurement in bytes.

$$\frac{17\,458\,000\,000\,000}{1\,000\,000\,000\,000} = 17.458$$

Thus, there are 17.458 terabytes in 17 458 000 000 000 bytes.

Now work margin exercise **17.**

Practice Problems

Change the following units as indicated.

1. 500 mg = _____ g   **2.** 43 g = _____ mg

3. 62 mg = _____ kg   **4.** 18 cm^3 = _____ mm^3

5. 7.9 dm^3 = _____ m^3 **6.** 2 mL = _____ L

7. 500 mL = _____ L   **8.** 16 mL = _____ cm^3

Practice Problem Answers

1. 0.5 g   **2.** 43 000 mg   **3.** 0.000 062 kg   **4.** 18 000 mm^3

5. 0.0079 m^3 **6.** 0.002 L   **7.** 0.5 L   **8.** 16 cm^3

Exercises 4.1

1. 3 m = _____ cm

2. 0.8 m = _____ cm

3. 1.5 m = _____ mm

4. 1.9 cm = _____ mm

5. 45 cm = _____ mm

6. 27 dm = _____ cm

7. 13.6 km = _____ m

8. 4.38 km = _____ m

9. 18.25 m = _____ dm

10. 60 m = _____ dm

11. 4.8 mm = _____ cm

12. 36 mm = _____ cm

13. 6.5 cm = _____ m

14. 82 cm = _____ m

15. 1300 mm = _____ m

16. 750 mm = _____ m

17. 5.25 cm = _____ m

18. 185 m = _____ km

19. 5500 m = _____ km

20. 140 cm = _____ dm

21. Change 245 mm to meters.

22. Convert 23 cm to meters.

23. How many kilometers are in 10 000 m?

24. How many meters are in 1.5 km?

25. What number of meters is equivalent to 20 000 cm?

26. Express 4.73 m in centimeters.

27. How many centimeters are in 3.2 mm?

28. Change 87 mm to meters.

29. Express 17.35 m in millimeters.

30. What number of kilometers is equivalent to 140 000 m?

31. $13 \text{ cm}^2 =$ _____ mm^2

32. $6.5 \text{ cm}^2 =$ _____ mm^2

33. $9.6 \text{ cm}^2 =$ _____ mm^2

34. $4.52 \text{ cm}^2 =$ _____ mm^2

35. $500 \text{ mm}^2 =$ _____ cm^2

36. $39 \text{ mm}^2 =$ _____ cm^2

37. $14 \text{ dm}^2 = \underline{\hspace{1.5cm}} \text{ cm}^2$

38. $19.5 \text{ dm}^2 = \underline{\hspace{1.5cm}} \text{ cm}^2$

39. $0.5 \text{ dm}^2 = \underline{\hspace{1.5cm}} \text{ mm}^2$

40. $3 \text{ dm}^2 = \underline{\hspace{1.5cm}} \text{ mm}^2$

41. $13 \text{ dm}^2 = \underline{\hspace{1.5cm}} \text{ cm}^2 = \underline{\hspace{1.5cm}} \text{ mm}^2$

42. $6.4 \text{ dm}^2 = \underline{\hspace{1.5cm}} \text{ cm}^2 = \underline{\hspace{1.5cm}} \text{ mm}^2$

43. $11.5 \text{ m}^2 = \underline{\hspace{1.5cm}} \text{ dm}^2 = \underline{\hspace{1.5cm}} \text{ cm}^2 = \underline{\hspace{1.5cm}} \text{ mm}^2$

44. $3.6 \text{ m}^2 = \underline{\hspace{1.5cm}} \text{ dm}^2 = \underline{\hspace{1.5cm}} \text{ cm}^2 = \underline{\hspace{1.5cm}} \text{ mm}^2$

45. $0.04 \text{ m}^2 = \underline{\hspace{1.5cm}} \text{ dm}^2 = \underline{\hspace{1.5cm}} \text{ cm}^2 = \underline{\hspace{1.5cm}} \text{ mm}^2$

46. $0.6 \text{ m}^2 = \underline{\hspace{1.5cm}} \text{ dm}^2 = \underline{\hspace{1.5cm}} \text{ cm}^2 = \underline{\hspace{1.5cm}} \text{ mm}^2$

47. $6.7 \text{ a} = \underline{\hspace{1.5cm}} \text{ m}^2$

48. $0.45 \text{ a} = \underline{\hspace{1.5cm}} \text{ m}^2$

49. $200 \text{ a} = \underline{\hspace{1.5cm}} \text{ m}^2$

50. $0.8 \text{ a} = \underline{\hspace{1.5cm}} \text{ m}^2$

51. How many hectares are in 2 a?

52. Change 5.75 km^2 to hectares.

53. How many hectares are in 150 a?

54. Change 0.4 km^2 to hectares.

55. How many ares are in 750 ha?

56. $9.56 \text{ ha} = \underline{\hspace{1.5cm}} \text{ a} = \underline{\hspace{1.5cm}} \text{ m}^2$

57. $0.27 \text{ ha} = \underline{\hspace{1.5cm}} \text{ a} = \underline{\hspace{1.5cm}} \text{ m}^2$

58. $6.25 \text{ km}^2 = \underline{\hspace{1.5cm}} \text{ a} = \underline{\hspace{1.5cm}} \text{ ha}$

59. $35 \text{ km}^2 = \underline{\hspace{1.5cm}} \text{ a} = \underline{\hspace{1.5cm}} \text{ ha}$

60. $15 \text{ km}^2 = \underline{\hspace{1.5cm}} \text{ m}^2$

Change the following units of mass (weight) as indicated. See Examples 9 and 10.

61. $2 \text{ g} = \underline{\hspace{1.5cm}} \text{ mg}$

62. $7 \text{ kg} = \underline{\hspace{1.5cm}} \text{ g}$

63. $3700 \text{ kg} = \underline{\hspace{1.5cm}} \text{ t}$

64. $34.5 \text{ mg} = \underline{\hspace{1.5cm}} \text{ g}$

65. $5600 \text{ g} = \underline{\hspace{1.5cm}} \text{ kg}$

66. $4000 \text{ kg} = \underline{\hspace{1.5cm}} \text{ t}$

67. $91 \text{ kg} = \underline{\hspace{1.5cm}} \text{ t}$

68. $73 \text{ kg} = \underline{\hspace{1.5cm}} \text{ mg}$

69. $0.7 \text{ g} = \underline{\hspace{1.5cm}} \text{ mg}$

70. $0.54 \text{ g} = \underline{\hspace{1.5cm}} \text{ mg}$

71. How many kilograms are there in 5 tons?

72. How many kilograms are there in 17 tons?

73. Change 2 tons to kilograms.

74. Change 896 mg to grams.

75. Express 896 g in milligrams.

76. Express 342 kg in grams.

77. Convert 75 000 g to kilograms.

78. Convert 3000 mg to grams.

79. Convert 7 tons to grams.

80. Convert 0.4 t to grams.

81. Change 0.34 g to kilograms.

82. Change 0.78 g to milligrams.

83. How many grams are in 16 mg?

84. How many milligrams are in 2.5 g?

85. 92.3 g = _____ kg

86. 3.94 g = _____ mg

87. 7.58 t = _____ kg

88. 5.6 t = _____ kg

89. 2963 kg = _____ t

90. 3547 kg = _____ t

Fill in the blanks below.

91. 1 cm^3 = _____ mm^3

1 dm^3 = _____ cm^3

1 m^3 = _____ dm^3

1 km^3 = _____ m^3

Change the following units of volume as indicated. See Examples 11 through 16.

92. 73 m^3 = _____ dm^3

93. 0.9 m^3 = _____ dm^3

94. 525 cm^3 = _____ m^3

95. 400 m^3 = _____ cm^3

96. 8.7 m^3 = _____ cm^3

97. 63 dm^3 = _____ m^3

98. How many cm^3 are in 45 mm^3?

99. How many mm^3 are in 3.1 cm^3?

100. Change 19 mm^3 to dm^3.

101. Change 5 cm^3 to mm^3.

102. Convert 2 dm^3 to cm^3.

103. Convert 76.4 mL to liters.

104. Change 5.3 L to milliliters.

105. Change 30 cm^3 to milliliters.

106. Change 30 cm^3 to liters.

107. Change 5.3 mL to liters.

108. 48 kL = _____ L

109. 72 000 L = _____ kL

110. 290 L = _____ kL

111. 569 mL = _____ L

112. 80 L = _____ mL = _____ cm^3

113. 7.3 L = _____ mL = _____ cm^3

Change the following units as indicated. See Example 17.

114. 6.5 megahertz = _____ hertz

115. 150 300 000 000 bytes = _____ gigabytes

116. How many gigahertz are in a computer processor with a speed of with 6 000 000 000 hertz?

117. How many metric tons are there in 5 megatons of dynamite?

118. How many gigabytes are in 0.5 terabytes? (**Hint:** Convert terabytes to bytes, then to gigabytes.)

119. A camera with 14 megapixels has how many more pixels than a camera with 12 megapixels?

4.2 U.S. Customary and Metric Equivalents

Objective A **Temperature**

We begin the following discussion of equivalent measures between the U.S. customary and metric systems with measures of temperature.

Temperature

U.S. customary measure is in **degrees Fahrenheit** (°F).

Metric measure is in **degrees Celsius** (°C).

The two scales are shown here on thermometers. Approximate conversions can be found by reading along a ruler or the edge of a piece of paper held horizontally across the page.

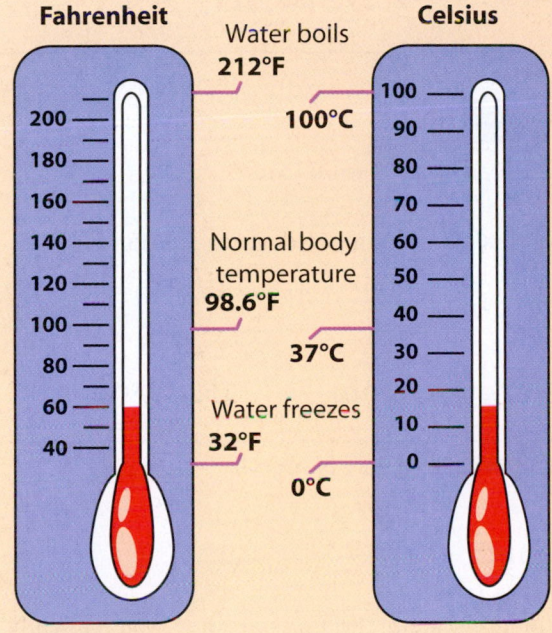

Example 1

Equivalent Measures of Temperature

Hold a straight edge horizontally across the two thermometers and you will read the following.

100°C = 212°F	water boils at sea level
40°C = 104°F	hot day in the desert
20°C = 68°F	comfortable room temperature

Objectives

A   Learn the U.S. customary and metric equivalents for measures of temperature.

B   Learn the U.S. customary and metric equivalents for measures of length.

C   Learn the U.S. customary and metric equivalents for measures of area.

D   Learn the U.S. customary and metric equivalents for measures of volume.

E   Learn the U.S. customary and metric equivalents for measures of mass.

Two formulas that give exact conversions are given here.

F = Fahrenheit temperature and C = Celsius temperature.

$$C = \frac{5(F-32)}{9} \qquad F = \frac{9 \cdot C}{5} + 32$$

A calculator will give answers accurate to eight digits. Answers that are not exact may be rounded to whatever place of accuracy you choose.

Example 2

Converting Measures of Temperature

Let $F = 86°$ and find the equivalent measure in Celsius.

Solution

$$C = \frac{5(86-32)}{9} = \frac{5(54)}{9} = 30 \quad \text{Thus } 86°F = 30°C.$$

Example 3

Converting Measures of Temperature

Let $C = 40°$ and convert this to degrees Fahrenheit.

Solution

$$F = \frac{9 \cdot 40}{5} + 32 = 72 + 32 = 104 \qquad \text{Thus } 40°C = 104°F.$$

Now work margin exercises 1 through 3.

Use a straight edge across the two thermometers to convert each temperature.

1. a. 0°C

b. 50°F

Use the conversion formulas given to convert each temperature.

2. 77°F

3. 45°C

Objective B **Length**

In the U.S. customary system, the units are not systematically related as are the units in the metric system. Historically, some of the units were associated with parts of the body, which would vary from person to person. For example, a foot was the length of a person's foot and a yard was the distance from the tip of one's nose to the tip of one's fingers with the arm outstretched.

There is considerably more consistency now because the official weights and measures are monitored by the government. Table 1 shows some of the basic units of length in the U.S. customary system.

U.S. Customary Units of Length		
1 foot (ft)	=	12 inches (in.)
1 yard (yd)	=	3 ft
1 mile (mi)	=	5280 ft

Table 1

We will use the equivalent measures (rounded) in Table 2 to convert units of length from the U.S. customary system to the metric system, and vice versa. Examples 4 through 7 illustrate how to make conversions using these equivalents and multiplying by the number of given units.

U.S. Customary and Metric Length Equivalents	
U.S. to Metric	**Metric to U.S.**
1 in. = 2.54 cm (exact)	1 cm = 0.394 in.
1 ft = 0.305 m	1 m = 3.28 ft
1 yd = 0.914 m	1 m = 1.09 yd
1 mi = 1.61 km	1 km = 0.62 mi

Table 2

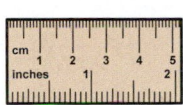

1 in. = 2.54 cm

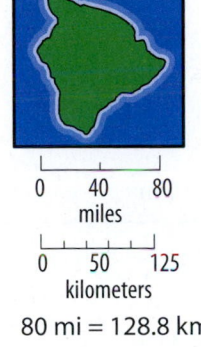

80 mi = 128.8 km

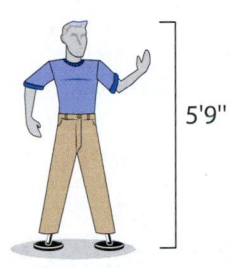

5 ft 9 in. = 175.26 cm

Figure 1

notes

- **Important Note About U.S. and Metric Equivalents**

- Most of the conversion units are not exact and slightly different answers are possible, depending on the conversion units used. Generally, the conversions will be accurate to at least the first three digits. Therefore, we will give conversions rounded to the nearest thousandth.

In Examples 4 – 7, use Table 2 to convert measurements as indicated.

Example 4

Converting Measures of Length

6 ft = _____ cm

Solution

6 ft = 72 in. = 72(2.54 cm) = 182.88 cm

or

6 ft = 6(0.305 m) = 1.83 m = 183 cm

Example 5

Converting Measures of Length

25 mi = _____ km

Solution

25 mi = 25(1.61 km) = 40.25 km

Example 6

Converting Measures of Length

30 m = _____ ft

Solution

30 m = 30(3.28 ft) = 98.4 ft

Example 7

Converting Measures of Length

10 km = _____ mi

Solution

10 km = 10(0.62 mi) = 6.2 mi

Now work margin exercises 4 through 7.

Use Table 2 to convert each of the following measurements as indicated.

4. 84 in. = _____ m

5. 100 yd = _____ m

6. 75 cm = _____ in.

7. 500 m = _____ ft

Objective C **Area**

As with units of length, the units of area in the U.S. customary system are not systematically related, as are the units of area in the metric system. Table 3 shows some of the basic units of area in the U.S. customary system.

U.S. Customary Units of Area
$1 \text{ ft}^2 = 144 \text{ in.}^2$
$1 \text{ yd}^2 = 9 \text{ ft}^2$
$1 \text{ acre} = 4840 \text{ yd}^2 = 43{,}560 \text{ ft}^2$

Table 3

In order to find U.S. Customary and metric equivalents for area, we can use the length equivalents as follows.

$$1 \text{ in.}^2 = 1 \text{ in.} \times 1 \text{ in.} = 2.54 \text{ cm} \times 2.54 \text{ cm} = 6.4516 \text{ cm}^2 \approx 6.45 \text{ cm}^2$$

In a similar manner, we obtain Table 4.

U.S. Customary and Metric Area Equivalents	
U.S. to Metric	**Metric to U.S.**
$1 \text{ in.}^2 = 6.45 \text{ cm}^2$	$1 \text{ cm}^2 = 0.155 \text{ in.}^2$
$1 \text{ ft}^2 = 0.093 \text{ m}^2$	$1 \text{ m}^2 = 10.764 \text{ ft}^2$
$1 \text{ yd}^2 = 0.836 \text{ m}^2$	$1 \text{ m}^2 = 1.196 \text{ yd}^2$
$1 \text{ acre} = 0.405 \text{ ha}$	$1 \text{ ha} = 2.47 \text{ acres}$

Table 4

We will use the equivalent measures (rounded) in Table 4 to convert units of area from the U.S. customary system to the metric system, and vice versa. Examples 8 through 11 illustrate how to make conversions using these equivalents and multiplying by the given number of units.

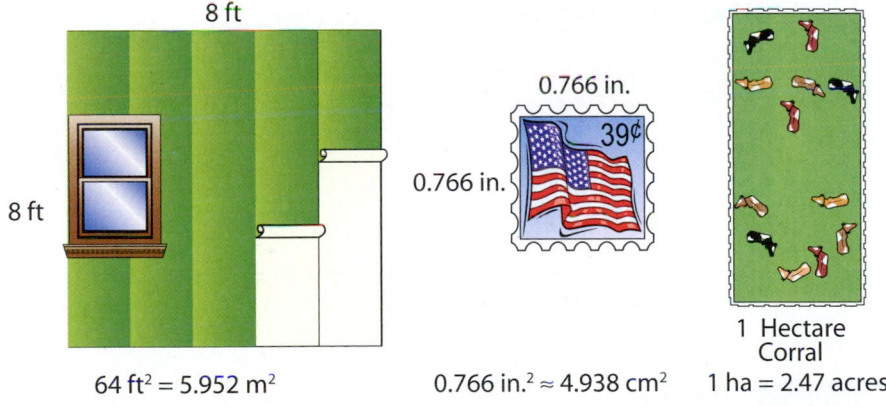

8 ft

8 ft

$64 \text{ ft}^2 = 5.952 \text{ m}^2$

0.766 in.

0.766 in.

39¢

$0.766 \text{ in.}^2 \approx 4.938 \text{ cm}^2$

1 Hectare Corral

$1 \text{ ha} = 2.47 \text{ acres}$

Figure 2

In Examples 8 – 11, use Table 4 to convert the measures as indicated.

Example 8

Converting Measures of Area

$$40 \text{ yd}^2 = \underline{\hspace{1cm}} \text{ m}^2$$

Solution

$$40 \text{ yd}^2 = 40\left(0.836 \text{ m}^2\right) = 33.44 \text{ m}^2$$

Example 9

Converting Measures of Area

$$5 \text{ acres} = \underline{\hspace{1cm}} \text{ ha}$$

Solution

$$5 \text{ acres} = 5(0.405 \text{ ha}) = 2.025 \text{ ha}$$

Example 10

Converting Measures of Area

$$5 \text{ ha} = \underline{\hspace{1cm}} \text{ acres}$$

Solution

$$5 \text{ ha} = 5(2.47 \text{ acres}) = 12.35 \text{ acres}$$

Example 11

Converting Measures of Area

$$100 \text{ cm}^2 = \underline{\hspace{1cm}} \text{ in.}^2$$

Solution

$$100 \text{ cm}^2 = 100\left(0.155 \text{ in.}^2\right) = 15.5 \text{ in.}^2$$

Now work margin exercises 8 through 11.

Use Table 4 to convert each of the following measurements as indicated.

8. $53 \text{ in.}^2 = \underline{\hspace{1cm}} \text{ cm}^2$

9. $16 \text{ acres} = \underline{\hspace{1cm}} \text{ ha}$

10. $3 \text{ ha} = \underline{\hspace{1cm}} \text{ acres}$

11. $38 \text{ m}^2 = \underline{\hspace{1cm}} \text{ ft}^2$

Objective D **Volume**

As with units of length and area, the units of volume in the U.S. customary system are not systematically related, as are the units of volume in the metric system. Table 5 shows some of the basic units of volume in the U.S. customary system.

U.S. Customary Units of Liquid Volume				
1 pint (pt)	=	16 fluid ounces (fl oz)		
1 quart (qt)	=	2 pt	=	32 fl oz
1 gallon (gal)	=	4 qt		

Table 5

In order to find U.S. Customary and metric equivalents for volume, we can use the length equivalents as follows.

$1 \text{ in.}^3 = 1 \text{ in.} \times 1 \text{ in.} \times 1 \text{ in.}$

$= 2.54 \text{ cm} \times 2.54 \text{ cm} \times 2.54 \text{ cm} = 16.387 \ 064 \text{ cm}^3 \approx 16.387 \text{ cm}^3$

In a similar manner, we obtain the following table of volume equivalents.

U.S. Customary and Metric Volume Equivalents	
U.S. to Metric	**Metric to U.S.**
$1 \text{ in.}^3 = 16.387 \text{ cm}^3$	$1 \text{ cm}^3 = 0.06 \text{ in.}^3$
$1 \text{ ft}^3 = 0.028 \text{ m}^3$	$1 \text{ m}^3 = 35.315 \text{ ft}^3$
$1 \text{ qt} = 0.946 \text{ L}$	$1 \text{ L} = 1.06 \text{ qt}$
$1 \text{ gal} = 3.785 \text{ L}$	$1 \text{ L} = 0.264 \text{ gal}$

Table 6

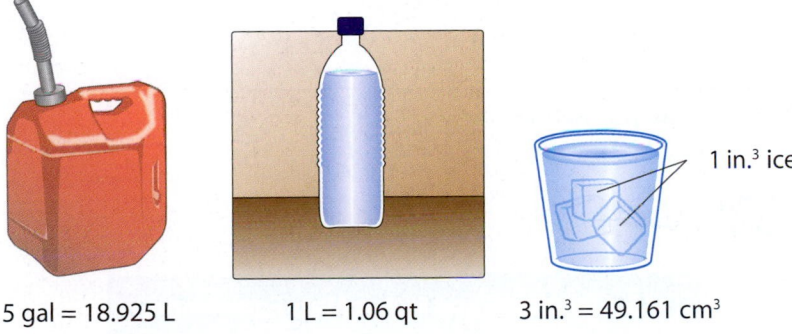

5 gal = 18.925 L   1 L = 1.06 qt   3 in.³ = 49.161 cm³

1 in.³ ice

Figure 3

In Examples 12 – 15, use Table 6 to convert the measures as indicated.

Example 12

Converting Measures of Liquid Volume

20 gal = _____ L

Solution

20 gal = 20(3.785 L) = 75.7 L

Example 13

Converting Measures of Liquid Volume

42 L = _____ gal

Solution

42 L = 42(0.264 gal) = 11.088 gal

Use Table 6 to convert each of the following measurements as indicated.

12. 4 gal = _____ L

13. 16 L = _____ qt

14. 28 qt = _____ L

15. 12 m³ = _____ ft³

Example 14

Converting Measures of Liquid Volume

6 qt = _____ L

Solution

6 qt = 6(0.946 L) = 5.676 L

Now work margin exercises 12 through 14.

Example 15

Converting Measures of Volume

$10 \text{ cm}^3 = $ _____ in.^3

Solution

$10 \text{ cm}^3 = 10\left(0.06 \text{ in.}^3\right) = 0.6 \text{ in.}^3$

Now work margin exercise 15.

Objective E **Mass**

U.S. Customary and Metric Mass Equivalents	
U.S. to Metric	**Metric to U.S.**
1 oz = 28.35 g	1 g = 0.035 oz
1 lb = 0.454 kg	1 kg = 2.205 lb

Table 7

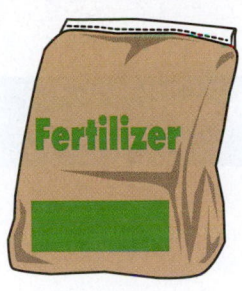

Fertilizer

25 lb = 11.35 kg

9 kg = 19.85 lb

Figure 4

In Examples 16 and 17, use Table 7 to convert the measures as indicated.

Example 16

Converting Measures of Mass

5 lb = _____ kg

Solution

5 lb = 5(0.454 kg) = 2.27 kg

Example 17

Converting Measures of Mass

15 kg = _____ lb

Solution

15 kg = 15(2.205 lb) = 33.075 lb

Now work margin exercises 16 and 17.

16. 3 oz = _____ g

17. 42 g = _____ oz

Practice Problems

Convert the following measures as indicated. You may need to refer to the tables throughout this lesson.

1. 23°F = _____ °C **2.** 14 in. = _____ cm

3. 36 ft^2 = _____ m^2 **4.** 24 cm^3 = _____ in.^{3}

5. 33 kg = _____ lb

Practice Problem Answers
1. −5°C **2.** 35.56 cm **3.** 3.348 m^2
4. 1.44 in.^{3} **5.** 72.765 lb

Exercises 4.2

Convert the following measures as indicated, rounding to the nearest hundredth when necessary. Use the tables in this section as a reference.

1. $25° \text{ C} = $ _____ °F

2. $80° \text{ C} = $ _____ °F

3. $50° \text{ C} = $ _____ °F

4. $35° \text{ C} = $ _____ °F

5. $50° \text{ F} = $ _____ °C

6. $100° \text{ F} = $ _____ °C

7. Change $32°\text{F}$ to degrees Celsius.

8. Change $41°\text{F}$ to degrees Celsius.

9. How many meters are in 3 yds?

10. How many meters are in 5 yds?

11. Change 60 miles to kilometers.

12. Change 100 miles to kilometers.

13. Convert 200 kilometers to miles.

14. Convert 65 kilometers to miles.

15. How many inches are in 50 cm?

16. How many inches are in 100 cm?

17. $3 \text{ in.}^2 = $ _____ cm^2

18. $16 \text{ in.}^2 = $ _____ cm^2

19. $600 \text{ ft}^2 = $ _____ m^2

20. $300 \text{ ft}^2 = $ _____ m^2

21. $100 \text{ yd}^2 = $ _____ m^2

22. $250 \text{ yd}^2 = $ _____ m^2

23. $1000 \text{ acres} = $ _____ ha

24. $250 \text{ acres} = $ _____ ha

25. How many acres are in 300 ha?

26. How many acres are in 400 ha?

27. Change 5 m^2 to square feet.

28. Change 10 m^2 to square feet.

29. Change 30 cm^2 to square inches.

30. Change 50 cm^2 to square inches.

31. $10 \text{ qt} = $ _____ L

32. $20 \text{ qt} = $ _____ L

33. $10 \text{ L} = $ _____ qt

34. $25 \text{ L} = $ _____ qt

35. $42 \text{ L} = $ _____ gal

36. $50 \text{ L} = $ _____ gal

37. $10 \text{ lb} = $ _____ kg

38. $500 \text{ kg} = $ _____ lb

39. $16 \text{ oz} = $ _____ g

40. $100 \text{ g} = $ _____ oz

41. **Area:** Suppose that the home you are considering buying sits on a rectangular shaped lot that is 270 feet by 121 feet. Convert this area to square meters.

42. **Area:** A new manufacturing building covers an area of 3 acres. How many hectares of ground does the new building cover?

43. A painting of a landscape is on a rectangular canvas that measures 3 feet by 4 feet.

    a. How many square centimeters of wall space will the painting cover when it is hanging?

    b. How many square meters?

A Learn the definitions of the basic geometric concepts point, line, and plane.

B Know the definition of an angle and learn how to classify angles.

C Know the definition of a triangle and learn how to classify triangles.

4.3 Introduction to Geometry

Objective A Introduction to Geometry

Several geometry concepts have been discussed throughout this text, including the following topics.

Perimeter	Area	Volume
Angles	Triangles	Circles

These discussions were brief and were based on generally known ideas or specific directions given in a problem rather than detailed definitions and formulas. They were designed to show applications of the topics in the corresponding chapters (such as decimal numbers, proportions, and square roots) and to ensure that some basic geometric concepts are known before a more in-depth course in algebra is attempted.

This section provides a more formal introduction to geometry and introduces special terminology related to angles and triangles. The remaining sections of this chapter emphasize formulas for perimeter, area, and volume for geometric figures, as well as reading graphs.

Plane geometry is the study of the properties of figures in a plane. The most basic ideas in plane geometry are **point**, **line,** and **plane**. These terms are considered so fundamental that any attempt to define their meaning would use terms more complicated and more difficult to understand than the terms themselves. Thus they are simply not defined and are called **undefined terms**. These undefined terms provide the foundation for the study of geometry and are used in the definitions of higher-level ideas such as rays, angles, triangles, circles, and so on.

Points, Lines, and Planes		
Undefined Term	**Representation**	**Discussion**
1. Point	$A \bullet$	A dot represents a point. Points are labeled with capital letters.
2. Line	line l or line $\overleftrightarrow{AB}$	A line has no beginning or end. Lines are labeled with small letters or by two points on the line.
3. Plane	plane P	Flat surfaces, such as table tops or walls, represent planes. Planes are labeled with capital letters.

Here, we give a formal definition of an angle by first defining a **ray** by using the two undefined terms **point** and **line**.

Ray, Angle, Vertex, Sides

A **ray** consists of a point (called the **endpoint**) and all the points on a line on one side of that point.

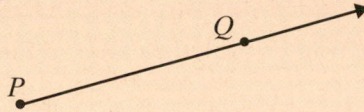

Ray $\overrightarrow{PQ}$ with endpoint P

An **angle** consists of two rays with a common endpoint. The common endpoint is called the **vertex** of the angle.

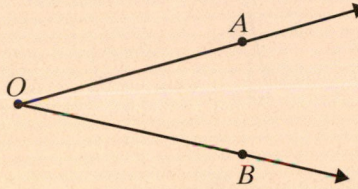

$\angle AOB$ with vertex O

In an angle, the two rays are called the sides of the angle. Thus, as shown in the figure, $\angle AOB$ has sides $\overrightarrow{OA}$ and $\overrightarrow{OB}$.

Every angle has a **measure** (or **measurement**) associated with it. If a circle is divided into 360 equal arcs and two rays are drawn from the center of the circle through two successive points of division on the circle, then that angle is said to **measure one degree** (symbolized 1°).

Figure 1 illustrates the use of a device (called a protractor) that shows the **measure** of $\angle AOB = 60°$. We write m $\angle AOB = 60°$.

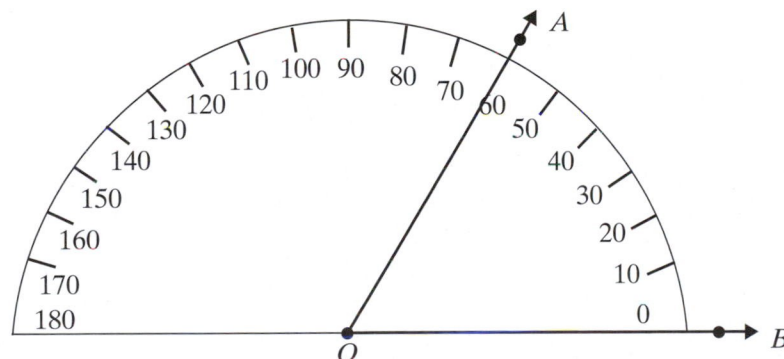

The protractor shows m $\angle AOB = 60°$.

Figure 1

To measure an angle with a protractor, lay the bottom edge of the protractor along one side of the angle with the vertex at the marked center point. Then read the measure from the protractor where the other side of the angle crosses it. You may need to extend one side of the angle for it to intersect the protractor.

Angles can be classified (or named) according to their measures.

Three Common Ways of Labeling Angles

1. Using three capital letters with the vertex as the middle letter.

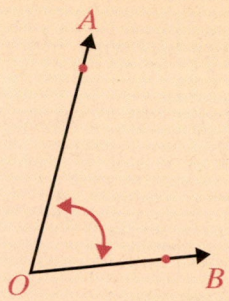

$\angle AOB$
or
$\angle BOA$

2. Using single numbers such as 1, 2, 3.

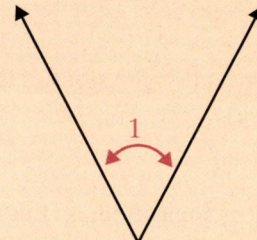

$\angle 1$

3. Using the single capital letter at the vertex when the meaning is clear.

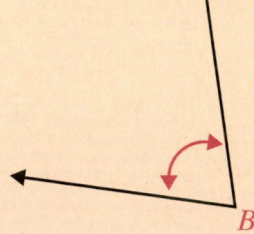

$\angle B$

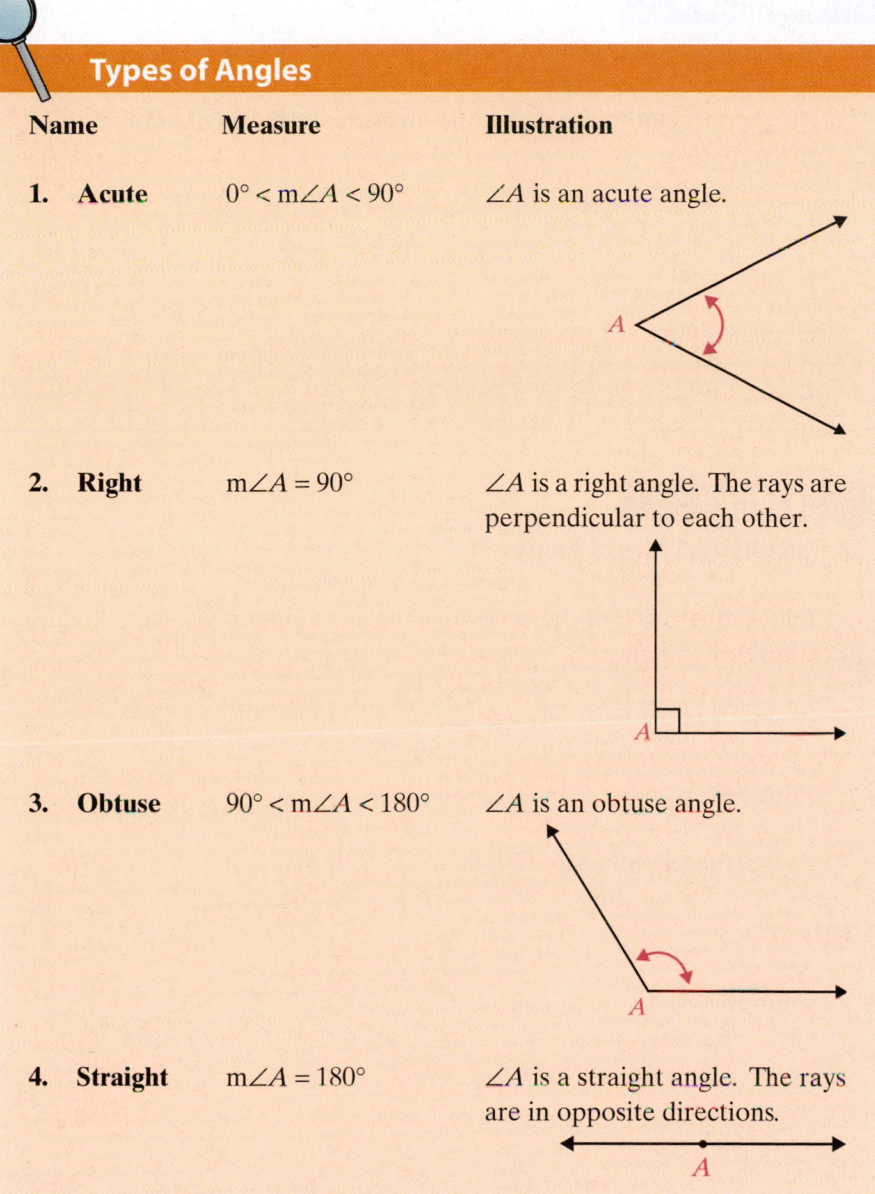

Types of Angles

Name	Measure	Illustration
1. Acute	$0° < m\angle A < 90°$	$\angle A$ is an acute angle.
2. Right	$m\angle A = 90°$	$\angle A$ is a right angle. The rays are perpendicular to each other.
3. Obtuse	$90° < m\angle A < 180°$	$\angle A$ is an obtuse angle.
4. Straight	$m\angle A = 180°$	$\angle A$ is a straight angle. The rays are in opposite directions.

The following figure is used for Examples 1 and 2.

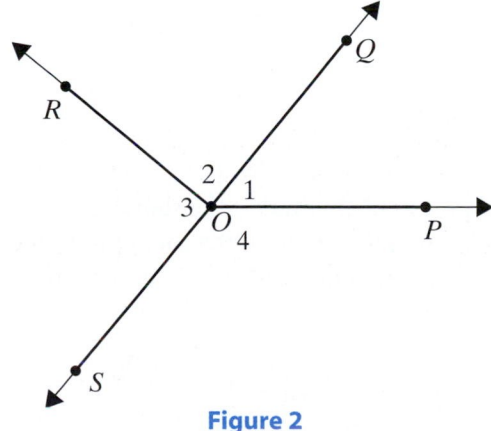

Figure 2

Example 1

Measuring Angles Using a Protractor

Use a protractor to check that the measures of the angles in Figure 2 are as follows.

a. m∠1 = 45°

b. m∠2 = 90°

c. m∠3 = 90°

d. m∠4 = 135°

1. Use a protractor to find the measure of ∠*SOQ* in Figure 2.

2. Tell whether each of the following angles in Figure 2 is acute, right, obtuse, or straight.

a. ∠3

b. ∠*ROP*

Example 2

Identifying Types of Angles

Tell whether each of the following angles in Figure 2 is acute, right, obtuse, or straight.

a. ∠1

Solution

∠1 is acute since 0° < m ∠1 < 90°.

b. ∠2

Solution

∠2 is a right angle since m ∠2 = 90°.

c. ∠*POR*

Solution

∠*POR* is obtuse since m∠*POR* = 45° + 90° = 135° > 90°.

Now work margin exercises 1 and 2.

Complementary, Supplementary, Congruent Angles

1. Two angles are **complementary** if the sum of their measures is 90°.
2. Two angles are **supplementary** if the sum of their measures is 180°.
3. Two angles are **congruent** if they have the same measure.

Example 3

Complementary, Supplementary, and Congruent Angles

In the following figure,

a. ∠1 and ∠2 are complementary because m∠1 + m∠2 = 90°.
b. ∠COD and ∠COA are supplementary because
m∠COD + m∠COA = 70° + 110° = 180°.
c. ∠AOD is a straight angle because m∠AOD = 180°.
d. ∠BOA and ∠BOD are supplementary, and in this case
m∠BOA = m∠BOD = 90°.

m∠1 = 70°
m∠2 = 20°
m∠3 = 90°

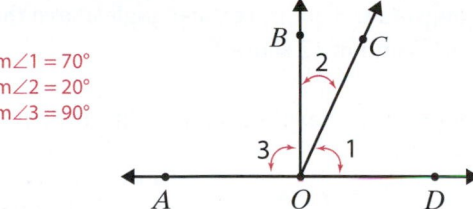

Example 4

Finding Measures of Angles

In the figure below, $\overleftrightarrow{PS}$ is a line and m∠QOP = 30°.

a. Find the measure of the angle ∠QOS.

Solution

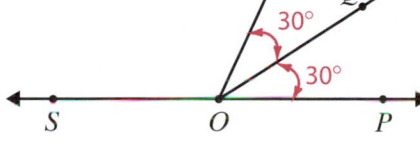

m∠QOS = 150°.

b. Find the measure of the angle ∠SOP.

Solution

m∠SOP = 180°.

c. Are any pairs complementary?

Solution

No pairs are complementary. No two angles have a total measure of 90°.

d. Are any pairs supplementary?

Solution

Yes. ∠QOP and ∠QOS are supplementary, and ∠ROP and ∠ROS are supplementary. The sum of the measures of each pair is 180°.

Now work margin exercises 3 and 4.

3. In the following figure, identify the complementary, supplementary, and straight angles.

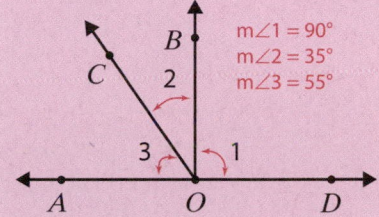

m∠1 = 90°
m∠2 = 35°
m∠3 = 55°

4. In the figure below, m∠TOU = 75°. Also, m∠VOT = 90°, and $\overleftrightarrow{WT}$ is a straight line. Find the measures of

a. ∠UOV

b. ∠WOV

c. Are any pairs complementary?

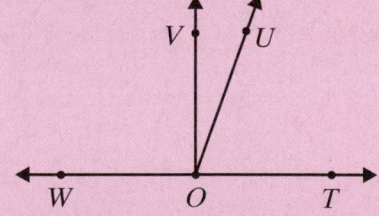

If two lines intersect, then two pairs of **vertical angles** are formed. **Vertical angles** are also called **opposite angles**. (See Figure 3.)

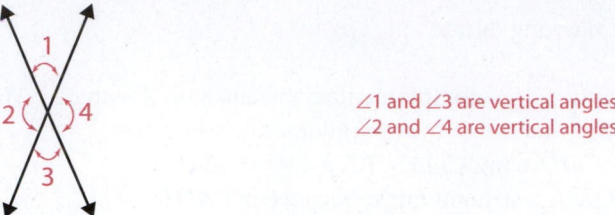

∠1 and ∠3 are vertical angles.
∠2 and ∠4 are vertical angles.

Figure 3

Vertical angles are congruent. That is, **vertical angles have the same measure.** (In Figure 3, m∠1 = m∠3 and m∠2 = m∠4.)

Two angles are **adjacent** if they have a common side. (See Figure 4.)

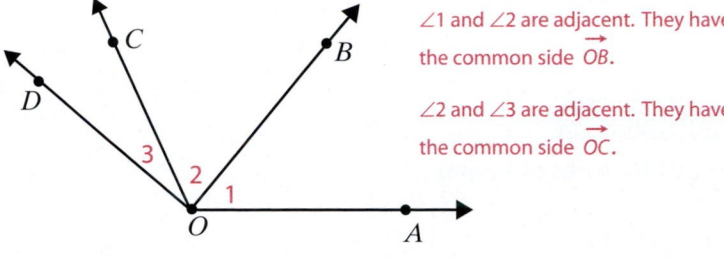

∠1 and ∠2 are adjacent. They have the common side $\overrightarrow{OB}$.

∠2 and ∠3 are adjacent. They have the common side $\overrightarrow{OC}$.

Figure 4

Example 5

Adjacent Angles

In the figure below, $\overleftrightarrow{AC}$ and $\overleftrightarrow{BD}$ are lines.

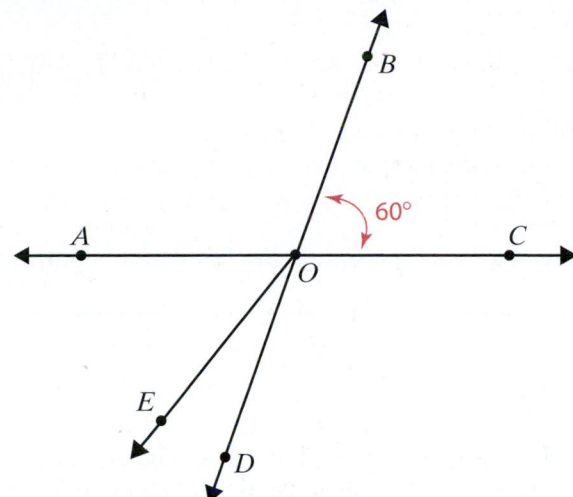

5. a. In the figure drawn in Example 5, name an angle adjacent to $\angle BOC$.

b. What is the measure of angle $\angle AOB$? $\angle COD$?

a. Name an angle adjacent to $\angle EOD$.

Solution

Four angles adjacent to $\angle EOD$ are $\angle AOE$, $\angle BOE$, $\angle BOD$, and $\angle COD$.

b. What is $m\angle AOD$?

Solution

Since $\angle BOC$ and $\angle AOD$ are vertical angles, they have the same measure. So $m\angle AOD = 60°$.

Now work margin exercise 5.

Objective C Triangles

A **triangle** consists of three line segments which join three points that do not lie on a line. The line segments are called the **sides** of the triangle, and the points are called the **vertices** of the triangle. If the points are labeled $A, B,$ and C, the triangle is symbolized $\triangle ABC$. (See Figure 5.)

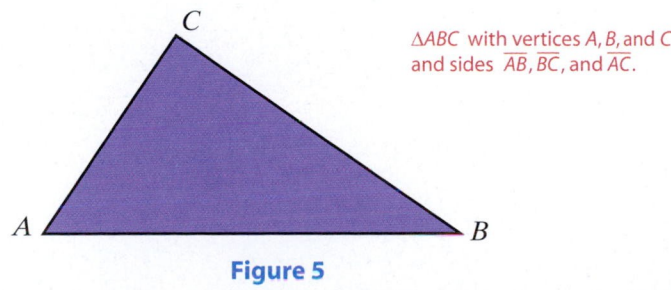

$\triangle ABC$ with vertices $A, B,$ and C and sides $\overline{AB}, \overline{BC},$ and $\overline{AC}$.

Figure 5

The sides of a triangle are said to determine three angles, and these angles are labeled by the vertices. Thus the angles of $\triangle ABC$ are $\angle A$, $\angle B$, and $\angle C$. (Since the definition of an angle involves rays, we can think of the sides of the triangle extended as rays that form these angles.)

Triangles are Classified in Two Ways

1. According to the lengths of their sides, and
2. According to the measures of their angles.

The corresponding names and properties are listed in the following tables.

notes

■ The line segment with endpoints A and B is indicated by placing a bar over the letters, as in $\overline{AB}$. The length of the segment is indicated by writing only the letters, as in AB.

Triangles Classified by Sides

Name	Property	Illustration
1. **Scalene**	No two sides are equal.	$\triangle ABC$ is scalene since no two sides are equal.

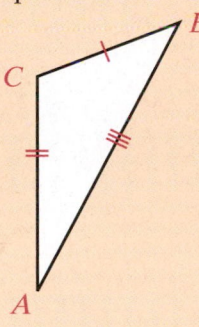

Name	Property	Illustration
2. **Isosceles**	Two sides are equal.	$\triangle PQR$ is isosceles since $PR = QR$.

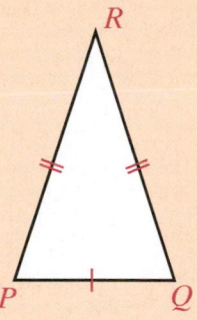

Name	Property	Illustration
3. **Equilateral**	All three sides are equal.	$\triangle XYZ$ is equilateral since $XY = XZ = YZ$.

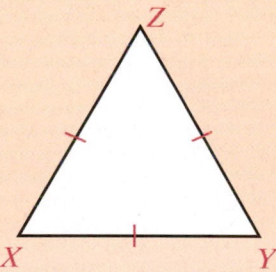

Note: The different number of tick marks on more than one side of a triangle indicate unequal sides, while the same amount of tick marks on more than one side indicate equal sides. For example, on the scalene triangle, each side has a different number of tick marks, indicating no sides are equal, while the equilateral triangle has a single tick mark on each side, indicating all sides are equal.

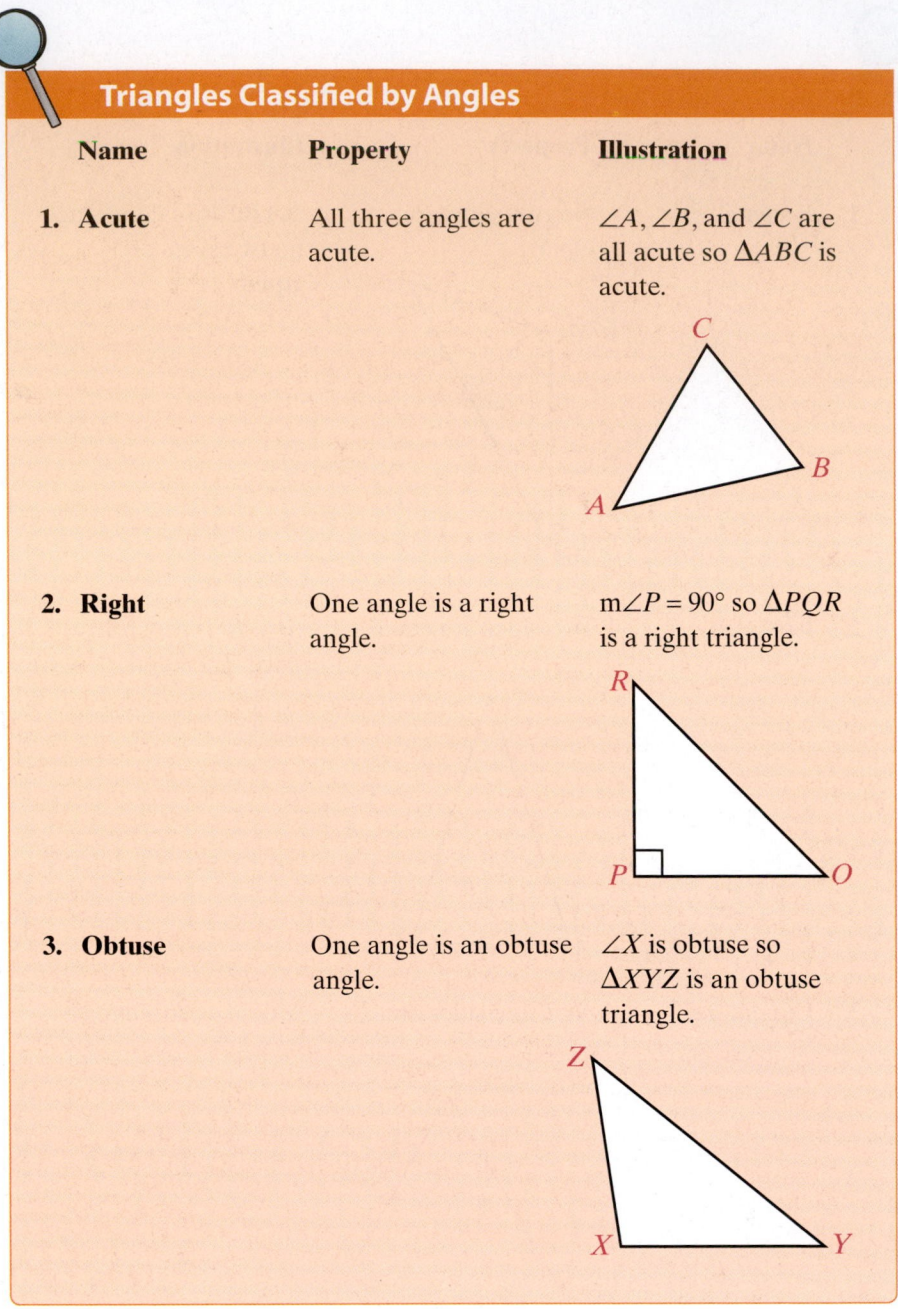

Triangles Classified by Angles

Name	Property	Illustration
1. **Acute**	All three angles are acute.	$\angle A$, $\angle B$, and $\angle C$ are all acute so $\triangle ABC$ is acute.
2. **Right**	One angle is a right angle.	$m\angle P = 90°$ so $\triangle PQR$ is a right triangle.
3. **Obtuse**	One angle is an obtuse angle.	$\angle X$ is obtuse so $\triangle XYZ$ is an obtuse triangle.

Every triangle is said to have six parts; namely, three angles and three sides. Two sides of a triangle are said to **include** the angle at their common endpoint or vertex. The third side is said to be **opposite** this angle. In a triangle, **congruent sides are opposite congruent angles.**

Example 6

Types of Triangles

In $\triangle ABC$ below, $AB = AC$. What kind of triangle is $\triangle ABC$?

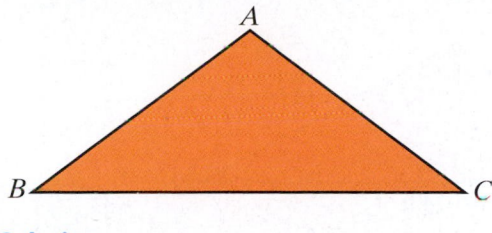

Solution

$\triangle ABC$ can be classified in two ways. $\triangle ABC$ is **isosceles** because two sides are equal. $\triangle ABC$ is also **obtuse** because m$\angle A > 90°$.

Now work margin exercise 6.

The sides in a right triangle have special names. The longest side, opposite the right angle, is called the **hypotenuse**, and the other two sides are called **legs.** (See Figure 6.)

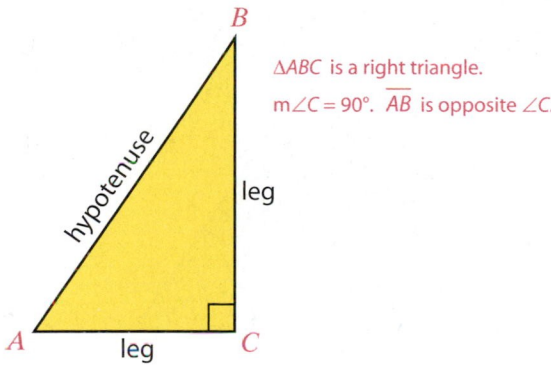

$\triangle ABC$ is a right triangle.

m$\angle C = 90°$. $\overline{AB}$ is opposite $\angle C$.

Figure 6

Two Important Properties of Any Triangle

1. The sum of the measures of its angles is 180°.
2. The sum of the lengths of any two sides must be greater than the length of the third side.

7. Suppose the lengths of the sides of a triangle are 48, 14, and 50. Is this triangle possible?

Example 7

Properties of Triangles

Suppose the lengths of the sides of $\triangle PQR$ are as shown in the figure below. Is this possible?

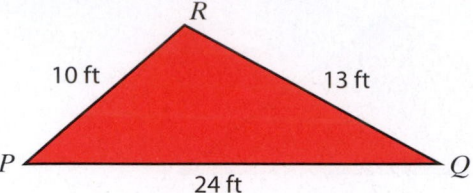

Solution

This is **not** possible because $PR + QR = 10\text{ ft} + 13\text{ ft} = 23\text{ ft}$ and $PQ = 24\text{ ft}$, which is longer than the sum of the other two sides. In a triangle, the sum of the lengths of any two sides must be greater than the length of the third side.

Now work margin exercise 7.

Example 8

Describing Triangles

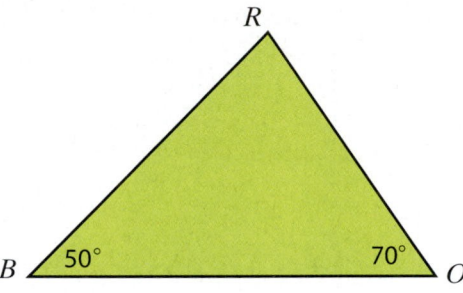

In $\triangle BOR$ below, $m\angle B = 50°$ and $m\angle O = 70°$.

a. What is $m\angle R$?

Solution

The sum of the measures of the angles must be $180°$, so $50° + 70° = 120°$ and $m\angle R = 180° - 120° = 60°$.

b. What kind of triangle is $\triangle BOR$?

Solution

$\triangle BOR$ is an acute triangle because all the angles are acute. Also, $\triangle BOR$ is a scalene triangle because no two sides are equal.

c. Which side is opposite $\angle R$?

Solution

$\overline{BO}$ is opposite $\angle R$.

d. Which sides include $\angle R$?

Solution

$\overline{RB}$ and $\overline{RO}$ include $\angle R$.

e. Is $\triangle BOR$ a right triangle? Why or why not?

Solution

$\triangle BOR$ is not a right triangle because none of the angles are right angles.

Now work margin exercise 8.

8. Suppose that $\triangle BOR$ in the figure in Example 8, $m\angle B = 50°$ and $m\angle R = 108°$.

   a. What is $m\angle O$?

   b. What kind of triangle is $\triangle BOR$?

   c. Which side is opposite $\angle O$?

   d. Which sides include $\angle O$?

   e. Is $\triangle BOR$ a right triangle? Why or why not?

Exercises 4.3

1. **a.** If ∠1 and ∠2 are complementary and m∠1 = 15°, what is m∠2?
    b. If m∠1 = 3°, what is m∠2?
    c. If m∠1 = 45°, what is m∠2?
    d. If m∠1 = 75°, what is m∠2?

2. **a.** If ∠3 and ∠4 are supplementary and m∠3 = 45°, what is m∠4?
    b. If m∠3 = 90°, what is m∠4?
    c. If m∠3 = 110°, what is m∠4?
    d. If m∠3 = 135°, what is m∠4?

3. The supplement of an acute angle is an obtuse angle.
    a. What is the supplement of a right angle?
    b. What is the supplement of an obtuse angle?
    c. What is the complement of an acute angle?
    d. What is the measure of the complement of a 45° angle?

4. In the figure shown below, $\overleftrightarrow{DC}$ is a line and m∠BOA = 90°.

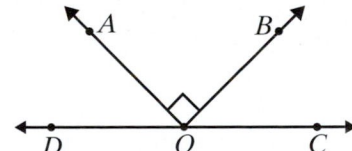

    a. What type of angle is ∠AOC?
    b. What type of angle is ∠BOC?
    c. What type of angle is ∠BOA?
    d. Name a pair of complementary angles.
    e. Name two pairs of supplementary angles.

5. An **angle bisector** is a ray that divides an angle into two angles with equal measures. If $\overrightarrow{OX}$ bisects ∠COD and m∠COD = 50°, what is the measure of each of the congruent angles formed?

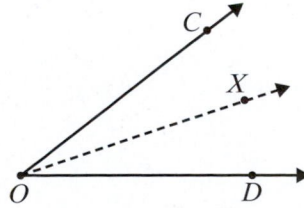

In Exercises 6 – 11, given that m∠AOB = 30° and m∠BOC = 80° and $\overrightarrow{OX}$ and $\overrightarrow{OY}$ are angle bisectors, find the measures of the following angles.

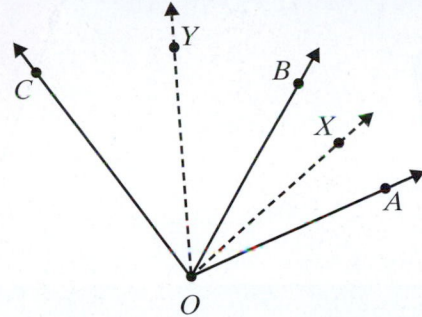

6. ∠AOX		**7.** ∠BOY	
8. ∠COX		**9.** ∠YOX	
10. ∠AOY		**11.** ∠AOC	

12. Use the figure to the right to answer the following questions:

    a. Name all the pairs of supplementary angles.

    b. Name all the pairs of complementary angles.

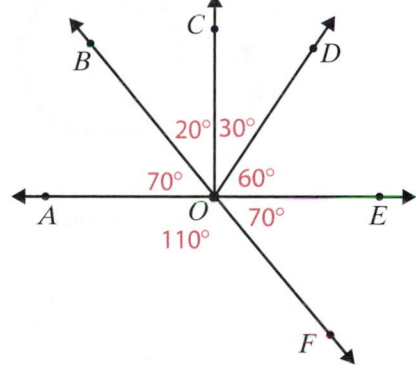

In Exercises 13 – 16, use a protractor to measure all the angles in each figure. Each line segment may be extended as a ray to form the side of an angle. See Example 1.

13.

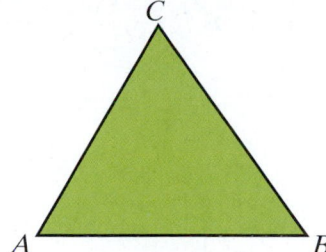

14.

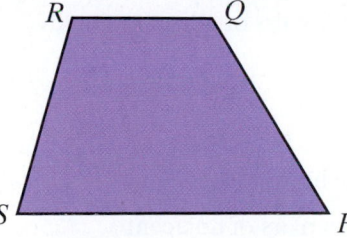

15.

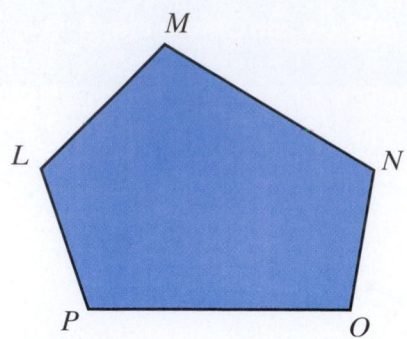

16.

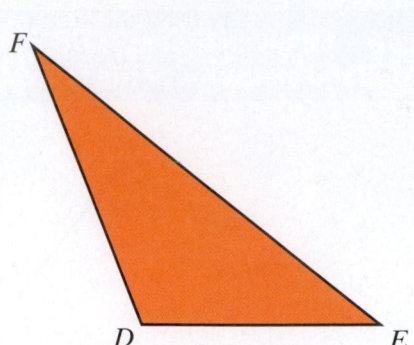

17. Name the type of angle formed by the hands on a clock.

    a. at six o'clock   **b.** at three o'clock

    c. at one o' clock   **d.** at five o'clock

 **a.**    **b.**    **c.**    **d.**

18. What is the measure of each angle formed by the hands of the clock in Exercise 17?

19. The figure at right shows two intersecting lines.

    a. If m∠1 = 30°, what is m∠2?

    b. Is m∠3 = 30°? Give a reason for your answer other

       than the fact that ∠1 and ∠3 are vertical angles.

    c. Name four pairs of adjacent angles.

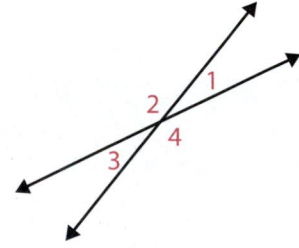

20. In the figure here, $\overleftrightarrow{AB}$ is a line.

    a. Name two pairs of adjacent
       angles.

    b. Name two vertical angles if there are any.

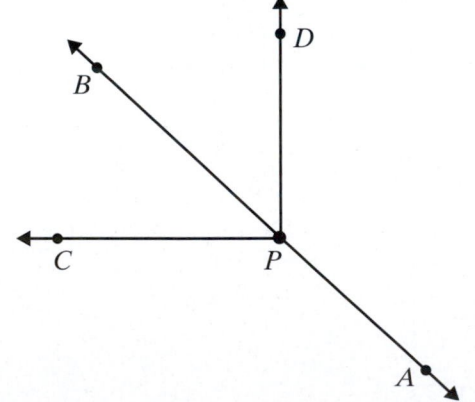

21. Given that m∠1 = 30° in the figure shown here, find the measures of the other three angles.

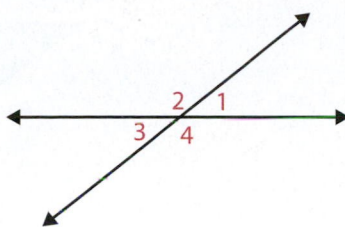

22. In the figure shown below, *l*, *m*, and *n* are straight lines with m∠1 = 20° and m∠6 = 90°.

    a. Find the measures of the other four angles.

    b. Which angle is supplementary to ∠6?

    c. Which angles are complementary to ∠1?

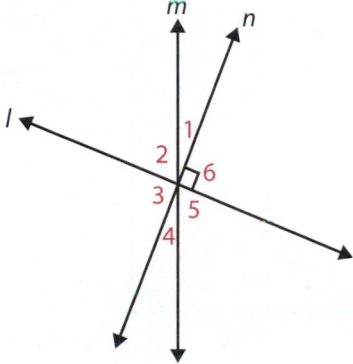

23. In the figure shown below, m∠2 = m∠3 = 40°. Find all other pairs of angles that have equal measures.

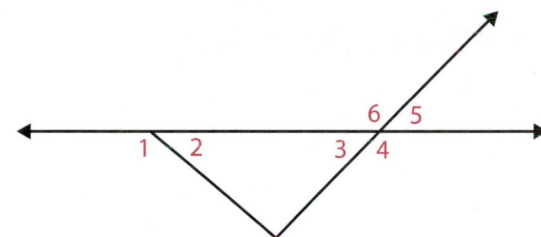

In Exercises 24 – 32, classify each of the following triangles. See Example 6.

24.

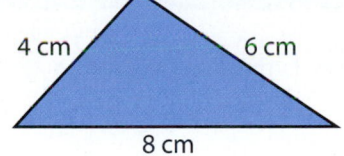

25.

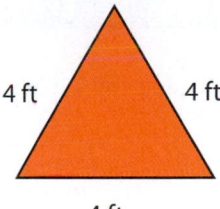

26.

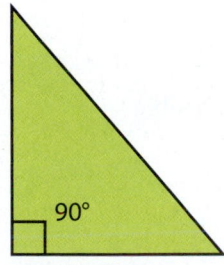

27.

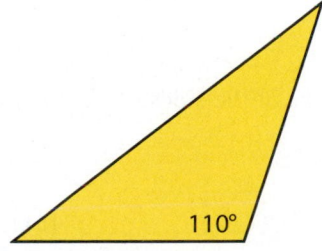

28.

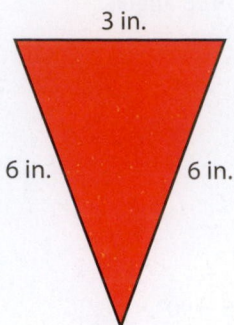

3 in.

6 in.   6 in.

29.

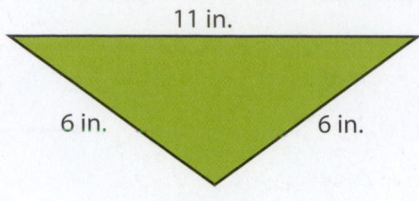

11 in.

6 in.   6 in.

30.

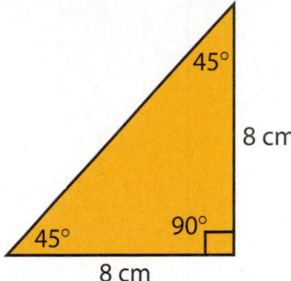

45°

8 cm

45°   90°

8 cm

31.

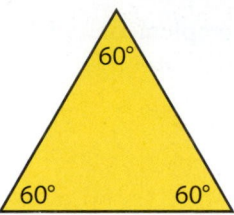

60°

60°   60°

32.

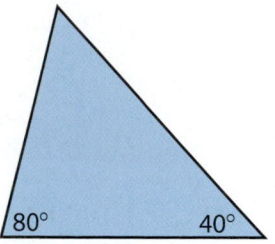

80°   40°

Solve the following problems. See Example 7.

33. Suppose the lengths of the sides of $\triangle ABC$ are as shown in the figure below. Is this possible?

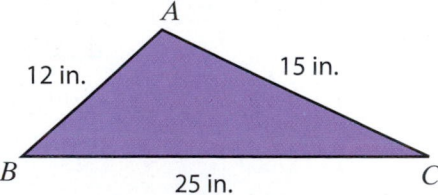

A

12 in.   15 in.

B   25 in.   C

34. Suppose the lengths of the sides of $\triangle DEF$ are as shown in the figure below. Is this possible?

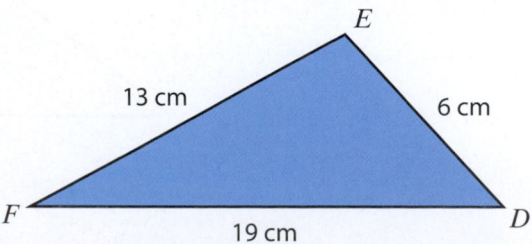

E

13 cm   6 cm

F   19 cm   D

35. In $\triangle XYZ$ below, m$\angle X = 30°$ and m$\angle Y = 70°$.

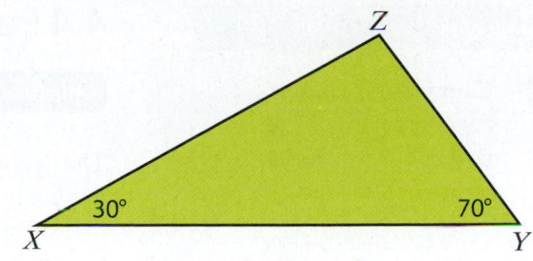

    a. What is m$\angle Z$?

    b. What kind of triangle is $\triangle XYZ$?

    c. Which side is opposite $\angle X$?

    d. Which sides include $\angle X$?

    e. Is $\triangle XYZ$ a right triangle? Why or why not?

36. In $\triangle STU$ below, m$\angle T = 50°$ and m$\angle U = 40°$.

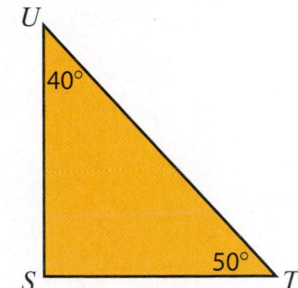

    a. What is m$\angle S$?

    b. What kind of triangle is $\triangle STU$?

    c. Which side is opposite $\angle T$?

    d. Which sides include $\angle T$?

    e. Is $\triangle STU$ a right triangle? Why or why not?

Objectives

A Know the formulas for the perimeter of several geometric figures.

4.4 Perimeter

Objective A **Geometry: Formulas for Perimeter**

The geometric figures and formulas discussed in this section illustrate applications of measures of length. The formulas are independent of any measurement system, and the units used are clearly labeled in each figure. Some of the exercises ask that the answers be given in both U.S. customary units and metric units. For these exercises, refer to Table 1 below.

U.S. and Metric Length Equivalents	
U.S. to Metric	**Metric to U.S.**
1 in. = 2.54 cm (exact)	1 cm = 0.394 in.
1 ft = 0.305 m	1 m = 3.28 ft
1 yd = 0.914 m	1 m = 1.09 yd
1 mi = 1.61 km	1 km = 0.62 mi

Table 1

Important Terms	
Perimeter:	Total distance around a plane geometric figure
Circumference:	Perimeter of a circle
Radius:	Distance from the center of a circle to a point on the circle
Diameter:	Distance from one point on a circle to another point on the circle measured through the center

notes

▪ We will use $\pi = 3.14$, but you should remember that 3.14 is only an approximation to π.
▪

Example 1

Finding the Perimeter of a Rectangle

Find the perimeter of a rectangle with length 20 inches and width 12 inches. Write the answer in both inches and centimeters.

Solution

20 in.

12 in.

$P = 2l + 2w$
$P = 2 \cdot 20 + 2 \cdot 12$
$P = 40 + 24 = 64$ in.
64 in. = 64 (1 in.) = 64 (2.54 cm) = 162.56 cm

The perimeter is 64 in., or 162.56 cm.

Now work margin exercise 1.

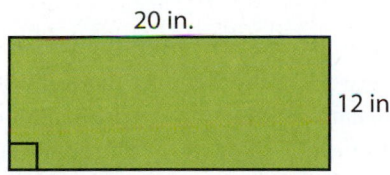

1. Find the perimeter of a rectangle with length 5 inches and width 3 inches. Write the answer in both inches and centimeters.

5 in.

3 in.

Geometric Figures and the Formulas for Finding their Perimeters

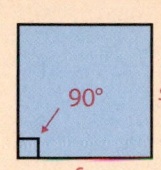

90°   s

s

Square
$P = 4s$

90°   w

l

Rectangle
$P = 2l + 2w$

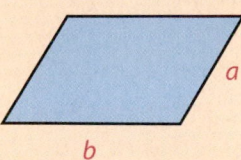

a

b

Parallelogram
$P = 2b + 2a$

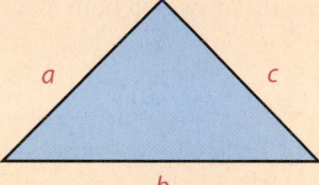

a   c

b

Triangle
$P = a + b + c$

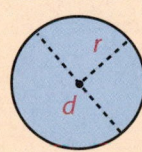

r

d

Circle
$C = 2\pi r$
$C = \pi d$

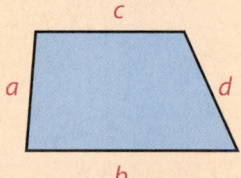

c

a   d

b

Trapezoid
$P = a + b + c + d$

2. Find the perimeter of a triangle with sides of 0.8 cm, 16 mm, and 0.12 dm. Find the perimeter in both millimeters and centimeters.

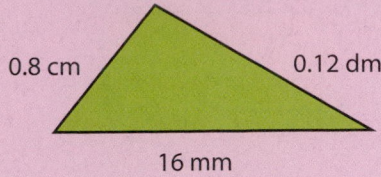

0.8 cm   0.12 dm

16 mm

Example 2

Finding the Perimeter of a Triangle

Find the perimeter of a triangle with sides of 4 cm, 80 mm, and 0.6 dm. Find the perimeter in both millimeters and centimeters.

Solution

Sketch the figure carefully so that your drawing shows how the lengths of the sides are related. (That is, the longest side must be labeled as 80 mm.) Change all the units to the same unit of measure. In this case, any of the three units of measure will do. Millimeters have been chosen.

$$4 \text{ cm} = 40 \text{ mm}, \quad 80 \text{ mm} = 80 \text{ mm}, \quad 0.6 \text{ dm} = 60 \text{ mm}$$

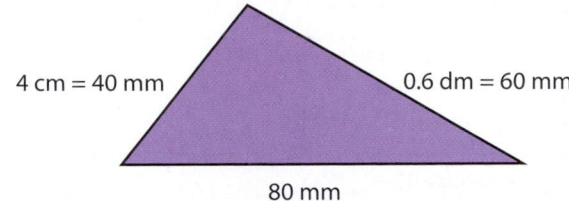

4 cm = 40 mm   0.6 dm = 60 mm

80 mm

$$P = a + b + c$$

$$P = 40 + 60 + 80 = 180 \text{ mm}$$

$$180 \text{ mm} = \frac{180}{10} = 18 \text{ cm}$$

So, the perimeter is 180 mm, or 18 cm.

Now work margin exercise **2.**

3. Find the circumference of a circle with a diameter of 4.6 meters. Write the answer in both meters and feet.

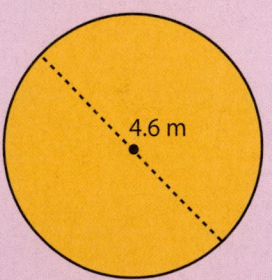

4.6 m

Example 3

Finding the Circumference of a Circle

Find the circumference of a circle with a diameter of 1.5 meters. Write the answer in both meters and feet.

Solution

Sketch the circle and label a diameter.

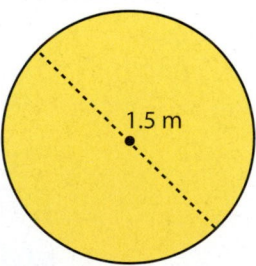

1.5 m

$$C = \pi d$$
$$C = 3.14 \cdot 1.5 = 4.71 \text{ m}$$
$$4.71 \text{ m} = 4.71 \, (1 \text{ m}) = 4.71 \, (3.28 \text{ ft}) = 15.449 \text{ ft (rounded)}$$
So, the circumference is 4.71 m, or 15.449 ft.

Now work margin exercise **3.**

Example 4

Finding the Perimeter of a Semicircle

Find the perimeter of a figure that is a semicircle (half of a circle) with a diameter of 100 mm. Find the perimeter in both millimeters and meters.

Solution

A sketch of the figure will help.

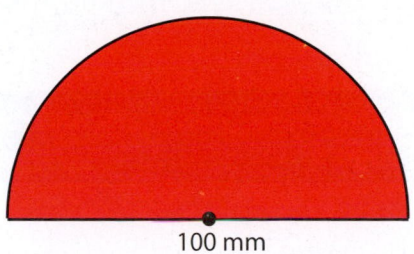

100 mm

Find the perimeter of the figure by adding the length of the diameter (100 mm) to the length of the arc of the semicircle.

Length of arc of semicircle $= \frac{1}{2}\pi d = \frac{1}{2}(3.14) \cdot 100 = 157$ mm

Perimeter of figure $= 157 + 100 = 257$ mm

257 mm $= \dfrac{257}{1000}$ m $= 0.257$ m

So, the perimeter of the figure is 257 mm, or 0.257 m.

Now work margin exercise 4.

4. Find the perimeter of a figure that is a semicircle (half of a circle) with a diameter of 25 cm. Find the perimeter in both centimeters and meters.

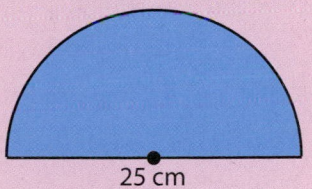

25 cm

Exercises 4.4

1. _____ **a.** square   A. $P = 2l + 2w$

   _____ **b.** parallelogram B. $P = 4s$

   _____ **c.** circle   C. $P = 2b + 2a$

   _____ **d.** rectangle   D. $C = 2\pi r$

   _____ **e.** trapezoid   E. $P = a + b + c$

   _____ **f.** triangle   F. $P = a + b + c + d$

Find the perimeter of the indicated figure. Use Table 1 for reference when necessary, and round your answers to the nearest thousandth. See Examples 1 through 3.

2. Find the perimeter of a triangle with sides of 4 cm, 8 cm, and 6 cm, in both centimeters and inches.

3. Find the perimeter of a rectangle with length 35 mm and width 17 mm, in both millimeters and inches.

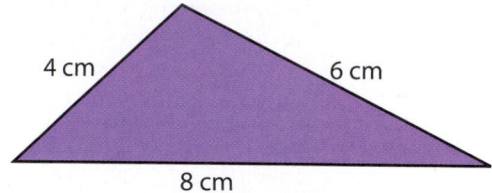

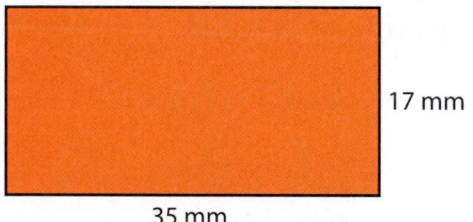

4. Find the circumference of a circle with a radius 5 ft, in both feet and meters.

5. Find the circumference of a circle with diameter 6 yd, in both yards and meters.

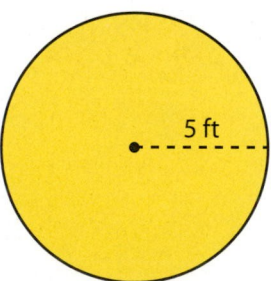

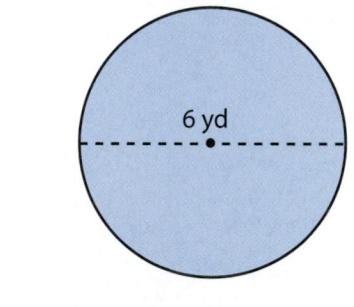

6. Find the perimeter of a triangle with sides of 5 cm, 40 mm, and 0.03 m. Write the answer in equivalent measures in meters, centimeters, and millimeters.

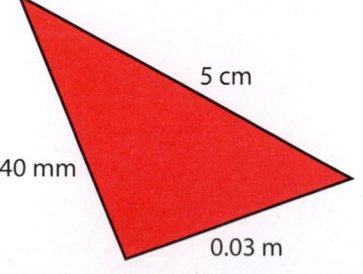

Find the perimeter of each of the following figures with indicated dimensions. Write the answers in equivalent measures in centimeters, millimeters, and inches regardless of the given units. See Example 4.

7.

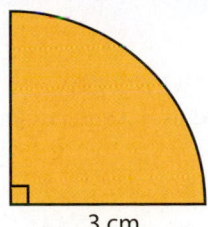

3 cm

8.

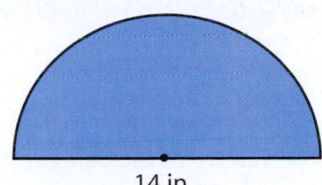

14 in.

9.

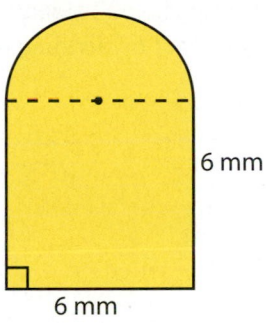

6 mm

6 mm

10.

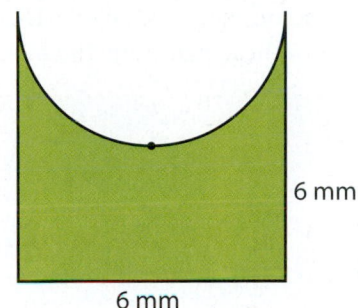

6 mm

6 mm

11.

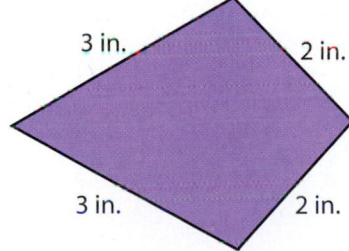

3 in.   2 in.

3 in.   2 in.

12.

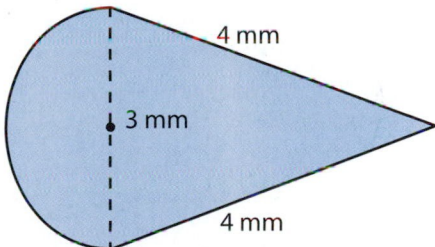

4 mm

3 mm

4 mm

13.

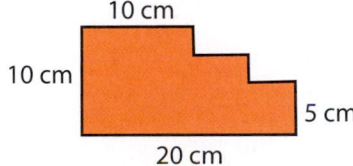

10 cm

10 cm

5 cm

20 cm

14.

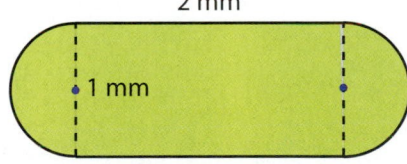

2 mm

1 mm

Writing & Thinking

15. First draw a circle with a diameter of 10 cm (radius 5 cm). Next, draw a square outside the circle so that each side of the square just touches the circle. (The square is said to be *circumscribed* about the circle, and the circle is said to be *inscribed* in the square.)

    a. Divide the perimeter of the square by the diameter of the circle. What number did you get? Next (on the same figure) draw an octagon (8-sided figure) that is circumscribed about the circle. Measure the perimeter of the octagon as accurately as you can.

    b. Divide the perimeter of the octagon by the diameter of the circle. What number did you get?

    c. Consider drawing circumscribed figures with more sides (such as a 16-sided figure, a 32-sided figure, and so on) about the circle. Describe the numbers you think you would get by dividing the perimeters of these figures by the diameter of the circle.

4.5 Area

Objectives

A Know the formulas for the area of several geometric figures.

Objective A **Geometry: Formulas for Area**

The following formulas for the areas of the geometric figures shown are valid regardless of the system of measurement used. **Be able to label your answers with the correct units.** Labeling the units is a good way, particularly in word problems, to remind yourself whether you are dealing with length, area, or volume. Some of the exercises ask that the answers be given in both U.S. customary units and metric units. For these exercises, refer to Table 1 below.

U.S. Customary and Metric Area Equivalents	
U.S. to Metric	**Metric to U.S.**
$1 \text{ in.}^2 = 6.45 \text{ cm}^2$	$1 \text{ cm}^2 = 0.155 \text{ in.}^2$
$1 \text{ ft}^2 = 0.093 \text{ m}^2$	$1 \text{ m}^2 = 10.764 \text{ ft}^2$
$1 \text{ yd}^2 = 0.836 \text{ m}^2$	$1 \text{ m}^2 = 1.196 \text{ yd}^2$
$1 \text{ acre} = 0.405 \text{ ha}$	$1 \text{ ha} = 2.47 \text{ acres}$

Table 1

notes

■ In triangles and other figures, the letter h is used to represent the **height** of the figure. The height is also called the **altitude**.
■

Six Geometric Figures and the Formulas for Finding their Areas

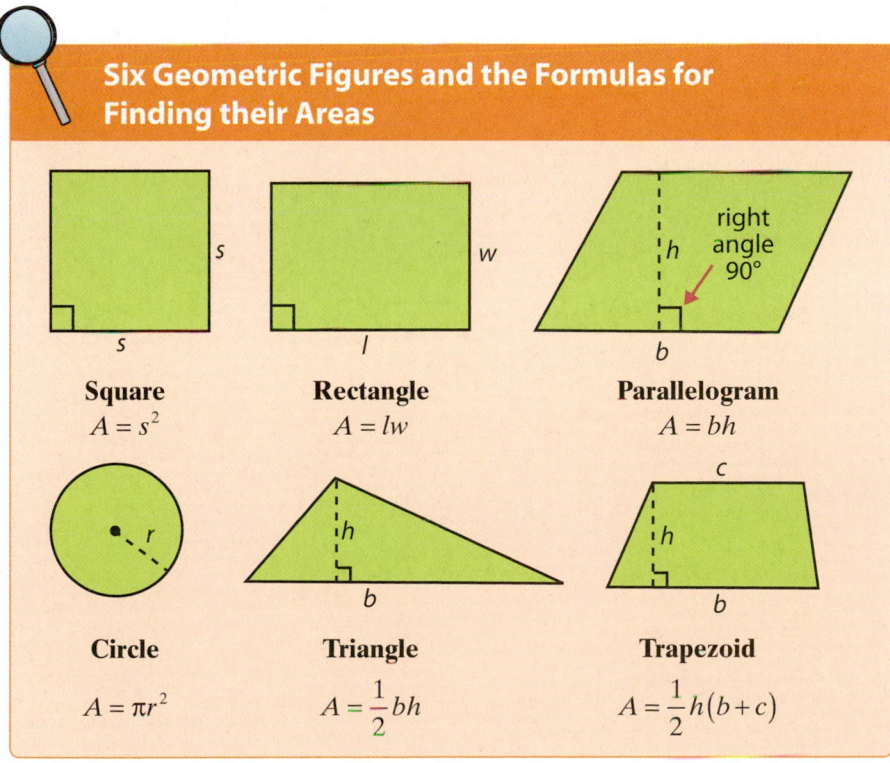

Square
$A = s^2$

Rectangle
$A = lw$

Parallelogram
$A = bh$

Circle
$A = \pi r^2$

Triangle
$A = \dfrac{1}{2}bh$

Trapezoid
$A = \dfrac{1}{2}h(b+c)$

Relationship of the Area of a Triangle to the Area of a Rectangle

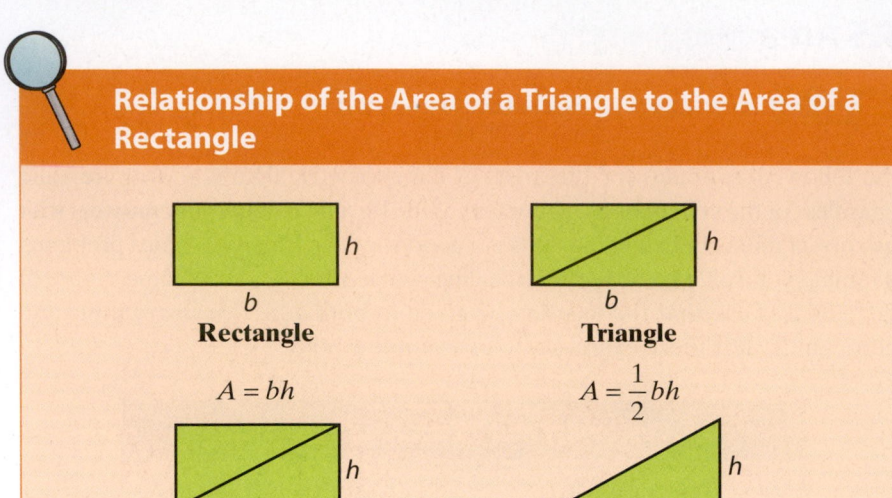

Rectangle

$A = bh$

Triangle

$A = \frac{1}{2}bh$

Relationship of the Area of a Parallelogram to the Area of a Rectangle

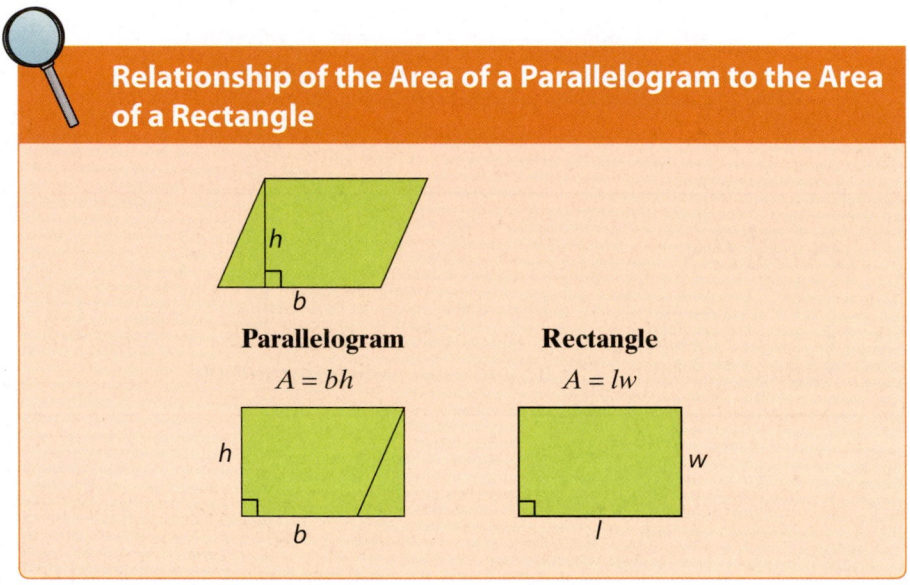

Parallelogram

$A = bh$

Rectangle

$A = lw$

Example 1

Finding the Area

Find the area of the figure shown here with the indicated dimensions. Write the answer in both square centimeters and square inches.

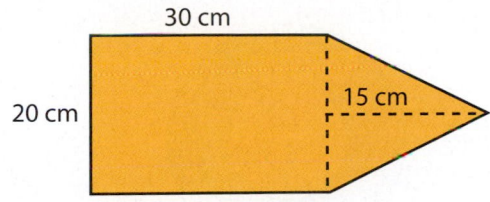

30 cm

20 cm

15 cm

Solution

Find the area of the rectangle and the area of the triangle, and then add the two areas.

Rectangle	**Triangle**
$A = lw$	$A = \dfrac{1}{2}bh$
$A = 20 \cdot 30$	$A = \dfrac{1}{2} \cdot 20 \cdot 15$
$= 600 \text{ cm}^2$	$= 150 \text{ cm}^2$

total area $= 600 \text{ cm}^2 + 150 \text{ cm}^2 = 750 \text{ cm}^2$

in square inches:

$$750 \text{ cm}^2 = 750\left(1 \text{ cm}^2\right) = 750\left(0.155 \text{ in.}^2\right) = 116.25 \text{ in.}^2$$

Now work margin exercise 1.

1. Find the area of the figure shown here with the indicated dimensions. Write the answer in both square centimeters and square inches.

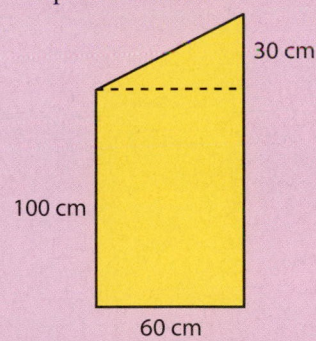

30 cm

100 cm

60 cm

Example 2

Finding the Area of a Washer

Find the area of the washer (shaded portion) with dimensions as shown. Write the answer in square millimeters and square centimeters.

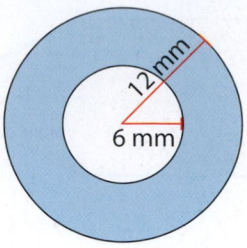

Solution

Subtract the area of the inside (smaller) circle from the area of the outside (larger) circle. This difference will be the area of the washer (shaded portion).

Larger Circle

$A = \pi r^2$

$A = 3.14(12)^2$

$\quad = 3.14(144)$

$\quad = 452.16 \text{ mm}^2$

Smaller Circle

$A = \pi r^2$

$A = 3.14(6)^2$

$\quad = 3.14(36)$

$\quad = 113.04 \text{ mm}^2$

area of shaded portion $= 452.16 \text{ mm}^2 - 113.04 \text{ mm}^2 = 339.12 \text{ mm}^2$
in square centimeters:

$$339.12 \text{ mm}^2 = \frac{339.12}{100} \text{ cm}^2 = 3.3912 \text{ cm}^2$$

An alternate approach is to leave π in the first answers and substitute later as follows.

$$A = \pi(12)^2 = 144\pi \text{ mm}^2 \quad \text{and} \quad A = \pi(6)^2 = 36\pi \text{ mm}^2$$

$$\text{total area} = (144\pi - 36\pi)\text{mm}^2 = 108\pi \text{ mm}^2$$

$$= 108(3.14) \text{ mm}^2$$

$$= 339.12 \text{ mm}^2$$

This approach is more algebraic in nature.

Example 3

Finding the Area of a Semicircle

Find the area enclosed by a semicircle with a diameter of 12 feet. Write the answer in both square feet and square meters.

Solution

First, sketch the figure. (A semicircle is half of a circle.)

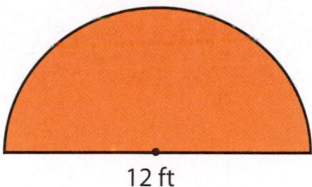

12 ft

For a circle, $A = \pi r^2$. So, for a semicircle, $A = \dfrac{1}{2}\pi r^2$.

For this semicircle, $d = 12$ ft, so $r = 6$ ft.

$$A = \frac{1}{2}\pi \cdot 6^2 = \frac{1}{2}(3.14)(36) = 56.52 \text{ ft}^2$$

in square meters:

$$56.52 \text{ ft}^2 = 56.52(0.093 \text{ m}^2) = 5.2564 \text{ m}^2 \text{ (rounded)}$$

Now work margin exercises **2** *and* **3**.

2. Find the area of the washer with the dimensions shown. Write the answer in square millimeters and square centimeters.

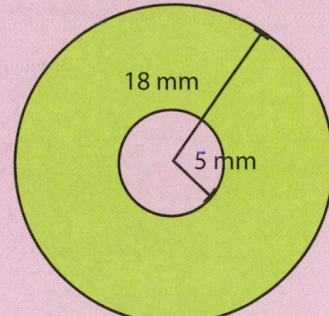

18 mm

5 mm

3. Find the area enclosed by a semicircle with a diameter of 45 feet long. Write the answer in both square feet and square meters.

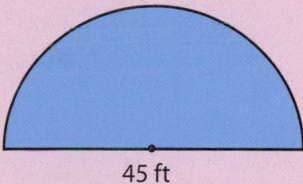

45 ft

Exercises 4.5

1. Match each formula for or to its corresponding geometric figure.

   _____ **a.** square   **A.** A = *lw*

   _____ **b.** parallelogram   **B.** A = *bh*

   _____ **c.** circle   **C.** A = *s*²

   _____ **d.** rectangle   **D.** A = π*r*²

   _____ **e.** trapezoid   **E.** $A = \frac{1}{2}bh$

   _____ **f.** triangle   **F.** $A = \frac{1}{2}h(b+c)$

Find the area of the indicated figure. Use Table 1 for reference when necessary, and round your answers to the nearest thousandth.

2. Find the area of a rectangle 35 inches long and 25 inches wide, in both square inches and square centimeters.

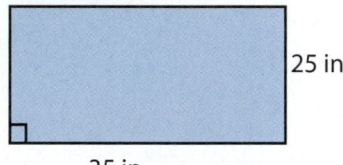

35 in.

25 in.

3. Find the area of a circle with radius 5 yards, in both square yards and square meters.

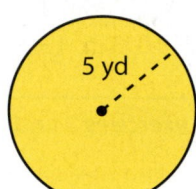

5 yd

4. Find the area of a triangle with base 4 centimeters and altitude 8 centimeters, in both square centimeters and square inches.

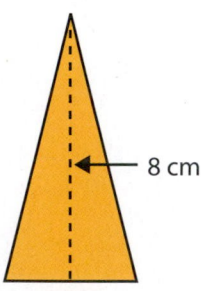

8 cm

4 cm

5. Find the area of a trapezoid with parallel sides of 3.5 mm and 4.2 mm and altitude of 10 mm, in both square millimeters and square centimeters.

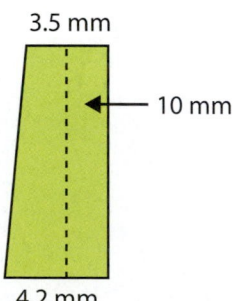

3.5 mm

10 mm

4.2 mm

6. Find the area of a parallelogram with height 1 m and base 50 cm, in both square meters and square centimeters.

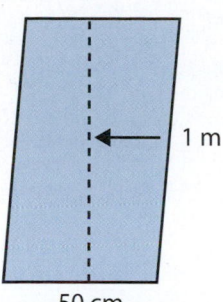

1 m

50 cm

Find the area of each of the following figures with the indicated dimensions. Write the answers in equivalent measures in square centimeters, square millimeters, and square inches, regardless of the given units. See Examples 1 through 3.

7.

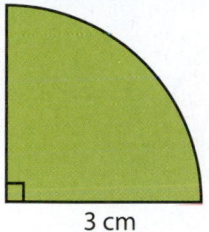

3 cm

8.

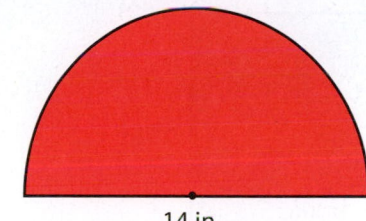

14 in.

9.

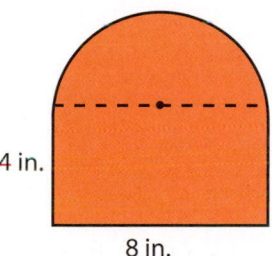

4 in.

8 in.

10.

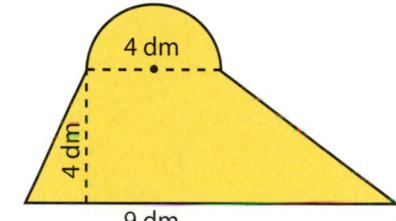

4 dm

4 dm

9 dm

11.

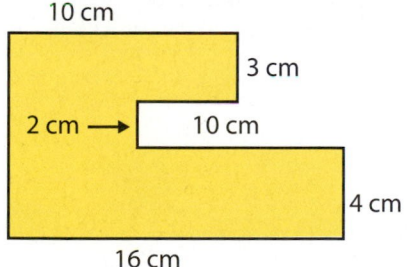

10 cm

3 cm

2 cm

10 cm

4 cm

16 cm

12.

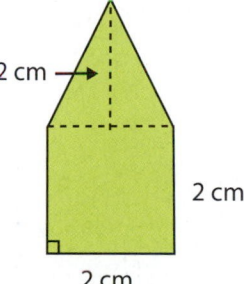

2 cm

2 cm

2 cm

13.

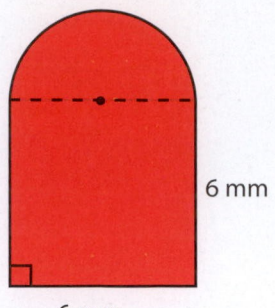

6 mm

6 mm

14.

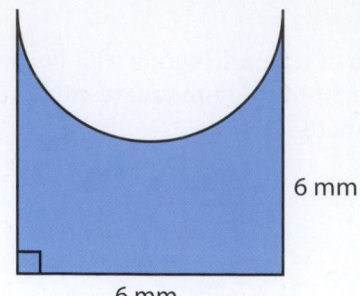

6 mm

6 mm

Find the areas of the shaded regions. Write the answers in equivalent measures in square centimeters, square millimeters, and square inches, regardless of the given units. See Example 2.

15.

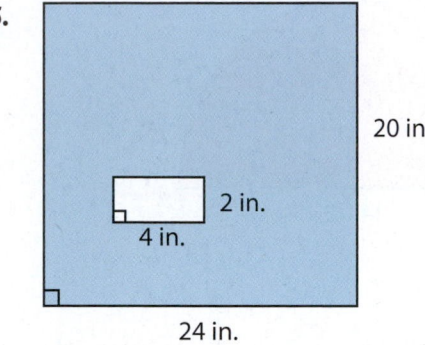

20 in.

2 in.

4 in.

24 in.

16.

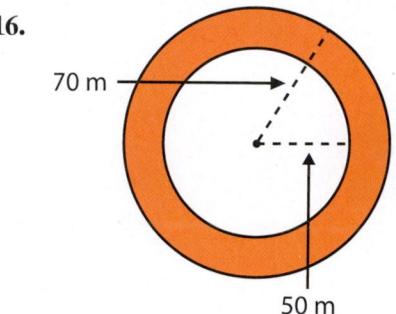

70 m

50 m

4.6 Volume

Objectives

A   Know the formulas for the volume of several geometric figures.

Objective A   **Geometry: Formulas for Volume**

The following formulas for the volumes of common geometric solids shown are valid regardless of the measurement system used. Always be sure to label your answers with the correct units of measure. Some of the exercises ask that the answers be given in both U.S. customary units and metric units. For these exercises, refer to Table 1 below.

U.S. Customary and Metric Volume Equivalents	
U.S. to Metric	**Metric to U.S.**
1 in.^{3} = 16.387 cm^3	1 cm^3 = 0.06 in.^{3}
1 ft^3 = 0.028 m^3	1 m^3 = 35.315 ft^3
1 qt = 0.946 L	1 L = 1.06 qt
1 gal = 3.785 L	1 L = 0.264 gal

Table 1

Five Geometric Solids and the Formulas for their Volumes

Rectangular Solid
$V = lwh$

Rectangular Pyramid
$V = \dfrac{1}{3} lwh$

Right Circular Cylinder
$V = \pi r^2 h$

Right Circular Cone
$V = \dfrac{1}{3} \pi r^2 h$

Sphere
$V = \dfrac{4}{3} \pi r^3$

1. Find the volume of a rectangular solid with length 5 cm, width 4 cm, and height 11 cm. Write the answer in both cubic centimeters and milliliters.

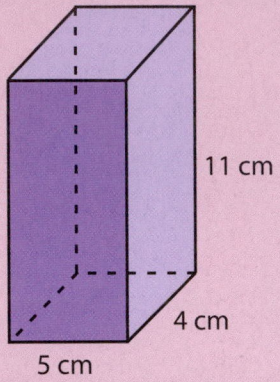

11 cm

4 cm

5 cm

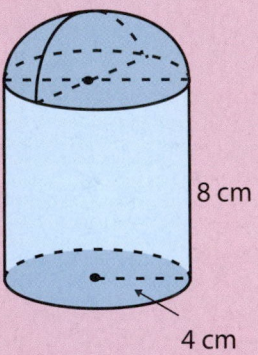

2. Find the volume of the solid with the dimensions indicated. Write the answer in both cubic centimeters and liters.

8 cm

4 cm

Example 1

Finding the Volume of a Rectangular Solid

Find the volume of the rectangular solid with length 8 cm, width 4 cm, and height 12 cm. Write the answer in both cubic centimeters and milliliters.

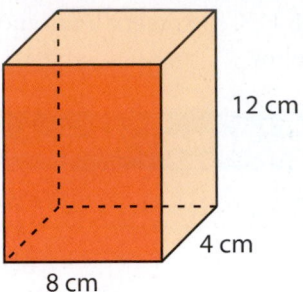

12 cm

4 cm

8 cm

Solution

$V = lwh$

$V = 8 \cdot 4 \cdot 12$

$ = 384 \text{ cm}^3$

Since $1 \text{ cm}^3 = 1 \text{ mL}$, $384 \text{ cm}^3 = 384 \text{ mL}$.

Now work margin exercise 1.

Example 2

Finding the Volume of a Solid

Find the volume of the solid with the dimensions indicated. Write the answer in both cubic centimeters and liters.

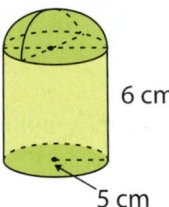

6 cm

5 cm

Solution

From the illustration, the solid is a hemisphere resting on top of a cylinder. Thus, the volume of the solid will be sum of the volumes of these two figures.

Cylinder	**Hemisphere**	**Total Volume**
$V = \pi r^2 h$	$V = \dfrac{1}{2} \cdot \dfrac{4}{3} \pi r^3$	471.00 cm^3
$V = 3.14(5^2)(6)$	$V = \dfrac{2}{3}(3.14)(5^3)$	$+\ 261.67 \text{ cm}^3$
$ = 471 \text{ cm}^3$		732.67 cm^3
	$ = 261.67 \text{ cm}^3$	

Since $1 \text{ cm}^3 = 1 \text{ mL} = 0.001 \text{ L}$, the total volume in liters is $732.67 \text{ cm}^3 = 0.73267 \text{ L}$ or 0.733 L (rounded).

Now work margin exercise 2.

Exercises 4.6

Match each formula for volume to its corresponding geometric figure.

1. _____ **a.** rectangular solid

   _____ **b.** rectangular pyramid

   _____ **c.** right circular cylinder

   _____ **d.** right circular cone

   _____ **e.** sphere

   A. $V = \dfrac{4}{3}\pi r^3$

   B. $V = \dfrac{1}{3}\pi r^2 h$

   C. $V = lwh$

   D. $V = \pi r^2 h$

   E. $V = \dfrac{1}{3}lwh$

2. Find the volume of a rectangular solid with length 5 in., width 2 in., and height 7 in., in both cubic inches and cubic centimeters.

 3. Find the volume of right circular cylinder 1.5 ft in height and 1 ft in diameter, in both cubic feet and cubic meters.

 4. Find the volume of a Christmas tree ornament in the shape of a sphere with radius 4.5 cm, in both cubic centimeters and cubic inches.

5. Find the volume of a right circular cone 3 dm high with a 2 dm radius, in both cubic decimeters and cubic meters.

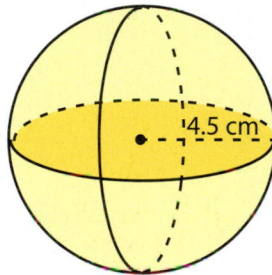

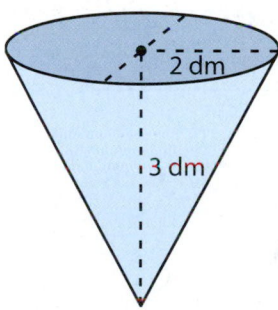

6. Find the volume of a rectangular pyramid with length 18 cm, width 10 cm, and altitude 3 cm, in both cubic centimeters and cubic millimeters.

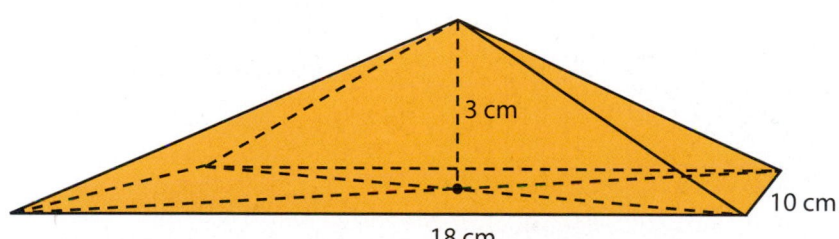

7. Disposable paper drinking cups like those used at water coolers are often cone-shaped. Find the volume of such a cup that is 9 cm high with a 3.2 cm radius. Express the answer to the nearest milliliter.

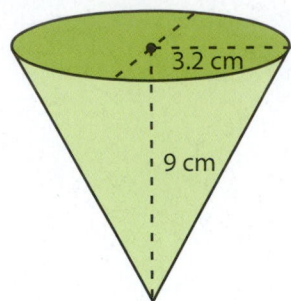

8. A manufacturer is to design a can in the shape of a right circular cylinder to hold 0.5 L of juice concentrate. If the can must have a diameter of 7 cm, how tall will the can be (to the nearest centimeter)?

Find the volume of each of the following figures with the indicated dimensions. Write the answers in equivalent measures in cubic centimeters, cubic millimeters, and cubic inches. See Example 2.

9.

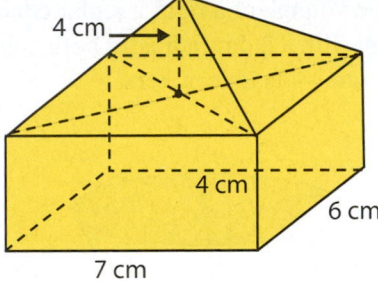

10.

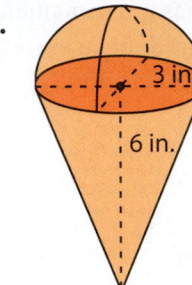

11.

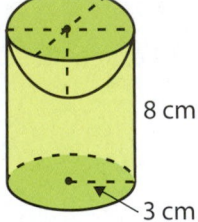

12.

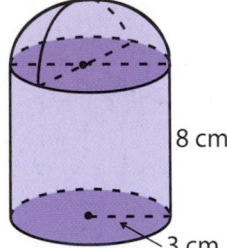

13.

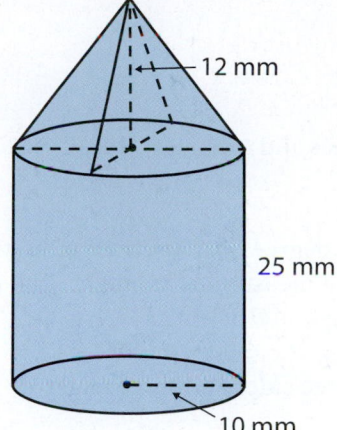

12 mm

25 mm

10 mm

14.

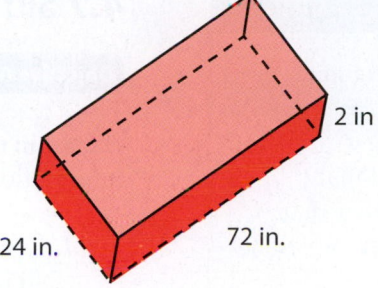

2 in

24 in.   72 in.

Objectives

A   Understand similar triangles.

B   Understand that polygons other than triangles can also be similar.

4.7 Similarity

Objective A   **Similar Triangles**

Earlier in this chapter, we discussed angles and triangles. Recall the following information:

1. An angle can be measured in degrees. A protractor, as shown in Figure 1, can be used to find the measure of an angle. The notation m∠*AOB* is read "measure of angle *AOB*."

2. Every triangle has six parts – three sides and three angles.

3. The sum of the measures of the angles of every triangle is 180°.

4. Each endpoint of the sides of a triangle is called a vertex of the triangle. Capital letters are used to label the vertices, and these letters can be used to name the angles and the sides.

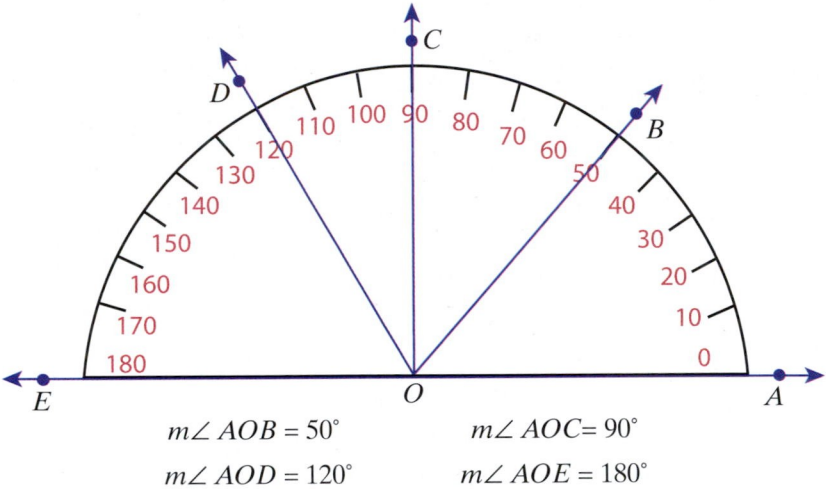

$$m\angle AOB = 50° \qquad m\angle AOC = 90°$$
$$m\angle AOD = 120° \qquad m\angle AOE = 180°$$

Figure 1

To measure an angle with a protractor, lay the bottom edge of the protractor along one side of the angle with the vertex at the marked centerpoint on the protractor. Then, read the measure from the protractor where the other side crosses the arch part of the protractor.

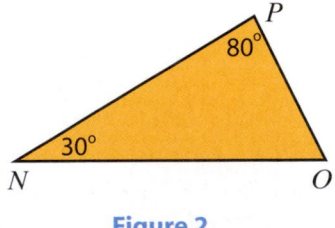

Figure 2

To illustrate the notation and information about triangles, consider triangle *NOP* (symbolized ∆*NOP*) in Figure 2. In this triangle,

$$m\angle N = 30° \quad \text{and} \quad m\angle P = 80°.$$

The sum of the measures of the three angles must be 180°. Therefore, we can set up and solve an equation for the unknown m∠O as follows:

$$m\angle N + m\angle P + m\angle O = 180°$$
$$30° + 80° + m\angle O = 180°$$
$$110° + m\angle O = 180°$$
$$m\angle O = 180° - 110°$$
$$m\angle O = 70°.$$

In this manner, if we know the measures of two angles of a triangle, we can always find the measure of the third angle.

Two triangles are said to be **similar triangles** if they have the same "shape." They may or may not have the same "size." More formally, two triangles are similar if they have the following two properties.

Two Triangles are Similar if:

1. Their **corresponding angles are congruent**. (The corresponding angles have the same measure.)
2. Their **corresponding sides are proportional**.

In similar triangles, **corresponding sides** are those sides opposite the congruent angles in the respective triangles. (See Figure 3.)

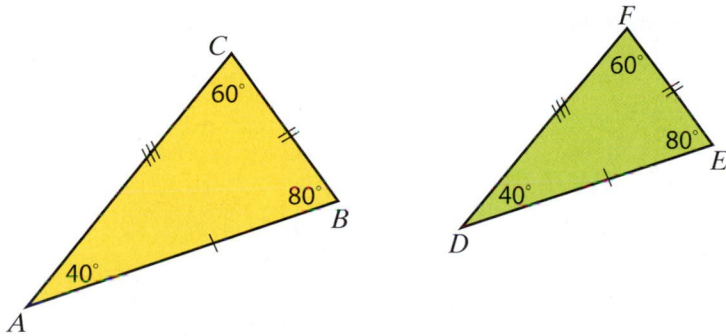

Figure 3

We write ∆**ABC** ~ ∆**DEF** (**~ is read "is similar to."**). The corresponding sides are proportional, so the ratios of corresponding sides are equal and

$$\frac{AB}{DE} = \frac{BC}{EF} \quad \text{and} \quad \frac{AB}{DE} = \frac{AC}{DF} \quad \text{and} \quad \frac{BC}{EF} = \frac{AC}{DF}.$$

To say that corresponding sides are proportional means that we can set up a proportion to solve for one of the unknown sides of two similar triangles. Example 1 illustrates how this can be done.

1. Given $\triangle DEF \sim \triangle TUV$, use the fact that corresponding sides are proportional and find the values of x and y.

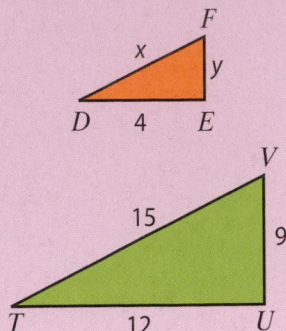

Example 1

Finding Unknown Lengths in Similar Triangles

Given that $\triangle ABC \sim \triangle PQR$, use the fact that corresponding sides are proportional and find the values of x and y.

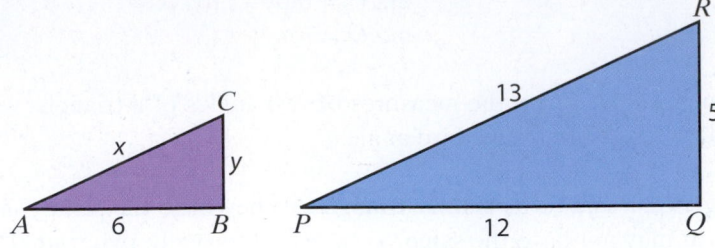

Solution

Set up two proportions and solve for the unknown terms.

$$\frac{x}{13} = \frac{6}{12} \qquad\qquad \frac{y}{5} = \frac{6}{12}$$

$$\frac{\cancel{13}}{1} \cdot \frac{x}{\cancel{13}} = \frac{13}{1} \cdot \frac{6}{12} \qquad\qquad \frac{\cancel{5}}{1} \cdot \frac{y}{\cancel{5}} = \frac{5}{1} \cdot \frac{6}{12}$$

$$x = \frac{13}{2} \left(\text{or } x = 6\frac{1}{2} \right) \qquad y = \frac{5}{2} \left(\text{or } y = 2\frac{1}{2} \right)$$

OR: Reducing first may lead to equations that are easier to solve. The answers will be the same.

$$\frac{x}{13} = \frac{6}{12} \qquad\qquad \frac{y}{5} = \frac{6}{12}$$

$$\frac{x}{13} = \frac{1}{2} \qquad\qquad \frac{y}{5} = \frac{1}{2}$$

$$\frac{\cancel{13}}{1} \cdot \frac{x}{\cancel{13}} = \frac{13}{1} \cdot \frac{1}{2} \qquad\qquad \frac{\cancel{5}}{1} \cdot \frac{y}{\cancel{5}} = \frac{5}{1} \cdot \frac{1}{2}$$

$$x = \frac{13}{2} \left(\text{or } x = 6\frac{1}{2} \right) \qquad y = \frac{5}{2} \left(\text{or } y = 2\frac{1}{2} \right)$$

Now work margin exercise 1.

Example 2

Height of a Christmas Tree

A very tall, magnificently decorated, tree was on display in an outside patio area at the mall during the month before Christmas. Joann knew that the height of a nearby lamppost was 10 feet. At 2 p.m. on a Tuesday afternoon, she measured the shadow of the lamppost to be $3\frac{1}{2}$ feet long and the shadow of the tree to be 21 feet long. With her understanding of similar triangles, Joann was able to calculate the height of the tree. What was the height of the tree?

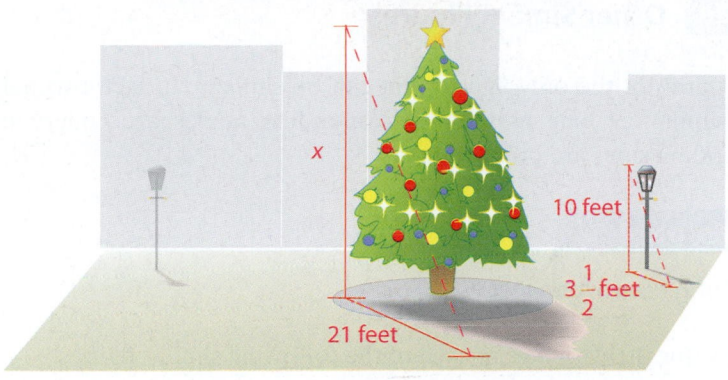

2. Two similar triangles have hypotenuses of lengths 5 and 15 meters, respectively. The shortest leg on the larger triangle is 9 meters. What is the length of the shortest leg on the smaller of the two triangles?

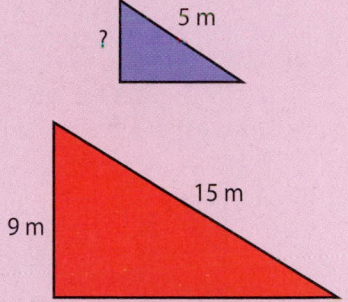

Solution

By letting x represent the height of the tree, Joann set up and solved the following proportion.

$$\frac{x \text{ ft (\textbf{height of tree})}}{21 \text{ ft (\textbf{length of tree shadow})}} = \frac{10 \text{ ft (\textbf{height of lamppost})}}{3\frac{1}{2} \text{ ft (\textbf{length of post shadow})}}$$

$$21 \cdot \frac{x}{21} = 21 \cdot \frac{10}{\frac{7}{2}}$$

$$x = 21^{\,3} \cdot 10 \cdot \frac{2}{7}$$

$$x = 60$$

The tree was 60 feet tall.

Now work margin exercise 2.

3. Find the values of x and y in the similar triangles $\triangle MNP$ and $\triangle MQR$.

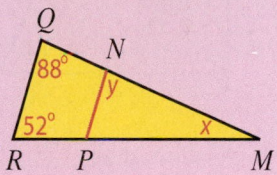

To say that corresponding angles are congruent means that the angles in the same relative positions in the two triangles are congruent. Example 3 illustrates corresponding angles.

Example 3

Finding Unknown Values in Similar Triangles

Find the values of x and y in triangles $\triangle ABC$ and $\triangle ADE$ where $\triangle ABC \sim \triangle ADE$.

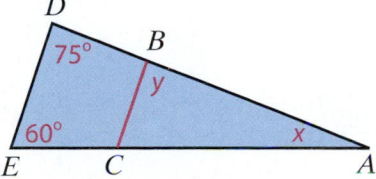

Solution

Let x be the measure of $\angle A$ and y be the measure of $\angle B$ in $\triangle ABC$ as illustrated. Since y is in the same relative position as $75°$ in $\triangle ADE$ ($\angle B$ and $\angle D$ are corresponding angles), we have $y = 75°$. In $\triangle ADE$, using the fact that the sum of the measures of the angles of a triangle must be $180°$ gives the equation $x + 75 + 60 = 180$. Solving for x gives $x = 45°$.

Now work margin exercise 3.

Objective B Other Similar Figures

Triangles are not the only figures that can be similar. In fact, two polygons can be similar as long as their corresponding angles are congruent and corresponding sides are proportional.

4. Solve for the variables in each of the following similar figures.

a. Assume that $LMNO \sim WXYZ$ and solve for a and b.

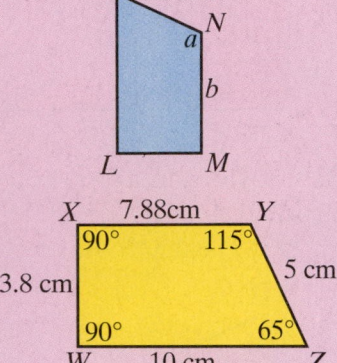

b. Assume that $UVWXYZ \sim LGHIJK$ and solve for a and b.

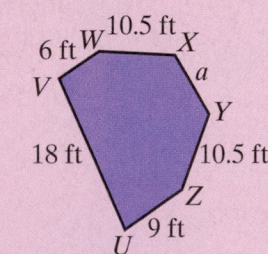

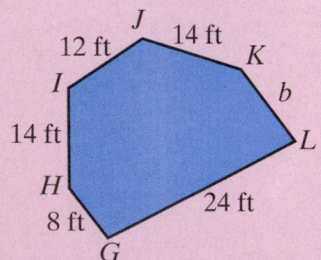

Example 4

Solving Proportions

Solve for the variables in each of the following similar figures.

a. Assume that $ABCD \sim STUR$ and solve for x and y.

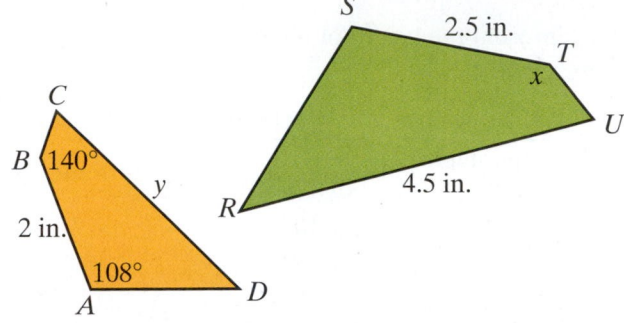

Solution

Because corresponding angles in similar figures have the same measure, $m\angle T = m\angle B = 140°$. Thus $x = 140°$.

Also recall that corresponding sides in similar figures are proportional. Thus, in order to solve for y, we set up and solve the following proportion.

$$\frac{2.5}{2} = \frac{4.5}{y}$$

$$2.5y = 9$$

$$y = 3.6 \text{ in.}$$

b. Assume that $ABCDE \sim HIJKL$ and solve for x and y.

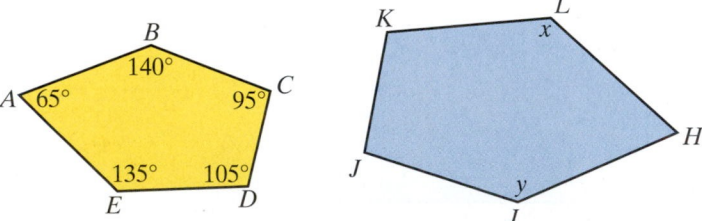

Solution

Because corresponding angles in similar figures have the same measure, $m\angle L = m\angle E = 135°$. Thus $x = 135°$.

Similarly, $m\angle I = m\angle B = 140°$. Thus $y = 140°$.

Now work margin exercise 4.

Exercises 4.7

Exercises 1 – 12 each illustrate a pair of similar triangles. Find the values of *x* and *y* in each of these exercises. See Examples 1 through 3.

1.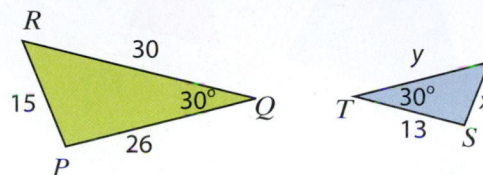

2.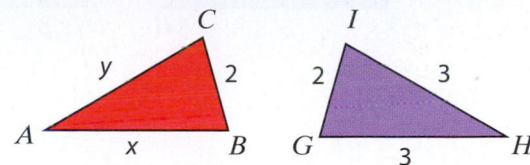

3.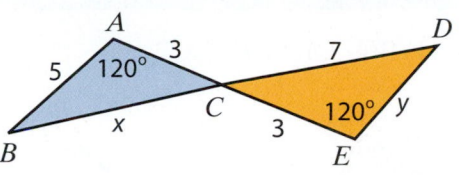

4.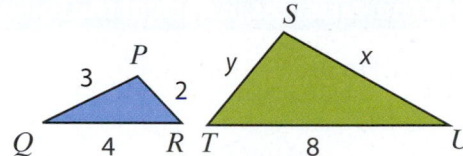

5.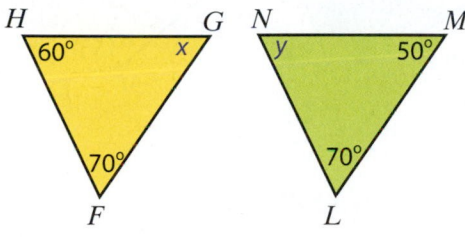

6.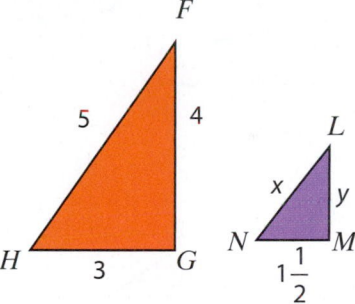

7.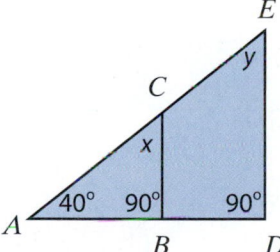

8.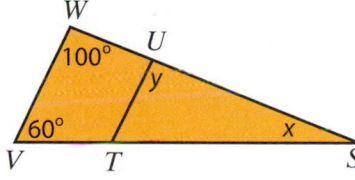

9.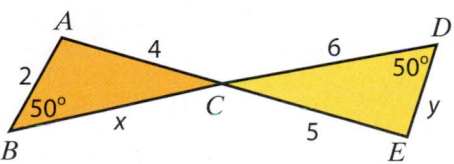

10.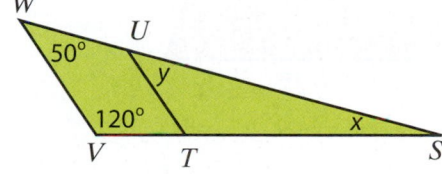

11.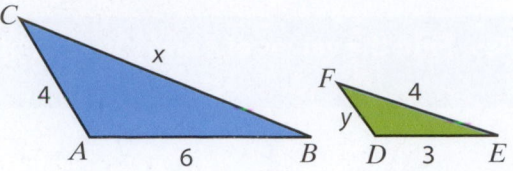

12.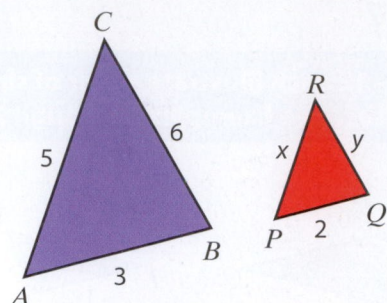

Solve for the variables in each of the following similar figures. See Example 4.

13 *HIJK ~ ABCD*

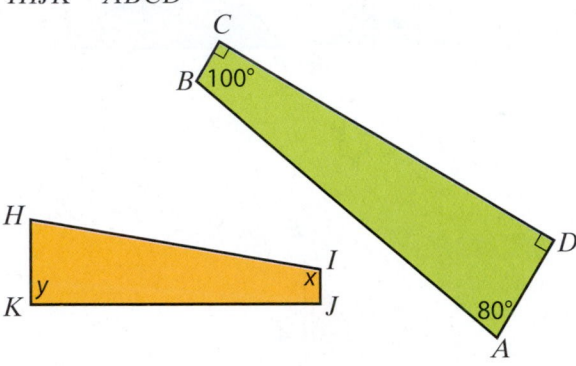

14. *WXYZV ~ LMNOP*

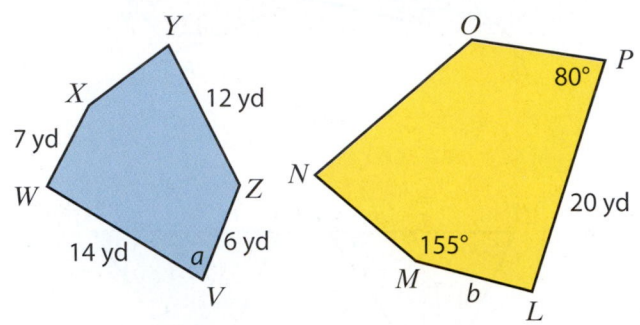

15. *ABCD ~ RSTU*

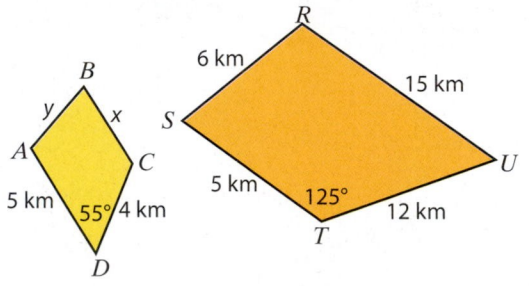

16. *ABCD ~ WXYZ*

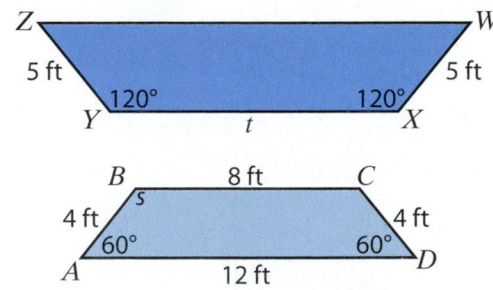

17. *RSTU ~ JIHG*

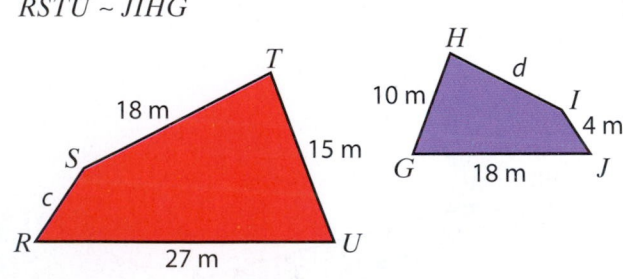

18. *ABCDEF ~ RSTUVW*

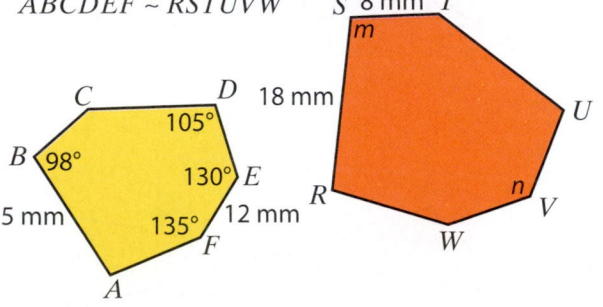

4.8 The Pythagorean Theorem

Objective A The Pythagorean Theorem

The following discussion involves right triangles.

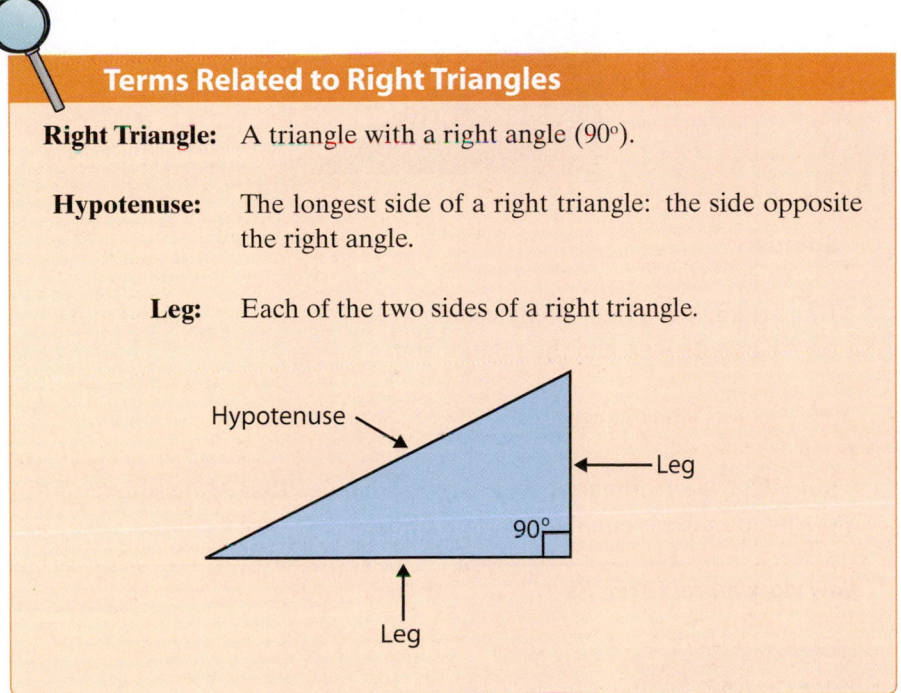

Terms Related to Right Triangles

Right Triangle: A triangle with a right angle (90°).

Hypotenuse: The longest side of a right triangle: the side opposite the right angle.

Leg: Each of the two sides of a right triangle.

Pythagoras (c. 585 – 501 B.C.), a famous Greek mathematician, is given credit for discovering the following theorem (even though historians have found that the facts of the theorem were known before the time of Pythagoras).

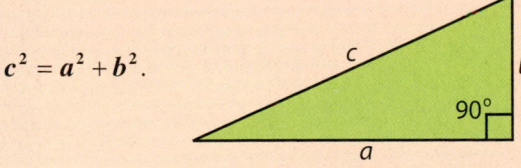

The Pythagorean Theorem

In a right triangle, the square of the hypotenuse is equal to the sum of the squares of the two legs:

$$c^2 = a^2 + b^2.$$

1. Show that a triangle with sides of 5 feet, 12 feet, and 13 feet must be a right triangle.

Example 1

Right Triangles

Show that a triangle with sides of 3 inches, 4 inches, and 5 inches must be a right triangle.

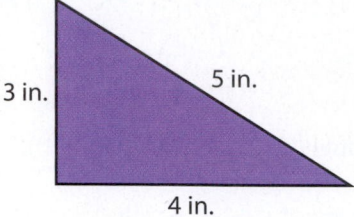

Solution

If the triangle is a right triangle, then the longest side (5 inches) must be the hypotenuse, and the relationship

$5^2 = 3^2 + 4^2$ must be true.

Since $25 = 9 + 16$, the triangle is a right triangle. That is, the square of the hypotenuse is equal to the sum of the squares of the legs.

Now work margin exercise 1.

 2. Find the length of the hypotenuse of a right triangle with legs of length 4 centimeters and 6 centimeters.

Example 2

Finding the Length of the Hypotenuse

Use a calculator and the Pythagorean theorem to find the length of the hypotenuse (accurate to the nearest hundredth) of a right triangle with legs 1 centimeter long and 3 centimeters long.

Solution

Let c = the length of the hypotenuse. Then, by the Pythagorean theorem,

$c^2 = 1^2 + 3^2$

$c^2 = 1 + 9$

$c^2 = 10$

$c = \sqrt{10}$ Recall that if $a^2 = b$, we write $a = \sqrt{b}$.

$c \approx 3.16$ cm.

The length of the hypotenuse is about 3.16 cm long.

Now work margin exercise 2.

Example 3

 Finding the Length of a Leg

Use a calculator and the Pythagorean theorem to find the length of a leg (accurate to the nearest hundredth) of a right triangle with hypotenuse 20 inches long and the other leg 13 inches long.

Solution

In this case, the length of the hypotenuse is known and the unknown is one of the legs.

Let x = the length of the unknown leg.

Then, by the Pythagorean theorem,

$$x^2 + 13^2 = 20^2$$
$$x^2 + 169 = 400$$
$$x^2 + 169 - 169 = 400 - 169$$
$$x^2 = 231$$
$$x = \sqrt{231}$$
$$x \approx 15.20.$$

So the leg is 15.20 inches long (accurate to the nearest hundredth).

Now work margin exercise 3.

Example 4

 **Finding the Length of a Guy Wire**

A guy wire is attached to the top of a telephone pole and anchored in the ground 20 feet from the base of the pole. If the pole is 40 feet high, what is the length of the guy wire (accurate to the nearest tenth)?

Solution

Let x = the length of the guy wire.

Then, by the Pythagorean theorem,

$$x^2 = 20^2 + 40^2$$
$$x^2 = 400 + 1600$$
$$x^2 = 2000$$
$$x = \sqrt{2000}$$
$$x \approx 44.7 \text{ ft.}$$

The length of the guy wire is about 44.7 ft long.

Now work margin exercise 4.

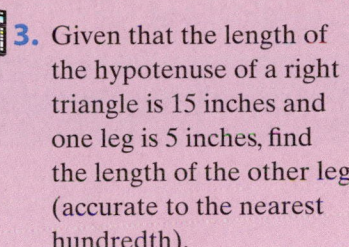

3. Given that the length of the hypotenuse of a right triangle is 15 inches and one leg is 5 inches, find the length of the other leg (accurate to the nearest hundredth).

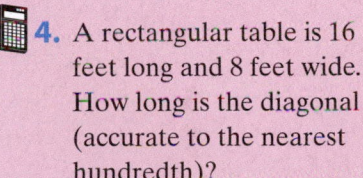

4. A rectangular table is 16 feet long and 8 feet wide. How long is the diagonal (accurate to the nearest hundredth)?

Exercises 4.8

Use the Pythagorean theorem to determine whether or not each of the following triangles is a right triangle. See Example 1.

1.

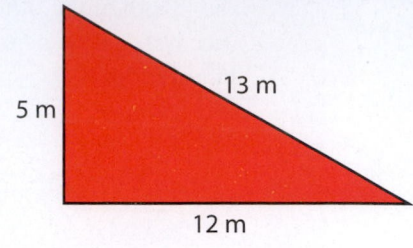

13 m

5 m

12 m

2.

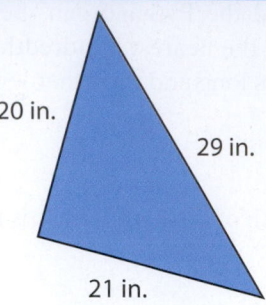

20 in.

29 in.

21 in.

3.

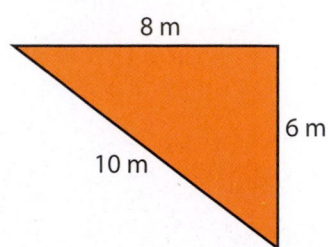

8 m

6 m

10 m

4.

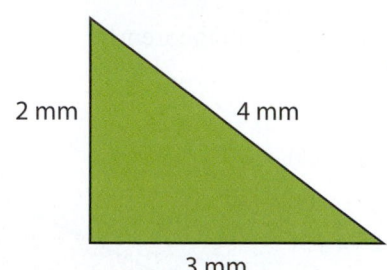

2 mm

4 mm

3 mm

Find the length of the hypotenuse (accurate to the nearest hundredth) of each of the following right triangles. See Example 2.

5.

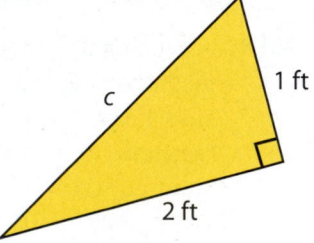

c

1 ft

2 ft

6.

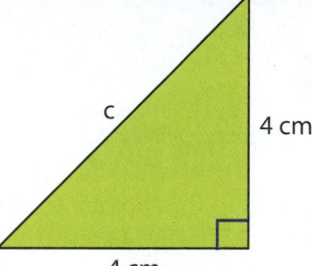

c

4 cm

4 cm

7.

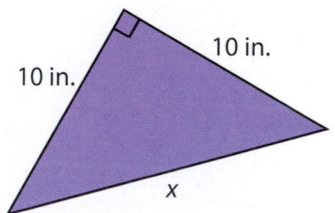

10 in.

10 in.

x

8.

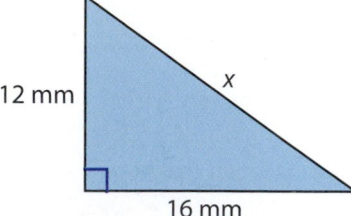

12 mm

x

16 mm

9.

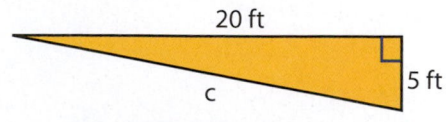

20 ft

5 ft

c

10.

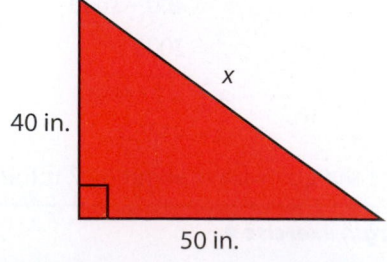

x

40 in.

50 in.

11.

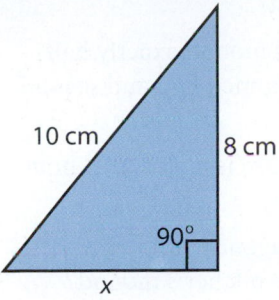

10 cm   8 cm

90°

x

12.

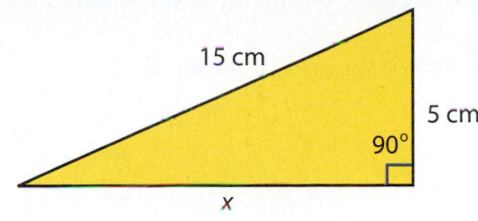

15 cm   5 cm

90°

x

13.

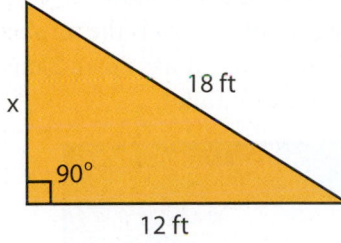

18 ft

x

90°

12 ft

14.

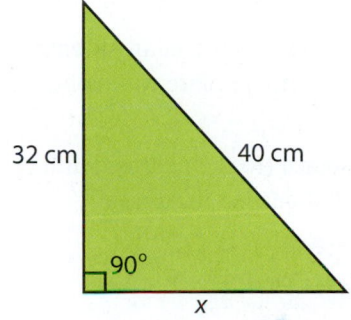

32 cm   40 cm

90°

x

Solve the following problems.

15. Find the lengths (accurate to the nearest hundredth) of the line segments $\overline{BD}$ and $\overline{AD}$ in the figure shown.

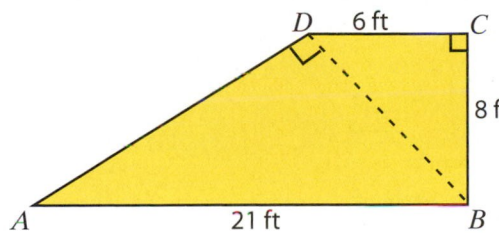

D   6 ft   C

8 ft

A   21 ft   B

16. Find the length of one side of the square in the figure shown with the square inscribed in a circle with radius 20 meters. (Each corner of the square lies on the circle.) (Use $\pi = 3.14$.)

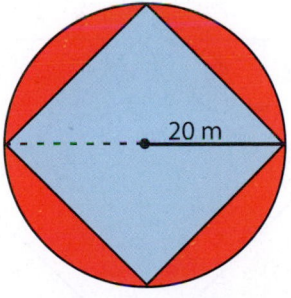

20 m

17. A square is inscribed in a circle with diameter 30 feet as shown in the figure. (Use $\pi = 3.14$.)

a. Find the circumference of the circle.

b. Find the area of the circle.

c. Find the perimeter of the square.

d. Find the area of the square.

e. Find the area of the shaded region in the figure.

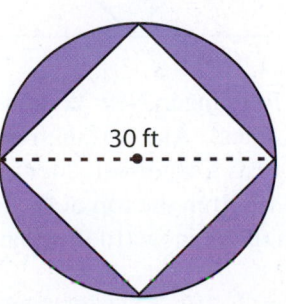

30 ft

18. A square has a diagonal of length 18 yd.

   a. What is the perimeter of the square?

   b. What is the area of the square?

19. The distance from home plate to the pitcher's mound is 60.5 feet.

   a. Is the pitcher's mound exactly half way between home plate and second base?

   b. If not, which base is it closer to, home plate or second base?

   c. Do the two diagonals of the square intersect at the pitcher's mound?

20. The shape of a baseball infield is a square with sides 90 feet long.

   a. Find the distance (to the nearest tenth of a foot) from home plate to second base.

   b. Find the distance (to the nearest tenth of a foot) from first base to third base.

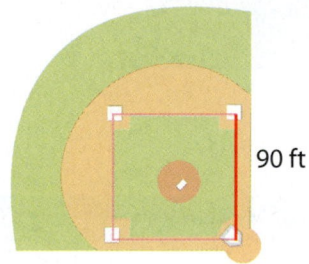

90 ft

21. If an airplane passes directly over your head at an altitude of 1 mile, how far (to the nearest hundredth of a mile) is the airplane from your position after it has flown 2 miles farther at the same altitude?

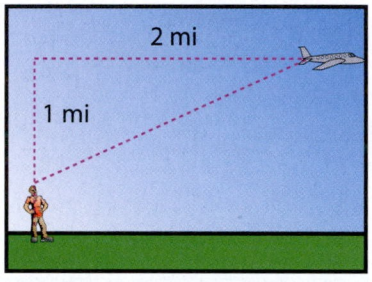

2 mi

1 mi

22. To create a square inside a square, a quilting pattern requires four triangular pieces like the one shaded in the figure shown here. If the square in the center measures 12 centimeters on a side, and the two legs of each triangle are of equal length, how long are the legs of each triangle, to the nearest tenth of a centimeter?

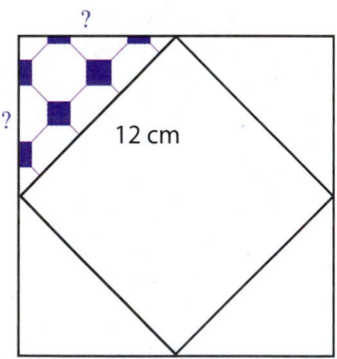

?

?

12 cm

23. The GE Building in New York is 850 feet tall (70 stories). At a certain time of day, the building casts a shadow 100 feet long. Find the distance from the top of the building to the tip of the shadow (to the nearest tenth of a foot).

24. The base of a fire engine ladder is 30 feet from a building and reaches to a third floor window 50 feet above ground level, how long (to the nearest hundredth of a foot) is the ladder extended?

4.9 Reading Graphs

Objective A **Introduction to Graphs**

Graphs are **pictures** of numerical information. Graphs appear almost daily in newspapers and magazines, and frequently in textbooks and corporate reports. Well-drawn graphs can organize and communicate information accurately, effectively, and quickly. Most computers can be programmed to draw graphs, and anyone whose work involves a computer in any way will probably be expected to understand graphs and even to create graphs.

There are many different types of graphs, each particularly well suited to the display and clarification of certain types of information. There have been various graphs used throughout the text. In this section we will discuss in more detail the uses of bar graphs, circle graphs, line graphs, and histograms.

The Purpose of the Four Types of Graphs

Bar Graphs: to emphasize comparative amounts

Circle Graphs: to help in understanding percents or parts of a whole.

Line Graphs: to indicate tendencies or trends over a period of time

Histograms: to indicate data in **classes** (a range or interval of numbers)

A common characteristic of all graphs is that they are intended to communicate information about numerical data quickly and easily. With this in mind, note the following three properties of all graphs:

1. They should be clearly labeled.
2. They should be easy to read.
3. They should have appropriate titles.

Example 1

Bar Graph

Figure 1 shows a bar graph. Note that the scale on the left and the data on the bottom (months) are clearly labeled and the graph itself has a title. The following questions can be easily answered by looking at Figure 1.

1. Use the bar graph in Figure 1 to answer parts **a.**, **b.**, and **c.**.

a. What were the sales during the highest sales month?

b. What was the amount of decrease in sales between January and February?

c. What was the percentage of decrease in sales between January and February?

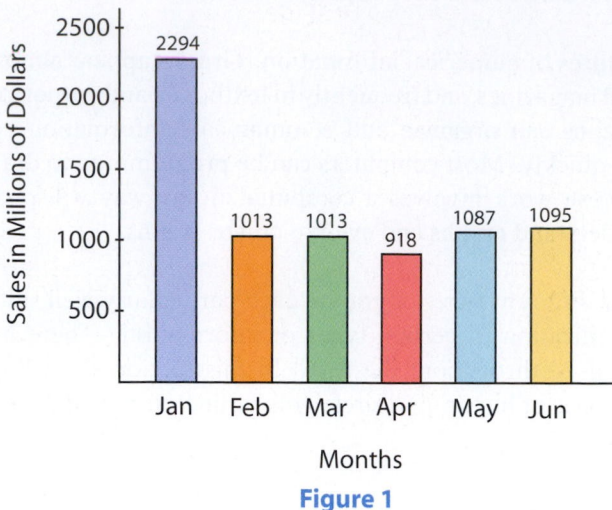

Monthly U.S. Book Sales 2010
Source: U.S. Census Bureau

Figure 1

a. What were the sales in February? __$1013 million__

b. During what month were sales lowest? __April__

c. During what month were sales highest? __January__

d. What were sales during the highest sales month? __$2294 million__

e. What were the sales in April? __$918 million__

f. What was the amount of increase in sales between April and May?

$1087   May sales
− $918   April sales
$169   increase in sales

g. What was the percent increase?

$\dfrac{\$169}{\$918}$ increase in April sales / April sales $\approx 0.184096 = 18.4\%$ increase

Now work margin exercise **1.**

Example 2

Circle Graph

Figure 2, on the next page, shows percents budgeted for various items in a home for one year. Suppose a person has an annual income of $15,000. Using Figure 2, calculate how much will be allocated to each item indicated in the graph.

Solution

Item			Amount
Housing:	$0.25 \times \$15,000$	=	$3750.00
Food:	$0.20 \times \$15,000$	=	3000.00
Taxes:	$0.05 \times \$15,000$	=	750.00
Clothing:	$0.07 \times \$15,000$	=	1050.00
Savings:	$0.10 \times \$15,000$	=	1500.00
Education:	$0.15 \times \$15,000$	=	2250.00
Entertainment:	$0.05 \times \$15,000$	=	750.00
Transportation & Maintenance:	$0.13 \times \$15,000$	=	1950.00

What is the total of all amounts? $15,000.00

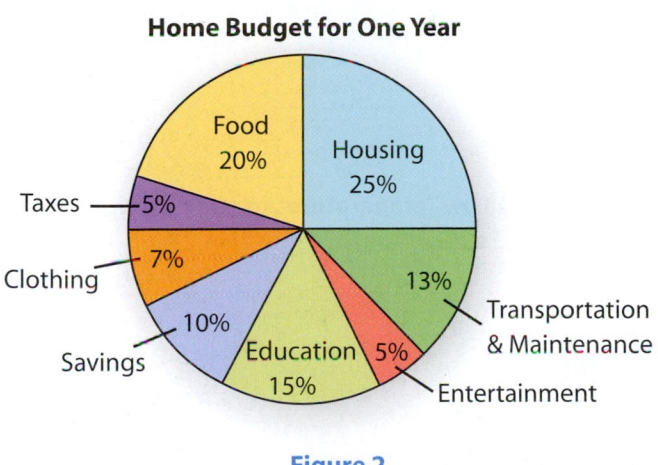

Home Budget for One Year

Figure 2

Now work margin exercise 2.

Example 3

Line Graph

Figure 3, on the next page, is a line graph that shows the relationships between daily high and low temperatures. You can see that temperatures tended to rise during the week but fell sharply on Saturday.

---

2. Use the circle graph in Figure 2 to answer the following question. If the person's annual income increases from $15,000 to $20,000 and the percentage allocated to each item does not change, how much will the person put into savings over the year?

3. Use the line graph in Figure 3 to answer parts **a.**, **b.**, and **c.**.

a. What was the high temperature for the week (highest daily high)?

b. What was the low temperature for the week (lowest daily low)?

c. Find the maximum daily difference between daily high and low temperatures in a single day for the week.

What was the lowest high temperature? ___66°___
On what day did this occur? ___Sunday___
What was the highest low temperature? ___70°___
On what day did this occur? ___Friday___

Find the mean difference between the daily high and low temperatures for the week shown.

Solution

First, find the differences; then divide to find the mean of these differences.

Sunday:	$66 - 60 =$	___6°___
Monday:	$70 - 62 =$	___8°___
Tuesday:	$76 - 66 =$	___10°___
Wednesday:	$72 - 66 =$	___6°___
Thursday:	$80 - 68 =$	___12°___
Friday:	$80 - 70 =$	___10°___
Saturday:	$74 - 62 =$	___12°___
		___64°___

$$\begin{array}{r} 9.1° \\ 7\overline{)64.0} \\ \underline{63} \\ 1\,0 \\ \underline{7} \\ 3 \end{array}$$ mean difference

64° total of differences

High & Low Temperatures for One Week

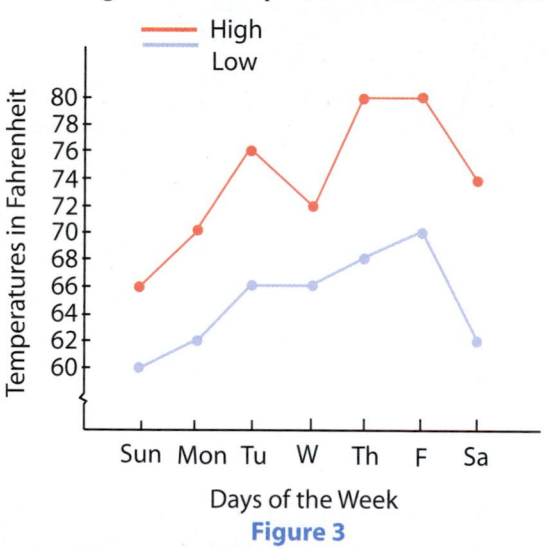

Figure 3

Now work margin exercise 3.

For bar graphs (see Example 1), we label the base line for each bar with individual names for people, months, days of the week, or other categories. Now we introduce a type of bar graph called **histogram**. In a histogram, the base line is labeled with numbers that indicate the boundaries of a range of numbers called a **class**. The bars are placed next to each other with no space between them.

Terms Related to Histograms

Class:	a range (or interval) of numbers that contain data items
Lower class limit:	the smallest number that belongs to a class
Upper class limit:	the largest number that belongs to a class
Class boundaries:	numbers that are halfway between the upper limit of one class and the lower limit of the next
Class width:	the difference between the class boundaries of a class (the width of each bar)
Frequency:	the number of data items in a class

4. Use the histogram in Figure 4 to answer parts **a.**, **b.**, and **c.**

a. Which class has the lowest frequency?

b. What percentage of the scores are below 300.5?

c. Which two classes appear to be equal?

Example 4

Histogram

Figure 4 shows a histogram that summarizes the scores of 50 students on an English placement test. Answer the following questions by referring to the graph.

a. How many classes are represented? __6__

b. What are the class limits of the first class? __201 and 250__

c. What are the class boundaries of the second class? __250.5 and 300.5__

d. What is the width of each class? __50__

e. Which class has the greatest frequency? __second class__

f. What is this frequency? __16__

g. What percentage of the scores are between 200.5 and 250.5?
$$\frac{2}{50} = 4\%$$

h. What percentage of the scores are above 400? $\dfrac{12}{50} = 24\%$

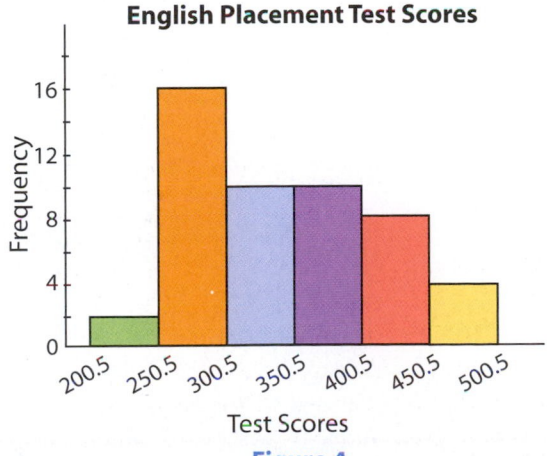

Figure 4

Now work margin exercise 4.

1. Using the bar graph in Example 1, what was the amount of decrease in sales between January and February?

2. Using the circle graph in Example 2, if the person's income increases to $30,000 and the percentage spent on each item does not change, what is the amount spent on entertainment over the year?

3. Using the line graph in Example 3, find the minimum difference between the high and low temperatures of a single day in the week shown.

4. Using the histogram in Example 4, what is the frequency of the class with the least frequency?

Practice Problem Answers

1. $1281 million **2.** $1500   **3.** 6°   **4.** 2

Exercises 4.9

Answer the questions related to each of the graphs. Some questions can be answered directly from the graphs; others may require some calculations. See Examples 1 through 4.

1. **Fields of Study:** The following bar graph shows the numbers of students in five fields of study at a university.

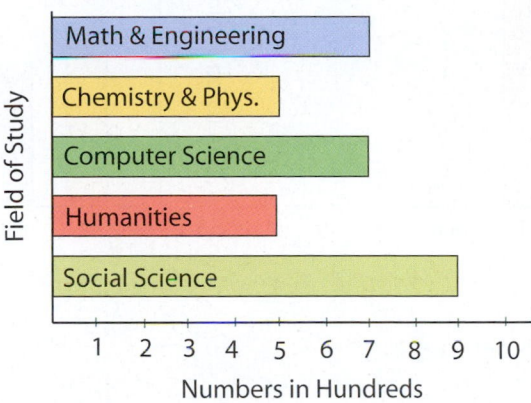

Declared College Majors at Downstate University

a. Which field(s) of study has the largest number of declared majors?

b. Which field(s) of study has the smallest number of declared majors?

c. How many declared majors are indicated in the entire graph?

d. What percent are computer science majors? Round your answer to the nearest tenth of a percent.

2. **Traffic:** The following bar graph shows the number of vehicles that crossed one intersection during a two-week period.

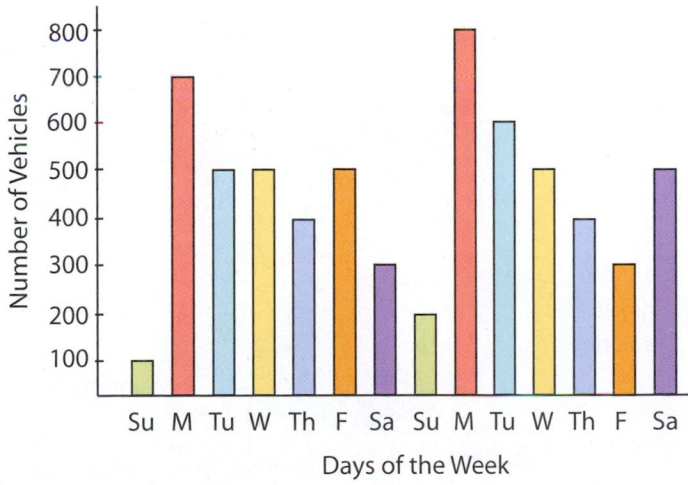

Traffic at One Intersection over a Two Week Period

a. On which day did the highest number of vehicles cross the intersection? How many crossed that day?

b. What was the mean number of vehicles that crossed the intersection on the two Sundays?

c. What was the total number of vehicles that crossed the intersection during the two weeks?

d. About what percent of the total traffic was counted on Saturdays? Round your answer to the nearest tenth of a percent.

3. College Life: The following bar graphs show the number of hours worked each week and the GPAs of five college students. When comparing the following two graphs, assume that all five students graduated with comparable grades from the same high school.

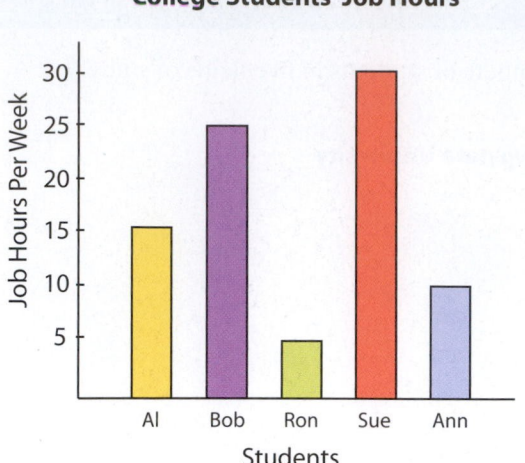

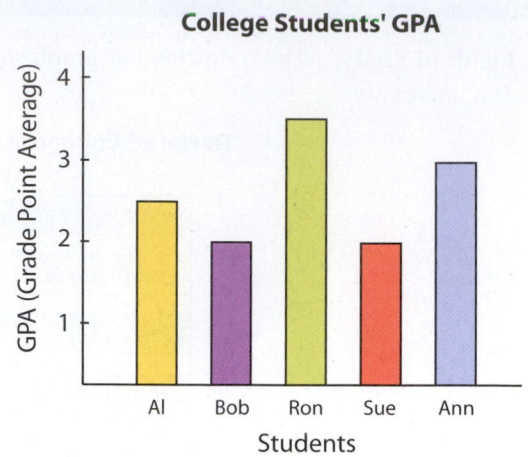

a. Who worked the most hours per week?

b. Who had the lowest GPA?

c. If Ron spent 30 hours per week studying for his classes, then the length of his total work week is the sum of the time he spent studying and the time he spent working. What percent of his work week did he spent studying? Round your answer to the nearest tenth of a percent.

d. Which two students worked the most hours? Which two students had the lowest GPA's? Do you think that this is typical?

e. Do you think that the two graphs shown here could be set as one graph? If so, show how you might do this.

4. Budgeting: The following circle graph represents the various areas of spending for a school with a total budget of $34,500,000.

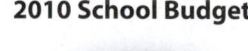

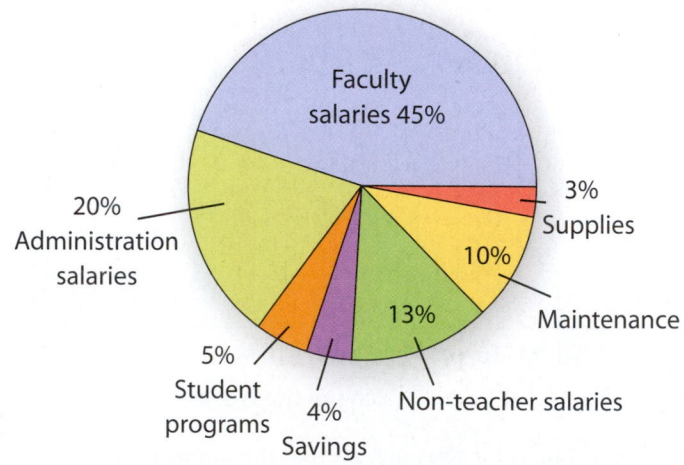

a. What amount will be allocated to each category?

b. What percent will be for expenditures other than salaries?

c. How much will be spent on maintenance and supplies?

d. How much more will be spent on teachers' salaries than on administration salaries?

5. Television Programming: The following circle graph represents the types of shows broadcast on television station KCBA. The station is off the air from 2 A.M. to 6 A.M., so there are only 20 hours of daily programming. Sports are not shown in the graph below because they are considered special events.

20-Hour TV Programming at Station KCBA

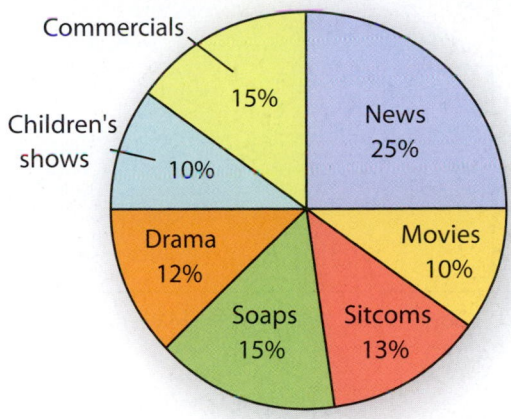

a. In the 20-hour period shown, how much time (in minutes) is devoted daily to each category?

b. What category has the most time devoted to it?

c. How much total time (in minutes) is devoted to drama, soaps, and sitcoms?

6. Budgeting: Mike just graduated from college and decided that he should try to live within a budget. The circle graph shows the categories he chose and the percents he allowed. His beginning take home salary is $24,000.

Mike's Budget

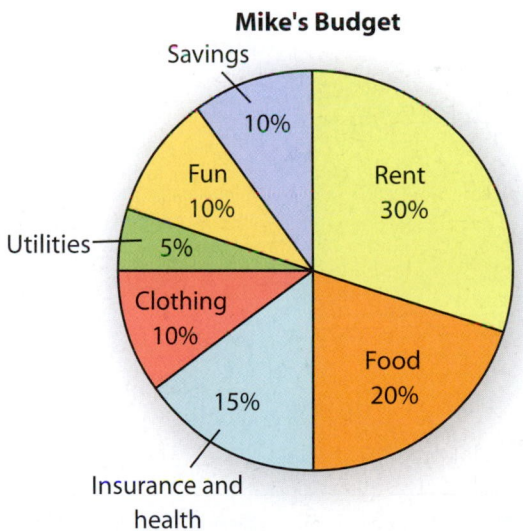

a. How much did he budget for each category?

b. What category was smallest in his budget?

c. What total amount did he budget for food, clothing, and rent?

7. Rainfall: The following line graph shows the total monthly rainfall in Tampa, Florida for the first 6 months of 2010. **Source:** weather.gov

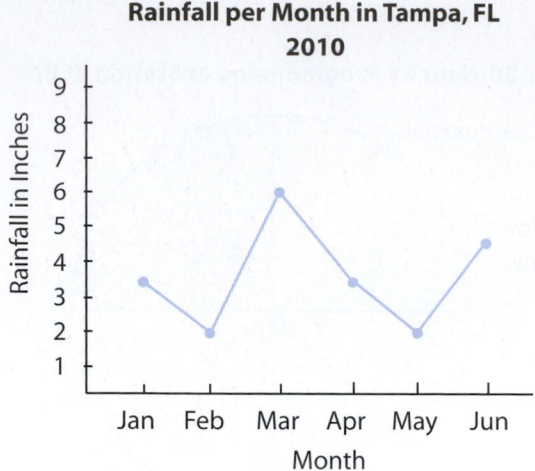

a. Which months had the least rainfall?

b. What was the most rainfall in a month?

c. What month had the most rainfall?

d. What was the mean rainfall over the six-month period (to the nearest hundredth)?

8. Mortgage Rates: The following line graph shows the average monthly mortgage rates for June and December for 2006-2010. **Source:** http://www.mortgage-x.com/trends.htm/

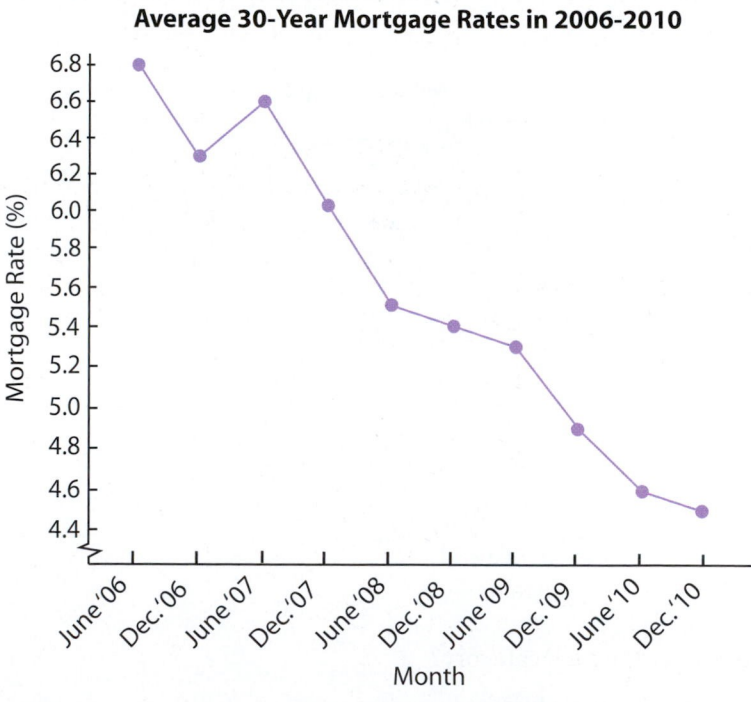

a. During what month or months were mortgage rates highest?

b. Lowest?

c. What was the mean of the interest rates given over the 5 year period? Round each value to the nearest tenth of a percent.

9. Baseball: The following line graph shows the number of home runs hit each month of the 2010 season by José Bautista and Albert Pujols. **Source:** mlb.com

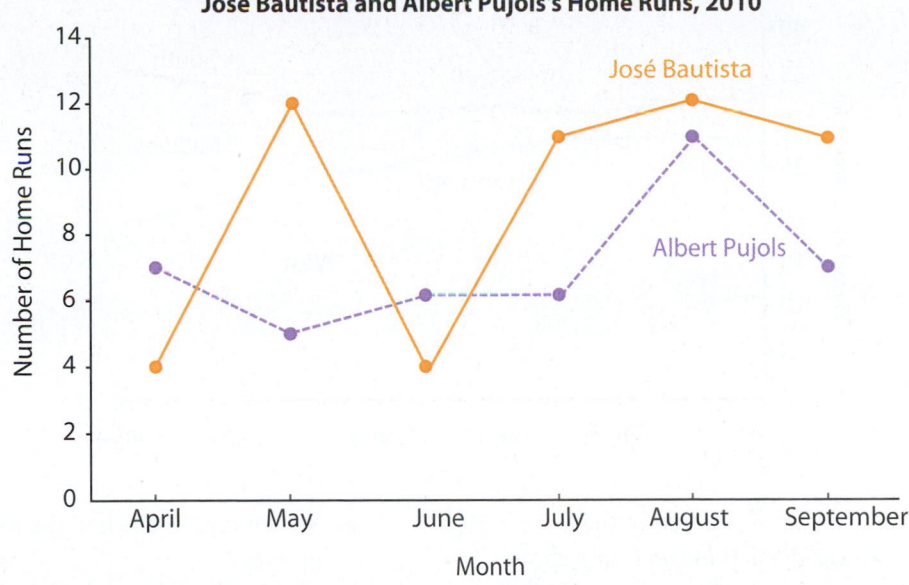

José Bautista and Albert Pujols's Home Runs, 2010

a. During which month did Albert hit the most home runs?

b. How much higher was his total for that month than for the lowest month?

c. In what months did José hit less home runs than Albert?

d. What was the difference between José and Albert's home runs in July?

e. What percent of José's total home runs did he hit in May? Round your answer to the nearest percent.

f. What percent of Albert's home runs did he hit in April? Round your answer to the nearest percent.

10. Investing: The following line graphs shows the stock market prices for oil, steel, and wheat over the course of a week. Assume that on Monday morning you had 100 shares of each of the three stocks shown.

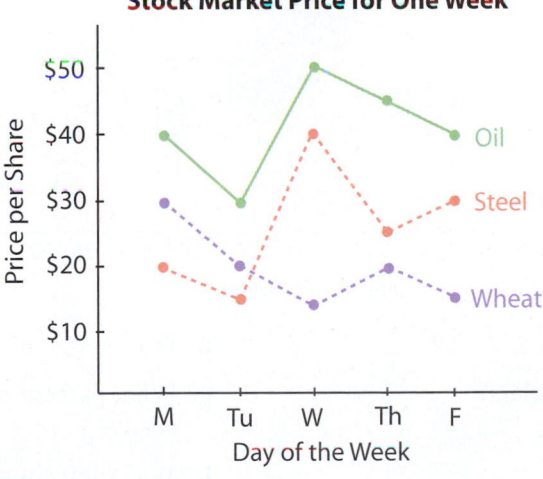

Stock Market Price for One Week

a. If you held the stock all week, on which stock would you have lost money?

b. How much would you have lost?

c. On which stock would you have gained money?

d. How much would you have gained?

e. On which stock could you have made the most money if you had sold at the best time?

f. How much could you have made?

11. Population: The following line graph shows the change in the percent of the U.S. population living in each of four major regions over the last century. **Source:** U.S. Census Bureau, decennial census of population, 1900 to 2000.

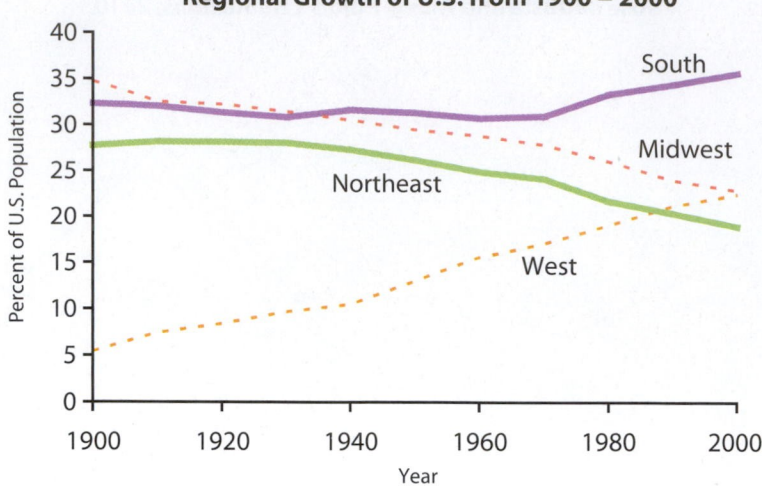

Regional Growth of U.S. from 1900 – 2000

a. Approximately what percent of the population was in each of the four regions in 1900?

b. In 2000?

c. Which region seems to have had the most stable percent of the population between 1900 and 2000?

d. What is the difference between the highest and lowest percents for this region?

e. Which region has had the most growth?

f. What was its lowest percent and when?

g. What was its highest percent and when?

h. Which region has had the most decline?

12. Tires: The following histogram summarizes the tread life for 100 types of new tires.

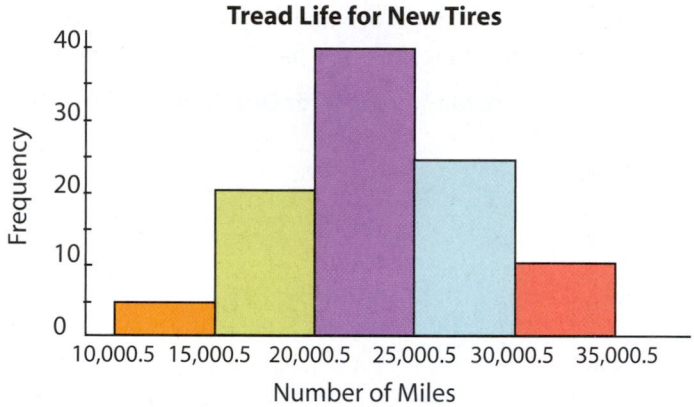

Tread Life for New Tires

a. How many classes are represented?

b. What is the width of each class?

c. Which class has the highest frequency?

d. What is this frequency?

e. What are the class boundaries of the second class?

f. How many tires were tested?

g. What percent of the tires were in the first class?

h. What percent of the tires lasted more than 25,000 miles?

13. **Fuel Efficiency:** A certain number of new cars were evaluated to find how many miles per gallon could be driven with a gallon of gas. The data is summarized in the following histogram.

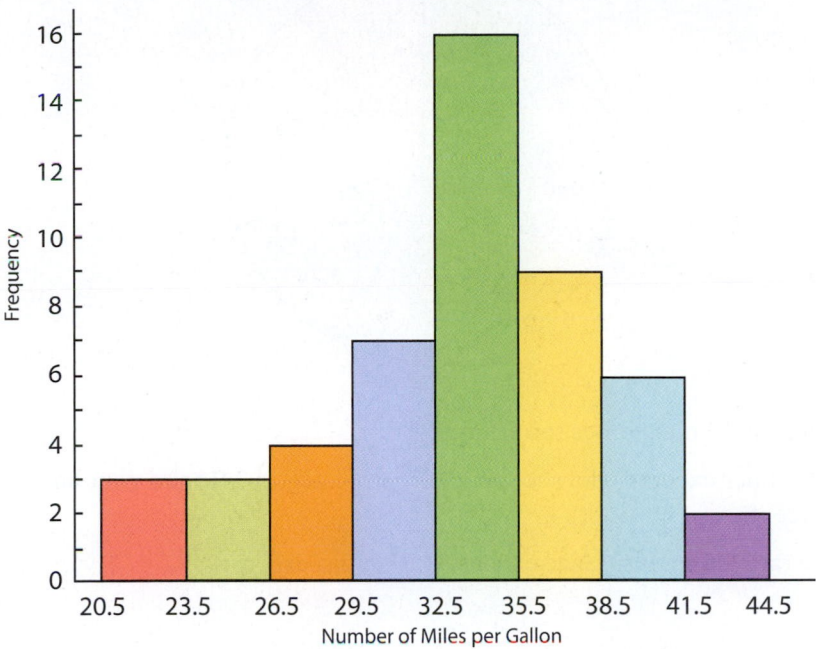

Miles per Gallon for New Cars Tested

a. How many classes are represented?

b. What is the class width?

c. Which class has the smallest frequency?

d. What is this frequency?

e. What are the class limits for the third class?

f. How many cars were tested?

g. How many cars tested below 30 miles per gallon?

h. What percent of the cars tested above 38 miles per gallon?

14. **Government:** The following circle graph represents the various sources of income for a city government with a total income of $100,000,000.

Sources of City Revenues

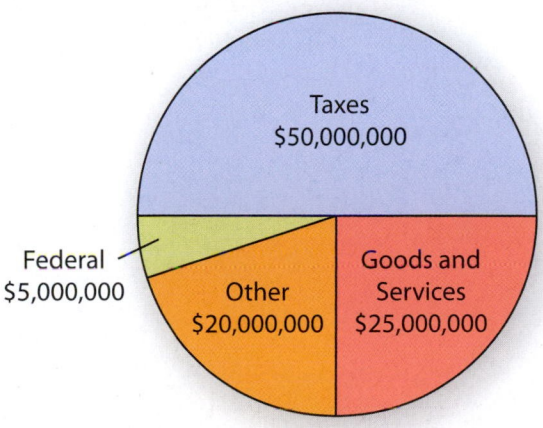

a. What is the city's largest source of income?

b. What percent of income comes from goods and services?

c. What is the ratio of income from taxes to the total income?

15. Cars: The following circle graph shows Sally's car expenses for the month of June.

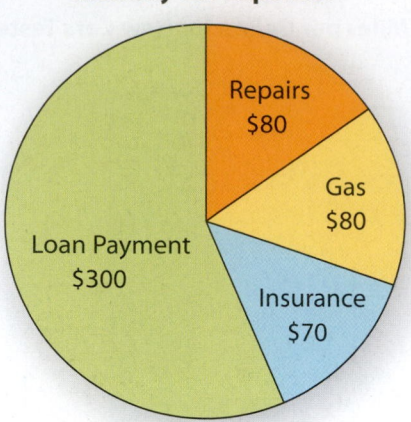

Monthly Car Expenses

Repairs $80

Gas $80

Insurance $70

Loan Payment $300

a. What were her total car expenses for the month?

b. What percent of her expenses did she spend on each category? Round your answer to the nearest tenth of a percent.

c. What was the ratio of her insurance expenses to her gas expenses?

Section 4.1: The Metric System

Meter   page 447

The meter is the basic unit of length in the metric system. Smaller and larger units are named by putting a prefix in front of the basic unit.

Metric Measures of Length   page 447

1 **milli**meter (mm)	= 0.001 meter	1 m = 10 dm
1 **centi**meter (cm)	= 0.01 meter	1 m = 100 cm
1 **deci**meter (dm)	= 0.1 meter	1 m = 1000 mm
1 meter (m)	= 1.0 meter (the basic unit)	
1 **deca**meter (dam)	= 10 meters	
1 **hecto**meter (hm)	= 100 meters	
1 **kilo**meter (km)	= 1000 meters	

Changing Metric Measures of Length   pages 448 – 449

Procedure	Example
To change to a measure of length that is	**Larger to Smaller**
one unit smaller, multiply by 10	6 cm = 60 mm
two units smaller, multiply by 100	3 m = 300 cm
three units smaller, multiply by 1000	12 m = 12 000 mm
and so on.	
To change to a measure of length that is	**Smaller to Larger**
one unit larger, divide by 10	35 mm = 3.5 cm
two units larger, divide by 100	250 cm = 2.5 m
three units larger, divide by 1000	32 m = 0.032 km
and so on.	

Area   pages 450 – 452

Area is a measure of the interior, or enclosure, of a surface.

Metric Measures of Small Areas   page 451

$1 \text{ cm}^2 = 100 \text{ mm}^2$

$1 \text{ dm}^2 = 100 \text{ cm}^2 = 10\,000 \text{ mm}^2$

$1 \text{ m}^2 = 100 \text{ dm}^2 = 10\,000 \text{ cm}^2 = 1\,000\,000 \text{ mm}^2$

Changing Metric Measures of Area   pages 451 – 452

Procedure	Example
To change to a measure of area that is	**Larger to Smaller**
one unit smaller, multiply by 100	$8.2 \text{ cm}^2 = 820 \text{ mm}^2$
two units smaller, multiply by 10 000	$3.56 \text{ m}^2 = 35\,600 \text{ cm}^2$
three units smaller, multiply by 1 000 000	$43 \text{ km}^2 = 43\,000\,000 \text{ m}^2$
and so on.	
To change to a measure of area that is	**Smaller to Larger**
one unit larger, divide by 100	$75 \text{ mm}^2 = 0.75 \text{ cm}^2$
two units larger, divide by 10 000	$52 \text{ cm}^2 = 0.0052 \text{ m}^2$
three units larger, divide by 1 000 000	$65 \text{ m}^2 = 0.000\,065 \text{ km}^2$
and so on.	

Metric Measurements of Land Area   page 453

$$1 \text{ a} = 100 \text{ m}^2$$
$$1 \text{ ha} = 100 \text{ a} = 10\ 000 \text{ m}^2$$

Mass   pages 453 – 455

Mass is the amount of material in an object. Regardless of where the object is in space, its mass remains the same.

Metric Measures of Mass   page 554

1 **milli**gram (mg)	= 0.001 gram
1 **centi**gram (cg)	= 0.01 gram
1 **deci**gram (dg)	= 0.1 gram
1 gram (g)	= 1.0 gram
1 **deka**gram (dag)	= 10 grams
1 **hecto**gram (hg)	= 100 grams
1 **kilo**gram (kg)	= 1000 grams
1 metric ton (t)	= 1000 kilograms

Units of Mass   page 455

A chart can be used to change from one unit of mass to another.

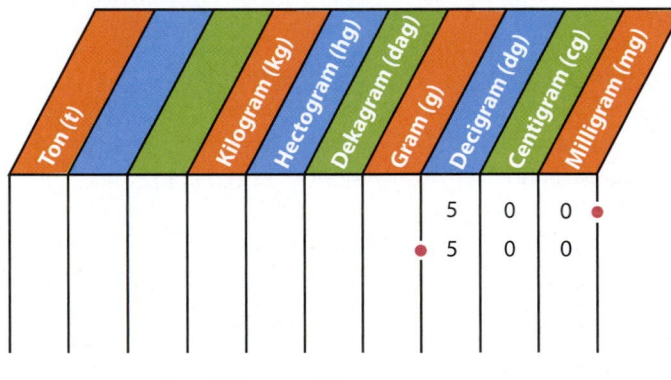

500 mg = 0.5 g

1. List each unit across the top.
2. Enter the given number so that there is one digit in each column with the decimal point on the given unit line.
3. Move the decimal point to the desired unit line.
4. Fill in the spaces with 0s using one digit per column.

Volume   page 456

Volume is a measure of the space enclosed by a three-dimensional figure and is measured in **cubic units**.

Metric Measures of Volume   page 457

$$1 \text{ cm}^3 = 1000 \text{ mm}^3$$
$$1 \text{ dm}^3 = 1000 \text{ cm}^3 = 1\ 000\ 000 \text{ mm}^3$$
$$1 \text{ m}^3 = 1000 \text{ dm}^3 = 1\ 000\ 000 \text{ cm}^3 = 1\ 000\ 000\ 000 \text{ mm}^3$$

Changing Metric Measures of Volume   page 458

Procedure	**Example**

To change to a measure of volume that is   **Larger to Smaller**

one unit smaller, multiply by 1000 6 cm^3 = 6000 mm^3

two units smaller, multiply by 1 000 000 3 m^3 = $3\ 000\ 000 \text{ m}^3$

three units smaller, multiply by 1 000 000 000 12 km^3 = $12\ 000\ 000\ 000 \text{ m}^3$

and so on.

To change to a measure of volume that is   **Smaller to Larger**

one unit larger, divide by 1000 35 mm^3 = 0.035 cm^3

two units larger, divide by 1 000 000 250 cm^3 = $0.000\ 25 \text{ m}^3$

three units larger, divide by 1 000 000 000 32 m^3 = $0.000\ 000\ 032 \text{ km}^3$

and so on.

Units of Volume   page 459

A chart can be used to change from one unit of volume to another.

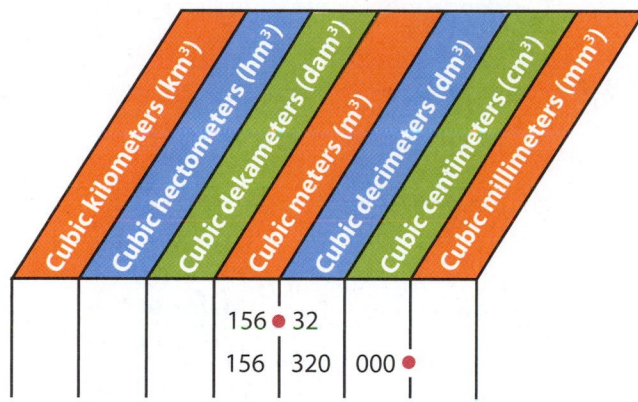

$156.32 \text{ m}^3 = 156\ 320\ 000 \text{ cm}^3$

1. List each volume unit across the top. (Abbreviations will do.)
2. Enter the given number so that there are three digits in each column with the decimal point on the given unit line.
3. Move the decimal point to the desired unit line.
4. Fill in the spaces with 0s using three digits per column.

Liquid Volume   pages 460 – 461

Liquid volume is measured in **liters** (abbreviated **L**). You are probably familiar with 1 L and 2 L bottles of soda on your grocer's shelf. A **liter** is the volume enclosed in a cube that is 10 cm on each edge.

1 milliliter (mL)   = 0.001 liter

1 liter (L)   = 1.0 liter

1 hectoliter (hL)   = 100 liters

1 kiloliter (kL)   = 1000 liters

Mega-, Giga-, and Tera-   pages 462 – 463

Prefix	Multiplier	Example
Mega-	1 million	1 megabyte = 1,000,000 bytes
Giga-	1 billion	1 gigabyte = 1,000,000,000 bytes
Tera-	1 trillion	1 terabyte = 1,000,000,000,000 bytes

Temperature   pages 468 – 469

U.S. customary measure is in **degrees Fahrenheit** ($°F$). Metric measure is in **degrees Celsius** ($°C$).

U.S. Customary Units of Length   page 469

1 foot (ft) = 12 inches (in.)
1 yard (yd) = 3 ft
1 mile (mi) = 5280 ft

Length Equivalents   page 470

U.S. to Metric	Metric to U.S.
1 in. = 2.54 cm (exact)	1 cm = 0.394 in.
1 ft = 0.305 m	1 m = 3.28 ft
1 yd = 0.914 m	1 m = 1.09 yd
1 mi = 1.61 km	1 km = 0.62 mi

U.S. Customary Units of Area   page 472

$1\,\text{ft}^2 = 144\,\text{in.}^2$
$1\,\text{yd}^2 = 9\,\text{ft}^2$
$1\,\text{acre} = 4840\,\text{yd}^2 = 43{,}560\,\text{ft}^2$

Area Equivalents   page 472

U.S. to Metric	Metric to U.S.
$1\,\text{in.}^2 = 6.45\,\text{cm}^2$	$1\,\text{cm}^2 = 0.155\,\text{in.}^2$
$1\,\text{ft}^2 = 0.093\,\text{m}^2$	$1\,\text{m}^2 = 10.764\,\text{ft}^2$
$1\,\text{yd}^2 = 0.836\,\text{m}^2$	$1\,\text{m}^2 = 1.196\,\text{yd}^2$
1 acre = 0.405 ha	1 ha = 2.47 acres

U.S. Customary Units of Liquid Volume

1 pint (pt)   = 16 fluid ounces (fl oz)
1 quart (qt) = 2 pt   = 32 fl oz
1 gallon (gal) = 4 qt

Volume Equivalents   page 474

U.S. to Metric	Metric to U.S.
$1\,\text{in.}^3 = 16.387\,\text{cm}^3$	$1\,\text{cm}^3 = 0.06\,\text{in.}^3$
$1\,\text{ft}^3 = 0.028\,\text{m}^3$	$1\,\text{m}^3 = 35.315\,\text{ft}^3$
1 qt = 0.946 L	1 L = 1.06 qt
1 gal = 3.785 L	1 L = 0.264 gal

Mass Equivalents   page 475

U.S. to Metric	Metric to U.S.
1 oz = 28.35 g	1 g = 0.035 oz
1 lb = 0.454 kg	1 kg = 2.205 lb

Plane Geometry page 479

Plane geometry is the study of the properties of figures in a plane.

Undefined Terms page 479

Point: A dot represents a point. Points are labeled with capital letters.

Line: A line has no beginning or end. Lines are labeled with small letters or by two points on the plane.

Plane: Flat surfaces, such as table tops or walls, represent planes. Planes are labeled with capital letters.

Angles page 480

Ray: A **ray** consists of a point (called the **endpoint**) and all the points on a line on one side of that point.

Angle: An **angle** consists of two rays with a common endpoint. The common endpoint is called the vertex of the angle.

In an angle, the two rays are called the **sides** of the angle.

Types of Angles page 482

Name	Measure	Illustration	
Acute	$0° < m\angle A < 90°$		
Right	$m\angle A = 90°$	The rays are perpendicular to each other.	
Obtuse	$90° < m\angle A < 180°$		
Straight	$m\angle A = 180°$	The rays are in opposite directions.	

Complementary, Supplementary, and Congruent Angles pages 483 – 484

1. Two angles are **complementary** if the sum of their measures is $90°$.
2. Two angles are **supplementary** if the sum of their measures is $180°$.
3. Two angles are **congruent** if they have the same measure.

Vertical and Adjacent Angles   pages 485 – 486

If two lines intersect, then two pairs of **vertical angles** are formed. Vertical angles are also called **opposite angles**. **Vertical angles are equal**. That is, **vertical angles have the same measure**.

Two angles are **adjacent** if they have a common side.

Triangles   page 486

A **triangle** consists of three line segments which join three points that do not lie on a straight line. The line segments are called the **sides** of the triangle, and the points are called the **vertices** of the triangle.

Triangles Classified by Sides   page 487

Name	Property
1. Scalene	No two sides are equal.
2. Isosceles	Two sides are equal.
3. Equilateral	All three sides are equal.

Triangles Classified by Angles   page 488

Name	Property
1. Acute	All three angles are acute.
2. Right	One angle is a right angle.
3. Obtuse	One angle is an obtuse angle.

Special Names for the Sides of a Right Triangle   page 490

The longest side, opposite the right angle is called the **hypotenuse**, and the other two sides are called **legs**.

Two Important Properties of Any Triangle   page 490

1. The sum of the measures of its angles is 180°.
2. The sum of the lengths of any two sides must be greater than the length of the third side.

Section 4.4: Perimeter

Important Terms   pages 499 – 500

Perimeter:	Total distance around a plane geometric figure
Circumference:	Perimeter of a circle
Radius:	Distance from the center of a circle to a point on the circle
Diameter:	Distance from one point on a circle to another point on the circle measured through the center

Geometric Figures and the Formulas for Finding Their Perimeters   pages 500 – 502

| **Square** | **Rectangle** | **Parallelogram** |
| $P = 4s$ | $P = 2l + 2w$ | $P = 2b + 2a$ |

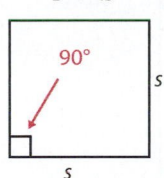
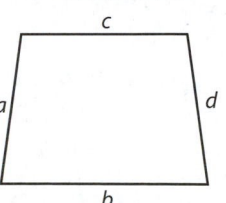

Triangle	**Circle**	**Trapezoid**
$P = a + b + c$	$C = 2\pi r$	$P = a + b + c + d$
	$C = \pi d$	

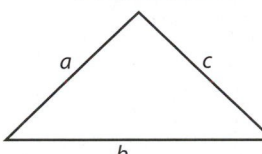
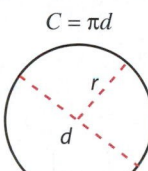

Six Geometric Figures and the Formulas for Finding Their Areas   page 506

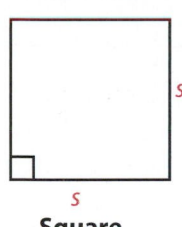

Square
$A = s^2$

Rectangle
$A = lw$

Parallelogram
$A = bh$

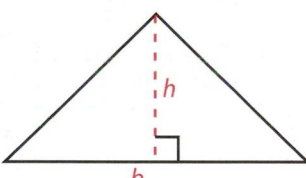
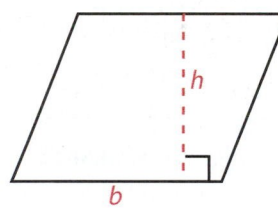
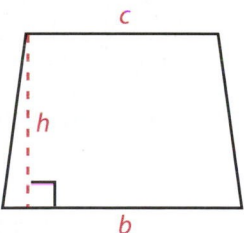

Circle
$A = \pi r^2$

Triangle
$A = \dfrac{1}{2}bh$

Trapezoid
$A = \dfrac{1}{2}h(b + c)$

Five Geometric Solids and the Formulas for their Volumes page 514

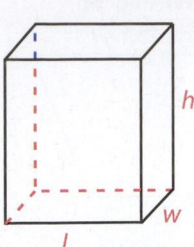

Rectangle Solid

$V = lwh$

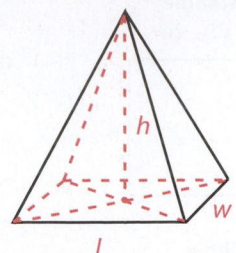

Rectangular Pyramid

$V = \dfrac{1}{3}lwh$

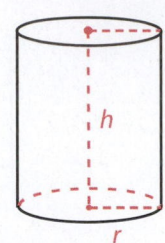

Right Circular Cylinder

$V = \pi r^2 h$

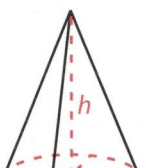

Right Circular Cone

$V = \dfrac{1}{3}\pi r^2 h$

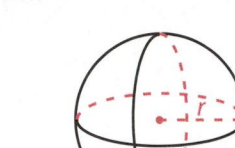

Sphere

$V = \dfrac{4}{3}\pi r^3$

Section 4.7: Similarity

Two Triangles are Similar if: pages 520 – 521

1. Their **corresponding angles are congruent**. (The corresponding angles have the same measure.)
2. Their **corresponding sides are proportional**.

Section 4.8: The Pythagorean Theorem

Terms Related to Right Triangles page 526

Right Triangle: A triangle with a right angle $(90°)$

Hypotenuse: The longest side of a right triangle: the side opposite the right angle

Leg: Each of the other two sides of a right triangle

The Pythagorean Theorem pages 526 – 527

In a right triangle, the square of the hypotenuse is equal to the sum of the squares of the two legs: $c^2 = a^2 + b^2$.

The Purpose of the Four Types of Graphs   page 532 – 533

1. **Bar Graphs:** to emphasize comparative amounts
2. **Circle Graphs:** to help in understanding percents or parts of a whole
3. **Line Graphs:** to indicate tendencies or trends over a period of time
4. **Histograms:** to indicate data in classes (a range or interval of numbers)

Terms Related to Histograms   page 536

Class: A range (or interval) of numbers that contain data items.

Lower class limit: The smallest number that belongs to a class.

Upper class limit: The largest number that belongs to a class.

Class boundaries: Numbers that are halfway between the upper limit of one class and the lower limit of the next class.

Class width: The difference between the class boundaries of a class (the width of each bar).

Frequency: The number of data items in a class.

Chapter 4: Test

Solve the following problems.

1. **a.** If ∠1 and ∠2 are complementary and m∠1 = 35°, what is m∠2?

    b. If ∠3 and ∠4 are supplementary and m∠3 = 15°, what is m∠4?

2. In the figure shown here, $\overrightarrow{AB}$ is a line with m∠*DOC* = 120°.

    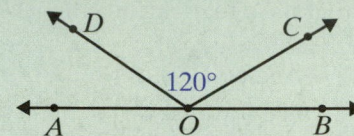

    a. What type of angle is ∠*BOD*?

    b. What type of angle is ∠*AOB*?

    c. Name two pairs of supplementary angles.

3. **a.** Find the measure of ∠*x*.

    b. What kind of triangle is Δ*RST*?

    c. Which side is opposite ∠*S*?

    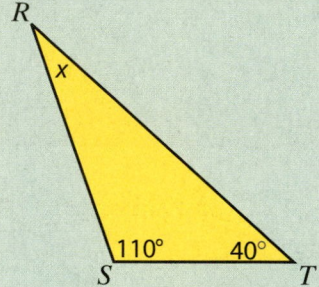

4. Classify each triangle.

    a.

    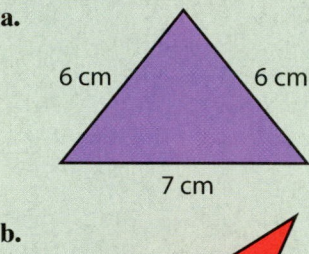

    b.

    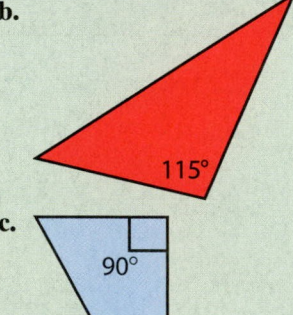

    c.

5. Which is longer, 10 mm or 10 cm? How much longer?

6. Which has the greater volume, 10 mL or 10 cm³ ? How much greater?

Change the following units of measure in the metric system as indicated.

7. 56 cm = _____ m

8. 33 m = _____ cm

9. 11 000 mm = ____ m

10. 83.5 mL = _____ L

11. 960 mm² = _____ cm²

12. 15 m² = _____ cm²

13. 75 a = _____ ha

14. 75 ha = _____ a

15. 140 cm³ = _____ mm³

16. 32 000 cm³ = _____ m³

17. 4000 mg = _____ g

18. 22 g = _____ kg

19. 50 cm = _____ inches

20. 25 L = _____ gal

21. 200 km = _____ mi

22. 7.8 in.² = _____ cm²

23. 30 qt = _____ L

24. 39 ft³ = _____ m³

Solve the following geometric problems.

25. Find

   a. the circumference and

   b. the area of a circle with diameter 30 cm.

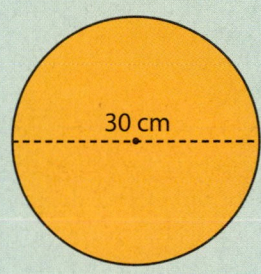

26. Find

   a. the perimeter and

   b. the area of a rectangle that is 8.5 in. wide and 11.6 in. long.

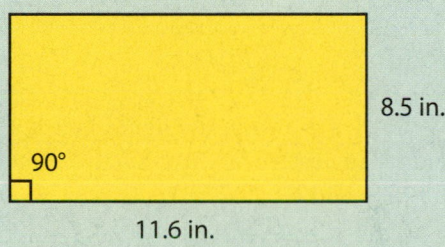

27. Find the volume of a sphere with diameter 12 in., in both cubic inches and cubic centimeters.

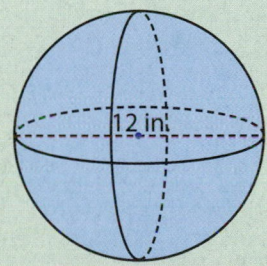

28. Find

   a. the perimeter and

   b. the area of the triangle with dimensions as shown here.

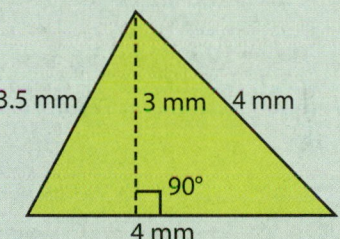

29. Find

   a. the perimeter and

   b. the area of the figure with dimensions as shown here.

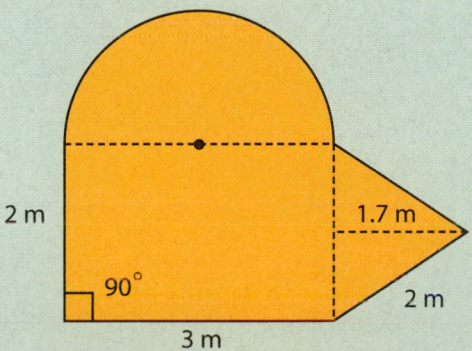

30. Find

   a. the perimeter and

   b. the area of a right triangle with sides of length 5 ft, 12 ft, and 13 ft.

31. A rectangular swimming pool is surrounded by a concrete border. If the swimming pool is 50 meters long and 25 meters wide and the concrete border is 3 meters wide all around the pool, what is the area of the concrete border?

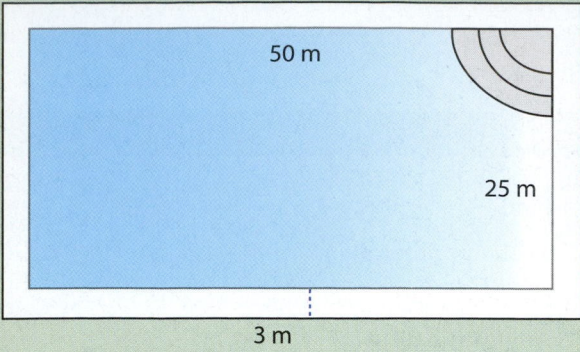

50 m

25 m

3 m

32. Draw a sketch of rectangular solid with length 5 cm, width 3 cm, and height 6 cm. Find the volume of this solid in both cubic centimeters and cubic inches.

33. The triangles shown in the figure below are similar triangles: $\triangle LMN \sim \triangle LKJ$. Find the values of x and y.

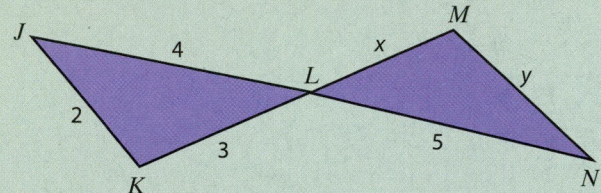

J

4

M

x

L

2

3

K

5

y

N

34.
a. What theorem do you use to determine whether or not a triangle is a right triangle?

b. If a triangle has sides of 24 inches, 26 inches and 10 inches, is it a right triangle? Why or why not?

35. Find the area of the red region (accurate to the nearest hundredth). [**HINT**: First find the value of x, then use the formula for area of a triangle, $A = \dfrac{1}{2}bh$.]

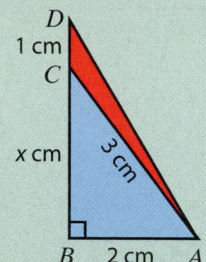

D

1 cm

C

x cm

3 cm

B   2 cm   *A*

36. A flagpole is 30 feet high. A rope is to be stretched from the top of the pole to a point on the ground 15 feet from the base of the pole. If one extra foot is needed at each end of the rope to tie knots, how long should the rope be (to the nearest tenth of a foot)?

37. The local athletic club wants to build a circular practice track that has an inner diameter of 50 yards.

a. What will be the inner circumference of the track?

b. If the track is to be 5 yards wide (all the way around), what will be the outer circumference of the track?

c. How many square yards of land will be covered by the surface of the track?

Budget for Apartment Complex

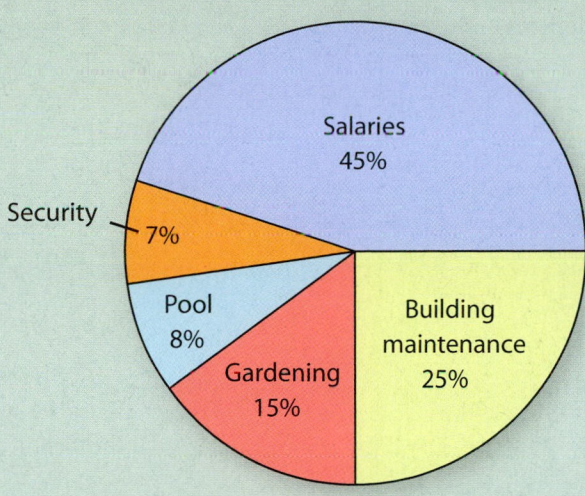

38. The budget for an apartment complex is shown in the graph. What fractional part of the budget is spent for salaries and security combined?

39. How much will be spent for building maintenance in one year if the budget is $250,000 per month?

40. How much will be spent for salaries and security combined in 6 months if the monthly budget is $150,000?

Ace Manufacturing Company Sales

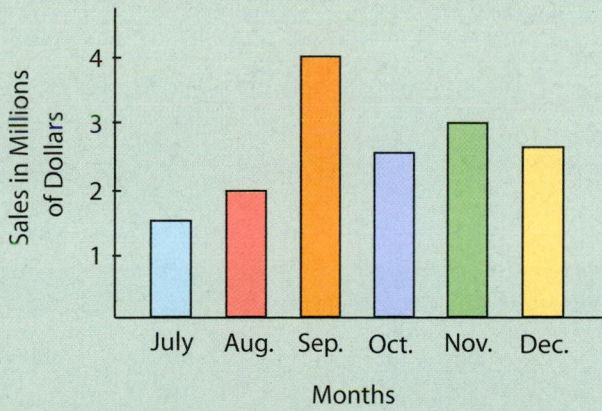

41. What were the total sales for the 6-month period?

42 What percent of the total sales for the 6 months were the July sales (to the nearest percent)?

43. What was the percent growth between August and September?

Cumulative Review: Chapters 1 - 4

Solve the following equations.

1. Name the property illustrated, and find the value of the variable.

   a. $21 + x = 15 + 21$

   b. $5(2 \cdot y) = (5 \cdot 2) \cdot 3$

   c. $6 + (3 + 17) = (a + 3) + 17$

2. Evaluate each expression.

   a. $(8 - 10)(15 - 10)$

   b. $(1.2)^2 - (1.5)^2$

   c. $\left(-\dfrac{2}{3}\right) \div \left(-\dfrac{7}{15}\right) \div (-2)$

3. Estimate the product of 845 and 3200.

4. Write the number five hundred twenty and sixty-five thousandths in decimal form.

5. Find the quotient of 89.5 and 17.6 to the nearest tenth.

6. Use a calculator to find the value of $\sqrt{63}$ to the nearest thousandth.

7. Find the LCM of 35, 49, and 105.

8. Evaluate the expression:
   $-16 + (12 \cdot 3 + 2^3) \div 4 - 20$

9. Evaluate the expression: $(0.5)^2 + 2\dfrac{1}{5} \div \dfrac{11}{6} - 0.3.$

10. Subtract 605 from the sum of 843 and 467; then find the quotient if the difference is divided by 5.

Perform the indicated operations. Reduce all answers.

11. $5\dfrac{3}{8} + 7\dfrac{5}{12}$

12. $|-52.3| + |16.1| - |-6.75|$

13. $\dfrac{3\dfrac{1}{2} + 4\dfrac{5}{8}}{2 - \dfrac{2}{3}}$

Solve the following equations.

14. $x - 17 = -25$

15. $-7x = -63$

16. $3x - 2x + 7 = 10 + 3$

17. $-52 + 48 = 6y - 2y$

18. If an investment pays 10% interest compounded daily, what will be the value of $6000 in 4 years?

19. If a savings account earned $75 simple interest at 5% for 9 months, what was the original principal in the account?

20. The circle graph shows a family budget for one year. What amount will be spent in each category if the family income is $45,000?

Family Budget

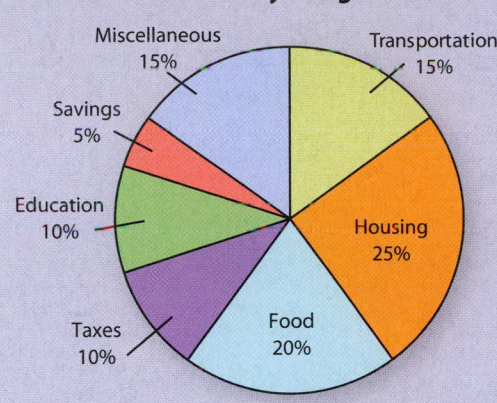

21. The Numero Uno Pizza Restaurant offers a "New York Style" thin pizza with the works for $16.95 for a large pizza with a 15 in. diameter. The same pizza with a diameter of 10 in. is priced at $11.95.

  a. What is the price per sq in. for the large pizza?

  b. Which is the better buy? Why?

22. For a circle with diameter 2.5 feet, find:

  a. the circumference, and

  b. the area.

23. Your friend plans to buy a used car for $16,000. He can either save up and pay cash or make a down payment of $1600 and 48 monthly payments of $420. What would you recommend that he do? Why?

24. Graphing calculators are on sale at a discount of 28%, and the original price was $62.50.

  a. What is the approximate amount of the discount?

  b. What is the sale price?

25. Solve each of the following proportions.

  a. $\dfrac{x}{7.2} = \dfrac{3.6}{14.4}$

  b. $\dfrac{4\frac{1}{2}}{y} = \dfrac{9}{3\frac{1}{3}}$

26. Find

  a. the mean,

  b. the median,

  c. the mode, and

  d. the range for the following data:

  26 51 49 61 41 46 41

27. Write each of the following numbers in scientific notation.

  a. 193,000,000

  b. 0.000386

28. **a.** Find the length of the hypotenuse (in radical notation) for the right triangle shown here.

  b. Use a calculator to find this value to the nearest hundredth.

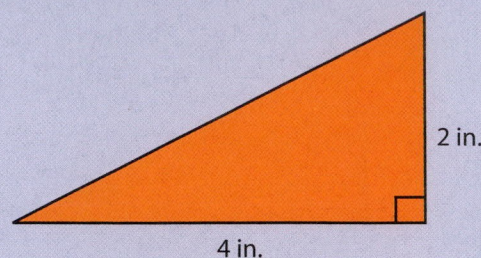

2 in.

4 in.

29. Find the value of x and the value of y if the two triangles shown are similar triangles.

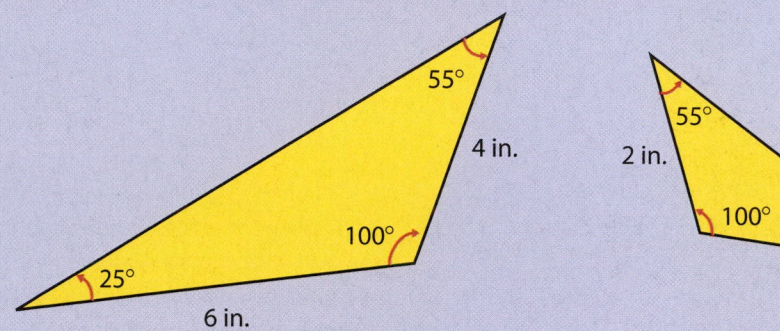

55°

4 in.

100°

25°

6 in.

55°

2 in.

100°

x

y

Simplify each of the following polynomials by combining like terms, and then tell what type of polynomial it is and state its degree.

30. $7x^3 + 8x - 15 + 9$

31. $9y^2 + 4y^2 + y^2 + 3y - y - 1$

32. $2(11x^2 + 3x - 1) + 3(5x^2 - 2x + 4)$

Mathematics @ Work!

Every day you are required to make decisions. Some decisions, like what you eat for breakfast or what you wear, will likely affect your life for only a day. However, other decisions, like how much to spend on your first house and which job offer to take, can affect the rest of your life. The skills you learn while studying algebra give you reasoning and problem-solving skills that can increase your confidence as you make some of life's big decisions, especially decisions relating to money. Consider the following situation.

Suppose that you are a traveling salesperson and your company has allowed you $65 per day for a car rental. You rent a car for $45 per day plus $0.05 per mile. How many miles can you drive for the $65 allowed?

Plan: Set up an equation relating the amount of money spent and the money allowed in your budget and solve the equation.

Solution: Let x = the number of miles driven.

Algebraic Expressions, Linear Equations, and Applications

5.1 Simplifying and Evaluating Algebraic Expressions

Objective A **Simplifying Algebraic Expressions**

A single number is called a **constant**. Any constant or variable or the indicated product and/or quotient of constants and variables is called a **term**. Examples of terms are

$$16, \quad 3x, \quad -5.2, \quad 1.3xy, \quad -5x^2, \quad 14b^3, \quad \text{and} \quad -\frac{x}{y}.$$

Note: As discussed in Section 1.6, an expression with 2 as the exponent is read "squared" and an expression with 3 as the exponent is read "cubed." Thus $7x^2$ is read "seven x squared" and $-4y^3$ is read "negative four y cubed."

The number written next to a variable is called the **coefficient** (or the **numerical coefficient**) of the variable. For example, in

$$5x^2, \quad 5 \text{ is the coefficient of } x^2.$$

Similarly, in $1.3xy$, 1.3 is the coefficient of xy.

> ## notes
>
> - If no number is written next to a variable, the coefficient is understood to be 1. If a negative sign (−) is next to a variable, the coefficient is
> - understood to be −1. For example,
> $$x = 1 \cdot x, \quad a^3 = 1 \cdot a^3, \quad -x = -1 \cdot x, \quad \text{and} \quad -y^5 = -1 \cdot y^5.$$
> -

Like Terms

Like terms (or **similar terms**) are terms that are constants or terms that contain the same variables raised to the same exponents.

Note: The sum of the exponents on the variables is the **degree** of the term.

Like Terms

$-6, \; 1.84, \; 145, \; \dfrac{3}{4}$ are like terms because each term is a constant.

$-3a, \; 15a, \; 2.6a, \; \dfrac{2}{3}a$ are like terms because each term contains the same variable, a, raised to the same exponent, 1. (Remember that $a = a^1$.) These terms are first-degree in a.

$5xy^2$ and $-3.2xy^2$ are like terms because each term contains the same two variables, x and y, with x first-degree in both terms and y second-degree in both terms.

Unlike Terms

$8x$ and $-9x^2$ are unlike terms (not like terms) because the variable x is not of the same power in both terms. $8x$ is first-degree in x and $-9x^2$ is second-degree in x.

Example 1

Like Terms

From the following list of terms, pick out the like terms.

$$-7, \ 2x, \ 4.1, \ -x, \ 3x^2 y, \ 5x, \ -6x^2 y, \ 0$$

Solution

$-7, 4.1$, and 0 are like terms. (All are constants.)
$2x, -x$, and $5x$ are like terms.
$3x^2 y$ and $-6x^2 y$ are like terms.

Now work margin exercise 1.

1. From the following list of terms, pick out the like terms: $4x, -3.9, 3xy^2, 5, 0, -2x, -6xy^2, x.$

An **algebraic expression** is a combination of variables and numbers using any of the operations of addition, subtraction, multiplication, or division, as well as exponents. Examples of algebraic expressions are

$$x^2 - 14, \quad \frac{2xy}{3z^3} + y^2, \quad \frac{12x - 9}{3x - 4}, \quad \text{and} \quad 10x^3 - 4x^2 - 11x + 1.$$

To simplify expressions that contain like terms we want to **combine like terms**.

Combining Like Terms

To **combine like terms**, add (or subtract) the coefficients and keep the common variable expression.

The procedure for combining like terms uses the distributive property in the form $ba + ca = (b + c)a$.

In particular, with numerical coefficients, we can combine like terms as follows.

$$9x + 6x = (9+6)x$$ by the distributive property

$$= 15x$$ Add the coefficients algebraically.

$$3x^2 - 5x^2 = (3-5)x^2$$ by the distributive property

$$= -2x^2$$ Subtract the coefficients algebraically.

$$xy - 1.6xy = (1.0 - 1.6)xy$$ by the distributive property
Note: $xy = 1xy = 1.0xy$

$$= -0.6xy$$ Subtract the coefficients algebraically.

Example 2

Combining Like Terms

a. $8x + 10x$

Solution

$$8x + 10x = (8+10)x = 18x$$ by the distributive property

b. $6.5y - 2.3y$

Solution

$$6.5y - 2.3y = (6.5 - 2.3)y = 4.2y$$ by the distributive property

c. $2x^2 + 3a + x^2 - a$

Solution

$$2x^2 + 3a + x^2 - a = 2x^2 + x^2 + 3a - a$$ by the commutative property of addition

$$= (2+1)x^2 + (3-1)a$$ **Note:** $+x^2 = +1x^2$ and $-a = -1a$

$$= 3x^2 + 2a$$

d. $4(n-7) + 5(n+1)$

Solution

$$4(n-7) + 5(n+1) = 4n - 28 + 5n + 5$$ Simplify by using the distributive property twice.

$$= 4n + 5n - 28 + 5$$ by the commutative property of addition

$$= (4+5)n + (-28+5)$$ by the distributive property

$$= 9n - 23$$ Combine like terms.

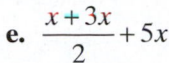

e. $\dfrac{x+3x}{2}+5x$

Solution

A fraction bar is a grouping symbol, similar to parentheses. So combine like terms in the numerator first.

$$\dfrac{x+3x}{2}+5x = \dfrac{4x}{2}+5x$$

$$= \dfrac{4}{2}\cdot x+5x$$

$$= 2x+5x \qquad \text{\color{red}Reduce the fraction.}$$

$$= 7x \qquad \text{\color{red}Combine like terms.}$$

*Now work margin exercise **2**.*

---

2. Combine like terms whenever possible.

a. $3y+8y$

b. $4.6x-2.5x$

c. $4x^2+6z-3x^2+4z$

d. $2(m+4)+9(m-2)$

e. $\dfrac{6x+2x}{4}+4x$

Objective B **Evaluating Algebraic Expressions**

In most cases, if an expression is to be evaluated, like terms should be combined first and then the resulting expression evaluated by following the rules for order of operations.

Parentheses must be used around negative numbers when substituting.

Without parentheses, an evaluation can be dramatically changed and lead to wrong answers, particularly when even exponents are involved. We analyze with the exponent 2 as follows.

In general, except for $x=0$,

1. $-x^2$ **is negative** $\quad \left[-6^2 = -1\cdot6^2 = -1\cdot36 = -36\right]$

2. $(-x)^2$ **is positive** $\quad \left[(-6)^2 = (-6)(-6) = 36\right]$

3. $-x^2 \neq (-x)^2$ $\quad \left[-36 \neq 36\right]$

To Evaluate an Algebraic Expression

1. Combine like terms, if possible.
2. Substitute the values given for any variables.
3. Follow the rules for order of operations.

3. a. Evaluate x^2 for $x = 2$ and for $x = -3$.

b. Evaluate $-x^2$ for $x = 2$ and for $x = -3$.

Example 3

Evaluating Algebraic Expressions

a. Evaluate x^2 for $x = 3$ and for $x = -4$.

Solution

For $x = 3$, $x^2 = (3)^2 = 9$.

For $x = -4$, $x^2 = (-4)^2 = 16$.

b. Evaluate $-x^2$ for $x = 3$ and for $x = -4$.

Solution

For $x = 3$, $-x^2 = -(3)^2 = -1(9) = -9$.

For $x = -4$, $-x^2 = -(-4)^2 = -1(16) = -16$.

Now work margin exercise 3.

Example 4

Simplifying and Evaluating Algebraic Expressions

Simplify each expression below by combining like terms. Then evaluate the resulting expression using the given values for the variables.

a. Simplify and evaluate $2x + 5 + 7x$ for $x = -3$.

Solution

Simplify first.
$$2x + 5 + 7x = 2x + 7x + 5$$
$$= 9x + 5$$

Now evaluate.
$$9x + 5 = 9(-3) + 5$$
$$= -27 + 5$$
$$= -22$$

b. Simplify and evaluate $3ab - 4ab + 6a - a$ for $a = 2, b = -1$.

Solution

Simplify first. $3ab - 4ab + 6a - a = -ab + 5a$

Now evaluate. $-ab + 5a = -1(2)(-1) + 5(2)$ **Note:** $-ab = -1ab$
$$= 2 + 10$$
$$= 12$$

c. Simplify and evaluate $\dfrac{5x+3x}{4}+2(x+1)$ for $x = 5$.

Solution

Simplify first.

$$\dfrac{5x+3x}{4}+2(x+1) = \dfrac{8x}{4}+2x+2$$
$$= 2x+2x+2$$
$$= 4x+2$$

Now evaluate.

$$4x+2 = 4(5)+2$$
$$= 20+2$$
$$= 22$$

Now work margin exercise 4.

4. a. Simplify and evaluate $-3x + 5x + 4$ for $x = -5$.

b. Simplify and evaluate $7ab - 3ab + 4a - a$ for $a = 3$ and for $b = -2$.

c. Simplify and evaluate $\dfrac{4x+2x}{2}+3(x+3)$ for $x = 3$.

Practice Problems

Simplify the following expressions by combining like terms.

1. $-2x - 5x$ **2.** $12y + 6 - y + 10$

3. $5(x-1)+4x$ **4.** $2b^2 - a + b^2 + a$

Simplify the expression. Then evaluate the resulting expression for $x = 3$ and $y = -2$.

5. $2(x+3y)+4(x-y)$

Practice Problem Answers

1. $-7x$ **2.** $11y + 16$ **3.** $9x - 5$
4. $3b^2$ **5.** $6x + 2y;\ 14$

Exercises 5.1

Pick out the like terms in each list of terms. See Example 1.

1. $-5,\ \dfrac{1}{6},\ 7x,\ 8,\ 9x,\ 3y$

2. $-2x^2,\ -13x^3,\ 5x^2,\ 14x^2,\ 10x^3$

3. $5xy,\ -x^2,\ -6xy,\ 3x^2y,\ 5x^2y,\ 2x^2$

4. $3ab^2,\ -ab^2,\ 8ab,\ 9a^2b,\ -10a^2b,\ ab,\ 12a^2$

5. $24,\ 8.3,\ 1.5xyz,\ -1.4xyz,\ -6,\ xyz,\ 5xy^2z,\ 2xyz^2$

6. $-35y,\ 1.62,\ -y^2,\ -y,\ 3y^2,\ \dfrac{1}{2},\ 75y,\ 2.5y^2$

Find the value of each numerical expression. See Example 3.

7. $(-8)^2$

8. -8^2

9. -11^2

10. $(-6)^2$

Simplify each expression by combining like terms. See Example 2.

11. $8x + 7x$

12. $3y + 8y$

13. $5x + (-2x)$

14. $7x + (-3x)$

15. $-n - n$

16. $-x - x$

17. $6y^2 - y^2$

18. $16z^2 - 5z^2$

19. $23x^2 + 11x^2$

20. $18x^3 + 7x^3$

21. $4x + 2 + 3x$

22. $3x - 1 + x$

23. $2x - 3y - x - y$

24. $x + y + x - 2y$

25. $2x^2 - 2y + 5x^2 + 6x^2$

26. $4a + 2a - 3b - a$

27. $3(n + 1) - n$

28. $2(n - 4) + n + 1$

29. $5(a - b) + 2a - 3b$

30. $4a - 3b + 2(a + 2b)$

31. $3(2x + y) + 2(x - y)$

32. $4(x + 5y) + 3(2x - 7y)$

33. $2x + 3x^2 - 3x - x^2$

34. $2y^2 + 4y - y^2 - 3y$

35. $2n^2 - 6n + 1 - 4n^2 + 8n - 3$

36. $3n^2 + 2n - 5 - n^2 + n - 4$

37. $3x^2 + 4xy - 5xy + y^2$

38. $2x^2 - 5xy + 11xy + 3$

39. $\dfrac{x + 5x}{6} + x$

40. $\dfrac{2y + 3y}{5} - 2y$

41. $y - \dfrac{2y + 4y}{3}$

42. $z - \dfrac{3z + 5z}{4}$

43. $5x + 4 - 2x$

44. $7x - 17 - x$

45. $5x - 2[x + 5(x - 3)]$

46. $x - 10 - 3x + 2$

47. $6a + 5a - a + 13$

48. $3(y - 1) + 2(y + 2)$

49. $4(y + 3) + 5(y - 2)$

50. $-5(x + y) + 2(x - y)$

51. $-2(a + b) + 3(b - a)$

52. $8.3x^2 - 5.7x^2 + x^2 + 2$

53. $3.1a^2 - 0.9a^2 + 4a - 5.3a^2$

54. $5ab + b^2 - 2ab + b^3$

55. $5a + ab^2 - 2ab^2 + 3a$

56. $2.4(x + 1) + 1.3(x - 1)$

57. $1.3(y + 2) - 2.6(8 - y)$

58. $\dfrac{3a + 5a}{-2} + 12a$

59. $8a + \dfrac{5a + 4a}{9}$

60. $\dfrac{-4b - 2b}{-3} + \dfrac{2b + 5b}{7}$

61. $\dfrac{5b + 3b}{4} + \dfrac{-4b - b}{-5}$

62. $2x + 3[x - 2(9 + x)]$

Writing & Thinking

63. Explain the difference between -5^2 and $(-5)^2$.

64. The text recommends simplifying an expression (combining like terms) before evaluating. Do you think this is necessary?

Evaluate the expression $4x^2 - 5(x + 2) + 3x + 10 + 2x$ for $x = 3$:

a. by substituting and then evaluating.

b. by first simplifying and then evaluating.

Which method would you recommend? Why?

5.2 Translating English Phrases and Algebraic Expressions

Objective A **Translating English Phrases into Algebraic Expressions**

Algebra is a language of mathematicians, and to understand mathematics, you must understand the language. We want to be able to change English phrases into their "algebraic" equivalents and vice versa. So if a problem is stated in English, we can translate the phrases into algebraic symbols and proceed to solve the problem according to the rules and methods developed for algebra.

Certain words are the keys to the basic operations. Some of these words are listed in Table 1 and highlighted in boldface in Example 1.

Key Words to Look For When Translating Phrases				
Addition	**Subtraction**	**Multiplication**	**Division**	**Exponent (Powers)**
add	subtract (from)	multiply	divide	square of
sum	difference	product	quotient	cube of
plus	minus	times		
more than	less than	twice		
increased by	decreased by	of (with fractions and percent)		
total	less			

Table 1

The following examples illustrate how these key words used in English phrases can be translated into algebraic expressions. **Note that in each case "a number" or "the number" implies the use of a variable (an unknown quantity).**

Example 1

Translating English Phrases into Algebraic Expressions

English Phrase	Algebraic Expression
a.	
the **product** of 3 and x	
3 **times** x	$3x$
3 **multiplied by** the number represented by x	
b.	
3 **added to** a number	
the **sum** of z and 3	
z **plus** 3	$z + 3$
3 **more than** z	
z **increased by** 3	
c.	
twice the **sum** of x and 1	
the **product** of 2 with the **sum** of x and 1	$2(x+1)$
2 **times** the quantity found by **adding** a number to 1	
d.	
twice x **plus** 1	
the **sum** of **twice** x and 1	$2x + 1$
2 **times** x **increased** by 1	
1 **more than** the **product** of 2 and a number	
e.	
the **difference** between 5 **times** a number and 3	
3 **less than** the **product** of a number and 5	
5 **times** a number **minus** 3	$5n - 3$
3 **subtracted from** the product of a number and 5	
5 **multiplied by** a number, **less** 3	
f.	
the **square of** a number	x^2
a number **squared**	
g. the cost of renting a truck for one day and driving x miles if the rate is $30 per day plus $0.25 per mile	$30 + 0.25x$

Now work margin exercise **1.**

1. Write the algebraic expression for each of the following English phrases.

a. 5 times y

b. the sum of x and 1

c. the product of 3 with the sum of z and 2

d. 3 more than the product of 8 and a number

e. 4 subtracted from 6 times x

f. the cube of a number

g. The cost of playing golf at a country club for a year if the membership dues are $5000 a year plus $10 per round, if r rounds of golf are played in a year.

notes

In Example **1b.**, the phrase "the sum of z and 3" was translated as $z + 3$. If the expression had been translated as $3 + z$, there would have been no mathematical error because addition is commutative. That is, $z + 3 = 3 + z$. However, in part **e.,** the phrase "3 less than the product of a number and 5" must be translated as it was because subtraction is **not** commutative. Thus

"3 less than 5 times a number" means $5n - 3$
while "5 times a number less than 3" means $3 - 5n$
and "3 less 5 times a number" means $3 - 5n$.

Therefore, be very careful when writing and/or interpreting expressions indicating subtraction. Be sure that the subtraction is in the order indicated by the wording in the problem. The same is true with expressions involving division.

The words **quotient** and **difference** deserve special mention because their use implies that the numbers given are to be operated on in the order given. That is, division and subtraction are done with the values in the same order that they are given in the problem. For example:

the quotient of y and 5 $\longrightarrow \dfrac{y}{5}$

the quotient of 5 and y $\longrightarrow \dfrac{5}{y}$

the difference between 6 and x $\longrightarrow 6 - x$

the difference between x and 6 $\longrightarrow x - 6$

If we did not have these agreements concerning subtraction and division, then the phrases just illustrated might have more than one interpretation and be considered **ambiguous**.

An **ambiguous phrase** is one whose meaning is not clear or for which there may be two or more interpretations. This is a common occurrence in everyday language, and misunderstandings occur frequently. Imagine the difficulties diplomats have in communicating ideas from one language to another trying to avoid ambiguities. Even the order of subjects, verbs, and adjectives may not be the same from one language to another. Translating grammatical phrases in any language into mathematical expressions is quite similar. To avoid ambiguous phrases in mathematics, we try to be precise in the use of terminology, to be careful with grammatical construction, and to follow the rules for order of operations.

Objective B **Translating Algebraic Expressions into English Phrases**

Consider the three expressions to be translated into English:

$$7(n+1), \; 6(n-3), \; \text{and} \; 7n+1.$$

In the first two expressions, we indicate the parentheses with a phrase such as "the quantity" or "the sum of" or "the difference between." **Without the parentheses, we agree that the operations used in the expression are to be indicated in the order given.** Thus

$7(n+1)$ can be translated as "seven times the sum of a number and 1," while $7n+1$ can be translated as "seven times a number plus 1."

$6(n-3)$ can be translated as "six times the difference between a number and 3."

Example 2

Translating Algebraic Expressions to Phrases

Write an English phrase that indicates the meaning of each algebraic expression.

Algebraic Expression	**Possible English Phrase**
a. $5x$	the product of 5 and a number
b. $2n + 8$	twice a number increased by 8
c. $3(a-2)$	three times the difference between a number and 2

Now work margin exercises 2.

Practice Problems

Change the following phrases to algebraic expressions.

1. 7 less than a number   **2.** the quotient of y and 5

3. 14 more than 3 times a number

Change the following algebraic expressions into English phrases. (There may be more than one correct translation.)

4. $10 - x$ **5.** $2(y-3)$ **6.** $5n + 3n$

Practice Problem Answers

1. $x - 7$ **2.** $\dfrac{y}{5}$ **3.** $3y + 14$

4. 10 decreased by a number

5. twice the difference between a number and 3

6. 5 times a number plus 3 times the same number

2. Write an English phrase for each of the following algebraic expressions.

a. $3x$

b. $7x + 6$

c. $2(n - 7)$

Exercises 5.2

Translate each algebraic expression into an equivalent English phrase. (There may be more than one correct translation.) See Example 2.

1. $4x$

2. $-9x$

3. $x + 5$

4. $4x - 7$

5. $7(x + 1.1)$

6. $3.2(x + 2.5)$

7. $-2(x - 8)$

8. $10(x + 4)$

9. $\dfrac{6}{(x-1)}$

10. $\dfrac{9}{(x+3)}$

11. $5(2x + 3)$

12. $3(4x - 5)$

Write each pair of expressions in words. Notice the differences between the algebraic expressions and the corresponding English phrases.

13. $3x + 7$; $3(x + 7)$

14. $4x - 1$; $4(x - 1)$

15. $7x - 3$; $7(x - 3)$

16. $5(x + 6)$; $5x + 6$

Write the algebraic expressions described by the English phrases. Choose your own variable. See Example 1.

17. 6 added to a number

18. 7 more than a number

19. 4 less than a number

20. a number decreased by 13

21. the quotient of twice a number and 10

22. the difference between a number and 3, all divided by 7

23. 5 subtracted from three times a number

24. the sum of twice a number and four times the number

25. 8 minus twice a number

26. the sum of a number and 9 times the number

27. twenty decreased by 4.8 times a number

28. the difference between three times a number and five times the same number

29. 9 times the sum of a number and 2

30. 3 times the difference between a number and 8

31. 13 less than the product of 4 with the sum of a number and 1

32. 4 more than the product of 8 with the difference between a number and 6

33. eight more than the product of 3 and the sum of a number and 6

34. six less than twice the difference between a number and 7

35. four less than 3 times the difference between 7 and a number

36. nine more than twice the sum of 17 and a number

37. **a.** 6 less than a number
b. 6 less a number

38. **a.** 20 less than a number
b. 20 less a number

39. **a.** 5 less than 3 times a number
b. 5 less 3 times a number

40. **a.** 6 less than 4 times a number
b. 6 less 4 times a number

Write the algebraic expressions described by the English phrases using the given variable. See Example 1.

41. **Time:** the number of hours in *d* days

42. **Graphing Calculators:** the cost of *x* graphing calculators if one calculator costs $115

43. **Gas Prices:** the cost of *x* gallons of gasoline if the cost of one gallon is $3.15

44. **Time:** the number of seconds in *m* minutes

45. **Time:** the number of days in *y* years (Assume 365 days in a year.)

46. **Candy:** the cost of *x* pounds of candy at $4.95 a pound

47. **Time:** the number of days in *t* weeks and 3 days

48. **Time:** the number of minutes in *h* hours and 20 minutes

49. **Football:** the points scored by a football team on *t* touchdowns (7 points) and 1 field goal (3 points)

50. **Vacation Time:** the amount of vacation days an employee has after *w* weeks if she gets 0.2 vacation days for every week she works

51. **Car Rentals:** the cost of renting a car for one day and driving *m* miles if the rate is $20 per day plus 15 cents per mile

52. **Fishing:** the cost of purchasing a fishing rod and reel if the rod costs *x* dollars and the reel costs $8 more than twice the cost of the rod

53. **Rectangles:** the perimeter of a rectangle if the width is *w* centimeters and the length is 3 cm less than twice the width

54. **Squares:** the area of a square with side *c* centimeters

---

Writing & Thinking

55. Discuss the meaning of the term "ambiguous phrase."

5.3 Solving Linear Equations: $x + b = c$ and $ax = c$

Objectives

A Define the term linear equation.

B Learn how to determine whether or not a number is a solution to a given equation.

C Solve equations of the form $x + b = c$.

D Solve equations of the form $ax = c$.

Objective A | **Linear Equations**

In this section, we will discuss solving linear equations in the following two forms:

$$x + b = c \quad \text{and} \quad ax = c.$$

In these equations we treat a, b, and c as constants and x as the unknown quantity.

In the following sections, we will combine the techniques developed here and discuss solving linear equations in the forms:

$$ax + b = c \quad \text{and} \quad ax + b = cx + d.$$

An **equation** is a statement that two algebraic expressions are equal. That is, both expressions represent the same number. If an equation contains a variable, any number that gives a true statement when substituted for the variable is called a **solution** to the equation. The solutions to an equation form a **solution set**. The process of finding the solution set is called **solving the equation**.

Linear Equation in x

If a, b, and c are **constants** and $a \neq 0$ then a **linear equation in x** is an equation that can be written in the form

$$ax + b = c.$$

Note: A linear equation in x is also called a **first-degree equation in x** because the variable x can be written with the exponent 1. That is, $x = x^1$.

Objective B | **Determining Possible Solutions**

We begin by determining whether or not a particular number satisfies an equation. A number is said to be a **solution** or to **satisfy an equation** if it gives a true statement when substituted for the variable.

Example 1

Determining a Possible Solution

Determine whether or not the given real number is a solution to the given equation by substituting for the variable and checking to see if the resulting equation is true or false.

a. $x + 5 = -2$ given that $x = -7$

Solution

$(-7) + 5 = -2$ is true, so -7 **is** a solution.

b. $1.4 + z = 0.5$ given that $z = -1.1$

Solution

$1.4 + (-1.1) = 0.5$ is false because $1.4 + (-1.1) = 0.3 \neq 0.5$.
So -1.1 **is not** a solution.

c. $5.6 - y = 2.9$ given that $y = 2.7$

Solution

$5.6 - 2.7 = 2.9$ is true. So, 2.7 **is** a solution.

d. $|z| - 14 = -3$ given that $z = -10$

Solution

$|(-10)| - 14 = -3$ is false because $|-10| - 14 = 10 - 14 = -4 \neq -3$.
So, -10 **is not** a solution.

Note: Notice that parentheses were used around negative numbers in the substitutions. This should be done to keep operations properly separated, particularly when negative numbers are involved.

Now work margin exercise 1.

Objective C **Solving Equations of the Form $x + b = c$**

To begin, we need the **addition principle of equality**.

> ### Addition Principle of Equality
>
> If the same algebraic expression is added to both sides of an equation, the new equation has the same solutions as the original equation. Symbolically, if A, B, and C are algebraic expressions, then the equations
>
> $$A = B$$
>
> and
>
> $$A + C = B + C$$
>
> have the same solutions.

1. Determine whether or not the given real number is a solution to the given equation by substituting for the variable and checking to see if the resulting equation is true or false.

a. $x - 2 = -9$ given that $x = -7$

b. $2.8 + z = 1.3$ given that $z = -1.7$

c. $6.7 - y = 3$ given that $y = 3.7$

d. $|z| - 7 = -4$ given that $x = -12$

Equations with the same solutions are said to be **equivalent equations**.

The objective of solving linear (or first-degree) equations is to get the variable by itself (with a coefficient of +1) on one side of the equation and any constants on the other side. The following procedure will help in solving linear equations such as

$$x - 3 = 7, \quad -11 = y + 5, \quad 3z - 2z + 2.5 = 3.2 + 0.8, \quad \text{and} \quad x - \frac{2}{5} = \frac{3}{10}.$$

Procedure for Solving Linear Equations that Simplify to the Form $x + b = c$

1. Combine any like terms on each side of the equation.
2. Use the **addition principle of equality** and add the opposite of the constant b to both sides. The objective is to isolate the variable on one side of the equation (either the left side or the right side) with a coefficient of +1.
3. Check your answer by substituting it for the variable in the original equation.

Every linear equation has exactly one solution. This means that once a solution has been found, there is no need to search for another solution.
Note that, as illustrated in Examples 1 through 3, variables other than x may be used.

Example 2

Solving $x + b = c$

a. $x - 3 = 7$

Solution

$x - 3 = 7$	Write the equation.
$x - 3 + 3 = 7 + 3$	Add 3 (the opposite of -3) to both sides.
$x = 10$	Simplify.

Check:

$x - 3 = 7$	
$(10) - 3 \overset{?}{=} 7$	Substitute $x = 10$.
$7 = 7$	true statement

b. $-11 = y + 5$

Solution

$-11 = y + 5$	Write the equation. Note that the variable can be on the right side.
$-11 - 5 = y + 5 - 5$	Add -5 (the opposite of $+5$) to both sides.
$-16 = y$	Simplify.

Check:

$$-11 = y + 5$$
$$-11 \stackrel{?}{=} (-16) + 5 \qquad \text{Substitute } y = -16.$$
$$-11 = -11 \qquad \text{true statement}$$

c. $\quad x - \dfrac{2}{5} = \dfrac{3}{10}$

Solution

$$x - \frac{2}{5} = \frac{3}{10} \qquad \text{Write the equation.}$$

$$x - \frac{2}{5} + \frac{2}{5} = \frac{3}{10} + \frac{2}{5} \qquad \text{Add } \frac{2}{5} \left(\text{the opposite of } -\frac{2}{5} \right) \text{ to both sides.}$$

$$x = \frac{3}{10} + \frac{4}{10} \qquad \text{(The common denominator is 10.)}$$

$$x = \frac{7}{10} \qquad \text{Simplify.}$$

Check:

$$x - \frac{2}{5} = \frac{3}{10}$$

$$\left(\frac{7}{10} \right) - \frac{2}{5} \stackrel{?}{=} \frac{3}{10} \qquad \text{Substitute } x = \frac{7}{10}.$$

$$\frac{7}{10} - \frac{4}{10} \stackrel{?}{=} \frac{3}{10} \qquad \text{The common denominator is 10.}$$

$$\frac{3}{10} = \frac{3}{10} \qquad \text{true statement}$$

Example 3

Simplfying and Solving Equations

$$3z - 2z + 2.5 = 3.2 + 0.8$$

Solution

$$3z - 2z + 2.5 = 3.2 + 0.8 \qquad \text{Write the equation.}$$

$$z + 2.5 = 4.0 \qquad \begin{array}{l}\text{Combine like terms on}\\ \text{both sides of the equation.}\end{array}$$

$$z + 2.5 - 2.5 = 4.0 - 2.5 \qquad \begin{array}{l}\text{Add } -2.5 \left(\text{the opposite of } +2.5 \right)\\ \text{to both sides.}\end{array}$$

$$z = 1.5 \qquad \text{Simplify.}$$

Check:

$$3z - 2z + 2.5 = 3.2 + 0.8$$

$$3(1.5) - 2(1.5) + 2.5 \stackrel{?}{=} 3.2 + 0.8 \qquad \text{Substitute } z = 1.5.$$

$$4.5 - 3.0 + 2.5 \stackrel{?}{=} 3.2 + 0.8 \qquad \text{Simplify.}$$

$$4.0 = 4.0 \qquad \text{true statement}$$

Now work margin exercises 2 and 3.

Solve each of the following linear equations.

2. a. $x - 5 = 12$

    b. $-8 = x + 4$

    c. $x - \dfrac{3}{8} = \dfrac{3}{4}$

3. $5z - 3z + 1.6 = 4.1 + 0.9$

Objective D **Solving Equations of the Form** *ax = c*

To solve equations of the form $ax = c$, where $a \neq 0$, we can use the idea of the reciprocal of the coefficient a. For example, as we studied earlier, the reciprocal of $\frac{3}{4}$ is $\frac{4}{3}$ and $\frac{3}{4} \cdot \frac{4}{3} = 1$. Also, we need the **multiplication** (or **division**) **principle of equality** as stated below.

Multiplication (or Division) Principle of Equality

If both sides of an equation are multiplied by (or divided by) the same nonzero constant, the new equation has the same solutions as the original equation. Symbolically, if A and B are algebraic expressions and C is any nonzero constant, then the equations

$$A = B,$$

$$AC = BC \text{ where } C \neq 0,$$

$$\text{and } \frac{A}{C} = \frac{B}{C} \text{ where } C \neq 0$$

have the same solutions. We say that the equations are equivalent.

Remember that the objective is to get the variable by itself on one side of the equation. That is, we want the variable to have +1 as its coefficient. The following procedure will accomplish this.

Procedure for Solving Linear Equations that Simplify to the Form *ax = c*

1. Combine any like terms on each side of the equation.
2. Use the **multiplication** (or **division**) **principle of equality** and multiply both sides of the equation by the reciprocal of the coefficient of the variable. (**Note:** This is the same as dividing both sides of the equation by the coefficient.) Thus the coefficient of the variable will become +1.
3. Check your answer by substituting it for the variable in the original equation.

Example 4

Solving $ax = c$

a. $5x = 20$

Solution

$$5x = 20 \qquad \text{Write the equation.}$$

$$\frac{1}{5} \cdot (5x) = \frac{1}{5} \cdot 20 \qquad \text{Multiply by } \frac{1}{5}, \text{ the reciprocal of 5.}$$

$$\left(\frac{1}{5} \cdot 5\right) x = \frac{1}{5} \cdot \frac{20}{1} \qquad \text{Use the associative property of multiplication.}$$

$$1 \cdot x = 4 \qquad \text{Simplify.}$$

$$x = 4$$

Multiplying by the reciprocal of the coefficient is the same as **dividing** by the coefficient itself. So, we can multiply both sides by $\frac{1}{5}$, as we did, or we can divide both sides by 5. In either case, the coefficient of x becomes $+1$.

$$5x = 20$$

$$\frac{5x}{5} = \frac{20}{5} \qquad \text{Divide both sides by 5.}$$

$$x = 4 \qquad \text{Simplify.}$$

Check:

$$5x = 20$$

$$5 \cdot (4) \overset{?}{=} 20 \qquad \text{Substitute } x = 4.$$

$$20 = 20 \qquad \text{true statement}$$

b. $1.1x + 0.2x = 12.2 - 3.1$

Solution

When decimal coefficients or constants are involved, you might want to use a calculator to perform some of the arithmetic.

$$1.1x + 0.2x = 12.2 - 3.1 \qquad \text{Write the equation.}$$

$$1.3x = 9.1 \qquad \text{Combine like terms.}$$

$$\frac{1.3x}{1.3} = \frac{9.1}{1.3} \qquad \begin{array}{l}\text{Use a calculator or pencil}\\\text{and paper to divide.}\end{array}$$

$$x = 7.0$$

4. Solve the following equations.

a. $3x = 33$

b. $2.5x + 0.4x = 15.2 - 3.6$

c. $\dfrac{2x}{3} = \dfrac{4}{5}$

d. $-x = -15$

Check:

$$1.1x + 0.2x = 12.2 - 3.1$$

$$1.1(7) + 0.2(7) \overset{?}{=} 12.2 - 3.1 \qquad \text{Substitute } x = 7.$$

$$7.7 + 1.4 \overset{?}{=} 9.1 \qquad \text{Simplify.}$$

$$9.1 = 9.1 \qquad \text{true statement}$$

c. $\dfrac{4x}{5} = \dfrac{3}{10}$ (This could be written $\dfrac{4}{5}x = \dfrac{3}{10}$ because $\dfrac{4}{5}x$ is the same as $\dfrac{4x}{5}$.)

Solution

$$\dfrac{4x}{5} = \dfrac{3}{10} \qquad \text{Write the equation.}$$

$$\dfrac{5}{4} \cdot \dfrac{4}{5} x = \dfrac{5}{4} \cdot \dfrac{3}{10} \qquad \text{Multiply both sides by } \dfrac{5}{4}.$$

$$1 \cdot x = \dfrac{1 \cdot \cancel{5}}{4} \cdot \dfrac{3}{2 \cdot \cancel{5}} \qquad \text{Simplify.}$$

$$x = \dfrac{3}{8}$$

Check:

$$\dfrac{4x}{5} = \dfrac{3}{10}$$

$$\dfrac{4}{5} \cdot \dfrac{3}{8} \overset{?}{=} \dfrac{3}{10} \qquad \text{Substitute } x = \dfrac{3}{8}.$$

$$\dfrac{3}{10} = \dfrac{3}{10} \qquad \text{true statement}$$

d. $-x = 4$

Solution

$$-x = 4 \qquad \text{Write the equation.}$$

$$-1x = 4 \qquad \text{-1 is the coefficient of x.}$$

$$\dfrac{-1x}{-1} = \dfrac{4}{-1} \qquad \text{Divide by -1 so that the coefficient will become $+1$.}$$

$$x = -4$$

Check:

$$-x = 4$$

$$-(-4) \overset{?}{=} 4 \qquad \text{Substitute } x = -4.$$

$$4 = 4 \qquad \text{true statement}$$

Now work margin exercise 4.

Example 5

Application

The original price of a Blu-Ray player was reduced by $45.50. The sale price was $165.90. Solve the equation $y - 45.50 = 165.90$ to determine the original price of the Blu-Ray player.

Solution

$y - 45.50 = 165.90$

$y - 45.50 + 45.50 = 165.90 + 45.50$ Use the addition principle by adding 45.50 to both sides.

$y = 211.40$ Simplify.

The original price of the Blu-Ray player was $211.40.

Now work margin exercise 5.

Practice Problems

Solve the following equations.

1. $-16 = x + 5$

2. $6y - 1.5 = 7.5$

3. $4x = -20$

4. $\dfrac{3y}{5} = 33$

5. $1.7z + 2.4z = 8.2$

6. $-x = -8$

7. $3x = 10$

8. $5x - 4x + 1.6 = -2.7$

9. $\dfrac{4}{5} = 3x$

5. The original price of a wool coat was reduced by $15.80. The reduced price was $84.79. Solve the equation $y - 15.80 = 84.79$ to determine the original price of the coat.

Practice Problem Answers

1. $x = -21$

2. $y = 1.5$

3. $x = -5$

4. $y = 55$

5. $z = 2$

6. $x = 8$

7. $x = \dfrac{10}{3}$

8. $x = -4.3$

9. $\dfrac{4}{15} = x$

Exercises 5.3

1. $x + 4 = 2$ given that $x = -2$

2. $z + (-12) = 6$ given that $z = 18$

3. $x - 3 = -7$ given that $x = 4$

4. $x - 2 = -3$ given that $x = 1$

5. $-10 + x = -14$ given that $x = -4$

6. $-9 - x = -14$ given that $x = 5$

7. $-26 + |x| = -8$ given that $x = -18$

8. $42 + |z| = -30$ given that $z = -72$

9. $|x| - |-3| = 25$ given that $x = -28$

10. $|-2| + |x| = 13$ given that $x = -11$

11. $x - 6 = 1$

12. $x - 10 = 9$

13. $y + 7 = 3$

14. $y + 12 = 5$

15. $x + 15 = -4$

16. $x + 17 = -10$

17. $22 = n - 15$

18. $36 = n - 20$

19. $6 = z + 12$

20. $18 = z + 1$

21. $x - 20 = -15$

22. $x - 10 = -11$

23. $y + 3.4 = -2.5$

24. $y + 1.6 = -3.7$

25. $x + 3.6 = 2.4$

26. $x + 2.7 = 3.8$

27. $x + \dfrac{1}{20} = \dfrac{3}{5}$

28. $n - \dfrac{2}{7} = \dfrac{3}{14}$

29. $5x = 45$

30. $9x = 108$

31. $32 = 4y$

32. $51 = 17y$

33. $\dfrac{3x}{4} = 15$

34. $\dfrac{5x}{7} = 65$

35. $4 = \dfrac{2y}{5}$

36. $46 = \dfrac{46y}{5}$

37. $4x - 3x = 10 - 12$

38. $7x - 6x = 13 + 15$

39. $7x - 8x = 13 - 25$

40. $10n - 11n = 20 - 14$

41. $3n - 2n + 6 = 14$

42. $7n - 6n + 13 = 22$

43. $1.7y + 1.3y = 6.3$

44. $2.5y + 7.5y = 4.2$

45. $\dfrac{3}{4}x = \dfrac{5}{3}$

46. $\dfrac{5}{6}x = \dfrac{5}{3}$

47. $7.5x = -99.75$

48. $-14 = 0.7x$

49. $1.5y - 0.5y + 6.7 = -5.3$

50. $2.6y - 1.6y - 5.1 = -2.9$

51. $10x - 9x - \dfrac{1}{2} = -\dfrac{9}{10}$

52. $6x - 5x + \dfrac{3}{4} = -\dfrac{1}{12}$

53. $1.4x - 0.4x + 2.7 = -1.3$

54. $3.5y - 2.5y - 6.3 = -1.0 - 2.5$

55. $\dfrac{7x}{4} - \dfrac{3x}{4} + \dfrac{7}{8} = \dfrac{3}{2}$

56. $\dfrac{5n}{2} - \dfrac{3n}{2} + \dfrac{4}{5} = \dfrac{7}{5} - \dfrac{1}{10}$

57. $6.2 = -3.5 + 7n - 6n$

58. $-7.2 = 1.3n - 0.3n - 1.0$

59. $1.7x = -5.1 - 1.7$

60. $3.2x = 2.8 - 9.2$

Solve the following word problems. See Example 5.

61. World Languages: The Japanese writing system consists of three sets of characters, two with 81 characters (which all Japanese students must know), and a third, kanji, with over 50,000 characters (of which only some are used in everyday writing). If a Japanese student knows 2107 total characters, solve the equation $x + 2(81) = 2107$ to determine the number of kanji characters the student knows.

62. Astronomy: 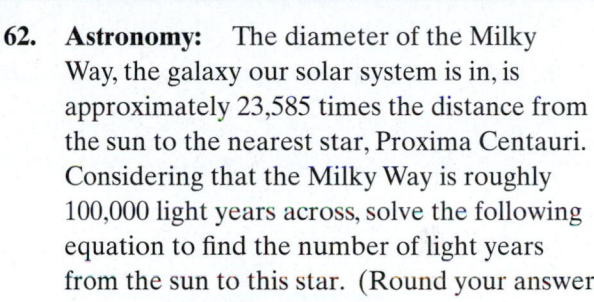 The diameter of the Milky Way, the galaxy our solar system is in, is approximately 23,585 times the distance from the sun to the nearest star, Proxima Centauri. Considering that the Milky Way is roughly 100,000 light years across, solve the following equation to find the number of light years from the sun to this star. (Round your answer to the nearest hundredth.)

$$23,585x = 100,000$$

63. Writing: An author is determined to have his first novel published by the publisher of George Orwell's 1984, his favorite book. However, his contract with the publisher requires his novel to be at least 75,000 words, and he has only written 63,500. Solve the following equation to determine how many more words he must write.

$$63,500 + x = 75,000$$

64. Sculpture: A sculptor has decided to begin a project to make scale models of famous landmarks out of stone. His first model will be of one of the moai, giant human figures carved from stone on Easter Island. If his model is to be $\dfrac{1}{12}$ scale and the original moai weighs 75 tons, solve the equation $12x = 75$ to determine how many tons his completed sculpture will weigh.

Use a graphing calculator to help solve the following equations.

65. $y + 32.861 = -17.892$

66. $x - 41.625 = 59.354$

67. $17.61x - 16.61x + 27.059 = 9.845$

68. $14.83y - 8.65 - 13.83y = 17.437 + 1.0$

69. $2.637x = 648.702$

70. $-0.3057y = 316.7052$

71. $-x = 145.6 + 17.89 - 10.32$

72. $-y = 143.5 + 178.462 - 200$

Writing & Thinking

73. **a.** Is the expression $6 + 3 = 9$ an equation? Explain.

   b. Is 4 a solution to the equation $5 + x = 10$? Explain.

5.4 Solving Linear Equations: $ax + b = c$

Objective A **Solving Equations of the Form $ax + b = c$**

Earlier we learned the equations to be solved were in one of two forms: $x + b = c$ or $ax = c$. In the first type, the coefficient of the variable x was always +1 and we used the addition principle to add or subtract constants to solve the equation. (We used the addition principle.) In the second type, we had to multiply both sides by the reciprocal of the coefficient of the variable (or divide both sides by the coefficient itself). Now we will apply both of these techniques in solving the same equation. That is, in the form $ax + b = c$, the coefficient a may be a number other than 1. The **general procedure for solving linear equations** is now a combination of the procedures stated in Section 5.3.

> ### Procedure for Solving Linear Equations that Simplify to the Form $ax + b = c$
>
> 1. Combine like terms on both sides of the equation.
> 2. Use the **addition principle of equality** and add the opposite of the constant b to both sides.
> 3. Use the **multiplication** (or **division**) **principle of equality** to multiply both sides by the reciprocal of the coefficient of the variable (or divide both sides by the coefficient itself). The coefficient of the variable will become +1.
> 4. Check your answer by substituting it for the variable in the original equation.

Example 1

Solving Linear Equations

a. $3x + 3 = -18$

Solution

$$3x + 3 = -18 \qquad \text{Write the equation.}$$
$$3x + 3 - 3 = -18 - 3 \qquad \text{Add } -3 \text{ to both sides.}$$
$$3x = -21 \qquad \text{Simplify.}$$
$$\frac{3x}{3} = \frac{-21}{3} \qquad \text{Divide both sides by 3.}$$
$$x = -7 \qquad \text{Simplify.}$$

Check:

$$3x + 3 = -18$$
$$3(-7) + 3 \overset{?}{=} -18 \qquad \text{Substitute } x = -7.$$
$$-21 + 3 \overset{?}{=} -18 \qquad \text{Simplify.}$$
$$-18 = -18 \qquad \text{true statement}$$

b. $-26 = 2y - 14 - 4y$

Solution

$-26 = 2y - 14 - 4y$	Write the equation.
$-26 = -2y - 14$	Combine like terms.
$-26 + 14 = -2y - 14 + 14$	Add 14 to both sides.
$-12 = -2y$	Simplify.
$\dfrac{-12}{-2} = \dfrac{-2y}{-2}$	Divide both sides by −2.
$6 = y$	Simplify.

Check:

$-26 = 2y - 14 - 4y$

$-26 \overset{?}{=} 2(6) - 14 - 4(6)$ Substitute $y = 6$.

$-26 \overset{?}{=} 12 - 14 - 24$ Simplify.

$-26 = -26$ true statement

Examples **2a.** and **2b.** illustrate solving equations with decimal coefficients. You may choose to work with the decimal coefficients as they are. However, another approach, shown here, is to multiply both sides in such a way to give integer coefficients. Generally, integers are easier to work with than decimals.

Example 2

Solving Linear Equations with Decimal Coefficients

a. $16.53 - 18.2z - 7.43 = 0$

Solution

$16.53 - 18.2z - 7.43 = 0$	Write the equation.
$100(16.53 - 18.2z - 7.43) = 100(0)$	Multiply both sides by 100. (This results in integer coefficients.)
$1653 - 1820z - 743 = 0$	Simplify.
$910 - 1820z = 0$	Combine like terms.
$910 - 1820z - 910 = 0 - 910$	Add −910 to both sides.
$-1820z = -910$	Simplify.
$\dfrac{-1820z}{-1820} = \dfrac{-910}{-1820}$	Divide both sides by −1820.
$z = 0.5 \left(\text{or } z = \dfrac{1}{2}\right)$	Simplify.

Check:

$$16.53 - 18.2z - 7.43 = 0$$

$$16.53 - 18.2(0.5) - 7.43 \stackrel{?}{=} 0 \qquad \text{Substitute } z = 0.5.$$

$$16.53 - 9.10 - 7.43 \stackrel{?}{=} 0 \qquad \text{Simplify.}$$

$$0 = 0 \qquad \text{true statement}$$

b. $5.1x + 7.4 - 1.8x = -9.1$

Solution

$5.1x + 7.4 - 1.8x = -9.1$	Write the equation.
$10(5.1x + 7.4 - 1.8x) = 10(-9.1)$	Multiply both sides by 10. (This results in integer coefficients.)
$51x + 74 - 18x = -91$	Simplify.
$33x + 74 = -91$	Combine like terms.
$33x + 74 - 74 = -91 - 74$	Add −74 to both sides.
$33x = -165$	Simplify.
$\dfrac{33x}{33} = \dfrac{-165}{33}$	Divide both sides by the coefficient 33.
$x = -5$	Simplify. Checking will show that −5 is a solution.

Examples **3a.** and **3b.** illustrate solving equations with coefficients that are fractions. You may choose to work with the coefficients as they are. However, another approach, shown here, is to multiply both sides by the LCM of the denominators which will give integer coefficients Generally, integers are easier to work with than fractions.

Example 3

Solving Linear Equations with Fractional Coefficients

a. $\dfrac{5}{6}x - \dfrac{5}{2} = -\dfrac{10}{9}$

Solution

$\dfrac{5}{6}x - \dfrac{5}{2} = -\dfrac{10}{9}$	Write the equation.
$18\left(\dfrac{5}{6}x - \dfrac{5}{2}\right) = 18\left(-\dfrac{10}{9}\right)$	Multiply both sides by 18 (the LCM of the denominators).
$18\left(\dfrac{5}{6}x\right) - 18\left(\dfrac{5}{2}\right) = 18\left(-\dfrac{10}{9}\right)$	Apply the distributive property.
$15x - 45 = -20$	Simplify.

Solve each of the following equations.

1. a. $2x + 5 = -21$

b. $-18 = 2y - 8 - 7y$

2. a. $3.69 + 12.80\,z - 6.25 = 0$

b. $0.7x + 9.5 - 2.4x = -4.1$

3. a. $\dfrac{3}{5}x - \dfrac{7}{3} = \dfrac{-17}{15}$

b. $\dfrac{1}{3}x + \dfrac{3}{8}x + \dfrac{9}{2} - \dfrac{8}{12}x = 0$

$$15x - 45 + 45 = -20 + 45 \qquad \text{Add 45 to both sides.}$$
$$15x = 25 \qquad \text{Simplify.}$$
$$\frac{15x}{15} = \frac{25}{15} \qquad \text{Divide both sides by 15.}$$
$$x = \frac{5}{3} \qquad \text{Simplify.}$$

Check:

$$\frac{5}{6}x - \frac{5}{2} = -\frac{10}{9}$$

$$\frac{5}{6}\left(\frac{5}{3}\right) - \frac{5}{2} \overset{?}{=} -\frac{10}{9} \qquad \text{Substitute } y = \frac{5}{3}.$$

$$\frac{25}{18} - \frac{45}{18} \overset{?}{=} -\frac{20}{18} \qquad \text{Simplify.}$$

$$-\frac{20}{18} = -\frac{20}{18} \qquad \text{true statement}$$

b. $\dfrac{1}{2}x + \dfrac{3}{4}x + \dfrac{7}{2} - \dfrac{2}{3}x = 0$

Solution

$$\frac{1}{2}x + \frac{3}{4}x + \frac{7}{2} - \frac{2}{3}x = 0 \qquad \text{Write the equation.}$$

$$12\left(\frac{1}{2}x + \frac{3}{4}x + \frac{7}{2} - \frac{2}{3}x\right) = 12(0) \qquad \begin{array}{l}\text{Multiply both sides by 12 (the}\\ \text{LCM of the denominators).}\end{array}$$

$$12\left(\frac{1}{2}x\right) + 12\left(\frac{3}{4}x\right) + 12\left(\frac{7}{2}\right) - 12\left(\frac{2}{3}x\right) = 12(0) \qquad \begin{array}{l}\text{Apply the distributive property.}\\ \text{Simplify.}\end{array}$$

$$6x + 9x + 42 - 8x = 0 \qquad \text{Combine like terms.}$$

$$7x + 42 = 0$$

$$7x + 42 - 42 = 0 - 42 \qquad \text{Add } -42 \text{ to both sides.}$$

$$7x = -42 \qquad \text{Simplify.}$$

$$\frac{7x}{7} = \frac{-42}{7} \qquad \text{Divide both sides by 7.}$$

$$x = -6 \qquad \begin{array}{l}\text{Simplify. Checking will show}\\ \text{that } -6 \text{ is the solution.}\end{array}$$

Now work margin exercises 1 through 3.

notes

■ **About Checking:** Checking can be quite time-consuming and need not be done for every problem. This is particularly important on exams. You should check only if you have time after the entire exam is completed.

■

Solve the following linear equations.

1. $x + 14 - 8x = -7$

2. $2.4 = 2.6y - 5.9y - 0.9$

3. $n - \dfrac{2n}{3} - \dfrac{1}{2} = \dfrac{1}{6}$

4. $\dfrac{3x}{14} + \dfrac{1}{2} - \dfrac{x}{7} = 0$

Practice Problem Answers

1. $x = 3$ 2. $y = -1$ 3. $n = 2$ 4. $x = -7$

Exercises 5.4

Solve each of the following linear equations. See Examples 1 through 3.

1. $3x + 11 = 2$

2. $3x + 10 = -5$

3. $5x - 4 = 6$

4. $4y - 8 = -12$

5. $6x + 10 = 22$

6. $3n + 7 = 19$

7. $9x - 5 = 13$

8. $2x - 4 = 12$

9. $1 - 3y = 4$

10. $5 - 2x = 9$

11. $14 + 9t = 5$

12. $5 + 2x = -7$

13. $-5x + 2.9 = 3.5$

14. $3x + 2.7 = -2.7$

15. $10 + 3x - 4 = 18$

16. $5 + 5x - 6 = 9$

17. $15 = 7x + 7 + 8$

18. $14 = 9x + 5 + 8$

19. $5y - 3y + 2 = 2$

20. $6y + 8y - 7 = -7$

21. $x - 4x + 25 = 31$

22. $3y + 9y - 13 = 11$

23. $-20 = 7y - 3y + 4$

24. $-20 = 5y + y + 16$

25. $4n - 10n + 35 = 1 - 2$

26. $-5n - 3n + 2 = 34$

27. $3n - 15 - n = 1$

28. $2n + 12 + n = 0$

29. $5.4x - 0.2x = 0$

30. $0 = 5.1x + 0.3x$

31. $\dfrac{1}{2}x + 7 = \dfrac{7}{2}$

32. $\dfrac{3}{5}x + 4 = \dfrac{9}{5}$

33. $\dfrac{1}{2} - \dfrac{8}{3}x = \dfrac{5}{6}$

34. $\dfrac{2}{5} - \dfrac{1}{2}x = \dfrac{7}{4}$

35. $\dfrac{3}{2} = \dfrac{1}{3}x + \dfrac{11}{3}$

36. $\dfrac{11}{8} = \dfrac{1}{5}x + \dfrac{4}{5}$

37. $\dfrac{7}{2} - 5 - \dfrac{5}{2}x = 9$

38. $\dfrac{8}{3} + 2 - \dfrac{7}{3}x = 6$

39. $\dfrac{5}{8}x - \dfrac{1}{4}x + \dfrac{1}{2} = \dfrac{3}{10}$

40. $\dfrac{1}{2}x + \dfrac{3}{4}x - \dfrac{5}{3} = \dfrac{5}{6}$

41. $\dfrac{y}{2} + \dfrac{1}{5} = 3$

42. $\dfrac{y}{3} - \dfrac{2}{3} = 7$

43. $\dfrac{7}{8} = \dfrac{3}{4}x - \dfrac{5}{8}$

44. $\dfrac{1}{10} = \dfrac{4}{5}x + \dfrac{3}{10}$

45. $\dfrac{y}{7} + \dfrac{y}{28} + \dfrac{1}{2} = \dfrac{3}{4}$

46. $\dfrac{5y}{6} - \dfrac{7y}{8} - \dfrac{1}{12} = \dfrac{1}{3}$

47. $x + 1.2x + 6.9 = -3.0$

48. $3x - 0.75x - 1.72 = 3.23$

49. $10 = x - 0.5x + 32$

50. $33 = y + 3 - 0.4y$

51. $2.5x + 0.5x - 3.5 = 2.5$

52. $4.7 - 0.5x - 0.3x = -0.1$ **53.** $6.4 + 1.2x + 0.3x = 0.4$ **54.** $5.2 - 1.3x - 1.5x = -0.4$

55. $-12.13 = 2.42y + 0.6y - 13.64$ **56.** $-7.01 = 1.75x + 3.05x - 8.45$ **57.** $-0.4x + x + 17.2 = 18.1$

58. $y - 0.75y + 13.76 = 14.66$ **59.** $0 = 17.3x - 15.02x - 0.456$ **60.** $0 = 20.5x - 16.35x + 0.1245$

Solve the following word problems.

61. **Music:** The tickets for a concert featuring the new hit band, Flying Sailor, sold out in 2.5 hours. If there were 35,000 tickets sold, solve the equation $35,000 - 2.5x = 0$ to find the number of tickets sold per hour.

62. **Parking Lots:** A rectangular shaped parking lot is to have a perimeter of 450 yards. If the width must be 90 yards because of a building code, solve the equation $2l + 2(90) = 450$ to determine the length of the parking lot.

63. **Temperature:** Jeff, who lives in England, is reading a letter from his pen pal in the United States. His pen pal says that the temperature in his city was 97.7° Fahrenheit one day, so it was too hot to play soccer outside. Jeff doesn't know how hot this is, because he is used to temperatures in Celsius. Help Jeff solve the equation below to determine the temperature in degrees Celsius.

$$1.8C + 32 = 97.7$$

64. **Height:** The tallest man-made structure in the world is the Burj Khalifa in Dubai, which stands at 2717 feet tall. The tallest tree in the world is a Mendocino tree in California. If 7 of these trees were stacked on top of each other, they would still be 144.5 feet shorter than the Burj Khalifa. Solve the equation below to determine the height of the tree.

$$7x + 144.5 = 2717$$

Use a graphing calculator to help solve the linear equations.

65. $0.15x + 5.23x - 17.815 = 15.003$ **66.** $15.97y - 12.34y + 16.95 = 8.601$

67. $13.45x - 20x - 17.36 = -24.696$ **68.** $26.75y - 30y + 23.28 = 4.4625$

5.5 Solving Linear Equations: $ax + b = cx + d$

Objectives

A Solve equations of the form $ax + b = cx + d$.

B Understand the terms conditional equations, identities, and contradictions.

Objective A **Solving Equations of the Form** $ax + b = cx + d$

Now we are ready to solve linear equations of the most general form $ax + b = cx + d$ where constants and variables may be on both sides. There may also be parentheses or other symbols of inclusion. **Remember that the objective is to get the variable on one side of the equation by itself with a coefficient of +1.**

> **General Procedure for Solving Linear Equations that Simplify to the Form** $ax + b = cx + d$
>
> 1. Simplify each side of the equation by removing any grouping symbols and combining like terms on both sides of the equation.
> 2. Use the **addition principle of equality** and add the opposite of a constant term and/or variable term to both sides so that variables are on one side and constants are on the other side.
> 3. Use the **multiplication** (or **division**) **principle of equality** to multiply both sides by the reciprocal of the coefficient of the variable (or divide both sides by the coefficient itself). The coefficient of the variable will become +1.
> 4. Check your answer by substituting it for the variable in the original equation.

Example 1

Solving Equations with Variables on Both Sides

a. $5x + 3 = 2x - 18$

Solution

$5x + 3 = 2x - 18$	Write the equation.
$5x + 3 - 3 = 2x - 18 - 3$	Add -3 to both sides.
$5x = 2x - 21$	Simplify.
$5x - 2x = 2x - 21 - 2x$	Add $-2x$ to both sides.
$3x = -21$	Simplify.
$\dfrac{3x}{3} = \dfrac{-21}{3}$	Divide both sides by 3.
$x = -7$	Simplify.

Check:

$$5x + 3 = 2x - 18$$

$$5(-7) + 3 \overset{?}{=} 2(-7) - 18 \qquad \text{Substitute } x = -7.$$

$$-35 + 3 \overset{?}{=} -14 - 18 \qquad \text{Simplify.}$$

$$-32 = -32 \qquad \text{true statement}$$

b. $4x + 1 - x = 2x - 13 + 5$

Solution

$$4x + 1 - x = 2x - 13 + 5 \qquad \text{Write the equation.}$$

$$3x + 1 = 2x - 8 \qquad \text{Combine like terms.}$$

$$3x + 1 - 1 = 2x - 8 - 1 \qquad \text{Add } -1 \text{ to both sides.}$$

$$3x = 2x - 9 \qquad \text{Simplify.}$$

$$3x - 2x = 2x - 9 - 2x \qquad \text{Add } -2x \text{ to both sides.}$$

$$x = -9 \qquad \text{Simplify.}$$

Check:

$$4x + 1 - x = 2x - 13 + 5$$

$$4(-9) + 1 - (-9) \overset{?}{=} 2(-9) - 13 + 5 \qquad \text{Substitute } x = -9.$$

$$-36 + 1 + 9 \overset{?}{=} -18 - 13 + 5 \qquad \text{Simplify.}$$

$$-26 = -26 \qquad \text{true statement}$$

Example 2

Solving Linear Equations Involving Decimals

$$6y + 2.5 = 7y - 3.6$$

Solution

$$6y + 2.5 = 7y - 3.6 \qquad \text{Write the equation.}$$

$$6y + 2.5 + 3.6 = 7y - 3.6 + 3.6 \qquad \text{Add 3.6 to both sides.}$$

$$6y + 6.1 = 7y \qquad \text{Simplify.}$$

$$6y + 6.1 - 6y = 7y - 6y \qquad \text{Add } -6y \text{ to both sides.}$$

$$6.1 = y \qquad \text{Simplify.}$$

Check:

$$6y + 2.5 = 7y - 3.6$$

$$6(6.1) + 2.5 \overset{?}{=} 7(6.1) - 3.6 \qquad \text{Substitute } y = 6.1.$$

$$36.6 + 2.5 \overset{?}{=} 42.7 - 3.6 \qquad \text{Simplify.}$$

$$39.1 = 39.1 \qquad \text{true statement}$$

Example 3

Solving Linear Equations with Fractional Coefficients

$$\frac{1}{3}x + \frac{1}{6} = \frac{2}{5}x - \frac{7}{10}$$

Solution

$\frac{1}{3}x + \frac{1}{6} = \frac{2}{5}x - \frac{7}{10}$	Write the equation.
$30\left(\frac{1}{3}x + \frac{1}{6}\right) = 30\left(\frac{2}{5}x - \frac{7}{10}\right)$	Multiply both sides by the LCM, 30.
$30\left(\frac{1}{3}x\right) + 30\left(\frac{1}{6}\right) = 30\left(\frac{2}{5}x\right) - 30\left(\frac{7}{10}\right)$	Apply the distributive property.
$10x + 5 = 12x - 21$	Simplify.
$10x + 5 - 5 = 12x - 21 - 5$	Add –5 to both sides.
$10x = 12x - 26$	Simplify.
$10x - 12x = 12x - 26 - 12x$	Add –12x to both sides.
$-2x = -26$	Simplify.
$\frac{-2x}{-2} = \frac{-26}{-2}$	Divide both sides by –2.
$x = 13$	Simplify. Checking will show that 13 is the solution.

Example 4

Solving Equations with Parentheses

a. $2(y - 7) = 4(y + 1) - 26$

Solution

$2(y - 7) = 4(y + 1) - 26$	Write the equation.
$2y - 14 = 4y + 4 - 26$	Use the distributive property.
$2y - 14 = 4y - 22$	Combine like terms.
$2y - 14 + 22 = 4y - 22 + 22$	Add 22 to both sides. Here we will put the variable on the right side to get a positive coefficient of y.
$2y + 8 = 4y$	Simplify.
$2y + 8 - 2y = 4y - 2y$	Add –2y to both sides.
$8 = 2y$	Simplify.
$\frac{8}{2} = \frac{2y}{2}$	Divide both sides by 2.
$4 = y$	Simplify. Checking will show that 4 is the solution.

b. $-2(5x+13)-2=-6(3x-2)-41$

Solution

$-2(5x+13)-2=-6(3x-2)-41$	Write the equation.
$-10x-26-2=-18x+12-41$	Use the distributive property. Be careful with the signs.
$-10x-28=-18x-29$	Combine like terms.
$-10x-28+18x=-18x-29+18x$	Add $18x$ to both sides.
$8x-28=-29$	Simplify.
$8x-28+28=-29+28$	Add 28 to both sides.
$8x=-1$	Simplify.
$\dfrac{8x}{8}=\dfrac{-1}{8}$	Divide both sides by 8.
$x=-\dfrac{1}{8}$	Simplify. Checking will show that $-\dfrac{1}{8}$ is the solution.

Now work margin exercises 1 through 4.

Solve each of the following equations.

1. a. $7x+4=3x-20$

 b. $6x+3-2x=2x-15+4$

2. $5y+1.4=7y-2.8.$

3. $\dfrac{1}{4}x+\dfrac{2}{5}=\dfrac{1}{2}x-\dfrac{6}{10}$

4. a. $3(y-5)=8(y+1)-3$

 b. $-5(3x+7)-11=-7(x-1)-5$

Objective B **Conditional Equations, Identities, and Contradictions**

When solving equations, there are times when we are concerned with the number of solutions that an equation has. If an equation has a finite number of solutions (the number of solutions is a countable number), the equation is said to be a **conditional equation**. As stated earlier, every linear equation has exactly one solution. Thus **every linear equation is a conditional equation**. However, in some cases, simplifying an equation will lead to a statement that is always true, such as $0=0$. In these cases the original equation is called an **identity** and has an infinite number of solutions which can be written as all real numbers or $\mathbb{R}$. If the equation simplifies to a statement that is never true, such as $0=2$, then the original equation is called a **contradiction** and there is no solution. Table 1 summarizes these ideas.

Type of Equation	Number of Solutions
conditional	finite number of solutions
identity	infinite number of solutions
contradiction	no solution

Table 1

5. Determine whether each of the following equations is a conditional equation, an identity, or a contradiction.

a. $2x + 9 = -13$

b. $4(x - 17) + 3x = 7(x + 7)$

c. $-3(x - 4) + x = 12 - 2x$

Example 5

Solutions of Equations

Determine whether each of the following equations is a conditional equation, an identity, or a contradiction.

a. $3x + 16 = -11$

Solution

$3x + 16 = -11$	Write the equation.
$3x + 16 - 16 = -11 - 16$	Add -16 to both sides.
$3x = -27$	Simplify.
$\dfrac{3x}{3} = \dfrac{-27}{3}$	Divide both sides by 3.
$x = -9$	Simplify.

The equation has one solution and it is a conditional equation.

b. $3(x - 25) + 3x = 6(x + 10)$

Solution

$3(x - 25) + 3x = 6(x + 10)$	Write the equation.
$3x - 75 + 3x = 6x + 60$	Use the distributive property.
$6x - 75 = 6x + 60$	Combine like terms.
$6x - 75 - 6x = 6x + 60 - 6x$	Add $-6x$ to both sides.
$-75 = 60$	Simplify.

The last equation is never true. Therefore, the original equation is a contradiction and has no solution.

c. $-2(x - 7) + x = 14 - x$

Solution

$-2(x - 7) + x = 14 - x$	Write the equation.
$-2x + 14 + x = 14 - x$	Use the distributive property.
$14 - x = 14 - x$	Combine like terms.
$14 - x + x = 14 - x + x$	Add x to both sides.
$14 = 14$	Simplify.

The last equation is always true. Therefore, the original equation is an identity and has an infinite number of solutions. Every real number is a solution.

Now work margin exercise 5.

Practice Problems

Solve the following linear equations.

1. $x + 14 - 6x = 2x - 7$

2. $6.4x + 2.1 = 3.1x - 1.2$

3. $\dfrac{2x}{3} - \dfrac{1}{2} = x + \dfrac{1}{6}$

4. $\dfrac{3}{14}n + \dfrac{1}{4} = \dfrac{1}{7}n - \dfrac{1}{4}$

5. $5 - (y - 3) = 14 - 4(y + 2)$

Determine whether each of the following equations is a conditional equation, an identity, or a contradiction.

6. $7(x - 3) + 42 = 7x + 21$

7. $-2x + 14 = -2(x + 1) + 10$

Practice Problem Answers

1. $x = 3$ **2.** $x = -1$ **3.** $x = -2$ **4.** $n = -7$

5. $y = -\dfrac{2}{3}$ **6.** identity **7.** contradiction

Exercises 5.5

Solve each of the following linear equations. See Examples 2 through 4.

1. $3x + 2 = x - 8$

2. $5x + 1 = 2x - 5$

3. $4n - 3 = n + 6$

4. $6y + 3 = y - 7$

5. $3y + 18 = 7y - 6$

6. $2y + 5 = 8y + 10$

7. $3x + 11 = 8x - 4$

8. $9x + 3 = 5x - 9$

9. $14n = 3n$

10. $1.6x = 0.8x$

11. $6y - 2.1 = y - 2.1$

12. $13x + 5 = 2x + 5$

13. $2(z + 1) = 3z + 3$

14. $6x - 3 = 3(x + 2)$

15. $16y + 23y - 3 = 16y - 2y + 2$

16. $5x - 2x + 4 = 3x + x - 1$

17. $0.25 + 3x + 6.5 = 0.75x$

18. $0.9y + 3 = 0.4y + 1.5$

19. $6.5 + 1.2x = 0.5 - 0.3x$

20. $x - 0.1x + 0.8 = 0.2x + 0.1$

21. $\dfrac{2}{3}x + 1 = \dfrac{1}{3}x - 6$

22. $\dfrac{4}{5}n + 2 = \dfrac{2}{5}n - 4$

23. $\dfrac{y}{5} + \dfrac{3}{4} = \dfrac{y}{2} + \dfrac{3}{4}$

24. $\dfrac{5n}{6} + \dfrac{1}{9} = \dfrac{3n}{2} + \dfrac{1}{9}$

25. $\dfrac{3}{8}\left(y - \dfrac{1}{2}\right) = \dfrac{1}{8}\left(y + \dfrac{1}{2}\right)$

26. $\dfrac{1}{2}\left(\dfrac{x}{2} + 1\right) = \dfrac{1}{3}\left(\dfrac{x}{2} - 1\right)$

27. $\dfrac{2x}{3} + \dfrac{x}{3} = -\dfrac{3}{4} + \dfrac{x}{2}$

28. $\dfrac{3}{4}x + \dfrac{1}{5}x = \dfrac{1}{2}x - \dfrac{3}{10}$

29. $x + \dfrac{2}{3}x - 2x = \dfrac{x}{6} - \dfrac{1}{8}$

30. $3x + \dfrac{1}{2}x - \dfrac{2}{5}x = \dfrac{x}{10} + \dfrac{7}{20}$

31. $3(1 + 9x) = 6(2 - 4x)$

32. $4(5 - x) = 8(3x + 10)$

33. $3(4x - 1) = 4(2x - 3) + 8$

34. $7(2x - 1) = 5(x + 6) - 13$

35. $5 - 3(2x + 1) = 4(x - 5) + 6$

36. $-2(y + 5) - 4 = 6(y - 2) + 2$

37. $8 + 4(2x - 3) = 5 - (x + 3)$

38. $8(3x + 5) - 9 = 9(x - 2) + 14$

39. $4.7 - 0.3x = 0.5x - 0.1$

40. $5.8 - 0.1x = 0.2x - 0.2$

41. $0.2(x + 3) = 0.1(x - 5)$

42. $0.4(x + 3) = 0.3(x - 6)$

43. $\dfrac{1}{2}(4-8x)=\dfrac{1}{3}(4x+7)-3$

44. $3+\dfrac{1}{4}(x-4)=\dfrac{2}{5}(2+3x)$

45. $0.6x-22.9=1.5x-18.4$

46. $0.1y+3.8=5.72-0.3y$

47. $0.12n+0.25n-5.895=4.3n$

48. $0.15n+32n-21.0005=10.5n$

49. $0.7(x+14.1)=0.3(x+32.9)$

50. $0.8(x-6.21)=0.2(x-24.84)$

Determine whether each of the following equations is a conditional equation, an identity, or a contradiction. See Example 5.

51. $2(3x-1)+5=3$

52. $-2x+13=-2(x-7)$

53. $5x+13=-2(x-7)+3$

54. $3x+9=-3(x-3)+6x$

55. $7(x-1)=-3(3-x)+4x$

56. $3(x-2)+4x=6(x-1)+x$

57. $5(x+1)=3(x+1)+2(x+1)$

58. $8x-20+x=-3(5-2x)+3(x-4)$

59. $2x+3x=5.2(3-x)$

60. $5.2x+3.4x=0.2(x-0.42)$

Solve each of the following word problems.

61. Construction: A farmer is putting a shed on his property. He has two designs. One uses wood and would cost $2 per square foot plus an extra $8400 in materials. The other design is metal and would cost $4 per square foot plus an additional $8800. Both sheds are the same size, and the wood shed costs $\dfrac{3}{4}$ what the metal shed costs. Solve the following equation to determine how many square feet the shed will be.

$$2x+8400=\dfrac{3}{4}(4x+8800)$$

62. Renting Buildings: Two rival shoe companies want to rent the same empty building for their office and shipping space. Schulster's Shoes needs 1000 square feet for offices, 600 for shipping, and another 6 square feet for every packaged box of shoes. Shoes, Shoes, Shoes! needs 750 square feet for offices, 400 for shipping, and 9 square feet per packaged box of shoes. If only one company will fit exactly into the empty building and they both plan to have the same amount of inventory, solve the following equation to determine how many boxes of shoes both companies hope to have at any given time.

$$1000+600+6x=750+400+9x$$

63. Ice Cream: An ice cream shop is having a special "Ice Cream Sunday" event in which they are giving away giant mixed sundaes of 3 scoops of vanilla ice cream and 2 scoops of chocolate. If they have 24 gallons of chocolate and 36 gallons of vanilla to start with, solve the given equation to determine how many sundaes they will have made when they run out of ice cream. (For this problem, we assume a gallon = 20 scoops.)

$$36 - \frac{1}{20}(3x) = 24 - \frac{1}{20}(2x)$$

64. Music: A guitarist and a drummer are getting ready for a gig. The length of the gig will depend on how much material they have prepared. For every hour of the show, the guitarist must practice for 5 days and the drummer for 3 days. Since the guitarist knows some of the songs, he saves 3 days of practice time. The drummer hurts his hand and loses 3 days of practice. If they plan to start and finish practicing at the same time, solve the following equation to determine how long the show will be.

$$5x - 3 = 3x + 3$$

 Use a graphing calculator to help solve the linear equations.

65. $0.17x - 23.0138 = 1.35x + 36.234$

66. $48.512 - 1.63x = 2.58x + 87.63553$

67. $0.32(x + 14.1) = 2.47x + 2.21795$

68. $1.6(9.3 + 2x) = 0.2(3x + 133.94)$

5.6 Applications: Number Problems and Consecutive Integers

Objectives

A Solve word problems involving translating number problems.

B Solve word problems involving translating consecutive integers.

C Learn how to solve word problems involving a variety of other applications.

Objective A **Number Problems**

In earlier chapters, we discussed translating English phrases into algebraic expressions, and learned the Pólya four-step process as an approach to problem solving. Now we will use those skills to read number problems and translate the sentences and phrases in the problem into a related equation. The solution of this equation will be the solution to the problem.

Example 1

Number Problems

a. If a number is decreased by 36 and the result is 76 less than twice the number, what is the number?

Solution

Let n = the unknown number.

a number is decreased by 36	the result is	76 less than twice the number

$$n - 36 \quad = \quad 2n - 76$$
$$n - 36 - n = 2n - 76 - n$$
$$-36 = n - 76$$
$$-36 + 76 = n - 76 + 76$$
$$40 = n \qquad \text{The number is 40.}$$

b. Three times the sum of a number and 5 is equal to twice the number plus 5. Find the number.

Solution

Let x = the unknown number.

3 times the sum of a number and 5	is equal to	twice the number plus 5

$$3(x + 5) \quad = \quad 2x + 5$$
$$3x + 15 = 2x + 5$$
$$3x + 15 - 2x = 2x + 5 - 2x$$
$$x + 15 = 5$$
$$x + 15 - 15 = 5 - 15$$
$$x = -10 \qquad \text{The number is } -10.$$

1. a. If a number is decreased by 28 and the result is 92 less than twice the number, what is the number?

b. 6 times the sum of a number and 3 is equal to 3 times the number plus 3. Find the number.

c. One integer is 3 more than 4 times a second integer. Their sum is 28. What are the two integers?

c. One integer is 4 more than three times a second integer. Their sum is 24. What are the two integers?

Solution

Let n = the second integer, then $3n + 4$ = the first integer.

$$\left(1\text{st integer}\right) + \left(2\text{nd integer}\right) = 24$$
$$\left(3n + 4\right) + n = 24 \qquad \text{Their sum is 24.}$$
$$4n + 4 = 24$$
$$4n + 4 - 4 = 24 - 4$$
$$4n = 20$$
$$\frac{4n}{4} = \frac{20}{4}$$
$$n = 5$$
$$3n + 4 = 19 \qquad \text{The two integers are 5 and 19.}$$

Now work margin exercise 1.

Objective B   **Consecutive Integers**

Remember that the set of **integers** consists of the whole numbers and their opposites.

$$\mathbb{Z} = \left\{ \ldots, -4, -3, -2, -1, 0, 1, 2, 3, 4, \ldots \right\}$$

Even integers are integers that are divisible by 2. The even integers are

$$\left\{ \ldots, -6, -4, -2, 0, 2, 4, 6, \ldots \right\}.$$

Odd integers are integers that are not even. If an odd integer is divided by 2, the remainder will be 1. The odd integers are

$$\left\{ \ldots, -5, -3, -1, 1, 3, 5, \ldots \right\}.$$

notes

In this discussion we will be dealing only with integers. Therefore, if you get a result that has a fraction or decimal number (not an integer), you will know that an error has been made and you should correct some part of your work.

The following terms and the ways of representing the integers must be understood before attempting the problems.

Consecutive Integers

Integers are **consecutive** if each is 1 more than the previous integer. Three consecutive integers can be represented as n, $n + 1$, and $n + 2$.

For example: 5, 6, 7

Consecutive Even Integers

Even integers are consecutive if each is 2 more than the previous even integer. Three consecutive even integers can be represented as $2n$, $2n + 2$, and $2n + 4$, where n is an integer.

For example: 24, 26, 28

Consecutive Odd Integers

Odd integers are consecutive if each is 2 more than the previous odd integer. Three consecutive odd integers can be represented as $2n + 1$, $2n + 3$, and $2n + 5$, where n is an integer.

For example: 41, 43, 45.

2. Three consecutive odd integers are such that their sum is –9. What are the integers?

Example 2

Consecutive Integers

Three consecutive **odd** integers are such that their sum is –3. What are the integers?

Solution

Let $2n + 1 =$ the first odd integer,
  $2n + 3 =$ the second odd integer, and
  $2n + 5 =$ the third odd integer.
Set up and solve the related equation.

$$\left(1\text{st integer}\right) + \left(2\text{nd integer}\right) + \left(3\text{rd integer}\right) = -3$$
$$2n + 1 + \left(2n + 3\right) + \left(2n + 5\right) = -3$$
$$6n + 9 = -3$$
$$6n + 9 - 9 = -3 - 9$$
$$6n = -12$$
$$\frac{6n}{6} = \frac{-12}{6}$$
$$n = -2$$
$$2n + 1 = -3$$
$$2n + 3 = -1$$
$$2n + 5 = 1$$

The three consecutive odd integers are $-3, -1,$ and 1.

Now work margin exercise 2.

Objective C **Other Applications**

As you learn more abstract mathematical ideas, you will find that you will use these ideas and the related processes to solve a variety of everyday problems as well as problems in specialized fields of study. Generally, you may not even be aware of the fact that you are using your mathematical knowledge. However, these skills and ideas will be part of your thinking and problem solving techniques for the rest of your life.

Example 3

Applications

a. Joe pays $800 per month to rent an apartment. If this is $\frac{2}{5}$ of his monthly income, what is his monthly income?

Solution

Let x = Joe's monthly income and $\frac{2}{5}x$ = rent.

$$\frac{2}{5}x = 800$$

$$\frac{5}{2} \cdot \frac{2}{5}x = \frac{5}{2} \cdot \frac{800}{1}$$

$$x = 2000$$

Joe's monthly income is $2000.

b. A student bought a calculator and a textbook for a total of $200.80 (including tax). If the textbook cost $20.50 more than the calculator, what was the cost of each item?

Solution

Let x = cost of the calculator, then $x + 20.50$ = cost of the textbook. The equation to be solved is:

$$\overbrace{x + 20.50}^{\substack{\text{cost of} \\ \text{text book}}} + \overbrace{x}^{\substack{\text{cost of} \\ \text{calculator}}} = \overbrace{200.80}^{\substack{\text{total} \\ \text{spent}}}$$

$$2x + 20.50 = 200.80$$

$$2x + 20.50 - 20.50 = 200.80 - 20.50$$

$$2x = 180.30$$

$$\frac{2x}{2} = \frac{180.30}{2}$$

$$x = 90.15 \qquad \text{cost of calculator}$$

$$x + 20.50 = 110.65 \qquad \text{cost of textbook}$$

The calculator costs $90.15 and the textbook costs $90.15 + $20.50 = $110.65, with tax included in each price.

Now work margin exercise 3.

3. a. Jim pays $1200 per month to send his son Taylor to basketball camp. If this is $\frac{3}{7}$ of his monthly income, what is his monthly income?

b. Graeme bought bagpipes and a soccer net for a total of $507.30 (including tax). If the bagpipes cost $70.40 more than the soccer net, what was the cost of each item?

Exercises 5.6

1. Five less than a number is equal to 13 decreased by the number. Find the number.

2. Three less than twice a number is equal to the number. What is the number?

3. Thirty-six is 4 more than twice a certain number. Find the number.

4. Fifteen decreased by twice a number is 27. Find the number.

5. Seven times a certain number is equal to the sum of twice the number and 35. What is the number?

6. The difference between twice a number and 3 is equal to 6 decreased by the number. Find the number.

7. Fourteen more than three times a number is equal to 6 decreased by the number. Find the number.

8. Two added to the quotient of a number and 7 is equal to −3. What is the number?

9. The quotient of twice a number and 5 is equal to the number increased by 6. What is the number?

10. Three times the sum of a number and 4 is equal to −9. Find the number.

11. Four times the difference between a number and 5 is equal to the number increased by 4. What is the number?

12. When 17 is added to six times a number, the result is equal to 1 plus twice the number. What is the number?

13. If the sum of twice a number and 5 is divided by 11, the result is equal to the difference between 4 and the number. Find the number.

14. If the difference between a number and 21 is divided by 2, the result is 4 times the number. What is the number?

15. Twice a number increased by three times the number is equal to 4 times the sum of the number and 3. Find the number.

16. Twice the difference between a number and 10 is equal to 6 times the number plus 16. What is the number?

17. The sum of two consecutive odd integers is 60. What are the integers?

18. The sum of two consecutive even integers is 78. What are the integers?

19. Find three consecutive integers whose sum is 69.

20. Find three consecutive integers whose sum is 93.

21. The sum of four consecutive integers is 74. What are the integers?

22. Find four consecutive integers whose sum is 90.

23. 171 minus the first of three consecutive integers is equal to the sum of the second and third. What are the integers?

24. If the first of three consecutive integers is subtracted from 120, the result is the sum of the second and third. What are the integers?

25. Four consecutive integers are such that if 3 times the first is subtracted from 208, the result is 50 less than the sum of the other three. What are the integers?

26. Find two consecutive integers such that twice the first plus three times the second equals 83.

27. Find three consecutive even integers such that the first plus twice the second is 54 less than four times the third.

28. Find three consecutive odd integers such that 4 times the first is 44 more than the sum of the second and third.

29. Find three consecutive even integers such that if the first is subtracted from the sum of the second and third, the result is 66.

30. Find three consecutive even integers such that their sum is 168 more than the second.

31. Find three consecutive odd integers such that the sum of twice the first and three times the second is 7 more than twice the third.

32. Find three consecutive even integers such that the sum of three times the first and twice the third is twenty less than six times the second.

Solve the following word problems. See Example 3.

33. School Supplies: A mathematics student bought a graphing calculator and a textbook for a course in statistics. If the text costs $49.50 more than the calculator, and the total cost for both was $125.74, what was the cost of each item?

34. Electronics: The total cost of a computer flash drive and a color printer was $225.50, including tax. If the cost of the flash drive was $170.70 less than the printer, what was the cost of each item?

35. Real Estate: A real estate agent says that the current value of a home is $90,000 more than twice its value when it was new. If the current value is $310,000, what was the value of the home when it was new? The home is 25 years old.

36. Guitars: Texas has a mean of 91,399 guitars sold each year, which is seven times the mean number of guitars sold each year in Wyoming. What is the mean number of electric guitars, sold each year in Wyoming?

37. Classic Cars: A classic car is now selling for $1500 more than three times its original price. If the selling price is now $12,000, what was the car's original price?

38. Mail: On August 24, the Fernandez family received 19 pieces of mail, consisting of magazines, bills, letters, and ads. If they received the same number of magazines as letters, three more bills than letters, and five more ads than bills, how many magazines did they receive?

39. Class Size: A high school graduating class is made up of 542 students. If there are 56 more girls than boys, how many boys are in the class?

40. Golfing: Lucinda bought two boxes of golf balls. She gave the pro-shop clerk a 50-dollar bill and received $10.50 in change. What was the cost of one box of golf balls? (Tax was included.)

41. Guitars: A guitar manufacturer spent $158 million on the production of acoustic and electric guitars last year. If the amount the company spent producing acoustic guitars was $68 million more than it spent on producing electric guitars, how much did the company spend producing electric guitars?

42. Telephone Calls: For a long-distance call, the telephone company charges 35¢ for each of the first three minutes and 15¢ for each additional minute. If the cost of a call was $12.30, how many minutes did the call last?

43. Rental Cars: If the rental price on a car is a fixed price per day plus $0.28 per mile, what was the fixed price per day if the total paid was $140 and the car was driven for 250 miles in two days?

44. Rental Cars: The U-Drive Company charges $20 per day plus 22¢ per mile driven. For a one-day trip, Louis paid a rental fee of $66.20. How many miles did he drive?

45. Carpentry: A 29 foot board is cut into three pieces at a sawmill. The second piece is 2 feet longer than the first and the third piece is 4 feet longer than the second. What are the lengths of the three pieces?

46. Triangles: The three sides of a triangle are $n, 4n - 4, 2n + 7$ (as shown in the figure below). If the perimeter of the triangle is 59 cm, what is the length of each side?

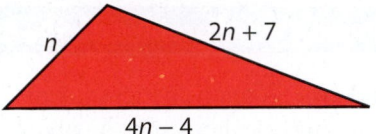

47. Construction: Joe Johnson decided to buy a lot and build a house on the lot. He knew that the cost of constructing the house was going to be $25,000 more than the cost of the lot. He told a friend that the total cost was going to be $275,000. As a test to see if his friend remembered the algebra they had together in school, he challenged his friend to calculate what he paid for the lot and what he was going to pay for the house. What was the cost of the lot and the cost of the house?

48. Triangles: The three sides of a triangle are $x, 3x - 1$, and $2x + 5$ (as shown in the figure below). If the perimeter of the triangle is 64 inches, what is the length of each side? (**Reminder:** The perimeter of a triangle is the sum of the lengths of the sides.)

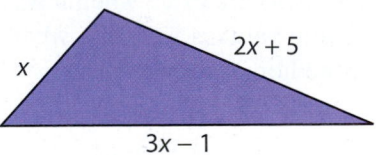

49. $5x - x = 8$

50. $2x + 3 = 9$

51. $n + (n + 1) = 33$

52. $n + (n + 4) = 3(n + 2)$

53. $3(n + 1) = n + 53$

54. $2(n + 2) - 1 = n + 4 - n$

Writing & Thinking

55. **a.** How would you represent four consecutive odd integers?

b. How would you represent four consecutive even integers?

c. Are these representations the same? Explain.

Objectives

A   Briefly review the concepts of decimal numbers and percentages.

B   Understand the concept of the basic percent equation.

C   Use the basic percent equation to solve application problems involving percentages.

5.7 Applications: Percent Problems

Our daily lives are filled with decimal numbers and percents: stock market reports, batting averages, win-loss records, salary raises, measures of pollution, taxes, interest on savings, home loans, discounts on clothes and cars, and on and on. Since decimal numbers and percents play such a prominent role in everyday life, we need to understand how to operate with them and apply them correctly in a variety of practical situations.

Objective A   A Brief Review of Decimals and Percentages

The word percent comes from the Latin *per centum,* meaning "per hundred." So, **percent means hundredths**. Thus

$$72\%, \quad \frac{72}{100}, \quad \text{and } 0.72 \text{ all have the same meaning.}$$

change a decimal number to percent:

$$0.036 = 3.6\% \qquad \text{Move the decimal point two places to the right and add the \% sign.}$$

change a percent to a decimal number:

$$56\% = 0.56 \qquad \text{Move the decimal point two places to the left and drop the \% sign.}$$

change a percent to a fraction (or mixed number):

$$30\% = \frac{30}{100} = \frac{3}{10} \qquad \text{Write the percent in hundredths form and reduce.}$$

change a fraction (or mixed number) to a percent:

$$\frac{1}{4} = 0.25 = 25\% \qquad \text{Write the fraction in decimal form and then change the decimal to a percent.}$$

Consider the statement

"15% of 80 is 12."

This statement has three numbers in it. In general, solving a percent problem involves knowing two of these numbers and trying to find the third. That is, **there are three basic types of percent problems**. We can break down the sentence in the following way.

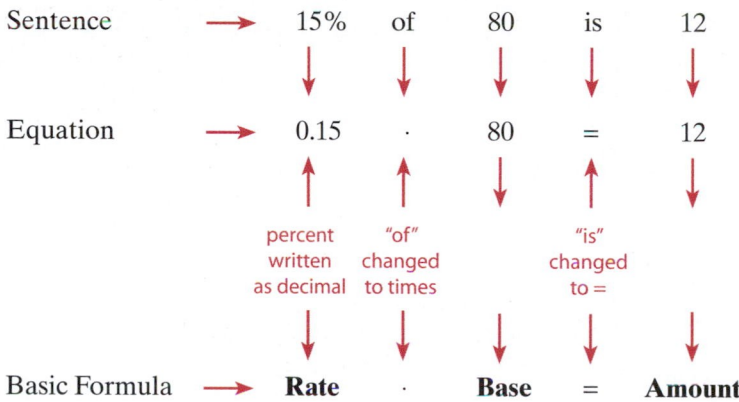

Sentence	→	15%	of	80	is	12
Equation	→	0.15	·	80	=	12

percent written as decimal   "of" changed to times   "is" changed to =

Basic Formula	→	**Rate**	·	**Base**	=	**Amount**

The terms that we have just discussed are explained in detail in the following box.

The Basic Formula $R \cdot B = A$

R = **RATE** or percent (as a decimal or fraction)

B = **BASE** (number we are finding the percent of)

A = **AMOUNT** (a part of the base)

"of" means to multiply.

"is" means equal (=).

The relationship among $R, B,$ and A is given in the equation

$R \cdot B = A$ (or $A = R \cdot B$).

Even though there are just three basic types of percent problems, many people have difficulty deciding whether to multiply or divide in a particular problem. Using the equation $R \cdot B = A$ helps to avoid these difficulties. If the values of any two of the quantities in the formula are known, they can be substituted in the equation and then the missing number can be found by solving the equation. **The process of solving the equation for the unknown quantity determines whether multiplication or division is needed.**

The following examples illustrate how to substitute into the equation and how to solve the resulting equations.

1. a. What is 30% of 660?

b. 63% of what number is 212.184?

c. What percent of 300 is 60?

Example 1

Percent of a Number

a. What is 72% of 800?

Solution

$R = 0.72$, $B = 800$, and A is unknown.

$$R \cdot B = A$$

$$0.72 \cdot 800 = A \qquad \text{Here, simply multiply to find } A.$$
$$576 = A$$

So **576** is 72% of 800.

b. 57% of what number is 163.191?

Solution

$R = 0.57$, B is unknown, and $A = 163.191$.

$$R \cdot B = A$$

$$0.57 \cdot B = 163.191$$
$$\frac{0.57 \cdot B}{0.57} = \frac{163.191}{0.57} \qquad \text{Now divide both sides by 0.57 to find } B.$$
$$B = 286.3$$

So 57% of **286.3** is 163.191.

c. What percent of 180 is 45?

Solution

R is unknown, $B = 180$, and $A = 45$.

$$R \cdot B = A$$

$$R \cdot 180 = 45$$
$$\frac{R \cdot 180}{180} = \frac{45}{180} \qquad \text{Now divide both sides by 180 to find } R.$$
$$R = 0.25$$
$$R = 25\% \qquad \text{Write the decimal in percent form.}$$

So **25%** of 180 is 45.

Now work margin exercise 1.

Objective C   Applications with Percentages

To sell goods that have been in stock for some time or simply to attract new customers, retailers and manufacturers sometimes offer a **discount**, a reduction in the selling price usually stated as a percent of the original price. The new, reduced price is called the **sale price**.

Example 2

Discounts

A bicycle was purchased at a discount of 25% of its original price of $1600. What was the sale price?

Solution

There are two ways to approach this problem. One way is to find the discount and then subtract this amount from $1600. Another way is to subtract 25% from 100% to get 75% and then find 75% of $1600. Here, we will calculate the discount and then subtract from the original price.

$$R \ \cdot \ B \ = \ A$$

$$0.25 \cdot 1600 = 400 \qquad \text{discount}$$

Sale Price =
original price − discount = $1600 − $400 = $1200
The sale price for the bicycle was $1200.

Now work margin exercise 2.

Sales tax is a tax charged on goods sold by retailers, and it is assessed by states and cities for income to operate various services. The **rate of sales tax** (a percent) varies by location.

Example 3

Sales Tax

Suppose 6% sales tax was added to the sale price $1200 of the bicycle in Example 2. What was the total paid for the bicycle?

Solution

The sales tax must be calculated on the sale price and then added to the sale price.

$$R \ \cdot \ B \ = \ A$$

$$0.06 \cdot 1200 = 72 \qquad \text{sales tax}$$

total paid = sale price + sales tax = $1200 + $72 = $1272
The total paid for the bicycle, including sales tax, was $1272.

Now work margin exercise 3.

2. A ping pong table was purchased at a discount of 40% of its original price of $1040. What was the sale price?

3. Suppose a 4% sales tax was added to the sale price of the ping pong table in margin exercise 2. What was the total paid for the ping pong table?

A **commission** is a fee paid to an agent or salesperson for a service. Commissions are usually a percent of a negotiated contract (as to a real estate agent) or a percent of sales.

4. A real estate agent earns a salary of $2300 a month plus a commission of 4% on the houses he sells after he has sold $1,000,000 of housing. What did he earn the month he sold $1,200,000 of housing?

Example 4

Commission

A saleswoman earns a salary of $1200 a month plus a commission of 8% on whatever she sells after she has sold $8000 in furniture. What did she earn the month she sold $25,000 worth of furniture?

Solution

First subtract $8000 from $25,000 to find the amount on which the commission is based.

$$\$25,000 - \$8000 = \$17,000$$

Now find the amount of the commission.

$$R \quad \cdot \quad B \quad = \quad A$$

$$0.08 \quad \cdot \quad 17{,}000 \quad = \quad 1360 \quad \text{commission}$$

Now add the commission to her salary to find what she earned that month.

$$\$1200 + \$1360 = \$2560 \qquad \text{earned for the month}$$

Now work margin exercise 4.

Problems involving percent come in a variety of forms. The next example illustrates another situation you may encounter in your own life.

Example 5

Commission

The Berrys sold their house. After paying the real estate agent a commission of 6% of the selling price and then paying $1486 in other costs and $90,000 on the mortgage, they received $49,514. What was the selling price of the house?

Solution

Use the relationship:

$$\text{selling price} \quad - \quad \text{cost} = \text{profit.}$$
$$S \qquad - \quad C = \quad P$$

Let S = selling price.

selling price − cost = profit

$$S - (0.06S + 1486 + 90{,}000) = 49{,}514$$
$$S - 0.06S - 1486 - 90{,}000 = 49{,}514$$
$$0.94S = 141{,}000$$
$$S = 150{,}000$$

Check:

$150{,}000$	selling price		9000	commission		$150{,}000$	selling price
× 0.06	commission %		1486	other costs		−$100{,}486$	cost
9000	commission		+ $90{,}000$	mortgage		$49{,}514$	profit
			$100{,}486$	cost			

The selling price was $150,000.

Percent of profit for an investment is the ratio (a fraction) that compares the money made to the money invested. If you make two investments of different amounts of money, then the amount of money you make on each investment is not a fair comparison. In comparing such investments, the investment with the greater percent of profit is considered the better investment. **To find the percent of profit, form the fraction (or ratio) of profit divided by the investment and change the fraction to a percent.**

Example 6

Percent of Profit

Calculate the percent of profit for both **a.** and **b.** and tell which is the better investment.

a. $300 profit on an investment of $2400 or
b. $500 profit on an investment of $5000

Solution

Set up ratios and find the corresponding percents.

a.

$$\frac{\$300 \text{ profit}}{\$2400 \text{ invested}} = \frac{300 \cdot 1}{300 \cdot 8} = \frac{1}{8} = 0.125 = 12.5\% \qquad \text{percent of profit}$$

b.

$$\frac{\$500 \text{ profit}}{\$5000 \text{ invested}} = \frac{500 \cdot 1}{500 \cdot 10} = \frac{1}{10} = 0.1 = 10\% \qquad \text{percent of profit}$$

Clearly, $500 is more than $300, but 12.5% is greater than 10% and investment **a.** is the better investment.

Now work margin exercises 5 and 6.

---

5. The Rotbergs sold the delivery truck from their family cookie business on eBay. After paying eBay a commission of 5% of the selling price and then paying $300 in other costs, they received $7965. What was the selling price of the cookie truck?

6. Calculate the percent of profit for both **a.** and **b.** and tell which is the better investment.

a. $900 profit on an investment of $2700

b. $400 profit on an investment of $1100

Practice Problems

1. Write 0.3 as a percent.

2. Write 6.4% as a decimal.

3. Write $\dfrac{1}{5}$ as a percent.

4. Find 12% of 200.

5. Find the percent of profit if $480 is made on an investment of $1500.

Practice Problem Answers

1. 30% **2.** 0.064 **3.** 20% **4.** 24 **5.** 32%

Exercises 5.7

Write each of the following numbers in percent form.

1. 0.91   **2.** 0.625   **3.** 1.37   **4.** 0.0075

5. $\dfrac{5}{8}$ **6.** $\dfrac{87}{100}$ **7.** $\dfrac{3}{4}$ **8.** $1\dfrac{1}{2}$

Write each of the following percents in decimal form.

9. 69%   **10.** 82%   **11.** 162%   **12.** 235%

13. 7.5%   **14.** 11.3%   **15.** 0.5%   **16.** 31.4%

Write each of the following percents in fraction form (reduced).

17. 35%   **18.** 72%   **19.** 130%   **20.** 40%

Solve the following word problems.

21. **Budgeting:** Heather's monthly income is $2500. Using the given circle graph answer the following questions.

     a. What percent of her income does Heather spend on rent each month?

     b. What percent of her income is spent on food and entertainment each month?

     c. What percent of her income does Heather save each month?

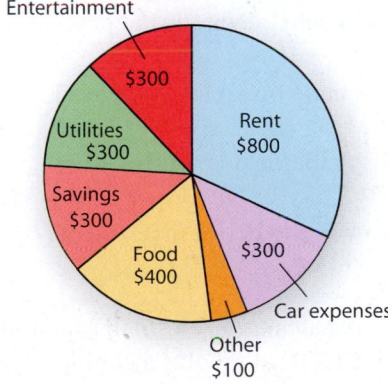

Monthly Expenses

22. **Earthquakes:** There were 4257 registered earthquakes in the U.S. in 2009. Using the given graph answer the following questions. (Round answers to the nearest tenth of a percent.)

     a. What percent of U.S. earthquakes in 2009 were less than a magnitude of 3.0?

     b. What percent of earthquakes were of magnitude 3.0 – 3.9?

     c. What percent of earthquakes were 3.0 and above?

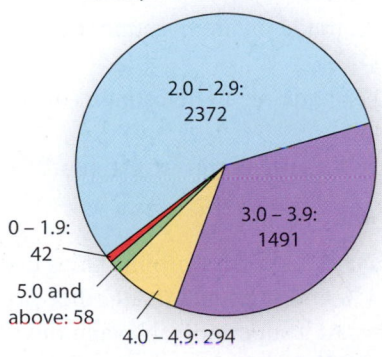

Magnitude of U.S. Earthquakes in 2009
Source: Data from the U.S. National Earthquake Information Center

23. Apple's Revenue: In 2009, Apple's revenue was $36 billion. (Round answers to the nearest tenth of a percent.)

   a. What percent of Apple's revenue came from computer sales?

   b. What percent of Apple's revenue came from mobile phones?

   c. What percent of Apple's revenue did not come from music related products?

24. Music Sales: In July, Daddy Fat's Record Shop sold 3900 CDs.

   a. What percent of their sales came from classical CDs?

   b. What percent of their sales came from R&B and rap CDs?

   c. What percent of their sales was neither rock nor country CDs?

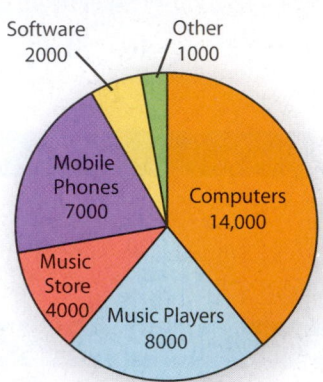

Apple's Revenue in 2009
(Data is in millions of dollars)
Source: http://tech.fortune.cnn.com

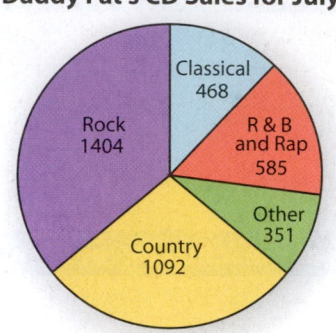

Daddy Fat's CD Sales for July

Find the missing rate, base, or amount. See Example 1.

25. 81% of 76 is _____ .

26. 102% of 87 is _____ .

27. _____% of 150 is 60.

28. _____% of 160 is 240.

29. 3% of _____ is 65.4.

30. 20% of _____ is 45.

31. What percent of 32 is 40?

32. 1250 is 50% of what number?

33. 100 is 125% of what number?

34. Find 24% of 244.

Solve the following word problems. See Example 2 through 5.

35. Percent of Profit What is the percent of profit if
   a. $400 is made on an investment of $5000?
   b. $350 is made on an investment of $3500?
   c. Which was the better investment?

36. Percent of Profit What is the percent of profit if
   a. $400 is made on an investment of $2000?
   b. $510 is made on an investment of $3000?
   c. Which was the better investment?

37. Sales Tax: In North Carolina sales tax on food is 4% and sales tax on all other products is 5.75%. At a superstore, Nevaeh buys $72 of food and $51 of other products. What will be the total cost with tax?

38. Salary: A calculator salesman's monthly income is $1500 salary plus a commission of 7% of his sales over $5000. What was his income the month that he sold $11,470 in calculators?

39. Salary Bonus: A computer programmer was told that he would be given a bonus of 5% of any money his programs could save the company. How much would he have to save the company to earn a bonus of $1000?

40. Property Tax: The property tax on a home was $2550. What was the tax rate if the home was valued at $170,000?

41. Basketball: In one season a basketball player missed 14% of her free throws. How many free throws did she make if she attempted 150 free throws?

42. Test Taking: A student missed 6 problems on a statistics test and received a grade of 85%. If all the problems were of equal value, how many problems were on the test?

43. Buying Textbooks: If sales tax is figured at 7.25%, how much tax will be added to the total purchase price of three textbooks, priced at $35.00, $55.00, and $70.00? What will be the total amount paid for the books?

44. Buying a Car: A car dealer bought a used car for $2500. He marked up the price so that he would make a profit of 25% of his cost.
   a. What was the selling price?
   b. If the customer paid 8% of the selling price in taxes and fees, what was the total cost of the car to the customer? The car was 6 years old.

45. Shopping: At a department store sale, sheets and pillowcases were discounted 25%. Sheets were originally marked $30.00 and pillowcases were originally marked $12.00.
   a. What was the sale price of the sheets?
   b. What was the sale price of the pillowcases?

46. Golfing: The Golf Pro Shop had a set of golf clubs that were marked on sale for $560. This was a discount of 20% off the original selling price.
   a. What was the original selling price?
   b. If the clubs cost the shop $420, what was the percent of profit based on cost?

47. Losing Weight: A man weighed 200 pounds. He lost 20 pounds in 3 months. Then he gained back 20 pounds 2 months later.
   a. What percent of his weight did he lose in the first 3 months?
   b. What percent of his weight did he gain back?
   c. The loss and the gain are the same amount, but the two percents are different. Explain.

48. Technology: Data cartridges for computer backup were on sale for $27.00 (a package of two cartridges). This price was a discount of 10% off the original price.
   a. Was the original price more than $27 or less than $27?
   b. What was the original price?

49. Farming: A citrus farmer figures that his fruit costs 96¢ a pound to grow. If he lost 20% of the crop he produced due to a frost and he sold the remaining 80% at $1.80 a pound, how many pounds did he produce to make a profit of $30,000?

50. Buying a Motorcycle: The dealer's discount to the buyer of a new motorcycle was $499.40. The manufacturer gave a rebate of $1000 on this model.
   a. What was the original price if the dealer discount was 7% of the original price?
   b. What would a customer pay for the motorcycle if taxes were 5% of the final selling price (after the rebate) and license fees were $642.60?

51. Farming: Farmer McGregor raises strawberries. They cost him $0.80 a basket to produce. He is able to sell only 85% of those he produces. If he sells his strawberries at $2.40 a basket, how many baskets must he produce to make a profit of $2480?

Writing & Thinking

52. A man and his wife wanted to sell their house and contacted a realtor. The realtor said that the price for the services (listing, advertising, selling, and paperwork) was 6% of the selling price. The realtor asked the couple how much cash they wanted after the realtor fee was deducted from the selling price. They said $141,000 and that therefore, the asking price should be 106% of $141,000 or a total of $149,460. The realtor said that this was the wrong figure and that the percent used should be 94% and not 106%.

   a. Explain why 106% is the wrong percent and $149,460 is not the right selling price.

   b. Explain why and how 94% is the correct percent and just what the selling price should be for the couple to receive $141,000 after the realtor's fee is deducted.

Collaborative Learning

With the class separated into teams of two to four students, each team is to analyze the following problem and decide how to answer the related questions. Then each team leader is to present the team's answers and related ideas to the class for general discussion.

53. **Business:** Sam works at a picture framing store and gets a salary of $500 per month plus a commission of 3% on sales he makes over $2000. Maria works at the same store, but she has decided to work on a straight commission of 8% of her sales.

   a. At what amount of sales will Sam and Maria make the same amount of money?

   b. Up to that point, who would be making more?

   c. After that point, who would be making more? Explain briefly. (If you were offered a job at that store, which method of payment would you choose?)

5.8 Working with Formulas

Objective A **Formulas**

Formulas are general rules or principles stated mathematically. There are many formulas in such fields of study as business, economics, medicine, physics, and chemistry as well as mathematics. Some of these formulas and their meanings are shown here.

> ## notes
>
> ■ Be sure to use the letters just as they are given in the formulas. In mathematics, there is little or no flexibility between capital and small
> ■ letters as they are used in formulas. In general, capital letters have special meanings that are different from corresponding small letters.
> ■ For example, capital *A* may mean the area of a triangle and small *a* may mean the length of one side, two completely different ideas.
> ■

Listed here are a few formulas with a description of the meaning of each formula. There are of course many more formulas.

Formula	Meaning
$I = Prt$	The **simple interest** I earned by investing money is equal to the product of the principal P times the rate of interest r times the time t in one year or part of a year.
$C = \dfrac{5}{9}(F - 32)$	**Temperature** in degrees Celsius C equals $\dfrac{5}{9}$ times the difference between the Fahrenheit temperature F and 32.
$d = rt$	The **distance traveled** d equals the product of the rate of speed r and the time t.
$P = 2l + 2w$	The **perimeter, P, of a rectangle** is equal to twice the length l plus twice the width w.
$L = 2\pi rh$	The **lateral surface area** L (top and bottom not included) of a cylinder is equal to 2π times the radius r of the base times the height h.
$F = ma$	In physics, the **force** F acting on an object is equal to its mass m times its acceleration a.
$\alpha + \beta + \gamma = 180°$	The **sum of the angles** (α, β, and γ) **of a triangle** is $180°$. **Note:** α, β, and γ are the Greek lowercase letters alpha, beta, and gamma, respectively.
$P = a + b + c$	The **perimeter, P, of a triangle** is equal to the sum of the lengths of the three sides a, b, and c.

Table 1

Objectives

A   Understand the concept of a formula, and learn several common formulas.

B   Evaluate formulas for given values of the variables.

C   Solve formulas for specified variables in terms of the other variables.

If you know values for all but one variable in a formula, you can substitute those values and find the value of the unknown variable using the techniques for solving equations discussed earlier. This section presents a variety of formulas from real-life situations, with complete descriptions of the meanings of these formulas. Working with these formulas will help you become familiar with a wide range of applications of algebra and provide practice in solving equations.

Example 1

Evaluating Formulas

a. A **note** is a loan for a period of 1 year or less, and the interest earned (or paid) is called **simple interest**. A note involves only one payment at the end of the term of the note and includes both principal and interest. The formula for calculating simple interest is:

$I = Prt$

where

I	=	**interest** (earned or paid)
P	=	**principal** (the amount invested or borrowed)
r	=	**rate** of interest (stated as an annual or yearly rate)
t	=	**time** (one year or part of a year)

Note: The rate of interest is usually given in percent form and converted to decimal or fraction form for calculations.

Maribel loaned $5000 to a friend for 3 months at an annual interest rate of 8%. How much will her friend pay her at the end of the 3 months?

Solution

Here,

$P = \$5000,$

$r = 8\% = 0.08,$ and

$t = 3 \text{ months} = \dfrac{3}{12} \text{ year} = \dfrac{1}{4} \text{ year}.$

Find the interest by substituting in the formula $I = Prt$ and evaluating.

$$I = 5000 \cdot \overset{0.02}{\cancel{0.08}} \cdot \frac{1}{\cancel{4}}$$

$$= 5000 \cdot 0.02$$

$$= \$100.00$$

The interest is $100 and the amount to be paid at the end of 3 months is principal + interest = $5000 + $100 = $5100.

b. Given the formula $C = \dfrac{5}{9}(F - 32)$, first find C if $F = 212°$ (212 degrees Fahrenheit) and then find F if $C = 20°$.

Solution

$F = 212°$, so substitute 212 for F in the formula.

$$C = \frac{5}{9}(212 - 32)$$

$$= \frac{5}{9}(180)$$

$$= 100$$

That is, 212°F is the same as 100°C. Water will boil at 212°F at sea level. This means that if the temperature is measured in degrees Celsius instead of degrees Fahrenheit, water will boil at 100°C at sea level.

$C = 20°$, so substitute 20 for C in the formula.

$$20 = \frac{5}{9}(F - 32) \qquad \text{Now solve for } F.$$

$$\frac{9}{5} \cdot 20 = \frac{9}{5} \cdot \frac{5}{9}(F - 32) \qquad \text{Multiply both sides by } \frac{9}{5}.$$

$$36 = F - 32 \qquad \text{Simplify.}$$

$$36 + 32 = F - 32 + 32 \qquad \text{Add 32 to both sides.}$$

$$68 = F \qquad \text{Simplify.}$$

That is, a temperature of 20°C is the same as a comfortable spring day temperature of 68°F.

c. The lifting force F exerted on an airplane wing is found by multiplying some constant k by the area A of the wing's surface and by the square of the plane's velocity v. The formula is $F = kAv^2$. Find the force on a plane's wing during take off if the area of the wing is 120 ft², k is $\dfrac{4}{3}$, and the plane is traveling 80 miles per hour during take off.

Solution

We know that $k = \dfrac{4}{3}$, $A = 120$, and $v = 80$.

1. a. You get a loan for $2000 for 2 months at an annual interest rate of 6%. How much do you have to pay at the end of the loan period, 2 months?

b. Given $C = \dfrac{5}{9}(F - 32)$, find F if $C = 50°$.

c. Given $F = kAv^2$ find the force exerted if $k = \dfrac{5}{4}$, $A = 100 \text{ ft}^2$, and $v = 10$ mph.

Substitution gives

$$F = \dfrac{4}{3} \cdot 120 \cdot 80^2$$
$$= \dfrac{4}{3} \cdot 120 \cdot 6400$$
$$= 160 \cdot 6400$$
$$F = 1{,}024{,}000 \text{ lb}$$

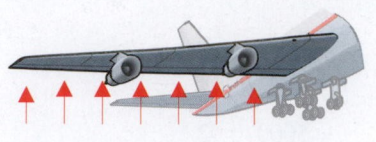

The force is measured in pounds.

d. The perimeter of a triangle is 38 feet. One side is 5 feet long and a second side is 18 feet long. How long is the third side?

Solution

Using the formula $P = a + b + c$, substitute $P = 38$, $a = 5$, and $b = 18$. Then solve for the third side.

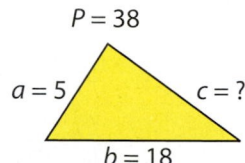

$P = 38$

$a = 5$ $c = ?$

$b = 18$

$$P = a + b + c$$
$$38 = 5 + 18 + c$$
$$38 = 23 + c$$
$$38 - 23 = 23 + c - 23$$
$$15 = c$$

The third side is 15 feet long.

Now work margin exercise 1.

Objective C **Solving Formulas for Different Variables**

We say that the formula $d = rt$ is "solved for" d in terms of r and t. Similarly, the formula $A = \dfrac{1}{2}bh$ is solved for A in terms of b and h, and the formula $P = R - C$ (profit is equal to revenue minus cost) is solved for P in terms of R and C. Many times we want to use a certain formula in another form. We want the formula "solved for" some variable other than the one given in terms of the remaining variables. **Treat the variables just as you would constants in solving linear equations.** Study the following examples carefully.

Example 2

Solving for Different Variables

a. Given $d = rt$, solve for t in terms of d and r. We want to represent the time in terms of distance and rate. We will use this concept later in word problems.

Solution

$$d = rt$$ Treat r and d as if they were constants.

$$\dfrac{d}{r} = \dfrac{rt}{r}$$ Divide both sides by r.

$$\dfrac{d}{r} = t$$ Simplify.

b. Given $V = \dfrac{k}{P}$, solve for P in terms of V and k.

Solution

$$V = \frac{k}{P}$$

$$P \cdot V = P \cdot \frac{k}{P} \qquad \text{Multiply both sides by } P.$$

$$PV = k \qquad \text{Simplify.}$$

$$\frac{PV}{V} = \frac{k}{V} \qquad \text{Divide both sides by } V.$$

$$P = \frac{k}{V} \qquad \text{Simplify.}$$

c. Given $C = \dfrac{5}{9}(F - 32)$ as in Example **1b.**, solve for F in terms of C. This would give a formula for finding Fahrenheit temperature given a Celsius temperature value.

Solution

$$C = \frac{5}{9}(F - 32) \qquad \text{Treat } C \text{ as a constant.}$$

$$\frac{9}{5} \cdot C = \frac{9}{5} \cdot \frac{5}{9}(F - 32) \qquad \text{Multiply both sides by } \frac{9}{5}.$$

$$\frac{9}{5}C = F - 32 \qquad \text{Simplify.}$$

$$\frac{9}{5}C + 32 = F - 32 + 32 \qquad \text{Add 32 to both sides.}$$

$$\frac{9}{5}C + 32 = F \qquad \text{Simplify.}$$

Thus $F = \dfrac{9}{5}C + 32$ is solved for F, and $C = \dfrac{5}{9}(F - 32)$ is solved for C.

These are two forms of the same formula.

d. Given the equation $2x + 4y = 10$,
  1. solve first for x in terms of y, and then
  2. solve for y in terms of x.

This equation is typical of the algebraic equations that we will discuss in later sections.

2. a. Given $F = kAv^2$ solve for k in terms of F, A and v.

b. Given the equation $16y + 25z = 400$,

  1. solve for y in terms of z, and then

  2. solve for z in terms of y.

c. Given $4x + 8y + 12z = 16$, solve for x.

Solution

1. Solving for x yields the following.

$$2x + 4y = 10 \qquad \text{Treat } 4y \text{ as a constant.}$$
$$2x + 4y - 4y = 10 - 4y \qquad \text{Subtract } 4y \text{ from both sides.}$$
$$\qquad\qquad\qquad\qquad \text{(This is the same as adding } -4y.\text{)}$$
$$2x = 10 - 4y \qquad \text{Simplify.}$$
$$\frac{2x}{2} = \frac{10 - 4y}{2} \qquad \text{Divide both sides by 2.}$$
$$x = \frac{10}{2} - \frac{4y}{2}$$
$$x = 5 - 2y \qquad \text{Simplify.}$$

2. Solving for y yields the following.

$$2x + 4y = 10 \qquad \text{Treat } 2x \text{ as a constant.}$$
$$2x + 4y - 2x = 10 - 2x \qquad \text{Subtract } 2x \text{ from both sides.}$$
$$4y = 10 - 2x \qquad \text{Simplify.}$$
$$\frac{4y}{4} = \frac{10 - 2x}{4} \qquad \text{Divide both sides by 4.}$$
$$y = \frac{10}{4} - \frac{2x}{4}$$
$$y = \frac{5}{2} - \frac{x}{2} \qquad \text{Simplify.}$$

or we can write

$$y = \frac{5 - x}{2} \quad \text{or} \quad y = -\frac{1}{2}x + \frac{5}{2}. \qquad \text{Both forms are correct.}$$

e. Given $3x - y = 15$, solve for y in terms of x.

Solution

Solving for y gives

$$3x - y = 15$$
$$3x - y - 3x = 15 - 3x \qquad \text{Subtract } 3x \text{ from both sides.}$$
$$-y = 15 - 3x$$
$$-1(-y) = -1(15 - 3x) \qquad \text{Multiply both sides by } -1 \text{ (or divide both sides by } -1\text{).}$$
$$y = -15 + 3x \qquad \text{Simplify using the distributive property.}$$
$$\text{or} \qquad y = 3x - 15$$

Now work margin exercise 2.

Practice Problems

1. $2x - y = 5$; solve for y.
2. $2x - y = 5$; solve for x.
3. $A = \dfrac{1}{2}bh$; solve for h.
4. $L = 2\pi rh$; solve for r.
5. $P = 2l + 2w$; solve for w.

Practice Problem Answers

1. $y = 2x - 5$
2. $x = \dfrac{y+5}{2}$
3. $h = \dfrac{2A}{b}$
4. $r = \dfrac{L}{2\pi h}$
5. $w = \dfrac{P - 2l}{2}$

Exercises 5.8

Refer to Example 1a for information concerning simple interest and the related formula $I = Prt$.

Simple Interest

1. You want to borrow $4000 at 12% for 3 months. How much interest would you pay?

2. For how many years must you leave $1000 in a savings account at 5.5% interest to earn $220 in interest?

3. What principal would you need to invest to earn $450 in simple interest in 6 months if the interest rate was 9%?

4. After 1 month, Gustav received $25 in simple interest on his savings account of $12,000. What was the interest rate?

5. A savings account of $3500 is left for 9 months and draws simple interest at a rate of 7%.
   a. How much interest is earned?
   b. What is the balance in the account at the end of the 9 months?

6. Tim just deposited $2562.50 to pay off a 3 month loan of $2500.
   a. How much of what he deposited was interest on the loan?
   b. What rate of interest was he charged?

In the following application problems, read the descriptive information carefully and then substitute the values given in the problem for the corresponding variables in the formulas. Evaluate the resulting expression for the unknown variable. See Example 1

Velocity

If an object is shot upward with an initial velocity v_0 in feet per second, the velocity v in feet per second is given by the formula $v = v_0 - 32t$, where t is time in seconds. (v_0 is read "v sub zero." The 0 is called a **subscript**.)

7. An object projected upward with an initial velocity of 106 feet per second has a velocity of 42 feet per second. How many seconds have passed?

8. Find the initial velocity of an object if the velocity after 4 seconds is 48 feet per second.

Medicine

In nursing, one procedure for determining the dosage for a child is

$$\text{child's dosage} = \frac{\text{age of child in years}}{\text{age of child} + 12} \cdot \text{adult dosage.}$$

9. If the adult dosage of a drug is 20 milliliters, how much should a 3-year-old child receive?

10. If the adult dosage of a drug is 340 milligrams, how much should a 5-year-old child receive?

Investment

The total amount of money in an account with P dollars invested in it is given by the formula $A = P + Prt$, where r is the rate expressed as a decimal and t is time (one year or part of a year).

11. If $1000 is invested at 6% interest, find the total amount in the account after 6 months.

12. How long will it take an investment of $600, at an annual rate of 5%, to be worth $615?

Carpentry

The number N of rafters in a roof or studs in a wall can be found by the formula $N = \dfrac{L}{d} + 1$, where L is the length of the roof or wall and d is the center-to-center distance from one rafter or stud to the next. Note that L and d must be in the same units.

13. How many rafters will be needed to build a roof 26 ft long if they are placed 2 ft on center?

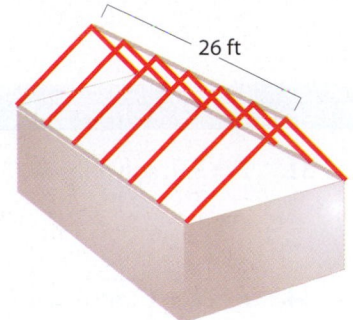

26 ft

14. A wall has studs placed 16 in. on center. If the wall is 20 ft long, how many studs are in the wall?

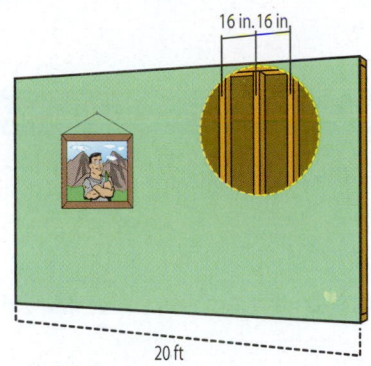

16 in. 16 in.

20 ft

15. How long is a wall if it requires 22 studs placed 16 in. on center?

16. What should the center-to-center distance be if you are building a 33 ft long roof using 12 rafters?

Cost

The total cost C of producing x items can be found by the formula $C = ax + k$, where a is the cost per item and k is the fixed costs (rent, utilities, and so on).

17. Find the total cost of producing 30 items if each costs $15 and the fixed costs are $580.

18. The total cost to produce 80 dolls is $1097.50. If each doll costs $9.50 to produce, find the fixed costs.

19. It costs a company $3.60 to produce a calculator. Last week the total costs were $1308. If the fixed costs are $480 weekly, how many calculators were produced last week?

20. Each week an electronics company builds 60 mp3 players for a total cost of $5340. If the fixed costs for a week are $750, what is the cost to produce each mp3 player?

Profit

The profit P is given by the formula $P = R - C$, where R is the revenue and C is the cost.

21. Find the revenue (income) of a company that shows a profit of $3.2 million and costs of $1.8 million.

22. Find the revenue of a company that shows a profit of $3.2 million and costs of $5.7 million.

Depreciation

Many items decrease in value as time passes. This decrease in value is called **depreciation**. One type of depreciation is called linear depreciation. The value V of an item after t years is given by $V = C - Crt$, where C is the original cost and r is the annual rate of depreciation expressed as a decimal.

23. If you buy a car for $6000 and it depreciates linearly at a rate of 10% per year, what will be its value after 6 years?

24. A contractor buys a 4 year-old piece of heavy equipment valued at $20,000. If the original cost of this equipment was $25,000, find the annual rate of depreciation.

Distance, Rate, and Time

The distance traveled d is given by the formula $d = rt$, where r is the rate of speed and t is the time it takes.

25. How long will a truck driver take to travel 350 miles if his mean speed is 50 mph?

26. What is the mean rate of speed of a biker who bikes 21.92 miles in 68.5 min?

27. What is Jonathan's mean rate of speed if he hikes 10.4 miles in 6.4 hours?

28. How long will it take a train traveling at 40 mph to go 140 miles?

Solve each formula for the indicated variable.

29. $P = a + b + c$; solve for b.

30. $P = 3s$; solve for s.

31. $F = ma$; solve for m.

32. $C = \pi d$; solve for d.

33. $A = lw$; solve for w.

34. $P = R - C$; solve for C.

35. $R = np$; solve for n.

36. $v = k + gt$; solve for k.

37. $I = A - P$; solve for P.

38. $L = 2\pi rh$; solve for h.

39. $A = \dfrac{m+n}{2}$; solve for m.

40. $P = a + 2b$; solve for a.

41. $I = Prt$; solve for t.

42. $R = \dfrac{E}{I}$; solve for E.

43. $P = a + 2b$; solve for b.

44. $c^2 = a^2 + b^2$; solve for b^2.

45. $\alpha + \beta + \gamma = 180°$; solve for β.

46. $y = mx + b$; solve for x.

47. $V = lwh$; solve for h.

48. $v = -gt + v_0$; solve for t.

49. $A = \dfrac{1}{2}bh$; solve for b.

50. $R = \dfrac{E}{I}$; solve for I.

51. $A = \pi r^2$; solve for π.

52. $A = \dfrac{R}{2L}$; solve for L.

53. $K = \dfrac{mv^2}{2g}$; solve for g.

54. $x + 4y = 4$; solve for y.

55. $2x + 3y = 6$; solve for y.

56. $3x - y = 14$; solve for y.

57. $5x + 2y = 11$; solve for x.

58. $-2x + 2y = 5$; solve for x.

59. $A = \frac{1}{2}h(b+c)$; solve for b. **60.** $A = P(1+rt)$; solve for r. **61.** $R = \frac{3(x-12)}{8}$ solve for x.

62. $-2x - 5 = -3(x+y)$; solve for x.

63. $3y - 2 = x + 4y + 10$; solve for y.

64. $V = \frac{1}{3}\pi r^2 h$; solve for h.

Make up a formula for each of the following situations.

65. Concert Tickets: Each ticket for a concert costs $\$t$ per person and parking costs $\$9.00$. What is the total cost per car C if there are n people in a car?

66. Car Rentals: ABC Car Rental charges $\$25$ per day plus $\$0.12$ per mile. What would you pay per day for renting a car from ABC if you were to drive the car x miles in one day?

67. Computers: Top-of-the-Line computer company knows that the cost (labor and materials) of producing a computer is $\$325$ per computer per week and the fixed overhead costs (lighting, rent, etc.) are $\$5400$ per week. What are the company's weekly costs of producing n computers per week?

68. Computers: If the Top-of-the-Line computer company (see Exercise 67) sells its computers for $\$683$ each, what is its profit per week if it sells the same number n that it produces? (Remember that profit is equal to revenue minus costs, or $P = R - C$.)

Writing & Thinking

69. The formula $z = \dfrac{x - \bar{x}}{s}$ is used extensively in statistics. In this formula, x represents one of a set of numbers, $\bar{x}$ represents the mean of those numbers in the set, and s represents a value called the standard deviation of the numbers. (The standard deviation is a positive number and is a measure of how "spread out" the numbers are.) The values for z are called z-scores, and they measure the number of standard deviation units a number x is from the mean $\bar{x}$.

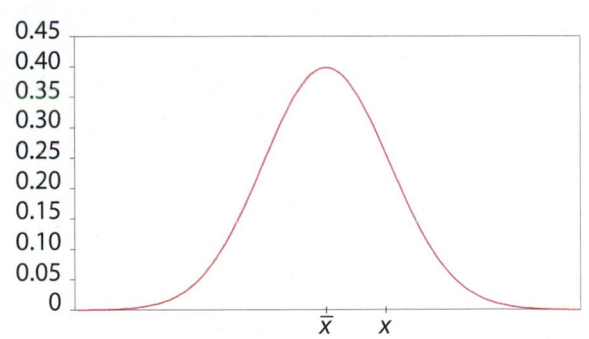

a. If $\bar{x} = 70$, what will be the z-score for $x = 70$? Does this z-score depend on the value of s? Explain.

b. If $\bar{x} = 70$, for what values of x will the corresponding z-scores be negative?

c. Calculate your z-score on each of the last two test scores in this class. (Your instructor will give you the mean and standard deviation for each test.) What do these scores tell you about your performance on the two exams?

70. Suppose that, for a particular set of exam scores, $\bar{x} = 72$ and $s = 6$. Find the z-score that corresponds to a score of
  a. 78 b. 66 c. 81 d. 60

5.9 Applications: Distance-Rate-Time, Interest, Mean, Cost

Objective A   Distance-Rate-Time

Word problems (or applications) are designed to teach you to read carefully, to organize, and to think clearly. Whether or not a particular problem is easy for you depends a great deal on your personal experiences and general reasoning abilities. The problems generally do not give specific directions to add, subtract, multiply, or divide. You must decide what relationships are indicated through careful analysis of the problem.

Problems involving distance usually make use of the relationship indicated by the formula $d = rt$, where d = distance, r = rate, and t = time. A chart or table showing the known and unknown values is quite helpful and is illustrated in the next example.

Example 1

Distance-Rate-Time

A motorist has a mean of 45 mph for the first part of a trip and 54 mph for the last part of the trip. If the total trip of 303 miles took 6 hours, what was the time for each part?

Solution

Analysis of strategy:

What is being asked for?

Total time minus time for 1st part of trip gives time for 2nd part of trip.

Let t = time for 1st part of trip and
$6 - t$ = time for 2nd part of trip.

	rate	·	time	=	distance
1st Part	45		t		$45t$
2nd Part	54		$6 - t$		$54(6 - t)$

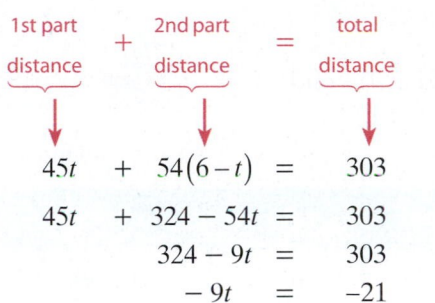

45 mph   54 mph

t hr $6 - t$ hr

303 miles

1st part distance	+	2nd part distance	=	total distance	Form an equation relating the given information.

$$45t + 54(6 - t) = 303 \qquad \text{Solve the equation.}$$

$$45t + 324 - 54t = 303$$

$$324 - 9t = 303$$

$$-9t = -21$$

$$t = \frac{-21}{-9} = \frac{7}{3} \text{ hr} \qquad \text{1st part of the trip}$$

$$6 - t = 6 - \frac{7}{3} = \frac{11}{3} \text{ hr} \qquad \text{2nd part of the trip}$$

Check:

$$45 \cdot \frac{7}{3} = 15 \cdot 7 = 105 \text{ miles (1st part)}$$

$$54 \cdot \frac{11}{3} = 18 \cdot 11 = 198 \text{ miles (2nd part)}$$

$$105 + 198 = 303 \text{ miles in total}$$

The first part took $\frac{7}{3}$ hr or $2\frac{1}{3}$ hr.

The second part took $\frac{11}{3}$ hr or $3\frac{2}{3}$ hr.

Now work margin exercise 1.

1. What is the time for each part of a trip if the speed of the first part is 60 mph and the speed of the second part is 45 mph and the total distance of 200 miles is covered in 4 hours?

Objective B Interest

To solve problems related to interest on money invested for one year, you need to know the basic relationship among the principal P (amount invested), the annual rate of interest r, and the amount of interest I (money earned). This relationship is described in the formula $P \cdot r = I$. (This is the formula for simple interest, $I = Prt$, with $t = 1$.) We use this relationship in Example 2.

2. Mark has had $15,000 invested for one year, some in a low-risk stock yielding 4% and the rest in a high-risk stock yielding 15%. If his investment has earned $1975 in interest, how much is in the low-risk stock and how much is in the high-risk stock?

Example 2

Interest

Kara has had $40,000 invested for one year, some in a savings account which paid 7% and the rest in a high-risk stock which yielded 12% for the year. If her interest income last year was $3550, how much did she have in the savings account and how much did she invest in the stock?

Solution

Let $x =$ amount invested at 7% and
$40,000 - x =$ amount invested at 12%.

Total amount invested minus amount invested at 7% represents amount invested at 12%.

	principal $\cdot$	rate $=$	interest
Savings Account	x	0.07	$0.07(x)$
Stock	$40,000 - x$	0.12	$0.12(40,000 - x)$

interest at 7% $+$ interest at 12% $=$ total interest

$$0.07(x) + 0.12(40,000 - x) = 3550$$

Multiply both sides of the equation by 100 to eliminate the decimal.

$$7x + 12(40,000 - x) = 355,000$$
$$7x + 480,000 - 12x = 355,000$$
$$-5x = -125,000$$
$$x = 25,000 \quad \text{amount invested at 7\%}$$
$$40,000 - x = 15,000 \quad \text{amount invested at 12\%}$$

Check:

$25,000(0.07) = 1750$ and $15,000(0.12) = 1800$
and $1750 + $1800 = $3550.

Kara had $25,000 in the savings account at 7% interest and invested $15,000 in the stock at 12% interest.

Now work margin exercise 2.

The **mean** of a set of numbers was defined and discussed earlier in the book. In this section we use the concept of mean to find unknown numbers and to read and understand graphs in more detail.

Example 3

Mean

Suppose that you have scores of 85, 92, 82 and 88 on four exams in your English class. What score will you need on the fifth exam to have a mean of 90?

Solution

Let x = your score on the fifth exam.

The sum of all the scores, including the unknown fifth exam, divided by 5 must equal 90.

$$\frac{85 + 92 + 82 + 88 + x}{5} = 90$$

$$\frac{347 + x}{5} = 90$$

$$5 \cdot \frac{347 + x}{5} = 5 \cdot 90$$

$$347 + x = 450$$

$$x = 103$$

Assuming that each exam is worth 100 points, you cannot attain an mean of 90 on the five exams.

Now work margin exercise 3.

3. Suppose on four of your five biology tests you scored 90, 95, 97, and 90 respectively. You can exempt the final if your mean on the 5 tests is greater than or equal to an 85. What do you need to make on your last test to have a mean of 85?

4. Using the graph in Example 4, find the difference in enrollment between the largest university and the smallest university.

Example 4

Bar Graphs

The bar graph shows the enrollment at the main campuses at six Big Ten universities. Use the graph to find the following (note that the units on the graph are in thousands):

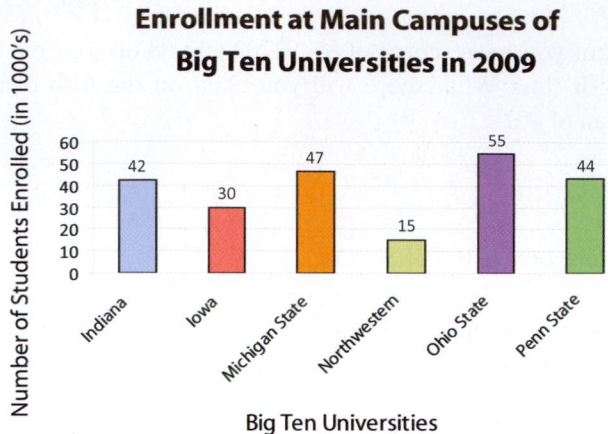

a. Find the mean enrollment over the six schools. (Round to the nearest thousand.)

Solution

Find the sum: $42 + 30 + 47 + 15 + 55 + 44 = 233$.
Divide by 6: $233 \div 6 \approx 39$.

The mean enrollment is approximately 39,000 students.

b. Which university has the lowest enrollment?

Solution

Northwestern has the lowest enrollment: 15,000 students.

c. Find the difference in enrollment between Ohio State and Penn State.

Solution

The difference is $55,000 - 44,000 = 11,000$ students.

Now work margin exercise 4.

Example 5

A jeweler paid $350 for a ring. He wants to price the ring for sale so that he can give a 30% discount on the marked selling price and still make a profit of 20% on his cost. What should be the marked selling price of the ring?

Solution

Again, we make use of the relationship:

$$S - C = P \text{ (selling price – cost = profit).}$$

Let x = marked selling price then
$x - 0.30x$ = actual selling price, and
        350 = cost.

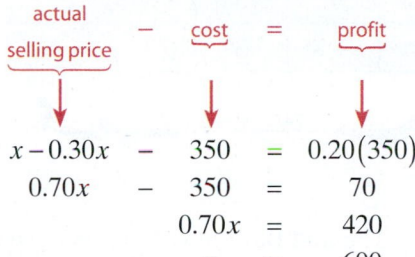

$$
\begin{array}{ccccc}
x - 0.30x & - & 350 & = & 0.20(350) \\
0.70x & - & 350 & = & 70 \\
 & & 0.70x & = & 420 \\
 & & x & = & 600
\end{array}
$$

The profit is 20% of what he paid originally.

Check:

Step 1:

  600 marked selling price
$\times\ 0.30$ discount %
  180 discount

Step 2:

  600 marked selling price
$-\ \$180$ discount
  420 actual selling price

Step 3:

  420 actual selling price
$-\ \$350$ cost
  70 profit

As a double check, 350 cost
               $\times\ 0.20$ profit %
                  70 profit.

The jeweler should set the selling price at $600.

Now work margin exercise 5.

5. A used car salesman paid $1500 for a car. He plans to sell the car at a 25% discount on marked selling price, but still wants to make a profit of 15% on his cost. What should be the maked selling price of the car?

Exercises 5.9

Solve each of the following application problems.

1. **Hiking:** What is Nathan's mean rate of speed if he hikes 12.6 miles in 7.5 hours?

2. **Biking:** What is Amy's mean rate of speed if she bikes 39.6 miles in 2.2 hours?

3. **Traveling by Car:** Jamie plans to take the scenic route from Los Angeles to San Francisco. Her GPS tells her it is a 420 mile trip. If she figures her mean speed will be 48 mph, how long will the trip take her?

4. **Traveling by Car:** Scott's mean speed on his drive from Memphis, TN is 60 mph. If the total trip is 285 miles, how long should he expect the drive to take?

5. **Biking:** Jane rides her bike to Lake Junaluska. Going to the lake, her mean speed is 12 mph. On the return trip, her mean speed is 10 mph. If the round trip takes a total of 5.5 hours, how long does the return trip take?

rate	· time	= distance
Going	12	$5.5 - t$	
Returning	10	t	

6. **Plane Speeds:** Two planes which are 2475 miles apart fly toward each other. Their speeds differ by 75 mph. If they pass each other in 3 hours, what is the speed of each?

	rate	· time	= distance
1st plane	r	3	
2nd plane	$r + 75$	3	

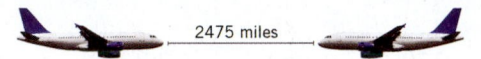

2475 miles

7. **Traveling by Car:** Marcus drives from Chicago to Detroit in 6 hours. On the return trip, his speed is increased by 10 mph and the trip takes 5 hours. Find his rate on the return trip. How far apart are the towns?

	rate	· time	= distance
Going	r	6	
Returning	$r + 10$	5	

8. **Hiking:** Tim and Barb have 8 hours to spend on a mountain hike. They can walk up the trail at a mean rate of 2 mph and can walk down at a mean of 3 mph. How long should they plan to hike uphill before turning around?

	rate	· time	= distance
Uphill	2	x	
Downhill	3	$8 - x$	

9. **Traveling by Car:** The Reeds are moving across Texas. Mr. Reed leaves $3\frac{1}{2}$ hours before Mrs. Reed. If his mean speed is 40 mph and her mean speed is 60 mph, how long will Mrs. Reed have to drive before she overtakes Mr. Reed?

10. **Traveling by Car:** After traveling for 40 minutes, Mr. Koole had to slow to $\frac{2}{3}$ his original speed for the rest of the trip due to heavy traffic. The total trip of 84 miles took 2 hours. Find his original speed.

11. **Train Speeds:** A train leaves Cincinnati at 2:00 P.M. A second train leaves the same station in the same direction at 4:00 P.M. The second train travels 24 mph faster than the first. If the second train overtakes the first at 7:00 P.M., what is the speed of each of the two trains?

12. **Running:** Maria runs through the countryside at a rate of 10 mph. She returns along the same route at 6 mph. If the total trip took 1 hour 36 minutes, how far did she run in total?

13. Traveling by Car: The distance from Atlanta, Georgia to Washington, DC is 620 miles. Driving in the middle of the night it takes about 9 hours to get to Washington. Due to higher traffic volume, it takes 2 more hours to travel there during the day. What is the mean rate of the driver during the day and during the night? (Round to the nearest whole number.)

14. Traveling by Car: Mr. Kent drove to a conference. The first half of the trip took 3 hours due to traffic. Traffic let up for the second half of the trip and he was able to increase his speed by 20 mph to make sure he got there on time. Find his rates of speed if he traveled 2 hours at the second rate.

15. Exercising: Jayden walked to his friend's house at a rate of 4 mph to borrow his friend's bicycle. Coming back home, he rode the bicycle at a mean rate of 12 mph. The total time for the round trip was 1 hour 30 minutes. How far away does Jayden's friend live?

16. Exercising: Once a week Felicia walks/runs for a total of 6 miles. Felicia spends twice as much time walking as she does running. If she walks at a rate of 4 mph and runs three times faster than she walks, what is the time for each part?

17. Investing: Amanda invests $25,000, part at 5% and the rest at 6%. The annual return on the 5% investment exceeds the annual return on the 6% investment by $40. How much did she invest at each rate?

18. Investing: Mr. Hill invests $10,000, part at 5.5% and part at 6%. The annual interest from the 5.5% investment exceeds the annual interest from the 6% investment by $251. How much did he invest at each rate?

19. Investing: The annual interest earned on a $6000 investment was $120 less than the interest earned on $10,000 invested at 1% less interest per year. What was the rate of interest on each amount?

20. Investing: Two investments totaling $16,000 produce an annual income of $1140. One investment yields 6% a year, while the other yields 8% per year. How much is invested at each rate?

21. Investing: The annual interest on a $4000 investment exceeds the interest earned on a $3000 investment by $80. The $4000 is invested at a 0.5% higher rate of interest than the $3000. What is the interest rate of each investment?

22. Investing: D'Andra makes two investments that total $12,000. One investment yields 8% per year and the other 10% per year. The total interest for one year is $1090. Find the amount invested at each rate.

23. Investing: A company is planning to invest $42,000 into two simple interest accounts. The annual interest rate on one of the accounts is 4.5% while the rate on the other is 6%. How much should the company invest in each account so that the two accounts will produce an equal annual interest income?

24. Shopping: A particular style of shoe costs the dealer $81 per pair. At what price should the dealer mark them so he can sell them at a 10% discount off the selling price and still make a 25% profit?

25. Investing: Sebastian would like both of his investments, a total of $12,000, to bring him the same annual interest income. One of his investments is at 5.5% annual interest rate and the other at 7%. Find the amount of money that Sebastian should invest in each account.

26. Used Car Sales: A car dealer paid $2850 for a used car. For his upcoming Labor Day sale, he wants to offer a 5% discount off the posted selling price, but would still like to make a 40% profit. What price should he advertise for that car?

27. Investing: Gabriella got some money from her grandparents as a graduation present. She decided to invest all of it. Part of the money was invested at a 2.5% interest rate, and the rest at a 4% interest rate. She invested $200 more in the 4% account than the 2.5% account. If her annual interest income was $47, how much did she invest at each rate?

28. Investing: Jordan is earning 1.5% interest from money invested in a savings account and 4% interest on a mutual bond fund. If the total of his investments is $18,000 and the annual interest from the savings account is less than the annual interest from the bond by $60, how much has Jordan invested at each rate?

29. Investing: A small company invested $20,500 such that a part of the money is in an account with a 4% interest rate and the rest at a 5% rate. The annual interest from the 5% account is $35 more than the interest earned from the 4% account. Find the amount of money the company invested at each rate.

30. Shopping: During the month of January, a department store would like to have a sale of 40% off of women's long boots. The store purchased the boots for $33 per pair. How much should the selling price be, if the manager of the store wants to make a 10% profit per pair?

31. Shopping: Carla plans on buying a new pair of sandals for the summer. They are on sale for 20% off of the original price. What was the original price of the sandals, if she pays $34.83, with 7.5% sales tax?

32. Investing: Last year an individual invested some money at a 5% interest rate and $2200 less than that amount at a 6% interest rate. If his interest income was $880, how much did he invest at each rate?

33. Shopping: A store purchased a certain style of leather jacket at $70 per jacket. If the store wants to sell the jackets at a 20% discount and still make a profit of 30%, what should be the marked selling price for each jacket?

34. Electronics: After receiving a 10% off coupon in the mail, Mark decided that it was time to buy a Blu-Ray player. Using the coupon and paying 8% sales tax, the final price came to $213.84. What was the listed price of the Blu-Ray player?

35. Exam Scores: Marissa has five exam scores of 75, 82, 90, 85, and 77 in her chemistry class. What score does she need on the final exam to have a mean grade of 80 (and thus earn a grade of B)? (All exams have a maximum of 100 points.)

36. Exam Scores: Gerald had scores of 80, 92, 89, and 95 on four exams in his algebra class. What score will he need on his fifth exam to have an overall mean grade of 90? (All exams have a maximum of 100 points.)

37. Biking: While riding her bike to the park and back home five times, Stacey timed herself at 60 min, 62 min, 55 min (the wind was helping), 58 min, and 63 min. She had set a goal of having a mean time of 60 minutes for her rides. How many minutes will she need on her sixth ride to attain her goal?

38. Cell Phones: For every 4 week period, Lauren wants to make a mean of 6 phone calls per week from her prepaid cell phone. The first week she made 9 phone calls, the second week 6 phone calls, and the third week 5 phone calls. How many phone calls does Lauren need to make in the fourth week to make sure she stays on track with her goal?

39. TV Watching: While growing up, Jason was allowed to watch TV a mean of 3 hours a day over a one week period. One particular week he watched 1 hour, 2 hours, 1 hour, 3 hours, 3 hours, and 5 hours. How many hours could Jason watch the seventh and last day of the week and still obey his parents?

40. Budgeting: A college student realized that he was spending too much money on video games. For the remaining 5 months of the year his goal is to spend a mean of $50 a month towards his hobby. How much can he spend in December, taking into consideration that the other 4 months he spent $70, $25, $105, $30 respectively?

41. Exam Scores: Wade has scores of 59, 68, 76, 84 and 69 on the first five tests in his social studies class. He knows that the final exam counts as two tests. What score will he need on the final to have a mean of 70? (All tests and exams have a maximum of 100 points.)

42. Exam Scores: A statistics student has grades of 86, 91, 95, and 76 on four hour-long exams. What score must he receive on the final exam to have a mean grade of 90 if:

a. the final is equivalent to a single hour-long exam (100 points maximum)?

b. the final is equivalent to two hour-long exams (200 points maximum)?

43. Temperature: Given the monthly temperatures over a year for Christchurch, New Zealand: **Source:** http://www.weather.com

44. Mean Rainfall: Given the means of monthly rainfall over a year for Visakhapatnam, India: **Source:** http://www.weather.com

Monthly Temperatures in Christchurch, New Zealand

Months of the Year

Mean Monthly Rainfall in Visakhapatnam, India

Months of the Year

a. Find the mean temperature for the year. (Round to the nearest tenth.)

b. Find the minimum mean temperature for the year.

c. Find the difference in mean temperature between the months of June and December.

a. Find the mean rainfall for the year. (Round to the nearest tenth.)

b. Find the maximum mean rainfall for the year.

c. Find the difference in mean rainfall between the months of October and December.

45. Company Profits: Given the yearly profits of the world's largest companies in 2009: **Source:** http://money.cnn.com/magazines/fortune/global500/2009/full_list/

Profits of World's Largest Companies in 2009

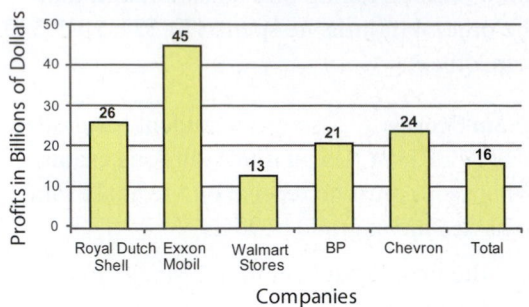

a. Find the mean profits of the companies in the year 2009. (Round to the nearest tenth of a billion.)

b. Find the yearly profits of Royal Dutch Shell in the year 2009.

c. What was the difference in profits between Exxon Mobil and BP?

46. Airport Usage: Given the number of passengers at the following airports: **Source:** Airports Council International – North America

Number of Passengers at World Airports in 2009

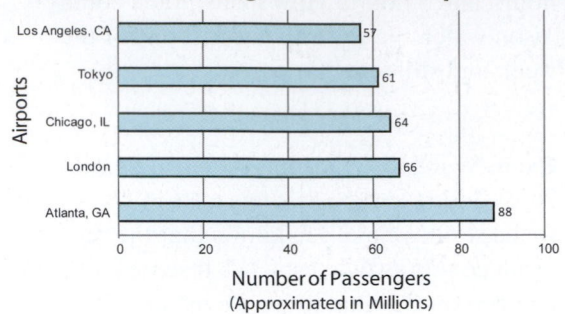

a. Find the mean number of passengers.

b. Find the difference in passengers between Atlanta, GA and Tokyo.

c. What was the total number of passengers to go through London?

Chapter 5: Index of Key Terms and Ideas

Section 5.1: Simplifying and Evaluating Algebraic Expressions

Algebraic Vocabulary   page 565
   Constant: A single number
   Term: Any constant or variable or the indicated product and/or quotient
   of constants and variables
   Numerical coefficient: The number written next to a variable
   Algebraic expression: A combination of variables and numbers using any
   of the operations of addition, subtraction, multiplication, or division, as
   well as exponents

Like Terms   pages 565 – 566
   Like terms (or **similar terms**) are terms that are constants or terms that
   contain the same variables that are raised to the same exponents.

Combining Like Terms   pages 566 – 568
   To **combine like terms**, add (or subtract) the coefficients and keep the
   common variable expression.

Evaluating Algebraic Expressions   pages 568 – 569
   1. Combine like terms, if possible.
   2. Substitute the values given for any variables.
   3. Follow the rules for order of operations.

Section 5.2: Translating English Phrases and Algebraic Expressions

Translating English Phrases and Algebraic Expressions   pages 573 – 574

Addition	Subtraction	Multiplication	Division	Exponent Powers
add	subtract (from)	multiply	divide	square of
sum	difference	product	quotient	cube of
plus	minus	times		
more than	less than	twice		
increased by	decreased by	of (with fractions and percent)		
total	less			

Section 5.3: Solving Linear Equations: $x + b = c$ and $ax = c$

Solution   page 580
   A solution to an equation is a number that gives a true statement when
   substituted for the variable in the equation.

Linear Equations in x   page 580
   If a, b, and c are **constants** and $a \neq 0$, then a **linear equation in x** (or **first-**
   degree equation in x) is an equation that can be written in the form $ax + b = c$.

Addition Principle of Equality   pages 581 – 582

If the same algebraic expression is added to both sides of an equation, the new equation has the same solutions as the original equation. Symbolically, if A, B, and C are algebraic expressions, then the equations $A = B$ and $A + C = B + C$ have the same solutions.

Equivalent Equations   pages 581 – 582

Equations with the same solutions are said to be **equivalent equations**.

Procedure for Solving Linear Equations that Simplify to the Form $x + b = c$ pages 582 –583

1. Combine any like terms on each side of the equation.
2. Use the **addition principle of equality** and add the opposite of the constant b to both sides. The objective is to isolate the variable on one side of the equation (either the left side or the right side) with a coefficient of +1.
3. Check your answer by substituting it for the variable in the original equation.

Multiplication (or Division) Principle of Equality   page 584

If both sides of an equation are multiplied by (or divided by) the same nonzero constant, the new equation has the same solutions as the original equation. Symbolically, if A and B are algebraic expressions and C is any nonzero constant, then the following equations have the same solutions.

$$A = B \qquad \text{and} \qquad AC = BC \qquad \text{where } C \neq 0$$

$$\text{and} \qquad \frac{A}{C} = \frac{B}{C} \qquad \text{where } C \neq 0$$

Procedure for Solving Linear Equations that Simplify to the Form $ax = c$ page 584 – 586

1. Combine any like terms on each side of the equation.
2. Use the **multiplication (or division) principle of equality** and multiply both sides of the equation by the reciprocal of the coefficient of the variable. (**Note:** This is the same as dividing both sides of the equation by the coefficient.) Thus the coefficient of the variable will become +1.
3. Check your answer by substituting it for the variable in the original equation.

Section 5.4: Solving Linear Equations: $ax + b = c$

Procedure for Solving Linear Equations that Simplify to the Form $ax + b = c$ pages 591 – 592

1. Combine like terms on both sides of the equation.
2. Use the **addition principle of equality** and add the opposite of the constant b to both sides.
3. Use the **multiplication (or division) principle of equality** to multiply both sides by the reciprocal of the coefficient of the variable (or divide both sides by the coefficient itself). The coefficient of the variable will become +1.
4. Check your answer by substituting it for the variable in the original equation.

Procedure for Solving Linear Equations that Simplify to the Form $ax + b = cx + d$ pages 598 – 599

1. Simplify each side of the equation by removing any grouping symbols and combining like terms on both sides of the equation.
2. Use the **addition principle of equality** and add the opposite of a constant term and/or variable term to both sides so that variables are on one side and constants are on the other side.
3. Use the **multiplication** (or **division**) **principle of equality** to multiply both sides by the reciprocal of the coefficient of the variable (or divide both sides by the coefficient itself). The coefficient of the variable will become +1.
4. Check your answer by substituting it for the variable in the original equation.

Types of Equations and Their Solutions   pages 601– 602

Type of Equation	Number of Solutions
Conditional	Finite number of solutions
Identity	Infinite number of solutions
Contradiction	No solutions

Section 5.6: Applications: Number Problems and Consecutive Integers

Consecutive Integers   pages 608 – 609

Integers are **consecutive** if each is 1 more than the previous integer. Three consecutive integers can be represented as n, $n + 1$, and $n + 2$.

Consecutive Even Integers   page 609

Even integers are consecutive if each is 2 more than the previous even integer. Three consecutive even integers can be represented as $2n$, $2n + 2$, and $2n + 4$, where n is an integer.

Consecutive Odd Integers   page 609

Odd integers are consecutive if each is 2 more than the previous odd integer. Three consecutive odd integers can be represented as $2n + 1$, $2n + 3$, and $2n + 5$, where n is an integer.

Section 5.7: Applications: Percent Problems

The Basic Percent Formula   page 617

$R \cdot B = A$ (or $A = R \cdot B$)
$R = RATE$ or percent (as a decimal or fraction)
$B = BASE$ (number we are finding the percent of)
$A = AMOUNT$ (a part of the base)
"**of**" means to multiply.
"**is**" means equal $(=)$.

Discount   page 619

A reduction in the selling price stated as a percent of the original price.

Sales Tax   page 619

A tax charged on goods sold by retailers (the rate, a percent, varies by location).

Commission   page 620 – 621

A fee paid to an agent or salesperson for a service.

Percent of Profit   page 621

A fraction (or ratio) of profit divided by the investment, changed to a percent.

Section 5.8: Working with Formulas

Lists of Common Formulas   page 627

Formula	Meaning
$I = Prt$	The **simple interest** I earned by investing money is equal to the product of the principal P times the rate of interest r times the time t in one year or part of a year.
$C = \dfrac{5}{9}(F - 32)$	**Temperature** in degrees Celsius C equals $\dfrac{5}{9}$ times the difference between the Fahrenheit temperature F and 32.
$d = rt$	The **distance traveled** d equals the product of the rate of speed r and the time t.
$P = 2l + 2w$	The **perimeter, P, of a rectangle** is equal to twice the length l plus twice the width w.
$L = 2\pi rh$	The **lateral surface area, L,** (top and bottom not included) of a cylinder is equal to 2π times the radius r of the base times the height h.
$F = ma$	In physics, the **force, F,** acting on an object is equal to its mass m times its acceleration a.
$\alpha + \beta + \gamma = 180°$	The **sum of the angles** ($\alpha, \beta,$ and γ) **of a triangle** is $180°$.
$P = a + b + c$	The **perimeter, P, of a triangle** is equal to the sum of the lengths of the three sides a, b, and c.

Evaluating Formulas   pages 628 – 630

If you know values for all but one variable in a formula, you can substitute those values and find the value of the unknown variable using the techniques for solving equations.

Solving Formulas for Different Variables   pages 630 – 632

We want the formula "solved for" some variable other than the one given in terms of the remaining variables. Treat the variables just as you would constants in solving linear equations.

Section 5.9: Applications: Distance-Rate-Time, Interest, Mean, Cost

Mean Problems   page 641

Use the concept of mean to find unknown numbers and to read and understand graphs.

Cost Problems   page 643

$S - C = P$ (selling price – cost = profit).

Chapter 5: Test

For the following expressions, a. simplify, and b. evaluate the simplified expression for $x = -2$ and $y = 3$.

1. $7x + 8x^2 - 3x^2$

2. $5y - y - 6 + 2y$

3. $2(x - 5) + 3(x + 4)$

4. $3y^2 + 6y - y^2 + y$

Translate each problem as directed.

5. Translate each algebraic expression into an equivalent English phrase. (There may be more than one correct translation.)

   a. $5n + 18$

   c. $42 - 7y$

   b. $3(x + 6)$

   d. $6x - 11$

6. Write the algebraic expression described by each English phrase.

   a. the product of a number and 6 decreased by 3

   c. 4 less than twice the sum of a number and 15

   b. twice the sum of a number and 5

   d. the quotient of a number and 10 increased by twice the number

Solve each of the following linear equations.

7. $0 = 2x + 8$

8. $y + 13 = -16$

9. $\dfrac{5}{3}x + 1 = -4$

10. $3 + 6x - 8x = 15$

11. $4x - 5 - x = 2x + 5 - x$

12. $8(3 + y) = -4(2y - 6)$

13. $\dfrac{3}{2}a + \dfrac{1}{2}a = -3 + \dfrac{3}{4}$

14. $6 + \dfrac{3}{2}x = 9 + \dfrac{1}{3}x - \dfrac{5}{6}x$

15. $0.7x + 2 = 0.4x + 8$

16. $4(2x - 1) + 3 = 2(x - 4) - 5$

Determine whether each of the following equations is a conditional equation, an identity, or a contradiction.

17. $9x + x - 4(x + 3) = 6(x - 2)$

18. $\dfrac{1}{4}(4x + 1) = -\dfrac{1}{2}(2x - 1)$

Set up an equation for each word problem and solve the equation.

19. One number is 5 more than twice another number. Their sum is -22. Find the numbers.

20. Find two consecutive integers such that twice the first added to three times the second is equal to 83.

21. Find three consecutive odd integers such that three times the second is equal to 27 more than the sum of the first and the third.

22. **Basketball:** Brandon scored 21 points in his basketball game last night. If he made two free throws (1 point each) and one 3-point shot, how many 2-point shots did he make?

Find the missing rate, base, or amount.

23. 62% of 180 is _____.

24. 48 is _____% of 150.

Solve each of the following application problems.

25. **Investing:** Which is the better investment, an investment of $6000 that earns a profit of $360 or an investment of $10,000 that earns a profit of $500? Explain your answer in terms of percent.

26. **Investing:** Julie made an investment of $2000 and lost 25% in one year. In the next year she made 25%. How much money did she have at the end of the two years?

27. **Triangles:** The triangle shown here indicates that the sides can be represented as x, $x + 1$, and $2x - 1$. What is the length of each side if the perimeter is 12 meters?

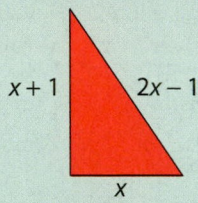

$x + 1$ $2x - 1$

x

28. **Shopping:** A man's suit was on sale for $547.50. If this price was a discount of 25% from the original price, what was the original price?

Solve each formula for the indicated variable.

29. $N = mrt + p$; solve for m.

30. $5x + 3y - 7 = 0$; solve for y.

Solve each of the following application problems.

31. **Text Messaging:** The cost to send/receive t text messages in a given month on AT&T's Messaging 1500 plan is $C = 0.05(t - 1500) + 15.00$ when $t \geq 1500$. How much did Jena spend the month she sent/received 1780 text messages.

32. **Traveling by Bus:** A bus leaves Kansas City headed for Phoenix traveling at a rate of 48 mph. Thirty minutes later, a second bus follows, traveling at 54 mph. How long will it take the second bus to overtake the first?

33. **Investing:** George and Marie have investments in two accounts totaling $25,000. One investment yields 5% a year and the other yields 3.5%. If their annual income from the 5% account yields $145 more each year than the other account, what amount do they have in each account?

34. **Shopping:** The manager of a jewelry store purchased a selection of watches from the manufacturer for $190 each. Next weekend he plans to have a 20% off sale. What price should he mark on each watch so that he will still make a profit of 40% on his cost?

35. Test Scores: In his calculus class a student has the following test scores: 77, 82, and 73. If his last test will have a maximum of 100 points, what score does he need to finish the class with a mean of 80?

36. Payment Plans: When purchasing an item on an installment plan, the total cost C equals the down payment d plus the product of the monthly payment p and the number of months t. ($C = d + pt$). Use this information to answer the following questions.

    a. A refrigerator costs $857.60 if purchased on the installment plan. If the monthly payments are $42.50 and the down payment is $92.60, how long will it take to pay for the refrigerator?

    b. A used car will cost $3250 if purchased on an installment plan. If the monthly payments are $115 for 24 months, what will be the down payment?

37. Cell Phone Usage: Given the amount of U.S. cellular telephone subscribers: **Source:** International Telecommunications Union

    a. Find the mean number of subscribers from 2002-2005, inclusive.

    b. Find the difference in the number of subscribers from 1999 and 2008.

    c. How many subscribers were there in 2001?

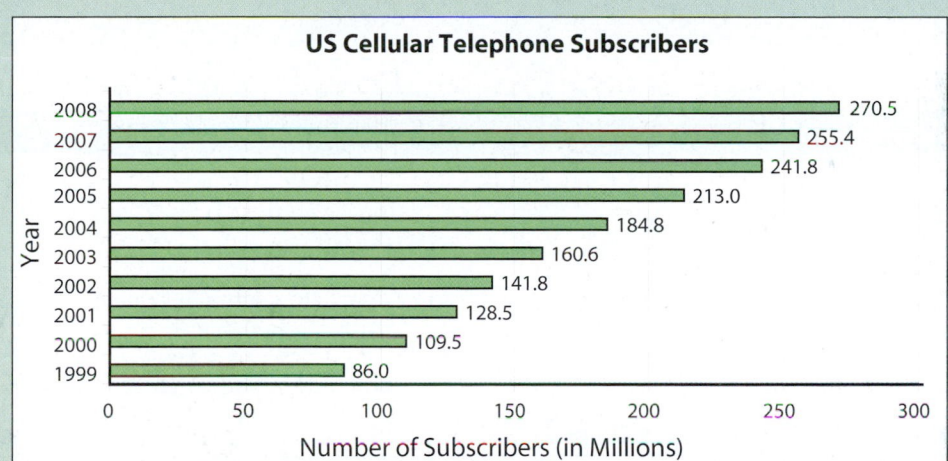

Cumulative Review: Chapters 1 - 5

Find the prime factorizaton.

1. 380

Simplify the expressions using the rules for order of operations. Reduce all answers to lowest terms.

2. $\dfrac{1}{2} + \dfrac{3}{8} \div \dfrac{3}{4} - \dfrac{5}{6}$

3. $\dfrac{1}{3} + \dfrac{2}{5} \cdot \dfrac{5}{8} \div \dfrac{3}{2} - \dfrac{1}{2}$

4. $\left(\dfrac{2}{3}\right)^2 \div \dfrac{5}{18} - \dfrac{3}{8}$

5. $\left(\dfrac{9}{10} + \dfrac{2}{15}\right) \div \left(\dfrac{1}{2} - \dfrac{2}{3}\right)$

For the following expressions, a. simplify, and b. evaluate the simplified expression for $x = -2$ and $y = 3$.

6. $-4(y+3)+2y$

7. $3(x+4)-5+x$

8. $2(x^2+4x)-(x^2-5x)$

9. $\dfrac{3(5x-x)}{4} - 2x - 7$

Solve the following problems.

10. $6x + 3x - 5 + 4x$

11. $\dfrac{a}{7} + \dfrac{a}{14} - \dfrac{a}{21}$

12. Translate each English phrase into an algebraic expression:

    a. twice a number decreased by 3

    b. the quotient of a number and 6 increased by 3 times the number

    c. thirteen less than twice the sum of a number and 10

13. Translate each algebraic expression into an English phrase:

    a. $6 + 4n$

    b. $18(n-5)$

Solve each of the following linear equations.

14. $x + 4.2 = -5.6$

15. $-5 + x = 13$

16. $17y = -51$

17. $-16y = -96$

18. $9x - 8 = 4x - 13$

19. $6 - (x-2) = 5$

20. $10 + 4x = 5x - 3(x+4)$

21. $4.4 + 0.6x = 1.2 - 0.2x$

22. $-2(5x+12) - 5 = -6(3x-2) - 10$

23. $\dfrac{3}{4}y - \dfrac{1}{2} = \dfrac{5}{8}y + \dfrac{1}{10}$

24. $\dfrac{1}{2} + \dfrac{1}{5}y = \dfrac{2}{15}y + 1$

25. $-\dfrac{3}{4}y + \dfrac{3}{8}y = \dfrac{1}{6}y$

Determine whether each of the following equations is a conditional equation, an identity, or a contradiction.

26. $0.2x + 25 = -1.6 + 0.1x$

27. $3(x-5) + 4 = x + 2(x-5)$

28. $4x - 3(x+6) = x - 12$

29. $-x + 7(x+3) = 3x + 3(x+7)$

30. $8 - 4(3-x) + 2x = 6(x+2) - 16$

Solve each of the following problems.

31. The two triangles shown here are similar triangles. Find the values of x and y.

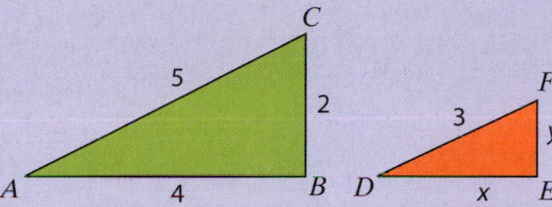

32. If the difference between $\dfrac{1}{2}$ and $\dfrac{11}{18}$ is divided by the sum of $\dfrac{2}{3}$ and $\dfrac{7}{15}$, what is the quotient?

33. From the sum of -2 and -11, subtract the product of -4 and 2.

34. **Entertaining:** Lucia is making punch for a party. She is making 5 gallons. The recipe calls for $\dfrac{2}{3}$ cup of sugar per gallon of punch. How many cups of sugar does she need?

35. **College Admission:** The percentages of applicants accepted into the eight Ivy League colleges are as follows: $8\%, 19\%, 10\%, 7\%,$ $18\%, 11\%, 13\%,$ and 10%. What is the mean percentage of applicants accepted to an Ivy League college? **Source:** College Board

Set up an equation for each problem and solve for the unknown quantity.

36. If twice a certain number is increased by 3, the result is 8 less than three times the number. Find the number.

37. **Computers:** Computers are on sale at a 30% discount. If you know that you will pay 6.5% in sales taxes, what will you pay for a computer that is priced at $680?

38. Find three consecutive even integers such that the sum of the first and twice the second is equal to 14 more than twice the third.

39. **Commission:** Barry makes $500 a month plus commission working at a clothing store on the weekends. If he receives a 3% commission on all clothes he sells, what was his income the month he sold $1860 worth of clothes?

40. Find four consecutive integers such that the sum of the first, second, and fourth integers is 2 less than the third.

41. Investing: Which is the better investment: **a.** a profit of $800 on an investment of $10,000 or **b.** a profit of $1200 on an investment of $15,000? Explain in terms of percents.

42. Electronics: The total cost of a portable hard drive and a spare laptop battery was $130.75. If the hard drive cost $38.45 more than the battery, what was the cost of each item?

Solve for the indicated variable.

43. $h = vt - 16t^2$; solve for v.

44. $A = P + Prt$; solve for r.

45. $5x - 3y = 14$; solve for y.

46. $V = \frac{1}{3}\pi r^2 h$; solve for h.

47. $C = \pi d$; solve for d.

48. $3x + 5y = 10$; solve for y.

Solve the following word problems.

49. Traveling by Train: Diego took the train into New York City to see a concert. On the way to the concert the train had a mean of 90 miles per hour. On the way home from the concert the train made a lot more stops and had a mean of only 50 miles per hour. If Diego spent a total of 2.8 hours on the train, how long did the return trip take?

50. Investing: Two investments totaling $22,000 produce an annual income of $742.50. One investment yields 4.5% a year, while the other yields 3% per year. How much is invested at each rate?

Mathematics @ Work!

Linear equations (or linear functions) can be used to represent many real-world applications. They play an especially large role in the business world, where they may be used to represent supply functions, demand functions, cost functions, or profit functions. Consider the relationship between supply and demand. When a company raises prices, less people may be inclined to buy the product, but the profit on each product sold may be higher. The federal government uses this tactic with special taxes on items like cigarettes to encourage more people to quit smoking by making a pack of cigarettes cost more. The problem given below provides another example of this phenomenon.

A pharmaceutical company is examining the demand of a certain drug. When the price was set at $55 per unit, sales were 1.2 million units. When the price was set at $40 per unit, sales increased to 1.8 million units. Using the price as the independent variable x and the quantity demanded (in millions of units) as the dependent variable y, find the linear equation that models this relationship. According to this equation, at what price will sales drop to 0?

For more problems like these, see Section 6.4.

Graphing Linear Equations

Objectives

A Understand the meaning of an equation in two variables.

B Graph and label ordered pairs of real numbers as points on a plane.

C Find ordered pairs of real numbers that satisfy a given equation.

D Locate points on a given graph of a line.

6.1 The Cartesian Coordinate System

René Descartes (1596 – 1650), a famous French mathematician, developed a system for solving geometric problems using algebra. This system is called the **Cartesian coordinate system** in his honor. Descartes based his system on a relationship between points in a plane and **ordered pairs** of real numbers. This section begins by relating algebraic formulas with ordered pairs and then shows how these ideas can be related to geometry.

Objective A **Equations in Two Variables**

Equations such as $d = 60t$, $I = 0.05P$, and $y = 2x + 3$ represent relationships between pairs of variables. For example, in the first equation, if $t = 3$, then $d = 60 \cdot 3 = 180$. With the understanding that t is first and d is second, we can represent $t = 3$ and $d = 180$ in the form of an ordered pair $(3, 180)$. In general, if t is the first number and d is the second number, then solutions to the equation $d = 60t$ can be written in the form of ordered pairs (t, d). Thus we see that $(180, 3)$ is different from $(3, 180)$. **The order of the numbers in an ordered pair is critical**.

We say that $(3, 180)$ **is a solution of** (or **satisfies**) the equation $d = 60t$. Similarly, $(5, 300)$ represents $t = 5$ and $d = 300$, and satisfies the equation $d = 60t$. In the same way, $(100, 5)$ satisfies $I = 0.05P$, where $P = 100$ and $I = 0.05 \cdot 100 = 5$. In this equation, solutions are ordered pairs in the form (P, I).

For the equation $y = 2x + 3$, ordered pairs are in the form (x, y), and $(2, 7)$ satisfies the equation. If $x = 2$, then substituting in the equation gives $y = 2 \cdot 2 + 3 = 7$. In the ordered pair (x, y), x is called the **first coordinate** and y is called the **second coordinate**. To find ordered pairs that satisfy an equation in two variables, we can **choose any value** for one variable and find the corresponding value for the other variable by substituting into the equation. For example, **for the equation $y = 2x + 3$**, we can find the following ordered pairs.

Choices for *x*	Substitution	Ordered Pairs
$x = 1$	$y = 2(1) + 3 = 5$	$(1, 5)$
$x = -2$	$y = 2(-2) + 3 = -1$	$(-2, -1)$
$x = \dfrac{1}{2}$	$y = 2\left(\dfrac{1}{2}\right) + 3 = 4$	$\left(\dfrac{1}{2}, 4\right)$

Table 1

All the ordered pairs $(1, 5)$, $(-2, -1)$, and $\left(\dfrac{1}{2}, 4\right)$ satisfy the equation $y = 2x + 3$.

There are an infinite number of such ordered pairs. Any real number could have been chosen for *x* and the corresponding value for *y* calculated.

Since the equation $y = 2x + 3$ is solved for y, we say that the value of y "depends" on the choice of x. Thus, in an ordered pair of the form (x, y), the first coordinate x is called the **independent variable** and the second coordinate y is called the **dependent variable**.

In the following table, the first variable in each case is the independent variable and the second variable is the dependent variable. Corresponding ordered pairs would be of the form $(t, d), (P, I)$, and (x, y). The choices for the values of the independent variables are arbitrary. There are an infinite number of other values that could have just as easily been chosen.

$d = 60t$			$I = 0.05P$			$y = 2x + 3$		
t	d	(t, d)	P	I	(P, I)	x	y	(x, y)
5	$60(5)$	$(5, 300)$	100	$0.05(100)$	$(100, 5)$	-2	$2(-2)+3$	$(-2, -1)$
10	$60(10)$	$(10, 600)$	200	$0.05(200)$	$(200, 10)$	-1	$2(-1)+3$	$(-1, 1)$
12	$60(12)$	$(12, 720)$	500	$0.05(500)$	$(500, 25)$	0	$2(0)+3$	$(0, 3)$
15	$60(15)$	$(15, 900)$	1000	$0.05(1000)$	$(1000, 50)$	3	$2(3)+3$	$(3, 9)$

Table 2

Objective B **Graphing Ordered Pairs**

The Cartesian coordinate system relates algebraic equations and ordered pairs to geometry. In this system, two number lines intersect at right angles and separate the plane into four **quadrants**. The **origin**, designated by the ordered pair $(0, 0)$, is the point of intersection of the two lines. The horizontal number line is called the **horizontal axis** or **x-axis**. The vertical number line is called the **vertical axis** or **y-axis**. Points that lie on either axis are not in any quadrant. They are simply on an axis. (See Figure 1).

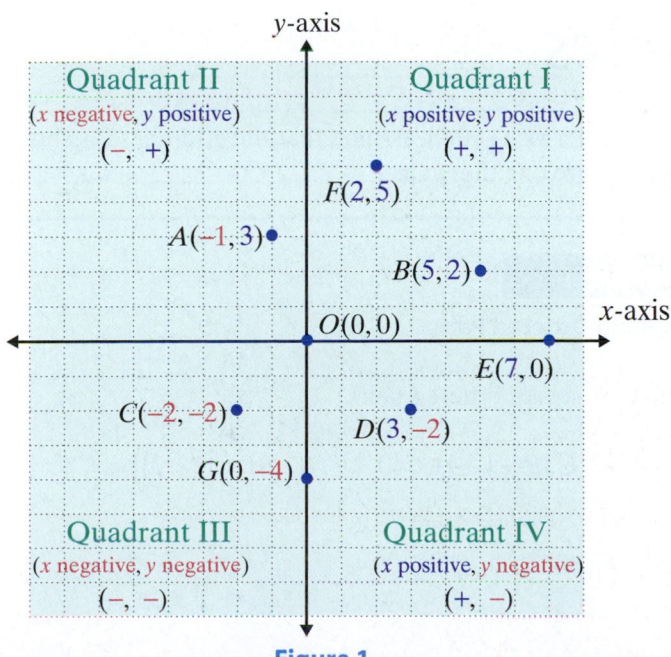

Figure 1

The following important relationship between ordered pairs of real numbers and points in a plane is the cornerstone of the Cartesian coordinate system.

One-to-One Correspondence

There is a **one-to-one correspondence** between points in a plane and ordered pairs of real numbers.

In other words, for each point in a plane there is one and only one corresponding ordered pair of real numbers, and for each ordered pair of real numbers there is one and only one corresponding point in the plane.

The **graphs of the points** $A(2,1), B(-2,3), C(-3,-2), D(1,-2),$ and $E(3,0)$ are shown in Figure 2. (**Note:** An ordered pair of real numbers and the corresponding point on the graph are frequently used to refer to each other. Thus, the ordered pair $(2,1)$ and the point $(2,1)$ are interchangeable ideas.)

Point	Quadrant
$A(2,1)$	I
$B(-2,3)$	II
$C(-3,-2)$	III
$D(1,-2)$	IV
$E(3,0)$	x-axis

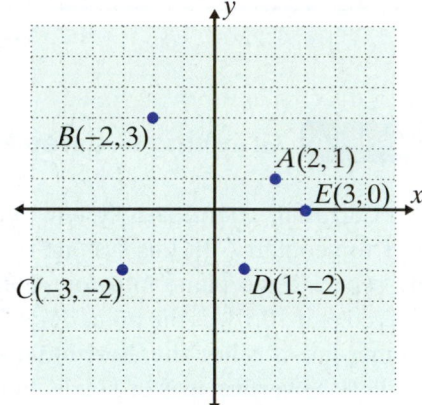

Figure 2

notes

▪ Unless otherwise stated, assume that the grid lines (as shown here in Figure 2) are one unit apart.

▪

Example 1

Graphing Ordered Pairs

Graph the sets of ordered pairs.

a. $\{A(-2,1), B(-1,-4), C(0,2), D(1,3), E(2,-3)\}$

Note: The listing of ordered pairs within the braces can be in any order.

Solution

To locate points: start at the **origin** $(0,0)$, move left or right for the x-coordinate and up or down for the y-coordinate.

For $A(-2, 1)$, move 2 units left and 1 unit up.

For $B(-1, -4)$, move 1 unit left and 4 units down.

For $C(0, 2)$, move no units left or right and 2 units up.

For $D(1, 3)$, move 1 unit right and 3 units up.

For $E(2, -3)$, move 2 units right and 3 units down.

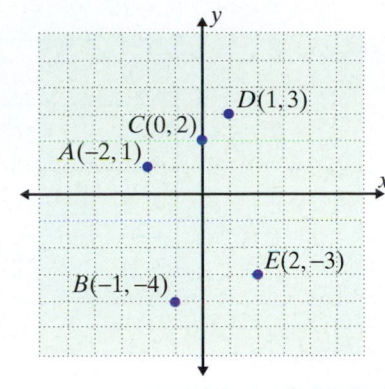

1. Graph the sets of ordered pairs.

a. $\{A(3,-1), B(-1,-4), C(4,0), D(-3,2), E(3,3)\}$

b. $\{A(0,2), B(-3,-1), C(-2,0), D(4,-3), E(3,1)\}$

b. $\{A(-1, 3), B(0, 1), C(1, -1), D(2, -3), E(3, -5)\}$

Solution

To locate each point, start at the **origin**, and move as follows:

For $A(-1, 3)$, move 1 unit left and 3 units up.

For $B(0, 1)$, move no units left or right and 1 unit up.

For $C(1, -1)$, move 1 unit right and 1 unit down.

For $D(2, -3)$, move 2 units right and 3 units down.

For $E(3, -5)$, move 3 units right and 5 units down.

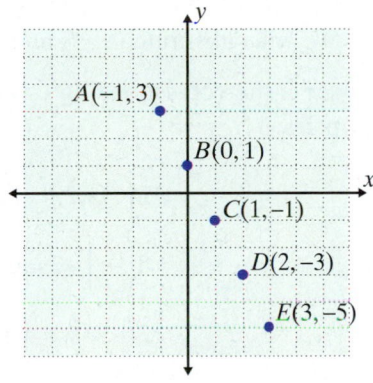

Now work margin exercise 1.

Objective C **Finding Ordered Pairs that Satisfy a Given Equation**

The points (ordered pairs) in Example **1b.** can be shown to satisfy the equation $y = -2x + 1$. For example, using $x = -1$ in the equation yields,

$$y = -2(-1) + 1 = 2 + 1 = 3,$$

and the ordered pair $(-1, 3)$ satisfies the equation. Similarly, letting $y = 1$ gives,

$$1 = -2x + 1$$
$$0 = -2x$$
$$0 = x$$

and the ordered pair $(0, 1)$ satisfies the equation.

We can write all the ordered pairs in Example **1b.** in table form.

x	−2x + 1 = y	(x,y)
−1	$-2(-1)+1=3$	$(-1, 3)$
0	$-2(0)+1=1$	$(0, 1)$
1	$-2(1)+1=-1$	$(1, -1)$
2	$-2(2)+1=-3$	$(2, -3)$
3	$-2(3)+1=-5$	$(3, -5)$

Table 3

Example 2

Determining Ordered Pairs

a. Determine which, if any, of the ordered pairs $(0, -2)$, $\left(\frac{2}{3}, 0\right)$, and $(2, 5)$ satisfy the equation $y = 3x - 2$.

Solution

We will substitute 0, $\frac{2}{3}$, and 2 for x in the equation $y = 3x - 2$ and see if the corresponding y-values match those in the given ordered pairs.

$x = 0$: $\quad y = 3(0) - 2 = -2$ so, $(0, -2)$ satisfies the equation.

$x = \frac{2}{3}$: $\quad y = 3\left(\frac{2}{3}\right) - 2 = 0$ so, $\left(\frac{2}{3}, 0\right)$ satisfies the equation.

$x = 2$: $\quad y = 3(2) - 2 = 4$ so, $(2, 4)$ satisfies the equation.

The point $(2, 5)$ does **not** satisfy the equation $y = 3x - 2$ because, as just illustrated, $y = 4$ when $x = 2$, not 5.

b. Determine the missing coordinate in each of the following ordered pairs so that the points will satisfy the equation $2x + 3y = 12$:
$(0, \quad)$, $(3, \quad)$, $(\quad, 0)$, $(\quad, -2)$.

Solution

The missing values can be found by substituting the given values for x (or for y) into the equation $2x + 3y = 12$ and solving for the other variable.

For $(0, \;)$, let $x = 0$:

$$2(0) + 3y = 12$$
$$3y = 12$$
$$y = 4.$$

The ordered pair is $(0, 4)$.

For $(3, \;)$, let $x = 3$:

$$2(3) + 3y = 12$$
$$6 + 3y = 12$$
$$3y = 6$$
$$y = 2.$$

The ordered pair is $(3, 2)$.

For $(\;, 0)$, let $y = 0$:

$$2x + 3(0) = 12$$
$$2x = 12$$
$$x = 6.$$

The ordered pair is $(6, 0)$.

For $(\;, -2)$, let $y = -2$:

$$2x + 3(-2) = 12$$
$$2x - 6 = 12$$
$$2x = 18$$
$$x = 9.$$

The ordered pair is $(9, -2)$.

c. Complete the table below so that each ordered pair will satisfy the equation $y = 1 - 2x$.

x	y = 1 - 2x	y	(x,y)
0			
		3	
$\frac{1}{2}$			
5			

Solution

Substituting each given value for x or y into the equation $y = 1 - 2x$ gives the following table of ordered pairs.

x	y = 1 - 2x	y	(x,y)
0	$1 = 1 - 2(0)$	1	$(0, 1)$
−1	$3 = 1 - 2(-1)$	3	$(-1, 3)$
$\frac{1}{2}$	$0 = 1 - 2\left(\frac{1}{2}\right)$	0	$\left(\frac{1}{2}, 0\right)$
5	$-9 = 1 - 2(5)$	−9	$(5, -9)$

For $x = 0$:
$$y = 1 - 2(0) = 1$$

For $y = 3$:
$$(3) = 1 - 2x$$
$$2 = -2x$$
$$-1 = x$$

For $x = \frac{1}{2}$:
$$y = 1 - 2\left(\frac{1}{2}\right) = 0$$

For $x = 5$:
$$y = 1 - 2(5) = -9$$

Now work margin exercise 2.

2. **a.** Determine which, if any, of the ordered pairs $(0, -3), (3, 6), (2, 5)$ satisfy the equation

$$y = 4x - 3.$$

b. Determine the missing coordinate in each of the following ordered pairs so that the points will satisfy the equation $x + 5y = 15$.

$$(0, \;), (5, \;)$$
$$(\;, 0), (\;, -3)$$

c. Complete the table below so that each ordered pair will satisfy the equation $y = 2 - 3x$.

x	y	(x,y)
0		
	1	
−2		
$\frac{2}{3}$		

notes

Although this discussion is related to ordered pairs of real numbers, most of the examples use ordered pairs of **integers**. This is because ordered pairs of integers are relatively easy to locate on a graph and relatively easy to read from a graph. Ordered pairs with fractions, decimals, or radicals must be located by estimating the positions of the points. The precise coordinates intended for such points can be difficult or impossible to read because large dots must be used so the points can be seen. **Even with these difficulties, you should understand that we are discussing ordered pairs of real numbers and that points with fractions, decimals, and radicals as coordinates do exist and should be plotted by estimating their positions.**

Objective D **Locating Points on a Line**

Example 3

Reading Points on a Graph

The graphs of two lines are given. Each line contains an infinite number of points. Use the grid to help you locate (or estimate) three points on each line.

a.

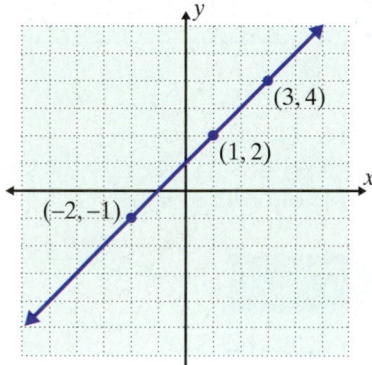

Solution

Three points on this graph are $(-2, -1)$, $(1, 2)$, and $(3, 4)$. (Of course there is more than one correct answer to this type of question. Use your own judgement.)

b.

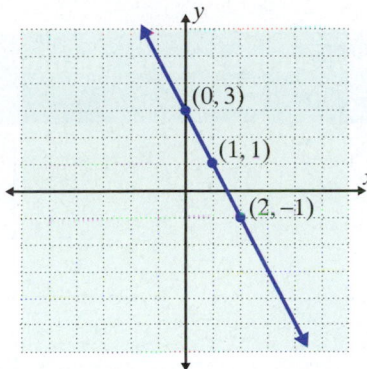

(0, 3)
(1, 1)
(2, –1)

Solution

Three points on this graph are $(0, 3)$, $(1, 1)$, and $(2, –1)$. (You may also estimate with fractions. For example, one point appears to be approximately $\left(\dfrac{1}{2}, 2\right)$.)

Now work margin exercise 3.

Practice Problems

1. Determine which ordered pairs satisfy the equation $3x + y = 14$.

   a. $(5, –1)$ **b.** $(4, 2)$ **c.** $(–1, 17)$

2. Given $3x + y = 5$, find the missing coordinate of each ordered pair so that it will satisfy the equation.

   a. $(0, \ )$ **b.** $\left(\dfrac{1}{3}, \ \right)$ **c.** $(\ , 2)$

3. Complete the table so that each ordered pair will satisfy the equation $y = \dfrac{2}{3}x + 1$.

x	y
0	
	–2
–3	
6	

4. List the sets of ordered pairs corresponding to the points on the graph.

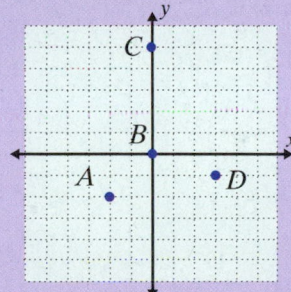

Practice Problem Answers

   1. All satisfy the equation.   **2. a.** $(0, 5)$ **b.** $\left(\dfrac{1}{3}, 4\right)$ **c.** $(1, 2)$

   3. $(0, 1), \left(-\dfrac{9}{2}, -2\right), (-3, -1), (6, 5)$

   4. $\{A(-2, -2), B(0, 0), C(0, 5), D(3, -1)\}$

3. The graphs of two lines are given. Use the grid to locate three points on each line.

a.

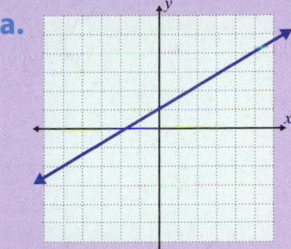

b.

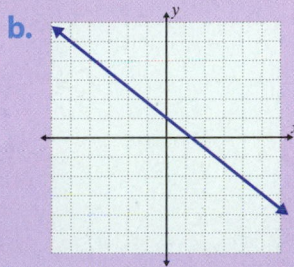

Exercises 6.1

1.

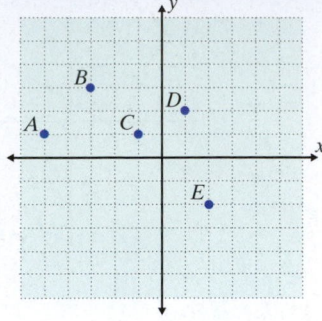

2.

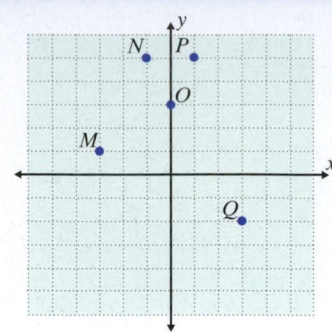

3.

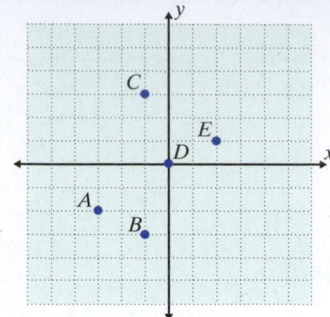

4.

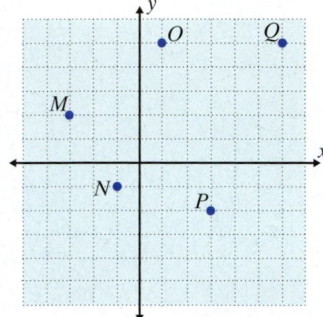

5.

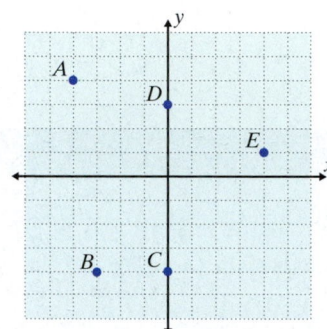

6.

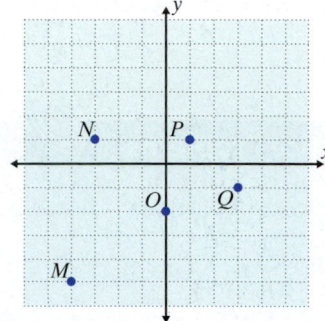

7.

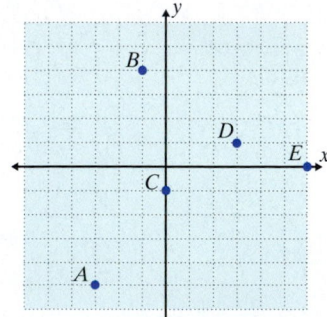

8.

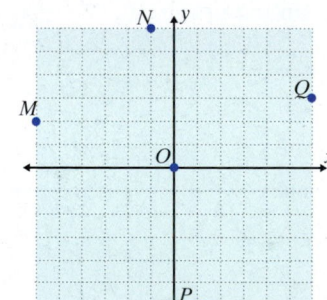

9.

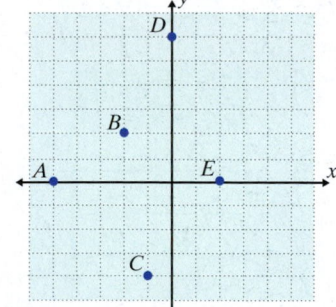

10.

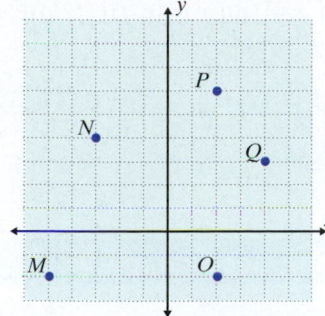

11. $\{A(4,-1), B(3,2), C(0,5), D(1,-1), E(1,4)\}$

12. $\{A(-1,-1), B(-3,-2), C(1,3), D(0,0), E(2,5)\}$

13. $\{A(1,2), B(0,2), C(-1,2), D(2,2), E(-3,2)\}$

14. $\{A(-1,4), B(0,-3), C(2,-1), D(4,1), E(-1,-1)\}$

15. $\{A(1,0), B(3,0), C(-2,1), D(-1,1), E(0,0)\}$

16. $\{A(-1,-1), B(0,1), C(1,3), D(2,5), E(3,10)\}$

17. $\{A(4,1), B(0,-3), C(1,-2), D(2,-1), E(-4,2)\}$

18. $\{A(0,1), B(1,0), C(2,-1), D(3,-2), E(4,-3)\}$

19. $\{A(1,4), B(-1,-2), C(0,1), D(2,7), E(-2,-5)\}$

20. $\{A(0,0), B(-1,3), C(3,-2), D(0,4), E(-7,0)\}$

21. $\left\{A(1,-3), B\left(-4,\frac{3}{4}\right), C\left(2,-2\frac{1}{2}\right), D\left(\frac{1}{2},4\right)\right\}$

22. $\left\{A\left(\frac{3}{4},\frac{1}{2}\right), B\left(2,-\frac{5}{4}\right), C\left(\frac{1}{3},-2\right), D\left(-\frac{5}{3},2\right)\right\}$

23. $\{A(1.6, -2), B(3, 2.5), C(-1, 1.5), D(0, -2.3)\}$　　**24.** $\{A(-2, 2), B(-3, 1.6), C(3, 0.5), D(1.4, 0)\}$

Determine which, if any, of the ordered pairs satisfy the given equations. See Example 2.

25. $2x - y = 4$
   a. $(1, 1)$
   b. $(2, 0)$
   c. $(1, -2)$
   d. $(3, 2)$

26. $x + 2y = -1$
   a. $(1, -1)$
   b. $(1, 0)$
   c. $(2, 1)$
   d. $(3, -2)$

27. $4x + y = 5$
   a. $\left(\dfrac{3}{4}, 2\right)$
   b. $(4, 0)$
   c. $(1, 1)$
   d. $(0, 3)$

28. $2x - 3y = 7$
   a. $(1, 3)$
   b. $\left(\dfrac{1}{2}, -2\right)$
   c. $\left(\dfrac{7}{2}, 0\right)$
   d. $(2, 1)$

29. $2x + 5y = 8$
   a. $(4, 0)$
   b. $(2, 1)$
   c. $(1, 1.2)$
   d. $(1.5, 1)$

30. $3x + 4y = 10$
   a. $(-2, 3)$
   b. $(0, 2.5)$
   c. $(4, -2)$
   d. $(1.2, 1.6)$

Determine the missing coordinate in each of the ordered pairs so that the point will satisfy the equation given. See Example 2.

31. $x - y = 4$
   a. $(0, \quad)$
   b. $(2, \quad)$
   c. $(\quad, 0)$
   d. $(\quad, -3)$

32. $x + y = 7$
   a. $(0, \quad)$
   b. $(-1, \quad)$
   c. $(\quad, 0)$
   d. $(\quad, 3)$

33. $x + 2y = 6$
   a. $(0, \quad)$
   b. $(2, \quad)$
   c. $(\quad, 0)$
   d. $(\quad, 4)$

34. $3x + y = 9$
   a. $(0, \quad)$
   b. $(4, \quad)$
   c. $(\quad, 0)$
   d. $(\quad, 3)$

35. $4x - y = 8$
   a. $(0, \quad)$
   b. $(1, \quad)$
   c. $(\quad, 0)$
   d. $(\quad, 4)$

36. $x - 2y = 2$
   a. $(0, \quad)$
   b. $(4, \quad)$
   c. $(\quad, 0)$
   d. $(\quad, 3)$

37. $2x + 3y = 6$
   a. $(0, \quad)$
   b. $\left(-1, \quad\right)$
   c. $(\quad, 0)$
   d. $(\quad, -2)$

38. $5x + 3y = 15$
   a. $(0, \quad)$
   b. $\left(2, \quad\right)$
   c. $(\quad, 0)$
   d. $\left(\quad, 4\right)$

39. $3x - 4y = 7$
   a. $\left(0, \quad\right)$
   b. $(1, \quad)$
   c. $\left(\quad, 0\right)$
   d. $\left(\quad, \dfrac{1}{2}\right)$

40. $2x + 5y = 6$
   a. $\left(0, \quad\right)$
   b. $\left(\dfrac{1}{2}, \quad\right)$
   c. $(\quad, 0)$
   d. $(\quad, 2)$

41. $y = 3x$

x	y
0	
	−3
−2	
	6

42. $y = -2x$

x	y
0	
	4
3	
	−2

43. $y = 2x - 3$

x	y
0	
	−1
−2	
	3

44. $y = 3x + 5$

x	y
0	
	−4
−2	
	2

45. $y = 9 - 3x$

x	y
0	
	0
1	
	−3

46. $y = 6 - 2x$

x	y
0	
	0
−2	
	−2

47. $y = \dfrac{3}{4}x + 2$

x	y
0	
	5
−4	
	$\dfrac{5}{4}$

48. $y = \dfrac{3}{2}x - 1$

x	y
0	
	2
−2	
	$-\dfrac{5}{2}$

49. $3x - 5y = 9$

x	y
0	
	0
−2	
	−1

50. $4x + 3y = 6$

x	y
0	
	0
3	
	−1

51. $5x - 2y = 10$

x	y
0	
	0
−1	
	5

52. $3x - 2y = 12$

x	y
	0
0	
	−3
6	

53. $2x + 3.2y = 6.4$

x	y
0	
3.2	
	0.8
	−0.2

54. $3x + y = -2.4$

x	y
	0
0	
	0.6
1.6	

55.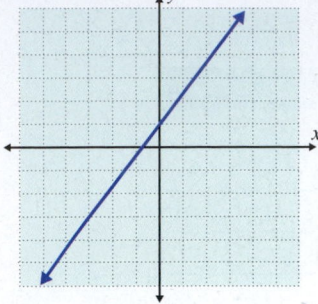

56.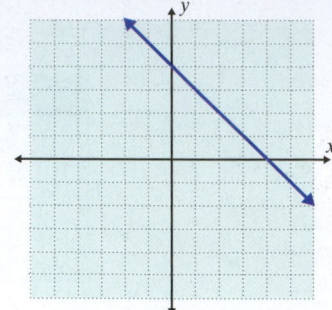

57.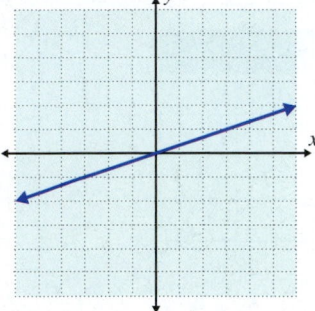

58.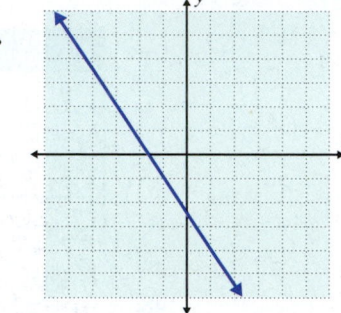

59.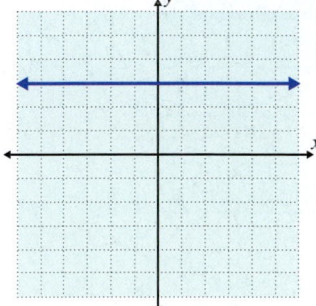

60.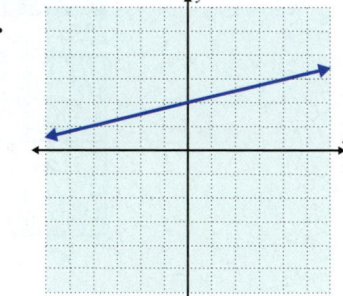

61.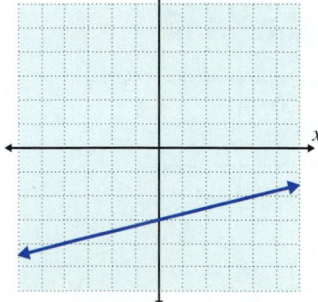

62.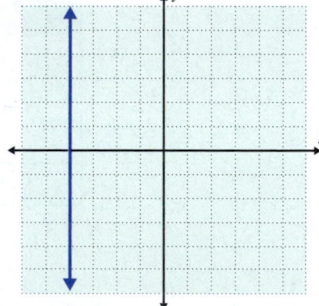

63.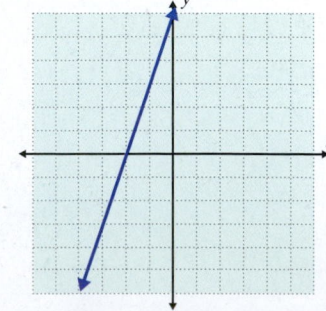

64.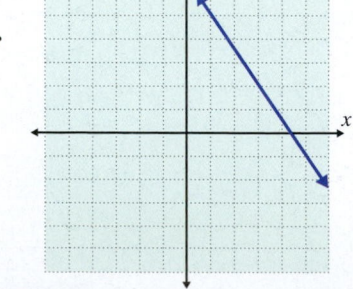

65. **Exchange Rate:** At one point in 2010, the current exchange rate from U.S. dollars to Euros was $E = 0.7802D$ where E is Euros and D is dollars.

a. Make a table of ordered pairs for the values of D and E if D has the values \$100, \$200, \$300, \$400, and \$500.

b. Graph the points corresponding to the ordered pairs.

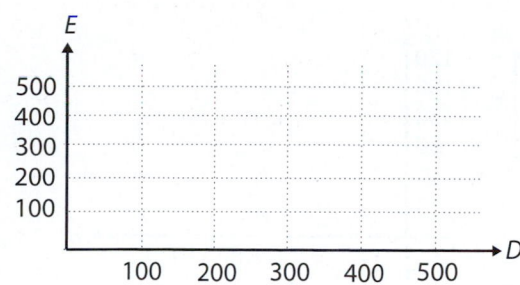

D	E
100	
200	
300	
400	
500	

66. **Temperature:** Given the equation $F = \dfrac{9}{5}C + 32$ where C is temperature in degrees Celsius and F is the corresponding temperature in degrees Fahrenheit:

a. Make a table of ordered pairs for the values of C and F if C has the values $-20°, -10°, -5°, 0°, 5°, 10°$, and $15°$.

b. Graph the points corresponding to the ordered pairs.

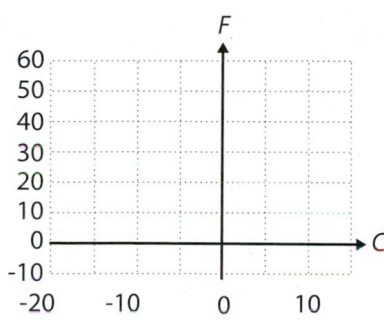

C	F
−20	
−10	
−5	
0	
5	
10	
15	

67. **Falling Objects:** Given the equation $d = 16t^2$, where d is the distance an object falls in feet and t is the time in seconds that the object falls:

a. Make a table of ordered pairs for the values of t and d with the values of $1, 2, 3.5, 4, 4.5$, and 5 for t seconds.

b. Graph the points corresponding to the ordered pairs.

c. These points do not lie on a straight line. What feature of the equation might indicate to you that the graph is not a straight line?

t	d
1	
2	
3.5	
4	
4.5	
5	

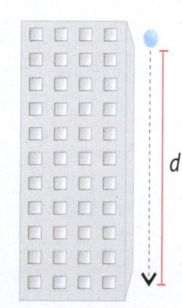

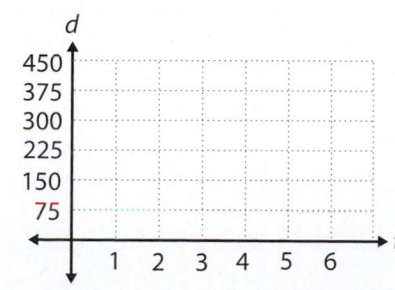

68. **Volume:** Given the equation $V = 9h$, where V is the volume (in cubic centimeters) of a box with a variable height h in centimeters and a fixed base of area 9 cm^2.

a. Make a table of ordered pairs for the values of h and V with h as the values 2 cm, 3 cm, 5 cm, 8 cm, 9 cm, and 10 cm.

b. Graph the points corresponding to the ordered pairs.

h	V
2	
3	
5	
8	
9	
10	

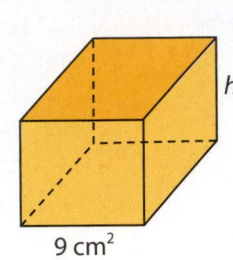

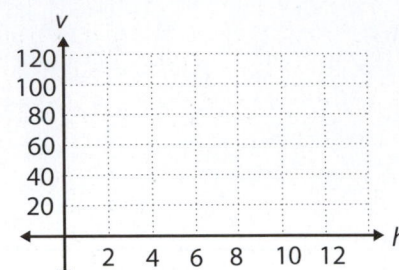

Writing & Thinking

In statistics, data is sometimes given in the form of ordered pairs where each ordered pair represents two pieces of information about one person. For example, ordered pairs might represent the height and weight of a person or the person's number of years of education and that person's annual income. The ordered pairs are plotted on a graph and the graph is called a **scatter diagram** (or **scatter plot**). Such scatter diagrams are used to see if there is any pattern to the data and, if there is, then the diagram is used to predict the value for one of the variables if the value of the other is known. For example, if you know that a person's height is 5 ft 6 in., then his or her weight might be predicted from information indicated in a scatter diagram that has several points of known information about height and weight.

69. a. The following table of values indicates the number of push-ups and the number of sit-ups that ten students did in a physical education class. Plot these points in a scatter diagram.

Person	#1	#2	#3	#4	#5	#6	#7	#8	#9	#10
x (push-ups)	20	15	25	23	35	30	42	40	25	35
y (sit-ups)	25	20	20	30	32	36	40	45	18	40

b. Does there seem to be a pattern in the relationship between push-ups and sit-ups? What is this pattern?

c. Using the scatter diagram in part **a.**, predict the number of sit-ups that a student might be able to do if he or she has just done each of the following numbers of push-ups: 22, 32, 35, and 45. (**Note:** In each case, there is no one correct answer. The answers are only estimates based on the diagram.)

70. Ask ten friends or fellow students what their height and shoe size is. (You may want to ask all boys or all girls since the scale for boys and girls' shoe sizes is different.) Organize the data in table form and then plot the corresponding scatter diagram. Knowing your own height, does the pattern indicated in the scatter diagram seem to predict your shoe size?

71. Ask ten friends or fellow students what their height and age is. Organize the data in table form and then plot the corresponding scatter diagram. Knowing your own height, does the pattern indicated in the scatter diagram seem to predict your age? Do you think that all scatter diagrams can be used to predict information related to the two variables graphed? Explain.

6.2 Graphing Linear Equations in Two Variables

Objectives

A Recognize the standard form of a linear equation in two variables: $Ax + By = C$.

B Plot points that satisfy a linear equation and graph the corresponding line.

C Find the y-intercept and x-intercept of a line and graph the corresponding line.

Objective A **The Standard Form:** $Ax + By = C$

In Section 6.1, we discussed ordered pairs of real numbers and graphed a few points (ordered pairs) that satisfied particular equations. Now, suppose we want to graph all the points that satisfy an equation such as:

$$3 - 3x = y.$$

The **solution set** for equations of this type (in the two variables x and y) consists of an infinite set of ordered pairs in the form (x, y) that satisfy the equation.

To find some of the solutions of the equation $3 - 3x = y$, we form a table (as we did in Section 6.1) by:

1. choosing arbitrary values for x and
2. finding the corresponding values for y by substituting into the equation.

In Figure 1, we have found five ordered pairs that satisfy the equation and graphed the corresponding points.

Choices x	Substitutions $3 - 3x = y$	Results (x, y)
-1	$3 - 3(-1) = 6$	$(-1, 6)$
0	$3 - 3(0) = 3$	$(0, 3)$
$\frac{2}{3}$	$3 - 3\left(\frac{2}{3}\right) = 1$	$\left(\frac{2}{3}, 1\right)$
2	$3 - 3(2) = -3$	$(2, -3)$
3	$3 - 3(3) = -6$	$(3, -6)$

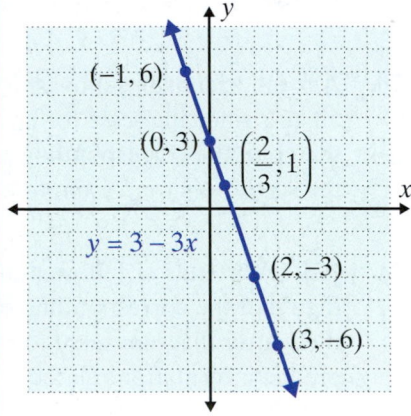

Figure 1

The five points in Figure 1 appear to lie on a line. They do, in fact, lie on a line, and any ordered pair that satisfies the equation $y = 3 - 3x$ will also lie on that same line.

Just as we use the terms **ordered pair** and **point** (the graph of an ordered pair) interchangeably, we use the terms **equation** and **graph of an equation** interchangeably. The equations

$$2x + 3y = 4, \quad y = -5, \quad x = 1.4, \quad \text{and} \quad y = 3x + 2$$

are called **linear equations**, and their graphs are lines on the Cartesian plane.

Standard Form of a Linear Equation

Any equation of the form

$$Ax + By = C,$$

where A, B, and C are real numbers and A and B are not both equal to 0, is called the **standard form of a linear equation**.

notes

Note that in the standard form $Ax + By = C$, A and B may be positive, negative, or 0, but A and B cannot **both** equal 0.

Objective B ### Graphing Linear Equations

Every line corresponds to some linear equation, and the graph of every linear equation is a line. We know from geometry that **two points determine a line**. This means that the graph of a linear equation can be found by locating any two points that satisfy the equation.

To Graph a Linear Equation in Two Variables

1. Locate any two points that satisfy the equation. (Choose values for x and y that lead to simple solutions. Remember that there is an infinite number of choices for either x or y. But, once a value for x or y is chosen, the corresponding value for the other variable is found by substituting into the equation.)
2. Plot these two points on a Cartesian coordinate system.
3. Draw a line through these two points. (**Note:** Every point on that line will satisfy the equation.)
4. **To check:** Locate a third point that satisfies the equation and check to see that it does indeed lie on the line.

1. Graph each of the following linear equations.

   a. $3x + 2y = 6$

   b. $2x - y = 2$

   c. $y = 3x$

Example 1

Graphing a Linear Equation in Two Variables

Graph each of the following linear equations.

a. $2x + 3y = 6$

Solution

Make a table with headings x and y and, whenever possible, **choose values for x or y that lead to simple solutions for the other variable**. (Values chosen for x and y are shown in red.)

x	$2x + 3y = 6$	y
0	$2(0) + 3y = 6$	2
−3	$2(-3) + 3y = 6$	4
3	$2x + 3(0) = 6$	0
$\dfrac{5}{2}$	$2x + 3\left(\dfrac{1}{3}\right) = 6$	$\dfrac{1}{3}$

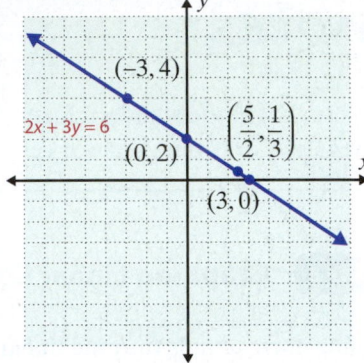

b. $x - 2y = 1$

Solution

Solve the equation for x ($x = 2y + 1$) and substitute 0, 1, and 2 for y.

x	$x = 2y + 1$	y
1	$x = 2(0) + 1$	0
3	$x = 2(1) + 1$	1
5	$x = 2(2) + 1$	2

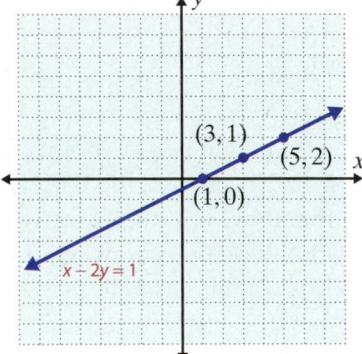

c. $y = 2x$

Solution

Substitute −1, 0, and 1 for x.

x	$y = 2x$	y
−1	$y = 2(-1)$	−2
0	$y = 2(0)$	0
1	$y = 2(1)$	2

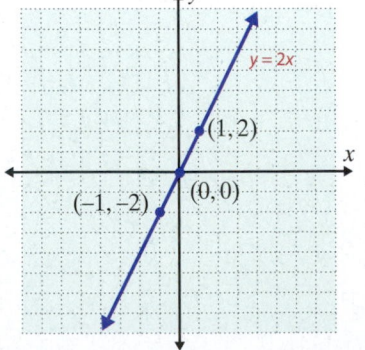

Now work margin exercise 1.

Locating the *y*-intercept and *x*-intercept

While the choice of the values for *x* or *y* can be arbitrary, letting $x = 0$ will locate the point on the graph where the line crosses (or intercepts) the *y*-axis. This point is called the **y-intercept**, and is of the form $(0, y)$. The **x-intercept** is the point found by letting $y = 0$. This is the point where the line crosses (or intercepts) the *x*-axis, and is of the form $(x, 0)$. These two points are generally easy to locate and are frequently used as the two points for drawing the graph of a linear equation. If a line passes through the point $(0, 0)$, then the *y*-intercept and the *x*-intercept are the same point, namely the origin. In this case, you will need to locate some other point to draw the graph.

Intercepts

1. To find the **y-intercept** (where the line crosses the *y*-axis), substitute $x = 0$ and solve for *y*.
2. To find the **x-intercept** (where the line crosses the *x*-axis), substitute $y = 0$ and solve for *x*.

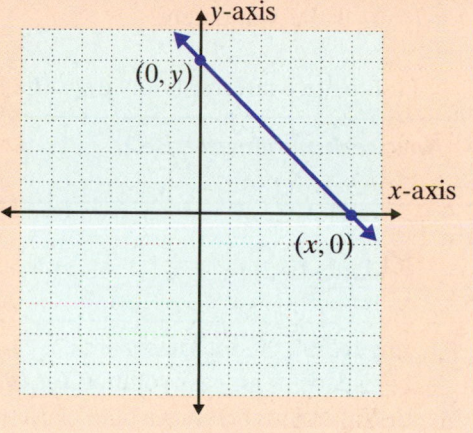

Example 2

y- and x- intercepts

Graph the following linear equations by locating the *y*-intercept and the *x*-intercept.

a. $x + 3y = 9$

Solution

$x = 0$: $(0) + 3y = 9$
$$3y = 9$$
$$y = 3$$

$(0, 3)$ is the *y*-intercept.

$y = 0$: $x + 3(0) = 9$
$$x = 9$$

$(9, 0)$ is the *x*-intercept.

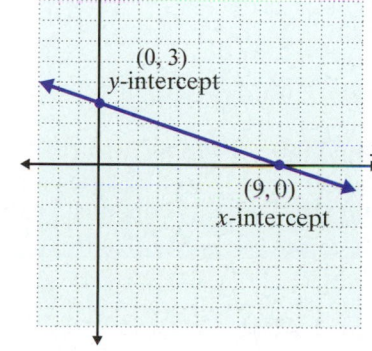

Plot the two intercepts and draw the line that contains them.

2. Graph the following linear equations by locating the x-intercept and the y-intercept.

a. $3x + 6y = 18$

b. $5x - 3y = 15$

b. $3x - 2y = 12$

Solution

$x = 0$: $3(0) - 2y = 12$
$$-2y = 12$$
$$y = -6$$

$(0, -6)$ is the y-intercept.

$y = 0$: $3x - 2(0) = 12$
$$3x = 12$$
$$x = 4$$

$(4, 0)$ is the x-intercept.

Plot the two intercepts and draw the line that contains them.

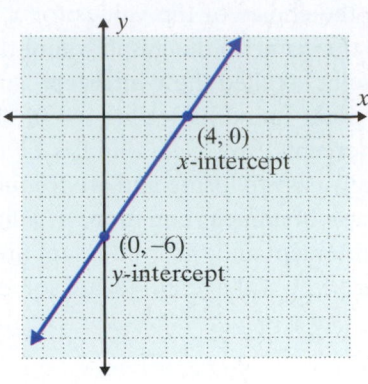

Now work margin exercise 2.

notes

■ In general, the intercepts are easy to find because substituting 0 for x or y leads to an easy solution for the other variable. However, when ■ the intercepts result in a point with fractional (or decimal) coordinates and estimation is involved, then a third point that satisfies the equation ■ should be found to verify that the line is graphed correctly.

Practice Problems

1. Find the missing coordinate of each ordered pair so that it belongs to the solution set of the equation $2x + y = 4$.
$$(0, \ ), (\ , 0), (\ , 8), (-1, \ )$$

2. Does the ordered pair $\left(1, \dfrac{3}{2}\right)$ satisfy the equation $3x + 2y = 6$?

3. Find the x-intercept and y-intercept of the equation $-3x + y = 9$.

4. Graph the linear equation $x - 2y = 3$.

Practice Problem Answers

1. $(0, 4), (2, 0), (-2, 8), (-1, 6)$ **2.** yes

3. x-intercept $= (-3, 0)$, y-intercept $= (0, 9)$ **4.**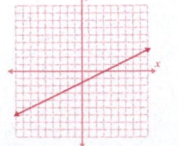

Exercises 6.2

Use your knowledge of *y*-intercepts and *x*-intercepts to match each of the following equations with its graph.

1. $4x + 3y = 12$

2. $4x - 3y = 12$

3. $x + 2y = 8$

4. $-x + 2y = 8$

5. $x + 4y = 0$

6. $5x - y = 10$

a.
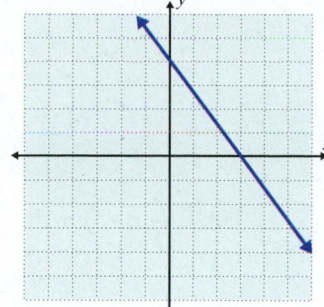

b.
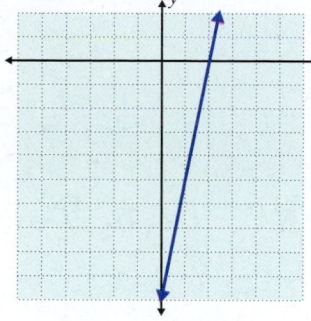

c.
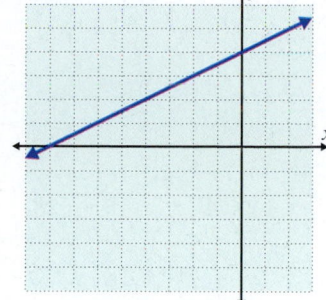

d.
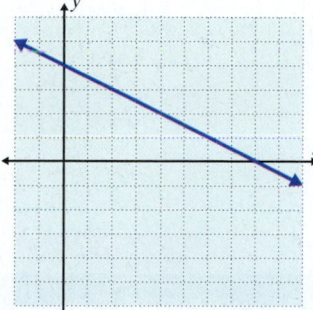

e.
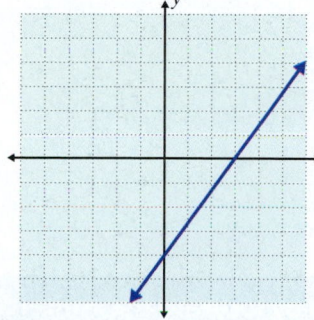

f.
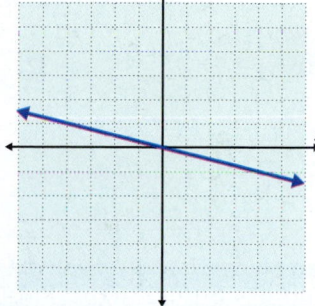

Locate at least two ordered pairs of real numbers that satisfy each of the linear equations and graph the corresponding line in the Cartesian coordinate system. See Example 1.

7. $x + y = 3$

8. $x + y = 4$

9. $y = x$

10. $2y = x$

11. $2x + y = 0$

12. $3x + 2y = 0$

13. $2x + 3y = 7$

14. $4x + 3y = 11$

15. $3x - 4y = 12$

16. $2x - 5y = 10$

17. $y = 4x + 4$

18. $y = x + 2$

19. $3y = 2x - 4$ **20.** $4x = 3y + 8$ **21.** $3x + 5y = 6$ **22.** $2x + 7y = -4$

23. $2x + 3y = 1$ **24.** $5x - 3y = -1$ **25.** $5x - 2y = 7$ **26.** $3x + 4y = 7$

27. $\dfrac{2}{3}x - y = 4$ **28.** $x + \dfrac{3}{4}y = 6$ **29.** $2x + \dfrac{1}{2}y = 3$ **30.** $\dfrac{2}{5}x - 3y = 5$

31. $5x = y + 2$ **32.** $4x = 3y - 5$

Graph the following linear equations by locating the *x*-intercept and the *y*-intercept. See Example 2

33. $x + y = 6$ **34.** $x + y = 4$ **35.** $x - 2y = 8$ **36.** $x - 3y = 6$

37. $4x + y = 8$ **38.** $x + 3y = 9$ **39.** $x - 4y = -6$ **40.** $x - 6y = 3$

41. $y = 4x - 10$ **42.** $y = 2x - 9$ **43.** $3x - 2y = 6$ **44.** $5x + 2y = 10$

45. $2x + 3y = 12$ **46.** $3x + 7y = -21$ **47.** $3x - 7y = -21$ **48.** $3x + 2y = 15$

49. $5x + 3y = 7$

50. $2x + 3y = 5$

51. $y = \dfrac{1}{2}x - 4$

52. $y = -\dfrac{1}{3}x + 3$

53. $\dfrac{2}{3}x - 3y = 4$

54. $\dfrac{1}{2}x + 2y = 3$

55. $\dfrac{1}{2}x - \dfrac{3}{4}y = 6$

56. $\dfrac{2}{3}x + \dfrac{4}{3}y = 8$

Writing & Thinking

57. Explain, in your own words, why it is sufficient to find the x-intercept and y-intercept to graph a line (assuming that they are not the same point).

58. Explain, in your own words, how you can determine if an ordered pair is a solution to an equation.

6.3 The Slope-Intercept Form

Objective A **The Meaning of Slope**

If you ride a bicycle up a mountain road, you certainly know when the **slope** (a measure of steepness called the **grade** for roads) increases because you have to pedal harder. The contractor who built the road was aware of the **slope** because trucks traveling the road must be able to control their downhill speed and be able to stop in a safe manner. A carpenter given a set of house plans calling for a roof with a **pitch** of 7 : 12 knows that for every 7 feet of rise (vertical distance) there are 12 feet of run (horizontal distance). That is, the ratio of rise to run is $\dfrac{rise}{run} = \dfrac{7}{12}$.

Objectives

A Interpret the slope of a line as a rate of change.

B Calculate the slope of a line given two points that lie on the line.

C Find the slopes of and graph horizontal and vertical lines.

D Recognize the slope-intercept form for a linear equation in two variables: $y = mx + b$.

Figure 1

Note that this ratio can be in units other than feet, such as inches or meters. (See Figure 2.)

$$\frac{rise}{run} = \frac{7 \text{ inches}}{12 \text{ inches}} = \frac{3.5 \text{ feet}}{6 \text{ feet}} = \frac{14 \text{ feet}}{24 \text{ feet}}$$

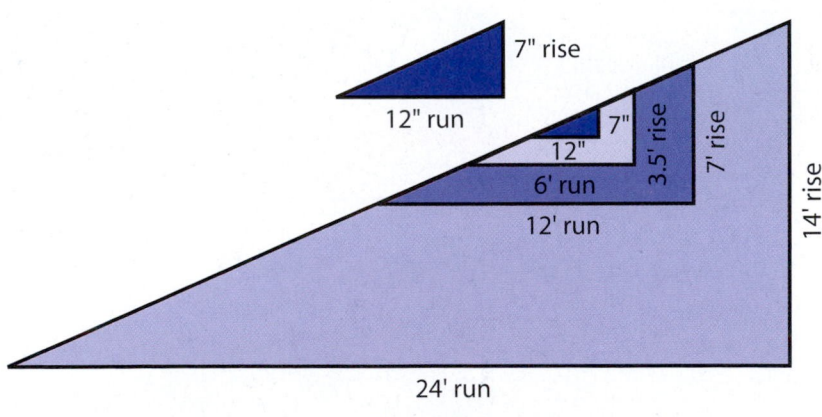

Figure 2

For a line, the **ratio of rise to run** is called the **slope of the line**. The graph of the linear equation $y = \dfrac{1}{3}x + 2$ is shown in Figure 3. What do you think is the slope of the line? Do you think that the slope is positive or negative? Do you think the slope might be $\dfrac{1}{3}$? $\dfrac{3}{1}$? 2?

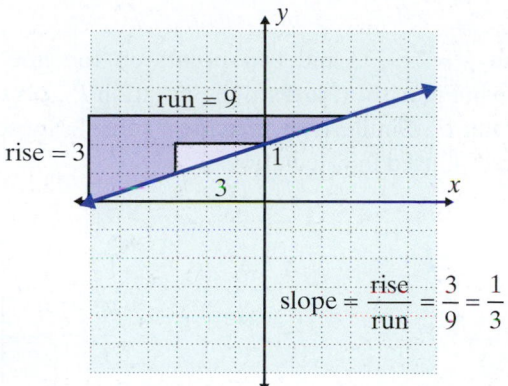

Figure 3

The concept of slope also relates to situations that involve **rate of change**. For example, the graphs in Figure 4 illustrate slope as miles per hour that a car travels and as pages per minute that a printer prints.

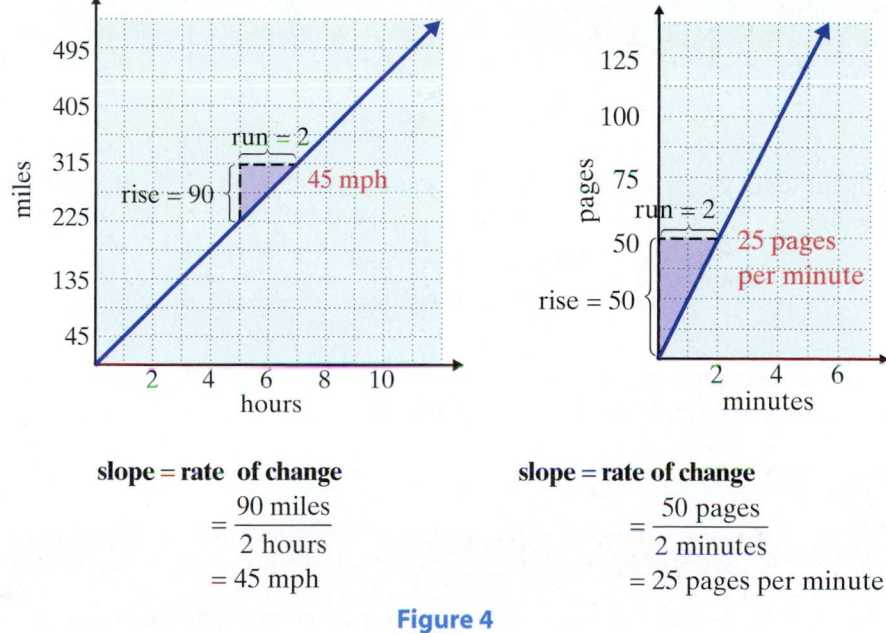

slope = rate of change

$$= \frac{90 \text{ miles}}{2 \text{ hours}}$$

$$= 45 \text{ mph}$$

slope = rate of change

$$= \frac{50 \text{ pages}}{2 \text{ minutes}}$$

$$= 25 \text{ pages per minute}$$

Figure 4

In general, the ratio of a change in one variable (say *y*) to a change in another variable (say *x*) is called the **rate of change of *y* with respect to *x***. Figure 5 shows how the rate of change (the slope) can change over periods of time.

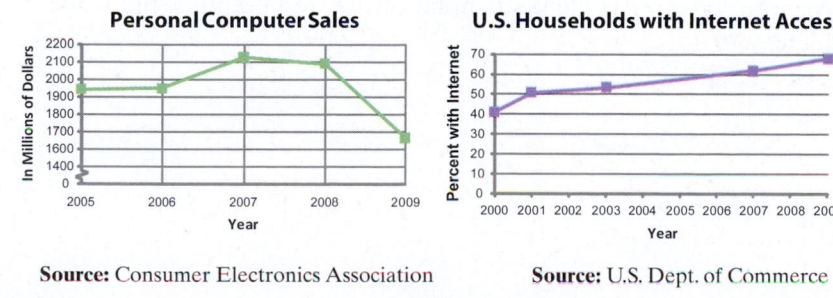

Source: Consumer Electronics Association

Source: U.S. Dept. of Commerce

Figure 5

Objective B **Calculating the Slope**

Consider the line $y = 2x + 3$, and two points on the line $P_1(-2, -1)$ and $P_2(2, 7)$ as shown in Figure 6. (**Note:** In the notation P_1, 1 is called a **subscript** and P_1 is read "P sub 1". Similarly, P_2 is read "P sub 2." Subscripts are used in "labeling" and are not used in calculations.)

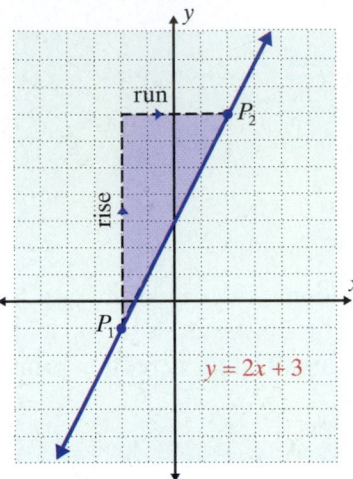 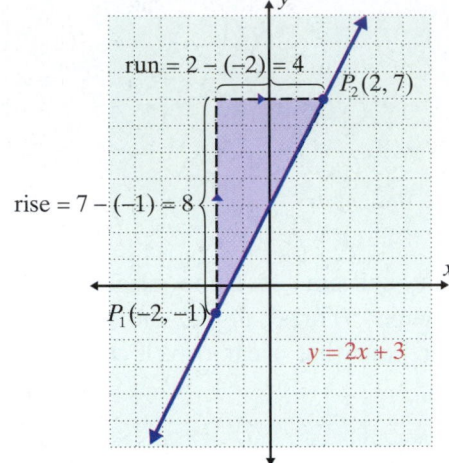

Figure 6

For the line $y = 2x + 3$ and using the points $(-2, -1)$ and $(2, 7)$ that are on the line,

$$\text{slope} = \frac{\text{rise}}{\text{run}} = \frac{\text{difference in } y\text{-values}}{\text{difference in } x\text{-values}} = \frac{7 - (-1)}{2 - (-2)} = \frac{8}{4} = 2.$$

From similar illustrations and the use of subscript notation, we can develop the following formula for the slope of any line.

Slope

Let $P_1(x_1, y_1)$ and $P_2(x_2, y_2)$ be two points on a line. The **slope** can be calculated as follows.

$$\text{slope} = m = \frac{\text{rise}}{\text{run}} = \frac{y_2 - y_1}{x_2 - x_1}$$

Note: The letter m is standard notation for representing the slope of a line.

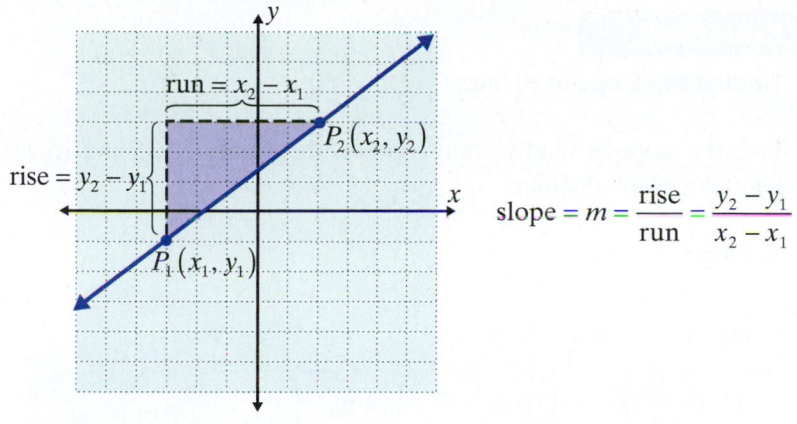

Figure 7

$$\text{slope} = m = \frac{\text{rise}}{\text{run}} = \frac{y_2 - y_1}{x_2 - x_1}$$

Example 1

Finding the Slope of a Line

Find the slope of the line that contains the points $(-1, 2)$ and $(3, 5)$, and then graph the line.

Solution

Using $(-1, 2)$ and $(3, 5)$:
(x_1, y_1) (x_2, y_2)

$$\text{slope} = m = \frac{y_2 - y_1}{x_2 - x_1}$$

$$= \frac{5 - 2}{3 - (-1)}$$

$$= \frac{3}{4}$$

Or, using $(3, 5)$ and $(-1, 2)$:
(x_1, y_1) (x_2, y_2)

$$\text{slope} = m = \frac{y_2 - y_1}{x_2 - x_1}$$

$$= \frac{2 - 5}{-1 - 3}$$

$$= \frac{-3}{-4}$$

$$= \frac{3}{4}$$

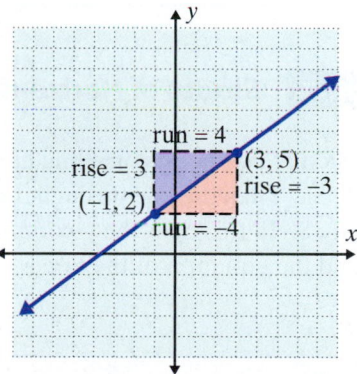

Now work margin exercise 1.

As we see in Example 1, **the slope is the same even if the order of the points is reversed**. The important part of the procedure is that **the coordinates must be subtracted in the same order in both the numerator and the denominator**.

In general,

$$\text{slope} = \frac{y_2 - y_1}{x_2 - x_1} = \frac{y_1 - y_2}{x_1 - x_2}.$$

1. Find the slope of the line that contains the points $(3, -1)$ and $(4, 2)$, and then graph the line.

The Slope–Intercept Form   **Section 6.3**   **688**

2. Find the slope of the line that contains the points $(0, 5)$ and $(4, 2)$.

Example 2

Finding the Slope of a Line

Find the slope of the line that contains the points $(1, 3)$ and $(5, 1)$, and then graph the line.

Solution

Using $(1, 3)$ and $(5, 1)$,
(x_1, y_1) (x_2, y_2)

$$\text{slope} = m = \frac{1-3}{5-1}$$

$$= \frac{-2}{4}$$

$$= -\frac{1}{2}$$

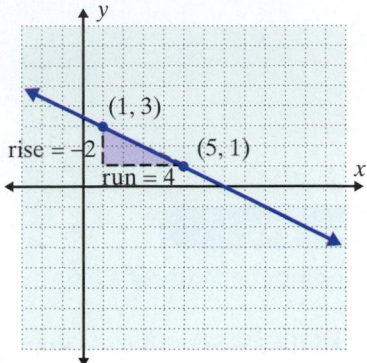

Now work margin exercise **2.**

notes

- Lines with **positive slope go up** (increase) as we move along the line from left to right.

- Lines with **negative slope go down** (decrease) as we move along the line from left to right.

Objective C **Slopes of Horizontal and Vertical Lines**

Suppose that two points on a line have the same *y*-coordinate, such as $(-2, 3)$ and $(5, 3)$. Then the line through these two points will be **horizontal** as shown in Figure 8. In this case, the *y*-coordinates of points on the horizontal line are all 3, and the equation of the line is simply $y = 3$. The slope is

$$m = \frac{3-3}{5-(-2)} = \frac{0}{7} = 0.$$

For any horizontal line, all of the *y*-values will be the same. Consequently, the formula for slope will always have 0 in the numerator. Therefore, **the slope of every horizontal line is 0**.

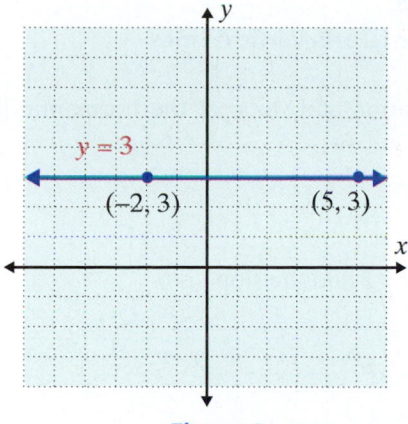

Figure 8

If two points have the same x-coordinates, such as $(1, 3)$ and $(1, -2)$, then the line through these two points will be **vertical** as in Figure 9. The x-coordinates for every point on the vertical line are all 1, and the equation of the line is simply $x = 1$. The slope is

$$m = \frac{-2 - 3}{1 - 1} = \frac{-5}{0}, \text{ which is } \textbf{undefined.}$$

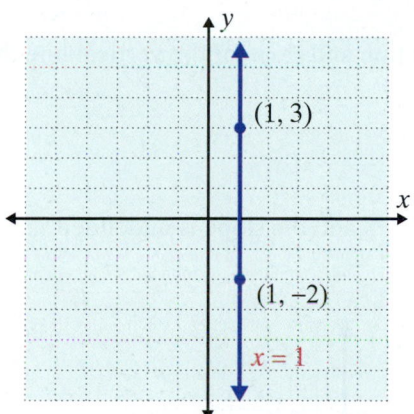

Figure 9

Horizontal and Vertical Lines

The following two general statements are true for horizontal and vertical lines.

1. For **horizontal lines** (of the form $y = b$), the **slope is 0**.
2. For **vertical lines** (of the form $x = a$), the **slope is undefined**.

3. Find the equation and slope of the indicated line.

a. horizonal line through $(3, -2)$

b. vertical line through $(2, 4)$

Example 3

Slopes of Horizontal and Vertical Lines

a. Find the equation and slope of the horizontal line through the point $(-2, 5)$.

Solution

The equation is $y = 5$ and the slope is 0.

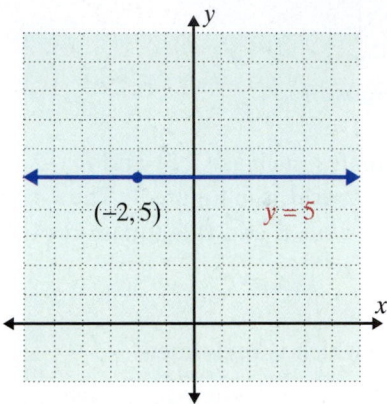

b. Find the equation and slope of the vertical line through the point $(3, 2)$.

Solution

The equation is $x = 3$ and the slope is undefined.

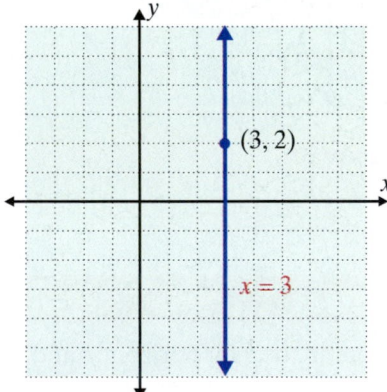

Now work margin exercise 3.

Objective D **Slope-Intercept Form**

There are certain relationships between the coefficients in the equation of a line and the graph of that line. For example, consider the equation

$$y = 5x - 7.$$

First, find two points on the line and calculate the slope. $(0, -7)$ and $(2, 3)$ both satisfy the equation.

$$\text{slope} = m = \frac{3 - (-7)}{2 - 0} = \frac{10}{2} = 5$$

Observe that the slope, $m = 5$, is the same as the coefficient of x in the equation $y = 5x - 7$. This is not just a coincidence. In fact, if a linear equation is solved for y, then the coefficient of x will always be the slope of the line.

For $y = mx + b$, m is the Slope

For an equation in the form $y = \mathbf{mx + b}$, the slope of the line is m.

For the line $y = mx + b$, the point where $x = 0$ is the point where the line crosses the y-axis. Recall that this point is called the **y-intercept**. By letting $x = 0$, we get

$$y = mx + b$$
$$y = m \cdot 0 + b$$
$$y = b.$$

Thus the point $(0, b)$ is the y-intercept. The concepts of slope and y-intercept lead to the following definition.

Slope-Intercept Form

$y = mx + b$ is called the **slope-intercept form** for the equation of a line, where m is the **slope** and $(0, b)$ is the **y-intercept**.

As illustrated in Example 4, an equation in the **standard form**

$$Ax + By = C \text{ with } B \neq 0$$

can be written in the slope-intercept form by solving for y.

Example 4

Using the Form $y = mx + b$

a. Find the slope and y-intercept of $-2x + 3y = 6$ and graph the line.

Solution

Solve for y.

$$-2x + 3y = 6$$
$$3y = 2x + 6$$
$$\frac{3y}{3} = \frac{2x}{3} + \frac{6}{3}$$
$$y = \frac{2}{3}x + 2$$

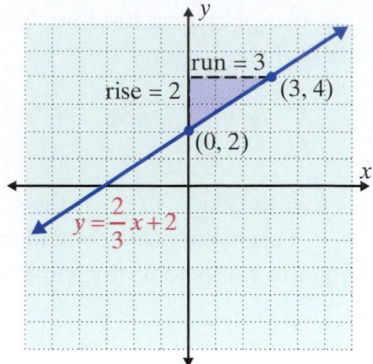

Thus $m = \dfrac{2}{3}$, which is the slope, and b is 2, making the y-intercept equal to $(0, 2)$.

As shown in the graph, if we "rise" 2 units up and "run" 3 units to the right **from the y-intercept** $(0, 2)$, we locate another point $(3, 4)$. The line can be drawn through these two points. **Note:** As shown in the graph on the right, we could also first "run" 3 units right and "rise" 2 units up from the y-intercept to locate the point $(3, 4)$ on the graph.

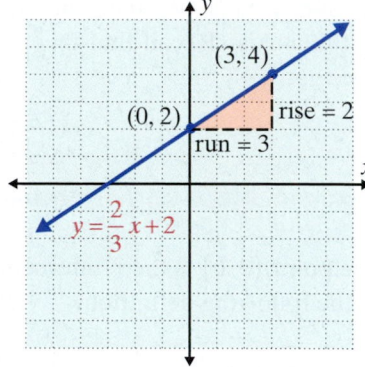

b. Find the slope and y-intercept of $x + 2y = -6$ and graph the line.

Solution

Solve for y.

$$x + 2y = -6$$
$$2y = -x - 6$$
$$\frac{2y}{2} = \frac{-x}{2} - \frac{6}{2}$$
$$y = -\frac{1}{2}x - 3$$

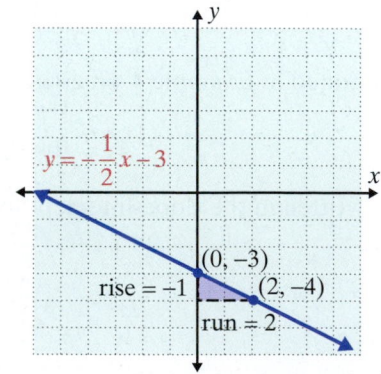

Thus $m = -\dfrac{1}{2}$, which is the slope, and b is -3, making the y-intercept equal to $(0, -3)$.

We can treat $m = -\dfrac{1}{2}$ as $m = \dfrac{-1}{2}$ and the "rise" as -1 and the "run" as 2. Moving from $(0, -3)$ as shown in the graph above, we locate another point $(2, -4)$ on the graph and draw the line.

c. Find the equation of the line through the point $(0, -2)$ with slope $\dfrac{1}{2}$.

Solution

Because the x-coordinate is 0, we know that the point $(0, -2)$ is the y-intercept. So $b = -2$. The slope is $\dfrac{1}{2}$. So $m = \dfrac{1}{2}$. Substituting in slope-intercept form $y = mx + b$ gives the result: $y = \dfrac{1}{2}x - 2$.

Now work margin exercise 4.

4. a. Find the slope and y-intercept of $-4x + 2y = 12$ and graph the line.

b. Find the slope and y-intercept of $3x + 2y = -10$ and graph the line.

c. Find the equation of the line through the point $(0, -3)$ with a slope of $\dfrac{2}{3}$.

Practice Problems

1. Find the slope of the line through the two points $(1, 3)$ and $(4, 6)$. Graph the line.

2. Find the equation of the line through the point $(0, 5)$ with slope $-\dfrac{1}{3}$.

3. Find the slope and y-intercept for the line $2x + y = 7$.

4. Write the equation for the horizontal line through the point $(-1, 3)$. What is the slope of this line?

5. Write the equation for the vertical line through the point $(-1, 3)$. What is the slope of this line?

Practice Problem Answers

1. $m = 1$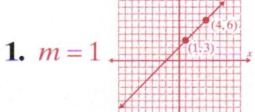

2. $y = -\dfrac{1}{3}x + 5$

3. $m = -2$; y-intercept $= (0, 7)$

4. $y = 3$; slope is 0

5. $x = -1$; slope is undefined

Exercises 6.3

1. $(2, 4); (1, -1)$

2. $(1, -2); (1, 4)$

3. $(-6, 3); (1, 2)$

4. $(-3, 7); (4, -1)$

5. $(-5, 8); (3, 8)$

6. $(-2, 3); (-2, -1)$

7. $(5, 1); (3, 0)$

8. $(0, 0); (-2, -3)$

9. $\left(\dfrac{3}{4}, \dfrac{3}{2}\right); (1, 2)$

10. $\left(4, \dfrac{1}{2}\right); (-1, 2)$

11. $\left(\dfrac{3}{2}, \dfrac{4}{5}\right); \left(-2, \dfrac{1}{10}\right)$

12. $\left(\dfrac{7}{2}, \dfrac{3}{4}\right); \left(\dfrac{1}{2}, -3\right)$

Determine whether each equation represents a horizontal line or vertical line and give its slope. Graph the line. See Example 3.

13. $y = 5$

14. $y = -2$

15. $x = -3$

16. $x = 1.7$

17. $3y = -18$

18. $4x = 2.4$

19. $-3x + 21 = 0$

20. $2y + 5 = 0$

21. $y = 2x - 1$

22. $y = 3x - 4$

23. $y = 5 - 4x$

24. $y = 4 - x$

25. $y = \dfrac{2}{3}x - 3$

26. $y = \dfrac{2}{5}x + 2$

27. $x + y = 5$

28. $x - 2y = 6$

29. $x + 5y = 10$

30. $4x + y = 0$

31. $4x + y + 3 = 0$

32. $2x + 7y + 7 = 0$

33. $2y - 8 = 0$

34. $3y - 9 = 0$

35. $2x = 3y$

36. $4x = y$

37. $3x + 9 = 0$

38. $4x + 7 = 0$

39. $5x - 6y = 18$

40. $3x + 6 = 6y$

41. $5 - 3x = 4y$

42. $5x = 11 - 2y$

43. $6x + 4y = -8$

44. $7x + 2y = 4$

45. $6y = -6 + 3x$

46. $4x = 3y - 7$

47. $5x - 2y + 5 = 0$

48. $6x + 5y = -15$

In reference to the equation $y = mx + b$, sketch the graph of three lines for each of the two characteristics listed below.

49. $m > 0$ and $b > 0$ **50.** $m < 0$ and $b > 0$ **51.** $m > 0$ and $b < 0$ **52.** $m < 0$ and $b < 0$

Find an equation in slope-intercept form for the line passing through the given point with the given slope. See Example 4.

53. $(0, 3); m = -\dfrac{1}{2}$ **54.** $(0, 2); m = \dfrac{1}{3}$ **55.** $(0, -3); m = \dfrac{2}{5}$ **56.** $(0, -6); m = \dfrac{4}{3}$

57. $(0, -5); m = 4$ **58.** $(0, 9); m = -1$ **59.** $(0, -4); m = 1$ **60.** $(0, 6); m = -5$

61. $(0, -3); m = -\dfrac{5}{6}$ **62.** $(0, -1); m = -\dfrac{3}{2}$

The graph of a line is shown with two points highlighted. Find a. the slope, b. the y-intercept (if there is one), and c. the equation of the line in slope-intercept form.

63.
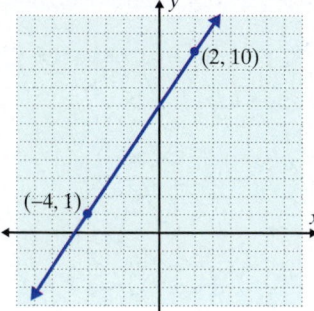

64.
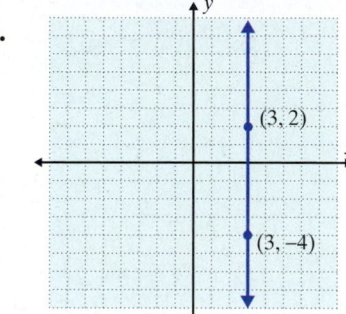

65.
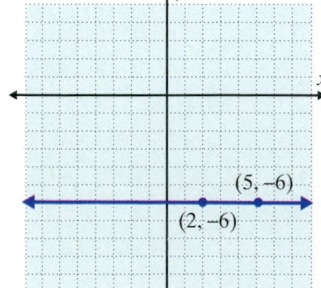

66.
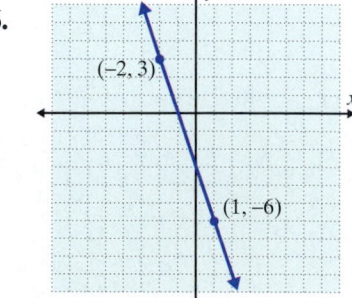

67.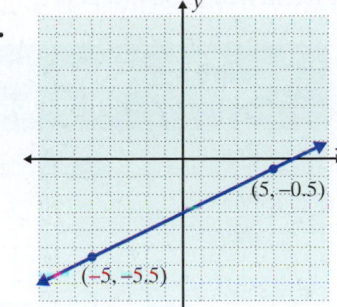

68.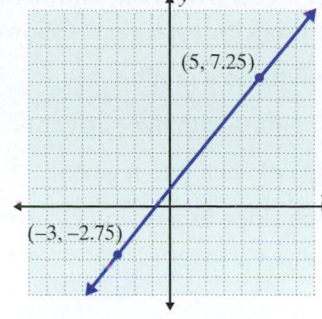

69.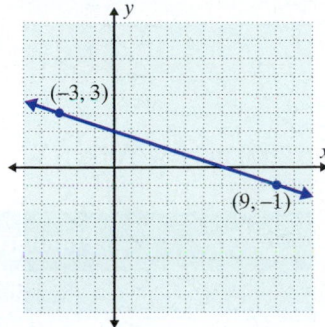

70.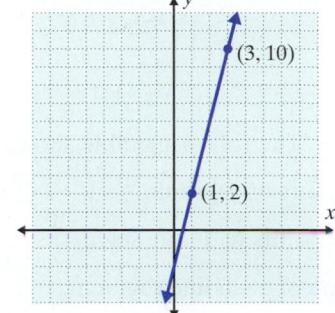

71. $\{(-1, 3), (0, 1), (5, -9)\}$ **72.** $\{(-2, -4), (0, 2), (3, 11)\}$ **73.** $\{(-2, 0), (0, 30), (1.5, 5.25)\}$

74. $\{(-1, -7), (1, 1), (2.5, 7)\}$ **75.** $\left\{\left(\frac{2}{3}, \frac{1}{2}\right), \left(0, \frac{5}{6}\right), \left(-\frac{3}{4}, \frac{29}{24}\right)\right\}$ **76.** $\left\{\left(\frac{3}{2}, -\frac{1}{3}\right), \left(0, \frac{1}{6}\right), \left(-\frac{1}{2}, \frac{3}{4}\right)\right\}$

Solve the following word problems.

77. **Buying a New Car:** John bought his new car for $35,000 in the year 2007. He knows that the value of his car has depreciated linearly. If the value of the car in 2010 was $23,000, what was the annual rate of depreciation of his car? Show this information on a graph. (When graphing, use years as the *x*-coordinates and the corresponding values of the car as the *y*-coordinates.)

 78. **Cell Phone Usage:** The number of people in the United States with mobile cellular phones was about 180 million in 2004 and about 286 million in 2009. If the growth in mobile cellular phones was linear, what was the approximate rate of growth per year from 2004 to 2009. Show this information on a graph. (When graphing, use years as the *x*-coordinates and the corresponding number of users as the *y*-coordinates.)

79. Internet Usage: The given table shows the estimated number of internet users from 2004 to 2008. The number of users for each year is shown in millions.

a. Plot these points on a graph.

b. Connect the points with line segments.

c. Find the slope of each line segment.

d. Interpret each slope as a rate of change.

Source: International Telecommunications Union Yearbook of Statistics

Year	Internet Users (in millions)
2004	185
2005	198
2006	210
2007	220
2008	231

80. Urban Growth: The following table shows the urban growth from 1850 to 2000 in New York, NY.

a. Plot these points on a graph.

b. Connect the points with line segments.

c. Find the slope of each line segment.

d. Interpret each slope as a rate of change.

Source: U.S. Census Bureau

Year	Population
1850	515,547
1900	3,437,202
1950	7,891,957
2000	8,008,278

81. Military: The following graph shows the number of female active duty military personnel over a span from 1945 to 2009. The number of women listed includes both officers and enlisted personnel from the Army, the Navy, the Marine Corps, and the Air Force.

a. Plot these points on a graph.

b. Connect the points with line segments.

c. Find the slope of each line segment.

d. Interpret each slope as a rate of change.

Source: U.S. Dept. of Defense

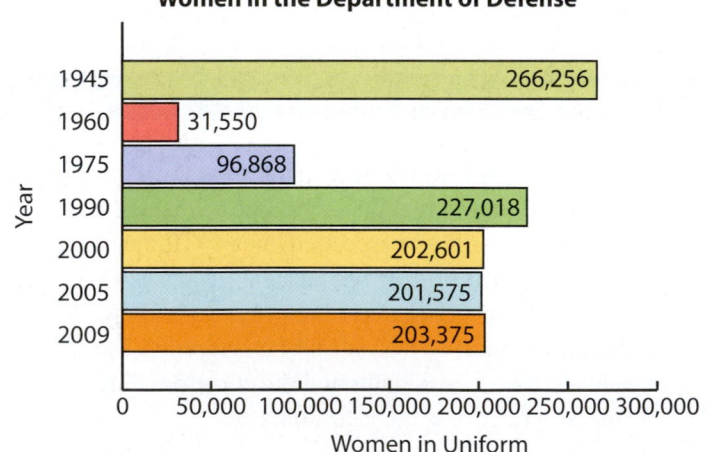

Women in the Department of Defense

82. Marriage: The following graph shows the rates of marriage per 1000 people in the U.S., over a span from 1920 to 2008.

a. Plot these points on a graph.

b. Connect the points with line segments.

c. Find the slope of each line segment.

d. Interpret each slope as a rate of change.

Source : U.S. National Center for Health Statistics

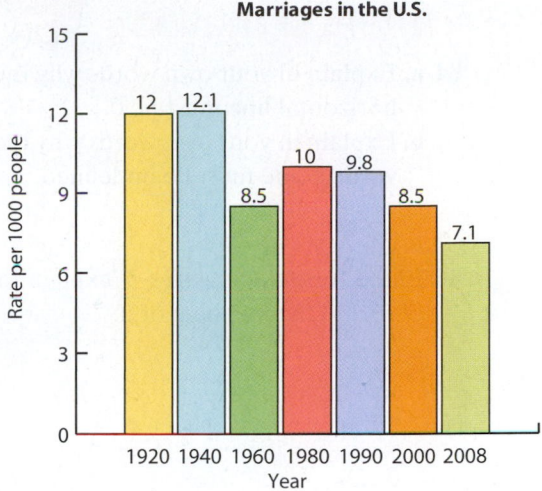

Collaborative Learning

83. The class should be divided into teams of 2 or 3 students. Each team will need access to a digital camera, a printer, and a ruler.

a. Take pictures of 8 things with a defined slope. (**Suggestions:** A roof, a stair railing, a beach umbrella, a crooked tree, etc. Be creative!)

b. Print each picture.

c. Use a ruler to draw a coordinate system on top of each picture. You will probably want to use increments of in. or cm, depending on the size of your picture.

d. Identify the line in each picture whose slope you are calculating and then use the coordinate systems you created to identify the coordinates of two points on each line.

e. Use the points you just found to calculate the slope of the line in each picture.

f. Share your findings with the class.

84. a. Explain in your own words why the slope of a horizontal line must be 0.

b. Explain in your own words why the slope of a vertical line must be undefined.

86. In the formula $y = mx + b$ explain the meaning of m and the meaning of b.

85. a. Describe the graph of the line $y = 0$.
b. Describe the graph of the line $x = 0$.

87. The slope of a road is called a **grade**. A steep grade is cause for truck drivers to have slow speed limits in mountains. What do you think that a "grade of 12%" means? Draw a picture of a right triangle that would indicate a grade of 12%.

6.4 The Point-Slope Form

Objective A   **Finding the Equations of Lines Given Two Points**

Given two points that lie on a line, the equation of the line can be found using the following method.

Finding the Equation of a Line Given Two Points

To find the equation of a line given two points on the line:

1. Use the formula $m = \dfrac{y_2 - y_1}{x_2 - x_1}$ to find the slope.

2. Use this slope, m, and either point in the point-slope formula $y - y_1 = m(x - x_1)$ to find the equation.

Example 1

Using Two Points to Find the Equation of a Line

Find the equation of the line containing the two points $(-1, 2)$ and $(4, -2)$.

Solution

First, find the slope.

$$m = \frac{y_2 - y_1}{x_2 - x_1}$$

$$= \frac{-2 - 2}{4 - (-1)}$$

$$= \frac{-4}{5}$$

$$= -\frac{4}{5}$$

Now use one of the given points and the point-slope form for the equation of a line. (**Note:** $(-1, 2)$ and $(4, -2)$ are used on the next page to illustrate that either point may be used.)

Objectives

A   Find the equations of a line given two points on the line.

B   Recognize and know how to find lines that are parallel and perpendicular.

C   Graph a line given its slope and one point on the line.

D   Recognize the point-slope form for a linear equation in two variables: $y - y_1 = m(x - x_1)$.

1. Find the equation of the line containing the two points $(-2, 4)$ and $(0, -1)$.

Using $(-1, 2)$

$y - y_1 = m(x - x_1)$	point-slope form
$y - 2 = -\dfrac{4}{5}\left[x - (-1)\right]$	Substitute.
$y - 2 = -\dfrac{4}{5}x - \dfrac{4}{5}$	distributive property
$y = -\dfrac{4}{5}x - \dfrac{4}{5} + 2$	
$y = -\dfrac{4}{5}x + \dfrac{6}{5}$	slope-intercept form
$4x + 5y = 6$	standard form

Using $(4, -2)$

$y - y_2 = m(x - x_2)$

$y - (-2) = -\dfrac{4}{5}(x - 4)$

$y + 2 = -\dfrac{4}{5}x + \dfrac{16}{5}$

$y = -\dfrac{4}{5}x + \dfrac{16}{5} - 2$

$y = -\dfrac{4}{5}x + \dfrac{6}{5}$

$4x + 5y = 6$

Now work margin exercise 1.

Objective B **Parallel Lines and Perpendicular Lines**

Parallel and Perpendicular Lines

Parallel lines are lines that never intersect (never cross each other) and these lines have the **same slope**. **Note:** All vertical lines (undefined slopes) are parallel to one another. All horizontal lines (zero slopes) are also parallel to one another.

Perpendicular lines are lines that intersect at 90° (right) angles and whose slopes are **negative reciprocals** of each other. Horizontal lines are perpendicular to vertical lines.

As illustrated in Figure 1, the lines $y = 2x + 1$ and $y = 2x - 3$ are **parallel**. They have the same slope, 2. The lines $y = \dfrac{2}{3}x + 1$ and $y = -\dfrac{3}{2}x - 2$ are **perpendicular**. Their slopes are negative reciprocals of each other, $\dfrac{2}{3}$ and $-\dfrac{3}{2}$.

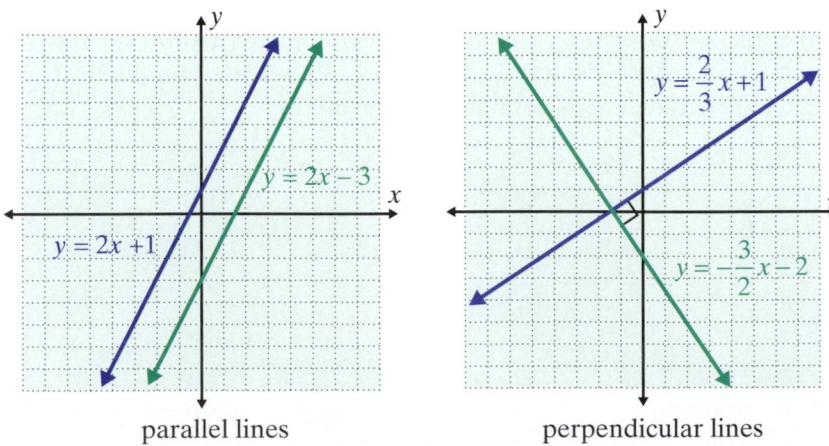

parallel lines   perpendicular lines

Figure 1

Example 2

Finding the Equations of Parallel Lines

Find the equation of the line through the point $(2, 3)$ and parallel to the line $5x + 3y = 1$. Graph both lines.

Solution

First, solve for y to find the slope of the given line.

$5x + 3y = 1$

$3y = -5x + 1$

$y = -\dfrac{5}{3}x + \dfrac{1}{3}$ Thus any line parallel to this line has slope $-\dfrac{5}{3}$.

Now use the point-slope form $y - y_1 = m(x - x_1)$ with $m = -\dfrac{5}{3}$ and $(x_1, y_1) = (2, 3)$.

$$y - 3 = -\dfrac{5}{3}(x - 2)$$ point-slope form

$$3(y - 3) = -5(x - 2)$$ Multiply both sides by the LCD, 3.

$$3y - 9 = -5x + 10$$ Simplify.

$$5x + 3y = 19$$ standard form

$$y = -\dfrac{5}{3}x + \dfrac{19}{3}$$ slope-intercept form

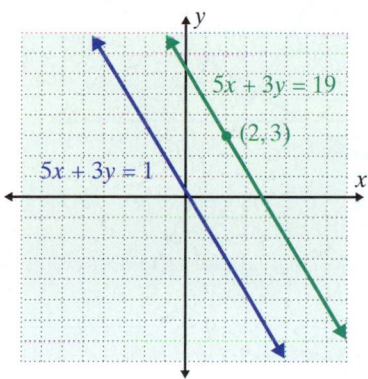

2. Find the equation of the line through the point $(0,2)$ and parallel to the line $3x + 2y = -2$. Graph both lines.

3. Find the equation of the line through the point $(0,2)$ and perpendicular to the line $3x + 2y = -2$. Graph both lines.

Example 3

Finding the Equations of Perpendicular Lines

Find the equation of the line through the point $(2,3)$ and perpendicular to the line $5x + 3y = 1$. Graph both lines.

Solution

We know from Example 2 that the slope of the line $5x + 3y = 1$ is $-\dfrac{5}{3}$. Thus, any line perpendicular to this line must have slope $m = \dfrac{3}{5}$ (the negative reciprocal of $-\dfrac{5}{3}$).

Now, using the point-slope form $y - y_1 = m(x - x_1)$ with $m = \dfrac{3}{5}$, and $(x_1, y_1) = (2, 3)$, we have the following.

$$y - 3 = \frac{3}{5}(x - 2) \qquad \text{Point-slope form}$$

$$5(y - 3) = 3(x - 2) \qquad \text{Multiply both sides by the LCD, 5.}$$

$$5y - 15 = 3x - 6 \qquad \text{Simplify.}$$

$$3x - 5y = -9 \qquad \text{Standard form}$$

or $\qquad y = \dfrac{3}{5}x + \dfrac{9}{5} \qquad$ Slope-intercept form

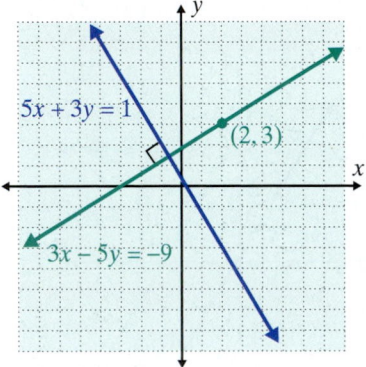

Now work margin exercises 2 and 3.

Objective C **Graphing a Line Given a Point and the Slope**

Lines represented by equations in the **standard form** $Ax + By = C$ and in the **slope-intercept form** $y = mx + b$ have been discussed in previous sections. Previously, we graphed lines using the y-intercept $(0, b)$ and the slope by moving vertically and then horizontally (or by moving horizontally and then vertically) from the y-intercept. This same technique can be used to graph lines if the given point is on the line but is not the y-intercept. Consider the following example.

Example 4

Graphing a Line Given a Point and the Slope

Graph the line with slope $m = -\dfrac{3}{4}$ and which passes through the point $(2, 5)$.

Solution

Start from the point $(2, 5)$ and locate another point on the line using the slope as $\dfrac{rise}{run} = \dfrac{-3}{4}$ or $\dfrac{3}{-4}$. There are many ways to proceed.

Here are two:

  1. Move 4 units right and 3 units down, or
  2. Move 3 units down and 4 units right.

Either way, you arrive at the same point $(6, 2)$.

This means that we can move from the given point either with the rise first or the run first.

Note: Any numbers in the ratio of –3 to 4 can be used for the moves, such as –6 to 8 or 9 to –12.

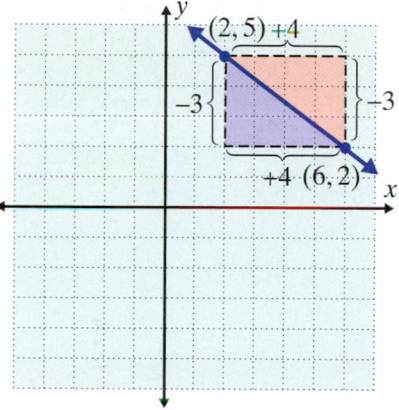

Now work margin exercise 4.

4. Graph the line with the slope $m = \dfrac{-1}{3}$ and which passes through the point $(3, 4)$.

Objective D ## Point-Slope Form

Now consider finding the equation of the line given the point (x_1, y_1) on the line and the slope m. If (x, y) is **any other point** on the line, then the slope formula gives the equation

$$\frac{y - y_1}{x - x_1} = m.$$

Multiplying both sides of this equation by the denominator (assuming the denominator is not 0) gives

$$y - y_1 = m(x - x_1), \qquad \text{which is called the \textbf{point-slope form}.}$$

For example, suppose that a point $(x_1, y_1) = (8, 3)$ and the slope $m = -\dfrac{3}{4}$ are given.

If (x, y) represents any point on the line other than $(8, 3)$, then substituting into the formula for slope gives the following

$$\frac{y - y_1}{x - x_1} = m$$ formula for slope

$$\frac{y - 3}{x - 8} = -\frac{3}{4}$$ Substitute the given information.

$$(x - 8)\left(\frac{y - 3}{x - 8}\right) = -\frac{3}{4}(x - 8)$$ Multiply both sides by $(x - 8)$.

$$y - 3 = -\frac{3}{4}(x - 8)$$ point-slope form: $y - y_1 = m(x - x_1)$

From this point-slope form, we can manipulate the equation to get the other two forms.

$$y - 3 = -\frac{3}{4}(x - 8)$$

$$y - 3 = -\frac{3}{4}x + 6$$

$$y = -\frac{3}{4}x + 9$$ slope-intercept form: $y = mx + b$

$$4(y) = 4\left(-\frac{3}{4}x + 9\right)$$

$$4y = -3x + 36$$

$$3x + 4y = 36$$ standard form: $Ax + By = C$

Point-Slope Form

An equation of the form

$$y - y_1 = m(x - x_1)$$

is called the **point-slope form** for the equation of a line that contains the point (x_1, y_1) and has slope m.

Example 5

Finding Equations of Lines Using the Slope and a Point

Find the equation of the line with a slope of $-\dfrac{1}{2}$ and which passes through the point $(2, 3)$. Graph the line using the point and slope.

Solution

Substitute the values into the point-slope form.

$$y - y_1 = m(x - x_1)$$

$$y - 3 = -\frac{1}{2}(x - 2) \qquad \text{point-slope form}$$

$$y - 3 = -\frac{1}{2}x + 1$$

$$y = -\frac{1}{2}x + 4 \qquad \text{slope-intercept form}$$

$$2y = -x + 8$$

$$x + 2y = 8 \qquad \text{standard form}$$

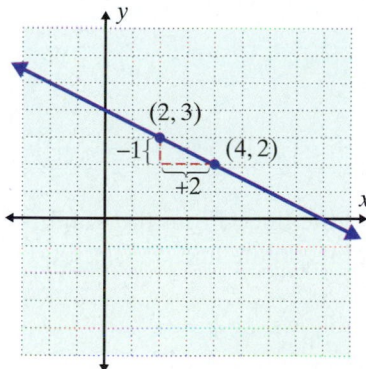

The point one unit down and two units right from $(2, 3)$ will be on the line because the slope is

$$m = \frac{\text{rise}}{\text{run}} = \frac{-1}{2} = -\frac{1}{2}.$$

With a negative slope, either the rise is negative and the run is positive, or the rise is positive and the run is negative. In either case, as the previous figure and the following figure illustrate, the line is the same.

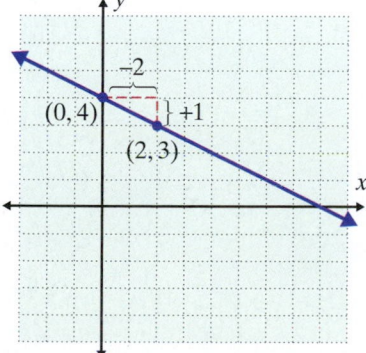

The point one unit up and two units to the left from $(2, 3)$ is on the line because the slope is

$$m = \frac{\text{rise}}{\text{run}} = \frac{1}{-2} = -\frac{1}{2}.$$

Now work margin exercise 5.

5. Find the equation of the line with a slope of $\dfrac{-2}{3}$ and which passes through the point $(3, 1)$. Graph the line using the point and slope.

In Example 5, the equation of the line is written in all three forms: point-slope form, slope-intercept form, and standard form. Generally, any one of these forms is sufficient.

However, there are situations in which one form is preferred over the others. Therefore, manipulation among the forms is an important skill. Also, if the answer in the text is in one form and your answer is in another form, you should be able to recognize that the answers are equivalent.

6. Write an equation in standard form for the following line.

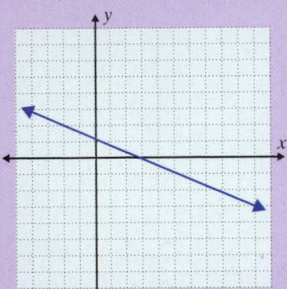

Example 6

Finding Equations of Lines Using a Graph

Write an equation in standard form for the following line.

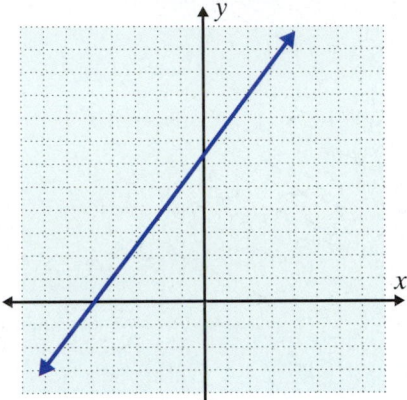

Solution

First, we must identify two points on the line. From the graph, we can see that $(2, 9)$ and $(-1, 5)$ are points on the line.

Next, we find the slope of the line as follows.

$$m = \frac{9-5}{2-(-1)} = \frac{4}{3}$$

Now, we substitute the values into point-slope form and then rewrite the equation in standard form.

$$y - y_1 = m\left(x - x_1\right)$$

$$y - 5 = \frac{4}{3}(x + 1)$$

$$y - 5 = \frac{4}{3}x + \frac{4}{3}$$

$$y = \frac{4}{3}x + \frac{19}{3}$$

$$3(y) = 3\left(\frac{4}{3}x + \frac{19}{3}\right)$$

$$3y = 4x + 19$$

$$-4x + 3y = 19$$

Now work margin exercise 6.

For easy reference, the following table summarizes what we know about lines.

Summary of Formulas and Properties of Lines

1. $y = b$ horizontal line, slope 0
2. $x = a$ vertical line, undefined slope
3. parallel lines   have the same slope
4. perpendicular lines   have slopes that are negative reciprocals of each other
5. $Ax + By = C$ standard form
6. $m = \dfrac{y_2 - y_1}{x_2 - x_1}$ slope of a line
7. $y = mx + b$ slope-intercept form
8. $y - y_1 = m(x - x_1)$ point-slope form

Practice Problems

Find a linear equation in standard form whose graph satisfies the given conditions.

1. passes through the point $(4, -1)$ with $m = 2$

2. parallel to $y = -3x + 4$ and contains the point $(-1, 5)$

3. perpendicular to $2x + y = 1$ and passes through the origin $(0, 0)$

4. contains the two points $(6, -2)$ and $(2, 0)$

Practice Problem Answers

1. $2x - y = 9$ **2.** $3x + y = 2$ **3.** $x - 2y = 0$ **4.** $x + 2y = 2$

Exercises 6.4

Find a. the slope, b. a point on the line, and c. the graph of the line for the given equations in point-slope form.

1. $y - 1 = 2(x - 3)$

2. $y - 4 = \frac{1}{2}(x - 1)$

3. $y + 2 = -5(x)$

4. $y = -(x + 8)$

5. $y - 3 = -\frac{1}{4}(x + 2)$

6. $y + 6 = \frac{1}{3}(x - 7)$

Find an equation in standard form for the line passing through the given point with the given slope. Graph the line. See Examples 4 and 5.

7. $(-2, 1)$; $m = -2$

8. $(3, 4)$; $m = 3$

9. $(5, -2)$; $m = 0$

10. $(0, 0)$; $m = -3$

11. $(-3, 6)$; $m = \frac{1}{2}$

12. $(-3, -1)$; m is undefined

13. $(7, 10)$; $m = \frac{3}{5}$

14. $(-1, -1)$; $m = -\frac{1}{4}$

15. $\left(-2, \frac{1}{3}\right)$; $m = \frac{2}{3}$

16. $\left(\frac{5}{2}, \frac{1}{2}\right)$; $m = -\frac{4}{3}$

711 **Chapter 6** *Graphing Linear Equations*

© Hawkes Learning Systems. All rights reserved.

Find an equation in slope-intercept form for the line passing through the two given points. See Example 1.

17. $(-5, 2); (3, 6)$ **18.** $(-3, 4); (2, 1)$ **19.** $(-5, 1); (2, 0)$ **20.** $(-4, -4); (3, 1)$

21. $(0, 2); \left(1, \dfrac{3}{4}\right)$ **22.** $\left(\dfrac{5}{2}, 0\right); \left(2, -\dfrac{1}{3}\right)$ **23.** $(2, -5); (4, -5)$ **24.** $(0, 4); \left(1, \dfrac{1}{2}\right)$

25. $(-2, 6); (3, 1)$ **26.** $(8, 2); (0, 0)$

Find an equation in standard form for each graph shown. See Example 6.

27. 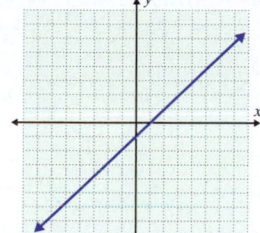   **28.** 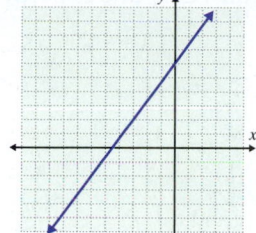   **29.** 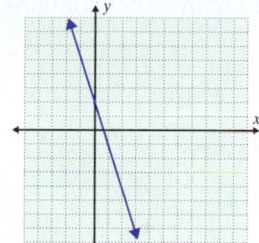   **30.**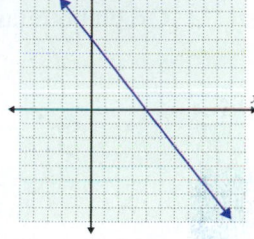

31. 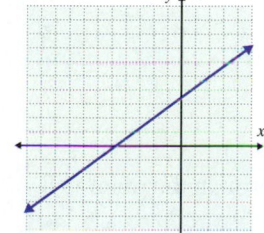   **32.** 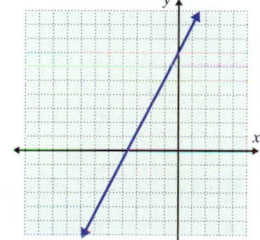   **33.** 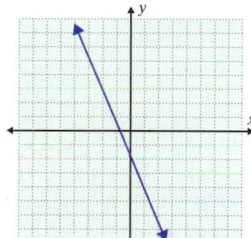   **34.**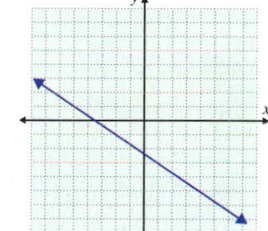

Find an equation in slope-intercept form that satisfies each set of conditions. See Examples 1 and 3.

35. Find an equation for the horizontal line through the point $(-2, 6)$.

36. Find an equation for the vertical line through the point $(-1, -4)$.

37. Write an equation for the line parallel to the x-axis and containing the point $(2, 7)$.

38. Find an equation for the line parallel to the y-axis and containing the point $(2, -4)$.

39. Find an equation for the line perpendicular to $x = 4$ and that passes through $(-1, 7)$.

40. Find an equation for the line parallel to the line $-6y = 1$ and containing the point $(-3, 2)$.

The Point–Slope Form   **Section 6.4**   712

41. Write an equation for the line parallel to the line $2x - y = 4$ and containing the origin. Graph both lines.

42. Find an equation for the line parallel to $7x - 3y = 1$ and containing the point $(1, 0)$. Graph both lines.

43. Write an equation for the line parallel to $5x = 7 + y$ and through the point $(-1, -3)$. Graph both lines.

44. Write an equation for the line that contains the point $(2, 2)$ and is perpendicular to the line $4x + 3y = 4$. Graph both lines.

45. Find an equation for the line that passes through the point $(4, -1)$ and is perpendicular to the line $5x - 3y + 4 = 0$. Graph both lines.

46. Write an equation for the line that is perpendicular to $8 - 3x - 2y = 0$ and passes through the point $(-4, -2)$.

47. Write an equation for the line through the origin that is perpendicular to $3x - y = 4$.

48. Find an equation for the line that is perpendicular to $2x + y = 5$ and that passes through $(6, -1)$.

49. Write an equation for the line that is perpendicular to $2x - y = 7$ and has the same y-intercept as $x - 3y = 6$.

50. Find an equation for the line with the same y-intercept as $5x + 4y = 12$ and that is perpendicular to $3x - 2y = 4$.

51. Show that the points $A(-2, 4)$, $B(0, 0)$, $C(6, 3)$, and $D(4, 7)$ are the vertices of a rectangle. (Plot the points and show that opposite sides are parallel and that adjacent sides are perpendicular.)

52. Show that the points $A(0, -1)$, $B(3, -4)$, $C(6, 3)$, and $D(9, 0)$ are the vertices of a parallelogram. (Plot the points and show that opposite sides are parallel.)

Determine whether each pair of lines is a. parallel, b. perpendicular, or c. neither. Graph both lines. (Hint: write the equations in slope-intercept form and then compare slopes and y-intercepts.)

53. $\begin{cases} y = -2x + 3 \\ y = -2x - 1 \end{cases}$

54. $\begin{cases} y = 3x + 2 \\ y = -\dfrac{1}{3}x + 6 \end{cases}$

55. $\begin{cases} 4x + y = 4 \\ x - 4y = 8 \end{cases}$

56. $\begin{cases} 2x + 3y = 5 \\ 3x + 2y = 10 \end{cases}$

57. $\begin{cases} 2x + 2y = 9 \\ 2x - y = 6 \end{cases}$

58. $\begin{cases} 3x - 4y = 16 \\ 4x + 3y = 15 \end{cases}$

Writing & Thinking

59. Handicapped Access: Ramps for persons in wheelchairs or otherwise handicapped are now built into most buildings and walkways. (If ramps are not present in a building, then there must be elevators.) What do you think that the slope of a ramp should be for handicapped access? Look in your library or contact your local building permit office to find the recommended slope for such ramps.

60. Discuss the difference between the concepts of a line having slope 0 and a line having undefined slope.

6.5 Linear Inequalities in One or Two Variables

Objectives

A Understand and use set-builder notation.

B Understand and use interval notation.

C Solve linear inequalities.

D Solve compound inequalities.

E Learn how to apply inequalities to solve word problems.

F Graph linear inequalities.

G Graph linear inequalities using a graphing calculator.

Objective A   Sets and Set-Builder Notation

A **set** is a collection of objects or numbers. The items in the set are called **elements**, and sets are indicated with braces $\{\ \}$, and named with capital letters. If the elements are listed within the braces, the set is said to be in **roster form**. For example,

$$A = \left\{ \frac{3}{4}, 1.7, 2, 4, 6, \sqrt{89} \right\}$$

$$\mathbb{W} = \{0, 1, 2, 3, 4, 5, ...\}$$

$$\mathbb{Z} = \{..., -4, -3, -2, -1, 0, 1, 2, 3, 4, ...\}$$

are all in roster form.

The symbol $\in$ is read "is an element of" and is used to indicate that a particular number belongs to a set. For example, in the previous illustrations, $1.7 \in A$, $0 \in \mathbb{W}$, and $-3 \in \mathbb{Z}$.

If the elements in a set can be counted, as in A above, the set is said to be **finite**. If the elements cannot be counted, as in $\mathbb{W}$ and $\mathbb{Z}$ above, the set is said to be **infinite**. If a set has absolutely no elements, it is called the **empty set** or **null set**, and is written in the form $\{\ \}$ or with the special symbol $\varnothing$. For example, the set of all people over 15 feet tall is the empty set, $\varnothing$.

The notation $\{x|\quad\}$ is read "the set of all x such that ..." and is called **set-builder notation**. The vertical bar $\left(\,|\,\right)$ is read "such that." A statement following the bar gives a condition (or restriction) for the variable x. For example,

$\{x|x \text{ is an even integer}\}$ is read "the set of all x such that x is an even integer."

The following note highlights the two important set concepts **union** and **intersection**.

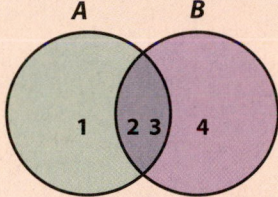

Special Comments about Union and Intersection

The concepts of union and intersection are part of set theory, which is very useful in a variety of courses including abstract algebra, probability, and statistics. These concepts are also used in analyzing inequalities and analyzing relationships among sets in general.

The **union** (symbolized $\cup$, as in $A \cup B$) of two (or more) sets is the set of all elements that belong to either one set or the other set or to both sets. The **intersection** (symbolized $\cap$, as in $A \cap B$) of two (or more) sets is the set of all elements that belong to both sets. The word **or** is used to indicate union and the word **and** is used to indicate intersection. For example, if $A = \{1, 2, 3\}$ and $B = \{2, 3, 4\}$, then the numbers that belong to A **or** B is the set $A \cup B = \{1, 2, 3, 4\}$. The set of numbers that belong to A **and** B is the set $A \cap B = \{2, 3\}$. These relationships can be illustrated using the following Venn diagram.

A B

1 2 3 4

Similarly, union and intersection notation can be used for sets with inequalities.

For example, $\{x \mid x < a \text{ **or** } x > b\}$ can be written in the form

$$\{x \mid x < a\} \cup \{x \mid x > b\}.$$

Also, $\{x \mid x > a \text{ **and** } x < b\}$ can be written in the form

$$\{x \mid x > a\} \cap \{x \mid x < b\} \text{ or } \{x \mid a < x < b\}.$$

Objective B **Intervals of Real Numbers**

Suppose that a and b are two real numbers and that $a < b$. The set of all real numbers between a and b is called an **interval of real numbers**. Intervals of real numbers are used in relationship to everyday concepts such as the length of time you talk on your cell phone, the shelf life of chemicals or medicines, and the speed of an airplane.

As an aid in reading inequalities and graphing inequalities correctly, note that an inequality may be read either from left to right or right to left. Because we are concerned about which numbers satisfy an inequality, we **read the variable first:**

$$x > 7 \quad \text{is read from } \textbf{left to right} \text{ as "}\textbf{\textit{x} is greater than 7.}\text{"}$$
$$\text{and} \quad 7 < x. \quad \text{is read from } \textbf{right to left} \text{ as "}\textbf{\textit{x} is greater than 7.}\text{"}$$

A compound interval, such as $-3 < x < 6$, is read

"**x is greater than –3 and x is less than 6.**"

Again, note that **the variable is read first**.

Various types of intervals and their corresponding **interval notation** are listed in Table 1.

Type of Interval	Algebraic Notation	Interval Notation	Graph
open interval	$a < x < b$	(a, b)	
closed interval	$a \leq x \leq b$	$[a, b]$	
half-open interval	$\begin{cases} a \leq x < b \\ a < x \leq b \end{cases}$	$[a, b)$ $(a, b]$	
open interval	$\begin{cases} x > a \\ x < b \end{cases}$	(a, ∞) $(-\infty, b)$	
half-open interval	$\begin{cases} x \geq a \\ x \leq b \end{cases}$	$[a, \infty)$ $(-\infty, b]$	

Table 1

notes

■ The symbol for infinity ∞ (or $-\infty$) is not a number. It is used to indicate that the interval is to include all real numbers from some point on (either in the positive direction or the negative direction) without end.

■

Example 1

Graphing Intervals

a. Graph the open interval $(3, \infty)$.

Solution

b. Graph the half-open interval $0 < x \leq 4$.

Solution

c. Represent the following graph using algebraic notation, and state what kind of interval it is.

Solution $x \geq 1$ is a half-open interval.

d. Represent the following graph using interval notation, and state what kind of interval it is.

Solution $(-3, 1)$ is an open interval.

Now work margin exercise 1.

Example 2 and Example 3 illustrate how the concepts of **union** (indicated by **or**) and **intersection** (indicated by **and**) are related to the graphs of intervals of real numbers on a real number line.

Example 2

Sets of Real Numbers Illustrating Union

Graph the set $\{x \mid x > 5 \ \textbf{or} \ x \leq 4\}$. The word **or** implies those values of x that satisfy **at least one** of the inequalities.

Solution

$x > 5$

$x \leq 4$

$x > 5$ or $x \leq 4$

The solution graph shows the union, $\cup$, of the first two graphs.

---

1. a. Graph the half-open interval $(-\infty, 6]$.

b. Graph the open interval $-1 < x < 1$.

c. Represent the following graph using algebraic notation, and state what kind of interval it is.

d. Represent the following graph using interval notation, and state what kind of interval it is.

2. Graph the set $\{t \mid t \le 8 \text{ or } t > 10\}$.

3. Graph the set $\{y \mid y \ge -1 \text{ and } y < 3\}$.

Example 3

Sets of Real Numbers Illustrating Intersection

Graph the set $\{x \mid x \le 2 \text{ and } x \ge 0\}$. The word **and** implies those values of x that satisfy **both** inequalities.

Solution

$x \le 2$

$x \ge 0$

$x \le 2 \text{ and } x \ge 0$

The solution graph shows the intersection, $\cap$, of the first two graphs. In other words, the third graph shows the points in common between the first two graphs in this example.

This set can also be indicated as $\{x \mid 0 \le x \le 2\}$.

Now work margin exercises 2 and 3.

Objective C **Solving Linear Inequalities**

In this section, we will solve **linear inequalities**, such as $6x + 5 \le -7$, and write the solution in **interval notation**, such as $(-\infty, -2]$ to indicate all real numbers x less than or equal to –2. Note that this can also be written in set-builder notation as $\{x \mid x \in (-\infty, -2]\}$ or $\{x \mid x \le -2\}$.

Linear Inequalities

Inequalities of the given form, where a, b, and c are real numbers and $a \ne 0$,

$$ax + b < c \quad \text{and} \quad ax + b \le c$$

$$ax + b > c \quad \text{and} \quad ax + b \ge c$$

are called **linear inequalities**.

The inequalities $c < ax + b < d$ and $c \le ax + b \le d$ are called **compound linear inequalities**. (This includes $c < ax + b \le d$ and $c \le ax + b < d$ as well.)

The solutions to linear inequalities are intervals of real numbers, and the methods for solving linear inequalities are similar to those used to solve linear equations. There is only one important exception.

Multiplying or dividing both sides of an inequality by a negative number causes the "direction" of the inequality to be reversed.

By the direction of the inequality, we mean "less than" or "greater than." Consider the following examples.

We know that $6 < 10$.

Add 5 to both sides	**Add –7 to both sides**	**Multiply both sides by 3**
$6 < 10$	$6 < 10$	$6 < 10$
$6+5$? $10+5$	$6+(-7)$? $10+(-7)$	$3 \cdot 6$? $3 \cdot 10$
$11 < 15$	$-1 < 3$	$18 < 30$

In the three cases just illustrated, addition, subtraction, and multiplication by a positive number, the direction of the inequality stayed the same. It remained <. Now we will see that multiplying or dividing each side by a negative number will reverse the direction of the inequality, from < to > or from > to <. This concept also applies to ≤ and ≥.

Multiply both sides by –3	**Divide both sides by –2**
$6 < 10$	$6 < 10$
$-3 \cdot 6$? $-3 \cdot 10$	$\dfrac{6}{-2}$? $\dfrac{10}{-2}$
$-18 > -30$	$-3 > -5$

In each of these last two examples, the direction of the inequality is changed from < to >. While two examples do not prove a rule to be true, this particular rule is true and is included in the following rules for solving inequalities.

> ### Rules for Solving Linear Inequalities
>
> 1. Simplify each side of the inequality by removing any grouping symbols and combining like terms.
> 2. Use the addition property of equality to add the opposites of constants or variable expressions so that variable expressions are on one side of the inequality and constants are on the other.
> 3. Use the multiplication property of equality to multiply both sides by the reciprocal of the coefficient of the variable (that is, divide both sides by the coefficient) so that the new coefficient is 1. **If this coefficient is negative, reverse the direction of the inequality.**
> 4. A quick (and generally satisfactory) check is to select any one number in your solution and substitute it into the original inequality.

As with solving equations, the object of solving an inequality is to find equivalent inequalities of simpler form that have the same solution set. We want the variable with a coefficient of +1 on one side of the inequality and any constants on the other side. One key difference between solving linear equations and solving linear inequalities is that linear equations have only one solution while linear inequalities generally have an infinite number of solutions.

Example 4

Solving Linear Inequalities

Solve the following linear inequalities and graph the solution set. Write the solution set using interval notation. Assume that x is a real number.

a. $6x + 5 \leq -1$

Solution

$6x + 5 \leq -1$	Write the inequality.
$6x + 5 - 5 \leq -1 - 5$	Add -5 to both sides.
$6x \leq -6$	Simplify.
$\dfrac{6x}{6} \leq \dfrac{-6}{6}$	Divide both sides by 6.
$x \leq -1$	Simplify.

x is in $\left(-\infty, -1\right]$ Use interval notation. Note that the interval $\left(-\infty, -1\right]$ is a half-open interval.

b. $x - 3 > 3x + 4$

Solution

$x - 3 > 3x + 4$	Write the inequality.
$x - 3 - x > 3x + 4 - x$	Add $-x$ to both sides.
$-3 > 2x + 4$	Simplify.
$-3 - 4 > 2x + 4 - 4$	Add -4 to both sides.
$-7 > 2x$	Simplify.
$\dfrac{-7}{2} > \dfrac{2x}{2}$	Divide both sides by 2.
$\dfrac{-7}{2} > x$	Simplify.

or $\qquad x < -\dfrac{7}{2}$

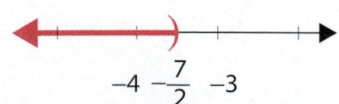

x is in $\left(-\infty, -\dfrac{7}{2}\right)$ Use interval notation. Note that the interval $\left(-\infty, -\dfrac{7}{2}\right)$ is an open interval.

4. Solve the following linear inequalities and graph the solution sets. Write each solution set using interval notation. Assume that x is a real number.

c. $6 - 4x \le x + 1$

Solution

$6 - 4x \le x + 1$	Write the inequality.
$6 - 4x - x \le x + 1 - x$	Add $-x$ to both sides.
$6 - 5x \le 1$	Simplify.
$6 - 5x - 6 \le 1 - 6$	Add -6 to both sides.
$-5x \le -5$	Simplify.
$\dfrac{-5x}{-5} \ge \dfrac{-5}{-5}$	Divide both sides by -5. **Note the reversal of the inequality sign!**
$x \ge 1$	Simplify.

a. $3x + 8 > -10$

b. $7 - x \le 9x - 1$

c. $5x - 9 \ge 2 - 4(x + 3)$

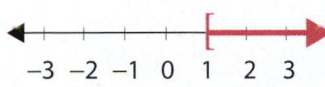

x is in $[1, \infty)$ Use interval notation. Note that the interval $[1, \infty)$ is a half-open interval.

d. $2x + 5 < 3x - (7 - x)$

Solution

$2x + 5 < 3x - (7 - x)$	Write the inequality.
$2x + 5 < 3x - 7 + x$	Distribute the negative sign.
$2x + 5 < 4x - 7$	Combine like terms.
$2x + 5 - 2x < 4x - 7 - 2x$	Add $-2x$ to both sides.
$5 < 2x - 7$	Simplify.
$5 + 7 < 2x - 7 + 7$	Add 7 to both sides.
$12 < 2x$	Simplify.
$\dfrac{12}{2} < \dfrac{2x}{2}$	Divide both sides by 2.
$6 < x$	Simplify.

x is in $(6, \infty)$ Use interval notation. Note that the interval $(6, \infty)$ is an open interval.

Now work margin exercise 4.

Solving Compound Inequalities

Compound inequalities have three parts, and can arise when a variable or variable expression is to be between two numbers. For example, the inequality

$$5 < x + 3 < 10$$

indicates that the values for the expression $x + 3$ are to be between 5 and 10. To solve this inequality, subtract 3 (or add −3) from each part.

$$5 < x + 3 < 10$$
$$5 - 3 < x + 3 - 3 < 10 - 3$$
$$2 < x < 7$$

Thus the variable x is isolated with coefficient +1, and we see that the solution set is the interval of real numbers $(2, 7)$. The graph of the solution set is the following.

Example 5

Solving Compound Inequalities

a. Solve the compound inequality $-5 \leq 4x - 1 < 11$ and graph the solution set. Write the solution set using interval notation. Assume that x is a real number.

Solution

$-5 \leq$	$4x - 1$	< 11	Write the inequality.
$-5 + 1 \leq$	$4x - 1 + 1$	$< 11 + 1$	Add 1 to each part.
$-4 \leq$	$4x$	< 12	Simplify.
$\dfrac{-4}{4} \leq$	$\dfrac{4x}{4}$	$< \dfrac{12}{4}$	Divide each part by 4.
$-1 \leq$	x	< 3	Simplify.

The solution set is the half-open interval $[-1, 3)$.

b. Solve the compound inequality $5 \leq -3 - 2x \leq 13$ and graph the solution set. Write the solution set using interval notation. Assume that x is a real number.

Solution

$5 \leq$	$-3 - 2x$	≤ 13	Write the inequality.
$5 + 3 \leq$	$-3 - 2x + 3$	$\leq 13 + 3$	Add 3 to each part.
$8 \leq$	$-2x$	≤ 16	Simplify.

$$\frac{8}{-2} \geq \frac{-2x}{-2} \geq \frac{16}{-2}$$ Divide each part by –2. **Note that the inequalities change direction.**

$$-4 \geq \quad x \quad \geq -8$$ Simplify.

$$(\text{or } -8 \leq \quad x \quad \leq -4)$$

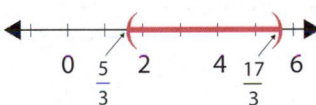

−9 −8 −7 −6 −5 −4 −3

The solution set is the closed interval $[-8, -4]$.

c. Solve the compound inequality $0 < \dfrac{3x-5}{4} < 3$ and graph the solution set. Write the solution set using interval notation. Assume that x is a real number.

Solution

$$0 < \frac{3x-5}{4} < 3$$ Write the inequality.

$$0 \cdot 4 < \frac{3x-5}{4} \cdot 4 < 3 \cdot 4$$ Multiply each part by 4.

$$0 < \quad 3x-5 \quad < 12$$ Simplify each part.

$$0+5 < 3x-5+5 < 12+5$$ Add 5 to each part.

$$5 < \quad 3x \quad < 17$$ Simplify.

$$\frac{5}{3} < \quad \frac{3x}{3} \quad < \frac{17}{3}$$ Divide each part by 3.

$$\frac{5}{3} < \quad x \quad < \frac{17}{3}$$ Simplify.

0 $\frac{5}{3}$ 2   4 $\frac{17}{3}$ 6

The solution set is the open interval $\left(\dfrac{5}{3}, \dfrac{17}{3}\right)$.

Now work margin exercise 5.

5. Solve the following compound inequalities and graph the solution sets. Write each solution set using interval notation. Assume that x is a real number.

a. $-12 < 5x + 3 \leq 8$

b. $-3 < 5 - 4x < 17$

c. $7 \leq \dfrac{5x-1}{2} \leq 13$

6. A certain anesthetic must be administered in a way that requires the mean dosage to remain under 450 milligrams per hour in order to guarantee patient safety. This anesthetic is being used during a 4-hour surgery. For the first three hours, the dosages were as follows: 500 milligrams the first hour, 380 milligrams during the second hour, and 520 milligrams the third hour. What are the dosages that can be administed safely to the patient for the fourth dosage?

In the following two examples, we show how inequalities can be related to real-world problems.

Example 6

Application with an Inequality

A math student has grades of 85, 98, 93, and 90 on four examinations. If he must have a mean score of 90 or better to receive an A for the course, what scores can he receive on the final exam and earn an A? (Assume that the final exam counts the same as the other exams.)

Solution

Let x = score on final exam.

The mean is found by adding the scores and dividing by 5.

$$\frac{85+98+93+90+x}{5} \geq 90$$

$$\frac{366+x}{5} \geq 90 \qquad \text{Simplify the numerator.}$$

$$5\left(\frac{366+x}{5}\right) \geq 5 \cdot 90 \qquad \text{Multiply both sides by 5.}$$

$$366+x \geq 450 \qquad \text{Simplify.}$$

$$366+x-366 \geq 450-366 \qquad \text{Add } -366 \text{ to each side.}$$

$$x \geq 84$$

If the student scores 84 or more on the final exam, he will have a mean score of 90 or more and receive an A in math.

Now work margin exercise 6.

Example 7

Application with an Inequality

Ellen is going to buy 30 stamps, some 28-cent stamps, and some 44-cent stamps. If she has $9.68, what is the maximum number of 44-cent stamps she can buy?

Solution

Let x = number of 44-cent stamps and
$30 - x$ = number of 28-cent stamps.

Ellen cannot spend more than $9.68.

$$0.44x + 0.28(30 - x) \leq 9.68$$
$$0.44x + 8.40 - 0.28x \leq 9.68$$
$$0.16x + 8.40 \leq 9.68$$
$$0.16x + 8.40 - 8.40 \leq 9.68 - 8.40$$
$$0.16x \leq 1.28$$
$$\frac{0.16x}{0.16} \leq \frac{1.28}{0.16}$$
$$x \leq 8$$

Ellen can buy at most eight 44-cent stamps if she buys a total of 30 stamps.

Now work margin exercise 7.

7. Ashley is planning her wedding, and is ordering flowered centerpieces. She wants some rose centerpieces and some lily centerpieces. The rose centerpieces cost $225, and the lily centerpieces cost $175. What is the maximum number of rose centerpieces she can buy on a budget of $3900?

Objective F **Graphing Linear Inequalities:** $y < mx + b$

In this section, we will develop techniques for analyzing and graphing linear inequalities. We need the following terminology.

half-plane	A straight line separates a plane into two **half-planes**. The points on one side of the line are in one of the half-planes, and the points on the other side of the line are in the other half-plane.
boundary line	The line itself is called the **boundary line**.
closed half-plane	If the boundary line is included in the solution set, then the half-plane is said to be **closed**.
open half-plane	If the boundary line is not included in the solution set, then the half-plane is said to be **open**.

Figure 1 shows both an open half-plane and a closed half-plane with the line $2x - 3y = 10$ as the boundary line. The solution set is the shaded region.

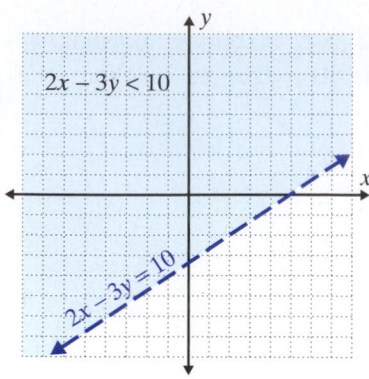

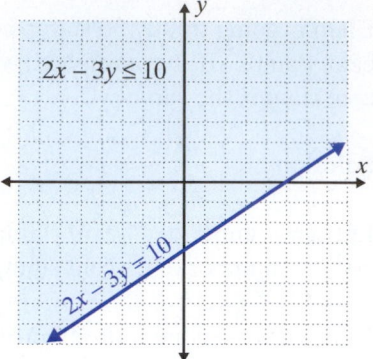

The points on the line $2x - 3y = 10$ are not included in the solution set so the line is dashed. The half-plane is open.

The points on the line $2x - 3y = 10$ are included in the solution set so the line is solid. The half-plane is closed.

Figure 1

Graphing Linear Inequalities

1. First, graph the boundary line (dashed if the inequality is < or >, solid if the inequality is ≤ or ≥).

2. Next, determine which side of the line to shade using one of the following methods.

   Method 1
   a. Test any one point obviously on one side of the line.
   b. If the test-point satisfies the inequality, shade the half-plane on that side of the line. Otherwise, shade the other half-plane.

   Note: The point $(0, 0)$, if it is not on the boundary line, is usually the easiest point to test.

   Method 2
   a. Solve the inequality for y (assuming that the line is not vertical).
   b. If the solution shows $y <$ or $y \leq$, then shade the half-plane below the line.
   c. If the solution shows $y >$ or $y \geq$, then shade the half-plane above the line.

   Note: If the boundary line is vertical, then solve for x. If the solution shows $x >$ or $x \geq$, then shade the half-plane to the right. If the solution shows $x <$ or $x \leq$, then shade the half-plane to the left.

3. The shaded half-plane (and the line if it is solid) is the solution to the inequality.

Example 8

Graphing Linear Inequalities

a. Graph the half-plane that satisfies the inequality $2x + y \leq 6$.

Solution

Method 1 is used in this example.

Step 1: Graph the boundary line $2x + y = 6$ as a solid line because the inequality is $\leq$ (less than or **equal to**).

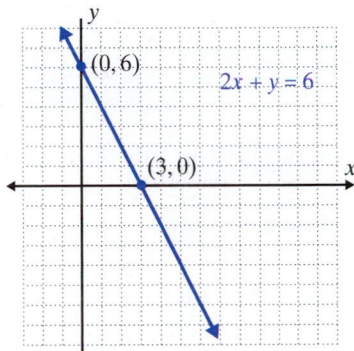

Step 2: Test any point on one side of the line. In this example, we have chosen $(0, 0)$.

$$2(0) + (0) \leq 6$$
$$0 \leq 6$$

This is a true statement.

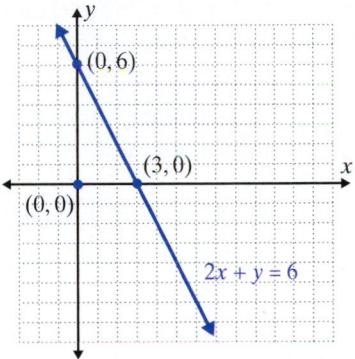

Step 3: Shade the half-plane on the same side of the line as the point $(0, 0)$. (The shaded half-plane and the boundary line is the solution set to the inequality.)

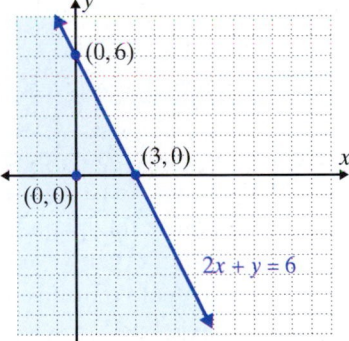

b. Graph the solution set to the inequality $y > 2x$.

Solution

Since the inequality is already solved for y, Method 2 is easy to apply.

Step 1: Graph the boundary line $y = 2x$ as a dashed line because the inequality is $>$.

8. Graph the solution set to the following inequalities.

a. $9x - 3y \le 21$

b. $x > -2$

c. $y < 4$

Step 2: The solution shows >, so by Method 2, the graph consists of those points above the line. Shade the half-plane above the line.

Note: As a check, we see that the point $(-3, 0)$ gives $0 > -6$, a true statement. Thus we know we have shaded the correct half-plane.

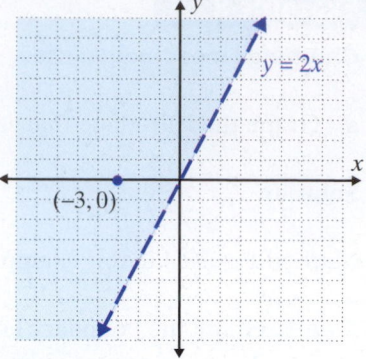

c. Graph the half-plane that satisfies the inequality $y > 1$.

Solution

Again, the inequality is already solved for y and Method 2 is used.

Step 1: Graph the boundary line $y = 1$ as a dashed line because the inequality is >. (The boundary line is a horizontal line.)

Step 2: By Method 2, shade the half-plane above the line.

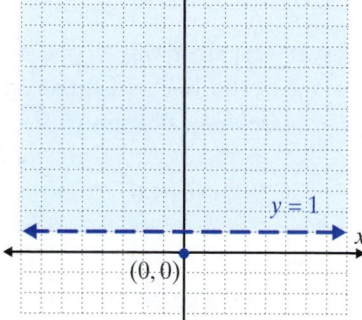

d. Graph the solution set to the inequality $x \le 0$.

Solution

The boundary line is a vertical line and Method 1 is used.

Step 1: Graph the boundary line $x = 0$ as a solid line because the inequality is $\le$ (less than or **equal to**). Note that this is the y-axis.

Step 2: Test the point $(-2, 1)$.

$$-2 \le 0$$

This statement is true.

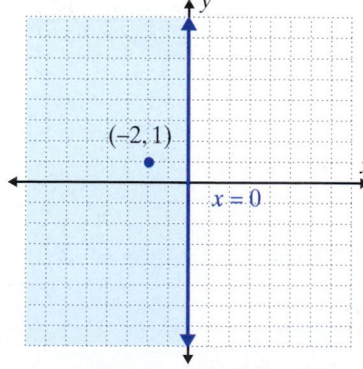

Step 3: Shade the half-plane on the same side of the line as $(-2, 1)$. This half-plane consists of the points with x-coordinate 0 or negative.

Now work margin exercise 8.

Objective G Using a TI-84 Plus Graphing Calculator to Graph Linear Inequalities

The first step in using the TI-84 Plus (or any other graphing calculator) to graph a linear inequality is to solve the inequality for y. This is necessary because this is the way that the boundary line equation can be graphed as a function. Thus Method 2 for graphing the correct half-plane is appropriate.

Note that when you press the ⌁Y=⌁ key on the calculator, a slash (\ ) appears to the left of the Y expression as in $\backslash Y_1$ =. This slash is actually a command to the calculator to graph the corresponding function as a solid line or curve. If you move the cursor to position it over the slash and hit ENTER repeatedly, the following options will appear.

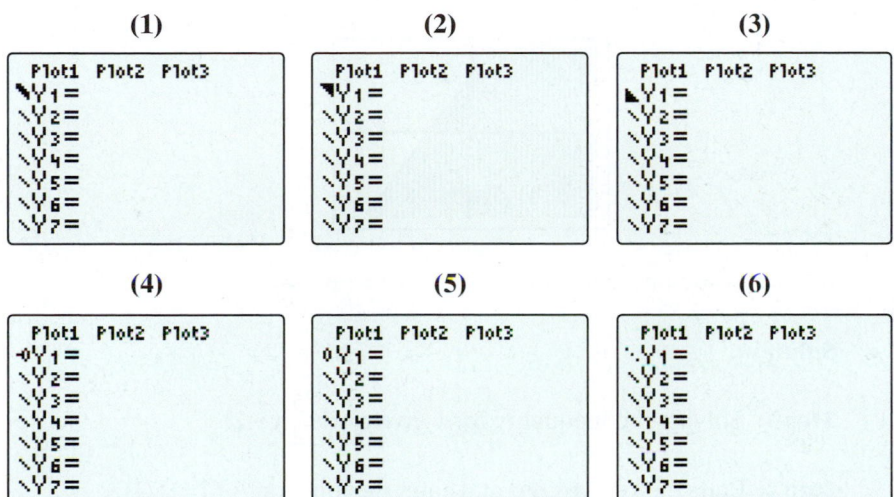

Figure 2

If the slash (which is actually four dots if you look closely) becomes a set of three dots **(6)**, then the corresponding graph of the function will be dotted. By setting the shading above the slash **(2)**, the corresponding graph on the display will show shading above the line or curve. By setting the shading below the slash **(3)**, the corresponding graph on the display will show shading below the line or curve. (The solid line occurs only when the slash is four dots, so the calculator is not good for determining whether the boundary curve is included or not.) Options 1, 4, and 5 are not used when graphing linear inequalities. The following examples illustrate two situations.

Example 9

 Graphing Linear Inequalities Using a Calculator

a. Graph the linear inequality $2x + y \leq 7$.

Solution

Step 1: Solving the inequality for y gives: $y \leq -2x + 7$.

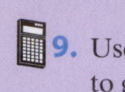

9. Use a graphing calculator to graph the following inequalities.

a. $5x - 8 > y - 6$

b. $6x - 3y \le 7$

Step 2: Press the key [Y=] and enter the function: $\backslash Y1 = -2X + 7$.

Step 3: Go to the \ and hit [ENTER] three times so that the display appears as follows.

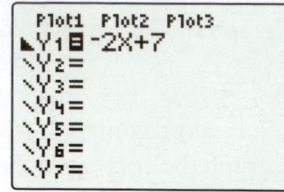

Step 4: Press [GRAPH] and (using the standard WINDOW settings) the following graph should appear on the display.

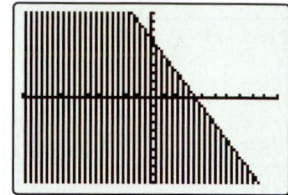

b. Graph the linear inequality $-5x + 4y > -8$.

Solution

Step 1: Solving the inequality for y gives: $y > \dfrac{5}{4}x - 2$.

Step 2: Press the key [Y=] and enter the function: $\backslash Y1 = (5/4)X - 2$.

Step 3: Go to the \ and hit [ENTER] two times so that the display appears as follows.

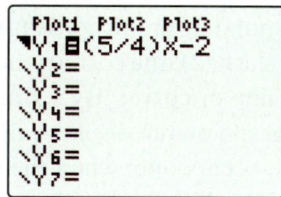

Step 4: Press [GRAPH] and (using the standard WINDOW settings) the following graph should appear on the display. (**Note:** The boundary line should actually be dotted.)

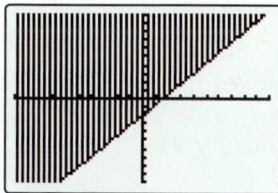

Now work margin exercise 9.

Practice Problems

Graph each set of real numbers on a real number line.

1. $\{x \mid x \le 3 \text{ and } x > 0\}$ **2.** $\{x \mid x \ge -1.5\}$ **3.** $\{x \mid x > 2 \text{ or } x < -4\}$

Solve each of the following inequalities and graph the solution sets. Write each solution set in interval notation. Assume that x is a real number.

4. $7 + x < 3$ **5.** $\dfrac{x}{2} + 1 \le \dfrac{1}{3}$ **6.** $-5 \le 2x + 1 < 9$

7. Which of the following points satisfy the inequality $x + y < 3$?

   a. $(2, 1)$ **b.** $\left(\dfrac{1}{2}, 3\right)$ **c.** $(0, 5)$ **d.** $(-5, 2)$

8. Which of the following points satisfy the inequality $x - 2y \ge 0$?

   a. $(2, 1)$ **b.** $(1, 3)$ **c.** $(4, 2)$ **d.** $(3, 1)$

9. Which of the following points satisfy the inequality $x < 3$?

   a. $(1, 0)$ **b.** $(0, 1)$ **c.** $(4, -1)$ **d.** $(2, 3)$

Practice Problem Answers

1. (number line: open circle at 0, open circle at 3, segment between)

2. (number line: bracket at −1.5, ray to the right)

3. (number line: open circle at −4 ray left, open circle at 2 ray right)

4. $(-\infty, -4)$ (number line: open circle at −4, ray to the left)

5. $\left(-\infty, -\dfrac{4}{3}\right]$ (number line: bracket at $-\dfrac{4}{3}$, ray to the left)

6. $[-3, 4)$ (number line: bracket at −3, open circle at 4)

7. d.

8. a., c., d.

9. a., b., d.

Exercises 6.5

Graph each set of indicated numbers on a real number line.

1. $\{x \mid x \text{ is a whole number less than } 3\}$

2. $\{x \mid x \text{ is an integer with } |x| \leq 3\}$

3. $\{x \mid x \text{ is a prime number less than } 20\}$

4. $\{x \mid x \text{ is a positive whole number divisible by } 0\}$

5. $\{x \mid x \text{ is a composite number between 2 and } 10\}$

6. $\{x \mid x \geq 4, x \text{ is an integer}\}$

7. $\{x \mid -8 < x < 0, x \text{ is a whole number}\}$

8. $\{x \mid -2 < x < 12, x \text{ is an integer}\}$

Use set-builder notation to indicate each set of numbers as described.

9. The set of all real numbers between 3 and 5, including 3

10. The set of all real numbers between −4 and 4

11. The set of all real numbers greater than or equal to −2.5

12. The set of all real numbers between −1.8 and 5, including both of these numbers

The graphs of sets of real numbers are given. a. Use set-builder notation to indicate the set of numbers shown in each graph. b. Use interval notation to represent the graph. c. Tell what type of interval is illustrated. See Example 1.

13.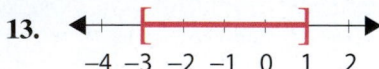

14.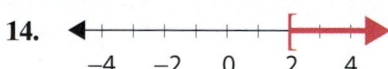

15.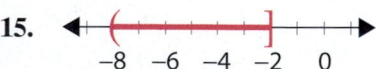

16.

17.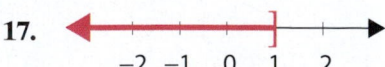

18.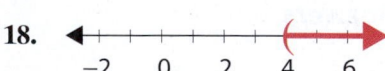

19.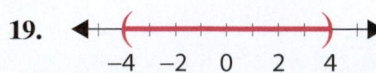

20.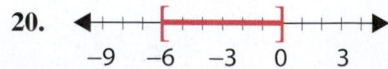

21. $x \leq -3$

22. $x \geq -0.5$

23. $x > 4$

24. $x < -\dfrac{1}{10}$

25. $0 < x \leq 2.5$

26. $-1.5 \leq x < 3.2$

27. $-2 \leq x \leq 0$

28. $-1 \leq x \leq 1$

29. $4 > x \geq 2$

30. $0 > x \geq -5$

31. $2x + 3 < 5$

32. $4x - 7 \geq 9$

33. $14 - 5x < 4$

34. $23 < 7x - 5$

35. $6x - 15 > 1$

36. $9 - 2x < 8$

37. $5.6 + 3x \geq 4.4$

38. $12x - 8.3 < 6.1$

39. $1.5x + 9.6 < 12.6$

40. $0.8x - 2.1 \geq 1.1$

41. $2 + 3x \geq x + 8$

42. $x - 6 \leq 4 - x$

43. $3x - 1 \leq 11 - 3x$

44. $5x + 6 \geq 2x - 2$

45. $4 - 2x < 5 + x$

46. $4 + x > 1 - x$ **47.** $x - 6 > 3x + 5$ **48.** $4 + 7x \leq 4x - 8$

49. $\dfrac{x}{2} - 1 \leq \dfrac{5x}{2} - 3$ **50.** $\dfrac{x}{4} + 1 \leq 5 - \dfrac{x}{4}$ **51.** $\dfrac{x}{3} - 2 > 1 - \dfrac{x}{3}$

52. $\dfrac{5x}{3} + 2 > \dfrac{x}{3} - 1$ **53.** $6x + 5.91 < 1.11 - 2x$ **54.** $4.3x + 21.5 \geq 1.7x + 0.7$

55. $6.2x - 5.9 > 4.8x + 3.2$ **56.** $0.9x - 11.3 < 3.1 - 0.7x$ **57.** $4(6 - x) < -2(3x + 1)$

58. $-3(2x - 5) \leq 3(x - 1)$ **59.** $-(3x + 8) \geq 2(3x + 1)$ **60.** $6(3x + 1) < 5(1 - 2x)$

61. $11x + 8 - 5x \geq 2x - (4 - x)$ **62.** $1 - (2x + 8) < (9 + x) - 4x$

63. $5 - 3(4 - x) + x \leq -2(3 - 2x) - x$ **64.** $x - 2(x + 3) \geq 7 - (4 - x) + 11$

65. $\dfrac{2(x - 1)}{3} < \dfrac{3(x + 1)}{4}$ **66.** $\dfrac{3(x - 2)}{2} \geq \dfrac{4(x - 1)}{3}$

67. $\dfrac{x - 2}{4} > \dfrac{x + 2}{2} + 6$ **68.** $\dfrac{x + 4}{9} \leq \dfrac{x}{3} - 2$

69. $\dfrac{2x + 7}{4} \leq \dfrac{x + 1}{3} - 1$ **70.** $\dfrac{4x}{7} - 3 > \dfrac{x - 6}{2} - 4$

71. $-4 < x + 5 < 6$

72. $2 \leq -x + 2 \leq 6$

73. $3 \geq 4x - 3 \geq -1$

74. $13 > 3x + 4 > -2$

75. $1 \leq \dfrac{2}{3}x - 1 \leq 9$

76. $-2 \leq \dfrac{1}{2}x - 5 \leq -1$

77. $14 > -2x - 6 > 4$

78. $-11 \geq -3x + 2 > -20$

79. $-1.5 < 2x + 4.1 < 3.5$

80. $0.9 < 3x + 2.4 < 6.9$

81. Test Scores: A statistics student has grades of 82, 95, 93, and 78 on four hour-long exams. He must have a mean score of 90 or higher to receive an A for the course. What scores can he receive on the final exam and earn an A if:

a. The final is equivalent to a single hour-long exam (100 points maximum)?

b. The final is equivalent to two hourly exams (200 points maximum)?

82. Test Scores: To receive a grade of B in a chemistry class, Melissa must have a mean score of 80 or more but less than 90. If her five hour-long exam scores were 75, 82, 90, 85, and 77, what score does she need on the final exam (100 points maximum) to earn a grade of B?

83. Car Sales: A car salesman makes $1000 each day that he works and makes approximately $250 commission for each car he sells. If a car salesman wants to make at least $3500 in one day, how many cars does he need to sell?

84. Postage: Allison is going to the post office to buy 38¢ stamps and 2¢ adjustment stamps. Since the current postage rate is 44¢, she will need 3 times as many 2¢ adjustment stamps as 38¢ stamps. If she has $11 to spend, what is the largest number of 38¢ stamps she can buy?

Objectives

A Understand the concept of a function.

B Find the domain and range of a relation or function.

C Determine whether a relation is a function or not.

D Use the vertical line test to determine whether a graph is or is not the graph of a function.

E Understand that all non-vertical lines represent functions.

F Determine the domain of nonlinear functions.

G Write a function using function notation.

H Use a graphing calculator to graph functions.

6.6 Introduction to Functions and Function Notation

Objective A **Functions**

Everyday use of the term **function** is not far from the technical use in mathematics. For example, distance traveled is a function of time; profit is a function of sales; heart rate is a function of exertion; and interest earned is a function of principal invested. In this sense, one variable "depends on" (or "is a function of") another.

Mathematicians distinguish between graphs of ordered pairs of real numbers as those that represent **functions** and those that do not. For example, every equation of the form $y = mx + b$ represents a function and we say that y "is a function of" x. Thus, lines that are not vertical are the graphs of functions. As the following discussion indicates, vertical lines do not represent functions.

notes

- The ordered pairs discussed in this text are ordered pairs of real numbers. However, more generally, ordered pairs might be other types of pairs such as (child, mother), (city, state), or (name, batting average).

Objective B **Domain and Range**

Relation, Domain, and Range

A **relation** is a set of ordered pairs of real numbers.

The **domain**, D, of a relation is the set of all first coordinates in the relation.

The **range**, R, of a relation is the set of all second coordinates in the relation.

In the graph of a relation, the horizontal axis (the x-axis) is called the **domain axis**, and the vertical axis (the y-axis) is called the **range axis**.

Example 1

Finding the Domain and Range

Find the domain and range for each of the following relations.

a. $g = \{(5, 7), (6, 2), (6, 3), (-1, 2)\}$

Solution

$$D = \{5, 6, -1\} \qquad \text{the set of all the first coordinates in } g$$
$$R = \{7, 2, 3\} \qquad \text{the set of all the second coordinates in } g$$

Note that 6 is written only once in the domain and 2 is written only once in the range, even though each appears more than once in the relation.

b. $f = \{(-1, 1), (1, 5), (0, 3)\}$

Solution

$$D = \{-1, 1, 0\} \qquad \text{the set of all the first coordinates in } f$$
$$R = \{1, 5, 3\} \qquad \text{the set of all the second coordinates in } f$$

Example 2

Reading the Domain and Range from the Graph of a Relation

Identify the domain and range from the graph of each relation.

Solution

a. The domain consists of the set of x-values for all points on the graph. In this case the domain is the interval $[-1, 3]$. The range consists of the set of y-values for all points on the graph. In this case, the range is the interval $[0, 6]$.

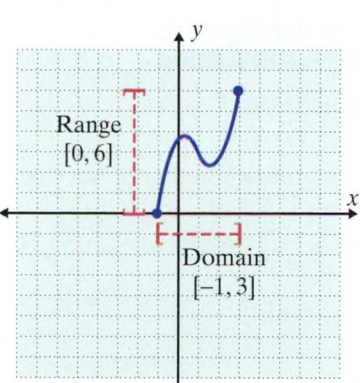

Solution

b. There is no restriction on the x-values which means that for every real number there is a point on the graph with that number as its x-value. Thus the domain is the interval $(-\infty, \infty)$. The y-values begin at -2 and then increase to infinity. The range is the interval $[-2, \infty)$.

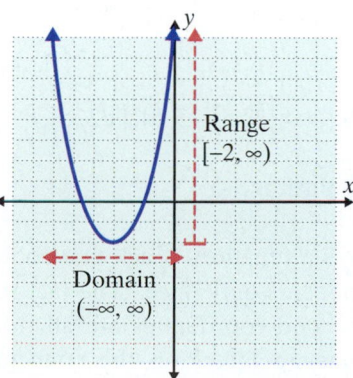

Now work margin exercises 1 and 2.

---

1. Find the domain and range of each of the following relations.

a. $g = \left\{ \begin{matrix} (4, 5), (7, 3), \\ (3, 6), (7, 5) \end{matrix} \right\}$

b. $f = \left\{ \begin{matrix} (-2, 3), (-4, -3), \\ (0, 0) \end{matrix} \right\}$

2. Identify the domain and range from the graph of each relation.

a.

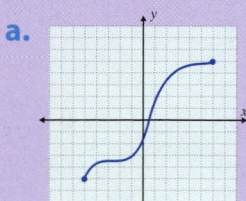

b.

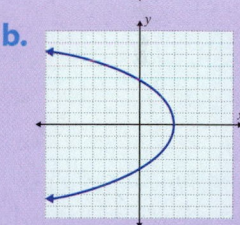

Introduction to Functions and Function Notation   **Section 6.6**

The relation $f = \{(-1, 1), (1, 5), (0, 3)\}$, used in Example **1b.**, meets a particular condition in that each first coordinate has a unique corresponding second coordinate. Such a relation is called a **function**. Notice that g in Example **1a.** is **not** a function because the first coordinate, 6, has two corresponding second coordinates, 2 and 3. Also, for ease in discussion and understanding, the relations illustrated in Examples 1 and 3 have only a finite number of ordered pairs. The graphs of these relations are isolated dots or points. As we will see, the graphs of most relations and functions have an infinite number of points, and their graphs are smooth curves. (**Note:** Lines are also deemed to be curves in mathematics.)

Functions

A **function** is a relation in which each domain element has exactly one corresponding range element.

The definition can also be stated in the following ways

1. A function is a relation in which each first coordinate appears only once.
2. A function is a relation in which no two ordered pairs have the same first coordinate.

3. Determine whether or not each of the following relations is a function.

a. $s = \left\{ \begin{matrix} (4,2), (5,7), \\ (3,8), (4,\sqrt{6}) \end{matrix} \right\}$

b. $t = \left\{ \begin{matrix} (3,\sqrt{4}), (7,4), \\ (-3,\sqrt{4}), (\sqrt{3},4) \end{matrix} \right\}$

Example 3

Functions

Determine whether or not each of the following relations is a function.

a. $s = \left\{(2, 3), (1, 6), (2, \sqrt{5}), (0, -1)\right\}$

Solution

s is not a function. The number 2 appears as a first coordinate more than once.

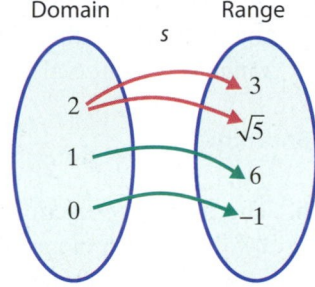

b. $t = \left\{(1, 5), (3, 5), (\sqrt{2}, 5), (-1, 5), (-4, 5)\right\}$

Solution

t is a function. Each first coordinate appears only once. The fact that the second coordinates are all the same has no effect on the concept of a function.

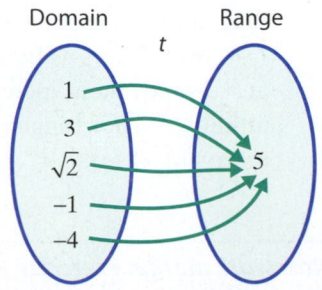

*Now work margin exercise **3.***

Objective D Vertical Line Test

If one point on the graph of a relation is directly above or below another point on the graph, then these points have the same first coordinate (or *x*-coordinate). Such a relation is **not** a function. Therefore, the **vertical line test** can be used to tell whether or not a graph represents a function. (See Figures 1 and 2.)

Vertical Line Test

If **any** vertical line intersects the graph of a relation at more than one point, then the relation is **not** a function.

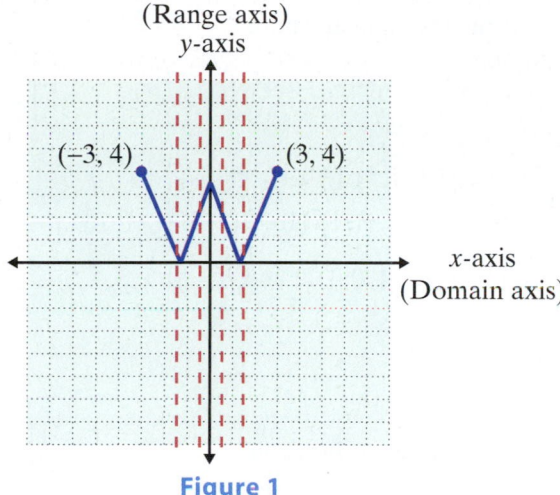

Figure 1

The vertical lines in Figure 1 indicate that this graph represents a function. From the graph, we see that the domain of the function is the interval of real numbers $[-3, 3]$ and the range of the function is the interval of real numbers $[0, 4]$.

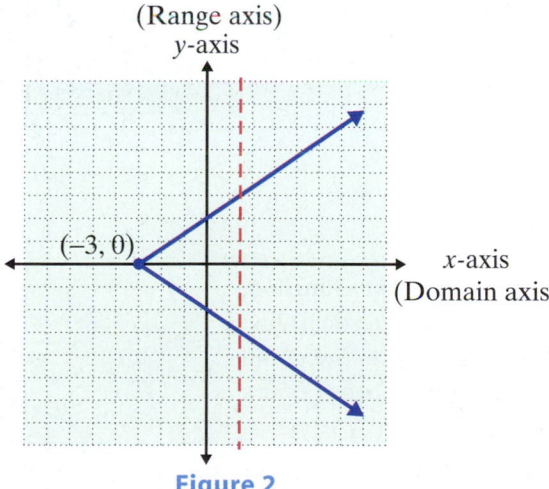

Figure 2

The relation in Figure 2 is **not** a function because the vertical line drawn intersects the graph at more than one point. Thus for that *x*-value, there is more than one corresponding *y*-value. Here $D = [-3, \infty)$ and $R = (-\infty, \infty)$.

Example 4

Vertical Line Test

Use the vertical line test to determine whether or not each of the following graphs represents a function. Then list the domain and range of each graph.

Solution

a. The relation is **not a function**, since a vertical line can be drawn that intersects the graph at more than one point. Listing the ordered pairs shows that several x-coordinates appear more than once.

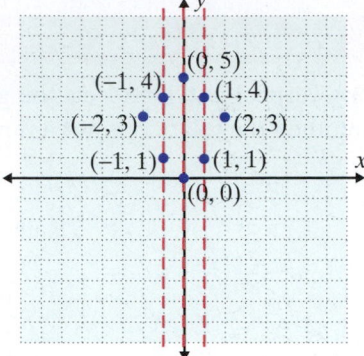

$$r = \left\{ \begin{array}{l} (-2, 3), (-1, 1), (-1, 4), \\ (0, 0), (0, 5), (1, 1), (1, 4), (2, 3) \end{array} \right\}$$

Here $D = \{-2, -1, 0, 1, 2\}$ and $R = \{0, 1, 3, 4, 5\}$.

Solution

b. The relation **is a function.** No vertical line will intersect the graph at more than one point. Several vertical lines are drawn to illustrate this.

For this function, we see from the graph that $D = [-2, 2]$ and $R = [0, 2]$.

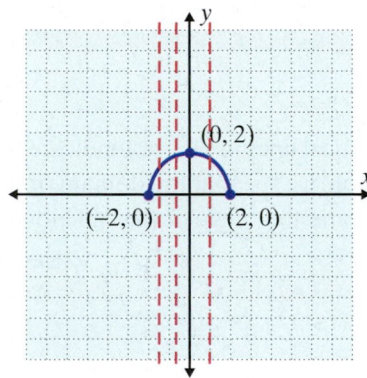

Solution

c. The relation is **not a function**. At least one vertical line (drawn) intersects the graph at more than one point.

Here $D = [-3, \infty)$ and $R = (-\infty, \infty)$.

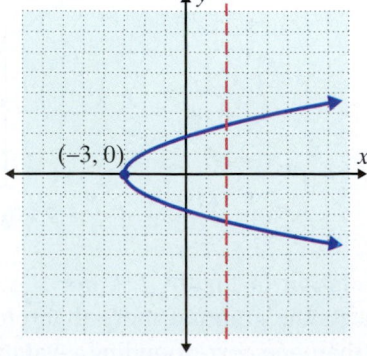

Solution

d. The relation is **not a function**. At least one vertical line intersects the graph at more than one point.

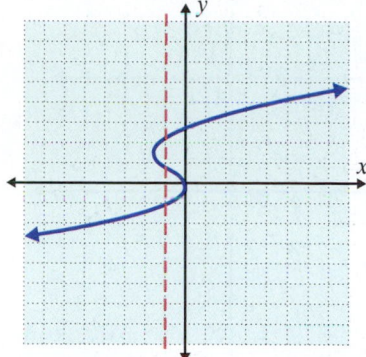

Here $D = (-\infty, \infty)$ and $R = (-\infty, \infty)$.

Now work margin exercise 4.

Objective E **Linear Functions**

All non-vertical lines represent functions. Thus we have the following definition for a linear function.

> ### Linear Function
>
> A **linear function** is a function represented by an equation of the form
>
> $$y = mx + b.$$
>
> The domain of a linear function is the set of all real numbers:
>
> $$D = (-\infty, \infty).$$

If the graph of a linear function is not a horizontal line, then the range is also the set of all real numbers. If the line is horizontal, then the domain is still the set of all real numbers; however, the range is a set containing just a single number. For example, the graph of the linear equation $y = 5$ is a horizontal line. The domain of the function is the set of all real numbers and the range is the set containing only the number 5 (written $\{5\}$). Figures 3 and 4 show two linear functions and the domain and range of each function.

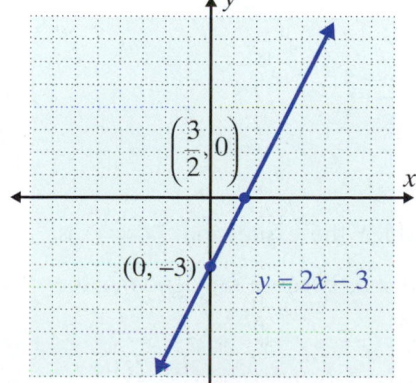

The graph of the linear function $y = 2x - 3$ is shown here.

$D = \{\text{all real numbers}\} = (-\infty, \infty)$

$R = \{\text{all real numbers}\} = (-\infty, \infty)$

Figure 3

4. Use the vertical line test to determine whether or not each of the following graphs represents a function. Then list the domain and range of each graph.

a.

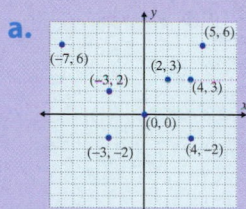

b.

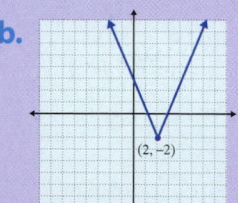

c.

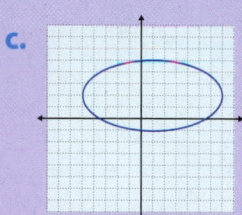

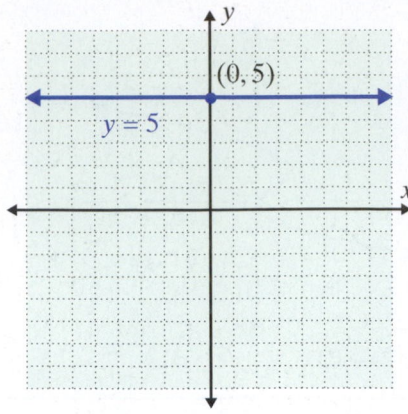

Figure 4

The graph of the linear function $y = 5$ is shown here.

$D = \{\text{all real numbers}\} = (-\infty, \infty)$

$R = \{5\}$

Domains of Nonlinear Functions

As we have seen, the domain for a linear function is the set of all real numbers. Now for the nonlinear function

$$y = \frac{2}{x-1},$$

we say that the domain (all possible values for x) is every real number for which the expression $\frac{2}{x-1}$ is **defined**. Because the denominator cannot be 0, the domain consists of all real numbers except 1. That is, $D = (-\infty, 1) \cup (1, \infty)$ or simply $x \neq 1$. We adopt the following rule concerning equations and domains.

Unless a finite domain is explicitly stated, the domain will be implied to be the set of all real x-values for which the given function is defined. That is, the domain consists of all values of x that give real values for y.

notes

- In determining the domain of a function, one fact to remember at this stage is that **no denominator can equal 0**. In future chapters, we will
- discuss other nonlinear functions with limited domains.

5. Find the domain for the function $f(x) = \dfrac{3x-2}{x+3}$.

Example 5

Domain

Find the domain for the function $y = \dfrac{2x+1}{x-5}$.

Solution

The domain is all real numbers for which the expression $\dfrac{2x+1}{x-5}$ is defined. Thus $D = (-\infty, 5) \cup (5, \infty)$ or $x \neq 5$, because the denominator is 0 when $x = 5$.

Note: Here, interval notation tells us that x can be any real number except 5.

Now work margin exercise 5.

We have used the ordered pair notation (x, y) to represent points in relations and functions. As the vertical line test will show, linear equations of the form

$$y = mx + b,$$

where the equation is solved for y, represent **linear functions**. Another notation, called **function notation**, is more convenient for indicating calculations of values of a function and indicating operations performed with functions. In function notation, instead of writing **y**, we write **$f(x)$**, read "**f of x.**"

The letter f is the name of the function. The letters f, g, h, F, G, and H are commonly used in mathematics, but any letter other than x will do. We have used r, s, and t in previous examples.

The linear equation $y = -3x + 2$ represents a linear function and we can replace y with $f(x)$ as follows:

$$f(x) = -3x + 2.$$

Now, in function notation, $f(4)$ means to replace x with 4 in the function.

$$f(4) = -3 \cdot (4) + 2 = -12 + 2 = -10$$

Thus, the ordered pair $(4, -10)$ can be written as $(4, f(4))$.

Example 6

Function Evaluation

For the function $g(x) = 4x + 5$, find:

a. $g(2)$

Solution

$$g(2) = 4(2) + 5 = 13$$

b. $g(-1)$

Solution

$$g(-1) = 4(-1) + 5 = 1$$

c. $g(0)$

Solution

$$g(0) = 4(0) + 5 = 5$$

Function notation is valid for a wide variety of types of functions. Example 7 illustrates the use of function notation with a nonlinear function.

6. For the function $g(x) = 3x - 2$, find the following.

a. $g(3)$

b. $g(-2)$

c. $g(0)$

7. For the function $f(y) = y^3 - 2y + 5$, find the following.

a. $f(1)$

b. $f(0)$

c. $f(-3)$

Example 7

Nonlinear Function Evaluation

For the function $h(x) = x^2 - 3x + 2$, find the following.

a. $h(4)$

Solution

$$h(4) = (4)^2 - 3(4) + 2 = 16 - 12 + 2 = 6$$

b. $h(0)$

Solution

$$h(0) = (0)^2 - 3(0) + 2 = 0 - 0 + 2 = 2$$

c. $h(-3)$

Solution

$$h(-3) = (-3)^2 - 3(-3) + 2 = 9 + 9 + 2 = 20$$

Now work margin exercises 6 and 7.

Example 8

Function Evaluation From a Graph

Using the graph of $g(x)$ below, find each of the following values.

a. $g(-4)$ **b.** $g(2)$ **c.** $g(-7)$

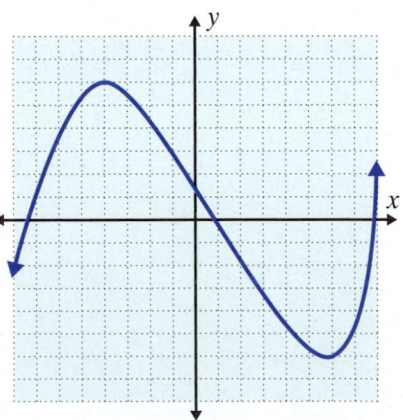

Solution

Using the graph, we need to find the *y*-value that corresponds to each *x*-value.

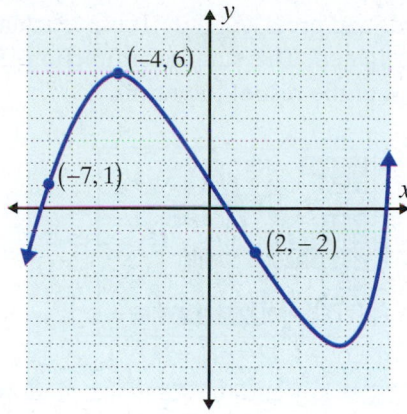

Thus, the function values are as follows:

a. $g(-4)=6$ **b.** $g(2)=-2$ **c.** $g(-7)=1$

Now work margin exercise 8.

Objective H **Using a TI-84 Plus Graphing Calculator to Graph Functions**

There are many types and brands of graphing calculators available. For convenience and so that directions can be specific, only the TI-84 Plus graphing calculator is used in the related discussions in this text. Other graphing calculators may be used, but the steps required may be different from those indicated in the text. If you choose to use another calculator, be sure to read the manual for your calculator and follow the relevant directions.

In any case, remember that a calculator is just a tool to allow for fast calculations and to help in understanding some abstract concepts. A calculator does not replace your ability to think and reason or the need for algebraic knowledge and skills.

You should practice and experiment with your calculator until you feel comfortable with the results. **Do not be afraid of making mistakes. Note that** **CLEAR** **or** **2ND** **QUIT** **will get you out of most trouble and allow you to start over.**

8. Using the graph of $f(x)$ below, find each of the following values.

a. $f(4)$

b. $f(-6)$

c. $f(-2)$

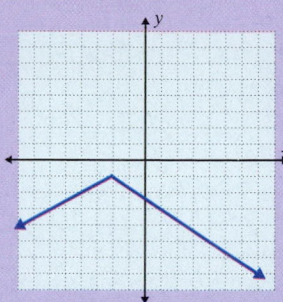

Some Basics about the TI-84 Plus

1. **MODE**: Turn the calculator **ON** and press the **MODE** key. The screen should be highlighted as shown below. If it is not, use the arrow keys in the upper right corner of the keyboard to highlight the correct words and press **ENTER**. It is particularly important that Func is highlighted. This stands for function. See the manual for the meanings of the rest of the terms.

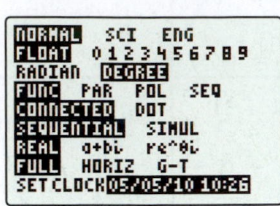

2. **WINDOW**: Press the **WINDOW** key and the standard window will be displayed. By default, the standard window displays a graph with x-values and y-values ranging from -10 to 10 with tic marks on the axis every 1 unit.

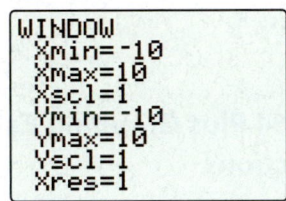

This window can be changed at any time by changing the individual numbers or pressing the **ZOOM** key and selecting an option from the menu displayed. To return the screen back to the standard dimensions with x-values and y-values ranging from -10 to 10, press **ZOOM** and 5: ZStandard. Because of the shape of the display screen, the standard screen is not a square screen (one unit along the x-axis looks longer than one unit along the y-axis). Be aware that the slopes of lines are not truly depicted unless the screen is in a scale of about $3:2$. A square screen can be attained by pressing zoom and 5: ZSquare or by pressing the window key and setting Xmin $= -15$ and Xmax $= 15$ to give the x-axis a length of 30 and the y-axis a length of 20 (a ratio of $3:2$).

3. **Y=**: The **Y=** key is in the upper left corner of the keyboard. This key will allow ten different functions to be entered. These functions are labeled as $Y_1, \ldots, Y_{10}$. The variable x may be entered using the **X,T,θ,n** key. The **^** key is used to indicate exponents. (Also note that the negative sign $(-)$ is next to the **ENTER** key.) For example, the equation $y = x^2 + 3x$ would be entered as:

$$Y_1 = X^2 + 3X.$$

To change an entry, practice with the keys **DEL** (delete), **CLEAR**, and **2ND** **INS** (insert).

```
Plot1 Plot2 Plot3
\Y1◼X^2+3X
\Y2=
\Y3=
\Y4=
\Y5=
\Y6=
\Y7=
```

4. **GRAPH** : If this key is pressed, then the screen will display the graph of whatever functions are indicated in the **Y=** list with the = sign highlighted using the minimum and maximum values indicated in the current WINDOW. In many cases the WINDOW must be changed to accommodate the domain and range of the function or to show a point where two functions intersect.

5. **TRACE** : The **TRACE** key will display the current graph even if it is not already displayed and give the *x*- and *y*- coordinates of a point highlighted on the graph. The curve may be traced by pressing the left and right arrow keys. At each point on the graph, the corresponding *x*- and *y*- coordinates are indicated at the bottom of the screen. **(Remember that because of the limitations of the pixels (lighted dots) on the screen, these *x*- and *y*- coordinates are generally only approximations.)**

6. **CALC:** The **CALC** key (press **2ND** **TRACE**) gives a menu with seven items. Items 1 – 5 are used with graphs.

```
CALCULATE
1◼value
2:zero
3:minimum
4:maximum
5:intersect
6:dy/dx
7:∫f(x)dx
```

After displaying a graph, select **CALC**. Then press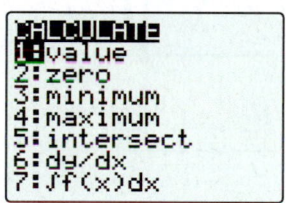

and follow the steps outlined below to locate the point where the graph crosses the *x*-axis (the *x*-intercept). The graph must actually cross the axis. The *x*-value of this point is called a **zero** of the function because the corresponding *y*-value will be 0.

Step 1: With the left arrow, move the cursor to the left of the *x*-intercept on the graph. Press ENTER in response to the question "LeftBound?".

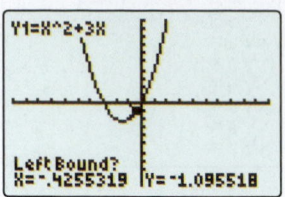

Step 2: With the right arrow, move the cursor to the right of the *x*-intercept on the graph. Press ENTER in response to the question "RightBound?".

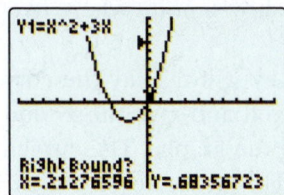

Step 3: With the left arrow, move the cursor near the *x*-intercept. Press ENTER in response to the question "Guess?".

The calculator's estimate of the zero will appear at the bottom of the display.

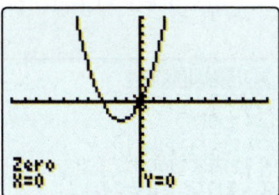

Example 9

Graphing Functions with a TI-84 Plus

Use a TI-84 Plus graphing calculator to find the graphs of each of the following functions. Use the CALC key to find the point where each graph intersects the *x*-axis. Changing the WINDOW may help you get a "better" or "more complete" picture of the function. This is a judgement call on your part.

a. $3x + y = -1$

Solution

To have the calculator graph a nonvertical straight line, you must first solve the equation for y. Solving for y gives

$$y = -3x - 1.$$

(It is important that the **(−)** key be used to indicate the negative sign in front of $3x$. This is not the same as the subtraction key.)

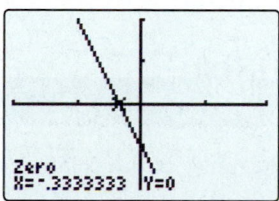

Note: Vertical lines are not functions and cannot be graphed by the calculator in function mode.

b. $y = x^2 + 3x$

Solution

Since the graph of this function has two x-intercepts, we have shown the graph twice. Each graph shows the coordinates of a distinct x-intercept.

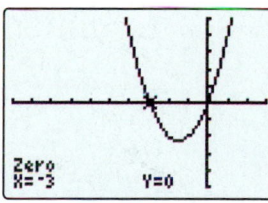   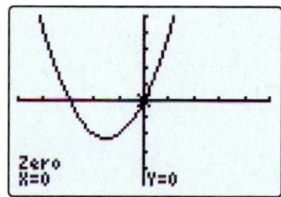

c. $y = 2x - 1; y = 2x + 1; y = 2x + 3$

Solution

$y = 2x - 1$	$y = 2x + 1$	$y = 2x + 3$

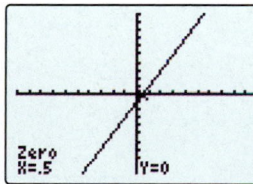   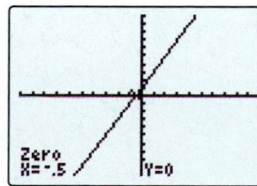   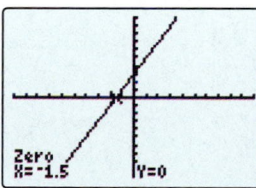

Now work margin exercise 9.

9. Use a TI-84 Plus graphing calculator to find the graphs of each of the following functions. Use the CALC key to find the point where each graph intersects the x-axis.

a. $2x + 3y = -8$

b. $y = x^3 + 2x$

c. $y = 4x - 2; y = 4x + 3;$
   $y = 4x + 7$

Practice Problems

1. State the domain and range of the relation

   a. $\{(5,6),(7,8),(9,0.5),(11,0.3)\}$.

   b. Is the relation a function? Explain briefly.

2. Use the vertical line test to determine whether the graph on the right represents a function.

3. For the function $f(x) = 3x^2 + x - 4$,

   find **a.** $f(2)$ **b.** $f(0)$ **c.** $f(-1)$.

Practice Problem Answers

1. **a.** $D = \{5, 7, 9, 11\}$; $R = \{0.3, 0.5, 6, 8\}$. **b.** Yes, the relation is a function because each *x*-coordinate appears only once.

2. not a function   3 . **a.** 10 **b.** −4 **c.** −2

Exercises 6.6

List the sets of ordered pairs that correspond to the points. State the domain and range and indicate which of the relations are also functions.

1.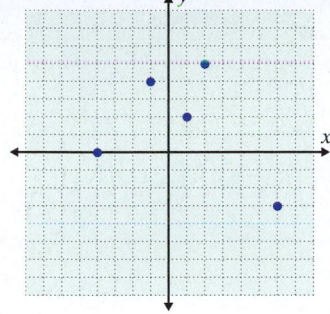

2.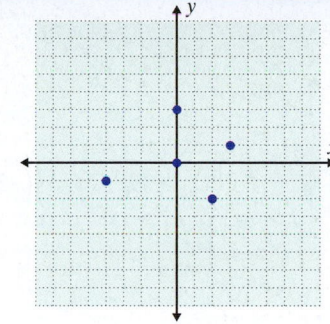

3.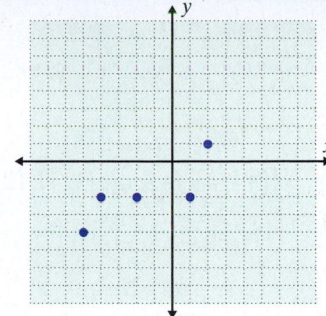

4.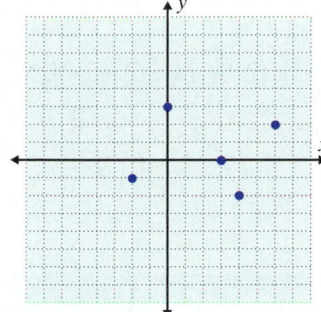

5.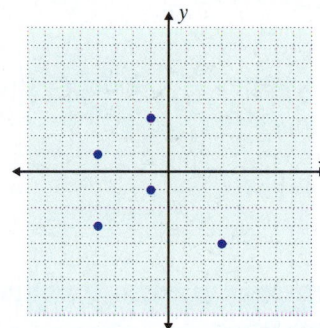

6.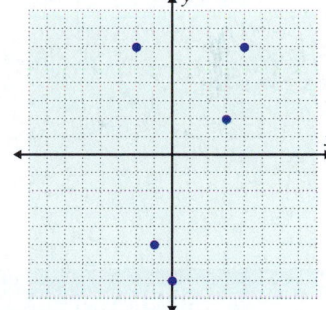

7.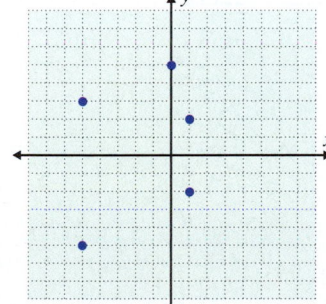

8.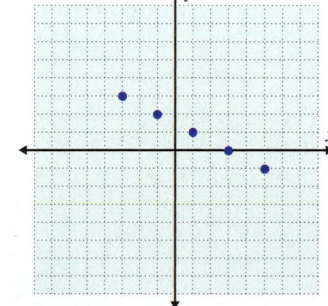

9. $f = \{(0,0),(1,6),(4,-2),(-3,5),(2,-1)\}$

10. $h = \{(1,-5),(2,-3),(-1,-3),(0,2),(4,3)\}$

11. $g = \{(-4,4),(-3,4),(1,4),(2,4),(3,4)\}$

12. $f = \{(-3,-3),(0,1),(-2,1),(3,1),(5,1)\}$

13. $s = \{(0,2),(-1,1),(2,4),(3,5),(-3,5)\}$

14. $t = \{(-1,-4),(0,-3),(2,-1),(4,1),(1,1)\}$

15. $f = \{(-1,4),(-1,2),(-1,0),(-1,6),(-1,-2)\}$

16. $g = \{(0,0),(-2,-5),(2,0),(4,-6),(5,2)\}$

17.

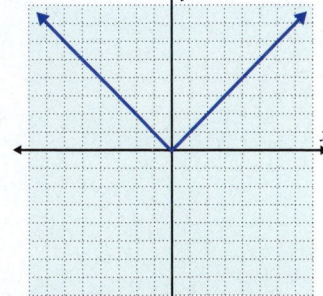

18.

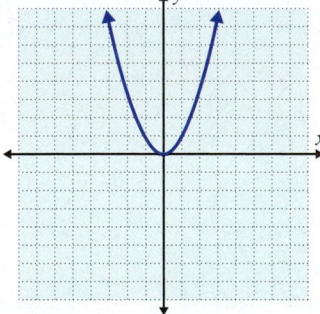

19.

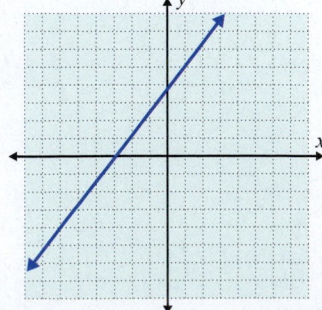

20.

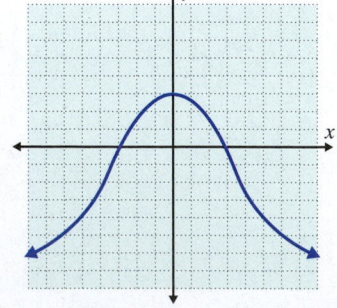

21.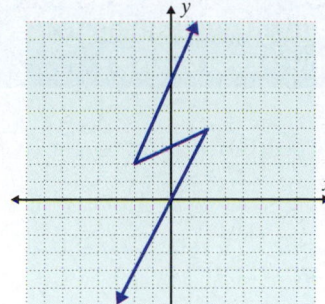

22.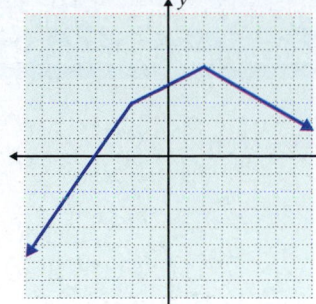

23.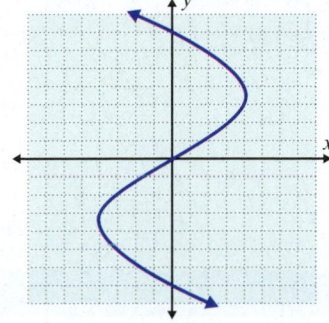

24.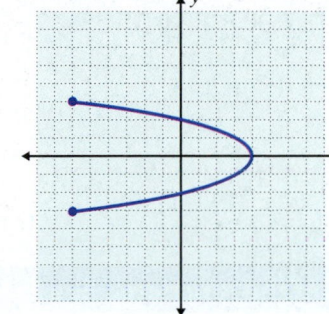

25.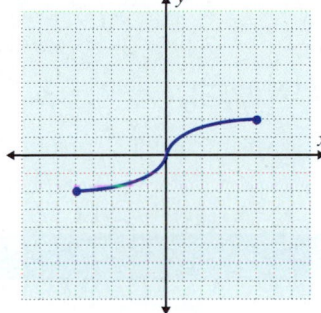

26.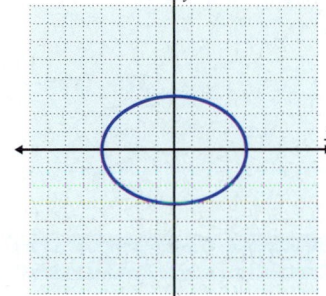

27.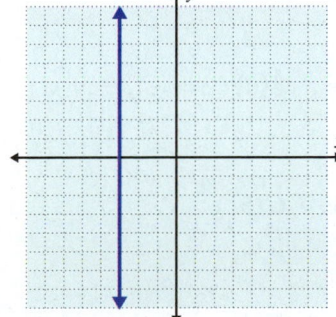

28.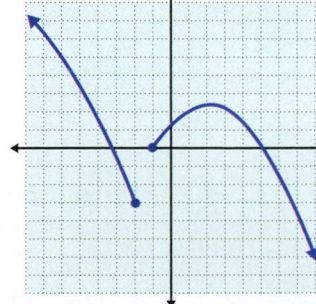

29. $y = 3x + 1; D = \left\{ -9, -\dfrac{1}{3}, 0, \dfrac{4}{3}, 2 \right\}$

30. $y = -\dfrac{3}{4}x + 2; D = \{-4, -2, 0, 3, 4\}$

31. $y = 1 - 3x^2; D = \{-2, -1, 0, 1, 2\}$

32. $y = x^3 - 4x; D = \left\{ -1, 0, \dfrac{1}{2}, 1, 2 \right\}$

State the domains of the functions. See Example 5.

33. $y = -5x + 10$

34. $2x + y = 14$

35. $g(x) = \dfrac{8}{x}$

36. $h(x) = \dfrac{7}{3x}$

37. $y = \dfrac{13x^2 - 5x + 8}{x - 3}$

38. $f(x) = \dfrac{35}{x - 6}$

Find the values of the functions as indicated. See Examples 6 and 7.

39. $f(x) = 3x - 10$

    a. $f(2)$ **b.** $f(-2)$ **c.** $f(0)$

40. $g(x) = -4x + 7$

    a. $g(-3)$ **b.** $g(6)$ **c** $g(0)$

41. $G(x) = x^2 + 5x + 6$

    a. $G(-2)$ **b.** $G(1)$ **c.** $G(5)$

42. $F(x) = 6x^2 - 10$

    a. $F(0)$ **b.** $F(-4)$ **c.** $F(4)$

43. $h(x) = x^3 - 8x$

    a. $h(-3)$ **b.** $h(0)$ **c.** $h(3)$

44. $P(x) = x^2 + 4x + 4$

    a. $P(-2)$ **b.** $P(10)$ **c.** $P(-5)$

45.   **a.** $f(-3)$ **b.** $f(0)$ **c.** $f(6)$ **46.**   **a.** $f(-5)$ **b.** $f(-1)$ **c.** $f(4)$

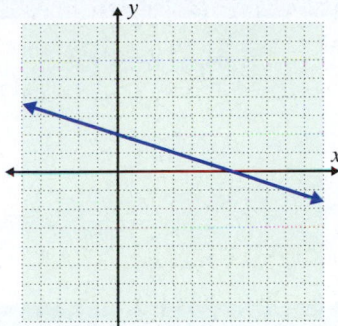 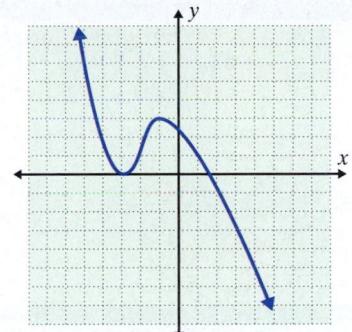

47.   **a.** $f(-5)$ **b.** $f(4)$ **c.** $f(0)$ **48.**   **a.** $f(-1)$ **b.** $f(1)$ **c.** $f(-6)$

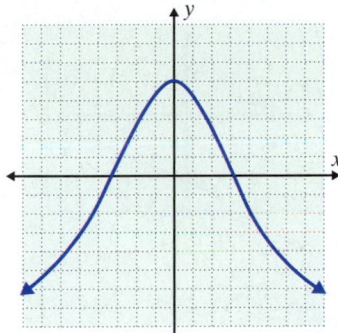 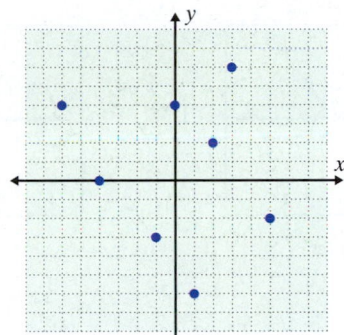

 Use a graphing calculator to graph the functions. Use the CALC features to find *x*-intercepts, if any. (The value of *y* will be 0 at those points.) For absolute value functions, select the MATH menu, then the NUM menu, and then 1 : abs ((Remember to press) after entering the absolute value.)

49. $y = 6$ **50.** $y = 4x$ **51.** $y = x + 5$ **52.** $y = -2x + 3$

53. $y = x^2 - 4x$ **54.** $y = 1 + 2x - x^2$ **55.** $y = -|3x|$ **56.** $y = |x + 2|$

57. $y = |x^2 - 3x|$ **58.** $y = 2x^3 - 5x^2 + 1$ **59.** $y = -x^3 + 3x - 1$ **60.** $y = x^4 - 10x^2 + 9$

Use the CALC features of the calculator to find the coordinates of any points of intersection of the graphs. (Hint: Item 5 on the CALC menu 5: intersect will help in finding that point (or points) of intersetion of two functions, if there is one.) In the Y = menu use both Y_1 = and Y_2 = to be able to graph both functions at the same time.

61. $y = 3x + 2$
$y = 4 - x$

62. $y = 2 - x$
$y = x$

63. $y = 2x - 1$
$y = x^2$

64. $y = x + 3$
$y = -x^2 + x + 7$

The calculator display shows an incorrect graph for the corresponding equation. Explain how you know, by just looking at the graph, that a mistake has been made.

65. $y = 2x + 5$

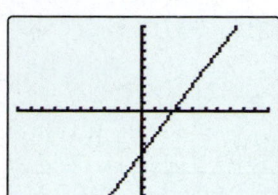

66. $y = -3x + 4$

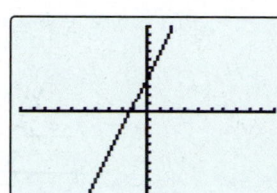

67. $y = \dfrac{2}{3}x - 2$

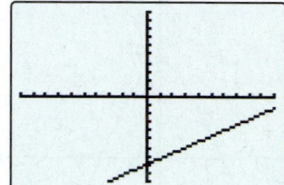

68. $y = -4x$

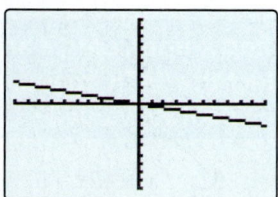

69. $y = -\dfrac{1}{3}x$

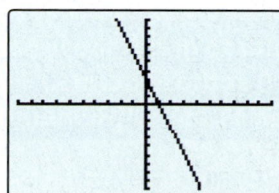

 **Writing & Thinking**

70. **Just for Fun:** Enter a variety of functions in your calculator, investigate your findings, and report these to your class. Certainly, interesting discussions will follow!

Chapter 6: Index of Key Terms and Ideas

Section 6.1: The Cartesian Coordinate System

Cartesian Coordinate System   page 661
   The Cartesian Coordinate System is a system for solving geometric
   problems using algebra.

Ordered Pair   page 662
   Pair of numbers of the form (x, y) where the order of the numbers is
   critical.

One-To-One Correspondence   pages 663 – 664
   There is a **one-to-one correspondence** between points in a
   plane and ordered pairs of real numbers.

Section 6.2: Graphing Linear Equations in Two Variables

Standard Form of a Linear Equation   pages 677 – 678
   Any equation of the form $Ax + By = C$, where A, B, and C are real
   numbers and where A and B are not both 0, is called the **standard form**
   of a **linear equation**.

To Graph a Linear Equation in Two Variables   pages 678 – 679
   1. Locate any two points that satisfy the equation.
   2. Plot these two points on a Cartesian coordinate system.
   3. Draw a line through these two points.
      Note: Every point on that line will satisfy the equation.
   4. To check, locate a third point that satisfies the equation and check to
      see that it does indeed lie on the line

y-intercept   pages 680 – 681
   The **y-intercept** is the point on the graph where the line crosses the y-axis.
   The x-coordinate will be 0, which makes the y-intercept of the form $(0, y)$.

x-intercept   pages 680 – 681
   The **x-intercept** is the point on the graph where the line crosses the x-axis.
   The y-coordinate will be 0, which makes the x-intercept of the form $(x, 0)$.

Section 6.3: The Slope-Intercept Form

Slope as a Rate of Change   pages 685 – 686

Slope of a Line   page 687

Let $P_1(x_1, y_1)$ and $P_2(x_2, y_2)$ be two points on a line. The **slope** can be calculated as follows:

$$\text{slope} = m = \frac{\text{rise}}{\text{run}} = \frac{y_2 - y_1}{x_2 - x_1}.$$

Positive and Negative Slopes   page 689

Lines with **positive slope** go up as we move along the line from left to right.

Lines with **negative slope** go down as we move along the line from left to right.

Horizontal and Vertical Lines   pages 689 – 691

The following two general statements are true for horizontal and vertical lines:

1. For **horizontal lines** (of the form $y = b$), the **slope is 0**.
2. For **vertical lines** (of the form $x = a$), the **slope is undefined**.

Slope-Intercept Form   pages 692 – 694

Any equation of the form $y = mx + b$ is called the **slope-intercept form** for the equation of a line. m is the slope and $(0, b)$ is the y-intercept.

Section 6.4: The Point-Slope Form

Finding the Equation of a Line Given Two Points   pages 702 – 703

To find the equation of a line given two points on the line:

1.   Use the formula $m = \dfrac{y_2 - y_1}{x_2 - x_1}$ to find the slope.

2.   Use this slope, m, and either point in the point-slope formula $y - y_1 = m(x - x_1)$ to find the equation.

Parallel Lines   pages 703 – 704

Parallel lines are lines that never intersect (cross each other) and these lines have the same slope. All vertical lines (undefined slopes) are parallel to one another. All horizontal lines (zero slopes) are also parallel to one another.

Perpendicular Lines   page 705

Perpendicular lines are lines that intersect at 90° (right) angles and whose slopes are negative reciprocals of each other. Horizontal lines are perpendicular to vertical lines.

Point-Slope Form

pages 706 – 708

An equation of the form $y - y_1 = m(x - x_1)$ is called the **point-slope form** for the equation of a line that contains the point (x_1, y_1) and has slope m.

Summary of Formulas and Properties of Lines

page 710

1. $y = b$ horizontal line, slope 0
2. $x = a$ vertical line, undefined slope
3. parallel lines   have the same slope
4. perpendicular lines   have slopes that are negative reciprocals of each other
5. $Ax + By = C$ standard form
6. $m = \dfrac{y_2 - y_1}{x_2 - x_1}$ slope of a line
7. $y = mx + b$ slope-intercept form
8. $y - y_1 = m(x - x_1)$ point-slope form

Section 6.5: Linear Inequalities in One or Two Variables

Set

page 715

A set is a collection of objects or numbers.

Union

page 716

The **union** (symbolized $\cup$, as in A $\cup$ B) of two (or more) sets is the set of all elements that belong to either one set or the other set or to both sets.

Intersection

page 716

The **intersection** (symbolized $\cap$, as in A $\cap$ B) of two (or more) sets is the set of all elements that belong to both sets.

Interval Notation

pages 716 – 717

Type of Interval Notation

Open $(a, b), (a, \infty), (-\infty, b)$

Closed $[a, b]$

Half-open $[a, \infty), (-\infty, b], [a, b), (a, b]$

Linear Inequalities

page 719

Inequalities of the given form, where a, b, and c are real numbers and $a \neq 0$, are called **linear inequalities**.

$$ax + b < c \quad \text{and} \quad ax + b \leq c$$
$$ax + b > c \quad \text{and} \quad ax + b \geq c$$

Compound Inequalities   pages 723 – 724

The inequalities $c < ax + b < d$ and $c \leq ax + b \leq d$ are called **compound linear inequalities**. (This includes $c < ax + b \leq d$ and $c \leq ax + b < d$ as well.)

Rules for Solving Linear Inequalities   page 720

1. Simplify each side of the inequality by removing any grouping symbols and combining like terms.
2. Use the addition property of equality to add the opposites of constants or variable expressions so that variable expressions are on one side of the inequality and constants are on the other.
3. Use the multiplication property of equality to multiply both sides by the reciprocal of the coefficient of the variable (that is, divide both sides by the coefficient) so that the new coefficient is 1. If this coefficient is negative, reverse the direction of the inequality.
4. A quick check is to select any one number in your solution and substitute it into the original inequality.

Terminology for Graphing Linear Inequalities   page 726

Half-plane
Boundary line
Closed Half-plane
Open Half-plane

Graphing Linear Inequalities   pages 726 – 728

1. First, graph the boundary line (dashed if the inequality is < or >, solid if the inequality is ≤ or ≥).
2. Next, determine which side of the line to shade using one of the following methods.

Method 1
   a. Test any one point obviously on one side of the line.
   b. If the test-point satisfies the inequality, shade the half-plane on that side of the line. Otherwise, shade the other half-plane.

Method 2
   a. Solve the inequality for y (assuming that the line is not vertical).
   b. If the solution shows $y <$ or $y \leq$, then shade the half-plane below the line.
   c. If the solution shows $y >$ or $y \geq$, then shade the half-plane above the line.

3. The shaded half-plane (and the line if it is solid) is the solution to the inequality.

Relation, Domain, and Range   page 737 – 738

A **relation** is a set of ordered pairs of real numbers. The **domain**, D, of a relation is the set of all first coordinates in the relation. The **range**, R, of a relation is the set of all second coordinates in the relation.

Functions   page 739

A **function** is a relation in which each domain element has exactly one corresponding range element. The definition can also be stated in the following ways:

1. A function is a relation in which each first coordinate appears only once.

2. A function is a relation in which no two ordered pairs have the same first coordinate.

Vertical Line Test

If **any** vertical line intersects the graph of a relation at more than one point, then the relation graphed is **not** a function.

Linear Functions   page 740 – 742

A **linear function** is a function represented by an equation of the form $y = mx + b$. The domain of a linear function is the set of all real numbers, $D = (-\infty, \infty)$.

Domains of Non-Linear Functions   page 743

Unless a finite domain is explicitly stated, the domain will be implied to be the set of all real x-values for which the given function is defined. That is, the domain consists of all values of x that give real values for y.

Function Notation   pages 743 – 744

In **function notation**, instead of writing y, write $f(x)$, read "f of x." The letter f is the name of the function. The notation $f(4)$ means to replace x with 4 in the function.

Chapter 6: Test

1. List the set of ordered pairs corresponding to the points on the graph. Assume that the grid lines are marked one unit apart.

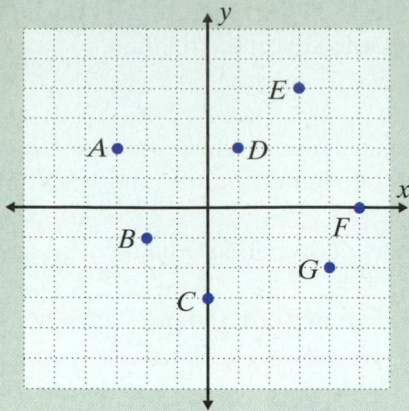

2. Graph the following set of ordered pairs and label the points.

$$\left\{ L(0, 2), M(4, -1), N(-3, 2), O(-1, -5), P(2, 1.5) \right\}$$

Determine the missing coordinate in each of the ordered pairs so that the point will satisfy the equation given.

3. $3x + y = 2$
    a. $(0, \quad)$
    b. $(\quad, 0)$
    c. $(-2, \quad)$
    d. $(\quad, -7)$

4. $x - 5y = 6$
    a. $(0, \quad)$
    b. $(\quad, 0)$
    c. $(11, \quad)$
    d. $(\quad, -2)$

Locate at least two ordered pairs of real numbers that satisfy each of the linear equations and graph the corresponding line in the Cartesian coordinate system.

5. $x + 4y = 5$

6. $2x - 5y = 1$

Graph the following linear equation by locating the x-intercept and the y-intercept.

7. $5x - 3y = 9$

8. $\dfrac{4}{3}x + 2y = 8$

9. $(1, -2), (9, 7)$

10. $(-2, 5), (8, 3)$

Write each equation in slope-intercept form. Find the slope and *y*-intercept, and then use them to draw the graph.

11. $x - 3y = 4$

12. $4x + 3y = 3$

Solve the following word problems.

13. **Bicycling:** The following table shows the number of miles Sam rode his bicycle from one hour to another.

    a. Plot these points on a graph.

    b. Connect the points with line segments.

    c. Find the slope of each line segment.

    d. Interpret the slope as a rate of change.

Hour	Miles
First	20
Second	31
Third	17

14. Find an equation in standard form for the line passing through the point $(3, 7)$ and with the slope $m = -\dfrac{5}{3}$. Graph the line.

15. Find an equation in standard form for the line passing through the points $(-4, 6)$ and $(3, -2)$. Graph the line.

Find an equation in slope-intercept form for each of the equations described.

16. Horizontal and passing through $(-1, 6)$

17. Parallel to $3x + 2y = -1$ and passing through $(2, 4)$

18. Perpendicular to the y-axis and passing through the point $(3, -2)$

19. Write the definition of a function.

20. List the set of ordered pairs that corresponds to the points. State the domain and range and indicate whether or not the relation is a function.

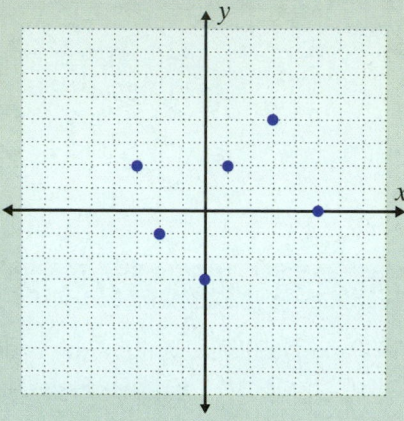

21. Use the vertical line test to determine whether or not the following graph represents a function. State the domain and range.

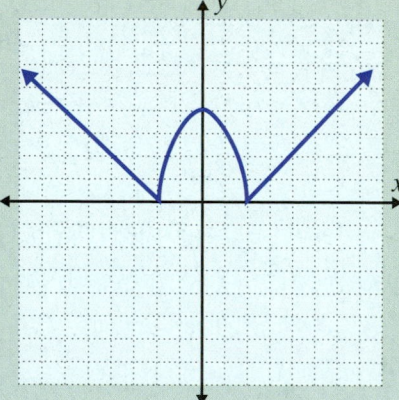

22. Given that $f(x) = x^2 - 2x + 5$, find

    a. $f(-2)$ **b.** $f(0)$ **c.** $f(1)$.

23. **a.** Use a graphing calculator to graph the function $y = 5 + 3x - 2x^2$.

    b. Use the CALC features of the calculator to find x-intercepts, if any.

Graph the solution set of each of the linear inequalities.

24. $3x - 5y \leq 10$ **25.** $3x + 4y > 7$

26. Use a graphing calculator to graph the following linear inequality: $y - 2x \leq -5$.

The graphs of sets of real numbers are given. a. Use set-builder notation to indicate the set of numbers shown in each graph. b. Use interval notation to represent the graph. c. Tell what type of interval is illustrated.

27.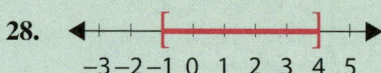
−2 −1 0 1 2 3 4 5 6

28.
−3 −2 −1 0 1 2 3 4 5

Solve the inequalities and graph the solution sets. Write each solution in interval notation. Assume that x is a real number.

29. $3x + 7 < 4(x + 3)$

30. $\dfrac{2x + 5}{4} > x + 3$

31. $-1 < 3x + 2 < 17$

Cumulative Review: Chapters 1 - 6

Find the LCM of each set of numbers or algebraic expressions.

1. $\{12, 15, 45\}$

2. $\{6, 20, 25, 35\}$

3. $\{6a^2, 24ab^2, 30ab, 40a^2b\}$

4. Find the mean of the set of numbers: $\{3.6, 8.9, 14.7, 25.3\}$

5. Fernando scored $80, 88,$ and 82 on three exams in statistics. What was his mean score on these exams (to the nearest tenth)?

Perform the indicated operations.

6. $|-9| + |-2|$

7. $17 - (-5)$

8. $6.5 + (-4.2) - 3.1$

9. $\dfrac{3}{4} - \dfrac{2}{3} + \left(-\dfrac{1}{6}\right)$

10. $8(-5)$

11. $(-4) \div (-6)$

12. $8 \div 0$

13. $0 \div \dfrac{3}{5}$

Simplify each expression by combining like terms.

14. $-4(x + 3) + 2x$

15. $x + \dfrac{x - 5x}{4}$

16. $(x^3 + 4x - 1) - (-2x^3 + x^2)$

17. $-2[7x - (2x + 5) + 3]$

Solve each of the equations.

18. $9x - 11 = x + 5$

19. $5(1 - 2x) = 3x + 57$

20. $5(2x + 3) = 3(x - 4) - 1$

21. $\dfrac{7x}{8} + 5 = \dfrac{x}{4}$

Determine whether each of the following equations is a conditional equation, an identity, or a contradiction.

22. $12x - 7 = -4(4x - 1) - 3$

23. $2(4z + 5) + 16 = 8z + 26$

24. Solve each equation for the indicated variable.

    a. Solve for n: $A = \dfrac{m + n}{2}$

    b. Solve for f: $\omega = 2\pi f$

Solve each inequality and graph the solution set on a real number line. Write each solution in interval notation. Assume that x is a real number.

25. $5x - 7 > x + 9$

26. $5x + 10 \leq 6(x + 3.8)$

27. $x + 8 - 5x \geq 2(x - 2)$

28. $\dfrac{2x + 1}{3} \leq \dfrac{3x}{5}$

Use the rules for order of operation to evaluate each expression.

29. Write 32,000,000 in scientific notation.

30. Solve for x: $\dfrac{1\frac{2}{3}}{x} = \dfrac{10}{2.25}$.

Mathematics
@ Work!

Physics is a natural science that deals with matter and its motion through space, as well as the concepts of energy and force. Within physics, scientists often use polynomials to define the rules that govern nature and all its effects. Consider the formula for force, $F = ma$ (force equals mass times acceleration). This formula defines why a basketball bounces higher when slammed into the ground than when given a simple dribble, and why a baseball may dent a car, while a ball of paper thrown at the same speed will not. Knowing how to operate with and factor polynomials is necessary to understand the information provided by these rules of nature. Consider the following problem related to physics.

Suppose a bowling ball has a mass of 10 lb and the ball's acceleration down the lane is represented by the polynomial $a = 4t^2 - 12t + 7$ m/s² where t represents time in seconds. (Assume that this is the ball's acceleration only until it strikes the pins.) Using the formula $F = ma$, find a polynomial that represents the force of the ball at time t. (**Note:** The force is measured in Newtons, N.)

For more problems like these, see Section 7.4.

Exponents, Polynomials, and Factoring Techniques

7.1 Exponents

The Product Rule

In earlier sections, an **exponent** was defined as a number that tells how many times a number (called the **base**) is used in multiplication. This definition is limited because it is valid only if the exponents are positive integers. In this section, we will develop four properties of exponents that will help in simplifying algebraic expressions and expand your understanding of exponents to include variable bases and negative exponents. (In later sections you will study fractional exponents.)

From earlier, we know that

Exponent

$$6^2 = 6 \cdot 6 = 36$$

Base

and

$$6^3 = 6 \cdot 6 \cdot 6 = 216.$$

Also, the base may be a variable so that

$$x^3 = x \cdot x \cdot x \quad \text{and} \quad x^5 = x \cdot x \cdot x \cdot x \cdot x.$$

Now to find the products of expressions such as $6^2 \cdot 6^3$ or $x^3 \cdot x^5$ and to simplify these products, we can write down all the factors as follows:

$$6^2 \cdot 6^3 = (6 \cdot 6) \cdot (6 \cdot 6 \cdot 6) = 6^5$$

and

$$x^3 \cdot x^5 = (x \cdot x \cdot x) \cdot (x \cdot x \cdot x \cdot x \cdot x) = x^8.$$

With these examples in mind, what do you think would be a simplified form for the product $3^4 \cdot 3^3$? You were right if you thought 3^7. That is, $3^4 \cdot 3^3 = 3^7$. Notice that in each case, **the base stays the same**.

The preceding discussion, along with the basic concept of whole-number exponents, leads to the following **product rule for exponents**.

The Product Rule for Exponents

If a is a nonzero real number and m and n are integers, then

$$a^m \cdot a^n = a^{m+n}.$$

In words, to multiply two powers with the same base, keep the base and add the exponents.

notes

Remember, If a variable or constant has no exponent written, the exponent is understood to be 1.

For example, $y = y^1$ and $7 = 7^1$.

In general, for any real number a, $\boldsymbol{a = a^1}$.

Example 1

Product Rule for Exponents

Use the product rule for exponents to simplify the following expressions.

a. $x^2 \cdot x^4$

Solution

$x^2 \cdot x^4 = x^{2+4} = x^6$

b. $y \cdot y^6$

Solution

$y \cdot y^6 = y^1 \cdot y^6 = y^{1+6} = y^7$

c. $4^2 \cdot 4$

Solution

$4^2 \cdot 4 = 4^{2+1} = 4^3 = 64$
      Note that the base stays 4.
      That is, the bases are not multiplied.

d. $2^3 \cdot 2^2$

Solution

$2^3 \cdot 2^2 = 2^{3+2} = 2^5 = 32$
      Note that the base stays 2.
      That is, the bases are not multiplied.

e. $(-2)^4 (-2)^3$

Solution

$(-2)^4 (-2)^3 = (-2)^{4+3} = (-2)^7 = -128$

Use the product rule for exponents to simplify the following expressions.

1. a. $x^3 \cdot x^2$

b. $y^5 \cdot y^4$

c. $5^2 \cdot 5$

d. $3^2 \cdot 3^3$

e. $(-3)^5 (-3)^2$

2. a. $4y^3 \cdot 5y^6$

b. $(-2x^4)(5x^3)$

c. $(-5ab^3)(7ab^4)$

To multiply terms that have numerical coefficients and variables with exponents, **the coefficients are multiplied** as usual and **the exponents are added by using the product rule**. Generally variables and constants may need to be rearranged using the commutative and associative properties of multiplication. Example 2 illustrates these concepts.

Example 2

Product Rule for Exponents

Use the product rule for exponents to simplify the following expressions.

a. $2y^2 \cdot 3y^9$

Solution

$$
\begin{aligned}
2y^2 \cdot 3y^9 &= 2 \cdot 3 \cdot y^2 \cdot y^9 \\
&= 6y^{2+9} \\
&= 6y^{11}
\end{aligned}
$$

Coefficients 2 and 3 are multiplied and exponents 2 and 9 are added.

b. $(-3x^3)(-4x^3)$

Solution

$$
\begin{aligned}
(-3x^3)(-4x^3) &= (-3)(-4) \cdot x^3 \cdot x^3 \\
&= 12x^{3+3} \\
&= 12x^6
\end{aligned}
$$

Coefficients −3 and −4 are multiplied and exponents 3 and 3 are added.

c. $(-6ab^2)(8ab^3)$

Solution

$$
\begin{aligned}
(-6ab^2)(8ab^3) &= (-6) \cdot 8 \cdot a^1 \cdot a^1 \cdot b^2 \cdot b^3 \\
&= -48 \cdot a^{1+1} \cdot b^{2+3} \\
&= -48a^2b^5
\end{aligned}
$$

Coefficients −6 and 8 are multiplied and exponents on each variable are added.

Now work margin exercises 1 and 2.

Objective B **The Exponent 0**

The product rule is stated for m and n as **integer** exponents. This means that the rule is also valid for 0 and for negative exponents. As an aid for understanding 0 as an exponent, consider the following patterns of exponents for powers of 2, 3, and 10.

Powers of 2	Powers of 3	Powers of 10
$2^5 = 32$	$3^5 = 243$	$10^5 = 100,000$
$2^4 = 16$	$3^4 = 81$	$10^4 = 10,000$
$2^3 = 8$	$3^3 = 27$	$10^3 = 1000$
$2^2 = 4$	$3^2 = 9$	$10^2 = 100$
$2^1 = 2$	$3^1 = 3$	$10^1 = 10$
$2^0 = ?$	$3^0 = ?$	$10^0 = ?$

Do you notice that the patterns indicate that the exponent 0 gives the same value for the last number in each column? That is, $2^0 = 1$, $3^0 = 1$, and $10^0 = 1$.

Another approach to understanding 0 as an exponent is to consider the product rule. Remember that the product rule is stated for **integer exponents**. Applying this rule with the exponent 0 and the fact that 1 is the multiplicative identity gives results such as the following equations.

$$5^0 \cdot 5^2 = 5^{0+2} = 5^2 \qquad \text{This implies that } 5^0 = 1.$$
$$4^0 \cdot 4^3 = 4^{0+3} = 4^3 \qquad \text{This implies that } 4^0 = 1.$$

This discussion leads directly to the **rule for 0 as an exponent**.

The Exponent 0

If a is a nonzero real number, then

$$a^0 = 1.$$

The expression 0^0 is undefined.

notes

Throughout this text, unless specifically stated otherwise, we will assume that the bases of exponents are nonzero.

Example 3

The Exponent 0

Simplify the following expressions using the rule for 0 as an exponent.

a. 10^0

Solution

$10^0 = 1$

3. Simplify the following expressions using the rule for 0 as an exponent.

a. 1587^0

b. $x^0 \cdot x^2$

c. $(-17)^0$

b. $x^0 \cdot x^3$

Solution

$x^0 \cdot x^3 = x^{0+3} = x^3$ or $x^0 \cdot x^3 = 1 \cdot x^3 = x^3$

c. $(-6)^0$

Solution

$(-6)^0 = 1$

Now work margin exercise 3.

Objective C **The Quotient Rule**

Now consider a fraction in which the numerator and denominator are powers with the same base, such as $\dfrac{5^4}{5^2}$ or $\dfrac{x^5}{x^2}$. We can write:

$$\frac{5^4}{5^2} = \frac{\cancel{5} \cdot \cancel{5} \cdot 5 \cdot 5}{\cancel{5} \cdot \cancel{5} \cdot 1} = \frac{5^2}{1} = 25 \quad \text{or} \quad \frac{5^4}{5^2} = 5^{4-2} = 5^2 = 25$$

and

$$\frac{x^5}{x^2} = \frac{\cancel{x} \cdot \cancel{x} \cdot x \cdot x \cdot x}{\cancel{x} \cdot \cancel{x} \cdot 1} = \frac{x^3}{1} = x^3 \quad \text{or} \quad \frac{x^5}{x^2} = x^{5-2} = x^3.$$

In fractions, as just illustrated, the exponents can be subtracted. Again, the base remains the same. We now have the following **quotient rule for exponents**.

Quotient Rule for Exponents

If a is a nonzero real number and m and n are integers, then

$$\frac{a^m}{a^n} = a^{m-n}.$$

In words, to divide two powers with the same base, keep the base and subtract the exponents. (Subtract the denominator exponent from the numerator exponent.)

Example 4

Quotient Rule for Exponents

Use the quotient rule for exponents to simplify the following expressions.

a. $\dfrac{x^6}{x}$

Solution

$$\frac{x^6}{x} = x^{6-1} = x^5$$

b. $\dfrac{y^8}{y^2}$

Solution

$$\frac{y^8}{y^2} = y^{8-2} = y^6$$

c. $\dfrac{x^2}{x^2}$

Solution

$$\frac{x^2}{x^2} = x^{2-2} = x^0 = 1$$

Note how this example shows another way to justify the idea that $a^0 = 1$. Since the numerator and denominator are the same and not 0, it makes sense that the fraction is equal to 1.

Now work margin exercise 4.

In division, with terms that have numerical coefficients, the **coefficients are divided** as usual and any **exponents are subtracted** by using the quotient rule. These ideas are illustrated in Example 5.

Example 5

Dividing Terms with Coefficients

Use the quotient rule for exponents to simplify the following expressions.

a. $\dfrac{15x^{15}}{3x^3}$

Solution

$$\frac{15x^{15}}{3x^3} = \frac{15}{3} \cdot \frac{x^{15}}{x^3}$$

Coefficients 15 and 3 are divided and exponents 15 and 3 are subtracted.

$$= 5 \cdot x^{15-3}$$

$$= 5x^{12}$$

b. $\dfrac{20x^{10}y^6}{2x^2y^3}$

Solution

$$\frac{20x^{10}y^6}{2x^2y^3} = \frac{20}{2} \cdot \frac{x^{10}}{x^2} \cdot \frac{y^6}{y^3}$$

Coefficients 20 and 2 are divided and exponents on each variable are subtracted.

$$= 10 \cdot x^{10-2} \cdot y^{6-3}$$

$$= 10x^8y^3$$

Now work margin exercise 5.

4. Use the quotient rule for exponents to simplify the following expressions.

a. $\dfrac{x^5}{x^3}$

b. $\dfrac{y^7}{y^4}$

c. $\dfrac{x^5}{x^5}$

5. Use the quotient rule for exponents to simplify the following expressions.

a. $\dfrac{12x^{12}}{4x^4}$

b. $\dfrac{21x^8y^7}{3x^2y^3}$

Objective D **Negative Exponents**

The quotient rule for exponents leads directly to the development of an understanding of negative exponents. In Examples 4 and 5, for each base, the exponent in the numerator was larger than or equal to the exponent in the denominator. Therefore, when the exponents were subtracted using the quotient rule, the result was either a positive exponent or the exponent 0. But, what if the larger exponent is in the denominator and we still apply the quotient rule?

For example, applying the quotient rule to $\dfrac{4^3}{4^5}$ gives

$$\frac{4^3}{4^5} = 4^{3-5} = 4^{-2}$$

which results in a negative exponent.

But, simply reducing $\dfrac{4^3}{4^5}$ gives

$$\frac{4^3}{4^5} = \frac{\cancel{4} \cdot \cancel{4} \cdot \cancel{4} \cdot 1}{\cancel{4} \cdot \cancel{4} \cdot \cancel{4} \cdot 4 \cdot 4} = \frac{1}{4 \cdot 4} = \frac{1}{4^2}.$$

This means that $4^{-2} = \dfrac{1}{4^2}$.

Similar discussions will show that $2^{-1} = \dfrac{1}{2}$, $5^{-2} = \dfrac{1}{5^2}$, and $x^{-3} = \dfrac{1}{x^3}$.

The **rule for negative exponents** follows.

Rule for Negative Exponents

If a is a nonzero real number and n is an integer, then

$$a^{-n} = \frac{1}{a^n}.$$

Example 6

Negative Exponents

Use the rule for negative exponents to simplify each expression so that it contains only positive exponents.

a. 5^{-1}

Solution

$$5^{-1} = \frac{1}{5^1} = \frac{1}{5} \qquad \textcolor{red}{\text{using the rule for negative exponents}}$$

b. x^{-3}

Solution

$$x^{-3} = \frac{1}{x^3}$$ using the rule for negative exponents

c. $x^{-9} \cdot x^7$

Solution

Here we use the product rule first and then the rule for negative exponents.

$$x^{-9} \cdot x^7 = x^{-9+7} = x^{-2} = \frac{1}{x^2}$$

Now work margin exercise 6.

> ## notes
>
> ■ There is nothing wrong with negative exponents. In fact, negative exponents are preferred in some courses in mathematics and science.
> ■ However, so that all answers are the same, in this course we will consider expressions to be simplified if:
> ■ **1.** all exponents are positive and
>   **2.** each base appears only once.
> ■

Example 7

Combining Rules for Exponents

Simplify each expression so that it contains only positive exponents.

a. $2^{-5} \cdot 2^8$

Solution

$$2^{-5} \cdot 2^8 = 2^{-5+8}$$ using the product rule with positive and negative exponents

$$= 2^3$$

$$= 8$$

b. $\dfrac{x^6}{x^{-1}}$

Solution

$$\frac{x^6}{x^{-1}} = x^{6-(-1)}$$ using the quotient rule with positive and negative exponents

$$= x^{6+1}$$

$$= x^7$$

6. Use the rule for negative exponents to simplify each expression so that it contains only positive exponents.

a. 7^{-1}

b. x^{-7}

c. $x^{-11} \cdot x^6$

7. Simplify each expression so that it contains only positive exponents.

a. $3^{-7} \cdot 3^{10}$

b. $\dfrac{x^5}{x^{-1}}$

c. $\dfrac{8^{-4}}{8^{-2}}$

d. $\dfrac{x^7 y^3}{x^4 y^6}$

e. $\dfrac{8x^9 \cdot 3x^4}{6x^{16}}$

c. $\dfrac{10^{-5}}{10^{-2}}$

Solution

$$\dfrac{10^{-5}}{10^{-2}} = 10^{-5-(-2)} \qquad \text{using the quotient rule with negative exponents}$$

$$= 10^{-5+2}$$

$$= 10^{-3}$$

$$= \dfrac{1}{10^3} \text{ or } \dfrac{1}{1000} \qquad \text{using the rule for negative exponents}$$

d. $\dfrac{x^6 y^3}{x^2 y^5}$

Solution

$$\dfrac{x^6 y^3}{x^2 y^5} = x^{6-2} y^{3-5} \qquad \text{using the quotient rule with two variables}$$

$$= x^4 y^{-2}$$

$$= \dfrac{x^4}{y^2} \qquad \text{using the rule for negative exponents}$$

e. $\dfrac{15x^{10} \cdot 2x^2}{3x^{15}}$

Solution

$$\dfrac{15x^{10} \cdot 2x^2}{3x^{15}} = \dfrac{(15 \cdot 2) x^{10+2}}{3x^{15}} \qquad \text{using the product rule}$$

$$= \dfrac{30x^{12}}{3x^{15}}$$

$$= 10x^{12-15} \qquad \text{using the quotient rule}$$

$$= 10x^{-3}$$

$$= \dfrac{10}{x^3} \qquad \text{using the rule for negative exponents}$$

Now work margin exercise 7.

Objective E **Using a TI-84 Plus Graphing Calculator to Evaluate Expressions with Exponents**

On a TI-84 Plus graphing calculator, the caret key [⌃] is used to indicate an exponent. Example 8 illustrates the use of a graphing calculator in evaluating expressions with exponents.

Example 8

 Evaluating Expressions with Exponents

Use a graphing calculator to evaluate each expression.

a. 2^{-3} **b.** 23.18^0 **c.** $(-3.2)^3 (1.5)^2$

Solutions

The following solutions show how the caret key 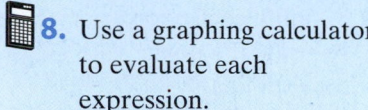 is used to
indicate exponents. Be careful to use the negative sign key **(–)**
(and not the minus sign key) for negative numbers and negative
exponents.

```
2^-3
              .125
23.18^0
                1
(-3.2)^3(1.5)^2
          -73.728
```

Now work margin exercise 8.

Objective F **Power Rule**

Now consider what happens when a power is raised to a power. For example,
to simplify the expressions $(x^2)^3$ and $(2^5)^2$, we can write

$$(x^2)^3 = x^2 \cdot x^2 \cdot x^2 = x^{2+2+2} = x^6 \text{ and } (2^5)^2 = 2^5 \cdot 2^5 = 2^{5+5} = 2^{10}.$$

However, this technique can be quite time-consuming when the exponent is
large such as in $(3y^3)^{17}$. The **power rule for exponents** gives a convenient way
to handle powers raised to powers.

8. Use a graphing calculator
to evaluate each
expression.

a. 3^{-3}

b. 17.06^0

c. $(-4.7)^2 (1.2)^3$

Power Rule for Exponents

If a is a nonzero real number and m and n are integers, then

$$\left(a^m\right)^n = a^{mn}.$$

In other words, the value of a power raised to a power can be found by multiplying the exponents and keeping the base.

9. Simplify each expression by using the power rule for exponents.

a. $\left(x^3\right)^5$

b. $\left(x^4\right)^{-3}$

c. $\left(y^{-5}\right)^3$

d. $\left(3^2\right)^{-3}$

Example 9

Power Rule for Exponents

Simplify each expression by using the power rule for exponents.

a. $\left(x^2\right)^4$

Solution

$$\left(x^2\right)^4 = x^{2(4)} = x^8$$

b. $\left(x^5\right)^{-2}$

Solution

$$\left(x^5\right)^{-2} = x^{5(-2)} = x^{-10} = \frac{1}{x^{10}} \quad \text{or} \quad \left(x^5\right)^{-2} = \frac{1}{\left(x^5\right)^2} = \frac{1}{x^{5(2)}} = \frac{1}{x^{10}}$$

c. $\left(y^{-7}\right)^2$

Solution

$$\left(y^{-7}\right)^2 = y^{(-7)(2)} = y^{-14} = \frac{1}{y^{14}}$$

d. $\left(2^3\right)^{-2}$

Solution

$$\left(2^3\right)^{-2} = 2^{3(-2)} = 2^{-6} = \frac{1}{2^6} \qquad \text{Evaluating gives } \frac{1}{2^6} = \frac{1}{64}.$$

Another approach, because we have a numerical base, would be

$$\left(2^3\right)^{-2} = (8)^{-2} = \frac{1}{8^2}. \qquad \text{Evaluating gives } \frac{1}{8^2} = \frac{1}{64}.$$

We see that while the base and exponent may be different, the value is the same.

Now work margin exercise 9.

If the base of an exponent is a product, we will see that each factor in the product can be raised to the power indicated by the exponent. For example, $(10x)^3$ indicates that the product of 10 and x is to be raised to the 3rd power and $(-2x^2y)^5$ indicates that the product of -2, x^2, and y is to be raised to the 5th power. We can simplify these expressions as follows.

$$(10x)^3 = 10x \cdot 10x \cdot 10x$$
$$= 10 \cdot 10 \cdot 10 \cdot x \cdot x \cdot x$$
$$= 10^3 \cdot x^3$$
$$= 1000x^3$$

$$(-2x^2y)^5 = (-2x^2y)(-2x^2y)(-2x^2y)(-2x^2y)(-2x^2y)$$
$$= (-2)(-2)(-2)(-2)(-2) \cdot x^2 \cdot x^2 \cdot x^2 \cdot x^2 \cdot x^2 \cdot y \cdot y \cdot y \cdot y \cdot y$$
$$= (-2)^5 (x^2)^5 \cdot y^5$$
$$= -32x^{10}y^5$$

We can simplify expressions such as these in a much easier fashion by using the following **power of a product rule for exponents**.

Rule for Power of a Product

If a and b are nonzero real numbers and n is an integer then

$$(ab)^n = a^n b^n.$$

In words, a power of a product is found by raising each factor to that power.

10. Simplify each expression by using the rule for power of a product.

a. $(4x)^2$

b. $(xy)^7$

c. $(-9ab)^2$

d. $(ab)^{-3}$

e. $\left(x^3 y^{-4}\right)^3$

Example 10

Rule for Power of a Product

Simplify each expression by using the rule for power of a product.

a. $(5x)^2$

Solution

$$(5x)^2 = 5^2 \cdot x^2 = 25x^2$$

b. $(xy)^3$

Solution

$$(xy)^3 = x^3 \cdot y^3 = x^3 y^3$$

c. $(-7ab)^2$

Solution

$$(-7ab)^2 = (-7)^2 a^2 b^2 = 49a^2 b^2$$

d. $(ab)^{-5}$

Solution

$$(ab)^{-5} = a^{-5} \cdot b^{-5} = \frac{1}{a^5} \cdot \frac{1}{b^5} = \frac{1}{a^5 b^5}$$

Alternatively, using the rule of negative exponents first and then the rule for the power of a product, we can simplify the expression as follows.

$$(ab)^{-5} = \frac{1}{(ab)^5} = \frac{1}{a^5 b^5}$$

e. $\left(x^2 y^{-3}\right)^4$

Solution

$$\left(x^2 y^{-3}\right)^4 = \left(x^2\right)^4 \left(y^{-3}\right)^4 = x^8 \cdot y^{-12} = \frac{x^8}{y^{12}}$$

Now work margin exercise 10.

notes

■ **Special Note about Negative Numbers and Exponents**

■ In an expression such as $-x^2$, we know that -1 is understood to be the coefficient of x^2. That is,

$$-x^2 = -1 \cdot x^2.$$

■ The same is true for expressions with numbers such as -7^2. That is,

$$-7^2 = -1 \cdot 7^2 = -1 \cdot 49 = -49.$$

■ We see that the exponent refers to 7 and not to -7. For the exponent to refer to -7 as the base, -7 **must be in parentheses** as follows:

$$(-7)^2 = (-7)(-7) = +49.$$

■ As another example,

$$-2^0 = -1 \cdot 2^0 = -1 \cdot 1 = -1 \quad \text{and} \quad (-2)^0 = 1.$$

Objective H **Rule for Power of a Quotient**

If the base of an exponent is a quotient (in fraction form), we will see that both the numerator and denominator are raised to the power indicated by the exponent. For example, in the expression $\left(\dfrac{2}{x}\right)^3$ the quotient (or fraction) $\dfrac{2}{x}$ is raised to the 3rd power.

We can simplify this expression as follows:

$$\left(\frac{2}{x}\right)^3 = \frac{2}{x} \cdot \frac{2}{x} \cdot \frac{2}{x} = \frac{2 \cdot 2 \cdot 2}{x \cdot x \cdot x} = \frac{2^3}{x^3} = \frac{8}{x^3}.$$

Or, we can simplify the expression in a much easier manner by applying the following **rule for the power of a quotient**.

Rule for Power of a Quotient

If a and b are nonzero real numbers and n is an integer, then

$$\left(\frac{a}{b}\right)^n = \frac{a^n}{b^n}.$$

In words, a power of a quotient (in fraction form) is found by raising both the numerator and the denominator to that power.

11. Simplify each expression by using the rule for the power of a quotient.

a. $\left(\dfrac{x}{y}\right)^7$

b. $\left(\dfrac{5}{6}\right)^2$

c. $\left(\dfrac{3}{a}\right)^3$

d. $\left(\dfrac{x}{6}\right)^3$

Example 11

Rule for Power of a Quotient

Simplify each expression by using the rule for the power of a quotient.

a. $\left(\dfrac{y}{x}\right)^5$ 　　　　　　**b.** $\left(\dfrac{3}{4}\right)^3$

Solution: $\left(\dfrac{y}{x}\right)^5 = \dfrac{y^5}{x^5}$ 　　**Solution:** $\left(\dfrac{3}{4}\right)^3 = \dfrac{3^3}{4^3} = \dfrac{27}{64}$

c. $\left(\dfrac{2}{a}\right)^4$ 　　　　　　**d.** $\left(\dfrac{x}{7}\right)^2$

Solution: $\left(\dfrac{2}{a}\right)^4 = \dfrac{2^4}{a^4} = \dfrac{16}{a^4}$ 　　**Solution:** $\left(\dfrac{x}{7}\right)^2 = \dfrac{x^2}{7^2} = \dfrac{x^2}{49}$

Now work margin exercise 11.

Objective I　**Using Combinations of Rules for Exponents**

There may be more than one way to apply the various rules for exponents. As illustrated in the following examples, if you apply the rules correctly (even in a different sequence), the answer will be the same in every case.

Example 12

Using Combinations of Rules for Exponents

Simplify each expression by using the appropriate rules for exponents.

a. $\left(\dfrac{-2x}{y^2}\right)^3$

Solution

$$\left(\dfrac{-2x}{y^2}\right)^3 = \dfrac{(-2x)^3}{\left(y^2\right)^3} = \dfrac{(-2)^3\, x^3}{y^6} = \dfrac{-8x^3}{y^6}$$

b. $\left(\dfrac{3a^2 b}{a^3 b^2}\right)^2$

Solution

Method 1: Simplify inside the parentheses first.

$$\left(\dfrac{3a^2 b}{a^3 b^2}\right)^2 = \left(3a^{2-3}b^{1-2}\right)^2 = \left(3a^{-1}b^{-1}\right)^2 = 3^2 a^{-2}b^{-2} = \dfrac{9}{a^2 b^2}$$

Method 2: Apply the power of a quotient rule first.

$$\left(\frac{3a^2b}{a^3b^2}\right)^2 = \frac{\left(3a^2b\right)^2}{\left(a^3b^2\right)^2} = \frac{3^2 a^{2(2)}b^2}{a^{3(2)}b^{2(2)}} = \frac{9a^4b^2}{a^6b^4} = 9a^{4-6}b^{2-4} = 9a^{-2}b^{-2} = \frac{9}{a^2b^2}$$

Note that the answer is the same even though the rules were applied in a different order.

Now work margin exercise 12.

Another general approach with fractions involving negative exponents is to note that

$$\left(\frac{a}{b}\right)^{-n} = \frac{a^{-n}}{b^{-n}} = \frac{b^n}{a^n} = \left(\frac{b}{a}\right)^n.$$

notes

In effect, there are two basic shortcuts with negative exponents and fractions:

1. Taking the reciprocal of a fraction changes the sign of any exponent in the fraction.
2. Moving any term from numerator to denominator or vice versa, changes the sign of the corresponding exponent.

Example 13

Two Approaches with Fractional Expressions and Negative Exponents

Simplify: $\left(\dfrac{x^3}{y^5}\right)^{-4}$

Solution

Method 1: Use the ideas of reciprocals first.

$$\left(\frac{x^3}{y^5}\right)^{-4} = \left(\frac{y^5}{x^3}\right)^4 = \frac{y^{5(4)}}{x^{3(4)}} = \frac{y^{20}}{x^{12}}$$

Method 2: Apply the power of a quotient rule first.

$$\left(\frac{x^3}{y^5}\right)^{-4} = \frac{\left(x^3\right)^{-4}}{\left(y^5\right)^{-4}} = \frac{x^{3(-4)}}{y^{5(-4)}} = \frac{x^{-12}}{y^{-20}} = \frac{y^{20}}{x^{12}}$$

Now work margin exercise 13.

12. Simplify each expression by using the appropriate rules for exponents.

a. $\left(\dfrac{-3x}{y^3}\right)^3$

b. $\left(\dfrac{4a^3b^2}{a^4b}\right)^2$

13. Simplify.

$$\left(\frac{x^6}{y^3}\right)^{-5}$$

14. Simplify.

$$\left(\frac{3x^3y^2}{4xy^{-3}}\right)^{-2}\left(\frac{5x^3y^{-2}}{2x^{-4}y^2}\right)^{-2}$$

Example 14

A More Complex Example

This example involves the application of a variety of steps. Study it carefully and see if you can get the same result by following a different sequence of steps.

Simplify: $\left(\dfrac{2x^2y^3}{3xy^{-2}}\right)^{-2}\left(\dfrac{4x^2y^{-1}}{3x^{-5}y^3}\right)^{-1}$

Solution

$$\left(\frac{2x^2y^3}{3xy^{-2}}\right)^{-2}\left(\frac{4x^2y^{-1}}{3x^{-5}y^3}\right)^{-1}=\left(\frac{3xy^{-2}}{2x^2y^3}\right)^{2}\left(\frac{3x^{-5}y^3}{4x^2y^{-1}}\right)$$

$$=\frac{3^2x^2y^{-2(2)}}{2^2x^{2(2)}y^{3(2)}}\cdot\frac{3x^{-5}y^3}{4x^2y^{-1}}$$

$$=\frac{9x^2y^{-4}}{4x^4y^6}\cdot\frac{3x^{-5}y^3}{4x^2y^{-1}}=\frac{9\cdot3x^{2-5}y^{-4+3}}{4\cdot4x^{4+2}y^{6-1}}$$

$$=\frac{27x^{-3}y^{-1}}{16x^6y^5}=\frac{27x^{-3-6}y^{-1-5}}{16}$$

$$=\frac{27x^{-9}y^{-6}}{16}=\frac{27}{16x^9y^6}$$

Now work margin exercise 14.

A complete summary of the rules for exponents includes the following eight rules.

Summary of the Rules for Exponents

For any nonzero real number a and integers m and n, the following rules hold.

1. The exponent 1: $a = a^1$

2. The exponent 0: $a^0 = 1$

3. The product rule: $a^m \cdot a^n = a^{m+n}$

4. The quotient rule: $\dfrac{a^m}{a^n} = a^{m-n}$

5. Negative exponents: $a^{-n} = \dfrac{1}{a^n}$

6. Power rule: $\left(a^m\right)^n = a^{mn}$

7. Power of a product: $(ab)^n = a^n b^n$

8. Power of a quotient: $\left(\dfrac{a}{b}\right)^n = \dfrac{a^n}{b^n}$

Practice Problems

Simplify each expression.

1. $2^3 \cdot 2^4$

2. $\dfrac{2^3}{2^4}$

3. $\dfrac{x^7 \cdot x^{-3}}{x^{-2}}$

4. $\dfrac{10^{-8} \cdot 10^2}{10^{-7}}$

5. $\dfrac{14x^{-3}y^2}{2x^{-3}y^{-2}}$

6. $\left(9x^4\right)^0$

7. $\dfrac{x^{-2}x^5}{x^{-7}}$

8. $\left(\dfrac{a^2 b^3}{4}\right)^0$

9. $\dfrac{-3^2 \cdot 5}{2 \cdot (-3)^2}$

10. $\left(\dfrac{5x}{3b}\right)^{-2}$

Use a graphing calculator to evaluate each expression.

11. 72^4

12. $(19.3)^0 (15)^3$

Practice Problem Answers

1. $2^7 = 128$

2. $\dfrac{1}{2}$

3. x^6

4. 10

5. $7y^4$

6. 1

7. x^{10}

8. 1

9. $-\dfrac{5}{2}$

10. $\dfrac{9b^2}{25x^2}$

11. $26{,}873{,}856$

12. 3375

Exercises 7.1

Simplify each expression. The final form of the expressions with variables should contain only positive exponents. Assume that all variables represent nonzero numbers.

1. $3^2 \cdot 3$

2. $7^2 \cdot 7^3$

3. $8^3 \cdot 8^0$

4. $5^0 \cdot 5^2$

5. 3^{-1}

6. 4^{-2}

7. 5^{-2}

8. 6^{-3}

9. $(-2)^4 (-2)^0$

10. $(-4)^3 (-4)^0$

11. $3(2^3)$

12. $6(3^2)$

13. $-4(5^3)$

14. $-2(3^3)$

15. $3(2^{-3})$

16. $4(3^{-2})$

17. $-3(5^{-2})$

18. $-5(2^{-2})$

19. $x^2 \cdot x^3$

20. $x^3 \cdot x$

21. $y^2 \cdot y^0$

22. $y^3 \cdot y^8$

23. x^{-3}

24. y^{-2}

25. $2x^{-1}$

26. $5y^{-4}$

27. $-8y^{-2}$

28. $-10x^{-3}$

29. $5x^6 y^{-4}$

30. $x^0 y^{-2}$

31. $3x^0 + y^0$

32. $5y^0 - 3x^0$

33. $\dfrac{7^3}{7}$

34. $\dfrac{9^5}{9^2}$

35. $\dfrac{10^3}{10^4}$

36. $\dfrac{10}{10^5}$

37. $\dfrac{2^3}{2^6}$

38. $\dfrac{5^7}{5^4}$

39. $\dfrac{x^4}{x^2}$

40. $\dfrac{x^6}{x^3}$

41. $\dfrac{x^3}{x}$

42. $\dfrac{y^7}{y^2}$

43. $\dfrac{x^7}{x^3}$

44. $\dfrac{x^8}{x^3}$

45. $\dfrac{x^{-2}}{x^2}$

46. $\dfrac{x^{-3}}{x}$

47. $\dfrac{x^4}{x^{-2}}$

48. $\dfrac{x^5}{x^{-1}}$

49. $\dfrac{x^{-3}}{x^{-5}}$

50. $\dfrac{x^{-4}}{x^{-1}}$

51. $\dfrac{y^{-2}}{y^{-4}}$

52. $\dfrac{y^3}{y^{-3}}$

53. $3x^3 \cdot x^0$

54. $3y \cdot y^4$

55. $x^3 \cdot x^2 \cdot x^{-1}$

56. $x^{-3} \cdot x^0 \cdot x^2$

57. $\left(4x^3\right)\left(9x^0\right)$

58. $\left(5x^2\right)\left(3x^4\right)$

59. $\left(-2x^2\right)\left(7x^3\right)$

60. $\left(3y^3\right)\left(-6y^2\right)$

61. $\left(-4x^5\right)(3x)$

62. $\left(6y^4\right)\left(5y^5\right)$

63. $\dfrac{8y^3}{2y^2}$

64. $\dfrac{12x^4}{3x}$

65. $\dfrac{9y^5}{3y^3}$

66. $\dfrac{-10x^5}{2x}$

67. $\dfrac{-8y^4}{4y^2}$

68. $\dfrac{12x^6}{-3x^3}$

69. $\dfrac{x^{-1}x^2}{x^3}$

70. $\dfrac{x \cdot x^3}{x^{-3}}$

71. $\dfrac{10^4 \cdot 10^{-3}}{10^{-2}}$

72. $\dfrac{10 \cdot 10^{-1}}{10^2}$

73. $\left(9x^2\right)^0$

74. $\left(-2x^{-3}y^5\right)^0$

75. $\left(9x^2y^3\right)\left(-2x^3y^4\right)$

76. $(-3xy)\left(-5x^2y^{-3}\right)$

77. $\dfrac{-8x^2y^4}{4x^3y^2}$

78. $\dfrac{-8x^{-2}y^4}{4x^2y^{-2}}$

79. $\left(3a^2b^4\right)\left(4ab^5c\right)$

80. $\left(-6a^3b^4\right)\left(4a^{-2}b^8\right)$

81. $\dfrac{36a^5b^0c}{-9a^{-5}b^{-3}}$

82. $\dfrac{7x^2y^{-2}}{28x^0yz^{-2}}$

83. $\dfrac{25y^6 \cdot 3y^{-2}}{15xy^4}$

84. $\dfrac{12a^{-2} \cdot 18a^4}{36a^2b^{-5}}$

85. $(2.16)^0$

86. $(-5.06)^2$

87. $(1.6)^{-2}$

88. $(2.1)^{-3}$

89. $(6.4)^4 (2.3)^2$

90. $(-14.8)^2 (21.3)^2$

Use the rules for exponents to simplify each of the expressions. Assume that all variables represent nonzero real numbers.

91. -3^4

92. -5^2

93. -2^4

94. -20^2

95. $(-10)^6$

96. $(-4)^6$

97. $(6x^3)^2$

98. $(-3x^4)^2$

99. $4(-3x^2)^3$

100. $7(2y^{-2})^4$

101. $5(x^2 y^{-1})$

102. $-3(7xy^2)^0$

103. $-2(3x^5 y^{-2})^{-3}$

104. $-4(5x^{-3} y)^{-1}$

105. $\left(\dfrac{3x}{y}\right)^3$

106. $\left(\dfrac{-4x}{y^2}\right)^2$

107. $\left(\dfrac{6m^3}{n^5}\right)^0$

108. $\left(\dfrac{3x^2}{y^3}\right)^2$

109. $\left(\dfrac{-2x^2}{y^{-2}}\right)^2$

110. $\left(\dfrac{2x}{y^5}\right)^{-2}$

111. $\left(\dfrac{x}{y}\right)^{-2}$

112. $\left(\dfrac{2a}{b}\right)^{-1}$

113. $\left(\dfrac{3x}{y^{-2}}\right)^{-1}$

114. $\left(\dfrac{4a^2}{b^{-3}}\right)^{-3}$

115. $\left(\dfrac{-3}{xy^2}\right)^{-3}$

116. $\left(\dfrac{5xy^3}{y}\right)^2$

117. $\left(\dfrac{m^2 n^3}{mn}\right)^2$

118. $\left(\dfrac{2ab^3}{b^2}\right)^4$

119. $\left(\dfrac{-7^2 x^2 y}{y^3}\right)^{-1}$

120. $\left(\dfrac{2ab^4}{b^2}\right)^{-3}$

121. $\left(\dfrac{5x^3 y}{y^2}\right)^2$

122. $\left(\dfrac{2x^2 y}{y^3}\right)^{-4}$

123. $\left(\dfrac{x^3 y^{-1}}{y^2}\right)^2$

124. $\left(\dfrac{2a^2 b^{-1}}{b^2}\right)^3$

125. $\left(\dfrac{6y^5}{x^2 y^{-2}}\right)^2$

126. $\left(\dfrac{3x^4}{x^{-2} y^{-4}}\right)^3$

127. $\dfrac{\left(7x^{-2} y\right)^2}{\left(xy^{-1}\right)^2}$

128. $\dfrac{\left(-5x^3 y^4\right)^2}{\left(3x^{-3} y\right)^2}$

129. $\dfrac{\left(3x^2 y^{-1}\right)^{-2}}{\left(6x^{-1} y\right)^{-3}}$

130. $\dfrac{\left(2x^{-3}\right)^{-3}}{\left(5y^{-2}\right)^{-2}}$

131. $\dfrac{\left(4x^{-2}\right)\left(6x^5\right)}{\left(9y\right)\left(2y^{-1}\right)}$

132. $\dfrac{\left(5x^2\right)\left(3x^{-1}\right)^2}{\left(25y^3\right)\left(6y^{-2}\right)}$

133. $\left(\dfrac{3xy^3}{4x^2 y^{-3}}\right)^{-1}\left(\dfrac{2x^3 y^{-1}}{9x^{-3} y^{-1}}\right)^2$

134. $\left(\dfrac{5a^4 b^{-2}}{6a^{-4} b^3}\right)^{-2}\left(\dfrac{5a^3 b^4}{2^{-2} a^{-2} b^{-2}}\right)^3$

135. $\left(\dfrac{6x^{-4} yz^{-2}}{4^{-1} x^{-4} y^3 z^{-2}}\right)^{-1}\left(\dfrac{2^{-2} xyz^{-3}}{12x^2 y^2 z^{-1}}\right)^{-2}$

136. $\left(\dfrac{3^{-5} a^5 b^3 c^{-1}}{3^{-2} abc}\right)^{-2}\left(\dfrac{7^{-1} a^{-4} bc^2}{7^{-2} a^{-3} bc^{-2}}\right)^{-2}$

7.2 Introduction to Polynomials

Objectives

A Define a polynomial, and learn how to classify polynomials.

B Evaluate a polynomial for given values of the variable.

Objective A **Definition of a Polynomial**

A **term** is an expression that involves only multiplication and/or division with constants and/or variables. Remember that a number written next to a variable indicates multiplication, and the number is called the **numerical coefficient** (or **coefficient**) of the variable. For example,

$$3x, \quad -5y^2, \quad 17, \quad \text{and} \quad \frac{x}{y}$$

are all algebraic terms.

In the term

$3x$, 3 is the coefficient of x,

$-5y^2$, -5 is the coefficient of y^2, and

$\frac{x}{y}$, 1 is the coefficient of $\frac{x}{y}$.

A term that consists of only a number, such as 17, is also called a **constant** or a **constant term**.

Monomial

A **monomial in** x is a term of the form

$$kx^n$$

where k is a real number and n is a whole number.

n is called the **degree** of the term, and k is called the **coefficient**.

A **monomial** may have more than one variable, and the **degree of such a monomial is the sum of the degrees of its variables**. For example, $4x^2y^3$ is a 5th degree monomial in x and y.

However, in this chapter, only monomials of one variable (note that any variable may be used in place of x) will be discussed.

Since $x^0 = 1$, a nonzero constant can be multiplied by x^0 without changing its value. Thus we say that a **nonzero constant is a monomial of degree 0**. For example,

$$17 = 17x^0 \quad \text{and} \quad -6 = -6x^0$$

which means that the constants 17 and -6 are monomials of degree 0. However, for the special number 0, we can write

$$0 = 0x^2 = 0x^5 = 0x^{13}$$

and we say that **the constant 0 is a monomial of no degree**.

Monomials may have fractional or negative coefficients; however, **monomials may not have fractional or negative exponents**. These facts are part of the definition since in the expression kx^n, k (the coefficient) can be any real number, but n (the exponent) must be a whole number.

expressions that **are not** monomials: $3\sqrt{x}$, $-15x^{\frac{2}{3}}$, $4a^{-2}$

expressions that **are** monomials: 17, $3x$, $5y^2$, $\frac{2}{7}a^4$, πx^2, $\sqrt{2}x^3$

A **polynomial** is a monomial or the indicated sum or difference of monomials. Examples of polynomials are

$$3x, \quad y+5, \quad 4x^2 - 7x + 1, \quad \text{and} \quad a^{10} + 5a^3 - 2a^2 + 6.$$

Polynomial

A **polynomial** is a monomial or the indicated sum or difference of monomials.

The **degree of a polynomial** is the largest of the degrees of its terms.

The coefficient of the term of the largest degree is called the **leading coefficient**.

Special Terminology for Polynomials

Term	Definition	Examples
Monomial:	polynomial with one term	$-2x^3$ and $4a^5$
Binomial:	polynomial with two terms	$3x+5$ and a^2+3
Trinomial:	polynomial with three terms	x^2+6x-7 and a^3-8a^2+12a

Polynomials with four or more terms are simply referred to as **polynomials**.

Examples of polynomials are:

$-1.4x^5$ a fifth-degree monomial in x (leading coefficient is -1.4),

$4z^3 - 7.5z^2 - 5z$ a third-degree trinomial in z (leading coefficient is 4),

$\frac{3}{4}y^4 - 2y^3 + 4y - 6$ a fourth-degree polynomial in y (leading coefficient is $\frac{3}{4}$).

In each of these examples, the terms have been written so that the exponents on the variables decrease in order from left to right. We say that the terms are written in **descending order**. If the exponents on the terms increase in order from left to right, we say that the terms are written in **ascending order**. **As a general rule, for consistency and style in operating with polynomials, the polynomials in this chapter will be written in descending order.**

1. Simplify each of the following polynomials by combining like terms. Write the polynomial in descending order and state the degree and type of the polynomial.

a. $4x^2 + 5x^2$

b. $4x^2 + 5x^2 - 5x$

c. $\frac{1}{3}y^2 + 4y^3 - 3y^2 + 8$

d. $3x^3 + 4x^2 - 8 - 3x^3$

e. $-4y^{-3} + 2y^{-1} + 3y^2$

Example 1

Simplifying Polynomials

Simplify each of the following polynomials by combining **like terms**. (To review the definition of like terms, see Section 5.1.) Write the polynomial in descending order and state the degree and type of the polynomial.

a. $5x^3 + 7x^3$

Solution

$5x^3 + 7x^3 = (5+7)x^3 = 12x^3$ third-degree monomial

b. $5x^3 + 7x^3 - 2x$

Solution

$5x^3 + 7x^3 - 2x = 12x^3 - 2x$ third-degree binomial

c. $\frac{1}{2}y + 3y - \frac{2}{3}y^2 - 7$

Solution

$\frac{1}{2}y + 3y - \frac{2}{3}y^2 - 7 = -\frac{2}{3}y^2 + \frac{7}{2}y - 7$ second-degree trinomial

d. $x^2 + 8x - 15 - x^2$

Solution

$x^2 + 8x - 15 - x^2 = 8x - 15$ first-degree binomial

e. $-3y^4 + 2y^2 + y^{-1}$

Solution

This expression is not a polynomial since y has a negative exponent.

Now work margin exercise 1.

To evaluate a polynomial for a given value of the variable, substitute the value for the variable wherever it occurs in the polynomial and follow the rules for order of operations. A convenient notation for evaluating polynomials is the function notation, $p(x)$ (read "p of x") discussed in section 6.6.

For example,

$$\text{if} \quad p(x) = x^2 - 4x + 13,$$

$$\text{then} \quad p(5) = (5)^2 - 4(5) + 13 = 25 - 20 + 13 = 18.$$

Example 2

Evaluating Polynomials

a. Given $p(x) = 4x^2 + 5x - 15$, find $p(3)$.

Solution

For $p(x) = 4x^2 + 5x - 15$,   Substitute 3 for x.

$$p(3) = 4(3)^2 + 5(3) - 15$$
$$= 4(9) + 15 - 15$$
$$= 36 + 15 - 15$$
$$= 36$$

b. Given $p(y) = 5y^3 + y^2 - 3y + 8$, find $p(-2)$.

Solution

For $p(y) = 5y^3 + y^2 - 3y + 8$,

$$p(-2) = 5(-2)^3 + (-2)^2 - 3(-2) + 8 \quad \text{Note the use of parentheses around } -2.$$
$$= 5(-8) + 4 - 3(-2) + 8$$
$$= -40 + 4 + 6 + 8$$
$$= -22$$

c. Rewrite the polynomial expression $f(x) = 6x + 13$ by substituting for x as indicated by the function notation $f(2a + 1)$.

Solution

$$f(2a + 1) = 6(2a + 1) + 13$$
$$= 12a + 6 + 13$$
$$= 12a + 19$$

Now work margin exercise 2.

2. a. Given $p(x) = 3x^3 - 5x^2 - 13$, find $p(3)$.

b. Given $p(y) = 4y^3 + 2y^2 - 8y + 2$ find, $p(-2)$.

c. Rewrite the polynomial expression $f(x) = 8x - 27$ by substituting for x as indicated by the function notation $f(3a - 1)$.

Practice Problems

Combine like terms and state the degree and type of the polynomial.

1. $8x^3 - 3x^2 - x^3 + 5 + 3x^2$ **2.** $5y^4 + y^4 - 3y^2 + 2y^2 + 4$

3. For the polynomial $p(x) = x^2 - 5x - 5$, find **a.** $p(3)$ and **b.** $p(-1)$.

Practice Problem Answers

1. $7x^3 + 5$; third-degree binomial

2. $6y^4 - y^2 + 4$; fourth-degree trinomial

3. a. −11   **b.** 1

Exercises 7.2

Identify the expression as a monomial, binomial, trinomial, or not a polynomial.

1. $3x^4$

2. $5y^2 - 2y + 1$

3. $8x^3 - 7$

4. $-2x^{-2}$

5. $14a^7 - 2a - 6$

6. $17x^{\frac{2}{3}} + 5x^2$

7. $6a^3 + 5a^2 - a^{-3}$

8. $-3y^4 + 2y^2 - 9$

9. $\frac{1}{2}x^3 - \frac{2}{5}x$

10. $\frac{5}{8}x^5 + \frac{2}{3}x^4$

Simplify the polynomials. Write the polynomial in descending order and state the degree and type of the simplified polynomial. For each polynomial, also state the leading coefficient.

11. $y + 3y$

12. $4x^2 - x + x^2$

13. $x^3 + 3x^2 - 2x$

14. $3x^2 - 8x + 8x$

15. $x^4 - 4x^2 + 2x^2 - x^4$

16. $2 - 6y + 5y - 2$

17. $-x^3 + 6x + x^3 - 6x$

18. $11x^2 - 3x + 2 - 7x^2$

19. $6a^5 + 2a^2 - 7a^3 - 3a^2$

20. $2x^2 - 3x^2 + 2 - 4x^2 - 2 + 5x^2$

21. $4y - 8y^2 + 2y^3 + 8y^2$

22. $2x + 9 - x + 1 - 2x$

23. $5y^2 + 3 - 2y^2 + 1 - 3y^2$

24. $13x^2 - 6x - 9x^2 - 4x$

25. $7x^3 + 3x^2 - 2x + x - 5x^3 + 1$

26. $-3y^5 + 7y - 2y^3 - 5 + 4y^2 + y^2$

27. $x^4 + 3x^4 - 2x + 5x - 10 - x^2 + x$

28. $a^3 + 2a^2 - 6a + 3a^3 + 2a^2 + 7a + 3$

29. $2x + 4x^2 + 6x + 9x^3$

30. $15y - y^3 + 2y^2 - 10y^2 + 2y - 16$

Evaluate the given polynomial as indicated.

31. Given $p(x) = x^2 + 14x - 3$, find $p(-1)$.

32. Given $p(x) = -5x^2 - 8x + 7$, find $p(-3)$.

33. Given $p(x) = 3x^3 - 9x^2 - 10x - 11$, find $p(3)$.

34. Given $p(y) = y^3 - 5y^2 + 6y + 2$, find $p(2)$.

35. Given $p(y) = -4y^3 + 5y^2 + 12y - 1$, find $p(-10)$.

36. Given $p(a) = a^3 + 4a^2 + a + 2$, find $p(-5)$.

37. Given $p(a) = 2a^4 + 3a^2 - 8a$, find $p(-1)$.

38. Given $p(x) = 8x^4 + 2x^3 - 6x^2 - 7$, find $p(-2)$.

39. Given $p(x) = x^5 - x^3 + x - 2$, find $p(-2)$.

40. Given $p(x) = 3x^6 - 2x^5 + x^4 - x^3 - 3x^2 + 2x - 1$, find $p(1)$.

A polynomial is given. Rewrite the polynomial by substituting for the variable as indicated with function notation.

41. Given $p(x) = 3x^4 + 5x^3 - 8x^2 - 9x$, find $p(a)$.

42. Given $p(x) = 6x^5 + 5x^2 - 10x + 3$, find $p(c)$.

43. Given $f(x) = 3x + 5$, find $f(a + 2)$.

44. Given $f(x) = -4x + 6$, find $f(a - 2)$.

45. Given $g(x) = 5x - 10$, find $g(2a + 7)$.

46. Given $g(x) = -4x - 8$, find $g(3a + 1)$.

First-degree polynomials are also called **linear polynomials**, second-degree polynomials are called **quadratic polynomials**, and third-degree polynomials are called **cubic polynomials**. The related functions are called linear functions, quadratic functions, and cubic functions, respectively.

47. Use a graphing calculator to graph the following linear functions.

a. $p(x) = 2x + 3$ **b.** $p(x) = -3x + 1$ **c.** $p(x) = \dfrac{1}{2}x$

48. Use a graphing calculator to graph the following quadratic functions.

a. $p(x) = x^2$ **b.** $p(x) = x^2 + 6x + 9$ **c.** $p(x) = -x^2 + 2$

49. Use a graphing calculator to graph the following cubic functions.

a. $p(x) = x^3$ **b.** $p(x) = x^3 - 4x$ **c.** $p(x) = x^3 + 2x^2 - 5$

50. Make up a few of your own linear, quadratic, and cubic functions and graph these functions with your calculator. Using the results from Exercises 47, 48, and 49, and your own functions, describe in your own words:

a. the general shape of the graphs of linear functions.

b. the general shape of the graphs of quadratic functions.

c. the general shape of the graphs of cubic functions.

7.3 Addition and Subtraction with Polynomials

Objectives

A Add polynomials.

B Subtract polynomials.

C Simplify algebraic expressions by removing grouping symbols and combining like terms.

Objective A **Addition with Polynomials**

The **sum** of two or more polynomials is found by combining **like terms**. Remember that like terms (or similar terms) are constants or terms that contain the same variables raised to the same powers. For example,

$3x^2$, $-10x^2$, and $1.4x^2$ are all **like terms**. Each has the same variable raised to the same power.

$5a^3$ and $-7a$ are **not** like terms. Each has the same variable, but the powers are different.

When adding polynomials, the polynomials may be written horizontally or vertically. For example,

$$\left(x^2 - 5x + 3\right) + \left(2x^2 - 8x - 4\right) + \left(3x^3 + x^2 - 5\right)$$
$$= 3x^3 + \left(x^2 + 2x^2 + x^2\right) + \left(-5x - 8x\right) + \left(3 - 4 - 5\right)$$
$$= 3x^3 + 4x^2 - 13x - 6.$$

If the polynomials are written in a vertical format, we align like terms, one beneath the other, in a column format and combine like terms in each column.

$$
\begin{array}{r}
x^2 - 5x + 3 \\
2x^2 - 8x - 4 \\
3x^3 + \ x^2 \qquad - 5 \\
\hline
3x^3 + 4x^2 - 13x - 6
\end{array}
$$

1. a. Add as indicated.

$$\left(6x^3 + 3x^2 - 8x + 4\right)$$
$$+ \left(2x^3 - x^2 + 2x - 8\right)$$
$$+ \left(2x^2 - 7\right)$$

b. Find the sum.

$$\left(7x^3 + 2x^2 + 9\right)$$
$$+ \left(-3x^3 - 5x^2 + 7x - 7\right)$$

Example 1

Adding Polynomials

a. Add as indicated.

$$\left(5x^3 - 8x^2 + 12x + 13\right) + \left(-2x^2 - 8\right) + \left(4x^3 - 5x + 14\right)$$

Solution

$$\left(5x^3 - 8x^2 + 12x + 13\right) + \left(-2x^2 - 8\right) + \left(4x^3 - 5x + 14\right)$$
$$= \left(5x^3 + 4x^3\right) + \left(-8x^2 - 2x^2\right) + \left(12x - 5x\right) + \left(13 - 8 + 14\right)$$
$$= 9x^3 - 10x^2 + 7x + 19$$

b. Find the sum.

$$\left(x^3 - x^2 + 5x\right) + \left(4x^3 + 5x^2 - 8x + 9\right)$$

Solution

$$
\begin{array}{r}
x^3 - \ x^2 + 5x \\
4x^3 + 5x^2 - 8x + 9 \\
\hline
5x^3 + 4x^2 - 3x + 9
\end{array}
$$

Now work margin exercise 1.

Subtraction with Polynomials

A negative sign written in front of a polynomial in parentheses indicates the **opposite of the entire polynomial**. The opposite can be found by changing the sign of every term in the polynomial.

$$-\left(2x^2 + 3x - 7\right) = -2x^2 - 3x + 7$$

We can also think of the opposite of a polynomial as -1 times the polynomial, applying the distributive property as follows.

$$\begin{aligned}
-\left(2x^2 + 3x - 7\right) &= -1\left(2x^2 + 3x - 7\right) \\
&= -1\left(2x^2\right) - 1\left(3x\right) - 1\left(-7\right) \\
&= -2x^2 - 3x + 7
\end{aligned}$$

The result is the same with either approach. So the **difference** between two polynomials can be found by changing the sign of each term of the second polynomial and then combining like terms.

$$\begin{aligned}
\left(5x^2 - 3x - 7\right) - \left(2x^2 + 5x - 8\right) &= 5x^2 - 3x - 7 - 2x^2 - 5x + 8 \\
&= \left(5x^2 - 2x^2\right) + \left(-3x - 5x\right) + \left(-7 + 8\right) \\
&= 3x^2 - 8x + 1
\end{aligned}$$

If the polynomials are written in a vertical format, one beneath the other, we change the signs of the terms of the polynomial being subtracted and then combine like terms.

Subtract.

$$\begin{array}{c}
5x^2 - 3x - 7 \\
-\left(2x^2 + 5x - 8\right) \\
\hline
\end{array}
\quad \longrightarrow \quad
\begin{array}{c}
5x^2 - 3x - 7 \\
-2x^2 - 5x + 8 \\
\hline
3x^2 - 8x + 1
\end{array}$$

Example 2

Subtracting Polynomials

a. Subtract as indicated.

$$\left(9x^4 - 22x^3 + 3x^2 + 10\right) - \left(5x^4 - 2x^3 - 5x^2 + x\right)$$

Solution

$$\begin{aligned}
\left(9x^4 - 22x^3 + 3x^2 + 10\right) &- \left(5x^4 - 2x^3 - 5x^2 + x\right) \\
&= 9x^4 - 22x^3 + 3x^2 + 10 - 5x^4 + 2x^3 + 5x^2 - x \\
&= \left(9x^4 - 5x^4\right) + \left(-22x^3 + 2x^3\right) + \left(3x^2 + 5x^2\right) - x + 10 \\
&= 4x^4 - 20x^3 + 8x^2 - x + 10
\end{aligned}$$

2. a. Subtract as indicated.

$$\left(7x^4 - 18x^3 + 8x^2 - 11\right)$$
$$-\left(-8x^4 - 15x^3 + 3x^2 - 12\right)$$

b. Find the difference.

$$9x^3 - 6x^2 + 1$$
$$-\left(3x^3 + 3x^2 + 8x\right)$$

b. Find the difference. $\qquad 8x^3 + 5x^2 - 14$
$$\qquad\qquad\qquad -\left(-2x^3 + x^2 + 6x\right)$$

Solution

$$8x^3 + 5x^2 - 14 \qquad\qquad 8x^3 + 5x^2 + 0x - 14$$
$$-\left(-2x^3 + x^2 + 6x\right) \longrightarrow \underline{2x^3 - x^2 - 6x + 0}$$
$$\qquad\qquad\qquad\qquad\qquad 10x^3 + 4x^2 - 6x - 14$$

Write in 0s for missing powers to help with alignment of like terms.

Now work margin exercise 2.

Objective C **Simplifying Algebraic Expressions**

If an algebraic expression contains more than one pair of grouping symbols, such as parentheses ( ), brackets [ ], braces { }, radicals signs $\sqrt{}$, absolute value bars $|\ |$, or fraction bars, simplify by working to remove the innermost pair of symbols first. **Apply the rules for order of operations** just as if the variables were numbers and proceed to combine like terms.

3. Simplify each of the following expressions.

a. $3x - \left[1 + 4\left(5 - 2x\right) + 3x\right]$

b. $-5\left(x - 3\right) + 3\left[x + 2\left(x - 3\right)\right]$

Example 3

Simplifying Algebraic Expressions

Simplify each of the following expressions.

a. $5x - \left[2x + 3\left(4 - x\right) + 1\right] - 9$

Solution

$$5x - \left[2x + 3\left(4 - x\right) + 1\right] - 9$$
$$= 5x - \left[2x + 12 - 3x + 1\right] - 9$$
$$= 5x - \left[-x + 13\right] - 9$$
$$= 5x + x - 13 - 9$$
$$= 6x - 22$$

Work with the parentheses first since they are included inside the brackets.

b. $-3\left(x - 4\right) + 2\left[x + 3\left(x - 3\right)\right]$

Solution

$$-3\left(x - 4\right) + 2\left[x + 3\left(x - 3\right)\right]$$
$$= -3\left(x - 4\right) + 2\left[x + 3x - 9\right]$$
$$= -3\left(x - 4\right) + 2\left[4x - 9\right]$$
$$= -3x + 12 + 8x - 18$$
$$= 5x - 6$$

Work with the parentheses first since they are included inside the brackets.

Now work margin exercise 3.

Practice Problems

1. Add. $(15x+4)+(3x^2-9x-5)$

2. Subtract. $(-5x^3-3x+4)-(3x^3-x^2+4x-7)$

3. Simplify. $2-\left[3a-(4-7a)+2a\right]$

Practice Problem Answers

1. $3x^2+6x-1$ **2.** $-8x^3+x^2-7x+11$ **3.** $-12a+6$

Exercises 7.3

Find the indicated sum.

1. $(2x^2 + 5x - 1) + (x^2 + 2x + 3)$

2. $(x^2 + 3x - 8) + (3x^2 - 2x + 4)$

3. $(x^2 + 7x - 7) + (x^2 + 4x)$

4. $(x^2 + 2x - 3) + (x^2 + 5)$

5. $(2x^2 - x - 1) + (x^2 + x + 1)$

6. $(3x^2 + 5x - 4) + (2x^2 + x - 6)$

7. $(-2x^2 - 3x + 9) + (3x^2 - 2x + 8)$

8. $(x^2 + 6x - 7) + (3x^2 + x - 1)$

9. $(-4x^2 + 2x - 1) + (3x^2 - x + 2) + (x - 8)$

10. $(8x^2 + 5x + 2) + (-3x^2 + 9x - 4) + (2x^2 + 6)$

11. $(x^2 + 2x - 1) + (3x^2 - x + 2) + (2x^3 - 4x - 8)$

12. $(x^3 + 2x - 9) + (x^2 - 5x + 2) + (x^3 - 4x^2 + 1)$

13. $\begin{array}{r} x^2 + 4x - 4 \\ -2x^2 + 3x + 1 \\ \hline \end{array}$

14. $\begin{array}{r} 2x^2 + 4x - 3 \\ 3x^2 - 9x + 2 \\ \hline \end{array}$

15. $\begin{array}{r} x^3 + 3x^2 + x \\ -2x^3 - x^2 + 2x - 4 \\ \hline \end{array}$

16. $\begin{array}{r} 4x^3 + 5x^2 \quad\ + 11 \\ 2x^3 - 2x^2 - 3x - 6 \\ \hline \end{array}$

17. $\begin{array}{r} 7x^3 + 5x^2 + x - 6 \\ -3x^2 + 4x + 11 \\ -3x^3 - x^2 - 5x + 2 \\ \hline \end{array}$

18. $\begin{array}{r} x^3 + 5x^2 + 7x - 3 \\ 4x^2 + 3x - 9 \\ 4x^3 + 2x^2 \quad\ - 2 \\ \hline \end{array}$

19. $\begin{array}{r} x^3 + 3x^2 \quad\ - 4 \\ 7x^2 + 2x + 1 \\ x^3 + x^2 - 6x \\ \hline \end{array}$

20. $\begin{array}{r} x^3 + 2x^2 \quad\ - 5 \\ -2x^3 \quad\quad + x - 9 \\ x^3 - 2x^2 \quad\ + 14 \\ \hline \end{array}$

Find the indicated difference.

21. $(2x^2 + 4x + 8) - (x^2 + 3x + 2)$

22. $(3x^2 + 7x - 6) - (x^2 + 2x + 5)$

23. $(x^2 - 9x + 2) - (4x^2 - 3x + 4)$

24. $(6x^2 + 11x + 2) - (4x^2 - 2x - 7)$

25. $\left(2x^2 - x - 10\right) - \left(-x^2 + 3x - 2\right)$

26. $\left(7x^2 + 4x - 9\right) - \left(-2x^2 + x - 9\right)$

27. $\left(x^4 + 8x^3 - 2x^2 - 5\right) - \left(2x^4 + 10x^3 - 2x^2 + 11\right)$

28. $\left(x^3 + 4x^2 - 3x - 7\right) - \left(3x^3 + x^2 + 2x + 1\right)$

29. $\left(-3x^4 + 2x^3 - 7x^2 + 6x + 12\right) - \left(x^4 + 9x^3 + 4x^2 + x - 1\right)$

30. $\left(2x^5 + 3x^3 - 2x^2 + x - 5\right) - \left(3x^5 - 2x^3 + 5x^2 + 6x - 1\right)$

31. $\left(9x^2 - 5\right) - \left(13x^2 - 6x - 6\right)$

32. $\left(8x^2 + 9\right) - \left(4x^2 - 3x - 2\right)$

33. $\left(3x^4 - 2x^3 - 8x - 1\right) - \left(5x^3 - 3x^2 - 3x - 10\right)$

34. $\left(x^5 + 6x^3 - 3x^2 - 5\right) - \left(2x^5 + 8x^3 + 5x + 17\right)$

35.
$$14x^2 - 6x + 9$$
$$-\left(8x^2 + x - 9\right)$$

36.
$$9x^2 - 3x + 2$$
$$-\left(4x^2 - 5x - 1\right)$$

37.
$$5x^4 + 8x^2 + 11$$
$$-\left(-3x^4 + 2x^2 - 4\right)$$

38.
$$11x^2 + 5x - 13$$
$$-\left(-3x^2 + 5x + 2\right)$$

39.
$$x^3 + 6x^2 - 3$$
$$-\left(-x^3 + 2x^2 - 3x + 7\right)$$

40.
$$3x^3 + 9x - 17$$
$$-\left(x^3 + 5x^2 - 2x - 6\right)$$

Simplify the algebraic expression and write the polynomial in descending order.

41. $5x + 2(x - 3) - (3x + 7)$

42. $-4(x - 6) - (8x + 2) - 3x$

43. $11 + \left[3x - 2(1 + 5x)\right]$

44. $2 + \left[9x - 4(3x + 2)\right]$

45. $8x - \left[2x + 4(x - 3) - 5\right]$

46. $17 - \left[-3x + 6(2x - 3) + 9\right]$

47. $3x^3 - \left[5 - 7\left(x^2 + 2\right) - 6x^2\right]$

48. $10x^3 - \left[8 - 5\left(3 - 2x^2\right) - 7x^2\right]$

49. $\left(2x^2 + 4\right) - \left[-8 + 2\left(7 - 3x^2\right) + x\right]$

50. $-\left[6x^2 - 3(4 + 2x) + 9\right] - \left(x^2 + 5\right)$

51. $2\left[3x + (x - 8) - (2x + 5)\right] - (x - 7)$

52. $-3\left[-x + (10 - 3x) - (8 - 3x)\right] + (2x - 1)$

53. $\left(x^2 - 1\right) + 2\left[4 + (3 - x)\right]$

54. $\left(4 - x^2\right) + 3\left[(2x - 3) - 5\right]$

55. $-(x - 5) + \left[6x - 2(4 - x)\right]$

56. $2(2x + 1) - \left[5x - (2x + 3)\right]$

57. Find the sum of $4x^2 - 3x$ and $6x + 5$.

58. Subtract $2x^2 - 4x$ from $7x^3 + 5x$.

59. Subtract $3(x+1)$ from $5(2x-3)$.

60. Find the sum of $10x-2(3x+5)$ and $3(x-4)+16$.

61. Subtract $3x^2-4x+2$ from the sum of $4x^2+x-1$ and $6x-5$.

62. Subtract $-2x^2+6x+12$ from the sum of $2x^2+3x-1$ and $x^2-13x+2$.

63. Add $5x^3-8x+1$ to the difference between $2x^3+14x-3$ and x^2+6x+5.

64. Add $2x^3+4x^2+1$ to the difference between $-x^2+10x-3$ and x^3+2x^2+4x.

Writing & Thinking

65. Write the definition of a polynomial.

66. Explain, in your own words, how to subtract one polynomial from another.

67. Describe what is meant by the degree of a polynomial in x.

68. Give two examples that show how the sum of two binomials might not be a binomial.

7.4 Multiplication with Polynomials

Objectives

A Multiply a polynomial by a monomial.

B Multiply two polynomials.

Up to this point, we have multiplied terms such as $5x^2 \cdot 3x^4 = 15x^6$ by using the product rule for exponents. Also, we have applied the distributive property to expressions such as $5(2x+3) = 10x + 15$.

Now we will use both the product rule for exponents and the distributive property to multiply polynomials. We discuss three cases here:

   a. the product of a monomial with a polynomial of two or more terms,
   b. the product of two binomials, and
   c. the product of a binomial with a polynomial of more than two terms.

Objective A Multiplying a Polynomial by a Monomial

Using the distributive property $a(b+c) = ab + ac$ with multiplication indicated on the left, we can find the product of a monomial with a polynomial of two or more terms as follows.

$$5x(2x+3) = 5x \cdot 2x + 5x \cdot 3 = 10x^2 + 15x$$

$$3x^2(4x-1) = 3x^2 \cdot 4x + 3x^2(-1) = 12x^3 - 3x^2$$

$$-4a^5(a^2 - 8a + 5) = -4a^5 \cdot a^2 - 4a^5(-8a) - 4a^5(5) = -4a^7 + 32a^6 - 20a^5$$

Objective B Multiplying Two Polynomials

Now suppose that we want to multiply two binomials, say, $(x+3)(x+7)$. We will apply the distributive property in the following way with multiplication indicated on the right of the parentheses.

Compare $(x+3)(x+7)$ to $(a+b)c = ac + bc$.

Think of $(x+7)$ as taking the place of c. Thus

$$(a+b)c = ac + bc$$

takes the form $(x+3)(x+7) = x(x+7) + 3(x+7)$.

Completing the products on the right, using the distributive property twice again, gives

$$
\begin{aligned}
(x+3)(x+7) &= x(x+7) + 3(x+7) \\
&= x \cdot x + x \cdot 7 + 3 \cdot x + 3 \cdot 7 \\
&= x^2 + 7x + 3x + 21 \\
&= x^2 + 10x + 21.
\end{aligned}
$$

In the same manner,

$$(x+2)(3x+4) = x(3x+4) + 2(3x+4)$$
$$= x \cdot 3x + x \cdot 4 + 2 \cdot 3x + 2 \cdot 4$$
$$= 3x^2 + 4x + 6x + 8$$
$$= 3x^2 + 10x + 8.$$

Similarly,

$$(2x-1)(x^2+x-5) = 2x(x^2+x-5) - 1(x^2+x-5)$$
$$= 2x \cdot x^2 + 2x \cdot x + 2x \cdot (-5) - 1 \cdot x^2 - 1 \cdot x - 1(-5)$$
$$= 2x^3 + 2x^2 - 10x - x^2 - x + 5$$
$$= 2x^3 + x^2 - 11x + 5.$$

The product of two polynomials can also be found by writing one polynomial under the other. **The distributive property is applied by multiplying each term of one polynomial by each term of the other.** Consider the product $(2x^2 + 3x - 4)(3x + 7)$. Now writing one polynomial under the other and applying the distributive property, we obtain the following.

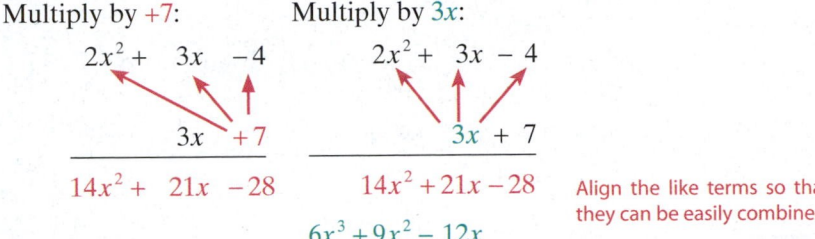

Multiply by $+7$:

Multiply by $3x$:

Align the like terms so that they can be easily combined.

Finally, combine like terms:

$$2x^2 + 3x - 4$$
$$3x + 7$$
$$\overline{ 14x^2 + 21x - 28}$$
$$\underline{6x^3 + 9x^2 - 12x}$$
$$6x^3 + 23x^2 + 9x - 28 \qquad \text{Combine like terms.}$$

Example 1

Multiplying Polynomials

Find each product.

a. $-4x(x^2 - 3x + 12)$

Solution

$$-4x(x^2 - 3x + 12) = -4x \cdot x^2 - 4x(-3x) - 4x \cdot 12$$
$$= -4x^3 + 12x^2 - 48x$$

b. $(2x-4)(5x+3)$

Solution

$$(2x-4)(5x+3) = 2x(5x+3)-4(5x+3)$$
$$= 2x\cdot 5x + 2x\cdot 3 - 4\cdot 5x - 4\cdot 3$$
$$= 10x^2 + 6x - 20x - 12$$
$$= 10x^2 - 14x - 12$$

c. $7y^2 - 3y + 2$
    $\underline{\qquad 2y + 3}$

Solution

$$\begin{array}{r} 7y^2 - 3y + 2 \\ \underline{2y + 3} \\ 21y^2 - 9y + 6 \\ \underline{14y^3 - 6y^2 + 4y \qquad} \\ 14y^3 + 15y^2 - 5y + 6 \end{array}$$

Multiply by 3.

Multiply by 2y.

Combine like terms.

d. $x^2 + 3x - 1$
    $\underline{x^2 - 3x + 1}$

Solution

$$\begin{array}{r} x^2 + 3x - 1 \\ \underline{x^2 - 3x + 1} \\ x^2 + 3x - 1 \\ -3x^3 - 9x^2 + 3x \\ \underline{x^4 + 3x^3 - x^2 \qquad\qquad} \\ x^4 \qquad -9x^2 + 6x - 1 \end{array}$$

Multiply by 1.

Multiply by –3x.

Multiply by x^2.

Combine like terms.

e. $(x-5)(x+2)(x-1)$

Solution

First multiply $(x-5)(x+2)$, then multiply this result by $(x-1)$.

$$(x-5)(x+2) = x(x+2) - 5(x+2)$$
$$= x^2 + 2x - 5x - 10$$
$$= x^2 - 3x - 10$$

$$(x^2 - 3x - 10)(x-1) = (x^2 - 3x - 10)x + (x^2 - 3x - 10)(-1)$$
$$= x^3 - 3x^2 - 10x - x^2 + 3x + 10$$
$$= x^3 - 4x^2 - 7x + 10$$

Now work margin exercise 1.

1. Find each product.

a. $-6x(3x^2 - x - 3)$

b. $(3x+3)(4x-3)$

c. $4x^2 + 8x - 2$
    $\underline{3x + 1}$

d. $x^2 + 4x - 2$
    $\underline{x^2 - 4x + 2}$

e. $(x-2)(x+3)(x-6)$

Practice Problems

Find each product.

1. $2x\left(3x^2 + x - 1\right)$ **2.** $(x+3)(x-7)$

3. $(x-1)\left(x^2 + x - 4\right)$ **4.** $(x+1)(x-2)(x+2)$

Practice Problem Answers

1. $6x^3 + 2x^2 - 2x$ **2.** $x^2 - 4x - 21$

3. $x^3 - 5x + 4$ **4.** $x^3 + x^2 - 4x - 4$

Exercises 7.4

Multiply as indicated and simplify if possible.

1. $-3x^2\left(2x^3+5x\right)$

2. $5x^2\left(-4x^2+6\right)$

3. $4x^5\left(x^2-3x+1\right)$

4. $9x^3\left(2x^3-x^2+5x\right)$

5. $-1\left(y^5-8y+2\right)$

6. $-7\left(2y^4+3y^2+1\right)$

7. $-4x^3\left(x^5-2x^4+3x\right)$

8. $-2x^4\left(x^3-x^2+2x\right)$

9. $5x^3\left(5x^2-x+2\right)$

10. $-2x^2\left(x^3+5x-4\right)$

11. $a^2\left(a^5+2a^4-5a+1\right)$

12. $7t^3\left(-t^3+5t^2+2t+1\right)$

13. $3x\left(2x+1\right)-2\left(2x+1\right)$

14. $x\left(3x+4\right)+7\left(3x+4\right)$

15. $3a\left(3a-5\right)+5\left(3a-5\right)$

16. $6x\left(x-1\right)+5\left(x-1\right)$

17. $5x\left(-2x+7\right)-2\left(-2x+7\right)$

18. $y\left(y^2+1\right)-1\left(y^2+1\right)$

19. $x\left(x^2+3x+2\right)+2\left(x^2+3x+2\right)$

20. $4x\left(x^2-x+1\right)+3\left(x^2-x+1\right)$

21. $\left(x+4\right)\left(x-3\right)$

22. $\left(x+7\right)\left(x-5\right)$

23. $\left(a+6\right)\left(a-8\right)$

24. $\left(x+2\right)\left(x-4\right)$

25. $\left(x-2\right)\left(x-1\right)$

26. $\left(x-7\right)\left(x-8\right)$

27. $3\left(t+4\right)\left(t-5\right)$

28. $-4\left(x+6\right)\left(x-7\right)$

29. $x\left(x+3\right)\left(x+8\right)$

30. $t\left(t-4\right)\left(t-7\right)$

31. $\left(2x+1\right)\left(x-4\right)$

32. $\left(3x-1\right)\left(x+4\right)$

33. $\left(6x-1\right)\left(x+3\right)$

34. $\left(8x+15\right)\left(x+1\right)$

35. $\left(2x+3\right)\left(2x-3\right)$

36. $(3t+5)(3t-5)$

37. $(4x+1)(4x+1)$

38. $(5x-2)(5x-2)$

39. $(y+3)(y^2-y+4)$

40. $(2x+1)(x^2-7x+2)$

41. $\begin{array}{r} 3x+7 \\ \underline{x-5} \end{array}$

42. $\begin{array}{r} 2x+6 \\ \underline{x+3} \end{array}$

43. $\begin{array}{r} x^2+3x+1 \\ \underline{5x-9} \end{array}$

44. $\begin{array}{r} 8x^2+3x-2 \\ \underline{-2x+7} \end{array}$

45. $\begin{array}{r} 2x^2+3x+5 \\ \underline{x^2+2x-3} \end{array}$

46. $\begin{array}{r} 6x^2-x+8 \\ \underline{2x^2+5x+6} \end{array}$

Find the product and simplify if possible.

47. $(3x-4)(x+2)$

48. $(t+6)(4t-7)$

49. $(2x+5)(x-1)$

50. $(5a-3)(a+4)$

51. $(7x+1)(x-2)$

52. $(x-2)(3x+8)$

53. $(2x+1)(3x-8)$

54. $(3x+7)(2x-5)$

55. $(2x+3)(2x+3)$

56. $(5y+2)(5y+2)$

57. $(x+3)(x^2-4)$

58. $(y^2+2)(y-4)$

59. $(2x+7)(2x-7)$

60. $(3x-4)(3x+4)$

61. $(x+1)(x^2-x+1)$

62. $(x-2)(x^2+2x+4)$

63. $(7a-2)(7a-2)$

64. $(5a-6)(5a-6)$

65. $(2x+3)(x^2-x-1)$

66. $(3x+1)(x^2-x+9)$

67. $(x+1)(x+2)(x+3)$

68. $(t-1)(t-2)(t-3)$

69. $(a^2+a-1)(a^2-a+1)$

70. $(y^2+y+2)(y^2+y-2)$

71. $(t^2+3t+2)^2$

72. $(a^2-4a+1)^2$

73. $(y+6)(y-6)+(y+5)(y-5)$

74. $(y-2)(y+2)+(y-1)(y+1)$

75. $(2a+1)(a-5)+(a-4)(a-4)$

76. $(x+4)(2x+1)+(x-3)(x-2)$

77. $(x-3)(x+5)-(x+3)(x+2)$

78. $(t+3)(t+3)-(t-2)(t-2)$

79. $(2a+3)(a+1)-(a-2)(a-2)$

80. $(4t-3)(t+4)-(t-2)(3t+1)$

Writing & Thinking

81. We have seen how the distributive property is used to multiply polynomials.

    a.   Show how the distributive property can be used to find the product
$$\begin{array}{r} 75 \\ \times\,93 \\ \hline \end{array}$$
(**Hint:** 75 = 70 + 5 and 93 = 90 + 3)

    b.   In the multiplication algorithm for multiplying whole numbers (as in the product above), we are told to "move to the left" when multiplying. For example

$$\begin{array}{r} 75 \\ \times\,93 \\ \hline 15 \\ 21 \\ 45 \\ 63 \\ \hline \end{array}$$

Why are the 21 and 45 moved one place to the left in the alignment?

When 9 and 7 are multiplied, we move the 63 two places left. Why?

7.5 Special Products of Polynomials

Objective A The FOIL Method

In the case of the **product of two binomials** such as $(2x+5)(3x-7)$, the **FOIL** method is useful. **F-O-I-L** is a mnemonic device (memory aid) to help in remembering which terms of the binomials to multiply together. First, by using the distributive property we can see how the terms are multiplied.

$$(2x+5)(3x-7) = 2x(3x-7)+5(3x-7)$$

$$= 2x \cdot 3x + 2x \cdot (-7) + 5 \cdot 3x + 5 \cdot (-7)$$

First terms	Outside terms	Inside terms	Last terms
F	O	I	L

Now we can use the FOIL method and then combine like terms to go directly to the answer.

$$(2x+5)(3x-7)$$

$$(2x+5)(3x-7) = 6x^2 - 14x + 15x - 35$$
$$= 6x^2 + x - 35$$

Example 1

FOIL Method

Use the FOIL method to find the products of the given binomials.

a. $(x+3)(2x+8)$

Solution

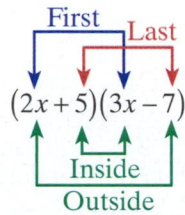

$$(x+3)(2x+8) = 2x^2 + 8x + 6x + 24$$
$$= 2x^2 + 14x + 24$$

b. $(2x-3)(3x-5)$

Solution

$$(2x-3)(3x-5) = 6x^2 - 10x - 9x + 15$$
$$= 6x^2 - 19x + 15$$

c. $(x+7)(x-7)$

Solution

$$(x+7)(x-7) = x^2 - 7x + 7x - 49 \qquad \text{Apply the FOIL method mentally.}$$
$$= x^2 - 49$$

Note that in this special case the two middle terms are opposites and their sum is 0.

Now work margin exercise 1.

Objective B **The Difference of Two Squares**

In Example **1c.**, the middle terms, $-7x$ and $+7x$, are opposites of each other and their sum is 0. Therefore the resulting product has only two terms and each term is a square.

$$(x+7)(x-7) = x^2 - 49$$

In fact, when two binomials are in the form of the sum and difference of the same two terms, the product will always be the difference of the squares of the terms. The product is called the **difference of two squares**.

Difference of Two Squares

$$(x+a)(x-a) = x^2 - a^2$$

In order to quickly recognize the difference of two squares, you should memorize the following squares of the positive integers from 1 to 20. The squares of integers are called **perfect squares.**

Perfect Squares from 1 to 400

1, 4, 9, 16, 25, 36, 49, 64, 81, 100, 121, 144, 169, 196, 225, 256, 289, 324, 361, 400

---

1. Use the FOIL method to find the products of the given binomials.

a. $(x+7)(3x+4)$

b. $(4x-5)(2x-2)$

c. $(x+4)(x-4)$

2. Find the following products.

a. $(x+6)(x-6)$

b. $(4y+3)(4y-3)$

c. $(x^4-3)(x^4+3)$

Example 2

Difference of Two Squares

Find the following products.

a. $(x+4)(x-4)$

Solution

The two binomials represent the sum and difference of x and 4. So, the product is the difference of their squares.

$(x+4)(x-4) = x^2 - 4^2 = x^2 - 16$ difference of two squares

b. $(3y+7)(3y-7)$

Solution

$(3y+7)(3y-7) = (3y)^2 - 7^2 = 9y^2 - 49$ difference of two squares

c. $(x^3-6)(x^3+6)$

Solution

$(x^3-6)(x^3+6) = (x^3)^2 - 6^2 = x^6 - 36$ difference of two squares

Now work margin exercise 2.

Objective C   **Squares of Binomials**

Now we consider the case where the two binomials being multiplied are the same. That is, we want to consider the **square of a binomial**. The following examples, using the distributive property, illustrate two patterns that, after some practice, allow us to go directly to the products.

$(x+3)^2 = (x+3)(x+3) = x^2 + 3x + 3x + 9$

$= x^2 + 2 \cdot 3x + 9$ The middle term is doubled.
$3x + 3x = 2 \cdot 3x$

$= x^2 + 6x + 9$ perfect square trinomial

$(x-11)^2 = (x-11)(x-11) = x^2 - 11x - 11x + 121$

$= x^2 - 2 \cdot 11x + 121$ The middle term is doubled.
$-11x - 11x = 2(-11x) = -2 \cdot 11x$

$= x^2 - 22x + 121$ perfect square trinomial

Note that in each case **the result of squaring the binomial is a trinomial**. These trinomials are called **perfect square trinomials**.

Squares of Binomials (Perfect Square Trinomials)

$(x+a)^2 = x^2 + 2ax + a^2$ **Square of a Binomial Sum**

$(x-a)^2 = x^2 - 2ax + a^2$ **Square of a Binomial Difference**

Example 3

Squares of Binomials

Find the following products.

a. $(2x+3)^2$

Solution

The pattern for squaring a binomial gives the following

$(2x+3)^2 = (2x)^2 + 2\cdot 3\cdot 2x + (3)^2$ **Note:** $(2x)^2 = 2x\cdot 2x = 4x^2$

$\qquad\qquad = 4x^2 + 12x + 9$

b. $(5x-1)^2$

Solution

$(5x-1)^2 = (5x)^2 - 2(1)(5x) + (1)^2$

$\qquad\qquad = 25x^2 - 10x + 1$

c. $(9-x)^2$

Solution

$(9-x)^2 = (9)^2 - 2(x)(9) + x^2$

$\qquad\qquad = 81 - 18x + x^2$

$\qquad\qquad = x^2 - 18x + 81$

d. $(y^3+1)^2$

Solution

$(y^3+1)^2 = (y^3)^2 + 2(1)(y^3) + 1^2$ **Note:** $(y^3)^2 = y^3\cdot y^3 = y^{3+3} = y^6$

$\qquad\qquad = y^6 + 2y^3 + 1$

Now work margin exercise 3.

3. Find the following products.

a. $(3x+5)^2$

b. $(4x-2)^2$

c. $(8-2x)^2$

d. $(2y^3-2)^2$

Practice Problems

Find the indicated products.

1. $(x+10)(x-10)$ **2.** $(x+3)^2$ **3.** $(2x-1)(x+3)$

4. $(2x-5)^2$ **5.** $(x^2+4)(x^2-3)$

Practice Problem Answers

1. $x^2 - 100$ **2.** $x^2 + 6x + 9$ **3.** $2x^2 + 5x - 3$

4. $4x^2 - 20x + 25$ **5.** $x^4 + x^2 - 12$

Exercises 7.5

Find each product and identify any that is either the difference of two squares or a perfect square trinomial.

1. $(x-7)^2$

2. $(x-5)^2$

3. $(x+4)(x+4)$

4. $(x+8)(x+8)$

5. $(x+3)(x-3)$

6. $(x-6)(x+6)$

7. $(x+9)(x-9)$

8. $(x+12)(x-12)$

9. $(2x+3)(x-1)$

10. $(3x+1)(2x+5)$

11. $(3x-4)^2$

12. $(3x+1)^2$

13. $(5x+2)(5x-2)$

14. $(2x+1)(2x-1)$

15. $(3x-2)(3x-2)$

16. $(3+x)^2$

17. $(8-x)(8-x)$

18. $(5-x)(5-x)$

19. $(4x+5)(4x-5)$

20. $(11-x)(11+x)$

21. $(5x-9)(5x+9)$

22. $(9x+2)(9x-2)$

23. $(4-x)^2$

24. $(3x+2)^2$

25. $(2x+7)(2x-7)$

26. $(6x+5)(6x-5)$

27. $(5x^2+2)(2x^2-3)$

28. $(4x^2+7)(2x^2+1)$

29. $(1+7x)^2$

30. $(2-5x)^2$

31. $(x+2)(5x+1)$

32. $(7x-2)(x-3)$

33. $(4x-3)(x+4)$

34. $(x+11)(x-8)$

35. $(3x-7)(x-6)$

36. $(x+7)(2x+9)$

37. $(5+x)(5+x)$

38. $(3-x)(3-x)$

39. $(x^2+1)(x^2-1)$

40. $(x^2+5)(x^2-5)$

41. $(x^2+3)(x^2+3)$

42. $(x^3+8)(x^3+8)$

43. $(x^3-2)^2$

44. $(x^2-4)^2$

45. $(x^2-6)(x^2+9)$

46. $(x^2+3)(x^2-5)$

47. $\left(x+\dfrac{2}{3}\right)\left(x-\dfrac{2}{3}\right)$

48. $\left(x-\dfrac{1}{2}\right)\left(x+\dfrac{1}{2}\right)$

49. $\left(x+\dfrac{3}{4}\right)\left(x-\dfrac{3}{4}\right)$

50. $\left(x+\dfrac{3}{8}\right)\left(x-\dfrac{3}{8}\right)$

51. $\left(x+\dfrac{3}{5}\right)\left(x+\dfrac{3}{5}\right)$

52. $\left(x+\dfrac{4}{3}\right)\left(x+\dfrac{4}{3}\right)$

53. $\left(x-\dfrac{5}{6}\right)^2$

54. $\left(x-\dfrac{2}{7}\right)^2$

55. $\left(x+\dfrac{1}{4}\right)\left(x-\dfrac{1}{2}\right)$

56. $\left(x-\dfrac{1}{5}\right)\left(x+\dfrac{2}{3}\right)$

57. $\left(x+\dfrac{1}{3}\right)\left(x+\dfrac{1}{2}\right)$

58. $\left(x-\dfrac{4}{5}\right)\left(x-\dfrac{3}{10}\right)$

 Use a calculator as an aid in multiplying the binomials.

59. $(x+1.4)(x-1.4)$

60. $(x-2.1)(x+2.1)$

61. $(x - 2.5)^2$

62. $(x + 1.7)^2$

63. $(x + 2.15)(x - 2.15)$

64. $(x + 1.36)(x - 1.36)$

65. $(x + 1.24)^2$

66. $(x - 1.45)^2$

67. $(1.42x + 9.6)^2$

68. $(0.46x - 0.71)^2$

69. $(11.4x + 3.5)(11.4x - 3.5)$

70. $(2.5x + 11.4)(1.3x - 16.9)$

71. $(12.6x - 6.8)(7.4x + 15.3)$

72. $(3.4x + 6)(3.4x - 6)$

Solve the following word problems.

73. A square is 20 inches on each side. A square x inches on each side is cut from each corner of the square.

a. Represent the area of the remaining portion of the square in the form of a polynomial function $A(x)$.

b. Represent the perimeter of the remaining portion of the square in the form of a polynomial function $P(x)$.

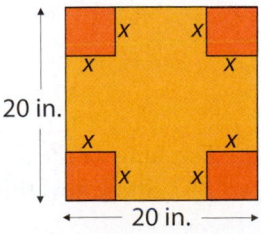

74. In the case of binomial probabilities, if x is the probability of success in one trial of an event, then the expression $f(x) = 15x^4(1 - x)^2$ is the probability of 4 successes in 6 trials where $0 \le x \le 1$.

a. Represent the expression $f(x)$ as a single polynomial by multiplying the polynomials.

b. If a fair coin is tossed, the probability of heads occurring is $\dfrac{1}{2}$. That is, $x = \dfrac{1}{2}$. Find the probability of 4 heads occurring in 6 tosses.

75. A rectangle has sides $(x + 3)$ ft and $(x + 5)$ ft. If a square x ft on a side is cut from the rectangle, represent the remaining area in the form of a polynomial function $A(x)$.

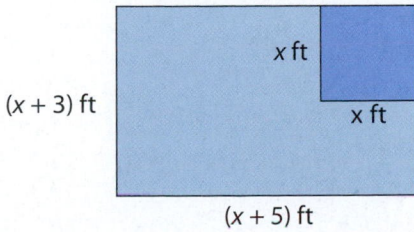

76. Architecture: The Americans with Disabilities Act requires sidewalks to be x feet wide in order for wheelchairs to fit on them. At the bottom, the Empire State Building is 425 feet long and 190 feet wide and a regulation sidewalk surrounds the building.

a. Represent the area covered by the building and the sidewalk in the form of a polynomial function.

b. Represent the area covered by the sidewalk only in the form of a polynomial function.

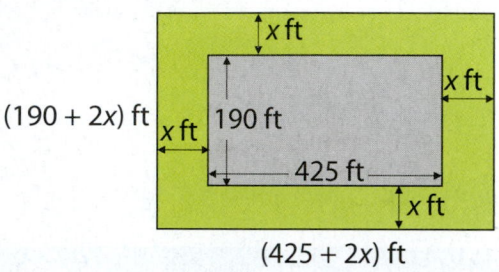

77. A rectangular piece of cardboard that is 10 inches by 15 inches has squares of length x inches on a side cut from each corner. (Assume that $0 < x < 5$.)

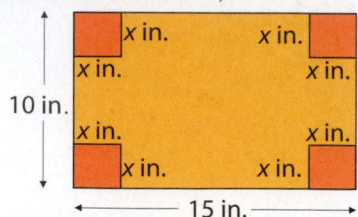

a. Represent the remaining area in the form of a polynomial function $A(x)$.

b. Represent the perimeter of the remaining figure in the form of a polynomial function $P(x)$.

c. If the flaps of the cardboard are folded up, an open box is formed. Represent the volume of this box in the form of a polynomial function $V(x)$.

78. Museums: The world's largest single aquarium habitat at the Georgia Aquarium is 284 feet long, 126 feet wide, and 30 feet deep. Another aquarium is attempting to make a tank that is x feet longer, wider, and deeper. Represent the volume of the new tank as a polynomial function $V(x)$.

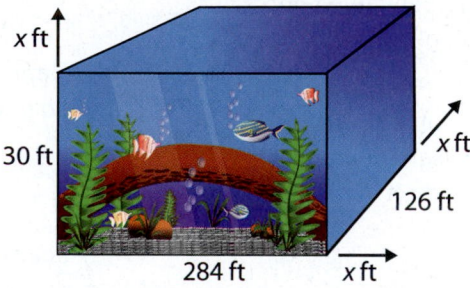

Writing & Thinking

79. A square with sides of length $(x + 5)$ can be broken up as shown in the diagram. The sums of the areas of the interior rectangles and squares is equal to the total area of the square: $(x + 5)^2$. Show how this fits with the formula for the square of a sum.

7.6 Greatest Common Factor (GCF) and Factoring by Grouping

Fractions, such as $\frac{127}{2}$ and $\frac{1}{8}$, in which the numerator and denominator are integers are called **rational numbers**. Fractions in which the numerator and denominator are polynomials are called **rational expressions**. (Recall that no denominator can be 0.)

rational expressions: $\dfrac{x^3 - 6x^2 + 2x}{3x}$, $\dfrac{6x^2 - 7x - 2}{2x - 1}$, $\dfrac{5x^3 - 3x + 1}{x + 3}$, and $\dfrac{1}{10x}$

In this section, we will treat a rational expression as a division problem. With this basis, there are two situations to consider:

1. the denominator (divisor) is a monomial, or
2. the denominator (divisor) is not a monomial.

Objective A   **Dividing by a Monomial**

We know that the sum of fractions with the same denominator can be written as a single fraction by adding the numerators and using the common denominator. For example,

$$\frac{5}{a} + \frac{3b}{a} + \frac{2c}{a} = \frac{5 + 3b + 2c}{a}.$$

If instead of adding the fractions, we start with the sum and want to divide the numerator by the denominator (with a monomial in the denominator), we divide each term in the numerator by the monomial denominator and simplify each fraction.

$$\frac{4x^3 + 8x^2 - 12x}{4x} = \frac{4x^3}{4x} + \frac{8x^2}{4x} - \frac{12x}{4x} = x^2 + 2x - 3$$

Example 1

Dividing by a Monomial

Divide each polynomial by the **monomial denominator** by writing each fraction as the sum (or difference) of fractions. Simplify each fraction, if possible.

a. $\dfrac{x^3 - 6x^2 + 2x}{3x}$

Solution

$$\frac{x^3 - 6x^2 + 2x}{3x} = \frac{x^3}{3x} - \frac{6x^2}{3x} + \frac{2x}{3x} = \frac{x^2}{3} - 2x + \frac{2}{3}$$

1. Divide each polynomial by the monomial denominator. Simplify each fraction, if possible.

a. $\dfrac{2x^3 - 8x^2 + 16x}{4x}$

b. $\dfrac{21y^4 - 14y^3 + 7y^2}{7y^2}$

b. $\dfrac{15y^4 - 20y^3 + 5y^2}{5y^2}$

Solution

$$\frac{15y^4 - 20y^3 + 5y^2}{5y^2} = \frac{15y^4}{5y^2} - \frac{20y^3}{5y^2} + \frac{5y^2}{5y^2} = 3y^2 - 4y + 1$$

Now work margin exercise **1.**

Factoring is the reverse of multiplication. That is, to factor polynomials, you need to remember how you multiplied them. In this way, the concept of factoring is built on your previous knowledge and skills with multiplication. For example, if you are given a product, such as $x^2 - a^2$ (the difference of two squares), you must recall the factors as $(x + a)$ and $(x - a)$ from your work in multiplying polynomials earlier in this chapter.

Studying mathematics is a building process with each topic dependent on previous topics with a few new ideas added each time. The equations and applications in the rest of this chapter involve many of the concepts studied earlier, yet you will find them more interesting and more challenging.

The result of multiplication is called the **product** and the numbers or expressions being multiplied are called **factors** of the product. The reverse of multiplication is called **factoring**. That is, given a product, we want to find the factors.

Multiplying Polynomials	**Factoring Polynomials**
$3x(x + 5) = 3x^2 + 15x$	$3x^2 + 15x = 3x(x + 5)$
factors product	product factors

Factoring polynomials relies heavily on the multiplication techniques developed in Section 7.4. You must remember how to multiply in order to be able to factor. Furthermore, you will find that the skills used in factoring polynomials are necessary when simplifying rational expressions (in a later chapter) and when solving equations. In other words, study this section and Sections 7.7, 7.8, and 7.9 with extra care.

Objective B **Greatest Common Factor of a Set of Terms**

The **greatest common factor** (GCF) of two or more integers is the largest integer that is a factor (or divisor) of all of the integers. For example, the GCF of 30 and 40 is 10. Note that 5 is also a common factor of 30 and 40, but 5 is not the **greatest** common factor. The number 10 is the largest number that will divide into both 30 and 40.

One way of finding the GCF is to use the prime factorization of each number. For example, to find the GCF for 36 and 60, we can write

$$36 = 4 \cdot 9 = 2 \cdot 2 \cdot 3 \cdot 3$$
$$60 = 4 \cdot 15 = 2 \cdot 2 \cdot 3 \cdot 5.$$

The common factors are 2, 2, and 3 and their product is the GCF.

$$GCF = 2 \cdot 2 \cdot 3 = 12$$

Writing the prime factorizations using exponents gives

$$36 = 2^2 \cdot 3^2 \quad \text{and} \quad 60 = 2^2 \cdot 3 \cdot 5.$$

We can see that the GCF is the product of the greatest power of each prime factor that is common to both numbers. That is,

$$GCF = 2^2 \cdot 3 = 12.$$

This procedure can be used to find the GCF for any set of integers or algebraic terms with integer exponents.

Procedure for Finding the GCF of a Set of Terms
1. Find the prime factorization of all integers and integer coefficients.
2. List all the factors that are common to all terms, including variables.
3. Choose the greatest power of each factor that is common to all terms.
4. Multiply these powers to find the GCF.
Note: If there is no common prime factor or variable, then the GCF is 1.

Example 2

Finding the GCF

Find the GCF for each of the following sets of algebraic terms.

a. $\{30, 45, 75\}$

Solution

Find the prime factorization of each number:
$30 = 2 \cdot 3 \cdot 5$, $45 = 3^2 \cdot 5$, and $75 = 3 \cdot 5^2$.

The common factors are 3 and 5 and the greatest power of each that is common to all numbers is 3^1 and 5^1.
Thus $GCF = 3^1 \cdot 5^1 = 15$.

2. Find the GCF for each of the following sets of algebraic terms.

a. $\{20, 5, 10\}$

b. $\{150xy, 250y^2x^3, 100x^2y^2\}$

b. $\{20x^4y, 15x^3y, 10x^5y^2\}$

Solution

Writing each integer coefficient in prime factored form gives:

$20x^4y = 2^2 \cdot 5 \cdot x^4 \cdot y,$
$15x^3y = 3 \cdot 5 \cdot x^3 \cdot y,$
$10x^5y^2 = 2 \cdot 5 \cdot x^5 \cdot y^2.$

The common factors are $5, x,$ and y and after finding the greatest power of each that is common to all three terms, we have 5^1, x^3, and y^1. Thus $GCF = 5^1 \cdot x^3 \cdot y^1 = 5x^3y.$

Now work margin exercise 2.

Objective C **Factoring Out the Greatest Common Monomial Factor**

Now consider the polynomial $3n + 15$. We want to write this polynomial as a product of two factors. Since

$$3n + 15 = 3 \cdot n + 3 \cdot 5,$$

we see that 3 is a common factor of the two terms in the polynomial. By using the distributive property, we can write

$$3n + 15 = 3 \cdot n + 3 \cdot 5 = 3(n + 5).$$

In this way, the polynomial $3n + 15$ has been **factored** into the product of 3 and $(n + 5)$.

For a more general approach to finding factors of a polynomial, we can use the quotient rule for exponents.

$$\frac{a^m}{a^n} = a^{m-n}$$

This property is used when dividing terms. For example,

$$\frac{35x^8}{5x^2} = 7x^6 \quad \text{and} \quad \frac{16a^5}{-8a} = -2a^4.$$

To divide a polynomial by a monomial each term in the polynomial is divided by the monomial. For example,

$$\frac{8x^3 - 14x^2 + 10x}{2x} = \frac{8x^3}{2x} - \frac{14x^2}{2x} + \frac{10x}{2x} = 4x^2 - 7x + 5.$$

Note: With practice, this division can be done mentally.

This concept of dividing each term by a monomial is part of finding a monomial factor of a polynomial. Finding the greatest common monomial factor (GCF) of a polynomial means to **find the GCF of the terms of the polynomial**. This monomial will be one factor, and the sum of the various quotients found by dividing each term by the GCF will be the other factor. Thus using the quotients just found and the fact that $2x$ is the GCF of the terms,

$$8x^3 - 14x^2 + 10x = 2x\left(4x^2 - 7x + 5\right).$$

We say that $2x$ is **factored out** and $2x$ and $\left(4x^2 - 7x + 5\right)$ are the **factors** of $8x^3 - 14x^2 + 10x$.

Factoring Out the GCF

To find a monomial that is the greatest common factor (GCF) of a polynomial:

1. Find the variable(s) of highest degree and the largest integer coefficient that are factors of each term of the polynomial. (The product of these is one factor.)
2. Divide this monomial factor into each term of the polynomial, resulting in another polynomial factor.

Now factor $24x^6 - 12x^4 - 18x^3$. The GCF is $6x^3$ and is factored out as follows:

$$24x^6 - 12x^4 - 18x^3 = 6x^3 \cdot 4x^3 + 6x^3\left(-2x\right) + 6x^3\left(-3\right)$$
$$= 6x^3\left(4x^3 - 2x - 3\right).$$

The factoring can be checked by multiplying $6x^3$ and $4x^3 - 2x - 3$. The product should be the expression we started with:

$$6x^3\left(4x^3 - 2x - 3\right) = 24x^6 - 12x^4 - 18x^3.$$

By definition, the GCF of a polynomial will have a positive coefficient. **However, if the leading coefficient is negative, we may choose to factor out the negative of the GCF (or $-1 \cdot \text{GCF}$).** This technique will leave a positive coefficient for the first term of the other polynomial factor. For example, the GCF for $-10a^4b + 15a^4$ is $5a^4$ and we can factor as follows.

$$-10a^4b + 15a^4 = 5a^4\left(-2b + 3\right)$$
$$\text{or} \qquad -10a^4b + 15a^4 = -5a^4\left(2b - 3\right).$$

Both answers are correct.

Suppose we factor $8n + 16$ as follows.

$$8n + 16 = 4\left(2n + 4\right)$$

But note that $2n + 4$ has 2 as a common factor. Therefore, we have not factored **completely**. **An expression is factored completely if none of its factors can be factored**.

Factoring polynomials means to find factors that are integers (other than +1) or polynomials with integer coefficients. If this cannot be done, we say that the polynomial is not factorable. For example, the polynomials $x^2 + 36$ and $3x + 17$ are **not factorable**.

Example 3

Factoring out the GCF of a polynomial

Factor each polynomial by factoring out the greatest common monomial factor.

a. $6n + 30$

Solution

$6n + 30 = 6 \cdot n + 6 \cdot 5 = 6(n + 5)$
Checking by multiplying gives $6(n + 5) = 6n + 30$, the original expression.

b. $x^3 + x$

Solution

$x^3 + x = x \cdot x^2 + x(+1) = x(x^2 + 1)$ +1 is the coefficient of x.

c. $5x^3 - 15x^2$

Solution

$5x^3 - 15x^2 = 5x^2 \cdot x + 5x^2(-3) = 5x^2(x - 3)$

d. $2x^4 - 3x^2 + 2$

Solution

$2x^4 - 3x^2 + 2$
This polynomial has no common monomial factor other than 1. In fact, this polynomial is **not factorable**.

e. $-4a^5 + 2a^3 - 6a^2$

Solution

The GCF is $2a^2$ and we can factor as follows.

$$-4a^5 + 2a^3 - 6a^2 = 2a^2\left(-2a^3 + a - 3\right)$$

However, the leading coefficient is negative and we can also factor as follows.

$$-4a^5 + 2a^3 - 6a^2 = -2a^2\left(2a^3 - a + 3\right)$$

Both answers are correct. However, we will see later that having a positive leading coefficient for the polynomial in parentheses may make that polynomial easier to factor.

A polynomial may be in more than one variable. For example, $5x^2y + 10xy^2$ is in the two variables x and y. Thus the GCF may have more than one variable.

$$5x^2y + 10xy^2 = 5xy \cdot x + 5xy \cdot 2y$$
$$= 5xy(x + 2y)$$

Similarly,

$$4xy^3 - 2x^2y^2 + 8xy^2 = 2xy^2 \cdot 2y + 2xy^2\left(-x\right) + 2xy^2 \cdot 4$$
$$= 2xy^2\left(2y - x + 4\right).$$

Example 4

Factoring out the GCF of a Polynomial

Factor each polynomial by finding the GCF (or $-1 \cdot \text{GCF}$).

a. $4ax^3 + 4ax$

Solution

$$4ax^3 + 4ax = 4ax\left(x^2 + 1\right) \qquad \text{Note that } 4ax = 1 \cdot 4ax.$$

Checking by multiplying gives $4ax\left(x^2 + 1\right) = 4ax^3 + 4ax$, the original expression.

b. $3x^2y^2 - 6xy^2$

Solution

$$3x^2y^2 - 6xy^2 = 3xy^2\left(x - 2\right)$$

Factor out the GCF of the following polynomials.

3. a. $10y + 30y^2$

   b. $-9b^4 + 9b^3 + 9b^2$

4. $-33x^2z - 55xz - 77xz^3$

c. $-14by^3 - 7b^2y + 21by^2$

Solution

$$-14by^3 - 7b^2y + 21by^2 = -7by\left(2y^2 + b - 3y\right)$$

Note that by factoring out a negative term, in this case $-7by$, the leading coefficient in parentheses is positive.

d. $13a^4b^5 - 3a^2b^9 - 4a^6b^3$

Solution

$$13a^4b^5 - 3a^2b^9 - 4a^6b^3 = a^2b^3\left(13a^2b^2 - 3b^6 - 4a^4\right)$$

Now work margin exercises 3 and 4.

Objective D   **Factoring by Grouping**

Consider the expression

$$y(x+4) + 2(x+4)$$

as the sum of two terms, $y(x+4)$ and $2(x+4)$. Each of these "terms" has the common **binomial** factor $(x+4)$. Factoring out this common binomial factor by using the distributive property gives

$$y(x+4) + 2(x+4) = (x+4)(y+2).$$

Similarly,

$$3(x-2) - a(x-2) = (x-2)(3-a).$$

Now consider the product

$$(x+3)(y+5) = (x+3)y + (x+3)5$$
$$= xy + 3y + 5x + 15,$$

which has four terms and no like terms. Yet the product has two factors, namely $(x+3)$ and $(y+5)$. Factoring polynomials with four or more terms can sometimes be accomplished by grouping the terms and using the distributive property, as in the above discussion and the following examples. Keep in mind that the common factor can be a binomial or other polynomial.

Example 5

Factoring Out a Common Binomial Factor

Factor each polynomial.

a. $3x^2(5x+1)-2(5x+1)$

Solution

$$3x^2(5x+1)-2(5x+1)=(5x+1)(3x^2-2)$$

b. $7a(2x-3)+(2x-3)$

Solution

$$7a(2x-3)+(2x-3)=7a(2x-3)+1\cdot(2x-3)$$
$$=(2x-3)(7a+1)$$

1 is the understood coefficient of $(2x-3)$.

Now work margin exercise 5.

In the examples just discussed, the common binomial factor was in parentheses. However, many times the expression to be factored is in a form with four or more terms. For example, by multiplying in Example **5a.**, we get the following expression.

$$3x^2(5x+1)-2(5x+1)=15x^3+3x^2-10x-2$$

The expression $15x^3+3x^2-10x-2$ has four terms with no common monomial factor; yet, we know that it has the two binomial factors $(5x+1)$ and $(3x^2-2)$. We can find the binomial factors by **grouping**. This means looking for common factors in each group and then looking for common binomial factors. The process is illustrated in Example 6.

5. Factor each polynomial.

a. $x^2(2x-y)-2(-2x+y)$

b. $x(6y+3v)-u(6y+3v)$

Example 6

Factor Each Polynomial by Grouping

a. $xy + 5x + 3y + 15$

Solution

$$xy + 5x + 3y + 15 = (xy + 5x) + (3y + 15)$$ Group terms that have a common monomial factor.

$$= x(y + 5) + 3(y + 5)$$ using the distributive property

$$= (y + 5)(x + 3)$$ $(y + 5)$ is a common binomial factor.

Checking gives $(y + 5)(x + 3) = xy + 5x + 3y + 15$.

b. $x^2 - xy - 5x + 5y$

Solution

$$x^2 - xy - 5x + 5y = (x^2 - xy) + (-5x + 5y)$$

$$= x(x - y) + 5(-x + y)$$

This does not work because $(x - y) \neq (-x + y)$. However, these two expressions are **opposites**. Thus we can find a common factor by factoring -5 instead of $+5$ from the last two terms.

$$x^2 - xy - 5x + 5y = (x^2 - xy) + (-5x + 5y)$$

$$= x(x - y) - 5(x - y)$$

$$= (x - y)(x - 5)$$ Success!

c. $x^2 + ax + 3x + 3y$

Solution

$$x^2 + ax + 3x + 3y = (x^2 + ax) + (3x + 3y)$$

$$= x(x + a) + 3(x + y)$$

But $x + a \neq x + y$ and there is no common factor.
So $x^2 + ax + 3x + 3y$ is **not factorable**.

d. $xy + 5x + y + 5$

6. Factor each polynomial by grouping.

a. $3xy - 3x + 2y - 2$

b. $2x^2 - x + 4y - 4$

Solution

$$xy + 5x + y + 5 = (xy + 5x) + (y + 5)$$
$$= x(y + 5) + 1 \cdot (y + 5)$$
$$= (y + 5)(x + 1)$$

1 is the understood coefficient of $(y + 5)$.

e. $5xy + 6uv - 3vy - 10ux$

Solution

In the expression $5xy + 6uv - 3vy - 10ux$ there is no common factor in the first two terms. However, the first and third terms have a common factor so we rearrange the terms as follows.

$$5xy + 6uv - 3vy - 10ux = (5xy - 3vy) + (6uv - 10ux)$$
$$= y(5x - 3v) + 2u(3v - 5x)$$

Now we see that $5x - 3v$ and $3v - 5x$ are opposites and we factor out $-2u$ from the last two terms. The result is as follows.

$$5xy + 6uv - 3vy - 10xu = (5xy - 3vy) + (6uv - 10ux)$$
$$= y(5x - 3v) - 2u(-3v + 5x)$$
$$= y(5x - 3v) - 2u(5x - 3v) \quad \textbf{Note:} \ 5x - 3v = -3v + 5x$$
$$= (5x - 3v)(y - 2u)$$

Now work margin exercise 6.

Practice Problems

Factor each expression completely.

1. $2x - 16$
2. $-5x^2 - 5x$
3. $7ax^2 - 7ax$
4. $a^4b^5 + 2a^3b^2 - a^3b^3$
5. $9x^2y^2 + 12x^2y - 6x^3$
6. $6a^3(x + 3) - (x + 3)$
7. $5x + 35 - xy - 7y$

Practice Problem Answers

1. $2(x - 8)$
2. $-5x(x + 1)$
3. $7ax(x - 1)$
4. $a^3b^2(ab^3 + 2 - b)$
5. $3x^2(3y^2 + 4y - 2x)$
6. $(x + 3)(6a^3 - 1)$
7. $(x + 7)(5 - y)$

Exercises 7.6

Find the GCF for each set of terms. See Example 2.

1. $\{10, 15, 20\}$

2. $\{25, 30, 75\}$

3. $\{16, 40, 56\}$

4. $\{30, 42, 54\}$

5. $\{9, 14, 22\}$

6. $\{44, 66, 88\}$

7. $\{30x^3, 40x^5\}$

8. $\{15y^4, 25y\}$

9. $\{8a^3, 16a^4, 20a^2\}$

10. $\{36xy, 48xy, 60xy\}$

11. $\{26ab^2, 39a^2b, 52a^2b^2\}$

12. $\{28c^2d^3, 14c^3d^2, 42cd^2\}$

13. $\{45x^2y^2z^2, 75xy^2z^3\}$

14. $\{21a^5b^4c^3, 28a^3b^4c^3, 35a^3b^4c^2\}$

Simplify the expressions.

15. $\dfrac{x^7}{x^3}$

16. $\dfrac{x^8}{x^3}$

17. $\dfrac{-8y^3}{2y^2}$

18. $\dfrac{12x^2}{2x}$

19. $\dfrac{9x^5}{3x^2}$

20. $\dfrac{-10x^5}{2x}$

21. $\dfrac{4x^3y^2}{2xy}$

22. $\dfrac{21x^4y^3}{-3xy^2}$

Complete the factoring of the polynomial as indicated.

23. $3m + 27 = 3(\quad)$

24. $2x + 18 = 2(\quad)$

25. $5x^2 - 30x = 5x(\quad)$

26. $6y^3 - 24y^2 = 6y^2(\quad)$

27. $13ab^2 + 13ab = 13ab(\quad)$

28. $8x^2y - 4xy = 4xy(\quad)$

29. $-15xy^2 - 20x^2y - 5xy = -5xy(\quad)$

30. $-9m^3 - 3m^2 - 6m = -3m(\quad)$

31. $11x - 121$

32. $14x + 21$

33. $16y^3 + 12y$

34. $-3x^2 + 6x$

35. $-6ax + 9ay$

36. $4ax - 8ay$

37. $10x^2y - 25xy$

38. $16x^4y - 14x^2y$

39. $-18y^2z^2 + 2yz$

40. $-14x^2y^3 - 14x^2y$

41. $8y^2 - 32y + 8$

42. $5x^2 - 15x - 5$

43. $2xy^2 - 3xy - x$

44. $ad^2 + 10ad + 25a$

45. $8m^2x^3 - 12m^2y + 4m^2z$

46. $36t^2x^4 - 45t^2x^3 + 24t^2x^2$

47. $-56x^4z^3 - 98x^3z^4 - 35x^2z^5$

48. $34x^4y^6 - 51x^3y^5 + 17x^5y^4$

49. $15x^4y^2 + 24x^6y^6 - 32x^7y^3$

50. $-3x^2y^4 - 6x^3y^4 - 9x^2y^3$

51. $7y^2(y+3) + 2(y+3)$

52. $6a(a-7) - 5(a-7)$

53. $3x(x-4) + (x-4)$

54. $2x^2(x+5) + (x+5)$

55. $4x^3(x-2) - (x-2)$

56. $9a(x+1) - (x+1)$

57. $10y(2y+3) - 7(2y+3)$

58. $a(x+5) + b(x+5)$

59. $a(x-2) - b(x-2)$

60. $3a(x-10) + 5b(x-10)$

61. $bx + b + cx + c$

62. $3x + 3y + ax + ay$

63. $x^3 + 3x^2 + 6x + 18$

64. $2z^3 - 14z^2 + 3z - 21$

65. $10a^2 - 5az + 2a + z$

66. $x^2 - 4x + 6xy - 24y$

67. $3x + 3y - bx - by$

68. $ax + 5ay + 3x + 15y$

69. $5xy + yz - 20x - 4z$

70. $x - 3xy + 2z - 6zy$

71. $z^2 + 3 + az^2 + 3a$

72. $x^2 - 5 + x^2y + 5y$

73. $6ax + 12x + a + 2$

74. $4xy + 3x - 4y - 3$

75. $xy + x + y + 1$

76. $xy + x - y - 1$

77. $10xy - 2y^2 + 7yz - 35xz$

78. $7xy - 3y + 2x^2 - 3x$

79. $3xy - 4uy - 6vx + 8uv$

80. $xy + 5vy + 6ux + 30uv$

81. $3ab + 4ac + 2b + 6c$

82. $24y - 3yz + 2xz - 16x$

83. $6ac - 9ad + 2bc - 3bd$

84. $2ac - 3bc + 6ad - 9bd$

Writing & Thinking

85. Explain why the GCF of $-3x^2 + 3$ is 3 and not −3.

7.7 Factoring Trinomials: $x^2 + bx + c$

In Section 7.5 we learned to use the FOIL method to multiply two binomials. In many cases the simplified form of the product was a trinomial. In this section we will learn to factor trinomials by reversing the FOIL method. In particular, we will focus on only factoring trinomials in one variable with leading coefficient 1.

Objectives

A   Factor trinomials with leading coefficients of 1 (of the form $x^2 + bx + c$).

B   Factor trinomials by first factoring out a common monomial factor.

Objective A   **Factoring Trinomials with Leading Coefficients of 1**

Using the FOIL method to multiply $(x+5)$ and $(x+3)$, we find

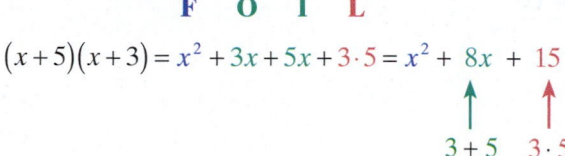

$$(x+5)(x+3) = x^2 + 3x + 5x + 3 \cdot 5 = x^2 + 8x + 15$$

$$3+5 \qquad 3 \cdot 5$$

We see that the leading coefficient (coefficient of x^2) is 1, the coefficient 8 is the sum of 3 and 5, and the constant term 15 is the product of 3 and 5.

More generally,

$$(x+a)(x+b) = x^2 + bx + ax + ab$$
$$= x^2 + (b+a)x + ab.$$

sum   product
of constants   of constants
a and b a and b

Now given a trinomial with leading coefficient 1, we want to find the binomial factors, if any. Reversing the relationship between a and b, as shown above, we can proceed as follows.

To factor a trinomial with leading coefficient 1, find two factors of the constant term whose sum is the coefficient of the middle term. (If these factors do not exist, the trinomial is **not factorable**.)

For example, to factor $x^2 + 11x + 30$, we need positive factors of +30 whose sum is +11.

Positive Factors of 30		**Sums of These Factors**
1 · 30	→	$1 + 30 = 31$
2 · 15	→	$2 + 15 = 17$
3 · 10	→	$3 + 10 = 13$
5 · 6	→	$5 + 6 = 11$

Now because $5 \cdot 6 = 30$ and $5 + 6 = 11$, we have
$$x^2 + 11x + 30 = (x+5)(x+6).$$

1. Factor the following trinomials.

a. $x^2 + 10x + 21$

b. $x^2 - x - 20$

Example 1

Factoring Trinomials with Leading Coefficient 1

Factor the following trinomials.

a. $x^2 + 8x + 12$

Solution

12 has three pairs of positive integer factors as illustrated in the following table. Of these 3 pairs, only $2 + 6$ is equal to 8.

Factors of 12

1	·	12
2	·	6	$\longrightarrow$ $2 + 6 = 8$
3	·	4

Note: If the middle term had been $-8x$, then we would have wanted pairs of negative integer factors to find a sum of -8.
Thus, $x^2 + 8x + 12 = (x+2)(x+6)$.

b. $y^2 - 8y - 20$

Solution

We want a pair of integer factors of -20 whose sum is -8. In this case, because the product is negative, one of the factors must be positive and the other negative.

Factors of -20

-1	·	20
1	·	-20
-2	·	10
2	·	-10	$\longrightarrow$ $2 + (-10) = -8$
-4	·	5
4	·	-5

We have listed all the pairs of integer factors of -20. You can see that 2 and -10 are the only two whose sum is -8. Thus listing all the pairs is not necessary. This stage is called the **trial-and-error stage**. That is, you can **try** different pairs (mentally or by making a list) until you find the correct pair. If such a pair does not exist, the polynomial is **not factorable**.

In this case, we have $y^2 - 8y - 20 = (y+2)(y-10)$.

Note that by the commutative property of multiplication, the order of the factors does not matter. That is, we can also write
$y^2 - 8y - 20 = (y-10)(y+2)$.

*Now work margin exercise **1.***

To Factor Trinomials of the Form $x^2 + bx + c$

To factor $x^2 + bx + c$, if possible, find an integer pair of factors of c whose sum is b.

1. If c is positive, then both factors must have the same sign.
   a. Both will be positive if b is positive.
      Example: $x^2 + 5x + 4 = (x + 4)(x + 1)$
   b. Both will be negative if b is negative.
      Example: $x^2 - 5x + 4 = (x - 4)(x - 1)$
2. If c is negative, then one factor must be positive and the other negative.
   Example: $x^2 + 6x - 7 = (x + 7)(x - 1)$ and $x^2 - 6x - 7 = (x - 7)(x + 1)$

Objective B   ## Finding a Common Monomial Factor First

If a trinomial does not have a leading coefficient of 1, then we look for a common monomial factor. **If there is a common monomial factor, factor out this common monomial factor first** and then factor the remaining trinomial factor, if possible. (We will discuss factoring trinomials with leading coefficients other than 1 in the next section.) A polynomial is **completely factored** if none of its factors can be factored.

Example 2

Finding a Common Monomial Factor

Completely factor the following trinomials by first factoring out the GCF in the form of a common monomial factor.

a. $5x^3 - 15x^2 + 10x$

Solution

First factor out the GCF, $5x$.

$5x^3 - 15x^2 + 10x = 5x(x^2 - 3x + 2)$ factored, but not completely factored

Now factor the trinomial $x^2 - 3x + 2$. Look for factors of $+2$ that add up to -3. Because $(-1)(-2) = +2$ and $(-1) + (-2) = -3$, we have

$$5x^3 - 15x^2 + 10x = 5x(x^2 - 3x + 2)$$
$$= 5x(x - 1)(x - 2). \quad \text{\color{red}completely factored}$$

Check: The factoring can be checked by multiplying the factors.

2. Completely factor the following trinomials by first factoring out the common monomial factor.

a. $7y^3 - 49y + 14y^2$

b. $11x^3y + 22x^2y - 33xy$

$$5x(x-1)(x-2) = 5x(x^2 - 2x - x + 2)$$
$$= 5x(x^2 - 3x + 2)$$
$$= 5x^3 - 15x^2 + 10x \qquad \text{the original expression}$$

b. $10y^5 - 20y^4 - 80y^3$

Solution

First factor out the GCF, $10y^3$.

$$10y^5 - 20y^4 - 80y^3 = 10y^3(y^2 - 2y - 8)$$

Now factor the trinomial $y^2 - 2y - 8$. Look for factors of -8 that add up to -2. Because $(-4)(+2) = -8$ and $(-4) + (+2) = -2$, we have

$$10y^5 - 20y^4 - 80y^3 = 10y^3(y^2 - 2y - 8) = 10y^3(y-4)(y+2).$$

The factoring can be checked by multiplying the factors.

Now work margin exercise 2.

---

notes

■ When factoring polynomials, always look for a common monomial factor first. Then, if there is one, remember to include this common monomial factor as part of the answer. Not all polynomials are factorable. For example, no matter what combinations are tried, $x^2 + 3x + 4$ does not have two binomial factors with integer coefficients. (There are no factors of $+4$ that will add to $+3$.) We say that the polynomial is **not factorable** (or **prime**). **A polynomial is not factorable if it cannot be factored as the product of polynomials with integer coefficients.**

Practice Problems

Completely factor each polynomial. If the polynomial cannot be factored, write "not factorable."

1. $x^2 - 3x - 28$ **2.** $x^2 + 4x + 3$ **3.** $y^2 - 12y + 35$

4. $a^2 - a + 1$ **5.** $2y^3 + 24y^2 + 64y$ **6.** $4a^4 + 48a^3 + 108a^2$

Practice Problem Answers

1. $(x-7)(x+4)$ **2.** $(x+3)(x+1)$ **3.** $(y-5)(y-7)$

4. not factorable   **5.** $2y(y+8)(y+4)$ **6.** $4a^2(a+3)(a+9)$

Exercises 7.7

List all pairs of integer factors for the given integer. Remember to include negative integers as well as positive integers.

1. 15

2. 12

3. 20

4. 30

5. −6

6. −7

7. 16

8. 18

9. −10

10. −25

Find the pair of integers whose product is the first integer and whose sum is the second integer.

11. $12, 7$

12. $25, 26$

13. $-14, -5$

14. $-30, -1$

15. $-8, 7$

16. $-40, 6$

17. $36, -12$

18. $16, -10$

19. $20, -9$

20. $4, -5$

Complete the factorization.

21. $x^2 + 6x + 5 = (x + 5)(\quad)$

22. $y^2 - 7y + 6 = (y - 1)(\quad)$

23. $p^2 - 9p - 10 = (p + 1)(\quad)$

24. $m^2 + 4m - 45 = (m - 5)(\quad)$

25. $a^2 + 12a + 36 = (a + 6)(\quad)$

26. $n^2 - 2n - 3 = (n - 3)(\quad)$

Completely factor the trinomials. If the trinomial cannot be factored, write "not factorable." See Example 1.

27. $x^2 - x - 12$

28. $x^2 - 6x - 27$

29. $y^2 + y - 30$

30. $x^2 + 6x - 36$

31. $m^2 + 3m - 1$

32. $x^2 + 3x - 18$

33. $x^2 - 8x + 16$

34. $a^2 + 10a + 25$

35. $x^2 + 7x + 12$

36. $a^2 + a + 2$

37. $y^2 - 3y + 2$

38. $y^2 - 14y + 24$

39. $x^2 + 3x + 5$ **40.** $y^2 + 12y + 35$ **41.** $x^2 - x - 72$

42. $y^2 + 8y + 7$ **43.** $z^2 - 15z + 54$ **44.** $a^2 + 4a - 21$

Completely factor the given polynomials. See Example 2.

45. $x^3 + 10x^2 + 21x$ **46.** $x^3 + 8x^2 + 15x$ **47.** $5x^2 - 5x - 60$

48. $6x^2 + 24x + 18$ **49.** $10y^3 - 10y^2 - 60y$ **50.** $7y^3 - 70y^2 + 168y$

51. $4p^4 + 36p^3 + 32p^2$ **52.** $15m^5 - 30m^4 + 15m^3$ **53.** $2x^4 - 14x^3 - 36x^2$

54. $3y^6 + 33y^5 + 90y^4$ **55.** $2x^2 - 2x - 72$ **56.** $3x^2 - 18x + 30$

57. $2a^4 - 8a^3 - 120a^2$ **58.** $2a^4 + 24a^3 + 54a^2$ **59.** $3y^5 - 21y^4 - 24y^3$

60. $4y^5 + 28y^4 + 24y^3$ **61.** $x^3 - 10x^2 + 16x$ **62.** $x^3 - 2x^2 - 3x$

63. $5a^2 + 10a - 30$ **64.** $6a^2 + 24a + 12$ **65.** $20a^4 + 40a^3 + 20a^2$

66. $6x^4 - 12x^3 + 6x^2$

Solve the following word problems.

67. **Triangles:** The area of a triangle is $\dfrac{1}{2}$ the product of its base and its height. If the area of the triangle shown is given by the function $A(x) = \dfrac{1}{2}x^2 + 24x$, find representations for the lengths of its base and its height (where the base is longer than the height).

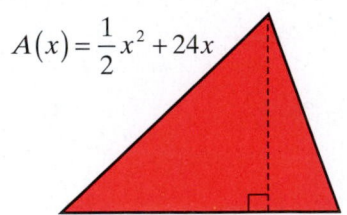

$A(x) = \dfrac{1}{2}x^2 + 24x$

68. Triangles: The area of a triangle is $\frac{1}{2}$ the product of its base and its height. If the area of the triangle shown is given by the function $A(x) = \frac{1}{2}x^2 + 14x$, find representations for the lengths of its base and its height (where the height is longer than the base).

$$A(x) = \frac{1}{2}x^2 + 14x$$

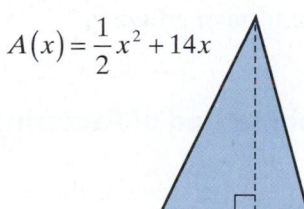

69. Rectangles: The area of the rectangle shown is given by the polynomial function $A(x) = 4x^2 + 20x$. If the width of the rectangle is $4x$, what is the length?

$$A(x) = 4x^2 + 20x \qquad 4x$$

70. Rectangles: The area of the rectangle shown is given by the polynomial function $A(x) = x^2 + 11x + 24$. If the length of the rectangle is $(x+8)$, what is the width?

$$A(x) = x^2 + 11x + 24$$

$$x + 8$$

Writing & Thinking

71. It is true that $2x^2 + 10x + 12 = (2x+6)(x+2) = (2x+4)(x+3)$. Explain how the trinomial can be factored in two ways. Is there some kind of error?

72. It is true that $5x^2 - 5x - 30 = (5x-15)(x+2)$. Explain why this is not the completely factored form of the trinomial.

7.8 Factoring Trinomials: $ax^2 + bx + c$

Second-degree trinomials in the variable x are of the general form

$$ax^2 + bx + c \quad \text{where the coefficients } a, b, \text{ and } c \text{ are real numbers and } a \neq 0.$$

In this section we will discuss two methods of factoring trinomials of the form $ax^2 + bx + c$ in which the coefficients are restricted to integers. These two methods are the **trial-and-error method** (a reverse of the **FOIL** method of multiplication) and the ***ac*-method** (a form of **grouping**).

> **Objective A** **The Trial-and-Error Method of Factoring**

To help in understanding this method we review the FOIL method of multiplying two binomials as follows.

$$(2x + 5)(3x + 1) = 6x^2 + 17x + 5$$

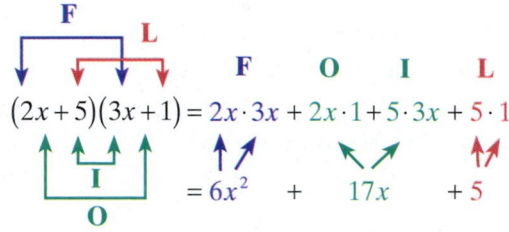

$$(2x + 5)(3x + 1) = 2x \cdot 3x + 2x \cdot 1 + 5 \cdot 3x + 5 \cdot 1$$
$$= 6x^2 + 17x + 5$$

F The product of the **first** two terms is $6x^2$.

O
I The sum of the **inner** and **outer** products is $17x$.

L The product of the **last** two terms is 5.

Now consider the problem of factoring $6x^2 + 23x + 7$

as the product of two binomials.

$$\mathbf{F} = 6x^2 \qquad \mathbf{L} = +7$$
$$6x^2 + 23x + 7 = (\quad)(\quad)$$

Now we use various combinations for **F** and **L** in the **trial-and-error method** as follows.

1. List all the possible combinations of factors of $6x^2$ and $+7$ in their respective **F** and **L** positions. (See the following list.)
2. Check the sum of the products in the **O** and **I** positions until you find the sum to be $+23x$.
3. If none of these sums is $+23x$, the trinomial is not factorable.

$$\overbrace{}^{\textbf{F}}\quad\underbrace{}_{\textbf{L}}$$

a. $(6x+1)(x+7)$

b. $(6x+7)(x+1)$

c. $(3x+1)(2x+7)$

d. $(3x+7)(2x+1)$

e. $(6x-1)(x-7)$

f. $(6x-7)(x-1)$

We really don't need to check these last four because the **O** and **I** would be negative, and we are looking for $+23x$. In this manner, the trial-and-error method is more efficient than it first appears to be.

g. $(3x-1)(2x-7)$

h. $(3x-7)(2x-1)$

Now investigating only the possibilities in the list with positive constants, we need to check the sums of the outer (**O**) and inner (**I**) products to find $+23x$.

a. $(6x+1)(x+7)$: $\mathbf{O}+\mathbf{I}= 42x+x = 43x$

b. $(6x+7)(x+1)$: $\mathbf{O}+\mathbf{I}= 6x+7x = 13x$

c. $(3x+1)(2x+7)$: $\mathbf{O}+\mathbf{I}= 21x+2x = \boxed{23x}$ ← We found 23x! With a little luck we could have found this first.

The correct factors, $(3x+1)$ and $(2x+7)$, have been found so we need not take the time to try the next product in the list, $(3x+7)(2x+1)$. Thus even though the list of possibilities of factors may be long, the actual time involved may be quite short if the correct factors are found early in the trial-and-error method. So we have

$$6x^2 +23x+7 =(3x+1)(2x+7).$$

Look at the constant term to determine what signs to use for the constants in the factors. The following guidelines will help limit the trial-and-error search.

Guidelines for the Trial-and-Error Method

1. If the sign of the constant term is positive (+), the signs in both factors will be the same, either both positive or both negative.
2. If the sign of the constant term is negative (−), the signs in the factors will be different, one positive and one negative.

Example 1

Using the Trial-and-Error Method

Factor by using the trial-and-error method.

a. $x^2 + 6x + 5$

Solution

Since the middle term is $+6x$ and the constant is 5, we know that the two factors of 5 must both be positive, +5 and +1.

$$x^2 + 6x + 5 = (x+5)(x+1)$$

$$5x$$
$$x$$

b. $4x^2 - 4x - 15$

Solution

For **F**: $4x^2 = 4x \cdot x$ and $4x^2 = 2x \cdot 2x$.
For **L**: $-15 = -15 \cdot 1$, $-15 = -1 \cdot 15$, $-15 = -3 \cdot 5$, and $-15 = -5 \cdot 3$.

Trials

$$(2x - 15)(2x + 1)$$ $2x - 30x = -28x$ is the wrong middle term.

$$-30x$$
$$2x$$

$$(2x - 3)(2x + 5)$$ $10x - 6x = +4x$ is the wrong middle term only because the sign is wrong. So just switch the signs and the factors will be right.

$$-6x$$
$$10x$$

$$(2x + 3)(2x - 5)$$ $-10x + 6x = -4x$ is the right middle term.

$$6x$$
$$-10x$$

Now that we have the answer, there is no need to try all the possibilities with $(4x \quad)(x \quad)$.

c. $6a^2 - 31a + 5$

Solution

Since the middle term is $-31a$ and the constant is $+5$, we know that the two factors of 5 must both be negative, -5 and -1. We try, $F = 6a^2 = 6a \cdot a$.

$$6a^2 - 31a + 5 = (6a - 1)(a - 5) \qquad {\color{red}-30a - a = -31a}$$

We found the correct factors on the first try.

Now work margin exercise **1.**

> ## notes
>
> ■ **Reminder: To factor completely** means to find factors of the polynomial, none of which are themselves factorable. Thus
> $$2x^2 + 12x + 10 = (2x + 10)(x + 1)$$
> ■ is not factored completely because $2x + 10 = 2(x + 5)$. We could write
> $$2x^2 + 12x + 10 = (2x + 10)(x + 1)$$
> $$= 2(x + 5)(x + 1).$$
> ■ This problem can be avoided by first factoring out the GCF (in this case, 2).

Example 2

Factoring Completely

Factor completely. Be sure to look first for the greatest common monomial factor.

a. $6x^3 - 8x^2 + 2x$

Solution

$$6x^3 - 8x^2 + 2x = 2x(3x^2 - 4x + 1) = 2x(3x - 1)(x - 1)$$

1. Use the trial-and-error method to factor.

a. $x^2 + 8x + 12$

b. $8u^2 - 2u - 21$

2. Factor completely.

a. $8x^3 - 12x^2 + 4x$

b. $21x^3 + 49x^2 - 7x$

Check:

Check the factorization by multiplying.

$$2x(3x-1)(x-1) = 2x(3x^2 - 3x - x + 1)$$
$$= 2x(3x^2 - 4x + 1)$$
$$= 6x^3 - 8x^2 + 2x \qquad \text{the original polynomial}$$

b. $-2x^2 - x + 6$

Solution

$$-2x^2 - x + 6 = -1(2x^2 + x - 6)$$
$$= -1(2x - 3)(x + 2)$$

Note that factoring out -1 gives a positive leading coefficient for the trinomial.

c. $10x^3 + 5x^2 + 5x$

Solution

$$10x^3 + 5x^2 + 5x = 5x(2x^2 + x + 1)$$

Now consider the trinomial: $2x^2 + x + 1 = (2x + ?)(x + ?)$.

The factors of $+1$ need to be $+1$ and $+1$, but $(2x + 1)(x + 1) = 2x^2 + \underline{3x} + 1$.

So there is no way to factor and get a middle term of $+x$ for the product. This trinomial, $2x^2 + x + 1$, is **not factorable**.

We have
$$10x^3 + 5x^2 + 5x = 5x(2x^2 + x + 1). \qquad \text{factored completely}$$

Now work margin exercise 2.

Objective B **The *ac*-Method of Factoring**

The *ac*-**method** of factoring is in reference to the coefficients a and c in the general form $ax^2 + bx + c$ and involves the method of factoring by grouping discussed in Section 7.6. The method is best explained by analyzing an example and explaining each step as follows.

We consider the problem of factoring the trinomial

$$2x^2 + 9x + 10 \text{ where } a = 2, b = 9, \text{ and } c = 10.$$

Analysis of Factoring by the *ac*-Method

General Method	Example
$ax^2 + bx + c$	$2x^2 + 9x + 10$

Step 1: Multiply $a \cdot c$.

Multiply $2 \cdot 10 = 20$.

Step 2: Find two integers whose product is ac and whose sum is b. If this is not possible, then the trinomial is **not factorable**.

Find two integers whose product is 20 and whose sum is 9. (In this case, $4 \cdot 5 = 20$ and $4 + 5 = 9$.)

Step 3: Rewrite the middle term (bx) using the two numbers found in Step 2 as coefficients.

Rewrite the middle term $(+9x)$ using $+4$ and $+5$ as coefficients.
$2x^2 + 9x + 10 = 2x^2 + 4x + 5x + 10$

Step 4: Factor by grouping the first two terms and the last two terms.

Factor by grouping the first two terms and the last two terms.
$$2x^2 + 4x + 5x + 10 = \left(2x^2 + 4x\right) + \left(5x + 10\right)$$
$$= 2x(x+2) + 5(x+2)$$

Step 5: Factor out the common binomial factor. This will give two binomial factors of the trinomial $ax^2 + bx + c$.

Factor out the common binomial factor $(x+2)$. Thus
$$2x^2 + 9x + 10 = 2x^2 + 4x + 5x + 10$$
$$= \left(2x^2 + 4x\right) + \left(5x + 10\right)$$
$$= 2x(x+2) + 5(x+2)$$
$$= (x+2)(2x+5).$$

Example 3

Using the *ac*-Method

a. Factor $4x^2 + 33x + 35$ using the *ac*-method.

Solution

$a = 4, b = 33, c = 35$

Step 1: Find the product ac: $4 \cdot 35 = 140$.

Step 2: Find two integers whose product is 140 and whose sum is 33.

$(+5)(+28) = 140$ and $(+5) + (+28) = +33$

Note: This step may take some experimenting with factors. You might try prime factoring.

For example $140 = 10 \cdot 14 = 2 \cdot 5 \cdot 2 \cdot 7$.

With combinations of these prime factors we can write the following.

Factors of 140	Sum
1 · 140	$1 + 140 = 141$
2 · 70	$2 + 70 = 72$
4 · 35	$4 + 35 = 39$
5 · 28	$5 + 28 = 33$ We can stop here!

5 and 28 are the desired coefficients.

Step 3: Rewrite $+33x$ as $+5x + 28x$, giving

$$4x^2 + 33x + 35 = 4x^2 + 5x + 28x + 35.$$

Step 4: Factor by grouping.

$$4x^2 + 33x + 35 = 4x^2 + 5x + 28x + 35$$
$$= x(4x + 5) + 7(4x + 5)$$

Step 5: Factor out the common binomial factor $(4x + 5)$.

$$4x^2 + 33x + 35 = 4x^2 + 5x + 28x + 35$$
$$= x(4x + 5) + 7(4x + 5)$$
$$= (4x + 5)(x + 7)$$

Note that in Step 3 we could have written $+33x$ as $+28x + 5x$. Try this to convince yourself that the result will be the same two factors.

b. Factor $12y^3 - 26y^2 + 12y$ using the *ac*-method.

Solution

First factor out the greatest common factor $2y$.

$$12y^3 - 26y^2 + 12y = 2y(6y^2 - 13y + 6)$$

Now factor the trinomial $6y^2 - 13y + 6$ with $a = 6$, $b = -13$, and $c = 6$.

Step 1: Find the product ac: $6(6) = 36$.

Step 2: Find two integers whose product is 36 and whose sum is −13.

Note: This may take some time and experimentation. We do know that both numbers must be negative because the product is positive and the sum is negative.

$$(-9)(-4) = +36 \text{ and } -9 + (-4) = -13$$

Steps 3 and 4: Factor by grouping.

$$6y^2 - 13y + 6 = 6y^2 - 9y - 4y + 6$$
$$= 3y(2y-3) - 2(2y-3)$$

Note: −2 is factored from the last two terms so that there will be a common binomial factor $(2y-3)$.

Step 5: Factor out the common binomial factor $(2y-3)$.

$$6y^2 - 13y + 6 = 6y^2 - 9y - 4y + 6$$
$$= 3y(2y-3) - 2(2y-3)$$
$$= (2y-3)(3y-2)$$

Thus for the original expression,

$$12y^3 - 26y^2 + 12y = 2y(6y^2 - 13y + 6)$$
$$= 2y(2y-3)(3y-2)$$

Do not forget to write the common monomial factor, 2y, in the answer.

c. Factor $4x^2 - 5x - 6$ using the *ac*-method.

Solution

$a = 4, b = -5, c = -6$

Step 1: Find the product *ac*: $4(-6) = -24$.

Step 2: Find two integers whose product is −24 and whose sum is −5.

Note: We know that one number must be positive and the other negative because the product is negative.

$$(+3)(-8) = -24 \text{ and } (+3)+(-8) = -5.$$

Steps 3 and 4: Factor by grouping.

$$4x^2 - 5x - 6 = 4x^2 + 3x - 8x - 6$$
$$= x(4x+3) - 2(4x+3)$$

Step 5: Factor out the common binomial factor $(4x+3)$.

$$4x^2 - 5x - 6 = 4x^2 + 3x - 8x - 6$$
$$= x(4x+3) - 2(4x+3)$$
$$= (4x+3)(x-2)$$

Now work margin exercise 3.

3. Use the *ac*-method to factor.

a. $5a^2 + 24a + 27$

b. $3b^2 + 6b - 24$

Tips to Keep in Mind while Factoring

1. When factoring polynomials, always look for the greatest common factor first. Then, if there is one, remember to include this common factor as part of the answer.

2. **To factor completely** means to find factors of the polynomial such that none of the factors are themselves factorable.

3. Not all polynomials are factorable. (See $2x^2 + x + 1$ in Example 2c.) **Any polynomial that cannot be factored as the product of polynomials with integer coefficients is not factorable.**

4. Factoring can be checked by multiplying the factors. The product should be the original expression.

notes

▪ No matter which method you use (the *ac*-method or the trial-and-error method), factoring trinomials takes time. With practice you will become more efficient with either method. Make sure to be patient and observant.

Practice Problems

Factor completely.

1. $3x^2 + 7x - 6$ **2.** $2x^2 + 6x - 8$ **3.** $3x^2 + 15x + 18$

4. $10x^2 - 41x - 18$ **5.** $x^2 + 11x + 28$ **6.** $-3x^3 + 6x^2 - 3x$

7. $3x^2 + 4x - 8$

Practice Problem Answers

1. $(3x - 2)(x + 3)$ **2.** $2(x - 1)(x + 4)$ **3.** $3(x + 2)(x + 3)$

4. $(5x + 2)(2x - 9)$ **5.** $(x + 7)(x + 4)$ **6.** $-3x(x - 1)(x - 1)$

7. not factorable

Exercises 7.8

1. $x^2 + 5x + 6$

2. $x^2 - 6x + 8$

3. $2x^2 - 3x - 5$

4. $3x^2 - 4x - 7$

5. $6x^2 + 11x + 5$

6. $4x^2 - 11x + 6$

7. $-x^2 + 3x - 2$

8. $-x^2 - 5x - 6$

9. $x^2 - 3x - 10$

10. $x^2 - 11x + 10$

11. $-x^2 + 13x + 14$

12. $-x^2 + 12x - 36$

13. $x^2 + 8x + 64$

14. $x^2 + 2x + 3$

15. $-2x^3 + x^2 + x$

16. $-2y^3 - 3y^2 - y$

17. $4t^2 - 3t - 1$

18. $2x^2 - 3x - 2$

19. $5a^2 - a - 6$

20. $3a^2 + 4a + 1$

21. $7x^2 + 5x - 2$

22. $4x^2 + 23x + 15$

23. $8x^2 - 10x - 3$

24. $6x^2 + 23x + 21$

25. $9x^2 - 3x - 20$

26. $4x^2 + 40x + 25$

27. $12x^2 - 38x + 20$

28. $12b^2 - 12b + 3$

29. $3x^2 - 7x + 2$

30. $7x^2 - 11x - 6$

31. $9x^2 - 6x + 1$

32. $4x^2 + 4x + 1$

33. $6y^2 + 7y + 2$

34. $12y^2 - 7y - 12$

35. $x^2 - 46x + 45$

36. $x^2 + 6x - 16$

37. $3x^2 + 9x + 5$

38. $5a^2 - 7a + 2$

39. $8a^2b - 22ab + 12b$

40. $12m^3n - 50m^2n + 8mn$

41. $x^2 + x + 1$

42. $x^2 + 2x + 2$

43. $16x^2 - 8x + 1$ **44.** $3x^2 - 11x - 4$ **45.** $64x^2 - 48x + 9$

46. $9x^2 - 12x + 4$ **47.** $6x^2 + 2x - 20$ **48.** $12y^2 - 15y + 3$

49. $10x^2 + 35x + 30$ **50.** $24y^2 + 4y - 4$ **51.** $-18x^2 + 72x - 8$

52. $7x^4 - 5x^3 + 3x^2$ **53.** $-45y^2 + 30y + 120$ **54.** $-12m^2 + 22m + 4$

55. $12x^2 - 60x + 75$ **56.** $32y^2 + 50$ **57.** $6x^3 + 9x^2 - 6x$

58. $-5y^2 + 40y - 60$ **59.** $12x^3 - 108x^2 + 243x$ **60.** $30a^3 + 51a^2 + 9a$

61. $9x^3y^3 + 9x^2y^3 + 9xy^3$ **62.** $48x^2y - 354xy + 126y$ **63.** $48xy^3 - 100xy^2 + 48xy$

64. $24a^2x^2 + 72a^2x + 243x$ **65.** $21y^4 - 98y^3 + 56y^2$ **66.** $72a^3 - 306a^2 + 189a$

Writing & Thinking

67. Discuss, in your own words, how the sign of the constant term determines what signs will be used in the factors when factoring trinomials.

68. **Volume of a Box:** The volume of an open box is found by cutting equal squares (x inches on a side) from a sheet of cardboard that is 5 inches by 25 inches. The function representing this volume is $V(x) = 4x^3 - 60x^2 + 125x$, where $0 < x < 2.5$. Factor this function and use the factors to explain, in your own words, how the function represents the volume.

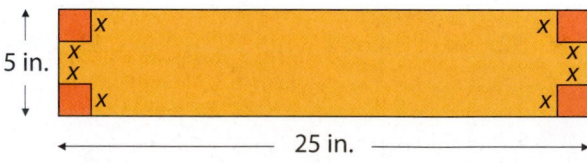

7.9 Special Factoring Techniques

In Section 7.5 we discussed the following three products.

I. $(x+a)(x-a) = x^2 - a^2$ — difference of two squares

II. $(x+a)^2 = x^2 + 2ax + a^2$ — square of a binomial sum

III. $(x-a)^2 = x^2 - 2ax + a^2$ — square of a binomial difference

Objectives

A Factor the difference of two squares.

B Factor perfect square trinomials.

C Factor the sums and differences of two cubes.

Two objectives in this section are to learn to factor products of these types (difference of two squares and squares of binomials) without using the trial-and-error method. That is, with practice you will learn to recognize the "form" of these products and go directly to the factors. **Memorize the products and their names listed above.**

For easy reference to squares, Table 1 lists the squares of the integers from 1 to 20.

Perfect Squares from 1 to 400									
Integer n	1	2	3	4	5	6	7	8	9	10
Square n^2	1	4	9	16	25	36	49	64	81	100
Integer n	11	12	13	14	15	16	17	18	19	20
Square n^2	121	144	169	196	225	256	289	324	361	400

Table 1

Objective A Difference of Two Squares

Consider the polynomial $x^2 - 25$. By recognizing this expression as the **difference of two squares**, we can go directly to the factors:

$$x^2 - 25 = (x)^2 - (5)^2 = (x+5)(x-5).$$

Similarly, we have

$$9 - y^2 = (3)^2 - (y)^2 = (3+y)(3-y)$$

and $\quad 49x^2 - 36 = (7x)^2 - (6)^2 = (7x+6)(7x-6).$

Remember to **look for a common monomial factor first**. For example,

$$6x^2y - 24y = 6y(x^2 - 4) = 6y(x+2)(x-2).$$

1. Factor completely.

a. $7yu^2 - 343y$

b. $v^6 - 100$

Example 1

Factoring the Difference of Two Squares

Factor completely.

a. $3a^2b - 3b$

Solution

$$3a^2b - 3b = 3b\left(a^2 - 1\right)$$

Factor out the GCF, $3b$.

$$= 3b(a+1)(a-1)$$

difference of two squares
Don't forget that $3b$ is a factor.

b. $x^6 - 400$

Solution

Even powers, such as x^6, can always be treated as squares: $x^6 = \left(x^3\right)^2$.

$$x^6 - 400 = \left(x^3\right)^2 - 20^2$$

$$= \left(x^3 + 20\right)\left(x^3 - 20\right)$$

difference of two squares

*Now work margin exercise **1**.*

Sum of Two Squares

The **sum of two squares** is an expression of the form $x^2 + a^2$ and is **not factorable**. For example, $x^2 + 36$ is the sum of two squares and is not factorable. There are no factors with integer coefficients whose product is $x^2 + 36$. To understand this situation, write

$$x^2 + 36 = x^2 + 0x + 36$$

and note that there are no factors of $+36$ that will add to 0.

Example 2

Using the Sum of Two Squares

Factor completely. Be sure to look first for the greatest common monomial factor.

a. $y^2 + 64$

Solution

$y^2 + 64$ is the **sum of two squares and is not factorable**.

b. $4x^2 + 100$

Solution

$$4x^2 + 100 = 4(x^2 + 25) \quad \text{\color{red}factored completely}$$

We see that 4 is the greatest common monomial factor and $x^2 + 25$ is the **sum of two squares and not factorable**.

Now work margin exercise 2.

2. Factor completely.

a. $x^2 + 49$

b. $45x^2 + 20$

Objective B **Perfect Square Trinomials**

We know that **squaring a binomial** leads to a **perfect square trinomial**. Therefore, factoring a perfect square trinomial gives the square of a binomial. We simply need to recognize whether or not a trinomial fits the "form."

In a perfect square trinomial, both the first and last terms of the trinomial must be perfect squares. If the first term is of the form x^2 and the last term is of the form a^2, then the middle term must be of the form $2ax$ or $-2ax$, as shown in the following examples.

$$x^2 + 8x + 16 = (x + 4)^2 \quad \text{\color{red}Here } 8x = 2 \cdot 4 \cdot x = 2ax \text{ and } 16 = 4^2 = a^2.$$

$$x^2 - 6x + 9 = (x - 3)^2 \quad \text{\color{red}Here } -6x = -2 \cdot 3 \cdot x = -2ax \text{ and } 9 = 3^2 = a^2.$$

Example 3

Factoring Perfect Square Trinomials

Factor completely.

a. $z^2 - 12z + 36$

Solution

In the form $x^2 - 2ax + a^2$ we have $x = z$ and $a = 6$.

$$z^2 - 12z + 36 = z^2 - 2 \cdot 6z + 6^2$$
$$= (z - 6)^2$$

b. $4y^2 + 12y + 9$

Solution

In the form $x^2 + 2ax + a^2$ we have $x = 2y$ and $a = 3$.

$$4y^2 + 12y + 9 = (2y)^2 + 2 \cdot 3 \cdot 2y + 3^2$$
$$= (2y + 3)^2$$

3. Factor completely.

a. $z^2 + 40z + 400$

b. $3u^2xy - 18uvxy + 27v^2xy$

c. $(v^2 + 8v + 16) - z^2$

c. $2x^3 - 8x^2y + 8xy^2$

Solution

Factor out the GCF first. Then factor the **perfect square trinomial**.

$$2x^3 - 8x^2y + 8xy^2 = 2x(x^2 - 4xy + 4y^2)$$
$$= 2x\left[x^2 - 2 \cdot 2y \cdot x + (2y)^2\right]$$
$$= 2x(x - 2y)^2$$

d. $(x^2 + 6x + 9) - y^2$

Solution

Treat $x^2 + 6x + 9$ as a perfect square trinomial, then factor the **difference of two squares**.

$$(x^2 + 6x + 9) - y^2 = (x + 3)^2 - y^2$$
$$= (x + 3 + y)(x + 3 - y)$$

Now work margin exercise 3.

Objective C **Sums and Differences of Two Cubes**

The formulas for the sums and differences of two cubes are new, and we can proceed to show that they are indeed true as follows.

$$(x + a)(x^2 - ax + a^2) = x \cdot x^2 - x \cdot ax + x \cdot a^2 + a \cdot x^2 - a \cdot ax + a \cdot a^2$$
$$= x^3 - ax^2 + a^2x + ax^2 - a^2x + a^3$$
$$= x^3 + a^3 \qquad \text{sum of two cubes}$$

$$(x - a)(x^2 + ax + a^2) = x \cdot x^2 + x \cdot ax + x \cdot a^2 - a \cdot x^2 - a \cdot ax - a \cdot a^2$$
$$= x^3 + ax^2 + a^2x - ax^2 - a^2x - a^3$$
$$= x^3 - a^3 \qquad \text{difference of two cubes}$$

notes

Important Notes about the Sum and Difference of Two Cubes
1. In each case, after multiplying the factors the middle terms drop out and only two terms are left.
2. The trinomials in parentheses are not perfect square trinomials. These trinomials are not factorable.
3. The sign in the binomial agrees with the sign in the result.

Because we know that factoring is the reverse of multiplication, we use these formulas to factor the sums and differences of two cubes. For example,

$$x^3 + 27 = (x)^3 + (3)^3$$
$$= (x+3)\left((x)^2 - (3)(x) + (3)^2\right)$$
$$= (x+3)(x^2 - 3x + 9)$$

and

$$x^6 - 125 = (x^2)^3 - (5)^3$$
$$= (x^2 - 5)\left[(x^2)^2 + (5)(x^2) + (5)^2\right]$$
$$= (x^2 - 5)(x^4 + 5x^2 + 25).$$

As an aid in factoring sums and differences of two cubes, Table 2 contains the cubes of the integers from 1 to 10. These cubes are called perfect cubes.

Perfect Cubes from 1 to 1000									
Integer n	1	2	3	4	5	6	7	8	9	10
Cube n^3	1	8	27	64	125	216	343	512	729	1000

Table 2

Example 4

Factoring Sums and Differences of Two Cubes

Factor completely.

a. $x^3 - 8$

Solution

$$x^3 - 8 = x^3 - 2^3$$
$$= (x-2)(x^2 + 2 \cdot x + 2^2)$$
$$= (x-2)(x^2 + 2x + 4)$$

Note: Notice that the second polynomial is not a perfect square trinomial and cannot be factored.

b. $x^6 + 64y^3$

Solution

$$x^6 + 64y^3 = (x^2)^3 + (4y)^3$$
$$= (x^2 + 4y)\left[(x^2)^2 - 4y \cdot x^2 + (4y)^2\right]$$
$$= (x^2 + 4y)(x^4 - 4x^2y + 16y^2)$$

4. Factor completely.

a. $v^3 - 27$

b. $48v^{12} - 750$

c. $16y^{12} - 250$

Solution

Factor out the GCF first. Then factor the **difference of two cubes**.

$$16y^{12} - 250 = 2\left(8y^{12} - 125\right)$$
$$= 2\left[\left(2y^4\right)^3 - 5^3\right]$$
$$= 2\left(2y^4 - 5\right)\left[\left(2y^4\right)^2 + (5)\left(2y^4\right) + 5^2\right]$$
$$= 2\left(2y^4 - 5\right)\left(4y^8 + 10y^4 + 25\right)$$

Now work margin exercise 4.

Practice Problems

Completely factor each of the following polynomials. If a polynomial cannot be factored, write "not factorable."

1. $5x^2 - 80$ **2.** $9x^2 - 12x + 4$ **3.** $4x^2 + 20x + 25$

4. $2x^3 - 250$ **5.** $\left(y^2 - 4y + 4\right) - x^4$ **6.** $x^3 + 8y^3$

Practice Problem Answers

1. $5(x-4)(x+4)$ **2.** $(3x-2)^2$

3. $(2x+5)^2$ **4.** $2(x-5)\left(x^2 + 5x + 25\right)$

5. $\left(y - 2 - x^2\right)\left(y - 2 + x^2\right)$ **6.** $(x+2y)\left(x^2 - 2xy + 4y^2\right)$

Exercises 7.9

Completely factor each of the given polynomials. If a polynomial cannot be factored, write "not factorable." See Examples 1 through 4.

1. $x^2 - 25$

2. $y^2 - 121$

3. $81 - y^2$

4. $25 - z^2$

5. $2x^2 - 128$

6. $3x^2 - 147$

7. $4x^4 - 64$

8. $5x^4 - 125$

9. $y^2 + 100$

10. $4x^2 + 49$

11. $y^2 - 16y + 64$

12. $z^2 + 18z + 81$

13. $-4x^2 + 100$

14. $-12x^4 + 3$

15. $9x^2 - 25$

16. $4x^2 - 49$

17. $y^2 - 10y + 25$

18. $x^2 + 12x + 36$

19. $4x^2 - 4x + 1$

20. $49x^2 - 14x + 1$

21. $25x^2 + 30x + 9$

22. $9y^2 + 12y + 4$

23. $16x^2 - 40x + 25$

24. $9x^2 - 12x + 4$

25. $4x^3 - 64x$

26. $50x^3 - 8x$

27. $2x^3y + 32x^2y + 128xy$

28. $3x^2y - 30xy + 75y$

29. $y^2 + 6y + 9$

30. $y^2 + 4y + 4$

31. $x^2 - 20x + 100$

32. $25x^2 - 10x + 1$

33. $x^4 + 10x^2y + 25y^2$

34. $16x^4 + 8x^2y + y^2$

35. $x^3 - 125$

36. $x^3 - 64$

37. $y^3 + 216$

38. $y^3 + 1$

39. $x^3 + 27y^3$

40. $8x^3 + 1$

41. $x^2 + 64y^2$

42. $3x^3 + 81$

43. $4x^3 - 32$

44. $64x^3 + 27y^3$

45. $54x^3 - 2y^3$

46. $3x^4 + 375xy^3$

47. $x^3y + y^4$

48. $x^4y^3 - x$

49. $x^2y^2 - x^2y^5$

50. $2x^2 - 16x^2y^3$

51. $24x^4y + 81xy^4$

52. $x^6 - 64y^3$

53. $x^6 - y^9$

54. $64x^2 + 1$

55. $27x^3 + y^6$

56. $x^3 + 64z^3$

57. $8x^3 + y^3$

58. $x^3 + 125y^3$

59. $8y^3 - 8$

60. $36x^3 + 36$

61. $9x^2 - y^2$

62. $x^2 - 4y^2$

63. $x^4 - 16y^4$

64. $81x^4 - 1$

65. $(x - y)^2 - 81$

66. $(x + 2y)^2 - 25$

67. $(x^2 - 2xy + y^2) - 36$

68. $(x^2 + 4xy + 4y^2) - 25$

69. $(16x^2 + 8x + 1) - y^2$

70. $x^2 - (y^2 + 6y + 9)$

Solve the following word problems.

71. **a.** Represent the area of the shaded region of the square shown below as the difference of two squares.

   b. Use the factors of the expression in part **a.** to draw (and label the sides of) a rectangle that has the same area as the shaded region.

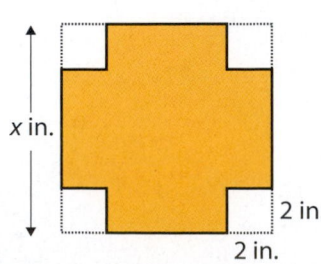

72. **a.** Use a polynomial function to represent the area of the shaded region of the square.

   b. Use a polynomial function to represent the perimeter of the shaded figure.

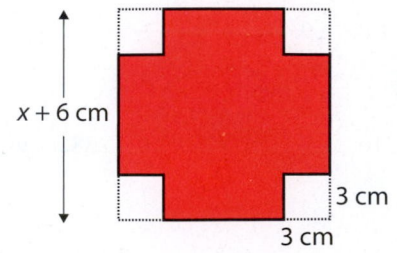

73. **a.** Show that the sum of the areas of the rectangles and squares in the figure is a perfect square trinomial.

     b. Rearrange the rectangles and squares in the form of a square and represent its area as the square of a binomial.

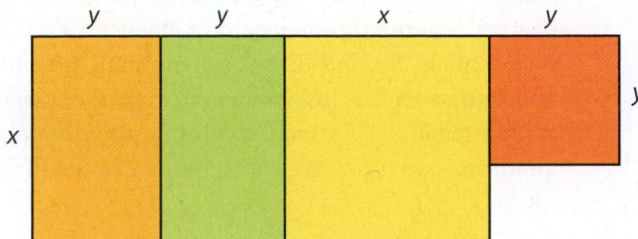

74. **Compound Interest:** Compound interest is interest earned on interest. If a principal P is invested and compounded annually (once a year) at a rate of r, then the amount, A_1 accumulated in one year is $A_1 = P + Pr$.

In factored form, we have: $A_1 = P + Pr = P(1 + r)$.

At the end of the second year the amount accumulated is $A_2 = (P + Pr) + (P + Pr)r$.

    a. Write the expression for A_2 in factored form similar to that for A_1.

    b. Write an expression for the amount accumulated in three years, A_3, in factored form.

    c. Write an expression for A_n the amount accumulated in n years.

    d. Use the formula you developed in part **c.** and your calculator to find the amount accumulated if $10,000 is invested at 6% and compounded annually for 20 years.

75. You may have heard of (or studied) the following rules for division of an integer by 3 and 9:

    1. An integer is divisible by 3 if the sum of its digits is divisible by 3.

    2. An integer is divisible by 9 if the sum of its digits is divisible by 9.

The proofs of both **1.** and **2.** can be started as follows.

Let abc represent a three-digit integer.

$$\begin{aligned} \text{Then } abc &= 100a + 10b + c \\ &= (99 + 1)a + (9 + 1)b + c \\ &= (\text{now you finish the proofs}) \end{aligned}$$

Use the pattern just shown and prove both **1.** and **2.** for a four-digit integer.

7.10 Solving Quadratic Equations by Factoring

In this chapter the emphasis has been on second-degree polynomials and functions. Such polynomials are particularly evident in physics with the path of thrown objects (or projectiles) affected by gravity, in mathematics involving area (area of a circle, rectangle, square, and triangle), and in many situations in higher-level mathematics. The methods of factoring we have studied (*ac*-method, trial-and-error method, difference of two squares, and perfect square trinomials) have been related, in general, to second-degree polynomials. Second-degree polynomials are called **quadratic polynomials** (or **quadratics**) and, as just discussed, they play a major role in many applications of mathematics. In fact, quadratic polynomials and techniques for solving quadratic equations are central topics in the first two courses in algebra.

Objective A **Solving Quadratic Equations by Factoring**

Quadratic Equations

Quadratic equations are equations that can be written in the form

$$ax^2 + bx + c = 0 \text{ where } a, b, \text{ and } c \text{ are real numbers and } a \neq 0.$$

The form $ax^2 + bx + c = 0$ is called the **standard form** (or **general form**) of a **quadratic equation**. In the standard form, a quadratic polynomial is on one side of the equation and 0 is on the other side. For example, $x^2 - 8x + 12 = 0$ is a quadratic equation in standard form, while $3x^2 - 2x = 27$ and $x^2 = 25x$ are quadratic equations, just not in standard form.

These last two equations can be manipulated algebraically so that 0 is on one side. When solving quadratic equations by factoring, having 0 on one side before we factor is necessary because of the following **zero-factor property**.

Zero-Factor Property

If the product of two (or more) factors is 0, then at least one of the factors must be 0. That is, for real numbers a and b,

if $a \cdot b = 0$, then $a = 0$ or $b = 0$ or both.

Example 1

Solving Factored Quadratic Equations

Solve the following quadratic equation.

$(x - 5)(2x - 7) = 0$

Solution

Since the quadratic is already factored and the other side of the equation is 0, we use the zero-factor property and set each factor equal to 0. This process yields two linear equations, which can then be solved.

$$x - 5 = 0 \quad \text{or} \quad 2x - 7 = 0$$
$$x = 5 \qquad\qquad 2x = 7$$
$$x = \frac{7}{2}$$

Thus the **two solutions** (or **roots**) to the original equation are $x = 5$ and $x = \frac{7}{2}$. Or, we can say that the solution set is $\left\{ \frac{7}{2}, 5 \right\}$.

The solutions can be **checked** by substituting them one at a time for x in the equation. That is, there will be two "checks."

Substituting $x = 5$ gives

$$(x - 5)(2x - 7) = (5 - 5)(2 \cdot 5 - 7)$$
$$= (0)(3)$$
$$= 0.$$

Substituting $x = \frac{7}{2}$ gives

$$(x - 5)(2x - 7) = \left(\frac{7}{2} - 5 \right)\left(2 \cdot \frac{7}{2} - 7 \right)$$
$$= \left(-\frac{3}{2} \right)(7 - 7)$$
$$= \left(-\frac{3}{2} \right)(0)$$
$$= 0.$$

Therefore, both 5 and $\frac{7}{2}$ are solutions to the original equation.

Now work margin exercise 1.

In Example 1, the polynomial was factored and the other side of the equation was 0. The equation was solved by setting each factor, in turn, equal to 0 and solving the resulting linear equations. These solutions tell us that the original equation has two solutions. **In general, a quadratic equation has two solutions.** In the special cases where the two factors are the same, there is only one solution and it is called a **double solution** (or **double root**). The following examples show how to solve quadratic equations by factoring. Remember that the equation must be in standard form with one side of the equation equal to 0. Study these examples carefully.

1. Find all possible solutions to the following equation.

$(y - 7)(3y - 5) = 0$

Example 2

Solving Quadratic Equations by Factoring

Solve the following equations by writing the equation in standard form with one side 0 and factoring the polynomial. Then set each factor equal to 0 and solve. Checking is left as an exercise for the student.

a. $3x^2 = 6x$

Solution

$$3x^2 = 6x$$

$$3x^2 - 6x = 0 \qquad$$ Write the equation in standard form with 0 on one side by subtracting $6x$ from both sides.

$$3x(x-2) = 0 \qquad$$ Factor out the common monomial, $3x$.

$$3x = 0 \quad \text{or} \quad x - 2 = 0 \qquad$$ Set each factor equal to 0.

$$x = 0 \quad \text{or} \qquad x = 2 \qquad$$ Solve each linear equation.

The solutions are 0 and 2. Or, we can say that the solution set is $\{0, 2\}$.

b. $x^2 - 8x + 16 = 0$

Solution

$$x^2 - 8x + 16 = 0$$

$$(x - 4)^2 = 0 \qquad$$ The trinomial is a perfect square.

$$x - 4 = 0 \quad \text{or} \quad x - 4 = 0 \qquad$$ Both factors are the same, so there is only one distinct solution.

$$x = 4 \quad \text{or} \qquad x = 4$$

The only solution is 4, and it is called a **double solution** (or **double root**). The solution set is $\{4\}$.

c. $4x^2 - 4x = 24$

Solution

$$4x^2 - 4x = 24$$

$$4x^2 - 4x - 24 = 0 \qquad$$ Add -24 to both sides so that one side is 0.

$$4(x^2 - x - 6) = 0 \qquad$$ Factor out the common monomial, 4.

$$4(x - 3)(x + 2) = 0$$

$$x - 3 = 0 \quad \text{or} \quad x + 2 = 0 \qquad$$ The constant factor 4 can never be 0 and does not affect the solution.

$$x = 3 \qquad\qquad x = -2$$

The solutions are 3 and -2. Or, we can say the solution set is $\{-2, 3\}$.

d. $(x + 5)^2 = 36$

Solution

$$(x + 5)^2 = 36 \qquad \text{Expand } (x + 5)^2.$$
$$x^2 + 10x + 25 = 36 \qquad \text{Add } -36 \text{ to both sides so that one side is 0.}$$
$$x^2 + 10x - 11 = 0 \qquad \text{Factor.}$$
$$(x + 11)(x - 1) = 0$$

$$x + 11 = 0 \qquad \text{or} \qquad x - 1 = 0$$
$$x = -11 \qquad\qquad x = 1$$

The solutions are −11 and 1.

e. $3x(x - 1) = 2(5 - x)$

Solution

$$3x(x - 1) = 2(5 - x)$$
$$3x^2 - 3x = 10 - 2x \qquad \text{Use the distributive property.}$$
$$3x^2 - 3x + 2x - 10 = 0 \qquad \text{Arrange terms so that 0 is on one side.}$$
$$3x^2 - x - 10 = 0 \qquad \text{Simplify.}$$
$$(3x + 5)(x - 2) = 0 \qquad \text{Factor.}$$

$$3x + 5 = 0 \qquad \text{or} \qquad x - 2 = 0$$
$$3x = -5 \qquad\qquad x = 2$$
$$x = -\frac{5}{3}$$

The solutions are $-\dfrac{5}{3}$ and 2.

f. $\dfrac{2x^2}{15} - \dfrac{x}{3} = -\dfrac{1}{5}$

Solution

$$\frac{2x^2}{15} - \frac{x}{3} = -\frac{1}{5}$$

$$15 \cdot \frac{2x^2}{15} - \frac{x}{3} \cdot 15 = -\frac{1}{5} \cdot 15 \qquad \text{Multiply each term by 15, the LCM of the denominators, to get integer coefficients.}$$

$$2x^2 - 5x = -3 \qquad \text{Simplify.}$$
$$2x^2 - 5x + 3 = 0 \qquad \text{Add 3 to both sides so that one side is 0.}$$
$$(2x - 3)(x - 1) = 0 \qquad \text{Factor by the trial-and-error method or the } ac\text{-method.}$$

2. Solve by factoring.

a. $4v^2 = 8v$

b. $x^2 - 6x + 9 = 0$

c. $9x^2 - 27x = 36$

d. $x(6x + 21) = 2(4x + 14)$

e. $\dfrac{x^2}{6} - \dfrac{x}{3} = \dfrac{1}{2}$

$$2x - 3 = 0 \quad \text{or} \quad x - 1 = 0$$
$$2x = 3 \qquad\qquad x = 1$$
$$x = \frac{3}{2}$$

The solutions are $\dfrac{3}{2}$ and 1.

Now work margin exercise **2.**

In the next example, we show that factoring can be used to solve equations of degrees higher than second-degree. There will be more than two factors, but just as with quadratics, if the product is 0, then the solutions are found by setting each factor equal to 0.

Example 3

Solving Higher Degree Equations

Solve the following equation: $2x^3 - 4x^2 - 6x = 0$.

Solution

$$2x^3 - 4x^2 - 6x = 0$$
$$2x(x^2 - 2x - 3) = 0 \qquad \text{Factor out the common monomial, } 2x.$$
$$2x(x - 3)(x + 1) = 0 \qquad \text{Factor the trinomial.}$$
$$2x = 0 \quad \text{or} \quad x - 3 = 0 \quad \text{or} \quad x + 1 = 0 \qquad \text{Set each factor equal to 0 and solve.}$$
$$x = 0 \qquad\qquad x = 3 \qquad\qquad x = -1$$

The solutions are $0, 3,$ and -1.
Or, we can say that the solution set is $\{-1, 0, 3\}$.

Now work margin exercise **3.**

3. Solve the following equation.

$$4x^3 - 12x^2 - 40x = 0$$

To Solve a Quadratic Equation by Factoring

1. Add or subtract terms as necessary so that 0 is on one side of the equation and the equation is in the standard form $\boldsymbol{ax^2 + bx + c = 0}$ where $a, b,$ and c are real numbers and $a \neq 0$.
2. Factor completely. (If there are any fractional coefficients, multiply each term by the least common denominator first so that all coefficients will be integers.)
3. Set each nonconstant factor equal to 0 and solve each linear equation for the unknown.
4. Check each solution, one at a time, in the original equation.

Special Comment: All of the quadratic equations in this section can be solved by factoring. That is, all of the quadratic polynomials are factorable. However, as we have seen in some of the previous sections, not all polynomials are factorable. In Chapter 12 we will develop techniques (other than factoring) for solving quadratic equations whether the quadratic polynomial is factorable or not.

Common Error

A **common error** is to divide both sides of an equation by the variable x. This error can be illustrated by using the equation in Example 2a.

$$3x^2 = 6x$$

$$\frac{3x^2}{x} = \frac{6x}{x}$$

Do not divide by x because you lose the solution $x = 0$.

INCORRECT

$$3x = 6$$

$$x = 2$$

Factoring is the method to use. By factoring, you will find all solutions as shown in the previous examples.

Objective B   ### Finding an Equation Given the Roots

To help develop a complete understanding of the relationships between factors, factoring, and solving equations, we reverse the process of finding solutions. That is, we want to **find an equation that has certain given solutions (or roots)**. For example, to find an equation that has the roots

$$x = 4 \quad \text{and} \quad x = -7,$$

we proceed as follows.

1. Rewrite the linear equations with 0 on one side:
   $x - 4 = 0$ and $x + 7 = 0$.

2. Form the product of the factors and set this product equal to 0:
   $(x - 4)(x + 7) = 0$.

3. Multiply the factors. The resulting quadratic equation must have the two given roots (because we know the factors):
   $x^2 + 3x - 28 = 0$.

notes

- This equation can be multiplied by any nonzero constant and a new equation will be formed, but it will still have the same solutions, namely 4 and −7. Thus technically, there are many equations with these two roots.

-

The formal reasoning is based on the following theorem called the **factor theorem**.

Factor Theorem

If $x = c$ is a root of a polynomial equation in the form $P(x) = 0$, then $x - c$ is a factor of the polynomial $P(x)$.

4. Use the factor theorem to find an equation with integer coefficients that has the given roots:

$$x = 4 \text{ and } x = \frac{3}{2}.$$

Example 4

Using the Factor Theorem to Find Equations with Given Roots

Find a polynomial equation with integer coefficients that has the given roots:

$$x = 3 \text{ and } x = -\frac{2}{3}.$$

Solution

Form the linear equations and then find the product of the factors.

$$x = 3 \qquad\qquad x = -\frac{2}{3}$$

$$x - 3 = 0 \qquad x + \frac{2}{3} = 0 \qquad \text{Rewrite the equations with 0 on one side.}$$

$$\qquad\qquad\qquad 3x + 2 = 0 \qquad \text{Multiply by 3 to get integer coefficients.}$$

Form the equation by setting the product of the factors equal to 0 and multiplying.

$$(x - 3)(3x + 2) = 0 \qquad \text{The equation has integer coefficients.}$$

$$3x^2 - 7x - 6 = 0 \qquad \text{This quadratic equation has the two given roots.}$$

Now work margin exercise 4.

Practice Problems

Solve each equation by factoring.

1. $x^2 - 6x = 0$ 　　　　　　　　 **2.** $6x^2 - x - 1 = 0$

3. $(x - 2)^2 - 25 = 0$ 　　　　　　 **4.** $x^3 - 8x^2 + 16x = 0$

5. Find a polynomial equation with integer coefficients that has the given roots:
$$x = -4 \text{ and } x = \frac{3}{5}.$$

Practice Problem Answers

1. $x = 0, x = 6$ 　　　 **2.** $x = \frac{1}{2}, x = -\frac{1}{3}$ 　　 **3.** $x = 7, x = -3$

4. $x = 0, x = 4$ 　　　 **5.** $5x^2 + 17x - 12 = 0$

Exercises 7.10

1. $(x-3)(x-2)=0$ **2.** $(x+5)(x-2)=0$ **3.** $(2x-9)(x+2)=0$ **4.** $(x+7)(3x-4)=0$

5. $0=(x+3)(x+3)$ **6.** $0=(x+10)(x-10)$ **7.** $(x+5)(x+5)=0$ **8.** $(x+5)(x-5)=0$

9. $2x(x-2)=0$ **10.** $3x(x+3)=0$ **11.** $(x+6)^2=0$ **12.** $5(x-9)^2=0$

Solve the equations by factoring. See Examples 2 and 3.

13. $x^2-3x-4=0$ **14.** $x^2+7x+12=0$ **15.** $x^2-x-12=0$

16. $x^2-11x+18=0$ **17.** $0=x^2+3x$ **18.** $0=x^2-3x$

19. $x^2+8=6x$ **20.** $x^2=x+30$ **21.** $2x^2+2x-24=0$

22. $9x^2+63x+90=0$ **23.** $0=2x^2-5x-3$ **24.** $0=2x^2-x-3$

25. $3x^2-4x-4=0$ **26.** $3x^2-8x+5=0$ **27.** $2x^2-7x=4$

28. $4x^2+8x=-3$ **29.** $-2x=3x^2-8$ **30.** $6x^2+2=-7x$

31. $4x^2-12x+9=0$ **32.** $25x^2-60x+36=0$ **33.** $8x=5x^2$

34. $15x=3x^2$ **35.** $9x^2-36=0$ **36.** $4x^2-16=0$

37. $5x^2=10x-5$ **38.** $2x^2=4x+6$ **39.** $8x^2+32=32x$

40. $6x^2=18x+24$ **41.** $\dfrac{x^2}{9}=1$ **42.** $\dfrac{x^2}{2}=8$

43. $\dfrac{x^2}{5}-x-10=0$ **44.** $\dfrac{2}{3}x^2+2x-\dfrac{20}{3}=0$ **45.** $\dfrac{x^2}{8}+x+\dfrac{3}{2}=0$

46. $\dfrac{x^2}{6} - \dfrac{1}{2}x - 3 = 0$

47. $x^2 - x + \dfrac{1}{4} = 0$

48. $\dfrac{x^2}{3} - 2x + 3 = 0$

49. $x^3 + 8x = 6x^2$

50. $x^3 = x^2 + 30x$

51. $6x^3 + 7x^2 = -2x$

52. $3x^3 = 8x - 2x^2$

53. $0 = x^2 - 100$

54. $0 = x^2 - 121$

55. $3x^2 - 75 = 0$

56. $5x^2 - 45 = 0$

57. $x^2 + 8x + 16 = 0$

58. $x^2 + 14x + 49 = 0$

59. $3x^2 = 18x - 27$

60. $5x^2 = 10x - 5$

61. $(x - 1)^2 = 4$

62. $(x - 3)^2 = 1$

63. $(x + 5)^2 = 9$

64. $(x + 4)^2 = 16$

65. $(x + 4)(x - 1) = 6$

66. $(x - 5)(x + 3) = 9$

67. $27 = (x + 2)(x - 4)$

68. $-1 = (x + 2)(x + 4)$

69. $x(x + 7) = 3(x + 4)$

70. $x(x + 9) = 6(x + 3)$

71. $3x(x + 1) = 2(x + 1)$

72. $2x(x - 1) = 3(x - 1)$

73. $x(2x + 1) = 6(x + 2)$

74. $3x(x + 3) = 2(2x - 1)$

Write a polynomial equation with integer coefficients that has the given roots. See Example 4.

75. $y = 3, y = -2$

76. $x = 5, x = 7$

77. $x = -5, x = -\dfrac{1}{2}$

78. $x = \dfrac{1}{4}, x = -1$

79. $x = \dfrac{1}{2}, x = \dfrac{3}{4}$

80. $y = \dfrac{2}{3}, y = \dfrac{1}{6}$

81. $x = 0, x = 3, x = -2$

82. $y = 0, y = -4, y = 1$

83. $y = -2, y = 3, y = 3$ (3 is a double root.)

84. $x = -1, x = -1, x = -1$ (−1 is a triple root.)

85. When solving equations by factoring, one side of the equation must be 0. Explain why this is so.

86. In solving the equation $(x+5)(x-4) = 6$, why can't we just put one factor equal to 3 and the other equal to 2? Certainly $3 \cdot 2 = 6$.

87. A ball is dropped from the top of a building that is 784 feet high. The height of the ball above ground level is given by the polynomial function $h(t) = -16t^2 + 784$ where t is measured in seconds.

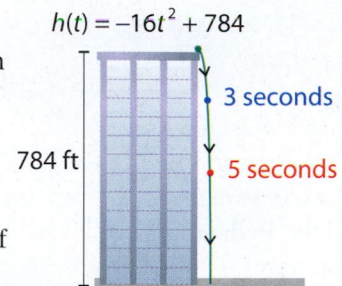

$h(t) = -16t^2 + 784$

784 ft

3 seconds

5 seconds

   a. How high is the ball after 3 seconds? 5 seconds?

   b. How far has the ball traveled in 3 seconds? 5 seconds?

   c. When will the ball hit the ground? Explain your reasoning in terms of factors.

7.11 Applications of Quadratic Equations

Whether or not application problems cause you difficulty depends a great deal on your personal experiences and general reasoning abilities. These abilities are developed over a long period of time. A problem that is easy for you, possibly because you have had experience in a particular situation, might be quite difficult for a friend and vice versa.

Most problems do not say specifically to add, subtract, multiply, or divide. You are to know from the nature of the problem what to do. You are to ask yourself, "What information is given? What am I trying to find? What tools, skills, and abilities do I need to use?"

Word problems should be approached in an orderly manner. You should have an "attack plan."

Attack Plan for Application Problems

1. Read the problem carefully at least twice.
2. Decide what is asked for and assign a variable or variable expression to the unknown quantities.
3. Organize a chart, table, or diagram relating all the information provided.
4. Form an equation. (A formula of some type may be necessary.)
5. Solve the equation.
6. Check your solution with the wording of the problem to be sure it makes sense.

Several types of problems lead to quadratic equations. The problems in this section are set up so that the equations can be solved by factoring. More general problems and approaches to solving quadratic equations are discussed in a later chapter.

Example 1

Applications of Quadratic Equations

a. One number is four more than another and the sum of their squares is 296. What are the numbers?

Solution

Let x = smaller number and
   $x + 4$ = larger number.

Set the sum of their squares equal to 296 and solve the equation.

$$x^2 + (x+4)^2 = 296 \quad \text{Add the squares.}$$

$$x^2 + x^2 + 8x + 16 = 296 \quad \text{Expand } (x+4)^2.$$

$$2x^2 + 8x - 280 = 0 \quad \text{Write the equation in standard form.}$$

$$2(x^2 + 4x - 140) = 0 \quad \text{Factor out the GCF.}$$

$$2(x+14)(x-10) = 0 \quad \text{Factor the trinomial.}$$

$$\begin{array}{lll} x + 14 = 0 & \text{or} & x - 10 = 0 \\ \quad x = -14 & & \quad x = 10 \\ x + 4 = -10 & & x + 4 = 14 \end{array}$$

There are two sets of answers to the problem: 10 and 14 or −14 and −10.

Check:

$$10^2 + 14^2 = 100 + 196 = 296$$

$$\text{and } (-14)^2 + (-10)^2 = 196 + 100 = 296$$

b. In an orange grove, there are 10 more trees in each row than there are rows. How many rows are there if there are 96 trees in the grove?

Solution

Let r = number of rows and
   $r + 10$ = number of trees per row.

Set up the equation and solve.

$$r(r+10) = 96$$

$$r^2 + 10r = 96$$

$$r^2 + 10r - 96 = 0$$

$$(r-6)(r+16) = 0$$

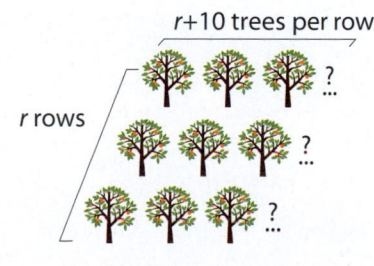

$r+10$ trees per row

r rows

$$\begin{array}{lll} r - 6 = 0 & \text{or} & r + 16 = 0 \\ \quad r = 6 & & \quad r = -16 \end{array}$$

There are 6 rows in the grove ($6 \cdot 16 = 96$ trees).

Note: While −16 is a solution to the equation, −16 does not fit the conditions of the problem and is discarded. You cannot have −16 rows.

c. A rectangle has an area of 135 square meters and a perimeter of 48 meters. What are the dimensions of the rectangle?

Solution

The area of a rectangle is the product of its length and width ($A = lw$). The perimeter of a rectangle is given by $P = 2l + 2w$. Since the perimeter is 48 meters, then the length plus the width must be 24 meters (one-half of the perimeter).

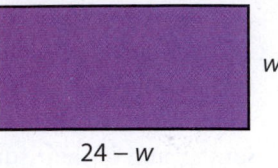

w

$24 - w$

Let w = width and
  $24 - w$ = length.

Set up an equation for the area and solve.

$$(24 - w) \cdot w = 135 \qquad \text{area} = lw = \text{length times width}$$
$$24w - w^2 = 135$$
$$0 = w^2 - 24w + 135$$
$$0 = (w - 9)(w - 15)$$

$$w - 9 = 0 \qquad \text{or} \qquad w - 15 = 0$$
$$w = 9 \qquad\qquad\qquad w = 15$$
$$l = 24 - 9 = 15 \qquad\qquad l = 24 - 15 = 9$$

The dimensions are 9 meters by 15 meters ($9 \cdot 15 = 135$).

d. A man wants to build a fence on three sides of a rectangular-shaped lot he owns. If 180 feet of fencing is needed and the area of the lot is 4000 square feet, what are the dimensions of the lot?

Solution

Let x = one of two equal sides and
$180 - 2x$ = third side.

Set up an equation for the area and solve.

$$x(180 - 2x) = 4000$$
$$180x - 2x^2 = 4000$$
$$0 = 2x^2 - 180x + 4000$$
$$0 = 2(x^2 - 90x + 2000)$$
$$0 = 2(x - 50)(x - 40)$$

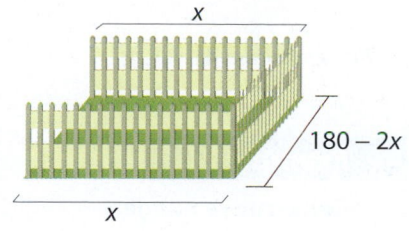

$$x - 50 = 0 \qquad \text{or} \qquad x - 40 = 0$$
$$x = 50 \qquad\qquad\qquad x = 40$$
$$180 - 2x = 180 - 2(50) = 80 \qquad 180 - 2x = 180 - 2(40) = 100 \quad \text{third side}$$

Thus there are two possible answers: the lot is 50 feet by 80 feet or the lot is 40 feet by 100 feet.

Now work margin exercise 1.

Objective B **Consecutive Integers**

In Section 5.6, we discussed applications with **consecutive integers**, **consecutive even integers**, and **consecutive odd integers**. Because the applications involved only addition and subtraction, the related equations were first degree. In this section the applications involve squaring expressions and solving quadratic equations. For convenience, the definitions and representations of the various types of integers are repeated here.

Consecutive Integers

Integers are **consecutive** if each is 1 more than the previous integer. Three consecutive integers can be represented as n, $n + 1$, and $n + 2$.

For example: 5, 6, 7

Consecutive Even Integers

Even integers are consecutive if each is 2 more than the previous even integer. Three consecutive even integers can be represented as $2n$, $2n + 2$, and $2n + 4$, where n is an integer.

For example: 24, 26, 28

1. a. One number is three more than another. The product of the two numbers is equal to 28. Find the numbers.

b. A subdivision of a house is set up on a grid with rows and columns. There are 5 more houses in each row than in each column. If there are a total of 84 houses in the subdivision, how many houses are in each row?

Consecutive Odd Integers

Odd integers are consecutive if each is 2 more than the previous odd integer. Three consecutive odd integers can be represented as **2n + 1**, **2n + 3**, and **2n + 5**, where *n* is an integer.

For example: 41, 43, 45.

Example 2

Consecutive Integers

a. Find two consecutive positive integers such that the sum of their squares is 265.

Solution

Let n = first integer.
Then $n + 1$ = next consecutive integer.
Set up and solve the related equation.

$$n^2 + (n+1)^2 = 265$$
$$n^2 + n^2 + 2n + 1 = 265$$
$$2n^2 + 2n - 264 = 0$$
$$2(n^2 + n - 132) = 0$$
$$2(n+12)(n-11) = 0$$

$$n + 12 = 0 \quad \text{or} \quad n - 11 = 0$$
$$n = -12 \qquad\qquad n = 11$$
$$n + 1 = -11 \qquad n + 1 = 12$$

Consider the solution $n = -12$. The next consecutive integer, $n + 1$, is -11. While it is true that the sum of their squares is 265, we must remember that the problem calls for **positive** consecutive integers. Therefore, we can only consider positive solutions. Hence, the two integers are 11 and 12.

b. Find three consecutive odd integers such that the product of the first and second is 68 more than the third.

Solution

Let $2n + 1$ = first odd integer
and $2n + 3$ = second consecutive odd integer
and $2n + 5$ = third consecutive odd integer.
Set up and solve the related equation.

$$(2n+1)(2n+3) = (2n+5)+68$$
$$4n^2 + 8n + 3 = 2n + 73$$
$$4n^2 + 6n - 70 = 0$$
$$2(2n^2 + 3n - 35) = 0$$
$$2(2n-7)(n+5) = 0$$

$$2n - 7 = 0 \quad \text{or} \quad n + 5 = 0$$
$$2n = 7 \qquad\qquad n = -5$$
$$n = \frac{7}{2} \qquad 2n + 1 = -9$$
$$2n + 3 = -7$$
$$2n + 5 = -5$$

The three consecutive odd integers are –9, –7, and –5. Note that, in the definition of consecutive odd integers, n must be an integer. Thus $n = \frac{7}{2}$ is not a possible value for n, and –9, –7, and –5 are the only solutions.

Now work margin exercise 2.

2. Find two consecutive negative odd integers such that the sum of their squares is 202.

Objective C ## The Pythagorean Theorem

A geometric topic that often generates quadratic equations is right triangles. In a **right triangle**, one of the angles is a right angle (measures 90°), and the side opposite this angle (the longest side) is called the **hypotenuse**. The other two sides are called **legs**. Pythagoras (c. 570 – 495 B.C.), a famous Greek mathematician, is given credit for proving the following very important and useful theorem (even though history indicates that the Babylonians knew of this theorem centuries before Pythagoras). Now there are entire books written that contain only proofs of the Pythagorean theorem developed by mathematicians since the time of Pythagoras. (You might want to visit the library!)

The Pythagorean Theorem

In a right triangle, if c is the length of the hypotenuse and a and b are the lengths of the legs, then

$$c^2 = a^2 + b^2.$$

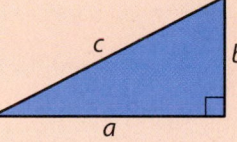

3. A kicker is attempting an extra point after an Eagles touchdown. The ball is held for the kicker 57 feet away from the goal post. The bar of the goal post is 10 feet high. Assuming the ball travels in a straight line when kicked, how far will the ball need to travel to score the extra point? Round your answer accurate to the nearest hundredth.

Example 3

The Pythagorean Theorem

A support wire is 25 feet long and stretches from a tree to a point on the ground. The point of attachment on the tree is 5 feet higher than the distance from the base of the tree to the point of attachment on the ground. How far up the tree is the point of attachment?

Solution

Let x = distance from base of tree to point of attachment on ground; then $x + 5$ = height of point of attachment on tree.
By the Pythagorean theorem, we have the following.

$$(x+5)^2 + x^2 = 25^2$$
$$x^2 + 10x + 25 + x^2 = 625$$
$$2x^2 + 10x - 600 = 0$$
$$2(x^2 + 5x - 300) = 0$$
$$2(x-15)(x+20) = 0$$
$$x = 15 \text{ or } x = -20$$

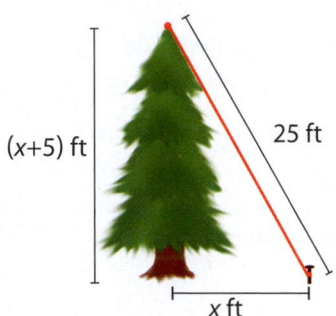

Because distance must be positive, −20 is not a possible solution. The solution is

$$x = 15$$
$$\text{and } x + 5 = 20.$$

Thus the point of attachment is 20 feet up the tree.

Now work margin exercise **3.**

Exercises 7.11

Solve the following word problems.

1. One number is eight more than another. Their product is −16. What are the numbers?

2. One number is 10 more than another. If their product is −25, find the numbers.

3. The square of an integer is equal to seven times the integer. Find the integer.

4. The square of an integer is equal to twice the integer. Find the integer.

5. If the square of a positive integer is added to three times the integer, the result is 28. Find the integer.

6. If the square of a positive integer is added to three times the integer, the result is 54. Find the integer.

7. One number is seven more than another. Their product is 78. Find the numbers.

8. One positive number is three more than twice another. If the product is 27, find the numbers.

9. One positive number is six more than another. The sum of their squares is 260. What are the numbers?

10. One number is five less than another. The sum of their squares is 97. Find the numbers.

11. The difference between two positive integers is 8. If the smaller is added to the square of the larger, the sum is 124. Find the integers.

12. One positive number is 3 more than twice another. If the square of the smaller is added to the larger, the sum is 51. Find the numbers.

13. The product of a negative integer and 5 less than twice the integer equals the integer plus 56. Find the integer.

14. Find a positive integer such that the product of the integer with a number three less than the integer is equal to the integer increased by 32.

15. The product of two consecutive positive integers is 72. Find the integers.

16. Find two consecutive integers whose product is 110.

17. Find two consecutive positive integers such that the sum of their squares is 85.

18. Find two consecutive positive integers such that the sum of their squares is 145.

19. The product of two consecutive odd integers is 63. Find the integers.

20. The product of two consecutive even integers is 168. Find the integers.

21. Find two consecutive positive integers such that the square of the second integer added to four times the first is equal to 41.

22. Find two consecutive negative integers such that 6 times the first plus the square of the second equals 34.

23. Find three consecutive positive integers such that twice the product of the two smaller integers is 88 more than the product of the two larger integers.

24. Find three consecutive odd integers such that the product of the first and third is 71 more than 10 times the second.

25. Four consecutive integers are such that if the product of the first and third is multiplied by 6, the result is equal to the sum of the second and the square of the fourth. What are the integers?

26. Find four consecutive even integers such that the square of the sum of the first and second is equal to 516 more than twice the product of the third and fourth.

27. Rectangles: The length of a rectangle is twice the width. The area is 72 square inches. Find the length and width of the rectangle.

28. Rectangles: The length of a rectangle is three times the width. If the area is 147 square centimeters, find the length and width of the rectangle.

29. Rectangles: The length of a rectangle is four times the width. If the area is 64 square feet, find the length and width of the rectangle.

30. Rectangles: The length of a rectangle is five times the width. If the area is 180 square inches, find the length and width of the rectangle.

31. Rectangles: The width of a rectangle is 4 feet less than the length. The area is 117 square feet. Find the length and width of the rectangle.

32. Rectangles: The length of a rectangular yard is 12 meters greater than the width. If the area of the yard is 85 square meters, find the length and width of the yard.

33. Triangles: The height of a triangle is 4 feet less than the base. The area of the triangle is 16 square feet. Find the length of the base and the height of the triangle.

34. Triangles: The base of a triangle exceeds the height by 5 meters. If the area is 42 square meters, find the length of the base and the height of the triangle.

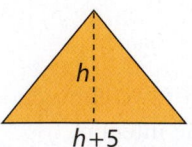

35. Triangles: The base of a triangle is 15 inches greater than the height. If the area is 63 square inches, find the length of the base.

36. Triangles: The base of a triangle is 6 feet less than the height. The area is 56 square feet. Find the height.

37. Rectangles: The perimeter of a rectangle is 32 inches. The area of the rectangle is 48 square inches. Find the dimensions of the rectangle.

38. Rectangles: The area of a rectangle is 24 square centimeters. If the perimeter is 20 centimeters, find the length and width of the rectangle.

39. Orchards: An orchard has 140 apple trees. The number of rows exceeds the number of trees per row by 13. How many trees are there in each row?

40. Military: One formation for a drill team is rectangular. The number of members in each row exceeds the number of rows by 3. If there is a total of 108 members in the formation, how many rows are there?

41. Theater: A theater can seat 144 people. The number of rows is 7 less than the number of seats in each row. How many rows of seats are there?

42. College Sports: An empty field on a college campus is being used for overflow parking for a football game. It currently has 187 cars in it. If the number of rows of cars is six less than the number of cars in each row, how many rows are there?

43. Parking: The parking garage at Baltimore-Washington International Airport contains 8400 parking spaces. The number of cars that can be parked on each floor exceeds the number of floors by 1675. How many floors are there in the parking garage?

44. Library Books: One bookshelf in the public library can hold 175 books. The number of books on each shelf exceeds the number of shelves by 18. How many books are on each shelf?

45. Rectangles: The length of a rectangle is 7 centimeters greater than the width. If 4 centimeters are added to both the length and width, the new area would be 98 square centimeters. Find the dimensions of the original rectangle.

46. Rectangles: The width of a rectangle is 5 meters less than the length. If 6 meters are added to both the length and width, the new area will be 300 square meters. Find the dimensions of the original rectangle.

47. Gardening: Susan is going to fence a rectangular flower garden in her back yard. She has 50 feet of fencing and she plans to use the house as the fence on one side of the garden. If the area is 300 square feet, what are the dimensions of the flower garden?

48. Ranching: A rancher is going to build a corral with 52 yards of fencing. He is planning to use the barn as one side of the corral. If the area is 320 square yards, what are the dimensions?

49. Communication: A telephone pole is to have a guy wire attached to its top and anchored to the ground at a point that is at a distance 34 feet less than the height of the pole from the base. If the wire is to be 2 feet longer than the height of the pole, what is the height of the pole?

50. Trees: Lucy is standing next to the the General Sherman tree in Sequoia National Park, home of some of the largest trees in the world. The distance from Lucy to the base of the tree is 71 m less than the height of the tree. If the distance from Lucy to the top of the tree is 1 m more than the height of the tree, how tall is the General Sherman?

51. Holiday Decorating: A Christmas tree is supported by a wire that is 1 foot longer than the height of the tree. The wire is anchored at a point whose distance from the base of the tree is 49 feet shorter than the height of the tree. What is the height of the tree?

52. Architecture: An architect wants to draw a rectangle with a diagonal of 13 inches. The length of the rectangle is to be 2 inches more than twice the width. What dimensions should she make the rectangle?

53. Gymnastics: Incline mats, or triangle mats, are offered with different levels of incline to help gymnasts learn basic moves. As the name may suggest, two sides of the mat are right triangles. If the height of the mat is 28 inches shorter than the length of the mat and the hypotenuse is 8 inches longer than the length of the mat, what is the length of the mat?

54. Laser Show: Bill uses mirrors to augment the "laser experience" at a laser show. At one show he places three mirrors, A, B, C, in a right triangular form. If the distance between A and B is 15 m more than the distance between A and C, and the distance between B and C is 15 m less than the distance between A and C, what is the distance between mirror A and mirror C?

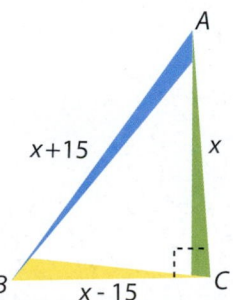

The demand for a product is the number of units of the product, x, that consumers are willing to buy when the market price is p dollars. The consumers' total expenditure for the product, S, is found by multiplying the price times the demand. $(S = px)$

55. During the summer at a local market, a farmer will sell $8p + 588$ pounds of peaches at p dollars per pound. If he sold $900 worth of peaches this summer, what was the price per pound of the peaches?

56. On a hot afternoon, fans at a stadium will buy $490 - 40p$ drinks for p dollars each. If the total sales after a game were $1225, what was the price per drink?

57. When fishing reels are priced at p dollars, local consumers will buy $36 - p$ fishing reels. What is the price if total sales were $320?

58. A manufacturer can sell $100 - 2p$ lamps at p dollars each. If the receipts from the lamps total $1200, what is the price of the lamps?

Writing & Thinking

59. If three positive integers satisfy the Pythagorean theorem, they are called a **Pythagorean triple**. For example, 3, 4, and 5 are a **Pythagorean triple** because $3^2 + 4^2 = 5^2$. There are an infinite number of such triples. To see how some triples can be found, fill out the following table and verify that the numbers in the rightmost three columns are indeed Pythagorean triples.

u	v	$2uv$	$u^2 - v^2$	$u^2 + v^2$
2	1	4	3	5
3	2	12	5	13
5	2			
4	3			
7	1			
6	5			

60. The pattern in Kara's linoleum flooring is in the shape of a square 8 inches on a side with right triangles of sides x inches placed on each side of the original square so that a new larger square is formed. What is the area of the new square? Explain why you do not need to find the value of x.

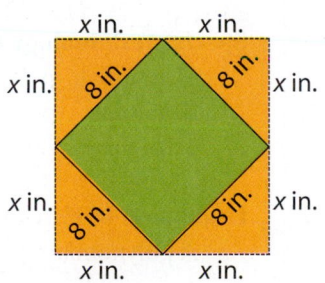

Chapter 7: Index of Key Terms and Ideas

Section 7.1: Exponents

Summary of the Rules for Exponents page 787

For any nonzero real number a and integers m and n:

1. The exponent 1: $a = a^1$

2. The exponent 0: $a^0 = 1 \quad (a \neq 0)$

3. The product rule: $a^m \cdot a^n = a^{m+n}$

4. The quotient rule: $\dfrac{a^m}{a^n} = a^{m-n}$

5. Negative exponents: $a^{-n} = \dfrac{1}{a^n}$

6. Power rule: $\left(a^m\right)^n = a^{mn}$

7. Power of a product: $(ab)^n = a^n b^n$

8. Power of a quotient: $\left(\dfrac{a}{b}\right)^n = \dfrac{a^n}{b^n}$

Section 7.2: Introduction to Polynomials

Monomial page 793

A **monomial in x** is a term of the form kx^n where k is a real number and n is a whole number. n is called the **degree** of the term, and k is called the **coefficient**.

Polynomial page 794

A **polynomial** is a monomial or the indicated sum or difference of monomials. The **degree of a polynomial** is the largest of the degrees of its terms. The coefficient of the term of the largest degree is called the **leading coefficient**.

Special Terminology for Some Polynomials page 794
  Monomial: polynomial with one term
  Binomial: polynomial with two terms
  Trinomial: polynomial with three terms

Evaluation of Polynomials page 796

A polynomial can be treated as a function and the notation $p(x)$ can be used.

Section 7.3: Addition and Subtraction with Polynomials

Addition with Polynomials page 801

The **sum** of two or more polynomials is found by combining **like terms**.

Subtraction with Polynomials pages 802 – 803

The **difference** between two polynomials can be found by changing the sign of each term of the second polynomial and then combining like terms.

Simplifying Algebraic Expressions   page 803

If an algebraic expression contains more than one pair of grouping symbols, such as parentheses ( ), brackets [ ], braces { }, radicals signs $\sqrt{\ }$, absolute value bars | |, or fraction bars, simplify by working to remove the innermost pair of symbols first. **Apply the rules for order of operations** just as if the variables were numbers and proceed to combine like terms.

Section 7.4: Multiplication with Polynomials

Multiplying a Polynomial by a Monomial   page 808

Using the distributive property $a(b+c) = ab + ac$ with multiplication indicated on the left, we can find the product of a monomial with a polynomial of two or more terms.

Multiplying Two Polynomials   pages 808 – 810

The **distributive property** is applied by multiplying each term of one polynomial by each term of the other.

Section 7.5: Special Products of Polynomials

The FOIL Method   pages 815 – 816

Multiply **F**irst Terms, **O**utside Terms, **I**nside Terms, **L**ast Terms

The Difference of Two Squares   page 816

$$(x+a)(x-a) = x^2 - a^2$$

Square of Binomials   pages 817 – 818

$$(x+a)^2 = (x+a)(x+a) = x^2 + 2ax + a^2$$
$$(x-a)^2 = (x-a)(x-a) = x^2 - 2ax + a^2$$

Section 7.6: Greatest Common Factor (GCF) and Factoring by Grouping

Dividing by a Monomial   pages 824 – 825

If we start with a polynomial in the numerator and want to divide the numerator by a monomial in the denominator, we divide each term in the numerator by the monomial denominator and simplify each fraction.

Greatest Common Factor (GCF)   pages 825 – 826

The **greatest common factor (GCF)** of two or more integers is the largest integer that is a factor (or divisor) of all of the integers.

Procedure for Finding the GCF   page 826

1. Find the prime factorization of all integers and integer coefficients.
2. List all the factors that are common to all terms, including variables.
3. Choose the greatest power of each factor common to all terms.
4. Multiply these powers to find the GCF.

Note: If there is no common prime factor or variable, then the GCF is 1.

Factoring Out the GCF   pages 827 – 828

To find a monomial that is the GCF of a polynomial:

1. Find the variable(s) of highest degree and the largest integer coefficient that is a factor of each term of the polynomial. (This is one factor.)
2. Divide this monomial factor into each term of the polynomial resulting, in another polynomial factor.

Factoring by Grouping   pages 831 – 832

To **factor by grouping** look for common factors in each group and then look for common binomial factors.

Factoring Trinomials with Leading Coefficient 1 ($x^2 + bx + c$)   pages 838 – 840

To factor $x^2 + bx + c$, if possible, find an integer pair of factors of c whose sum is b.

1. If c is positive, then both factors must have the same sign.
   a. Both will be positive if b is positive.
   b. Both will be negative if b is negative.
2. If c is negative, then one factor must be positive and the other negative.

Not Factorable (or Prime)   page 841

A polynomial is **not factorable** if it cannot be factored as the product of polynomials with integer coefficients.

Trial-and-Error Method   pages 847 – 848

1. If the sign of the constant term is positive (+), the signs in both factors will be the same, either both positive or both negative.
2. If the sign of the constant term is negative (−), the signs in the factors will be different, one positive and one negative.

Section 7.8: Factoring Trinomials: $ax^2 + bx + c$ (cont.)

The *ac*-Method (Grouping)   pages 849 – 850
1. Multiply $a \cdot c$.
2. Find two integers whose product is ac and whose sum is b. If this is not possible, then the trinomial is **not factorable**.
3. Rewrite the middle term (bx) using the two numbers found in Step 2 as coefficients.
4. Factor by grouping the first two terms and the last two terms.
5. Factor out the common binomial factor. This will give two binomial factors of the trinomial $ax^2 + bx + c$.

Tips to Keep in Mind while Factoring   page 853
1. When factoring polynomials, always look for the greatest common factor first. Then, if there is one, remember to include this common factor as part of the answer.
2. **To factor completely** means to find factors of the polynomial such that none of the factors are themselves factorable.
3. Not all polynomials are factorable. (See $2x^2 + x + 1$ in Example **2c.**) **Any polynomial that cannot be factored as the product of polynomials with integer coefficients is not factorable**.
4. Factoring can be checked by multiplying the factors. The product should be the original expression.

Section 7.9: Special Factoring Techniques

Special Factoring Techniques   page 856
1. $x^2 - a^2 = (x + a)(x - a)$:   difference of two squares
2. $x^2 + 2ax + a^2 = (x + a)^2$:   perfect square trinomial
3. $x^2 - 2ax + a^2 = (x - a)^2$:   perfect square trinomial
4. $x^3 + a^3 = (x + a)(x^2 - ax + a^2)$:   sum of two cubes
5. $x^3 - a^3 = (x - a)(x^2 + ax + a^2)$:   difference of two cubes

Sum of Two Squares   pages 857 – 858
The **sum of two squares** is an expression of the form $x^2 + a^2$ and is **not factorable**.

Section 7.10: Solving Quadratic Equations by Factoring

Quadratic Equations   page 865
Quadratic equations are equations that can be written in the form $ax^2 + bx + c = 0$ where a, b, and c are real numbers and $a \neq 0$.

Zero-Factor Property   page 865
If the product of two (or more) factors is 0, then at least one of the factors must be 0. That is, if a and b are real numbers, then if $a \cdot b = 0$, then $a = 0$ or $b = 0$ or both.

Section 7.10: Solving Quadratic Equations by Factoring

Solving Quadratic Equations by Factoring   pages 867 – 869
   1. Add or subtract terms as necessary so that 0 is on one side of the equation and the equation is in the standard form $ax^2 + bx + c = 0$, where $a, b,$ and c are real numbers and $a \neq 0$.
   2. Factor completely.
   3. Set each nonconstant factor equal to 0 and solve each linear equation for the unknown.
   4. Check each solution, one at a time, in the original equation.

Finding an Equation Given the Roots   page 870
   1. Rewrite the linear equations with 0 on one side.
   2. Form the product of the factors and set this product equal to 0:
   3. Multiply the factors. The resulting quadratic equation must have the two given roots (because we know the factors):

Factor Theorem   page 871
   If $x = c$ is a root of a polynomial equation in the form $P(x) = 0$, then $x - c$ is a factor of the polynomial $P(x)$.

Section 7.11: Applications of Quadratic Equations

Attack Plan for Word Problems   page 875
   1. Read the problem carefully at least twice.
   2. Decide what is asked for and assign a variable or variable expression to the unknown quantities.
   3. Organize a chart, table, or diagram relating all the information provided.
   4. Form an equation. (A formula of some type may be necessary.)
   5. Solve the equation.
   6. Check your solution with the wording of the problem to be sure it makes sense.

Consecutive Integers   page 878
   Integers are **consecutive** if each is 1 more than the previous integer. Three consecutive integers can be represented as **n, $n + 1$**, and **$n + 2$**.

Consecutive Even Integers   page 878
   Even integers are consecutive if each is 2 more than the previous even integer. Three consecutive even integers can be represented as **$2n$, $2n + 2$**, and **$2n + 4$**, where n is an integer.

Consecutive Odd Integers   page 879
   Odd integers are consecutive if each is 2 more than the previous odd integer. Three consecutive odd integers can be represented as **$2n + 1$**, **$2n + 3$**, and **$2n + 5$**, where n is an integer.

The Pythagorean Theorem   page 880
   In a right triangle, if c is the length of the hypotenuse and a and b are the lengths of the legs, then $c^2 = a^2 + b^2$.

Chapter 7: Test

Use the rules for exponents to simplify each expression. Each answer should have only positive exponents. Assume that all variables represent nonzero numbers.

1. $\left(5a^2b^5\right)\left(-2a^3b^{-5}\right)$

2. $\left(-7x^4y^{-3}\right)^0$

3. $\dfrac{\left(-8x^2y^{-3}\right)^2}{16xy}$

4. $\dfrac{3x^{-2}}{9x^{-3}y^2}$

5. $\left(\dfrac{4xy^2}{x^3}\right)^{-1}$

6. $\left(\dfrac{2x^0y^3}{x^{-1}y}\right)^2$

Simplify each polynomial. Write the polynomial in descending order and state the degree, type, and leading coefficient of the simplified polynomial.

7. $3x+4x^2-x^3+4x^2+x^3$

8. $2x^2+3x-x^3+x^2-1$

9. $17-14+4x^5-3x+2x^4+x^5-8x$

10. For the polynomial $P(x)=2x^3-5x^2+7x+10$, find

   a. $P(2)$

   b. $P(-3)$

Simplify the following expressions.

11. $7x+\left[2x-3(4x+1)+5\right]$

12. $12x-2\left[5-(7x+1)+3x\right]$

Perform the indicated operations and simplify each expression. Tell which, if any, answers are the difference of two squares and which, if any, are perfect square trinomials.

13. $\left(5x^3-2x+7\right)+\left(-x^2+8x-2\right)$

14. $\left(x^4+3x^2+9\right)-\left(-6x^4-11x^2+5\right)$

15. $5x^2\left(3x^5-4x^4+3x^3-8x^2-2\right)$

16. $(7x+3)(7x-3)$

17. $(4x+1)^2$

18. $(6x-5)(6x-5)$

19. $(2x+5)(6x-3)$

20. $3x(x-7)(2x-9)$

21. $(3x+1)(3x-1)-(2x+3)(x-5)$

22.
$$\begin{array}{r} 2x^3 - 3x - 7 \\ \times \quad\quad 5x + 2 \\ \hline \end{array}$$

Express each quotient as a sum (or difference) of fractions and simplify if possible.

23. $\dfrac{4x^3 + 3x^2 - 6x}{2x^2}$

24. $\dfrac{5a^3 + 6a^4 + 3a}{3a^2}$

25. Subtract $5x^2 - 3x + 4$ from the sum of $4x^2 + 2x + 1$ and $3x^2 - 8x - 10$.

Solve the following word problem.

26. Rectangles: A sheet of metal is in the shape of a rectangle with width 12 inches and length 20 inches. A slot (see the figure) of width x inches and length $x + 3$ inches is cut from the top of the rectangle.

  a. Write a polynomial function $A(x)$ that represents the area of the remaining figure.

  b. Write a polynomial function $P(x)$ that represents the perimeter of the remaining figure.

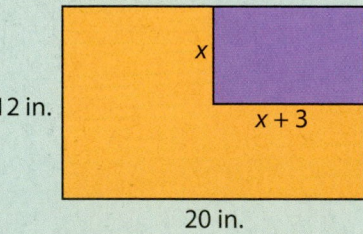

Factor each of the polynomials by finding the GCF (or $-1 \cdot$ GCF).

27. $28ab^2x - 21ab^2y$

28. $18yz^3 - 6y^2z^3 + 12yz^2$

Completely factor each of the given polynomials. If the polynomial cannot be factored, write "not factorable."

29. $x^2 - 9x + 20$

30. $-x^2 - 14x - 49$

31. $xy - 7x + 35 - 5y$

32. $6x^2 - 6$

33. $12x^2 + 2x - 10$

34. $3x^2 + x - 24$

35. $16x^2 - 25y^2$

36. $2x^3 - x^2 - 3x$

37. $6x^2 - 13x + 6$

38. $2xy - 3y + 14x - 21$

39. $4x^2 + 25$

40. $-3x^3 + 6x^2 - 6x$

Solve the equations.

41. $x^2 - 7x - 8 = 0$

42. $-3x^2 = 18x$

43. $0 = 4x^2 - 17x - 15$

44. $(2x - 7)(x + 1) = 6x - 19$

Solve the following word problems.

45. Find a polynomial equation with integer coefficients that has $x = 3$ and $x = -8$ as solutions.

46. Find a polynomial equation with integer coefficients that has $x = \dfrac{1}{2}$ as a double root.

47. One number is 10 less than five times another number. Their product is 120. Find the numbers.

48. The difference between two positive numbers is 9. If the smaller is added to the square of the larger, the result is 147. Find the numbers.

49. The product of two consecutive positive integers is 342. Find the two integers.

50. **Rectangles:** The length of a rectangle is 7 centimeters less than twice the width. If the area of the rectangle is 165 square centimeters, find the length and width.

51. **Architecture:** The average staircase has steps with a width that is 6 cm more than the height and a diagonal that is 12 cm more than the height. How tall is the average step?

52. **Squares:** The area of a square can be represented by the function $A(x) = 9x^2 + 30x + 25$. Write a polynomial function $P(x)$ that represents the perimeter of the square. (**Hint:** Factor the expression to find the length and the width.)

Cumulative Review: Chapters 1 - 7

Find the LCM for each set of terms.

1. $\{20, 12, 24\}$

2. $\{8x^2, 14x^2y, 21xy\}$

Find the value of each expression by using the rules for order of operations.

3. $2 \cdot 3^2 \div 6 \cdot 3 - 3$

4. $6 + 3\left[4 - 2\left(3^3 - 1\right)\right]$

Perform the indicated operation. Reduce all answers to lowest terms.

5. $\dfrac{7}{12} + \dfrac{9}{16}$

6. $\dfrac{11}{15a} - \dfrac{5}{12a}$

7. $\dfrac{6x}{25} \cdot \dfrac{5}{4x}$

8. $\dfrac{40}{92} \div \dfrac{2}{15x}$

Solve each of the equations.

9. $4(2x - 3) + 2 = 5 - (2x + 6)$

10. $\dfrac{4x}{7} - 3 = 9$

11. $\dfrac{2x + 3}{6} - \dfrac{x + 1}{4} = 2$

Determine whether each of the following equations is a conditional equation, an identity, or a contradiction.

12. $4(x + 2) + 5x - 5 = 2 - (x + 4)$

13. $5(3x - 2) + 1 = 9(x - 1) + 6x$

14. $4(2x + 4) - 3 = 7(x + 2) + x$

Solve each formula for the indicated variable.

15. $y = mx + b$; solve for x.

16. $3x + 5y = 10$; solve for y.

Solve the following inequalities and graph the solution sets. Write each solution in interval notation. Assume that x is a real number.

17. $5x + 3 \geq 2x - 15$

18. $-16 < 3x + 5 < 17$

19. Write the equation $3x + 7y = -14$ in slope-intercept form. Find the slope and the y-intercept, and then use them to draw the graph.

20. Find the equation for the line passing through the point $(-4, 7)$ with slope 0. Graph the line.

21. Find the equation in standard form of the line determined by the two points $(-5, 3)$ and $(2, -4)$. Graph the line.

22. Find the equation in slope-intercept form of the line parallel to the line $y = 3x - 7$ and passing through the point $(-1, 1)$. Graph both lines.

23. Find the equation in slope-intercept form of the line perpendicular to the line $y = 3x - 7$ and passing through the point $(-1, 1)$. Graph both lines.

24. Given the relation $r = \{(2, -3), (3, -2), (5, 0), (7.1, 3.2)\}$.

    a. Graph the relation.

    b. State the domain of the relation.

    c. State the range of the relation.

    d. Is the relation a function? Explain.

Use the vertical line test to determine whether each of the graphs does or does not represent a function. State the domain and range.

25.
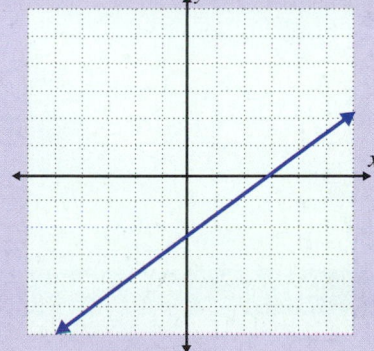

26.
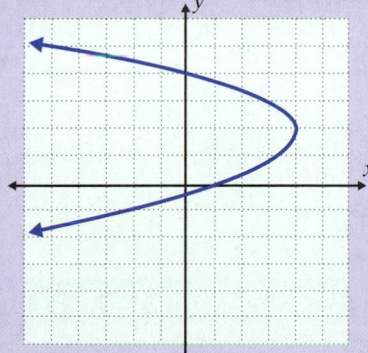

27. For the function $f(x) = x^2 - 3x + 4$, find

a. $f(-6)$ **b.** $f(0)$ **c.** $f\left(\dfrac{1}{2}\right)$

Graph the inequalities.

28. $y > 6x + 2$

29. $3x + 2y \geq 10$

Use the properties of exponents to simplify each of the expressions. Answers should contain only positive exponents. Assume that all variables represent nonzero real numbers.

30. $\left(4x^2 y\right)^3$

31. $\left(-2x^3 y^2\right)^{-3}$

32. $\dfrac{21xy^2}{3x^{-1}y}$

33. $\left(\dfrac{6x^2}{y^5}\right)^2$

34. $\dfrac{\left(xy^0\right)^3}{\left(x^3 y^{-1}\right)^2}$

35. $\left(\dfrac{x^2 y^{-2}}{3y^5}\right)^{-2}$

36. Write each number in the following expression in scientific notation and simplify. Show the steps you use. Do not use a calculator. Leave the answers in scientific notation.

     a. 0.00000056×0.0003

     b. $\dfrac{81,000 \times 6200}{0.003 \times 0.2}$

Perform the indicated operations and simplify by combining like terms.

37. $2(4x + 3) + 5(x - 1)$

38. $2x(x + 5) - (x + 1)(x - 3)$

39. $\left(2x^2 + 6x - 7\right) + \left(2x^2 - x - 1\right)$

40. $(x + 1) - \left(4x^2 + 3x - 2\right)$

41. $(2x - 7)(x + 4)$

42. $-(x + 6)(3x - 1)$

43. $(x + 6)^2$

44. $(2x - 7)(2x - 7)$

45. Express the following quotient as a sum of fractions and simplify, if possible.

$$\dfrac{8x^2 y^2 - 5xy^2 + 4xy}{4xy^2}$$

Factor each expression as completely as possible.

46. $8x - 20$

47. $6x - 96$

48. $xy + 3x + 2y + 6$

49. $ax - 2a - 2b + bx$

50. $x^2 - 9x + 18$

51. $6x^2 - x - 12$

52. $-12x^2 - 16x$

53. $16x^2 y - 24xy$

54. $5x^4 + 15x^3 - 200x^2$ **55.** $4x^2 - 1$ **56.** $3x^2 - 48y^2$ **57.** $x^2 + x + 3$

58. $3x^2 + 5x + 2$ **59.** $2x^3 - 20x^2 + 50x$ **60.** $4x^3 + 100x$ **61.** $5x^3 - 135y^3$

Solve the equations.

62. $(x-7)(x+1) = 0$ **63.** $21x - 3x^2 = 0$ **64.** $0 = x^2 + 8x + 12$

65. $x^2 = 3x + 28$ **66.** $x^3 + 14x^2 + 49x = 0$ **67.** $8x = 12x + 2x^2$

68. $(4x-3)(x+1) = 2$ **69.** $2x(x+5)(x-2) = 0$

Solve the following problems.

70. Find an equation that has $x = -10$ and $x = -5$ as roots.

71. Find an equation that has $x = 0$, $x = 4$ and $x = 13$ as roots.

 72. What is the percent of profit if:

   a. $520 is made on an investment of $4000?

   b. $385 is made on an investment of $3500?

   c. Which was the better investment based on return rate?

73. **Traveling by Car:** For winter break, Lindsey and Sloan decided to make the 510 mile trip from Charlotte, NC to Philadelphia, PA together. Sloan drove first at a mean speed of 63 mph for 4.5 hours. Lindsey drove for 3.5 hours. What was Lindsey's mean speed? Please round to the nearest mph.

74. **Investing:** Karl invested a total of $10,000 in two separate accounts. One account paid 6% interest and the other paid 8% interest. If the annual income from both accounts was $650, how much did he have invested in each account?

75. **Rectangles:** The perimeter of a rectangle is 60 inches. The area of the rectangle is 221 square inches. Find the dimensions of the rectangle. (**Hint:** After substituting and simplifying, you will have a quadratic equation.)

76. **Number Problem:** The difference between two positive numbers is 7. If the square of the smaller is added to the square of the larger, the result is 137. Find the numbers.

77. **Consecutive Integers:** Find two consecutive integers such that the sum of their squares is equal to 145.

78. **Water Slides:** A pool's straight water slide is 8 m longer than the height of the slide. If the distance from the ladder to the end of the slide is 1 m shorter than the length, what is the height of the slide?

Mathematics @ Work!

Many times, one equation does not provide enough information for us to find the best solution. When this occurs, we must construct more than one equation in more than one variable, and use information from both to find an appropriate solution. These sets of equations are called systems. The Chinese recorded methods for solving systems of two equations in two variables as early as 200 A.D., but it was not until the 18th and 19th centuries that European mathematicians Gabriel Cramer, Carl Frederick Gauss, and Wilhelm Jordan introduced methods for solving systems of equations to the Western world. Thanks in large part to their work, algorithms have now been derived for calculating solutions to systems with infinitely many equations and variables. With the invention of computers, the time it took to solve these systems was reduced drastically and now solving systems of equations plays a prominent role in engineering, physics, chemistry, computer science, and economics. The following scenario depicts a problem that must be solved using more than one equation and variable.

A farmer is building a fenced-in pasture for his cows, and has 900 feet of fencing to use. The farmer wants the length of the pasture to be twice the width. What should the dimensions of the pasture be? Recall that the perimeter of a rectangle can be found by using the equation $P = 2l + 2w$.

For more examples like these, see Section 8.4.

Systems of Linear Equations

8.1 Systems of Linear Equations: Solutions by Graphing

Many applications involve two (or more) quantities, and by using two (or more) variables, we can form linear equations using the information given. Such a set of equations is called a **system**, and in this chapter we will develop techniques for solving systems of linear equations.

Graphing systems of two equations in two variables is helpful in visualizing the relationships between the equations. However, this approach is somewhat limited in finding solutions since numbers might be quite large, or solutions might involve fractions or decimals that must be estimated on the graph. Therefore, algebraic techniques are necessary to accurately solve systems of linear equations. Graphing in three dimensions will be left to later courses.

Two ideas probably new to you are matrices and determinants. Matrices and determinants provide powerful general approaches to solving large systems of equations with many variables. Discussions in this chapter will be restricted to two linear equations in two variables and three linear equations in three variables.

Objective A Systems of Linear Equations

Determine if given points lie on both lines in a specified system of equations.

Many applications, as we will see in Sections 8.4 and 8.5, involve solving pairs of linear equations. Such pairs are said to form **systems of equations** or **sets of simultaneous equations**. For example,

$$\begin{cases} x - 2y = 0 \\ 3x + y = 7 \end{cases} \quad \text{and} \quad \begin{cases} 2x - 5y = 6 \\ 4x - 10y = -2 \end{cases}$$

are two systems of linear equations.

Each individual equation has an infinite number of solutions. That is, there are an infinite number of ordered pairs that satisfy each equation. But, we are interested in finding ordered pairs that satisfy **both** equations.

A **solution of a system** of linear equations is an ordered pair (or point) that satisfies **both** equations. To determine whether or not a particular ordered pair is a solution to a system, substitute the values for x and y in **both** equations. If the results for both equations are true statements, then the ordered pair is a solution to the system.

Example 1

Solution of a System

Show that $(2, 1)$ is a solution to the system $\begin{cases} x - 2y = 0 \\ 3x + y = 7 \end{cases}$.

Solution

Substitute $x = 2$ and $y = 1$ into **both** equations. In the first equation:

$$2 - 2(1) \overset{?}{=} 0$$
$$2 - 2 = 0 \quad \text{a true statement}$$

In the second equation:

$$3(2) + 1 \overset{?}{=} 7$$
$$6 + 1 = 7 \quad \text{a true statement}$$

Because $(2, 1)$ satisfies both equations, $(\mathbf{2, 1})$ **is a solution to the system**.

Now work margin exercise 1.

1. Show that $(6,4)$ is a solution to the following system.
$$\begin{cases} x - 2y = -2 \\ 3x + 2y = 26 \end{cases}$$

Example 2

Not a Solution of a System

Show that $(0, 4)$ is not a solution to the system $\begin{cases} y = -x + 4 \\ y = 2x + 1 \end{cases}$.

Solution

Substitute $x = 0$ and $y = 4$ into **both** equations. In the first equation:

$$4 \overset{?}{=} -(0) + 4$$
$$4 = 0 + 4 \quad \text{a true statement}$$

In the second equation:

$$4 \overset{?}{=} 2(0) + 1$$
$$4 = 0 + 1 \quad \text{a false statement}$$

Because $(0, 4)$ does not satisfy **both** equations, $(\mathbf{0, 4})$ **is NOT a solution to the system**.

Now work margin exercise 2.

2. Show that $(3,2)$ is not a solution to the following system.
$$\begin{cases} y = -2x + 8 \\ y = 3x - 8 \end{cases}$$

Objective B   **Solving Systems of Linear Equations by Graphing**

The following questions concerning systems of equations need to be addressed:

     How do we find the solution, if there is one?

     Will there always be a solution to a system of linear equations?

     Can there be more than one solution?

In this chapter, we will discuss four methods (graphing, substitution, addition, and using matrices) for solving a system of linear equations. The first method is to **solve by graphing**. In this method, both equations are graphed and the point of intersection (if there is one) is the solution to the system. Table 1 illustrates the three possibilities for a system of two linear equations. Each system will be one of the following:

1. **consistent**   (has exactly one solution)
2. **inconsistent**   (has no solution)
3. **dependent**   (has an infinite number of solutions)

System	Graph	Intersection	Classification
$\begin{cases} 2x + y = 5 \\ x - y = 1 \end{cases}$		There is exactly one solution: $(2, 1)$ (The lines intersect at one point.)	consistent
$\begin{cases} 3x - 2y = 2 \\ 6x - 4y = -4 \end{cases}$		There is no solution. (The lines are parallel.)	inconsistent
$\begin{cases} 2x - 4y = 6 \\ x - 2y = 3 \end{cases}$		There are an infinite number of solutions: every ordered pair that satisfies $x - 2y = 3$. (The lines are the same line.)	dependent

Table 1

To Solve a System of Linear Equations by Graphing

1. Graph both linear equations on the same set of axes.
2. Observe the point of intersection (if there is one).
   a. If the slopes of the two lines are different, then the lines intersect in one and only one point. The system has a single point as its solution.
   b. If the lines are distinct and have the same slope, then the lines are parallel. The system has no solution.
   c. If the lines are the same line, then all the points on the line constitute the solution. There is an infinite number of solutions.
3. Check the solution (if there is one) in both of the original equations.

Solving by graphing can involve estimating the solutions whenever the intersection of the two lines is at a point not represented by a pair of integers. (There is nothing wrong with this technique. Just be aware that at times it can lack accuracy. See Example 6.)

You might want to review graphing lines in an earlier chapter. Recall that lines can be graphed by plotting intercepts (as illustrated in Example 3). Other methods include finding a point on the line and using the slope to find another point (see Examples 4 and 5) or finding two points (see Example 6).

Example 3

Solving a System of Linear Equations by Graphing

Solve the following system of linear equations by graphing:
$$\begin{cases} x + y = 6 \\ y = x + 4 \end{cases}$$

Solution

The two lines intersect at the point $(1, 5)$. Thus the solution to the system is $(1, 5)$ or $x = 1$ and $y = 5$.

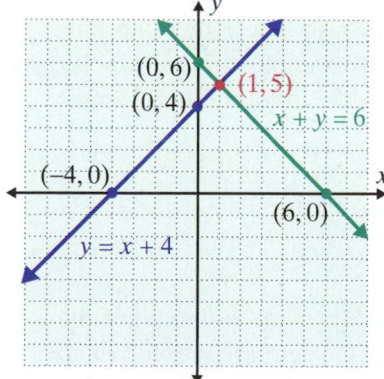

Check: Substitution shows that $(1, 5)$ satisfies **both** of the equations in the system.

$$x + y = 6 \qquad y = x + 4$$
$$\overset{?}{(1) + (5) = 6} \qquad \overset{?}{(5) = (1) + 4}$$
$$6 = 6 \qquad 5 = 5$$

Example 4

System with No Solution

Solve the following system of linear equations by graphing:

$$\begin{cases} y = 3x \\ y - 3x = -4 \end{cases}$$

Solution

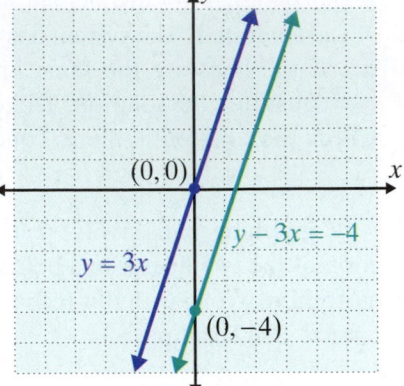

The lines are parallel with the same slope, 3, and there are no points of intersection.

Thus the system is inconsistent, and there is no solution to the system.

Example 5

A System with Infinitely Many Solutions

Solve the following system of linear equations by graphing:

$$\begin{cases} x + 2y = 6 \\ y = -\dfrac{1}{2}x + 3 \end{cases}$$

Solution

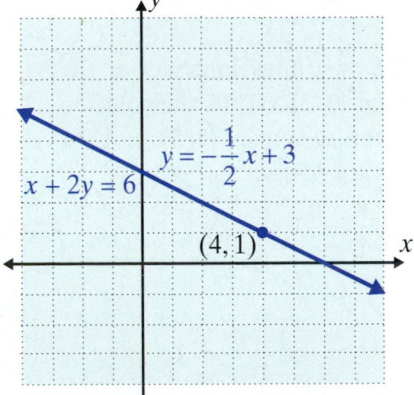

All points that lie on one line also lie on the other line. For example, $(4, 1)$ is a point on the line $x + 2y = 6$ since $4 + 2(1) = 6$. The point $(4, 1)$ is also on the line $y = -\dfrac{1}{2}x + 3$ since $1 = -\dfrac{1}{2}(4) + 3$.

Because all points that lie on the line are a solution to the system, we say that the system is dependent and that there are infinitely many solutions to the system.

Example 6

A System that Requires Estimation

Solve the following system of linear equations by graphing:

$$\begin{cases} x - 3y = 4 \\ 2x + y = 3 \end{cases}$$

Solution

The two lines intersect at one point. However, we can only estimate the point of intersection as $\left(2, -\dfrac{1}{2}\right)$. In this situation be aware that, although graphing gives a good "estimate," finding exact solutions to the system is not likely.

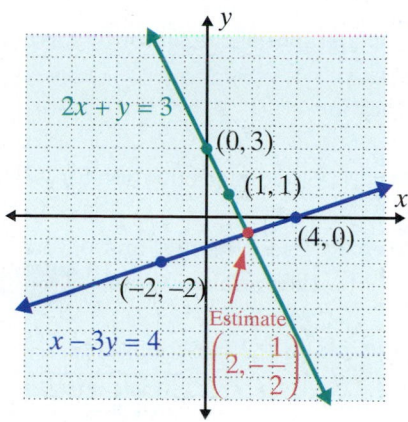

Check: Substitute $x = 2$ and $y = -\dfrac{1}{2}$.

$$2 - 3\left(-\frac{1}{2}\right) \overset{?}{=} 4 \qquad \text{and} \qquad 2(2) + \left(-\frac{1}{2}\right) \overset{?}{=} 3$$

$$\frac{7}{2} \neq 4 \qquad\qquad\qquad\qquad \frac{7}{2} \neq 3$$

Thus checking shows that the estimated solution $\left(2, -\dfrac{1}{2}\right)$ does not satisfy either equation. The estimated point of intersection is just that, an estimate. The following discussion provides a technique based on using of a graphing calculator that would give the exact solution as $\left(\dfrac{13}{7}, -\dfrac{5}{7}\right)$.

Now work margin exercises 3 through 6.

notes

1. To use the graphing method, graph the lines as accurately as you can.
2. Be sure to check your solution by substituting it back into both of the original equations. (Of course, fractional estimates may not check exactly.)

7. Use a graphing calculator to solve the following system.

$$\begin{cases} x + y = -5 \\ x - 3y = -1 \end{cases}$$

A graphing calculator can be used to locate (or estimate) the point of intersection of two lines (and therefore the solution to the system).

Example 7

 Using a Graphing Calculator

Use a graphing calculator to solve the system $\begin{cases} 2x + y = 8 \\ x - y = 1 \end{cases}$.

Solution

Step 1: Solve each equation for *y*. This system becomes $\begin{cases} y = -2x + 8 \\ y = x - 1 \end{cases}$.

Step 2: Press <kbd>Y=</kbd> and enter the two expressions for *y*. To get the variable *X*, press <kbd>X,T,θ,n</kbd>. The display screen will appear as follows.

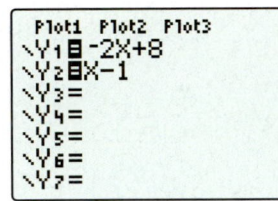

Step 3: Press <kbd>GRAPH</kbd>. (Both lines should appear. If not, you may need to adjust the <kbd>WINDOW</kbd>. For help, refer to Section 6.6).

Step 4: Press <kbd>2ND</kbd> and CALC. Select 5: intersect. The cursor will appear on one of the lines. Use the right or left arrow to get near the point of intersection and press <kbd>ENTER</kbd>.

Then move the up or down arrow to get to the other line.

Now use the right or left arrow to move closer to the point of intersection on this line and press <kbd>ENTER</kbd>. Follow the directions for Guess? by moving the cursor to the point of intersection and pressing <kbd>ENTER</kbd>.

Step 5: The answer *x* = 3 and *y* = 2 will appear at the bottom of the display screen.

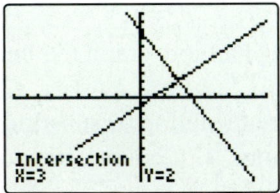

Note: Step 4 may seem somewhat complicated, but TRY IT. It is fun and accurate! (**Note:** If the lines are parallel (an inconsistent system) the calculator will give an error message.)

Now work margin exercise 7.

Practice Problems

1. Determine which of the given points, if any, lie on both of the lines in the given system of equations by substituting each point into both equations.

$$\begin{cases} x + 2y = 8 \\ 2x - y = 1 \end{cases}$$

   a. $(8, 0)$ **b.** $(2, 3)$ **c.** $(0, -1)$ **d.** $(4, 2)$

2. Show that the following system of equations is inconsistent by determining the slope of each line and the y-intercept. Explain your reasoning.

$$\begin{cases} 3x + y = 1 \\ 6x + 2y = 7 \end{cases}$$

3. Solve the following system graphically.

$$\begin{cases} y = 2x + 5 \\ 4x - 2y = -10 \end{cases}$$

Practice Problem Answers

1. b.

2. $m_1 = -3, b_1 = 1, m_2 = -3, b_2 = \dfrac{7}{2}$. The lines do not intersect because they are parallel, they have the same slope, but different y-intercepts.

3. The system is dependent and there are infinitely many solutions.

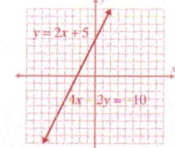

Exercises 8.1

Determine which of the given points, if any, lie on both of the lines in the systems of equations by substituting each point into both equations. See Examples 1 and 2.

1. $\begin{cases} x - y = 6 \\ 2x + y = 0 \end{cases}$

    a. $(1, -2)$
    b. $(4, -2)$
    c. $(2, -4)$
    d. $(-1, 2)$

2. $\begin{cases} x + 3y = 5 \\ 3y = 4 - x \end{cases}$

    a. $(2, 1)$
    b. $(2, -2)$
    c. $(-1, 2)$
    d. $(4, 0)$

3. $\begin{cases} 2x + 4y - 6 = 0 \\ 3x + 6y - 9 = 0 \end{cases}$

    a. $(1, 1)$
    b. $(2, 0)$
    c. $\left(0, \dfrac{3}{2}\right)$
    d. $(-1, 3)$

4. $\begin{cases} 5x - 2y - 5 = 0 \\ 5x = -3y \end{cases}$

    a. $(1, 0)$
    b. $\left(\dfrac{3}{5}, -1\right)$
    c. $(0, 0)$
    d. $(1, 4)$

The graphs of the lines represented by each system of equations are given. Determine the solution of the system by looking at the graph. Check your solution by substituting into both equations.

5. $\begin{cases} x + 2y = 4 \\ x - y = -2 \end{cases}$

6. $\begin{cases} x + 2y = 1 \\ 2x + y = -1 \end{cases}$

7. $\begin{cases} 2x - y = 6 \\ 3x + y = 14 \end{cases}$

8. $\begin{cases} x - y = 4 \\ 3x - y = 6 \end{cases}$

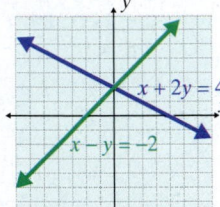
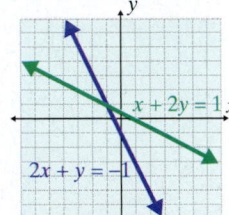
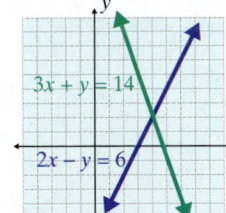
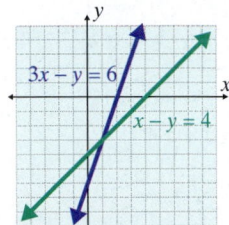

Show that each system of equations is inconsistent by determining the slope of each line and the *y*-intercept. (That is, show that the lines are parallel and do not intersect.)

9. $\begin{cases} 2x + y = 3 \\ 4x + 2y = 5 \end{cases}$

10. $\begin{cases} 3x - 5y = 1 \\ 6x - 10y = 4 \end{cases}$

11. $\begin{cases} y = \dfrac{1}{2}x + 3 \\ x - 2y = 1 \end{cases}$

12. $\begin{cases} 3x - y = 8 \\ x - \dfrac{1}{3}y = 2 \end{cases}$

Solve the following systems graphically. Classify each system as consistent, inconsistent, or dependent. See Examples 3 through 5.

13. $\begin{cases} y = x + 1 \\ y + x = -5 \end{cases}$

14. $\begin{cases} y = 2x + 5 \\ 4x - 2y = 7 \end{cases}$

15. $\begin{cases} 2x - y = 4 \\ 3x + y = 6 \end{cases}$

16. $\begin{cases} 2x + 3y = 6 \\ 4x + 6y = 12 \end{cases}$

17. $\begin{cases} x + y - 5 = 0 \\ \quad x = 5 - y \end{cases}$

18. $\begin{cases} y = 4x - 3 \\ x = 2y - 8 \end{cases}$

19. $\begin{cases} 3x - y = 6 \\ \quad y = 3x \end{cases}$

20. $\begin{cases} \quad y = 2x \\ 2x + y = 4 \end{cases}$

21. $\begin{cases} x - y = 5 \\ \quad x = -3 \end{cases}$

22. $\begin{cases} x - 2y = 4 \\ \quad x = 4 \end{cases}$

23. $\begin{cases} x + 2y = 8 \\ 3x - 2y = 0 \end{cases}$

24. $\begin{cases} 2x + y = 0 \\ 4x + 2y = -8 \end{cases}$

25. $\begin{cases} 5x - 4y = 5 \\ \quad 8y = 10x - 10 \end{cases}$

26. $\begin{cases} x + y = 8 \\ \quad 5y = 2x + 5 \end{cases}$

27. $\begin{cases} 4x + 3y + 7 = 0 \\ 5x - 2y + 3 = 0 \end{cases}$

28. $\begin{cases} 4x - 2y = 10 \\ -6x + 3y = -15 \end{cases}$

29. $\begin{cases} x = 5 \\ y = -1 \end{cases}$

30. $\begin{cases} y = 7 \\ x = 8 \end{cases}$

31. $\begin{cases} \dfrac{1}{2}x + 2y = 7 \\ \quad 2x = 4 - 8y \end{cases}$

32. $\begin{cases} 4x + y = 6 \\ 2x + \dfrac{1}{2}y = 3 \end{cases}$

33. $\begin{cases} 7x - 2y = 1 \\ \quad y = 3 \end{cases}$

34. $\begin{cases} \quad x = 1.5 \\ x - 3y = 9 \end{cases}$

35. $\begin{cases} \quad y = \dfrac{1}{2}x + 2 \\ x - 2y + 4 = 0 \end{cases}$

36. $\begin{cases} 2x - 5y = 6 \\ y = \dfrac{2}{5}x + 1 \end{cases}$

37. $\begin{cases} 2x + 3y = 5 \\ 3x - 2y = 1 \end{cases}$

38. $\begin{cases} \dfrac{2}{3}x + y = 2 \\ x - 4y = 3 \end{cases}$

39. $\begin{cases} x - y = 4 \\ 2y = 2x - 4 \end{cases}$

40. $\begin{cases} x + y = 4 \\ 2x - 3y = 3 \end{cases}$

41. $\begin{cases} \dfrac{1}{2}x + \dfrac{1}{3}y = \dfrac{1}{6} \\ \dfrac{1}{4}x + \dfrac{1}{4}y = 0 \end{cases}$

42. $\begin{cases} \dfrac{1}{4}x - y = \dfrac{13}{4} \\ \dfrac{1}{3}x + \dfrac{1}{6}y = -\dfrac{1}{6} \end{cases}$

Each of the following application problems has been modeled using a system of equations. Solve the system graphically.

43. The sum of two numbers is 25 and their difference is 15. What are the two numbers?

Let x = larger number and y = the other number.

The corresponding modeling system is $\begin{cases} x + y = 25 \\ x - y = 15 \end{cases}$

44. Rectangles: The perimeter of a rectangle is 50 m and the length is 5 m longer than the width. Find the dimensions of the rectangle.

Let x = the length and y = the width.

The corresponding modeling system is $\begin{cases} 2x + 2y = 50 \\ x - y = 5 \end{cases}$

45. Swimming Pools: OSHA recommends that swimming pool owners clean their pool decks with a solvent composed of a 12% chlorine solution and a 3% chlorine solution. Fifteen gallons of the solvent consists of 6% chlorine. How much of each of the mixing solutions were used?

Let x = the number of gallons of the 12% solution

and y = the number of gallons of the 3% solution.

The corresponding modeling system is $\begin{cases} x + y = 15 \\ 0.12x + 0.03y = 0.06(15) \end{cases}$

46. School Supplies: A student bought a calculator and a textbook for a course in algebra. He told his friend that the total cost was $170 (without tax) and that the calculator cost $20 more than twice the cost of the textbook. What was the cost of each item?

Let x = the cost of the calculator

and y = the cost of the textbook.

The corresponding modeling system is $\begin{cases} x + y = 170 \\ x = 2y + 20 \end{cases}$

Use a graphing calculator and the CALC and 5:intersect commands to find the solutions to the given systems of linear equations. If necessary, round values to four decimal places. (Remember to solve each equation for *y*. Use both Y1 and Y2 in the [Y=] menu.) See Example 7.

47. $\begin{cases} x + 2y = 9 \\ x - 2y = -7 \end{cases}$

48. $\begin{cases} x - 3y = 0 \\ 2x + y = 7 \end{cases}$

49. $\begin{cases} y = 2 \\ 2x - 3y = -3 \end{cases}$

50. $\begin{cases} 2x - 3y = 0 \\ 3x + 3y = \dfrac{5}{2} \end{cases}$

51. $\begin{cases} y = -3 \\ 2x + y = 0 \end{cases}$

52. $\begin{cases} 2x - 3y = 1.25 \\ x + 2y = 5 \end{cases}$

53. $\begin{cases} x + y = 3.5 \\ -2x + 5y = 7.7 \end{cases}$

54. $\begin{cases} 4x + y = -0.5 \\ x + 2y = -8 \end{cases}$

Writing & Thinking

55. Explain, in your own words, why the answer to a consistent system of linear equations can be written as an ordered pair.

8.2 Systems of Linear Equations: Solutions by Substitution

Objective A   **Solving Systems of Linear Equations by Substitution**

As we discussed in Section 8.1, solving systems of linear equations by graphing is somewhat limited in accuracy. The graphs must be drawn very carefully and even then the points of intersection (if there are any) can be difficult to estimate accurately.

In this section, we will develop an algebraic method called the **method of substitution**. The objective of the substitution method is to eliminate one of the variables, so that a new equation is formed with just one variable. If this new equation has one solution, then the solution to the system is a single point. If this new equation is never true, then the system has no solution. If this new equation is always true, then the system has an infinite number of solutions.

> ### To Solve a System of Linear Equations by Substitution
>
> 1. Solve one of the equations for one of the variables.
> 2. Substitute the resulting expression into the other equation.
> 3. Solve this new equation, if possible, and then substitute back into one of the original equations to find the value of the other variable. (This is known as **back substitution**.)
> 4. Check the solution in both of the original equations.

To illustrate the substitution method, consider the following system.

$$\begin{cases} y = -2x + 5 \\ x + 2y = 1 \end{cases}$$

How would you substitute? Since the first equation is already solved for y, a reasonable substitution would be to put $-2x + 5$ for y in the second equation. Try this and see what happens.

first equation already solved for y: $y = \boxed{-2x + 5}$

substitute into second equation: $x + 2y = 1$

$$x + 2(-2x + 5) = 1$$

We now have one equation in only one variable, namely x. The problem has been reduced from one of solving two equations in two variables to solving one equation in one variable. Solve this equation for x. Then find the corresponding y-value by substituting this x-value into **either of the two original equations**.

$$x + 2(-2x + 5) = 1$$
$$x - 4x + 10 = 1$$
$$-3x = -9$$
$$x = 3$$

Back substitute $x = 3$ into $y = -2x + 5$.

$$y = -2(3) + 5$$
$$= -6 + 5$$
$$= -1$$

Thus the solution to the system is the point $(3, -1)$.

Substitution is not the only algebraic technique for solving a system of linear equations. It does work in all cases but is most often used when one of the equations is easily solved for one variable.

In the following examples, note how the results in Example 2 indicate that the system has no solution and the results in Example 3 indicate that the system has an infinite number of solutions.

Example 1

A System with One Solution

Solve the following system of linear equations by using substitution.

$$\begin{cases} y = \dfrac{5}{6}x + 2 \\ \dfrac{1}{6}x + y = 8 \end{cases}$$

Solution

The first equation is already solved for y. Substituting $\dfrac{5}{6}x + 2$ for y in the second equation gives the following.

$$\frac{1}{6}x + \left(\frac{5}{6}x + 2\right) = 8$$

$$6 \cdot \frac{1}{6}x + 6 \cdot \left(\frac{5}{6}x + 2\right) = 6 \cdot 8 \qquad \text{Multiply both sides by 6, the LCD.}$$

$$x + 5x + 12 = 48$$
$$6x + 12 = 48$$
$$6x = 36$$
$$x = 6$$

Substituting 6 for x in the first equation gives the corresponding value for y.

$$y = \frac{5}{6}(6) + 2 = 5 + 2 = 7$$

The solution to the system is $(6, 7)$.

Check: Substitution shows that $(6, 7)$ satisfies **both** of the equations in the system.

$$\frac{1}{6}x + y = 8 \qquad\qquad y = \frac{5}{6}x + 2$$

$$\frac{1}{6}(6) + 7 \overset{?}{=} 8 \qquad 7 \overset{?}{=} \frac{5}{6}(6) + 2$$

$$8 = 8 \qquad\qquad 7 = 7$$

Solve the following system of linear equations by using substitution

1. $\begin{cases} y = \dfrac{2}{5}x + 3 \\ \dfrac{3}{5}x + y = 10 \end{cases}$

2. $\begin{cases} 2x + y = 1 \\ 10x + 5y = 4 \end{cases}$

3. $\begin{cases} 2x + y = 1 \\ 10x + 5y = 5 \end{cases}$

Example 2

A System with No Solution

Solve the following system of linear equations by using substitution.

$$\begin{cases} 3x + y = 1 \\ 6x + 2y = 3 \end{cases}$$

Solution

Solving the first equation for y gives $y = 1 - 3x$. Substituting $1 - 3x$ for y in the second equation gives the following.

$$6x + 2(1 - 3x) = 3$$
$$6x + 2 - 6x = 3$$
$$2 = 3$$

This last equation $(2 = 3)$ is **never true**. This tells us that the system is inconsistent and has **no solution**. Graphically, the lines are parallel and there is no intersection.

Example 3

A System with Infinitely Many Solutions

Solve the following system of linear equations by using substitution.

$$\begin{cases} x - 2y = 1 \\ 3x - 6y = 3 \end{cases}$$

Solution

Solving the first equation for x gives $x = 1 + 2y$. Substituting $1 + 2y$ for x in the second equation gives the following.

$$3(1 + 2y) - 6y = 3$$
$$3 + 6y - 6y = 3$$
$$3 = 3$$

This last equation $(3 = 3)$ is **always true**. This tells us that the system is dependent and has an **infinite number of solutions**

*Now work margin exercises **1** through **3**.*

notes

- The solution can take two forms for a system with an infinite number of solutions. In one form we can solve for x and then use this expression for x in the ordered pair format (x, y). For example, in Example 3, solving either equation for x gives $x = 1 + 2y$ and we can write $(1 + 2y, y)$ to represent all solutions.

- Alternatively, we can solve for y which gives $y = \dfrac{1}{2}x - \dfrac{1}{2}$ and we have the form $\left(x, \dfrac{1}{2}x - \dfrac{1}{2} \right)$ for all solutions. Try substituting various values for x and y in these expressions and you will see that all results satisfy both equations.

Example 4

A System with Decimal Numbers

Solve the following system of linear equations by using substitution.

$$\begin{cases} x + y = 5 \\ 0.2x + 0.3y = 0.9 \end{cases}$$

Solution

Solving the first equation for y gives $y = 5 - x$. Substituting $5 - x$ for y in the second equation gives the following.

$$0.2x + 0.3y = 0.9$$
$$0.2x + 0.3(5 - x) = 0.9$$
$$0.2x + 1.5 - 0.3x = 0.9$$
$$-0.1x = -0.6$$
$$\frac{-0.1x}{-0.1} = \frac{-0.6}{-0.1}$$
$$x = 6$$
$$y = 5 - x = 5 - 6 = -1$$

Note that you could multiply each term by 10 to eliminate the decimal.

The solution to the system is $(6, -1)$.

To check, substitute $x = 6$ and $y = -1$ in both of the original equations.

Now work margin exercise 4.

4. Solve the following system of linear equations by using substitution.

$$\begin{cases} x + y = 6 \\ 0.8x + 0.4y = 1.8 \end{cases}$$

Solve the following systems by using the method of substitution.

1. $\begin{cases} x + y = 3 \\ \quad y = 2x \end{cases}$

2. $\begin{cases} \quad y = -3 + 2x \\ 4x - 2y = 6 \end{cases}$

3. $\begin{cases} x + 2y = -1 \\ x - 4y = -4 \end{cases}$

4. $\begin{cases} y = 4 - 3x \\ y = -3x + 6 \end{cases}$

5. $\begin{cases} \quad y = 3x - 1 \\ 2x + y = 4 \end{cases}$

Practice Problem Answers

1. $(1, 2)$ **2.** dependent system **3.** $\left(-2, \dfrac{1}{2}\right)$ **4.** no solution **5.** $(1, 2)$

Exercises 8.2

Solve the following systems of linear equations by using the method of substitution.

1. $\begin{cases} x + y = 6 \\ \quad y = 2x \end{cases}$

2. $\begin{cases} 5x + 2y = 21 \\ \quad x = y \end{cases}$

3. $\begin{cases} 3x - 7 = y \\ 2y = 6x - 14 \end{cases}$

4. $\begin{cases} y = 3x + 4 \\ 2y = 3x + 5 \end{cases}$

5. $\begin{cases} x = 3y \\ 3y - 2x = 6 \end{cases}$

6. $\begin{cases} 4x = y \\ 4x - y = 7 \end{cases}$

7. $\begin{cases} x - 5y + 1 = 0 \\ \quad x = 7 - 3y \end{cases}$

8. $\begin{cases} 2x + 5y = 15 \\ \quad x = y - 3 \end{cases}$

9. $\begin{cases} 7x + y = 9 \\ \quad y = 4 - 7x \end{cases}$

10. $\begin{cases} 3y + 5x = 5 \\ \quad y = 3 - 2x \end{cases}$

11. $\begin{cases} 3x - y = 7 \\ x + y = 5 \end{cases}$

12. $\begin{cases} 4x - 2y = 5 \\ \quad y = 2x + 3 \end{cases}$

13. $\begin{cases} 3x + 5y = -13 \\ \quad y = 3 - 2x \end{cases}$

14. $\begin{cases} 15x + 5y = 20 \\ \quad y = -3x + 4 \end{cases}$

15. $\begin{cases} x - y = 5 \\ 2x + 3y = 0 \end{cases}$

16. $\begin{cases} 4x = 8 \\ 3x + y = 8 \end{cases}$

17. $\begin{cases} 2y = 5 \\ 3x - 4y = -4 \end{cases}$

18. $\begin{cases} x + y = 8 \\ 3x + 2y = 8 \end{cases}$

19. $\begin{cases} y = 2x - 5 \\ 2x + y = -3 \end{cases}$

20. $\begin{cases} 2x + 3y = 5 \\ x - 6y = 0 \end{cases}$

21. $\begin{cases} x + 5y = 1 \\ x - 3y = 5 \end{cases}$

22. $\begin{cases} 3x + 8y = -2 \\ x + 2y = -1 \end{cases}$

23. $\begin{cases} 9x + 3y = 6 \\ 3x = 2 - y \end{cases}$

24. $\begin{cases} 5x + 2y = -10 \\ 10x = -3 - 4y \end{cases}$

25. $\begin{cases} x - 2y = -4 \\ 3x + y = -5 \end{cases}$

26. $\begin{cases} x + 4y = 3 \\ 3x - 4y = 7 \end{cases}$

27. $\begin{cases} 3x - y = -1 \\ 7x - 4y = 0 \end{cases}$

28. $\begin{cases} x + 5y = -1 \\ 2x + 7y = 1 \end{cases}$

29. $\begin{cases} x + 3y = 5 \\ 3x + 2y = 7 \end{cases}$

30. $\begin{cases} 3x - 4y - 39 = 0 \\ 2x - y - 13 = 0 \end{cases}$

31. $\begin{cases} \dfrac{1}{4}x - \dfrac{3}{2}y = -5 \\ -x + 6y = 20 \end{cases}$

32. $\begin{cases} \dfrac{-4}{3}x + 2y = 7 \\ \dfrac{8}{3}x - 4y = -5 \end{cases}$

33. $\begin{cases} 6x - y = 15 \\ 0.2x + 0.5y = 2.1 \end{cases}$

34. $\begin{cases} x + 2y = 3 \\ 0.4x + y = 0.6 \end{cases}$

35. $\begin{cases} 0.2x - 0.1y = 0 \\ \quad y = x + 10 \end{cases}$

36. $\begin{cases} 0.1x - 0.2y = 1.4 \\ 3x + y = 14 \end{cases}$

37. $\begin{cases} 3x - 2y = 5 \\ \quad y = 1.5x + 2 \end{cases}$

38. $\begin{cases} \quad x = 2y - 7.5 \\ 2x + 4y = -15 \end{cases}$

39. $\begin{cases} \dfrac{1}{2}x + \dfrac{1}{3}y = 4 \\ 3x + 2y = 24 \end{cases}$

40. $\begin{cases} \dfrac{1}{3}x + \dfrac{1}{7}y = 2 \\ 7x + 3y = 42 \end{cases}$

41. $\begin{cases} \dfrac{x}{3} + \dfrac{y}{5} = 1 \\ x + 6y = 12 \end{cases}$

42. $\begin{cases} \dfrac{x}{5} + \dfrac{y}{4} - 3 = 0 \\ \dfrac{x}{10} - \dfrac{y}{2} + 1 = 0 \end{cases}$

A word problem is stated with equations given that represent a mathematical model for the problem. Solve the system by using the method of substitution.

43. The sum of two numbers is 25 and their difference is 15. What are the two numbers? Let x = larger number and y = the other number.

The corresponding modeling system is

$\begin{cases} x + y = 25 \\ x - y = 15 \end{cases}$

44. **Rectangles:** The perimeter of a rectangle is 50 meters and the length is 5 meters longer than the width. Find the dimensions of the rectangle.

Let x = the length and y = the width.

The corresponding modeling system is
$\begin{cases} 2x + 2y = 50 \\ \quad x - y = 5 \end{cases}$

45. **Swimming Pools:** OSHA recommends that swimming pool owners clean their pool decks with a solvent composed of a 12% chlorine solution and a 3% chlorine solution. Fifteen gallons of the solvent consists of 6% chlorine. How much of each of the mixing solutions were used?

Let x = the number of gallons of the 12% solution

and y = the number of gallons of the 3% solution.

The corresponding modeling system is

$\begin{cases} \quad\quad x + y = 15 \\ 0.12x + 0.03y = 0.06(15) \end{cases}$

46. **School Supplies:** A student bought a calculator and a textbook for a course in algebra. He told his friend that the total cost was \$170 (without tax) and that the calculator cost \$20 more than twice the cost of the textbook. What was the cost of each item?

Let x = the cost of the calculator

and y = the cost of the textbook.

The corresponding modeling system is

$\begin{cases} x + y = 170 \\ \quad x = 2y + 20 \end{cases}$

Writing & Thinking

47. Explain the advantages of solving a system of linear equations
   a. by graphing,
   b. by substitution.

8.3 Systems of Linear Equations: Solutions by Addition

Objective A **Solving Systems of Linear Equations by Addition**

Objectives

A Solve systems of linear equations using the method of addition.

B Use systems of equations to find the equation of a line through two points.

We have discussed two methods for solving systems of linear equations:

1. graphing and
2. substitution.

We know that solutions found by graphing are not necessarily exact (estimation may be involved); and, in some cases, the method of substitution can lead to complicated algebraic steps. In this section, we consider a third method:

3. **addition** (or **method of elimination**).

In the **method of addition**, as with the method of substitution, the objective is to eliminate one of the variables so that a new equation is found with just one variable, if possible. If this new equation:

1. has one solution, the solution to the system is a **single point**.
2. is never true, the system has **no solution**.
3. is always true, the system has **infinite solutions**.

Consider solving the following system where both equations are written in standard form.

$$\begin{cases} x - 2y = -9 \\ x + 2y = 11 \end{cases}$$

In the **method of addition**, we write one equation under the other so that like terms are aligned vertically. (Note that, in this example, the coefficients of y are opposites, namely -2 and $+2$.)

Then add like terms as follows.

$$\begin{aligned} x - 2y &= -9 \\ \underline{x + 2y} &= \underline{11} \\ 2x &= 2 \qquad \text{\textcolor{red}{The } } y \text{ \textcolor{red}{terms are eliminated because the coefficients are opposites.}} \\ x &= 1 \end{aligned}$$

Now substitute $x = 1$ into either of the original equations and solve for y.

$$\begin{aligned} 1 - 2y &= -9 & \text{OR} && 1 + 2y &= 11 \\ -2y &= -10 &&& 2y &= 10 \\ y &= 5 &&& y &= 5 \end{aligned}$$

Therefore the solution to the system is $(1, 5)$.

This example was relatively easy because the coefficients for y were opposites. The general procedure can be outlined as follows.

To Solve a System of Linear Equations by Addition

1. Write the equations in **standard form** one under the other so that **like terms are vertically aligned**.
2. Multiply all terms of one equation by a constant (and possibly all terms of the other equation by another constant) so that **two like terms have opposite coefficients**.
3. Add the two equations by **combining like terms** and solve the resulting equation, if possible.
4. **Back substitute into one of the original equations** to find the value of the other variable.
5. Check the solution (if there is one) in both of the original equations.

Example 1

A System with One Solution

Use the method of addition to solve the following system:

$$\begin{cases} 3x + 5y = -3 \\ -7x + 2y = 7 \end{cases}.$$

Solution

Multiply each term in the first equation by 2 and each term in the second equation by −5. This will result in the y-coefficients being opposites. Add the two equations by combining like terms which will eliminate y. Solve for x.

$$\begin{cases} 3x + 5y = -3 \\ -7x + 2y = 7 \end{cases}$$

$$\begin{cases} [2](3x + 5y = -3) \longrightarrow 6x + 10y = -6 \\ [-5](-7x + 2y = 7) \longrightarrow 35x - 10y = -35 \end{cases}$$

$$41x \qquad = -41 \qquad \text{\textcolor{red}{y is eliminated.}}$$
$$x \qquad = -1$$

Substitute $x = -1$ into either one of the original equations.

$3x + 5y = -3$	$-7x + 2y = 7$
$3(-1) + 5y = -3$	$-7(-1) + 2y = 7$
$-3 + 5y = -3$	$7 + 2y = 7$
$5y = 0$	$2y = 0$
$y = 0$	$y = 0$

OR

The solution is $(-1, 0)$.

Check: Substitution shows that $(-1, 0)$ satisfies **both** of the equations in the system.

$$3x + 5y = -3 \qquad\qquad -7x + 2y = 7$$
$$\overset{?}{3(-1) + 5(0) = -3} \qquad\qquad \overset{?}{-7(-1) + 2(0) = 7}$$
$$-3 = -3 \qquad\qquad\qquad 7 = 7$$

notes

In Example 1, we could have eliminated x instead of y by multiplying the terms in the first equation by 7 and the terms in the second equation by 3. The solution would be the same. Try this yourself to confirm this method.

Example 2

A System with Infinitely Many Solutions

Solve the following system by using the method of addition.

$$\begin{cases} 3x - \dfrac{1}{2}y = 6 \\ 6x - y = 12 \end{cases}$$

Solution

Multiply the first equation by -2 so that the y-coefficients will be opposites.

$$\begin{cases} [-2]\left(3x - \dfrac{1}{2}y = 6\right) \longrightarrow -6x + y = -12 \\ (6x - y = 12) \longrightarrow \underline{6x - y = 12} \\ 0 = 0 \end{cases}$$

Because this last equation, $0 = 0$, is **always true**, the system is dependent and has infinitely many solutions.

Use the method of addition
to solve the following systems.

1. $\begin{cases} 3x + 2y = 0 \\ 5x - 2y = 16 \end{cases}$

2. $\begin{cases} 6x + 3y = 15 \\ 4x + 2y = 10 \end{cases}$

3. $\begin{cases} x + 0.5y = 0.1 \\ 0.5x - 2y = 10.4 \end{cases}$

Example 3

A System with Decimal Numbers

Solve the following system by using the method of addition.

$$\begin{cases} x + 0.4y = 3.08 \\ 0.1x - y = 0.1 \end{cases}$$

Solution

Multiply the second equation by -10 so that the x-coefficients will be opposites.

$$\begin{cases} (x + 0.4y = 3.08) & \longrightarrow & 1.0x + 0.4y = 3.08 \\ [-10](0.1x - y = 0.1) & \longrightarrow & \underline{-1.0x + 10.0y = -1.0} \\ & & 10.4y = 2.08 \\ & & y = 0.2 \end{cases}$$

x is eliminated.
Substitute $y = 0.2$ into one of the original equations.

$$x + 0.4y = 3.08$$
$$x + 0.4(0.2) = 3.08$$
$$x + 0.08 = 3.08$$
$$x = 3$$

The solution is $(3, 0.2)$.

To check, substitute $x = 3$ and $y = 0.2$ in both of the original equations.

Now work margin exercises 1 through 3.

Objective B   **Using Systems of Equations to Find the Equation of a Line**

Example 4

Using Systems of Equations to Find the Equation of a Line

Using the formula $y = mx + b$, find the equation of the line determined by the two points $(3, 5)$ and $(-6, 2)$.

Solution

Write two equations in m and b by substituting the coordinates of the points for x and y. This gives the following system.

$$\begin{cases} 5 = 3m + b \\ 2 = -6m + b \end{cases}$$

Multiply the second equation by -1 so that the b-coefficients will be opposites.

$$\begin{cases} (5 = 3m + b) \\ [-1](2 = -6m + b) \end{cases} \longrightarrow \begin{aligned} 5 &= 3m + b \\ \underline{-2 &= 6m - b} \\ 3 &= 9m \\ \tfrac{1}{3} &= m \end{aligned}$$

4. Find the equation of the line determined by the two points $(4, -2)$ and $(6, 3)$.

Substitute $m = \dfrac{1}{3}$ into one of the original equations.

$$5 = 3m + b$$
$$5 = 3\left(\dfrac{1}{3}\right) + b$$
$$5 = 1 + b$$
$$4 = b$$

The equation of the line is $y = \dfrac{1}{3}x + 4$.

Now work margin exercise 4.

Guidelines for Deciding which Method to Use when Solving a System of Linear Equations

1. The graphing method is helpful in "seeing" the geometric relationship between the lines and finding approximate solutions. A calculator can be very helpful here.
2. Both the substitution method and the addition method give exact solutions.
3. The substitution method may be reasonable and efficient if one of the coefficients of one of the variables is 1.
4. The addition method is particularly efficient if the coefficients for one of the variables are opposites.

Practice Problems

Solve the following systems by using the method of addition.

1. $\begin{cases} 2x + 2y = 4 \\ x - y = -3 \end{cases}$
2. $\begin{cases} 3x + 4y = 12 \\ \dfrac{1}{3}x - 8y = -5 \end{cases}$
3. $\begin{cases} 2x - 5y = 6 \\ 4x - 10y = -2 \end{cases}$

4. $\begin{cases} 0.02x + 0.06y = 1.48 \\ 0.03x - 0.02y = 0.02 \end{cases}$
5. $\begin{cases} y = 3x + 15 \\ 6x - 2y = -30 \end{cases}$

Practice Problem Answers

1. $\left(-\dfrac{1}{2}, \dfrac{5}{2}\right)$
2. $\left(3, \dfrac{3}{4}\right)$
3. no solution
4. $(14, 20)$
5. infinitely many solutions

Exercises 8.3

Solve each system by using the addition method. Classify each system as consistent, inconsistent, or dependent.

1. $\begin{cases} 8x - y = 29 \\ 2x + y = 11 \end{cases}$

2. $\begin{cases} x + 3y = 9 \\ x - 7y = -1 \end{cases}$

3. $\begin{cases} 3x + 2y = 0 \\ 5x - 2y = 8 \end{cases}$

4. $\begin{cases} 12x - 3y = 21 \\ 4x - y = 7 \end{cases}$

5. $\begin{cases} 2x + 2y = 5 \\ x + y = 3 \end{cases}$

6. $\begin{cases} 2x - y = 7 \\ x + y = 2 \end{cases}$

7. $\begin{cases} 3x + 3y = 9 \\ x + y = 3 \end{cases}$

8. $\begin{cases} 9x + 2y = -42 \\ 5x - 6y = -2 \end{cases}$

9. $\begin{cases} \dfrac{1}{2}x + y = -4 \\ 3x - 4y = 6 \end{cases}$

10. $\begin{cases} x + y = 1 \\ x - \dfrac{1}{3}y = \dfrac{11}{3} \end{cases}$

11. $\begin{cases} x + y = 12 \\ 0.05x + 0.25y = 1.6 \end{cases}$

12. $\begin{cases} x + 0.1y = 8 \\ 0.1x + 0.01y = 0.64 \end{cases}$

Solve each system by using either the substitution method or the addition method (whichever seems better to you).

13. $\begin{cases} x = 11 + 2y \\ 2x - 3y = 17 \end{cases}$

14. $\begin{cases} 6x - 3y = 6 \\ y = 2x - 2 \end{cases}$

15. $\begin{cases} x - 2y = 4 \\ y = \dfrac{1}{2}x - 2 \end{cases}$

16. $\begin{cases} 2x + y = 3 \\ 4x + 2y = 7 \end{cases}$

17. $\begin{cases} x = 3y + 4 \\ y = 6 - 2x \end{cases}$

18. $\begin{cases} y = 2x + 14 \\ x = 14 - 3y \end{cases}$

19. $\begin{cases} 7x - y = 16 \\ 2y = 2 - 3x \end{cases}$

20. $\begin{cases} 3x + y = -10 \\ 2y - 1 = x \end{cases}$

21. $\begin{cases} 4x - 2y = 8 \\ 2x - y = 4 \end{cases}$

22. $\begin{cases} x + y = 6 \\ 2x + y = 16 \end{cases}$

23. $\begin{cases} 3x + 2y = 4 \\ x + 5y = -3 \end{cases}$

24. $\begin{cases} x + 2y = 0 \\ 2x = 4y \end{cases}$

25. $\begin{cases} 4x + 3y = 2 \\ 3x + 2y = 3 \end{cases}$

26. $\begin{cases} x - 3y = 4 \\ 3x - 9y = 10 \end{cases}$

27. $\begin{cases} 5x - 2y = 17 \\ 2x - 3y = 9 \end{cases}$

28. $\begin{cases} \dfrac{1}{2}x + 2y = 9 \\ 2x - 3y = 14 \end{cases}$

29. $\begin{cases} 3x + 2y = 14 \\ 7x + 3y = 26 \end{cases}$

30. $\begin{cases} 4x + 3y = 28 \\ 5x + 2y = 35 \end{cases}$

31. $\begin{cases} 2x + 7y = 2 \\ 5x + 3y = -24 \end{cases}$

32. $\begin{cases} 7x - 6y = -1 \\ 5x + 2y = 37 \end{cases}$

33. $\begin{cases} 10x + 4y = 7 \\ 5x + 2y = 15 \end{cases}$

34. $\begin{cases} 6x - 5y = -40 \\ 8x - 7y = -54 \end{cases}$

35. $\begin{cases} 0.5x - 0.3y = 7 \\ 0.3x - 0.4y = 2 \end{cases}$

36. $\begin{cases} 0.6x + 0.5y = 5.9 \\ 0.8x + 0.4y = 6 \end{cases}$

37. $\begin{cases} 2.5x + 1.8y = 7 \\ 3.5x - 2.7y = 4 \end{cases}$

38. $\begin{cases} 0.75x - 0.5y = 2 \\ 1.5x - 0.75y = 7.5 \end{cases}$

39. $\begin{cases} \dfrac{2}{3}x - \dfrac{1}{2}y = \dfrac{2}{3} \\ \dfrac{8}{3}x - 2y = \dfrac{17}{6} \end{cases}$

40. $\begin{cases} \dfrac{3}{4}x + \dfrac{1}{4}y = \dfrac{3}{8} \\ \dfrac{3}{2}x + \dfrac{1}{2}y = \dfrac{3}{4} \end{cases}$

41. $\begin{cases} \dfrac{1}{6}x - \dfrac{1}{12}y = -\dfrac{13}{6} \\ \dfrac{1}{5}x + \dfrac{1}{4}y = 2 \end{cases}$

42. $\begin{cases} \dfrac{5}{3}x - \dfrac{2}{3}y = -\dfrac{29}{30} \\ 2x + 5y = 0 \end{cases}$

Write an equation for the line determined by the two given points using the formula $y = mx + b$.

43. $(2, 3), (1, -2)$

44. $(0, 6), (-3, -3)$

45. $(1, -3), (5, -3)$

46. $(5, 3), (5, -4)$

47. $(1, 2), (-3, 0)$

48. $(-4, 2), (5, -1)$

A word problem is stated with equations given that represent a mathematical model for the problem. Solve the system by using either the method of substitution or the method of addition.

49. Investing: Georgia had \$10,000 to invest, and she put the money into two accounts. One of the accounts will pay 6% interest and the other will pay 10%. How much did she put in each account if the interest from the 10% account exceeded the interest from the 6% account by \$40?

Let x = amount in 10% account and y = amount in 6% account.

The system that models the problem is

$\begin{cases} x + y = 10,000 \\ 0.10x - 0.06y = 40 \end{cases}$

50. Baseball: A minor league baseball team has a game attendance of 4500 people. Tickets cost \$5 for children and \$8 for adults. The total revenue made at this game was \$26,100. How many adults and how many children attended the game?

Let x = adults and y = children.

The system that models the problem is

$\begin{cases} x + y = 4500 \\ 8x + 5y = 26,100 \end{cases}$

51. Acid Solutions: How many liters each of a 30% acid solution and a 40% acid solution must be used to produce 100 liters of a 36% acid solution?

Let x = amount of 30% solution and y = amount of 40% solution.

The system that models the problem is

$\begin{cases} x + y = 100 \\ 0.30x + 0.40y = 0.36(100) \end{cases}$

52. Traveling by Car: Two cars leave Denver at the same time traveling in opposite directions. One travels at a mean speed of 55 mph and the other at 65 mph. In how many hours will they be 420 miles apart?

Let x = time of travel for first car and y = time of travel for second car.

The system that models the problem is

$\begin{cases} x = y \\ 55x + 65y = 420 \end{cases}$

65 mph   55 mph   Denver

8.4 Applications: Distance-Rate-Time, Number Problems, Amounts and Costs

Objective A **Distance-Rate-Time Problems**

Systems of equations occur in many practical situations such as: supply and demand in business, velocity and acceleration in engineering, money and interest in investments, and mixture in physics and chemistry.

Many of the problems in earlier sections illustrated these ideas. However, in those exercises, the system of equations was given. As you study the applications in Sections 8.4 and 8.5, you may want to refer to some of those exercises, as well as the examples, as guides in solving the given applications. **For these applications you will need to create your own systems of equations.**

Remember that the emphasis in all application problems is to develop your reasoning as well as your reading skills. You must learn how to transfer English phrases into algebraic expressions.

Note: You should consider making tables similar to those illustrated in Examples 1, 2, and 4 when working with applications. These tables can help you organize the information in a more understandable form.

Example 1

Solving Proportions Distance-Rate-Time (Rates Unknown)

A small plane flew 300 miles in 2 hours flying with the wind. Then on the return trip, flying against the wind, it traveled only 200 miles in 2 hours. What were the wind speed and the speed of the plane? (**Note:** The "speed of the plane" means how fast the plane would be flying with no wind.)

Solution

Let s = speed of plane, then let,
   w = wind speed.

When flying with the wind, the plane's actual rate will increase to $s + w$. When flying against the wind, the plane's actual rate will decrease to $s - w$. (**Note:** If the wind had been strong enough, the plane could actually have been flying backward, away from its destination.)

	Rate $\times$	Time	= Distance
With the wind	$s + w$	2	$2(s + w)$
Against the wind	$s - w$	2	$2(s - w)$

The system of linear equations is,

with the wind $\longrightarrow$ $\begin{cases} 2(s+w) = 300 \\ 2(s-w) = 200 \end{cases}$ $\longrightarrow$ $\begin{array}{r} 2s+2w = 300 \\ 2s - 2w = 200 \\ \hline 4s = 500 \\ s = 125 \end{array}$ speed of plane

against the wind $\longrightarrow$

Back substitute $s = 125$ into one of the original equations.

$$2(125 + w) = 300$$
$$125 + w = 150$$
$$w = 25 \qquad \text{wind speed}$$

The speed of the plane was 125 mph, and the wind speed was 25 mph

Now work margin exercise 1.

Example 2

Distance-Rate-Time (Times Unknown)

Two buses leave a bus station traveling in opposite directions. One leaves at noon and the second leaves at 1 P.M. The first one travels at a mean speed of 55 mph and the second one at a mean speed of 59 mph. At what time will the buses be 226 miles apart?

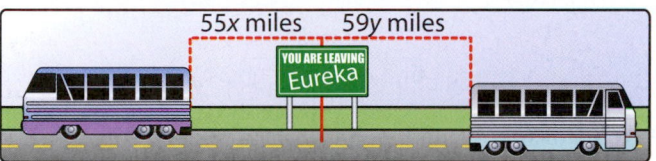

Solution

Let x = time of travel for first bus, and
y = time of travel for second bus.

	Rate	×	Time	=	Distance
Bus 1	55		x		$55x$
Bus 2	59		y		$59y$

The system of linear equations is

$$\begin{cases} x = y + 1 & \text{The first bus travels 1 hour longer than the second.} \\ 55x + 59y = 226 & \text{The sum of the distances is 226 miles.} \end{cases}$$

Substitution gives

$$55(y+1) + 59y = 226$$
$$55y + 55 + 59y = 226$$
$$114y = 171$$
$$y = 1.5$$

This gives $x = 1.5 + 1 = 2.5$.
Thus the first bus travels 2.5 hours and the second bus travels 1.5 hours. The buses will be 226 miles apart at 2:30 P.M.

Now work margin exercise 2.

1. Bob ran 5 miles in 30 minutes running with the wind, and then turned around. For the next 30 minutes running against the wind, he ran 4.2 miles. What were the wind speed, and the speed Bob was running? (**Hint:** Convert minutes to hours).

2. One man takes the eastbound line to go to work at 9 A.M., and his wife takes the westbound line at 9:30 A.M. The mean speed of the husband's train was 25 mph, while the mean of his wife's train was 30 mph. At what time will the husband and wife be 122.5 miles apart?

3. The sum of two numbers is 150 and their difference is 36. What are the two numbers?

Example 3

Number Problem

The sum of two numbers is 80 and their difference is 10. What are the two numbers?

Solution

Let x = one number and
    y = the other number.
The system of linear equations is as follows.

$$\begin{cases} x + y = 80 \\ x - y = 10 \end{cases}$$ The sum is 80.
    The difference is 10.

Solving by addition gives the following.

$$\begin{array}{rcl} x + y & = & 80 \\ x - y & = & 10 \\ \hline 2x & = & 90 \\ x & = & 45 \end{array}$$

Back substitute 45 for x in the first equation gives:

$$45 + y = 80$$
$$y = 80 - 45 = 35$$

The two numbers are 45 and 35.

Check:

$$45 + 35 = 80 \text{ and } 45 - 35 = 10$$

Now work margin exercise 3.

Example 4

Counting Coins

Mike has $1.05 worth of change in nickels and quarters. If he has twice as many nickels as quarters, how many of each type of coin does he have?

Solution

We use two equations – one relating the number of coins and the other relating the value of the coins. (The value of each nickel is 5 cents and the value of each quarter is 25 cents.)

Let n = number of nickels and
    q = number of quarters.

Coins	# of coins	Value	Total value
Nickels	n	0.05	$0.05n$
Quarters	q	0.25	$0.25q$

The system of linear equations is the following.

$$\begin{cases} n = 2q & \text{There are two times as many nickels as quarters.} \\ 0.05n + 0.25q = 1.05 & \text{The total amount of money is \$1.05.} \end{cases}$$

Note carefully that in the first equation, q is multiplied by 2 because the number of nickels is twice the number of quarters. Therefore, n is bigger.

The first equation is already solved for n, so substitute $2q$ for n in the second equation.

$$0.05(2q) + 0.25q = 1.05$$
$$0.10q + 0.25q = 1.05$$
$$10q + 25q = 105 \qquad \text{Multiply the equation by 100.}$$
$$35q = 105$$
$$q = 3 \qquad \text{number of quarters}$$
$$n = 2q = 6 \qquad \text{number of nickels}$$

Mike has 3 quarters and 6 nickels.

Now work margin exercise 4.

Example 5

Calculating Age

Kathy is 6 years older than her sister, Sue. In 3 years, she will be twice as old as Sue. How old is each girl now?

Solution

We use two equations: one relating their ages now and the other relating their ages in 3 years.

Let K = Kathy's age now and
    S = Sue's age now.
Then the system of linear equations is

$$\begin{cases} K - S = 6 & \text{The difference in their ages is 6 years.} \\ K + 3 = 2(S + 3) & \text{In 3 years each age is increased by 3 and Kathy is twice as old as Sue.} \end{cases}$$

Rewrite the second equation in standard form.
$$K + 3 = 2(S + 3)$$
$$K + 3 = 2S + 6$$
$$K - 2S = 3$$

4. Seth has \$3.85 worth of change in nickels and dimes. If he has 3 times as many dimes as nickels, how many of each type of coin does he have?

5. Tom is 9 years older than his sister, Jane. In 5 years, he will be twice as old as Jane. How old are Tom and Jane now?

Solve the system using the method of addition.

$$\begin{cases} (K - S = 6) \\ [-1](K - 2S = 3) \end{cases} \longrightarrow \begin{array}{l} K - S = 6 \\ \underline{-K + 2S = -3} \\ \qquad S = 3 \quad \text{Sue's current age} \end{array}$$

Back substitute $S = 3$ into one of the original equations.

$$K - 3 = 6$$
$$K = 6 + 3$$
$$K = 9 \qquad \qquad \text{Kathy's current age}$$

Kathy is 9 years old; Sue is 3 years old.

Now work margin exercise 5.

Objective C **Amounts and Costs**

6. Seven sodas and 6 water bottles cost $13.55. Three sodas and four water bottles cost $6.95. What is the cost of a soda? What is the cost of a water bottle?

Example 6

Amounts and Costs

Three hot dogs and two orders of French fries cost $10.30. Four hot dogs and four orders of fries cost $15.60. What is the cost of a hot dog? What is the cost of an order of fries?

Solution

Let x = cost of one hot dog and
$\quad\ y$ = cost of one order of fries.
The system of linear equations is the following.

$$\begin{cases} 3x + 2y = 10.30 \quad \text{Three hot dogs and two orders of French fries cost \$10.30.} \\ 4x + 4y = 15.60 \quad \text{Four hot dogs and four orders of fries cost \$15.60.} \end{cases}$$

Both equations are in standard form. Solve using the addition method.

$$\begin{cases} [-2](3x + 2y = 10.30) \\ \ \ \ \ (4x + 4y = 15.60) \end{cases} \longrightarrow \begin{array}{l} -6x - 4y = -20.60 \\ \underline{4x + 4y = \ \ \ 15.60} \\ -2x \qquad\ = -5.00 \\ \ \ \ x \qquad\ = \ \ \ 2.50 \quad \text{cost of one hot dog} \end{array}$$

Back substitute $x = 2.50$ into one of the original equations.

$$3(2.50) + 2y = 10.30$$
$$7.50 + 2y = 10.30$$
$$2y = 2.80$$
$$y = 1.40 \qquad \text{cost of one order of fries}$$

One hot dog costs $2.50 and one order of fries costs $1.40.

Now work margin exercise 6.

Exercises 8.4

Solve each problem by setting up a system of two equations in two unknowns and solve.

1. The sum of two numbers is 56. Their difference is 10. Find the numbers.

2. The sum of two numbers is 40. The sum of twice the larger and 4 times the smaller is 108. Find the numbers.

3. The sum of two numbers is 36. Three times the smaller plus twice the larger is 87. Find the two numbers.

4. The sum of two integers is 102, and the larger number is 10 more than three times the smaller. Find the two integers.

5. The difference between two integers is 13, and their sum is 87. What are the two integers?

6. The difference between two numbers is 17. Four times the smaller is equal to 7 more than the larger. What are the numbers?

7. **Supplementary Angles:** Two angles are supplementary if the sum of their measures is 180°. Find two supplementary angles such that the smaller is 30° more than one-half of the larger.

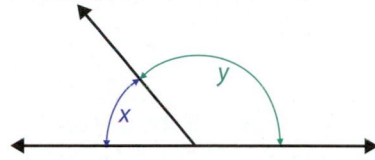

8. **Complementary Angles:** Two angles are complementary if the sum of their measures is 90°. Find two complementary angles such that one is 15° less than six times the other.

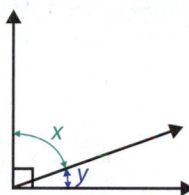

9. **Triangles:** The sum of the measures of the three angles of a triangle is 180°. In an isosceles triangle, two of the angles have the same measure. What are the measures of the angles of an isosceles triangle in which one angle measures 15° more than each of the other two congruent angles?

10. **Triangles:** The sum of the measures of the three angles of a triangle is 180°. In an isosceles triangle, two of the angles have the same measure. What are the measures of the angles of an isosceles triangle in which each of the two congruent angles measures 15° more than the third angle?

11. **Boating:** Liam makes a 4-mile motorboat trip downstream in 20 minutes $\left(\frac{1}{3}\ \text{hr}\right)$. The return trip takes 30 minutes $\left(\frac{1}{2}\ \text{hr}\right)$. Find the rate of the boat in still water and the rate of the current.

12. **Flying an Airplane:** Mr. McKelvey finds that flying with the wind he can travel 1188 miles in 6 hours. However, when flying against the wind, he travels only $\frac{2}{3}$ of the distance in the same amount of time. Find the speed of the plane in still air and the wind speed.

13. **Running:** Usain Bolt, the world-record holder in the 100 meter dash, ran 100 meters in 9.69 seconds with no wind. He later ran the same distance in 9.58 seconds with the wind. What was his speed and what was the wind speed?

14. **Boating:** Jessica drove her speedboat upriver this morning. It took her 1 hour going upriver and 54 minutes going down river. If she traveled 36 miles each way, what would have been the rate of the boat in still water and what was the rate of the current (in miles per hour)?

15. **Traveling by Car:** Randy made a business trip of 190 miles. His mean speed for the first part of the trip was 52 mph and 56 mph for the second part. If the total trip took $3\frac{1}{2}$ hours, how long did he travel at each rate?

16. **Traveling by Car:** Marian drove to a resort 335 miles from her home. Her mean speed was 60 mph for the first part of her trip and 55 mph for the second part. If her total driving time was $5\frac{3}{4}$ hours, how long did she travel at each rate?

17. **Traveling by Car:** Marcos lives 364 miles away from his cousin Cana. They start driving at the same time and travel toward each other. Cana's speed is 11 mph faster than Marcos' speed. If they meet in 4 hrs, find their speeds.

18. **Traveling by Car:** Naomi and Linda live 324 miles apart. They start at the same time and travel toward each other. Naomi's speed is 8 mph greater than Linda's. If they meet in 3 hours, find their speeds.

19. **Traveling by Car:** Steve travels 4 times as fast as Tim. Starting at the same point, but traveling in opposite directions, they are 105 miles apart after 3 hours. Find their rates of travel.

20. **Traveling by Car:** Bella travels 5 mph less than twice as fast as June. Starting at the same point and traveling in the same direction, they are 80 miles apart after 4 hours. Find their speeds.

21. **Traveling by Train:** Two trains leave Dallas at the same time. One train travels east and the other travels west. The speed of the westbound train is 5 mph greater than the speed of the eastbound train. After 6 hours, they are 510 miles a part. Find the rate of each train. Assume the trains travel in a straight line in opposite directions.

22. **Boating:** A boat left Dana Point Marina at 11:00 A.M. traveling at 10 knots (nautical miles per hour). Two hours later, a Coast Guard boat left the same marina traveling at 14 knots trying to catch the first boat. If both boats traveled the same course, at what time did the Coast Guard captain anticipate overtaking the first boat?

23. **Jogging:** A jogger runs into the countryside at a rate of 10 mph. He returns along the same route at 6 mph. If the total trip took 1 hour 36 minutes, how far did he jog?

24. **Biking:** A cyclist traveled to her destination at a mean rate of 15 mph. By traveling 3 mph faster, she took 30 minutes less to return. What distance did she travel each way?

25. Coin Collecting: Sonja has some nickels and dimes. If she has 30 coins worth a total of $2.00, how many of each type of coin does she have?

26. Coin Collecting: Conner has a total of 27 coins consisting of quarters and dimes. The total value of the coins is $5.40. How many of each type of coin does he have?

27. Coin Collecting: A bag contains pennies and nickels only. If there are 182 coins in all and their value is $3.90, how many pennies and how many nickels are in the bag?

28. Coin Collecting: Your friend challenges you to figure out how many dimes and quarters are in a cash register. He tells you that there are 65 coins and that their value is $11.90. How many dimes and how many quarters are in the register?

29. Basketball Admission: Tickets for the local high school basketball game were priced at $3.50 for adults and $2.50 for students. If the income for one game was $9550 and the attendance was 3500, how many adults and how many students attended that game?

30. Rectangles: The width of a rectangle is $\frac{3}{4}$ of its length. If the perimeter of the rectangle is 140 feet, what are the dimensions of the rectangle?

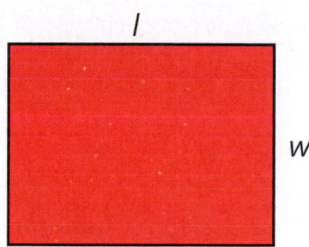

31. Rectangles: The length of a rectangle is 10 meters more than one-half the width. If the perimeter is 44 meters, what are the length and width?

32. Building a Fence: A farmer has 260 meters of fencing to build a rectangular corral. He wants the length to be 3 times as long as the width. What dimensions should he make his corral?

33. Soccer: At present, the length of a rectangular soccer field is 55 yards longer than the width. The city wants to rearrange the area containing the soccer field into two square playing fields. A math teacher on the council told them that if the width of the current field were to be increased by 5 yards and the length cut in half, the resulting field would be a square. What are the dimensions of the field currently?

34. Perimeter: Consider a square and a regular hexagon (a six-sided figure with sides of equal length). One side of the square is 5 feet longer than a side of the hexagon, and the two figures have the same perimeter. What are the lengths of the sides of each figure?

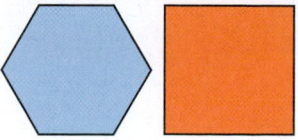

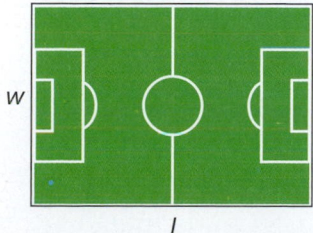

35. Rectangles: The length of a rectangle is 1 meter less than twice the width. If each side is increased by 4 meters, the perimeter will be 116 meters. Find the length and the width of the original rectangle.

36. Age: Ava is 8 years older than her brother Curt. Four years from now, Ava will be twice as old as Curt. How old is each at the present time?

37. Age: When they got married, Elvis Presley was 11 years older than his wife Priscilla. One year later, Priscilla was two-thirds of Elvis' age. How old was each of them when they got married?

38. Charity Admission: A Christmas charity party sold tickets for $45.00 for adults and $25.00 for children. The total number of tickets sold was 320 and the total for the ticket sales was $13,000. How many adult and how many children's tickets were sold?

39. Buying Books: Joan went to a book sale on campus and bought paperback books for $0.25 each and hardback books for $1.75 each. If she bought a total of 15 books for $11.25, how many of each type of book did she buy?

40. Recycling: Morton took some old newspapers and aluminum cans to the recycling center. Their total weight was 180 pounds. He received 1.5¢ per pound for the newspapers and 30¢ per pound for the cans. The total received was $14.10. How many pounds of each did Morton have?

41. Baseball Admission: Admission to the baseball game is $2.00 for general admission and $3.50 for reserved seats. The receipts were $36,250 for 12,500 paid admissions. How many of each ticket, general and reserved, were sold?

42. Going to the Theater: 70 children and 160 adults attended a play. The total receipts were $620. One adult ticket and 2 children's tickets cost $7. Find the price of each type of ticket.

43. Surfing: Last summer, Ernie sold surfboards. One style sold for $625 and the other sold for $550. He sold a total of 47 surfboards. How many of each style did he sell if the sales from each style were equal?

44. Selling Candy: The Candy Shack sells a particular candy in two different size packages. One size sells for $1.25 and the other sells for $1.75. If the store received $65.50 for 42 packages of candy, how many of each size were sold?

8.5 Applications: Interest and Mixture

Objectives

A Use systems of linear equations to solve interest application problems.

B Use systems of linear equations to solve mixture application problems.

If one of the numbers in a proportion is unknown, it can be replaced with a variable such as x. For example, in this section we will study two more types of applications that can be solved using systems of linear equations; interest on money invested and mixture. These applications can be "wordy" and you will need to read carefully and analyze the information thoroughly to be able to translate it into a system involving two variables.

Objective A **Interest Problems**

People in business and banking know several formulas for calculating interest. The formula used depends on the method of payment (monthly or yearly) and the type of interest (simple or compound). Also, penalties for late payments and even penalties for early payments might be involved. In any case, standard notation is the following:

$P \longrightarrow$ the principal (amount of money invested or borrowed),
$r \longrightarrow$ the rate of interest (an annual rate),
$t \longrightarrow$ the time (in one year or part of a year), and
$I \longrightarrow$ the interest (paid or earned).

In this section, we will use only the basic formula for simple interest

$$I = Prt,$$

with interest calculated on an annual basis. In this special case, we have $t = 1$ and the formula becomes

$$I = Pr.$$

Example 1

Interest

James has two investment accounts, one pays 6% interest and the other pays 10% interest. He has $1000 more in the 10% account than he has in the 6% account. Each year, the interest from the 10% account is $260 more than the interest from the 6% account. How much does he have in each account?

Solution

Careful reading indicates two types of information:
1. He has two accounts.
2. He earns two amounts of interest.

Let x = amount (principal) invested at 6%
   y = amount (principal) invested at 10%.
Then let,
   $0.06x$ = interest earned on first account and
   $0.10y$ = interest earned on second account.

Now set up two equations.

$$\begin{cases} y = x + 1000 \\ 0.10y - 0.06x = 260 \end{cases}$$

y is larger than *x* by $1000. Interest from the 10% account is $260 more than interest from the 6% account.

Because the first equation is already solved for *y*, we use the substitution method and substitute for *y* in the second equation.

$$0.10(x + 1000) - 0.06x = 260$$
$$10(x + 1000) - 6x = 26{,}000 \qquad \text{Multiply by 100.}$$
$$10x + 10{,}000 - 6x = 26{,}000$$
$$4x = 16{,}000$$
$$x = 4000 \qquad \text{amount at 6\%}$$

Back substitute $x = 4000$ into one of the original equations to find *y*.

$$y = x + 1000 = 4000 + 1000 = 5000 \qquad \text{amount at 10\%}$$

James has $4000 invested at 6% and $5000 invested at 10%.

Example 2

Interest

Lila has $7000 to invest. She decides to separate her funds into two accounts. One yields interest at the rate of 7% and the other at 12%. If she wants a total annual interest from both accounts to be $690, how should she split the money? (**Note:** The higher interest account is considered more risky. Otherwise, she would put the entire $7000 into that account.)

Solution

Again, careful reading indicates two types of information:
1. She has two accounts.
2. She earns two amounts of interest.

Let x = amount (principal) invested at 7%
    y = amount (principal) invested at 12%.
Then let
    $0.07x$ = interest earned on first account and
    $0.12y$ = interest earned on second account.

Now set up two equations.

$$\begin{cases} x + y = 7000 \\ 0.07x + 0.12y = 690 \end{cases}$$

The sum of both accounts is $7000.

The total interest from both accounts is $690.

Both equations are in standard form. Solve by addition. Multiply the first equation by –7 and the second by 100 to get opposite coefficients for x as follows:

$$\begin{cases} [-7] & (x+ \quad y = 7000) \longrightarrow \\ [100] & (0.07x + 0.12y = 690) \longrightarrow \end{cases}$$

$$\begin{array}{r} -7x - 7y = -49{,}000 \\ 7x + 12y = 69{,}000 \\ \hline 5y = 20{,}000 \\ y = 4000 \quad \text{amount at 12\%} \end{array}$$

Substitute $y = 4000$ into one of the original equations.

$x + 4000 = 7000$

$\qquad x = 3000$ amount at 7%

She should invest $3000 at 7% and $4000 at 12%.

Now work margin exercises 1 and 2.

Objective B **Mixture Problems**

Problems involving mixtures occur in physics and chemistry and in such places as candy stores or coffee shops. These problems arise when two or more items of a different percentage of concentration of a chemical such as salt, chlorine, or antifreeze are to be mixed; or two or more types of food such as coffee, nuts, or candy are to be mixed to form a final mixture that satisfies certain conditions of percentage of concentration.

The basic plan is to write an equation that deals with only one part of the mixture (such as the salt in the mixture). The following examples explain how this can be accomplished.

Example 3

Mixture

How many ounces each of a 10% salt solution and a 15% salt solution must be used to produce 50 ounces of a 12% salt solution?

Solution

Let $x =$ amount of 10% solution and
$\quad y =$ amount of 15% solution.

	amount of solution	×	percent of salt	=	amount of salt
10% solution	x		0.10		$0.10x$
15% solution	y		0.15		$0.15y$
12% solution	50		0.12		$0.12(50)$

1. Fergus has two investment accounts for his toupee company. One pays 9% interest and the other pays 12% interest. He has $800 more in the 12% account than he has in the 9% account. Each year, the interest from the 12% account is $246 more than the interest from the 9% account. How much does Fergus have in each account?

2. Darnell has $9000 to invest. He decides to separate his money into two accounts. One yields interest at the rate of 5% and the other at 9%. If he wants a total annual income from both accounts to be $550, how should he split up the money?

Then the system of linear equations is the following.

$$\begin{cases} x + y = 50 \\ 0.10x + 0.15y = 0.12(50) \end{cases}$$

The sum of the two amounts must be 50 ounces.

The sum of the amounts of salt from the two solutions equals the total amount of salt in the final solution.

Multiplying the first equation by -10 and the second by 100 gives

$$\begin{cases} [-10] \quad (x + \quad y = 50) \\ [100](0.10x + 0.15y = 0.12(50)) \end{cases} \longrightarrow \begin{array}{l} -10x - 10y = -500 \\ \underline{10x + 15y = 600} \\ 5y = 100 \\ y = 20 \quad \text{amount of 15\%} \end{array}$$

Substitute $y = 20$ into one of the original equations.

$$x + 20 = 50$$
$$x = 30 \quad \text{amount of 10\%}$$

Use 30 ounces of the 10% solution and 20 ounces of the 15% solution.

Example 4

Mixture

How many gallons of a 20% acid solution should be mixed with a 30% acid solution to produce 100 gallons of a 23% solution?

Solution

Let x = amount of 20% solution and
$\quad\;\; y$ = amount of 30% solution.

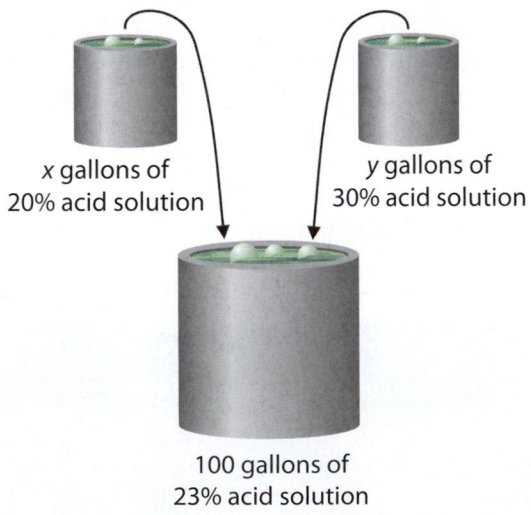

x gallons of
20% acid solution

y gallons of
30% acid solution

100 gallons of
23% acid solution

	amount of solution	×	percent of acid	=	amount of acid
20% solution	x		0.20		$0.20x$
30% solution	y		0.30		$0.30y$
23% solution	100		0.23		$0.23(100)$

Then the system of linear equations is

$$\begin{cases} x + y = 100 \\ 0.20x + 0.30y = 0.23(100) \end{cases}$$

The sum of the two amounts must be 100 gallons.

The sum of the amounts of acid from the two solutions equals the total amount of acid in the final solution.

Multiplying the first equation by -20 and the second by 100 gives,

$$\begin{cases} [-20] & (x + y = 100) \\ [100] & (0.20x + 0.30y = 0.23(100)) \end{cases}$$

$\longrightarrow$ $-20x - 20y = -2000$

$\longrightarrow$ $\underline{20x + 30y = 2300}$

$10y = 300$

$y = 30$ amount of 30%

Substitute $y = 30$ into one of the original equations.

$x + 30 = 100$

$x = 70$ amount of 20%

70 gallons of the 20% solution should be added to 30 gallons of the 30% solution. This will produce 100 gallons of a 23% solution.

Now work margin exercises 3 and 4.

3. How many ounces each of a 12% chlorine solution and a 18% chlorine solution must be used to produce 150 ounces of a 14% chlorine solution?

4. How many gallons of a 17% soap solution should be mixed with a 22% soap solution to produce 120 gallons of a 21% solution?

Exercises 8.5

Solve each problem by setting up a system of two equations in two unknowns and solving the system.

1. **Investing:** Carmen invested $9000, part in a 6% passbook account and the rest in a 10% certificate account. If her annual interest was $680, how much did she invest at each rate?

2. **Investing:** Mr. Brown has $12,000 invested. Part is invested at 6% and the remainder at 8%. If the interest from the 6% investment is $230 more than the interest from the 8% investment, how much is invested at each rate?

3. **Investing:** Ten thousand dollars is invested, part at 5.5% and part at 6%. The interest from the 5.5% investment is $251 more than the interest from the 6% investment. How much is invested at each rate?

4. **Investing:** On two investments totaling $9500, Darius lost 3% on one and earned 6% on the other. If his net annual receipts were $282, how much was each investment?

5. **Investing:** Meredith has money in two savings accounts. One rate is 8% and the other is 10%. If she has $200 more in the 10% account, how much is invested at 8% if the total interest is $101?

6. **Investing:** Money is invested at two rates. One rate is 9% and the other is 13%. If there is $700 more invested at 9%, find the amount invested at each rate if the total annual interest is $239.

7. **Investing:** Ethan has half of his investments in stock paying an 11% dividend and the other half in debentures paying 13% interest. If his total annual interest is $840, how much does he have invested?

8. **Investing:** Betty invested some of her money at 12% interest. She invested $300 more than twice that amount at 10%. How much is invested at each rate if her interest income is $318 annually?

9. **Investing:** GFA invested some money in a development yielding 24% and $9000 less in a development yielding 18%. If the first investment produces $2820 more per year than the second, how much is invested in each development?

10. **Investing:** Victoria invests a certain amount of money at 7% annual interest and three times that amount at 8%. If her annual income is $232.50, how much does she have invested at each rate?

11. **Investing:** Jamal has a certain amount of money invested at 5% annual interest and $500 more than twice that amount invested in bonds yielding 7%. His total income from interest is $187. How much does he have invested at each rate?

12. **Investing:** A total of $6000 is invested, part at 8% and the remainder at 12%. How much is invested at each rate if the annual interest is $620?

13. **Investing:** Ms. Merriman has $12,000 invested. Part is invested at 9% and the remainder at 11%. If the interest from the 9% investment is $380 more than the interest from the 11% investment, how much is invested at each rate?

14. **Investing:** Eight thousand dollars is invested, part at 15% and the remainder at 12%. If the annual interest income from the 15% investment is $66 more than the annual interest income from the 12% investment, how much is invested at each rate?

15. **Investing:** Morgan inherited $124,000 from her Uncle Edward. She invested a portion in bonds and the remainder in a long-term certificate account. The amount invested in bonds was $24,000 less than 3 times the amount invested in certificates. How much was invested in bonds and how much in certificates?

16. **Investing:** Sang has invested $48,000, part at 6% and the rest in a higher risk investment at 10%. How much did she invest at each rate to receive $4000 in interest after one year?

17. **Manufacturing:** A metallurgist has one alloy containing 20% copper and another containing 70% copper. How many pounds of each alloy must he use to make 50 pounds of a third alloy containing 50% copper?

18. **Manufacturing:** A manufacturer has received an order for 24 tons of a 60% copper alloy. His stock contains only alloys of 80% copper and 50% copper. How much of each will he need to fill the order?

19. **Tobacco:** A tobacco shop wants 50 ounces of tobacco that is 24% rare Turkish blend. How much each of a 30% Turkish blend and a 20% Turkish blend will be needed?

20. **Acid Solutions:** How many liters each of a 40% acid solution and a 55% acid solution must be used to produce 60 liters of a 45% acid solution?

21. **Dairy Production:** A dairy man wants to mix a 35% protein supplement and a standard 15% protein ration to make 1800 pounds of a high-grade 20% protein ration. How many pounds of each should he use?

22. **Manufacturing:** To meet the government's specifications, a certain alloy must be 65% aluminum. How many pounds each of a 70% aluminum alloy and a 54% aluminum alloy will be needed to produce 640 pounds of the 65% aluminum alloy?

23. **Food Science:** A meat market has ground beef that is 40% fat and extra lean ground beef that is only 15% fat. How many pounds of each will be needed to obtain 50 pounds of lean ground beef that is 25% fat?

24. **Gasoline:** George decides to mix grades of gasoline in his truck. He puts in 8 gallons of regular and 12 gallons of premium for a total cost of $55.80. If premium gasoline costs $0.15 more per gallon than regular, what was the price of each grade of gasoline?

25. **Acid Solutions:** How many grams of pure acid (100% acid) and how many grams of a 40% solution should be mixed together to get a total of 30 grams of a 60% solution?

26. **Coffee:** Pure dark coffee beans are to be mixed with a mixture that is 60% dark beans. How much of each (pure dark and 60% dark beans) should be used to get a mixture of 50 pounds that contains 70% of the dark beans?

27. **Salt Solutions:** Salt is to be added to a 4% salt solution. How many ounces of salt and how many ounces of the 4% solution should be mixed together to get 60 ounces of a 20% salt solution?

28. **Chemistry:** How many liters each of a 12% iodine solution and a 30% iodine solution must be used to produce a total mixture of 90 liters of a 22% iodine solution?

29. Food Science: A candy maker is making truffles using a mixture of a melted dark chocolate that is 72% cocoa and milk chocolate that is 42% cocoa. If she wants 6 pounds of melted chocolate that is 52% cocoa, how much of each type of chocolate does she need?

30. Dairy Farming: A dairy needs 360 gallons of milk containing 4% butterfat. How many gallons each of milk containing 5% butterfat and milk containing 2% butterfat must be used to obtain the desired 360 gallons?

31. Body Piercing: It is recommended that one cleans new body piercing with a 1% salt solution for the first few weeks with the new piercing. You have a 0.5% solution and a 5% solution and you need to make 8 ounces of the 1% solution. How much of the 0.5% and 5% solutions will you need? (Round your answers to the nearest hundredth.)

32. Pharmacy: A druggist has two solutions of alcohol. One is 25% alcohol. The other is 45% alcohol. He wants to mix these two solutions to get 36 ounces that will be 30% alcohol. How many ounces of each of these two solutions should he mix together?

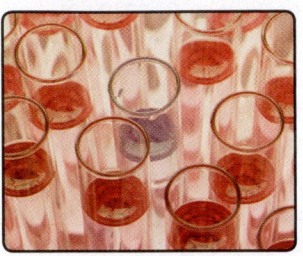

Writing & Thinking

33. Your friend has $20,000 to invest and decided to invest part at 4% interest and the rest at 10% interest. Why might you advise him (or her) to invest all of it at

    a. 4%?

    b. 10%?

8.6 Systems of Linear Equations: Three Variables

Objective A **Solving Systems of Linear Equations in Three Variables**

The equation $2x + 3y - z = 16$ is called a **linear equation in three variables**. The general form is

$$Ax + By + Cz = D \text{ where } A, B, \text{ and } C \text{ are not all equal to } 0.$$

The solutions to such equations are called **ordered triples** and are of the form (x_0, y_0, z_0). One ordered triple that satisfies the equation $2x + 3y - z = 16$ is $(1, 4, -2)$. To check this, substitute $x = 1$, $y = 4$, and $z = -2$ into the equation to see if the result is 16:

$$2(1) + 3(4) - (-2) = 2 + 12 + 2$$
$$= 16.$$

There are an infinite number of ordered triples that satisfy any linear equation in three variables in which at least two of the coefficients are nonzero. Any two values may be substituted for two of the variables, and then the value for the third variable can be calculated. For example, by letting $x = -1$ and $y = 5$ in the equation $2x + 3y - z = 16$, we find:

$$2(-1) + 3(5) - z = 16$$
$$-2 + 15 - z = 16$$
$$-z = 3$$
$$z = -3.$$

Hence, the ordered triple $(-1, 5, -3)$ also satisfies the equation $2x + 3y - z = 16$. Graphs can be drawn in three dimensions by using a coordinate system involving three mutually perpendicular number lines labeled as the x-axis, y-axis, and z-axis. Three planes are formed: the xy-plane, the xz-plane, and the yz-plane. The three axes separate space into eight regions called octants. You can "picture" the first octant as the region bounded by the floor of a room and two walls with the axes meeting in a corner. The floor is the xy-plane. The axes can be ordered in a "right-hand" or "left-hand" format. Figure 1 shows the point represented by the ordered triple $(2, 3, 1)$ in a right-hand system.

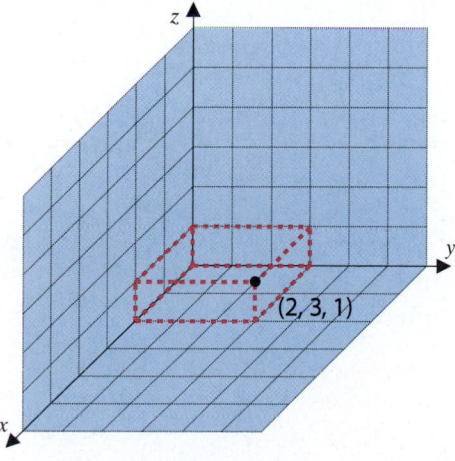

Figure 1

The graphs of linear equations in three variables are planes in three dimensions. A portion of the graph of $2x + 3y - z = 16$ appears in Figure 2.

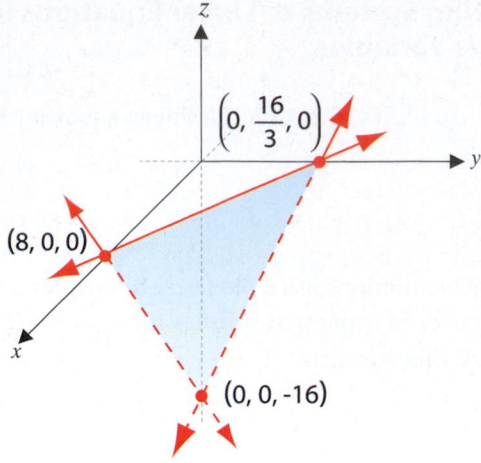

Figure 2

Two distinct planes will either be parallel or they will intersect. If they intersect, their intersection will be a line. If three distinct planes intersect, they will intersect in a line or in a single point represented by an ordered triple.

The graphs of systems of three linear equations in three variables can be both interesting and informative, but they can be difficult to sketch and points of intersection difficult to estimate. Also, most graphing calculators are limited to graphs in two dimensions, so they are not useful in graphically analyzing systems of linear equations in three variables.

Therefore, in this text, only algebraic techniques for solving these systems will be discussed. Figure 3 illustrates four different possibilities for the relative positions of three planes.

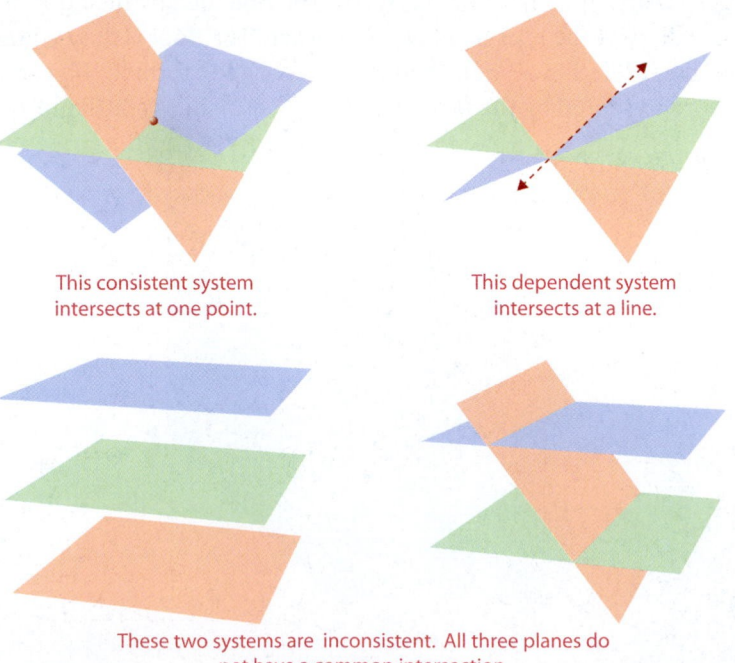

This consistent system intersects at one point.

This dependent system intersects at a line.

These two systems are inconsistent. All three planes do not have a common intersection.

Figure 3

To Solve a System of Three Linear Equations in Three Variables

1. Select two equations and eliminate one variable by using the addition method.
2. Select a different pair of equations and eliminate the **same** variable.
3. Steps 1 and 2 give **two** linear equations in **two** variables. Solve these equations by either addition or substitution as discussed in Sections 8.2 and 8.3.
4. Back substitute the values found in Step 3 into any one of the original equations to find the value of the third variable.
5. Check the solution (if one exists) in all three of the original equations.

The solution possibilities for a system of three equations in three variables are as follows:

1. There will be exactly one ordered triple solution. (Graphically, the three planes intersect at one point.)
2. There will be an infinite number of solutions. (Graphically, the three planes intersect in a line or are the same plane.)
3. There will be no solutions. (Graphically, there are no points common to all 3 planes.)

The technique is illustrated with the following system.

$$\begin{cases} 2x + 3y - z = 16 & \text{(I)} \\ x - y + 3z = -9 & \text{(II)} \\ 5x + 2y - z = 15 & \text{(III)} \end{cases}$$

Step 1: Using equations (I) and (II), eliminate y.

Note: We could just as easily have chosen to begin by eliminating x or z. To be sure that you understand the process you might want to solve the system by first eliminating x and then again by first eliminating z. In all cases, the answer will be the same.

$$\begin{array}{ll} \text{(I)} & (2x + 3y - z = 16) \longrightarrow \quad 2x + 3y - z = 16 \\ \text{(II)} \quad [3] & (x - y + 3z = -9) \longrightarrow \quad \underline{3x - 3y + 9z = -27} \\ & \qquad\qquad\qquad\qquad\qquad\qquad\quad 5x \qquad\; + 8z = -11 \;\text{(IV)} \end{array}$$

Step 2: Using a different pair of equations, (II) and (III), eliminate the **same** variable, y.

$$\begin{array}{ll} \text{(II)} \quad [2] & (x - y + 3z = -9) \longrightarrow \quad 2x - 2y + 6z = -18 \\ \text{(III)} & (5x + 2y - z = 15) \longrightarrow \quad \underline{5x + 2y - z = 15} \\ & \qquad\qquad\qquad\qquad\qquad\qquad\quad 7x \qquad\; + 5z = -3 \;\text{(V)} \end{array}$$

Step 3: Using the results of Steps 1 and 2, solve the two equations for x and z.

$$(IV) \begin{cases} [-7] & (5x + 8z = -11) \\ (V) & [5] & (7x + 5z = -3) \end{cases} \longrightarrow \begin{array}{r} -35x - 56z = 77 \\ 35x + 25z = -15 \\ \hline -31z = 62 \\ z = -2 \end{array}$$

Back substitute $z = -2$ into equation (IV) to find x.

$5x + 8(-2) = -11$ use equation (IV)

$\qquad 5x - 16 = -11$

$\qquad\qquad 5x = 5$

$\qquad\qquad x = 1$

Step 4: Using $x = 1$ and $z = -2$, back substitute to find y.

$\qquad 1 - y + 3(-2) = -9$ use equation (II)

$\qquad\qquad -y - 5 = -9$

$\qquad\qquad -y = -4$

$\qquad\qquad y = 4$

The solution is $(1, 4, -2)$. The solution can be checked by substituting the results into **all three** of the original equations.

$$\begin{cases} 2(1) + 3(4) - (-2) = 16 & (I) \\ (1) - (4) + 3(-2) = -9 & (II) \\ 5(1) + 2(4) - (-2) = 15 & (III) \end{cases}$$

Example 1

Three Variables (A System with One Solution)

Solve the following system of linear equations.

$$\begin{cases} x - y + 2z = -4 & (I) \\ 2x + 3y + z = \dfrac{1}{2} & (II) \\ x + 4y - 2z = 4 & (III) \end{cases}$$

Solution

Using equations (I) and (III), eliminate z.

$$\begin{array}{ll} (I) & x - y + 2z = -4 \\ (III) & x + 4y - 2z = 4 \\ \hline & 2x + 3y = 0 \quad (IV) \end{array}$$

Using equations (I) and (II), eliminate z.

$$(I) \begin{cases} (x - y + 2z = -4) \\ (II) \ [-2] \left(2x + 3y + z = \dfrac{1}{2}\right) \end{cases} \longrightarrow \begin{array}{r} x - y + 2z = -4 \\ -4x - 6y - 2z = -1 \\ \hline -3x - 7y = -5 \ (V) \end{array}$$

Eliminate the variable x using the two equations in x and y.

$$\begin{array}{lll} \text{(IV)} & [3] & (2x + 3y = 0) \longrightarrow & 6x + 9y = 0 \\ \text{(V)} & [2] & (-3x - 7y = -5) \longrightarrow & \underline{-6x - 14y = -10} \\ & & & -5y = -10 \\ & & & y = 2 \end{array}$$

Back substituting into equation (IV) to find x yields:

$$2x + 3(2) = 0$$
$$2x + 6 = 0$$
$$2x = -6$$
$$x = -3.$$

Finally, using $x = -3$ and $y = 2$, back substitute into (I).

$$(-3) - (2) + 2z = -4$$
$$-5 + 2z = -4$$
$$2z = 1$$
$$z = \frac{1}{2}$$

The solution is $\left(-3, 2, \dfrac{1}{2}\right)$.

Check: The solution can be checked by substituting $\left(-3, 2, \dfrac{1}{2}\right)$ into all three of the original equations.

$$\begin{cases} (-3) - (2) + 2\left(\dfrac{1}{2}\right) = -4 & \text{(I)} \\[2mm] 2(-3) + 3(2) + \left(\dfrac{1}{2}\right) = \dfrac{1}{2} & \text{(II)} \\[2mm] (-3) + 4(2) - 2\left(\dfrac{1}{2}\right) = 4 & \text{(III)} \end{cases}$$

Example 2

Three Variables (A System with No Solution)

Solve the following system of linear equations.

$$\begin{cases} 3x - 5y + z = 6 & \text{(I)} \\ x - y + 3z = -1 & \text{(II)} \\ 2x - 2y + 6z = 5 & \text{(III)} \end{cases}$$

Solve the following system of linear equations.

1. $\begin{cases} 2x + y + z = 4 \\ x + 2y + z = 1 \\ 3x + y - z = -3 \end{cases}$

2. $\begin{cases} 7x + 2y - z = 7 \\ 3x - y - 2z = 1 \\ 9x - 3y - 6z = -3 \end{cases}$

3. $\begin{cases} y + 3z = 1 \\ x + y = 3 \\ x + 2y + 3z = 4 \end{cases}$

Solution

Using equations (I) and (II), eliminate z.

$\begin{array}{ll} \text{(I)} & [-3]\,(3x - 5y + z = 6) \\ \text{(II)} & \quad\ \ (x - y + 3z = -1) \end{array}$ ⟶ $\begin{array}{r} -9x + 15y - 3z = -18 \\ x - y + 3z = -1 \\ \hline -8x + 14y \quad\quad = -19 \ \text{(IV)} \end{array}$

Using equations (II) and (III), eliminate z.

$\begin{array}{ll} \text{(II)} & [-2]\,(x - y + 3z = -1) \\ \text{(III)} & \quad\ \ (2x - 2y + 6z = 5) \end{array}$ ⟶ $\begin{array}{r} -2x + 2y - 6z = 2 \\ 2x - 2y + 6z = 5 \\ \hline 0 = 7 \ \text{(V)} \end{array}$

This last equation is **false**. Thus the system has **no solution**.

Example 3

Three Variables (A System with Infinitely Many Solutions)

Solve the following system of linear equations.

$\begin{cases} 3x - 2y + 4z = 1 & \text{(I)} \\ 3x - y + 7z = -1 & \text{(II)} \\ 6x - 5y + 5z = 4 & \text{(III)} \end{cases}$

Solution

Using equations (I) and (II), eliminate x.

$\begin{array}{ll} \text{(I)} & \quad\ \ (3x - 2y + 4z = 1) \\ \text{(II)} & [-1]\,(3x - y + 7z = -1) \end{array}$ ⟶ $\begin{array}{r} 3x - 2y + 4z = 1 \\ -3x + y - 7z = 1 \\ \hline -y - 3z = 2 \ \text{(IV)} \end{array}$

Using equations (I) and (III), eliminate x.

$\begin{array}{ll} \text{(I)} & [-2]\,(3x - 2y + 4z = 1) \\ \text{(III)} & \quad\ \ (6x - 5y + 5z = 4) \end{array}$ ⟶ $\begin{array}{r} -6x + 4y - 8z = -2 \\ 6x - 5y + 5z = 4 \\ \hline -y - 3z = 2 \ \text{(V)} \end{array}$

Using equations (IV) and (V), eliminate y.

$\begin{array}{ll} \text{(IV)} & \quad\ \ (-y - 3z = 2) \\ \text{(V)} & [-1]\,(-y - 3z = 2) \end{array}$ ⟶ $\begin{array}{r} -y - 3z = 2 \\ y + 3z = -2 \\ \hline 0 = 0 \end{array}$

Because this last equation, $0 = 0$, is **always true**, the system has an **infinite number of solutions**.

Now work margin exercises 1 through 3.

Example 4

Three Variables (Application)

A cash register contains $341 in $20, $5, and $2 bills. There are twenty-eight bills in all and three more $2 bills than $5 bills. How many bills of each kind are there?

Solution

Let x = number of $20 bills,

y = number of $5 bills, and

z = number of $2 bills.

$$\begin{cases} x + y + z = 28 & \text{(I)} \\ 20x + 5y + 2z = 341 & \text{(II)} \\ z = y + 3 & \text{(III)} \end{cases}$$

There are twenty-eight total bills.

The total value is $341.

There are three more $2 bills than $5 bills.

Using equations (I) and (II), eliminate x.

$$\begin{array}{rl} \text{(I)} & [-20] \quad (x + y + z = 28) \rightarrow -20x - 20y - 20z = -560 \\ \text{(II)} & \quad (20x + 5y + 2z = 341) \rightarrow \underline{20x + 5y + 2z = 341} \\ & \qquad\qquad\qquad\qquad\qquad\qquad\quad -15y - 18z = -219 \ \text{(IV)} \end{array}$$

As equation (III) already has no x term, we rewrite it in the form $y - z = -3$ and use this equation along with the equation just found.

$$\begin{array}{rl} \text{(III)} & [15] \quad (y - z = -3) \longrightarrow 15y - 15z = -45 \\ \text{(IV)} & \quad (-15y - 18z = -219) \longrightarrow \underline{-15y - 18z = -219} \\ & \qquad\qquad\qquad\qquad\qquad\qquad -33z = -264 \\ & \qquad\qquad\qquad\qquad\qquad\qquad\quad z = 8 \end{array}$$

Back substituting into equation (III) to solve for y gives:

$8 = y + 3$

$5 = y$.

Now we can substitute the values $z = 8$ and $y = 5$ into equation (I).

$x + 5 + 8 = 28$

$x + 13 = 28$

$x = 15$

There are fifteen $20 bills, five $5 bills, and eight $2 bills.

Now work margin exercise 4.

Practice Problems

Solve the following system of linear equations. $\begin{cases} 2x + y + z = 4 \\ x + 2y + z = 1 \\ 3x + y - z = -3 \end{cases}$

Practice Problem Answers

1. $(1, -2, 4)$

---

4. A piggy bank contains $21.70 in $1 coins, quarters and dimes. There are 37 coins in all and 5 more quarters than dimes. How many coins of each kind are there?

Exercises 8.6

Solve each system of equations. State which systems, if any, have no solution or an infinite number of solutions. See Examples 1 through 3.

1. $\begin{cases} x + y - z = 0 \\ 3x + 2y + z = 4 \\ x - 3y + 4z = 5 \end{cases}$

2. $\begin{cases} x - y + 2z = 3 \\ -6x + y + 3z = 7 \\ x + 2y - 5z = -4 \end{cases}$

3. $\begin{cases} 2x - y - z = 1 \\ 2x - 3y - 4z = 0 \\ x + y - z = 4 \end{cases}$

4. $\begin{cases} y + z = 6 \\ x + 5y - 4z = 4 \\ x - 3y + 5z = 7 \end{cases}$

5. $\begin{cases} x + y - 2z = 4 \\ 2x + y = 1 \\ 5x + 3y - 2z = 6 \end{cases}$

6. $\begin{cases} x - y + 5z = -6 \\ x + 2z = 0 \\ 6x + y + 3z = 0 \end{cases}$

7. $\begin{cases} y + z = 2 \\ x + z = 5 \\ x + y = 5 \end{cases}$

8. $\begin{cases} 2y + z = -4 \\ 3x + 4z = 11 \\ x + y = -2 \end{cases}$

9. $\begin{cases} x - y + 2z = -3 \\ 2x + y - z = 5 \\ 3x - 2y + 2z = -3 \end{cases}$

10. $\begin{cases} x - y - 2z = 3 \\ x + 2y + z = 1 \\ 3y + 3z = -2 \end{cases}$

11. $\begin{cases} 2x - y + 5z = -2 \\ x + 3y - z = 6 \\ 4x + y + 3z = -2 \end{cases}$

12. $\begin{cases} 2x - y + 5z = 5 \\ x - 2y + 3z = 0 \\ x + y + 4z = 7 \end{cases}$

13. $\begin{cases} 3x + y + 4z = -6 \\ 2x + 3y - z = 2 \\ 5x + 4y + 3z = 2 \end{cases}$

14. $\begin{cases} 2x + y - z = -3 \\ -x + 2y + z = 5 \\ 2x + 3y - 2z = -3 \end{cases}$

15. $\begin{cases} x - 2y + z = 7 \\ x - y - 4z = -4 \\ x + 4y - 2z = -5 \end{cases}$

16. $\begin{cases} 2x - 2y + 3z = 4 \\ x - 3y + 2z = 2 \\ x + y + z = 1 \end{cases}$

17. $\begin{cases} 2x + 3y + z = 4 \\ 3x - 5y + 2z = -5 \\ 4x - 6y + 3z = -7 \end{cases}$

18. $\begin{cases} x + y + z = 3 \\ 2x - y - 2z = -3 \\ 3x + 2y + z = 4 \end{cases}$

19. $\begin{cases} 2x - 3y + z = -1 \\ 6x - 9y - 4z = 4 \\ 4x + 6y - z = 5 \end{cases}$

20. $\begin{cases} x + 6y + z = 6 \\ 2x + 3y - 2z = 8 \\ 2x + 4z = 3 \end{cases}$

Solve each word problem. See Example 4.

21. The sum of three integers is 67. The sum of the first and second integers is 13 more than the third integer. The third integer is 7 less than the first. Find the three integers.

22. The sum of three integers is 189. The first integer is 28 less than the second. The second integer is 21 less than the sum of the first and third integers. Find the three integers.

23. Money: A wallet contains $218 in $10, $5, and $1 bills. There are forty-six bills in all and four more fives than tens. How many bills of each kind are there?

24. Money: Sally is trying to get her brother Robert to learn to think algebraically. She tells him that she has 23 coins in her purse, including nickels, dimes, and quarters. She has two more dimes than quarters, and the total value of the coins is $2.50. How many of each kind of coin does she have?

25. Perimeter of a Triangle: The perimeter of a triangle is 73 cm. The longest side is 13 cm less than the sum of the other two sides. The shortest side is 11 cm less than the longest side. Find the lengths of the three sides.

26. Triangles: The sum of the measures of the three angles of a triangle is 180°. In one particular triangle, the largest angle is 10° more than three times the smallest angle, and the third angle is one-half the largest angle. What are the measures of the three angles?

27. Theater: A theater has three types of seats for Broadway plays: main floor, balcony, and mezzanine. Main floor tickets are $60, balcony tickets are $45, and mezzanine tickets are $30. Opening night the sales totaled $27,600. Main floor sales were 20 more than the total of balcony and mezzanine sales. Mezzanine sales were 40 more than two times balcony sales. How many of each type of ticket was sold?

28. Fruit Stand: At Steve's Fruit Stand, 4 pounds of bananas, 2 pounds of apples, and 3 pounds of grapes cost $16.40. Five pounds of bananas, 4 pounds of apples, and 2 pounds of grapes cost $16.60. Two pounds of bananas, 3 pounds of apples, and 1 pound of grapes cost $9.60. Find the price per pound of each kind of fruit.

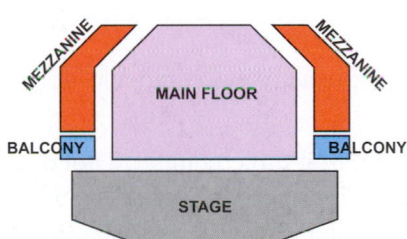

29. Flower Arranging: A florist is creating bridesmaids' bouquets for a wedding. Each 16 flower bouquet will have a mixture of lilies, roses, and daisies and cost $92. Lilies cost $10, roses cost $6, and daisies cost $4 a stem. If each bouquet will have as many daisies as it will roses and lilies combined, how many of each type of flower will be in the bouquet?

30. Construction: The Tates are having a house built. The cost is split up into three parts: the house, the lot, and the improvements. The cost of building the house is $16,000 more than three times the cost of the lot. The cost of the improvements (the landscaping, sidewalks, and upgrades) is one-third the cost of the lot. If the total cost is $159,000, what is the cost of each part of the construction?

31. Investing: Kirk inherited $100,000 from his aunt and decided to invest in three different accounts: savings, bonds, and stocks. The amount in his bond account was $10,000 more than three times the amount in his stock account. At the end of the first year, the savings account returned 5%, the bond 8%, and the stocks 10% for total interest of $7400. How much did he invest in each account?

32. Stock Market: Melissa has saved a total of $30,000 and wants to invest in three different stocks: PepsiCo, IBM, and Microsoft. She wants the PepsiCo amount to be $1000 less than twice the IBM amount and the Microsoft amount to be $2000 more than the total in the other two stocks. How much should she invest in each stock?

33. Chemistry: A chemist wants to mix 9 liters of a 25% acid solution. Because of limited amounts on hand, the mixture is to come from three different solutions, one with 10% acid, another with 30% acid, and a third with 40% acid. The amount of the 10% solution must be twice the amount of the 40% solution, and the amount of the 30% solution must equal the total amount of the other two solutions. How much of each solution must be used?

34. Manufacturing: An appliance company makes three versions of their popular stand mixer. The standard model has production costs of $100, the deluxe model of $175, and the premium model of $250. On any given day the factory line makes 140 mixers and spends $21,500 on production costs. If the number of standard mixers made equals the sum of the premium and deluxe models made, how many of each kind of mixer are made each day?

Writing & Thinking

35. Is it possible for three linear equations in three unknowns to have exactly two solutions? Explain your reasoning in detail.

36. In geometry, three non-collinear points determine a plane. (That is, if three points are not on a line, then there is a unique plane that contains all three points.) Find the values of A, B, and C (and therefore the equation of the plane) given $Ax + By + Cz = 3$ and the three points on the plane $(0, 3, 2)$, $(0, 0, 1)$ and $(-3, 0, 3)$. Sketch the plane in three dimensions as best you can by locating the three given points.

37. As stated in Exercise 36, three non-collinear points determine a plane. Find the values of A, B, and C (and therefore the equation of the plane) given $Ax + By + Cz = 10$ and the three points on the plane $(2, 0, -2)$, $(3, -1, 0)$ and $(-1, 5, -4)$. Sketch the plane in three dimensions as best you can by locating the three given points.

8.7 Matrices and Gaussian Elimination

Objective A   **Matrices**

A rectangular array of numbers is called a **matrix** (plural **matrices**). Matrices are usually named with capital letters, and each number in the matrix is called an **entry**. Entries written horizontally are said to form a **row**, and entries written vertically are said to form a **column**. The matrix A shown below has two rows and three columns and is a **2 × 3 matrix** (read "two by three matrix"). We say that the **dimension** of the matrix is the number of rows by the number of columns, which for this matrix is two by three (or 2×3). Similarly, if a matrix has three rows and two columns then its dimension is 3×2.

$$A = \begin{bmatrix} 3 & 4 & 0 \\ 7 & -2 & 5 \end{bmatrix} \qquad \begin{bmatrix} 3 & 4 & 0 \\ 7 & -2 & 5 \end{bmatrix} \leftarrow \text{row 1} \atop \leftarrow \text{row 2} \qquad \begin{matrix} \text{column 1} \quad \text{column 2} \quad \text{column 3} \\ \begin{bmatrix} 3 & 4 & 0 \\ 7 & -2 & 5 \end{bmatrix} \end{matrix}$$

Three more examples are:

$$B = \begin{bmatrix} 5 & -1 \\ 2 & 3 \end{bmatrix} \qquad C = \begin{bmatrix} 5 & -1 & 0 & 7 \\ 2 & 3 & 2 & 8 \\ 1 & -3 & 0 & 6 \end{bmatrix} \qquad D = \begin{bmatrix} 0 & 4 \\ 1 & 6 \\ -1 & 3 \end{bmatrix}$$

$$2 \times 2 \text{ matrix} \qquad 3 \times 4 \text{ matrix} \qquad 3 \times 2 \text{ matrix}$$

A matrix with the same number of rows as columns is called a **square matrix**. Matrix B (shown above) is a square 2×2 matrix.

Objective B   **Coefficient Matrices and Augmented Matrices**

Matrices have many uses and are generated from various types of problems because they allow data to be presented in a systematic and orderly manner. (Business majors may want to look up a topic called Markov chains.)

Also, matrices can sometimes be added, subtracted, and multiplied. Some square matrices have inverses, much the same as multiplicative inverses for real numbers. Matrix methods of solving systems of linear equations can be done manually or with graphing calculators and computers. These topics are presented in courses such as finite mathematics and linear algebra. In this text, we will see that matrices can be used to solve systems of linear equations in which the equations are written in standard form. The two matrices derived from such a system are the **coefficient matrix** (made up of the coefficients of the variables) and the **augmented matrix** (including the coefficients and the constant terms). For example:

System	**Coefficient Matrix**	**Augmented Matrix**
$\begin{cases} x - y + z = -6 \\ 2x + 3y = 17 \\ x + 2y + 2z = 7 \end{cases}$	$\begin{bmatrix} 1 & -1 & 1 \\ 2 & 3 & 0 \\ 1 & 2 & 2 \end{bmatrix}$	$\begin{bmatrix} 1 & -1 & 1 & -6 \\ 2 & 3 & 0 & 17 \\ 1 & 2 & 2 & 7 \end{bmatrix}$

coefficients   constants

Objectives

A   Understand the concept of a matrix.

B   Write a system of equations as a coefficient matrix and an augmented matrix.

C   Become familiar with elementary row operations.

D   Become familiar with the general notation for a matrix and a system of equations.

E   Solve systems of linear equations by using the Gaussian elimination method.

F   Use a graphing calculator to solve a system of linear equations with matrices.

Note that 0 is the entry in the second row, third column of both matrices. This 0 corresponds to the missing z-variable in the second equation. The second equation could have been written $2x + 3y + 0z = 17$.

As discussed in an earlier section, notation with a small number to the right and below a variable is called **subscript** notation. For example, a_1 is read "a sub one," b_3 is read "b sub three" and R_2 is read "R sub two."

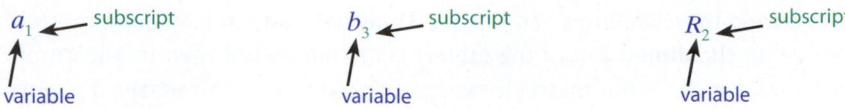

When working with matrices, R_1 represents row 1, R_2 represents row 2, and so on.

Objective C   Elementary Row Operations

In solving these systems, we can make any of the following three manipulations **without changing the solution set of the system**.

The system $\begin{cases} x - y + z = -6 \\ 2x + 3y + 0z = 17 \\ x + 2y + 2z = 7 \end{cases}$ is used here to illustrate some possibilities.

1. Any two equations may be interchanged.

$$\begin{cases} x - y + z = -6 \\ 2x + 3y + 0z = 17 \\ x + 2y + 2z = 7 \end{cases} \xrightarrow{R_1 \leftrightarrow R_2} \begin{cases} 2x + 3y + 0z = 17 \\ x - y + z = -6 \\ x + 2y + 2z = 7 \end{cases}$$

Here we have interchanged the first two equations.

2. All terms of any equation may be multiplied by a constant.

$$\begin{cases} x - y + z = -6 \\ 2x + 3y + 0z = 17 \\ x + 2y + 2z = 7 \end{cases} \xrightarrow{-2R_1} \begin{cases} -2x + 2y - 2z = 12 \\ 2x + 3y + 0z = 17 \\ x + 2y + 2z = 7 \end{cases}$$

Here we have multiplied each term of the first equation by –2.

3. All terms of any equation may be multiplied by a constant and these new terms may be added to like terms of another equation. **The original equation remains unchanged.**

$$\begin{cases} x - y + z = -6 \\ 2x + 3y + 0z = 17 \\ x + 2y + 2z = 7 \end{cases} \xrightarrow{R_3 - R_1} \begin{cases} x - y + z = -6 \\ 2x + 3y + 0z = 17 \\ 0x + 3y + z = 13 \end{cases}$$

Here we have multiplied the first equation by –1 (mentally) and added the results to the third equation.

When dealing with matrices, the three operations mentioned above are called **elementary row operations**. These operations are listed below and illustrated in Example 1. Follow the steps outlined in Example 1 carefully, and note how the row operations are indicated, such as $\frac{1}{2}R_3$ to indicate that all numbers in row 3 are multiplied by $\frac{1}{2}$. (Reasons for using these row operations are discussed under **Gaussian elimination** in Objective E.)

Elementary Row Operations

1. Interchange two rows.
2. Multiply a row by a nonzero constant.
3. Add a multiple of a row to another row.

If any elementary row operation is applied to a matrix, the new matrix is said to be **row-equivalent** to the original matrix.

Example 1

Coefficient and Augmented Matrices

a. For the system $\begin{cases} y + z = 6 \\ x + 5y - 4z = 4 \\ 2x - 6y + 10z = 14 \end{cases}$, write the corresponding coefficient matrix and the corresponding augmented matrix.

Solution

Coefficient Matrix

$$\begin{bmatrix} 0 & 1 & 1 \\ 1 & 5 & -4 \\ 2 & -6 & 10 \end{bmatrix}$$

Augmented Matrix

$$\left[\begin{array}{ccc|c} 0 & 1 & 1 & 6 \\ 1 & 5 & -4 & 4 \\ 2 & -6 & 10 & 14 \end{array}\right]$$

b. In the augmented matrix in Example **1a.**, interchange rows 1 and 2 and multiply row 3 by $\dfrac{1}{2}$.

Solution

$$\left[\begin{array}{ccc|c} 0 & 1 & 1 & 6 \\ 1 & 5 & -4 & 4 \\ 2 & -6 & 10 & 14 \end{array}\right] \xrightarrow[\tfrac{1}{2}R_3]{R_1 \leftrightarrow R_2} \left[\begin{array}{ccc|c} 1 & 5 & -4 & 4 \\ 0 & 1 & 1 & 6 \\ 1 & -3 & 5 & 7 \end{array}\right]$$

c. For the system $\begin{cases} x - y = 5 \\ 3x + 4y = 29 \end{cases}$, write the corresponding coefficient matrix and the corresponding augmented matrix.

Solution

Coefficient Matrix

$$\begin{bmatrix} 1 & -1 \\ 3 & 4 \end{bmatrix}$$

Augmented Matrix

$$\left[\begin{array}{cc|c} 1 & -1 & 5 \\ 3 & 4 & 29 \end{array}\right]$$

1. a. Write the coefficient matrix and the augmented matrix for the following system.

$$\begin{cases} 2x + y + z = 4 \\ x + 2y + z = 1 \\ 3x + y - z = -3 \end{cases}$$

b. In the augmented matrix you found in part **a.**, interchange rows 1 and 3 and multiply row 2 by 3.

c. In the augmented matrix you found in part **a.**, add 2 times row 2 to row 1.

d. In the augmented matrix in Example **1c.**, add −3 times row 1 to row 2. (**Note:** Row 1 is unchanged in the resulting matrix. Only row 2 is changed.)

Solution

$$\begin{bmatrix} 1 & -1 & | & 5 \\ 3 & 4 & | & 29 \end{bmatrix} \xrightarrow{R_2 - 3R_1} \begin{bmatrix} 1 & -1 & | & 5 \\ 0 & 7 & | & 14 \end{bmatrix}$$

Now work margin exercise 1.

Objective D **General Notation for a Matrix and a System of Equations**

With matrices we use capital letters to name a matrix and **double subscript** notation with corresponding lower case letters to indicate both the row and column location of an entry. For example, in a matrix A, the entries will be designated as described below and as shown on the following page.

a_{11} is read "a sub one one" and indicates the entry in the first row and first column; a_{12} is read "a sub one two" and indicates the entry in the first row and second column; a_{13} is read "a sub one three" and indicates the entry in the first row and third column; a_{21} is read "a sub two one" and indicates the entry in the second row and first column; and so on.

notes

- We will see in dealing with polynomials later that a_{11} can be read simply as "a sub eleven." However, with matrices, we need to indicate the row and column corresponding to the entry. If there are more than nine rows or columns, then commas are used to separate the numbers as $a_{10,10}$. You will see the commas in use on your calculator.

With double subscript notation we can write the general form of a 2×3 matrix A and a 3×3 matrix B as follows.

$$A = \begin{bmatrix} a_{11} & a_{12} & a_{13} \\ a_{21} & a_{22} & a_{23} \end{bmatrix} \qquad B = \begin{bmatrix} b_{11} & b_{12} & b_{13} \\ b_{21} & b_{22} & b_{23} \\ b_{31} & b_{32} & b_{33} \end{bmatrix}$$

We will use this notation when discussing the use of calculators to define matrices and to operate with matrices. The general form of a system of three linear equations might use this notation in the following way.

$$\begin{cases} a_{11}x + a_{12}y + a_{13}z = k_1 \\ a_{21}x + a_{22}y + a_{23}z = k_2 \\ a_{31}x + a_{32}y + a_{33}z = k_3 \end{cases}$$

A matrix is in **upper triangular form** (or just **triangular form** for our purposes) if its entries in the lower left triangular region are all 0's. The **main diagonal** consists of the entries in the positions of b_{11}, b_{22}, and b_{33}. Note that in the main diagonal the column and row indices are equal. Thus if all the entries below the main diagonal of a matrix are all 0's, the matrix is in triangular form, as shown below.

$$B = \begin{bmatrix} b_{11} & b_{12} & b_{13} \\ 0 & b_{22} & b_{23} \\ 0 & 0 & b_{33} \end{bmatrix}$$

The upper triangular form of a matrix with all 1's in the main diagonal is called the **row echelon form** (or **ref**). We will see that a graphing calculator can be used to change a matrix into the row echelon form.

Objective E ## Gaussian Elimination Method

Another method of solving a system of linear equations is the **Gaussian elimination** method (named after the famous German mathematician Carl Friedrich Gauss, 1777 – 1855). This method makes use of augmented matrices and elementary row operations. The objective is to transform an augmented matrix into row echelon form and then use back substitution to find the values of the variables. The method is outlined as follows:

Strategy for Gaussian Elimination

1. Write the augmented matrix for the system.
2. Use elementary row operations to transform the matrix into row echelon form.
3. Solve the corresponding system of equations by using back substitution.

The following examples illustrate the method. Study the steps and the corresponding comments carefully.

2. Solve the following system of linear equations using the Gaussian elimination method with back substitution.

$$\begin{cases} 4x - 4y = 20 \\ 3x + 6y = -3 \end{cases}$$

Example 2

Gaussian Elimination

Solve the following system of linear equations by using the Gaussian elimination method with back substitution.

$$\begin{cases} 2x + 4y = -6 \\ 5x - y = 7 \end{cases}$$

Solution

Step 1: Write the augmented matrix.

$$\begin{bmatrix} 2 & 4 & | & -6 \\ 5 & -1 & | & 7 \end{bmatrix}$$

(The following steps show how to use elementary row operations to get the matrix in row echelon form.)

Step 2: Multiply row 1 by $\dfrac{1}{2}$ so that the entry in the upper left corner will be 1. This will help to get 0 below the 1 in the next step.

$$\begin{bmatrix} 2 & 4 & | & -6 \\ 5 & -1 & | & 7 \end{bmatrix} \xrightarrow{\frac{1}{2}R_1} \begin{bmatrix} 1 & 2 & | & -3 \\ 5 & -1 & | & 7 \end{bmatrix}$$

Step 3: To get 0 in the lower left corner, add -5 times row 1 to row 2.

$$\begin{bmatrix} 1 & 2 & | & -3 \\ 5 & -1 & | & 7 \end{bmatrix} \xrightarrow{R_2 - 5R_1} \begin{bmatrix} 1 & 2 & | & -3 \\ 0 & -11 & | & 22 \end{bmatrix}$$

Step 4: Now multiply row 2 by $-\dfrac{1}{11}$.

$$\begin{bmatrix} 1 & 2 & | & -3 \\ 0 & -11 & | & 22 \end{bmatrix} \xrightarrow{-\frac{1}{11}R_2} \begin{bmatrix} 1 & 2 & | & -3 \\ 0 & 1 & | & -2 \end{bmatrix}$$

Step 5: The last matrix (in triangular form) in Step 4 represents the following system of linear equations.

$$\begin{cases} x + 2y = -3 \\ 0x + y = -2 \end{cases}$$

The last equation gives $y = -2$.
Back substitute to find the value for x.

$$x + 2(-2) = -3$$
$$x - 4 = -3$$
$$x = 1$$

Thus the solution is $(1, -2)$.

Now work margin exercise 2.

Example 3

Gaussian Elimination

Solve the following system of linear equations by using the Gaussian elimination method with back substitution.

$$\begin{cases} 2x - 3y - z = -4 \\ -x + 2y + z = 6 \\ x - y + 2z = 14 \end{cases}$$

Solution

Step 1: Write the augmented matrix.

$$\begin{bmatrix} 2 & -3 & -1 & | & -4 \\ -1 & 2 & 1 & | & 6 \\ 1 & -1 & 2 & | & 14 \end{bmatrix}$$

Step 2: Exchange row 1 and row 3 so that the entry in the upper left corner will be 1.

$$\begin{bmatrix} 2 & -3 & -1 & | & -4 \\ -1 & 2 & 1 & | & 6 \\ 1 & -1 & 2 & | & 14 \end{bmatrix} \xrightarrow{R_1 \leftrightarrow R_3} \begin{bmatrix} 1 & -1 & 2 & | & 14 \\ -1 & 2 & 1 & | & 6 \\ 2 & -3 & -1 & | & -4 \end{bmatrix}$$

Step 3: To get a 0 under the 1 in Column 1, add row 1 to row 2.

$$\begin{bmatrix} 1 & -1 & 2 & | & 14 \\ -1 & 2 & 1 & | & 6 \\ 2 & -3 & -1 & | & -4 \end{bmatrix} \xrightarrow{R_2 + R_1} \begin{bmatrix} 1 & -1 & 2 & | & 14 \\ 0 & 1 & 3 & | & 20 \\ 2 & -3 & -1 & | & -4 \end{bmatrix}$$

Step 4: To get a_{31} to be 0, add -2 times row 1 to row 3.

$$\begin{bmatrix} 1 & -1 & 2 & | & 14 \\ 0 & 1 & 3 & | & 20 \\ 2 & -3 & -1 & | & -4 \end{bmatrix} \xrightarrow{R_3 - 2R_1} \begin{bmatrix} 1 & -1 & 2 & | & 14 \\ 0 & 1 & 3 & | & 20 \\ 0 & -1 & -5 & | & -32 \end{bmatrix}$$

Step 5: Add row 2 to row 3 to arrive at triangular form.

$$\begin{bmatrix} 1 & -1 & 2 & | & 14 \\ 0 & 1 & 3 & | & 20 \\ 0 & -1 & -5 & | & -32 \end{bmatrix} \xrightarrow{R_3 + R_2} \begin{bmatrix} 1 & -1 & 2 & | & 14 \\ 0 & 1 & 3 & | & 20 \\ 0 & 0 & -2 & | & -12 \end{bmatrix}$$

Step 6: Then multiply row 3 by $-\dfrac{1}{2}$.

$$\begin{bmatrix} 1 & -1 & 2 & | & 14 \\ 0 & 1 & 3 & | & 20 \\ 0 & 0 & -2 & | & -12 \end{bmatrix} \xrightarrow{-\frac{1}{2}R_3} \begin{bmatrix} 1 & -1 & 2 & | & 14 \\ 0 & 1 & 3 & | & 20 \\ 0 & 0 & 1 & | & 6 \end{bmatrix}$$

$$\begin{cases} x - y + 2z = 14 \\ \quad\quad y + 3z = 20 \\ \quad\quad\quad\quad z = 6 \end{cases}$$

From the last equation, we have $z = 6$.
Back substitution into the equation $y + 3z = 20$ gives:

$$y + 3(6) = 20$$
$$y = 2.$$

Back substitution into the equation $x - y + 2z = 14$ gives:

$$x - 2 + 2(6) = 14$$
$$x = 4.$$

Thus the solution is $(4, 2, 6)$.

Now work margin exercise 3.

3. Solve the following system of linear equations using the Gaussian elimination method.

$$\begin{cases} x + 3y - 2z = -3 \\ -2x - y - 3z = -2 \\ 3x + y + z = 8 \end{cases}$$

If the final matrix, in triangular form, has a row with all entries 0, then the system has an infinite number of solutions.

For example, solving the system $\begin{cases} x + 3y = 8 \\ 2x + 6y = 16 \end{cases}$

will result in the matrix $\begin{bmatrix} 1 & 3 & | & 8 \\ 0 & 0 & | & 0 \end{bmatrix}$.

The last line indicates that $0x + 0y = 0$ which is always true. Therefore, the solution to the system is the set of all solutions of the equation $x + 3y = 8$.

If the triangular form of the augmented matrix shows the coefficient entries in one or more rows to be all 0's and the constant not 0, then the system has no solution.

For example, the last row of the augmented matrix $\begin{bmatrix} 1 & 2 & 2 & | & 7 \\ 0 & 1 & 3 & | & 6 \\ 0 & 0 & 0 & | & 15 \end{bmatrix}$ indicates

that $0x + 0y + 0z = 15$. Since this is never true, the system has no solution.

Using a TI-84 Plus Graphing Calculator to Solve a System of Linear Equations

The TI-84 Plus calculator can be used to define a matrix and perform matrix operations. (Matrices can be added, subtracted, and multiplied under special restrictions. These operations are saved for another course.) In particular, the Gaussian elimination method is used by the calculator to solve a system of linear equations by reducing the corresponding augmented matrix to **row echelon form (ref)**.

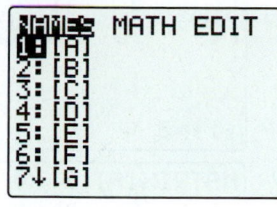

Pressing the MATRIX key (found by pressing 2ND x⁻¹) will give the menu shown here. This menu will allow you to choose existing matrices to operate with them.

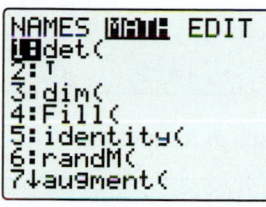

Pressing the right arrow and moving to MATH will give the shown choices. This menu will give you a list of operations you can do on matrices we will use the reduced echelon form (ref) option here.

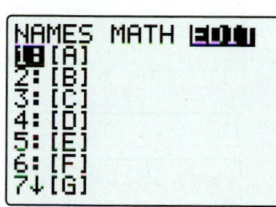

Pressing the right arrow again and moving to EDIT will give the shown choices. This menu allows you to edit existing matrices and create new ones.

Example 4 shows how to use the calculator to solve a system of three linear equations in three variables. Study each step carefully.

Example 4

 Using a Graphing Calculator to Solve Systems of Equations

Use a TI-84 Plus calculator to solve the following system of linear equations.

$$\begin{cases} x + 2y + z = 1 \\ -x + y + z = -6 \\ 4x - y + 3z = -1 \end{cases}$$

Solution

Step 1: Press 2ND > MATRIX and move to the EDIT menu.

```
MATRIX[A] 1 ×1
[ 0          ]
```

Press ENTER. The following display will appear.

4. Use a TI-84 Plus calculator to solve the following system of linear equations.

$$\begin{cases} x - 3y + 2z = 10 \\ 2x + 5y - 2z = -9 \\ -x + 6y + 3z = 2 \end{cases}$$

Step 2: The augmented matrix is a 3×4 matrix. So, in the top line enter **3**, press **ENTER**, enter **4**, press **ENTER** and the display will appear as shown.

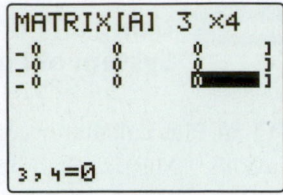

Note: If other numbers are already present on the display, just type over them. The calculator will adjust automatically.

Step 3: Move the cursor to the upper left entry position and enter the coefficients and constants in the matrix. As you enter each number press **ENTER** and the cursor will automatically move to the next position in the matrix. Note that the double subscripts appear at the bottom of the display as each number is entered. The final display for matrix [A] should appear as shown.

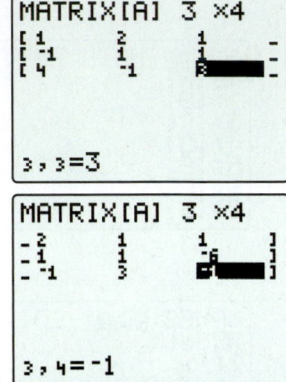

Note: The display only shows three columns at a time.

Step 4: Press **2ND** > QUIT, press **2ND** > MATRIX again, go to MATH; move the cursor down to A:ref(; press **ENTER**. The display will appear as shown.

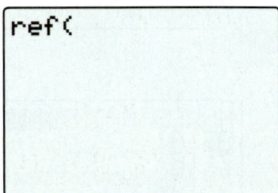

Step 5: Press **2ND** > MATRIX again; press **ENTER** (this selects the matrix A); enter a right parenthesis) ; press the **MATH** key; and choose 1:>Frac by pressing **ENTER**. The display will appear as shown.

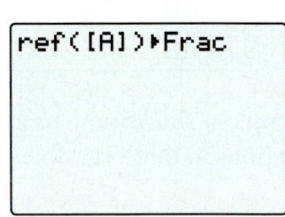

Note: You must select the matrix from the matrix menu. The calculator will not recognize the matrix if you manually type in [A].

Step 6: Press **ENTER** and the row echelon form of matrix will appear as follows.

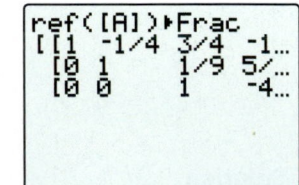

Note: To see the rest of the matrix move the cursor on the screen to the right.

With back substitution we get the following solution $(3, 1, -4)$.

Now work margin exercise 4.

Practice Problems

Solve the following system of linear equations by using the Gaussian elimination method with back substitution.

$$\begin{cases} x - 2y + 3z = 4 \\ 2x + y = 0 \\ 3x + y - z = -4 \end{cases}$$

Practice Problem Answers

$(-1, 2, 3)$

Exercises 8.7

1. $\begin{cases} 2x + 2y = 13 \\ 5x - y = 10 \end{cases}$

2. $\begin{cases} x + 4y = -1 \\ 2x - 3y = 7 \end{cases}$

3. $\begin{cases} 7x - 2y + 7z = 2 \\ -5x + 3y = 2 \\ 4y + 11z = 8 \end{cases}$

4. $\begin{cases} -8x + 2y - z = 6 \\ 2x + 3z = -3 \\ -4x - 2y + 5z = 13 \end{cases}$

5. $\begin{cases} 3w + x - y + 2z = 6 \\ w - x + 2y - z = -8 \\ 2x + 5y + z = 2 \\ w + 3x + 3z = 14 \end{cases}$

6. $\begin{cases} 4w + x + 3y - 2z = 13 \\ w - 2x + y - 4z = -3 \\ w + x + 4y + 2z = 12 \\ -2w + 3x - y - 3z = 5 \end{cases}$

Write the system of linear equations represented by each of the augmented matrices. Use x, y, and z as the variables.

7. $\begin{bmatrix} -3 & 5 & | & 1 \\ -1 & 3 & | & 2 \end{bmatrix}$

8. $\begin{bmatrix} 3 & -1 & | & 5 \\ -2 & 10 & | & 9 \end{bmatrix}$

9. $\begin{bmatrix} 1 & 3 & 4 & | & 1 \\ 2 & -3 & -2 & | & 0 \\ 1 & 1 & 0 & | & -4 \end{bmatrix}$

10. $\begin{bmatrix} 2 & -9 & 14 & | & 0 \\ -3 & 0 & -8 & | & 5 \\ 2 & -6 & 1 & | & 3 \end{bmatrix}$

Perform the indicated row operations on the given matrix. See Example 1.

11. $\begin{bmatrix} 1 & 4 \\ -1 & 7 \end{bmatrix}$

    a. interchange rows 1 and 2

    b. multiply row 2 by -2

12. $\begin{bmatrix} 3 & -2 & | & 8 \\ 1 & 5 & | & 9 \end{bmatrix}$

    a. multiply row 1 by $\dfrac{1}{2}$

    b. add -3 times row 2 to row 1

13. $\begin{bmatrix} 1 & 3 & 7 \\ -8 & -2 & 5 \\ 4 & -1 & 6 \end{bmatrix}$

    a. interchange rows 2 and 3

    b. add −2 times row 3 to row 2

14. $\left[\begin{array}{ccc|c} 1 & 2 & -3 & 7 \\ 0 & -2 & -1 & 9 \\ 0 & 1 & 1 & 4 \end{array}\right]$

    a. interchange rows 2 and 3

    b. Using your answer from part **a.**, add 2 times row 2 to row 3.

Use the Gaussian elimination method to solve the given system of linear equations. See Example 3.

15. $\begin{cases} x + 2y = 3 \\ 2x - y = -4 \end{cases}$

16. $\begin{cases} 4x + 3y = 5 \\ -x - 2y = 0 \end{cases}$

17. $\begin{cases} -8x + 2y = 6 \\ x - 2y = 1 \end{cases}$

18. $\begin{cases} 2x + y = -2 \\ 4x + 3y = -2 \end{cases}$

19. $\begin{cases} x - 3y + 2z = 11 \\ -2x + 4y + z = -3 \\ x - 2y + 3z = 12 \end{cases}$

20. $\begin{cases} x + 2y - z = 6 \\ x + 3y - 3z = 3 \\ x + y + z = 6 \end{cases}$

21. $\begin{cases} x + 2y + 3z = 4 \\ x - y - z = 0 \\ 4x - 3y + z = 5 \end{cases}$

22. $\begin{cases} x + y - 2z = -1 \\ 3x + 4y - 2z = 0 \\ x - y + z = 4 \end{cases}$

23. $\begin{cases} x - y - 2z = 3 \\ x + 2y - z = 5 \\ 2x - 3y - 2z = 3 \end{cases}$

24. $\begin{cases} x + y + 3z = 2 \\ 2x - y + z = 1 \\ 4x + y + 7z = 5 \end{cases}$

25. $\begin{cases} x - y + 5z = -6 \\ x + 2z = 0 \\ 6x + y + 3z = 0 \end{cases}$

26. $\begin{cases} x - 3y - z = -4 \\ 3x - 2y + z = 1 \\ -2x + y + 2z = 13 \end{cases}$

27. $\begin{cases} x - y + 2z = 5 \\ 2x - 2y + 4z = 5 \\ 3x - 3y + 6z = 8 \end{cases}$

28. $\begin{cases} 2x - y - 5z = -9 \\ x - 3y + 2z = 0 \\ 3x + 2y + 10z = 4 \end{cases}$

29. $\begin{cases} x - 5y + 2z = -7 \\ x + y + 4z = 7 \\ 2x - y + 7z = 7 \end{cases}$

30. $\begin{cases} 2x - y + 5z = -2 \\ 4x + y + 3z = -2 \\ x + 3y - z = 6 \end{cases}$

31. $\begin{cases} 3x + 4z = 11 \\ x + y = -2 \\ 2y + z = -4 \end{cases}$

32. $\begin{cases} y + z = 2 \\ x + y = 5 \\ x + z = 5 \end{cases}$

Set up a system of linear equations that represents the information and solve the system using Gaussian elimination. See Example 4.

33. The sum of three integers is 169. The first integer is twelve more than the second integer. The third integer is fifteen less than the sum of the first and second integers. What are the integers?

34. **Pizza:** A pizzeria sells three sizes of pizzas: small, medium, and large. The pizzas sell for $6.00, $8.00, and $9.50, respectively. One evening they sold 68 pizzas for a total of $528.00. If they sold twice as many medium-sized pizzas as large-sized pizzas, how many of each size did they sell?

35. Grocery Shopping: Caroline bought a pound of bacon, a dozen eggs, and a loaf of bread. The total cost was $8.52. The eggs cost $0.94 more than the bacon. The combined cost of the bread and eggs was $2.34 more than the cost of the bacon. Find the cost of each item.

36. Investments: An investment firm is responsible for investing $250,000 from an estate according to three conditions in the will of the deceased. The money is to be invested in three accounts paying 6%, 8%, and 11% interest. The amount invested in the 6% account is to be $5000 more than the total invested in the other two accounts, and the total annual interest for the first year is to be $19,250. How much is the firm supposed to invest in each account?

Use your graphing calculator to solve the systems of linear equations. See Example 4.

37. $\begin{cases} x + y = -4 \\ 2x + 3y = -12 \end{cases}$

38. $\begin{cases} 2x + 3y = 1 \\ x - 5y = -19 \end{cases}$

39. $\begin{cases} x + y + z = 10 \\ 2x - y + z = 10 \\ -x + 2y + 2z = 14 \end{cases}$

40. $\begin{cases} 2x + y + 2z = 2 \\ x - y + 4z = 1 \\ 3x - y + z = -4 \end{cases}$

41. $\begin{cases} 2y - 3z = -18 \\ 4x + 5y = -7 \\ 6x - z = 8 \end{cases}$

42. $\begin{cases} w + x + y + z = 0 \\ w + x - y - z = -2 \\ -w + 3x + 3y - z = 11 \\ y - 2z = 6 \end{cases}$

43. $\begin{cases} x - 3y + z = 0 \\ 2x + 2y - z = 2 \\ x + y + z = 5 \end{cases}$

44. $\begin{cases} x - 2y - 2z = -13 \\ 2x + y - z = -5 \\ x + y + z = 6 \end{cases}$

Writing & Thinking

45. Suppose that Gaussian elimination with a system of three linear equations in three unknowns results in the following triangular matrix. Discuss how you can use back substitution to find that the system has an infinite number of solutions and these solutions satisfy the equation $x + 5y = 6$. (**Hint:** Solve the second equation for z.)

$$\begin{bmatrix} 1 & 2 & -1 & | & 4 \\ 0 & 3 & 1 & | & 2 \\ 0 & 0 & 0 & | & 0 \end{bmatrix}$$

8.8 Systems of Linear Inequalities

Objectives

A Solve systems of linear inequalities graphically.

B Use a graphing calculator to graph systems of linear inequalities.

In some branches of mathematics, in particular a topic called game theory, the solution to a very sophisticated problem can involve the set of points that satisfies a system of several **linear inequalities**. In business these ideas relate to problems such as minimizing the cost of shipping goods from several warehouses to distribution outlets. In this section we will consider graphing the solution sets to only two inequalities. We will leave the problem solving techniques to another course.

In an earlier section we graphed linear inequalities of the form $y < mx + b$ (or $y \le mx + b$ or $y > mx + b$ or $y \ge mx + b$). These graphs are called **half-planes**. The line $y = mx + b$ is called the **boundary line** and the half-planes are **open** (the boundary line is not included) or **closed** (the boundary line is included).

In this section, we will develop techniques for graphing (and therefore solving) **systems of two linear inequalities**. The **solution set** (if there are any solutions) to such a system of linear inequalities consists of the points in the **intersection** of two half-planes and portions of boundary lines indicated by the inequalities. The following procedure may be used to solve a system of linear inequalities.

To Solve a System of Two Linear Inequalities

1. For each inequality, graph the boundary line and shade the appropriate half-plane.
2. Determine the region of the graph that is common to both half-planes (the region where the shading overlaps).
   (This region is called the **intersection** of the two half-planes and is the **solution set** of the system.)
3. To check, pick one test-point in the intersection and verify that it satisfies both inequalities.

Note: If there is no intersection, then the system has no solution.

Example 1

Graphing Systems of Linear Inequalities

a. Graph the points that satisfy the following system of inequalities.

$$\begin{cases} x \le 2 \\ y \ge -x + 1 \end{cases}$$

Solution

Step 1: For $x \le 2$, the points are to the left of and on the line $x = 2$.
For $y \ge -x + 1$, the points are above and on the line $y = -x + 1$.

Step 2: Determine the region that is common to both half-planes.

Step 3: In this case, we test the point $(0, 3)$. On the graph below, the solution is the purple-shaded region and its boundary lines.

$0 \leq 2$ a true statement

$3 \geq -0 + 1$ a true statement

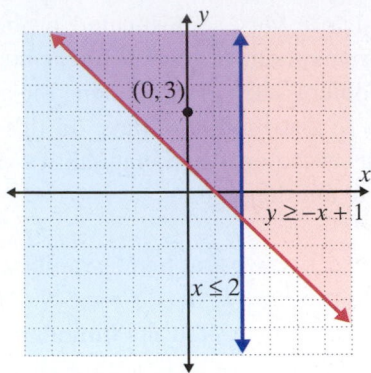

b. Solve the system of linear inequalities graphically: $\begin{cases} 2x + y \leq 6 \\ x + y < 4 \end{cases}$.

Solution

Frist solve each inequality for y: $\begin{cases} y \leq -2x + 6 \\ y < -x + 4 \end{cases}$.

Step 1: For $y \leq -2x + 6$, the points are below and on the line $y = -2x + 6$. For $y < -x + 4$, the points are below but not on the line $y = -x + 4$.

Step 2: Determine the region that is common to both half-planes. Note that the line $y = -x + 4$ is dashed to indicate that the points on the line are not included.

Step 3: In this case, we test the point $(0, 0)$. On the graph below, the solution is the purple-shaded region and its boundary lines.

$2 \cdot 0 + 0 \leq 6$ a true statement

$0 + 0 < 4$ a true statement

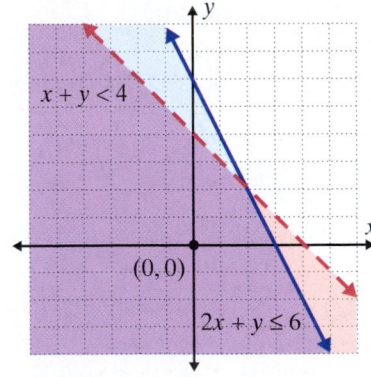

c. Solve the system of linear inequalities graphically: $\begin{cases} y \geq x \\ y \leq x+2 \end{cases}$

1. Solve the following systems of inequalities graphically.

Solution

Step 1: For $y \geq x$, the points are above and on the line $y = x$. For $y \leq x + 2$, the points are below and on the line $y = x + 2$.

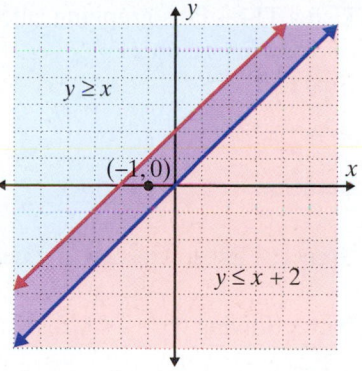

Step 2: Determine the region that is common to both half-planes.

Step 3: We test point $(-1, 0)$. The solution set consists of the boundary lines and the region between them.

Now work margin exercise **1.**

a. $\begin{cases} y \geq 2 \\ x - y < 4 \end{cases}$

b. $\begin{cases} x + 2y < 3 \\ 3x + y \geq -6 \end{cases}$

c. $\begin{cases} y < -3x \\ y > -3x + 4 \end{cases}$

notes

▪ When the boundary lines are parallel there are three possibilities:
   1. The common region will be in the form of a strip between two lines (as in Example **1c.**).
   2. The common region will be a half-plane, as the solution to one inequality will be entirely contained within the solution of the other inequality.
   3. There will be no common region which means there is no solution.

Objective B ### Using a TI-84 Plus Graphing Calculator to Graph Systems of Linear Inequalities

To graph (and therefore solve) a system of linear inequalities (with no vertical line) with a TI-84 Plus graphing calculator, proceed as follows:

1. Solve each inequality for y.
2. Press the ⬚ Y= key and enter the two expressions for Y_1 and Y_2.
3. Move to the left of Y_1 and Y_2 and press **ENTER** until the desired graphing symbol appears.
4. Press ⬚ GRAPH and the desired region will be graphed as a cross-hatched area. (You may need to reset the ⬚ WINDOW so that both regions appear. If you need assistance with this refer back to Section 6.6)

Example 2 illustrates how this can be done.

2. Use a TI-84 Plus graphing calculator to graph the following system of linear inequalities.

$$\begin{cases} y - 2x \geq -5 \\ y + 4x < 7 \end{cases}$$

 ## Example 2

Graphing Systems of Linear Inequalities

Use a TI-84 Plus graphing calculator to graph the following system of linear inequalities.

$$\begin{cases} 2x + y < 4 \\ 2x - y \leq 0 \end{cases}$$

Solution

Step 1: First solve each inequality for y: $\begin{cases} y < -2x + 4 \\ y \geq 2x \end{cases}$.

Note: Solving $2x - y \leq 0$ for y can be written as $2x \leq y$ and then as $y \geq 2x$.

Step 2: Press the [Y=] key and enter both functions and the corresponding symbols as they appear here.

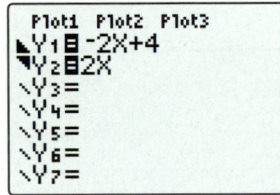

Remember: To shade your graphs, position the cursor over the slash next to Y_1 (or Y_2) and hit **ENTER** repeatedly until the appropriate shading is displayed.

Step 3: Press [GRAPH]. The display should appear as follows. The solution is the cross-hatched region and the points on the line $2x - y = 0$.

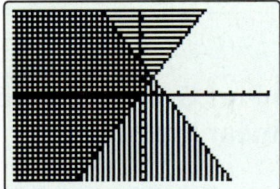

*Now work margin exercise **2**.*

Practice Problems

Solve the systems of two linear inequalities graphically.

1. $\begin{cases} y \leq -3 \\ y \geq x - 3 \end{cases}$ **2.** $\begin{cases} y > 2 \\ x < 2 \end{cases}$ **3.** $\begin{cases} x + y \geq 0 \\ x - 2y \geq 4 \end{cases}$

Practice Problem Answers

1. 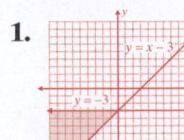   **2.**   **3.**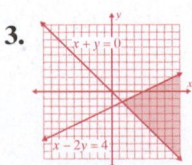

Exercises 8.8

Solve the systems of two linear inequalities graphically. See Example 1.

1. $\begin{cases} y > 2 \\ x \ge -3 \end{cases}$

2. $\begin{cases} 2x + 5 < 0 \\ y \ge 2 \end{cases}$

3. $\begin{cases} x < 3 \\ y > -x + 2 \end{cases}$

4. $\begin{cases} y \le -5 \\ y \ge x - 5 \end{cases}$

5. $\begin{cases} x \le 3 \\ 2x + y > 7 \end{cases}$

6. $\begin{cases} 2x - y > 4 \\ y < -1 \end{cases}$

7. $\begin{cases} x - 3y \le 3 \\ x < 5 \end{cases}$

8. $\begin{cases} 3x - 2y \ge 8 \\ y \ge 0 \end{cases}$

9. $\begin{cases} x - y \ge 0 \\ 3x - 2y \ge 4 \end{cases}$

10. $\begin{cases} y \ge x - 2 \\ x + y \ge -2 \end{cases}$

11. $\begin{cases} 3x + y \le 10 \\ 5x - y \ge 6 \end{cases}$

12. $\begin{cases} y > 3x + 1 \\ -3x + y < -1 \end{cases}$

13. $\begin{cases} 3x + 4y \ge -7 \\ y < 2x + 1 \end{cases}$

14. $\begin{cases} 2x - 3y \ge 0 \\ 8x - 3y < 36 \end{cases}$

15. $\begin{cases} x + y < 4 \\ 2x - 3y < 3 \end{cases}$

16. $\begin{cases} 2x + 3y < 12 \\ 3x + 2y > 13 \end{cases}$

17. $\begin{cases} x + y \ge 0 \\ x - 2y \ge 6 \end{cases}$

18. $\begin{cases} y \ge 2x + 3 \\ y \le x - 2 \end{cases}$

19. $\begin{cases} x + 3y \le 9 \\ x - y \ge 5 \end{cases}$

20. $\begin{cases} x - y \ge -2 \\ x + 2y < -1 \end{cases}$

21. $\begin{cases} y \le x + 3 \\ x - y \le -5 \end{cases}$

22. $\begin{cases} y \ge 2x - 5 \\ 3x + 2y > -3 \end{cases}$

23. $\begin{cases} y \le -2x \\ y > -2x - 6 \end{cases}$

24. $\begin{cases} y > x - 4 \\ y < x + 2 \end{cases}$

25. $\begin{cases} y \geq 0 \\ 3x - 5y \leq 10 \end{cases}$ **26.** $\begin{cases} y \leq 0 \\ 3x + y \leq 11 \end{cases}$ **27.** $\begin{cases} 4x - 3y \geq 6 \\ 3x - y \leq 3 \end{cases}$ **28.** $\begin{cases} 3x + 2y \leq 15 \\ 2x + 5y \geq 10 \end{cases}$

29. $\begin{cases} 3x - 4y \geq -6 \\ 3x + 2y \leq 12 \end{cases}$ **30.** $\begin{cases} 3y \leq 2x + 2 \\ x + 2y \leq 11 \end{cases}$ **31.** $\begin{cases} x + y \leq 8 \\ 3x - 2y \geq -6 \end{cases}$ **32.** $\begin{cases} x + y \leq 7 \\ 2x - y \leq 8 \end{cases}$

33. $\begin{cases} y \leq x \\ y < 2x + 1 \end{cases}$ **34.** $\begin{cases} x - y \geq -2 \\ 4x - y < 16 \end{cases}$

Writing & Thinking

35. Graph the inequalities and explain how you can tell that there is no solution.

$$\begin{cases} y \leq 2x - 5 \\ y \geq 2x + 3 \end{cases}$$

Chapter 8: Index of Key Terms and Ideas

Section 8.3: Systems of Linear Equations: Solutions by Addition

To Solve a System of Linear Equations by Addition   pages 920 – 921
1. Write the equations in **standard form** one under the other so that **like terms are aligned**.
2. Multiply all terms of one equation by a constant (and possibly all terms of the other equation by another constant) so that **two like terms have opposite coefficients**.
3. Add the two equations by **combining like terms** and solve the resulting equation, if possible.
4. **Back substitute into one of the original equations** to find the value of the other variable.
5. Check the solution (if there is one) in both of the original equations.

Guidelines for Deciding which Method to Use when Solving a System of Linear Equations   page 924
1. The graphing method is helpful in "seeing" the geometric relationship between the lines and finding approximate solutions. A calculator can be very helpful here.
2. Both the substitution method and the addition method give exact solutions.
3. The substitution method may be reasonable and efficient if one of the coefficients of one of the variables is 1.
4. The addition method is particularly efficient if the coefficients for one of the variables are opposites.

Section 8.4: Applications: Distance-Rate-Time, Number Problems, Amounts and Costs

Applications   pages 927 – 931
Distance-Rate-Time, Number Problems, Amounts and Costs
1. For all these applications, you will need to create your own system of linear equations.
2. Creating tables can help you organize the information into a more understandable form.

Section 8.5: Applications: Interest and Mixture

Applications   pages 936 – 940
1. Interest: In many of these problems you will find two types of information: more than one account and more than one amount of interest. Use these types of information to write equations.
2. Mixture: The basic plan for these problems is to write an equation that deals with only one part of the mixture.

Section 8.6: Systems of Linear Equations: Three Variables

Linear Equation in Three Variables   pages 944 – 948

The general form of an equation in three variables is
$Ax + By + Cz = D$ where A, B, and C are not all equal to 0.

Ordered Triples   page 944

An **ordered triple** is an ordering of three real numbers in the form
(x_0, y_0, z_0) and is used to represent the solution to a linear equation in
three variables.

Graphs in Three Dimensions   page 945

Three mutually perpendicular number lines labeled as the x-axis, y-axis,
and z-axis are used to separate space into eight regions called **octants**.

To Solve a System of Three Linear Equations in Three Variables   page 946

1. Select two equations and eliminate one variable by using the addition
   method.
2. Select a different pair of equations and eliminate the same variable.
3. Steps 1 and 2 give two linear equations in two variables. Solve these
   equations by either addition or substitution as discussed in Sections 7.2 and
   7.3.
4. Back substitute the values found in Step 3 into any one of the original
   equations to find the value of the third variable.
5. Check the solution (if one exists) in all three of the original equations.

Section 8.7: Matrices and Gaussian Elimination

Matrices   page 954

Matrix – a rectangular array of numbers.
Entry – a number in a matrix
Row – a group of entries written horizontally
Column – a group of entries written vertically
Dimension of a matrix – the number of rows by the number of columns,
such as 2×3
Square matrix – a matrix with the same number of rows as columns, such
as 3×3

Coefficient Matrix   page 954

A matrix formed from the coefficients of the variables in a system of
linear equations is called a **coefficient matrix**.

Augmented Matrix   page 954

A matrix formed from the coefficients of the variables and the constant
terms in a system of linear equations is called an **augmented matrix**.

Elementary Row Operations   page 956

There are three elementary row operations with matrices:
1. Interchange two rows.
2. Multiply a row by a nonzero constant.
3. Add a multiple of a row to another row.

If any elementary row operation is applied to a matrix, the new matrix is said to be **row-equivalent** to the original matrix.

Upper Triangular Form   page 958

A matrix is said to be in **upper triangular form** if all the entries in the lower left triangular region are 0's.

Row Echelon Form   page 958

The upper triangular form of a matrix with all 1's in the main diagonal is called the **row echelon form** (or **ref**).

Gaussian Elimination   pages 958 – 959

To solve a system of linear equations using Gaussian elimination:
1. Write the augmented matrix for the system.
2. Use elementary row operations to transform the matrix into row echelon form.
3. Solve the corresponding system of equations by using back substitution.

Section 8.8: Systems of Linear Inequalities

To Solve a System of Two Linear Inequalities   page 968
1. For each inequality, graph the boundary line and shade the appropriate half-plane. (Refer to a previous section to review this process.)
2. Determine the region of the graph that is common to both half-planes (the region where the shading overlaps). (This region is called the **intersection** of the two half-planes and is the **solution set** of the system.)
3. To check, pick one test-point in the intersection and verify that it satisfies both inequalities.

Note: If there is no intersection, then the system is inconsistent and has no solution.

Chapter 8: Test

Determine which of the given points, if any, lie on both of the lines in the system of equations by substituting each point into both equations.

1. $\begin{cases} 3x - 7y = 5 \\ 5x - 2y = -11 \end{cases}$ **a.** $(1, 8)$ **b.** $(4, 1)$ **c.** $(-3, -2)$ **d.** $(13, 10)$

Solve the systems by graphing.

2. $\begin{cases} 2x = -3y + 9 \\ 4x + 6y = 18 \end{cases}$

3. $\begin{cases} x - y = 3 \\ x + 2y = 6 \end{cases}$

Solve the systems by using the method of substitution.

4. $\begin{cases} 5x - 2y = 0 \\ y = 3x + 4 \end{cases}$

5. $\begin{cases} 3x = y - 12 \\ 4x + 3y = 10 \end{cases}$

Solve the systems by using the method of addition.

6. $\begin{cases} 3x + y = 5 \\ -12x - 4y = -7 \end{cases}$

7. $\begin{cases} 3x + 4y = 10 \\ x + 6y = 1 \end{cases}$

Solve the systems by using any method.

8. $\begin{cases} x + y = 2 \\ y = -2x - 1 \end{cases}$

9. $\begin{cases} 6x + 2y - 8 = 0 \\ y = -3x \end{cases}$

10. $\begin{cases} 7x + 5y = -9 \\ 6x + 2y = 6 \end{cases}$

11. $\begin{cases} x + 3y = -2 \\ 2x = -6y - 4 \end{cases}$

12. Solve the following system of equations algebraically: $\begin{cases} x - 2y - 3z = 3 \\ x + y - z = 2 \\ 2x - 3y - 5z = 5 \end{cases}$

13. Solve the following system of equations algebraically: $\begin{cases} x + 2y - 2z = 0 \\ x - y + z = 2 \\ -x + 4y - 4z = -8 \end{cases}$

Solve the following system of equations.

14. For the following system of equations:

a. Write the coefficient matrix and the augmented matrix, and

b. Solve the system using the Gaussian elimination method with back substitution.

$$\begin{cases} x + 2y - 3z = -11 \\ x - y - z = 2 \\ x + 3y + 2z = -4 \end{cases}$$

Solve the systems of equations graphically.

15. Solve the following system of linear inequalities graphically: $\begin{cases} x \geq -3 \\ y \geq 3x \end{cases}$

16. Solve the following system of linear inequalities graphically: $\begin{cases} y < 3x + 4 \\ 2x + y \geq 1 \end{cases}$

Solve the following word problems.

17. Coffee: Every time Tish goes to a coffee shop she either buys a soy latte or an iced coffee. In one month she spent a total of $79.10. If the money she spent on soy lattes is $3.50 plus three times the money she spent on iced coffees, how much money did she spend on each drink?

18. Boating: Pete's boat can travel 48 miles upstream in 4 hours. The return trip takes 3 hours. Find the speed of the boat in still water and the speed of the current.

19. School Supplies: Eight pencils and two pens cost $2.22. Three pens and four pencils cost $2.69. What is the price of each pen and each pencil?

20. Investing: Gary has two investments yielding a total annual interest of $185.60. The amount invested at 8% is $320 less than twice the amount invested at 6%. How much is invested at each rate?

21. Rectangles: The perimeter of a rectangle is 60 inches and the length is 4 inches longer than the width. Find the dimensions of the rectangle.

22. Manufacturing: A metallurgist needs 2000 pounds of an alloy that is 80% copper. In stock, he has only alloys of 83% copper and 68% copper. How many pounds of each must be used?

23. Making Change: Sonia has a bag of coins with only nickels and quarters. She wants you to figure out how many of each type of coin she has and tells you that she has 105 coins and that the value of the coins is $17.25. Show her that you know algebra and determine how many nickels and how many quarters she has.

24. Stamp Collecting: Kimberly bought 90 stamps in denominations of 41¢, 58¢, and 75¢. To test her daughter, who is taking an algebra class, she said that she bought three times as many 41¢ stamps as 58¢ stamps and that the total cost of the stamps was $43.70. How many stamps of each denomination did she buy?

Cumulative Review: Chapters 1 - 8

Working with decimals.

1. Write the number two hundred thousand, sixteen and four hundredths in decimal notation.

2. Round 17.986 to the nearest hundredth.

3. Find the decimal equivalent of $\dfrac{14}{35}$.

4. Write $\dfrac{9}{5}$ as a percent.

5. Write $1\dfrac{1}{2}\%$ in decimal form.

Find the value of each expression by performing the indicated operations.

6. $\dfrac{2}{15}+\dfrac{11}{15}+\dfrac{7}{15}$

7. $4-\dfrac{10}{11}$

8. $\left(\dfrac{2}{5}\right)\left(-\dfrac{5}{6}\right)\left(\dfrac{4}{7}\right)$

9. $2\dfrac{4}{15}+3\dfrac{1}{6}+4\dfrac{7}{10}$

10. $70\dfrac{1}{4}-23\dfrac{5}{6}$

11. $\left(-4\dfrac{5}{7}\right)\left(-2\dfrac{6}{11}\right)$

12. $6\div3\dfrac{1}{3}$

13. $(700)(-800)$

14. $40.3-67.2$

15. $71+\left|-0.35\right|+4.39$

16. $(0.27)(0.043)^2$

17. $27.404\div(-0.34)$

Use the rules for order of operations to evaluate each expression.

18. $\left(36\div3^2\cdot2\right)+12\div4-2^2$

19. $1.7^2-3\dfrac{1}{2}\div\dfrac{1}{4}$

Solve the following problems.

20. Evaluate $3x^2-5x-17$ for $x=-1.5$.

21. Find the prime factorization of 396.

22. Find $\left(\text{Use } \pi=3.14.\right)$

    a. the circumference, and

    b. the area of a circle with a radius of 10 feet.

23. Find the length of the hypotenuse of a right triangle if one of its legs is 5 inches long and the other leg is 10 inches long.

Solve each of the equations.

24. $5x-4=11$

25. $3x=7x-16$

26. $4(3x-1)=2(2x-5)-3$

27. $\dfrac{4x-1}{3}+\dfrac{x-5}{2}=2$

Solve the inequalities and graph the solution sets. Write each solution using interval notation.

28. $x+4-3x\geq2x+5$

29. $\dfrac{x}{5}-3.4>\dfrac{x}{2}+1.6$

30. $4x = 8$

31. $3x - 2 = 5y$

32. $4x - y = 1$

33. $3 - 2y = 0$

Solve the following equations as indicated.

34 Find an equation in standard form for the line that passes through the point $(3, -4)$ and is parallel to the line $2x - y = 5$. Graph both lines.

35. Find an equation in slope-intercept form for the line passing through the points $(1, -4)$ and $(4, 2)$. Graph the line.

36. Three graphs are shown below. Use the vertical line test to determine whether or not each graph represents a function. State each graph's domain and range.

a.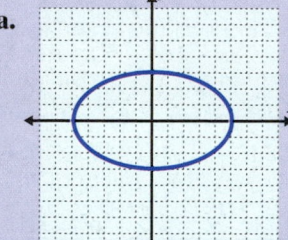

b.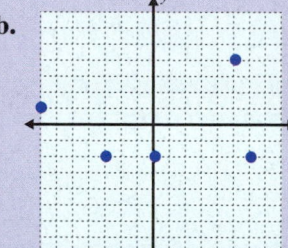

c.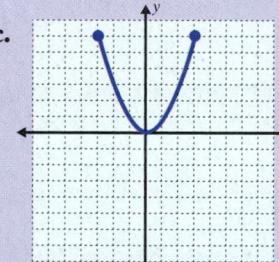

Solve the systems of linear equations by the stated method.

37. Determine which of the given points, if any, lie on both of the lines in the systems of equations by substituting each point into both equations.

$$\begin{cases} x - 2y = 6 \\ y = \dfrac{1}{2}x - 3 \end{cases}$$

a. $(0, 6)$ **b.** $(6, 0)$ **c.** $(2, -2)$ **d.** $\left(3, -\dfrac{3}{2}\right)$

38. Solve the system of linear equations by graphing: $\begin{cases} 2x + y = 6 \\ 3x - 2y = -5 \end{cases}$

39. Solve the system of linear equations by the addition method: $\begin{cases} x + 3y = 10 \\ 5x - y = 2 \end{cases}$

40. Solve the system of linear equations by the substitution method: $\begin{cases} x + y = -4 \\ 2x + 7y = 2 \end{cases}$

Solve the systems of linear equations using any method.

41. $\begin{cases} 5x + 2y = 3 \\ y = 4 \end{cases}$

42. $\begin{cases} 2x + 4y = 9 \\ 3x + 6y = 8 \end{cases}$

43. $\begin{cases} x + 5y = 10 \\ y - 2 = -\dfrac{1}{5}x \end{cases}$

44. $\begin{cases} 2x + y = 7 \\ 2x - y = 1 \end{cases}$

Solve the systems of linear equations using Gaussian elimination.

45. $\begin{cases} x - 3y + 2z = -1 \\ -2x + y + 3z = 1 \\ x - y + 4z = 9 \end{cases}$

46. $\begin{cases} x + y + z = -2 \\ 2x - 3y - z = 15 \\ -x + y - 3z = -12 \end{cases}$

Solve the systems of linear inequalities graphically.

47. $\begin{cases} x + 2y > 4 \\ x - y > 7 \end{cases}$

48. $\begin{cases} y \le x - 5 \\ 3x + y \ge 2 \end{cases}$

49. The sum of two numbers is 14. Twice the larger number added to three times the smaller is equal to 31. Find both numbers.

50. Investing: Alicia has $7000 invested, some at 7% and the remainder at 8%. After one year, the interest from the 7% investment exceeds the interest from the 8% investment by $70. How much is invested at each rate?

Mathematics @ Work!

When you watch a sunset over the ocean, how far are you from the horizon? On a clear day, a person looking over the ocean can see an island on the horizon up to 80 miles away (the distance from San Diego, California to San Clemente Island, for example). Yet the visible distance varies due to the clarity and temperature of the air. In 1923 N. Korzenewsky, leader of an expedition in the deserts of Turkestan, recorded seeing snow-capped mountains 750 km (466 miles) away. A more reliable recorded observation is one that was made in 1933 by Commander C. L. Garner of the Coast and Geodetic Survey. He says that instrumental measurements were made in both directions "between Mt. Shasta and Mt. St. Helena in California", a distance of 192 miles. A general rule of thumb for calculating the distance in miles between you and the horizon is the formula $d = 1.32 \sqrt{h}$ where h is your eye's height above the ground in feet. Use this equation to answer the question below.

If you are watching the sunset from the roof of a two-story building, and your eye is 28 feet above the ground, how far away is the horizon?

For more problems like these, see Section 9.2.

Roots and Radicals

9.1 Radical Expressions

Objectives

A Evaluate square roots.

B Evaluate cube roots.

C Evaluate squares and cubes of sums and differences.

D Use a calculator to evaluate square and cube roots.

You are probably familiar with the concept of **square roots** and the square root symbol $\left(\text{or } \textbf{radical sign } \sqrt{\ }\right)$ from your previous work in algebra. For example, $\sqrt{3}$ represents the square root of 3 and is the number whose square is 3.

Objective A **Square Roots**

A number is **squared** when it is multiplied by itself. For example,

$$6^2 = 6 \cdot 6 = 36 \qquad \text{and} \qquad (-1.5)^2 = (-1.5)(-1.5) = 2.25.$$

If an integer is squared, the result is called a **perfect square**. The squares for the integers from 1 to 20 are shown in Table 1 for easy reference.

Squares of Integers from 1 to 20 (Perfect Squares)										
Integers (n)	1	2	3	4	5	6	7	8	9	10
Perfect Squares (n^2)	1	4	9	16	25	36	49	64	81	100
Integers (n)	11	12	13	14	15	16	17	18	19	20
Perfect Squares (n^2)	121	144	169	196	225	256	289	324	361	400

Table 1

Now we want to reverse the process of squaring. That is, given a number, we want to find a number that when squared will result in the given number. This is called **finding a square root** of the given number. In general,

if $b^2 = a$, then b is a square root of a.

For example,

- because $5^2 = 25$, 5 is a **square root** of 25 and we write $\sqrt{25} = 5$.

- because $9^2 = 81$, 9 is a **square root** of 81 and we write $\sqrt{81} = 9$.

Radical Terminology

The symbol $\sqrt{\ }$ is called a **radical sign**.

The number under the radical sign is called the **radicand**.

The complete expression, such as $\sqrt{64}$, is called a **radical** or **radical expression**.

Every positive real number has two square roots, one positive and one negative. The positive square root is called the **principal square root**. For example,

- because $(8)^2 = 64$, $\sqrt{64} = 8$. ← the **principal square root**

- because $(-8)^2 = 64$, $-\sqrt{64} = -8$. ← the **negative square root**

The number 0 has only one square root, namely 0.

Square Root

If a is a nonnegative real number, then

$$\sqrt{a} \text{ is the \textbf{principal square root} of } a,$$

and

$$-\sqrt{a} \text{ is the \textbf{negative square root} of } a.$$

---

notes

■ Square roots of negative numbers are not real numbers. For example, $\sqrt{-4}$ is not a real number. There is no real number whose square is
■ -4. Numbers of this type will be discussed in a future chapter.

Example 1

Evaluating Square Roots

a. 64 has two square roots, one positive and one negative. The $\sqrt{}$ sign is understood to represent the positive square root (or the principal square root) and $-\sqrt{}$ represents the negative square root. Therefore, we have

$$\sqrt{64} = 8 \qquad \text{and} \qquad -\sqrt{64} = -8.$$

b. Because $11^2 = 121$, we have $\sqrt{121} = 11$ and $-\sqrt{121} = -11$.

c. Because $0^2 = 0$, $\sqrt{0^2} = 0$.

d. $\sqrt{-25}$ is not a real number.

Now work margin exercise 1.

1. If real square roots exist, state them (state both the principal square root and the negative square root).

a. 81

b. 196

c. 49

d. −144

2. Evaluate the square roots.

a. $\sqrt{\dfrac{9}{64}}$

b. $\sqrt{0.0016}$

Example 2

Evaluating Square Roots

a. Because $\left(\dfrac{4}{5}\right)^2 = \dfrac{16}{25}$, we know that $\sqrt{\dfrac{16}{25}} = \dfrac{4}{5}$.

b. $-\sqrt{0.0009} = -0.03$ because $(0.03)^2 = 0.0009$.

Now work margin exercise 2.

Recall that **rational numbers** are numbers that can be expressed as the quotient of integers and are of the form $\dfrac{a}{b}$ where a and b are integers and $b \neq 0$. In decimal form, rational numbers are of the form of terminating decimals or repeating infinite decimals. Square roots of perfect square radicands simplify to rational numbers.

No rational number will square to give 2. So, $\sqrt{2}$ is an **irrational number** and a decimal representation is an infinite nonrepeating decimal. Your calculator will show the following.

$$\sqrt{2} = 1.414213562\ldots \quad \text{accurate to the nearest billionth}$$

$$\sqrt{2} = 1.4142 \quad \text{rounded to the nearest ten-thousandth}$$

To get a better idea of $\sqrt{2}$ we can compare as follows:

$$\sqrt{1} < \sqrt{2} < \sqrt{4} \quad \text{Note that 1 and 4 are perfect squares.}$$

which gives $1 < \sqrt{2} < 2$

and we see that $\sqrt{2}$ is between 1 and 2 and the value 1.4142 is reasonable.

3. The square root of 67 is approximately 8.1854. State why or why not this is a reasonable estimate.

Example 3

Estimating Square Roots

Using a calculator, we find that $\sqrt{30} \approx 5.4772$, accurate to the nearest ten-thousandth. Check that this is a reasonable estimate.

Solution

Because $25 < 30 < 36$, we have $\sqrt{25} < \sqrt{30} < \sqrt{36}$ and $5 < \sqrt{30} < 6$. The approximation 5.4772 is between 5 and 6 and is reasonable.

Another approach is to square as follows:
$$(5.4772)^2 = 29.99971984 \quad \text{which is close to 30.}$$

Now work margin exercise 3.

A number is **cubed** when it is used as a factor 3 times. For example,

$$5^3 = 5 \cdot 5 \cdot 5 = 125 \qquad \text{and} \qquad (-30)^3 = (-30) \cdot (-30) \cdot (-30) = -27{,}000.$$

If an integer is cubed, the result is called a **perfect cube**. The cubes for the integers from 1 to 10 are shown here for easy reference.

Cubes of Integers from 1 to 10 (Perfect Cubes)				
$1^3 = 1$	$2^3 = 8$	$3^3 = 27$	$4^3 = 64$	$5^3 = 125$
$6^3 = 216$	$7^3 = 343$	$8^3 = 512$	$9^3 = 729$	$10^3 = 1000$

Table 2

The reverse of cubing is finding the **cube root**, symbolized $\sqrt[3]{}$. For example, the cube root of 125 is 5 and we write $\sqrt[3]{125} = 5$.

Cube Root

If a is a real number, then $\sqrt[3]{a}$ is the **cube root** of a.

> ## notes
>
> In the cube root expression, $\sqrt[3]{a}$, the number 3 is called the **index**. In a square root expression such as $\sqrt{a}$, the index is understood to be 2 and **is not written.** Expressions with square roots and cube roots (as well as other roots) are called **radical expressions.**

Example 4

Evaluating Cube Roots

a. Because $2^3 = 8$, $\sqrt[3]{8} = 2$.

b. Because $(-6)^3 = -216$, $\sqrt[3]{-216} = -6$.

Note that the cube root of a negative number is a real number and is negative.

c. Because $\left(\dfrac{1}{3}\right)^3 = \dfrac{1}{27}$, $\sqrt[3]{\dfrac{1}{27}} = \dfrac{1}{3}$.

Now work margin exercise 4.

4. Evaluate the following radical expressions without a calculator

a. $\sqrt[3]{64}$

b. $\sqrt[3]{-125}$

c. $\sqrt[3]{\dfrac{1}{1000}}$

Objective C Roots of Sums and Differences

If a radical contains a sum or difference inside the radical sign, we must simplify the expression inside the radical before evaluating it.

Common Error

Wrong Solution

DO NOT distribute the radical to sums and differences inside it.

$$\sqrt{2+7} = \sqrt{2} + \sqrt{7}$$

Correct Solution

DO simplify the radicand, then evaluate the radical expression.

$$\sqrt{2+7} = \sqrt{9} = 3$$

5. Simplify the following radicals.

a. $\sqrt[3]{3+5}$

b. $\sqrt{4+12}$

Example 5

Roots of Sums and Differences

Simplify the following radicals.

a. $\sqrt{9+16}$

Solution

First, perform the addition in the radicand, then evaluate the radical.

$$\sqrt{9+16} = \sqrt{25} = 5$$

b. $\sqrt[3]{8-35}$

Solution

$$\sqrt[3]{8-35} = \sqrt[3]{-27} = -3$$

Now work margin exercise 5.

Using a TI-84 Plus Graphing Calculator to Evaluate Expressions with Radicals

A TI-84 Plus graphing calculator can be used to find decimal approximations for radicals and expressions containing radicals. The displays in Example 6 illustrate the advantage of being able to see the entire expression being evaluated.

Example 6

Evaluating Radical Expressions with a Calculator

The following radical expressions are evaluated by using a TI-84 Plus graphing calculator. In each example, the steps (or keys to press) are shown. The TI-84 Plus gives answers accurate up to nine decimal places. You may choose (through the **MODE** key) to have answers accurate to fewer than nine places.

a. $\sqrt{17}$

Solution

Step 1: Press **2ND** **x^2** to get the square root symbol $\sqrt{\ }$.

Note: When the $\sqrt{\ }$ symbol appears, it will appear with a left-hand parenthesis. You should press the right-hand parenthesis to close the square root operation.

Step 2: Enter **1** **7** and the right-hand parenthesis **)**.

Step 3: Press **ENTER**.

The display will appear as follows:

b. $3\sqrt{20}$ Note: This expression represents 3 times $\sqrt{20}$.

Solution

Step 1: Enter **3**.

Step 2: Press **2ND** **x^2**. This gives the $\sqrt{\ }$ symbol.

Step 3: Enter **2** **0** and the right-hand parenthesis **)**.

Step 4: Press **ENTER**.

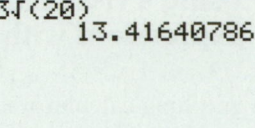

6. Evaluate the following radical expressions with a calculator.

a. $\sqrt{45}$

b. $5\sqrt{30}$

c. $8+2\sqrt{7}$

The display will appear as follows:

```
3√(20)
        13.41640786
```

c. $1+2\sqrt{3}$

Note: The calculator is programmed to follow the rules for order of operations.

Solution

Step 1: Enter **1**.

Step 2: Press the **+** key.

Step 3: Enter **2**.

Step 4: Press **2ND** **x²** (This gives the $\sqrt{}$ symbol.)

Step 5: Enter **3** and the right-hand parenthesis **)**.

Step 6: Press **ENTER**.

```
1+2√(3)
        4.464101615
```

The display will appear as follows:

Now work margin exercise 6.

Practice Problems

Simplify the following square roots and cube roots.

1. $\sqrt{49}$ **2.** $\sqrt[3]{125}$ **3.** $-\sqrt{25}$ **4.** $\sqrt{196}$

Use your knowledge of square roots and cube roots to determine whether each number is rational, irrational, or nonreal.

5. $\sqrt{\dfrac{9}{16}}$ **6.** $\sqrt{324}$ **7.** $\sqrt{-25}$ **8.** $\sqrt{3}$

Use your calculator to find the value (accurate to the nearest ten-thousandth) of each of the following radical expressions.

9. $5\sqrt{2}$ **10.** $6-4\sqrt{7}$

Practice Problem Answers

1. 7   **2.** 5   **3.** −5   **4.** 14

5. rational   **6.** rational   **7.** nonreal   **8.** irrational

9. 7.0711   **10.** −4.5830

Exercises 9.1

Simplify the following square roots and cube roots. See Examples 1, 2, and 4.

1. $\sqrt{9}$ **2.** $\sqrt{49}$ **3.** $\sqrt{81}$ **4.** $\sqrt{36}$ **5.** $\sqrt{289}$

6. $\sqrt{121}$ **7.** $\sqrt{169}$ **8.** $\sqrt{361}$ **9.** $\sqrt[3]{1}$ **10.** $\sqrt[3]{1000}$

11. $\sqrt[3]{125}$ **12.** $\sqrt[3]{343}$ **13.** $\sqrt[3]{216}$ **14.** $\sqrt[3]{512}$ **15.** $\sqrt{\dfrac{1}{4}}$

16. $\sqrt{\dfrac{9}{16}}$ **17.** $\sqrt[3]{\dfrac{27}{64}}$ **18.** $\sqrt[3]{\dfrac{1}{8}}$ **19.** $\sqrt{0.04}$ **20.** $\sqrt{0.0081}$

21. $-\sqrt{100}$ **22.** $-\sqrt{144}$ **23.** $-\sqrt{0.0016}$ **24.** $-\sqrt{0.000004}$ **25.** $\sqrt[3]{-27}$

26. $\sqrt[3]{-64}$ **27.** $\sqrt[3]{-125}$ **28.** $\sqrt[3]{-1000}$ **29.** $\sqrt{\dfrac{9}{25}}$ **30.** $\sqrt{\dfrac{25}{81}}$

Estimates (accurate to the nearest ten-thousandth) of radicals are given. Show that these are reasonable estimates. See Example 3.

31. $\sqrt{74} \approx 8.6023$ **32.** $\sqrt{18} \approx 4.2426$

33. $\sqrt{32} \approx 5.6569$ **34.** $\sqrt{110} \approx 10.4881$

Use your knowledge of square roots and cube roots to determine whether each number is rational, irrational, or nonreal.

35. $\sqrt{4}$ **36.** $\sqrt{17}$ **37.** $\sqrt{169}$ **38.** $\sqrt[3]{8}$

39. $\sqrt{\dfrac{2}{9}}$ **40.** $-\sqrt{\dfrac{1}{4}}$ **41.** $\sqrt{-36}$ **42.** $\sqrt[3]{-27}$

43. $-\sqrt[3]{125}$ **44.** $\sqrt{-10}$ **45.** $\sqrt{1.68}$ **46.** $\sqrt{5.29}$

47. $\sqrt{1+15}$ **48.** $\sqrt[3]{9-1}$ **49.** $\sqrt[3]{8-16}$ **50.** $\sqrt{6+30}$

51. $\sqrt[3]{27+37}$ **52.** $\sqrt{80-16}$ **53.** $\sqrt{144+25}$ **54.** $\sqrt[3]{64-189}$

55. $\sqrt{39}$ **56.** $\sqrt{150}$ **57.** $\sqrt{6.23}$ **58.** $\sqrt{9.6}$

59. $\sqrt{\dfrac{1}{5}}$ **60.** $\sqrt{\dfrac{3}{8}}$ **61.** $4\sqrt{5}$ **62.** $6\sqrt{3}$

63. $-2\sqrt{17}$ **64.** $-3\sqrt{6}$ **65.** $3+2\sqrt{6}$ **66.** $4-3\sqrt{5}$

67. $-2-\sqrt{12}$ **68.** $5+\sqrt{90}$

Writing & Thinking

69. Discuss, in your own words, why the square root of a negative number is not a real number.

70. Discuss, in your own words, why the cube root of a negative number is a negative number.

9.2 Simplifying Radical Expressions

Objective A **Simplifying Algebraic Expressions with Square Roots**

Objectives

A Simplify algebraic expressions that contain square roots.

B Simplify radical expressions that contain variables.

C Simplify algebraic expressions that contain cube roots.

Various roots can be related to solutions of equations, and we want such numbers to be in a **simplified form** for easier calculations and algebraic manipulations. We need the two properties of radicals stated here for square roots.

Properties of Square Roots

If a and b are **positive** real numbers, then

1. $\sqrt{ab} = \sqrt{a}\sqrt{b}$

2. $\sqrt{\dfrac{a}{b}} = \dfrac{\sqrt{a}}{\sqrt{b}}$

As an example, we know that $\sqrt{144} = 12$. However, in a situation where you may have forgotten this, you can proceed as follows using property 1 of square roots:

$$\sqrt{144} = \sqrt{36} \cdot \sqrt{4} = 6 \cdot 2 = 12.$$

Similarly, using property 2, we can write

$$\sqrt{\frac{49}{36}} = \frac{\sqrt{49}}{\sqrt{36}} = \frac{7}{6}.$$

Simplest Form for Square Roots

A square root is considered to be in **simplest form** when the radicand has no perfect square as a factor.

The number 200 is not a perfect square. Now to simplify $\sqrt{200}$, we can use property 1 of square roots and any of the following three approaches.

Approach 1: Factor 200 as $4 \cdot 50$ because 4 is a perfect square. This gives,

$$\sqrt{200} = \sqrt{4 \cdot 50} = \sqrt{4} \cdot \sqrt{50} = 2\sqrt{50}.$$

However, $2\sqrt{50}$ is **not in simplest form** because 50 has a perfect square factor, 25. Thus to complete the process, we have

$$\sqrt{200} = 2\sqrt{50} = 2\sqrt{25 \cdot 2} = 2 \cdot \sqrt{25} \cdot \sqrt{2} = 2 \cdot 5 \cdot \sqrt{2} = 10\sqrt{2}.$$

Approach 2:   Note that 100 is a perfect square factor of 200 and $200 = 100 \cdot 2$.

$$\sqrt{200} = \sqrt{100 \cdot 2} = \sqrt{100} \cdot \sqrt{2} = 10\sqrt{2}$$

Approach 3:   Use prime factors.

$$\begin{aligned}
\sqrt{200} &= \sqrt{2 \cdot 2 \cdot 2 \cdot 5 \cdot 5} \\
&= \sqrt{2 \cdot 2 \cdot 5 \cdot 5} \cdot \sqrt{2} \\
&= \sqrt{2 \cdot 2} \cdot \sqrt{5 \cdot 5} \cdot \sqrt{2} \\
&= 2 \cdot 5 \cdot \sqrt{2} \\
&= 10\sqrt{2}
\end{aligned}$$

notes

Of these three approaches, the second appears to be the easiest because it has the fewest steps. However, "seeing" the largest perfect square factor may be difficult. If you do not immediately see a perfect square factor, proceed by finding other factors or prime factors as illustrated.

---

1. Simplify each numerical expression so that there are no perfect square numbers in the radicand.

a. $\sqrt{98}$

b. $\sqrt{45}$

c. $\sqrt{\dfrac{12}{25}}$

Example 1

Simplifying Numerical Expressions with Square Roots

Simplify each numerical expression so that there are no perfect square numbers in the radicand.

a. $\sqrt{48}$

Solution

$$\sqrt{48} = \sqrt{16 \cdot 3} = \sqrt{16} \cdot \sqrt{3} = 4\sqrt{3} \qquad \text{16 is the largest perfect square factor.}$$

b. $\sqrt{63}$

Solution

$$\sqrt{63} = \sqrt{9 \cdot 7} = \sqrt{9} \cdot \sqrt{7} = 3\sqrt{7} \qquad \text{9 is the largest perfect square factor.}$$

c. $\sqrt{\dfrac{75}{16}}$

Solution

$$\sqrt{\frac{75}{16}} = \frac{\sqrt{75}}{\sqrt{16}} = \frac{\sqrt{25 \cdot 3}}{\sqrt{16}} = \frac{\sqrt{25} \cdot \sqrt{3}}{\sqrt{16}} = \frac{5\sqrt{3}}{4}$$

Now work margin exercise 1.

To simplify square root expressions that contain variables, such as $\sqrt{x^2}$, we must be aware of whether the variable represents a positive number $(x > 0)$, zero $(x = 0)$, or a negative number $(x < 0)$.

For example,

$$\text{if } x = 0, \text{ then } \sqrt{x^2} = \sqrt{0^2} = \sqrt{0} = 0 = x.$$

$$\text{If } x = 5, \text{ then } \sqrt{x^2} = \sqrt{5^2} = \sqrt{25} = 5 = x.$$

$$\text{But, if } x = -5, \text{ then } \sqrt{x^2} = \sqrt{(-5)^2} = \sqrt{25} = 5 \neq x.$$

$$\text{In fact, if } x = -5, \text{ then } \sqrt{x^2} = \sqrt{(-5)^2} = \sqrt{25} = 5 = |-5| = |x|.$$

Thus, simplifying radical expressions with variables involves more detailed analysis than simplifying radical expressions with only constants. The following definition indicates the correct way to simplify $\sqrt{x^2}$.

Square Root of x^2

If x is a real number, then $\sqrt{x^2} = |x|$.

Note: If $x \geq 0$ is given, then we can write $\sqrt{x^2} = x$.

Although using the absolute value when simplifying square roots is mathematically correct, we can avoid some confusion by assuming that the variable under the radical sign represents only positive numbers or 0. This eliminates the need for absolute value signs.

Therefore, for the remainder of this text, we will assume that $x > 0$ and write $\sqrt{x^2} = x$, in all square root expressions, unless specifically stated otherwise.

Example 2

Simplifying Square Roots with Variables

Simplify each of the following radical expressions. Assume that all variables represent positive numbers. (Note that, by making this assumption, we need not be concerned about the absolute value sign.)

a. $\sqrt{16y^2}$

Solution

$$\sqrt{16y^2} = 4y$$

2. Simplify each of the following radical expressions. Assume that all variables represent positive numbers.

a. $\sqrt{36z^2}$

b. $\sqrt{75b^2}$

c. $\sqrt{45c^2d^2}$

b. $\sqrt{72a^2}$

Solution

$$\sqrt{72a^2} = \sqrt{36a^2} \cdot \sqrt{2} = 6a\sqrt{2}$$

c. $\sqrt{12x^2y^2}$

Solution

$$\sqrt{12x^2y^2} = \sqrt{4x^2y^2} \cdot \sqrt{3} = 2xy\sqrt{3}$$

Now work margin exercise 2.

To find the square root of an expression with even exponents, divide the exponents by 2. For example,

$$x^2 \cdot x^2 = x^4 \qquad a^3 \cdot a^3 = a^6 \qquad y^5 \cdot y^5 = y^{10}$$

$$\sqrt{x^4} = x^2 \qquad \sqrt{a^6} = a^3 \qquad \sqrt{y^{10}} = y^5.$$

To find the square root of an expression with odd exponents, factor the expression into two terms, one with exponent 1 and the other with an even exponent. For example,

$$x^3 = x^2 \cdot x \qquad \text{and} \qquad y^9 = y^8 \cdot y,$$

which means that

$$\sqrt{x^3} = \sqrt{x^2 \cdot x} = \sqrt{x^2} \cdot \sqrt{x} = x \cdot \sqrt{x} \qquad \text{and} \qquad \sqrt{y^9} = \sqrt{y^8 \cdot y} = \sqrt{y^8} \cdot \sqrt{y} = y^4 \sqrt{y}.$$

Example 3

Simplifying Square Roots

Simplify each of the following radical expressions. Look for perfect square factors and even powers of the variables. Assume that all variables represent positive numbers.

a. $\sqrt{81x^4}$

Solution

$$\sqrt{81x^4} = 9x^2 \qquad\qquad \text{The exponent 4 is divided by 2.}$$

b. $\sqrt{64x^5y}$

Solution

$$\sqrt{64x^5y} = \sqrt{64x^4} \cdot \sqrt{xy} = 8x^2\sqrt{xy}$$

c. $\sqrt{18a^4b^6}$

Solution

$$\sqrt{18a^4b^6} = \sqrt{9a^4b^6} \cdot \sqrt{2} = 3a^2b^3\sqrt{2}$$

Each exponent is divided by 2.

d. $\sqrt{\dfrac{9a^{13}}{b^4}}$

Solution

$$\sqrt{\frac{9a^{13}}{b^4}} = \frac{\sqrt{9a^{13}}}{\sqrt{b^4}} = \frac{\sqrt{9a^{12}} \cdot \sqrt{a}}{\sqrt{b^4}} = \frac{3a^6\sqrt{a}}{b^2}$$

recall $a, b > 0$

Now work margin exercise 3.

3. Simplify each of the following radical expressions. Assume that all variables represent positive numbers.

a. $\sqrt{16x^8}$

b. $\sqrt{50b^5}$

c. $\sqrt{12x^8y^{12}}$

d. $\sqrt{\dfrac{25z^{18}}{y^8}}$

Objective C ## Simplifying Algebraic Expressions with Cube Roots

When simplifying expressions with cube roots, we need to be aware of perfect cube numbers and variables with exponents that are multiples of 3. (Multiples of 3 are 3, 6, 9, 12, 15, and so on.) **Thus exponents are divided by 3 in simplifying cube root expressions.** For example,

$$x^2 \cdot x^2 \cdot x^2 = x^6 \qquad a^3 \cdot a^3 \cdot a^3 = a^9 \qquad y^5 \cdot y^5 \cdot y^5 = y^{15}$$

and $\qquad \sqrt[3]{x^6} = x^2 \qquad \sqrt[3]{a^9} = a^3 \qquad \sqrt[3]{y^{15}} = y^5.$

Simplest Form for Cube Roots

A cube root is considered to be in **simplest form** when the radicand has no perfect cube as a factor.

Example 4

Simplifying Expressions with Cube Roots

Simplify each of the following radical expressions. Look for perfect cube factors and powers of the variables that are multiples of 3.

a. $\sqrt[3]{54x^6}$

Solution

$$\sqrt[3]{54x^6} = \sqrt[3]{27x^6} \cdot \sqrt[3]{2} = 3x^2\sqrt[3]{2}$$

27 is a perfect cube and the exponent 6 is divisible by 3.

4. Simplify each of the following radical expressions.

a. $\sqrt[3]{48z^3}$

b. $\sqrt[3]{-81a^8b^{12}}$

c. $\sqrt[3]{686x^6y^{11}}$

b. $\sqrt[3]{-40x^4y^{13}}$

Solution

$$\sqrt[3]{-40x^4y^{13}} = \sqrt[3]{-8x^3y^{12}} \cdot \sqrt[3]{5xy} = -2xy^4\sqrt[3]{5xy}$$

−8 is a perfect cube and each exponent is divided by 3.

c. $\sqrt[3]{250a^8b^{11}}$

Solution

$$\sqrt[3]{250a^8b^{11}} = \sqrt[3]{125a^6b^9} \cdot \sqrt[3]{2a^2b^2} = 5a^2b^3\sqrt[3]{2a^2b^2}$$

125 is a perfect cube and each exponent is divided by 3..

Now work margin exercise 4.

Practice Problems

Simplify each of the radical expressions. Assume that all variables represent positive real numbers.

1. $\sqrt{192}$

2. $\sqrt{32x^2}$

3. $\sqrt{\dfrac{3}{4}}$

4. $\sqrt{\dfrac{54y^5}{25x^2}}$

5. $\sqrt[3]{120}$

6. $\sqrt[3]{32x^5}$

Practice Problem Answers

1. $8\sqrt{3}$

2. $4x\sqrt{2}$

3. $\dfrac{\sqrt{3}}{2}$

4. $\dfrac{3y^2\sqrt{6y}}{5x}$

5. $2\sqrt[3]{15}$

6. $2x\sqrt[3]{4x^2}$

Exercises 9.2

Simplify each of the radical expressions. Assume all variables represent positive real numbers. See Examples 1 through 4.

1. $\sqrt{12}$

2. $-\sqrt{45}$

3. $\sqrt{288}$

4. $-\sqrt{63}$

5. $-\sqrt{72}$

6. $\sqrt{98}$

7. $-\sqrt{56}$

8. $\sqrt{162}$

9. $-\sqrt{125}$

10. $-\sqrt{121}$

11. $\sqrt{\dfrac{1}{4}}$

12. $\sqrt{\dfrac{32}{49}}$

13. $-\sqrt{\dfrac{11}{64}}$

14. $-\sqrt{\dfrac{125}{100}}$

15. $\sqrt{\dfrac{28}{25}}$

16. $\sqrt{\dfrac{147}{100}}$

17. $\sqrt{36x^2}$

18. $\sqrt{49y^2}$

19. $\sqrt{8x^3}$

20. $\sqrt{18a^5}$

21. $\sqrt{24x^{11}y^2}$

22. $\sqrt{20x^{15}y^3}$

23. $\sqrt{125x^3y^6}$

24. $\sqrt{8x^5y^4}$

25. $-\sqrt{18x^2y^2}$

26. $-\sqrt{32x^4y^8}$

27. $\sqrt{12ab^2c^3}$

28. $\sqrt{45a^2b^3c^4}$

29. $\sqrt{75x^4y^6z^8}$

30. $\sqrt{200x^2y^2z^2}$

31. $\sqrt{\dfrac{5x^4}{9}}$

32. $-\sqrt{\dfrac{7y^6}{16x^4}}$

33. $\sqrt{\dfrac{32a^5}{81b^{16}}}$

34. $\sqrt{\dfrac{75x^8}{121y^{12}}}$

35. $\sqrt{\dfrac{200x^8}{289}}$

36. $\sqrt{\dfrac{32x^{15}y^{10}}{169}}$

37. $\sqrt[3]{216}$

38. $\sqrt[3]{1}$

39. $\sqrt[3]{56}$

40. $\sqrt[3]{72}$

41. $\sqrt[3]{-1}$

42. $\sqrt[3]{-125}$

43. $\sqrt[3]{-128}$

44. $\sqrt[3]{-250}$

45. $\sqrt[3]{125x^4}$

46. $\sqrt[3]{64a^{12}}$

47. $\sqrt[3]{-8x^8}$

48. $\sqrt[3]{-512a^5}$

49. $\sqrt[3]{72a^6b^4}$

50. $\sqrt[3]{108ab^9}$

51. $\sqrt[3]{216x^6y^5}$

52. $\sqrt[3]{64x^9y^2}$

53. $\sqrt[3]{24x^5y^7z^9}$

54. $\sqrt[3]{250x^6y^9z^{15}}$

55. $\dfrac{\sqrt[3]{81}}{6}$

56. $\dfrac{\sqrt[3]{192}}{10}$

57. $\sqrt[3]{\dfrac{375}{8}}$

58. $\sqrt[3]{\dfrac{-48}{125}}$

59. $\sqrt[3]{\dfrac{125y^{12}}{27x^6}}$

60. $\sqrt[3]{\dfrac{x^6z^3}{64y^9}}$

Use the formula $s = \sqrt[3]{V}$, **which relates the length of the sides of a cube and the volume V, to answer the questions below.**

61. **Puzzle cube:** The volume of a puzzle cube is 343 cubic inches. What is the length of one side?

62. **Building Blocks:** Three cubic blocks of different volumes were stacked on top of each other. The top block was 216 cubic centimeters. The middle block was 343 cubic centimeters, and the bottom block was 512 cubic centimeters. How tall was the stack of blocks?

Use the following two formulas used in electricity to answer the questions below. Round your answer to the nearest hundredth.

$$I = \sqrt{\frac{P}{R}}$$

$$E = \sqrt{PR}$$

P = power (in watts)
I = current (in amperes)
E = voltage (in volts)
R = resistance (in ohms, Ω)

63. Electricity: What is the current in amperes of a light bulb that produces 150 watts of power and has a 25 Ω resistance?

64. Electricity: If a light bulb has a resistance of 30 Ω and produces 90 watts of power, what is its current in amperes?

65. Electricity: How many volts of electricity would Meghan need to produce 48 Ω of resistance from a 300 watt lamp?

66. Electricity: A 5000 Ω resistor is rated at 2.5 watts. What is the maximum voltage of electricity that should be connected across it?

Writing & Thinking

67. Under what conditions is the expression $\sqrt{a}$ not a real number?

68. Explain why the expression $\sqrt[3]{y}$ is a real number regardless of whether $y > 0$, $y < 0$, or $y = 0$.

9.3 Rational Exponents

Objective A nth **Roots**

Objectives

A Understand the meaning of nth root.

B Translate expressions using radicals into expressions using rational exponents and translate expressions using rational exponents into expressions using radicals.

C Simplify expressions using the properties of rational exponents.

D Evaluate expressions of the form $a^{\frac{m}{n}}$ with a calculator.

In Section 9.1, we restricted our discussions to radicals involving square roots and cube roots. In this section, we will expand on those ideas by discussing radicals indicating n^{th} roots in general and how to relate radical expressions to expressions with rational (fractional) exponents. For example, the fifth root of x can be written in radical form as $\sqrt[5]{x}$ or with a fractional exponent as $x^{\frac{1}{5}}$.

To understand roots in general, consider the following notation. (**Note:** In this discussion, we assume that n is a positive integer.)

Type of Root	Radical Notation and Exponential Notation
square roots	If $b = \sqrt{a}$, then $b = a^{\frac{1}{2}}$.
cube roots	If $b = \sqrt[3]{a}$, then $b = a^{\frac{1}{3}}$.
fourth roots	If $b = \sqrt[4]{a}$, then $b = a^{\frac{1}{4}}$.
nth roots	If $b = \sqrt[n]{a}$, then $b = a^{\frac{1}{n}}$.

Table 1

For example,

- because $\sqrt[4]{16} = 2$, we can say that $2 = 16^{\frac{1}{4}}$.

- because $\sqrt[5]{243} = 3$, we can say that $3 = 243^{\frac{1}{5}}$.

The following notation is used for all radical expressions.

Radical Notation

If n is a positive integer, then $\sqrt[n]{a} = a^{\frac{1}{n}}$ (assuming $\sqrt[n]{a}$ is a real number).

The expression $\sqrt[n]{a}$ is called a **radical**.

The symbol $\sqrt[n]{}$ is called a **radical sign**.

n is called the **index**.

a is called the **radicand**.

Note: If no index is given, it is understood to be 2. For example, $\sqrt{3} = \sqrt[2]{3} = 3^{\frac{1}{2}}$.

notes

Special Notes about the Index n:

For the expression $\sqrt[n]{a}$ (or $a^{\frac{1}{n}}$) to be a real number:
1. when a is nonnegative, n can be any index, and
2. when a is negative, n must be odd.
   (If a is negative and n is even, then $\sqrt[n]{a}$ is nonreal.)

Example 1

Evaluating n^{th} Roots

a. $49^{\frac{1}{2}} = \sqrt{49} = 7$, because $7^2 = 49$.

b. $81^{\frac{1}{4}} = \sqrt[4]{81} = 3$, because $3^4 = 81$.

c. $(-8)^{\frac{1}{3}} = \sqrt[3]{-8} = -2$, because $(-2)^3 = -8$.

d. $(0.00001)^{\frac{1}{5}} = \sqrt[5]{0.00001} = 0.1$, because $(0.1)^5 = 0.00001$.

e. $(-16)^{\frac{1}{2}} = \sqrt{-16}$ is not a real number.   Any even root of a negative number is nonreal.

Now work margin exercise 1.

Objective B   **Rational Exponents**

Earlier we discussed the rules of exponents using only integer exponents. These same rules of exponents apply to rational exponents (fractional exponents) as well and are repeated here for easy reference.

1. Evaluate the nth root.

a. $36^{\frac{1}{2}}$

b. $16^{\frac{1}{4}}$

c. $27^{-\frac{1}{3}}$

d. $(0.0016)^{\frac{1}{4}}$

e. $(-25)^{\frac{1}{2}}$

Now consider the problem of evaluating the expression $8^{\frac{2}{3}}$ where the exponent, $\frac{2}{3}$, is of the form $\frac{m}{n}$. Using the power rule for exponents, we can write

$$8^{\frac{2}{3}} = \left(8^{\frac{1}{3}}\right)^2 = \left(2\right)^2 = 4 \qquad \text{or,} \qquad 8^{\frac{2}{3}} = \left(8^2\right)^{\frac{1}{3}} = \left(64\right)^{\frac{1}{3}} = 4.$$

The result is the same with either approach. That is, we can take the cube root first and then square the answer. Or, we can square first and then take the cube root. In general, for an exponent of the form $\dfrac{m}{n}$, taking the nth root first and then raising this root to the exponent m is easier because the numbers are smaller.

For example,

$$81^{\frac{3}{4}} = \left(81^{\frac{1}{4}}\right)^3 = \left(3\right)^3 = 27$$

is easier to calculate and work with than

$$81^{\frac{3}{4}} = \left(81^3\right)^{\frac{1}{4}} = \left(531{,}441\right)^{\frac{1}{4}} = 27.$$

The fourth root of 81 is more commonly known than the fourth root of 531,441.

The General Form $a^{\frac{m}{n}}$

If n is a positive integer, m is any integer, and $a^{\frac{1}{n}}$ is a real number, then

$$a^{\frac{m}{n}} = \left(a^{\frac{1}{n}}\right)^m = \left(a^m\right)^{\frac{1}{n}}.$$

In radical notation: $a^{\frac{m}{n}} = \left(\sqrt[n]{a}\right)^m = \sqrt[n]{a^m}.$

Example 2

Conversion from Exponential Notation to Radical Notation

Assume that each variable represents a positive real number. Each expression is changed to an equivalent expression in either radical or exponential notation.

a. $x^{\frac{2}{3}} = \sqrt[3]{x^2}$ The index, 3, is the denominator in the rational exponent.

b. $3x^{\frac{4}{5}} = 3\sqrt[5]{x^4}$ The coefficient, 3, is not affected by the exponent.

c. $-a^{\frac{3}{2}} = -\sqrt{a^3}$ −1 is the understood coefficient.

d. $\sqrt[6]{a^5} = a^{\frac{5}{6}}$ The index, 6, is the denominator of the rational exponent.

e. $5\sqrt{x} = 5x^{\frac{1}{2}}$ In a square root, the index is understood to be 2.

f. $-\sqrt[3]{4} = -4^{\frac{1}{3}}$ The coefficient, −1, is not affected by the exponent.

Also, we could write $-4^{\frac{1}{3}} = -1 \cdot 4^{\frac{1}{3}}.$

Now work margin exercise 2.

Objective C   **Simplifying Expressions with Rational Exponents**

Expressions with rational exponents, such as

$$x^{\frac{2}{3}} \cdot x^{\frac{1}{6}}, \quad \frac{x^{\frac{3}{4}}}{x^{\frac{1}{3}}}, \quad \text{and} \quad \left(2a^{\frac{1}{4}}\right)^3,$$

can be simplified using the rules for exponents.

notes

For the remainder of this chapter we will assume that all variables represent positive real numbers, unless otherwise stated.

---

2. Convert each expression to an equivalent radical or exponential form. Assume all variables represent positive, real numbers.

a. $x^{\frac{2}{5}}$

b. $8z^{\frac{6}{7}}$

c. $-b^{\frac{5}{4}}$

d. $\sqrt[7]{x^5}$

e. $3\sqrt{s}$

f. $-\sqrt[4]{5}$

3. Simplify each expression using one or more of the rules for exponents.

a. $x^{\frac{1}{3}} \cdot x^{\frac{1}{4}}$

b. $\dfrac{a^{\frac{4}{9}}}{a^{\frac{2}{3}}}$

c. $\left(3b^{\frac{1}{5}}\right)^4$

d. $\left(64z^{-\frac{4}{9}}\right)^{-\frac{1}{2}}$

e. $(-49)^{-\frac{2}{4}}$

f. $16^{\frac{1}{2}}$

Example 3

Simplifying Expressions with Rational Exponents

Each expression is simplified using one or more of the rules for exponents.

a. $x^{\frac{2}{3}} \cdot x^{\frac{1}{6}} = x^{\frac{2}{3}+\frac{1}{6}}$

$\qquad = x^{\frac{4}{6}+\frac{1}{6}} = x^{\frac{5}{6}}$ Find a common denominator and add the exponents.

b. $\dfrac{x^{\frac{3}{4}}}{x^{\frac{1}{3}}} = x^{\frac{3}{4}-\frac{1}{3}}$

$\qquad = x^{\frac{9}{12}-\frac{4}{12}} = x^{\frac{5}{12}}$ Find a common denominator and subtract the exponents.

c. $\left(2a^{\frac{1}{4}}\right)^3 = 2^3 \cdot a^{\frac{1}{4}(3)} = 8a^{\frac{3}{4}}$

d. $\left(27y^{-\frac{9}{10}}\right)^{-\frac{1}{3}} = 27^{-\frac{1}{3}} \cdot y^{-\frac{9}{10}\left(-\frac{1}{3}\right)}$

$\qquad = \dfrac{y^{\frac{3}{10}}}{27^{\frac{1}{3}}} = \dfrac{y^{\frac{3}{10}}}{3}$ Multiply the exponents of *y* and reduce.

e. $(-36)^{-\frac{1}{2}} = \dfrac{1}{(-36)^{\frac{1}{2}}}$ This is not a real number because $(-36)^{\frac{1}{2}} = \sqrt{-36}$ is not real.

f. $9^{\frac{2}{4}} = 9^{\frac{1}{2}} = 3$ The exponent can be reduced as long as the expression is real.

g. $\left(\dfrac{49x^6y^{-2}}{z^{-4}}\right)^{\frac{1}{2}} = \dfrac{\left(49x^6y^{-2}\right)^{\frac{1}{2}}}{\left(z^{-4}\right)^{\frac{1}{2}}}$

$\qquad = \dfrac{49^{\frac{1}{2}}x^{6\left(\frac{1}{2}\right)}y^{-2\left(\frac{1}{2}\right)}}{z^{-4\left(\frac{1}{2}\right)}}$ Use the power rule four times.

$\qquad = \dfrac{7x^3y^{-1}}{z^{-2}}$ Simplify the exponents.

$\qquad = \dfrac{7x^3z^2}{y}$ Use the properties of negative exponents.

Now work margin exercise 3.

Example 4 shows how to use fractional exponents to simplify rather complicated looking radical expressions. The results may seem surprising at first.

Example 4

Simplifying Radical Notation by Changing to Exponential Notation

Simplify each expression by first changing it into an equivalent expression with rational exponents. Then rewrite the answer in simplified radical form.

a. $\sqrt[4]{\sqrt[3]{x}} = \left(\sqrt[3]{x}\right)^{\frac{1}{4}} = \left(x^{\frac{1}{3}}\right)^{\frac{1}{4}}$ Note that $\dfrac{1}{3}\cdot\dfrac{1}{4}=\dfrac{1}{12}$.

$\qquad = x^{\frac{1}{12}} = \sqrt[12]{x}$

b. $\sqrt[3]{a}\sqrt{a} = a^{\frac{1}{3}}\cdot a^{\frac{1}{2}}$

$\qquad = a^{\frac{1}{3}+\frac{1}{2}} = a^{\frac{2}{6}+\frac{3}{6}}$

$\qquad = a^{\frac{5}{6}} = \sqrt[6]{a^5}$

c. $\dfrac{\sqrt{x^3}\sqrt[3]{x^2}}{\sqrt[5]{x^2}} = \dfrac{x^{\frac{3}{2}}\cdot x^{\frac{2}{3}}}{x^{\frac{2}{5}}} = \dfrac{x^{\frac{3}{2}+\frac{2}{3}}}{x^{\frac{2}{5}}} = \dfrac{x^{\frac{9}{6}+\frac{4}{6}}}{x^{\frac{2}{5}}} = \dfrac{x^{\frac{13}{6}}}{x^{\frac{2}{5}}}$

$\qquad = x^{\frac{13}{6}-\frac{2}{5}} = x^{\frac{65}{30}-\frac{12}{30}} = x^{\frac{53}{30}}$ Note that

$\qquad = x^{\frac{30}{30}}\cdot x^{\frac{23}{30}} = x\cdot x^{\frac{23}{30}} = x\sqrt[30]{x^{23}}$ $\dfrac{53}{30}=\dfrac{30}{30}+\dfrac{23}{30}=1+\dfrac{23}{30}.$

Now work margin exercise 4.

Objective D **Evaluating Roots with a TI-84 Plus Calculator (The ⌃ key)**

The caret key ⌃ on the TI-84 Plus calculator (and most graphing calculators) is used to indicate exponents. Using this key, roots of real numbers can be calculated with up to nine-digit accuracy. To set the number of decimal places you wish in any calculations, press the **MODE** key and highlight the digit opposite the word FLOAT that indicates the desired accuracy. If no digit is highlighted, then the accuracy will be to nine decimal places (in some cases ten decimal places).

4. Rewrite the expression in simplified radical form.

a. $\sqrt[5]{\sqrt[2]{x}}$

b. $\sqrt[4]{x}\cdot\sqrt{x}$

c. $\dfrac{\sqrt[3]{x^2}\cdot\sqrt{x^3}}{\sqrt[6]{x^5}}$

 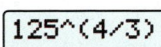

5. Evaluate the following expressions using a TI-84 Plus graphing calculator.

a. $64^{\frac{4}{3}}$

b. $42^{\frac{5}{6}}$

Example 5

 Evaluating Rational Exponents with a Calculator

Evaluate the following expressions using a TI-84 Plus graphing calculator.

a. $125^{\frac{4}{3}}$

Solution

To find $125^{\frac{4}{3}}$ proceed as follows.

Step 1: Enter the base, 125.

Step 2: Press the caret key .

Step 3: Enter the exponent in parentheses, $\frac{4}{3}$.

Step 4: Press **ENTER**.

The display should read as follows.

```
125^(4/3)
                625
```

b. $36^{\frac{3}{5}}$

Solution

To find $36^{\frac{3}{5}}$ proceed as follows.

Step 1: Enter the base, 36.

Step 2: Press the caret key .

Step 3: Enter the exponent in parentheses, $\frac{3}{5}$.

Step 4: Press **ENTER**.

The display should read as follows.

```
36^(3/5)
        8.585814487
```

Now work margin exercise 5.

Practice Problems

Simplify each of the following expressions. Leave the answers with rational exponents.

1. $64^{\frac{2}{3}}$

2. $x^{\frac{3}{4}} \cdot x^{\frac{1}{5}} \cdot x^{\frac{1}{2}}$

3. $\dfrac{x^{\frac{1}{6}} \cdot y^{\frac{1}{2}}}{x^{\frac{1}{3}} \cdot y^{\frac{1}{4}}}$

4. $\left(16^{\frac{3}{4}}\right)^{-2}$

5. $-81^{\frac{1}{4}}$

Simplify each expression by first changing to an equivalent expression with rational exponents. Rewrite the answer in simplified radical form.

6. $\sqrt[4]{x} \cdot \sqrt{x}$

7. $\sqrt[5]{\sqrt[3]{a^2}}$

8. $\dfrac{\sqrt{36x^5}}{\sqrt[3]{8x^3}}$

Use a graphing calculator to find the following values accurate to the nearest ten-thousandth.

9. $128^{\frac{1}{5}}$

10. $100^{-\frac{1}{4}}$

Practice Problem Answers

1. 16

2. $x^{\frac{29}{20}}$

3. $\dfrac{y^{\frac{1}{4}}}{x^{\frac{1}{6}}}$

4. $\dfrac{1}{64}$

5. -3

6. $\sqrt[4]{x^3}$

7. $\sqrt[15]{a^2}$

8. $3x\sqrt{x}$

9. 2.6390

10. 0.3162

Exercises 9.3

Write an equivalent expression using radical notation. See Example 2.

1. $8^{\frac{1}{3}}$

2. $5^{\frac{1}{2}}$

3. $-x^{\frac{1}{6}}$

4. $4y^{\frac{3}{4}}$

5. $(2z)^{\frac{2}{5}}$

Write an equivalent expression using exponential notation. See Example 2.

6. $\sqrt{3}$

7. $\sqrt[7]{13}$

8. $4\sqrt[3]{x^2}$

9. $\sqrt[3]{-9}$

10. $\sqrt[5]{16x^2}$

Simplify each numerical expression. See Example 1.

11. $9^{\frac{1}{2}}$

12. $121^{\frac{1}{2}}$

13. $100^{-\frac{1}{2}}$

14. $25^{-\frac{1}{2}}$

15. $-64^{\frac{3}{2}}$

16. $(-64)^{\frac{3}{2}}$

17. $(-64)^{\frac{1}{3}}$

18. $-(64)^{\frac{1}{3}}$

19. $\left(-\dfrac{4}{25}\right)^{\frac{1}{2}}$

20. $-\left(\dfrac{4}{25}\right)^{\frac{1}{2}}$

21. $\left(\dfrac{9}{49}\right)^{\frac{1}{2}}$

22. $\left(\dfrac{225}{144}\right)^{\frac{1}{2}}$

23. $64^{\frac{2}{3}}$

24. $8^{-\frac{2}{3}}$

25. $(-216)^{-\frac{1}{3}}$

26. $(-125)^{\frac{1}{3}}$

27. $\left(\dfrac{8}{125}\right)^{-\frac{1}{3}}$

28. $-\left(\dfrac{16}{81}\right)^{-\frac{3}{4}}$

29. $\left(-\dfrac{1}{32}\right)^{\frac{2}{5}}$

30. $\left(\dfrac{27}{64}\right)^{\frac{2}{3}}$

31. $3\cdot 16^{-\frac{3}{4}}$

32. $2\cdot 25^{-\frac{1}{2}}$

33. $-100^{-\frac{3}{2}}$

34. $-49^{-\frac{5}{2}}$

35. $\left[\left(\dfrac{1}{32}\right)^{\frac{2}{5}}\right]^{-3}$

36. $\left[(-27)^{\frac{2}{3}}\right]^{-2}$

Use a graphing calculator to find the value of each numerical expression accurate to the nearest ten-thousandth, if necessary. See Example 5.

37. $25^{\frac{2}{3}}$

38. $81^{\frac{7}{4}}$

39. $100^{\frac{7}{2}}$

40. $100^{\frac{1}{3}}$

41. $250^{\frac{5}{6}}$

42. $2000^{\frac{2}{3}}$

43. $24^{-\frac{3}{4}}$

44. $18^{-\frac{3}{2}}$

45. $\sqrt[9]{72}$

46. $\sqrt[8]{63}$

47. $\sqrt[4]{0.0025}$ **48.** $\sqrt[5]{0.00032}$ **49.** $\sqrt[4]{3600}$ **50.** $\sqrt[6]{4500}$ **51.** $\sqrt[5]{35.4}$

52. $\sqrt[10]{1.8}$

53. $\left(2x^{\frac{1}{3}}\right)^3$ **54.** $\left(3x^{\frac{1}{2}}\right)^4$ **55.** $\left(9a^4\right)^{-\frac{1}{2}}$ **56.** $\left(16a^3\right)^{-\frac{1}{4}}$

57. $8x^2 \cdot x^{\frac{1}{2}}$ **58.** $3x^3 \cdot x^{\frac{2}{3}}$ **59.** $5a^2 \cdot a^{-\frac{1}{3}} \cdot a^{\frac{1}{2}}$ **60.** $a^{\frac{2}{3}} \cdot a^{-\frac{3}{5}} \cdot a^0$

61. $\dfrac{x^{\frac{3}{4}}}{x^{\frac{1}{6}}}$ **62.** $\dfrac{a^{\frac{2}{3}}}{a^{\frac{1}{9}}}$ **63.** $\dfrac{x^{\frac{2}{5}}}{x^{-\frac{1}{10}}}$ **64.** $\dfrac{a^{\frac{1}{2}}}{a^{-\frac{2}{3}}}$

65. $\dfrac{a^{\frac{3}{4}} \cdot a^{\frac{1}{8}}}{a^2}$ **66.** $\dfrac{x^{\frac{2}{3}} \cdot x^{\frac{4}{3}}}{x^2}$ **67.** $\dfrac{a^{\frac{1}{2}} \cdot a^{-\frac{3}{4}}}{a^{-\frac{1}{2}}}$ **68.** $\dfrac{x^{\frac{2}{3}} x^{-1}}{x^{-\frac{3}{2}}}$

69. $\dfrac{a^{\frac{3}{2}} b^{\frac{4}{5}}}{a^{-\frac{1}{2}} b^2}$ **70.** $\dfrac{a^{\frac{3}{4}} b^{-\frac{1}{3}}}{a^{\frac{3}{2}} b^{\frac{1}{6}}}$ **71.** $\left(2x^{\frac{1}{2}} y^{\frac{1}{3}}\right)^3$ **72.** $\left(a^{\frac{1}{2}} a^{\frac{1}{3}}\right)^6$

73. $\left(4x^{-\frac{3}{4}} y^{\frac{1}{5}}\right)^{-2}$ **74.** $\left(81a^{-8} b^2\right)^{-\frac{1}{4}}$ **75.** $\left(-x^3 y^6 z^{-6}\right)^{\frac{2}{3}}$ **76.** $\left(9x^2 y^{-4} z^{-3}\right)^{\frac{3}{2}}$

77. $\left(\dfrac{x^2 y^{-3}}{z^4}\right)^{-\frac{1}{2}}$ **78.** $\left(\dfrac{27a^3 b^6}{c^9}\right)^{-\frac{1}{3}}$ **79.** $\left(\dfrac{16a^{-4} b^3}{c^4}\right)^{\frac{3}{4}}$ **80.** $\left(\dfrac{-27a^2 b^3}{c^{-3}}\right)^{\frac{1}{3}}$

81. $\dfrac{\left(x^{\frac{1}{4}}y^{\frac{1}{2}}\right)^3}{x^{\frac{1}{2}}y^{\frac{1}{4}}}$

82. $\dfrac{\left(x^{\frac{1}{2}}y\right)^{-\frac{1}{3}}}{x^{\frac{2}{3}}y^{-1}}$

83. $\dfrac{\left(8x^2y\right)^{-\frac{1}{3}}}{\left(5x^{\frac{1}{3}}y^{-\frac{1}{2}}\right)^2}$

84. $\dfrac{\left(25a^4b^{-1}\right)^{\frac{1}{2}}}{\left(2a^{\frac{1}{5}}b^{\frac{3}{5}}\right)^3}$

85. $\left(\dfrac{a^{-3}b^{\frac{1}{3}}}{a^{\frac{1}{2}}b}\right)^{\frac{1}{2}}\cdot\left(\dfrac{ab^{\frac{1}{2}}}{a^{-\frac{2}{3}}b^{-1}}\right)^{\frac{1}{2}}$

86. $\left(\dfrac{x^2y^{\frac{1}{3}}}{x^{\frac{1}{2}}y^{\frac{3}{2}}}\right)^{\frac{1}{2}}\cdot\left(\dfrac{x^{-\frac{1}{2}}y^{\frac{2}{3}}}{x^{-1}y^{\frac{3}{4}}}\right)^2$

87. $\dfrac{\left(27xy^{\frac{1}{2}}\right)^{\frac{1}{3}}\left(x^{\frac{1}{2}}y\right)^{\frac{1}{6}}}{\left(25x^{-\frac{1}{2}}y\right)^{\frac{1}{2}}\left(16x^{\frac{1}{3}}y\right)^{\frac{1}{2}}}$

88. $\dfrac{\left(4a^{-6}b\right)^{\frac{1}{2}}}{\left(7a^2b^3\right)^{-1}}\cdot\dfrac{\left(49a^4b^3\right)^{-\frac{1}{2}}}{\left(64a^{-3}b^6\right)^{\frac{2}{3}}}$

Simplify each expression by first changing it into an equivalent expression with rational exponents. Rewrite the answer in simplified radical form. Assume that all variables represent positive real numbers. See Example 4.

89. $\sqrt[3]{a}\cdot\sqrt{a}$

90. $\sqrt[3]{x^2}\cdot\sqrt[5]{x^3}$

91. $\dfrac{\sqrt[4]{y^3}}{\sqrt[6]{y}}$

92. $\dfrac{\sqrt[3]{x^4}}{\sqrt[4]{x}}$

93. $\dfrac{\sqrt[3]{x^2}\sqrt[5]{x^6}}{\sqrt{x^3}}$

94. $\dfrac{a\sqrt[4]{a}}{\sqrt[3]{a}\sqrt{a}}$

95. $\sqrt{\sqrt[3]{y}}$

96. $\sqrt[5]{\sqrt{x}}$

97. $\sqrt[3]{\sqrt[3]{x}}$

98. $\sqrt{\sqrt{a}}$

99. $\sqrt[15]{(7a)^5}$

100. $\sqrt[21]{(3x)^7}$

101. $\sqrt[4]{\sqrt[3]{\sqrt{x}}}$

102. $\sqrt[5]{\sqrt[4]{\sqrt[3]{x}}}$

103. $\left(\sqrt[3]{a^4bc^2}\right)^{15}$

104. $\left(\sqrt[4]{a^3b^6c}\right)^{12}$

Writing & Thinking

105. Is $\sqrt[5]{a}\cdot\sqrt{a}$ the same as $\sqrt[5]{a^2}$? Explain why or why not.

106. Assume that x represents a positive real number. Describe what kind of number the exponent n must be for x^n to mean

    a. a product.

    b. a quotient.

    c. 1.

    d. a radical state.

9.4 Operating with Radicals

Objective A **Addition and Subtraction with Radical Expressions**

Recall that to find the sum $2x^2 + 3x^2 - 8x^2$, you can use the distributive property and write

$$2x^2 + 3x^2 - 8x^2 = (2 + 3 - 8)x^2$$
$$= -3x^2.$$

Recall that the terms $2x^2, 3x^2,$ and $-8x^2$ are called **like terms** because each term contains the same variable expression, x^2. Similarly,

$$2\sqrt{5} + 3\sqrt{5} - 8\sqrt{5} = (2 + 3 - 8)\sqrt{5}$$
$$= -3\sqrt{5}.$$

In the example above, $2\sqrt{5}, 3\sqrt{5},$ and $-8\sqrt{5}$ are called **like radicals** because each term contains the same radical expression, $\sqrt{5}$. **Like radicals** have the same index and radicand or they can be simplified so that they have the same index and radicand.

The terms $2\sqrt{3}$ and $2\sqrt{7}$ are **not** like radicals because the radicands are not the same, and neither expression can be simplified. Therefore, a sum such as $2\sqrt{3} + 2\sqrt{7}$ cannot be simplified. That is, the terms cannot be combined.

In some cases, radicals that are not like radicals can be simplified, and the results may lead to like radicals. For example, $4\sqrt{12}, \sqrt{75},$ and $-\sqrt{108}$ are not like radicals. However, simplification of each radical allows the sum of these radicals to be found as follows.

$$4\sqrt{12} + \sqrt{75} - \sqrt{108} = 4\sqrt{4 \cdot 3} + \sqrt{25 \cdot 3} - \sqrt{36 \cdot 3}$$
$$= 4 \cdot 2\sqrt{3} + 5\sqrt{3} - 6\sqrt{3}$$
$$= 8\sqrt{3} + 5\sqrt{3} - 6\sqrt{3}$$
$$= (8 + 5 - 6)\sqrt{3}$$
$$= 7\sqrt{3}$$

Objectives

A Perform addition and subtraction involving radical expressions.

B Perform multiplication involving radical expressions.

C Rationalize the denominators of rational expressions.

D Use a graphing calculator to evaluate radical expressions.

1. Perform the indicated operation and simplify, if possible. Assume that all variables are positive.

a. $\sqrt{75a} + \sqrt{48a}$

b. $\sqrt{45} + \sqrt{20} + \sqrt{27}$

c. $\sqrt[3]{9x} - \sqrt[3]{72x}$

Example 1

Addition and Subtraction with Radicals

Perform the indicated operation and simplify, if possible. Assume that all variables are positive.

a. $\sqrt{32x} + \sqrt{18x}$

Solution

$$\sqrt{32x} + \sqrt{18x} = \sqrt{16 \cdot 2x} + \sqrt{9 \cdot 2x}$$
$$= 4\sqrt{2x} + 3\sqrt{2x}$$
$$= (4+3)\sqrt{2x}$$
$$= 7\sqrt{2x}$$

b. $\sqrt{12} + \sqrt{18} + \sqrt{27}$

Solution

$$\sqrt{12} + \sqrt{18} + \sqrt{27}$$
$$= \sqrt{4 \cdot 3} + \sqrt{9 \cdot 2} + \sqrt{9 \cdot 3}$$
$$= 2\sqrt{3} + 3\sqrt{2} + 3\sqrt{3}$$
$$= (2+3)\sqrt{3} + 3\sqrt{2}$$
$$= 5\sqrt{3} + 3\sqrt{2}$$

Note that $\sqrt{3}$ and $\sqrt{2}$ are **not** like radicals. Therefore, the last expression is already fully simplified.

c. $\sqrt[3]{5x} - \sqrt[3]{40x}$

Solution

$$\sqrt[3]{5x} - \sqrt[3]{40x} = \sqrt[3]{5x} - \sqrt[3]{8 \cdot 5x}$$
$$= \sqrt[3]{5x} - 2\sqrt[3]{5x}$$
$$= (1-2)\sqrt[3]{5x}$$
$$= -\sqrt[3]{5x}$$

d. $x\sqrt{4y^3} - 5\sqrt{x^2 y^3}$

Solution

$$x\sqrt{4y^3} - 5\sqrt{x^2 y^3} = x\sqrt{4y^2}\sqrt{y} - 5\sqrt{x^2 y^2}\sqrt{y}$$
$$= 2xy\sqrt{y} - 5xy\sqrt{y}$$
$$= -3xy\sqrt{y}$$

Now work margin exercise 1.

To find a product such as $\left(\sqrt{3}+5\right)\left(\sqrt{3}-7\right)$ treat the two expressions as two binomials and multiply just as with polynomials. For example, using the FOIL method, we get

$$\left(\sqrt{3}+5\right)\left(\sqrt{3}-7\right)=\left(\sqrt{3}\right)^2-7\sqrt{3}+5\sqrt{3}+5(-7)$$
$$=3-7\sqrt{3}+5\sqrt{3}-35$$
$$=3-35+(-7+5)\sqrt{3}$$
$$=-32-2\sqrt{3}.$$

Example 2

Multiplication of Radicals

Multiply and simplify the following expressions.

a. $\sqrt{5}\cdot\sqrt{15}$

Solution

$$\sqrt{5}\cdot\sqrt{15}=\sqrt{5}\cdot\sqrt{5}\cdot\sqrt{3}=\left(\sqrt{5}\right)^2\cdot\sqrt{3}=5\sqrt{3}$$

b. $\sqrt{7}\left(\sqrt{7}-\sqrt{14}\right)$

Solution

$$\sqrt{7}\left(\sqrt{7}-\sqrt{14}\right)=\sqrt{7}\cdot\sqrt{7}-\sqrt{7}\cdot\sqrt{14}$$
$$=\left(\sqrt{7}\right)^2-\sqrt{7}\left(\sqrt{7}\cdot\sqrt{2}\right)$$
$$=\left(\sqrt{7}\right)^2-\left(\sqrt{7}\right)^2\sqrt{2}$$
$$=7-7\sqrt{2}$$

c. $\left(3\sqrt{7}-2\right)\left(\sqrt{7}+3\right)$

Solution

$$\left(3\sqrt{7}-2\right)\left(\sqrt{7}+3\right)=3\left(\sqrt{7}\right)^2+3\cdot3\sqrt{7}-2\sqrt{7}-2\cdot3$$
$$=3\cdot7+9\sqrt{7}-2\sqrt{7}-6$$
$$=21-6+(9-2)\sqrt{7}$$
$$=15+7\sqrt{7}$$

2. Multiply and simplify the following expressions

a. $\sqrt{3} \cdot \sqrt{75}$

b. $\sqrt{13}\left(\sqrt{26} - \sqrt{13}\right)$

c. $\left(2\sqrt{5} - 1\right)\left(\sqrt{5} + 2\right)$

d. $\left(\sqrt{7} - \sqrt{5}\right)^2$

e. $\left(\sqrt{3z} - \sqrt{5}\right)\left(\sqrt{3z} + \sqrt{5}\right)$

d. $\left(\sqrt{6} + \sqrt{2}\right)^2$

Solution

$$\left(\sqrt{6} + \sqrt{2}\right)^2 = \left(\sqrt{6}\right)^2 + 2\sqrt{6}\sqrt{2} + \left(\sqrt{2}\right)^2 \qquad \textcolor{red}{(a+b)^2 = a^2 + 2ab + b^2}$$

$$= 6 + 2\sqrt{12} + 2$$

$$= 8 + 2\sqrt{4} \cdot \sqrt{3}$$

$$= 8 + 2 \cdot 2\sqrt{3}$$

$$= 8 + 4\sqrt{3}$$

Now work margin exercise 2.

Objective C Rationalizing Denominators of Rational Expressions

An expression with a radical in the denominator may not be in the most usable form for further algebraic manipulation or operations. In many situations, we may want to **rationalize the denominator**, to make calculations easier. That is, we want to find an equivalent fraction in which the denominator does not contain a radical. The numerator may still have a radical in it, but a rational expression with no radicals in the denominator definitely makes arithmetic with radicals much easier. (**Note:** In this section we will deal only with radicals that involve square roots or cube roots, however the ideas discussed here also apply to n^{th} roots in general.)

> **To Rationalize a Denominator Containing a Square Root or a Cube Root**
>
> 1. If the denominator contains a square root, multiply both the numerator and denominator by an expression that will give a denominator with no square roots.
> 2. If the denominator contains a cube root, multiply both the numerator and denominator by an expression that will give a denominator with no cube roots.

Example 3

Rationalizing the Denominator

Simplify each radical expression so that the denominator contains no radicals. Assume all variables represent positive numbers.

a. $\dfrac{\sqrt{5}}{\sqrt{2}}$

Solution

Multiply the numerator and denominator by $\sqrt{2}$ because $2 \cdot 2 = 4$, a perfect square.

$$\frac{\sqrt{5}}{\sqrt{2}} = \frac{\sqrt{5}}{\sqrt{2}} \cdot \frac{\sqrt{2}}{\sqrt{2}} = \frac{\sqrt{10}}{\sqrt{4}} = \frac{\sqrt{10}}{2}$$

b. $\sqrt{\dfrac{5}{12x}}$

Solution

Multiply the numerator and denominator by $\sqrt{3x}$ because $12x \cdot 3x = 36x^2$, a perfect square expression.

$$\sqrt{\frac{5}{12x}} = \frac{\sqrt{5}}{\sqrt{12x}} = \frac{\sqrt{5}}{\sqrt{12x}} \cdot \frac{\sqrt{3x}}{\sqrt{3x}} = \frac{\sqrt{15x}}{\sqrt{36x^2}} = \frac{\sqrt{15x}}{6x}$$

c. $\dfrac{7}{\sqrt[3]{32y}}$

Solution

Multiply the numerator and denominator by $\sqrt[3]{2y^2}$ because $32y \cdot 2y^2 = 64y^3$, a perfect cube expression.

$$\frac{7}{\sqrt[3]{32y}} = \frac{7}{\sqrt[3]{32y}} \cdot \frac{\sqrt[3]{2y^2}}{\sqrt[3]{2y^2}} = \frac{7\sqrt[3]{2y^2}}{\sqrt[3]{64y^3}} = \frac{7\sqrt[3]{2y^2}}{4y}$$

Now work margin exercise 3.

If the denominator has a radical expression with **a sum or difference involving square roots** such as

$$\frac{2}{4 - \sqrt{2}} \quad \text{or} \quad \frac{12}{3 + \sqrt{5}},$$

then a different method is used for rationalizing the denominator. In this method, we think of the denominator in the form of $a - b$ or $a + b$. Thus

$$\text{if } a - b = 4 - \sqrt{2}, \text{ then } a + b = 4 + \sqrt{2}$$
$$\text{and}$$
$$\text{if } a + b = 3 + \sqrt{5}, \text{ then } a - b = 3 - \sqrt{5}.$$

The two expressions $(a - b)$ and $(a + b)$ are called **conjugates** of each other, and, as we know, their product $(a - b)(a + b)$ results in the **difference of two squares**:

$$(a - b)(a + b) = a^2 - b^2.$$

3. Simplify each radical expression so that the denominator contains no radicals. Assume all variables represent positive numbers.

a. $\dfrac{\sqrt{7}}{\sqrt{3}}$

b. $\sqrt{\dfrac{7}{18z}}$

c. $\dfrac{11}{\sqrt[3]{9x}}$

To Rationalize a Denominator Containing a Sum or Difference Involving Square Roots

If the denominator of a fraction contains a sum or difference involving a square root, rationalize the denominator by multiplying both the numerator and denominator by the **conjugate of the denominator**.

1. If the denominator is of the form $a - b$, multiply both the numerator and denominator by $a + b$.
2. If the denominator is of the form $a + b$, multiply both the numerator and denominator by $a - b$.

The new denominator will be the difference of two squares and therefore will not contain a radical term.

Example 4

Rationalizing a Denominator with a Sum or Difference Involving a Square Root

Simplify each expression by rationalizing the denominator.

a. $\dfrac{2}{4 - \sqrt{2}}$

Solution

Multiply the numerator and denominator by $4 + \sqrt{2}$.

$$\frac{2}{4 - \sqrt{2}} = \frac{2\left(4 + \sqrt{2}\right)}{\left(4 - \sqrt{2}\right)\left(4 + \sqrt{2}\right)}$$

If $a - b = 4 - \sqrt{2}$, then $a + b = 4 + \sqrt{2}$.

$$= \frac{2\left(4 + \sqrt{2}\right)}{4^2 - \left(\sqrt{2}\right)^2}$$

The denominator is the difference of two squares.

$$= \frac{2\left(4 + \sqrt{2}\right)}{16 - 2}$$

$$= \frac{\cancel{2}\left(4 + \sqrt{2}\right)}{\cancel{14}_{7}}$$

$$= \frac{4 + \sqrt{2}}{7}$$

The denominator is a rational number. Note that the numerator is now irrational. However, this is generally preferred over having an irrational denominator.

b. $\dfrac{31}{6 + \sqrt{5}}$

Solution

Multiply the numerator and denominator by $6 - \sqrt{5}$.

$$\frac{31}{6+\sqrt{5}} = \frac{31(6-\sqrt{5})}{(6+\sqrt{5})(6-\sqrt{5})}$$

$$= \frac{31(6-\sqrt{5})}{36-5}$$

$$= \frac{\cancel{31}(6-\sqrt{5})}{\cancel{31}}$$

$$= 6-\sqrt{5}$$

4. Simplify each expression by rationalizing the denominator.

a. $\dfrac{5}{\sqrt{3}-5}$

b. $\dfrac{15}{7+\sqrt{6}}$

c. $\dfrac{1}{\sqrt{8}+\sqrt{3}}$

d. $\dfrac{8}{1-\sqrt{z}}$

e. $\dfrac{x+y}{\sqrt{x}+\sqrt{y}}$

c. $\dfrac{1}{\sqrt{7}-\sqrt{2}}$

Solution

Multiply the numerator and denominator by $\sqrt{7}+\sqrt{2}$.

$$\frac{1}{\sqrt{7}-\sqrt{2}} = \frac{1(\sqrt{7}+\sqrt{2})}{(\sqrt{7}-\sqrt{2})(\sqrt{7}+\sqrt{2})}$$

$$= \frac{\sqrt{7}+\sqrt{2}}{7-2} = \frac{\sqrt{7}+\sqrt{2}}{5}$$

d. $\dfrac{6}{1+\sqrt{x}}$

Solution

Multiply the numerator and denominator by $1-\sqrt{x}$.

$$\frac{6}{1+\sqrt{x}} = \frac{6(1-\sqrt{x})}{(1+\sqrt{x})(1-\sqrt{x})}$$

$$= \frac{6(1-\sqrt{x})}{1-x}$$

$$= \frac{6-6\sqrt{x}}{1-x}$$

e. $\dfrac{x-y}{\sqrt{x}-\sqrt{y}}$

Solution

Multiply the numerator and denominator by $\sqrt{x}+\sqrt{y}$.

$$\frac{x-y}{\sqrt{x}-\sqrt{y}} = \frac{(x-y)(\sqrt{x}+\sqrt{y})}{(\sqrt{x}-\sqrt{y})(\sqrt{x}+\sqrt{y})}$$

$$= \frac{\cancel{(x-y)}(\sqrt{x}+\sqrt{y})}{\cancel{x-y}}$$

$$= \sqrt{x}+\sqrt{y}$$

Now work margin exercise 4.

Evaluating Radical Expressions with a TI-84 Plus Graphing Calculator

Techniques for using a TI-84 Plus graphing calculator to evaluate radical expressions and expressions with rational exponents were illustrated in Sections 9.1, 9.2, and 9.3. These same basic techniques are used to evaluate numerical expressions that contain sums, differences, products, and quotients of radicals. Be careful to use parentheses to ensure that the rules for order of operations are maintained. In particular, sums and differences in numerators and denominators of fractions must be enclosed in parentheses. Study the following examples carefully.

Example 5

 Using a TI-84 Graphing Calculator to Evaluate Radical Expressions

Use a TI-84 Plus graphing calculator to evaluate each expression. Round answers to the nearest ten-thousandth.

a. $3 + 2\sqrt{5}$

Solution

The display should appear as follows.

```
3+2√(5)
        7.472135955
```

Thus $3 + 2\sqrt{5} = 7.4721359... \approx 7.4721$.   *rounded to 4 decimal places*

b. $\left(\sqrt{2} + 5\right)\left(\sqrt{2} - 5\right)$

Solution

The display should appear as follows.

```
(√(2)+5)(√(2)-5)
             -23
```

Note that the right parenthesis on 2 must be included. Otherwise, the calculator will interpret the expression as $\sqrt{(2+5)}$ (or $\sqrt{7}$) which is not intended.

Thus, $\left(\sqrt{2} + 5\right)\left(\sqrt{2} - 5\right) = -23$.

c. $\dfrac{3}{\sqrt{6}-\sqrt{2}}$

Solution

The display should appear as follows.

```
3/(√(6)−√(2))
        2.897777479
```
Count the parentheses in pairs.

Thus $\dfrac{3}{\sqrt{6}-\sqrt{2}} = 2.8977774... \approx 2.8978.$ *rounded to the nearest ten-thousandth*

Now work margin exercise 5.

Practice Problems

Simplify each expression. Assume that all variables represent positive numbers.

1. $2\sqrt{10}-6\sqrt{10}$ **2.** $\sqrt{5}+\sqrt{45}-\sqrt{15}$ **3.** $\sqrt{8x}-3\sqrt{2x}+\sqrt{18x}$

4. $\sqrt[3]{x^5}+x\sqrt[3]{27x^2}$ **5.** $\sqrt{\dfrac{3}{8a^2}}$ **6.** $\dfrac{\sqrt[3]{4ab}}{\sqrt[3]{2a^2b^4}}$

7. $\dfrac{4}{\sqrt{2}+\sqrt{6}}$ **8.** $\dfrac{x-5}{\sqrt{x}-\sqrt{5}}$ **9.** $\left(\sqrt{3}+\sqrt{2}\right)^2$

10. $\left(3+\sqrt{2}\right)^2$ **11.** $\left(\sqrt{3}+\sqrt{8}\right)^2$ **12.** $\dfrac{\sqrt{5}-3\sqrt{2}}{\sqrt{6}+\sqrt{10}}$

Practice Problem Answers

1. $-4\sqrt{10}$ **2.** $4\sqrt{5}-\sqrt{15}$ **3.** $2\sqrt{2x}$

4. $4x\sqrt[3]{x^2}$ **5.** $\dfrac{\sqrt{6}}{4a}$ **6.** $\dfrac{\sqrt[3]{2a^2}}{ab}$

7. $\sqrt{6}-\sqrt{2}$ **8.** $\sqrt{x}+\sqrt{5}$ **9.** $5+2\sqrt{6}$

10. $11+6\sqrt{2}$ **11.** $11+4\sqrt{6}$

12. $\dfrac{5\sqrt{2}+6\sqrt{3}-6\sqrt{5}-\sqrt{30}}{4}$

5. Use a TI-84 Plus graphing calculator to evaluate each expression. Round answers to the nearest ten-thousandth.

a. $4+5\sqrt{2}$

b. $\left(\sqrt{8}-1\right)\left(\sqrt{8}+1\right)$

c. $\dfrac{8}{\sqrt{5}-\sqrt{10}}$

Exercises 9.4

Perform the indicated operations and simplify. Assume all variables represent positive numbers. See Examples 1 and 2.

1. $\sqrt{2} - 7\sqrt{2}$

2. $3\sqrt{5} - 6\sqrt{5}$

3. $6\sqrt{11} + 4\sqrt{11} - 3\sqrt{11}$

4. $2\sqrt{x} + 4\sqrt{x} - \sqrt{x}$

5. $9\sqrt[3]{7x^2} - 4\sqrt[3]{7x^2} - 8\sqrt[3]{7x^2}$

6. $12\sqrt[3]{4x} - 10\sqrt[3]{4x} - 6\sqrt[3]{4x}$

7. $2\sqrt{3} + 4\sqrt{12}$

8. $2\sqrt{48} - 3\sqrt{75}$

9. $2\sqrt{18} + \sqrt{8} - 3\sqrt{50}$

10. $5\sqrt{48} + 2\sqrt{45} - 3\sqrt{20}$

11. $2\sqrt{12} + \sqrt{72} - \sqrt{75}$

12. $3\sqrt{28} - \sqrt{63} + 8\sqrt{10}$

13. $7\sqrt{12x} - 4\sqrt{27x} + \sqrt{108x}$

14. $\sqrt{32x} + 7\sqrt{12x} + \sqrt{98x}$

15. $6x\sqrt{45x} + \sqrt{80x^3} - \sqrt{20x^3}$

16. $2\sqrt{18xy^2} + \sqrt{8xy^2} - 3y\sqrt{50x}$

17. $\sqrt{125} - \sqrt{63} + 3\sqrt{45}$

18. $5\sqrt{48} + 2\sqrt{24} - \sqrt{75}$

19. $\sqrt[3]{81x^2} - 5\sqrt[3]{48x^2} - 5\sqrt[3]{24x^2}$

20. $\sqrt[3]{16x} - 5\sqrt[3]{54x} + 2\sqrt[3]{40x}$

21. $x\sqrt{y} + \sqrt{x^2 y} - \sqrt{xy^3}$

22. $x\sqrt{y^3} - 2\sqrt{x^2 y^3} - y\sqrt{x^2 y}$

23. $x\sqrt{2x^3 y^2} - 3y\sqrt{8x^5} + xy\sqrt{72x^3}$

24. $x\sqrt{9x^3 y^2} - 5x^2\sqrt{xy^2} + 6y\sqrt{x^5}$

25. $\left(3 + \sqrt{2}\right)\left(5 - \sqrt{2}\right)$

26. $\left(\sqrt{6} + 2\right)\left(\sqrt{6} - 2\right)$

27. $\left(\sqrt{3x} - 8\right)\left(\sqrt{3x} - 1\right)$

28. $\left(6 + \sqrt{2x}\right)\left(4 + \sqrt{2x}\right)$

29. $\left(2\sqrt{7} + 4\right)\left(\sqrt{7} - 3\right)$

30. $\left(5\sqrt{3} - 2\right)\left(2\sqrt{3} - 7\right)$

31. $\left(3\sqrt{5} + 4\sqrt{3}\right)\left(3\sqrt{5} - 4\sqrt{3}\right)$

32. $\left(3\sqrt{2} + \sqrt{5}\right)\left(\sqrt{2} + \sqrt{5}\right)$

33. $\left(\sqrt{5} + 2\sqrt{2}\right)^2$

34. $\left(2\sqrt{5} + 3\sqrt{2}\right)^2$

35. $\left(\sqrt{2}+\sqrt{3}\right)\left(\sqrt{5}-\sqrt{3}\right)$

36. $\left(\sqrt{6}+\sqrt{5}\right)\left(\sqrt{6}-\sqrt{2}\right)$

37. $\left(\sqrt{x}+\sqrt{6}\right)\left(\sqrt{x}-3\sqrt{6}\right)$

38. $\left(\sqrt{11}+\sqrt{3}\right)\left(\sqrt{11}-2\sqrt{3}\right)$

39. $\left(3\sqrt{7}+\sqrt{5}\right)\left(3\sqrt{7}-\sqrt{5}\right)$

40. $\left(7\sqrt{x}+\sqrt{2}\right)\left(7\sqrt{x}-\sqrt{2}\right)$

41. $\left(\sqrt{x}+5\sqrt{y}\right)^2$

42. $\left(3\sqrt{x}+\sqrt{y}\right)^2$

43. $\left(4\sqrt{x}+3\sqrt{y}\right)\left(\sqrt{x}-3\sqrt{y}\right)$

44. $\left(2\sqrt{2x}+\sqrt{y}\right)\left(3\sqrt{2x}+2\sqrt{y}\right)$

Rationalize the denominator and simplify if possible. Assume that all variables represent positive numbers. See Examples 3 and 4.

45. $\dfrac{\sqrt{14}}{\sqrt{6}}$

46. $\dfrac{\sqrt{12}}{\sqrt{8}}$

47. $\dfrac{\sqrt[3]{35}}{\sqrt[3]{4}}$

48. $\dfrac{\sqrt[3]{10}}{\sqrt[3]{9}}$

49. $-\sqrt{\dfrac{2}{3y}}$

50. $-\sqrt{\dfrac{25}{x^3}}$

51. $\dfrac{\sqrt{8x}}{\sqrt{5y^2}}$

52. $\dfrac{\sqrt{4x}}{\sqrt{3y^2}}$

53. $\dfrac{\sqrt{16y^2}}{\sqrt{2y^3}}$

54. $\dfrac{\sqrt{24b}}{\sqrt{6b^2}}$

55. $\sqrt[3]{\dfrac{2y^3}{27x^2}}$

56. $\sqrt[3]{\dfrac{7x}{2y^4}}$

57. $\dfrac{\sqrt[3]{6a^4}}{\sqrt[3]{25a^2b^4}}$

58. $\dfrac{\sqrt[3]{x^5}}{\sqrt[3]{9xy}}$

59. $\dfrac{1}{\sqrt{2}+1}$

60. $\dfrac{3}{\sqrt{3}-5}$

61. $\dfrac{\sqrt{3}}{\sqrt{5}-4}$

62. $\dfrac{\sqrt{6}}{\sqrt{7}+3}$

63. $\dfrac{2}{\sqrt{2}+\sqrt{3}}$

64. $\dfrac{8}{\sqrt{5}-\sqrt{3}}$

65. $\dfrac{\sqrt{5}}{\sqrt{7}-\sqrt{3}}$

66. $\dfrac{\sqrt{10}}{\sqrt{5}-2\sqrt{2}}$

67. $\dfrac{2-\sqrt{6}}{\sqrt{6}-3}$

68. $\dfrac{\sqrt{10}-5}{2-\sqrt{10}}$

69. $\dfrac{\sqrt{3}+\sqrt{7}}{\sqrt{3}-\sqrt{7}}$

70. $\dfrac{2\sqrt{3}+\sqrt{2}}{\sqrt{3}-\sqrt{2}}$

71. $\dfrac{2\sqrt{x}-y}{\sqrt{x}-y}$

72. $\dfrac{x+2\sqrt{y}}{x-2\sqrt{y}}$

Use a graphing calculator to evaluate each expression. Round to the nearest ten-thousandth if necessary. See Example 5.

73. $13-\sqrt{75}$

74. $5-\sqrt{67}$

75. $\sqrt{900}+\sqrt{2.56}$

76. $\sqrt{1600}-\sqrt{1.69}$

77. $\left(\sqrt{7}+8\right)\left(\sqrt{7}-8\right)$

78. $\left(\sqrt{8}-\sqrt{5}\right)\left(\sqrt{8}+\sqrt{5}\right)$

79. $\left(2\sqrt{3}+5\sqrt{2}\right)\left(\sqrt{10}-3\sqrt{5}\right)$

80. $\left(6\sqrt{5}+5\sqrt{7}\right)\left(3\sqrt{2}-\sqrt{6}\right)$

81. $\dfrac{\sqrt{6}}{16-3\sqrt{6}}$

82. $\dfrac{\sqrt{5}}{1+2\sqrt{5}}$

83. $\dfrac{\sqrt{10}-\sqrt{2}}{\sqrt{10}+\sqrt{2}}$

84. $\dfrac{\sqrt{5}+\sqrt{2}}{\sqrt{5}-\sqrt{2}}$

Solve the following word problems.

85. Radio Circuits: For a complete radio circuit, $d=\sqrt{2g}+\sqrt{2h}$, where d equals the visual horizon distance and g and h are the heights of the radio antennas at the respective stations. What is d when $g=75$ ft and $h=85$ ft?

86. Tile Patterns: Mary is making a tile decoration for her wall. Using square tiles of different sizes, Mary created one decoration that is five tiles across, with sides touching. The first tile is 10 in.^{2}, the second is 20 in.^{2}, the third is 30 in.^{2}, the fourth is 20 in.^{2}, and the fifth is 10 in.^{2} What is the length of the decoration?

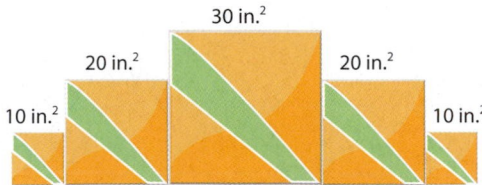

Writing & Thinking

87. In your own words, explain how to rationalize the denominator of a fraction containing the sum or difference of square roots in the denominator. Why does this work?

9.5 Solving Radical Equations

Objective A   **Solving Equations that Contain One or More Radical Expressions**

Each of the following equations involves at least one radical expression:

$$x + 3 = \sqrt{x + 5}, \qquad \sqrt{x} - \sqrt{2x - 14} = 1, \quad \text{and} \quad \sqrt[3]{x + 1} = 5.$$

If the radicals are square roots, we solve by squaring both sides of the equations. If the radical is some other root and this root can be isolated on one side of the equation, we solve by raising both sides of the equation to the integer power corresponding to the index of the radical. For example, with a cube root both sides are raised to the third power.

Squaring both sides of an equation may introduce new solutions. For example, the first-degree equation $x = -3$ has only one solution, namely, -3. However, squaring both sides gives the quadratic equation

$$x^2 = (-3)^2 \qquad \text{or} \qquad x^2 = 9.$$

The quadratic equation $x^2 = 9$ has two solutions, 3 and -3. Thus a new solution that is not a solution to the original equation has been introduced. Such a solution is called an **extraneous solution**.

When both sides of an equation are raised to a power, an extraneous solution may be introduced. Be sure to check all solutions in the original equation.

The following examples illustrate a variety of situations involving radicals. The steps used are related to the following general method.

Method for Solving Equations with Radicals

1. Isolate one of the radicals on one side of the equation. (An equation may have more than one radical.)
2. Raise both sides of the equation to the power corresponding to the index of the radical.
3. If the equation still contains a radical, repeat steps 1 and 2.
4. Solve the equation after all the radicals have been eliminated.
5. Be sure to check all possible solutions in the original equation and eliminate any extraneous solutions.

Example 1

Equations with One Radical

Solve the following equations.

a. $\sqrt{x^2 + 13} = 7$

Solution

The radical is by itself on one side of the equation, so square both sides.

$$\sqrt{x^2 + 13} = 7$$
$$\left(\sqrt{x^2 + 13}\right)^2 = 7^2 \qquad \text{Square both sides.}$$
$$x^2 + 13 = 49 \qquad \text{This new equation contains no radical.}$$
$$x^2 - 36 = 0$$
$$(x + 6)(x - 6) = 0 \qquad \text{Solve by factoring.}$$
$$x = -6 \quad \text{or} \quad x = 6$$

Check: **Check both answers** in the original equation.

$$\sqrt{(-6)^2 + 13} \overset{?}{=} 7 \qquad \sqrt{(6)^2 + 13} \overset{?}{=} 7$$
$$\sqrt{36 + 13} \overset{?}{=} 7 \qquad \sqrt{36 + 13} \overset{?}{=} 7$$
$$\sqrt{49} \overset{?}{=} 7 \qquad \sqrt{49} \overset{?}{=} 7$$
$$7 = 7 \qquad 7 = 7$$

Both −6 and 6 are solutions.

b. $\sqrt{y^2 - 10y - 11} = 1 + y$

Solution

Since there is only one radical and it is by itself on one side of the equation, square both sides.

$$\sqrt{y^2 - 10y - 11} = 1 + y$$
$$\left(\sqrt{y^2 - 10y - 11}\right)^2 = (1 + y)^2 \qquad \text{Square both sides.}$$
$$y^2 - 10y - 11 = 1 + 2y + y^2$$
$$-12y - 12 = 0 \qquad \text{Simplifying gives a first-degree equation.}$$
$$-12y = 12$$
$$y = -1$$

Check:

$$\sqrt{(-1)^2 - 10(-1) - 11} \overset{?}{=} 1 + (-1)$$

$$\sqrt{1 + 10 - 11} \overset{?}{=} 0$$

$$\sqrt{0} \overset{?}{=} 0$$

$$0 = 0$$

There is one solution, -1.

c. $\sqrt{3x + 13} + 3 = 2x$

Solution

$$\sqrt{3x + 13} + 3 = 2x$$

$$\sqrt{3x + 13} = 2x - 3 \qquad \text{Isolate the radical.}$$

$$\left(\sqrt{3x + 13}\right)^2 = (2x - 3)^2 \qquad \text{Square both sides.}$$

$$3x + 13 = 4x^2 - 12x + 9$$

$$0 = 4x^2 - 15x - 4$$

$$0 = (4x + 1)(x - 4) \qquad \text{Solve by factoring.}$$

$$x = -\frac{1}{4} \quad \text{or} \quad x = 4$$

Check: **Check both answers** in the original equation.

$$\sqrt{3\left(-\frac{1}{4}\right) + 13} + 3 \overset{?}{=} 2\left(-\frac{1}{4}\right) \qquad\qquad \sqrt{3(4) + 13} + 3 \overset{?}{=} 2(4)$$

$$\sqrt{\frac{49}{4}} + 3 \overset{?}{=} -\frac{1}{2} \qquad\qquad\qquad \sqrt{25} + 3 \overset{?}{=} 8$$

$$\frac{7}{2} + 3 \overset{?}{=} -\frac{1}{2} \qquad\qquad\qquad 5 + 3 \overset{?}{=} 8$$

$$\frac{13}{2} \neq -\frac{1}{2} \qquad\qquad\qquad\qquad 8 = 8$$

$-\frac{1}{4}$ is **not** a solution. The only solution is 4.

d. $\sqrt{x + 1} = -3$

Solution

We could stop right here. There is **no real solution** to this equation because the radical on the left is nonnegative and cannot possibly equal -3, a negative number. Suppose we did not notice this relationship. Then proceeding as usual, we will find an answer that does not check.

1. Solve the following equations.

a. $\sqrt{x^2 - 48} = 4$

b. $\sqrt{y^2 + y - 1} = y + 3$

c. $\sqrt{-x^2 + x + 12} - 3 = x$

d. $\sqrt{x + 5} = -9$

$$\sqrt{x + 1} = -3$$
$$\left(\sqrt{x + 1}\right)^2 = (-3)^2 \qquad \text{Square both sides.}$$
$$x + 1 = 9$$
$$x = 8$$

Check:

$$\sqrt{(8) + 1} \overset{?}{=} -3$$
$$\sqrt{9} \overset{?}{=} -3$$
$$3 \neq -3$$

So, 8 does **not** check, and there is **no solution**.

Now work margin exercise 1.

notes

■ It is possible that after checking the answers you may find that none of the answers are solutions. In this case the answer is **no solution**.

■

Example 2

Equations with Two Radicals

Solve the following equations with two radicals. You may need to rearrange terms and to square twice.

a. $\sqrt{x + 4} = \sqrt{3x - 2}$

Solution

There are two radicals on opposite sides of the equation. Squaring both sides will give a new equation with no radicals.

$$\sqrt{x + 4} = \sqrt{3x - 2}$$
$$\left(\sqrt{x + 4}\right)^2 = \left(\sqrt{3x - 2}\right)^2$$
$$x + 4 = 3x - 2$$
$$6 = 2x \qquad \text{Simplifying results in a first-degree equation.}$$
$$3 = x$$

Check: $\quad \sqrt{(3) + 4} \overset{?}{=} \sqrt{3(3) - 2}$
$$\sqrt{7} = \sqrt{7}$$

There is one solution, 3.

b. $\sqrt{x} - \sqrt{2x - 14} = 1$

Solution

2. Solve the following equations with two radicals. You may need to rearrange terms to square twice.

a. $\sqrt{2x + 5} = \sqrt{x + 7}$

b. $2\sqrt{x} - \sqrt{3x - 11} = 2$

Where there is a sum or difference of radicals, squaring is easier if the radicals are on different sides of the equation. Also, squaring both sides of the equation is easier if one of the radicals is by itself on one side of the equation.

$$\sqrt{x} - \sqrt{2x - 14} = 1$$

$$\sqrt{x} = 1 + \sqrt{2x - 14} \quad \text{Isolate one of the radicals.}$$

$$\left(\sqrt{x}\right)^2 = \left(1 + \sqrt{2x - 14}\right)^2 \quad \text{Square both sides.}$$

$$x = 1 + 2\sqrt{2x - 14} + (2x - 14) \quad \text{Treat the right-hand side as the square of a binomial.}$$

$$x = 2\sqrt{2x - 14} + 2x - 13$$

$$-x + 13 = 2\sqrt{2x - 14} \quad \text{Simplify so that the radical is on one side by itself.}$$

$$\left(-x + 13\right)^2 = \left(2\sqrt{2x - 14}\right)^2 \quad \text{Square both sides again.}$$

$$x^2 - 26x + 169 = 4(2x - 14)$$

$$x^2 - 26x + 169 = 8x - 56$$

$$x^2 - 34x + 225 = 0$$

$$(x - 9)(x - 25) = 0 \quad \text{Solve by factoring.}$$

$$x = 9 \quad \text{or} \quad x = 25$$

Check: **Check both answers** in the original equation.

$$\sqrt{(9)} - \sqrt{2(9) - 14} \overset{?}{=} 1 \qquad \sqrt{(25)} - \sqrt{2(25) - 14} \overset{?}{=} 1$$

$$3 - \sqrt{4} \overset{?}{=} 1 \qquad\qquad 5 - \sqrt{36} \overset{?}{=} 1$$

$$3 - 2 \overset{?}{=} 1 \qquad\qquad 5 - 6 \overset{?}{=} 1$$

$$1 = 1 \qquad\qquad\qquad -1 \neq 1$$

25 is **not** a solution. The only solution is 9.

Now work margin exercise 2.

3. Solve the following equation containing a cube root.

$$\sqrt[3]{3x+3}+2=5$$

Example 3

Equations Containing a Cube Root

Solve the following equation containing a cube root: $\sqrt[3]{2x+1}+1=3$.

Solution

First, get the radical by itself on one side of the equation. Then since this radical is a cube root, cube both sides of the equation.

$$\sqrt[3]{2x+1}+1=3$$
$$\sqrt[3]{2x+1}=2 \qquad \text{Add } -1 \text{ to both sides.}$$
$$\left(\sqrt[3]{2x+1}\right)^3=2^3 \qquad \text{Cube both sides.}$$
$$2x+1=8 \qquad \text{Solve the equation.}$$
$$x=\frac{7}{2}$$

Check: **Check** in the original equation.

$$\sqrt[3]{2\left(\frac{7}{2}\right)+1}+1 \overset{?}{=} 3$$
$$\sqrt[3]{7+1}+1 \overset{?}{=} 3$$
$$\sqrt[3]{8}+1 \overset{?}{=} 3$$
$$2+1 \overset{?}{=} 3$$
$$3=3$$

There is one solution, $\frac{7}{2}$.

Now work margin exercise 3.

Practice Problems

Solve the following equations.

1. $2\sqrt{x+4}=x+1$ **2.** $\sqrt{3x+1}+1=\sqrt{x}$

3. $\sqrt[3]{2x-9}+4=3$ **4.** $\sqrt{2x-5}=-1$

Practice Problem Answers

1. $x=5$ **2.** no solution   **3.** $x=4$ **4.** no solution

Exercises 9.5

Solve the following equations. Be sure to check your answers in the original equation. See Examples 1 and 2.

1. $\sqrt{8x+1} = 5$

2. $\sqrt{7x+1} = 6$

3. $\sqrt{3x+4} = -5$

4. $\sqrt{4x-3} = 7$

5. $\sqrt{6-x} = 3$

6. $\sqrt{11-x} = 5$

7. $\sqrt{5x-6} = 8$

8. $\sqrt{2x-5} = -1$

9. $\sqrt{5x+4} = 7$

10. $\sqrt{3x-2} = 4$

11. $\sqrt{x-4} + 6 = 2$

12. $\sqrt{6x+4} + 2 = 10$

13. $\sqrt{4x+1} + 4 = 9$

14. $\sqrt{2x-7} + 5 = 3$

15. $\sqrt{x(x+3)} = 2$

16. $\sqrt{x(x-5)} = 6$

17. $\sqrt{x(2x+5)} = 5$

18. $\sqrt{x(3x-14)} = 7$

19. $\sqrt{x+6} = x+4$

20. $\sqrt{x+7} = 2x-1$

21. $\sqrt{x-2} = x-2$

22. $\sqrt{x+3} = x+3$

23. $x-2 = \sqrt{3x-6}$

24. $x+6 = \sqrt{2x+12}$

25. $\sqrt{x^2-16} = 3$

26. $\sqrt{x^2-25} = 12$

27. $5 + \sqrt{x+5} - 2x = 0$

28. $x-2 - \sqrt{x+4} = 0$

29. $2x = \sqrt{7x-3} + 3$

30. $x - \sqrt{3x-8} = 4$

31. $\sqrt{2x+5} = \sqrt{4x-1}$

32. $\sqrt{5x-1} = \sqrt{x+7}$

33. $\sqrt{3x+2} = \sqrt{9x-10}$

34. $\sqrt{2-x} = \sqrt{2x-7}$

35. $\sqrt{2x-1} = \sqrt{x+1}$

36. $\sqrt{3x+2} = \sqrt{x+4}$

37. $\sqrt{x+2} = \sqrt{2x-5}$

38. $\sqrt{2x-5} = \sqrt{3x-9}$

39. $\sqrt{4x-3} = \sqrt{2x+5}$

40. $\sqrt{4x-6} = \sqrt{3x-1}$

41. $\sqrt{3x+1} = 1 - \sqrt{x}$

42. $\sqrt{x} = \sqrt{x+16} - 2$

43. $\sqrt{x+4} = \sqrt{x+11} - 1$

44. $\sqrt{1-x} + 2 = \sqrt{13-x}$

45. $\sqrt{x+1} = \sqrt{x+6} + 1$

46. $\sqrt{x+4} = \sqrt{x+20} - 2$

47. $\sqrt{x+5} + \sqrt{x} = 5$

48. $\sqrt{x} + \sqrt{x-3} = 3$

49. $\sqrt{2x+3} = 1 + \sqrt{x+1}$

50. $\sqrt{5x-18} - 4 = \sqrt{5x+6}$

51. $\sqrt{3x+1} - \sqrt{x+4} = 1$

52. $\sqrt{3x+4} - \sqrt{x+5} = 1$

53. $\sqrt{5x-1} = 4 - \sqrt{x-1}$

54. $\sqrt{2x-5} - 2 = \sqrt{x-2}$

55. $\sqrt{2x-1} + \sqrt{x+3} = 3$

56. $\sqrt{2x+3} - \sqrt{x+5} = 1$

57. $\sqrt[3]{4+3x} = -2$

58. $\sqrt[3]{2+9x} = 9$

59. $\sqrt[3]{5x+4} = 4$

60. $\sqrt[3]{7x+1} = -5$

61. $\sqrt[4]{2x+1} = 3$

62. $\sqrt[4]{x-6} = 2$

Writing & Thinking

63. Explain, in your own words, why $(a+b)^2 \neq a^2 + b^2$. (Assume $a \neq 0$ and $b \neq 0$.)

9.6 Functions with Radicals

Objective A **Review of Functions and Function Notation**

The concept of functions is among the most important and useful ideas in all of mathematics. Functions were introduced earlier along with function notation, such as $f(x)$ (read "f of x"). We also discussed **linear functions** and the use of function notation in evaluating functions. The function notation $p(x)$ was used again to represent polynomials. In this section, the function concept is expanded to include **radical functions** (functions with radicals). The definitions of relations and functions and the vertical line test are restated here for review and easy reference.

Relation, Domain, and Range

A **relation** is a set of ordered pairs of real numbers.

The **domain D** of a relation is the set of all first coordinates in the relation.

The **range R** of a relation is the set of all second coordinates in the relation.

Function

A **function** is a relation in which each domain element has exactly one corresponding range element.

The definition can also be stated in the following ways:

1. A function is a relation in which each first coordinate appears only once.
2. A function is a relation in which no two ordered pairs have the same first coordinate.

Vertical Line Test

If **any** vertical line intersects the graph of a relation at more than one point, then the relation is **not** a function.

Objectives

A Briefly review the concepts of functions and function notation.

B Find the domain and range of radical functions.

C Evaluate radical functions.

D Graph radical functions.

E Use a graphing calculator to graph radical functions.

We have used the ordered pair notation, (x, y), to represent points on the graphs of relations and functions. For example, $y = 2x - 5$ represents a linear function and its graph is a line (as shown in the figure below).

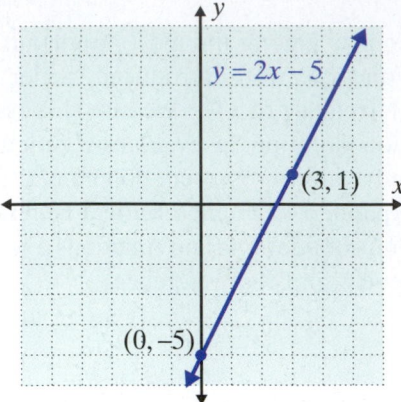

Figure 1

Finding the Domain and Range of Radical Functions

We define **radical functions** (functions with radical expressions) as follows.

Radical Function

A **radical function** is a function of the form $y = \sqrt[n]{g(x)}$ in which the radicand contains a variable expression.

The **domain** of such a function depends on the index, n.

1. If n is an even number, the domain is the set of all x such that $g(x) \geq 0$.
2. If n is an odd number, the domain is the set of all real numbers, $(-\infty, \infty)$.

Examples of radical functions are

$$y = 3\sqrt{x}, \quad f(x) = \sqrt{2x + 3}, \quad \text{and} \quad y = \sqrt[3]{x - 7}.$$

Example 1

Finding the Domain of a Radical Function

Determine the domain of each radical function.

a. $f(x) = \sqrt{2x + 3}$

Solution

Because the index is 2, the radicand must be nonnegative. (That is, the expression under the radical sign cannot be negative.)

Thus, we must have

$$2x + 3 \geq 0$$
$$2x \geq -3$$
$$x \geq -\frac{3}{2}$$

and the domain is the interval of real numbers $\left[-\frac{3}{2}, \infty\right)$.

b. $y = \sqrt[3]{x - 7}$

Solution

Because the index is 3, an odd number, the radicand may be any real number. Thus the domain is the set of all real numbers $(-\infty, \infty)$.

Now work margin exercise 1.

1. Determine the domain of
each radical function.

a. $f(x) = \sqrt{5x + 10}$

b. $y = \sqrt[3]{-x - 9}$

Objective C **Evaluating Radical Functions**

As illustrated earlier, function notation is particularly useful when evaluating functions for specific values of the variable. For example,

$$\text{if } f(x) = \sqrt{2x + 3}, \text{ then } f\left(\frac{1}{2}\right) = \sqrt{2\left(\frac{1}{2}\right) + 3} = \sqrt{1 + 3} = \sqrt{4} = 2$$
$$\text{and } f(0) = \sqrt{2(0) + 3} = \sqrt{0 + 3} = \sqrt{3}.$$

A calculator can be used to find decimal approximations. Such approximations are helpful when estimating the locations of points on a graph. For example,

if $f(x) = \sqrt{x - 5}$,

then $f(8) = \sqrt{(8) - 5} = \sqrt{3} \approx 1.7321$ accurate to the nearest ten-thousandth

and $f(25) = \sqrt{(25) - 5} = \sqrt{20} \approx 4.4721.$ accurate to the nearest ten-thousandth

2. Evaluate each function for the given values of x.

a. $f(x) = 5\sqrt{2x}$; find $f(x)$ values for $x = 0, 2, 8$

b. $f(x) = \sqrt[3]{2x - 3}$; find $f(x)$ values for $x = 1, 2, 15$

Example 2

Evaluating Radical Functions

Complete each table by finding the corresponding $f(x)$ values for the given values of x.

a. $f(x) = \sqrt[3]{x - 7}$

x	f(x)
7	?
6	?
−1	?

Solution

x	f(x)
7	$\sqrt[3]{7-7} = \sqrt[3]{0} = 0$
6	$\sqrt[3]{6-7} = \sqrt[3]{-1} = -1$
−1	$\sqrt[3]{-1-7} = \sqrt[3]{-8} = -2$

b. $f(x) = 3\sqrt{x}$

x	f(x)
0	?
4	?
6	?

Solution

x	f(x)
0	$3\sqrt{0} = 0$
4	$3\sqrt{4} = 3 \cdot 2 = 6$
6	$3\sqrt{6} \approx 7.3485$

Now work margin exercise 2.

Graphing Radical Functions

To graph a radical function, we need to be aware of its domain and plot at least a few points to see the nature of the resulting curve. Example 3 shows how to begin to graph the radical function $y = \sqrt{x+5}$.

Example 3

Graphing a Radical Function

Graph the function $y = \sqrt{x+5}$.

Solution

For the domain we have $x + 5 \geq 0$, so $x \geq -5$.

To see the nature of the graph, we select a few values for x in the domain and find the corresponding values of y. Then we plot the points on a graph.

x	y
−5	0
−4	1
−3	$\sqrt{2} \approx 1.41$
0	$\sqrt{5} \approx 2.24$
4	3

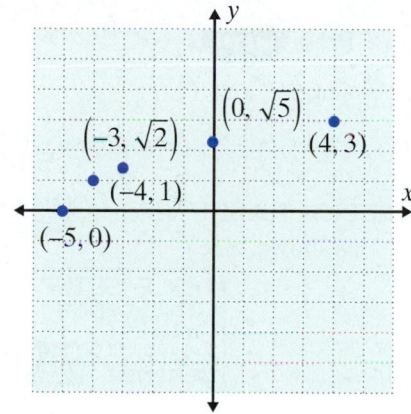

Note that $\sqrt{x+5}$ is the principal square root. This means that $y \geq 0$. Thus the point $(-5, 0)$ is on the x-axis and the remaining points on the graph are above the x-axis. So, we can complete the graph by drawing a smooth curve that passes through the selected points. The graph of the function is shown here.

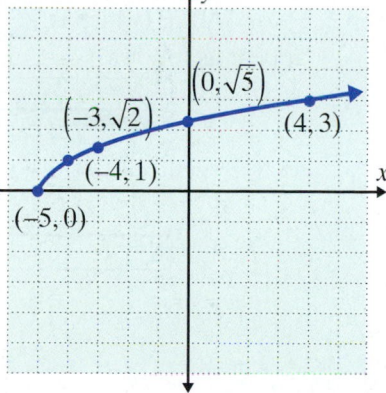

We see that the domain is $[-5, \infty)$ and the range is $[0, \infty)$.

Now work margin exercise 3.

3. Graph the function $y = -\sqrt{2x+1}$.

Objective E **Using a TI-84 Plus Graphing Calculator to Graph Radical Functions**

Example 4 shows how to use a TI-84 Plus graphing calculator to find many points on the graph of a radical function and then how to graph the function.

Example 4

Using a TI-84 Plus to Graph a Radical Function

a. Use the **TABLE** feature of a TI-84 Plus graphing calculator to locate many points on the graph of the function $y = \sqrt[3]{2x - 3}$.

Solution

Using the **TABLE** feature of a TI-84 Plus:

Step 1: Press ⬤ Y= ⬤ and enter the function as follows:

    a. Press **MATH**.

    b. Choose 4: $\sqrt[3]{}$ (.

    c. Enter $2x - 3$) and press **ENTER**.

Step 2: Press TBLSET (which is **2ND** **WINDOW**) and set the display as shown here.

Step 3: Press TABLE (which is **2ND** **GRAPH**) and the display will appear as follows:

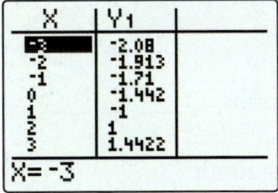

b. Plot several points (approximately) on a graph and then connect them with a smooth curve.

Once your calculator is displaying the table, you may scroll up and down the display to find as many points as you like. A few are shown here to see the nature of the graph.

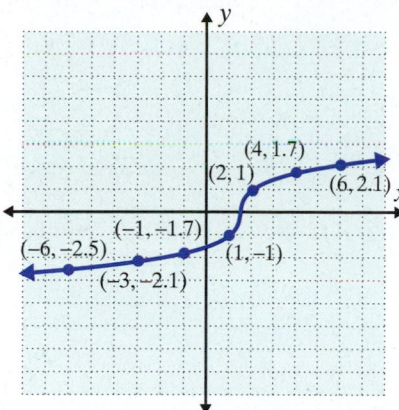

c. Use a TI-84 Plus graphing calculator to graph the function.

Solution

Press ⬛GRAPH and the display will appear with the curve as follows.

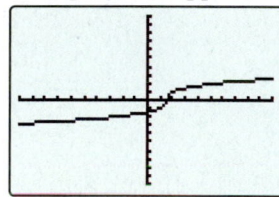

Note that you may need to adjust the window. Refer to section 6.6 if you need help with this.

Press the ⬛WINDOW key and the standard window will be displayed. By default, the standard window displays a graph with x-values and y-values ranging from -10 to 10 with tic marks on the axis every 1 unit. This window can be changed at any time by changing the individual numbers or pressing the ⬛ZOOM key and selecting an option from the menu displayed. To return the screen back to the standard dimensions with x-values and y-values ranging from -10 to 10, press zoom and **5: ZStandard**. A square screen can be attained by pressing zoom and **5: ZSquare** or by pressing the window key and setting $Xmin = -15$ and $Xmax = 15$ to give the x-axis a length of 30 and the y-axis a length of 20 (a ratio of 3:2).

Now work margin exercise 4.

4. a. Use the **TABLE** feature of a TI-84 Plus graphing calculator to locate five points on the graph of the function $y = \sqrt[3]{2 - 4x}$.

x	y_1

b. Use a TI-84 Plus graphing calculator to graph the function.

Practice Problems

Find the indicated value of the function.

1. For $f(x) = \sqrt{x+7}$, find $f(5)$.

2. For $g(x) = \sqrt{3x-1}$, find $g\left(\dfrac{2}{3}\right)$.

3. For $h(x) = \sqrt[3]{x+5}$, find $h(-13)$.

Use a calculator to estimate the value of the function accurate to the nearest ten-thousandth.

4. Estimate $f(2)$ for $f(x) = \sqrt{4x-1}$.

5. Estimate $g(-3)$ for $g(x) = \sqrt[3]{1-3x}$.

Practice Problem Answers

1. $2\sqrt{3}$ **2.** 1 **3.** -2

4. 2.6458 **5.** 2.1544

Exercises 9.6

Find each function value as indicated and round decimal values to the nearest ten-thousandth, if necessary. See Example 2.

1. Given $f(x) = \sqrt{2x+1}$, find

   a. $f(2)$ b. $f(4)$ c. $f(24.5)$ d. $f(1.5)$

2. Given $f(x) = \sqrt{5-3x}$, find

   a. $f(0)$ b. $f(-2)$ c. $f\left(-\dfrac{20}{3}\right)$ d. $f(-2.4)$

3. Given $g(x) = \sqrt[3]{x+6}$, find

   a. $g(21)$ b. $g(-7)$ c. $g(-14)$ d. $g(18)$

4. Given $h(x) = \sqrt[3]{4-x}$, find

   a. $h(4)$ b. $h(-4)$ c. $h(3.999)$ d. $h(-2.5)$

Use interval notation to indicate the domain of each radical function. See Example 1.

5. $y = \sqrt{x+8}$ 6. $y = \sqrt{2x-1}$ 7. $y = \sqrt{2.5-5x}$ 8. $y = \sqrt{1-3x}$

9. $f(x) = \sqrt[3]{x+4}$ 10. $f(x) = \sqrt[3]{6x}$ 11. $g(x) = \sqrt[4]{x}$ 12. $g(x) = \sqrt[4]{7-x}$

13. $y = \sqrt[5]{4x-1}$ 14. $y = \sqrt[5]{8+x}$

15. $y = \sqrt{x-2}$

16. $y = \sqrt{2-x}$

17. $y = -\sqrt{x-3}$

18. $y = -\sqrt{3-x}$

19. $y = \sqrt{x+4}$

20. $y = \sqrt{x-4}$

A.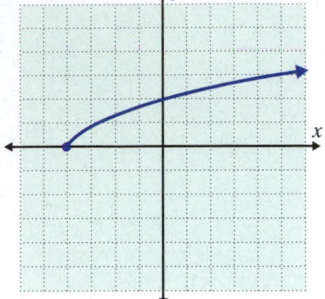

B.

C.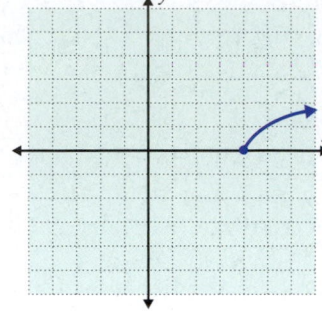

D.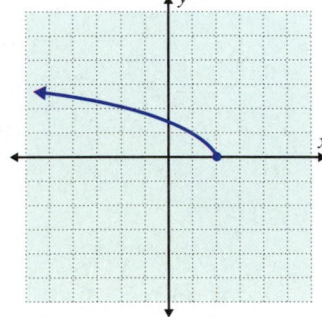

E.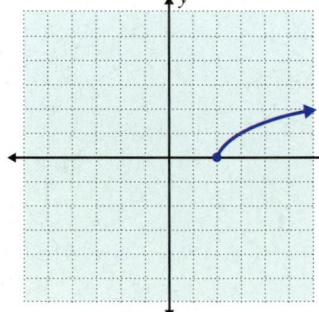

F.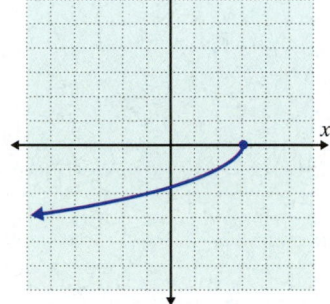

Determine at least 5 points for the given function and then sketch the graph. See Example 3.

21. $y = \sqrt{x-1}$

22. $y = \sqrt{2x+6}$

23. $f(x) = -\sqrt{3x+3}$

24. $h(x) = -\sqrt{x+1}$

25. $f(x) = \sqrt[3]{x+2}$

26. $g(x) = \sqrt[3]{x-6}$

27. $y = \sqrt[3]{3x+6}$

28. $y = \sqrt[3]{2x-4}$

29. $y = 3\sqrt{x+2}$

30. $y = 2\sqrt{3-x}$

31. $g(x) = -\sqrt{2x}$

32. $f(x) = \sqrt{3x}$

33. $f(x) = -\sqrt{x+4}$

34. $f(x) = -\sqrt{5-x}$

35. $y = -\sqrt[3]{x+2}$

36. $y = -\sqrt[3]{3x+4}$

37. $g(x) = -\sqrt[4]{x+5}$

38. $y = \sqrt[4]{2x+6}$

39. $y = \sqrt[5]{2x+1}$

40. $y = \sqrt[5]{x+7}$

Writing & Thinking

41. The graph of the radical function $f(x) = \sqrt{x}$ is shown with two values of x on the x-axis, 3 and $3 + h$.

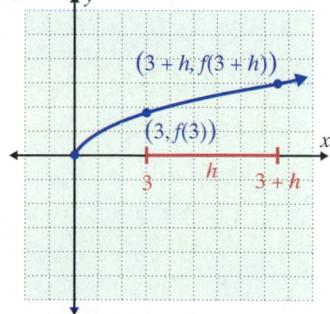

a. Rationalize the numerator of the expression $\dfrac{f(3+h) - f(3)}{h} = \dfrac{\sqrt{3+h} - \sqrt{3}}{h}$ by multiplying both the numerator and denominator by the conjugate of the numerator. Then simplify the resulting expression.

b. What do you think this expression represents graphically?
(**Hint:** Two points determine a line.)

c. Using your results from parts **a.** and **b.**, what do you see happening on the graph if the value of h shrinks slowly to 0?

d. Using your analysis from part **c.**, what happens to the value of your simplified expression in part **a.** and what do you think this value represents?

Section 9.1: Radical Expressions

Perfect Squares   page 985

The square of an integer is called a **perfect square**.

Radical Terminology   page 985

The symbol $\sqrt{}$ is called a **radical sign**.
The number under the radical sign is called the **radicand**.
The complete expression, such as $\sqrt{64}$, is called a **radical** or **radical expression**.

Square Root   pages 985 – 987

If a is a nonnegative real number, then $\sqrt{a}$ is the **principal square root** of a and $-\sqrt{a}$ is the **negative square root** of a.

Rational Number   page 987

Rational numbers are numbers that can be expressed as the quotient of integers and are of the form $\dfrac{a}{b}$ where a and b are integers and $b \neq 0$.

Irrational Number   page 987

An **irrational number** is a number whose decimal representation is an infinite nonrepeating decimal.

Perfect Cube   page 988

If an integer is cubed, the result is called a **perfect cube**.

Cube Root   page 988

If a is a real number, then $\sqrt[3]{a}$ is the **cube root** of a.

Roots of Sums and Differences   page 989

If a radical contains a sum or difference inside the radical sign, we must simplify the expression inside the radical before evaluating it.

Section 9.2: Simplifying Radical Expressions

Properties of Square Roots   page 994

If a and b are positive real numbers, then

1. $\sqrt{ab} = \sqrt{a}\sqrt{b}$

2. $\sqrt{\dfrac{a}{b}} = \dfrac{\sqrt{a}}{\sqrt{b}}$

Simplest Form   page 994

A square root is considered to be in simplest form when the radicand has no perfect square as a factor.
A cube root is considered to be in simplest form when the radicand has no perfect cube as a factor.

Square Root of x^2 page 996

If x is a real number, then $\sqrt{x^2} = |x|$.

Note: If $x \geq 0$ is given, then we can write $\sqrt{x^2} = x$.

Section 9.3: Rational Exponents

Radical Notation page 1003

The expression $\sqrt[n]{a}$ is called a **radical** or **radical expression**.
The symbol $\sqrt[n]{}$ is called a **radical sign**.
n is called the **index**.
a is called the **radicand**.
(**Note:** If no index is given, the index is 2.)

Special Notes about the Index n page 1004

For the expression $\sqrt[n]{a}$ or $a^{\frac{1}{n}}$ to be a real number:

1. when a is nonnegative, n can be any index, and
2. when a is negative, n must be odd.
   (If a is negative and n is even, then $\sqrt[n]{a}$ is nonreal.)

The General Form $a^{\frac{m}{n}}$ page 1006

If n is a positive integer, m is any integer, and $a^{\frac{1}{n}}$ is a real number, then
$$a^{\frac{m}{n}} = \left(a^{\frac{1}{n}}\right)^m = \left(a^m\right)^{\frac{1}{n}}.$$

In radical notation:
$$a^{\frac{m}{n}} = \left(\sqrt[n]{a}\right)^m = \sqrt[n]{a^m}.$$

Section 9.4: Operating with Radicals

Like Radicals page 1014

Like radicals have the same index and radicand or they can be simplified so that they have the same index and radicand.

Addition and Subtraction with Radical Expressions page 1015

To add and subtract radical expressions, combine like radicals.

Multiplication with Radical Expressions pages 1016 – 1017

To find the product of two radicals, treat the expressions as binomials and multiply just as with polynomials.

To Rationalize a Denominator Containing a Square Root   pages 1017 – 1018
or a Cube Root
    1. If the denominator contains a square root, multiply both the numerator and denominator by an expression that will give a denominator with no square roots.
    2. If the denominator contains a cube root, multiply both the numerator and denominator by an expression that will give a denominator with no cube roots.

Conjugates   page 1018

    The two expressions $(a - b)$ and $(a + b)$ are called **conjugates** of each other, and their product, $(a - b)(a + b)$, results in the difference of two squares.

To Rationalize a Denominator Containing a Sum or Difference   pages 1019 – 1020
Involving Square Roots
    If the denominator of a fraction contains a sum or difference involving a square root, rationalize the denominator by multiplying both the numerator and denominator by the conjugate of the denominator.
    1. If the denominator is of the form $a - b$, multiply both the numerator and denominator by $a + b$.
    2. If the denominator is of the form $a + b$, multiply both the numerator and denominator by $a - b$.
    The new denominator will be the difference of two squares and therefore will not contain a radical term.

Section 9.5: Solving Radical Equations

Extraneous Solution   page 1026

    An **extraneous solution** is a number that is found when solving an equation that does not satisfy the original equation. They can be unintentionally introduced by raising both sides of an equation to a power.

Method for Solving Equations with Radicals   page 1026
    1. Isolate one of the radicals on one side of the equation. (An equation may have more than one radical.)
    2. Raise both sides of the equation to the power corresponding to the index of the radical.
    3. If the equation still contains a radical, repeat Steps 1 and 2.
    4. Solve the equation after all the radicals have been eliminated.
    5. Be sure to check all possible solutions in the original equation and eliminate any extraneous solutions.

Radical Function   pages 1035 – 1036

A **radical function** is a function of the form $y = \sqrt[n]{g(x)}$ in which the radicand contains a variable expression.

The **domain** of such a function depends on the index, n:

1. If n is an even number, the domain is the set of all x such that $g(x) \geq 0$.
2. If n is an odd number, the domain is the set of all real numbers $(-\infty, \infty)$.

Chapter 9: Test

Simplify the expressions. Assume that all variables represent positive real numbers.

1. $\sqrt{112}$

2. $\sqrt{120xy^4}$

3. $\sqrt[3]{48x^2y^5}$

4. $\sqrt[3]{\dfrac{343x^{18}y^4}{8z^6}}$

5. Write $(2x)^{\frac{2}{3}}$ in radical notation.

6. Write $\sqrt[6]{8x^2y^4}$ as an equivalent, simplified expression with rational exponents.

Simplify each numerical expression. Assume that all variables represent positive real numbers. Leave the answers in rational exponent form.

7. $(-8)^{\frac{2}{3}}$

8. $4x^{\frac{1}{2}} \cdot x^{\frac{2}{3}}$

9. $\left(\dfrac{16x^{-4}y}{y^{-1}}\right)^{\frac{3}{4}}$

10. Simplify the expression, $\sqrt[3]{x^2} \cdot \sqrt[4]{x}$, by first changing it into an equivalent expression with rational exponents. Rewrite the answer in simplified radical form.

Perform the indicated operations and simplify. Assume that all variables represent positive real numbers.

11. $2\sqrt{75} + 3\sqrt{27} - \sqrt{12}$

12. $5x\sqrt{y^3} - 2\sqrt{x^2y^3} - 4y\sqrt{x^2y}$

13. $\left(\sqrt{3} - \sqrt{2}\right)^2$

14. $\left(6 + \sqrt{3x}\right)\left(5 - 2\sqrt{3x}\right)$

15. Rationalize the denominator and simplify, if possible.

    a. $\sqrt{\dfrac{5y^2}{8x^3}}$

    b. $\dfrac{1-x}{1-\sqrt{x}}$

Solve the following equations. Be sure to check your answers in the original equation.

16. $\sqrt{9x-5} - 3 = 8$

17. $\sqrt[3]{3x+4} = -2$

18. $\sqrt{x+8} - 2 = x$

19. $\sqrt{5x+1} = 1 + \sqrt{3x+4}$

20. Find the domain of each of the following radical functions. Write the answer in interval notation.

    a. $y = \sqrt{3x+4}$

    b. $f(x) = \sqrt[3]{2x+5}$

21. Use a graphing calculator to graph each of the following radical functions. Sketch the graph and label 3 points on the graph.

    a. $f(x) = -\sqrt{x+3}$ **b.** $y = \sqrt[3]{x-4}$

Use a calculator to find the value of each expression accurate to the nearest ten-thousandth.

22. $\sqrt[5]{119}$ **23.** $32^{-\frac{3}{5}}$ **24.** $\left(\sqrt{2}+6\right)\left(\sqrt{2}-1\right)$ **25.** $\dfrac{\sqrt{3}-\sqrt{5}}{\sqrt{7}-\sqrt{10}}$

Cumulative Review: Chapters 1 - 9

All answers should be in simplest form. Reduce all fractions to lowest terms, and express all improper fractions as mixed numbers.

1. $\dfrac{2}{3} + \dfrac{5}{6} + \dfrac{2}{9}$

2. $17\dfrac{5}{8} + 12\dfrac{7}{10}$

3. $6.09 + 10.6 + 7$

4. $(-2.03) + (16.7) + (-5.602)$

Solve each of the following equations.

5. $15x + 6(x + 1) = 7$

6. $4(5x - 1) = 10(2x + 3)$

Solve the inequality and graph the solution set. Write the solution using interval notation. Assume that x is a real number.

7. $6x - 2 < 4x + 10$

Solve the following problems.

8. Graph the linear equation $4x - 2y = -8$ by locating the y-intercept and x-intercept.

9. Write the equation $x + 6y = 18$ in slope-intercept form. Then find the slope and the y-intercept, and use them to draw the graph.

10. Find an equation for the vertical line through the point $(5, -7)$.

11. Find an equation in standard form for the line passing through the point $\left(\dfrac{1}{4}, -3\right)$ with slope $m = -4$. Graph the line.

12. The graphs of three curves are shown. Use the vertical line test to determine whether or not each graph represents a function. Then state the domain and range of each graph.

a.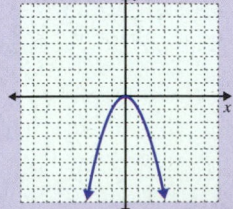

b.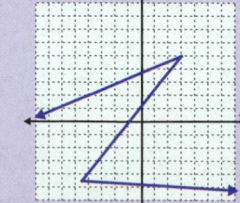

c.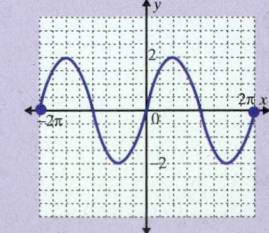

13. Write the expression $\dfrac{0.008 \times 40{,}000}{320 \times 0.001}$ in scientific notation and simplify.

14. Given $P(x) = x^3 - 8x^2 + 19x - 12,$ find

    a. $P(0)$ **b.** $P(4)$ **c.** $P(-3)$

Perform the indicated operations.

15. $\left(x^2 + 7x - 5\right) - \left(-2x^3 + 5x^2 - x - 1\right)$ **16.** $(2x - 7)(3x - 1)$

17. $(5x + 2)(4 - x)$ **18.** $(7x - 2)^2$

Factor completely.

19. $28 + x - 2x^2$ **20.** $64y^4 + 32y^3 + 4y^2$

21. $5x^3 - 320$ **22.** $x^3 + 4x^2 - x - 4$

Solve each quadratic equation by factoring.

23. $x^2 - 13x - 48 = 0$ **24.** $x = 2x^2 - 6$ **25.** $0 = 15x^2 - 11x + 2$

26. **a.** Find an equation that has $x = 4$ and $x = 7$ as roots.

    b. Find an equation that has $x = -2$, $x = 12$, and $x = 1$ as roots.

27. Solve the following system of linear equations by graphing both equations and locating the point of intersection.

$$\begin{cases} 3x - 2y = 7 \\ x + 3y = -5 \end{cases}$$

Solve each system by the method given.

Solve each system by using either the substitution method or the addition method.

28. $\begin{cases} -2y = x - 26 \\ \quad y = 2x - 22 \end{cases}$ **29.** $\begin{cases} x + y = -5 \\ \dfrac{1}{2}x + \dfrac{1}{2}y = \dfrac{7}{2} \end{cases}$

Solve the following system of linear equations using Gaussian elimination. You may use your calculator.

30. $\begin{cases} x + y - z = 1 \\ 3x - y + z = 3 \\ -x + 2y + 2z = 3 \end{cases}$

Solve the following system of inequalities graphically.

31. $\begin{cases} x + y \le 4 \\ 3x - y \le 2 \end{cases}$

Simplify the expressions. Assume that all variables represent positive real numbers.

32. $8^{-\frac{4}{3}}$

33. $\left(x^{\frac{1}{2}} \cdot x^{\frac{2}{3}} \right)^2$

34. $\sqrt{288}$

35. $\sqrt[3]{16x^6 y^{10}}$

Rationalize the denominator and simplify if possible.

36. $\dfrac{\sqrt{5}}{\sqrt{2y}}$

37. $\dfrac{\sqrt{5} - \sqrt{6}}{\sqrt{5} + \sqrt{6}}$

Solve the radical equations given. Be sure to check your answers in the original equation.

38. $\sqrt{x + 10} + 1 = x + 5$

39. $\sqrt[3]{x - 7} + 3 = 1$

40. Use interval notation to indicate the domain of the radical function $\sqrt[4]{6 - 2x}$.

41. Find

   a. the circumference and

   b. the area of a circle with a radius of 8 feet.

42. Find the length of the hypotenuse of a right triangle with legs of length 10 meters and 20 meters. Write the answer in both simplified radical form and decimal form (accurate to the nearest thousandth).

43. a. If $1550 is deposited in an account paying 8% compounded monthly, what will be the total amount in the account at the end of 3 years (to the nearest dollar)?

   b. How much interest will be earned?

44. Find two consecutive integers such that 3 more than twice the smaller is equal to 13 less than the larger.

45. The sum of the squares of two consecutive positive odd integers is 514. What are the integers?

46. Investing: Harold has $50,000 that he wants to invest in two accounts. One pays 6% interest, and the other (at a higher risk) pays 10% interest. If he wants a $3600 annual return on these two investments, how much should he put into each account?

47. Eating Out: For lunch, Jason had one burrito and 2 tacos. Matt had 2 burritos and 3 tacos. If Jason spent $5.30 and Matt spent $9.25, find the price of one burrito and the price of one taco.

48. Selling Candy: A grocer plans to make up a special mix of two popular kinds of candy for Halloween. He wants to mix a total of 100 pounds to sell for $1.75 per pound. Individually, the two types sell for $1.25 and $2.50 per pound. How many pounds of each of the two kinds should he put in the mix?

49. Pizza: The Local Pizza Hotspot Restaurant offers a cheese pizza in three sizes: 7 in. diameter for $3.25, 12 in. diameter for $8.95, 14 in. diameter for $10.95

   a. What is the price per sq in. for each size pizza?

   b. Which is the best buy? Why?

50. Docking a Boat: A boat is being pulled to the shore from a dock. When the rope to the boat is 40 meters long, the boat is 30 meters from the dock. What is the height of the dock (to the nearest tenth of a meter) above the deck of the boat?

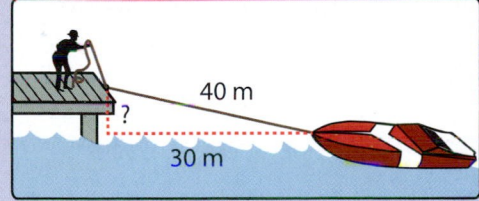

Mathematics @ Work!

The notes of your favorite song relate mathematically using a function called a logarithm. Western tonal music is written using a set of notes that are each a semitone, or half-step, apart. An octave consists of 12 semitones, where each note has its own frequency or pitch. The frequency of each consecutive note does not increase by a fixed value, as it would in a linear relationship. Instead, the frequency of each note is $\sqrt[12]{2}$ times the previous note. For functions like this, we say the frequency grows exponentially. See the following problem.

In music, certain combinations of notes have been given names. Some of these include major thirds, minor thirds, and tritones. If r is the frequency ratio between two notes, the equation $S = 12\log_2(r)$ tells you the number of semitones, S, separating the notes. Find the number of semitones between two notes of the following type.

    a. A major third, which has a frequency ratio of $\sqrt[3]{2}$

    b. A minor third, which has a frequency ratio of $\sqrt[4]{2}$

    c. A tritone, which has a frequency ratio of $\sqrt{2}$

For more problems like these, see Section 10.5.

Exponential and Logarithmic Functions

10.1 Algebra of Functions

As discussed throughout this text, functions are an important topic in mathematics. Function notation $f(x)$ is particularly helpful in evaluating functions and indicating graphical relationships. Operating algebraically with functions as well as understanding and finding the **composition** and **inverses** of functions rely heavily on function notation. The concepts of composite and inverse functions form the basis of the relationship between logarithmic and exponential functions.

Logarithms are exponents. Traditionally, logarithmic and exponential values were calculated with the extensive use of printed tables and techniques for estimating values not found in the tables. As some of your "older" teachers will tell you, this was a long and detailed process. These tables are no longer printed in textbooks. Now hand-held calculators have programs stored in their electronic memories that calculate the values in these tables with even greater accuracy, and complicated expressions can be evaluated by pressing a few keys.

Of all the topics discussed in algebra, logarithmic functions and exponential functions probably have the most value in terms of applied problems. Learning curves, important in business and education, can be described with logarithmic and exponential functions. Exponential growth and decay are basic concepts in biology and medicine. (Cancer cells grow exponentially and radium decays exponentially.) Computers use logarithmic and exponential concepts in their design and implementation. These concepts are likely to be encountered in almost any field of study.

Objective A   **Algebraic Operations with Functions**

If two (or more) functions have the same domain, then we can perform the operations of addition, subtraction, multiplication, and division with these functions. For example, consider the following quadratic functions:

$$f(x) = 2x^2 - 1 \quad \text{and} \quad g(x) = x^2 + 2x - 5.$$

Both have the same domain: $\mathbb{R} = $ all real numbers $= (-\infty, \infty)$. This means that we can

    a. choose any value for x from the common domain,
    b. evaluate each function for that value of x, and
    c. perform operations with those functional values.

The functional values are the y-values. For example, if we choose $x = 3$, then

$$f(3) = 2(3)^2 - 1 = 17 \quad \text{and} \quad g(3) = (3)^2 + 2(3) - 5 = 10.$$

Now we can easily find the sum and difference

$$f(3) + g(3) = 17 + 10 = 27 \quad \text{and} \quad f(3) - g(3) = 17 - 10 = 7.$$

However, if we want to find, say $f(5)+g(5)$ and $f(5)-g(5)$, we would need to again evaluate both functions, this time at $x=5$. Another way to find sums and differences of functions is to find the algebraic sum (or difference) of the two expressions. Then these new expressions will allow us to find the sum (or difference) directly for any value of x that is in the original domain of both functions. For example, we find the sum $f+g$ as follows.

$$
\begin{aligned}
(f+g)(x) &= f(x)+g(x) \\
&= (2x^2-1)+(x^2+2x-5) \\
&= 3x^2+2x-6
\end{aligned}
$$

With this new function, we find directly that $(f+g)(3)$ directly.

$$(f+g)(3)=3(3)^2+2(3)-6=27$$

Similarly,

$$
\begin{aligned}
(f-g)(x) &= f(x)-g(x) \\
&= (2x^2-1)-(x^2+2x-5) \\
&= x^2-2x+4,
\end{aligned}
$$

and we have

$$(f-g)(3)=(3)^2-2(3)+4=7.$$

Similar notation is used for the product and quotient of two functions. One important condition is that both functions **must have the same domain**. If not, then the algebraic sums, differences, products, and quotients are **restricted to portions of the domains that are in common. Also, in the case of quotients, no denominator can be 0.**

Algebraic Operations with Functions

If $f(x)$ and $g(x)$ represent two functions and x is a value in the **domain of both functions**, then we define the following operations.

1. Sum of two functions: $(f+g)(x)=f(x)+g(x)$

2. Difference of two functions: $(f-g)(x)=f(x)-g(x)$

3. Product of two functions: $(f\times g)(x)=f(x)\times g(x)$

4. Quotient of two functions: $\left(\dfrac{f}{g}\right)(x)=\dfrac{f(x)}{g(x)}$ where $g(x)\neq 0$

Example 1

Algebraic Operations with Functions

Let $f(x) = 3x^2 + x - 4$ and $g(x) = x - 6$. Find the following functions.

a. $(f + g)(x)$

Solution

$$(f + g)(x) = (3x^2 + x - 4) + (x - 6) = 3x^2 + 2x - 10$$

b. $(f - g)(x)$

Solution

$$(f - g)(x) = (3x^2 + x - 4) - (x - 6) = 3x^2 + 2$$

c. $(f \cdot g)(x)$

Solution

$$(f \cdot g)(x) = (3x^2 + x - 4)(x - 6) = 3x^3 - 17x^2 - 10x + 24$$

d. Evaluate each of the functions found in parts **a.** - **c.** at $x = 2$.

Solution

Evaluating each of these functions at $x = 2$ gives the following results.

$$(f + g)(2) = 3(2)^2 + 2(2) - 10 = 6$$

$$(f - g)(2) = 3(2)^2 + 2 = 14$$

$$(f \cdot g)(2) = 3(2)^3 - 17(2)^2 - 10(2) + 24 = -40$$

Example 2

Algebraic Operations with Functions

Let $f(x) = x^2 - x$ and $g(x) = 2x + 1$. Find the following functions.

a. $(f + g)(x)$

Solution

$$(f + g)(x) = (x^2 - x) + (2x + 1) = x^2 + x + 1$$

b. $(g-f)(x)$

Solution

$$(g-f)(x) = (2x+1) - (x^2 - x) = -x^2 + 3x + 1$$

c. $\left(\dfrac{f}{g}\right)(x)$

Solution

$$\left(\dfrac{f}{g}\right)(x) = \dfrac{x^2 - x}{2x + 1} \qquad \text{where } 2x + 1 \neq 0 \left(\text{or } x \neq -\dfrac{1}{2}\right)$$

d. Evaluate each of the functions found in parts **a.–c.** at $x = 3$.

Solution

Evaluating each of these functions at $x = 3$ gives the following results.

a. $(f + g)(3) = (3)^2 + (3) + 1 = 13$

b. $(g - f)(3) = -(3)^2 + 3(3) + 1 = 1$

c. $\left(\dfrac{f}{g}\right)(3) = \dfrac{(3)^2 - (3)}{2(3) + 1} = \dfrac{6}{7}$

Now work margin exercises 1 and 2.

Note that, except for Example **2c.**, the domain for all of the functions discussed in Examples 1 and 2 is the set of all real numbers, $(-\infty, \infty)$. In Example **2c.** we noted that the denominator cannot equal 0. In Example 3 we show how the domain may need to be limited before performing algebra with functions that contain radical expressions.

Example 3

Algebraic Operations with Functions with Limited Domains

Let $f(x) = x + 5$ and $g(x) = \sqrt{x - 2}$. Find the following functions and state the domain of each function.

a. $(f + g)(x)$

Solution

$$(f + g)(x) = x + 5 + \sqrt{x - 2}$$

---

1. Let $f(x) = 2x^2 + 3x - 9$ and $g(x) = x + 3$. Find the following functions.

a. $(f + g)(x)$

b. $(f - g)(x)$

c. $(f \cdot g)(x)$

2. Let $f(x) = 2x^2 - x$ and $g(x) = x - 1$. Find the following functions.

a. $(g - f)(x)$

b. $\left(\dfrac{g}{f}\right)(x)$

3. Let $f(x) = x - 1$ and
$g(x) = \sqrt{x+3}$. Find the
following functions.

a. $(f+g)(x)$

b. $\left(\dfrac{g}{f}\right)(x)$

4. The dimensions of a
rectangle are constantly
changing. The length
of the rectangle can
be represented by the
formula $2x^2 - 4x + 5$ after
x seconds have gone by.
The width of the rectangle
can be represented by
$x^2 - 6x + 10$ after x seconds
have gone by. Find an
expression that represents
the perimeter of the
rectangle after x seconds
have gone by.

The domain of f is the set of all real numbers. However, the domain of the sum is restricted to the domain of g, the radical function. In this case we must have $x - 2 \geq 0$. Thus in interval notation, the domain is $[2, \infty)$.

b. $\left(\dfrac{f}{g}\right)(x)$

Solution

$$\left(\dfrac{f}{g}\right)(x) = \dfrac{x+5}{\sqrt{x-2}}$$

For this function, the denominator cannot be 0, so $x \neq 2$. Therefore, we must have $x - 2 > 0$ and the domain, in interval notation, is $(2, \infty)$.

Note: The domain can become smaller than, but never larger than, the domain of the two original functions.

Now work margin exercise 3.

Example 4

Algebraic Operations with Functions

Sally is analyzing the finances for her bakery. She finds that the bakery's revenue can be represented by the function $23x + 18$ when they sell x cakes. The cost function can be represented by $8x - 4$ when they make x cakes. Using the formula profit = revenue − cost, find an expression that represents the bakery's profits when they sell x cakes.

Solution

Plugging the algebraic expressions into the formula profit = revenue − cost and simplifying, we obtain the following.

$$\begin{aligned}
\text{profit} &= (23x + 18) - (8x - 4) \\
&= 23x + 18 - 8x + 4 \\
&= 15x + 22
\end{aligned}$$

Thus, the profit function for the bakery can be represented by $15x + 22$ when they sell x cakes.

Now work margin exercise 4.

Objective B **Graphing the Sum of Two Functions**

In this section we discuss how to graph the sum of two functions. (Graphing the difference, product, and quotient can be accomplished in a similar manner.) Remember that, in any case, algebraic operations with functions can be performed only over a common domain.

We begin with two functions f and g that have the same domain and only a finite number of ordered pairs.

$$f = \{(-2, 1), (0, 4), (3, 5)\}$$

$$g = \{(-2, 3), (0, 1), (3, 2)\}$$

Note that the domain of both functions is $\{-2, 0, 3\}$.

Both functions have only three points and are graphed in Figure 1.

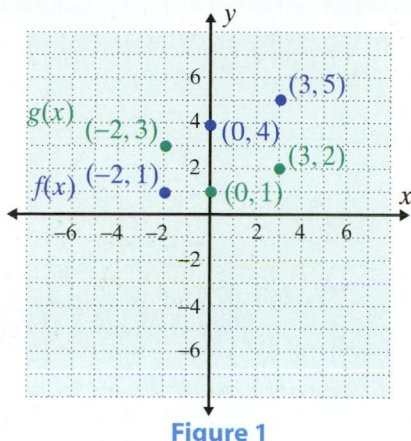

Figure 1

To add the two functions graphically, look at each of the x-values on the x-axis and the points directly above (or below) them and add the corresponding y-values. This process will give new points, and these new points represent a function that is the sum of the two original functions.

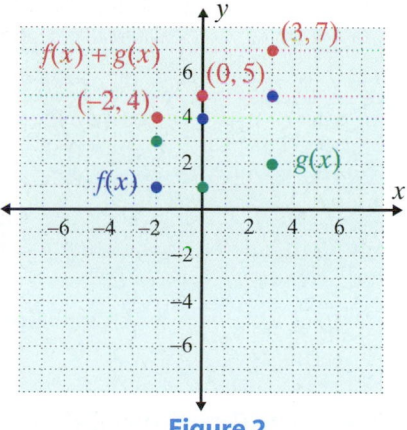

Figure 2

From the points in the Figure 2 we see that
$$(f + g)(x) = f(x) + g(x)$$
$$= \{(-2, 1+3), (0, 4+1), (3, 5+2)\}$$
$$= \{(-2, 4), (0, 5), (3, 7)\}.$$

In general, graphing the sum (difference, product, or quotient) of two functions will involve an infinite number of points. Certainly not all of these points can be plotted one at a time. However, by making a table of a few key points and joining these points with smooth curves or line segments, the general nature of the result can be found. In fact, if the two functions consist of line segments, then the sum will also consist of line segments. Table 1 and Figures 3 and 4 illustrate such a case.

Remember that when operating with functions, the operations are performed with the y-values for each value of x in the common domain. (Don't mess with the x-values!)

Points Illustrated in Figure 3 and Figure 4

x	f(x)	g(x)	f(x) + g(x)
−3	1	−3	$1 + (-3) = -2$
−2	1	0	$1 + 0 = 1$
−1	1	3	$1 + 3 = 4$
0	−2	3	$(-2) + 3 = 1$
1	1	3	$1 + 3 = 4$
2	1	4	$1 + 4 = 5$
3	0	5	$0 + 5 = 5$

Table 1

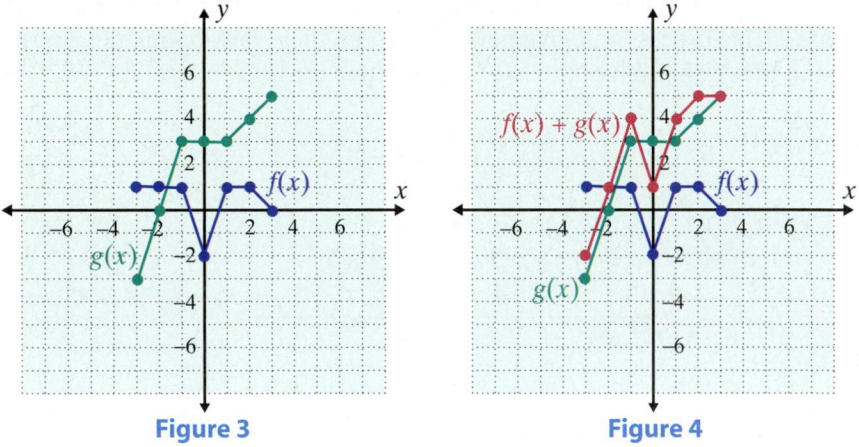

Figure 3   **Figure 4**

Objective C   **Using a TI-84 Plus Graphing Calculator to Graph the Sum of Two Functions**

A graphing calculator can be used to graph the sum (difference, product, or quotient) of two functions. An interesting way to do this is to first press ▬Y=▬ and enter the functions as Y_1 and Y_2; then assign Y_3 to be the sum of the first two. Figures **5a** and **5b** show the displays for entering two functions and their graphs.

Using the TI-84 Plus graphing calculator, we enter:
$$Y_1 = x^2 - 3 \quad \text{(a parabola)}$$
and
$$Y_2 = 5x - 1 \quad \text{(a straight line)}.$$

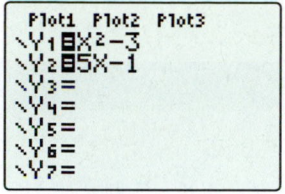

Figure 5a:
The two functions
entered

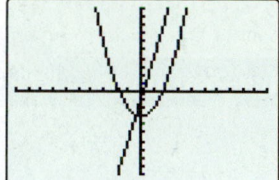

Figure 5b:
The two functions
graphed

Now we can graph the sum of the two functions, $f(x) + g(x)$, (or $Y_1 + Y_2$ on the calculator) by assigning $Y_3 = Y_1 + Y_2$ as follows. (This saves performing the actual algebraic operations. The calculator does it for us.)

Step 1: Press ⬭ᵧ₌ and select Y_3.

Step 2: Press the **VARS** key.

Step 3: Move the cursor to the Y-VARS at the top of the display.

Step 4: Select 1:FUNCTION... and press **ENTER**.

Step 5: Select Y_1 and press **ENTER**.

Step 6: Press the ➕ key.

Step 7: Go to Y-VARS again, select 1:FUNCTION... and select Y_2.

The display will now appear as follows.

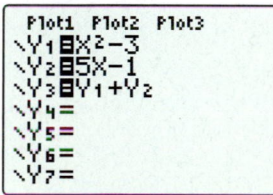

Figure 6

Now press **GRAPH** and the display will show all three curves, Y_1, Y_2, and the sum $Y_3 = Y_1 + Y_2$. You may need to reset the **WINDOW** to allow a more complete display of all of the graphs.

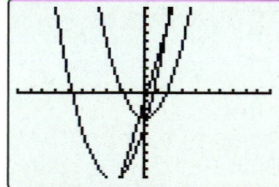

Figure 7

If this graph is somewhat confusing (because three graphs are shown), you can turn off the first two graphs and show just the sum. To turn off a graph, go to the highlighted equal sign, ▇, next to the equation and hit **ENTER**. You should see the following screens.

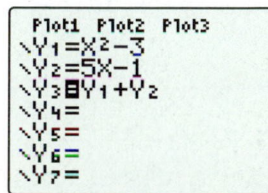

Figure 8a

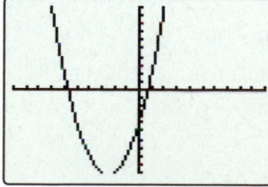

Figure 8b

1. For $f(x) = x^2 - 2x - 3$ and $g(x) = x - 3$ find:

   a. $(f + g)(x)$ **b.** $(f - g)(x)$ **c.** $(f \cdot g)(x)$ **d.** $\left(\dfrac{f}{g}\right)(x)$

2. Evaluate each of the functions **a.-d.** in Practice Problem 1 for $x = 4$.

3. For $f(x) = 2x - 5$ and $h(x) = \sqrt{x + 6}$,

   a. Find $f(x) + h(x)$ and state the domain of the sum function.

   b. Find $f(x) \cdot h(x)$ and state the domain of the product function.

4. Given $f(x) = \{(-3, 2), (-1, 1), (0, -1), (2, 3)\}$ and
   $g(x) = \{(-3, 4), (-1, 2), (0, 1), (2, 1)\}$ graph the sum of
   the two functions.

Practice Problem Answers

1. **a.** $x^2 - x - 6$ **b.** $x^2 - 3x$ **c.** $x^3 - 5x^2 + 3x + 9$ **d.** $x + 1, x \neq 3$

2. **a.** 6 **b.** 4 **c.** 5 **d.** 5

3. **a.** $2x - 5 + \sqrt{x + 6}$, $D = [-6, \infty)$ **b.** $2x\sqrt{x + 6} - 5\sqrt{x + 6}$, $D = [-6, \infty)$

4.

Exercises 10.1

For the following pairs of functions find, a. $(f+g)(x)$, b. $(f-g)(x)$, c. $(f \cdot g)(x)$ and d. $\left(\dfrac{f}{g}\right)(x)$. See Examples 1 and 2.

1. $f(x) = x+2,\ g(x) = x-5$

2. $f(x) = 2x,\ g(x) = x+4$

3. $f(x) = x^2,\ g(x) = 3x-4$

4. $f(x) = x-3,\ g(x) = x^2+1$

5. $f(x) = x^2-9,\ g(x) = x-3$

6. $f(x) = x^2-25,\ g(x) = x+5$

7. $f(x) = 2x^2+x,\ g(x) = x^2+2$

8. $f(x) = x^3+6x,\ g(x) = x^2+6$

9. $f(x) = x^2+4x+1,\ g(x) = x^2-4x+1$

10. $f(x) = x^3-x^2,\ g(x) = 6-x^2$

Let $f(x) = x^2+4$ and $g(x) = -x+3$. Find the values of the indicated expressions. See Examples 1 and 2.

11. $f(2)+g(2)$

12. $f(2) \cdot g(2)$

13. $g(a)-f(a)$

14. $\dfrac{g(a)}{f(a)}$

15. $(f+g)(-4)$

16. $(f-g)(0.5)$

17. $\left(\dfrac{f}{g}\right)(-2)$

18. $(f \cdot g)(-3)$

19. $(g-f)(-6)$

20. $\left(\dfrac{g}{f}\right)(-1)$

21. If $f(x) = \sqrt{2x-6}$ and $g(x) = x+4$, find $(f+g)(x)$.

22. If $f(x) = x^2 - 2x + 1$ and $g(x) = x - 1$, find $\left(\dfrac{f}{g}\right)(x)$.

23. Find $f(x) \cdot g(x)$ given that $f(x) = 3x + 2$ and $g(x) = x - 7$.

24. Find $f(x) - g(x)$ given that $f(x) = x^2$ and $g(x) = x^2 - 2$.

25. For $f(x) = x - 5$ and $g(x) = \sqrt{x+3}$, find $\dfrac{f(x)}{g(x)}$.

26. For $f(x) = 2x - 8$ and $g(x) = \sqrt{2-x}$, find $f(x) \cdot g(x)$.

27. If $f(x) = -\sqrt{x-3}$ and $g(x) = 3x$, find $(f \cdot g)(x)$.

28. If $f(x) = -\sqrt{4-x}$ and $g(x) = 5 - x$, find $(g - f)(x)$.

29. If $f(x) = \sqrt[3]{x+3}$ and $g(x) = \sqrt{5+x}$, find $f(x) + g(x)$.

30. If $f(x) = \sqrt{x-1}$ and $g(x) = \sqrt[3]{2x+1}$, find $f(x) - g(x)$.

For the following pairs of functions, graph a. the sum $(f+g)$ and b. the difference $(f-g)$ on two different graphs.

31. $f = \{(-2, 5), (0, -3), (2, 1)\}$
$g = \{(-2, -2), (0, -1), (2, -3)\}$

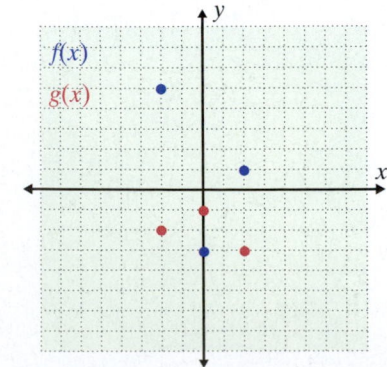

32. $f = \{(0, 2), (1, 1), (2, -1)\}$
$g = \{(0, 4), (1, 2), (2, -2)\}$

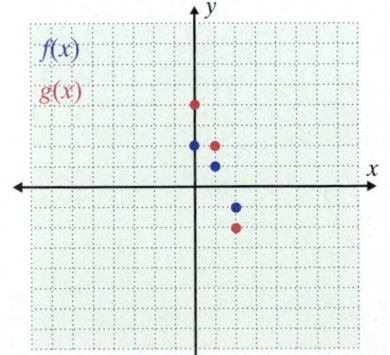

33. $f = \{(-3, -1), (-1, 0), (2, -4)\}$

$g = \{(-3, -3), (-1, 5), (2, 1)\}$

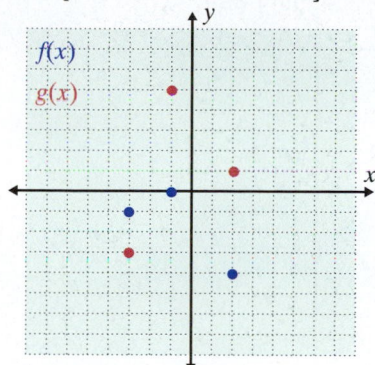

34. $f = \{(-4, 4), (-2, -5), (1, 3)\}$

$g = \{(-4, 1), (-2, 0), (1, -2)\}$

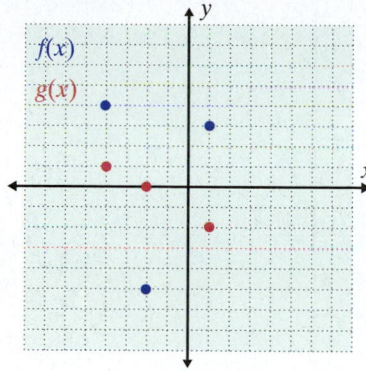

35. $f = \{(-2, -5), (-1, 4), (0, -1), (1, 5)\}$

$g = \{(-2, -1), (-1, 0), (0, -4), (1, 1)\}$

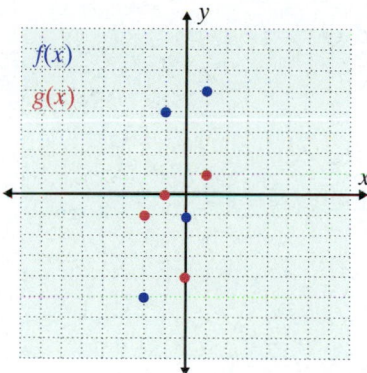

36. $f = \{(-2, 3), (1, 2), (2, 1), (3, 0)\}$

$g = \{(-2, -4), (1, -3), (2, -2), (3, -1)\}$

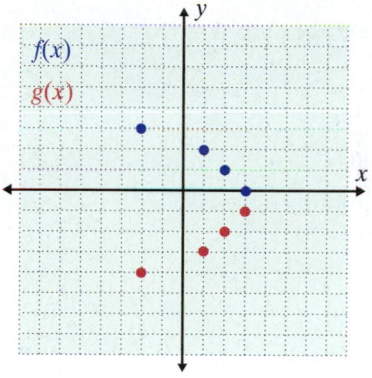

37. $f = \{(-3, 1), (-1, 2), (1, 1), (3, 2)\}$

$g = \{(-3, -3), (-1, -4), (1, 3), (3, 4)\}$

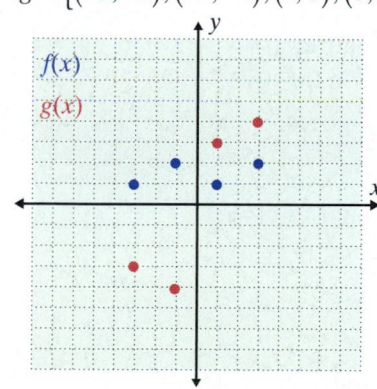

38. $f = \{(-4, 6), (-2, 0), (0, -2), (3, -3)\}$

$g = \{(-4, 0), (-2, -4), (0, 1), (3, 3)\}$

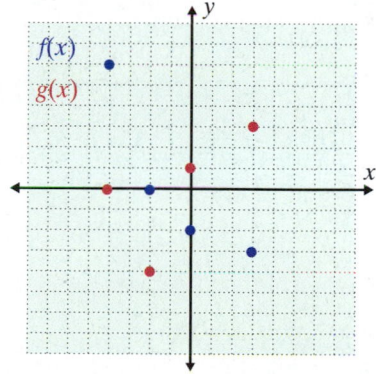

39. $f = \{(-5, 1), (-1, 2), (2, 3), (3, 2)\}$
$g = \{(-5, 2), (-1, 1), (2, 0), (3, -1)\}$

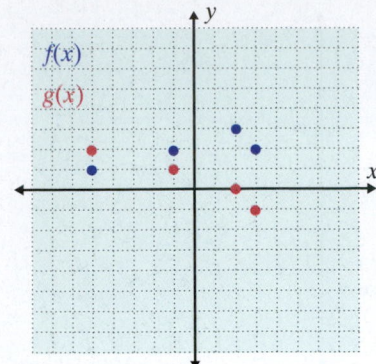

40. $f = \{(-3, 2), (0, 0), (3, -1), (4, 1)\}$
$g = \{(-3, -4), (0, -3), (3, 2), (4, -1)\}$

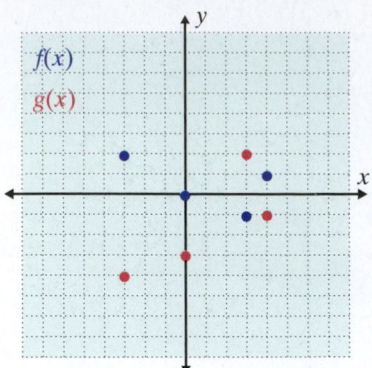

Graph each pair of functions and the sum of these functions on the same set of axes.

41. $f(x) = x^2$ and $g(x) = -1$

42. $f(x) = x^2$ and $g(x) = 2$

43. $f(x) = x + 1$ and $g(x) = 2x$

44. $f(x) = x + 5$ and $g(x) = x - 5$

45. $f(x) = x + 4$ and $g(x) = -x$

46. $f(x) = 2 - x$ and $g(x) = x$

47. $f(x) = x + 1$ and $g(x) = x^2 - 1$

48. $f(x) = x^2 + 2$ and $g(x) = x^2 - 2$

49. $f(x) = \sqrt{x - 6}$ and $g(x) = 2$

50. $f(x) = \sqrt{3 - x}$ and $g(x) = -1$

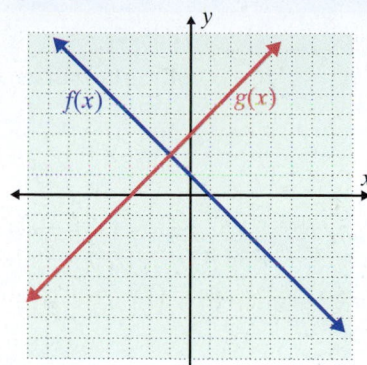

51. $(f+g)(-2)$

52. $(f-g)(2)$

53. $(f \cdot g)(3)$

54. $(g-f)(0)$

55. $\left(\dfrac{f}{g}\right)(4)$

56. $(g \cdot f)(4)$

Graph the sum of each function.

57.

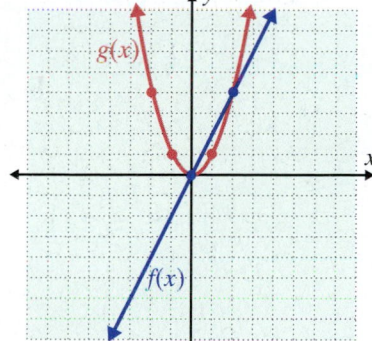

58.

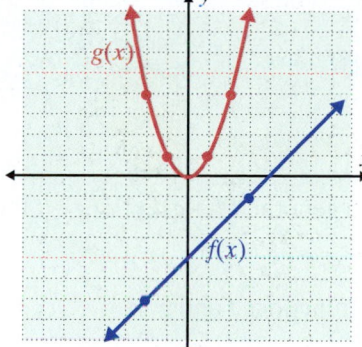

59.

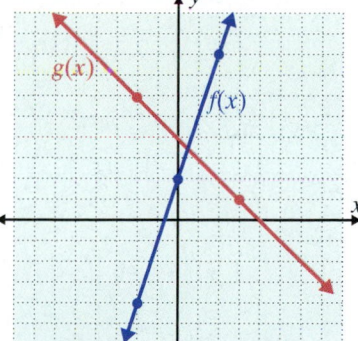

60.

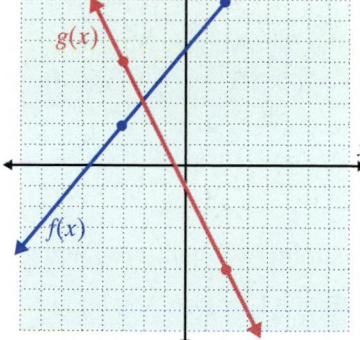

61.

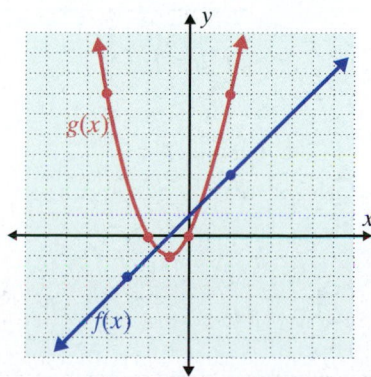

62.

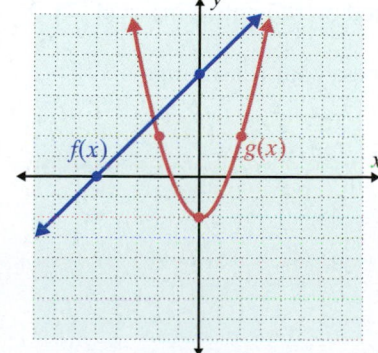

Use a graphing calculator to graph each pair of functions and the sum of these functions on the same set of axes.

63. $f(x) = x^2$ and $h(x) = 2x + 1$

64. $g(x) = x^2 + x$ and $h(x) = 3x + 4$

65. $f(x) = \sqrt{x+4}$ and $g(x) = -2$

66. $f(x) = -\sqrt{x-1}$ and $g(x) = 3$

67. $f(x) = \sqrt[3]{x+5}$ and $h(x) = 2x$

68. $h(x) = \sqrt[3]{x-1}$ and $g(x) = x - 1$

69. $g(x) = 7 - x^2$ and $h(x) = x^2 - 3$

70. $f(x) = x^2 + 5$ and $g(x) = 4 - x^2$

Writing & Thinking

71. Explain why, in general, $(f - g)(x) \neq (g - f)(x)$ if $f(x) \neq g(x)$.

72. Given the two functions f and g,
$$f = \{(-2, 0), (-1, 1), (0, 4), (2, 4), (3, 5), (4, 1)\}$$
$$g = \{(-2, 3), (-1, 4), (0, 1), (2, -1), (3, 2), (4, 6)\}$$
find and graph the following.
   a. $f - g$
   b. $f \cdot g$
   c. $\dfrac{f}{g}$

73. Use the graphs of the two functions f and g shown in the graph.
   a. Sketch the graph of $f - g$.
   b. Sketch the graph of $f \cdot g$.
   c. Is $\dfrac{f}{g}$ defined on the entire interval $[-3, 3]$?

   Briefly discuss your reasoning.

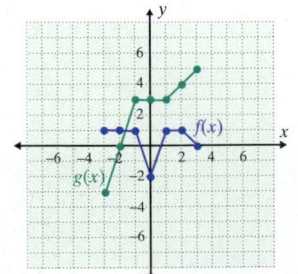

10.2 Composite and Inverse Functions

Objective A **Evaluating a Function at an Algebraic Expression**

We have previously studied how to evaluate a function at a given value of x. However, we have always input a number into the function to evaluate it. We can expand our thinking to input an algebraic expression into the function to evaluate it. The process works the same as with numbers. Replace the variable with the expression given everywhere the variable appears.

Example 1

Evaluating a Function at an Algebraic Expression

Find the following function values when $f(x) = x^2 - 5x + 8$.

a. $f(2a)$

Solution

Substitute $2a$ in for x everywhere it appears and simplify.
$$f(2a) = (2a)^2 - 5(2a) + 8$$
$$= 4a^2 - 10a + 8$$

b. $f(g+3)$

Solution

$$f(g+3) = (g+3)^2 - 5(g+3) + 8$$
$$= g^2 + 6g + 9 - 5g - 15 + 8$$
$$= g^2 + g + 2$$

Now work margin exercise 1.

Objective B **Composition of Two Functions**

Previously, function notation, $f(x)$, has proven useful in evaluating functions and in indicating arithmetic operations with functions. This same notation is needed in developing the concept of **a function of a function**, called the **composition** of two functions. For example, suppose that,

$$g(x) = 2x - 4 \qquad \text{and} \qquad f(x) = x^2 - 4x + 1.$$

Now for $x = 3$, we have

$$g(3) = 2 \cdot (3) - 4 = 2 \qquad \text{and} \qquad f(2) = (2)^2 - 4 \cdot (2) + 1 = -3.$$

Objectives

A Know how to input an algebraic expression into a function.

B Form the composition of two functions.

C Determine if a function is one-to-one using the horizontal line test.

D Understand the concept of an inverse function, and determine if two functions are inverses.

E Find the inverse of a one-to-one function.

1. Find the following function values when $g(x) = x - 8x^2$.

a. $g(r+1)$

b. $g(b^2)$

For the **composition** $f(g(3))$ (read "f of g of 3"), we have

$$f(\overbrace{g(3)}) = f(2) = -3.$$

We see that $g(3)$ which equals 2, replaces x in the function $f(x)$.

More generally, given two functions $f(x)$ and $g(x)$, a new function $f(g(x))$ called the **composition** (or **composite**) of f and g, is found by substituting the expression for $g(x)$ in place of x in the function f. In this case the value of $g(x)$ must be in the domain of f. Thus for,

$$g(x) = 2x - 4 \quad \text{and} \quad f(x) = x^2 - 4x + 1, \quad \text{the composition}$$

$$f(g(x)) \quad \text{(read "f of g of x")} \quad \text{is found as follows:}$$

$$f(x) = x^2 - 4x + 1$$
$$f(g(x)) = (g(x))^2 - 4(g(x)) + 1 \qquad \text{Replace the x in $f(x)$ with $g(x)$.}$$
$$= (2x - 4)^2 - 4(2x - 4) + 1 \qquad \text{Replace $g(x)$ with $2x - 4$.}$$
$$= 4x^2 - 16x + 16 - 8x + 16 + 1 \qquad \text{Simplify.}$$
$$= 4x^2 - 24x + 33.$$

The composition of g and f (reversing the order of f and g) is indicated by

$$g(f(x)) \qquad \text{(read "g of f of x")}$$

and is found by substituting the expression for $f(x)$ in place of x in the function g. Thus,

$$g(x) = 2x - 4$$
$$g(f(x)) = 2(f(x)) - 4 \qquad \text{Replace the x in $g(x)$ with $f(x)$.}$$
$$= 2(x^2 - 4x + 1) - 4 \qquad \text{Replace $f(x)$ with $x^2 - 4x + 1$.}$$
$$= 2x^2 - 8x + 2 - 4 \qquad \text{simplify}$$
$$= 2x^2 - 8x - 2.$$

As we can see with these examples, in general, $f(g(x)) \neq g(f(x))$ and substitutions must be done carefully and accurately. The following definition shows another notation (a small raised circle) often used to indicate the composition of functions.

Composite Function

For two functions f and g, the **composite function** $f \circ g$ is defined as follows:

$$(f \circ g)(x) = f(g(x)).$$

The domain of $f \circ g$ consists of those values of x in the domain of g for which $g(x)$ is in the domain of f.

Example 2

Compositions

a. Form the compositions $(f \circ g)(x)$ and $(g \circ f)(x)$ if $f(x) = 5x + 2$ and $g(x) = 3x - 7$.

Solution

$(f \circ g)(x) = f(g(x)) = 5 \cdot g(x) + 2 = 5(3x - 7) + 2 = 15x - 33$

$(g \circ f)(x) = g(f(x)) = 3 \cdot f(x) - 7 = 3(5x + 2) - 7 = 15x - 1$

Note: Both $(f \circ g)(x)$ and $(g \circ f)(x)$ are defined for all real numbers.

b. Form the composite functions $(f \circ g)(x)$ and $(g \circ f)(x)$ if $f(x) = \sqrt{x - 3}$ and $g(x) = x^2 + 4$.

Solution

$$(f \circ g)(x) = \sqrt{g(x) - 3}$$
$$= \sqrt{(x^2 + 4) - 3} = \sqrt{x^2 + 1}$$

Note: For the expression under the radical to be defined, we must have $g(x) \geq 3$. Because $x^2 + 4 \geq 3$ for all real numbers, the domain of $(f \circ g)(x)$ is all real numbers.

$$(g \circ f)(x) = (f(x))^2 + 4$$
$$= (\sqrt{x - 3})^2 + 4 = x - 3 + 4 = x + 1$$

Note: $(g \circ f)(x)$ is defined only for $x \geq 3$. (Here the domain comes from f before the simplification.)

c. Find $f(g(x))$ and $g(f(x))$ if $f(x) = \sqrt{x + 3}$ and $g(x) = 2x - 5$.

Solution

$$f(g(x)) = \sqrt{g(x) + 3}$$
$$= \sqrt{(2x - 5) + 3} = \sqrt{2x - 2}$$

Note: $f(g(x))$ is defined only for $2x - 2 \geq 0$ or $x \geq 1$.

$$g(f(x)) = 2(f(x)) - 5$$
$$= 2(\sqrt{x + 3}) - 5 = 2\sqrt{x + 3} - 5$$

Note: $g(f(x))$ is defined only for $x \geq -3$.

Now work margin exercise 2.

2. a. If $f(x) = x^2$ and $g(x) = x^2 + 2$ form the compositions $(f \circ g)(x)$ and $(g \circ f)(x)$.

b. If $f(x) = 3x + 3$ and $g(x) = \sqrt{x + 3}$ form the compositions $f(g(x))$ and $g(f(x))$.

By the definition of a function, there is only one corresponding y-value for each x-value in a function's domain. Graphically, the **vertical line test** can be used to help determine whether or not a graph represents a function. Now in order to develop the concept of **inverse functions**, we need to study functions that have only one x-value for each y-value in the range. Such functions are said to be **one-to-one functions** (or **1-1 functions**).

Consider the following **functions**:

$$f = \left\{(1, 2), (2, 4), (3, 6), (4, 8), (5, 10)\right\}$$
$$g = \left\{(-2, 6), (0, 6), (1, 5), (2, 4), (4, 1)\right\}.$$

Both sets of ordered pairs are functions because each value of x appears only once. In the function f, each y-value appears only once. But, in function g, the y-value 6 appears twice: in $(-2, 6)$ and $(0, 6)$. Figure 1 illustrates both functions.

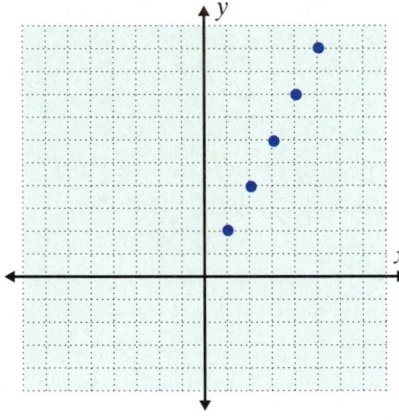

Figure 1a

Function f is a one-to-one function.

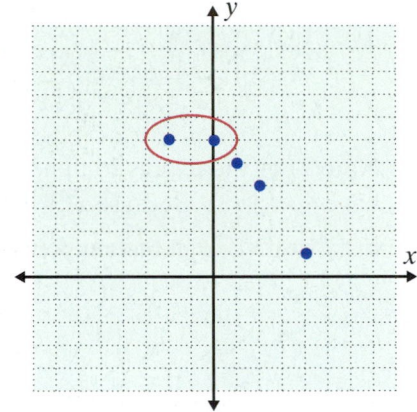

Figure 1b

Function g is not a one-to-one function.

One-to-One Functions

A function is a **one-to-one function** (or **1-1 function**) if for each value of y in the range there is only one corresponding value of x in the domain.

Graphically, as illustrated in Figure **1b**, if a horizontal line intersects the graph of a function in more than one point then it is **not** one-to-one. This is, in effect, the **horizontal line test.**

Horizontal Line Test

A function is one-to-one if no **horizontal line** intersects the graph of the function at more than one point.

The graphs in Figure 2 (below) illustrate the concept of one-to-one functions. (Note that each graph passes the vertical line test and is indeed a function.)

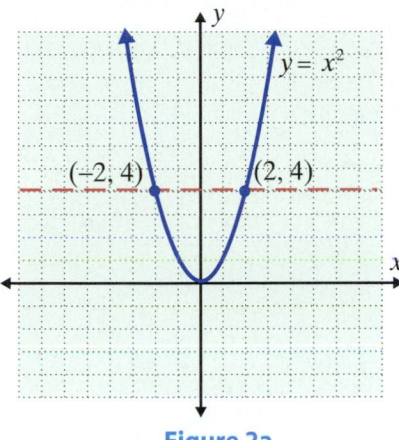

Figure 2a

not one-to-one

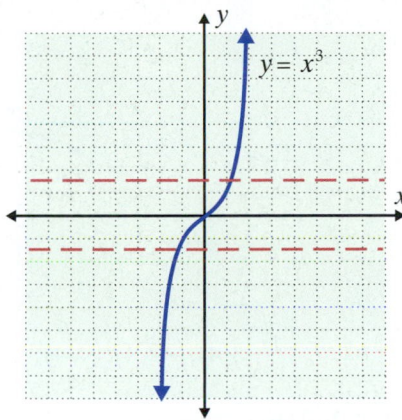

Figure 2b

one-to-one

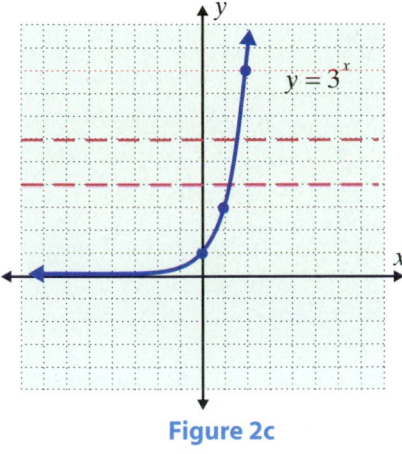

Figure 2c

one-to-one

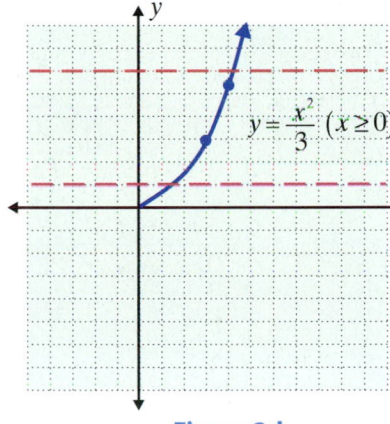

Figure 2d

one-to-one

Example 3

One-to-One Functions

Determine whether each function is or is not one-to-one.

a. $y = \sqrt{x+5}$

Solution

The horizontal line test shows that this function is one-to-one.

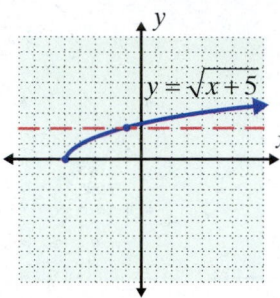

b. $f = \{(-3, 4), (-2, 1), (0, 4), (3, 1)\}$

Solution

This function is not one-to-one. Both y-values, 4 and 1, have more than one corresponding x-value.

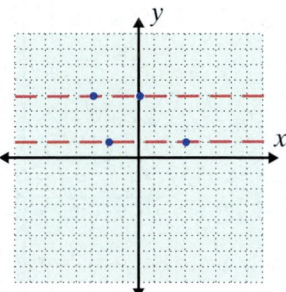

c. $y = 2x - 1$

Solution

The graph of the function $y = 2x - 1$ is a straight line. Straight lines that are not vertical and not horizontal represent one-to-one functions. (Vertical lines are not functions in the first place and horizontal lines fail the horizontal line test.)

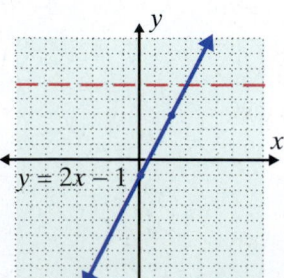

d. $y = -x^2 + 1$

3. Determine whether the following functions are one-to-one.

a. $y = \sqrt{3x + 3}$

b. $y = x^2 + x + 17$

Solution

The graph of the function $y = -x^2 + 1$ is a parabola and the horizontal line test shows that the function is not one-to-one.

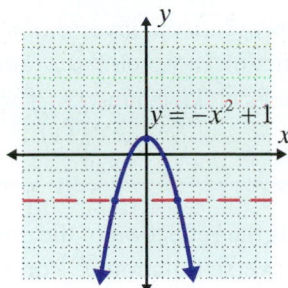

Now work margin exercise 3.

Objective D **Inverse Functions**

Now that we have discussed one-to-one functions, we can develop the concept of **inverse functions**. To find the inverse of a one-to-one function represented by a set of ordered pairs, exchange x and y in each ordered pair. That is, if (x, y) is in the original one-to-one function, then (y, x) is in the inverse function.

For example,

if, $f = \{(-1, 1), (0, 2), (1, 4)\}$, then,

interchanging the coordinates in each ordered pair gives

$g = \{(1, -1), (2, 0), (4, 1)\}.$

The functions f and g are called **inverses** of each other. If g is the inverse of f, we write

f^{-1} (read "f inverse") rather than use g.

Thus in this example we can write $f^{-1} = \{(1, -1), (2, 0), (4, 1)\}.$

Inverse Functions

If f is a one-to-one function with ordered pairs of the form (x, y), then its **inverse function**, denoted as f^{-1}, is also a one-to-one function with ordered pairs of the form (y, x).

Only one-to-one functions have inverse functions. If a function is not one-to-one, then interchanging the x- and y-values would yield a relation that is not a function. Graphically, the function must satisfy the horizontal line test. If it did not, then the inverse would not pass the vertical line test and would not be a function.

notes

- The notation $f^{-1}(x)$ represents the inverse of a one-to-one function. This inverse is a new function in which the x- and y-values have been interchanged. $f^{-1}(x)$ does NOT mean $\dfrac{1}{f(x)}$ because the -1 is NOT an exponent.

The graph of any point (b, a) is the reflection of the point (a, b) across the line $y = x$. Thus the points of the inverse function f^{-1} are reflections of the points of the function f across the line $y = x$. We say that the graphs are **symmetric about the line** $y = x$. Figure 3 illustrates these reflections and the symmetry.

We see that the domain, denoted D_f, and range, denoted R_f, of the two functions are interchanged.

That is,

$$D_f = R_{f^{-1}} = \{-1, 0, 1\}$$
$$R_f = D_{f^{-1}} = \{1, 2, 4\}.$$

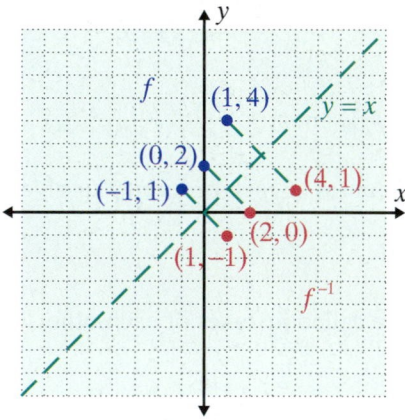

Figure 3

In general, every one-to-one function has an inverse function (or just inverse), and the graph of the inverse function of any one-to-one function f can be found by reflecting the graph of f across the line $y = x$. Figure 4 shows two more illustrations of this concept.

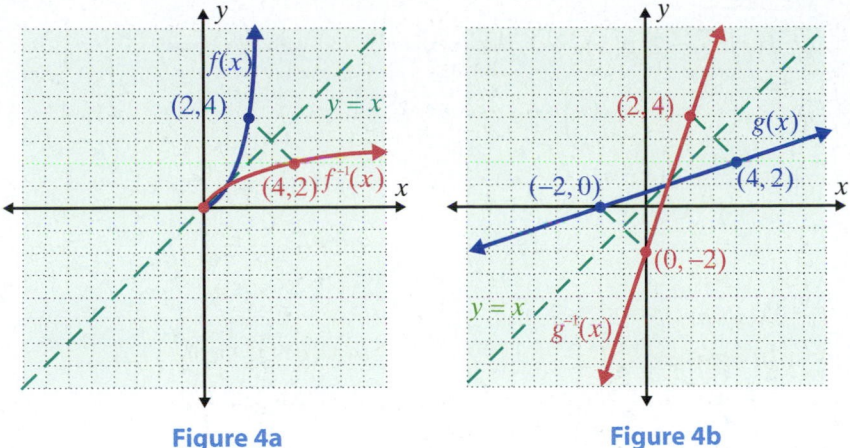

| Figure 4a | Figure 4b |

The following definition of inverse functions helps to determine whether or not two functions are inverses of each other.

To Determine whether Two Functions are Inverses

If f and g are one-to-one functions and

$$f(g(x)) = x \qquad \text{for all } x \text{ in } D_g \text{ and}$$

$$g(f(x)) = x \qquad \text{for all } x \text{ in } D_f,$$

then f and g are **inverse functions.**

That is, $g = f^{-1}$ and $f = g^{-1}$.

Example 4

Inverse Functions

Use the definition of inverse functions to show that f and g are inverse functions.

a. $f(x) = 2x + 6$ and $g(x) = \dfrac{x-6}{2}$.

Solution

$$f(x) = 2x + 6 \text{ and } g(x) = \dfrac{x-6}{2}.$$

The domain of both functions is the set of all real numbers.

We have

$$f\big(g(x)\big) = 2 \cdot g(x) + 6 \qquad \text{Replace the } x \text{ in } f(x) \text{ with } g(x) \text{ and simplify.}$$

$$= 2\left(\frac{x-6}{2}\right) + 6$$

$$= (x-6) + 6$$

$$= x.$$

Also,

$$g\big(f(x)\big) = \frac{f(x) - 6}{2} \qquad \text{Replace the } x \text{ in } g(x) \text{ with } f(x) \text{ and simplify.}$$

$$= \frac{(2x+6) - 6}{2}$$

$$= \frac{2x}{2}$$

$$= x.$$

Therefore, $g = f^{-1}$ and $f = g^{-1}$.

The graph shows that the line $y = x$ is a line of symmetry for the graphs of the inverse functions.

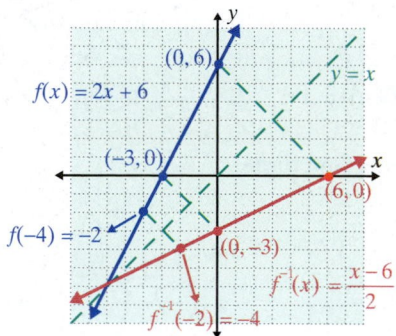

b. $f(x) = \sqrt{x-3}$ and $g(x) = x^2 + 3$ for $x \ge 0$.

Solution

$f(x) = \sqrt{x-3}$ and $g(x) = x^2 + 3$ for $x \ge 0$. The domain of f is the interval $[3, \infty)$ and the domain of g is the interval $[0, \infty)$.

We have,

$$f\big(g(x)\big) = \sqrt{g(x) - 3} \qquad \text{Replace the } x \text{ in } f(x) \text{ with } g(x) \text{ and simplify.}$$

$$= \sqrt{(x^2 + 3) - 3}$$

$$= \sqrt{x^2}$$

$$= x \qquad \text{for } x \ge 0.$$

Also,

$$g\big(f(x)\big) = \big(f(x)\big)^2 + 3 \qquad \text{Replace the } x \text{ in } g(x) \text{ with } f(x) \text{ and simplify.}$$

$$= \big(\sqrt{x-3}\big)^2 + 3$$

$$= x - 3 + 3$$

$$= x \qquad \text{for } x \geq 3.$$

Therefore, $g = f^{-1}$ and $f = g^{-1}$.

The graph shows that the line $y = x$ is a line of symmetry for the graphs of the inverse functions. Note that these graphs are only parts of parabolas and the domains have been restricted accordingly.

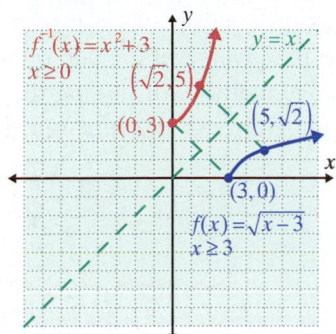

Now work margin exercise 4.

4. Show that f and g are inverse functions.

$$f(x) = x^2 - 3 \text{ and}$$

$$g(x) = \sqrt{x+3} \text{ for } x \geq 0$$

Example 5

Evaluating Compositions of Inverses

Given $f(x) = \sqrt{x+4}$ for $x \geq -4$ and $f^{-1}(x) = x^2 - 4$ for $x \geq 0$, evaluate the indicated compositions.

a. $f\big(f^{-1}(2)\big)$

Solution

$$f^{-1}(2) = (2)^2 - 4 = 0 \text{ so } f\big(f^{-1}(2)\big) = f(0) = \sqrt{(0)+4} = \sqrt{4} = 2$$

b. $f\big(f^{-1}(5)\big)$

Solution

$$f^{-1}(5) = (5)^2 - 4 = 21 \text{ so } f\big(f^{-1}(5)\big) = f(21) = \sqrt{(21)+4} = \sqrt{25} = 5$$

c. $f^{-1}\big(f(-2)\big)$

Solution

$$f(-2) = \sqrt{(-2)+4} = \sqrt{2} \text{ , so}$$

$$f^{-1}\big(f(-2)\big) = f^{-1}\big(\sqrt{2}\big) = \big(\sqrt{2}\big)^2 - 4 = 2 - 4 = -2$$

5. Given
$f(x) = \sqrt{x+2}$ for $x \geq -2$
and
$f^{-1}(x) = x^2 - 2$ for $x \geq 0$
evaluate the indicated
compositions.

a. $f\left(f^{-1}(4)\right)$

b. $f^{-1}\left(f(-1)\right)$

d. $f^{-1}\left(f(-5)\right)$

Solution

$f(-5)$ does not exist because -5 is not in the domain of f. Therefore $f^{-1}\left(f(-5)\right)$ does not exist.

Now work margin exercise 5.

Objective E **Finding the Inverse of a One-to-One Function**

In Examples 4 and 5 the given functions were inverses of each other. The next question is how to find the inverse of a given one-to-one function. The following procedure shows one method for finding the inverse using the fact that if an ordered pair (x, y) belongs to the function f, then (y, x) belongs to f^{-1}.

To Find the Inverse of a One-to-One Function

1. Let $y = f(x)$. (In effect, substitute y for $f(x)$.)
2. Interchange x and y.
3. In the new equation, solve for y in terms of x.
4. Substitute $f^{-1}(x)$ for y. (This new function is the inverse of f.)

Example 6

Finding the Inverse

a. Find $f^{-1}(x)$ if $f(x) = 5x - 7$.

Solution

$$f(x) = 5x - 7$$
$$y = 5x - 7 \qquad \text{Substitute } y \text{ for } f(x).$$
$$x = 5y - 7 \qquad \text{Interchange } x \text{ and } y.$$
$$x + 7 = 5y \qquad \text{Solve for } y \text{ in terms of } x.$$
$$\frac{x+7}{5} = y$$
$$f^{-1}(x) = \frac{x+7}{5} \qquad \text{Substitute } f^{-1}(x) \text{ for } y.$$

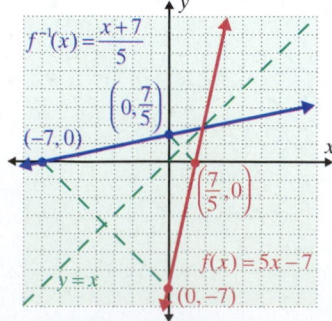

b. Find $g^{-1}(x)$ if $g(x) = x^2 - 2$ for $x \geq 0$.

Solution

$$g(x) = x^2 - 2 \qquad x \geq 0 \text{ (Note that } g \text{ is one-to-one for } x \geq 0.)$$

$$y = x^2 - 2 \qquad \text{Substitute } y \text{ for } g(x).$$

$$x = y^2 - 2 \qquad \text{Interchange } y \text{ and } x.$$

$$\pm\sqrt{x + 2} = y \qquad \text{Solve for } y \text{ in terms of } x.$$

$$g^{-1}(x) = \sqrt{x + 2} \qquad \text{Take the positive square root because we}$$

must have $y \geq 0$. (The domain of g is $x \geq 0$,

so the range of g^{-1} is $y \geq 0$.)

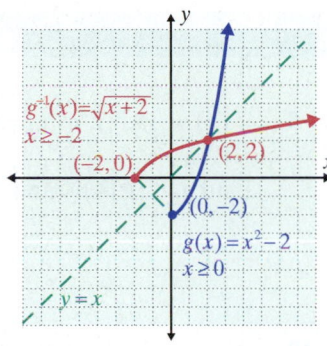

Now work margin exercise 6.

Practice Problems

1. For $f(x) = 2x - 1$ and $g(x) = x^2$, find:

   a. $f(g(x))$ **b.** $g(f(x))$

2. Given $f(x) = 3x + 4$ and $f^{-1}(x) = \dfrac{x - 4}{3}$, find the following:

   a. $f(f^{-1}(2))$ **b.** $f^{-1}(f(1))$

3. Given $g(x) = \sqrt{x - 2}$ and $f(x) = \dfrac{1}{x - 1}$, find $f(g(3))$.

4. Find the inverse of the function $f(x) = 3x - 1$.

5. Find $f^{-1}(x)$ if $f(x) = x^2 - 4$, $x \geq 0$.
   (Note that f is one-to-one for $x \geq 0$.)

Practice Problem Answers

1. a. $f(g(x)) = 2x^2 - 1$ **b.** $g(f(x)) = 4x^2 - 4x + 1$

2. a. 2 **b.** 1

3. undefined

4. $f^{-1}(x) = \dfrac{x + 1}{3}$ **5.** $f^{-1}(x) = \sqrt{x + 4}$, $x \geq -4$

6. Find the inverse.

   a. Find $f^{-1}(x)$ if $f(x) = 7x + 2$.

   b. Find $f(x)$ if $f^{-1}(x) = \sqrt{2x + 2}$
      for $x \geq -1$.

Exercises 10.2

1. $f(x) = 8x - 5$
   a. $f(r)$
   b. $f(3a - 1)$

2. $r(x) = 4x - 6$
   a. $r(g - 5)$
   b. $r(h^2 + 8)$

3. $g(y) = 5y^2 + 4$
   a. $g(x - 2)$
   b. $g(3n^2)$

4. $h(y) = y^4 + 8$
   a. $h(3p)$
   b. $h(2s^2)$

5. $f(c) = 3c^2 + 6c - 9$
   a. $f(n - 2)$
   b. $f(4y^3)$

6. $b(t) = t^2 - 2t + 7$
   a. $b(5k)$
   b. $b(x + 1)$

Find the following function compositions.

7. $f(x) = 3x + 5$, $g(x) = \dfrac{x + 4}{2}$

   Find **a.** $f(g(2))$ and **b.** $g(f(2))$.

8. $f(x) = \dfrac{1}{4}x + 1$, $g(x) = 6x - 7$

   Find **a.** $f(g(4))$ and **b.** $g(f(4))$.

9. $f(x) = x^2$, $g(x) = 2x + 3$

   Find **a.** $(f \circ g)(-5)$ and **b.** $(g \circ f)(-1)$.

10. $f(x) = x^2 + 1$, $g(x) = x - 6$

    Find **a.** $(f \circ g)(3)$ and **b.** $(g \circ f)(-2)$.

Form the compositions $f(g(x))$ and $g(f(x))$ for each pair of functions. See Example 2.

11. $f(x) = \sqrt{x}$, $g(x) = x^2$

12. $f(x) = \dfrac{1}{x}$, $g(x) = \dfrac{1}{x}$

13. $f(x) = \sqrt{x}$, $g(x) = x - 2$

14. $f(x) = \sqrt{x}$, $g(x) = x^2 - 9$

15. $f(x) = x - 1$, $g(x) = \dfrac{1}{x^2}$

16. $f(x) = \dfrac{1}{x^2}$, $g(x) = x^2 + 1$

17. $f(x) = x^3 + x + 1$, $g(x) = x + 1$

18. $f(x) = x^3$, $g(x) = 2x - 1$

19. $f(x) = \dfrac{1}{\sqrt{x}},\ g(x) = x^2$

20. $f(x) = \dfrac{1}{\sqrt{x}},\ g(x) = x^2 - 4$

21. $f(x) = \dfrac{1}{x},\ g(x) = x^2 + 7x - 8$

22. $f(x) = \dfrac{1}{x+1},\ g(x) = x^2 + x - 3$

23. $f(x) = x^{3n},\ g(x) = 2x - 6$

24. $f(x) = x^{\frac{1}{3}},\ g(x) = 4x + 7$

25. $f(x) = x^3,\ g(x) = \sqrt{x-8}$

26. $f(x) = x^3 + 1,\ g(x) = \dfrac{1}{x}$

27. For the functions $f(x) = 6x - 3$ and
$g(x) = \dfrac{1}{3}x + 3,$ find:

a. $f\big(g(3)\big)$

b. $g\big(f(0)\big)$

c. Does it appear that f and g are inverses of each other? Explain.

28. For the functions $h(x) = -2x + 4$ and
$g(x) = \dfrac{4-x}{2},$ find:

a. $h\big(g(6)\big)$

b. $g\big(h(-4)\big)$

c. Does it appear that h and g are inverses of each other? Explain.

29. Given $f(x) = \dfrac{1}{2x+1}$ and $g(x) = -\dfrac{1}{x},$ find:

a. $g\big(f(4)\big)$

b. $f\big(g(2)\big)$

c. Explain the different results of **a.** and **b.**

30. Given $f(x) = \sqrt{x-9}$ and $g(x) = x - 9,$ find:

a. $g\big(f(109)\big)$

b. $f\big(g(9)\big)$

c. Explain the different results of **a.** and **b.**

31. $f(x) = 3x + 1$ and $g(x) = \dfrac{x-1}{3}$

32. $f(x) = -2x + 3$ and $g(x) = \dfrac{3-x}{2}$

33. $f(x) = \sqrt[3]{x-1}$ and $g(x) = x^3 + 1$

34. $f(x) = x^3 - 4$ and $g(x) = \sqrt[3]{x+4}$

35. $f(x) = x^2$ for $x \geq 0$ and $g(x) = \sqrt{x}$

36. $f(x) = \sqrt{x+3}$ and $g(x) = x^2 - 3$ for $x \geq 0$

37. $f(x) = x^3 + 2$ and $g(x) = \sqrt[3]{x-2}$

38. $f(x) = \sqrt[5]{x+6}$ and $g(x) = x^5 - 6$

39. $f(x) = \dfrac{2}{x}$ and $g(x) = \dfrac{2}{x}$

40. $f(x) = \dfrac{3}{x}$ and $g(x) = \dfrac{3}{x}$

41. $f(x) = 2x - 3$

42. $f(x) = 2x - 5$

43. $g(x) = x$

44. $g(x) = 1 - 4x$

45. $f(x) = 5x + 1$

46. $g(x) = -3x + 1$

47. $g(x) = \dfrac{2}{3}x + 2$

48. $f(x) = -\dfrac{1}{2}x - 3$

49. $f(x) = -x - 2$

50. $f(x) = -2x + 4$

51. $f(x) = x^2 + 1,\ x \geq 0$

52. $f(x) = x^2 - 1,\ x \geq 0$

53. $f(x) = -\sqrt{x},\ x \geq 0$

54. $f(x) = -\sqrt{x-2},\ x \geq 2$

Using the horizontal line test, determine which of the graphs are graphs of one-to-one functions. If the graph represents a one-to-one function, graph its inverse by reflecting the graph of the function across the line $y = x$.

(**Hint:** If a function is one-to-one, label a few points on the graph and use the fact that the x- and y-coordinates are interchanged on the graph of the inverse.)

55.

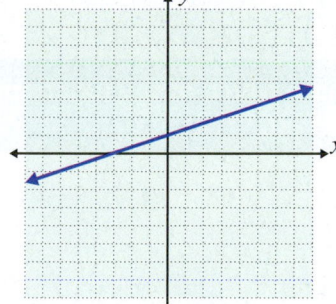

56.

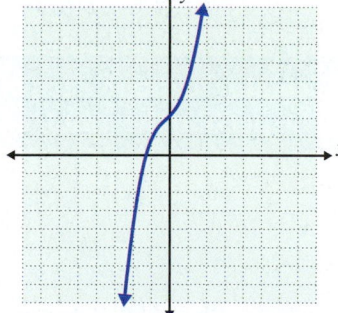

57.

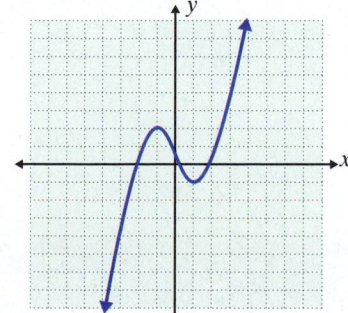

58.

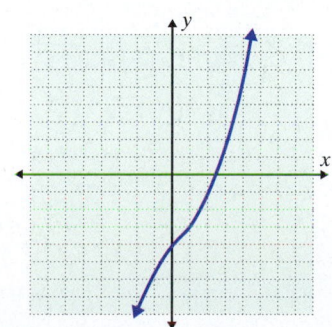

59.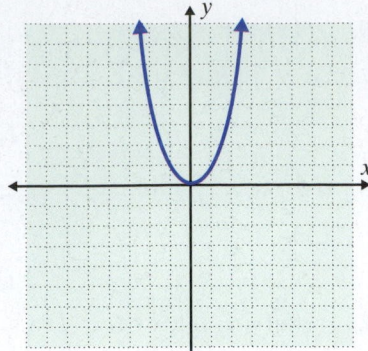

60.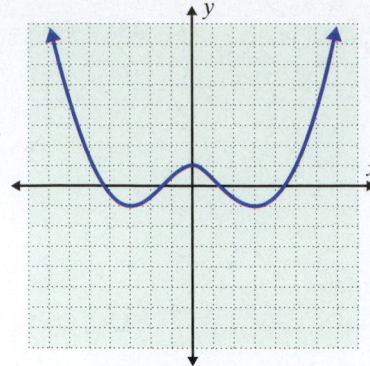

61.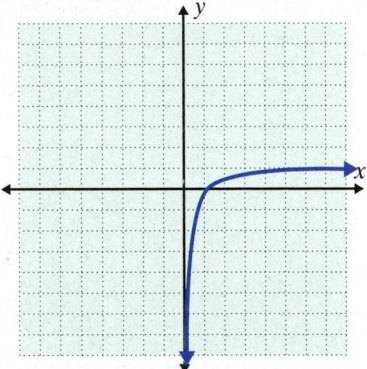

62.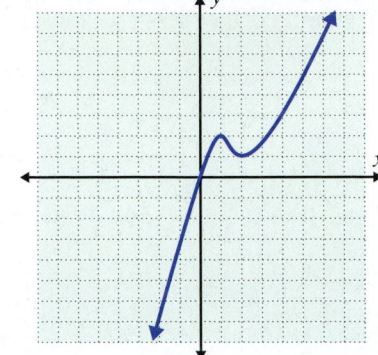

63.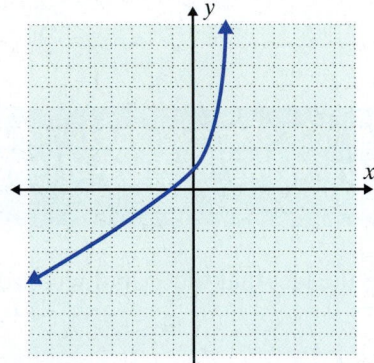

64.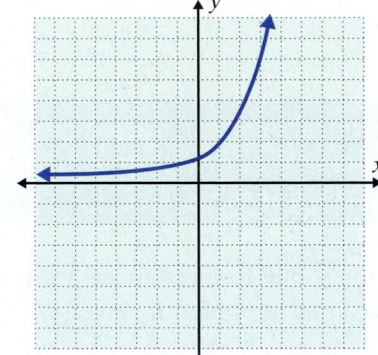

 Use a graphing calculator to graph each of the functions and determine which of the functions are one-to-one by inspecting the graph and using the horizontal line test.

65. $f(x) = 2x + 3$

66. $f(x) = 7 - 4x$

67. $g(x) = x^2 - 2$

68. $g(x) = 9 - x^2$

69. $f(x) = 4 - x^3$

70. $f(x) = x^3 + 2$

71. $f(x) = \dfrac{4}{x}$

72. $g(x) = \dfrac{1}{x}$

73. $g(x) = \sqrt{x - 3}$

74. $f(x) = \sqrt{x + 5}$

75. $f(x) = |x + 1|$

76. $f(x) = |x - 5|$

77. $f(x) = x^3$

78. $f(x) = (x+1)^3$

79. $f(x) = \dfrac{1}{x-3}$

80. $f(x) = \dfrac{1}{x}$

81. $f(x) = x^2, \; x \geq 0$

82. $f(x) = x^2 + 2, \; x \geq 0$

83. $g(x) = x^3 + 2$

84. $g(x) = 6 - x^3$

85. $f(x) = \sqrt{x+5}, \; x \geq -5$

86. $g(x) = \sqrt{x-3}, \; x \geq 3$

87. $f(x) = -x^2 + 1, \; x \geq 0$

88. $g(x) = -x^2 - 2, \; x \geq 0$

Writing & Thinking

89. Explain in your own words, why the domains of the two composite functions $f(g(x))$ and $g(f(x))$ might not be the same. Give an example of two functions that illustrate this possibility.

90. Explain briefly why a function must be one-to-one to have an inverse.

10.3 Exponential Functions

Exponential Functions

You may have read that the population of the world is growing exponentially or studied the exponential growth of bacteria in a biology class. Radioactive materials decay exponentially and never actually disappear. The graph in Figure 1 illustrates that exponential growth has a relatively slow beginning and then builds at an exceedingly rapid rate. This can be extremely important to a doctor trying to curb the growth of "bad" bacteria in a patient.

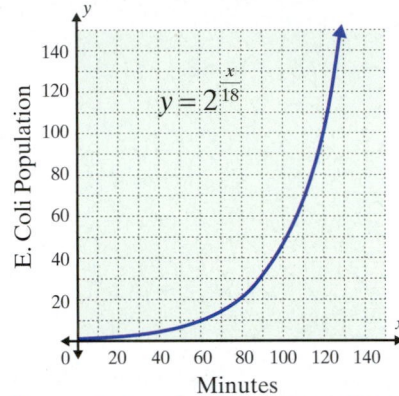

$$y = 2^{\frac{x}{18}}$$

The population of certain strains of E. coli doubles every 18 minutes under optimal conditions.

Figure 1

Quadratic functions have a variable base and a constant exponent, as in $f(x) = x^2$. However, in **exponential functions**, the base is constant and the exponent is a variable, as in $f(x) = 2^x$. As we will see, these two types of functions have major differences in their characteristics. Exponential functions are defined as follows.

> ## Exponential Functions
>
> An **exponential function** is a function of the form
>
> $$f(x) = b^x$$
>
> where $b > 0$, $b \neq 1$, and x is any real number.

Examples of exponential functions are:

$$f(x) = 2^x, \quad f(x) = 3^x, \quad \text{and} \quad y = \left(\frac{1}{3}\right)^x.$$

Objective B   Exponential Growth

The following table of values and the graphs of the corresponding points give a very good idea of what the graph of the **exponential growth** function $y = 2^x$ looks like (see Figure **2a.**). Because we know that 2^x is defined for all real exponents, points such as $\left(\sqrt{2}, 2^{\sqrt{2}}\right)$, $\left(\pi, 2^{\pi}\right)$, and $\left(\sqrt{5}, 2^{\sqrt{5}}\right)$ are on the graph, and the graph for $f(x) = 2^x$ is a smooth curve as shown in Figure **2b.**

x	$y = 2^x$
3	$2^3 = 8$
2	$2^2 = 4$
1	$2^1 = 2$
$\dfrac{1}{2}$	$2^{\frac{1}{2}} = \sqrt{2} \approx 1.4142$
0	$2^0 = 1$
$-\dfrac{1}{2}$	$2^{-\frac{1}{2}} = \dfrac{1}{\sqrt{2}} \approx 0.7071$
-1	$2^{-1} = \dfrac{1}{2}$
-2	$2^{-2} = \dfrac{1}{2^2} = \dfrac{1}{4}$
-3	$2^{-3} = \dfrac{1}{2^3} = \dfrac{1}{8}$
-4	$2^{(-4)} = \dfrac{1}{2^4} = \dfrac{1}{16}$

$\text{Domain} = (-\infty, \infty)$
$\text{Range} = (0, \infty)$

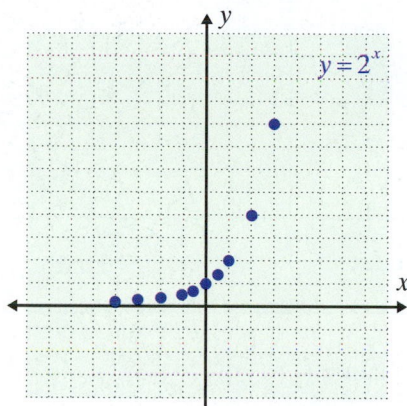

Figure 2a

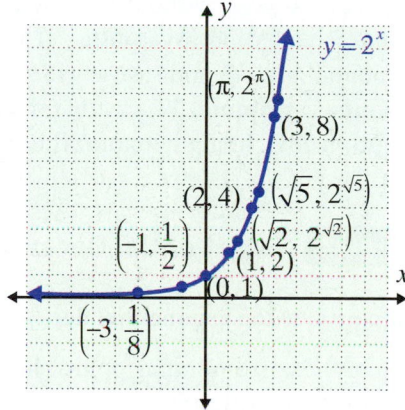

Figure 2b

Figure 3 shows a table of values and the graph of the function $y = 3^x$. Note that the graphs of $y = 2^x$ and $y = 3^x$ are quite similar, but that the graph of $y = 3^x$ rises faster. That is, the **exponential growth is faster if the base is larger**.

x	y = 3x
2	$3^2 = 9$
1	$3^1 = 3$
0	$3^0 = 1$
−1	$3^{-1} = \dfrac{1}{3} \approx 0.3333$
−3	$3^{-3} = \dfrac{1}{27} \approx 0.0370$

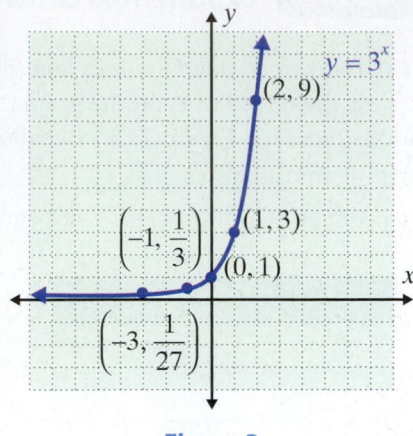

Figure 3

Domain = $(-\infty, \infty)$

Range = $(0, \infty)$

Notice that in both graphs the curves tend to get **very close** to the line $y = 0$ (the x-axis) without ever touching the x-axis. When this happens, the line is called an **asymptote**. If the line is horizontal, as in the cases of exponential growth and (as we will see) exponential decay, the line is called a **horizontal asymptote**. We say that the curve (or function) approaches the line **asymptotically**. In mathematics, this phenomenon happens frequently.

Objective C **Exponential Decay**

Now consider the **exponential decay** function $f(x) = \left(\dfrac{1}{2}\right)^x$. The table and the graph of the corresponding points shown in Figure 4 indicate the nature of the graph of this function.

x	$y = \left(\dfrac{1}{2}\right)^x = 2^{-x}$
-3	$2^{-(-3)} = 2^3 = 8$
-2	$2^{-(-2)} = 2^2 = 4$
-1	$2^{-(-1)} = 2^1 = 2$
$-\dfrac{1}{2}$	$2^{-\left(-\frac{1}{2}\right)} = 2^{\frac{1}{2}} = \sqrt{2} \approx 1.4142$
0	$2^{-0} = 2^0 = 1$
$\dfrac{1}{2}$	$2^{-\frac{1}{2}} = \dfrac{1}{\sqrt{2}} \approx 0.7071$
1	$2^{-1} = \dfrac{1}{2}$
2	$2^{-2} = \dfrac{1}{2^2} = \dfrac{1}{4}$
3	$2^{-3} = \dfrac{1}{2^3} = \dfrac{1}{8}$

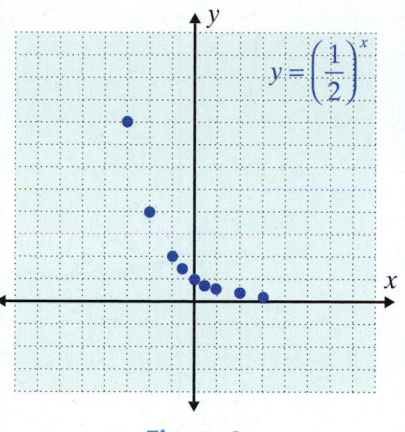

Figure 4

Domain = $(-\infty, \infty)$
Range = $(0, \infty)$

notes

- Because $\dfrac{1}{2} = 2^{-1}$, $\dfrac{1}{3} = 3^{-1}$, $\dfrac{1}{4} = 4^{-1}$, and so on, for fractions between

- 0 and 1, we can write an exponential function with a fractional base between 0 and 1 (these are exponential decay functions) in the form of an exponential function with a base greater than 1 and a negative exponent. Thus we write

$$y = \left(\frac{1}{2}\right)^x = \left(2^{-1}\right)^x = 2^{-x} \qquad \text{and} \qquad y = \left(\frac{1}{3}\right)^x = \left(3^{-1}\right)^x = 3^{-x}.$$

Figure 5 shows the complete graphs of the two exponential decay functions

$$y = \left(\frac{1}{2}\right)^x = 2^{-x} \qquad \text{and} \qquad y = \left(\frac{1}{3}\right)^x = 3^{-x}.$$

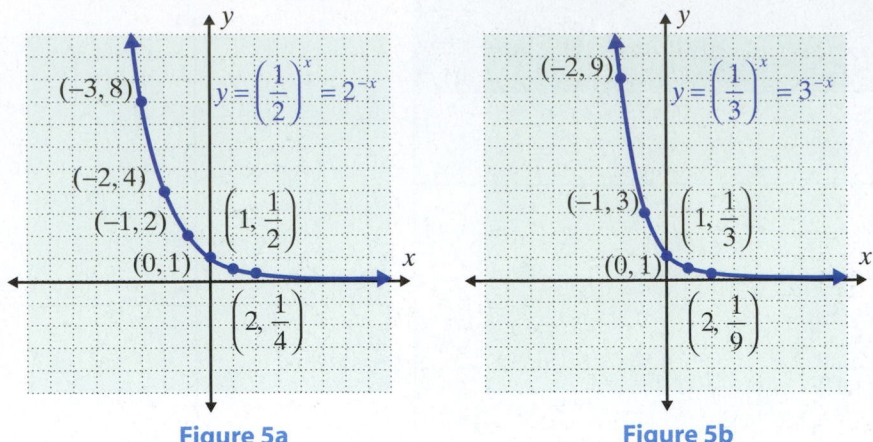

Figure 5a   **Figure 5b**

Note again, in Figure 5, that the line $y = 0$ (the x-axis) is a **horizontal asymptote** for both curves. For exponential growth the curves approach the asymptote as x moves further in the negative direction as in Figures 2b and 3. For exponential decay, the graph approaches the asymptote as x moves further in the positive direction as in Figure 5.

The following general concepts are helpful in understanding the graphs and the nature of exponential functions, both exponential growth and exponential decay.

General Concepts of Exponential Functions

For $b > 1$:

1. $b^x > 0$
2. b^x increases to the right and is called an **exponential growth function**.
3. $b^0 = 1$, so $(0, 1)$ is the y-intercept.
4. b^x approaches the x-axis for negative values of x. (The x-axis is a horizontal asymptote. **See Figure 3**.)

For $0 < b < 1$:

1. $b^x > 0$
2. b^x decreases to the right and is called an **exponential decay function**.
3. $b^0 = 1$, so $(0, 1)$ is the y-intercept.
4. b^x approaches the x-axis for positive values of x. (The x-axis is a horizontal asymptote. **See Figure 5**.)

As with all functions, exponential functions can be multiplied by constants, shifted horizontally, and shifted vertically. (**Note:** We will also do this with parabolas in a future chapter.) Thus,

$$y = a \cdot b^{r \cdot x}, \quad y = b^{x-h}, \quad \text{and} \quad y = b^x + k$$

and various combinations of these expressions are all exponential functions.

Objective D **Bacterial Growth**

Exponential functions are related to many practical applications, among which are bacterial growth, radioactive decay, compound interest, and light absorption. For example, a bacteria culture kept at a certain temperature may grow according to the exponential function

$$y = y_0 \cdot 2^{0.5t}$$

where t = time in hours and

y_0 = amount of bacteria present when $t = 0$.

(y_0 is called the initial value of y.)

Example 1

 **Bacterial Growth**

a. A scientist has 10,000 bacteria present when $t = 0$. She knows the bacteria grow according to the function $y = y_0 \cdot 2^{0.5t}$ where t is measured in hours. How many bacteria will be present at the end of one day?

Solution

Substitute $t = 24$ hours and $y_0 = 10,000$ into the function.

$$y = 10,000 \cdot 2^{0.5(24)} = 10,000 \cdot 2^{12}$$
$$= 10,000(4096)$$
$$= 40,960,000$$
$$= 4.096 \times 10^7 \text{ bacteria}$$

You could also use your TI-84 Plus graphing calculator by entering the numbers as shown in the following display. Press **ENTER** to get the result.

```
10000*2^(0.5*24)
            40960000
```

b. Use the formula for exponential growth, $y = y_0 b^t$, to determine the exponential function that fits the following information: $y_0 = 5000$ bacteria, and there are 135,000 bacteria present after 3 days.

Solution

Use $y = y_0 b^t$ where t is measured in days. Substitute 135,000 for y, 3 for t, and 5000 for y_0, then solve for b.

1. How many bacteria will be present at the end of 36 hours?

$t = 36$ hours and $y_0 = 10{,}000$

$$135{,}000 = 5000b^3$$
$$27 = b^3$$
$$\sqrt[3]{(27)} = \sqrt[3]{(b^3)}$$
$$3 = b$$

The function is $y = 5000 \cdot 3^t$.

Now work margin exercise 1.

Objective E **Compound Interest**

The topic of compound interest (interest paid on interest) leads to a particularly interesting (and useful) exponential function. The formula $A = P(1+r)^t$ can be used for finding the value (amount) accumulated when a principal P is invested and interest is compounded once a year. If compounding is performed more than once a year, we use the following formula to find A.

Compound Interest

Compound interest on a principal P invested at an annual interest rate r (in decimal form) for t years that is compounded n times per year can be calculated using the following formula.

$$A = P\left(1 + \frac{r}{n}\right)^{nt}$$

where A is the amount accumulated.

Example 2

 Compound Interest

a. If P dollars are invested at a rate of interest r (in decimal form) compounded annually (once a year, $n = 1$) for t years, the formula for the amount A becomes $A = P\left(1 + \dfrac{r}{1}\right)^{1 \cdot t} = P(1+r)^t$. Find the value of \$1000 invested at $r = 6\% = 0.06$ for 3 years.

Solution

We have $P = 1000$, $r = 0.06$, and $t = 3$.
$$A = 1000(1 + 0.06)^3$$
$$= 1000(1.06)^3$$
$$= 1000(1.191016) \approx \$1191.02$$

Note: To use your calculator to evaluate $(1.06)^3$, enter 1.06, press the key, enter 3, then press the  key.

b. Find the value of A if $1000 is invested at 6% for 3 years and interest is compounded quarterly. (**Note:** Banks and savings institutions often use 360 days per year.)

Solution

Use the formula for compound interest.
We have $P = 1000, r = 0.06, n = 4,$ and $t = 3$.

$$A = 1000\left(1 + \frac{0.06}{4}\right)^{4(3)}$$

$$= 1000(1.015)^{12}$$

$$= 1000(1.195618171...) \qquad \text{using a calculator}$$

$$\approx \$1195.62$$

c. What will be the value of a principal investment of $1000 invested at 6% for 3 years if interest is compounded monthly (12 times per year)?

Solution

Use the formula for compound interest.
We have $P = 1000, r = 0.06, n = 12,$ and $t = 3$.

$$A = 1000\left(1 + \frac{0.06}{12}\right)^{12(3)}$$

$$= 1000(1 + 0.005)^{36}$$

$$= 1000(1.005)^{36}$$

$$= 1000(1.196680524...) \qquad \text{using a calculator}$$

$$\approx \$1196.68$$

Now work margin exercise 2.

Examples **2a.**, **2b.**, and **2c.** illustrate the effects of compounding interest more frequently over 3 years. The formula gives the results

$$A = \$1191.02 \quad \text{for} \quad n = 1 \text{ (once a year)},$$
$$A = \$1195.62 \quad \text{for} \quad n = 4 \text{ (quarterly)},$$

These numbers might not seem very dramatic, only a difference of $5.66 for 3 years; but, if you use your calculator, in 20 years you will see a difference of $103.07 for a $1000 investment. An investment of $10,000 for 20 years at 9% will show a difference of $4047.41 The results show that more frequent compounding will result in higher income.

2. Find the compound interest.

a. Find the value of $2000 compounded annually invested at $r = 5\%$ for 2 years.

b. Find the value of $2000 invested at $r = 5\%$ for 2 years and is compounded monthly.

If interest is **compounded continuously** (even faster than every second), then the irrational number e $(e \approx 2.718)$ can be shown to be the base of the corresponding exponential function for calculating interest. Table 1 shows how the expression $\left(1+\dfrac{1}{n}\right)^n$ changes as n takes on larger and larger values. The number e is the **limit** (or **limiting value**) of the expression "as n approaches infinity" $(n \to \infty)$ and we write $e = \lim\limits_{n \to \infty}\left(1+\dfrac{1}{n}\right)^n$. Study the following table to help understand the ideas.

Values of the expression $\left(1+\dfrac{1}{n}\right)^n$ as n approaches infinity $(n \to \infty)$		
n	$\left(1+\dfrac{1}{n}\right)$	$\left(1+\dfrac{1}{n}\right)^n$
1	$\left(1+\dfrac{1}{1}\right)=2$	$2^1 = 2$
2	$\left(1+\dfrac{1}{2}\right)=1.5$	$(1.5)^2 = 2.25$
5	$\left(1+\dfrac{1}{5}\right)=1.2$	$(1.2)^5 = 2.48832$
10	$\left(1+\dfrac{1}{10}\right)=1.1$	$(1.1)^{10} \approx 2.59374246$
100	$\left(1+\dfrac{1}{100}\right)=1.01$	$(1.01)^{100} \approx 2.704813829$
1000	$\left(1+\dfrac{1}{1000}\right)=1.001$	$(1.001)^{1000} \approx 2.716923932$
10,000	$\left(1+\dfrac{1}{10,000}\right)=1.0001$	$(1.0001)^{10,000} \approx 2.718145927$
100,000	$\left(1+\dfrac{1}{100,000}\right)=1.00001$	$(1.00001)^{100,000} \approx 2.718268237$
$\downarrow$		$\downarrow$
∞		$e = 2.718281828459...$

Table 1

This gives us the following definition of e. (Be aware that these are very sophisticated mathematical concepts and it will take some careful reading to understand how to arrive at e and the formula $A = Pe^{rt}$.)

The Number e

The number e is defined to be

$$e = 2.718281828459 \ldots$$

Note: The number e is on the TI-84 Plus calculator above the divide key.

As we know, the formula for compound interest is

$$A = P\left(1 + \frac{r}{n}\right)^{nt}.$$

This formula for compound interest takes a different form when interest is **compounded continuously**. The new form involves the number e and is stated below.

Continuously Compounded Interest

Continuously compounded interest on a principal P invested at an annual interest rate r for t years can be calculated using the following formula:

$$A = Pe^{rt}$$

where A is the amount accumulated.

As illustrated in Example 3, a calculator is needed to use the formula for continuously compounded interest.

3. Find the value of $1500 invested at 5.5% for 4 years if interest is compounded continuously.

Example 3

 Calculating Continuously Compounded Interest Using a TI-84 Plus Calculator

Find the value of $1000 invested at 6% for 3 years if interest is compounded continuously. (In this case, $P = \$1000$, $r = 6\% = 0.06$, and $t = 3$.)

Solution

Press $\boxed{\text{2ND}}$ and $\boxed{\text{LN}}$ and $e^{\wedge}($ will appear on the display.

To find the value of $A = Pe^{rt} = 1000e^{0.06 \cdot 3}$
enter the numbers as shown and press $\boxed{\text{ENTER}}$ to get the result.

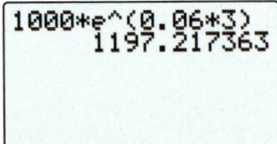

```
1000*e^(0.06*3)
            1197.217363
```
The entire exponent must be in parentheses.

Thus the value of $1000 compounded continuously at 6% for 3 years will be $1197.22. (Note that from Example 2c there is only a 54 cent gain in A when $1000 is compounded continuously instead of monthly at 6% for 3 years.)

Now work margin exercise **3.**

Practice Problems

1. Sketch the graph of the exponential function $f(x) = 2 \cdot 3^x$ and label 3 points on the graph.

2. Sketch the graph of the exponential decay function $y = 0.5 \cdot 2^{-x}$ and label 3 points on the graph.

 3. Find the value of $5000 invested at 8% for 10 years if interest is **a.** compounded monthly, **b.** compounded continuously.

Practice Problem Answers

1. 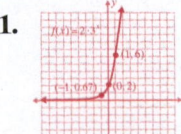 **2.** 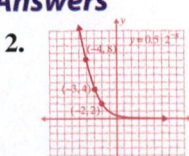 **3. a.** $11,098.20, **b.** $11,127.70

Exercises 10.3

1. $y = 4^x$

2. $y = 5^x$

3. $y = \left(\dfrac{1}{3}\right)^x$

4. $y = \left(\dfrac{1}{5}\right)^x$

5. $y = \left(\dfrac{2}{3}\right)^x$

6. $y = \left(\dfrac{5}{2}\right)^x$

7. $y = \left(\dfrac{1}{2}\right)^{-x}$

8. $y = \left(\dfrac{3}{4}\right)^{-x}$

9. $y = 2^{x-1}$

10. $y = 3^{x+1}$

11. $f(x) = 2^x + 1$

12. $f(x) = 3^x - 1$

13. $f(x) = -4^{-x}$

14. $g(x) = -2^{-x}$

15. $f(x) = 2^{0.5x}$

16. $g(x) = 10^{0.5x}$

17. $f(x) = 4^{-x} - 1$

18. $g(x) = 10^{-x} - 3$

19. $f(x) = 3 \cdot \left(\dfrac{1}{2}\right)^{x+1}$

20. $y = -4 \cdot \left(\dfrac{1}{3}\right)^{x-1}$

21. If $f(t) = 3 \cdot 4^t$ what is the value of $f(2)$?

22. For $f(x) = 3 \cdot 10^{2x}$, find the value of $f(0.5)$.

 Use your calculator to find the value as indicated. Round your answer to the nearest hundredth.

23. Find $f(2)$ if $f(x) = 27.3 \cdot e^{-0.4x}$.

24. Find $f(3)$ if $f(x) = 41.2 \cdot e^{-0.3x}$.

25. Find $f(9)$ if $f(t) = 2000 \cdot e^{0.08t}$.

26. Find $f(22)$ if $f(t) = 2000 \cdot e^{0.05t}$.

 Solve the following word problems. See Examples 1 through 3.

27. Bacteria Growth: A biologist knows that in the laboratory, bacteria in a culture grow according to the function $y = y_0 \cdot 5^{0.2t}$, where y_0 is the initial number of bacteria present and t is time measured in hours. How many bacteria will be present in a culture at the end of 5 hours if there were 5000 present initially?

28. Bacteria Growth: Referring to Exercise 27, how many bacteria were present initially if at the end of 15 hours, there were 2,500,000 bacteria present?

29. Savings Accounts: Four thousand dollars is deposited into a savings account with a rate of 8% per year. Find the total amount A on deposit at the end of 5 years if the interest is compounded

a. annually.

b. semiannually.

c. quarterly.

d. continuously.

30. Savings Accounts: Find the amount A in a savings account if $2000 is invested at 7% for 4 years and the interest is compounded

a. annually.

b. semiannually.

c. quarterly.

d. continuously.

31. Investing: Find the value of $1800 invested at 6% for 3 years if the interest is compounded continuously.

32. Investing: Find the value of $2500 invested at 5% for 5 years if the interest is compounded continuously.

33. Sales Revenue: The revenue function is given by $R(x) = x \cdot p(x)$ dollars, where x is the number of units sold and $p(x)$ is the unit price. If $p(x) = 25(2)^{\frac{-x}{5}}$, find the revenue if 15 units are sold.

34. Sales Revenue: Referring to Exercise 33, if $p(x) = 40(3)^{\frac{-x}{6}}$, find the revenue if 12 units are sold.

35. Broadcasting: A radio station knows that during an intense advertising campaign, the number of people N who will hear a commercial is given by $N = A\left(1 - 2^{-0.05t}\right)$, where A is the number of people in the broadcasting area and t is the number of hours the commercial has been run. If there are 500,000 people in the area, how many will hear a commercial during the first 20 hours?

36. Investing: Bethany invested $45,000 in a retirement fund that earns 8% interest and is compounded continuously, how much money will the account be worth after:

a. 10 years?

b. 20 years?

c. 40 years?

37. Battery Reliability: Statistics show that the fractional part of flashlight batteries f that are still good after t hours of use is given by $f = 4^{-0.02t}$. What fractional part of the batteries are still operating after 150 hours of use?

$t = 0$

$t = 150$

38. Investing: If a principal P is invested at a rate r compounded continuously, the interest earned is given by $I = A - P$.

a. Find the interest earned in 20 years on $10,000 invested at 10% and compounded continuously?

b. Find the interest earned in 20 years on $10,000 invested at 5% and compounded continuously.

c. Explain why the interest earned at 5% is not just one-half of the interest earned at 10% in parts **a.** and **b.**

39. Machine Value: The value V of a machine at the end of t years is given by $V = C\left(1 - r\right)^t$, where C is the original cost and r is the rate of depreciation. Find the value of a machine at the end of 4 years if the original cost was $1200 and $r = 0.20$.

40. Machine Value: Referring to Exercise 39, find the value of a machine at the end of 3 years if the original cost was $2000 and $r = 0.15$.

41. Drug Absorption: A cancer patient is given a dose of 50 mg of a particular drug. In five days the amount of the drug in her system is reduced to 1.5625 mg. If the drug decays (or is absorbed) at an exponential rate, find the function that represents the amount of the drug. (**Hint:** Use the formula $y = y_0 b^{-t}$ and solve for b.)

42. Growth of Cancer Cells: Determine the exponential function that fits the following information concerning exponential growth of cancer cells: $y_0 = 10,000$ cancer cells, and there are 160,000 cancer cells present after 4 days. (**Hint:** Use the formula $y = y_0 b^t$ and solve for b.)

Use a graphing calculator to graph each of the following functions. In each case the x-axis is a horizontal asymptote.

43.   **a.** $y = e^x$

b. $y = e^{-x}$

c. $y = e^{-x^2}$

10.4 Logarithmic Functions

Objectives

A Understand the concept of a logarithm, and understand the relationship between logarithms and exponential functions.

B Use the basic properties of logarithms to evaluate logarithms.

C Graph logarithmic functions and exponential functions on the same set of axes.

Objective A Logarithms

You have learned to solve many different algebraic expressions in this textbook. However, you still cannot solve $y = 3^x$ for x. In order to solve equations of this form, we need a function called a **logarithm**.

Definition of Logarithm (base b)

For all positive values of x and for all positive values of b (except $b \neq 1$), the expression $b^y = x$ can be written as $y = \log_b x$. The function $f(x) = \log_b x$ is the **logarithmic function** with base b.

$\log_b x = y$ is read "the log, base b, of x equals y."

---

notes

■ Two very important things to remember:

■ 1. A logarithm is an exponent.
   2. You cannot take the logarithm of a negative number.

■

---

1. Write $2^3 = 8$ in logarithmic form.

Example 1

Writing an Exponential Expression as a Logarithmic Expression

a. $5^2 = 25$ becomes $\log_5 25 = 2$

b. $3^4 = 81$ becomes $\log_3 81 = 4$

c. $6^0 = 1$ becomes $\log_6 1 = 0$

d. $4^a = 16$ becomes $\log_4 16 = a$

Now work margin exercise 1.

---

2. Write $\log_2 32 = 5$ in exponential form.

Example 2

Writing a Logarithmic Expression as an Exponential Expression

a. $\log_3 9 = 2$ is written as $3^2 = 9$ Is this a true statement? Always do a mental check.

b. $\log_2 16 = 4$ is written as $2^4 = 16$

c. $\log_e 1 = 0$ is written as $e^0 = 1$

d. $\log_{10} 100 = 2$ is written as $10^2 = 100$

Now work margin exercise 2.

Evaluating Logarithms

Example 3

Evaluating Logarithms

Evaluate $\log_4 4$.

Solution

Step 1: Set the logarithmic expression equal to x.

$$\log_4 4 = x$$

Step 2: Rewrite the expression in exponential form.

$$4^x = 4$$

Step 3: Ask yourself, 4 raised to what power is equal to 4?

$$x = 1 \text{ because } 4^1 = 4$$

Step 4: So $\log_4 4 = 1$ (because the logarithm is an exponent).

Example 4

Evaluating Logarithms

Evaluate $\log_2 8$.

Solution

Step 1: Set the logarithmic expression equal to x.

$$\log_2 8 = x$$

Step 2: Rewrite the expression in exponential form.

$$2^x = 8$$

Step 3: Ask yourself, 2 raised to what power is equal to 8?

$$x = 3 \text{ because } 2^3 = 8$$

Step 4: So $\log_2 8 = 3$ (because the logarithm is an exponent).

Evaluate the following logarithms.

3. $\log_3 27$

4. $\log_5 1$

5. $\log_{36} 6$

6. Solve by first changing the equation to exponential form.

$$\log_4 x = \frac{1}{2}$$

Example 5

Evaluating Logarithms

Evaluate $\log_{25} 5$.

Solution

Step 1: Set the logarithmic expression equal to x.

$$\log_{25} 5 = x$$

Step 2: Rewrite the expression in exponential form.

$$25^x = 5$$

Step 3: Ask yourself, what power of 25 is equal to 5?

$$x = \frac{1}{2}, \text{ because } \sqrt{25} = 5$$

Step 4: So $\log_{25} 5 = \frac{1}{2}$ (because the logarithm is an exponent).

Now work margin exercises 3 through 5.

Example 6

Using the Exponential Form to Solve Logarithmic Equations

Solve the following equation by first changing the equation to exponential form.

$$\log_{16} x = \frac{3}{4}$$

Solution

$$\log_{16} x = \frac{3}{4} \qquad \text{Write the equation in exponential form and solve for } x.$$

$$x = 16^{\frac{3}{4}}$$

$$x = \left(16^{\frac{1}{4}}\right)^3 = 2^3 = 8$$

Thus $\log_{16} 8 = \frac{3}{4}$.

Now work margin exercise 6.

Objective C **Graphs of Logarithmic Functions**

As illustrated in Figure 1, because logarithmic functions are the inverses of exponential functions, the graphs of logarithmic functions can be found by reflecting the corresponding exponential functions across the line $y = x$. Figure **1a** shows how the graphs of $y = 2^x$ and $y = \log_2 x$ are related. Figure **1b** shows how the graphs of $y = 10^x$ and $y = \log_{10} x$ are related. Note that in the graphs of both logarithmic functions the values of y are negative when x is between 0 and 1 $(0 < x < 1)$.

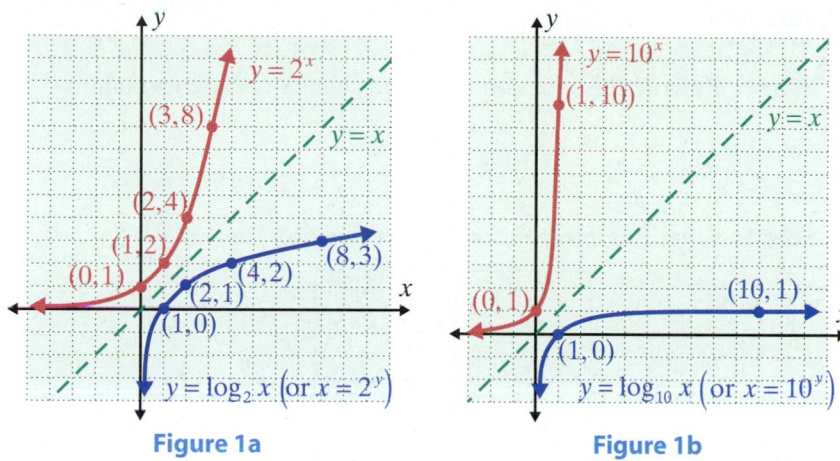

Figure 1a   **Figure 1b**

Recall that points on the graphs of inverse functions can be found by reversing the coordinates of ordered pairs. This means that the domain and range of a function and its inverse are interchanged. Thus for exponential functions and logarithmic functions, we have the following.

- For the exponential function $y = b^x$,

  the domain is all real x, and
  the range is all $y > 0$.   (The graph is above the x-axis.)
  There is a horizontal asymptote at $y = 0$.

- For the logarithmic function $y = \log_b x$ $\left(\text{or } x = b^y\right)$,

  the domain is all $x > 0$, and   (The graph is to the right of the y-axis.)
  the range is all real y.
  There is a vertical asymptote at $x = 0$.

In order to graph logarithmic functions with bases other than 10 or e, we need to use the **change-of-base formula**.

Change-of-Base Formula

For $a, b, x > 0$ and $a, b \neq 1$,

$$\log_b x = \frac{\log_a x}{\log_a b}.$$

Because graphing calculators are designed to graph base 10 or base e logarithms, we will do examples changing to these bases. The log key on your calculator is base 10. Base 10 logarithms are referred to as common logarithms.

Example 7

Graphing Logarithms Using the Change-of-Base Formula

Graph $y = \log_5 x$ using the change-of-base formula.

Solution

Step 1: Rewrite the equation using the change-of-base formula.

$$y = \log_5 x = \frac{\log x}{\log 5}$$

Step 2: Enter this into your graphing calculators as $y = \frac{\log x}{\log 5}$.

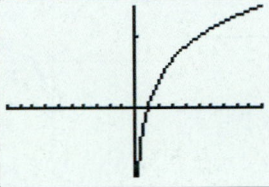

Graph the equation $y = \log_9 x$ using the change-of-base formula as indicated.

7. Use the common logarithm, log, in the change-of-base formula.

8. Use the natural logarithm, ln, in the change-of-base formula.

Example 8

Graphing Logarithms Using the Change-of-Base Formula

Graph $y = \log_5 x$ using the ln in change-of-base formula. The ln key uses the natural base, e, and results in the same graph.

Solution

Step 1: Rewrite the equation using the change-of-base formula.

$$y = \log_5 x = \frac{\ln x}{\ln 5}$$

Step 2: Enter this into your graphing calculators as $y = \frac{\ln x}{\ln 5}$.

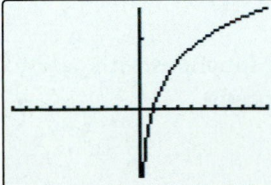

Now work margin exercises 7 and 8.

Practice Problems

Express the given equation in logarithmic form.

1. $4^2 = 16$ **2.** $5^3 = 125$ **3.** $10^{-1} = \dfrac{1}{10}$

Express the given equation in exponential form.

4. $\log_2 x = -1$ **5.** $\log_5 x = 2$ **6.** $\log_3 x = 0$

Find the value of each expression.

7. $\log_2 64$ **8.** $\log_3 27$ **9.** $\log_7 7$

Practice Problem Answers

1. $\log_4 16 = 2$ **2.** $\log_5 125 = 3$ **3.** $\log_{10} \dfrac{1}{10} = -1$ **4.** $x = 2^{-1}$

5. $x = 5^2$ **6.** $x = 3^0$ **7.** 6 **8.** 3 **9.** 1

Exercises 10.4

1. $7^2 = 49$

2. $3^3 = 27$

3. $5^{-2} = \dfrac{1}{25}$

4. $2^{-5} = \dfrac{1}{32}$

5. $\pi^0 = 1$

6. $6^0 = 1$

7. $10^2 = 100$

8. $10^1 = 10$

9. $10^k = 23$

10. $4^k = 11.6$

11. $\left(\dfrac{2}{3}\right)^2 = \dfrac{4}{9}$

12. $\left(\dfrac{3}{4}\right)^2 = \dfrac{9}{16}$

Express each equation in exponential form. See Example 2.

13. $\log_3 9 = 2$

14. $\log_5 125 = 3$

15. $\log_9 3 = \dfrac{1}{2}$

16. $\log_b 4 = \dfrac{2}{3}$

17. $\log_7 \dfrac{1}{7} = -1$

18. $\log_{1/2} 8 = -3$

19. $\log_{10} N = 1.74$

20. $\log_2 42.3 = x$

21. $\log_b 18 = 4$

22. $\log_b 39 = 10$

23. $\log_n y^2 = x$

24. $\log_b a = x^2$

Solve by first changing each equation to exponential form. See Example 6.

25. $\log_4 x = 2$

26. $\log_3 x = 4$

27. $\log_{14} 196 = x$

28. $\log_{25} 125 = x$

29. $\log_5 \dfrac{1}{125} = x$

30. $\log_3 \dfrac{1}{9} = x$

31. $\log_{36} x = -\dfrac{1}{2}$

32. $\log_{81} x = -\dfrac{3}{4}$

33. $\log_x 32 = 5$

34. $\log_x 121 = 2$

35. $\log_8 x = \dfrac{5}{3}$

36. $\log_{16} x = \dfrac{3}{4}$

37. $\log_8 8^{3.7} = x$

38. $\log_{10} 10^{1.52} = x$

39. $\log_5 5^{\log_5 25} = x$

40. $\log_4 4^{\log_2 8} = x$

41. $f(x) = 6^x$

42. $f(x) = 2^x$

43. $y = \left(\dfrac{2}{3}\right)^x$

44. $y = \left(\dfrac{1}{4}\right)^x$

45. $f(x) = \log_4 x$

46. $f(x) = \log_5 x$

47. $y = \log_{1/2} x$

48. $y = \log_{1/3} x$

49. $y = \log_8 x$

50. $y = \log_7 x$

Writing & Thinking

51. Discuss, in your own words, the symmetrical relationship of the graphs of the two functions $y = 10^x$ and $y = \log_{10} x$.

52. Discuss, in your own words, the symmetrical relationship of the graphs of the two logarithmic functions $y = \log_{10} x$ and $y = -\log_{10} x$.

10.5 Properties of Logarithms

Objectives

A Understand the basic properties of logarithms.

B Understand the Product Rule of Logarithms.

C Understand the Quotient Rule of Logarithms.

D Understand the Power Rule of Logarithms.

Objective A **Basic Properties of Logarithms**

We know that **exponents are logarithms** and from our previous knowledge of exponents, we can make the following equivalent statements for logarithms, base b.

$$b^0 = 1 \quad \Leftrightarrow \quad \log_b 1 = 0$$
$$b^1 = b \quad \Leftrightarrow \quad \log_b b = 1$$

Also, directly from the definition of $y = \log_b x$, we can make two more general statements:

$$x = b^{\log_b x} \quad \text{and} \quad \log_b b^x = x.$$

In summary, we have the following four basic properties of logarithms.

Basic Properties of Logarithms

For $b > 0$ and $b \neq 1$,

1. $\log_b 1 = 0$ Regardless of the base, the logarithm of 1 is 0.
2. $\log_b b = 1$ The logarithm of the base is always 1.
3. $x = b^{\log_b x}$ for $x > 0$
4. $\log_b b^x = x$

1. Simplify the following logarithms.

a. $\log_5 25$

b. $\ln e^5$

c. $\log 100$

Example 1

Using Basic Properties of Logarithms

Use the basic properties of logarithms to simplify the following logarithms.

a. $\log_2 8$

Solution

Step 1: Rewrite with an exponential expression: $\log_2 2^3$.
Step 2: Use rule #4 to simplify the expression: $\log_2 2^3 = 3$.

b. $\ln e^a$

Solution

Step 1: Remember that natural logs always have base e: $\ln_e e^a$.
Step 2: Use rule #4 to simplify the expression: $\ln_e e^a = a$.

c. $\log 1000$

Solution

Step 1: Recognize that this is a common logarithm (since it has base 10) and that 1000 can be written exponentially: $\log_{10} 10^3$.
Step 2: Use rule #4 to simplify the expression: $\log_{10} 10^3 = 3$.

Now work margin exercise 1.

Because logarithms are exponents, their properties are similar to those of exponents. In fact, the properties of exponents are used to prove the rules for logarithms. Study the developments carefully and you will see that logarithms are handled just as exponents.

> ## notes
>
> ■ Using a calculator to find the values of logarithms is discussed in the next section. To illustrate some of the properties of logarithms in
> ■ this section, the following base 10 logarithmic values, accurate to the nearest ten-thousandth.
>
> ■
> $$\log_{10} 2 \approx 0.3010 \quad \log_{10} 3 \approx 0.4771$$
> $$\log_{10} 5 \approx 0.6990 \quad \log_{10} 6 \approx 0.7782$$
> ■
>
> ■ Also, because the logarithms are rounded to the nearest ten-thousandth, a slight difference may occur in the answers if a calculator
> ■ is used to find the logarithms in some of the illustrations.
>
> ■

Product Rule for Logarithms

For $b > 0$, $b \neq 1$, and $x, y > 0$,

$$\log_b xy = \log_b x + \log_b y.$$

In words, **the logarithm of a product is equal to the sum of the logarithms of the factors.**

Example 2

Using the Product Rule for Logarithms

Expand each of the following expressions using the product rule.

a. $\log_2 3y$

Solution

$\log_2 3y = \log_2 3 + \log_2 y$

b. $\log_5 3 \cdot 4$

Solution

$\log_5 3 \cdot 4 = \log_5 3 + \log_5 4$

Now work margin exercise 2.

2. Expand the following expressions using the product rule.

a. $\log_8 15k$

b. $\log_9 5 \cdot 8$

Quotient Rule for Logarithms

For $b > 0$, $b \neq 1$, and $x, y > 0$,

$$\log_b \frac{x}{y} = \log_b x - \log_b y.$$

In words, **the logarithm of a quotient is equal to the difference between the logarithm of the numerator and the logarithm of the denominator.**

3. Expand each of the following expressions using the quotient rule.

a. $\log_3 \dfrac{4}{t}$

b. $\log_{10} \dfrac{5}{2}$

Example 3

Using the Quotient Rule for Logarithms

Expand each of the following expressions using the quotient rule.

a. $\log_{10} \dfrac{5}{x}$

Solution

$$\log_{10} \frac{5}{x} = \log_{10} 5 - \log_{10} x$$
$$= \log 5 - \log x$$

Logarithms with base 10 are called common logs, with the base of 10 understood, but not written.

b. $\log_e \dfrac{x}{y}$

Solution

$$\log_e \frac{x}{y} = \log_e x - \log_e y$$
$$= \ln x - \ln y$$

Logarithms with base e are called natural logs, with the base of e understood, but not written. ln is read as "the natural log of."

Now work margin exercise **3.**

Objective D **The Power Rule**

Power Rule for Logarithms

For $b > 0$, $b \neq 1$, $x > 0$, and any real number r,

$$\log_b x^r = r \cdot \log_b x.$$

In words, **the logarithm of a number raised to a power is equal to the product of the exponent and the logarithm of the number.**

Example 4

Using the Power Rule

a. $\log_{10} \sqrt{2}$

Solution

$$\log_{10} \sqrt{2} = \log_{10} 2^{\frac{1}{2}} = \frac{1}{2} \cdot \log_{10} 2 \approx \frac{1}{2}(0.3010) = 0.1505$$

b. $\log_{10} 8$

Solution

$$\log_{10} 8 = \log_{10} 2^3 = 3 \cdot \log_{10} 2 \approx 3(0.3010) = 0.9030$$

c. $\log_{10} 25$

Solution

$$\log_{10} 25 = \log_{10} 5^2 = 2 \cdot \log_{10} 5 \approx 2(0.6990) = 1.3980$$

Now work margin exercise 4.

Summary of Properties of Logarithms

For $b > 0$, $b \neq 1$, $x, y > 0$, and any real number r,

1. $\log_b 1 = 0$
2. $\log_b b = 1$
3. $x = b^{\log_b x}$
4. $\log_b b^x = x$
5. $\log_b xy = \log_b x + \log_b y$ the product rule
6. $\log_b \dfrac{x}{y} = \log_b x - \log_b y$ the quotient rule
7. $\log_b x^r = r \cdot \log_b x$ the power rule

4. Simplify each of the expressions using the power rule.

a. $\log_{10} \sqrt{3}$

b. $\log_{10} 16$

c. $\log_{10} 9$

5. Simplify each of the expressions using the power rule.

a. $\log_b 4x^4$

b. $\log_b \sqrt{4x}$

Example 5

Expanding Logarithms

Use the properties of logarithms to expand each expression as much as possible.

a. $\log_b 2x^3$

Solution

$$\begin{aligned}
\log_b 2x^3 &= \log_b 2 + \log_b x^3 && \text{product rule}\\
&= \log_b 2 + 3\log_b x && \text{power rule}
\end{aligned}$$

b. $\log_b \dfrac{xy^2}{z}$

Solution

$$\begin{aligned}
\log_b \frac{xy^2}{z} &= \log_b xy^2 - \log_b z && \text{quotient rule}\\
&= \log_b x + \log_b y^2 - \log_b z && \text{product rule}\\
&= \log_b x + 2\log_b y - \log_b z && \text{power rule}
\end{aligned}$$

c. $\log_b (xy)^{-3}$

Solution

$$\begin{aligned}
\log_b (xy)^{-3} &= -3\log_b xy && \text{power rule}\\
&= -3\left(\log_b x + \log_b y\right) && \text{product rule}\\
&= -3\log_b x - 3\log_b y
\end{aligned}$$

d. $\log_b \sqrt{3x}$

Solution

$$\begin{aligned}
\log_b \sqrt{3x} &= \log_b (3x)^{\frac{1}{2}} = \frac{1}{2}\log_b 3x && \text{power rule}\\
&= \frac{1}{2}\left(\log_b 3 + \log_b x\right) && \text{product rule}\\
&= \frac{1}{2}\log_b 3 + \frac{1}{2}\log_b x
\end{aligned}$$

Now work margin exercise 5.

Example 6

Using the Properties of Logarithms to Write a Single Logarithmic Expression

Use the properties of logarithms to write each expression as a single logarithm. Note that the bases must be the same when simplifying sums and differences of logarithms.

a. $2\log_b x - 3\log_b y$

Solution

$$2\log_b x - 3\log_b y = \log_b x^2 - \log_b y^3 \qquad \text{power rule}$$

$$= \log_b\left(\frac{x^2}{y^3}\right) \qquad \text{quotient rule}$$

b. $\dfrac{1}{2}\log_a 4 - \log_a 5 - \log_a y$

Solution

$$\frac{1}{2}\log_a 4 - \log_a 5 - \log_a y$$

$$= \log_a 4^{\frac{1}{2}} - \log_a 5 - \log_a y \qquad \text{power rule}$$

$$= \log_a 2 - \left(\log_a 5 + \log_a y\right) \qquad 4^{\frac{1}{2}}=2;\ \text{Factor out } -1.$$

$$= \log_a 2 - \log_a 5y \qquad \text{product rule}$$

$$= \log_a\left(\frac{2}{5y}\right) \qquad \text{quotient rule}$$

c. $\log_a(x+1) + \log_a(x-1)$

Solution

$$\log_a(x+1) + \log_a(x-1) = \log_a\left[(x+1)(x-1)\right] \qquad \text{product rule}$$

$$= \log_a\left(x^2 - 1\right) \qquad \text{multiply}$$

d. $\log_b \sqrt{x} + \log_b \sqrt[3]{x}$

Solution

$$\log_b \sqrt{x} + \log_b \sqrt[3]{x} = \log_b x^{\frac{1}{2}} + \log_b x^{\frac{1}{3}}$$

$$= \frac{1}{2}\log_b x + \frac{1}{3}\log_b x \qquad \text{power rule}$$

$$= \left(\frac{1}{2} + \frac{1}{3}\right)\log_b x \qquad \text{distributive property}$$

$$= \frac{5}{6}\log_b x \qquad \frac{1}{2} + \frac{1}{3} = \frac{3}{6} + \frac{2}{6} = \frac{5}{6}$$

6. Use the properties of logarithms to write each expression as a single logarithm.

a. $4\log_b x - 2\log_b z$

b. $\log_z(x-1) + \log_z(x+2)$

c. $\log_b \sqrt{x} + \log_b x^2$

Or

$$\log_b \sqrt{x} + \log_b \sqrt[3]{x} = \log_b x^{\frac{1}{2}} + \log_b x^{\frac{1}{3}}$$

$$= \log_b \left(x^{\frac{1}{2}} \cdot x^{\frac{1}{3}}\right) \qquad \text{product rule}$$

$$= \log_b x^{\left(\frac{1}{2} + \frac{1}{3}\right)}$$

$$= \log_b x^{\frac{5}{6}}$$

$$= \frac{5}{6}\log_b x \qquad \text{power rule}$$

Now work margin exercise 6.

Common Misunderstandings

Common Misunderstandings about Logarithms

There is no logarithmic property for the logarithm of a sum or a difference.

$\log_b(x+y)$ cannot be simplified

$\log_b(x-y)$ cannot be simplified

Also,

$\log_b(xy) \neq \log_b x \cdot \log_b y$ The log of a product does not equal the product of the logs.

$\log_b \dfrac{x}{y} \neq \dfrac{\log_b x}{\log_b y}$ The log of a quotient does not equal the quotient of the logs.

Practice Problems

1. Write $\log_b(x^3 y)$ as a sum of logarithmic expressions.

2. Write the expression $2\log_b 5 + \log_b x - \log_b 3$ as a single logarithm.

3. Write the expression $2\log_b x - \log_b(x+1)$ as a single logarithm.

Practice Problem Answers

1. $3\log_b x + \log_b y$ **2.** $\log_b\left(\dfrac{25x}{3}\right)$ **3.** $\log_b\left(\dfrac{x^2}{x+1}\right)$

Exercises 10.5

Use the following logarithms (accurate to the nearest ten-thousandth) in exercises 1 and 2.

$$\log_{10} 2 \approx 0.3010 \qquad \log_{10} 3 \approx 0.4771$$
$$\log_{10} 5 \approx 0.6990 \qquad \log_{10} 6 \approx 0.7782$$

1. Find the value of the following expressions.

a. $10^{0.3010}$ **b.** $10^{0.4771}$ **c.** $10^{0.6990}$ **d.** $10^{0.7782}$

2. Find the value of the following expressions.

a. $10^{0.3010 + 0.7782}$ **b.** $10^{0.4771 + 0.7782}$ **c.** $10^{0.4771 + 0.6990}$ **d.** $10^{0.6990 - 0.3010}$

Use your knowledge of logarithms and exponents to find the value of each expression. See Example 1.

3. $\log_2 32$ **4.** $\log_3 9$ **5.** $\log_4 \dfrac{1}{16}$ **6.** $\log_5 \dfrac{1}{125}$

7. $\log_3 \sqrt{3}$ **8.** $\log_2 \sqrt{8}$ **9.** $5^{\log_5 10}$ **10.** $3^{\log_3 17}$

11. $6^{\log_6 \sqrt{3}}$ **12.** $5^{\log_5 5}$

Use the properties of logarithms to expand each expression as much as possible. See Example 5.

13. $\log_b 5x^4$ **14.** $\log_b 3x^2 y$ **15.** $\log_b 2x^{-3} y$ **16.** $\log_5 xy^2 z^{-1}$

17. $\log_6 \dfrac{2x}{y^3}$ **18.** $\log_3 \dfrac{xy}{4z}$ **19.** $\log_b \dfrac{x^2}{yz}$ **20.** $\log_3 \dfrac{xy^2}{z^2}$

21. $\log_5 (xy)^{-2}$ **22.** $\log_b (x^2 y)^4$ **23.** $\log_6 \sqrt[3]{xy^2}$ **24.** $\log_5 \sqrt{2x^3 y}$

25. $\log_3 \sqrt{\dfrac{xy}{z}}$ **26.** $\log_6 \sqrt[3]{\dfrac{x^2}{y}}$ **27.** $\log_5 21x^2 y^{\frac{2}{3}}$ **28.** $\log_b 15x^{-\frac{1}{2}} y^{\frac{1}{3}}$

29. $\log_6 \dfrac{x}{\sqrt{x^3 y^5}}$ **30.** $\log_3 \dfrac{1}{\sqrt{x^4 y}}$ **31.** $\log_b \left(\dfrac{x^3 y^2}{z}\right)^{-3}$ **32.** $\log_4 \left(\dfrac{x^{-\frac{1}{2}} y}{z^2}\right)^{-2}$

> **Use the properties of logarithms to write each expression as a single logarithm of a single expression. See Example 6.**

33. $2\log_b 3 + \log_b x - \log_b 5$ **34.** $\dfrac{1}{2}\log_b 25 + \log_b 3 - \log_b x$

35. $\log_2 7 + \log_2 9 + 2\log_2 x$ **36.** $\log_5 4 + \log_5 6 + \log_5 y$

37. $2\log_b x + \log_b y$ **38.** $\log_2 x + 3\log_2 y$

39. $3\log_5 y - \dfrac{1}{2}\log_5 x$ **40.** $3\log_{10} x - 2\log_{10} y$

41. $\dfrac{1}{2}\left(\log_5 x - \log_5 y\right)$ **42.** $\dfrac{1}{3}\left(\log_{10} x - 2\log_{10} y\right)$

43. $\log_2 x - \log_2 y + \log_2 z$ **44.** $\log_b x - 2\log_b y - 2\log_b z$

45. $\log_b x + 2\log_b y - \dfrac{1}{2}\log_b z$ **46.** $-\dfrac{2}{3}\log_2 x - \dfrac{1}{3}\log_2 y + \dfrac{2}{3}\log_2 z$

47. $2\log_5 x + \log_5 (2x+1)$ **48.** $\log_b (3x+1) + 2\log_b x$

49. $\log_2 (x-1) + \log_2 (x+3)$ **50.** $\log_{10} (x+3) + \log_{10} (x-3)$

51. $\log_b\left(x^2 - 2x - 3\right) - \log_b\left(x - 3\right)$

52. $\log_2\left(x - 4\right) - \log_2\left(x^2 - 2x - 8\right)$

53. $\log_{10}\left(x + 6\right) - \log_{10}\left(2x^2 + 9x - 18\right)$

54. $\log_5\left(3x^2 + 5x - 2\right) - \log_5\left(3x - 1\right)$

Writing & Thinking

55. Prove the quotient rule for logarithms: For $b > 0$, $b \neq 1$, and $x, y > 0$,

$$\log_b \frac{x}{y} = \log_b x - \log_b y.$$

56. Prove the following property of logarithms: For $b > 0$, $b \neq 1$, and $x > 0$,

$$\log_b b^x = x.$$

10.6 Common Logarithms and Natural Logarithms

Objectives

A Understand base 10 logarithms, or common logarithms.

B Know how to find the inverse of a common logarithm.

C Use a calculator to find the inverse logarithm of N.

D Understand base e logarithms, or natural logarithms.

E Know how to find the inverse of a natural logarithm.

Base 10 logarithms (called **common logarithms**) and base e logarithms (called **natural logarithms**) are the two most commonly used logarithms. Common logarithms occur frequently because of their close relationship with the decimal system and are used in chemistry, astronomy, and physical science. Natural logarithms are an important tool in calculus and appear "naturally" in growth and decay problems. Both common and natural logarithms have designated keys on calculators.

Objective A Common Logarithms (Base 10 Logarithms)

The abbreviated notation $\log x$ is used to indicate common logarithms (base 10 logarithms). That is, **whenever the base notation is omitted, the base is understood to be 10**:

$$\log_{10} x = \log x.$$

Example 1

Translations Between Exponential Form and Logarithmic Form Using Common Logarithms

	Exponential Form		Logarithmic Form	
a.	$10^4 = 10,000$	$\Leftrightarrow$	$\log 10,000 = 4$	This is a common logarithm that equals 4.
b.	$10^{-2} = 0.01$	$\Leftrightarrow$	$\log 0.01 = -2$	This is a common logarithm that equals -2.
c.	$10^x = 6.3$	$\Leftrightarrow$	$\log 6.3 = x$	This is a common logarithm that equals x.

Now work margin exercise 1.

1. Translate $10^2 = 100$ into logarithmic form.

The TI-84 Plus graphing calculator has the designated key **LOG** that can be used in the following 3-step process to find the values of common logarithms.

1123   Chapter 10   *Exponential and Logarithmic Functions*

Using a Calculator to Evaluate Common Logarithms

Using a Calculator to Evaluate Common Logarithms (Base 10)

To evaluate a **common logarithm** on a TI-84 Plus graphing calculator:

1. Press the **LOG** key. `log(` will appear on the display.
2. Enter the number and a right-hand parenthesis, .
3. Press **ENTER**.

Example 2

 Evaluating Common Logarithms Using a TI-84 Plus Graphing Calculator

Use a TI-84 Plus graphing calculator to find the values of the following common logarithms.

a. $\log 200$ b. $\log 50,000$ c. $\log 0.0006$

Solutions

From the display we see the results (accurate to the nearest billionth).

a. $\log 200 \approx 2.301029996$

b. $\log 50,000 \approx 4.698970004$

c. $\log 0.0006 \approx -3.22184875$

```
log(200)
        2.301029996
log(50000)
        4.698970004
log(0.0006)
        -3.22184875
```

Now work margin exercise 2.

Example **2c.** shows a negative value for the logarithm. This is because 0.0006 is between 0 and 1. **In fact, the logarithm (with base greater than 1) of any number between 0 and 1 will always be negative.** See Figure 1.

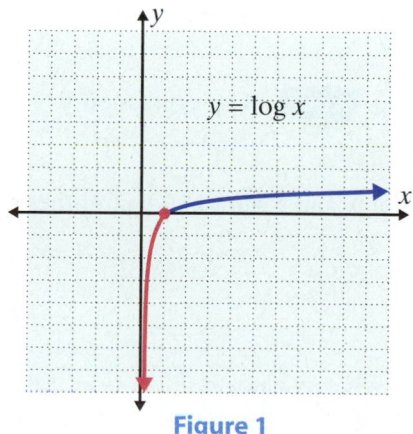

$y = \log x$

Figure 1

2. Use a TI-84 Plus graphing calculator to find the values of the following common logarithms, accurate to the nearest ten-thousandth

a. $\log 400$

b. $\log 5000$

Logarithms are exponents and they may be positive, negative, or 0. For example,

$$10^{-2} = \frac{1}{100} = 0.01 \qquad \text{and} \qquad \log 0.01 = -2.$$

However, there are no logarithms of negative numbers or 0. The domains of logarithmic functions are positive real numbers. The logarithm of a negative number is undefined. For example, $\log(-2)$ is **undefined.** That is, there is no real number x such that $10^x = -2$. If $\log(-2)$ is entered in a calculator, an error message will appear as shown here.

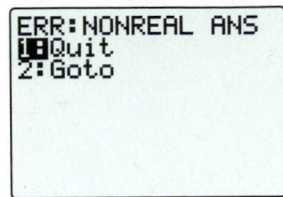

Figure 2

Objective B ## Finding the Inverse of a Logarithm (Base 10)

Because logarithms are exponents, if the value of a common logarithm is known, an exponent is known. Finding the value of the related exponential expression is called finding the **inverse of the logarithm** (or finding the **antilog**). For example,

if $\log x = N$, then $x = 10^N$ and the number x is called the **inverse log of** N.

Objective C ## Using a TI-84 Plus Graphing Calculator to Find the Inverse Log of N

Finding the inverse log of N is a three-step process.

Using a Calculator to Find the Inverse Log of N

To evaluate the **inverse log of N** on a TI-84 Plus graphing calculator:

1. Press **2ND** and **LOG**. The expression 10^(will appear on the display.

2. Enter the value of N and a right-hand parenthesis **)** .

3. Press **ENTER** .

Example 3

 Using a TI-84 Plus Graphing Calculator to Find the Inverse Log of N

Use a TI–84 Plus graphing calculator to find the **inverse log of N** for each expression. (That is, find the value of x.)

a. $\log x = 5$ **b.** $\log x = -3$

c. $\log x = 2.4142$ **d.** $\log x = 16.5$

Solutions

Thus the calculator gives,

a. $x = 10^5 = 100{,}000$

b. $x = 10^{-3} = 0.001$

c. $x = 10^{2.4142} \approx 259.5374301$

d. $x = 10^{16.5} \approx 3.16227766\text{E}16$

The letter E in the solution is the calculator version of scientific notation. Thus,

$3.16227766\text{E}16 = 3.16227766 \times 10^{16}$.

```
10^(5)
             100000
10^( -3)
                .001
```

```
10^(2.4142)
        259.5374301
10^(16.5)
      3.16227766E16
```

Now work margin exercise 3.

Objective D   **Natural Logarithms (Base e Logarithms)**

The number $e \approx 2.718281828$ is an irrational number that appears quite naturally in continuously compounded interest. Base e logarithms are called natural logarithms. **The notation for natural logarithms is shortened to ln x** (read "L N of x" or "natural log of x"). That is,

$$\log_e x = \ln x.$$

Figure 3 shows the graphs of the two functions $y = e^x$ and its inverse $y = \ln x$.

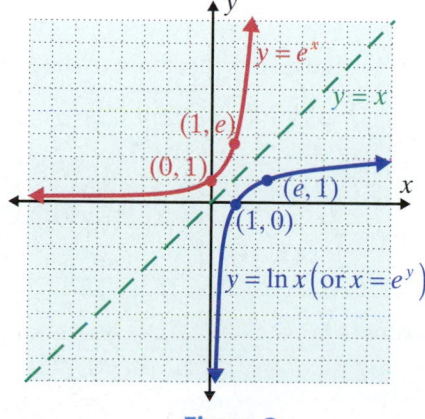

Figure 3

---

3. Use a TI–84 Plus graphing calculator to find the inverse log of N for each expression, accurate to the nearest ten-thousandth.

a. $\log x = 2.5$

b. $\log x = 3.333$

c. $\log x = \dfrac{1}{2}$

4. Translate each of the following expressions into logarithmic form using natural logarithms.

a. $e^x = 2$

b. $e^4 = c$

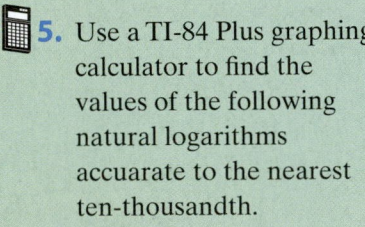 **5.** Use a TI-84 Plus graphing calculator to find the values of the following natural logarithms accuarate to the nearest ten-thousandth.

a. $\ln 2$

b. $\ln 0.64$

Example 4

Translations between Exponential Form and Logarithmic Form using Natural Logarithms

	Exponential Form		Logarithmic Form	
a.	$e^t = 3.21$	$\Leftrightarrow$	$\ln 3.21 = t$	This is a natural logarithm that equals t.
b.	$e^0 = 1$	$\Leftrightarrow$	$\ln 1 = 0$	This is a natural logarithm that equals 0.
c.	$e^2 = n$	$\Leftrightarrow$	$\ln n = 2$	This is a natural logarithm that equals 2.

*Now work margin exercise **4**.*

Using a Calculator to Evaluate Natural Logarithms

1. Press the **LN** key. `1n(` will appear on the display.

2. Enter the number and a right-hand parenthesis, **)** .

3. Press **ENTER** .

Example 5

 Evaluating Natural Logarithms Using a TI-84 Plus Graphing Calculator

Use a TI-84 Plus graphing calculator to find the values of the following natural logarithms.

a. $\ln 1$ **b.** $\ln 3$ **c.** $\ln 0.02$

Solutions

From the display we see the results (accurate to the nearest billionth).
a. $\ln 1 = 0$
b. $\ln 3 \approx 1.098612289$
c. $\ln 0.02 \approx -3.912023005$

```
ln(1)
                    0
ln(3)
        1.098612289
ln(0.02)
       -3.912023005
```

*Now work margin exercise **5**.*

Objective E Finding the Inverse of a Natural Logarithm (Base *e*)

Again, because logarithms are exponents, if we know the value of a natural logarithm, we know an exponent. Finding the value of the related exponential expression is called finding the **inverse of the logarithm**. For example,

if $\ln x = N$, then $x = e^N$ and the number x is called the **inverse ln of *N*.**

Using a Calculator to Find the Inverse ln of *N*

To evaluate the **inverse ln of *N*** on a TI-84 Plus graphing calculator:

1. Press **2ND** and **LN**. The expression e^(will appear on the display.

2. Enter the value of *N* and a right-hand parenthesis **)**.

3. Press **ENTER**.

Example 6

Using a TI-84 Plus Graphing Calculator to find the Inverse ln of *N*

Use a TI-84 Plus graphing calculator to find the inverse ln of *N* for each expression. (That is, find the value of *x*.)

a. $\ln x = 3$ **b.** $\ln x = -1$
c. $\ln x = -0.1$ **d.** $\ln x = 50$

Solutions

Thus the calculator gives,

a. $x = e^3 \approx 20.08553692$

b. $x = e^{-1} \approx 0.3678794412$

c. $x = e^{-0.1} \approx 0.904837418$

d. $x = e^{50} \approx 5.184705529\text{E}21$
$= 5.184705529 \times 10^{21}$

```
e^(3)
          20.08553692
e^(-1)
          .3678794412
```

```
e^(-0.1)
          .904837418
e^(50)
    5.184705529E21
```

Now work margin exercise 6.

---

6. Use a TI-84 Plus graphing calculator to find the inverse ln of *N* for each expression, accurate to the nearest ten-thousandth.

a. $\ln x = 40$

b. $\ln x = -\dfrac{1}{100}$

Practice Problems

Express each equation in logarithmic form.

1. $10^1 = 10$
2. $e^3 = x$
3. $10^{-4} = 0.0001$

Express each equation in exponential form.

4. $\log(0.1) = -1$
5. $\log \dfrac{1}{100} = -2$
6. $\ln x = 6$

Use a calculator to evaluate each logarithm (accurate to the nearest ten-thousandth).

7. $\log 175$
8. $\ln 52$
9. $\log(-6.1)$

Use a calculator to find the value of x in each equation (accurate to the nearest ten-thousandth).

10. $\ln x = -2$
11. $\log x = 0.5$
12. $\ln x = 30$

Practice Problem Answers

1. $\log 10 = 1$
2. $\ln x = 3$
3. $\log 0.0001 = -4$

4. $10^{-1} = 0.1$
5. $10^{-2} = \dfrac{1}{100}$
6. $e^6 = x$

7. 2.2430
8. 3.9512
9. error (undefined)

10. 0.1353
11. 3.1623
12. 1.0686×10^{13}

Exercises 10.6

Express each equation in logarithmic form. See Examples 1 and 4.

1. $10^{1.5} = x$

2. $10^k = 23$

3. $10^{-3} = \dfrac{1}{1000}$

4. $10^{-4} = 0.0001$

5. $e^x = 27$

6. $e^k = 12.4$

7. $e^0 = 1$

8. $e^4 = x$

9. $10^x = 3.2$

10. $10^y = x$

Express each equation in exponential form. See Examples 1 and 4.

11. $\log 1 = 0$

12. $\log 100 = 2$

13. $\log 5.4 = y$

14. $\ln 40.1 = x$

15. $\ln x = 1.54$

16. $\log x = 25.3$

17. $\ln e = 1$

18. $\log 10 = 1$

19. $\ln x = a$

20. $\log a = x$

Use a calculator to evaluate the logarithms accurate to the nearest ten-thousandths place. See Examples 2 and 5.

21. $\log 173$

22. $\log 396$

23. $\log 88.4$

24. $\log 0.0061$

25. $\log 0.0573$

26. $\log(-8.47)$

27. $\ln 37.5$

28. $\ln 96$

29. $\ln(-14.9)$

30. $\ln 157.6$

31. $\ln 0.00461$

32. $\ln 0.0139$

33. $\log x = 2.31$

34. $\log x = -3$

35. $\log x = -1.7$

36. $\log x = 4.1$

37. $2 \log x = -0.038$

38. $5 \log x = 9.4$

39. $\ln x = 5.17$

40. $\ln x = 4.9$

41. $\ln x = -8.3$

42. $\ln x = 6.74$

43. $0.2 \ln x = 0.0079$

44. $3 \ln x = -0.066$

Writing & Thinking

45. Explain the difference in the meaning of the expressions $\log x$ and $\ln x$.

46. The function $y = \log x$ is defined only for $x > 0$. Discuss the function $y = \log(-x)$. That is, does this function even exist? If it does exist, what is its domain? Sketch its graph and the graph of the function $y = \log x$.

47. What is the domain of the function $y = \ln|x|$? Graph the function.

10.7 Logarithmic and Exponential Equations and Change-of-Base

Objectives

A   Solve exponential equations in which the bases are the same.

B   Solve exponential equations in which the bases are not the same.

C   Solve equations with logarithms.

D   Use the change-of-base formula and a calculator to evaluate logarithmic expressions.

Objective A   **Solving Exponential Equations with the Same Base**

All the properties of exponents that have been discussed are valid for real exponents, not just integers or fractions. For example, $2^{\sqrt{2}} \cdot 2^3 = 2^{\sqrt{2}+3}$ is a valid operation and gives a real number. To be complete in our development of the properties of exponents, the properties for real exponents and positive real bases are stated here.

Properties of Real Exponents

If a and b are positive real numbers and x and y are any real numbers, then:

1. $b^0 = 1$ **2.** $b^{-x} = \dfrac{1}{b^x}$ **3.** $b^x \cdot b^y = b^{x+y}$

4. $\dfrac{b^x}{b^y} = b^{x-y}$ **5.** $\left(b^x\right)^y = b^{xy}$ **6.** $(ab)^x = a^x b^x$

7. $\left(\dfrac{a}{b}\right)^x = \dfrac{a^x}{b^x}$

The following properties are used in solving equations containing exponents and logarithms.

Properties of Equations with Exponents and Logarithms

For $b > 0$ and $b \neq 1$,

1. If $b^x = b^y$, then $x = y$.
2. If $x = y$, then $b^x = b^y$.
3. If $\log_b x = \log_b y$, then $x = y$ ($x > 0$ and $y > 0$).
4. If $x = y$, then $\log_b x = \log_b y$ ($x > 0$ and $y > 0$).

1. Solve each equation for x.

a. $4^{x^2+3} = 4^{4x}$

b. $4^{\frac{x}{2}} = 2^{x^2-2}$

Example 1

Solving Exponential Equations with the Same Base

Solve each equation for x.

a. $2^{x^2-7} = 2^{6x}$

Solution

$2^{x^2-7} = 2^{6x}$ Both bases are 2.

$x^2 - 7 = 6x$ The exponents are equal because the bases are the same.

$x^2 - 6x - 7 = 0$ Solve for x.

$(x-7)(x+1) = 0$

$x = 7$ or $x = -1$

b. $8^{4-2x} = 4^{x+2}$

Solution

$8^{4-2x} = 4^{x+2}$ Here the bases are different.

$\left(2^3\right)^{4-2x} = \left(2^2\right)^{x+2}$ Rewrite both sides so that the bases are the same.

$2^{12-6x} = 2^{2x+4}$ Use the property $\left(b^x\right)^y = b^{xy}$.

$12 - 6x = 2x + 4$ The exponents are equal because the bases are the same.

$8 = 8x$ Solve for x.

$1 = x$

*Now work margin exercise **1**.*

Objective B   **Solving Exponential Equations with Different Bases**

As Example 1 demonstrated, if the bases are the same in an exponential equation, there is no need to involve logarithms. However, **when the bases are not the same, equations are solved by taking the logarithm of both sides.** (See Property 4 of Equations with Exponents and Logorithms.)

Example 2

 Solving Exponential Equations with Different Bases

Solve each of the following exponential equations by taking the log (or ln) of both sides or using the definition of logarithm as an exponent.

a. $10^{3x} = 2.1$

Solution

First, take the log of both sides, and then use the definition of logarithm as an exponent as follows.

$$10^{3x} = 2.1$$

$$\log\left(10^{3x}\right) = \log 2.1 \qquad \text{Take the log of both sides.}$$

$$3x = \log 2.1 \qquad \text{by the definition of a common logarithm}$$

$$x = \frac{\log 2.1}{3}$$

Use a calculator to find a decimal approximation:

$$x = \frac{\log 2.1}{3} \approx \frac{0.3222}{3} = 0.1074$$

b. $6^x = 18$

Solution

The base is 6, not 10 or e, but we can solve by taking the log of both sides or by taking the ln of both sides. The result is the same.

Taking the log of both sides:	**Taking the ln of both sides:**

$$6^x = 18 \qquad\qquad\qquad 6^x = 18$$

$$\log 6^x = \log 18 \qquad\qquad \ln 6^x = \ln 18$$

$$x \cdot \log 6 = \log 18 \qquad\qquad x \cdot \ln 6 = \ln 18$$

$$x = \frac{\log 18}{\log 6} \qquad\qquad\qquad x = \frac{\ln 18}{\ln 6}$$

Using a calculator,   **Using a calculator,**

$$x = \frac{\log 18}{\log 6} \approx \frac{1.2553}{0.7782} \qquad\qquad x = \frac{\ln 18}{\ln 6} \approx \frac{2.8904}{1.7918}$$

$$\approx 1.6131. \qquad\qquad\qquad\qquad \approx 1.6131.$$

2. Solve each of the following exponential equations.

a. $4^x = 3$

b. $5^x = 20$

c. $5^{2x-1} = 10^x$

Solution

$$5^{2x-1} = 10^x$$
$$\log 5^{2x-1} = \log 10^x \qquad \text{Take the log of both sides.}$$
$$(2x-1)\log 5 = x\log 10 \qquad \text{power rule}$$
$$2x\cdot\log 5 - 1\cdot\log 5 = x\cdot 1 \qquad \log 10 = 1$$
$$2x\log 5 - x = \log 5 \qquad \text{Arrange } x-\text{terms on one side.}$$
$$x(2\log 5 - 1) = \log 5 \qquad \text{Factor out the } x.$$
$$x = \frac{\log 5}{2\log 5 - 1}$$

As a decimal approximation,

$$x = \frac{\log 5}{2\log 5 - 1} \approx \frac{0.6990}{2(0.6990) - 1} \approx 1.7563. \qquad \text{using rounded values}$$

Now work margin exercise 2.

Objective C **Solving Equations with Logarithms**

All the various properties of logarithms can be used to solve equations that involve logarithms. **Remember that logarithms are defined only for positive real numbers, so each answer should be checked in the original equation.**

Example 3

Solving Equations with Logarithms

Use the properties of logarithms to solve the following equations.

a. $\log 5x = 3$

Solution

$$\log 5x = 3$$
$$5x = 10^3 \qquad \text{definition of a common logarithm}$$
$$5x = 1000$$
$$x = 200 \qquad \text{Checking will show that the equation is defined at } x = 200.$$

b. $\log(x-1)+\log(x-4)=1$

Solution

$$\log(x-1)+\log(x-4)=1$$
$$\log[(x-1)(x-4)]=1 \qquad \text{product rule}$$
$$(x-1)(x-4)=10^1 \qquad \text{definition of a common logarithm}$$
$$x^2-5x+4=10$$
$$x^2-5x-6=0$$
$$(x-6)(x+1)=0 \qquad \text{Solve by factoring.}$$

$$x=6 \quad \text{or} \quad \cancel{x=-1} \qquad \begin{array}{l}\text{Checking } x=-1 \text{ yields} \\ \log(-1-1)=\log(-2), \\ \text{which is undefined.}\end{array}$$

c. $\log x - \log(x-1) = \log 3$

Solution

$$\log x - \log(x-1) = \log 3$$
$$\log\left(\frac{x}{x-1}\right) = \log 3 \qquad \text{quotient rule}$$
$$\frac{x}{x-1} = 3 \qquad \text{If } \log_b x = \log_b y, \text{ then } x = y.$$
$$x = 3(x-1) \qquad \text{Solve for } x.$$
$$x = 3x-3$$
$$3 = 2x$$
$$\frac{3}{2} = x$$

d. $\ln(x^2-x-6)-\ln(x+2)=2$

Solution

$$\ln(x^2-x-6)-\ln(x+2)=2$$
$$\ln\left(\frac{x^2-x-6}{x+2}\right)=2 \qquad \text{quotient rule}$$
$$\ln\left(\frac{(x+2)(x-3)}{(x+2)}\right)=2 \qquad \text{Factor the numerator.}$$
$$\ln(x-3)=2 \qquad \text{Simplify.}$$
$$x-3=e^2 \qquad \text{Change to exponential}$$
$$x=3+e^2 \qquad \text{form with base } e.$$

Or, using a calculator, $x = 3+e^2 \approx 3+7.3891 = 10.3891$.

Now work margin exercise 3.

3. Use the properties of logarithms to solve the following equations.

a. $\log 2x = 2$

b. $\log(x+2)+\log(x+2)=0$

Because a calculator can be used to evaluate common logarithms and natural logarithms, most of the examples have been restricted to base 10 or base e expressions. If an equation involves logarithms of other bases, the following discussion shows how to rewrite each logarithm using any base.

Change-of-Base Formula

For $a, b, x > 0$ and $a, b \neq 1$,

$$\log_b x = \frac{\log_a x}{\log_a b}.$$

The change-of-base formula can be derived using properties of logarithms as follows.

$$b^{\log_b x} = x$$

$$\log_a \left(b^{\log_b x} \right) = \log_a x \qquad \text{Take the } \log_a \text{ of both sides.}$$

$$\log_b x \left(\log_a b \right) = \log_a x \qquad \text{by the power rule using } \log_b x \text{ as}$$
$$\text{the exponent } r$$

$$\log_b x = \frac{\log_a x}{\log_a b}. \qquad \text{Divide both sides by } \log_a b \text{ to arrive}$$
$$\text{at the change-of-base formula.}$$

Example 4

 Change-of-Base

Use the change-of-base formula to evaluate the expressions in parts **a.** and **b.** and to solve the equation in part **c.**

a. $\log_2 3.42$

Solution

This expression can be evaluated using either base 10 or base e since both are easily available on a calculator.

$$\log_2 3.42 = \frac{\ln 3.42}{\ln 2} \qquad \text{change-of-base formula}$$

$$\approx \frac{1.2296}{0.6931} \approx 1.7741 \qquad \text{using rounded values}$$

(The student can show that $\dfrac{\log 3.42}{\log 2}$ gives the same result.)

In exponential form: $2^{1.7741} \approx 3.42$.

40. $3^{x-2} = 100$

41. $5^{2x} = \dfrac{1}{100}$

42. $7^{3x} = \dfrac{1}{10}$

43. $5^{1-x} = 1$

44. $12^{5x+2} = 1$

45. $4^{2x+5} = 0.01$

46. $4^{2-3x} = 0.1$

47. $14^{3x-1} = 10^3$

48. $12^{2x+7} = 10^4$

49. $7^x = 9$

50. $2^x = 20$

51. $3^{3x} = 23$

52. $5^{2x} = 23$

53. $6^{2x-1} = 14.8$

54. $4^{7-3x} = 26.3$

55. $5\log x = 7$

56. $3\log x = 13.2$

57. $4\log x - 6 = 0$

58. $2\log x - 15 = 0$

59. $4\log x^{\frac{1}{2}} + 8 = 0$

60. $\dfrac{2}{3}\log x^{\frac{2}{3}} + 9 = 0$

61. $5\ln x - 8 = 0$

62. $2\ln x + 3 = 0$

63. $\ln x^2 + 2.2 = 0$

64. $\ln x^2 - 41.6 = 0$

65. $\log x + \log 2x = \log 18$

66. $\log(x+4) + \log(x-4) = \log 9$

67. $\log x^2 - \log x = 2$

68. $\log x + \log x^2 = 3$

69. $\ln(x-3) + \ln x = \ln 18$

70. $\ln(x+5) + \ln(x-1) = \ln 16$

71. $\log(x-15) = 2 - \log x$

72. $\log(2x-17) = 2 - \log x$

73. $\log(3x-5) + \log(x-1) = 1$

74. $\log(2x-3) + \log(x+3) = 3$

75. $\log(x-3) - \log(x+1) = 1$

76. $\ln(x+1) + \ln(x-1) = 0$

77. $\log(x^2-9) - \log(x-3) = -2$

78. $\ln\left(x^2 + 4x - 5\right) - \ln\left(x + 5\right) = -2$

79. $\log\left(x^2 - 4x - 5\right) - \log\left(x + 1\right) = 2$

80. $\log\left(x^2 - x - 12\right) - \log\left(x - 4\right) = -2$

81. $\ln\left(x^2 - 4\right) = 3 + \ln\left(x + 2\right)$

82. $\ln\left(x^2 + 2x - 3\right) = 1 + \ln\left(x - 1\right)$

83. $\log \sqrt[3]{x^2 + 2x + 20} = \dfrac{2}{3}$

84. $\log \sqrt{x^2 - 24} = \dfrac{3}{2}$

Use the change-of-base formula to evaluate each of the expressions or solve the equations. Round answers to the nearest ten-thousandth. See Example 4.

85. $\log_3 12$

86. $\log_4 36$

87. $\log_5 1.68$

88. $\log_{11} 39.6$

89. $\log_8 0.271$

90. $\log_7 0.849$

91. $\log_{15} 739$

92. $\log_2 14.2$

93. $\log_{20} 0.0257$

94. $\log_9 2.384$

95. $2^x = 5$

96. $3^{2x} = 10$

97. $9^{2x-1} = 100$

98. $5^{x-1} = 30$

99. $4^{3-x} = 20$

100. $6^{4-3x} = 25$

Writing & Thinking

101. Solve the following equation for x two different ways: $a^{2x-1} = 1$.

102. Rewrite each of the following expressions as products.

    a. 5^{x+2} **b.** 3^{x-2}

103. Explain, in your own words, why $7 \cdot 7x \neq 49x$ when $x \neq 1$. Show each of the expressions $7 \cdot 7x$ and $49x$ as a single exponential expression with base 7.

Chapter 10: Index of Key Terms and Ideas

Section 10.1: Algebra of Functions

Algebraic Operations with Functions

pages 1058 – 1060

If $f(x)$ and $g(x)$ represent two functions and x is a value in the domain of both functions, then:

1. Sum of two functions: $(f+g)(x) = f(x) + g(x)$
2. Difference of two functions: $(f-g)(x) = f(x) - g(x)$
3. Product of two functions: $(f \cdot g)(x) = f(x) \cdot g(x)$
4. Quotient of two functions: $\left(\dfrac{f}{g}\right)(x) = \dfrac{f(x)}{g(x)}$ where $g(x) \neq 0$.

Graphing the Sum of Two Functions

pages 1061 – 1063

To add the two functions graphically, look at each of the x-values an the x-axis and the points directly above (or below) them and add the corresponding y-values.

Section 10.2: Composite and Inverse Functions

Composite Functions

pages 1072 – 1074

For two functions f and g, the **composite function** $f \circ g$ is defined as follows: $(f \circ g)(x) = f(g(x))$.

The **domain of** $f \circ g$ consists of those values of x in the domain of g for which $g(x)$ is in the domain of f.

One-to-One Functions

pages 1075 – 1076

A function is a **one-to-one function** (or **1-1 function**) if for each value of y in the range there is only one corresponding value of x in the domain.

Horizontal Line Test

page 1076

A function is one-to-one if no **horizontal line** intersects the graph of the function at more than one point.

Inverse Functions

pages 1078 – 1080

If f is a one-to-one function with ordered pairs of the form (x, y), then its **inverse function**, denoted as f^{-1}, is also a one-to-one function with ordered pairs of the form (y, x). (Only one-to-one functions have inverses.)

To Determine whether Two Functions are Inverses

page 1080

If f and g are one-to-one functions and
$f(g(x)) = x$ for all x in D_g, and
$g(f(x)) = x$ for all x in D_f

then f and g are **inverse functions**. That is, $g = f^{-1}$ and $f = g^{-1}$.

To Find the Inverse of a One-to-One Function

page 1083

1. Let $y = f(x)$. (In effect, substitute y for $f(x)$.)
2. Interchange x and y.
3. In the new equation, solve for y in terms of x.
4. Substitute $f^{-1}(x)$ for y. (This new function is the inverse of f.)

Section 10.3: Exponential Functions

Exponential Functions   page 1091

An **exponential function** is a function of the form $f(x) = b^x$ where $b > 0$, $b \neq 1$, and x is any real number.

Asymptotes   page 1093

When the graph of a function gets closer and closer to a line (for example the x- or y-axis) without ever touching it, the line is called an **asymptote**.

Exponential Growth and Decay   pages 1095 – 1096

1. If $b > 1$, then b^x is called an **exponential growth function**. This type of function increases to the right and approaches the x-axis for negative values of x.

2. If $0 < b < 1$, then b^x is called an **exponential decay function**. This type of function decreases to the right and approaches the x-axis for positive values of x.

Compound Interest   pages 1097 – 1098

Compound interest on a principal P invested at an annual interest rate r (in decimal form) for t years that is compounded n times per year can be calculated using the following formula: $A = P\left(1 + \dfrac{r}{n}\right)^{nt}$ where A is the amount accumulated.

The Number e   page 1100

The number e is defined to be $e = 2.718281828459\ldots$

Continuously Compounded Interest   pages 1100 – 1101

Continuously compounded interest on a principal P invested at an annual interest rate r for t years, can be calculated using the following formula: $A = Pe^{rt}$ where A is the amount accumulated.

Section 10.4: Logarithmic Functions

Logarithmic Functions   pages 1105 – 1107

For all positive values of x and for all positive values of b (except $b \neq 1$), the expression $b^y = x$ can be written as $y = \log_b x$. The function $f(x) = \log_b x$ is the **logarithmic function** with base b.

Graphs of Logarithmic Functions   pages 1108 – 1109

For the logarithmic function $y = \log_b x$ $\left(\text{or } x = b^y\right)$,

Change-of-Base Formula   pages 1108 – 1109

For $a, b, x > 0$ and $a, b \neq 1$, $\log_b x = \dfrac{\log_a x}{\log_a b}$.

Section 10.5 Properties of Logarithms

Properties of Logarithms   pages 1113 – 1119

For $b > 0$, $b \neq 1$, $x, y > 0$, and any real number r,

1. $\log_b 1 = 0$
2. $\log_b b = 1$
3. $x = b^{\log_b x}$
4. $\log_b b^x = x$
5. $\log_b xy = \log_b x + \log_b y$ the product rule
6. $\log_b \dfrac{x}{y} = \log_b x - \log_b y$ the quotient rule
7. $\log_b x^r = r \cdot \log_b x$ the power rule

Section 10.6: Common Logarithms and Natural Logarithms

Common Logarithms   page 1123

A **common logarithm** is a base 10 logarithm. The abbreviated notion $\log x$ is used to indicate common logorithms (base 10 logarithms).

Inverse Logarithms   page 1125

Natural Logarithms   pages 1126 – 1127

A **natural logarithm** is a base e logarithm. The notation for natural logarithms is shortened to $\ln x$.

Section 10.7: Logarithmic and Exponential Equations and Change-of-Base

Properties of Equations with Exponents and Logarithms   page 1123

For $b > 0$ and $b \neq 1$,

1. If $b^x = b^y$, then $x = y$.
2. If $x = y$, then $b^x = b^y$.
3. If $\log_b x = \log_b y$, then $x = y$ ($x > 0$ and $y > 0$).
4. If $x = y$, then $\log_b x = \log_b y$ ($x > 0$ and $y > 0$).

Chapter 10: Test

For the following pairs of functions, graph a. the sum and b. the difference on two different graphs.

1.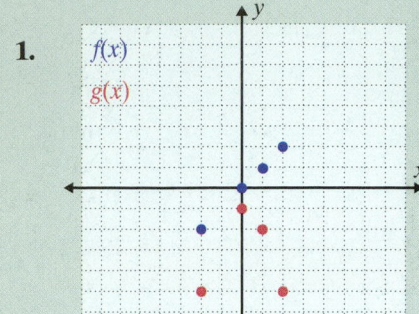

$f = \{(-2, -2), (0, 0), (1, 1), (2, 2)\}$

$g = \{(-2, -5), (0, -1), (1, -2), (2, -5)\}$

2.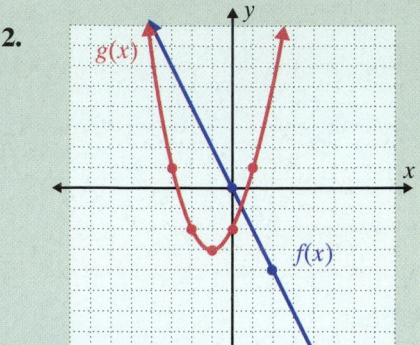

3. Given the two functions, $f(x) = \sqrt{x - 3}$ and $g(x) = x^2 + 1$, find:

    a. $(f + g)(x)$ **b.** $(f - g)(x)$ **c.** $(f \cdot g)(x)$ **d.** $\left(\dfrac{f}{g}\right)(x)$

4. If $f(x) = 2x - 5$ and $g(x) = 3 - 2x^2$, form the following compositions.

    a. $f(g(x))$ **b.** $g(f(x))$

5. Use the horizontal line test to determine whether the graph on the right is a one-to-one function. If the graph represents a one-to-one function, graph its inverse by reflecting the graph of the function across the line $y = x$.

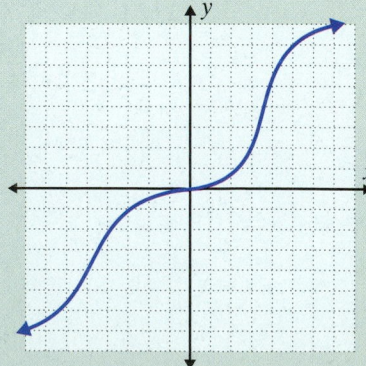

6. Determine algebraically whether or not each pair of functions are inverses of each other. Graph both functions on the same set of axes and show the line $y = x$ as a dotted line on each graph.

    a. $f(x) = x^2$ and $g(x) = -x^2$

    b. $f(x) = \dfrac{1}{x}$ and $g(x) = \dfrac{1}{x}$

7. Find $f^{-1}(x)$ if $f(x) = \dfrac{1}{x+1} - 3$. Graph both functions on the same set of axes and show the line $y = x$ as a dotted line on the graph.

8. Sketch the graphs of the exponential functions given in the following equations. Label three points on each graph.

    a. $y = -3^x$

    b. $y = 5^{x-2} - 1$

9. Express the following equations in logarithmic form.

    a. $10^5 = 100,000$

    b. $\left(\dfrac{1}{2}\right)^{-3} = 8$

    c. $e^x = 15$

10. Express the following equations in exponential form.

    a. $\log x = 4$

    b. $\log_3 \dfrac{1}{9} = -2$

    c. $\log_4 256 = x$

11. Find the inverse of the function $y = \left(\dfrac{1}{2}\right)^x$. Graph both functions on the same set of axes and show the line $y = x$ as a dotted line on each graph. Label two points on each graph.

12. Solve the following equations by first changing them to exponential form.

    a. $\log_7 x = 3$

    b. $\log_9 27 = x$

13. Use the properties of logarithms to expand the following expressions as much as possible.

    a. $\ln\left(x^2 - 25\right)$

    b. $\log \sqrt[3]{\dfrac{x^2}{y}}$

14. Use the properties of logarithms to write each expression as a logarithm of a single expression.

a. $\ln(x+5)+\ln(x-4)$

b. $\log\sqrt{x}+\log x^2-\log 5x$

Use a calculator to find the value of x accurate to the nearest ten-thousandth.

15. a. $x=\log 579$

b. $5\ln x = 9.35$

Use the properties of exponents and logarithms to solve each of the equations for x. If necessary, round answers to nearest ten-thousandth.

16. $7^3 \cdot 7^x = 7^{-1}$

17. $6^{x-1} = 36^{x+1}$

18. $10^{x+2} = 283$

19. $2e^{0.24x} = 26$

20. $4^x = 12$

21. $\log(2x+3)-\log(x+1)=0$

22. $\ln(x^2+3x-4)-\ln(x+4)=3$

23. Use the change-of-base formula to evaluate the following expression: $\log_6 25$. Round your answer to nearest ten-thousandth.

Cumulative Review: Chapters 1 - 10

Evaluate each of the following expressions. Reduce all fractions to lowest terms.

1. $\dfrac{5}{8} + \dfrac{3}{10}$

2. $\dfrac{1}{3} - \dfrac{7}{6}$

3. $\left(\dfrac{5}{6}\right)\left(\dfrac{4}{15}\right)\left(-\dfrac{7}{8}\right)$

4. $-\dfrac{15}{16} \div \dfrac{32}{35}$

Find the value of each expression using the rules for order of operations.

5. $8 + \left[5 \cdot 6 - (9 \div 3 + 3)\right]$

6. $2^3 \div 4 + 7 - 10 \div 5$

Simplify. Assume all variables are positive.

7. $\sqrt{80x^2 y^3}$

8. $\dfrac{\sqrt{8x^3 y^2}}{\sqrt{5xy^3}}$

9. $\dfrac{x^{\frac{2}{3}} y^{\frac{1}{3}}}{x^{\frac{1}{2}} y^{\frac{2}{3}}}$

10. $\left(x^2 y^{-1}\right)^{\frac{1}{2}} \left(4xy^3\right)^{-\frac{1}{2}}$

Change each expression to an equivalent exponential expression and simplify. Assume all variables are positive.

11. $\sqrt{x^4 y^3}$

12. $\sqrt[3]{x^2 y^3}$

Perform the indicated operations and simplify the results.

13. $(7x + 2)(x - 4)$

14. $-6(y + 5)(y - 2)$

Use your knowledge of logarithms and exponents to find the value of each expression.

15. $\log_2 \dfrac{1}{64}$

16. $7^{\log_7 12}$

17. Use the properties of logarithms to expand the expression $\log_b \left(x^3 \sqrt{y}\right)$ as much as possible.

18. Use the properties of logarithms to write the expression $2 \log_b x - \log_b y + \log_b 7$ as a single logarithm.

Solve each of the equations. Round answers to nearest ten-thousandth, if necessary.

19. $4x^2 - 28x + 49 = 0$

20. $x - 2 = \sqrt{x + 10}$

21. $6^{3x+5} = 55$

22. $\ln\left(x^2 - 7x + 10\right) - \ln(x - 2) = 2.5$

23. Solve the equation $\log_6 \dfrac{1}{216} = x$ by first changing it to exponential form.

24. Use the Gaussian elimination method to solve the system of linear equations:

$$\begin{cases} x + y + z = 2 \\ 2x + y - z = -1 \\ x + 3y + 2z = 2 \end{cases}$$

25. Find an equation in slope-intercept form for the line passing through the point $(-3, 4)$ with slope $m = -\dfrac{2}{3}$. Graph the line.

For the systems of equations: a. Determine whether each pair of lines is parallel, perpendicular, or neither. b. Solve the system of equations. c. Graph both lines.

26. $\begin{cases} y = 5x - 6 \\ y = 5x + 1 \end{cases}$

27. $\begin{cases} y = -2x + 5 \\ y = \dfrac{1}{2}x \end{cases}$

28. Solve the following system of two linear inequalities graphically. $\begin{cases} y \le 4x - 1 \\ x + y < 3 \end{cases}$.

29. Find and label at least 5 points on the graph of the function $\sqrt{4x - 3}$ and then sketch the graph of the function.

30. The graphs of three curves are shown. Use the vertical line test to determine whether or not each graph represents a function. Then state the domain and range of each graph.

a.

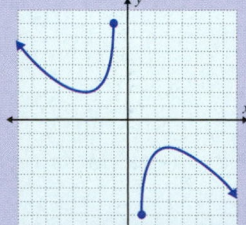

b.

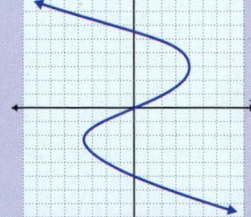

c.

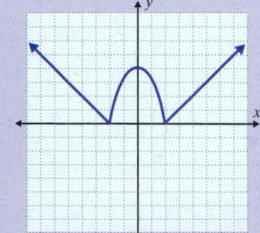

31. $y = x^2 - 2x + 5$
$D = \{-2, -1, 0, 1, 2\}$

32. $y = x^3 - 5x^2$
$D = \{-2, -1, 0, 1, 2\}$

33. Given the two functions, $f(x) = x^2 + x$ and $g(x) = \sqrt{x+2}$, find:

a. $(f+g)(x)$ **b.** $(f-g)(x)$ **c.** $(f \cdot g)(x)$ **d.** $\left(\dfrac{f}{g}\right)(x)$

34. Given the two functions, $f(x) = x^2 + 3$ for $x \geq 0$ and $g(x) = 2x + 1$ for $x \geq 0$, find:

a. $f(g(x))$ **b.** $g(f(x))$

Which of the functions are one-to-one functions? If the graph represents a one-to-one function, graph its inverse by reflecting the graph of the function across the line $y = x$.

35.

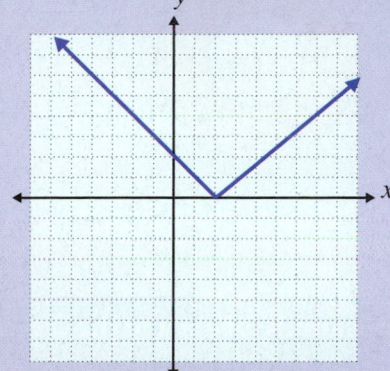

36.

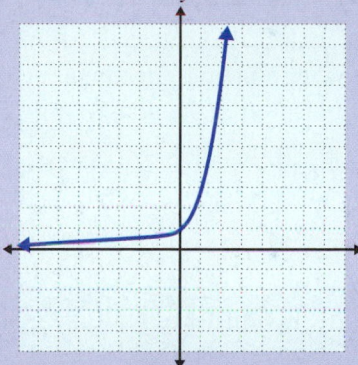

Find the inverse of each function and graph both the function and its inverse on the same set of axes. Include the graph of the line $y = x$ as a dotted line in each graph.

37. $f(x) = x^2 - 4$ for $x \geq 0$

38. $g(x) = (x-2)^3$

39. Find three consecutive odd integers such that 2 times the second plus 5 times the third is equal to 6 less than the first.

40. If the quotient of a number and 7 is decreased by 12, the result is the number plus −6. What is the number?

41. Find the simple interest earned on $5000 loaned at 18% for 10 months.

42. What will be the value of $10,000 invested at 5% and compounded monthly for 3 years?

43. Computer Screens: In order to display images properly, the width of a computer screen must be $\frac{4}{3}$ times the height of the screen. If a manufacturer wants the length of the diagonal of the screen to be 20 inches, find the width and height of the screen.

44. Publishing Books: A book is available in both hardback and paperback. A bookstore sold a total of 43 books during the week. The total receipts were $297.50. If hardback books sell for $12.50 and paperbacks sell for $4.50, how many of each were sold?

45. Mileage: Mr. John kept the mileage records for 5 months as shown in the following bar graph.

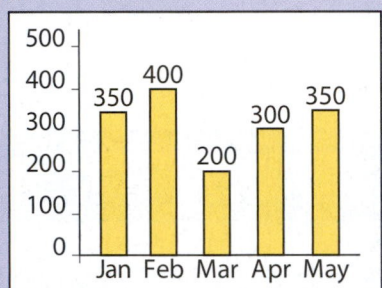

Find the following statistics for the information indicated in the graph.

a. Mean = _____   **b.** Median = _____

c. Mode = _____   **d.** Range = _____

46. In the figure shown, there are two concentric (have the same center) circles. (Use $\pi = 3.14$.)

a. Find the circumference of each circle.

b. Find the area (accurate to nearest hundredth) of the shaded region between the two circles.

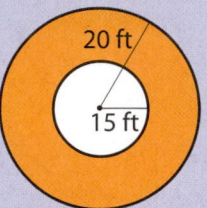

Use your calculator to find the value (accurate to nearest ten-thousandth) of each of the radical expressions.

47. $6 - 5\sqrt{7}$

48. $\dfrac{3 + 2\sqrt{11}}{5}$

Use a calculator to evaluate the logarithms accurate to nearest ten-thousandth.

49. $\log 54.6$

50. $\ln 10,000$

Mathematics @ Work!

Many people have heard the phrase "Many hands make light work," a quote attributed to John Heywood, an English playwright just before Shakespeare. The truth behind this statement has proven itself time and time again. Effects of collaboration are everywhere; collaboration helps companies meet deadlines, collaborating pistons make for faster cars, and multiple mirrors collaborating in a solar array create an enormous amount of energy. As companies look at how best to distribute their resources, they can use rational equations to determine exactly how much time each task will take a team of people to complete. All they need to know is the amount of time it takes each individual to complete the entire task and then they can find the time it should take the team to complete the task if they work together. By evaluating these times they can make sure each pending deadline has the manpower necessary to complete it in a timely fashion.

Sarah and Jeff are raking leaves. Sarah can rake the yard in 3 hours by herself, and Jeff can rake the yard in 2 5/8 hours working alone. How long will it take the two of them to rake the yard if they work together?

For more problems like these, see Section 11.5.

Rational Expressions

11.1 Rational Expressions

Objective A **Introduction to Rational Expressions**

The term **rational number** is the technical name for a fraction in which both the numerator and denominator are integers. Similarly, the term **rational expression** is the technical name for a fraction in which both the numerator and denominator are polynomials.

Rational Expressions
A **rational expression** is an algebraic expression that can be written in the form $$\frac{P}{Q} \text{ where } P \text{ and } Q \text{ are polynomials and } Q \neq 0.$$

Examples of rational expressions are

$$\frac{4x^2}{9}, \quad \frac{y^2 - 25}{y^2 + 25}, \quad \text{and} \quad \frac{x^2 + 7x - 6}{x^2 - 5x - 14}.$$

As with rational numbers, the denominators of rational expressions cannot be 0. If a numerical value is substituted for a variable in a rational expression and the denominator assumes a value of 0, we say that the expression is undefined for that value of the variable.

notes

■ **Remember, the denominator of a rational expression can never be 0.** Division by 0 is undefined.

■

Example 1

Finding Restrictions on the Variable

Determine what values of the variable, if any, will make the rational expression undefined. (These values are called restrictions on the variable.)

a. $\dfrac{5}{3x-1}$

Solution

$$3x - 1 = 0 \qquad \text{Set the denominator equal to 0.}$$
$$3x = 1 \qquad \text{Solve the equation.}$$
$$x = \frac{1}{3}$$

Thus, the expression $\dfrac{5}{3x-1}$ is undefined for $x = \dfrac{1}{3}$. Any other real number may be substituted for x in the expression. We write $x \neq \dfrac{1}{3}$ to indicate the restriction on the variable.

b. $\dfrac{x^2-4}{x^2-5x-6}$

Solution

$$x^2 - 5x - 6 = 0 \qquad \text{Set the denominator equal to 0.}$$
$$(x-6)(x+1) = 0 \qquad \text{Solve the equation by factoring.}$$
$$x - 6 = 0 \quad \text{or} \quad x + 1 = 0$$
$$x = 6 \qquad\qquad x = -1$$

Thus, there are two restrictions on the variable: 6 and −1. We write $x \neq -1, 6$.

c. $\dfrac{x+3}{x^2+36}$

Solution

$$x^2 + 36 = 0 \qquad \text{Set the denominator equal to 0.}$$
$$x^2 = -36 \qquad \text{Solve the equation.}$$

However, there is no real number whose square is −36. Thus there are no restrictions on the variable.

Now work margin exercise 1.

1. Determine which values of the variable make the expression undefined.

a. $\dfrac{9}{5x-1}$

b. $\dfrac{x^2-25}{x^2-7x+12}$

c. $\dfrac{x+8}{x^2+16}$

notes

■ **Comments about the Numerator being 0**

■ If the numerator of a rational expression has a value of 0 and the denominator is not 0 for that value of the variable, then the expression ■ is defined and has a value of 0. If both numerator and denominator are 0, then the expression is undefined just as in the case where only ■ the denominator is 0.

The rules for operating with rational expressions are essentially the same as those for operating with fractions in arithmetic. That is, simplifying, multiplying, and dividing rational expressions involve factoring and reducing. Addition and subtraction of rational expressions require common denominators. The basic rules for fractions were discussed in earlier sections and are summarized here for easy reference.

Summary of Arithmetic Rules for Rational Numbers (or Fractions)

A **fraction** (or **rational number**) is a number that can be written in the form $\frac{a}{b}$ where a and b are integers and $b \neq 0$. (Remember, no denominator can be 0.)

The Fundamental Principle: $\dfrac{a}{b} = \dfrac{a \cdot k}{b \cdot k}$ where $b, k \neq 0$

The reciprocal of $\dfrac{a}{b}$ is $\dfrac{b}{a}$ and $\dfrac{a}{b} \cdot \dfrac{b}{a} = 1$ where $a, b \neq 0$.

Multiplication: $\dfrac{a}{b} \cdot \dfrac{c}{d} = \dfrac{a \cdot c}{b \cdot d}$ where $b, d \neq 0$

Division: $\dfrac{a}{b} \div \dfrac{c}{d} = \dfrac{a}{b} \cdot \dfrac{d}{c}$ where $b, c, d \neq 0$

Addition: $\dfrac{a}{b} + \dfrac{c}{b} = \dfrac{a+c}{b}$ where $b \neq 0$

Subtraction: $\dfrac{a}{b} - \dfrac{c}{b} = \dfrac{a-c}{b}$ where $b \neq 0$

For rational expressions, each rule can be restated by replacing a and b with P and Q where P and Q represent polynomials. In particular, the fundamental principle can be restated as follows:

The Fundamental Principle of Rational Expressions

If $\dfrac{P}{Q}$ is a rational expression and P, Q, and K are polynomials where

$$Q, K \neq 0, \text{ then } \quad \frac{P}{Q} = \frac{P \cdot K}{Q \cdot K}.$$

Objective B Reducing (or Simplifying) Rational Expressions

The fundamental principle can be used to **reduce** (or **simplify**) a rational expression to **lower terms,** for multiplication, or division and to **build** a rational expression to **higher terms,** for addition or subtraction. Just as with rational numbers, a rational expression is said to be **reduced to lowest terms** if the numerator and denominator have no common factors other than 1 and –1.

Example 2

Reducing Rational Expressions

Use the fundamental principle to reduce each expression to lowest terms. State any restrictions on the variable by using the fact that no denominator can be 0. This restriction applies to denominators **before and after** a rational expression is reduced.

a. $\dfrac{2x - 10}{3x - 15}$

Solution

$$\frac{2x - 10}{3x - 15} = \frac{2(x - 5)}{3(x - 5)} = \frac{2}{3} \qquad (x \neq 5)$$

Note that $x - 5$ is a common factor. The key word here is factor. We reduce using factors only.

b. $\dfrac{x^3 - 64}{x^2 - 16}$

Solution

$$\frac{x^3 - 64}{x^2 - 16} = \frac{(x - 4)(x^2 + 4x + 16)}{(x + 4)(x - 4)}$$

Reduce. The common factor is $x - 4$. Note that $x^3 - 64$ is the difference of two cubes. Also, note that $x^2 + 4x + 16$ is not factorable.

$$= \frac{x^2 + 4x + 16}{x + 4} \qquad (x \neq -4, 4)$$

2. Reduce the expression to lowest terms. State any restrictions on the variable.

a. $\dfrac{2x-6}{5x-15}$

b. $\dfrac{x^3-125}{x^2-25}$

c. $\dfrac{x-5}{5-x}$

c. $\dfrac{y-10}{10-y}$

Solution

$$\dfrac{y-10}{10-y} = \dfrac{y-10}{-y+10}$$

$$= \dfrac{1\cancel{(y-10)}}{-1\cancel{(y-10)}}$$

$$= \dfrac{1}{-1} = -1 \qquad (y \neq 10)$$

> Note that the expression $10-y$ is the opposite of $y-10$. When nonzero opposites are divided, the quotient is always -1.

*Now work margin exercise **2**.*

In Example **2c.**, the result was -1. The expression $10-y$ is the opposite of $y-10$ for any value of y. That is,

$$10-y = -y+10 = -1(y-10) = -(y-10).$$

When nonzero opposites are divided, the quotient is always -1. For example,

$$\dfrac{-9}{9} = -1, \qquad \dfrac{23}{-23} = -1, \qquad \dfrac{x-5}{5-x} = \dfrac{\cancel{(x-5)}}{-1\cancel{(x-5)}} = \dfrac{1}{-1} = -1 \ (x \neq 5).$$

Opposites in Rational Expressions

For a polynomial P, $\dfrac{-P}{P} = -1$ where $P \neq 0$.

In particular, $\dfrac{a-x}{x-a} = \dfrac{-(x-a)}{x-a} = -1$ where $x \neq a$.

Remember that the key word when reducing is **factor.** Many students make mistakes similar to the following when working with rational expressions.

Common Error

"Divide out" only common factors.

Wrong Solution	Correct Solution

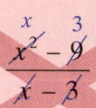

$$\frac{4x + 8}{8}$$

8 is not a common factor.

WRONG

$$\frac{4x+8}{8} = \frac{\cancel{4}(x+2)}{\underset{2}{\cancel{8}}}$$

4 is a common factor.

CORRECT

$$\frac{x^{\cancel{2}}-\overset{3}{\cancel{9}}}{\cancel{x}-\cancel{3}}$$

3 and x are not common factors.

WRONG

$$\frac{x^2-9}{x-3} = \frac{(x+3)\,\cancel{(x-3)}}{\cancel{(x-3)}}$$

$x - 3$ is a common factor.

CORRECT

Objective C **Multiplication with Rational Expressions**

To Multiply Rational Expressions

To multiply any two (or more) rational expressions,

1. Completely factor each numerator and denominator.
2. Multiply the numerators and multiply the denominators, keeping the expressions in factored form.
3. "Divide out" any common factors from the numerators and denominators. Remember that no denominator can have a value of 0.

No factoring is necessary in this example:

$$\frac{2x}{x-6} \cdot \frac{x+5}{x-4} = \frac{2x(x+5)}{(x-6)(x-4)} = \frac{2x^2+10x}{(x-6)(x-4)}. \qquad (x \neq 4, 6)$$

However, in this case the numerators and denominators must be factored.

$$\frac{y^2-4}{y^3}\cdot\frac{y^2-3y}{y^2-y-6}=\frac{(y+2)\,(y-2)\,(y)\,(y-3)}{y^3\,(y-3)\,(y+2)}=\frac{y-2}{y^2}\quad(y\neq2,0,3)$$

Multiplication with Rational Expressions

If $P, Q, R,$ and S are polynomials and $Q, S \neq 0,$ then

$$\frac{P}{Q}\cdot\frac{R}{S}=\frac{P\cdot R}{Q\cdot S}.$$

3. Multiply and reduce, if possible. State any restrictions on the variable(s).

a. $\dfrac{7x^4y}{9x^2y^6}\cdot\dfrac{3x^3y^3}{14xy^4}$

b. $\dfrac{x}{x-3}\cdot\dfrac{x^2-9}{x^3}$

c. $\dfrac{2x+2}{x^2-x}\cdot\dfrac{x^2-2x+1}{2x^2+4x+2}$

d. $\dfrac{x^2-8x+12}{3x+9}\cdot\dfrac{x^2-25}{x^2-7x+10}$

Example 3

Multiplication with Rational Expressions

Multiply and reduce if possible.

a.
$$\frac{5x^2y}{9xy^3}\cdot\frac{6x^3y^2}{15xy^4}=\frac{5\cdot2\cdot3\cdot x^5\cdot y^3}{3\cdot3\cdot3\cdot5\cdot x^2\cdot y^7}=\frac{2x^{5-2}y^{3-7}}{9}=\frac{2x^3y^{-4}}{9}=\frac{2x^3}{9y^4}$$

$$\left(x\neq0,y\neq0\right)$$

b.
$$\frac{x}{x-2}\cdot\frac{x^2-4}{x^2}=\frac{x\,(x+2)\,(x-2)}{(x-2)\,x^2}=\frac{x+2}{x}$$

$$\left(x\neq0,2\right)$$

c.
$$\frac{3x-3}{x^2+x}\cdot\frac{x^2+2x+1}{3x^2-6x+3}=\frac{3\,(x-1)\,(x+1)^2}{x\,(x+1)\cdot3\,(x-1)^2}=\frac{x+1}{x\,(x-1)}\text{ or }\frac{x+1}{x^2-x}$$

$$\left(x\neq-1,0,1\right)$$

d.
$$\frac{x^2-7x+12}{2x+6}\cdot\frac{x^2-4}{x^2-2x-8}=\frac{(x-4)\,(x-3)\,(x+2)\,(x-2)}{2\,(x+3)\,(x-4)\,(x+2)}$$

$$=\frac{(x-3)\,(x-2)}{2\,(x+3)}\text{ or }\frac{x^2-5x+6}{2\,(x+3)}\text{ or }\frac{x^2-5x+6}{2x+6}$$

$$\left(x\neq-3,-2,4\right)$$

Now work margin exercise 3.

notes

- As shown in Examples **3c.** and **3d.,** there may be more than one correct form of an answer. After a rational expression has been reduced, the numerator and denominator may be multiplied out or left in factored form. **Generally, the denominator will be left in factored form and the numerator multiplied out.** As we will see in the next section, this form makes the results easier to add and subtract. However, be aware that this form is just an option, and multiplying out the denominator is not an error.

Objective D **Division with Rational Expressions**

To divide any two rational expressions, multiply the first fraction by the **reciprocal** of the second fraction.

Division with Rational Expressions

If $P, Q, R,$ and S are polynomials with $Q, R, S \neq 0,$ then

$$\frac{P}{Q} \div \frac{R}{S} = \frac{P}{Q} \cdot \frac{S}{R}.$$

Note that $\dfrac{S}{R}$ is the reciprocal of $\dfrac{R}{S}.$

4. Divide and reduce, if possible. Assume that no denominator has a value of 0.

a. $\dfrac{15x^3y}{12xy^3} \div \dfrac{5x^5y}{xy^2}$

b. $\dfrac{x^3+y^3}{y^3} \div \dfrac{-x-y}{xy}$

c. $\dfrac{x^2-9x+18}{3x^2+19x+6} \div \dfrac{3x^2-17x-6}{x^2+x-30}$

Example 4

Division with Rational Expressions

Divide and reduce, if possible. Assume that no denominator has a value of 0.

a. $\dfrac{12x^2y}{10xy^2} \div \dfrac{3x^4y}{xy^3}$

Solution

$$\dfrac{12x^2y}{10xy^2} \div \dfrac{3x^4y}{xy^3} = \dfrac{12x^2y}{10xy^2} \cdot \dfrac{xy^3}{3x^4y}$$

$$= \dfrac{\cancel{2}\cdot 2 \cdot \cancel{3} \cdot x^3 \cdot y^4}{\cancel{2}\cdot 5 \cdot \cancel{3} \cdot x^5 \cdot y^3}$$

$$= \dfrac{2x^{3-5}y^{4-3}}{5}$$

$$= \dfrac{2x^{-2}y}{5} = \dfrac{2y}{5x^2}$$

Note that in this example we have used the quotient rule for exponents.

b. $\dfrac{x^3-y^3}{x^3} \div \dfrac{y-x}{xy}$

Solution

$$\dfrac{x^3-y^3}{x^3} \div \dfrac{y-x}{xy} = \dfrac{x^3-y^3}{x^3} \cdot \dfrac{xy}{y-x}$$

$$= \dfrac{\overset{-1}{\cancel{(x-y)}}\left(x^2+xy+y^2\right)\cancel{x}y}{\underset{x^2}{\cancel{x^3}}\,\cancel{(y-x)}}$$

Note that $\dfrac{x-y}{y-x} = -1$.

$$= \dfrac{-y\left(x^2+xy+y^2\right)}{x^2} = \dfrac{-x^2y-xy^2-y^3}{x^2}$$

c. $\dfrac{x^2-8x+15}{2x^2+11x+5} \div \dfrac{2x^2-5x-3}{4x^2-1}$

Solution

$$\dfrac{x^2-8x+15}{2x^2+11x+5} \div \dfrac{2x^2-5x-3}{4x^2-1} = \dfrac{x^2-8x+15}{2x^2+11x+5} \cdot \dfrac{4x^2-1}{2x^2-5x-3}$$

$$= \dfrac{\cancel{(x-3)}(x-5)(2x-1)\cancel{(2x+1)}}{(2x+1)(x+5)\cancel{(x-3)}\cancel{(2x+1)}}$$

$$= \dfrac{(x-5)(2x-1)}{(2x+1)(x+5)} = \dfrac{2x^2-11x+5}{(2x+1)(x+5)}$$

Remember that you have the option of leaving the numerator and/or denominator in factored form.

Now work margin exercise **4.**

Practice Problems

Reduce to lowest terms. State any restrictions on the variables.

1. $\dfrac{5x+20}{7x+28}$ **2.** $\dfrac{6-3x}{3x-6}$ **3.** $\dfrac{x^2+x-2}{x^2+3x+2}$

Perform the following operations and simplify the results. Assume that no denominator has a value of 0.

4. $\dfrac{x-7}{x^3} \cdot \dfrac{x^2}{49-x^2}$ **5.** $\dfrac{y^2-y-6}{y^2-5y+6} \cdot \dfrac{y^2-4}{y^2+4y+4}$

6. $\dfrac{x^3+3x}{2x+1} \div \dfrac{x^2+3}{x+1}$

7. $\dfrac{x^2+2x-3}{x^2-3x-10} \cdot \dfrac{2x^2-9x-5}{x^2-2x+1} \div \dfrac{4x+2}{x^2-x}$

Practice Problem Answers

1. $\dfrac{5}{7}$, $x \neq -4$ **2.** -1, $x \neq 2$ **3.** $\dfrac{x-1}{x+1}$, $x \neq -2, -1$

4. $\dfrac{-1}{x(x+7)}$ **5.** 1 **6.** $\dfrac{x^2+x}{2x+1}$ **7.** $\dfrac{x^2+3x}{2(x+2)}$

Exercises 11.1

1. $\dfrac{9x^2y^3}{12xy^4}$

2. $\dfrac{18xy^4}{27x^2y}$

3. $\dfrac{20x^5}{30x^2y^3}$

4. $\dfrac{15y^4}{20x^3y^2}$

5. $\dfrac{x}{x^2-3x}$

6. $\dfrac{3x}{x^2+5x}$

7. $\dfrac{7x-14}{x-2}$

8. $\dfrac{4-2x}{2x-4}$

9. $\dfrac{9-3x}{4x-12}$

10. $\dfrac{2x-8}{16-4x}$

11. $\dfrac{6x^2+4x}{3xy+2y}$

12. $\dfrac{1+3y}{4x+12xy}$

13. $\dfrac{x^2+6x}{x^2+5x-6}$

14. $\dfrac{x^2-y^2}{3x^2+3xy}$

15. $\dfrac{x^3+27}{x^2-9}$

16. $\dfrac{x^3-8}{x^2-4}$

17. $\dfrac{xy-3y+2x-6}{y^2-4}$

18. $\dfrac{3x^2+14x-24}{18-9x-2x^2}$

19. $\dfrac{x^3-8}{5x-2y+xy-10}$

20. $\dfrac{x^3+64}{2x^2+x-28}$

Perform the indicated operations and reduce to lowest terms. Assume that no denominator has a value of 0. See Examples 3 and 4.

21. $\dfrac{3ax^2}{4b}\cdot\dfrac{6b^2}{27x^2y}$

22. $\dfrac{18x^3}{5y^2}\cdot\dfrac{30y^3}{9x^4}$

23. $\dfrac{24x^3}{25y^2}\cdot\dfrac{10y^5}{18x}$

24. $\dfrac{16x^8}{3y^{11}}\cdot\dfrac{-21y^9}{10x^7}$

25. $\dfrac{x^2-9}{x^2+2x}\cdot\dfrac{x+2}{x-3}$

26. $\dfrac{16x^2-9}{3x^2-15x}\cdot\dfrac{6}{4x+3}$

27. $\dfrac{x^2+2x-3}{x^2+3x} \cdot \dfrac{x}{x+1}$

28. $\dfrac{4x+16}{x^2-16} \cdot \dfrac{x-4}{x}$

29. $\dfrac{x^2+6x-16}{x^2-64} \cdot \dfrac{1}{2-x}$

30. $\dfrac{4-x^2}{x^2-4x+4} \cdot \dfrac{3}{x+2}$

31. $\dfrac{x^2-5x+6}{x^2-4x} \cdot \dfrac{x-4}{x-3}$

32. $\dfrac{2x^2+x-3}{x^2+4x} \cdot \dfrac{2x+8}{x-1}$

33. $\dfrac{2x^2+10x}{3x^2+5x+2} \cdot \dfrac{6x+4}{x^2}$

34. $\dfrac{x+3}{x^2-16} \cdot \dfrac{x^2-3x-4}{x^2-1}$

35. $\dfrac{x}{x^2+7x+12} \cdot \dfrac{x^2-2x-24}{x^2-7x+6}$

36. $\dfrac{x^2-2x-3}{x+5} \cdot \dfrac{x^2-5x-14}{x^2-x-6}$

37. $\dfrac{8-2x-x^2}{x^2-2x} \cdot \dfrac{x-4}{x^2-3x-4}$

38. $\dfrac{3x^2+21x}{x^2-49} \cdot \dfrac{x^2-5x+4}{x^2+3x-4}$

39. $\dfrac{(x-2y)^2}{x^2-5xy+6y^2} \cdot \dfrac{x+2y}{x^2-4xy+4y^2}$

40. $\dfrac{4x^2+6x}{x^2+3x-10} \cdot \dfrac{x^2+4x-12}{x^2+5x-6}$

41. $\dfrac{2x^2+5x+2}{3x^2+8x+4} \cdot \dfrac{3x^2-x-2}{4x^3-x}$

42. $\dfrac{x^2+5x}{4x^2+12x+9} \cdot \dfrac{6x^2+7x-3}{x^2+10x+25}$

43. $\dfrac{x^2+x+1}{x^2-1} \cdot \dfrac{x^2-2x+1}{x^3-1}$

44. $\dfrac{x^2-9}{2x^2+4x+8} \cdot \dfrac{x^3-8}{x^2-5x+6}$

45. $\dfrac{x-2}{x^2-2x+4} \cdot \dfrac{x^3+8}{x^2-4x+4}$

46. $\dfrac{2x^2-7x+3}{x^2-9} \cdot \dfrac{3x^2+8x-3}{6x^2+x-1}$

47. $\dfrac{12x^2y}{9xy^9} \div \dfrac{4x^4y}{x^2y^3}$

48. $\dfrac{35xy^3}{24x^3y} \div \dfrac{15x^4y^3}{84xy^4}$

49. $\dfrac{45xy^4}{21x^2y^2} \div \dfrac{40x^4}{112xy^5}$

50. $\dfrac{x-3}{15x} \div \dfrac{4x-12}{5}$

51. $\dfrac{x-1}{6x+6} \div \dfrac{2x-2}{x^2+x}$

52. $\dfrac{7x-14}{x^2} \div \dfrac{x^2-4}{x^3}$

53. $\dfrac{6x^2-54}{x^4} \div \dfrac{x-3}{x^2}$

54. $\dfrac{x^2-25}{6x+30} \div \dfrac{x-5}{x}$

55. $\dfrac{2x-1}{x^2+2x} \div \dfrac{10x^2-5x}{6x^2+12x}$

56. $\dfrac{x+3}{x^2+3x-4} \div \dfrac{x+2}{x^2+x-2}$

57. $\dfrac{6x^2-7x-3}{x^2-1} \div \dfrac{2x-3}{x-1}$

58. $\dfrac{x^2-9}{2x^2+7x+3} \div \dfrac{x^2-3x}{2x^2+11x+5}$

59. $\dfrac{x^2-6x+9}{x^2-4x+3} \div \dfrac{2x^2-7x+3}{x^2-3x+2}$

60. $\dfrac{x^3+2x^2}{x^3+64} \div \dfrac{4x^2}{x^2-4x+16}$

61. $\dfrac{2x+1}{4x-x^2} \div \dfrac{4x^2-1}{x^2-16}$

62. $\dfrac{x^2-4x+4}{x^2+5x+6} \div \dfrac{x^2+2x-8}{x^2+7x+12}$

63. $\dfrac{x^2-x-6}{x^2+6x+8} \div \dfrac{x^2-4x+3}{x^2+5x+4}$

64. $\dfrac{x^2-x-12}{6x^2+x-9} \div \dfrac{x^2-6x+8}{3x^2-x-6}$

65. $\dfrac{6x^2+5x+1}{4x^3-3x^2} \div \dfrac{3x^2-2x-1}{3x^2-2x+1}$

66. $\dfrac{8x^2 + 2x - 15}{3x^2 + 13x + 4} \div \dfrac{2x^2 + 5x + 3}{6x^2 - x - 1}$

67. $\dfrac{3x^2 + 13x + 14}{4x^3 - 3x^2} \div \dfrac{6x^2 - x - 35}{4x^2 + 5x - 6}$

68. $\dfrac{3x^2 + 2x}{9x^2 - 4} \div \dfrac{27x^3 - 8}{9x^2 - 6x + 4}$

69. $\dfrac{x^2 - 8x + 15}{x^2 - 9x + 14} \div \dfrac{x^2 + 4x - 21}{x - 1}$

70. $\dfrac{6 - 11x - 10x^2}{2x^2 + x - 3} \div \dfrac{5x^3 - 2x^2}{3x^2 - 5x + 2}$

71. $\dfrac{x - 6}{x^2 - 7x + 6} \cdot \dfrac{x^2 - 3x}{x + 3} \cdot \dfrac{x^2 - 9}{x^2 - 4x + 3}$

72. $\dfrac{3x^2 + 11x + 10}{2x^2 + x - 6} \cdot \dfrac{x^2 + 2x - 3}{2x - 1} \cdot \dfrac{2x - 3}{3x^2 + 2x - 5}$

73. $\dfrac{x^3 + 3x^2}{x^2 + 7x + 12} \cdot \dfrac{2x^2 + 7x - 4}{2x^2 - x} \cdot \dfrac{x^2 + 4x - 5}{2x^2 - x - 1}$

74. $\dfrac{x^2 + 2x - 3}{x^2 + 10x + 21} \cdot \dfrac{x^2 + 6x + 5}{x^2 - 7x - 8} \cdot \dfrac{x^2 - x - 56}{x^2 - 3x - 40}$

75. $\dfrac{2x^2 - 5x + 2}{4xy - 2y + 6x - 3} \div \dfrac{xy - 2y + 3x - 6}{2y^2 + 9y + 9}$

76. $\dfrac{2xy - 12x + y - 6}{y^2 - 2y - 24} \div \dfrac{2x^2 + 11x + 5}{xy + 5y + 4x + 20}$

Answer the following word problems.

77. Rectangles: The area of a rectangle (in square feet) is represented by the polynomial function $A(x) = 4x^2 - 4x - 15$. If the length of the rectangle is $(2x + 3)$ feet, find a representation for the width.

$$A(x) = 4x^2 - 4x - 15$$

$2x + 3$

78. Rectangles: The area of a rectangle (in square feet) is represented by the polynomial function $A(x) = 3x^2 - x - 10$. If the length of the rectangle is $(3x + 5)$ feet, find a representation for the width.

$$A(x) = 3x^2 - x - 10$$

$3x + 5$

79. **a.** Define rational expression.

    b. Give an example of a rational expression that is undefined for $x = -2$ and $x = 3$ and has a value of 0 for $x = 1$. Explain how you determined this expression.

    c. Give an example of a rational expression that is undefined for $x = -5$ and never has a value of 0. Explain how you determined this expression.

80. Write the opposite of each of the following expressions.

    a. $3 - x$ **b.** $2x - 7$ **c.** $x + 5$ **d.** $-3x - 2$

81. Given the rational function $f(x) = \dfrac{x - 4}{x^2 - 100}$:

    a. For what values, if any, will $f(x) = 0$?

    b. For what values, if any, is $f(x)$ undefined?

11.2 Addition and Subtraction with Rational Expressions

Objectives

A Add rational expressions.

B Subtract rational expressions.

Objective A   **Addition with Rational Expressions**

To add rational expressions with a common denominator, proceed just as with fractions: add the numerators and keep the common denominator. For example,

$$\frac{5}{x+2}+\frac{6}{x+2}=\frac{5+6}{x+2}=\frac{11}{x+2}. \qquad (x\neq -2)$$

In some cases the sum can be reduced.

$$\frac{x^2+6}{x+2}+\frac{5x}{x+2}=\frac{x^2+5x+6}{x+2}=\frac{(x+2)(x+3)}{x+2}=x+3 \qquad (x\neq -2)$$

Addition with Rational Expressions

For polynomials $P, Q,$ and $R,$ with $Q\neq 0,$

$$\frac{P}{Q}+\frac{R}{Q}=\frac{P+R}{Q}.$$

Example 1

Adding Rational Expressions with a Common Denominator

Find each sum and reduce if possible. (Note the importance of the factoring techniques we studied earlier.) State any restrictions on the variable.

a. $\dfrac{x}{x^2-1}+\dfrac{1}{x^2-1}$

Solution

$$\frac{x}{x^2-1}+\frac{1}{x^2-1}=\frac{x+1}{x^2-1}$$

$$=\frac{x+1}{(x+1)(x-1)}$$

$$=\frac{1}{x-1} \qquad (x\neq -1, 1)$$

Remember, if we use the entire expression in the numerator (or denominator) to reduce, we are left with a factor of 1.

1. Find each sum and reduce if possible. State any restrictions on the variable.

a. $\dfrac{x}{x^2-25}+\dfrac{5}{x^2-25}$

b. $\dfrac{2}{x^2+8x+15}+\dfrac{3x+7}{x^2+8x+15}$

b. $\dfrac{1}{x^2+7x+10}+\dfrac{2x+3}{x^2+7x+10}$

Solution

$$\dfrac{1}{x^2+7x+10}+\dfrac{2x+3}{x^2+7x+10}=\dfrac{2x+4}{x^2+7x+10}$$

$$=\dfrac{2\cancel{(x+2)}}{(x+5)\cancel{(x+2)}}$$

$$=\dfrac{2}{x+5}\qquad (x\neq-5,-2)$$

*Now work margin exercise **1**.*

The rational expressions in Example 1 had common denominators. To add expressions with different denominators, we need to find the least common multiple (LCM) of the denominators. The LCM was discussed in an earlier chapter. The procedure is stated here for polynomials.

To Find the LCM for a Set of Polynomials

1. Completely factor each polynomial (including prime factors for numerical factors).
2. Form the product of all factors that appear, using each factor the most number of times it appears in any one polynomial.

The LCM of a set of denominators is called the **least common denominator (LCD)**. To add fractions with different denominators, begin by changing each fraction to an equivalent fraction with the LCD as the denominator. This is called **building the fraction to higher terms**.

Use the following procedure when adding rational expressions with different denominators.

Procedure for Adding Rational Expressions with Different Denominators

1. Find the LCD (the LCM of the denominators).
2. Rewrite each fraction in an equivalent form with the LCD as the denominator.
3. Add the numerators and keep the common denominator.
4. Reduce if possible.

Example 2

Adding Rational Expressions with Different Denominators

Find each sum and reduce if possible. Assume that no denominator has a value of 0.

a. $\dfrac{y}{y-3}+\dfrac{6}{y+4}$

Solution

In this case, neither denominator can be factored so the LCD is the product of these factors. That is, $\text{LCD}=(y-3)(y+4)$.
Now, using the fundamental principle, we have

$$\dfrac{y}{y-3}+\dfrac{6}{y+4}=\dfrac{y(y+4)}{(y-3)(y+4)}+\dfrac{6(y-3)}{(y+4)(y-3)}$$

$$=\dfrac{(y^2+4y)+(6y-18)}{(y-3)(y+4)}$$

$$=\dfrac{y^2+10y-18}{(y-3)(y+4)}. \qquad \text{The numerator is not factorable and the expression is reduced.}$$

b. $\dfrac{1}{x^2+6x+9}+\dfrac{1}{x^2-9}+\dfrac{1}{2x+6}$

Solution

First, find the LCD.

Step 1: Factor each expression completely.

$$x^2+6x+9=(x+3)^2$$
$$x^2-9=(x+3)(x-3)$$
$$2x+6=2(x+3)$$

Step 2: Form the product of 2, $(x+3)^2$, and $(x-3)$. That is, use each factor the most number of times it appears in any one factorization.

$$\text{LCD}=2(x+3)^2(x-3)$$

Now use the LCD and add as follows.

$$\dfrac{1}{x^2+6x+9}+\dfrac{1}{x^2-9}+\dfrac{1}{2x+6}$$

$$=\dfrac{1}{(x+3)^2}+\dfrac{1}{(x+3)(x-3)}+\dfrac{1}{2(x+3)}$$

$$=\dfrac{1\cdot 2(x-3)}{(x+3)^2\cdot 2(x-3)}+\dfrac{1\cdot 2(x+3)}{(x+3)(x-3)\cdot 2(x+3)}+\dfrac{1\cdot(x+3)(x-3)}{2(x+3)\cdot(x+3)(x-3)}$$

$$=\dfrac{(2x-6)+(2x+6)+(x^2-9)}{2(x+3)^2(x-3)}=\dfrac{x^2+4x-9}{2(x+3)^2(x-3)}$$

Now work margin exercise 2.

---

2. Find each sum and reduce if possible. Assume that no denominator has a value of 0.

a. $\dfrac{x}{x+2}+\dfrac{3}{x+3}$

b.

$$\dfrac{1}{x^2+10x+25}+\dfrac{1}{x^2-25}+\dfrac{1}{2x+10}$$

Addition and Subtraction with Rational Expressions **Section 11.2** **1172**

notes

Objective B   **Subtraction with Rational Expressions**

When subtracting fractions, the placement of negative signs can be critical. For example, note how −2 can be indicated in three different forms:

$$-\frac{6}{3} = -2, \quad \frac{-6}{3} = -2, \quad \text{and} \quad \frac{6}{-3} = -2.$$

Thus we have $-\dfrac{6}{3} = \dfrac{-6}{3} = \dfrac{6}{-3}$.

With polynomials we have the following statement about the placement of negative signs which can be **very useful in subtraction**. We seldom leave the negative sign in the denominator.

Placement of Negative Signs

If P and Q are polynomials and $Q \neq 0$, then

$$-\frac{P}{Q} = \frac{P}{-Q} = \frac{-P}{Q}.$$

To subtract rational expressions with a common denominator, proceed just as with fractions: subtract the numerators and keep the common denominator. For example,

$$\frac{17}{x+7} - \frac{23}{x+7} = \frac{17-23}{x+7} = \frac{-6}{x+7} \quad \left(\text{or } -\frac{6}{x+7}\right).$$

Subtraction with Rational Expressions

For polynomials P, Q, and R, with $Q \neq 0$,

$$\frac{P}{Q} - \frac{R}{Q} = \frac{P-R}{Q}.$$

Example 3

Subtracting Rational Expressions with a Common Denominator

Find each difference and reduce if possible. Assume that no denominator has a value of 0.

a. $\dfrac{2x-5y}{x+y} - \dfrac{3x-7y}{x+y}$

Solution

$$\dfrac{2x-5y}{x+y} - \dfrac{3x-7y}{x+y} = \dfrac{2x-5y-(3x-7y)}{x+y} \qquad \text{\color{red}Subtract the entire numerator.}$$

$$= \dfrac{2x-5y-3x+7y}{x+y}$$

$$= \dfrac{-x+2y}{x+y}$$

b. $\dfrac{x^2}{x^2+4x+4} - \dfrac{2x+8}{x^2+4x+4}$

Solution

$$\dfrac{x^2}{x^2+4x+4} - \dfrac{2x+8}{x^2+4x+4} = \dfrac{x^2-(2x+8)}{x^2+4x+4} \qquad \text{\color{red}Subtract the entire numerator.}$$

$$= \dfrac{x^2-2x-8}{x^2+4x+4}$$

$$= \dfrac{(x-4)\,\cancel{(x+2)}}{(x+2)\,\cancel{(x+2)}} \qquad \text{\color{red}Factor and reduce.}$$

$$= \dfrac{x-4}{x+2}$$

c. $\dfrac{x}{x-5} - \dfrac{3}{5-x}$

Solution

Each denominator is the **opposite** of the other. Multiply both the numerator and denominator of the second fraction by −1 so that both denominators will be the same, in this case $x-5$.

$$\dfrac{x}{x-5} - \dfrac{3}{5-x} = \dfrac{x}{x-5} - \dfrac{3}{(5-x)} \cdot \dfrac{\color{red}(-1)}{\color{red}(-1)}$$

$$= \dfrac{x}{x-5} - \dfrac{-3}{x-5}$$

$$= \dfrac{x-(-3)}{x-5}$$

$$= \dfrac{x+3}{x-5}$$

Now work margin exercise 3.

3. Find each difference and reduce if possible. Assume that no denominator has a value of 0.

a. $\dfrac{4x-7y}{3x-y} - \dfrac{3x-9y}{3x-y}$

b. $\dfrac{x^2}{x^2+6x+9} - \dfrac{2x+15}{x^2+6x+9}$

c. $\dfrac{x}{x-4} - \dfrac{5}{4-x}$

As with addition, if the rational expressions do not have the same denominator, find the LCM of the denominators (the LCD) and use the fundamental principle to **build each fraction to higher terms**, if necessary, so that each has the LCD as the denominator.

Example 4

Subtracting Rational Expressions with Different Denominators

Find each difference and reduce if possible. Assume that no denominator has a value of 0.

a. $\dfrac{x+5}{x-5} - \dfrac{100}{x^2-25}$

Solution

$$x - 5 = x - 5$$
$$x^2 - 25 = (x+5)(x-5)$$

$\text{LCD} = (x+5)(x-5)$

$$\frac{x+5}{x-5} - \frac{100}{x^2 - 25} = \frac{(x+5)(x+5)}{(x-5)(x+5)} - \frac{100}{(x+5)(x-5)}$$

$$= \frac{(x^2 + 10x + 25) - 100}{(x+5)(x-5)}$$

$$= \frac{x^2 + 10x + 25 - 100}{(x+5)(x-5)}$$

$$= \frac{x^2 + 10x - 75}{(x+5)(x-5)}$$

$$= \frac{(x+15)\,\cancel{(x-5)}}{(x+5)\,\cancel{(x-5)}}$$

$$= \frac{x+15}{x+5}$$

b. $\dfrac{x+y}{(x-y)^2} - \dfrac{x}{2x^2 - 2y^2}$

Solution

$$(x-y)^2 = (x-y)^2$$
$$2x^2 - 2y^2 = 2(x-y)(x+y)$$

$\text{LCD} = 2(x-y)^2(x+y)$

$$\frac{x+y}{(x-y)^2} - \frac{x}{2x^2 - 2y^2}$$

$$= \frac{(x+y) \cdot 2(x+y)}{(x-y)^2 \cdot 2(x+y)} - \frac{x(x-y)}{2(x-y)(x+y)(x-y)}$$

$$= \frac{2(x^2 + 2xy + y^2) - (x^2 - xy)}{2(x-y)^2(x+y)}$$

$$= \frac{2x^2 + 4xy + 2y^2 - x^2 + xy}{2(x-y)^2(x+y)}$$

$$= \frac{x^2 + 5xy + 2y^2}{2(x-y)^2(x+y)}$$

4. Find each difference and reduce if possible. Assume that no denominator has a value of 0.

a. $\dfrac{x+6}{x-6} - \dfrac{144}{x^2-36}$

b. $\dfrac{5x+y}{(x+y)^2} - \dfrac{y}{3x^2-3y^2}$

c. $\dfrac{4x-8}{x^2+4x-12} - \dfrac{x^2+6x}{x^2+9x+18}$

d.

$\dfrac{y+4}{xy-2y+x-2} - \dfrac{y-5}{xy-2y+3x-6}$

c. $\dfrac{3x-12}{x^2+x-20} - \dfrac{x^2+5x}{x^2+9x+20}$

Hint: In this problem, both expressions can be reduced before looking for the LCD.

Solution

$$\dfrac{3x-12}{x^2+x-20} - \dfrac{x^2+5x}{x^2+9x+20}$$

$$= \dfrac{3(x-4)}{(x+5)(x-4)} - \dfrac{x(x+5)}{(x+5)(x+4)}$$

$$= \dfrac{3}{x+5} - \dfrac{x}{x+4}$$

Now subtract these two expressions with $\text{LCD} = (x+5)(x+4)$.

$$\dfrac{3}{x+5} - \dfrac{x}{x+4} = \dfrac{3(x+4)}{(x+5)(x+4)} - \dfrac{x(x+5)}{(x+4)(x+5)}$$

$$= \dfrac{(3x+12) - (x^2+5x)}{(x+5)(x+4)}$$

$$= \dfrac{3x+12-x^2-5x}{(x+5)(x+4)}$$

$$= \dfrac{-x^2-2x+12}{(x+5)(x+4)}$$

d. $\dfrac{x+1}{xy-3y+4x-12} - \dfrac{x-3}{xy+6y+4x+24}$

Solution

$$xy-3y+4x-12 = y(x-3)+4(x-3)$$
$$= (x-3)(y+4)$$

$$xy+6y+4x+24 = y(x+6)+4(x+6)$$
$$= (x+6)(y+4)$$

$\text{LCD} = (x-3)(y+4)(x+6)$

$$\dfrac{x+1}{xy-3y+4x-12} - \dfrac{x-3}{xy+6y+4x+24}$$

$$= \dfrac{(x+1)(x+6)}{(y+4)(x-3)(x+6)} - \dfrac{(x-3)(x-3)}{(x+6)(y+4)(x-3)}$$

$$= \dfrac{x^2+7x+6 - (x^2-6x+9)}{(y+4)(x-3)(x+6)}$$

$$= \dfrac{x^2+7x+6-x^2+6x-9}{(y+4)(x-3)(x+6)}$$

$$= \dfrac{13x-3}{(y+4)(x-3)(x+6)}$$

Now work margin exercise 4.

Practice Problems

Perform the indicated operations and reduce if possible. Assume that no denominator is 0.

1. $\dfrac{1}{1-y}+\dfrac{2}{y^2-1}$

2. $\dfrac{x+3}{x^2+x-6}+\dfrac{x-2}{x^2+4x-12}$

3. $\dfrac{1}{y+2}-\dfrac{1}{y^3+8}$

4. $\dfrac{x}{x^2-1}-\dfrac{1}{x-1}$

Practice Problem Answers

1. $\dfrac{-1}{y+1}$

2. $\dfrac{2x+4}{(x-2)(x+6)}$

3. $\dfrac{y^2-2y+3}{(y+2)(y^2-2y+4)}$

4. $\dfrac{-1}{(x+1)(x-1)}$

Exercises 11.2

1. $\dfrac{3x}{x+4} + \dfrac{12}{x+4}$

2. $\dfrac{7x}{x+5} + \dfrac{35}{x+5}$

3. $\dfrac{x-1}{x+6} + \dfrac{x+13}{x+6}$

4. $\dfrac{3x-1}{2x-6} + \dfrac{x-11}{2x-6}$

5. $\dfrac{3x+1}{5x+2} + \dfrac{2x+1}{5x+2}$

6. $\dfrac{x^2+3}{x+1} + \dfrac{4x}{x+1}$

7. $\dfrac{x-5}{x^2-2x+1} + \dfrac{x+3}{x^2-2x+1}$

8. $\dfrac{2x^2+5}{x^2-4} + \dfrac{3x-1}{x^2-4}$

9. $\dfrac{13}{7-x} - \dfrac{1}{x-7}$

10. $\dfrac{6x}{x-6} + \dfrac{36}{6-x}$

11. $\dfrac{3x}{x-4} + \dfrac{16-x}{4-x}$

12. $\dfrac{20}{x-10} - \dfrac{3}{10-x}$

13. $\dfrac{x^2+2}{x^2+x-12} + \dfrac{x+1}{12-x-x^2}$

14. $\dfrac{10}{x^2-x-6} - \dfrac{5x}{6+x-x^2}$

15. $\dfrac{x^2+2}{x^2-4} - \dfrac{4x-2}{x^2-4}$

16. $\dfrac{2x+5}{2x^2-x-1} - \dfrac{4x+2}{2x^2-x-1}$

17. $\dfrac{x+3}{7x-2} + \dfrac{2x-1}{14x-4}$

18. $\dfrac{3x+1}{4x+10} + \dfrac{4-x}{2x+5}$

19. $\dfrac{5}{x-3} + \dfrac{x}{x^2-9}$

20. $\dfrac{x+1}{x^2-3x-10} + \dfrac{x}{x-5}$

21. $\dfrac{x}{x-1} - \dfrac{4}{x+2}$

22. $\dfrac{x-1}{3x-1} - \dfrac{8+4x}{x+2}$

23. $\dfrac{x+2}{x+3} - \dfrac{4}{3-x}$

24. $\dfrac{x-1}{4-x} + \dfrac{3x}{x+5}$

25. $\dfrac{x+2}{3x+9} + \dfrac{2x-1}{2x-6}$

26. $\dfrac{x}{4x-8} - \dfrac{3x+2}{3x+6}$

27. $\dfrac{3x}{6+x} - \dfrac{2x}{x^2-36}$

28. $\dfrac{4}{x+5} - \dfrac{2x+3}{x^2+4x-5}$

29. $\dfrac{4x+1}{7-x} + \dfrac{x-1}{x^2-8x+7}$

30. $\dfrac{3x-4}{x^2-x-20} - \dfrac{2}{5-x}$

31. $\dfrac{4x}{x^2+3x-28} + \dfrac{3}{x^2+6x-7}$

32. $\dfrac{3x}{x^2+2x+1} - \dfrac{x}{x^2+4x+4}$

33. $\dfrac{3x+4}{2x^2-23x+30} - \dfrac{x+5}{2x^2-19x+24}$

34. $\dfrac{x+1}{x^2-3x+2} + \dfrac{6}{x^2-6x+8}$

35. $\dfrac{4x-1}{x^2-5x+4} + \dfrac{2x+7}{x^2-11x+28}$

36. $\dfrac{7x+3}{5x^2+27x+36} + \dfrac{3x-2}{5x^2+22x+24}$

37. $\dfrac{x-6}{7x^2-3x-4} + \dfrac{7-x}{7x^2+18x+8}$

38. $\dfrac{x+10}{x^2+5x+4} - \dfrac{4}{x^2+6x+8}$

39. $\dfrac{x-3}{4x^2-5x-6} - \dfrac{4x+10}{2x^2+x-10}$

40. $\dfrac{2x+1}{8x^2-37x-15} + \dfrac{2-x}{8x^2+11x+3}$

41. $\dfrac{3x}{4-x} + \dfrac{7x}{x+4} - \dfrac{x-3}{x^2-16}$

42. $\dfrac{x}{x+3} + \dfrac{x+1}{3-x} + \dfrac{x^2+4}{x^2-9}$

43. $-\dfrac{1}{2} + \dfrac{x-5}{x-3} + \dfrac{x-1}{x^2-5x+6}$

44. $-4 + \dfrac{1-2x}{x+6} + \dfrac{x^2+1}{x^2+4x-12}$

45. $\dfrac{2}{x^2-4} - \dfrac{3}{x^2-3x+2} + \dfrac{x-1}{x^2+x-2}$

46. $\dfrac{5}{x^2+3x+2} + \dfrac{4}{x^2+6x+8} - \dfrac{6}{x^2+5x+4}$

47. $\dfrac{x}{x^2+4x-21} + \dfrac{1-x}{x^2+8x+7} + \dfrac{3x}{x^2-2x-3}$

48. $\dfrac{x+2}{9x^2-6x+4} + \dfrac{10x-5x^2}{27x^3+8} - \dfrac{2}{3x+2}$

49. $\dfrac{3x+9}{x^2-5x+4} + \dfrac{49}{12+x-x^2} + \dfrac{3x+21}{x^2+2x-3}$

50. $\dfrac{5x+22}{x^2+8x+15} + \dfrac{4}{x^2+4x+3} + \dfrac{6}{x^2+6x+5}$

51. $\dfrac{x}{xy+x-2y-2} + \dfrac{x+2}{xy+x+y+1}$

52. $\dfrac{4x}{xy-3x+y-3} + \dfrac{x+2}{xy+2y-3x-6}$

53. $\dfrac{3y}{xy+2x+3y+6} + \dfrac{x}{x^2-2x-15}$

54. $\dfrac{2}{xy-4x-2y+8} + \dfrac{5y}{y^2-3y-4}$

55. $\dfrac{x+6}{x^2+x+1} - \dfrac{3x^2+x-4}{x^3-1}$

56. $\dfrac{2x-5}{8x^2-4x+2} + \dfrac{x^2-2x+5}{8x^3+1}$

57. $\dfrac{x+1}{x^3-3x^2+x-3} + \dfrac{x^2-5x-8}{x^4-8x^2-9}$

58. $\dfrac{x+4}{x^3-5x^2+6x-30} - \dfrac{x-7}{x^3-2x^2+6x-12}$

59. $\dfrac{x+1}{2x^2-x-1} + \dfrac{2x}{2x^2+5x+2} - \dfrac{2x}{3x^2+4x-4}$

60. $\dfrac{x-6}{3x^2+10x+3} - \dfrac{2x}{5x^2-3x-2} + \dfrac{2x}{3x^2-2x-1}$

Writing & Thinking

61. Discuss the steps in the process you go through when adding two rational expressions with different denominators. That is, discuss how you find the least common denominator when adding rational expressions and how you use this LCD to find equivalent rational expressions that you can add.

11.3 Simplifying Complex Fractions

Objective A Simplifying Complex Fractions (First Method)

Objectives

A Simplify complex fractions using the first method.

B Simplify complex fractions using the second method.

C Simplify complex algebraic expressions.

A **complex fraction** is a fraction in which the numerator and/or denominator are themselves fractions or the sum or difference of fractions. Examples of complex fractions are:

$$\frac{\dfrac{1}{x+3}-\dfrac{1}{x}}{1+\dfrac{3}{x}} \qquad \text{and} \qquad \frac{x+y}{x^{-1}+y^{-1}}=\frac{x+y}{\dfrac{1}{x}+\dfrac{1}{y}}.$$

The objective here is to develop techniques for simplifying complex fractions so that they are written in the form of a single reduced rational expression.

In a complex fraction such as $\dfrac{\dfrac{1}{x+3}-\dfrac{1}{x}}{1+\dfrac{3}{x}}$, the large fraction bar is a symbol of inclusion.

The expression could also be written as follows.

$$\frac{\dfrac{1}{x+3}-\dfrac{1}{x}}{1+\dfrac{3}{x}}=\left(\frac{1}{x+3}-\frac{1}{x}\right)\div\left(1+\frac{3}{x}\right)$$

Thus a complex fraction indicates that the numerator is to be divided by the denominator.

To Simplify Complex Fractions (First Method)

1. Simplify the numerator so that it is a single rational expression.
2. Simplify the denominator so that it is a single rational expression.
3. Divide the numerator by the denominator and reduce to lowest terms.

This method is used to simplify the complex fractions in Examples 1 and 2. Study the examples closely so that you understand what happens at each step.

Example 1

First Method for Simplifying Complex Fractions

Simplify the complex fraction $\dfrac{\dfrac{3x}{y^2}}{\dfrac{12x}{7y}}$.

Solution

$$\frac{\dfrac{3x}{y^2}}{\dfrac{12x}{7y}} = \frac{\cancel{3}\,\cancel{x}}{y^{\cancel{2}}_{\,y}} \cdot \frac{7\,\cancel{y}}{\cancel{12}_{\,4}\,\cancel{x}}$$

To divide, multiply by the reciprocal of the denominator.

$$= \frac{7}{4y}$$

Example 2

First Method for Simplifying Complex Fractions

Simplify the following complex fractions.

a. $\dfrac{\dfrac{1}{x+3} - \dfrac{1}{x}}{1 + \dfrac{3}{x}}$

Solution

$$\frac{\dfrac{1}{x+3} - \dfrac{1}{x}}{1 + \dfrac{3}{x}} = \frac{\dfrac{1 \cdot x}{(x+3) \cdot x} - \dfrac{1(x+3)}{x(x+3)}}{\dfrac{x}{x} + \dfrac{3}{x}}$$

Combine the fractions in the numerator and in the denominator separately.

Note that $1 = \dfrac{x}{x}$.

$$= \frac{\dfrac{x - (x+3)}{x(x+3)}}{\dfrac{x+3}{x}}$$

$$= \frac{\dfrac{x - x - 3}{x(x+3)}}{\dfrac{x+3}{x}}$$

$$= \frac{\dfrac{-3}{x(x+3)}}{\dfrac{x+3}{x}}$$

$$= \frac{-3}{\cancel{x}(x+3)} \cdot \frac{\cancel{x}}{x+3}$$

To divide, multiply by the reciprocal of the denominator.

$$= \frac{-3}{(x+3)^2}$$

b. $\dfrac{x+y}{x^{-1}+y^{-1}}$

Solution

$$\frac{x+y}{x^{-1}+y^{-1}} = \frac{x+y}{\dfrac{1}{x}+\dfrac{1}{y}}$$

Recall that $x^{-1} = \dfrac{1}{x}$ and $y^{-1} = \dfrac{1}{y}$.

$$= \frac{\dfrac{x+y}{1}}{\dfrac{1}{x}\cdot\dfrac{y}{y}+\dfrac{1}{y}\cdot\dfrac{x}{x}}$$

Add the two fractions in the denominator.

$$= \frac{\dfrac{x+y}{1}}{\dfrac{y}{xy}+\dfrac{x}{xy}}$$

$$= \frac{\dfrac{x+y}{1}}{\dfrac{y+x}{xy}}$$

$$= \frac{\cancel{x+y}}{1}\cdot\frac{xy}{\cancel{y+x}}$$

Multiply by the reciprocal of the denominator.

$$= \frac{xy}{1}$$

$$= xy$$

Now work margin exercise 1 and 2.

Objective B **Simplifying Complex Fractions (Second Method)**

A second method is to find the LCM of the denominators in the fractions in both the original numerator and the original denominator and then multiply **both** the numerator and denominator by this LCM.

> **To Simplify Complex Fractions (Second Method)**
>
> 1. Find the LCM of all the denominators in the numerator and denominator of the complex fraction.
> 2. Multiply both the numerator and denominator of the complex fraction by this LCM.
> 3. Simplify both the numerator and denominator and reduce to lowest terms.

Simplify the following complex fractions.

1. $\dfrac{\dfrac{5x}{y^3}}{\dfrac{15x}{y^2}}$

2. a. $\dfrac{\dfrac{1}{x+6}-\dfrac{1}{x}}{1+\dfrac{6}{x}}$

b. $\dfrac{3x-3y}{(3x)^{-1}-(3y)^{-1}}$

3. Simplify the following complex fractions, using the second method for simplifying complex fractions.

a. $\dfrac{\dfrac{1}{x+6} - \dfrac{1}{x}}{1 + \dfrac{6}{x}}$

b. $\dfrac{3x - 3y}{(3x)^{-1} - (3y)^{-1}}$

Example 3

Second Method for Simplifying Complex Fractions

Simplify the following complex fractions.

a. $\dfrac{\dfrac{1}{x+3} - \dfrac{1}{x}}{1 + \dfrac{3}{x}}$

Solution

$$\dfrac{\dfrac{1}{x+3} - \dfrac{1}{x}}{1 + \dfrac{3}{x}} = \dfrac{\left(\dfrac{1}{x+3} - \dfrac{1}{x}\right) \cdot x(x+3)}{\left(1 + \dfrac{3}{x}\right) \cdot x(x+3)}$$

Multiply by $x(x+3)$, the LCM of $\{x, x+3\}$. This multiplication can be done because the net effect is that the fraction is multiplied by 1.

$$= \dfrac{\dfrac{1}{x+3} \cdot x(x+3) - \dfrac{1}{x} \cdot x(x+3)}{1 \cdot x(x+3) + \dfrac{3}{x} \cdot x(x+3)}$$

$$= \dfrac{x - (x+3)}{x(x+3) + 3(x+3)}$$

$$= \dfrac{x - x - 3}{(x+3)(x+3)}$$

$$= \dfrac{-3}{(x+3)^2}$$

Note that this matches the result found in Example **2a.** using Method 1.

b. $\dfrac{x+y}{x^{-1} + y^{-1}}$

Solution

$$\dfrac{x+y}{x^{-1} + y^{-1}} = \dfrac{\dfrac{x+y}{1}}{\dfrac{1}{x} + \dfrac{1}{y}}$$

$$= \dfrac{\left(\dfrac{x+y}{1}\right)xy}{\left(\dfrac{1}{x} + \dfrac{1}{y}\right)xy}$$

Multiply by xy, the LCM of $\{1, x, y\}$.

$$= \dfrac{(x+y)xy}{\dfrac{1}{x} \cdot xy + \dfrac{1}{y} \cdot xy}$$

$$= \dfrac{(x+y)xy}{y + x} = xy$$

*Now work margin exercise **3**.*

Objective C Simplifying Complex Algebraic Expressions

A **complex algebraic expression** is an expression that involves rational expressions and more than one operation. In simplifying such expressions, the rules for order of operations apply. As with complex fractions, the objective is to simplify the expression so that it is written in the form of a single reduced rational expression.

Example 4

Simplifying Complex Algebraic Expressions

Simplify the following expression.

$$\frac{4-x}{x+3} + \frac{x}{x+3} \div \frac{x}{x-3}$$

Solution

The rules for order of operations indicate that the division is to be done first followed by the addition.

$$\frac{4-x}{x+3} + \frac{x}{x+3} \div \frac{x}{x-3} = \frac{4-x}{x+3} + \frac{\cancel{x}}{x+3} \cdot \frac{x-3}{\cancel{x}}$$

$$= \frac{4-x}{x+3} + \frac{x-3}{x+3}$$

$$= \frac{4-x+x-3}{x+3}$$

$$= \frac{1}{x+3}$$

Now work margin exercise 4.

Practice Problems

Simplify each of the following expressions.

1. $\dfrac{\dfrac{1}{x}}{1+\dfrac{1}{x}}$

2. $\dfrac{\dfrac{1}{x+2}-\dfrac{1}{x}}{1+\dfrac{2}{x}}$

3. $\dfrac{1+\dfrac{3}{x-3}}{x-\dfrac{x^2}{x-3}}$

4. $\dfrac{\dfrac{1}{x+y}-\dfrac{1}{x-y}}{\dfrac{2y}{x^2-y^2}}$

5. $\dfrac{5}{x}-\dfrac{3}{x-2} \div \dfrac{x}{x-2}$

4. Simplify the following expression.

$$\frac{6}{x+4} + \frac{x}{x+4} \div \frac{x}{x-4}$$

Practice Problem Answers

1. $\dfrac{1}{x+1}$

2. $\dfrac{-2}{(x+2)^2}$

3. $-\dfrac{1}{3}$

4. -1

5. $\dfrac{2}{x}$

Exercises 11.3

1. $\dfrac{\dfrac{2x}{3y^2}}{\dfrac{5x^2}{6y}}$

2. $\dfrac{\dfrac{6x^2}{5y}}{\dfrac{x}{10y^2}}$

3. $\dfrac{\dfrac{12x^3}{7y^2}}{\dfrac{3x^5}{2y}}$

4. $\dfrac{\dfrac{9x^2}{7y^3}}{\dfrac{3xy}{14}}$

5. $\dfrac{\dfrac{x+3}{2x}}{\dfrac{2x-1}{4x^2}}$

6. $\dfrac{\dfrac{x-2}{6x}}{\dfrac{x+3}{3x^2}}$

7. $\dfrac{\dfrac{2x-1}{x}}{\dfrac{2}{x}+3}$

8. $\dfrac{2-\dfrac{3}{x}}{\dfrac{x^2-4}{x}}$

9. $\dfrac{\dfrac{3}{x}+\dfrac{1}{2x}}{1+\dfrac{2}{x}}$

10. $\dfrac{\dfrac{3}{x}+\dfrac{5}{2x}}{\dfrac{1}{x}+4}$

11. $\dfrac{1+\dfrac{1}{x}}{1-\dfrac{1}{x^2}}$

12. $\dfrac{\dfrac{2}{y}+1}{\dfrac{4}{y^2}-1}$

13. $\dfrac{\dfrac{1}{x}+\dfrac{1}{3x}}{\dfrac{x+6}{x^2}}$

14. $\dfrac{\dfrac{3}{x}-\dfrac{6}{x^2}}{\dfrac{x-2}{x^2}}$

15. $\dfrac{\dfrac{7}{x}-\dfrac{14}{x^2}}{\dfrac{1}{x}-\dfrac{4}{x^3}}$

16. $\dfrac{\dfrac{x}{3}+\dfrac{1}{9x^2}}{\dfrac{1}{27x^2}+\dfrac{x}{9}}$

17. $\dfrac{\dfrac{1}{3}+\dfrac{1}{x}}{\dfrac{1}{2}-\dfrac{1}{x}}$

18. $\dfrac{\dfrac{x}{y}-\dfrac{1}{3}}{\dfrac{6}{y}-\dfrac{2}{x}}$

19. $\dfrac{\dfrac{2}{x}+\dfrac{3}{4y}}{\dfrac{3}{2x}-\dfrac{5}{3y}}$

20. $\dfrac{\dfrac{4}{3x}-\dfrac{5}{y}}{\dfrac{1}{3}+\dfrac{3}{y}}$

21. $\dfrac{1+x^{-1}}{1-x^{-2}}$

22. $\dfrac{x^{-3}+1}{1-x^{-1}}$

23. $\dfrac{1}{x^{-1}+y^{-1}}$

24. $\dfrac{x-y}{x^{-2}-y^{-2}}$

25. $\dfrac{x^{-1}+y^{-1}}{x+y}$

26. $\dfrac{y^{-2}-x^{-2}}{x+y}$

27. $\dfrac{x^{-1}+y^{-1}}{x^{-1}-y^{-1}}$

28. $\dfrac{x^{-1}+y^{-1}}{x^{-2}-y^{-2}}$

29. $\dfrac{\dfrac{4}{x}-1}{1-\dfrac{1}{x-3}}$

30. $\dfrac{x+\dfrac{3}{x-4}}{1-\dfrac{1}{x}}$

31. $\dfrac{1-\dfrac{4}{x+3}}{1-\dfrac{2}{x+1}}$

32. $\dfrac{1+\dfrac{4}{2x-3}}{1+\dfrac{x}{x+1}}$

33. $\dfrac{\dfrac{1}{x+h}-\dfrac{1}{x}}{h}$

34. $\dfrac{\dfrac{1}{(x+h)^2}-\dfrac{1}{x^2}}{h}$

35. $\dfrac{\left(2+\dfrac{1}{x+h}\right)-\left(2+\dfrac{1}{x}\right)}{h}$

36. $\dfrac{\left(\dfrac{1}{(x+h)^2}-3\right)-\left(\dfrac{1}{x^2}-3\right)}{h}$

37. $\dfrac{x^2-4y^2}{1-\dfrac{2x+y}{x-y}}$

38. $\dfrac{8x^2-2y^2}{\dfrac{4x-1}{x-y}-2}$

39. $\dfrac{\dfrac{x+1}{x-1}-\dfrac{x-1}{x+1}}{\dfrac{x+1}{x-1}+\dfrac{x-1}{x+1}}$

40. $\dfrac{\dfrac{1}{x^2-1}-\dfrac{1}{x+1}}{\dfrac{1}{x-1}+\dfrac{1}{x^2-1}}$

41. $\dfrac{\dfrac{x}{x-4}-\dfrac{1}{x-1}}{\dfrac{x}{x-1}+\dfrac{2}{x-3}}$

42. $\dfrac{\dfrac{1}{x+1}-\dfrac{x}{x+2}}{\dfrac{x}{x+2}-\dfrac{2}{x-1}}$

Simplify the following complex algebraic expressions. See Example 4.

43. $\dfrac{1}{x+1}-\dfrac{3}{2x}\cdot\dfrac{4x}{x+1}$

44. $\dfrac{4}{x}-\dfrac{2}{x^2-2x}\cdot\dfrac{x-2}{5}$

45. $\left(\dfrac{8}{x}-\dfrac{3}{4x}\right)\div\dfrac{4x+5}{x}$

46. $\left(\dfrac{2}{x}+\dfrac{5}{x-3}\right)\div\dfrac{x}{2x-6}$

47. $\dfrac{x}{x-1}-\dfrac{3}{x-1}\cdot\dfrac{x+2}{x}$

48. $\dfrac{x+3}{x+2}+\dfrac{x}{x+2}\cdot\dfrac{x-3}{x^2}$

49. $\dfrac{x-1}{x+4}+\dfrac{x-6}{x^2+3x-4}\div\dfrac{x-4}{x-1}$

50. $\dfrac{x}{x+3}-\dfrac{3}{x+5}\div\dfrac{x-2}{x^2+3x-10}$

51. Some complex fractions involve the sum (or difference) of complex fractions. Beginning with the outermost denominator, simplify each of the following expressions.

a. $1 + \dfrac{1}{1 + \dfrac{1}{1 + \dfrac{1}{1+1}}}$

b. $2 - \dfrac{1}{2 - \dfrac{1}{2 - \dfrac{1}{2-1}}}$

c. $x + \dfrac{1}{x + \dfrac{1}{x + \dfrac{1}{x+1}}}$

11.4 Solving Rational Equations

Objective A **Proportions**

A **ratio** is a comparison of two numbers by division. Ratios are written in the form

$$a : b \quad \text{or} \quad \frac{a}{b} \quad \text{or} \quad a \text{ to } b.$$

For example, suppose the ratio of female to male faculty at the local community college is 3 to 2. We can also write this ratio as $\frac{3 \text{ females}}{2 \text{ males}}$ or as 3 females : 2 males. This ratio does not mean that there are only 5 teachers at the college. There are many ratios that reduce to $\frac{3}{2}$. There might be 35 teachers at the college, 21 females and 14 males, because $21 + 14 = 35$ and $\frac{21}{14} = \frac{3}{2}$. Until you know the total number of teachers at the college, you cannot know the exact numbers of female and male teachers. But you do know that there are 3 female teachers for every 2 male teachers.

A **proportion** is a special type of equation stating that two ratios are equal. For example,

$$\frac{21}{14} = \frac{3}{2} \quad \text{and} \quad \frac{3.5}{7} = \frac{3.75}{7.5} \quad \text{are proportions.}$$

Proportion

A **proportion** is an equation stating that two ratios are equal.

In symbols, $\dfrac{a}{b} = \dfrac{c}{d}$ is a proportion.

Proportions may involve only numbers as in $\dfrac{2}{3} = \dfrac{10}{15}$. However, proportions can also be used to find unknown quantities in applications, and in such cases, will involve variables.

One method of solving proportions with variables is to "clear" the equation of fractions by first multiplying both sides of the equation by the LCM of the denominators, the LCD. This method is illustrated in Example 1.

Objectives

A Solve proportions.

B Solve equations that involve rational expressions.

C Use the process of solving equations to manipulate formulas.

1. Solve the following proportions.

a. $\dfrac{x-7}{2x} = \dfrac{8}{4x}$

b. $\dfrac{4}{x-8} = \dfrac{8}{x}$

Example 1

Proportions

Solve the following proportions.

a. $\dfrac{x-5}{2x} = \dfrac{6}{3x}$ $\qquad$ LCD $= 6x$

Solution

$$\dfrac{x-5}{2x} = \dfrac{6}{3x} \qquad \text{Note: } x \neq 0.$$

$$6x \cdot \dfrac{x-5}{2x} = 6x \cdot \dfrac{6}{3x} \qquad \text{Multiply both sides by the LCD, } 6x.$$

$$3(x-5) = 2(6)$$

$$3x - 15 = 12$$

$$3x = 27$$

$$x = 9$$

Check: $\dfrac{9-5}{2\cdot 9} \overset{?}{=} \dfrac{6}{3\cdot 9}$

$$\dfrac{4}{18} \overset{?}{=} \dfrac{6}{27}$$

$$\dfrac{2}{9} = \dfrac{2}{9}$$

Thus, the solution is $x = 9$.

b. $\dfrac{3}{x-6} = \dfrac{5}{x}$ $\qquad$ LCD $= x(x-6)$

Solution

$$\dfrac{3}{x-6} = \dfrac{5}{x} \qquad \text{Note: } x \neq 0, 6.$$

$$x(x-6) \cdot \dfrac{3}{x-6} = x(x-6) \cdot \dfrac{5}{x} \qquad \text{Multiply both sides by the LCD, } x(x-6).$$

$$3x = 5x - 30$$

$$-2x = -30$$

$$x = 15$$

Check: $\dfrac{3}{15-6} \overset{?}{=} \dfrac{5}{15}$

$$\dfrac{3}{9} \overset{?}{=} \dfrac{5}{15}$$

$$\dfrac{1}{3} = \dfrac{1}{3}$$

Thus, the solution is $x = 15$.

Now work margin exercise 1.

Proportions can be used to solve many everyday types of word problems. Using the correct ratios of units on both sides of the proportion is critical. One of the following conditions must be true:

1. The numerators agree in type and the denominators agree in type.
2. The numerators correspond and the denominators correspond.

Example 2 illustrates one situation.

Example 2

Application of Proportions

On an architect's scale drawing of a home, $\frac{1}{2}$ inch represents 10 feet. What length does a measure of $2\frac{1}{2}$ inches represent?

Solution

Set up a proportion representing the information. In this example the numerators are the same type (inches) and the denominators are the same type (feet).

$$\frac{\frac{1}{2} \text{ inch}}{10 \text{ feet}} = \frac{2\frac{1}{2} \text{ inches}}{x \text{ feet}} \qquad \text{LCD} = 10x$$

$$10x \cdot \frac{\frac{1}{2}}{10} = 10x \cdot \frac{\frac{5}{2}}{x} \qquad \text{Multiply both sides by } 10x.$$

$$\frac{1}{2}x = 25 \qquad \text{Simplify.}$$

$$2 \cdot \frac{1}{2}x = 2 \cdot 25 \qquad \text{Multiply both sides by 2.}$$

$$x = 50 \qquad \text{Simplify.}$$

On this drawing, $2\frac{1}{2}$ inches represent 50 feet.

Now work margin exercise 2.

2. On a scale of a map of a highway, 1 inch represents 60 miles. What length does a measure of $3\frac{3}{4}$ inches represent?

Any of the following four equations could have been used to solve the problem in Example 2:

Agree in type

$$\frac{\frac{1}{2}\text{ inch}}{10\text{ feet}} = \frac{2\frac{1}{2}\text{ inches}}{x\text{ feet}} \qquad \frac{10\text{ feet}}{\frac{1}{2}\text{ inch}} = \frac{x\text{ feet}}{2\frac{1}{2}\text{ inches}}$$

Correspond

$$\frac{\frac{1}{2}\text{ inch}}{2\frac{1}{2}\text{ inches}} = \frac{10\text{ feet}}{x\text{ feet}} \qquad \frac{2\frac{1}{2}\text{ inches}}{\frac{1}{2}\text{ inch}} = \frac{x\text{ feet}}{10\text{ feet}}$$

Objective B **Solving Equations with Rational Expressions**

An equation such as $\dfrac{3}{x} + \dfrac{1}{8} = \dfrac{13}{4x}$ that involves the sum of rational expressions is **not a proportion**. However, the method of finding the solution to the equation is similar in that we "clear" the fractions by multiplying both sides of the equation by the LCD of the fractions. In this example, the LCD is $8x$ and we can proceed as follows:

$$\frac{3}{x} + \frac{1}{8} = \frac{13}{4x} \qquad \text{Note that } x \neq 0.$$

$$8x\left(\frac{3}{x} + \frac{1}{8}\right) = 8x \cdot \frac{13}{4x} \qquad \text{Multiply both sides by } 8x, \text{ the LCD.}$$

$$8x \cdot \frac{3}{x} + 8x \cdot \frac{1}{8} = \overset{2}{8x} \cdot \frac{13}{4x} \qquad \text{Use the distributive property and reduce.}$$

$$24 + x = 26 \qquad \text{Simplify.}$$

$$x = 2.$$

As we have seen, rational expressions may contain variables in either the numerator or denominator or both. In any case, a general approach to solving equations that contain rational expressions is as follows.

To Solve an Equation Containing Rational Expressions

1. Find the LCD of the fractions.
2. Multiply both sides of the equation by this LCD and simplify.
3. Solve the resulting equation. (This equation will have only polynomials on both sides.)
4. Check each solution in the **original equation**. (Remember that no denominator can be 0 and any solution that gives a 0 denominator is to be discarded.)

Checking is particularly important when equations have rational expressions. Multiplying by the LCD may introduce solutions that are not solutions to the original equation. Such solutions are called **extraneous solutions** or **extraneous roots** and occur because multiplication by a variable expression may, in effect, be multiplying the original equation by 0.

Example 3

Solving Equations Involving Rational Expressions

State any restrictions on the variable, and then solve each equation.

a. $\dfrac{1}{x-4} = \dfrac{3}{x^2-5x}$

Solution

First, find the LCD of the fractions, and then multiply both sides of the equation by the LCD.

$$\left.\begin{array}{l} x-4 = x-4 \\ x^2-5x = x(x-5) \end{array}\right\} \quad \text{LCD} = x(x-5)(x-4)$$

$$x(x-5)(x-4)\cdot\frac{1}{x-4} = x(x-5)(x-4)\cdot\frac{3}{x(x-5)} \qquad (x \neq 0, 4, 5)$$

$$x(x-5) = 3(x-4)$$
$$x^2 - 5x = 3x - 12$$
$$x^2 - 8x + 12 = 0$$
$$(x-6)(x-2) = 0$$

$$x-6 = 0 \quad \text{or} \quad x-2 = 0$$
$$x = 6 \qquad\qquad x = 2$$

Since 6 and 2 are not restrictions, there are two solutions, $x = 6$ and $x = 2$.

b. $\dfrac{x}{x-2} + \dfrac{x-6}{x(x-2)} = \dfrac{5x}{x-2} - \dfrac{10}{x-2}$

Solution

First, find the LCD of the fractions and then multiply each term on both sides of the equation by the LCD.

$$\left.\begin{array}{l} x-2 \\ x(x-2) \end{array}\right\} \quad \text{LCD} = x(x-2)$$

$$x(x-2)\cdot\frac{x}{x-2} + x(x-2)\cdot\frac{x-6}{x(x-2)} = x(x-2)\cdot\frac{5x}{x-2} - x(x-2)\cdot\frac{10}{x-2}$$

$$x^2 + x - 6 = 5x^2 - 10x \qquad (x \neq 0, 2)$$
$$0 = 4x^2 - 11x + 6$$
$$0 = (4x-3)(x-2)$$
$$4x - 3 = 0 \quad \text{or} \quad x-2 = 0$$
$$4x = 3 \qquad\qquad x = 2$$
$$x = \frac{3}{4}$$

3. State any restrictions on the variable, and then solve each equation.

a. $\dfrac{1}{x+2} = \dfrac{6}{x^2+7x}$

b.

$\dfrac{x}{x-2} + \dfrac{4x-4}{x(x-2)} = \dfrac{6x}{x-2} - \dfrac{8}{x-2}$

c. $\dfrac{4}{x^2-25} = \dfrac{2}{x^2} + \dfrac{2}{x^2-5x}$

The only solution is $x = \dfrac{3}{4}$, since 2 is a restricted value $(x \neq 0, 2)$ and thus **not** a solution. No denominator can be 0.

c. $\dfrac{2}{x^2-9} = \dfrac{1}{x^2} + \dfrac{1}{x^2-3x}$

Solution

First, find the LCD of the fractions and then multiply each term on both sides of the equation by the LCD.

$$\left.\begin{array}{l} x^2-9 = (x+3)(x-3) \\ x^2 = x^2 \\ x^2-3x = x(x-3) \end{array}\right\} \quad \text{LCD} = x^2(x+3)(x-3)$$

$$x^2 \cancel{(x+3)} \cancel{(x-3)} \cdot \dfrac{2}{\cancel{(x+3)}\cancel{(x-3)}}$$

$$= \cancel{x^2}(x+3)(x-3) \cdot \dfrac{1}{\cancel{x^2}} + x^{\cancel{2}}{}^{x}(x+3) \cancel{(x-3)} \cdot \dfrac{1}{\cancel{x}\,\cancel{(x-3)}}$$

$$2x^2 = (x+3)(x-3) + x(x+3) \qquad (x \neq 0, -3, 3)$$

$$2x^2 = x^2 - 9 + x^2 + 3x$$

$$2x^2 = 2x^2 + 3x - 9$$

$$9 = 3x$$

$$\cancel{3 = x} \qquad\qquad \text{Note that 3 is one of the restrictions.}$$

There is no solution. The solution set is the empty set, $\varnothing$. The original equation is a contradiction.

Now work margin exercise 3.

Objective C **Formulas**

As discussed in an earlier chapter, many formulas are equations relating more than one variable. The equation is solved for one variable in terms of another variable (maybe more than one). The process of solving equations can be used to manipulate the formula so that it is solved for one of the other variables. As illustrated in Example 4, the resulting equation may involve a rational expression.

Example 4

Solving a Formula for a Specified Variable

The formula $S = 2\pi r^2 + 2\pi rh$ is used to find the surface area (S) of a right circular cylinder, where r is the radius of the cylinder and h is the height of the cylinder. Solve the formula for h.

Solution

$$S = 2\pi r^2 + 2\pi rh \qquad \text{Write the formula.}$$

$$S - 2\pi r^2 = 2\pi rh \qquad \text{Add } -2\pi r^2 \text{ to both sides of the equation.}$$

$$\frac{S - 2\pi r^2}{2\pi r} = h \qquad \text{Divide both sides by } 2\pi r.$$

Thus, the formula solved for h is: $h = \dfrac{S - 2\pi r^2}{2\pi r} \left(\text{or } h = \dfrac{S}{2\pi r} - r \right)$.

Now work margin exercise 4.

4. The surface area of a rectangular box is given by the formula $SA = 2(lh) + 2(wh) + 2(lw)$. Solve the formula for l.

Practice Problems

Solve each proportion. State any restrictions on the variable.

1. $\dfrac{5x}{3} = \dfrac{9x + 4}{5}$

2. $\dfrac{16}{x} = \dfrac{4}{x - 5}$

3. On a road map, each inch represents 50 miles. What distance is represented by 4.5 inches on the map?

Solve each equation. State any restrictions on the variable.

4. $\dfrac{5}{3x + 2} = \dfrac{4}{3x + 1}$

5. $\dfrac{x}{x - 3} - \dfrac{2x + 3}{x^2 + x - 12} = \dfrac{x - 1}{x + 4}$

6. Solve the formula $A = P + Pr$ for r.

Practice Problem Answers

1. $x = -6$

2. $x = \dfrac{20}{3}; \ x \neq 0, 5$

3. 225 miles

4. $x = 1; \ x \neq -\dfrac{1}{3}, -\dfrac{2}{3}$

5. $x = 1; \ x \neq -4, 3$

6. $r = \dfrac{A - P}{P}$

Exercises 11.4

1. $\dfrac{4x}{7} = \dfrac{x+5}{3}$

2. $\dfrac{3x+1}{4} = \dfrac{2x+1}{3}$

3. $\dfrac{10}{x} = \dfrac{5}{x-2}$

4. $\dfrac{8}{x-3} = \dfrac{12}{2x-3}$

5. $\dfrac{4}{x-4} = \dfrac{2}{x+3}$

6. $\dfrac{3}{x+5} = \dfrac{6}{x-2}$

7. $\dfrac{x+2}{5x} = \dfrac{x-6}{3x}$

8. $\dfrac{x-4}{3x} = \dfrac{x-2}{5x}$

9. $\dfrac{5x+2}{x-6} = \dfrac{11}{4}$

10. $\dfrac{x+9}{3x+2} = \dfrac{5}{8}$

Solve the following word problems. See Example 2.

11. **Computers:** Making a statistical analysis, Ana found 3 defective computers in a sample of 20 computers. If this ratio is consistent, how many defective computers does she expect to find in a batch of 2400 computers?

12. **Manufacturing:** At the Bright-As-Day light bulb plant, 3 out of each 100 bulbs produced are defective. If the daily production is 4800 bulbs, how many are defective?

13. **Education:** The University of Arizona has a ratio of 1 professor for every 18 students. If there are 1600 faculty members at the university, how many students are enrolled there?

14. **Baseball:** New York Yankees player Alex Rodriguez records a hit a mean of 14 times for every 50 times at bat. If he maintains this mean, how many at-bats will he need to achieve 110 hits? (Round to the nearest whole number.)

15. **Cartography:** On a map of Maryland, one inch represents 4 miles. If there are 8.5 inches between Baltimore, MD and Washington, DC, how far are the two cities from each other?

16. **Architecture:** A floor plan is drawn to scale in which 1 inch represents 4 feet. What size will the drawing be for a room that is 30 feet by 40 feet? (**Hint:** Set up two proportions.)

17. **Baking:** The recipe for Nestle Tollhouse Chocolate Chip Cookies calls for 2 cups of chocolate chips to make 5 dozen cookies. If you want to bake 17 dozen cookies, how many cups of chocolate chips do you need?

18. **Car Maintenance:** In the instructions for Never-Ice Antifreeze it states that 4 quarts of antifreeze are needed for every 10 quarts of radiator capacity. If Sal's car has a 22-quart radiator, how many quarts of antifreeze will it need?

19. **Landscape Architecture:** An architect is to draw plans for a city park. He intends to use a scale of $\dfrac{1}{2}$ inch to represent 25 feet. How many inches will be needed to use for the length and width of a rectangular playing field that is 50 yards by 125 yards? (1 yard = 3 feet)

20. **Testing Cars:** A test driver wants to increase the speed of the car he is driving by 3 miles per hour every 2 seconds. But he can only check his speed every 5 seconds because he is busy with other items during the test drive.

    a. By how much should he increase his speed in 5 seconds?

    b. If he starts checking his speed at 40 miles per hour, how fast should he be going after 10 seconds?

21. $\dfrac{5x}{4} - \dfrac{1}{2} = -\dfrac{3}{16}$

22. $\dfrac{x}{6} - \dfrac{1}{42} = \dfrac{1}{7}$

23. $\dfrac{3x-1}{6} - \dfrac{x+3}{4} = \dfrac{7}{12}$

24. $\dfrac{x-2}{3} - \dfrac{x-3}{5} = \dfrac{13}{15}$

25. $\dfrac{2+x}{4} - \dfrac{5x-2}{12} = \dfrac{8-2x}{5}$

26. $\dfrac{4x+1}{5} = \dfrac{2x+3}{2} - \dfrac{x+2}{4}$

27. $\dfrac{2}{3x} = \dfrac{1}{4} - \dfrac{1}{6x}$

28. $\dfrac{1}{x} - \dfrac{8}{21} = \dfrac{3}{7x}$

29. $\dfrac{3}{5x} - \dfrac{1}{5} = \dfrac{3}{4x}$

30. $\dfrac{3}{8x} - \dfrac{7}{10} = \dfrac{1}{5x}$

31. $\dfrac{3}{4x} - \dfrac{1}{2} = \dfrac{7}{8x} + \dfrac{1}{6}$

32. $\dfrac{5}{3x} + \dfrac{1}{2} = \dfrac{7}{9x} - \dfrac{5}{6}$

33. $\dfrac{2}{4x+1} = \dfrac{4}{x^2+9x}$

34. $\dfrac{3}{4x-1} = \dfrac{4}{x^2+x}$

35. $\dfrac{9}{x^2-6x} = \dfrac{5}{2x-3}$

36. $\dfrac{-9}{x^2+5x} = \dfrac{8}{4-9x}$

37. $\dfrac{x}{x-4} - \dfrac{4}{2x-1} = 1$

38. $\dfrac{x}{x+3} + \dfrac{1}{x+2} = 1$

39. $\dfrac{x+2}{x+1} + \dfrac{x+2}{x+4} = 2$

40. $\dfrac{3x-2}{x+4} + \dfrac{2x+5}{x-1} = 5$

41. $\dfrac{2}{4x-1} + \dfrac{1}{x+1} = \dfrac{3}{x+1}$

42. $\dfrac{x-2}{x+4} - \dfrac{3}{2x+1} = \dfrac{x-7}{x+4}$

43. $\dfrac{x-2}{x-3} + \dfrac{x-3}{x-2} = \dfrac{2x^2}{x^2-5x+6}$

44. $\dfrac{x}{x-4} - \dfrac{12x}{x^2+x-20} = \dfrac{x-1}{x+5}$

45. $\dfrac{3x+5}{3x+2} + \dfrac{8x+16}{3x^2-4x-4} = \dfrac{x+2}{x-2}$

46. $\dfrac{3x+5}{3x+2} - \dfrac{4-2x}{3x^2+8x+4} = \dfrac{x+4}{x+2}$

47. $\dfrac{3}{3x-1} + \dfrac{1}{x+1} = \dfrac{4}{2x-1}$

48. $\dfrac{2}{x+1} + \dfrac{4}{2x-3} = \dfrac{4}{x-5}$

49. $S = \dfrac{a}{1-r}$; solve for r (formula for the sum of an infinite geometric sequence)

50. $z = \dfrac{x - \bar{x}}{s}$; solve for x (formula used in statistics)

51. $z = \dfrac{x - \bar{x}}{s}$; solve for s (formula used in statistics)

52. $a_n = a_1 + (n-1)d$; solve for d (formula for the nth term in an arithmetic sequence)

53. $m = \dfrac{y - y_1}{x - x_1}$; solve for y (formula for the slope of a line)

54. $v_{\text{avg}} = \dfrac{d_2 - d_1}{t_2 - t_1}$; solve for d_2 (formula for mean velocity)

55. $\dfrac{1}{R_{\text{total}}} = \dfrac{1}{R_1} + \dfrac{1}{R_2}$; solve for R_{total} (formula used in electronics)

56. $\dfrac{1}{x} = \dfrac{1}{t_1} + \dfrac{1}{t_2}$; solve for x (formula used in mathematics)

57. $A = P + Pr$; solve for P (formula used for compound interest)

58. $y = \dfrac{ax + b}{cx + d}$; solve for x (formula used in mathematics)

Writing & Thinking

In simplifying rational expressions, the result is a rational or polynomial expression. However, in solving equations with rational expressions, the goal is to find a value (or values) for the variable that will make the equation a true statement. Many students confuse these two ideas. To avoid confusing the techniques for adding and subtracting rational expressions with the techniques for solving equations, simplify the expression in part **a.** and solve the equation in part **b.** Explain, in your own words, the differences in your procedures. Assume no denominator has a value of 0.

59. **a.** $\dfrac{10}{x} + \dfrac{31}{x-1} + \dfrac{4x}{x-1}$ **b.** $\dfrac{10}{x} + \dfrac{31}{x-1} = \dfrac{4x}{x-1}$

60. **a.** $\dfrac{-4}{x^2 - 16} + \dfrac{x}{2x+8} - \dfrac{1}{4}$ **b.** $\dfrac{-4}{x^2 - 16} + \dfrac{x}{2x+8} = \dfrac{1}{4}$

61. **a.** $\dfrac{3x}{x^2 - 4} + \dfrac{5}{x+2} + \dfrac{2}{x-2}$ **b.** $\dfrac{3x}{x^2 - 4} + \dfrac{5}{x+2} = \dfrac{2}{x-2}$

62. **a.** $\dfrac{7}{5x} + \dfrac{2}{x-4} - \dfrac{3}{5x}$ **b.** $\dfrac{7}{5x} + \dfrac{2}{x-4} = \dfrac{3}{5x}$

63. **a.** $\dfrac{2}{x+9} - \dfrac{2}{x-9} + \dfrac{1}{2}$ **b.** $\dfrac{2}{x+9} - \dfrac{2}{x-9} = \dfrac{1}{2}$

11.5 Applications: Rational Expressions

The following strategy for solving word problems is valid for all word problems that involve algebraic equations.

Objectives

A   Solve applied problems related to fractions.

B   Solve applied problems related to work.

C   Solve applications involving distance, rate, and time.

To Solve a Word Problem Containing Rational Expressions

1. Read the problem carefully. Read it several times if necessary.
2. Decide what is asked for and assign a variable to the unknown quantity.
3. Draw a diagram or set up a chart whenever possible as a visual aid.
4. Form an equation that relates the information provided.
5. Solve the equation.
6. Check your solution with the wording of the problem to be sure it makes sense.

Objective A   Problems Related to Fractions

We now introduce word problems involving rational expressions with problems relating the numerator and denominator of a fraction.

Example 1

Fractions

The denominator of a fraction is 8 more than the numerator. If both the numerator and denominator are increased by 3, the new fraction is equal to $\frac{1}{2}$. Find the original fraction.

Solution

Reread the problem to be sure that you understand all terminology used. Assign variables to the unknown quantities.

Let, n = original numerator

     $n + 8$ = original denominator

     $\dfrac{n}{n+8}$ = original fraction

1. The denominator of a fraction is 3 more than the numerator. If both the numerator and denominator are increased by 4, the new fraction is equal to $\dfrac{3}{4}$. Find the original fraction.

$$\frac{n+3}{(n+8)+3} = \frac{1}{2}$$

$$\frac{n+3}{n+11} = \frac{1}{2}$$

The numerator and the denominator are each increased by 3, making a new fraction that is equal to $\dfrac{1}{2}$.

$$2(n+11) \cdot \frac{n+3}{n+11} = 2(n+11) \cdot \frac{1}{2}$$

$$2n+6 = n+11$$

$$n = 5 \quad \longleftarrow \text{ Original numerator}$$

$$n+8 = 13 \quad \longleftarrow \text{ Original denominator}$$

Check: $\dfrac{(5)+3}{(13)+3} = \dfrac{8}{16} = \dfrac{1}{2}$

The original fraction is $\dfrac{5}{13}$.

Now work margin exercise **1.**

Objective B **Problems Related to Work**

Problems involving work usually translate into equations involving rational expressions. The basic idea is to **represent what part of the work is done in one unit of time**. For example, if a man can dig a ditch in 3 hours, what part (of the ditch-digging job) can he do in one hour? The answer is $\dfrac{1}{3}$ of the work in one hour. If a fence was painted in 2 days, then $\dfrac{1}{2}$ of the work of painting the fence was done in 1 day. (These ideas assume a steady working pace.) In general, if the total work took x hours, then $\dfrac{1}{x}$ of the total work would be done in one hour.

Example 2

Work Problems

a. A carpenter can build a certain type of patio cover in 6 hours. His partner takes 8 hours to build the same cover. How long would it take them working together to build this type of patio cover?

Solution

Let x = number of hours to build the cover working together.

Person(s)	Time of Work (in Hours)	Part of Work Done in 1 Hour
Carpenter	6	$\dfrac{1}{6}$
Partner	8	$\dfrac{1}{8}$
Together	x	$\dfrac{1}{x}$

	Part done in		Part done in		Part done in
	1 hr by carpenter	+	1 hr by partner	=	1 hr together

$$\frac{1}{6} \quad + \quad \frac{1}{8} \quad = \quad \frac{1}{x}$$

$$\frac{1}{6}\left(\overset{4}{24}x\right)+\frac{1}{8}\left(\overset{3}{24}x\right)=\frac{1}{x}\left(24x\right)$$ Multiply each term on both sides by $24x$, the LCD of the fractions.

$$4x+3x=24$$

$$7x=24$$

$$x=\frac{24}{7}$$

Together, they can build the patio cover in $\frac{24}{7}$ hours, or $3\frac{3}{7}$ hours.

(Note that this answer is reasonable because the time is less than either person would take working alone.)

b. A man can wax his car three times faster than his daughter can. Together they can do the job in 4 hours. How long would it take each of them working alone?

Solution

Let t = number of hours for the man alone to wax the car and $3t$ = number of hours for the daughter alone to wax the car.

Person(s)	Time of Work (in Hours)	Part of Work Done in 1 Hour
Man	t	$\frac{1}{t}$
Daughter	$3t$	$\frac{1}{3t}$
Together	4	$\frac{1}{4}$

Part done by man alone in 1 hour | Part done by daughter alone in 1 hour | Part done working together in 1 hour

$$\underbrace{\frac{1}{t}}_{} + \underbrace{\frac{1}{3t}}_{} = \underbrace{\frac{1}{4}}_{}$$

$$\frac{1}{\cancel{t}}\left(12\cancel{t}\right) + \frac{1}{\cancel{3t}}\left(\cancel{12t}^{\,4}\right) = \frac{1}{\cancel{4}}\left(\cancel{12}^{\,3}t\right)$$

$$12 + 4 = 3t$$

$$16 = 3t$$

$$t = \frac{16}{3}$$

$$3t = (3)\left(\frac{16}{3}\right) = 16$$

Check:

man's part in 1 hr: $\dfrac{1}{t} = \dfrac{1}{\frac{16}{3}} = \dfrac{3}{16}$

daughter's part in 1 hr: $\dfrac{1}{3t} = \dfrac{1}{3 \cdot \frac{16}{3}} = \dfrac{1}{16}$

man's part in 4 hr: $\dfrac{3}{16} \cdot 4 = \dfrac{3}{4}$

daughter's part in 4 hr : $\dfrac{1}{16} \cdot 4 = \dfrac{1}{4}$

$\dfrac{3}{4} + \dfrac{1}{4} = 1$ car waxed in 4 hours

Working alone, the man takes $\dfrac{16}{3}$ hours, or $5\dfrac{1}{3}$ hours, and his daughter takes 16 hours.

c. A man was told that his new pool would fill through an inlet valve in 3 hours. He knew something was wrong when the pool took 8 hours to fill. He found he had left the drain valve open. How long would it take to drain the pool once it is completely filled and only the drain valve is open?

Solution

Let t = time to drain pool with only the drain valve open.

Note: We use the information gained when the pool was filled with both valves open. In that situation, the inlet and outlet valves worked against each other.

Valves	Hours to Fill or Drain	Part Filled or Drained in 1 Hour
Inlet	3	$\dfrac{1}{3}$
Outlet	t	$\dfrac{1}{t}$
Together	8	$\dfrac{1}{8}$

Part filled by inlet in 1 hour − Part emptied by outlet in 1 hour = Part filled together in 1 hour

$$\frac{1}{3} \quad - \quad \frac{1}{t} \quad = \quad \frac{1}{8}$$

$$\frac{1}{3}\left(\overset{8}{\cancel{24}}t\right) - \frac{1}{t}\left(24\cancel{t}\right) = \frac{1}{8}\left(\overset{3}{\cancel{24}}t\right)$$

$$8t - 24 = 3t$$

$$5t = 24$$

$$t = \frac{24}{5}$$

The pool would drain in $\dfrac{24}{5}$ hours, or $4\dfrac{4}{5}$ hours. (Note that this is more time than the inlet valve would take to fill the pool. If the outlet valve worked faster than the inlet valve, then the pool would never have filled in the first place.)

Now work margin exercise 2.

Objective C **Problems Related to Distance-Rate-Time:** $d = rt$

You may recall that the basic formula involving distance, rate, and time is $d = rt$. This relationship can also be stated in the forms $t = \dfrac{d}{r}$ and $r = \dfrac{d}{t}$.

If distance and rate are known or can be represented, then $t = \dfrac{d}{r}$ is the way to represent time. Similarly, if the distance and time are known or can be represented, then $r = \dfrac{d}{t}$ is the way to represent rate.

2. Solve the following work problems.

a. An electrician can wire a house in 12 hours. Her assistant takes 18 hours to wire the same size house. How long would it take them to wire the house together?

b. A mother can carve a pumpkin twice as fast as her eight year old son. Together they can carve the pumpkin in 3 hours. How long would it take each of them working alone?

c. Janet's pool normally takes 4 hours to fill. Last time it took 10 hours to fill and she found that she had left the drain open. How long would it take to drain the pool once it is completely filled and only the drain valve is open?

Example 3

Distance-Rate-Time

a. On Lake Itasca, a man can row his boat 5 miles per hour. On the nearby Mississippi River, it takes him the same time to row 5 miles downstream as it does to row 3 miles upstream. What is the speed of the river current in miles per hour?

Solution

Let c = the speed of the current.

Distance and rate are represented first in the table below. Then the time going downstream and coming back upstream is represented in terms of distance and rate. Since the rate is in miles per hour, the distance is in miles and the time is in hours.

	Distance d	Rate r	Time $t = \dfrac{d}{r}$
Downstream	5	$5 + c$	$\dfrac{5}{5+c}$
Upstream	3	$5 - c$	$\dfrac{3}{5-c}$

$$\frac{5}{5+c} = \frac{3}{5-c} \qquad \textcolor{red}{\text{The times are equal.}}$$

$$\textcolor{red}{(5+c)}\,(5-c) \cdot \frac{5}{5+c} = (5+c)\,\textcolor{red}{(5-c)} \cdot \frac{3}{5-c}$$

$$25 - 5c = 15 + 3c$$

$$-8c = -10$$

$$c = \frac{5}{4}$$

Check:

$$\text{time downstream} = \frac{5}{5 + \dfrac{5}{4}} = \frac{5}{\dfrac{20}{4} + \dfrac{5}{4}} = \frac{5}{\dfrac{25}{4}} = 5 \cdot \frac{4}{25} = \frac{4}{5} \text{ hours}$$

$$\text{time upstream} = \frac{3}{5 - \dfrac{5}{4}} = \frac{3}{\dfrac{20}{4} - \dfrac{5}{4}} = \frac{3}{\dfrac{15}{4}} = 3 \cdot \frac{4}{15} = \frac{4}{5} \text{ hours}$$

The times are equal. The rate of the river current is $\dfrac{5}{4}$ mph, or $1\dfrac{1}{4}$ mph.

b. If a passenger train travels three times as fast as a freight train, and the freight train takes 4 hours longer to travel 210 miles, what is the speed of each train?

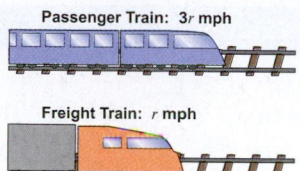

Passenger Train: $3r$ mph

Freight Train: r mph

Solution

Let $r =$ rate of freight train in miles per hour
and $3r =$ rate of passenger train in miles per hour.

	Distance d	Rate r	Time $t = \dfrac{d}{r}$
Freight	210	r	$\dfrac{210}{r}$
Passenger	210	$3r$	$\dfrac{210}{3r}$

Note: If the rate is faster, then the time is shorter. Thus the fraction $\dfrac{210}{3r}$ is smaller than the fraction $\dfrac{210}{r}$.

$$\frac{210}{r} - \frac{210}{3r} = 4 \qquad \text{\color{red}{The difference between their times is 4 hours.}}$$

$$\frac{210}{r} - \frac{70}{r} = 4$$

$$\frac{210}{r} \cdot r - \frac{70}{r} \cdot r = 4 \cdot r$$

$$210 - 70 = 4r$$

$$140 = 4r$$

$$35 = r$$

$$105 = 3r$$

Check:

$$\text{time for freight train} = \frac{210}{35} = 6 \text{ hours}$$

$$\text{time for passenger train} = \frac{210}{105} = 2 \text{ hours}$$

$$6 - 2 = 4 \text{ hours difference in time}$$

The freight train travels 35 mph, and the passenger train travels 105 mph.

Now work margin exercise 3.

3. Solve the following distance-rate-time problems.

a. On Lake Wespanee a man can paddle his kayak at a rate of 6 miles per hour. On the nearby Millitonka River it takes him the same time to paddle 6 miles downstream as it does to paddle 3 miles upstream. What is the speed of the current of the river?

b. If a commercial airplane travels twice as fast as a private aircraft, and the private aircraft takes $2\dfrac{1}{2}$ hours longer to travel 900 miles, what is the speed of each airplane?

Exercises 11.5

Solve the following problems. See Example 1.

1. The sum of two numbers is 117 and they are in the ratio of 8 to 5. Find the two numbers.

2. If 4 is subtracted from a certain number and the difference is divided by 2, the result is 1 more than $\frac{1}{5}$ of the original number. Find the original number.

3. What number must be added to both the numerator and denominator of $\frac{16}{21}$ to make the resulting fraction equal to $\frac{5}{6}$?

4. Find the number that can be subtracted from both the numerator and denominator of the fraction $\frac{69}{102}$ so that the result is $\frac{5}{8}$.

5. The denominator of a fraction exceeds the numerator by 7. If the numerator is increased by 3 and the denominator is increased by 5, the resulting fraction is equal to $\frac{1}{2}$. Find the original fraction.

6. The numerator of a fraction exceeds the denominator by 5. If the numerator is decreased by 4 and the denominator is increased by 3, the resulting fraction is equal to $\frac{4}{5}$. Find the original fraction.

7. One number is $\frac{3}{4}$ of another number. Their sum is 63. Find the numbers.

8. The sum of two numbers is 24. If $\frac{2}{5}$ of the larger number is equal to $\frac{2}{3}$ of the smaller number, find the numbers.

9. One number exceeds another by 5. The sum of their reciprocals is equal to 19 divided by the product of the two numbers. Find the numbers.

10. One number is 3 less than another. The sum of their reciprocals is equal to 7 divided by the product of the two numbers. Find the numbers.

Solve the following word problems. See Examples 2 and 3.

11. **Shirt Sales:** A manufacturer sold a group of shirts for $1026. One-fifth of the shirts were priced at $18 each and the remainder at $24 each. How many shirts were sold?

12. **Paying Bills:** Luis spent $\frac{1}{5}$ of his monthly salary for rent and $\frac{1}{6}$ of his monthly salary for his car payment. If $950 was left, what was his monthly salary?

13. **Painting:** Suppose that an artist expects that for every 9 special brushes she orders, 7 will be good and 2 will be defective. If she orders 54 brushes, how many will she expect to be defective?

14. Travel by Car: It takes Rosa, traveling at 30 mph, 30 minutes longer to go a certain distance than it takes Melody traveling at 50 mph. Find the distance traveled.

15. Travel by Plane: It takes a plane flying at 450 mph 25 minutes longer to travel a certain distance than it takes a second plane to fly the same distance at 500 mph. Find the distance.

16. Landscaping: Toni needs 4 hours to complete the yard work. Her husband, Sonny, needs 6 hours to do the work. How long will the job take if they work together?

17. Manufacturing: In 1921, automated wrapping machines were used to aid in the wrapping of Hershey Kisses in the Hershey chocolate factory. The machine could wrap the candies 100 times faster than a person could. Together they can wrap a crate full of Kisses in 5 minutes. How long would it take each of them working alone?

18. Mass Mailings: Ben's secretary can address the weekly newsletters in $4\frac{1}{2}$ hours. Charlie's secretary needs only 3 hours. How long will it take if they both work on the job?

19. Shoveling Snow: Working together, Greg and Cindy can clean the snow from the driveway in 20 minutes. It would have taken Cindy 36 minutes working alone. How long would it have taken Greg alone?

20. Carpentry: A carpenter and his partner can put up a patio cover in $3\frac{3}{7}$ hours. If the partner needs 8 hours to complete the patio alone, how long would it take the carpenter working alone?

21. Travel: Beth can travel 208 miles in the same length of time it takes Anna to travel 192 miles. If Beth's speed is 4 mph greater than Anna's, find both rates.

22. Biking: Kirk can bike 32 miles in the same amount of time that his twin brother Karl can bike 24 miles. If Kirk bikes 2 mph faster than Karl, how fast does each man bike?

23. Plane Speeds: A commercial airliner can travel 750 miles in the same amount of time that it takes a private plane to travel 300 miles. The speed of the airliner is 60 mph more than twice the speed of the private plane. Find the speed of each aircraft.

24. Car Speeds: Gabriela drives her car 350 miles and has a mean of a certain speed. If the mean speed had been 9 mph less, she could have traveled only 300 miles in the same length of time. What was her mean speed?

25. Boating: A family travels 18 miles down river and returns. It takes 8 hours to make the round trip. Their rate in still water is twice the rate of the river's current. How long will the return trip take?

26. Boat Speed: Cruise ships travel 5 times faster than sailboats (in optimal wind conditions). If it takes 16 hours longer for a sailboat to travel 100 miles from Charleston, SC to Savannah, GA, what is the speed of each boat?

27. Wind Speed: An airplane can fly 650 mph in still air. If it can travel 2800 miles with the wind in the same time it can travel 2400 miles against the wind, find the wind speed. (**Note**: A tailwind increases the speed of the plane and a headwind decreases the speed of the plane.)

28. Wind Speed: A one-engine plane can fly 120 mph in still air. If it can fly 490 miles with a tailwind in the same time that it can fly 350 miles against a headwind, what is the speed of the wind?

29. Filling a Pool: Using a small inlet pipe it takes 9 hours to fill a pool. Using a large inlet pipe it only takes 3 hours. If both are used simultaneously, how long will it take to fill the pool?

30. Filling a Pool: An inlet pipe on a swimming pool can be used to fill the pool in 36 hours. The drain pipe can be used to empty the pool in 40 hours. If the pool is $\frac{2}{3}$ filled using the inlet pipe and then the drain pipe is accidentally opened, how long from that time will it take to fill the pool?

31. Clearing Land: A contractor hires two bulldozers to clear the trees from a 20-acre tract of land. One works twice as fast as the other. It takes them 3 days to clear the tract working together. How long would it take each of them alone?

32. Store Maintenance: John, Ralph, and Denny, working together, can clean their bait and tackle store in 6 hours. Working alone, Ralph takes twice as long to clean the store as John does. Denny needs three times as long as John does. How long would it take each man working alone?

33. Boating: Francois rode his jet ski 36 miles downstream and then 36 miles back. The round trip took $5\frac{1}{4}$ hours. Find the speed of the jet ski in still water and the speed of the current if the speed of the current is $\frac{1}{7}$ the speed of the jet ski.

34. Boating: Momence, IL is 12 miles upstream on the same side of the river from Kankakee, IL on the Kankakee River. A motorboat that can travel 8 mph in still water leaves Momence and travels downstream toward Kankakee. At the same time, another boat that can travel 10 mph leaves Kankakee and travels upstream toward Momence. Each boat completes the trip in the same amount of time. Find the rate of the current.

35. Skiing: Samantha rides the ski lift to the top of Blue Mountain, a distance of $1\frac{3}{4}$ kilometers (a little more than 1 mile). She then skis directly down the slope. If she skis five times as fast as the lift travels and the total trip takes 45 minutes, find the rate at which she skis.

Writing & Thinking

36. If n is any integer, then $2n$ is an even integer and $2n + 1$ is an odd integer. Use these ideas to solve the following problems.

   a. Find two consecutive odd integers such that the sum of their reciprocals is $\frac{12}{35}$.

   b. Find two consecutive even integers such that the sum of the first and the reciprocal of the second is $\frac{9}{4}$.

11.6 Applications: Variation

Objective A **Direct Variation**

Suppose that you ride your bicycle at a steady rate of 15 miles per hour (not quite as fast as Lance Armstrong, but you are enjoying yourself). If you ride for 1 hour, the distance you travel would be 15 miles. If you ride for two hours, the distance you travel would be 30 miles. This relationship can be written in the form of the

15 mph

formula $d = 15t$ (or $\frac{d}{t} = 15$), where d is the distance traveled and t is the time in hours. We say that distance and time **vary directly** (or are in **direct variation** or are **directly proportional**). The term proportional implies that the ratio is constant. In this example, 15 is the constant and is called the **constant of variation**. When two variables vary directly, an **increase in the value of one variable indicates an increase in the other**, and the ratio of the two quantities is constant.

Objectives

A Solve problems related to direct variation.

B Solve problems related to inverse variation.

C Solve problems involving combined variation.

Direct Variation

A variable quantity y **varies directly** as (or is **directly proportional** to) a variable x if there is a constant k such that

$$\frac{y}{x} = k \text{ or } y = kx.$$

The constant k is called the **constant of variation**.

Example 1

Direct Variation

If y varies directly as x, and $y = 6$ when $x = 2$, find y if $x = 6$.

Solution

$y = kx$	general formula for direct variation
$6 = 2k$	Substitute the known values and solve for k.
$3 = k$	Use this value for k in the general formula.

So $y = 3x$. Thus, if $x = 6$ then, $y = 3 \cdot 6 = 18$.

1. If y varies directly as x, and $y = 10$ when $x = 5$, find y if $x = 2$.

2. If a 25 g weight is placed on the same spring mentioned in Example 2, calculate how far the spring will stretch.

Example 2

Direct Variation

A spring will stretch a greater distance as more weight is placed on the end of it. The distance (d) the spring stretches varies directly as the weight (w) placed at the end of the spring. This is a property of springs studied in physics and is known as Hooke's Law. If a weight of 10 g stretches a certain spring 6 cm, how far will the spring stretch with a weight of 15 g? (**Note:** We assume that the weight is not so great as to break the spring.)

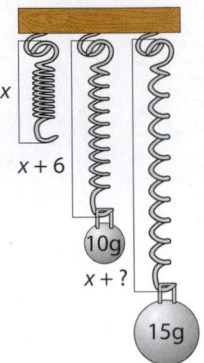

Solution

Because the two variables are directly proportional, the relationship can be indicated with the formula

$d = k \cdot w$ where d = distance spring stretches in cm,
                            w = weight in g, and
                            k = constant of variation.

First, substitute the given information to find the value for k. (The value of k will depend on the particular spring. Springs made of different material or which are wound more tightly will have different values for k.)

$d = k \cdot w$ Substitute the known values into the formula.

$6 = k \cdot 10$

$\dfrac{3}{5} = k$ The constant of variation is $\dfrac{3}{5}$ (or 0.6).

$d = \dfrac{3}{5}w.$ Use this value for k in the general formula.

If $w = 15$, we have $d = \dfrac{3}{5} \cdot 15 = 9$.

The spring will stretch 9 cm if a weight of 15 g is placed at its end.

Now work margin exercises **1 and 2.**

Listed here are several formulas involving direct variation.

$d = \dfrac{3}{5}w$ Hooke's Law for a spring where $k = \dfrac{3}{5}$.

$C = 2\pi r$ The circumference of a circle is directly proportional to the radius.

$A = \pi r^2$ The area of a circle varies directly as the radius squared.

$P = 625d$ Water pressure is proportional to the depth of the water.

When two variables vary in such a way that their product is constant, we say that the two variables **vary inversely** or are **inversely proportional**. For example, if a gas is placed in a container as in an automobile engine and pressure is increased on the gas, then the product of the pressure and the volume of gas will remain constant. That is, pressure and volume are related by the formula $V \cdot P = k$ or, $V = \dfrac{k}{P}$.

Note that if a product of two variables is to remain constant, then **an increase in the value of one variable must be accompanied by a decrease in the other**. Or, in the case of a fraction with a constant numerator, if the denominator increases in value, then the fraction decreases in value. For the gas in an engine, an increase in pressure indicates a decrease in the volume of gas.

Inverse Variation

A variable quantity y **varies inversely as** (or is **inversely proportional to**) a variable x if there is a constant k such that

$$x \cdot y = k \text{ or } y = \frac{k}{x}.$$

The constant k is called the **constant of variation**.

Example 3

Inverse Variation

If y varies inversely as the cube of x, and $y = -1$ when $x = 3$, find y if $x = -3$.

Solution

$y = \dfrac{k}{x^3}$ general formula for inverse variation

$-1 = \dfrac{k}{3^3}$ Substitute the known values and solve for k.

$-1 = \dfrac{k}{27}$

$-27 = k$

So $y = \dfrac{-27}{x^3}$. Use this value for k in the general formula.

Thus, if $x = -3$ then, $y = \dfrac{-27}{(-3)^3} = \dfrac{-27}{-27} = 1$.

3. If y varies inversely as x raised to the fifth power, and $y = -2$ when $x = 2$, find y when $x = -2$.

4. Assuming the same astronaut in Example 4 is on his way to the moon and is 300 miles from the surface of the Earth, calculate his weight rounded to the nearest pound.

Example 4

Inverse Variation

The gravitational force (F) between an object and the Earth is inversely proportional to the square of the distance (d) from the object to the center of the Earth. Hence we have the formula

$$F \cdot d^2 = k \text{ or } F = \frac{k}{d^2}, \text{ where } F = \text{force}, d = \text{distance}, \text{ and}$$

$k =$ constant variation.

(As the distance of an object from the Earth becomes larger, the gravitational force exerted by the Earth on the object becomes smaller.)

If an astronaut weighs 200 pounds on the surface of the Earth, what will he weigh 100 miles above the Earth? Assume that the radius of the Earth is 4000 miles.

Solution

We know $\quad F = 200 = 2 \times 10^2$ pounds when $\quad d = 4000 = 4 \times 10^3$ miles.

Use scientific notation to make values simpler to work with in the calculations.

$$2 \times 10^2 = \frac{k}{\left(4 \times 10^3\right)^2}$$

Substitute and solve for k.

$$k = 2 \times 10^2 \times 16 \times 10^6 = 32 \times 10^8 = 3.2 \times 10^9$$

So, $\quad F = \dfrac{3.2 \times 10^9}{d^2}$.

When the astronaut is 100 miles above the Earth, $d = 4100 = 4.1 \times 10^3$ miles. Then

$$F = \frac{3.2 \times 10^9}{16.81 \times 10^6} \approx 0.190 \times 10^3 = 190 \text{ pounds.}$$

That is, 100 miles above the Earth the astronaut will weigh about 190 pounds.

Now work margin exercises 3 and 4.

Objective C **Combined Variation**

If a variable varies either directly or inversely with more than one other variable, the variation is said to be **combined variation**. If the combined variation is all direct variation (the variables are multiplied), then it is called **joint variation**. For example, the volume of a cylinder varies jointly as its height and the square of its radius.

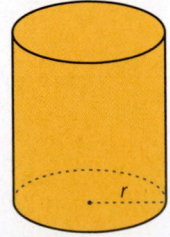

$V = kr^2h$

where r = radius, h = height, and k = constant of variation

Using this information, what is the value of k, the constant of variation, if a cylinder has the approximate measurements $V = 198$ cubic feet, $r = 3$ feet, and $h = 7$ feet?

$$V = k \cdot r^2 \cdot h \qquad \text{\textcolor{red}{V varies jointly as r^2 and h.}}$$

$$198 = k \cdot 3^2 \cdot 7 \qquad \text{\textcolor{red}{Substitute the known values.}}$$

$$\frac{198}{9 \cdot 7} = k$$

$$k = \frac{22}{7} \approx 3.14$$

We know from experience that $k = \pi$. Since the measurements are only approximate, the estimate for k is only approximate.

Substituting the constant of variation, the formula is $V = \pi r^2 h$.

5. If z varies jointly as x^3 and y^2, and $z = 108$ when $x = 2$ and $y = 3$, what is z when $x = 3$ and $y = 4$?

Example 5

Joint Variation

If z varies jointly as x^2 and y, and $z = 18$ when $x = 2$ and $y = 4$, what is z when $x = 4$ and $y = 3$?

Solution

$$z = k \cdot x^2 \cdot y$$

$$18 = k \cdot 2^2 \cdot 4 \qquad \text{Substitute the known values and solve for k.}$$

$$18 = k \cdot 16$$

$$\frac{9}{8} = k$$

So, $z = \frac{9}{8} x^2 y.$ \qquad Substitute $\frac{9}{8}$ for k in the general formula.

If, $x = 4$ and $y = 3$ then,

$$z = \frac{9}{8} \cdot 4^2 \cdot 3$$

$$z = 54.$$

Now work margin exercise 5.

Example 6

More Variation

a. The distance an object falls varies directly as the square of the time it falls (until it hits the ground and assuming little or no air resistance). If an object fell 64 feet in two seconds, how far would it have fallen by the end of 3 seconds?

Solution

$$d = k \cdot t^2 \qquad \text{where d = distance, t = time (in seconds), and k = constant of variation}$$

$$64 = k \cdot 2^2 \qquad \text{Substitute the known values and solve for k.}$$

$$16 = k$$

6. Solve the following problems on variation.

a. The distance an object falls varies directly as the square of the time it falls. If an object fell 64 feet in two seconds, how many feet will it fall in 5 seconds?

b. A volume of a gas varies inversely as the pressure on the gas. If the gas has a volume of 300 cubic inches under pressure of 8 pounds per square inch, what will be its volume if the pressure increases to 12 pounds per square inch?

c. The safe load a bridge support beam can carry varies jointly as the width w and the square of the depth d and inversely as the length l. A 5 inch wide steel beam that is 12 inches deep and 15 feet long can support a load of 18,000 pounds. What is the safe load of a beam of the same material that is 4 inches wide, 8 inches deep, and 12 feet long?

So, $d = 16t^2$.   Substitute 16 for k in the general formula.

If, $t = 3$ then,

$$d = 16 \cdot 3^2$$

$$d = 144.$$

The object would have fallen 144 feet in 3 seconds.

b. The volume of a gas in a container varies inversely as the pressure on the gas. If a gas has a volume of 200 cubic inches under pressure of 5 pounds per square inch, what will be its volume if the pressure is increased to 8 pounds per square inch?

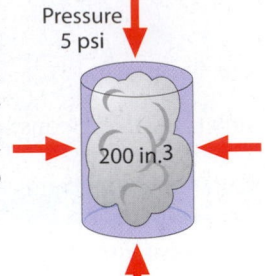

Pressure 5 psi

200 in.3

Solution

$$V = \frac{k}{P}$$ where V = volume, P = pressure, and k = constant variation

$$200 = \frac{k}{5}$$ Substitute the known values and solve for k.

$$k = 1000$$

So, $$V = \frac{1000}{P}$$ Substitute 1000 for k in the general formula.

$$V = \frac{1000}{8} = 125.$$

The volume will be 125 cubic inches.

c. The safe load L of a wooden beam supported at both ends varies jointly as the width w and the square of the depth d and inversely as the length l. A 3 in. wide by 10 in. deep beam that is 8 ft long supports a load of 9600 lb safely. What is the safe load of a beam of the same material that is 4 in. wide, 9 in. deep, and 12 ft long?

Solution

$$L = \frac{k \cdot w \cdot d^2}{l}$$ where L = safe load, w = width, d = depth, and l = length

$$9600 = \frac{k \cdot 3 \cdot 10^2}{8}$$ Substitute the known values and solve for k.

$$9600 = \frac{k \cdot 300}{8}$$

$$k = \frac{9600 \cdot 8}{300}$$

$$k = 256$$

So, $$L = \frac{256 \cdot w \cdot d^2}{l}$$ Substitute 256 for k in the general formula.

$$L = \frac{256 \cdot 4 \cdot 9^2}{12} = \frac{256 \cdot 4 \cdot 81}{12} = 6912.$$

The safe load will be 6912 lb.

Now work margin exercise 6.

Practice Problems

1. The length that a hanging spring stretches varies directly as the weight placed on the end of the spring. If a weight of 5 mg stretches a certain spring 3 cm, how far will the spring stretch with a weight of 6 mg?

2. The volume of propane in a container varies inversely as the pressure on the gas. If the propane has a volume of 200 in.^{3} under a pressure of 4 lb per in.^{2}, what will be its volume if the pressure is increased to 5 lb per in.^{2}?

Practice Problem Answers

1. $\dfrac{18}{5}$ cm

2. 160 in.^{3}

Exercises 11.6

1. If y varies directly as x, and $y = 3$ when $x = 9$, find y if $x = 7$.

2. If y is directly proportional to x^2, and $y = 3$ when $x = 2$, what is y when $x = 8$?

3. If y varies inversely as x, and $y = 5$ when $x = 8$, find y if $x = 20$.

4. If y is inversely proportional to x, and $y = 5$ when $x = 4$, what is y when $x = 2$?

5. If y varies inversely as x^2, and $y = -8$ when $x = 2$, find y if $x = 3$.

6. If y is inversely proportional to x^3, and $y = 40$ when $x = \dfrac{1}{2}$, what is y when $x = \dfrac{1}{3}$?

7. If y is directly proportional to the square root of x, and $y = 6$ when $x = \dfrac{1}{4}$, what is y when $x = 9$?

8. If y is directly proportional to the square of x, and $y = 80$ when $x = 4$, what is y when $x = 6$?

9. z varies jointly as x and y, and $z = 60$ when $x = 2$ and $y = 3$. Find z if $x = 3$ and $y = 4$.

10. z varies jointly as x and y, and $z = -6$ when $x = 5$ and $y = 8$. Find z if $x = 12$ and $y = 15$.

11. z varies jointly as x and y^2, and $z = 63$ when $x = 5$ and $y = 3$. Find z if $x = \dfrac{10}{3}$ and $y = 2$.

12. z varies jointly as x^2 and y, and $z = 20$ when $x = 2$ and $y = 3$. Find z if $x = 4$ and $y = \dfrac{7}{10}$.

13. z varies directly as x and inversely as y^2. If $z = 5$ when $x = 1$ and $y = 2$, find z if $x = 2$ and $y = 1$.

14. z varies directly as x^3 and inversely as y^2. If $z = 24$ when $x = 2$ and $y = 2$, find z if $x = 3$ and $y = 2$.

15. z varies directly as $\sqrt{x}$ and inversely as y. If $z = 24$ when $x = 4$ and $y = 3$, find z if $x = 9$ and $y = 2$.

16. z varies directly as x^2 and inversely as $\sqrt{y}$. If $z = 108$ when $x = 6$ and $y = 4$, find z if $x = 4$ and $y = 9$.

17. s varies directly as the sum of r and t and inversely as w. If $s = 24$ when $r = 7$ and $t = 8$ and $w = 9$, find s if $r = 9$ and $t = 3$ and $w = 18$.

18. s varies directly as r and inversely as the difference of t and u. If $s = 36$ when $r = 12$ and $t = 9$ and $u = 6$, find s if $r = 18$ and $t = 11$ and $u = 8$.

19. L varies jointly as m and n and inversely as p. If $L = 6$ when $m = 7$ and $n = 8$ and $p = 12$, find L if $m = 15$ and $n = 14$ and $p = 10$.

20. W varies jointly as x and y and inversely as z. If $W = 10$ when $x = 6$ and $y = 5$ and $z = 2$, find W if $x = 12$ and $y = 6$ and $z = 3$.

21. **Free Falling Object:** The distance a free falling object falls is directly proportional to the square of the time it falls (before it hits the ground). If an object fell 256 feet in 4 seconds, how far will it have fallen by the end of 5 seconds?

22. **Stretching a Spring:** The length a hanging spring stretches varies directly with the weight placed on the end. If a spring stretches 5 in. with a weight of 10 lb, how far will the spring stretch if the weight is increased to 12 lb?

23. Gas Prices: The total price (P) of gasoline purchased varies directly with the number of gallons purchased. If 10 gallons are purchased for $39.80, what will be the price of 15 gallons?

24. Economics: Research shows that the value of gold and the value of the dollar are inversely proportional. In 2008, gold cost $900 per ounce and the dollar had a rating of 75 on the US dollar index. In 2010, the cost of gold was $1100 per ounce. What was the current rating of the dollar? (Round your answer to the nearest hundredth.)

25. Pizza: The circumference of a circle varies directly as the diameter. A circular pizza pie with a diameter of 1 foot has a circumference of 3.14 feet. What will be the circumference of a pizza pie with a diameter of 1.5 feet?

26. Pizza: The area of a circle varies directly as the square of its radius. A circular pizza pie with a radius of 6 in. has an area of 113.04 in.^{2} What will be the area of a pizza pie with a radius of 9 in.?

27. Triangles: Several triangles have the same area. In this set of triangles the height and base are inversely proportional. In one such triangle the height is 5 m and the base is 12 m. Find the height of the triangle in this set with a base of 10 m.

28. Weight in Space: If an astronaut weighs 250 pounds on the surface of the earth, what will the astronaut weigh 150 miles above the earth? Assume that the radius of the earth is 4000 miles, and round to the nearest tenth. (See Example 4.)

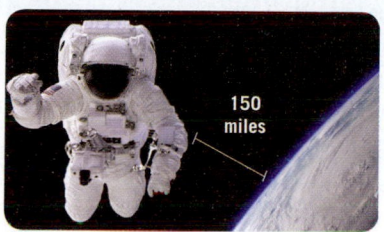

29. Elongation of a Wire: The elongation (E) in a wire when a mass (m) is hung at its free end varies jointly as the mass and the length (l) of the wire and inversely as the cross-sectional area (A) of the wire. The elongation is 0.0055 cm when a mass of 120 g is attached to a wire 330 cm long with a cross-sectional area of 0.4 cm^2. Find the elongation if a mass of 160 g is attached to the same wire.

30. Elongation of a Wire: When a mass of 240 oz is suspended by a wire 49 in. long whose cross-sectional area is 0.035 in.^{2}, the elongation of the wire is 0.016 in. Find the elongation if the same mass is suspended by a 28 in. wire of the same material with a cross-sectional area of 0.04 in.^{2} (See Exercise 29.)

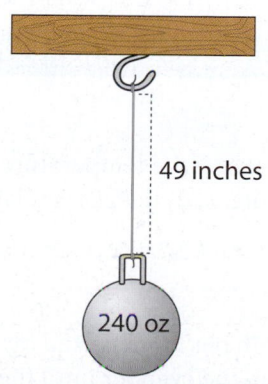

49 inches

240 oz

31. Safe Load of a Wooden Beam: The safe load (L) of a wooden beam supported at both ends varies jointly as the width (w) and the square of the depth (d) and inversely as the length (l). A beam 4 in. wide, 6 in. deep, and 12 ft long supports a load of 4800 lb safely. What is the safe load of a beam of the same material that is 6 in. wide 10 in. deep and 15 ft long? (See Example **6c.**)

32. Safe Load of a Wooden Beam: A wooden beam 2 in. wide, 8 in. deep, and 14 ft long holds up to 2400 lb. What load would a beam 3 in. wide 6 in. deep and 15 ft long, of the same material, support?

33. Gravitational Force: The gravitational force of attraction, F, between two bodies varies directly as the product of their masses, m_1 and m_2, and inversely as the square of the distance, d, between them. The gravitational force between a 5-kg mass and a 2-kg mass 1 m apart is 1.5×10^{-10} N. Find the force between a 24-kg mass and a 9-kg mass that are 6 m apart. (N represents a unit of force called a newton.)

34. Gravitational Force: In Exercise 33, what is the force if the distance between the 24-kg mass and the 9-kg mass is cut in half?

Solve the following lifting force problems.

Lifting Force

The lifting force (or lift), L, in pounds exerted by the atmosphere on the wings of an airplane is related to the area, A, of the wings in square feet and the speed (or velocity), v, of the plane in miles per hour by the formula $L = kAv^2$, where k is the constant of variation.

35. If the lift is 9600 lb for a wing area of 120 ft^2 and a speed of 80 mph, find the lift of the same airplane at a speed of 100 mph.

36. The lift for a wing of area 280 ft^2 is 34,300 lb when the plane is traveling at 210 mph. What is the lift if the speed is decreased to 180 mph?

37. The lift for a wing with an area of 144 ft^2 is 10,000 lb when the plane is traveling at 150 mph. What is the lift if the speed is decreased to 120 mph?

38. A plane traveling 140 mph with wing area 195 ft^2 has 12,500 lb of lift exerted on the wings. Find the lift for the same plane traveling at 168 mph.

Solve the following pressure problems.

Pressure

Boyle's Law states that if the temperature of a gas sample remains the same, the pressure, P, of the gas is related to the volume V by the formula $P = \dfrac{k}{V}$, where k is the constant of proportionality.

39. A pressure of 1600 lb per ft^2 is exerted by 2 ft^3 of air in a cylinder. If a piston is pushed into the cylinder until the pressure is 1800 lb per ft^2, what will be the volume of the air? Round to the nearest tenth.

40. The volume of gas in a container is 300 cm^3 when the pressure on the gas is 20 g per cm^2. What will be the volume if the pressure is increased to 30 g per cm^2?

41. The pressure in a canister of gas is 1360 g per in.^{2} when the volume of gas is 5 in.^{3} If the volume is reduced to 4 in.^{3}, what is the pressure?

42. A scuba diver is using a diving tank that can hold 6 liters of air. If the tank has a pressure rating of 220 bar when full, what is the pressure rating when the volume of gas is 4 liters?

Solve the following electricity problems.

Electricity

The resistance, R (in ohms), in a wire is given by the formula $R = \dfrac{kL}{d^2}$, where k is the constant of variation, L is the length of the wire and d is the diameter.

43. The resistance of a wire 500 ft long with a diameter of 0.01 in. is 20 ohms. What is the resistance of a wire 1500 ft long with a diameter of 0.02 in.?

44. The resistance is 2.6 ohms when the diameter of a wire is 0.02 in. and the wire is 10 ft long. Find the resistance of the same type of wire with a diameter of 0.01 in. and a length of 5 ft.

45. Tristan's car stereo uses a 5 ft audio wire with diameter 0.025 in. and resistance of 1.6 ohms. What is the resistance of 8 ft of the same type of audio wire?

46. Nicole purchased a spool of wire with diameter 0.01 in. for the speakers in her home audio system. If the resistance of 15 ft of this wire is 6 ohms, what is the resistance of 25 ft of the wire?

Solve the following lever problems.

Levers

If a lever is balanced with weight on opposite sides of its balance point, then the following proportion exists.

$$\dfrac{W_1}{W_2} = \dfrac{L_2}{L_1} \quad \text{or} \quad W_1 L_1 = W_2 L_2$$

where $L_1 + L_2 = L$, the total length of the lever

47. How much weight can be raised at one end of a bar 8 ft long by the downward force of 60 lb when the balance point is $\dfrac{1}{2}$ ft from the unknown weight?

48. Where should the balance point of a bar 12 ft long be located if a 120 lb force is to raise a load weighing 960 lb?

49. Find the location of the balance point of a 25 ft board that can raise a 300 lb package with a downward force of 75 lb.

50. How much weight can be raised on one end of a 17 meter board by 90 kilograms, if the balance point is 5 meters from the unknown weight?

51. Explain, in your own words, the meaning of the terms

    a. direct variation,

    b. inverse variation,

    c. joint variation, and

    d. combined variation.

Discuss an example of each type of variation that you have observed in your daily life.

11.7 Polynomial Division

Objective A **The Division Algorithm**

Objectives

A   Divide polynomials by using the division algorithm.

B   Divide polynomials by using synthetic division.

C   Understand the remainder theorem.

In arithmetic, the **division algorithm** (called **long division**) is the process (or series of steps) that we follow when dividing two numbers. By this division algorithm, we can find $64 \div 5$ as follows.

$$
\begin{array}{r}
12 \quad \leftarrow \text{quotient} \\
5\overline{)64} \quad \leftarrow \text{dividend} \\
\underline{5} \quad \leftarrow \text{subtraction} \\
14 \\
\underline{10} \quad \leftarrow \text{subtraction} \\
4 \quad \leftarrow \text{4 is the remainder (The remainder is always smaller than the divisor.)}
\end{array}
$$

divisor $\longrightarrow$

Check: $5 \cdot 12 + 4 = 60 + 4 = 64$ (Multiply the divisor times the quotient and add the remainder. The result should be the original dividend.)

We can also write the division in fraction form with the remainder over the divisor, giving a mixed number.

$$64 \div 5 = \frac{64}{5} = 12 + \frac{4}{5} = 12\frac{4}{5}$$

In algebra, **the division algorithm with polynomials** is quite similar. In dividing one polynomial by another, with the degree of the divisor smaller than the degree of the dividend, the quotient will be another polynomial with a remainder.

$\underline{\text{For } 64 \div 5}$	$\underline{\text{For } P \div D}$	where
$64 = 5 \cdot 12 + 4$	$P = D \cdot Q + R$	P is the dividend,
		D is the divisor,
or		Q is the quotient, and
$\frac{64}{5} = 12 + \frac{4}{5}$	$\frac{P}{D} = Q + \frac{R}{D}$	R is the remainder.

The remainder must be of smaller degree than the divisor. If the remainder is 0, then the divisor and quotient are factors of the dividend.

The Division Algorithm

For polynomials P and D, the division algorithm gives

$$\frac{P}{D} = Q + \frac{R}{D}, \; D \neq 0$$

where Q and R are polynomials and the **degree of R < degree of D**.

The actual process of long division is not clear from this abstract definition. Although you are familiar with the process of long division with decimal numbers, this same procedure appears more complicated with polynomials. The **division algorithm (long division)** is illustrated in a step-by-step form in the following example. Study it carefully.

Example 1

The Division Algorithm

Simplify $\dfrac{6x^2 - 7x - 2}{2x - 1}$ by using long division.

Solution

Step 1: $2x-1\overline{\smash{)}6x^2 - 7x - 2}$

Write both polynomials in order of descending powers. If any powers are missing, fill in with 0's.

Step 2: $2x-1\overline{\smash{)}\,\overset{\textstyle 3x}{6x^2 - \ 7x - 2}}$

Mentally divide $6x^2$ by $2x$: $\dfrac{6x^2}{2x} = 3x$. Write $3x$ above $6x^2$.

Step 3: $\begin{array}{r} 3x \\ 2x-1\overline{\smash{)}\,6x^2 - \ 7x - 2} \\ -\left(6x^2 - 3x\right) \end{array}$

Multiply $3x$ times $\left(2x - 1\right)$ and write the terms of the product, $6x^2 - 3x$ under the like terms in the dividend. Use a '−' sign to indicate that the product is to be subtracted.

Step 4: $\begin{array}{r} 3x \\ 2x-1\overline{\smash{)}\,6x^2 - 7x - 2} \\ -6x^2 + 3x \\ \hline -4x \end{array}$

Subtract $6x^2 - 3x$ by changing signs and adding.

Step 5: $\begin{array}{r} 3x \\ 2x-1\overline{\smash{)}\,6x^2 - 7x - 2} \\ -6x^2 + 3x \quad\downarrow \\ \hline -4x - 2 \end{array}$

Bring down the −2.

Step 6: $\begin{array}{r} 3x \ - \ 2 \\ 2x-1\overline{\smash{)}\,6x^2 - 7x - 2} \\ -6x^2 + 3x \\ \hline -4x - 2 \end{array}$

Mentally divide $-4x$ by $2x$: $\dfrac{-4x}{2x} = -2$. Write −2 in the quotient.

Step 7:

$$2x-1\overline{)6x^2-7x-2}$$ with quotient $3x - 2$

$$\underline{-6x^2+3x}$$
$$-4x-2$$
$$\underline{-(-4x+2)}$$

Multiply -2 times $(2x-1)$ and write the terms of the product, $-4x + 2$, under the like terms in the expression $-4x - 2$. Use a '$-$' sign to indicate that the product is to be subtracted.

Step 8:

$$2x-1\overline{)6x^2-7x-2}$$ with quotient $3x - 2$

$$\underline{-6x^2+3x}$$
$$-4x-2$$
$$\underline{+4x-2}$$
$$-4$$

Subtract $-4x + 2$ by changing signs and adding.

Thus the quotient is $3x - 2$ and the remainder is -4.

In the form $Q+\dfrac{R}{D}$ we can write $\dfrac{6x^2-7x-2}{2x-1}=3x-2-\dfrac{4}{2x-1}$.

Check: Show $Q\cdot D+R=P$.

$$(3x-2)(2x-1)-4=6x^2-3x-4x+2-4=6x^2-7x-2$$

Now work margin exercise 1.

Example 2

Long Division (Remainder 0)

Divide $(25x^3-5x^2+3x+1)\div(5x+1)$ by using long division.

Solution

$$5x+1\overline{)25x^3-5x^2+3x+1}$$ with quotient $5x^2-2x+1$

$$\underline{-(25x^3+5x^2)}$$
$$-10x^2+3x$$
$$\underline{-(-10x^2-2x)}$$
$$5x+1$$
$$\underline{-(5x+1)}$$
$$0$$

There is no remainder, thus the quotient is simply $5x^2-2x+1$.

Now work margin exercise 2.

1. Simplify.

$$\frac{12x^2-5x+3}{3x-2}$$

2. Perform the following division problem by using long division.

$$(8x^3-14x^2-7x+6)$$
$$\div(4x+3)$$

In Example 2, because the remainder is 0, both $(5x+1)$ and $(5x^2-2x+1)$ are **factors** of $25x^3-5x^2+3x+1$. That is,

$$(5x^2-2x+1)(5x+1) = 25x^3-5x^2+3x+1 \quad \text{or} \quad Q \cdot D = P.$$

Factoring polynomials was discussed in detail in earlier chapters.

3. Simplify.

$$\frac{x^4-x^2+8x+11}{x^2-2x+1}$$

Example 3

Long Division (Terms Missing)

Simplify $\dfrac{x^4+9x^2-3x+5}{x^2-x+2}$ using long division.

Solution

Note that 0 is written as a placeholder for any missing powers of the variable. In this way, like terms are easily aligned vertically.

$$
\begin{array}{r}
x^2 + x + 8 \\
x^2-x+2 \overline{) x^4 + 0x^3 + 9x^2 - 3x + 5} \\
\underline{-\left(x^4 - x^3 + 2x^2\right)} \\
x^3 + 7x^2 - 3x \\
\underline{-\left(x^3 - x^2 + 2x\right)} \\
8x^2 - 5x + 5 \\
\underline{-\left(8x^2 - 8x + 16\right)} \\
3x - 11
\end{array}
$$

Note that the remainder is of smaller degree than the divisor.

Thus the quotient is x^2+x+8 and the remainder is $3x-11$.

In the form $Q + \dfrac{R}{D}$ we can write $x^2+x+8+\dfrac{3x-11}{x^2-x+2}$.

Now work margin exercise 3.

In an earlier section, polynomials are divided by using the division algorithm (long division). In the special case **when the divisor of a rational expression is a first-degree binomial with leading coefficient 1**, long division can be simplified by omitting the variables entirely and writing only certain coefficients. The procedure is called **synthetic division**. The following analysis describes how the procedure works for $\dfrac{5x^3 + 11x^2 - 3x + 1}{x + 3}$. **Note:** $x + 3$ is first-degree with leading coefficient 1.

a. With Variables

$$
\begin{array}{r}
5x^2 - 4x + 9 \\
x+3\overline{)5x^3 + 11x^2 - 3x + 1} \\
\underline{5x^3 + 15x^2} \\
-4x^2 - 3x \\
\underline{-4x^2 - 12x} \\
9x + 1 \\
\underline{9x + 27} \\
-26
\end{array}
$$

b. Without Variables

$$
\begin{array}{r}
5 \quad -4 \quad +9 \\
1+3\overline{)5 \quad +11 \quad -3 \quad +1} \\
\boxed{5} \;+ 15 \\
-4 \quad \boxed{-3} \\
\boxed{-4} \quad -12 \\
9 \quad \boxed{+1} \\
\boxed{9} \quad +27 \\
-26
\end{array}
$$

The boxed numbers in step **b.** can be omitted since they are repetitions of the numbers directly above them.

c. Boxed numbers omitted

$$
\begin{array}{r}
5 \quad -4 \quad +9 \\
1+3\overline{)5 \quad +11 \quad -3 \quad +1} \\
+15 \\
-4 \\
-12 \\
9 \\
+27 \\
-26
\end{array}
$$

d. Numbers moved up to fill in spaces

$$
\begin{array}{r}
5 \quad -4 \quad +9 \\
1+3\overline{)5 \quad +11 \quad -3 \quad +1} \\
+15 \quad -12 \quad +27 \\
-4 \quad +9 \quad -26
\end{array}
$$

Next, omit the 1 in the divisor, change +3 to −3, and write the opposites of the boxed numbers (because the quotient coefficient will now be multiplied by −3 instead of +3), as shown in steps **e.** and **f.** This allows the numbers to be added instead of subtracted. The number 5 is written on the bottom line, and the top line is omitted. The quotient and remainder can now be read from the bottom line.

e.
$$
\begin{array}{r}
5 \quad -4 \quad +9 \\
1+3\overline{)5 \quad +11 \quad -3 \quad +1} \\
\boxed{+15} \; \boxed{-12} \; \boxed{+27} \\
-4 \quad +9 \quad -26
\end{array}
$$

f.
$$
\begin{array}{r}
-3\overline{)5 \quad +11 \quad -3 \quad +1} \\
\downarrow \; -15 + 12 - 27 \\
5 \quad -4 + \;9 \; -26
\end{array}
$$

This represents
$$
5x^2 - 4x + 9 + \frac{-26}{x + 3}.
$$

The numbers on the bottom now represent the coefficients of a polynomial of **one degree less than the dividend,** along with the remainder. The last number to the right is the remainder.

In summary, synthetic division can be accomplished as follows:

1. Write only the coefficients of the dividend and the opposite of the constant in the divisor.

$$-3\rfloor \quad 5 \quad\quad 11 \quad\quad -3 \quad\quad 1$$
_____

2. Rewrite the first coefficient (5) as the first coefficient in the quotient.

$$\begin{array}{r|rrrr} -3 & 5 & 11 & -3 & 1 \\ & \downarrow & & & \\ \hline & 5 & & & \end{array}$$

3. Multiply the coefficient (5) by the constant divisor (-3) and **add** this product (-15) to the second coefficient (11).

$$\begin{array}{r|rrrr} -3 & 5 & 11 & -3 & 1 \\ & \downarrow & -15 & & \\ \hline & 5 & -4 & & \end{array}$$

4. Continue to multiply each new coefficient by the constant divisor and add this product to the next coefficient in the dividend.

$$\begin{array}{r|rrrr} -3 & 5 & 11 & -3 & 1 \\ & \downarrow & -15 & 12 & -27 \\ \hline & 5 & -4 & 9 & -26 \end{array}$$

5. The constants on the bottom line are the coefficients of the quotient and the remainder.

$$\frac{5x^3 + 11x^2 - 3x + 1}{x+3} = 5x^2 - 4x + 9 + \frac{-26}{x+3}$$

$$= 5x^2 - 4x + 9 - \frac{26}{x+3}$$

Example 4

Synthetic Division

Use synthetic division to write each expression in the form $Q + \dfrac{R}{D}$.

a. $\dfrac{4x^3 + 10x^2 + 11}{x+5}$

Solution

$$\begin{array}{r|rrrr} -5 & 4 & 10 & 0 & 11 \\ & \downarrow & -20 & 50 & -250 \\ \hline & 4 & -10 & 50 & -239 \end{array}$$

$$\frac{4x^3 + 10x^2 + 11}{x+5} = 4x^2 - 10x + 50 + \frac{-239}{x+5}$$

$$= 4x^2 - 10x + 50 - \frac{239}{x+5}$$

b. $\dfrac{2x^4 - x^3 - 5x^2 - 2x + 7}{x - 2}$

Solution

$$\begin{array}{r|rrrrr} 2 & 2 & -1 & -5 & -2 & 7 \\ & \downarrow & 4 & 6 & 2 & 0 \\ \hline & 2 & 3 & 1 & 0 & 7 \end{array}$$

$$\dfrac{2x^4 - x^3 - 5x^2 - 2x + 7}{x - 2} = 2x^3 + 3x^2 + x + \dfrac{7}{x - 2}$$

Now work margin exercise 4.

Remember!

Remember that synthetic division is used only when the divisor is a first-degree polynomial of the form $(x + c)$ or $(x - c)$.

Example 5

When to Perform Synthetic Division

Which of the following problems can be performed by using synthetic division?

a. $\dfrac{x^3 - 3x^2 - 5x - 8}{x^2 + 2x + 3}$

b. $\dfrac{x^2 - 16}{-4 + x}$

Solution

Because the divisor of **a.** cannot be written in the form $(x + c)$ or $(x - c)$, **a.** cannot be performed by synthetic division.

However, the divisor of **b.** can be written as $x - 4$, and thus **b.** can be performed with synthetic division as follows.

$$\begin{array}{r|rrr} 4 & 1 & 0 & -16 \\ & \downarrow & 4 & 16 \\ \hline & 1 & 4 & 0 \end{array}$$

Thus, $\dfrac{x^2 - 16}{-4 + x} = x + 4.$

Now work margin exercise 5.

Objective C **The Remainder Theorem**

Synthetic division can be used for several purposes, one of which is to find the value of a polynomial for a particular value of x. For example, we know that if

$$P(x) = x^3 - 5x^2 + 7x - 10, \text{ then, } P(2) = 2^3 - 5 \cdot 2^2 + 7 \cdot 2 - 10 = -8.$$

---

4. Use synthetic division to write each expression in the form $Q + \dfrac{R}{D}$.

a. $\dfrac{3x^4 - 3x^2 + x + 7}{x - 1}$

b. $\dfrac{8x^3 + 10x^2 - 8x - 9}{x + 4}$

5. Can the following problems be performed by using synthetic division? If so, find the quotient.

a. $\dfrac{x^3 - 3x^2 + 3x - 1}{-1 + x}$

b. $\dfrac{x^4 + 5x - 9}{x^2 + 3}$

Using synthetic division to divide $x^3 - 5x^2 + 7x - 10$ by $x - 2$ we have

$$
\begin{array}{r|rrrr}
2 & 1 & -5 & 7 & -10 \\
  &   & 2 & -6 & 2 \\
\hline
  & 1 & -3 & 1 & -8
\end{array} \quad \longleftarrow \text{remainder}
$$

The fact that the remainder is the same as $P(2)$ is not an accident. In fact, as the following theorem states, the remainder when a polynomial is divided by a first-degree factor of the form $(x - c)$ will always be $P(c)$.

The Remainder Theorem

If a polynomial $P(x)$ is divided by $(x - c)$, then the remainder will be $P(c)$.

Proof:

By the division algorithm we know that $\dfrac{P(x)}{x - c} = Q(x) + \dfrac{R}{x - c}$ where R is a constant. Remember that the degree of the remainder must be less than the degree of the divisor.

Multiplying through by $(x - c)$, we have

$$P(x) = (x - c) \cdot Q(x) + R,$$

and substituting $x = c$ gives

$$
\begin{aligned}
P(c) &= (c - c) \cdot Q(c) + R \\
     &= 0 \cdot Q(c) + R \\
     &= 0 + R \\
     &= R.
\end{aligned}
$$

The proof is complete.

Example 6

The Remainder Theorem and Synthetic Division

a. Use synthetic division to find $P(5)$ given $P(x) = -2x^2 + 15x - 50$.

Solution

$$
\begin{array}{r|rrr}
5 & -2 & 15 & -50 \\
  &   & -10 & 25 \\
\hline
  & -2 & 5 & -25
\end{array} \quad \longleftarrow \text{remainder} = P(5)
$$

Thus $P(5) = -25$.
(Checking shows $P(5) = -2 \cdot 5^2 + 15 \cdot 5 - 50 = -50 + 75 - 50 = -25$.)

b. Use synthetic division to find $P(-3)$, given
$$P(x) = 3x^4 + 10x^3 - 5x^2 + 125.$$

Note: To evaluate $P(-3)$, think of the divisor in the form $(x+3) = (x-(-3))$. That is, in the form $(x-c)$, $c = -3$.

Solution

$$
\begin{array}{r|rrrrr}
-3 & 3 & 10 & -5 & 0 & 125 \\
   &   & -9 & -3 & 24 & -72 \\
\hline
   & 3 & 1 & -8 & 24 & 53
\end{array}
$$
← remainder $= P(-3)$

Thus $P(-3) = 53$.

c. Use synthetic division to show that $(x-6)$ is a factor of $P(x) = x^3 - 14x^2 + 53x - 30.$

Solution

$$
\begin{array}{r|rrrr}
6 & 1 & -14 & 53 & -30 \\
  &   & 6   & -48 & 30 \\
\hline
  & 1 & -8 & 5 & 0
\end{array}
$$
← remainder $= P(6)$

Thus the remainder is $P(6) = 0$ and $(x-6)$ **is a factor of** $P(x)$.
Note: The coefficients in the quotient tell us that $x^2 - 8x + 5$ is also a factor of $P(x)$.

Now work margin exercise 6.

6. a. Use synthetic division to find $J(3)$ given $J(y) = -4y^3 + 15y + 30.$

b. Use synthetic division to find $f(-2)$ given $f(x) = 3x^4 + 8x^3 - 9x^2 - 5.$

c. Use synthetic division to show that $(x-8)$ is a factor of $R(x) = x^3 - 6x^2 - 20x + 32.$

Practice Problems

1. Express the quotient as a sum of fractions in simplified form:
$$\frac{8x^2 + 6x + 1}{2x}.$$

Use the division algorithm to divide. Write the answer in the form
$$Q + \frac{R}{D}.$$

2. $\dfrac{3x^2 - 8x + 5}{x + 2}$

3. $(x^3 + 4x^2 - 10) \div (x^2 + x - 1)$

Practice Problem Answers

1. $4x + 3 + \dfrac{1}{2x}$ **2.** $3x - 14 + \dfrac{33}{x+2}$ **3.** $x + 3 + \dfrac{-2x - 7}{x^2 + x - 1}$

Exercises 11.7

1. $\dfrac{x^2 - 2x - 20}{x + 4}$

2. $\dfrac{x^2 + 9x - 5}{x - 1}$

3. $\dfrac{6x^2 - 11x - 3}{2x - 1}$

4. $\dfrac{10x^2 + 16x + 5}{5x + 3}$

5. $\dfrac{21x^2 + 25x - 3}{7x - 1}$

6. $\dfrac{15x^2 - 14x - 11}{3x - 4}$

7. $\dfrac{x^2 - 12x + 27}{x - 3}$

8. $\dfrac{x^2 - 12x + 35}{x - 5}$

9. $\dfrac{x^3 - 9x^2 + 8x - 3}{x - 8}$

10. $\dfrac{x^3 - 6x^2 + 8x - 5}{x - 2}$

11. $\dfrac{4x^3 + 2x^2 - 3x + 1}{x + 2}$

12. $\dfrac{3x^3 + 6x^2 + 8x - 5}{x + 1}$

13. $\dfrac{x^3 + 6x + 3}{x - 7}$

14. $\dfrac{2x^3 + 3x - 2}{x - 1}$

15. $\dfrac{2x^3 - 5x^2 + 6}{x + 2}$

16. $\dfrac{4x^3 - x^2 + 13}{x - 1}$

17. $\dfrac{21x^3 + 41x^2 + 13x + 5}{3x + 5}$

18. $\dfrac{6x^3 - 7x^2 + 14x - 8}{3x - 2}$

19. $\dfrac{2x^3 + 7x^2 + 10x - 6}{2x + 3}$

20. $\dfrac{6x^3 - 4x^2 + 5x - 7}{x - 2}$

21. $\dfrac{x^3 - x^2 - 10x - 10}{x - 4}$

22. $\dfrac{2x^3 - 3x^2 + 7x + 4}{2x - 1}$

23. $\dfrac{10x^3 + 11x^2 - 12x + 9}{5x + 3}$

24. $\dfrac{6x^3 + 19x^2 - 3x - 7}{6x + 1}$

25. $\dfrac{2x^3 - 7x + 2}{x + 4}$

26. $\dfrac{2x^3 + 4x^2 - 9}{x + 3}$

27. $\dfrac{9x^3 - 19x + 9}{3x - 2}$

28. $\dfrac{4x^3 - 8x^2 - 9x}{2x - 3}$

29. $\dfrac{6x^3 + 11x^2 + 25}{2x + 5}$

30. $\dfrac{16x^3 + 7x + 12}{4x + 3}$

31. $\dfrac{x^4 - 3x^3 + 2x^2 - x + 2}{x - 3}$

32. $\dfrac{x^4 + x^3 - 4x^2 + x - 3}{x + 6}$

33. $\dfrac{x^4 + 2x^2 - 3x + 5}{x - 2}$

34. $\dfrac{3x^4 + 2x^3 - 2x^2 - 1}{x + 1}$

35. $\dfrac{x^4 - x^2 + 3}{x - \dfrac{1}{2}}$

36. $\dfrac{x^3 + 2x^2 + 1}{x - \dfrac{2}{3}}$

37. $\dfrac{3x^3 + 5x^2 + 7x + 9}{x^2 + 2}$

38. $\dfrac{2x^4 + 2x^3 + 3x^2 + 6x - 1}{2x^2 + 3}$

39. $\dfrac{x^4 + x^3 - 4x + 1}{x^2 + 4}$

40. $\dfrac{2x^4 + x^3 - 8x^2 + 3x - 2}{x^2 - 5}$

41. $\dfrac{6x^3 + 5x^2 - 8x + 3}{3x^2 - 2x - 1}$

42. $\dfrac{x^3 - 9x^2 + 20x - 38}{x^2 - 3x + 5}$

43. $\dfrac{3x^4 - 7x^3 + 5x^2 + x - 2}{x^2 + x + 1}$

44. $\dfrac{2x^4 - x^3 - 10x^2 - 3x - 1}{x^2 - 3x + 1}$

45. $\dfrac{x^4 + 3x - 7}{x^2 + 2x - 3}$

46. $\dfrac{3x^4 - 4x^2 + 3}{x^2 + x - 1}$

47. $\dfrac{x^3 - 27}{x - 3}$

48. $\dfrac{x^3 + 125}{x + 5}$

49. $\dfrac{x^6 - 1}{x + 1}$

50. $\dfrac{x^6 - 1}{x - 1}$

51. $\dfrac{x^5 + 1}{x - 1}$

52. $\dfrac{x^6 + 1}{x + 1}$

53. $\dfrac{x^5 - x^3 + x}{x + \dfrac{1}{2}}$

54. $\dfrac{x^4 - 2x^3 + 4}{x + \dfrac{4}{5}}$

> **Divide the following expressions using synthetic division. a. Write the answer in the form** $Q + \dfrac{R}{D}$ **where R is a constant. b. In each exercise, $D = (x - c)$, state the value of c and the value of $P(c)$. See Examples 4 through 6.**

55. $\dfrac{x^2 - 12x + 27}{x - 3}$

56. $\dfrac{x^2 - 12x + 35}{x - 5}$

57. $\dfrac{x^3 + 4x^2 + x - 1}{x + 8}$

58. $\dfrac{x^3 - 6x^2 + 8x - 5}{x - 2}$

59. $\dfrac{4x^3 + 2x^2 - 3x + 1}{x + 2}$

60. $\dfrac{3x^3 + 6x^2 + 8x - 5}{x + 1}$

61. $\dfrac{x^3 + 6x + 3}{x - 7}$

62. $\dfrac{2x^3 - 7x + 2}{x + 4}$

63. $\dfrac{2x^3 + 4x^2 - 9}{x + 3}$

64. $\dfrac{4x^3 - x^2 + 13}{x - 1}$

65. $\dfrac{x^4 - 3x^3 + 2x^2 - x + 2}{x - 3}$

66. $\dfrac{x^4 + x^3 - 4x^2 + x - 3}{x + 6}$

67. $\dfrac{x^4 + 2x^2 - 3x + 5}{x - 2}$

68. $\dfrac{3x^4 + 2x^3 + 2x^2 + x - 1}{x + 1}$

69. $\dfrac{x^4 - x^2 + 3}{x - \dfrac{1}{2}}$

70. $\dfrac{x^3 + 2x^2 + 1}{x - \dfrac{2}{3}}$

71. $\dfrac{x^5 - 1}{x - 1}$

72. $\dfrac{x^5 - x^3 + x}{x + \dfrac{1}{2}}$

73. $\dfrac{x^4 - 2x^3 + 4}{x + \dfrac{4}{5}}$

74. $\dfrac{x^6 + 1}{x + 1}$

75. Suppose that a polynomial is divided by $(3x-2)$ and the answer is given as $x^2+2x+4+\dfrac{20}{3x-2}$. What was the original polynomial? Explain how you arrived at this conclusion.

76. Suppose that a polynomial is divided by $(x+5)$ and the answer is given as $x^2-3x+2-\dfrac{6}{x+5}$. What was the original polynomial? Explain how you arrived at this conclusion

77. Given that $P(x)=2x^3-8x^2+10x+15$.

   a. Find $P(2)$ then divide $P(x)$ by $x-2$.

   b. Find $P(-1)$ then divide $P(x)$ by $x+1$.

   c. Find $P(4)$ then divide $P(x)$ by $x-4$.

   Do you see any pattern in the values of $P(a)$ for $x=a$ and the remainders you found in the division process? (**Hint:** Check the appendix section on Synthetic Division and the Remainder Theorem.)

Collaborative Learning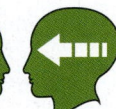

78. With the class divided into teams of 3 or 4 students, each team should develop answers to the following questions and be prepared to discuss the answers in class.

   a. First use long division to divide the polynomial $P(x)=2x^3-8x^2+10x+15$ by $2x-1$.

   Then use synthetic division to divide the same polynomial by $x-\dfrac{1}{2}$.

   Do the same process with two or three other polynomials and divisors. Next compare the corresponding long and synthetic division answers and explain how the answers are related.

   b. Use the results from part **a.** and explain algebraically the relationship of the answers when a polynomial is divided (using long division) by $ax-b$ and (using synthetic division) by $x-\dfrac{b}{a}$.

   c. Show how the remainder theorem should be restated if $x-c$ is replaced by $ax-b$.

Section 11.1: Rational Expressions

Rational Expression   page 1155

A **rational expression** is an algebraic expression of the form $\dfrac{P}{Q}$ where P and Q are polynomials and $Q \neq 0$.

Restrictions on a Variable   pages 1155 – 1156

Values of the variable that make a rational expression undefined are called **restrictions** on the variable.

The Fundamental Principle of Rational Expressions   page 1158

If $\dfrac{P}{Q}$ is a rational expression and P, Q, and K are polynomials where $Q, K \neq 0$, then $\dfrac{P}{Q} = \dfrac{P \cdot K}{Q \cdot K}$.

Opposites in Rational Expressions   page 1159

For a polynomial $\dfrac{-P}{P} = -1$ where $P \neq 0$.

In particular, $\dfrac{a-x}{x-a} = \dfrac{-(x-a)}{x-a} = -1$ where $x \neq a$.

To Multiply Rational Expressions   page 1160

1. Completely factor each numerator and denominator.
2. Multiply the numerators and multiply the denominators, keeping the expressions in factored form.
3. "Divide out" any common factors from the numerators and denominators. Remember that no denominator can have a value of 0.

Multiplication with Rational Expressions   page 1161

$\dfrac{P}{Q} \cdot \dfrac{R}{S} = \dfrac{P \cdot R}{Q \cdot S}$ where P, Q, R, and S are polynomials with $Q, S \neq 0$.

Division with Rational Expressions   pages 1162 – 1163

$\dfrac{P}{Q} \div \dfrac{R}{S} = \dfrac{P}{Q} \cdot \dfrac{S}{R}$ where P, Q, R, and S are polynomials with $Q, R, S \neq 0$.

Section 11.2: Addition and Subtraction with Rational Expressions

Addition and Subtraction with Rational Expressions   pages 1170 – 1177

$\dfrac{P}{Q} + \dfrac{R}{Q} = \dfrac{P+R}{Q}$ and $\dfrac{P}{Q} - \dfrac{R}{Q} = \dfrac{P-R}{Q}$ where $Q \neq 0$.

To Find the LCM for a Set of Polynomials   page 1171

1. Completely factor each polynomial.
2. Form the product of all factors that appear, using each factor the most number of times it appears in any one polynomial.

Section 11.2: Addition and Subtraction with Rational Expressions (cont.)

**Procedure for Adding (or Subtracting) Rational Expressions
with Different Denominators**   page 1171

1. Find the LCD (the LCM of the denominators).
2. Rewrite each fraction in an equivalent form with the LCD as the denominator.
3. Add (or subtract) the numerators and keep the common denominator.
4. Reduce if possible.

Placement of Negative Signs   page 1173

$$-\frac{P}{Q} = \frac{P}{-Q} = \frac{-P}{Q} \text{ where } Q \neq 0$$

Section 11.3: Simplifying Complex Fractions

Complex Fractions   page 1182

A **complex fraction** is a fraction in which the numerator and/or denominator are themselves fractions or the sum or difference of fractions.

To Simplify Complex Fractions   page 1182 – 1185

First Method

1. Simplify the numerator so that it is a single rational expression.
2. Simplify the denominator so that it is a single rational expression.
3. Divide the numerator by the denominator and reduce to lowest terms.

Second Method

1. Find the LCM of all the denominators in the numerator and denominator of the complex fraction.
2. Multiply both the numerator and denominator of the complex fraction by this LCM.
3. Simplify both the numerator and denominator and reduce to lowest terms.

Simplifying Complex Algebraic Expressions   page 1186

A complex algebraic expression is an expression that involves rational expressions and more than one operation. In simplifying such expressions, the rules for order of operations apply.

Section 11.4: Solving Rational Equations

Ratio   page 1190

A **ratio** is a comparison of two numbers by division.

Proportion   page 1190

A **proportion** is an equation stating that two ratios are equal. One of the
following conditions must be true:
1. The numerators agree in type and the denominators agree in type.
2. The numerators correspond and the denominators correspond.

Similar Figures (including Similar Triangles)   pages 1192 – 1193

Similar figures are figures that meet the following two conditions:
1. The measures of the corresponding angles are equal.
2. The lengths of the corresponding sides are proportional.

To Solve an Equation Containing Rational Expressions   pages 1193 – 1195
1. Find the LCD of the fractions.
2. Multiply both sides of the equation by this LCD and simplify.
3. Solve the resulting equation.
4. Check each solution in the original equation. (Remember that no
   denominator can equal 0.)

Checking for Extraneous Solutions   pages 1193 – 1194

Section 11.5: Applications: Rational Expressions

Strategy for Solving Word Problems   page 1200

Applications   pages 1200 – 1206

Number problems related to fractions
Work problems
Distance-Rate-Time problems

Section 11.6: Applications: Variation

Direct Variation   pages 1210 – 1211

A variable quantity y **varies directly** as (or is **directly proportional to**) a
variable x if there is a constant k such that

$$\frac{y}{x} = k \text{ or } y = kx.$$

The constant k is called the **constant of variation**.

Inverse Variation   pages 1212 – 1213

A variable quantity y **varies inversely as** (or is **inversely proportional to**) a variable x if there is a constant k such that

$$x \cdot y = k \text{ or } y = \frac{k}{x}.$$

The constant k is called the **constant of variation**.

Combined Variation   pages 1213 – 1214

If a variable varies either directly or inversely with more than one other variable, the variation is said to be **combined variation**.

Joint Variation   pages 1213 – 1214

If the combined variation is all direct variation (the variables are multiplied), then it is called **joint variation**.

Section 11.7: Polynomial Division

The Division Algorithm   pages 1222 – 1224

For polynomials P and D, the division algorithm gives

$$\frac{P}{D} = Q + \frac{R}{D}, \ D \neq 0$$

where Q and R are polynomials and the **degree of R < degree of D**.

Synthetic Division   pages 1226 – 1228

1. Write only the coefficients of the dividend and the opposite of the constant in the divisor.
2. Rewrite the first coefficient as the first coefficient in the quotient.
3. Multiply the coefficient by the constant divisor and **add** this product to the second coefficient.
4. Continue to multiply each new coefficient by the constant divisor and add this product to the next coefficient in the dividend.
5. The constants on the bottom line are the coefficients of the quotient and the remainder.

Remember that synthetic division is used only when the divisor is a first-degree polynomial of the form $(x + c)$ or $(x - c)$.

The Remainder Theorem   pages 1228 – 1230

If a polynomial $P(x)$ is divided by $(x - c)$, then the remainder will be $P(c)$.

Chapter 11: Test

Reduce to lowest terms. State any restrictions on the variable(s).

1. $\dfrac{x^2 + 3x}{x^2 + 7x + 12}$

2. $\dfrac{2x + 5}{4x^2 + 20x + 25}$

Perform the indicated operations and reduce the answers to lowest terms. Assume that no denominator has a value of 0.

3. $\dfrac{x + 3}{x^2 + 3x - 4} \cdot \dfrac{x^2 + x - 2}{x + 2}$

4. $\dfrac{6x^2 - x - 2}{12x^2 + 5x - 2} \div \dfrac{4x^2 - 1}{8x^2 - 6x + 1}$

5. $\dfrac{x}{x^2 + 3x - 10} + \dfrac{3x}{4 - x^2}$

6. $\dfrac{x - 4}{3x^2 + 5x + 2} - \dfrac{x - 1}{x^2 - 3x - 4}$

7. $\dfrac{x^2 - 16}{x^2 - 4x} \cdot \dfrac{x^2}{x + 4} \div \dfrac{x - 1}{2x^2 - 2x}$

8. $\dfrac{x}{x + 3} - \dfrac{x + 1}{x - 3} + \dfrac{x^2 + 4}{x^2 - 9}$

Simplify the complex fractions.

9. $\dfrac{\dfrac{5y}{3x^2}}{\dfrac{10y^4}{9x}}$

10. $\dfrac{\dfrac{4}{3x} + \dfrac{1}{6x}}{\dfrac{1}{x^2} - \dfrac{1}{2x}}$

11. $\dfrac{x^{-1} - y^{-1}}{x - y}$

12. $\dfrac{\dfrac{4}{x} + \dfrac{2}{x + 3}}{\dfrac{x^2 - 9}{2x}}$

13. Simplify the expression in part **a.** and solve the equation in part **b.**

a. $\dfrac{3}{x} - \dfrac{2}{x + 1} + \dfrac{5}{2x}$

b. $\dfrac{3}{x} - \dfrac{2}{x + 1} = \dfrac{5}{2x}$

State any restrictions on x and then solve the equations.

14. $\dfrac{x-1}{x+4} = \dfrac{4}{5}$

15. $\dfrac{4}{7} - \dfrac{1}{2x} = 1 + \dfrac{1}{x}$

16. $\dfrac{4}{x+4} + \dfrac{3}{x-1} = \dfrac{1}{x^2+3x-4}$

17. $\dfrac{x}{x+2} = \dfrac{1}{x+1} - \dfrac{x}{x^2+3x+2}$

Solve each of the formulas for the specified variables.

18. $S = \dfrac{n(a_1 + a_n)}{2}$; solve for n

19. $y = mx + b$; solve for x

Divide by using the division algorithm. Write the answers in the form $Q + \dfrac{R}{D}$ where the degree of $R <$ the degree of D.

20. $(2x^2 - 9x - 20) \div (2x + 3)$

21. $\dfrac{x^3 - 8x^2 + 3x + 15}{x^2 + x - 3}$

Divide the following expressions using synthetic division. a. Write the answer in the form $Q + \dfrac{R}{D}$ where R is a constant. b. State the value of c and the value of $P(c)$.

22. $\dfrac{x^3 + x^2 - 18x + 9}{x + 5}$

23. $\dfrac{x^4 + x + 1}{x - 1}$

Solve the following problems as indicated.

24. The denominator of a fraction is three more than twice the numerator. If eight is added to both the numerator and the denominator, the resulting fraction is equal to $\dfrac{2}{3}$. Find the original fraction.

Solve the following word problems.

25. House Cleaning: Sonya can clean the apartment in 6 hours. It takes Lucy 12 hours to clean it. If they work together, how long will it take them?

26. Travel: Mario can travel 228 miles in the same time that Carlos travels 168 miles. If Mario's speed is 15 mph faster than Carlos's, find their rates.

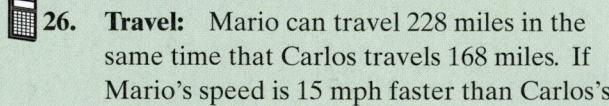

27. Boating: Bob travels 4 miles upstream. In the same time, he could have traveled 7 miles downstream. If the speed of the current is 3 mph, find the speed of the boat in still water.

28. z varies directly as x^3. If $z = 3$ when $x = 2$, find z if $x = 4$.

29. z varies directly as x^2 and inversely as $\sqrt{y}$. If $z = 24$ when $x = 3$ and $y = 4$, find z if $x = 5$ and $y = 9$.

30. Stretching a Spring: Hooke's Law states that the distance a spring will stretch vertically is directly proportional to the weight placed at its end. If a particular spring will stretch 5 cm when a weight of 4 g is placed at its end, how far will the spring stretch if a weight of 6 g is placed at its end?

31. Candy: The volume of a gumball is directly proportional to the radius cubed. A gumball with a diameter of 1.75 inches has an approximate volume of 2.805 inches cubed. What is the volume of a gumball that has a diameter of 2.14 inches? (Round your answer to the nearest hundredth.)

Cumulative Review: Chapters 1 - 11

Solve the following problems as indicated.

1. Graph the set of integers less than –5 on a number line.

2. Perform the indicated operations and reduce.

   a. $\dfrac{5}{6} + \dfrac{3}{4}$

   b. $\dfrac{1}{2} - \dfrac{5}{7}$

   c. $\dfrac{1}{2} \cdot \dfrac{1}{2}$

   d. $\dfrac{7}{32} \div \dfrac{21}{8}$

3. Write an algebraic expression for the following English phrase.

   "5 less than twice the sum of a number and 6"

Perform the indicated operations and simplify the expressions.

4. $(3x+7)(x-4)$

5. $(2x-5)^2$

6. $-4(x+2)+5(2x-3)$

7. $x(x+7)-(x+1)(x-4)$

8. Solve the inequality $\dfrac{x+5}{2} < \dfrac{3x}{4} + 1$ and graph the solution set. Write the solution using interval notation. Assume that x is a real number.

9. Given the relation $r = \{(-1,5),\ (0,2),\ (-1,0),\ (5,0),\ (6,0)\}$.

   a. Graph the relation.

   b. What is the domain of the relation?

   c. What is the range of the relation?

   d. Is the relation a function? Explain.

10. For the function $f(x) = x^3 - 2x^2 - 1$, find: **a.** $f(3)$ **b.** $f(0)$ **c.** $f\left(-\dfrac{2}{3}\right)$

11. Graph the equation $6x + 4y = 18$ by locating the x-intercept and the y-intercept.

12. Find the equation in slope-intercept form of the line that has slope $\frac{3}{4}$ and passes through the point $(-2, 4)$. Graph the line.

13. Find the equation in slope-intercept form of the line passing through the two points $(-5, -1)$ and $(4, 2)$. Graph the line.

14. Find the equation of the line passing through the point $(-3, 7)$ with undefined slope. Graph the line.

15. Find the equation in standard form of the line perpendicular to the line $x - 2y = 3$ and passing through the point $(-4, 0)$. Graph both lines.

Use the vertical line test to determine whether or not each graph represents a function. State the domain and range.

16.

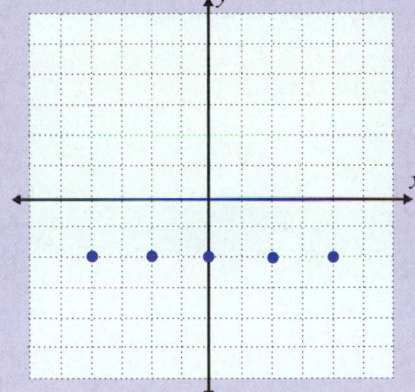

17.

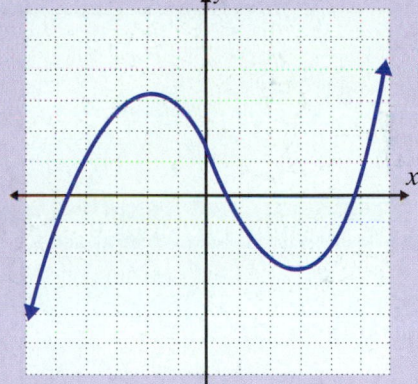

18. Graph the solution to the following inequality: $3x + y > 4$.

Simplify each expression. The final form of the expressions with variables should contain only positive exponents. Assume that all variables represent nonzero numbers.

19. $\dfrac{32x^3y^2}{4xy^2}$

20. $\dfrac{15xy^{-1}}{3x^{-2}y^{-3}}$

Completely factor each polynomial.

21. $3xy + y^2 + 3x + y$

22. $4x^2 - 4x - 15$

23. $6x^2 - 7x + 2$

24. $4x^2 + 16x + 15$

25. $2x^2 + 6x - 20$

26. $6x^3 - 22x^2 - 8x$

27. $9x^6 - 4y^2$

28. $8x^3 + 125$

29. Find an equation that has $x = 3$ and $x = -6$ as roots.

30. Find an equation that has $x = -1$, $x = 0$ and $x = -3$ as roots.

Express each quotient as a sum (or difference) of fractions and simplify if possible.

31. $\dfrac{3x^3y + 10x^2y^2 + 5xy^3}{5x^2y}$

32. $\dfrac{7x^2y^2 - 21x^2y^3 + 8x^2y^4}{7x^2y^2}$

Divide by using the long division algorithm.

33. $\left(x^2 - 14x - 16\right) \div (x + 1)$

34. $\dfrac{2x^3 + 5x^2 + 7}{x + 3}$

Divide the following expressions using synthetic division. a. Write the answer in the form $Q + \dfrac{R}{D}$ where R is a constant. b. State the value of c and the value of $P(c)$.

35. $\dfrac{x^3 + x + 2}{x - 2}$

36. $\dfrac{x^4 + 3x^3 + 3x^2 + 4x + 3}{x + 1}$

37. $\dfrac{8x^3 + 4x^2}{2x^2 - 5x - 3}$

38. $\dfrac{x^2 - 9}{x + 3}$

39. $\dfrac{x}{x^2 + x}$

40. $\dfrac{x^2 + 2x - 15}{2x^2 - 12x + 18}$

Perform the indicated operations and simplify. Assume that no denominator is 0.

41. $\dfrac{x^2}{x + y} - \dfrac{y^2}{x + y}$

42. $\dfrac{4x}{3x + 3} - \dfrac{x}{x + 1}$

43. $\dfrac{4x}{x - 4} \div \dfrac{12x^2}{x^2 - 16}$

44. $\dfrac{x^2 + 3x + 2}{x + 3} \cdot \dfrac{3x^2 + 6x}{x + 1}$

45. $\dfrac{2x + 1}{x^2 + 5x - 6} \cdot \dfrac{x^2 + 6x}{x}$

46. $\dfrac{8}{x^2 + x - 6} + \dfrac{2x}{x^2 - 3x + 2}$

47. $\dfrac{x}{x^2 + 3x - 4} + \dfrac{x + 1}{x^2 - 1}$

48. $\dfrac{x + 1}{x^2 + 4x + 4} \div \dfrac{x^2 - x - 2}{x^2 - 2x - 8}$

Simplify the complex algebraic fractions.

49. $\dfrac{\dfrac{3}{x} + \dfrac{1}{6x}}{\dfrac{7}{3x}}$

50. $\dfrac{\dfrac{1}{4x} + \dfrac{1}{x^2}}{\dfrac{1}{2x} + \dfrac{1}{x}}$

Solve each of the equations.

51. $4(x + 2) - 7 = -2(3x + 1) - 3$

52. $(x + 4)(x - 5) = 10$

53. $4x^2 + 20x + 25 = 0$

54. $x^3 - x^2 = 20x$

55. $0 = x^2 - 7x + 10$

56. $\dfrac{7}{2x - 1} = \dfrac{3}{x + 6}$

57. **a.** Simplify the following expression: $\dfrac{3}{x} - \dfrac{5}{x + 3}$.

b. Solve the following equation: $\dfrac{3}{x} = \dfrac{5}{x + 3}$.

58. **a.** Simplify the following expression: $\dfrac{3x}{x+2} + \dfrac{2}{x-4} + 3$.

  b. Solve the following equation: $\dfrac{3x}{x+2} + \dfrac{2}{x-4} = 3$.

59. In the formula $A = P + Prt$, solve for t.

Solve the following word problems.

60. **Investing:** How long will it take for an investment of $600 at a rate of 5% to be worth $615? (**Hint:** $I = Prt$)

61. **Consecutive Integers:** Find three consecutive integers such that half the product of the two smallest integers is equal to 19 plus the largest.

62. **Airplanes:** An airplane can travel 1035 miles in the same time that a train travels 270 miles. The speed of the plane is 50 mph more than three times the speed of the train. Find the speed of each.

63. **Waxing a Car:** A man can wax his car three times as fast as his daughter can. Together they can complete the job in 2 hours. How long would it take each of them working alone?

64. **Boating:** A family travels 18 miles downriver and returns. It takes 8 hours to make the round trip. Their rate in still water is twice the rate of the current. Find the rate of the current.

65. **Springs:** The weight on a spring varies directly as the length the spring stretches. If a hanging spring stretches 5 cm when a weight of 13 g is placed at its end, how far will the spring stretch if a weight of 20 g is placed at its end?

Mathematics @ Work!

In 2010, Walmart surged past Exxon Mobil to once again claim the title of America's largest corporation. Walmart's revenues were about $408 billion, while Exxon Mobil's revenues where only about $285 billion. For any business, generating revenue is a top priority. Without revenue, companies have no way to pay their workers or produce and improve their product lines. A key element in generating revenue requires creating a product that appeals to the consumer population. However, the product must also be priced such that people feel that the benefit they receive from the product is worth the expense. The formula for calculating revenue is $R = px$, where p is the price and x is the number of units being sold. The optimal price to charge to receive maximum revenue can be found by calculating the maximum value of the quadratic function derived using the revenue formula. Consider the following scenario:

A golf resort estimates that by charging x dollars to rent a set of golf clubs for the day, they will rent out $300 - x$ sets of clubs over the course of a week. What price will yield maximum revenue?

  a. What is the revenue function $R(x)$?
  b. What price will yield a maximum revenue?

For more problems like these, see Section 12.5.

Complex Numbers and Quadratic Equations

12.1 Complex Numbers

Objective A **Square Roots of Negative Numbers**

Objectives

A Simplify square roots of negative numbers.

B Identify the real parts and the imaginary parts of complex numbers.

C Solve equations with complex numbers by setting the real parts and the imaginary parts equal to each other.

D Add and subtract with complex numbers.

E Multiply with complex numbers.

F Divide with complex numbers.

G Simplify powers of i.

One of the properties of real numbers is that the square of any real number is nonnegative. That is, for any real number x, $x^2 \geq 0$. The square roots of negative numbers, such as $\sqrt{-4}$ and $\sqrt{-5}$, are not real numbers. However, they can be defined as part of the system of **complex numbers**.

Complex numbers include all the real numbers and the even roots of negative numbers. Earlier we saw how these numbers occur as solutions to quadratic equations. At first such numbers seem to be somewhat impractical because they are difficult to picture in any type of geometric setting and they are not solutions to the types of word problems that are familiar. However, complex numbers do occur quite naturally in trigonometry and higher level mathematics and have practical applications in such fields as electrical engineering.

The first step in the development of complex numbers is to define $\sqrt{-1}$.

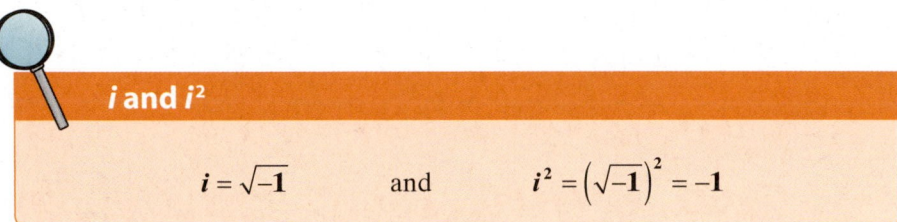

i and i^2

$$i = \sqrt{-1} \qquad \text{and} \qquad i^2 = \left(\sqrt{-1}\right)^2 = -1$$

Using the definition of $\sqrt{-1}$, the following definition for the square root of a negative number can be made.

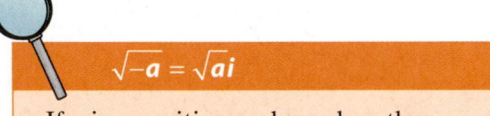

$\sqrt{-a} = \sqrt{a}\,i$

If a is a positive real number, then

$$\sqrt{-a} = \sqrt{a} \cdot \sqrt{-1} = \sqrt{a}\,i.$$

Note: The number i is not under the radical sign. To avoid confusion, we sometimes write $i\sqrt{a}$.

Example 1

Square Roots of Negative Numbers

Simplify the following radicals.

a. $\sqrt{-25} = \sqrt{-1}\sqrt{25} = i \cdot 5 = 5i$ $(5i)^2 = 5^2 i^2 = 25(-1) = -25$

b. $\sqrt{-36} = \sqrt{-1}\sqrt{36} = i \cdot 6 = 6i$

c. $\sqrt{-24} = \sqrt{-1}\sqrt{4 \cdot 6} = i \cdot 2 \cdot \sqrt{6} = 2\sqrt{6}\,i$ or $2i\sqrt{6}$

d. $\sqrt{-45} = \sqrt{-1}\sqrt{9 \cdot 5} = i \cdot 3 \cdot \sqrt{5} = 3\sqrt{5}\,i$ or $3i\sqrt{5}$

We can write $2\sqrt{6}\,i$ and $3\sqrt{5}\,i$ as long as we take care not to include the i under the radical sign.

Now work margin exercise 1.

1. Simplify the following radicals.

a. $\sqrt{-100}$

b. $\sqrt{-49}$

c. $\sqrt{-18}$

d. $\sqrt{-72}$

Objective B   **Real and Imaginary Parts of Complex Numbers**

Complex Numbers

The **standard form** of a **complex number** is $a + bi$, where a and b are real numbers. a is called the **real part** and b is called the **imaginary part**.

If $b = 0$, then $a + bi = a + 0i = a$ is a **real number**.

If $a = 0$, then $a + bi = 0 + bi = bi$ is called a **pure imaginary number** (or an **imaginary number**).

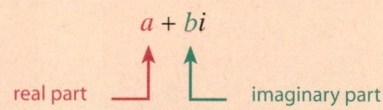

$$a + bi$$

real part   imaginary part

---

notes

■ The term "imaginary" is somewhat misleading. Complex numbers
and imaginary numbers are no more "imaginary" than any other type
■ of number. In fact, all the types of numbers that we have studied
(whole numbers, integers, rational numbers, irrational numbers, and
■ real numbers) are products of human imagination.

In general, if a is a real number, then we can write $a = a + 0i$. This means that a is a complex number. **Thus every real number is a complex number.** Figure 1 illustrates the relationships among the various types of numbers we have studied.

2. Identify the real and imaginary parts of each complex number.

a. $5i$

b. $14 + \sqrt{7}i$

c. $\dfrac{6 - 11i}{5}$

d. -13

Example 2

Real and Imaginary Parts

Identify the real and imaginary parts of each complex number.

a. $4 - 2i$ \qquad 4 is the real part; -2 is the imaginary part.

b. $\dfrac{5 + 2i}{3}$ \qquad $\dfrac{5 + 2i}{3} = \dfrac{5}{3} + \dfrac{2}{3}i$ in standard form.

Thus $\dfrac{5}{3}$ is the real part; $\dfrac{2}{3}$ is the imaginary part.

c. 7 \qquad $7 = 7 + 0i$ in standard form.
Thus 7 is the real part; 0 is the imaginary part. (Remember, if $b = 0$, the complex number is a real number.)

d. $-\sqrt{3}i$ \qquad $-\sqrt{3}i = 0 - \sqrt{3}i$ in standard form.

Thus 0 is the real part; $-\sqrt{3}$ is the imaginary part. (If $a = 0$ and $b \neq 0$, then the complex number is a pure imaginary number.)

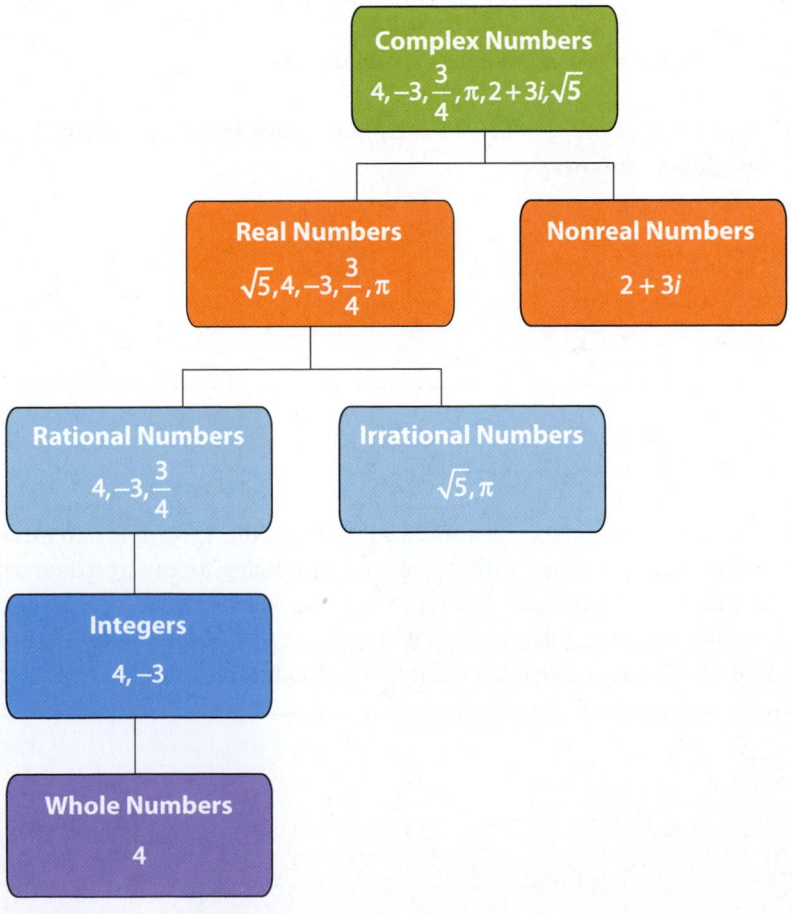

Figure 1

Solving Equations with Complex Numbers

If two complex numbers are equal, then the real parts are equal and the imaginary parts are equal. For example, if $x + yi = 7 + 2i$, then $x = 7$ and $y = 2$.

This relationship can be used to solve equations involving complex numbers.

> ### Equality of Complex Numbers
>
> For complex numbers $a + bi$ and $c + di$,
> if $\boldsymbol{a + bi = c + di}$, then $\boldsymbol{a = c}$ and $\boldsymbol{b = d}$.

Example 3

Solving Equations

Solve each equation for x and y.

a. $(x + 3) + 2yi = 7 - 6i$

Solution

Equate the real parts and the imaginary parts, and solve the resulting equations.

$$x + 3 = 7 \quad \text{and} \quad 2y = -6$$
$$x = 4 \qquad\qquad y = -3$$

b. $2y + 3 - 8i = 9 + 4xi$

Solution

Equate the real parts and the imaginary parts, and solve the resulting equations.

$$2y + 3 = 9 \quad \text{and} \quad -8 = 4x$$
$$2y = 6 \qquad\qquad -2 = x$$
$$y = 3$$

Now work margin exercise 3.

3. Solve each equation for x and y.

a. $(x - 7) + 4yi = 3 - 8i$

b. $4y + 7 - 5i = 11 + xi$

Addition and Subtraction with Complex Numbers

Adding and subtracting complex numbers is similar to adding and subtracting polynomials. Simply combine like terms. For example,

$$(2 + 3i) + (9 - 8i) = 2 + 9 + 3i - 8i$$
$$= (2 + 9) + (3 - 8)i$$
$$= 11 - 5i.$$

Similarly,

$$(5 - 2i) - (6 + 7i) = 5 - 2i - 6 - 7i$$
$$= (5 - 6) + (-2 - 7)i$$
$$= -1 - 9i.$$

Addition and Subtraction with Complex Numbers

For complex numbers $a + bi$ and $c + di$,

$$(a + bi) + (c + di) = (a + c) + (b + d)i$$
and
$$(a + bi) - (c + di) = (a - c) + (b - d)i.$$

4. Find each sum or difference as indicated.

a. $(5 - 3i) + (2 + 7i)$

b. $\left(-7 - \sqrt{2}i\right) - \left(-7 + 3\sqrt{2}i\right)$

c. $\left(4 - \sqrt{5}i\right) + \left(\sqrt{2} + 8i\right)$

Example 4

Addition and Subtraction with Complex Numbers

Find each sum or difference as indicated.

a. $(6 - 2i) + (1 - 2i)$

Solution

$$(6 - 2i) + (1 - 2i) = (6 + 1) + (-2 - 2)i$$
$$= 7 - 4i$$

b. $\left(-8 - \sqrt{2}\,i\right) - \left(-8 + \sqrt{2}\,i\right)$

Solution

$$\left(-8 - \sqrt{2}\,i\right) - \left(-8 + \sqrt{2}\,i\right) = \left(-8 - (-8)\right) + \left(-\sqrt{2} - \sqrt{2}\right)i$$
$$= (-8 + 8) + \left(-2\sqrt{2}\right)i$$
$$= 0 - 2\sqrt{2}i$$
$$= -2\sqrt{2}\,i \text{ or } -2i\sqrt{2}$$

c. $\left(\sqrt{3} - 2i\right) + \left(1 + \sqrt{5}\,i\right)$

Solution

$$\left(\sqrt{3} - 2i\right) + \left(1 + \sqrt{5}\,i\right) = \left(\sqrt{3} + 1\right) + \left(-2 + \sqrt{5}\right)i$$
$$= \left(\sqrt{3} + 1\right) + \left(\sqrt{5} - 2\right)i$$

Note: Here, the coefficients do not simplify. This means that the real part is $\sqrt{3} + 1$ and the imaginary part is $\sqrt{5} - 2$.

Now work margin exercise 4.

Multiplication with Complex Numbers

The product of two complex numbers can be found by using the FOIL method for multiplying two binomials (see previous section). **Remember that $i^2 = -1$.** For example,

$$(3+5i)(2+i) = 6+3i+10i+5i^2$$
$$= 6+13i-5 \qquad \textcolor{red}{5i^2 = 5(-1) = -5}$$
$$= 1+13i.$$

Example 5

Multiplication with Complex Numbers

Find the following products.

a. $(3i)(2-7i)$

Solution

$$(3i)(2-7i) = 6i - 21i^2$$
$$= 6i - 21(-1)$$
$$= 21 + 6i$$

b. $(5+i)(2+6i)$

Solution

$$(5+i)(2+6i) = 5(2+6i) + i(2+6i)$$
$$= 10 + 30i + 2i + 6i^2$$
$$= 10 + 32i - 6$$
$$= 4 + 32i$$

c. $(\sqrt{2} - i)(\sqrt{2} - i)$

Solution

$$(\sqrt{2}-i)(\sqrt{2}-i) = (\sqrt{2})^2 - \sqrt{2}\cdot i - \sqrt{2}\cdot i + i^2$$
$$= 2 - 2\sqrt{2}\,i - 1 \qquad \textcolor{red}{\text{Remember } i^2 = -1.}$$
$$= 1 - 2i\sqrt{2}$$

d. $(-1+i)(2-i)$

Solution

$$(-1+i)(2-i) = -2 + i + 2i - i^2$$
$$= -2 + 3i + 1$$
$$= -1 + 3i$$

Now work margin exercise 5.

5. Find the following products.

a. $(4i)(1-7i)$

b. $(4+2i)(3+5i)$

c. $(\sqrt{2}-i)(\sqrt{2}+i)$

Common Error

Remember that $\sqrt{a} \cdot \sqrt{b} = \sqrt{ab}$ **only** if a and b are nonnegative real numbers.

Applying this rule to negative real numbers can lead to an error. The error can be avoided by first changing the radicals to imaginary form.

Wrong Solution

$$\sqrt{-6} \cdot \sqrt{-2} = \sqrt{12}$$
$$= \sqrt{4} \cdot \sqrt{3}$$
$$= 2\sqrt{3}$$

Correct Solution

$$\sqrt{-6} \cdot \sqrt{-2} = \sqrt{6}\, i \cdot \sqrt{2}\, i$$
$$= \sqrt{12}\, i^2$$
$$= 2\sqrt{3}(-1)$$
$$= -2\sqrt{3}$$

Objective F — Division with Complex Numbers

The two complex numbers $a + bi$ and $a - bi$ are called **complex conjugates** or simply **conjugates** of each other. As the following steps show, **the product of two complex conjugates will always be a nonnegative real number**.

$$(a+bi)(a-bi) = a^2 - abi + abi - b^2 i^2$$
$$= a^2 - b^2 i^2$$
$$= a^2 + b^2$$

The resulting product, $a^2 + b^2$, is a real number, and it is nonnegative since it is the sum of the squares of real numbers.

Remember that the form $a + bi$ is called the **standard form** of a complex number. The standard form allows for easy identification of the real and imaginary parts. Thus,

$$\frac{1+3i}{5} = \frac{1}{5} + \frac{3}{5}i \text{ in standard form.}$$

The real part is $\frac{1}{5}$ and the imaginary part is $\frac{3}{5}$.

To write the fraction $\frac{1+i}{2-3i}$ in standard form, multiply both the numerator and denominator by $2 + 3i$, the complex conjugate of the denominator, and simplify. This will give a positive real number in the denominator.

$$\frac{1+i}{2-3i} = \frac{(1+i)(2+3i)}{(2-3i)(2+3i)} \qquad \text{\textcolor{red}{2+ 3i is the conjugate of the denominator.}}$$

$$= \frac{2+3i+2i+3i^2}{2^2+6i-6i-3^2i^2}$$

$$= \frac{2+5i-3}{4-(9)(-1)} \qquad \text{\textcolor{red}{Reminder: }} i^2 = -1.$$

$$= \frac{-1+5i}{13}$$

$$= -\frac{1}{13} + \frac{5}{13}i$$

Writing Fractions with Complex Numbers

1. Multiply both the numerator and denominator by the complex conjugate of the denominator.
2. Simplify the resulting products in both the numerator and denominator.
3. Write the simplified result in standard form.

Remember the following special product. We restate it here to emphasize its importance.

$$(a+bi)(a-bi) = a^2 + b^2$$

Example 6

Division with Complex Numbers

Write the following fractions in standard form.

a. $\dfrac{4}{-1-5i}$

Solution

$$\frac{4}{-1-5i} = \frac{4(-1+5i)}{(-1-5i)(-1+5i)}$$

$$= \frac{-4+20i}{(-1)^2-(5i)^2} = \frac{-4+20i}{1+25}$$

$$= \frac{-4+20i}{26} = -\frac{4}{26} + \frac{20}{26}i$$

$$= -\frac{2}{13} + \frac{10}{13}i$$

6. Write the following fractions in standard form.

a. $\dfrac{8}{1-2i}$

b. $\dfrac{\sqrt{5}-2i}{\sqrt{5}+2i}$

c. $\dfrac{4-i}{4i}$

d. $\dfrac{\sqrt{3}+4i}{-\sqrt{3}+4i}$

b. $\dfrac{\sqrt{3}+i}{\sqrt{3}-i}$

Solution

$$\frac{\sqrt{3}+i}{\sqrt{3}-i} = \frac{\left(\sqrt{3}+i\right)\left(\sqrt{3}+i\right)}{\left(\sqrt{3}-i\right)\left(\sqrt{3}+i\right)}$$

$$= \frac{3+2\sqrt{3}\,i+i^2}{\left(\sqrt{3}\right)^2-i^2} = \frac{2+2\sqrt{3}\,i}{3+1}$$

$$= \frac{2+2\sqrt{3}\,i}{4} = \frac{2}{4}+\frac{2\sqrt{3}}{4}\,i$$

$$= \frac{1}{2}+\frac{\sqrt{3}}{2}\,i$$

c. $\dfrac{6+i}{i}$

Solution

$$\frac{6+i}{i} = \frac{\left(6+i\right)\left(-i\right)}{i\left(-i\right)}$$

Since $i = 0+i$ and $-i = 0-i$, the number $-i$ is the conjugate of i.

$$= \frac{-6i-i^2}{-i^2} = \frac{-6i+1}{1}$$

$$= 1-6i$$

d. $\dfrac{\sqrt{2}+i}{-\sqrt{2}+i}$

Solution

$$\frac{\sqrt{2}+i}{-\sqrt{2}+i} = \frac{\left(\sqrt{2}+i\right)\left(-\sqrt{2}-i\right)}{\left(-\sqrt{2}+i\right)\left(-\sqrt{2}-i\right)}$$

$$= \frac{-\left(\sqrt{2}\right)^2-\sqrt{2}\,i-\sqrt{2}\,i-i^2}{\left(-\sqrt{2}\right)^2-i^2}$$

$$= \frac{-2-2\sqrt{2}\,i+1}{2+1} = \frac{-1-2\sqrt{2}\,i}{3}$$

$$= -\frac{1}{3}-\frac{2\sqrt{2}}{3}\,i$$

Now work margin exercise 6.

The powers of *i* form an interesting pattern. Regardless of the particular integer exponent, there are only four possible values for any power of *i*:

$$i, \quad -1, \quad -i, \quad \text{and} \quad 1.$$

The fact that these are the only four possibilities for powers of *i* becomes apparent from studying the following powers.

$$i^1 = i$$
$$i^2 = -1$$
$$i^3 = i^2 \cdot i = -1 \cdot i = -i$$
$$i^4 = i^2 \cdot i^2 = (-1)(-1) = 1$$
$$i^5 = i^4 \cdot i = 1 \cdot i = i$$
$$i^6 = i^4 \cdot i^2 = (1)(-1) = -1$$
$$i^7 = i^4 \cdot i^3 = (1)(-i) = -i$$
$$i^8 = i^4 \cdot i^4 = 1 \cdot 1 = 1$$

Higher powers of *i* can be simplified using the fact that when *i* is raised to a power that is a multiple of 4, the result is 1. Thus if *n* is a positive integer, then

$$i^{4n} = \left(i^4\right)^n = 1^n = 1$$
$$i^{4n+1} = i^{4n} \cdot i = 1 \cdot i = i$$
$$i^{4n+2} = i^{4n} \cdot i^2 = (1) \cdot (-1) = -1$$
$$i^{4n+3} = i^{4n} \cdot i^3 = (1) \cdot (-i) = -i.$$

Example 7

Powers of *i*

Simplify each power of *i*.

a. $i^{45} = i^{44} \cdot i = \left(i^4\right)^{11} \cdot i = 1^{11} \cdot i = i$

$i = 0 + i$ in standard form.

b. $i^{58} = i^{56} \cdot i^2 = \left(i^4\right)^{14} \cdot i^2 = 1^{14} \cdot (-1) = -1$

$1 = -1 + 0i$ in standard form.

c. $i^{-7} = \dfrac{1}{i^7} = \dfrac{1}{i^7} \cdot \dfrac{i}{i} = \dfrac{i}{i^8} = \dfrac{i}{1} = i$

$i = 0 + i$ in standard form.

Now work margin exercise 7.

7. Simplify each power of *i*.

a. i^{37}

b. i^{14}

c. i^{-8}

Practice Problems

1. Find the real part and the imaginary part of $2 - \sqrt{39}i$.

Add or subtract as indicated. Simplify your answers.

2. $\left(-7 + \sqrt{3}i\right) + (5 - 2i)$
    3. $(4 + i) - (5 + 2i)$

Solve for x and y.

4. $x + yi = \sqrt{2} - 7i$
    5. $3y + (x - 7)i = -9 + 2i$

Write each of the following numbers in standard form.

6. $-2i(3 - i)$
    7. $(2 + 4i)(1 + i)$
    8. i^{13}

9. i^{-2}
    10. $\dfrac{2}{1 + 5i}$
    11. $\dfrac{7 + i}{2 - i}$

Practice Problem Answers

1. Real part is 2, imaginary part is $-\sqrt{39}$.

2. $-2 + \left(\sqrt{3} - 2\right)i$
    3. $-1 - i$

4. $x = \sqrt{2}$ and $y = -7$
    5. $x = 9$ and $y = -3$

6. $-2 - 6i$
    7. $-2 + 6i$
    8. $0 + i$
    9. $-1 + 0i$

10. $\dfrac{1}{13} - \dfrac{5}{13}i$
    11. $\dfrac{13}{5} + \dfrac{9}{5}i$

Exercises 12.1

Find the real part and the imaginary part of each of the complex numbers. See Example 2.

1. $4 - 3i$

2. $\dfrac{3}{4} + i$

3. $-11 + \sqrt{2}\,i$

4. $6 + \sqrt{3}\,i$

5. $\dfrac{3}{8}$

6. $\dfrac{4}{7}i$

7. $\dfrac{4 + 7i}{5}$

8. $\dfrac{2 - i}{4}$

9. $\dfrac{2}{3} + \sqrt{17}\,i$

10. $-\sqrt{5} + \dfrac{\sqrt{2}}{2}i$

Simplify the following radicals. See Example 1

11. $\sqrt{-49}$

12. $\sqrt{-121}$

13. $-\sqrt{-64}$

14. $-\sqrt{-169}$

15. $\sqrt{128}$

16. $4\sqrt{-99}$

17. $-2\sqrt{-108}$

18. $\sqrt{242}$

19. $\sqrt{-1000}$

20. $\sqrt{-243}$

Solve the equations for x and y. See Example 3

21. $x + 3i = 6 - yi$

22. $2x - 8i = -2 + 4yi$

23. $\sqrt{5} - 2i = y + xi$

24. $\sqrt{2} - 2yi = 3x + 6i$

25. $2x + 3 + 6i = 7 - yi - 2i$

26. $x + yi + 8 = 2i + 4 - 3yi$

27. $x + 2i = 5 - yi - 3 - 4i$

28. $11i - 2x + 4 = 10 - 3i + 2yi$

29. $2x - 2yi + 6 = 6i - x + 2$

Find each sum or difference as indicated. See Example 4

30. $(2 + 3i) + (4 - i)$

31. $(-3 + 2i) - (6 + 2i)$

32. $(4 - 3i) + (2 - 3i)$

33. $(8 + 9i) - (8 - 5i)$

34. $(4 + 3i) - (\sqrt{2} + 3i)$

35. $(7 + \sqrt{6}\,i) + (-2 + i)$

36. $(\sqrt{11} + 2i) + (5 - 7i)$

37. $(\sqrt{3} + \sqrt{2}\,i) - (5 + \sqrt{2}\,i)$

38. $(\sqrt{5} + \sqrt{3}\,i) + (1 - i)$

39. $(5 + \sqrt{-25}) - (7 + \sqrt{-100})$

40. $(7 + \sqrt{-9}) - (3 - 2\sqrt{-25})$

41. $(4 + i) + (-3 - 2i) - (-1 - i)$

42. $(-2 - 3i) + (6 + i) - (2 + 5i)$

43. $(7 + 3i) + (2 - 4i) - (6 - 5i)$

44. $(-5 + 7i) + (4 - 2i) - (3 - 5i)$

45. $8(2+3i)$

46. $-3(7-4i)$

47. $\sqrt{3}\left(\sqrt{3}+2i\right)$

48. $-i\left(\sqrt{3}+i\right)$

49. $2i\left(\sqrt{5}+2i\right)$

50. $\sqrt{3}\,i\left(2-\sqrt{3}\,i\right)$

51. $5i\left(2-\sqrt{2}\,i\right)$

52. $(2+7i)(6+i)$

53. $(-3+5i)(-1+2i)$

54. $(2-3i)(2+3i)$

55. $(4+5i)(4-5i)$

56. $(5+7i)^2$

57. $\left(\sqrt{3}+i\right)\left(\sqrt{3}-2i\right)$

58. $\left(5-\sqrt{2}\,i\right)\left(5-\sqrt{2}\,i\right)$

59. $\left(\sqrt{7}+3i\right)\left(\sqrt{7}+i\right)$

60. $\left(4+\sqrt{5}\,i\right)\left(4-\sqrt{5}\,i\right)$

61. $\left(2\sqrt{3}+i\right)(4+3i)$

62. $\left(3+\sqrt{5}\,i\right)\left(3+\sqrt{6}\,i\right)$

63. $\left(2-\sqrt{3}\,i\right)\left(3-\sqrt{2}\,i\right)$

64. $\dfrac{-3}{i}$

65. $\dfrac{5}{4i}$

66. $\dfrac{2+i}{-4i}$

67. $\dfrac{-4}{1+2i}$

68. $\dfrac{2i}{5-i}$

69. $\dfrac{-4i}{1+3i}$

70. $\dfrac{6+i}{3-4i}$

71. $\dfrac{9-2i}{\sqrt{5}+i}$

72. $\dfrac{\sqrt{6}-3i}{\sqrt{6}+3i}$

73. i^{13}

74. i^{20}

75. i^{30}

76. i^{15}

77. i^{-3}

78. i^{-5}

79. i^{4k}

80. i^{4k+2}

81. i^{4k+3}

82. i^{4k+1}

83. $(x+3i)(x-3i)$

84. $(y+5i)(y-5i)$

85. $\left(x+\sqrt{2}\,i\right)\left(x-\sqrt{2}\,i\right)$

86. $\left(2x+\sqrt{7}\,i\right)\left(2x-\sqrt{7}\,i\right)$

87. $\left(\sqrt{5}y+2i\right)\left(\sqrt{5}y-2i\right)$

88. $\left(y-\sqrt{3}\,i\right)\left(y+\sqrt{3}\,i\right)$

89. $\left[(x+1)-\sqrt{8}\,i\right]\left[(x+1)+\sqrt{8}\,i\right]$ **90.** $\left[(y-3)+2i\right]\left[(y-3)-2i\right]$ **91.** $\left[(x-1)+5i\right]\left[(x-1)-5i\right]$

Writing & Thinking

92. List 5 numbers that do, and 5 numbers that do not fit each of the following categories (if possible).

    a. rational number   **b.** integer

    c. real number   **d.** pure imaginary number

    e. complex number   **f.** irrational number

93. Answer the following questions and give a brief explanation of your answer.

    a. Is every real number a complex number?

    b. Is every complex number a real number?

94. Explain why $\sqrt{-4} \cdot \sqrt{-4} \neq 4$. What is the correct value of $\sqrt{-4} \cdot \sqrt{-4}$?

95. Explain why the product of every complex number and its conjugate is a non-negative real number.

12.2 Completing the Square and the Square Root Property

Objective A **Solving Quadratic Equations by Factoring**

Objectives

A Solve quadratic equations by factoring.

B Solve quadratic equations by using the square root property.

C Solve quadratic equations by completing the square.

D Write quadratic equations with given roots.

Not every polynomial can be factored so that the factors have integer coefficients, and not every polynomial equation can be solved by factoring. However, when the solutions of a polynomial equation can be found by factoring, the method depends on the **zero-factor property**, which is restated here for easy reference.

Zero-Factor Property

If the product of two (or more) factors is 0, then at least one of the factors must be 0. That is, if a and b are real numbers and $a \cdot b = 0$, then

$$a = 0 \text{ or } b = 0 \text{ or both}.$$

Also, as discussed in a previous section, polynomial equations of second degree are called **quadratic equations** and, because these are the equations of interest in this chapter, the definition is restated here.

Quadratic Equations

Quadratic equations are equations that can be written in the form

$$ax^2 + bx + c = 0,$$

where a, b, and c are real numbers and $a \neq 0$.

The procedure for solving quadratic equations by factoring involves making sure that one side of the equation is 0 and then applying the zero-factor property. The following list of steps outlines the procedure.

Solving Quadratic Equations by Factoring

1. Add or subtract terms as necessary so that 0 is on one side of the equation and the equation is in the standard form $ax^2 + bx + c = 0$ where a, b, and c are real numbers and $a \neq 0$.
2. Factor completely. (If there are any fractional coefficients, multiply each term by the least common denominator so that all coefficients will be integers.)
3. Set each nonconstant factor equal to 0 and solve each linear equation for the unknown.
4. Check each solution, one at a time, in the original equation.

Example 1

Solving Quadratic Equations by Factoring

Solve the following quadratic equations by factoring.

a. $x^2 - 15x = -50$

Solution

$$x^2 - 15x = -50$$
$$x^2 - 15x + 50 = 0 \qquad \text{Add 50 to both sides. One side must be 0.}$$
$$(x - 5)(x - 10) = 0 \qquad \text{Factor the left-hand side.}$$

$$x - 5 = 0 \quad \text{or} \quad x - 10 = 0 \qquad \text{Set each factor equal to 0.}$$
$$x = 5 \qquad\qquad x = 10 \qquad \text{Solve each linear equation.}$$

Check:

$$(5)^2 - 15 \cdot (5) \overset{?}{=} -50 \qquad (10)^2 - 15 \cdot (10) \overset{?}{=} -50$$
$$25 - 75 \overset{?}{=} -50 \qquad\qquad 100 - 150 \overset{?}{=} -50$$
$$-50 = -50 \qquad\qquad\qquad -50 = -50$$

b. $3x^2 - 24x = -48$

Solution

$$3x^2 - 24x = -48$$
$$3x^2 - 24x + 48 = 0 \qquad \text{Add 48 to both sides. One side must be 0.}$$
$$3(x^2 - 8x + 16) = 0 \qquad \text{Factor out the GCF, 3.}$$
$$3(x - 4)^2 = 0 \qquad\qquad \text{The trinomial is a perfect square.}$$
$$x - 4 = 0 \qquad\qquad\quad \text{Two factors are the same.}$$
$$x = 4 \qquad\qquad\qquad \text{The solution is a } \textbf{double root}.$$

Check:

$$3 \cdot (4)^2 - 24 \cdot (4) \overset{?}{=} -48$$
$$48 - 96 \overset{?}{=} -48$$
$$-48 = -48$$

Now work margin exercise 1.

Now that complex numbers have been introduced and discussed, we will see that quadratic equations may have nonreal complex solutions. **In particular, the sum of two squares (previously declared "not factorable" because real factors were implied) can be factored as the product of complex conjugates.** For example,

$$x^2 + 9 = (x + 3i)(x - 3i).$$

As shown in Example 2, such factors can lead to nonreal solutions to a quadratic equation.

2. Solve the following quadratic equation by factoring: $x^2 + 25 = 0$.

Example 2

Quadratic Equations Involving the Sum of Two Squares

Solve the following quadratic equation by factoring: $x^2 + 4 = 0$.

Solution

$$x^2 + 4 = 0$$
$$(x + 2i)(x - 2i) = 0$$
$$x + 2i = 0 \quad \text{or} \quad x - 2i = 0$$
$$x = -2i \qquad\qquad x = 2i$$

Note that $x^2 + 4$ is the sum of two squares and can be factored into the product of conjugates of complex numbers.

Check:

$$(-2i)^2 + 4 \overset{?}{=} 0 \qquad (2i)^2 + 4 \overset{?}{=} 0$$
$$4i^2 + 4 \overset{?}{=} 0 \qquad 4i^2 + 4 \overset{?}{=} 0$$
$$-4 + 4 \overset{?}{=} 0 \qquad -4 + 4 \overset{?}{=} 0$$
$$0 = 0 \qquad\qquad 0 = 0$$

Now work margin exercise 2.

Objective B Solving Quadratic Equations Using the Square Root Property

Consider the equation $x^2 = 13$. Allowing that the variable x might be positive or negative, we use the definition of square root, $\sqrt{x^2} = |x|$. Taking the square root of both sides of the equation gives: $|x| = \sqrt{13}$. So, we have two solutions, $x = \sqrt{13}$ or $x = -\sqrt{13}$. To indicate both solutions, we write $x = \pm\sqrt{13}$. Similarly, for the equation $(x - 3)^2 = 5$, the same process gives $x - 3 = \pm\sqrt{5}$.

This leads to the two equations and the two solutions, as follows.

$$x - 3 = \sqrt{5} \qquad \text{or} \qquad x - 3 = -\sqrt{5}$$
$$x = 3 + \sqrt{5} \qquad\qquad x = 3 - \sqrt{5}$$

We can write the two solutions in the form $x = 3 \pm \sqrt{5}$.

This discussion shows how the definition of a square root can be used to solve quadratic equations in certain forms. In particular, if one side of the equation is a squared expression and the other side is a constant, we can simply take the square root of both sides. If the constant is negative, then the solutions will involve nonreal numbers. We are using the following **square root property**.

Square Root Property

If $x^2 = c$, then $x = \pm\sqrt{c}$.

If $(x-a)^2 = c$, then $x - a = \pm\sqrt{c}$ $\left(\text{or } x = a \pm \sqrt{c}\right)$.

Note: If c is negative $(c < 0)$, then the solutions will be nonreal.

Example 3

Using the Square Root Property to Solve Quadratic Equations

Solve the following quadratic equations using the square root property.

a. $(y+4)^2 = 8$

Solution

$$(y+4)^2 = 8 \qquad \text{Note that } 8 > 0.$$
$$y + 4 = \pm\sqrt{8}$$
$$y = -4 \pm 2\sqrt{2}$$

b. $x^2 = -25$

Solution

$$x^2 = -25 \qquad \text{Note that } -25 < 0.$$
$$x = \pm\sqrt{-25}$$
$$x = \pm 5i$$

Now work margin exercise 3.

3. Solve the following quadratic equations using the square root property.

a. $(x+6)^2 = 12$

b. $x^2 = -16$

Objective C **Completing the Square**

Recall that a perfect square trinomial is the result of squaring a binomial. Our objective here is to find the constant term of a perfect square trinomial when the first two terms are given. This is called **completing the square**. We will find this procedure useful in solving quadratic equations and in developing the quadratic formula.

The following examples illustrate the concept of a perfect square trinomial.

Perfect Square Trinomials		Equal Factors		Square of a Binomial
$x^2 - 8x + 16$	$=$	$(x-4)(x-4)$	$=$	$(x-4)^2$
$x^2 + 20x + 100$	$=$	$(x+10)(x+10)$	$=$	$(x+10)^2$
$x^2 - 9x + \dfrac{81}{4}$	$=$	$\left(x-\dfrac{9}{2}\right)\left(x-\dfrac{9}{2}\right)$	$=$	$\left(x-\dfrac{9}{2}\right)^2$
$x^2 - 2ax + a^2$	$=$	$(x-a)(x-a)$	$=$	$(x-a)^2$
$x^2 + 2ax + a^2$	$=$	$(x+a)(x+a)$	$=$	$(x+a)^2$

Table 1

The last two examples in Table 1 are in the form of formulas. We see two things in each case:

1. The leading coefficient (the coefficient of x^2) is 1.

2. The constant term is the square of $\dfrac{1}{2}$ of the coefficient of x.

For example, $\dfrac{1}{2}(2a) = a$ and the square of this result is the constant a^2.

What constant should be added to $x^2 - 16x$ to get a perfect square trinomial? By following the ideas just discussed, we find that $\dfrac{1}{2}(-16) = -8$ and $(-8)^2 = 64$. Therefore, to complete the square, we add 64. Thus

$$x^2 - 16x + 64 = (x-8)^2.$$

By adding 64, we have completed the square for $x^2 - 16x$.

Example 4

Completing the Square

Add the constant that will complete the square for each expression, and write the new expression as the square of a binomial.

a. $x^2 + 10x$

Solution

$$x^2 + 10x + \underline{\;\;?\;\;} = \left(\underline{\;\;?\;\;}\right)^2$$

$$\dfrac{1}{2}(10) = 5 \text{ and } (5)^2 = 25$$

Find $\dfrac{1}{2}$ of the coefficient of x and square the result. Add this square constant to complete the square. The resulting trinomial will equal the square of a binomial.

So, add 25: $x^2 + 10x + \underline{25} = (\underline{x+5})^2$

b. $x^2 - 7x$

Solution

$$x^2 - 7x + \underline{\ ?\ } = \left(\underline{\quad ?\quad}\right)^2$$

$$\frac{1}{2}(-7) = -\frac{7}{2} \text{ and } \left(-\frac{7}{2}\right)^2 = \frac{49}{4} \quad \text{Find } \frac{1}{2} \text{ of the coefficient of } x \text{ and square the result.}$$

So, add $\dfrac{49}{4}$: $x^2 - 7x + \dfrac{49}{4} = \left(x - \dfrac{7}{2}\right)^2$

Now work margin exercise 4.

Now we want to use the process of completing the square to help in solving quadratic equations. This technique involves the following steps.

Solve a Quadratic Equation by Completing the Square

1. If necessary, divide or multiply on both sides of the equation so that the leading coefficient (the coefficient of x^2) is 1.
2. If necessary, isolate the constant term on one side of the equation.
3. Find the constant that completes the square of the polynomial and **add this constant to both sides**. Remember that the constant term is the square of $\dfrac{1}{2}$ of the coefficient of x. Rewrite the polynomial as the square of a binomial.
4. Use the square root property to find the solutions of the equation.

Example 5

Solving Quadratic Equations by Completing the Square

Solve the following quadratic equations by completing the square.

a. $x^2 - 8x = 25$

Solution

$$x^2 - 8x = 25 \qquad \text{The coefficient of } x^2 \text{ is already 1 and the constant is isolated on one side of the equation.}$$

$$x^2 - 8x + 16 = 25 + 16 \qquad \text{Complete the square: } \frac{1}{2}(-8) = -4 \text{ and } (-4)^2 = 16.$$
$$\text{Therefore, add 16 to both sides.}$$

$$(x - 4)^2 = 41 \qquad \text{Factor the polynomial.}$$

4. Add the constant that will complete the square for each expression, and write the new expression as the square of a binomial.

a. $y^2 - 14y + \_ = (\_\_\_)^2$

b. $x^2 + 9x + \_ = (\_\_\_)^2$

$$x - 4 = \sqrt{41} \quad \text{or} \quad x - 4 = -\sqrt{41}$$
$$x = 4 + \sqrt{41} \qquad\qquad x = 4 - \sqrt{41}$$

Use the square root property.

There are two real solutions: $4 + \sqrt{41}$ and $4 - \sqrt{41}$.

We write $x = 4 \pm \sqrt{41}$.

b. $3x^2 + 6x - 15 = 0$

Solution

$$3x^2 + 6x - 15 = 0$$

$$\frac{3x^2}{3} + \frac{6x}{3} - \frac{15}{3} = \frac{0}{3}$$

Divide each term by 3. The leading coefficient must be 1.

$$x^2 + 2x - 5 = 0$$

$$x^2 + 2x = 5$$

Isolate the constant term.

$$x^2 + 2x + 1 = 5 + 1$$

Complete the square:
$\frac{1}{2}(2) = 1$ and $1^2 = 1$.
Therefore, add 1 to both sides.

$$(x + 1)^2 = 6$$

Factor the polynomial.

$$x + 1 = \pm\sqrt{6}$$

Use the square root property.

$$x = -1 \pm \sqrt{6}$$

c. $2x^2 + 2x - 7 = 0$

Solution

$$2x^2 + 2x - 7 = 0$$

Divide each term by 2 so that the leading coefficient will be 1.

$$x^2 + x - \frac{7}{2} = 0$$

$$x^2 + x = \frac{7}{2}$$

Isolate the constant term.

$$x^2 + x + \frac{1}{4} = \frac{7}{2} + \frac{1}{4}$$

Complete the square: the coefficient of x is 1 and $\frac{1}{2}(1) = \frac{1}{2}$ and $\left(\frac{1}{2}\right)^2 = \frac{1}{4}$.
Therefore, add $\frac{1}{4}$ to both sides.

$$\left(x + \frac{1}{2}\right)^2 = \frac{15}{4}$$

Factor the polynomial.

$$x + \frac{1}{2} = \pm\sqrt{\frac{15}{4}}$$

Use the square root property.

$$x = -\frac{1}{2} \pm \frac{\sqrt{15}}{2}$$

$$x = \frac{-1 \pm \sqrt{15}}{2}$$

d. $x^2 - 2x + 13 = 0$

Solution

$x^2 - 2x + 13 = 0$

$\quad x^2 - 2x = -13$ Isolate the constant term.

$x^2 - 2x + 1 = -13 + 1$ Complete the square: $\frac{1}{2}(-2) = -1$ and $(-1)^2 = 1$.

                                    Therefore, add 1 to both sides.

$\quad\quad (x-1)^2 = -12$ Factor the polynomial.

$\quad\quad\quad x - 1 = \pm\sqrt{-12}$ Use the square root property.

$\quad\quad\quad x - 1 = \pm i\sqrt{12} = \pm 2i\sqrt{3}$

$\quad\quad\quad\quad x = 1 \pm 2i\sqrt{3}$ The solutions are nonreal complex conjugates.

Now work margin exercise 5.

Objective D   **Writing Quadratic Equations with Known Roots**

In an earlier section, we found equations with known roots by setting the product of factors equal to 0 and simplifying. The same method is applied here with roots that are nonreal and roots that involve radicals.

Example 6

Quadratic Equations with Known Roots

Find quadratic equations that have the given roots.

a. $y = 3 + 2i$ and $y = 3 - 2i$

Solution

$\quad\quad y = 3 + 2i \quad\quad\quad y = 3 - 2i$

$y - 3 - 2i = 0 \quad\quad y - 3 + 2i = 0$ Get 0 on one side of each equation.

Set the product of the two factors equal to 0 and simplify.

$\quad (y - 3 - 2i)(y - 3 + 2i) = 0$ Regroup the terms to represent the product of complex conjugates. This makes the multiplication easier.

$\big[(y - 3) - 2i\big]\big[(y - 3) + 2i\big] = 0$

$\quad\quad\quad (y - 3)^2 - 4i^2 = 0$

$\quad\quad y^2 - 6y + 9 + 4 = 0$ Remember, $i^2 = -1$.

$\quad\quad\quad y^2 - 6y + 13 = 0$ This equation has two solutions: $y = 3 + 2i$ and $y = 3 - 2i$.

5. Solve the following quadratic equations by completing the square.

a. $x^2 - 10x = 31$

b. $2x^2 + 4x - 12 = 0$

c. $2x^2 + 2x - 9 = 0$

d. $x^2 + 4x + 11 = 0$

b. $x = 5 - \sqrt{2}$ and $x = 5 + \sqrt{2}$

Solution

$$x = 5 - \sqrt{2} \qquad x = 5 + \sqrt{2}$$
$$x - 5 + \sqrt{2} = 0 \qquad x - 5 - \sqrt{2} = 0 \qquad \text{Get 0 on one side of each equation.}$$

Set the product of the two factors equal to 0 and simplify.

$$\left(x - 5 + \sqrt{2}\right)\left(x - 5 - \sqrt{2}\right) = 0 \qquad \text{Regroup the terms to make the}$$
$$\left[(x - 5) + \sqrt{2}\right]\left[(x - 5) - \sqrt{2}\right] = 0 \qquad \text{multiplication easier.}$$
$$(x - 5)^2 - \left(\sqrt{2}\right)^2 = 0$$
$$x^2 - 10x + 25 - 2 = 0$$
$$x^2 - 10x + 23 = 0 \qquad \text{This equation has two solutions:}$$
$$\qquad\qquad\qquad\qquad x = 5 - \sqrt{2} \text{ and } x = 5 + \sqrt{2}.$$

6. Find polynomial equations that have the given roots.

a. $y = 4 + 3i$ and $y = 4 - 3i$

b. $x = 2 - 3\sqrt{2}$ and $x = 2 + 3\sqrt{2}$

c. $x = 1 + i\sqrt{5}$ and $x = 1 - i\sqrt{5}$

c. $x = 3 + i\sqrt{5}$ and $x = 3 - i\sqrt{5}$

Solution

$$x = 3 + i\sqrt{5} \qquad x = 3 - i\sqrt{5}$$
$$x - 3 - i\sqrt{5} = 0 \qquad x - 3 + i\sqrt{5} = 0 \qquad \text{Get 0 on one side of each equation.}$$

Set the product of the two factors equal to 0 and simplify.

$$\left(x - 3 - i\sqrt{5}\right)\left(x - 3 + i\sqrt{5}\right) = 0 \qquad \text{Regroup the terms to make the}$$
$$\left[(x - 3) - i\sqrt{5}\right]\left[(x - 3) + i\sqrt{5}\right] = 0 \qquad \text{multiplication easier.}$$
$$(x - 3)^2 - \left(i\sqrt{5}\right)^2 = 0$$
$$x^2 - 6x + 9 - i^2 \left(\sqrt{5}\right)^2 = 0$$
$$x^2 - 6x + 9 - (-1)(5) = 0 \qquad \text{Remember, } i^2 = -1.$$
$$x^2 - 6x + 14 = 0 \qquad \text{This equation has two solutions:}$$
$$\qquad\qquad\qquad\qquad x = 3 + i\sqrt{5} \text{ and } x = 3 - i\sqrt{5}$$

Now work margin exercise 6.

Solve each of the following quadratic equations by completing the square.

1. $x^2 + 2x + 2 = 0$

2. $3x^2 - 6x + 15 = 0$

3. $x^2 - 24x + 72 = 0$

4. $2x^2 + 5x - 3 = 0$

5. $x^2 - 3x + 1 = 0$

6. Find a quadratic equation that has the roots $2 + 3i$ and $2 - 3i$.

Practice Problem Answers

1. $x = -1 \pm i$

2. $x = 1 \pm 2i$

3. $x = 12 \pm 6\sqrt{2}$

4. $x = -3, \dfrac{1}{2}$

5. $x = \dfrac{3 \pm \sqrt{5}}{2}$

6. $x^2 - 4x + 13 = 0$

Exercises 12.2

Solve the following equations by factoring. See Example 2.

1. $x^2 - 18x + 45 = 0$

2. $12y^2 - 18y - 12 = 0$

3. $2x^2 = 34x - 120$

4. $7x^2 = 11x + 6$

5. $3y^2 = 363$

6. $x^4 = x^2$

7. $4z^3 + 16z^2 = 48z$

8. $5x^3 - 25x^2 = -30x$

9. $(x+8)(x+2) = -9$

10. $(z+3)^2 = 100$

Solve the equations using the square root property. See Example 3.

11. $x^2 - 144 = 0$

12. $x^2 - 169 = 0$

13. $x^2 + 25 = 0$

14. $x^2 + 81 = 0$

15. $x^2 + 24 = 0$

16. $x^2 + 18 = 0$

17. $(x-2)^2 = 9$

18. $(x-4)^2 = 25$

19. $x^2 = 5$

20. $x^2 = 12$

21. $2(x+3)^2 = 6$

22. $3(x-1)^2 = 15$

23. $(x-3)^2 = -4$

24. $(x+8)^2 = -9$

25. $(x+2)^2 = -7$

26. $(x-5)^2 = -10$

Add the correct constant to complete the square; then factor the trinomial as indicated. See Example 4.

27. $x^2 - 12x + \underline{} = (\underline{})^2$

28. $y^2 + 14y + \underline{} = (\underline{})^2$

29. $x^2 + 6x + \underline{} = (\underline{})^2$

30. $x^2 + 8x + \underline{} = (\underline{})^2$

31. $x^2 - 5x + \underline{} = (\underline{})^2$

32. $x^2 + 7x + \underline{} = (\underline{})^2$

33. $y^2 + y + \underline{} = (\underline{})^2$

34. $x^2 + \dfrac{1}{2}x + \underline{} = (\underline{})^2$

35. $x^2 + \dfrac{1}{3}x + \underline{} = (\underline{})^2$

36. $y^2 + \dfrac{3}{4}y + \underline{} = (\underline{})^2$

Solve the quadratic equations by completing the square. See Example 5.

37. $x^2 + 4x - 5 = 0$

38. $x^2 + 6x - 7 = 0$

39. $y^2 + 2y = 5$

40. $x^2 + 3 = 8x$

41. $x^2 + 3 = 10x$

42. $z^2 + 4z = 2$

43. $x^2 - 6x + 10 = 0$

44. $x^2 - 2x + 5 = 0$

45. $x^2 + 11 = 12x$

46. $x^2 = 6 - x$

47. $y^2 = 10y - 4$

48. $x^2 = 3 - 4x$

49. $z^2 + 3z - 5 = 0$

50. $x^2 - 5x + 5 = 0$

51. $x^2 + x + 2 = 0$

52. $y^2 + 3y + 3 = 0$

53. $x^2 + 5x + 2 = 0$

54. $4x^2 + 7x + 2 = 0$

55. $3x^2 - 10x + 5 = 0$

56. $3y^2 + 5y - 3 = 0$

57. $3x^2 + 6x + 18 = 0$

58. $4x^2 + 8x + 16 = 0$

59. $2x - 3 = 4x^2$

60. $2x + 2 = -6x^2$

61. $5y^2 + 15y + 25 = 0$

62. $4x^2 + 20x + 32 = 0$

63. $3y^2 = 4 - y$

64. $2x^2 + 4 = -9x$

65. $2x^2 - 8x + 4 = 0$

66. $3x^2 - 18x + 12 = 0$

67. $x = \sqrt{7},\, x = -\sqrt{7}$

68. $x = \sqrt{6},\, x = -\sqrt{6}$

69. $x = 1 + \sqrt{3},\, x = 1 - \sqrt{3}$

70. $z = 3 + \sqrt{2},\, z = 3 - \sqrt{2}$

71. $y = -2 + 2\sqrt{5},\, y = -2 - 2\sqrt{5}$

72. $x = 1 + 2\sqrt{3},\, x = 1 - 2\sqrt{3}$

73. $x = 4i,\, x = -4i$

74. $x = 7i,\, x = -7i$

75. $y = i\sqrt{6},\, y = -i\sqrt{6}$

76. $y = i\sqrt{5},\, y = -i\sqrt{5}$

77. $x = 2 + i,\, x = 2 - i$

78. $x = -3 + 2i,\, x = -3 - 2i$

79. $x = 1 + i\sqrt{2},\, x = 1 - i\sqrt{2}$

80. $x = 2 + i\sqrt{3},\, x = 2 - i\sqrt{3}$

81. $x = -5 + 2i\sqrt{6},\, x = -5 - 2i\sqrt{6}$

82. $y = 4 + 3i\sqrt{2},\, y = 4 - 3i\sqrt{2}$

Writing & Thinking

83. Explain, in your own words, the steps involved in the process of solving a quadratic equation by completing the square.

12.3 The Quadratic Formula

Objective A   **Developing the Quadratic Formula**

The **quadratic formula** gives the roots of any quadratic equation in terms of the coefficients a, b, and c of the **general quadratic equation**

$$ax^2 + bx + c = 0.$$

Therefore, if you memorize the quadratic formula, you can solve any quadratic equation by simply substituting the coefficients into the formula. To develop the quadratic formula, we solve the general quadratic equation by **completing the square**.

$$ax^2 + bx + c = 0$$

$$x^2 + \frac{b}{a}x + \frac{c}{a} = \frac{0}{a}$$
Divide each term of the equation by a. The leading coefficient must be 1.

$$x^2 + \frac{b}{a}x = -\frac{c}{a}$$
Add $-\dfrac{c}{a}$ to both sides of the equation.

$$x^2 + \frac{b}{a}x + \frac{b^2}{4a^2} = \frac{b^2}{4a^2} - \frac{c}{a}$$
Add $\dfrac{b^2}{4a^2}$ to both sides to complete the square.

$$\left(x + \frac{b}{2a}\right)^2 = \frac{b^2}{4a^2} - \frac{4ac}{4a^2}$$
Factor the left side. $4a^2$ is the common denominator on the right-hand side.

$$\left(x + \frac{b}{2a}\right)^2 = \frac{b^2 - 4ac}{4a^2}$$
Simplify.

$$x + \frac{b}{2a} = \pm\sqrt{\frac{b^2 - 4ac}{4a^2}}$$
Use the square root property.

$$x + \frac{b}{2a} = \pm\frac{\sqrt{b^2 - 4ac}}{2a}$$
Simplify.

$$x = -\frac{b}{2a} \pm \frac{\sqrt{b^2 - 4ac}}{2a}$$
Solve for x.

$$x = \frac{-b \pm \sqrt{b^2 - 4ac}}{2a}$$
This equation is called the **quadratic formula**.

Objectives

A   Develop the quadratic formula.

B   Use the quadratic formula to solve quadratic equations.

C   Calculate the discriminant and use it to determine the nature of the solutions of a quadratic equation.

---

notes

Note about the coefficient a

For convenience and without loss of generality, in the development of the quadratic formula (and in the examples and exercises) the leading coefficient a is positive. If a is a negative number, we can multiply both sides of the equation by -1. This will make the leading coefficient positive without changing any solutions of the original equation.

The Quadratic Formula

For the general quadratic equation $ax^2 + bx + c = 0$ where $a \neq 0$ the solutions are

$$x = \frac{-b \pm \sqrt{b^2 - 4ac}}{2a}.$$

The quadratic formula should be memorized.

Applications of quadratic equations are found in such fields as economics, business, computer science, and chemistry, and in almost all branches of mathematics. Most instructors assume that their students know the quadratic formula and how to apply it. You should recognize that the importance of the quadratic formula lies in the fact that **any** quadratic equation can be solved.

Example 1

The Quadratic Formula

Solve the following quadratic equations using the quadratic formula.

a. $x^2 - 5x + 3 = 0$

Solution

Substitute $a = 1$, $b = -5$, and $c = 3$ into the formula.

$$x = \frac{-b \pm \sqrt{b^2 - 4ac}}{2a} = \frac{-(-5) \pm \sqrt{(-5)^2 - 4 \cdot 1 \cdot 3}}{2 \cdot 1}$$

$$= \frac{5 \pm \sqrt{25 - 12}}{2}$$

$$= \frac{5 \pm \sqrt{13}}{2}$$

b. $7x^2 - 2x + 1 = 0$

Solution

Substitute $a = 7$, $b = -2$, and $c = 1$ into the formula.

$$x = \frac{-b \pm \sqrt{b^2 - 4ac}}{2a} = \frac{-(-2) \pm \sqrt{(-2)^2 - 4 \cdot 7 \cdot 1}}{2 \cdot 7}$$

$$= \frac{2 \pm \sqrt{4 - 28}}{14}$$

$$= \frac{2 \pm \sqrt{-24}}{14}$$

$$= \frac{2 \pm 2i\sqrt{6}}{14}$$

$$= \frac{2\left(1 \pm i\sqrt{6}\right)}{2 \cdot 7}$$

$$= \frac{1 \pm i\sqrt{6}}{7}$$

The solutions are nonreal complex conjugates. In standard form:

$$\frac{1}{7} + \frac{\sqrt{6}}{7}i \text{ and } \frac{1}{7} - \frac{\sqrt{6}}{7}i.$$

1. Solve the following equations using the quadratic formula.

a. $x^2 - 7x + 4 = 0$

b. $5x^2 - 3x + 2 = 0$

c. $x^2 + 9 = 0$

c. $\dfrac{3}{4}x^2 - \dfrac{1}{2}x = \dfrac{1}{3}$

Solution

The quadratic formula is easier to use with integer coefficients. Multiply each term by the LCD, 12, so that the coefficients will be integers.

$$12 \cdot \frac{3}{4}x^2 - 12 \cdot \frac{1}{2}x = 12 \cdot \frac{1}{3}$$

$$9x^2 - 6x = 4$$

$$9x^2 - 6x - 4 = 0$$

To apply the formula, one side must be 0.

$$x = \frac{-(-6) \pm \sqrt{(-6)^2 - 4 \cdot 9 \cdot -4}}{2 \cdot 9}$$

Substitute $a = 9$, $b = -6$, and $c = -4$ into the quadratic formula.

$$= \frac{6 \pm \sqrt{36 + 144}}{18}$$

$$= \frac{6 \pm \sqrt{180}}{18} = \frac{6 \pm 6\sqrt{5}}{18}$$

$$= \frac{6\left(1 \pm \sqrt{5}\right)}{6 \cdot 3} = \frac{1 \pm \sqrt{5}}{3}$$

Factor and reduce.

Now work margin exercise **1.**

Many students make a mistake when simplifying fractions by dividing the denominator into only one of the terms in the numerator.

Wrong Solution

$$\frac{4+\cancel{2}\sqrt{3}}{\cancel{2}} = 4+\sqrt{3}$$

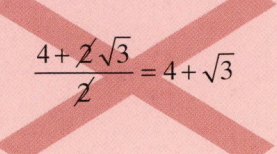

Correct Solution

$$\frac{4+2\sqrt{3}}{2} = \frac{4}{2} + \frac{2\sqrt{3}}{2} = 2+\sqrt{3}$$

$$\frac{4+2\sqrt{3}}{2} = \frac{\cancel{2}\left(2+\sqrt{3}\right)}{\cancel{2}} = 2+\sqrt{3}$$

In Example 2, the equation is third-degree (a cubic equation) and one of the factors is quadratic. The quadratic formula can be applied to this factor.

2. Solve the following cubic equation by factoring and using the quadratic equation.

$$3x^3 - 15x^2 + 6x = 0$$

Example 2

Cubic Equations

Solve the following cubic equation by factoring and using the quadratic formula.

$$2x^3 - 10x^2 + 6x = 0$$

Solution

$$2x^3 - 10x^2 + 6x = 0$$
$$2x\left(x^2 - 5x + 3\right) = 0 \qquad \text{Factor out } 2x.$$

$$2x = 0 \quad \text{or} \quad x^2 - 5x + 3 = 0 \qquad \text{Set each factor equal to 0.}$$

$$x = 0 \qquad\qquad x = \frac{5 \pm \sqrt{13}}{2} \qquad \text{Solve each equation. (The quadratic equation was solved in Example \textbf{1a.} using the quadratic formula.)}$$

There are 3 solutions: 0, $\dfrac{5+\sqrt{13}}{2}$, and $\dfrac{5-\sqrt{13}}{2}$.

Now work margin exercise 2.

Objective C The Discriminant

The expression $b^2 - 4ac$, the part of the quadratic formula that lies under the radical sign, is called the **discriminant**. The discriminant identifies the number and type of solutions to a quadratic equation. Assuming a, b, and c are all real numbers, there are three possibilities: the discriminant is either positive, negative, or zero.

In Example **1a.**, the discriminant was positive, $b^2 - 4ac = (-5)^2 - 4(1)(3) = 13$, and there were two real solutions: $x = \dfrac{5 \pm \sqrt{13}}{2}$. In Example **1b.**, the discriminant was negative, $b^2 - 4ac = (-2)^2 - 4(7)(1) = -24$, and there were two nonreal solutions: $x = \dfrac{1 \pm i\sqrt{6}}{7}$.

The discriminant gives the following information.

Discriminant	Nature of Solutions
$b^2 - 4ac > 0$	two real solutions
$b^2 - 4ac = 0$	one real solution, $x = \dfrac{-b \pm 0}{2a} = -\dfrac{b}{2a}$
$b^2 - 4ac < 0$	two nonreal solutions

Table 1

In the case where $b^2 - 4ac = 0$, we say $x = -\dfrac{b}{2a}$ is a **double root**. Additionally, if the discriminant is a perfect square, the equation is factorable.

Example 3

Finding the Discriminant

Find the discriminant and determine the nature of the solutions to each of the following quadratic equations.

a. $3x^2 + 11x - 7 = 0$

Solution

Substitute $a = 3$, $b = 11$, and $c = -7$ into the discriminant:

$$b^2 - 4ac = 11^2 - 4(3)(-7)$$
$$= 121 + 84$$
$$= 205 > 0 \qquad \text{There are two real solutions.}$$

3. Find the discriminant and determine the nature of the solutions to each of the following quadratic equations.

a. $x^2 + 15x - 7$

b. $x^2 + 8x + 16$

c. $x^2 + 9$

b. $x^2 + 6x + 9 = 0$

Solution

Substitute $a = 1$, $b = 6$, and $c = 9$ into the discriminant:

$$b^2 - 4ac = 6^2 - 4(1)(9)$$
$$= 36 - 36$$
$$= 0 \qquad \text{There is one real solution, a double root.}$$

c. $x^2 + 1 = 0$

Solution

Here $b = 0$. We could write $x^2 + 0x + 1 = 0$.

$$b^2 - 4ac = 0^2 - 4(1)(1)$$
$$= 0 - 4$$
$$= -4 \qquad \text{There are two nonreal solutions.}$$

Now work margin exercise 3.

Example 4

a. Determine the value(s) for c such that $x^2 + 8x + c = 0$ will have one real solution. (**Hint:** Set the discriminant equal to 0 and solve the equation for c.)

Solution

$$b^2 - 4ac = 0$$
$$8^2 - 4(1)(c) = 0$$
$$64 - 4c = 0$$
$$-4c = -64$$
$$c = 16$$

Check:

$$x^2 + 8x + (16) = 0$$
$$x^2 + 8x + 16 = 0$$
$$(x + 4)^2 = 0$$
$$x = -4$$

There is only one real solution. Thus -4 is a double root.

b. Determine the value(s) for a such that $ax^2 - 8x + 4 = 0$ will have two nonreal solutions. (**Hint:** Set the discriminant less than 0 and solve for a.)

Solution

$$b^2 - 4ac < 0$$

$$(-8)^2 - 4(a)(4) < 0$$

$$64 - 16a < 0$$

$$-16a < -64$$

$$a > 4$$

Thus, if a is any real number greater than 4, the discriminant will be negative and the equation will have two nonreal solutions.

Now work margin exercise 4.

4. a. Determine the value(s) for c such that $x^2 - 6x + c = 0$ will have one real solution.

b. Determine the value(s) for a such that $ax^2 - 8x + 1 = 0$ will have two non-real solutions.

Practice Problems

Solve each of the following quadratic equations using the quadratic formula.

1. $x^2 + 2x - 4 = 0$ **2.** $2x^2 - 3x + 4 = 0$ **3.** $5x^2 - x - 4 = 0$

4. $\dfrac{1}{4}x^2 - \dfrac{1}{2}x = -\dfrac{1}{4}$ **5.** $3x^2 + 5 = 0$

6. Determine the value(s) for c such that $x^2 - 6x + c = 0$ will have two real solutions.

Practice Problem Answers

1. $x = -1 \pm \sqrt{5}$ **2.** $x = \dfrac{3 \pm i\sqrt{23}}{4}$ **3.** $x = -\dfrac{4}{5}, 1$

4. $x = 1$ **5.** $x = \dfrac{\pm i\sqrt{15}}{3}$ **6.** $c < 9$

Exercises 12.3

Find the discriminant and determine the nature of the solutions of each quadratic equation. See Example 3.

1. $x^2 + 6x - 8 = 0$

2. $x^2 + 3x + 1 = 0$

3. $x^2 - 8x + 16 = 0$

4. $x^2 + 3x + 5 = 0$

5. $4x^2 + 2x + 3 = 0$

6. $3x^2 - x + 2 = 0$

7. $5x^2 + 8x + 3 = 0$

8. $4x^2 + 12x + 9 = 0$

9. $100x^2 - 49 = 0$

10. $9x^2 + 121 = 0$

11. $3x^2 + x + 1 = 0$

12. $5x^2 - 3x - 2 = 0$

Solve each of the quadratic equations using the quadratic formula.

13. $x^2 + 4x - 4 = 0$

14. $x^2 - 6x - 1 = 0$

15. $9x^2 + 12x + 4 = 0$

16. $4x^2 - 20x + 25 = 0$

17. $x^2 - 2x + 7 = 0$

18. $x^2 - 2x + 3 = 0$

19. $2x^2 + 5x - 3 = 0$

20. $3x^2 - 7x + 4 = 0$

21. $4x^2 + 6x + 1 = 0$

22. $2x^2 - 3x - 1 = 0$

23. $4x^2 + 6x + 3 = 0$

24. $x^2 - 5x + 7 = 0$

Solve the given equations using any of the techniques discussed for solving quadratic equations: factoring, completing the square, or using the quadratic formula.

25. $x^2 + 3x - 5 = 0$

26. $x^2 - 7x - 3 = 0$

27. $x^2 + 4x + 3 = 0$

28. $x^2 + 14x + 49 = 0$

29. $x^2 + 8 = 0$

30. $x^2 - 7 = 0$

31. $x^2 - 5x + 2 = 0$

32. $3x^2 + 2x - 2 = 0$

33. $16x^2 + 8x = -1$

34. $6x^2 = 5x + 1$

35. $3x^2 - 4 = 0$

36. $4x^2 + 9 = 0$

37. $9x^2 - 12x + 4 = 0$ **38.** $9x^2 - 6x + 1 = 0$ **39.** $2x^2 = -8x - 9$

40. $3x^2 = 6x - 4$ **41.** $5x^2 + 5 = 7x$ **42.** $4x^2 - 5x = -3$

43. $6x^2 + 2x = 20$ **44.** $10x^2 + 30 = -35x$ **45.** $3x^2 = 18x - 33$

46. $2x^2 = 16x - 36$ **47.** $x^2 + 4x = x - 2x^2$ **48.** $3x^2 + 4x = 0$

49. $x^3 - 9x^2 + 4x = 0$ **50.** $x^3 - 8x^2 = 3x^2 + 3x$ **51.** $x^3 + 3x^2 + x = 0$

52. $4x^3 + 10x^2 - 3x = 0$

First multiply each side of the equation by the LCM of the denominators to get integer coefficients and then solve the resulting equation. See Example 1c.

53. $3x^2 - 4x + \dfrac{1}{3} = 0$ **54.** $\dfrac{3}{4}x^2 - 2x + \dfrac{1}{8} = 0$ **55.** $2x^2 - \dfrac{2}{3}x + \dfrac{2}{9} = 0$

56. $2x^2 + 3x + \dfrac{5}{4} = 0$ **57.** $\dfrac{1}{2}x^2 - x + \dfrac{3}{4} = 0$ **58.** $\dfrac{2}{3}x^2 - \dfrac{1}{3}x + \dfrac{1}{2} = 0$

59. $\dfrac{1}{4}x^2 + \dfrac{7}{8}x + \dfrac{1}{2} = 0$ **60.** $\dfrac{5}{12}x^2 - \dfrac{1}{2}x - \dfrac{1}{4} = 0$

61. Determine the value(s) for c such that $x^2 - 8x + c = 0$ will have two real solutions.

62. Determine the value(s) for c such that $x^2 + 5x + c = 0$ will have two real solutions.

63. Determine the value(s) for c such that $x^2 + 9x + c = 0$ will have one real solution.

64. Determine the value(s) for c such that $x^2 - 7x + c = 0$ will have one real solution.

65. Determine the value(s) for a such that $ax^2 - 6x + 3 = 0$ will have two nonreal solutions.

66. Determine the value(s) for a such that $ax^2 + 4x - 2 = 0$ will have two nonreal solutions.

67. Determine the value(s) for a such that $ax^2 + x - 9 = 0$ will have two real solutions.

68. Determine the value(s) for a such that $ax^2 + 6x + 3 = 0$ will have two real solutions.

69. Determine the value(s) for a such that $ax^2 + 7x + 12 = 0$ will have one real solution.

70. Determine the value(s) for a such that $ax^2 - 2x + 8 = 0$ will have one real solution.

71. Determine the value(s) for c such that $3x^2 + 4x + c = 0$ will have two nonreal solutions.

72. Determine the value(s) for c such that $2x^2 + 3x + c = 0$ will have two nonreal solutions.

Solve the quadratic equations using the quadratic formula and your calculator. Write the solutions accurate to the ten-thousandth.

73. $0.02x^2 - 1.26x + 3.14 = 0$

74. $0.5x^2 + 0.07x - 5.6 = 0$

75. $\sqrt{2}x^2 - \sqrt{3}x - \sqrt{5} = 0$

76. $x^2 - 2\sqrt{10}x + 10 = 0$

77. $0.3x^2 + \sqrt{2}x + 0.72 = 0$

78. $\sqrt[3]{4}x^2 - \sqrt[4]{2}x - \sqrt{11} = 0$

79. $x^2 + 2\sqrt{15} - 15 = 0$

80. $0.05x^2 - \sqrt{30} = 0$

Writing & Thinking

81. Find an equation of the form $Ax^4 + Bx^2 + C = 0$ that has the four roots ± 2 and ± 3. Explain how you arrived at this equation.

82. The surface area of a right circular cylinder can be found using the following formula:

$S = 2\pi r^2 + 2\pi rh$, where r is the radius of the cylinder and h is the height.

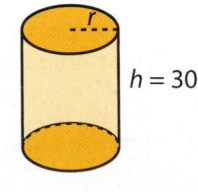

Estimate the radius of a circular cylinder of height 30 cm and surface area 300 cm². Explain how you used your knowledge of quadratic equations.

12.4 Applications

The following strategy for solving word problems, is a valid approach to solving word problems at all levels.

Strategy for Solving Word Problems

1. Understand the problem.
   a. Read the problem carefully. (Read it several times if necessary.)
   b. If it helps, restate the problem in your own words.
2. Devise a plan.
   a. Decide what is asked for; assign a variable to the unknown quantity. Label this variable so you know exactly what it represents.
   b. Draw a diagram or set up a chart whenever possible.
   c. Write an equation that relates the information provided.
3. Carry out the plan.
   a. Study your picture or diagram for insight into the solution.
   b. Solve the equation.
4. Look back over the results.
   a. Does your solution make sense in terms of the wording of the problem?
   b. Check your solution in the equation.

The problems in this section can be solved by setting up quadratic equations and then solving these equations by factoring, completing the square, or using the quadratic formula.

Objective A   The Pythagorean Theorem

The Pythagorean Theorem

In a right triangle, if c is the length of the hypotenuse and a and b are the lengths of the legs, then

$$c^2 = a^2 + b^2.$$

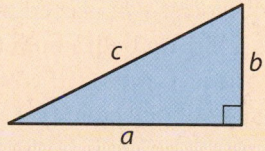

1. The length of a small parking lot is 14 meters longer than its width. If a diagonal is stretched across the parking lot is 26 meters long, what are the dimensions of the parking lot?

Example 1

The Pythagorean Theorem

The length of a rectangular field is 6 meters longer than its width. If a diagonal foot path stretching from one corner of the field to the opposite corner is 30 meters long, what are the dimensions of the field?

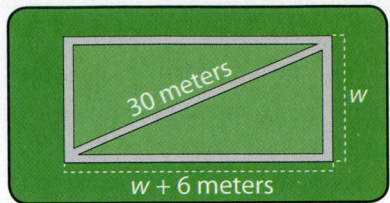

Solution

Let w = width, and
$w + 6$ = length.

$(w+6)^2 + w^2 = 30^2$ Use the Pythagorean theorem.

$w^2 + 12w + 36 + w^2 = 900$

$2w^2 + 12w - 864 = 0$

$w^2 + 6w - 432 = 0$ Divide both sides by 2.

$(w+24)(w-18) = 0$

~~$w = -24$~~   or $w = 18$ A negative number does not fit the conditions of the problem as width cannot be negative.

$w = 18$ meters
$w + 6 = 24$ meters

The length is 24 meters and the width is 18 meters.

Now work margin exercise 1.

Objective B   Projectiles

The formula $h = -16t^2 + v_0 t + h_0$ is used in physics and relates to the height of a projectile such as a thrown ball, a bullet, or a rocket.

h =   height of object, in feet
t =   time the object is in the air, in seconds
v_0 =   beginning velocity, in feet per second
h_0 =   beginning height   ($h_0 = 0$ if the object is initially at ground level.)

Example 2

Projectiles

A bullet is fired straight up from ground level with a muzzle velocity of 320 feet per second.

a. When will the bullet hit the ground?

Solution

In this problem, $v_0 = 320$ feet per second and
$h_0 = 0$ feet.

The bullet hits the ground when its height h equals 0.

$$h = -16t^2 + v_0 t + h_0$$
$$0 = -16t^2 + 320t + 0$$
$$0 = t^2 - 20t \qquad \text{Divide both sides by } -16.$$
$$0 = t(t - 20) \qquad \text{Factor.}$$
$$t = 0 \text{ or } t = 20$$

The bullet hits the ground in 20 seconds. The solution $t = 0$ confirms the fact that the bullet was fired from the ground.

b. When will the bullet be 1200 feet above the ground?

Solution

Let $h = 1200$.

$$1200 = -16t^2 + 320t$$
$$0 = -16t^2 + 320t - 1200$$
$$0 = t^2 - 20t + 75 \qquad \text{Divide both sides by } -16.$$
$$0 = (t - 5)(t - 15) \qquad \text{Factor.}$$
$$t = 5 \text{ or } t = 15$$

Both solutions are meaningful. The bullet is at a height of 1200 feet twice; once at 5 seconds while going up and once at 15 seconds while coming down.

Now work margin exercise 2.

2. A bullet is fired straight up from the ground level with a muzzle velocity of 480 feet per second.

a. When will the bullet hit the ground?

b. When will the bullet be 2000 feet above the ground?

3. A company is making gift boxes out of rectangular sheets of cardboard. The sheets are 8 in. longer than they are wide. Two inch by two inch squares are cut from each corner and the sides are folded up to form an open box. If the box has a volume of 616 in.^{3}, what were the dimensions of the original sheet of cardboard?

Example 3

A rectangular sheet of copper was 6 in. longer than it was wide. Three inch by three inch squares were cut from each corner and the sides were folded up to form an open box. If the box has a volume of 336 in.^{3}, what were the dimensions of the original sheet of copper?

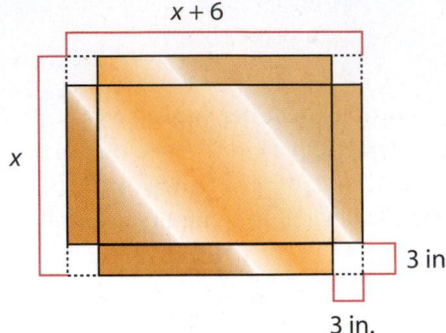

Solution

Let x = width of copper sheet (before the squares are cut out); and $x + 6$ = length of copper sheet (before the squares are cut out).
After the box has been folded:

$$3 = \text{height of box}$$
$$(x+6)-3-3 = x = \text{length of box} \quad \text{3 is subtracted twice.}$$
$$x-3-3 = x-6 = \text{width of box.}$$

The volume of the box is length × width × height: $V = lwh$.
Thus

length × width × height = volume

$$x \cdot (x-6) \cdot 3 = 336$$

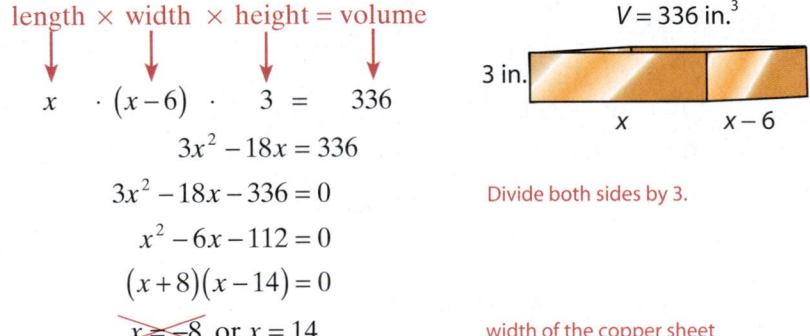

$V = 336$ in.^{3}

$$3x^2 - 18x = 336$$
$$3x^2 - 18x - 336 = 0 \quad \text{Divide both sides by 3.}$$
$$x^2 - 6x - 112 = 0$$
$$(x+8)(x-14) = 0$$
$$\cancel{x = -8} \text{ or } x = 14 \quad \text{width of the copper sheet}$$
$$x+6 = 20. \quad \text{length of the copper sheet}$$

The length of the sheet was 20 in. and the width was 14 in.

Now work margin exercise 3.

Cost per Person

In the following example, note that the **cost per person** is found by dividing the total cost by the number of people going to the tournament. The cost per person changes because the total cost remains fixed while the number of people changes.

Example 4

Cost per Person

The members of a bowling club were going to rent a bus to drive them to a tournament at a total cost of $2420, which was to be divided equally among the members. At the last minute, two of the members decided to drive their own cars. The cost to the remaining members increased $11 each. How many members rode the bus?

Solution

Let x = number of club members and
$x - 2$ = number of club members who rode the bus.

final cost per member		initial cost per member		difference in cost per member
$\dfrac{2420}{x-2}$	$-$	$\dfrac{2420}{x}$	$=$	11

$$x(x-2)\frac{2420}{x-2} - x(x-2)\frac{2420}{x} = x(x-2)\cdot 11 \qquad \text{LCD} = x(x-2)$$

$$2420x - 2420(x-2) = 11x(x-2)$$

$$2420x - 2420x + 4840 = 11x^2 - 22x$$

$$0 = 11x^2 - 22x - 4840$$

$$0 = x^2 - 2x - 440 \qquad \text{Divide both sides by 11.}$$

$$0 = (x-22)(x+20)$$

$$x = 22 \text{ or } x = -20 \qquad \text{-20 does not fit the conditions.}$$
$$\qquad\qquad\qquad\qquad \text{That is, the number of people}$$
$$x - 2 = 20 \qquad\qquad \text{in a club is a positive number.}$$

Check:

Final cost per member $= \dfrac{2420}{20} = \$121$

Initial cost per member $= \dfrac{2420}{22} = \$110$

$\$121 - \$110 = \$11$ difference in cost per member

Twenty members rode the bus.

Now work margin exercise 4.

4. The members and family of a curling team were going to the national championships and were going to rent a bus to drive them there at a total cost of $800, which was to be divided equally among the members. At the last minute, five members had an emergency and could not attend. The cost to the remaining members increased $8 each. How many members attended the national championship?

Exercises 12.4

1. A positive integer is one more than twice another. Their product is 78. Find the two integers.

2. One number is equal to the square of another. Find the numbers if their sum is 72.

3. Find two positive numbers whose difference is 10 and whose product is 56. (**Hint:** If x is one number, then the other is either $x + 10$ or $x - 10$. Try both and analyze your answers.)

4. One number is three more than twice a second number. Their product is 119. Find the numbers.

5. The sum of two numbers is -17. Their product is 66. Find the numbers. (**Hint:** If x is one number, then the other is $(-17 - x)$.)

6. Find a positive real number such that its square is equal to twice the number increased by 2.

7. Find a negative real number such that the square of the sum of the number and 5 is equal to 48.

8. The square of a negative real number is decreased by 2.25 and the result is equal to 3 times the number. What is the number?

9. Twice the square of a positive real number is equal to 4 more than the number. What is the number?

10. The sum of three positive integers is 68. The second is one more than twice the first. The third is three less than the square of the first. Find the integers.

11. **Right Triangles:** A right triangle has two equal sides. The hypotenuse is 14 cm. Find the length of the sides.

12. **Right Triangles:** The length of one leg of a right triangle is twice the length of the second leg. The hypotenuse is 15 meters. Find the lengths of the two legs.

13. **Mean Speed:** Mel and John leave Desert Point at the same time. Mel drives north and John drives east. Mel's mean speed is 10 mph slower than John's. At the end of one hour they are 50 miles apart. Find the mean speed of each driver.

14. **Height of a Telephone Pole:** A telephone pole was broken over at a point $\frac{4}{9}$ of the distance from its base to the top. The top of the pole reached a point on the ground 9 meters from the base of the pole (thus forming a triangle with the ground). What was the original height of the pole?

15. **Dimensions of a Rectangle:** The length of a rectangle is 2 feet less than three times the width. If the area of the rectangle is 40 square feet, find the length and width.

16. **Dimensions of a Rectangle:** A rectangle is 3 meters longer than it is wide. If the width is doubled and the length is decreased by 4 meters, the area is unchanged. Find the original length and width.

17. **Dimensions of a Rectangle:** The area of a rectangle is 102 square inches and the perimeter of the rectangle is 46 inches. Find the dimensions.

18. **Dimensions of a Box:** A rectangular piece of cardboard that is twice as long as it is wide has a small square 4 cm by 4 cm cut from each corner. The edges are then folded up to form an open box with a volume of 1536 cubic cm. What are the dimensions of the box? (See Example 3.)

19. **Size of an Orchard:** An orchard has 2030 trees. The number of trees in each row is 12 more than twice the number of rows. How many trees are in each row?

20. **Capacity of an Auditorium:** A rectangular auditorium seats 960 people. The number of seats in each row is 16 more than the number of rows. Find the number of seats in each row.

21. **Size of an Apartment Complex:** An apartment building has the same number of units on each floor. The building has five times as many units per floor as number of floors, and there are 405 units total. How many floors does the building have?

22. **Shipping:** A large U-Haul truck is 8 ft tall. The length of the truck is 4 ft longer than three times the width. What are the dimensions of the truck if the volume is 1590 ft^3?

23. **Dimensions of a Frame:** A photograph 9 in. wide and 12 in. long is surrounded by a frame of uniform thickness. The area of the frame itself, not including the center, is 162 in.^{2} Find the thickness of the frame.

Area of Frame - 162 sq. in.

9 in.

12 in.

24. **Art:** The *Mona Lisa* is a famous painting by Leonardo da Vinci. The painting is 30 in. by 21 in. It is surrounded by a frame of uniform thickness whose area (not including the center) is 756 in.^{2}. Find the thickness of the frame.

25. **Power of a Generator:** A 40-volt generator with a resistance of 4 ohms delivers power externally of $40I - 4I^2$ watts, where I is the current measured in amperes. Find the current needed for the generator to deliver 100 watts of power.

26. **Power of a Generator:** Find the current needed for the 40-volt generator in Exercise 25 to deliver 64 watts of power.

27. **Selling Signs:** Raymond operates a small sign-making business. He finds that if he charges x dollars for each sign, he sells $40 - x$ signs per week. What is the least number of signs he can sell to have an income of $336 in one week?

28. **Selling Picture Frames:** It costs Mrs. Snow $3 to build a picture frame. She estimates that if she charges x dollars each, she can sell $60 - x$ frames per week. What is the lowest price necessary to make a profit of $432 each week?

29. Selling Peanuts: Samuel operates a small peanut stand. He estimates that he can sell 600 bags of peanuts per day if he charges 50¢ for each bag. He determines that he can sell 20 more bags for each 1¢ reduction in price.

a. What would be his revenue for one day if he charged 48¢ per bag?

b. What should he charge in order to have receipts of $315?

30. Rock Climbing: The Piton Rock Climbing Club planned a climbing expedition. The total cost was $900, which was to be divided equally among the members going. While practicing, three members fell and were injured so they were unable to go. If the cost per person increased by $15, how many members went on the expedition?

31. Manufacturing: A manufacturing crew needs to assemble 1000 boxes per day, divided equally among the workers. One day, three workers call in sick, and the remaining members each need to assemble 75 more boxes than usual. How many workers are on the manufacturing crew?

32. Traveling by Car: Jim traveled 240 miles to a Bridge convention. Later, his wife Ann drove up to meet him. Ann's mean speed exceeded Jim's by 4 mph, and the trip took her 15 minutes less time. Find Ann's speed.

33. Boating: In two hours, a motorboat can travel 8 miles down a river and return 4 miles back. If the river flows at a rate of 2 miles per hour, how fast can the boat travel in still water?

34. Product Assembly: Two employees together can prepare a large order in 2 hrs. Working alone, one employee takes three hours longer than the other. How long does it take each person working alone?

35. Library Processing: Two librarians working together can catalog a new shipment of books in 3 hours. Working alone, one librarian takes 8 hours longer than the other. How long does it take each librarian working alone?

36. Traveling by Plane: Fern can fly her plane 240 miles against the wind in the same time it takes her to fly 360 miles with the wind. The speed of the plane in still air is 30 mph more than four times the speed of the wind. Find the speed of the plane in still air.

37. Selling Coffee: A grocer mixes $9.00 worth of Grade A coffee with $12.00 worth of Grade B coffee to obtain 30 pounds of a blend. If Grade A costs 30¢ a pound more than Grade B, how many pounds of each were used?

38. Theater: The Andersonville Little Theater Group sold 340 tickets to their spring production. Receipts from the sale of reserved tickets were $855. Receipts from general admission tickets were $375. How many of each type of ticket were sold if the cost of a reserved ticket is $2 more than a general admission ticket?

Use the formula for a projectile $h = -16t^2 + v_0 t + h_0$ where t is in seconds, v_0 is in ft/s and h_0 is in feet. See Example 2.

39. Throwing a Ball: A ball is thrown vertically from ground level with an initial speed of 108 ft/s.

a. When will the ball hit the ground?

b. When will the ball be 180 ft above the ground?

40. Throwing a Ball: After scoring the championship winning touchdown, a football player throws the ball directly upward from the ground with an initial velocity of 52 ft/s.

a. When will the ball strike the ground?

b. When will the ball be 40 ft above the ground?

41. **Shooting an Arrow:**   An arrow is shot vertically upward from a platform 40 ft high at a rate of 224 ft/s.

    a. When will the arrow be 824 ft above the ground?

    b. When will it be 424 ft above the ground?

42. **Dropping a Stone:**   A stone is dropped from a platform 196 ft high.
(**Hint:** Since the stone is dropped, $v_0 = 0$.)

    a. When will it hit the ground?

    b. How far has it fallen after 3 seconds?

> **Use your calculator to find the answers accurate to the nearest hundredth.**

43. **Geometry:**   If a triangle is inscribed in a circle so that one side of the triangle is a diameter of the circle, the triangle will be a right triangle (every time). If an isosceles triangle (two sides equal) is inscribed in this manner in a circle with diameter 20 cm, find the length of the two equal sides.

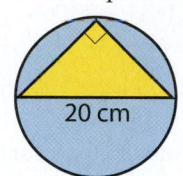

44. **Geometry:**   If a triangle is inscribed in a semicircle such that one side of the triangle is the diameter of a circle with radius 6 in., and one side of the triangle is 5 in., what is the length of the third side? (See Exercise 43.)

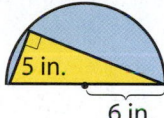

45. **Geometry:**   A square is said to be inscribed in a circle if each corner of the square lies on the circle. (Use $\pi = 3.14$.)

    a. Find the circumference and area of a circle with diameter 30 ft.

    b. Find the perimeter and area of the square inscribed in the circle.

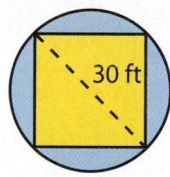

46. **Baseball:**   The shape of a baseball infield is a square with sides 90 ft long.

    a. Find the distance from home plate to second base.

    b. Find the distance from first base to third base.

47. **Baseball:**   The distance from home plate to the pitcher's mound for the field in Exercise 46 is 60.5 feet.

    a. Is the pitcher's mound exactly halfway between home plate and second base?

    b. If not, which base is it closer to, home plate or second base?

    c. Do the two diagonals of the square intersect at the pitcher's mound?

48. **Flying an Airplane:**   If an airplane passes directly over your head at an altitude of 1 mile, how far is the airplane from your position after it has flown 2 miles farther at the same altitude?

49. **Length of a Shadow:** The GE Building in New York is 850 feet tall (70 stories). At a certain time of day, the building casts a shadow 100 feet long. Find the distance from the top of the building to the tip of the shadow (to the nearest hundredth of a foot).

50. **Quilting:** To create a square inside a square, a quilting pattern requires four triangular pieces like the one shaded in the figure shown here. If the square in the center measures 12 cm on a side, and the two legs of each triangle are of equal length, how long are the legs of each triangle?

Writing & Thinking

51. Develop the quadratic formula using the technique of completing the square with the general quadratic equation $ax^2 + bx + c = 0$.

52. Suppose that you are to solve an applied problem and the solution leads to a quadratic equation. You decide to use the quadratic formula to solve the equation. Explain what restrictions you must be aware of when you use the formula.

12.5 Graphs of Quadratic Equations

Objective A **Introduction to Quadratic Functions**

We have studied various types of **functions** and their corresponding graphs: linear functions, polynomial functions, and functions with radicals. Related concepts include **domain**, **range**, and **zeros**. The vertical line test (restated here) can be used to tell whether or not a graph represents a function.

> ### Vertical Line Test
>
> If **any** vertical line intersects a graph in more than one point, then the relation is **not** a function.

In this section, we expand our interest in functions to include a detailed analysis of **quadratic functions**, functions that are represented by quadratic expressions. For example, consider the function

$$y = x^2 - 4x + 3.$$

What is the graph of this function? Since the equation is not linear, the graph will not be a straight line. The nature of the graph can be investigated by plotting several points (See Figure 1).

x	$x^2 - 4x + 3 = y$
-1	$(-1)^2 - 4(-1) + 3 = 8$
0	$(0)^2 - 4(0) + 3 = 3$
$\dfrac{1}{2}$	$\left(\dfrac{1}{2}\right)^2 - 4\left(\dfrac{1}{2}\right) + 3 = \dfrac{5}{4}$
1	$(1)^2 - 4(1) + 3 = 0$
2	$(2)^2 - 4(2) + 3 = -1$
3	$(3)^2 - 4(3) + 3 = 0$
$\dfrac{7}{2}$	$\left(\dfrac{7}{2}\right)^2 - 4\left(\dfrac{7}{2}\right) + 3 = \dfrac{5}{4}$
4	$(4)^2 - 4(4) + 3 = 3$
5	$(5)^2 - 4(5) + 3 = 8$

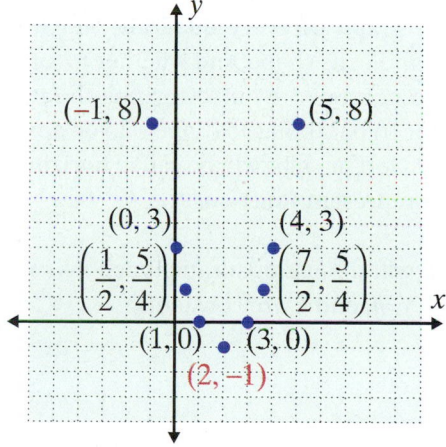

Figure 1

The complete graph of $y = x^2 - 4x + 3$ is shown in Figure 2. The curve is called a **parabola**. The point $(2, -1)$ is the "turning point" of the parabola and is called the **vertex** of the parabola. The line $x = 2$ is the **line of symmetry** or **axis of symmetry** for the parabola. That is, the curve is a "mirror image" of itself with respect to the line $x = 2$.

Objectives

A Become familiar with quadratic functions, and know that the graph of a quadratic function is a parabola.

B Graph quadratic functions of the form $y = ax^2$.

C Graph quadratic functions of the form $y = ax^2 + k$.

D Graph quadratic functions of the form $y = a(x - h)^2$.

E Graph quadratic functions of the form $y = a(x - h)^2 + k$ and $y = ax^2 + bx + c$.

F Find the zeros of a quadratic function.

G Solve applied problems using quadratic functions and the concept of maximum and minimum.

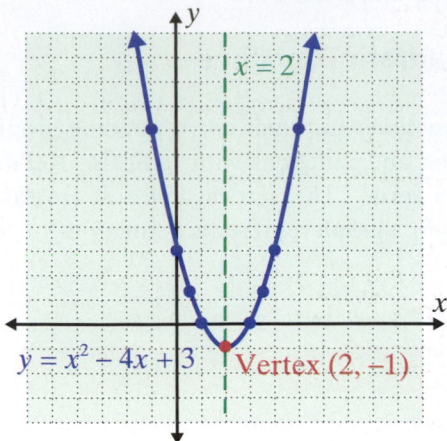

$y = x^2 - 4x + 3$ is a parabola.

Vertex is $(2, -1)$.

$x = 2$ is the line of symmetry.

Figure 2

Quadratic Functions

Any function that can be written in the form

$$y = ax^2 + bx + c$$

where a, b, and c are real numbers and $a \neq 0$ is a **quadratic function**.

The graph of every quadratic function is a parabola. The position of the parabola, its shape, and whether it "opens upward" or "opens downward" can be determined by investigating the function itself. For convenience, we will refer to parabolas that open up or down as **vertical parabolas**. Parabolas that open left or right will be called **horizontal parabolas**. **Horizontal parabolas do not represent functions.**

We will discuss quadratic functions in each of the following five forms where a, b, c, h, and k are constants:

$$y = ax^2$$
$$y = ax^2 + k$$
$$y = a(x - h)^2$$
$$y = a(x - h)^2 + k$$
$$y = ax^2 + bx + c.$$

Functions of the Form $y = ax^2$

For any real number x, $x^2 \geq 0$. So, $ax^2 \geq 0$ if $a > 0$ and $ax^2 \leq 0$ if $a < 0$. This means that the graph of $y = ax^2$ is "above" the x-axis if $a > 0$ and "below" the x-axis if $a < 0$. The **vertex** is at the origin $(0, 0)$ in either of these cases and is the one point where each graph touches (or is tangent to) the x-axis.

For all quadratic functions, the **domain** is the set of all real numbers. That is, x can be replaced by any real number and there will be one corresponding y-value. The **range** of the function $y = ax^2$ depends on the value of a. If $a > 0$, then $y \geq 0$. If $a < 0$, then $y \leq 0$.

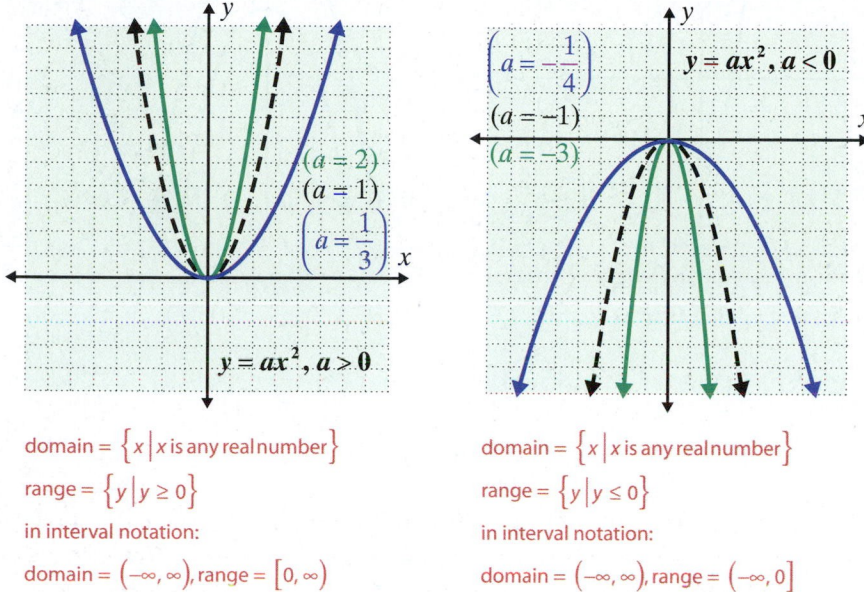

domain = $\{x \mid x \text{ is any real number}\}$

range = $\{y \mid y \geq 0\}$

in interval notation:

domain = $(-\infty, \infty)$, range = $[0, \infty)$

domain = $\{x \mid x \text{ is any real number}\}$

range = $\{y \mid y \leq 0\}$

in interval notation:

domain = $(-\infty, \infty)$, range = $(-\infty, 0]$

Figure 3

Figure 3 illustrates the following characteristics of quadratic functions of the form $y = ax^2$:

- **a.** The vertex is at $(0, 0)$.
- **b.** If $a > 0$, the parabola "opens upward."
- **c.** If $a < 0$, the parabola "opens downward."
- **d.** The bigger $|a|$ is, the narrower the opening.
- **e.** The smaller $|a|$ is, the wider the opening.
- **f.** The line $x = 0$ (the y-axis) is the line of symmetry.

Functions of the Form $y = ax^2 + k$

Adding k to ax^2 simply changes each y-value of $y = ax^2$ by k units (increase if k is positive, decrease if k is negative). That is, the graph of $y = ax^2 + k$ can be found by "sliding" or "shifting" the graph of $y = ax^2$ up k units if $k > 0$ or down $|k|$ units if $k < 0$ (Figure 4).

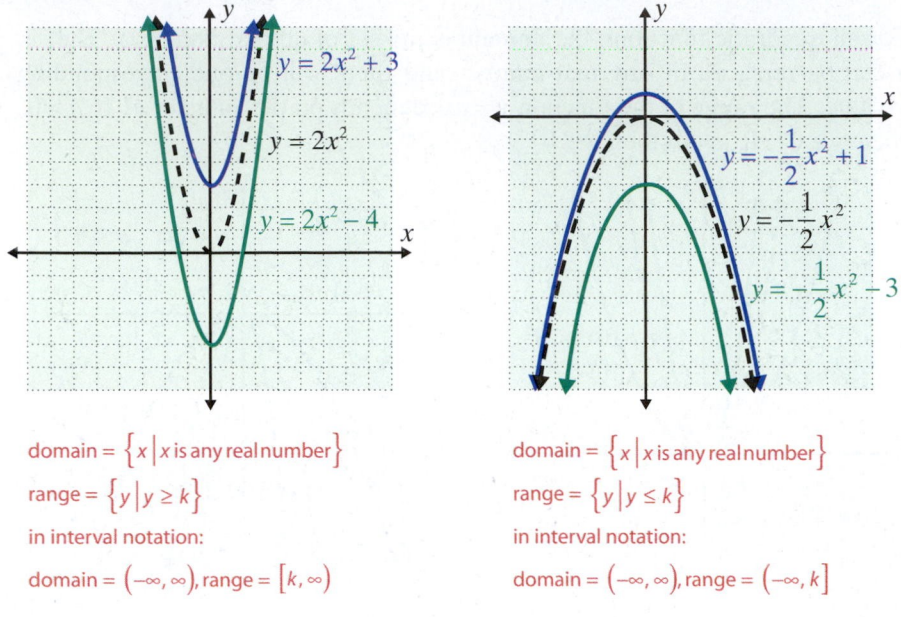

domain = $\left\{ x \,\middle|\, x \text{ is any real number} \right\}$

range = $\left\{ y \,\middle|\, y \geq k \right\}$

in interval notation:

domain = $(-\infty, \infty)$, range = $[k, \infty)$

domain = $\left\{ x \,\middle|\, x \text{ is any real number} \right\}$

range = $\left\{ y \,\middle|\, y \leq k \right\}$

in interval notation:

domain = $(-\infty, \infty)$, range = $(-\infty, k]$

Figure 4

The vertex of $y = ax^2 + k$ is at the point $(0, k)$. The graph of $y = ax^2 + k$ is a vertical shift (or vertical translation) of the graph of $y = ax^2$. The line $x = 0$ (the y-axis) is the line of symmetry just as with functions of the form $y = ax^2$.

Functions of the Form $y = a(x - h)^2$

We know that the square of a real number is nonnegative. Therefore, $(x - h)^2 \geq 0$. So, if $a > 0$, then $y = a(x - h)^2 \geq 0$. If $a < 0$, then $y = a(x - h)^2 \leq 0$. Also notice that $ax^2 = 0$ when $x = 0$, and $a(x - h)^2 = 0$ when $x = h$. Thus the vertex is at $(h, 0)$, and the parabola "opens upward" if $a > 0$ and "opens downward" if $a < 0$ (Figure 5).

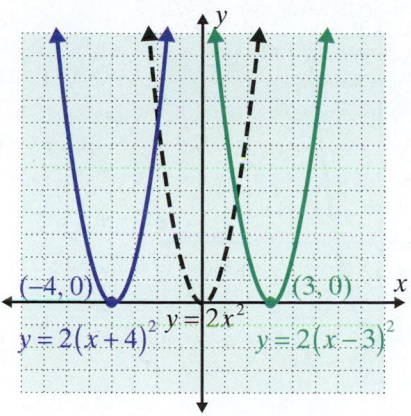

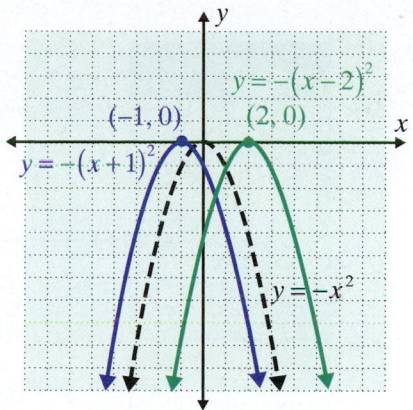

domain = $\left\{x \mid x \text{ is any real number}\right\}$

range = $\left\{y \mid y \geq 0\right\}$

in interval notation:

domain = $(-\infty, \infty)$, range = $[0, \infty)$

domain = $\left\{x \mid x \text{ is any real number}\right\}$

range = $\left\{y \mid y \leq 0\right\}$

in interval notation:

domain = $(-\infty, \infty)$, range = $(-\infty, 0]$

Figure 5

The graph of $y = a(x - h)^2$ is a **horizontal shift** (or **horizontal translation**) of the graph of $y = ax^2$. The shift is to the right if $h > 0$ and to the left if $h < 0$. As a special comment, note that if $h = -3$, then

$$y = a(x - h)^2 \text{ gives } y = a\left(x - (-3)\right)^2 \text{ or } y = a(x + 3)^2. \quad \text{a shift of 3 units to the left}$$

Thus if h is negative, the expression $(x - h)^2$ appears with a plus sign. If h is positive, the expression $(x - h)^2$ appears with a minus sign. In either case, the line $x = h$ is the line of symmetry.

Example 1

Quadratic Functions

Find the line of symmetry, the vertex, the domain, and the range of each quadratic function. Then graph the functions, setting up a table of values for x and y as an aid and choosing values of x on each side of the line of symmetry.

a. $y = 2x^2 - 3$

Solution

The line of symmetry is $x = 0$. (The parabola opens upward since a is positive.)

vertex = $(0, -3)$; domain = $(-\infty, \infty)$ or $\left\{x \mid x \text{ is any real number}\right\}$;

range = $[-3, \infty)$ or $\left\{y \mid y \geq -3\right\}$.

1. Find the line of symmetry, the vertex, the domain, and the range of each quadratic function.

a. $3x^2 + 5$

b. $(x-2)^2$

x	y
0	−3
$\frac{1}{2}$	$-\frac{5}{2}$
$-\frac{1}{2}$	$-\frac{5}{2}$
1	−1
−1	−1
2	5
−2	5

b. $y = -\left(x - \dfrac{5}{2}\right)^2$

Solution

The line of symmetry is $x = \dfrac{5}{2}$. (The parabola opens down since a is negative.)

vertex $= \left(\dfrac{5}{2}, 0\right)$; domain $= (-\infty, \infty)$ or $\{x \mid x \text{ is any real number}\}$;

range $= (-\infty, 0]$ or $\{y \mid y \le 0\}$

x	y
$\frac{5}{2}$	0
2	$-\frac{1}{4}$
3	$-\frac{1}{4}$
1	$-\frac{9}{4}$
4	$-\frac{9}{4}$
0	$-\frac{25}{4}$
5	$-\frac{25}{4}$

*Now work margin exercise **1**.*

Functions of the Form $y = a(x-h)^2 + k$ **and**
$y = ax^2 + bx + c$

The graphs of equations of the form

$$y = a(x-h)^2 + k$$

combine both the vertical shift of k units and the horizontal shift of h units. The **vertex is at** (h, k). For example, the graph of the function $y = -2(x-3)^2 + 5$ is a shift of the graph of $y = -2x^2$ up 5 units and to the right 3 units and has its vertex at $(3, 5)$.

The graph of $y = \left(x + \dfrac{1}{2}\right)^2 - 2$ is the same as the graph of $y = x^2$ but is shifted left $\dfrac{1}{2}$ unit and down 2 units. The vertex is at $\left(-\dfrac{1}{2}, -2\right)$ (Figure 6).

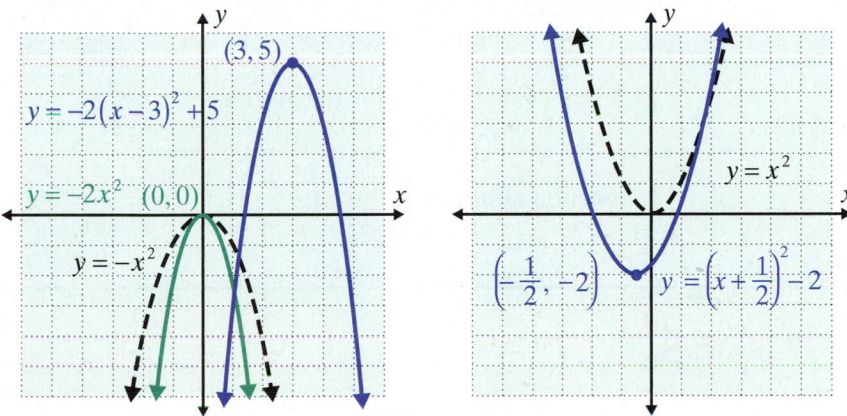

Figure 6

The general form of a quadratic function is $y = ax^2 + bx + c$. However, this form does not give as much information about the graph of the corresponding parabola as the form $y = a(x-h)^2 + k$. **Therefore, to easily find the vertex, line of symmetry, and range, and to graph the parabola, we want to change the general form $y = ax^2 + bx + c$ into the form $y = a(x-h)^2 + k$.** This can be accomplished by completing the square using the following technique, illustrated in Example 2. Be aware that you are not solving an equation. You do not "do something" to both sides. You are **changing the form** of a function. (**Note:** A graphing calculator will give the same graph regardless of the form of the function.)

$$y = ax^2 + bx + c \qquad \text{Write the function.}$$

$$= a\left(x^2 + \frac{b}{a}x\right) + c \qquad \text{Factor } a \text{ from just the first two terms.}$$

$$= a\left(x^2 + \frac{b}{a}x + \frac{b^2}{4a^2} - \frac{b^2}{4a^2}\right) + c \quad \text{Complete the square of } x^2 + \frac{b}{a}x.$$

$$\frac{1}{2}\left(\frac{b}{a}\right) = \frac{b}{2a} \text{ and } \left(\frac{b}{2a}\right)^2 = \frac{b^2}{4a^2}.$$

Add and subtract $\frac{b^2}{4a^2}$ inside the parentheses.

$$= a\left(x^2 + \frac{b}{a}x + \frac{b^2}{4a^2}\right) - \frac{b^2}{4a} + c \quad \text{Multiply } a\left(\frac{-b^2}{4a^2}\right) \text{ and write this term outside}$$

the parentheses.

$$= a\left(x + \frac{b}{2a}\right)^2 + \frac{4ac - b^2}{4a} \qquad \text{Write the square of the binomial and add the}$$

last two terms to get the form $y = a(x - h)^2 + k$.

In terms of the coefficients a, b, and c,

$$x = -\frac{b}{2a} \text{ is the \textbf{line of symmetry}}$$

and

$$(h, k) = \left(-\frac{b}{2a}, \frac{4ac - b^2}{4a}\right) \text{ is the \textbf{vertex}.}$$

$$\left(\text{In function notation, } \left(-\frac{b}{2a}, f\left(-\frac{b}{2a}\right)\right).\right)$$

notes

- Rather than memorize the formula for the coordinates of the vertex, you should just remember that the x-coordinate of the vertex is
- $x = -\dfrac{b}{2a}$. Substituting this value for x in the function will give the
- y-value for the vertex.

Objective F   **Zeros of a Quadratic Function**

The points where a parabola crosses the x-axis, if any, are the x-intercepts. This is where $y = 0$. These points are called the **zeros of the function**. We find these points by substituting 0 for y and solving the resulting quadratic equation.

$$y = ax^2 + bx + c \qquad \text{quadratic function}$$

$$0 = ax^2 + bx + c \qquad \text{quadratic equation}$$

If the solutions are nonreal complex numbers, then the graph does not cross the x-axis. It is either entirely above the x-axis or entirely below the x-axis.

The following examples illustrate how to apply all our knowledge about quadratic functions.

Example 2

Graphing Quadratic Functions

a. For $y = x^2 - 6x + 1$, find the zeros of the function, the line of symmetry, the vertex, the domain, the range, and graph the parabola.

Solution

$x^2 - 6x + 1 = 0$

$$x = \frac{6 \pm \sqrt{(-6)^2 - 4}}{2} \qquad \text{quadratic formula}$$

$$= \frac{6 \pm \sqrt{32}}{2}$$

$$= \frac{6 \pm 4\sqrt{2}}{2}$$

$$= 3 \pm 2\sqrt{2}$$

The zeros are $3 \pm 2\sqrt{2}$.

Change the form of the function for easier graphing.

$y = x^2 - 6x + 1$

$= \left(x^2 - 6x + 9 - 9\right) + 1 \qquad \text{Add } 0 = 9 - 9 \text{ inside the parentheses.}$

$= \left(x^2 - 6x + 9\right) - 9 + 1$

$= \left(x - 3\right)^2 - 8$

Summary:

zeros: $3 \pm 2\sqrt{2}$

line of symmetry is $x = 3$.

vertex: $(3, -8)$

domain: $(-\infty, \infty)$ or

$\qquad \left\{x \mid x \text{ is any real number}\right\}$

range: $[-8, \infty)$ or $\left\{y \mid y \geq -8\right\}$

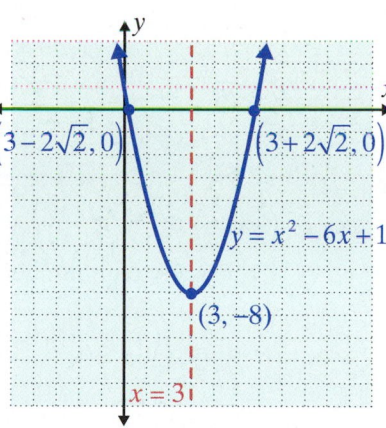

b. For $y = -x^2 - 4x + 2$, find the zeros of the function, the line of symmetry, the vertex, the domain, the range, and graph the parabola.

Solution

$-x^2 - 4x + 2 = 0$

$$x = \frac{4 \pm \sqrt{(-4)^2 - 4(-1)(2)}}{2(-1)} \quad \text{quadratic formula}$$

$$= \frac{4 \pm \sqrt{24}}{-2}$$

$$= \frac{4 \pm 2\sqrt{6}}{-2}$$

$$= -2 \pm \sqrt{6}$$

The zeros are $-2 \pm \sqrt{6}$.

Change the form of the function for easier graphing.

$y = -x^2 - 4x + 2$

$= -\left(x^2 + 4x\right) + 2$ Factor –1 from the first two terms only.

$= -\left(x^2 + 4x + 4 - 4\right) + 2$ Add $0 = 4 - 4$ inside the parentheses.

$= -\left(x^2 + 4x + 4\right) + 4 + 2$ Multiply –1(–4) and put this outside the parentheses.

$= -\left(x + 2\right)^2 + 6$

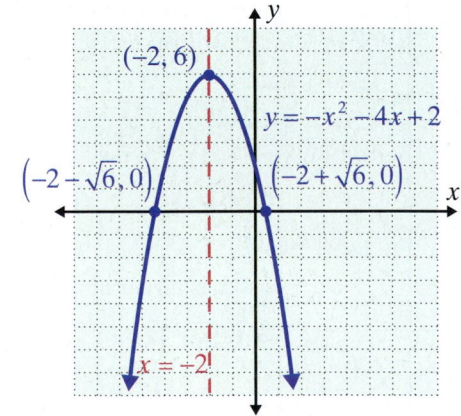

Summary:

zeros: $-2 \pm \sqrt{6}$

line of symmetry is $x = -2$.

vertex: $\left(-2, 6\right)$

domain: $\left(-\infty, \infty\right)$ or

$\left\{x \mid x \text{ is any real number}\right\}$

range: $\left(-\infty, 6\right]$ or $\left\{y \mid y \le 6\right\}$

c. For $y = 2x^2 - 6x + 5$, find the zeros of the function, the line of symmetry, the vertex, the domain, the range, and graph the parabola.

Solution

$2x^2 - 6x + 5 = 0$

$$x = \frac{6 \pm \sqrt{(-6)^2 - 4(2)(5)}}{2(2)} \quad \text{quadratic formula}$$

$$= \frac{6 \pm \sqrt{-4}}{4}$$

There are **no real zeros** because the discriminant is negative. The graph will not cross the x-axis. Now using another approach, the vertex is at

$$x = -\frac{b}{2a} = -\frac{-6}{2\cdot2} = \frac{3}{2}$$

and

$$y = 2\left(\frac{3}{2}\right)^2 - 6\left(\frac{3}{2}\right) + 5$$

$$= \frac{9}{2} - 9 + 5$$

$$= \frac{1}{2}.$$

So we have the vertex at $\left(\frac{3}{2},\frac{1}{2}\right)$ and the following results.

Summary:

zeros: no real zeros

line of symmetry is $x = \dfrac{3}{2}$.

vertex: $\left(\dfrac{3}{2},\dfrac{1}{2}\right)$

domain: $(-\infty, \infty)$ or
$\{x \mid x \text{ is any real number}\}$

range: $\left[\dfrac{1}{2},\infty\right)$ or $\left\{y \mid y \geq \dfrac{1}{2}\right\}$

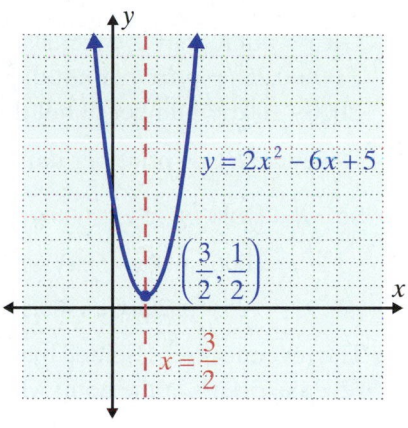

2. Find the zeros, the line of symmetry, the vertex, the domain, and the range of each quadratic function.

a. $y = x^2 - 10x + 21$

b. $y = -x^2 - 6x + 3$

c. $y = x^2 - 2x + 4$

Now work margin exercise 2.

Objective G **Applications with Maximum and Minimum Values**

The vertex of a vertical parabola is either the lowest point or the highest point on the parabola.

Minimum and Maximum Values

For a parabola with its equation in the form $y = a(x - h)^2 + k$,

1. If $a > 0$, then the parabola opens upward and (h, k) is the lowest point and the y-value k is called the **minimum value** of the function.
2. If $a < 0$, then the parabola opens downward and (h, k) is the highest point and y-value k is called the **maximum value** of the function.

If the function is in the general quadratic form $y = ax^2 + bx + c$, then the maximum or minimum value can be found by letting $x = -\dfrac{b}{2a}$ and solving for y.

The concepts of maximum and minimum values of a function help not only in graphing but also in solving many types of applications. Applications involving quadratic functions are discussed here. Other types of applications are discussed in more advanced courses in mathematics.

Example 3

Minimum and Maximum Values

a. A sandwich company sells hot dogs at the local baseball stadium for $3.00 each. Their mean sales per game is 2000 hot dogs. The company estimates that each time the price is raised by 25¢, they will sell 100 fewer hot dogs. What price should they charge to maximize their revenue (income) per game? What will be the maximum revenue?

Solution

Let x = number of 25¢ increases in price and
$3.00 + 0.25x$ = price per hot dog, and
$2000 - 100x$ = number of hot dogs sold.

$$\text{Revenue}(R) = (\text{price per unit}) \cdot (\text{number of units sold})$$

So, $R = (3.00 + 0.25x)(2000 - 100x)$
$$= 6000 - 300x + 500x - 25x^2$$
$$= -25x^2 + 200x + 6000.$$

The revenue is represented by a quadratic function and the maximum revenue occurs at the point where

$$x = -\frac{b}{2a} = -\frac{200}{-50} = 4.$$

For $x = 4$,
$$\text{price per hot dog} = 3.00 + 0.25(4) = \$4.00, \text{and}$$
$$\text{revenue} = (4)(2000 - 100 \cdot 4) = \$6400.$$

Thus the company will make its maximum revenue of $6400 by charging $4 per hot dog.

b. A rancher is going to build three sides of a rectangular corral next to a river. He has 240 feet of fencing and wants to enclose the maximum area possible. What are the dimensions of the fencing that gives the corral its maximum area?

240 feet of fencing in total

Solution

Let x = length of the two equal sides of the rectangular corral and $240 - 2x$ = length of third side of the rectangular corral.

Since area equals length times width, the area, A, of the corral is represented by the quadratic function $A = x(240 - 2x) = 240x - 2x^2$.

The maximum area occurs at the point where $x = -\dfrac{b}{2a} = -\dfrac{240}{-4} = 60$.

Two sides of the rectangle are 60 feet and the third side is $240 - 2(60) = 120$ feet.

The maximum area possible is $60(120) = 7200$ square feet.

Now work margin exercise 3.

Practice Problems

1. Write the function $y = 2x^2 - 4x + 3$ in the form $y = a(x-h)^2 + k$.

2. Find the zeros of the function $y = x^2 - 7x + 10$.

3. Find the vertex and the range of the function $y = -x^2 + 4x - 5$.

3. Solve the following minimum and maximum value equations.

a. A minor league hockey team sells tickets for games at a price of $8.00 per ticket. Their mean ticket sales per home game is 4000 tickets. The team estimates that for each $1.00 increase in ticket price they will sell 200 fewer tickets. What price should they charge to maximize their revenue (profit) per game? What will be their maximum revenue?

b. A towing company is going to build a rectangular fence against the side of a building to store impounded cars. The manager has 360 yards of fencing and wants to enclose the maximum area possible inside the lot. What are the dimensions of the lot with the maximum area and what is this area?

Practice Problem Answers

1. $y = 2(x-1)^2 + 1$ **2.** $x = 2, 5$ **3.** vertex: $(2, -1)$, range: $(-\infty, -1]$

Exercises 12.5

For each of the quadratic functions, determine the line of symmetry, the vertex, the domain, and the range. See Example 1.

1. $y = 3x^2 - 4$

2. $y = \dfrac{2}{3}x^2 + 7$

3. $y = 7x^2 + 9$

4. $y = 5x^2 - 1$

5. $y = -4x^2 + 1$

6. $y = -2x^2 - 6$

7. $y = -\dfrac{3}{4}x^2 + \dfrac{1}{5}$

8. $y = \dfrac{5}{3}x^2 + \dfrac{7}{8}$

9. $y = (x+1)^2$

10. $y = (x-1)^2$

11. $y = -\dfrac{2}{3}(x-4)^2$

12. $y = -5(x+2)^2$

13. $y = 2(x+3)^2 - 2$

14. $y = 4(x-5)^2 + 1$

15. $y = \dfrac{3}{4}(x+2)^2 - 6$

16. $y = -2(x+1)^2 - 4$

17. $y = -\dfrac{1}{2}\left(x - \dfrac{3}{2}\right)^2 + \dfrac{7}{2}$

18. $y = -\dfrac{5}{3}\left(x - \dfrac{9}{2}\right)^2 + \dfrac{3}{4}$

19. $y = \dfrac{1}{4}\left(x - \dfrac{4}{5}\right)^2 - \dfrac{11}{5}$

20. $y = \dfrac{10}{3}\left(x + \dfrac{7}{8}\right)^2 - \dfrac{9}{16}$

21. Graph the function $y = x^2$. Then without additional computation, graph the following translations.

   a. $y = x^2 - 2$

   b. $y = (x - 3)^2$

   c. $y = -(x - 1)^2$

   d. $y = 5 - (x + 1)^2$

22. Graph the function $y = 2x^2$. Then without additional computation, graph the following translations.

   a. $y = 2x^2 - 3$

   b. $y = 2(x - 4)^2$

   c. $y = -2(x + 1)^2$

   d. $y = -2(x + 2)^2 - 4$

23. Graph the function $y = \dfrac{1}{2}x^2$. Then without additional computation, graph the following translations.

   a. $y = \dfrac{1}{2}x^2 + 3$

   b. $y = \dfrac{1}{2}(x + 2)^2$

   c. $y = -\dfrac{1}{2}x^2$

   d. $y = \dfrac{1}{2}(x - 1)^2 - 4$

24. Graph the function $y = \dfrac{1}{4}x^2$. Then without additional computation, graph the following translations.

   a. $y = -\dfrac{1}{4}x^2$

   b. $y = \dfrac{1}{4}x^2 - 5$

   c. $y = \dfrac{1}{4}(x + 4)^2$

   d. $y = 2 - \dfrac{1}{4}(x + 2)^2$

25. $y = 2x^2 - 4x + 2$

26. $y = -3x^2 + 12x - 12$

27. $y = x^2 - 2x - 3$

28. $y = x^2 - 4x + 5$

29. $y = x^2 + 6x + 5$

30. $y = x^2 - 8x + 12$

31. $y = 2x^2 - 8x + 5$

32. $y = 2x^2 - 12x + 16$

33. $y = -3x^2 - 12x - 9$

34. $y = 3x^2 - 6x - 1$

35. $y = 5x^2 - 10x + 8$

36. $y = -4x^2 + 16x - 11$

37. $y = -x^2 - 5x - 2$ **38.** $y = x^2 + 3x - 1$ **39.** $y = 2x^2 + 7x + 5$

40. $y = 2x^2 + x - 3$

Graph the two given functions and answer the following questions:

a. Are the graphs the same?
b. Do the functions have the same zeros?
c. Briefly, discuss your interpretation of the results in parts a. and b.

41. $\begin{cases} y = x^2 - 3x - 10 \\ y = -x^2 + 3x + 10 \end{cases}$ **42.** $\begin{cases} y = x^2 - 5x + 6 \\ y = -x^2 + 5x - 6 \end{cases}$

43. $\begin{cases} y = 2x^2 - 5x - 3 \\ y = -2x^2 + 5x + 3 \end{cases}$ **44.** $\begin{cases} y = -4x^2 - 15x + 4 \\ y = 4x^2 + 15x - 4 \end{cases}$

 Use the function $h = -16t^2 + v_0 t + h_0$ **, where h is the height of the object after time t, v_0 is the initial velocity, and h_0 is the initial height.**

45. Throwing a Ball: A ball is thrown vertically upward from the ground with an initial velocity of 112 ft/s.

  a. When will the ball reach its maximum height?

  b. What will be the maximum height?

46. Launching a Rocket: A water rocket is launched from the ground and has an initial velocity of 104 ft/s.

  a. When will the rocket reach its maximum height?

  b. What will be the maximum height?

47. Throwing a Stone: A stone is projected vertically upward from a platform that is 20 feet high, at a rate of 160 feet per second.
  a. When will the stone reach its maximum height?
  b. What will be the maximum height?

48. Throwing a Stone: A cannonball is projected vertically upward from a platform that is 32 feet high, at a rate of 128 feet per second.
  a. When will the cannonball reach its maximum height?
  b. What will be the maximum height?

49. Selling Radios: A retailer sells radios. He estimates that by selling them for x dollars each, he will be able to sell $100 - x$ radios each month.
  a. What price will yield maximum revenue?
  b. What will be the maximum revenue?

50. Selling Picture Frames: Mrs. Richey can sell 72 picture frames each month if she charges $24 each. She estimates that for each $1 increase in price, she will sell 2 fewer frames.
  a. Find the price that will yield maximum revenue.
  b. What will be the maximum revenue?

51. Selling Lamps: A store owner estimates that by charging x dollars each for a certain lamp, he can sell $40 - x$ lamps each week. What price will give him maximum sales revenue?

52. Construction: A contractor is to build a brick wall 6 feet high to enclose a rectangular garden. The wall will be on three sides of the rectangle because the fourth side is a building. The owner wants to enclose the maximum area but wants to pay for only 150 feet of wall. What dimensions should the contractor make the garden?

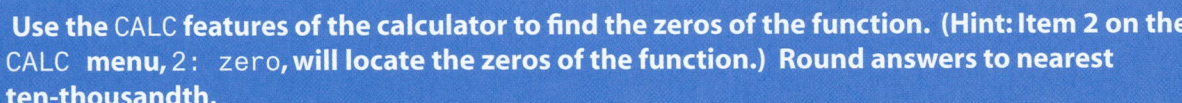

Use the CALC features of the calculator to find the zeros of the function. (Hint: Item 2 on the CALC menu, `2: zero`, will locate the zeros of the function.) Round answers to nearest ten-thousandth.

53. $y = x^2 - 2x - 2$

54. $y = 3x^2 + x - 1$

55. $y = -2x^2 + 2x + 5$

56. $y = -x^2 - 2x + 7$

57. $y = x^2 + 3x + 3$

58. $y = -4x^2 - x - 6$

Use a graphing calculator to graph each function by pressing ⬜ Y= ⬜ and entering the function. Find the coordinates of the maximum as follows. Round answers to nearest ten-thousandth.

Step 1: Press CALC (**2ND** **TRACE**).

Step 2: Press or choose `4: maximum`.

Step 3: Follow the directions for moving the cursor to `Left Bound?`, `Right Bound?`, and `Guess?`. (Press **ENTER** each time.)

59. $y = 4x - x^2$

60. $y = 1 - 2x - x^2$

61. $y = -8 + 4x - x^2$

62. $y = 3 - 2x - x^2$

Use a graphing calculator to graph each function by pressing Y= and entering the function. Find the coordinates of the minimum as follows. Round answers to nearest ten-thousandth.

Step 1: Press CALC (**2ND** **TRACE**).

Step 2: Press or choose 3: minimum.

Step 3: Follow the directions for moving the cursor to Left Bound?, Right Bound?, and Guess?. (Press **ENTER** each time.)

63. $y = x^2 - 8x + 15$ **64.** $y = x^2 + 10x + 22$ **65.** $y = 2x^2 + 4x + 3$ **66.** $y = 3x^2 - 6x + 5$

Writing & Thinking

67. Discuss the following features of the general quadratic function
$y = ax^2 + bx + c.$

    a. What type of curve is its graph?

    b. What is the value of x at its vertex?

    c. What is the equation of the line of symmetry?

    d. Does the graph always cross the x-axis? Explain.

68. Discuss the discriminant of the general quadratic equation $ax^2 + bx + c = 0$ and how the value of the discriminant is related to the graph of the corresponding quadratic function $y = ax^2 + bx + c.$

69. Discuss the domain and range of a quadratic function in the form $y = a(x - h)^2 + k.$

Section 12.1: Complex Numbers

Complex Numbers   page 1251

$i = \sqrt{-1}$ and $i^2 = -1$

If a is a positive real number, then $\sqrt{-a} = \sqrt{a} \cdot \sqrt{-1} = \sqrt{a}\, i$.

(To avoid confusion, we sometimes write $i\sqrt{a}$.)

Standard Form of Complex Numbers   page 1252

The **standard form** of a **complex number** is $a + bi$, where a and b are real numbers. a is called the **real part** and b is called the **imaginary part**.

If $b = 0$, then $a + bi = a + 0i = a$ is a real number.

If $a = 0$, then $a + bi = 0 + bi = bi$ is called a **pure imaginary number** (or an **imaginary number**).

Equality of Complex Numbers   page 1254

For complex numbers $a + bi$ and $c + di$, if $a + bi = c + di$, then $a = c$ and $b = d$.

Addition and Subtraction with Complex Numbers   pages 1254 – 1255

For complex numbers $a + bi$ and $c + di$,

$$(a + bi) + (c + di) = (a + c) + (b + d)i$$

and $(a + bi) - (c + di) = (a - c) + (b - d)i.$

Multiplication with Complex Numbers   page 1256

The product of two complex numbers can be found by using the FOIL method for multiplying two binomials.

Complex Conjugates   pages 1257 – 1258

$a + bi$ and $a - bi$ are **complex conjugates** of each other.

The product of two complex conjugates will always be a nonnegative real number: $(a + bi)(a - bi) = a^2 + b^2$.

To Write a Fraction with Complex Numbers in Standard Form   pages 1258 – 1259

1. Multiply both the numerator and denominator by the complex conjugate of the denominator.
2. Simplify the resulting products in both the numerator and denominator.
3. Write the simplified result in standard form.

Powers of i: i^n   page 1260

$$i^{4n} = \left(i^4\right)^n = 1^n = 1$$

$$i^{4n+1} = i^{4n} \cdot i = 1 \cdot i = i$$

$$i^{4n+2} = i^{4n} \cdot i^2 = 1 \cdot (-1) = -1$$

$$i^{4n+3} = i^{4n} \cdot i^3 = 1 \cdot (-i) = -i$$

Review of Solving Quadratic Equations by Factoring   page 1265

1. Add or subtract terms as necessary so that 0 is on one side of the equation and the equation is in the standard form $ax^2 + bx + c = 0$ where a, b, and c are real numbers and $a \neq 0$.
2. Factor completely. (If there are any fractional coefficients, multiply each term by the least common denominator so that all coefficients will be integers.)
3. Set each nonconstant factor equal to 0 and solve each linear equation for the unknown.
4. Check each solution, one at a time, in the original equation.

Square Root Property   pages 1267 – 1268

If $x^2 = c$, then $x = \pm\sqrt{c}$.

If $(x - a)^2 = c$, then $x - a = \pm\sqrt{c}$ (or $x = a \pm \sqrt{c}$).

Note: If c is negative $(c < 0)$, then the solutions will be nonreal.

To Solve a Quadratic Equation by Completing the Square   page 1270

1. If necessary, divide or multiply both sides of the equation so that the leading coefficient (the coefficient of x^2) is 1.
2. If necessary, isolate the constant term on one side of the equation.
3. Find the constant that completes the square of the polynomial and add this constant to both sides. Rewrite the polynomial as the square of a binomial.
4. Use the square root property to find the solutions of the equation.

Writing Equations with Known Roots   pages 1272 – 1273

Quadratic Formula   pages 1279 – 1280

For the general quadratic equation $ax^2 + bx + c = 0$, where $a \neq 0$, the solutions are $x = \dfrac{-b \pm \sqrt{b^2 - 4ac}}{2a}$.

Discriminant   pages 1281 – 1283

The expression $b^2 - 4ac$, the part of the quadratic formula that lies under the radical sign, is called the discriminant.
If $b^2 - 4ac > 0$, there are two real solutions.
If $b^2 - 4ac = 0$, there is one real solution.
If $b^2 - 4ac < 0$, there are two nonreal solutions.

Strategy for Solving Word Problems   page 1288
   1. Understand the problem.
      a. Read the problem carefully. (Read it several times if necessary.)
      b. If it helps, restate the problem in your own words.
   2. Devise a plan.
      a. Decide what is asked for; assign a variable to the unknown quantity. Label this variable so you know exactly what it represents.
      b. Draw a diagram or set up a chart whenever possible.
      c. Write an equation that relates the information provided.
   3. Carry out the plan.
      a. Study your picture or diagram for insight into the solution.
      b. Solve the equation.
   4. Look back over the results.
      a. Does your solution make sense in terms of the wording of the problem?
      b. Check your solution in the equation.

The Pythagorean Theorem   pages 1288 – 1289

   In a right triangle, if c is the length of the hypotenuse and a and b are the lengths of the legs, then $c^2 = a^2 + b^2$.

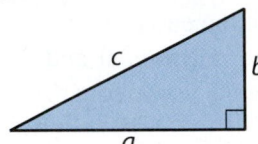

Projectiles   pages 1289 – 1290

   The formula $h = -16t^2 + v_0 t + h_0$ is used in physics and relates to the height of a projectile such as a thrown ball, a bullet, or a rocket.

   h = height of object, in feet
   t = time the object is in the air, in seconds
   v_0 = beginning velocity, in feet per second
   h_0 = beginning height

Cost Per Person   page 1292

   The cost per person is found by dividing the total cost by the number of people.

Quadratic Function

Any function that can be written in the form
$y = ax^2 + bx + c$, where a, b, and c are real numbers and
$a \neq 0$ is a quadratic function.

Parabolas

The graph of every quadratic function is a vertical parabola (a parabola
that opens up or down). Horizontal parabolas (parabolas that open left
or right) do not represent functions.

Forms of Quadratic Functions

$y = ax^2$
$y = ax^2 + k$
$y = a(x - h)^2$
$y = a(x - h)^2 + k$
$y = ax^2 + bx + c$

Line of Symmetry

The line $x = -\dfrac{b}{2a}$ is the line of symmetry (or **axis of symmetry**) of the
graph of the quadratic function $y = ax^2 + bx + c$. The curve is a "mirror
image" of itself with respect to the line of symmetry.

Vertex

The vertex of the graph of $y = ax^2 + bx + c$ is at the point $\left(-\dfrac{b}{2a}, \dfrac{4ac - b^2}{4a} \right)$.
The vertex is the "turning point" of the parabola.

Zeros of a Quadratic Function

The zeros of a quadratic function are the points where the parabola
crosses the x-axis.

Minimum and Maximum Values

For a parabola with equation of the form $y = a(x - h)^2 + k$,
1. If $a > 0$, then the parabola opens upward and (h, k) is the lowest point
   and the y-value, k, is called the minimum value of the function.
2. If $a < 0$, then the parabola opens downward and (h, k) is the highest
   point and the y-value, k, is called the maximum value of the function.

Chapter 12: Test

Find the real part and the imaginary part of each of the complex numbers.

1. $6 - \dfrac{1}{3}i$

2. $\dfrac{-2 + 3i}{5}$

3. Solve for x and y: $(2x + 3i) - (6 + 2yi) = 5 - 3i$.

Perform the indicated operations and write each result in the standard form.

4. $(5 + 8i) + (11 - 4i)$

5. $\left(2 + 3\sqrt{-4}\right) - \left(7 - 2\sqrt{-25}\right)$

6. $(4 + 3i)(2 - 5i)$

7. $(x + 2i)(x - 2i)$

8. $\dfrac{2 + i}{3 + 2i}$

Solve the following.

9. Simplify the following power of i: i^{23}.

10. Solve the following equation using the square root property: $(x + 1)^2 = 9$.

11. Add the correct constant to complete the square; then factor the trinomial as indicated.

    $x^2 - 30x + \underline{\quad} = \left(\underline{\quad}\right)^2$

12. Solve the following equation by completing the square: $x^2 + 4x + 1 = 0$

13. Solve the following equation by factoring: $4x^3 = -4x^2 - x$

14. Write a quadratic equation with integer coefficients that has the two numbers $1 \pm \sqrt{5}$ as roots.

15. Find the discriminant of the quadratic equation $4x^2 + 5x - 3 = 0$ and determine the nature of its solutions.

16. Using the discriminant, determine the value(s) for b such that the equation $2x^2 + bx + 3 = 0$ will have exactly one real root.

Solve the given equations using any of the techniques discussed for solving quadratic equations: factoring, completing the square, or the quadratic formula.

17. $2x^2 + x + 1 = 0$

18. $2x^2 - 3x - 4 = 0$

19. $2x^2 + 3 = 4x$

For each of the quadratic functions, determine the line of symmetry, the vertex, the domain, and the range.

20. $y = (x-3)^2 - 1$

21. $y = 2(x-3)^2 - 9$

Rewrite each of the quadratic functions in the form $y = a(x-h)^2 + k$. Find the vertex, range, and zero(s) of each function. Graph the function.

22. $y = x^2 + 4x - 6$

23. $y = x^2 - 6x + 8$

Solve the following word problems.

24. Right Triangles: The length of one leg of a right triangle is one meter less than twice the length of the second leg. The hypotenuse is 17 m. Find the lengths of the two legs.

25. Throwing a Ball: A man standing at the edge of a cliff 112 ft above the beach throws a ball straight up into the air with a velocity of 96 ft/s. Use the formula $h = -16t^2 + v_0 t + h_0$. (Round answers to the nearest hundredth.)
  a. When will the ball hit the beach?

  b. When will the ball be 64 ft above the beach?

26. Dimensions of a Rectangle: The length of a rectangle is 4 inches longer than the width. If the diagonal is 20 inches long, what are the dimensions of the rectangle?

27. Traveling by Car: Sandy made a business trip to a city 200 miles away and then returned home. Her mean speed on the return trip was 10 mph less than her mean speed going. If her total travel time was 9 hours, what was her mean rate in each direction?

28. Number Problem: One number is 10 more than another number. Form a quadratic function that will allow you to find the minimum product of the two numbers. Find the numbers and the minimum product.

29. Maximize Area: The perimeter of a rectangle is 22 in. Find the dimensions that will maximize the area.

Cumulative Review: Chapters 1 - 12

Simplify the expressions. Assume that all variables are positive.

1. $\left(4x^{-3}\right)\left(2x\right)^{-2}$

2. $\dfrac{x^{-2}y^4}{x^{-5}y^{-2}}$

3. $\left(27x^{-3}y^{\frac{3}{4}}\right)^{\frac{2}{3}}$

4. $\left(\dfrac{9x^{\frac{4}{3}}}{4y^{\frac{2}{3}}}\right)^{\frac{3}{2}}$

5. $\sqrt[3]{-27x^6y^8}$

6. $\sqrt[4]{32x^9y^{15}}$

7. $\dfrac{\sqrt{72}}{3}+5\sqrt{\dfrac{1}{2}}$

8. $\dfrac{1}{2}\sqrt{\dfrac{4}{3}}+3\sqrt{\dfrac{1}{3}}$

Rationalize the denominator and simplify each expression.

9. $\dfrac{5}{\sqrt{2}+\sqrt{3}}$

10. $\dfrac{x^2-81}{3+\sqrt{x}}$

Perform the indicated operations and write the results in the standard form $a+bi$.

11. $\left(2+5i\right)+\left(2-3i\right)$

12. $\left(2+5i\right)\left(2-3i\right)$

13. $\dfrac{2+5i}{2-3i}$

Completely factor each polynomial.

14. $xy+2x-5y-10$

15. $64x^2-81$

16. $2x^6-432y^3$

17. Given $P\left(x\right)=x^4-10x^3+20x^2-8x-2$, find **a.** $P\left(2\right)$ and **b.** $P\left(-1\right)$.

18. Use the division algorithm to divide and write the answer in the form $Q+\dfrac{R}{D}$ where the degree of $R<$ the degree of D.

$\dfrac{x^3-7x^2+2x-15}{x^2+2x-1}$

Perform the indicated operations and simplify the results.

19. $\dfrac{x+1}{2x^2+7x-4}+\dfrac{3x}{3x^2+10x-8}$

20. $\dfrac{x}{x^2-7x-8}-\dfrac{3x+1}{3x^2+x-2}$

21. $\dfrac{x^3+3x^2}{6x^2-36x+30}\cdot\dfrac{2x^2+2x-4}{x^3+2x^2}$

22. $\dfrac{2x^2+x}{x^2-2x+1}\div\dfrac{4x^2-6x}{x^2-1}$

Solve the equations.

23. $7(2x-5)=5(x+3)+4$

24. $(2x+1)(x-4)=(2x-3)(x+6)$

25. $10x^2+11x-6=0$

26. $4x^2+7x+1=0$

27. $\sqrt{x+5}-2=x+1$

28. For the following equations, find the discriminant and determine the nature of the solutions.

a. $4x^2-7x+6=0$

b. $x^2+10x+25=0$

Find an equation with integer coefficients that has the indicated roots.

29. $x=\dfrac{3}{4}, x=-5$

30. $x=1-2\sqrt{5}, x=1+2\sqrt{5}$

Solve the inequalities algebraically, write the answers in interval notation, and then graph each solution set on a real number line.

31. $4(x+3)-1\geq 2(x-4)$

32. $\dfrac{7}{2}x+3\leq x+\dfrac{13}{2}$

33. Graph the equation $4x+3y=5$ by locating the x-intercept and the y-intercept.

34. Find an equation in slope-intercept form for the line passing through the points $(-1,-8)$ and $(4,-3)$. Graph the line.

35. Write an equation in standard form for the line parallel to $y=3x-4$ that passes through the point $(4,0)$. Graph both lines.

36. What is the vertical line test for functions? Why does it work?

37. Solve the following system of two linear equations graphically: $\begin{cases} -2x+3y=-4 \\ x-2y=3 \end{cases}$

38. Solve the following system of two linear equations algebraically: $\begin{cases} 3x - 2y = 7 \\ -2x + y = -6 \end{cases}$

39. Use Gaussian elimination to solve the following system: $\begin{cases} x - 3y - z = 1 \\ 2x + y - 2z = -5 \\ 3x - y + 2z = 10 \end{cases}$

40. Solve the following system of linear inequalities graphically: $\begin{cases} x < 4 \\ x - y \le 3 \end{cases}$

For each of the quadratic functions, determine the line of symmetry, the vertex, the domain, the range, and the zeros. Graph the function.

41. $y = x^2 + x - 6$

42. $y = -x^2 + x + 12$

43. $y = 2x^2 - 9x - 5$

Solve for the indicated variable.

44. $3x + 2y = 6$ for y

45. $\dfrac{3}{4}x + \dfrac{1}{2}y = 5$ for y

Solve the following word problems.

46. Volume of a Gas: V (volume) varies inversely as P (pressure) when a gas is enclosed in a container. In a particular situation, the volume of gas is 25 in.^{3} when a pressure of 10 lb per in.^{2} is exerted. What would be the volume of gas if a pressure of 15 lb per in.^{2} were to be used?

47. Boating: Susan traveled 25 miles downstream. In the same length of time, she could have traveled 15 miles upstream. If the speed of the current is 2.5 mph, find the speed of Susan's boat in still water.

48. Rectangles: Find the dimensions of a rectangle that has an area of 520 m^2 and a perimeter of 92 m.

49. Integers: Find three consecutive even integers such that the square of the first added to the product of the second and third gives a result of 368.

---

50. **Catering:** Jennifer and Michael work at a local restaurant and must prepare enough mirepoix (a mixture of onions, celery, and carrots) for a community event. If it takes Michael 3 hours to cut enough mirepoix and 1.25 hours if Michael and Jennifer work together, how long would it take (to the nearest hundredth) for Jennifer to prepare everything by herself?

51. **Word Problem:** z varies directly as x^2 and inversely as $\sqrt{y}$. If $z = 50$ when $x = 2$ and $y = 25$, find z if $x = 4$ and $y = 100$.

52. **Investing:** Melissa plans to invest $10,000 into two accounts, one that returns 9% interest and another that returns 11% interest. If she invests three times as much in the account that returns 11% interest and expects a return of $1050, how much money does she invest in each account? Please round your answer to the nearest cent.

53. **Medicine:** A doctor needs to administer two drugs to a patient. The amount of Drug A administered must be three times the amount of Drug B, and the total amount of medication must equal 20 mg. How much of each drug should be given to the patient?

54. **Selling Golf Balls:** The profit function (in dollars) for a golf ball manufacturer is given by the function $P(x) = -0.01x^2 + 200x - 5000$ where x is the number of dozens of ball sold in one week.

   a. How many dozens of golf balls need to be sold to give the maximum profit in a week?

   b. What will be the maximum profit?

   c. How many golf balls is this?

 **Use a calculator to estimate the value of each number accurate to the nearest ten-thousandth.**

55. $\sqrt{6} + 2\sqrt{3}$

56. $\sqrt{2}\left(1 + 3\sqrt{10}\right)$

57. $\dfrac{\sqrt{2} + 2}{\sqrt{2} - 2}$

Use a graphing calculator to graph each of the functions. State the domain and range of each.

58. $f(x) = \sqrt{x - 3}$

59. $g(x) = \sqrt{1 - x}$

60. $y = -\sqrt{x + 1}$

Answers

Chapter 1: Whole Numbers, Integers, and Introduction to Solving Equations

Section 1.1: Introduction to Whole Numbers

Margin Examples **1. a.** thirty-two million, four hundred fifty thousand, ninety **b.** five thousand seven hundred eighty-four **2. a.** 6041 **b.** 9,000,483,000 **3. a.** 580 **b.** 300
c. 10,000 **d.** 9700 **4. a.** 7400 **b.** 30,000 **c.** 100,000 **d.** 28,000,000
Exercises **1.** eight hundred ninety-three **3.** seven thousand eight hundred thirty-one **5.** eight million, nine hundred fifty-seven thousand **7.** six million, six thousand, six **9.** twenty-five million **11.** sixty-seven million, nine hundred forty thousand, three hundred **13.** three hundred five thousand, four hundred forty-four **15.** nineteen billion, four hundred sixty-two million, five hundred forty-one thousand, three hundred **17.** 84 **19.** 606 **21.** 114,225 **23.** 200,016,053
25. 14,063,172,081 **27.** 14,014 **29.** 15,015,015 **31.** 80; **33.** 660; **35.** 500;
37. 2000; **39.** 4000; **41.** 780 **43.** 630 **45.** 300
47. 1000 **49.** 8000 **51.** 7000 **53.** 80,000 **55.** 1,070,000 **57.** 100,000 **59.** 3,600,000 **61.** 350,000 square miles; 910,000
square kilometers **63.** 564,000 people and 626,000 people

Section 1.2: Addition and Subtraction with Whole Numbers

Margin Examples **1.** 2149 **2. a.** 17; associative property of addition **b.** 29; commutative property of addition **c.** 39; additive identity property **3. a.** $y = 12$; commutative property of addition **b.** $n = 0$; additive identity property **c.** $x = 10$; associative property of addition **4.** 38cm **5.** 155 **6.** $3640 **7.** 360; 314 **8.** 2200; 1838 **9.** 2363 **10.** 9677
Exercises **1.** zero **3.** $(4 + 2) + 3 = 4 + (2 + 3)$; $(5 + 11) + 13 = 5 + (11 + 13)$. Answers will vary. **5.** commutative property of addition **7.** associative property of addition **9.** 150 **11.** 2127 **13.** 596 **15.** 4,301,692 **17.** 2209
19. 301,846 **21.** 755,915 **23.** 4676 **25.** 315,999 **27.** 230; 224 **29.** 2400; 2281 **31.** 8000; 7576 **33.** 1000; 1687 **35.** 2642
miles **37.** $847 **39.** $165 **41.** 24 in. **43.** 12 in. **45.** 32 cm **47.** 156 in. **49.** 1854 **51.** 677,543 **53.** 2,945,584 **55.** 2736

Section 1.3: Multiplication and Division with Whole Numbers

Margin Examples **1. a.** 96; commutative property of multiplication **b.** 0; zero-factor law **c.** 35; multiplicative identity property **d.** 42; associative property of multiplication **2. a.** x = 25; multiplicative identity property **b.** n = 14; commutative property of multiplication **c.** y = 1; associative property of multiplication **d.** t = 0; zero-factor law

3. a. 63,000 **b.** 150,000 **c.** 160,000 **d.** 4,500,000 **4.** 884 **5.** 480 **6.** 3796 square yards **7.** 257 R15 **8.**
$$\begin{array}{r} 68 \\ 26\overline{)1768} \\ \underline{156} \\ 208 \\ \underline{208} \\ 0 \end{array}$$
9. $15 **10.** 2800; 2952 **11.** 800; 813 **12.** 15; 21 R8; The difference between the estimate and the quotient is 6. **13.** 215,922 **14.** 151 with an undetermined remainder **15.** 9900 R12 **16.** 300
Exercises **1.** A factor of a number is divisor of that number. **3.** $y = 7$; associative property of multiplication **5.** $n = 1$; multiplicative identity **7.** $x = 8$; commutative property of multiplication
9. $4(7 + 9) = 4 \cdot 7 + 4 \cdot 9 = 28 + 36 = 64$ **11.** $7(3 + 4) = 7 \cdot 3 + 7 \cdot 4 = 21 + 28 = 49$ **13.** 140 **15.** 3813
17. 32,640 **19.** 6 **21.** 8 **23.** 1 **25.** 4 **27.** 11 **29.** 8 **31.** 7 **33.** undefined **35.** 8 R 1 **37.** 11 R 9 **39.** 9 R 0
41. $$\begin{array}{r} 45 \\ 35\overline{)1575} \\ \underline{140} \\ 175 \\ \underline{175} \\ 0 \end{array}$$ **43.** $$\begin{array}{r} 39 \\ 56\overline{)2184} \\ \underline{168} \\ 504 \\ \underline{504} \\ 0 \end{array}$$ **45. a.** 400 **b.** 480 **c.** 300 **d.** 720 **47. a.** 100 **b.** 6 **c.** 400 **d.** 200 **49.** 280; 264 **51.** 1000; 1176
53. 3200; 2916 **55.** 80,000; 85,680 **57.** 10; 12 R 0 **59.** 20; 20 R 13 **61.** 30; 31 R 80
63. 1600; 1542 R 8 **65. a.** $38,400 **b.** $82,800 **c.** $252,000 **67.** 648 sq ft **69.** 600 pages
71. 30 sq ft **73.** 384 sq in. **75.** 6,588,436 **77.** 780 **79.** 961,800 **81.** 950 **83.** $3000; $2548
85. 40,000; 36,750 sq ft **87.** 237 R 13 **89.** 728 R 3 **91.** 750 R 17 **93.** 13,799 R 83

Section 1.4: Problem Solving with Whole Numbers

Margin Examples 1. 51 **2.** 519 **3.** $25,125 **4.** $868 **5.** 28 square inches
Exercises 1. 785 **3.** 1741 **5.** 2305 **7.** $220 per month **9.** $1296 **11.** $150 **13.** $538 **15.** $7717 **17. a.** 24 ft **b.** 24 sq ft
19. 174 cm **21.** 380 sq in. **23.** 111 sq in. **25.** 299 sq ft **27.** 1. Read each problem carefully until you understand the problem and know what is being asked. 2. Draw any type of figure or diagram that might be helpful, and decide what operations are needed. 3. Perform these operations. 4. Mentally check to see if your answer is reasonable and see if you can think of another more efficient or more interesting way to do the same problem **29.** False, because $\dfrac{0}{0}$ is undefined and can never equal 1.

Section 1.5: Solving Equations with Whole Numbers

Margin Examples 1. a. Substituting 8 for x gives $9 + 8 = 17$, which is true. **b.** Substituting 11 for x gives $9 + 11 = 17$, which is false. **2.** 9 **3. a.** $x = 12$ **b.** $x = 16$ **4. a.** $y = 82$ **b.** $y = 47$ **5. a.** $y = 4$ **b.** $x = 6$ **c.** $n = 13$
Exercises 1. 16 **3.** 18 **5.** 15 **7.** 3 **9.** $x = 4$ **11.** $13 = x$ **13.** $y = 0$ **15.** $y = 18$ **17.** $10 = n$ **19.** $125 = n$ **21.** $x = 4$ **23.** $y = 4$
25. $4 = n$ **27.** $x = 6$ **29.** $47 = y$ **31.** $x = 7$ **33.** $x = 5$ **35.** $x = 12$ **37.** $3 = y$ **39.** $9 = x$ **41.** $n = 3$ **43.** $0 = x$ **45.** $0 = x$
47. $6 = x$ **49.** $n = 2$

Section 1.6: Exponents and Order of Operations

Margin Examples 1. a. $100^3 = 1,000,000$ **b.** $2^4 = 16$ **c.** $1^3 = 1$ **d.** $8^2 = 64$ **2. a.** Six cubed is equal to two hundred sixteen.
b. Nine squared is equal to eighty-one. **3. a.** 16,807 **b.** 175,616 **c.** 1369 **4.** 19 **5.** 29 **6.** 68 **7.** 40 **8.** 39
Exercises 1. a. 2 **b.** 3 **c.** 8 **3. a.** 4 **b.** 2 **c.** 16 **5. a.** 9 **b.** 2 **c.** 81 **7. a.** 11 **b.** 2 **c.** 121 **9. a.** 3 **b.** 5 **c.** 243 **11. a.** 19 **b.** 0 **c.** 1
13. a. 1 **b.** 6 **c.** 1 **15. a.** 24 **b.** 1 **c.** 24 **17. a.** 20 **b.** 3 **c.** 8000 **19. a.** 30 **b.** 2 **c.** 900 **21.** 2^4 or 4^2 **23.** 5^2 **25.** 7^2
27. 6^2 **29.** 10^3 **31.** 2^3 **33.** 6^3 **35.** 3^5 **37.** 10^5 **39.** 3^3 **41.** 6^5 **43.** 11^3 **45.** $2^3 \cdot 3^2$ **47.** $2 \cdot 3^2 \cdot 11^2$ **49.** $3^3 \cdot 7^3$
51. 64 **53.** 121 **55.** 400 **57.** 169 **59.** 900 **61.** 2704 **63.** 390,625 **65.** 15,625 **67.** 78,125 **69.** 21 **71.** 19 **73.** 28 **75.** 2
77. 18 **79.** 0 **81.** 0 **83.** 46 **85.** 9 **87.** 74 **89.** 20 **91.** 132 **93.** 176 **95.** 61 **97.** 1532 **99.** 132 **101.** error; Answers will vary.

Section 1.7: Introduction to Polynomials

Margin Examples 1. a. first-degree monomial **b.** eighteenth-degree binomial **c.** sixth-degree trinomial **2.** 72 **3.** 81 **4.** 60
Exercises 1. 2; trinomial **3.** 2; binomial **5.** 12; monomial **7.** 4; trinomial **9.** 5; trinomial **11.** 20 **13.** 8 **15.** 53,248 **17.** 9
19. 81 **21.** 0 **23.** 3 **25.** 0 **27.** 5 **29.** 245 **31.** 74 **33.** 706 **35.** 4, 10, 18, 28, 40, 54 **37.** 8, 13, 22, 35, 52, 73 **39.** 9
41. 38 **43.** 30 **45.** 15,525 **47.** 500 **49. a.** 331,969 **b.** 3650 **c.** 20,854 **d.** 1675 **51.** Answers will vary.

Section 1.8: Introduction to Integers

Margin Examples 1. a. $+10$ **b.** $+8$ **c.** -17 **2.** **3.**
4. a. false; $6 \le 7$ **b.** true **c.** true **d.** false; $2 \ge -2$ **5.** 1 **6.** 4 **7.** true **8.** $z = -3, 3$ **9.** There are no possible values.
10. all integers except zero. **11.** $-2, -1, 0, 1, 2$
Exercises 1. $>$ **3.** $>$ **5.** $<$ **7.** $<$ **9.** $=$ **11.** $>$ **13.** true **15.** true **17.** false; $|2| > -2$ **19.** true **21.** false; $|-3| = 3$
23. false; $-4 < |-4|$ **25.** **27.** **29.**
31. **33.** **35.** 5, -5 **37.** 2, -2 **39.** no solution **41.** 23, -23
43. **45.** **47.** **49.** Sometimes. Examples will
vary. **51.** Always. Examples will vary. **53.** Answers will vary. If y represents a negative number, then $-y$ will represent a positive number. **55.** Answers will vary. **a.** The absolute value of an integer is never negative. **b.** $|x| \ge 0$ for all integers.

Section 1.9: Addition with Integers

Margin Examples 1. a. 8 **b.** 9 **c.** -10 **2. a.** -9 **b.** 4 **c.** 28 **3. a.** -45 **b.** 51 **c.** -18 **4. a.** yes **b.** yes **c.** no
Exercises 1. -15 **3.** 40 **5.** 9 **7.** -2 **9.** -8 **11.** -17 **13.** 0 **15.** 0 **17.** -18 **19.** -45 **21.** 0 **23.** -32 **25.** 3 **27.** -19 **29.** -10
31. -24 **33.** 31 **35.** 76 **37.** -310 **39.** 27 **41.** 0 **43. a.** 6, -6 **b.** 15, 5 **c.** 12, -6 **45.** 18 **47.** 14 **49.** 35 **51.** no **53.** yes
55. yes **57.** yes **59.** no **61.** Sometimes. Examples will vary. **63.** Sometimes. Examples will vary. **65.** Never. Examples will vary. **67.** Never. Examples will vary. **69.** If they are opposites. **71.** -3405 **73.** -2575 **75.** $-23,094$ **77.** 495

Section 1.10: Subtraction with Integers

Margin Examples **1. a.** 0 **b.** −11 **c.** 5 **2. a.** −1 **b.** −18 **c.** −12 **d.** −6 **3.** −46 **4.** −107 **5.** 0 **6. a.** 3 is a solution **b.** 15 is a solution **c.** 3 is a solution

Exercises **1. a.** −8 **b.** −20 **3. a.** −29 **b.** −9 **5. a.** −8 **b.** −32 **7.** 6 **9.** 5 **11.** −20 **13.** −34 **15.** 4 **17.** −4 **19.** 6 **21.** −15 **23.** −22 **25.** −19 **27.** −7 **29.** −21 **31.** −2 **33.** 0 **35.** 0 **37.** −28 **39.** −45 **41.** −1 **43.** −19 **45.** −8 **47.** −11 **49.** 0 **51.** 127 **53.** 2, 0, −2, −4 **55.** −18, −20, −22, −24 **57.** > **59.** < **61.** > **63.** = **65.** yes **67.** yes **69.** yes **71.** no **73.** yes **75.** 3° **77.** 4238 points **79.** 85 yd **81. a.** −12,544 **b.** 57,456 **83. a.** −618,000 **b.** 1,618,000 **85. a.** −5 **b.** 19 **87.** −1 **89.** Answers will vary. Think of −x as the "opposite of x." If x is positive, then −x is negative. If x is negative, then −x is positive. Also, note that 0 is its own opposite.

Section 1.11: Multiplication, Division, and Order of Operations with Integers

Margin Examples **1. a.** −15 **b.** −32 **c.** −3 **2. a.** 24 **b.** 0 **c.** 20 **3. a.** −60 **b.** 126 **c.** −125 **4. a.** −3 **b.** −4 **c.** 4 **5. a.** 0 **b.** undefined **c.** undefined **6.** 23 **7.** 13 **8.** −12

Exercises **1.** −24 **3.** −24 **5.** 28 **7.** −18 **9.** −50 **11.** 0 **13.** −14 **15.** 105 **17.** −32 **19.** 0 **21.** −1 **23.** −50 **25.** 60 **27.** 0 **29. a.** 1 **b.** −1 **c.** 1 **d.** 1 **31.** −4 **33.** −2 **35.** −6 **37.** −4 **39.** −14 **41.** −3 **43.** 5 **45.** 8 **47.** 0 **49.** undefined **51.** −70 **53. a.** −17 **b.** −27 **55.** −14 **57.** −22 **59.** −9 **61.** 27 **63.** 31 **65.** 45 **67.** 1026 **69.** −9 **71.** 4624 **73.** −105 **75.** $x \div y = 0$ if $x = 0$ and $y \neq 0$.

Section 1.12: Applications: Change in Value and Average

Margin Examples **1.** −38° F **2.** 20,500 ft **3.** 74°F **4.** 85 **5.** 304,700 cars

Exercises **1.** 72 years old **3.** −18,500 ft **5.** 16,300 ft **7.** −105 points **9.** 0 **11.** −6 **13.** $78 per share **15.** 85 **17.** 8 degrees **19. a.** 68 **b.** 4 strokes under par **21.** $331,432,400 **23.** −22°C **25.** 100; 80 **27.** −17,500; −18,152 **29.** −1000; −804 **31. a.** −10°C **b.** −20°F **c.** −40°F

Section 1.13: Introduction to Like Terms and Polynomials

Margin Examples **1.** 5, −1, and 0 are like terms; $10x$, −$5x$, and −$3x$ are like terms; and $4x^2z$ and $2x^2z$ are like terms. **2. a.** $16x$ **b.** $6x − 8a$ **c.** $4x^2 − 3b$ **d.** −x^2 **3.** $8xy − x^2$; −52 **4.** $x + 5xy$; −32 **5.** $6x^2 + x$; 22

Exercises **1.** $6x + 18$ **3.** −$7x − 35$ **5.** −$11n − 33$ **7.** $5x$ **9.** $2y$ **11.** −$2n$ **13.** 0 **15.** −$3y$ **17.** $9x − 24$ **19.** $15x^2 + 5x − 1$ **21.** $4x^2 − 4x − 3$ **23.** $5x − 1$ **25.** $4x − 28$ **27.** −$xy + 11x − y$ **29.** $2x^2y + 2$ **31.** $8a^2b + 7ab^2 + 3ab$ **33.** $5x − a − 5$ **35.** −$2xyz − 6$ **37.** $4x^2 − 4x + 5$; 29 **39.** $2y^2 − 6y − 5$; 3 **41.** −$2x^3 + x^2 − x$; 22 **43.** $7y^3 + 5y^2 − 6$; −8 **45.** $3y^2 − 16y − 5$;14 **47.** −$a^3 − 2a − 7$; −4 **49.** $3ab − 3a + 3b$; 3 **51.** $9a − b − 15$; −22 **53.** $6abc − 5ab + 2bc$; 14 **55.** $a^2b^2c^2 + 4ab^2 + bc^2$; 2 **57.** $12a + 8b − 3c − 64$; −101 **59.** 0; 0 **61. a.** −16 **b.** 16 **c.** −64 **d.** −64 **63. a.** −216 **b.** −1 **c.** −343 **d.** −32 **65. a.** −2 **b.** 3 **c.** −10 **d.** 21

Section 1.14: Solving Equations with Integers

Margin Examples **1. a.** −10 is a solution. **b.** 2 is not a solution. **2. a.** $x = −7$ **b.** $x = 28$ **3. a.** $n = 8$ **b.** $n = −19$ **4. a.** $n = 1$ **b.** $y = −4$

Exercises **1.** −12 **3.** −25 **5.** −15 **7.** −3 **9.** $x = −14$ **11.** $x = −37$ **13.** $y = −18$ **15.** $y = −9$ **17.** $n = 0$ **19.** $n = −105$ **21.** $x = 4$ **23.** $y = −5$ **25.** $n = −21$ **27.** $x = −21$ **29.** $y = −96$ **31.** $x = 43$ **33.** $x = 5$ **35.** $x = −18$ **37.** $y = 8$ **39.** $x = 0$ **41.** $x = 0$ **43.** $y = 0$ **45.** $x = 6$ **47.** $n = −42$ **49.** $n = −11$ **51.** Answers will vary.

Chapter 1 Test

1. three million, seventy-five thousand, one hundred twenty-two **2.** 15,320,584 **3.** 680 **4.** 14,000 **5.** 16; commutative property of addition **6.** 3; associative property of multiplication **7.** 0; additive identity **8.** 4; associative property of addition **9.** 12,300; 12,007 **10.** 850,000; 870,526 **11.** 2000; 2484 **12.** 50,000; 51,170 **13.** 1500; 1175 **14.** 270,000; 222,998 **15.** 160; 170 R 240 **16.** 58,000 **17. a.** 150 m **b.** 174 m **c.** 486 sq m **18.** $39,080 **19.** $x = 42$ **20.** $x = 35$ **21.** $y = 3$ **22. a.** 4 **b.** 3 **c.** 64 **23. a.** 32,768 **b.** 6859 **c.** 1 **24.** 24 **25.** 5 **26.** 8 **27.** 47 **28.** 12 **29.** 17 **30.** second-degree; binomial **31.** fourth-degree; trinomial **32.** 24 **33.** 81 **34. a.** yes; $10^2 = 100$ **b.** no; no whole number multiplied by itself will give 80. **c.** yes; $15^2 = 225$ **35.** Answers may vary. **36.** 12 sq in. **37.** 13, −3; **38.** 3, −23; **39.** **40.** −2, −1, 0, 1, 2 **41.** **42.** −3 **43.** −165 **44.** −40 **45.** −2 **46.** −420 **47.** 0 **48.** 480 **49.** −8 **50.** 0 **51.** 4 **52.** −15 **53.** −35 **54.** 43 **55.** −27,000 ft **56.** −60 **57.** 5 yd gain **58.** 64 mph **59.** 0°F **60.** 28,224 ft; 8606 m **61.** 0 **62.** 76 **63.** $5x$ **64.** −$2x^2 + 8x + 15$ **65.** −$3y^3 + 3y^2 − 4y$ **66.** $5y^2 − 6$ **67.** −51 **68.** 96 **69.** $x = −2$ **70.** $y = 3$ **71.** $x = −28$ **72.** $y = 1$

Chapter 2: Fractions, Mixed Numbers, and Proportions

Section 2.1: Tests for Divisibility

Margin Examples 1. a. Yes, 3 divides the sum of the digits and the units digit is 4. **b.** No, 3 does not divide the sum of the digits. **c.** yes; no **d.** yes; no **2.** 2 **3.** 2, 5, 10 **4.** none **5.** 2, 3, 5, 6, and 10 are factors of 2400
6. $3 \cdot 5 \cdot 8 \cdot 13 = (3 \cdot 13) \cdot (5 \cdot 8) = 39 \cdot 40$ **7.** 7 does not divide the product
Exercises 1. 2 **3.** 3, 9 **5.** 3, 9 **7.** 2 **9.** 5 **11.** 3, 9 **13.** 2, 5, 10 **15.** none **17.** 2, 3, 6 **19.** 2, 3, 5, 6, 10 **21.** 3
23. 2, 3, 5, 6, 9, 10 **25.** 2 **27.** 2, 3, 6, 9 **29.** 2 **31.** 2, 3, 6 **33.** none **35.** 2 **37.** 2, 5, 10 **39.** none **41. a.** yes **b.** yes **c.** no
d. no **e.** yes **43. a. i.** 7 **ii.** 7 **iii.** 7 **iv.** 8 **b. i.** 1, 4, 7 **ii.** 4. **iii.** none **iv.** 2, 5, 8 **45.** Answers will vary. Not necessarily. To be divisible by 27, an integer must have 3 as a factor three times. Two counterexamples are 9 and 18. **47.** The product is 126, and 6 divides the product 21 times. **49.** The product is 270, and 10 divides the product 27 times. **51.** The product is 1050, and 25 divides the product 42 times. **53.** The product is 1890, and 45 divides the product 42 times. **55.** The product is 2520, and 40 divides the product 63 times. **57. a.** 30, 45; Answers will vary. **b.** 9, 12; Answers will vary. **c.** 10, 25; Answers will vary. **59.** Answers will vary. The question is designed to help the students become familiar with their calculators and to provide an early introduction to scientific notation.

Section 2.2: Prime Numbers

Margin Examples 1. Each has exactly two factors, 1 and itself. **2.** 1, 2, 7, and 14 are all factors of 14. 1, 2, 3, 6, 7, 14, 21, and 42 are all factors of 42. 1, 2, 3, 4, 6, and 12 are all factors of 12. **3.** No, since the sum of its digits is divisible by 9, 999 is divisible by 9. The sum of the digits is also divisible by 3. **4.** No, $17 \cdot 17 = 289$. **5.** 221 is composite. Tests for 2, 3, and 5 all fail. Next, divide by 7, 11, and 13. The first divisor of 221 greater than 1 is 13. **6.** 239 is prime. Tests for 2, 3, and 5 all fail. Trial division by 7, 11, 13, and 17 fail to yield a divisor, but when dividing by 17, the quotient is less than 17. **7.** 5 and 54
Exercises 1. 3, 6, 9, 12, 15 **3.** 8, 16, 24, 32, 40 **5.** 11, 22, 33, 44, 55 **7.** 20, 40, 60, 80, 100 **9.** 41, 82, 123, 164, 205 **11.** prime
13. prime **15.** composite; 1, 55; 5, 11 **17.** prime **19.** composite; 1, 205; 5, 41 **21.** prime **23.** composite; 1, 943; 23, 41
25. composite; 1, 551; 19, 29 **27.** 3 and 8 **29.** 2 and 8 **31.** 2 and 10 **33.** 4 and 9 **35.** 3 and 17 **37.** 5 and 5 **39.** 3 and 24
41. 5 and 11 **43.** 3 and 21 **45.** 3 and 25 **47.** −6 and −10 **49.** 2 and −10 **51.** −5 and 10 **53. a.** Answers will vary. The number 1 does not have two or more distinct factors. **b.** Answers will vary. Since 0 is not a counting number, it does not satisfy the definition of either prime or composite.

Section 2.3: Prime Factorization

Margin Examples 1. $74 = 2 \cdot 37$ **2.** $52 = 2 \cdot 13 \cdot 2 = 2^2 \cdot 13$ **3.** $460 = 2 \cdot 2 \cdot 5 \cdot 23 = 2^2 \cdot 5 \cdot 23$ **4.** $616 = 2 \cdot 2 \cdot 2 \cdot 7 \cdot 11$
$= 2^3 \cdot 7 \cdot 11$ **5.** 1, 3, 7, 9, 21, 63 **6.** 1, 2, 3, 6, 9, 18, 27, 54
Exercises 1. The prime number is a whole number greater than 1 that has exactly two different factors (or divisors), itself and 1. **3.** $2^2 \cdot 5$ **5.** 2^5 **7.** $2 \cdot 5^2$ **9.** $2 \cdot 5 \cdot 7$ **11.** $3 \cdot 17$ **13.** $2 \cdot 31$ **15.** $3^2 \cdot 11$ **17.** 37 is prime. **19.** $2^3 \cdot 3 \cdot 5$ **21.** $2^2 \cdot 7^2$ **23.** 19^2
25. $5 \cdot 13$ **27.** $2^3 \cdot 5^3$ **29.** $2^5 \cdot 5^5$ **31.** $2^3 \cdot 3 \cdot 5^2$ **33.** 107 is prime. **35.** $3 \cdot 103$ **37.** $3 \cdot 5 \cdot 11$ **39.** $3^3 \cdot 5^2$ **41.** $2^3 \cdot 3^3$ **43.** $2^2 \cdot 3^2 \cdot 47$
45. $2^2 \cdot 13^2$ **47. a.** $2 \cdot 3 \cdot 7$ **b.** 1, 2, 3, 6, 7, 14, 21, 42 **49. a.** $2^2 \cdot 3 \cdot 5^2$ **b.** 1, 2, 3, 4, 5, 6, 10, 12, 15, 20, 25, 30, 50, 60, 75, 100, 150, 300
51. a. $2 \cdot 3 \cdot 11$ **b.** 1, 2, 3, 6, 11, 22, 33, 66 **53. a.** $2 \cdot 3 \cdot 13$ **b.** 1, 2, 3, 6, 13, 26, 39, 78 **55. a.** $2 \cdot 3 \cdot 5^2$ **b.** 1, 2, 3, 5, 6, 10, 15, 25, 30, 50, 75, 150 **57. a.** $2 \cdot 3^2 \cdot 5$ **b.** 1, 2, 3, 5, 6, 9, 10, 15, 18, 30, 45, 90 **59. a.** $7 \cdot 11 \cdot 13$ **b.** 1, 7, 11, 13, 77, 91, 143, 1001
61. a. $3^2 \cdot 5 \cdot 13$ **b.** 1, 3, 5, 9, 13, 15, 39, 45, 65, 117, 195, 585 **63. a.** There are 12 factors of 500. These factors are: 1, 2, 4, 5, 10, 20, 25, 50, 100, 125, 250, and 500. **b.** There are 24 factors of 660. These factors are: 1, 2, 3, 4, 5, 6, 10, 11, 12, 15, 20, 22, 30, 33, 44, 55, 60, 66, 110, 132, 165, 220, 330, and 660. **c.** There are 18 factors of 450. These factors are: 1, 2, 3, 5, 6, 9, 10, 15, 18, 25, 30, 45, 50, 75, 90, 150, 225, and 450. **d.** There are 27 factors of 148,225. These factors are: 1; 5; 7; 11; 25; 35; 49; 55; 77; 121; 175; 245; 275; 385; 539; 605; 847; 1225; 1925; 2695; 3025; 4235; 5929; 13,475; 21,175; 29,645; and 148,225.

Section 2.4: Least Common Multiple (LCM)

Margin Examples 1. 840 **2.** 126 **3.** 840 **4.** 140; $140 = 20 \cdot 7$; $140 = 35 \cdot 4$; $140 = 70 \cdot 2$ **5.** $350xy^2$ **6.** $60a^3b^3$ **7.** $210ab^2$
8. a. 30 minutes **b.** 5, 10, and 6 **9.** Set of divisors for 45: {1, 3, 5, 9, 15, 45}; Set of divisors for 60: {1, 2, 3, 4, 5, 6, 10, 12, 15, 20, 30, 60}; Set of divisors for 90: {1, 2, 3, 5, 6, 9, 10, 15, 18, 30, 45, 90}; GCD = 15 **10.** GCD = 26 **11.** GCD = 49 **12.** GCD = 30
13. GCD = 1 **14.** GCD = 1

Exercises 1. a. 5, 10, 15, 20, 25, 30, 35, 40, 45, 50, 55, 60 **b.** 6, 12, 18, 24, 30, 36, 42, 48, 54, 60, 66, 72 **c.** 15, 30, 45, 60, 75, 90, 105, 120, 135, 150, 165, 180 **3.** 105 **5.** 30 **7.** 66 **9.** 140 **11.** 150 **13.** 180 **15.** 196 **17.** 60 **19.** 484 **21.** 56 **23.** 120 **25.** 1,192,464 **27.** 7,485,696 **29. a.** LCM = 120 **b.** $120 = 6 \cdot 20 = 24 \cdot 5 = 30 \cdot 4$ **31. a.** LCM = 108 **b.** $108 = 12 \cdot 9 = 18 \cdot 6 = 27 \cdot 4$ **33. a.** LCM = 14,157 **b.** $14,157 = 99 \cdot 143 = 121 \cdot 117 = 143 \cdot 99$ **35. a.** LCM = 7840 **b.** $7840 = 40 \cdot 196 = 56 \cdot 140 = 160 \cdot 49 = 196 \cdot 40$ **37. a.** LCM = 194,040 **b.** $194,040 = 45 \cdot 4312 = 56 \cdot 3465 = 98 \cdot 1980 = 99 \cdot 1960$ **39.** $600xy^2z$ **41.** $112a^3bc$ **43.** $210a^2b^3$ **45.** $84ab^2c^2$ **47.** $180m^2np^2$ **49.** $108x^2y^2z^2$ **51.** $726ab^3$ **53.** $120y^4$ **55.** $1140c^4$ **57.** $4590a^2bc^2x^2y$ **59. a.** 60 minutes **b.** 4, 3, and 2 trips, respectively **61. a.** every 24 days **b.** every 24 days **63.** every 180 days **65.** 4 **67.** 17 **69.** 10 **71.** 3 **73.** 6 **75.** 2 **77.** 8 **79.** 4 **81.** 12 **83.** 25 **85.** 1 **87.** 1 **89.** 35 **91.** relatively prime **93.** relatively prime **95.** relatively prime **97.** not relatively prime; GCD is 2 **99.** relatively prime **101.** Answers will vary. The multiples of any number are that number and other numbers that are larger. Therefore, the LCM of a set of numbers must be the largest number of the set or some number larger than the largest number.

Section 2.5: Introduction to Fractions

Margin Examples 1. a. $\dfrac{5}{9}$ **b.** $\dfrac{10}{11}$ **c.** $\dfrac{54}{6} = 9$ **2.** $-\dfrac{9}{80} = \dfrac{-9}{80} = \dfrac{9}{-80}$ **3.** $\dfrac{16}{33}$ **4.** $\dfrac{15}{32}$ **5. a.** $\dfrac{12}{35}$ **b.** 0 **c.** $\dfrac{28}{5}$ **6.** $\dfrac{27}{4}$; commutative property of multiplication **7.** $\dfrac{1}{24}$; associative property of multiplication **8.** $\dfrac{30}{36}$ **9.** $\dfrac{18x}{42x}$ **10.** $\dfrac{3}{5}$

11. $\dfrac{3x}{2y}$ **12.** $-\dfrac{7}{15}$ **13.** $\dfrac{35}{46}$ **14.** $\dfrac{3}{7}$ **15.** $\dfrac{xy}{6}$ **16.** $\dfrac{4}{15}$ **17.** $\dfrac{45a}{224b}$ **18.** -9 **19.** $\dfrac{65}{4}$ **20.** $\dfrac{5}{7}; \dfrac{2}{7}$

Exercises 1. a. $\dfrac{1}{2}$ **b.** $\dfrac{1}{2}$ **3. a.** $\dfrac{2}{3}$ **b.** $\dfrac{1}{3}$ **5. a.** $\dfrac{4}{7}$ **b.** $\dfrac{3}{7}$ **7. a.** $\dfrac{33}{100}$ **b.** $\dfrac{67}{100}$ **9. a.** 0 **b.** 0 **c.** undefined **d.** undefined **11.** $\dfrac{3}{25}$ **13.** $\dfrac{4}{9}$ **15.** $-\dfrac{18}{35}$ **17.** $\dfrac{1}{4}$ **19.** $\dfrac{9}{16}$ **21.** $\dfrac{4}{27}$ **23.** 0 **25.** $-\dfrac{12}{5}$ **27.** $\dfrac{1}{150}$ **29.** $\dfrac{125}{512}$ **31.** commutative property of multiplication **33.** associative property of multiplication **35.** $\dfrac{3}{4} \cdot \dfrac{3}{3} = \dfrac{9}{12}$ **37.** $\dfrac{6}{7} \cdot \dfrac{2}{2} = \dfrac{12}{14}$ **39.** $\dfrac{3n}{-8} \cdot \dfrac{-5}{-5} = \dfrac{-15n}{40}$ **41.** $\dfrac{-5x}{13} \cdot \dfrac{3y}{3y} = \dfrac{-15xy}{39y}$ **43.** $\dfrac{4a^2}{17b} \cdot \dfrac{2a}{2a} = \dfrac{8a^3}{34ab}$ **45.** $\dfrac{-9x^2}{10y^2} \cdot \dfrac{10y}{10y} = \dfrac{-90x^2y}{100y^3}$ **47.** $\dfrac{4}{5}$ **49.** $\dfrac{22}{65}$ **51.** $-\dfrac{2}{3}$ **53.** $-\dfrac{27}{56x^2}$ **55.** $-\dfrac{2}{3x}$ **57.** $\dfrac{17x}{2}$ **59.** $\dfrac{6a}{b}$ **61.** $\dfrac{11y}{14}$ **63.** $-\dfrac{1}{6}$ **65.** $\dfrac{21}{8}$ **67.** $-\dfrac{273a}{80}$ **69.** $\dfrac{21x^2}{2}$ **71.** $-\dfrac{77}{4}$ **73.** $\dfrac{7}{10}; \dfrac{3}{10}$ **75.** 6 in. **77.** $\dfrac{4}{5}; 60$ miles **79.** 15

Section 2.6: Division with Fractions

Margin Examples 1. $\dfrac{18}{11}$ **2.** $\dfrac{7x}{2}$ **3.** $\dfrac{8}{9}$ **4.** $\dfrac{9}{70}$ **5.** 1 **6.** $\dfrac{4}{5}$ **7.** $\dfrac{5x^2}{38}$ **8.** $\dfrac{45}{16}$ **9.** $\dfrac{1}{6}$ **10. a.** yes **b.** less than 3000 **c.** 4500 pounds

Exercises 1. $\dfrac{25}{13}$ **3.** 0 **5.** $10 \div 2 = 5$, while $10 \div \dfrac{1}{2} = 10 \cdot 2 = 20$ **7.** Answers will vary. **9.** $\dfrac{25}{24}$ **11.** $\dfrac{14}{11}$ **13.** 1 **15.** $\dfrac{9}{16}$ **17.** $-\dfrac{1}{4}$ **19.** $\dfrac{1}{16}$ **21.** undefined **23.** 0 **25.** $-\dfrac{8}{5}$ **27.** $\dfrac{9}{20}$ **29.** $\dfrac{4}{3a}$ **31.** $\dfrac{4x^2}{3y^2}$ **33.** $98x^2$ **35.** $-300a$ **37.** $\dfrac{29}{155}$ **39.** $-\dfrac{3x}{8}$ **41. a.** $\dfrac{2}{5}$ **b.** $\dfrac{12}{25}$ **43. a.** larger **b.** 315 **45. a.** more **b.** less **c.** 75 **47.** $\dfrac{128}{3}$ **49. a.** 35 **b.** 21 **c.** yes **d.** pass by 3 votes **51. a.** 1500 **b.** 2500 **c.** 1000 **d.** 2500 **53. a.** more **b.** less **c.** 8000 **55. a.** more **b.** less **c.** 900

Section 2.7: Addition and Subtraction with Fractions

Margin Examples 1. $\dfrac{4}{7}$ **2.** $\dfrac{8}{9}$ **3.** $\dfrac{7}{15}$ **4.** $\dfrac{53}{40}$ **5.** $\dfrac{537}{100}$ **6.** $\dfrac{2}{15}$ **7.** $-\dfrac{6}{13}$ **8.** $-\dfrac{23}{38}$ **9.** $\dfrac{11}{20}$ **10.** $-\dfrac{31}{36}$ **11.** $\dfrac{14}{c}$ **12.** $\dfrac{10+3x}{5x}$ **13.** $\dfrac{4y-5}{4}$ **14.** $\dfrac{3n+7}{6}$ **15.** $\dfrac{6x-5}{6}$ **16.** $13,000

Exercises 1. a. $\dfrac{1}{9}$ **b.** $\dfrac{2}{3}$ **3. a.** $\dfrac{4}{49}$ **b.** $\dfrac{4}{7}$ **5. a.** $\dfrac{25}{16}$ **b.** $\dfrac{5}{2}$ **7. a.** $\dfrac{64}{81}$ **b.** $\dfrac{16}{9}$ **9. a.** $\dfrac{9}{25}$ **b.** $\dfrac{6}{5}$ **11.** $\dfrac{3}{7}$ **13.** $\dfrac{4}{3}$ **15.** $\dfrac{8}{25}$ **17.** $-\dfrac{1}{5}$ **19.** $\dfrac{1}{16}$ **21.** $\dfrac{17}{21}$ **23.** $\dfrac{1}{3}$ **25.** $-\dfrac{15}{14}$ **27.** 0 **29.** $-\dfrac{2}{25}$ **31.** $\dfrac{7}{12}$ **33.** $\dfrac{2}{135}$ **35.** $\dfrac{1}{3}$ **37.** $-\dfrac{63}{50}$ **39.** $\dfrac{2}{15}$ **41.** $\dfrac{3}{16}$ **43.** $\dfrac{3}{8}$ **45.** $\dfrac{19}{10}$ **47.** $\dfrac{21}{16}$ **49.** $\dfrac{5x+3}{15}$ **51.** $\dfrac{y-2}{6}$ **53.** $\dfrac{3+4x}{3x}$ **55.** $\dfrac{8a-7}{8}$ **57.** $\dfrac{10a+3}{100}$ **59.** $\dfrac{4x+13}{20}$ **61.** $\dfrac{3x+35}{5x}$

63. $\dfrac{2n-9}{3n}$ **65.** $\dfrac{n-4}{5}$ **67.** $\dfrac{35+4x}{5x}$ **69.** $\dfrac{16}{15}$ **71. a.** $\dfrac{13}{30}$ **b.** $1430 **73.** $\dfrac{19}{20}$ in. by $\dfrac{7}{10}$ in.

Section 2.8: Introduction to Mixed Numbers

Margin Examples 1. $\dfrac{37}{8}$ **2.** $\dfrac{39}{4}$ **3.** $\dfrac{54}{5}$ **4.** $\dfrac{65}{9}$ **5.** $\dfrac{101}{32}$ **6.** $1\dfrac{9}{11}$ **7.** $24\dfrac{5}{6}$ **8.** $\dfrac{46}{10} = \dfrac{\cancel{2} \cdot 23}{\cancel{2} \cdot 5} = \dfrac{23}{5} = 4\dfrac{3}{5}$; $\dfrac{46}{10} = 4\dfrac{6}{10} = 4\dfrac{\cancel{2} \cdot 3}{\cancel{2} \cdot 5} = 4\dfrac{3}{5}$

Exercises 1. $5+\dfrac{1}{2}$ **3.** $10+\dfrac{11}{15}$ **5.** $7+\dfrac{1}{6}$ **7.** $32+\dfrac{1}{16}$ **9.** $\dfrac{14}{5}$ **11.** $\dfrac{13}{6}$ **13.** $\dfrac{16}{9}$ **15.** $\dfrac{17}{6}$ **17.** $\dfrac{29}{8}$ **19.** $\dfrac{28}{5}$ **21.** $\dfrac{46}{3}$

23. $\dfrac{25}{2}$ **25.** $\dfrac{156}{5}$ **27.** $\dfrac{37}{5}$ **29.** $\dfrac{53}{10}$ **31.** $\dfrac{39}{2}$ **33.** $\dfrac{3931}{1000}$ **35.** $\dfrac{707}{100}$ **37.** $1\dfrac{1}{2}$ **39.** $7\dfrac{1}{2}$ **41.** $1\dfrac{1}{4}$ **43.** $1\dfrac{12}{17}$ **45.** 5 **47.** $1\dfrac{1}{4}$

49. 2 in. **51.** $\dfrac{1799}{97}$ **53.** $\dfrac{755}{37}$ **55.** $\dfrac{918}{47}$ **57.** $\dfrac{8957}{100}$ **59.** $\dfrac{248,345}{412}$

Section 2.9: Multiplication and Division with Mixed Numbers

Margin Examples 1. $\dfrac{11}{2} = 5\dfrac{1}{2}$ **2.** $\dfrac{77}{4} = 19\dfrac{1}{4}$ **3.** $-\dfrac{7}{3} = -2\dfrac{1}{3}$ **4.** $1\dfrac{1}{5}$ **5.** 16 **6.** 12 **7.** $-\dfrac{5}{2}$ or $-2\dfrac{1}{2}$ **8.** $\dfrac{1651}{18} = 91\dfrac{13}{18}$ square inches

9. $13\dfrac{1}{5}$ in.^{2} **10.** $1160\dfrac{1}{4}$ in.^{3} **11.** 9 **12.** $\dfrac{42}{19} = 2\dfrac{4}{19}$ **13.** $-\dfrac{59}{10} = -5\dfrac{9}{10}$

Exercises 1. $3\dfrac{1}{5}$ **3.** $13\dfrac{1}{3}$ **5.** 35 **7.** 24 **9.** $12\dfrac{1}{4}$ **11.** $32\dfrac{32}{35}$ **13.** $40\dfrac{1}{4}$ **15.** $-72\dfrac{9}{20}$ **17.** $-15\dfrac{99}{160}$ **19.** $1\dfrac{3}{10}$ **21.** $851\dfrac{1}{30}$

23. $-1291\dfrac{5}{42}$ **25.** $640\dfrac{1}{16}$ **27.** 60 **29.** 60 **31.** $1\dfrac{7}{8}$ **33.** 4 **35.** $3\dfrac{1}{7}$ **37.** $-1\dfrac{1}{2}$ **39.** $-1\dfrac{5}{6}$ **41.** $-\dfrac{9}{32}$ **43.** $2\dfrac{2}{7}$ **45. a.** $\dfrac{11}{16}$

b. above ground **c.** 20 ft **d.** 44 ft **47. a.** $\dfrac{2}{5}$ **b.** 180 pages **c.** 15 hours **49. a.** $20\dfrac{3}{4}$ m **b.** $26\dfrac{7}{16}$ m^2 **51.** 10 cm^2

53. $2907\dfrac{44}{125}$ cm^3 **55. a.** less **b.** $320 **57. a.** more **b.** 60 passengers **59. a.** 20 gallons **b.** $57\dfrac{1}{2}$ (or $57.50) **61. a.** The product will always be smaller than the other number if the number is a positive fraction. Answers will vary. **b.** The product will be smaller if it is multiplied by a positive whole number. Examples will vary. **c.** If the number is negative, the product will always be greater than the other number. Examples will vary.

Section 2.10: Addition and Subtraction with Mixed Numbers

Margin Examples 1. $11\dfrac{4}{5}$ **2.** $8\dfrac{7}{10}$ **3.** $29\dfrac{5}{12}$ **4.** $15\dfrac{3}{11}$ **5.** $4\dfrac{1}{5}$ **6.** $7\dfrac{3}{14}$ **7.** $6\dfrac{4}{8} = 6\dfrac{1}{2}$ **8.** $3\dfrac{2}{9}$ **9.** $3\dfrac{7}{8}$ **10.** $6\dfrac{7}{10}$ **11.** $-8\dfrac{41}{55}$

12. $-8\dfrac{25}{36}$ **13.** $\dfrac{331}{24} = 13\dfrac{19}{24}$ **14.** $\dfrac{49}{4} = 12\dfrac{1}{4}$ **15.** $-\dfrac{153}{65} = -2\dfrac{23}{65}$

Exercises 1. 10 **3.** $11\dfrac{3}{10}$ **5.** $10\dfrac{11}{12}$ **7.** $17\dfrac{9}{28}$ **9.** $21\dfrac{1}{16}$ **11.** $13\dfrac{3}{4}$ **13.** $5\dfrac{5}{8}$ **15.** $30\dfrac{37}{60}$ **17.** $8\dfrac{2}{5}$ **19.** $8\dfrac{1}{6}$ **21.** 2 **23.** $3\dfrac{3}{10}$

25. $7\dfrac{1}{6}$ **27.** $2\dfrac{16}{21}$ **29.** $20\dfrac{19}{24}$ **31.** $53\dfrac{41}{60}$ **33.** $15\dfrac{4}{7}$ **35.** $\dfrac{11}{16}$ **37.** $1\dfrac{2}{5}$ **39.** $7\dfrac{1}{4}$ **41.** $-12\dfrac{1}{4}$ **43.** $-5\dfrac{3}{8}$ **45.** $-6\dfrac{19}{24}$

47. $13\dfrac{13}{30}$ **49.** $-3\dfrac{8}{15}$ **51.** $-50\dfrac{29}{30}$ **53.** $-45\dfrac{3}{20}$ **55.** $5\dfrac{3}{4}$ ft **57.** $10\dfrac{13}{24}$ inches **59.** $29\dfrac{1}{10}$ cm

Section 2.11: Complex Fractions and Order of Operations

Margin Examples 1. $\dfrac{10}{9} = 1\dfrac{1}{9}$ **2.** 4 **3.** $-\dfrac{1}{10}$ **4.** $8\dfrac{25}{38}$ **5.** $\dfrac{3x+2}{3}$ **6.** $4\dfrac{31}{108}$

Exercises 1. A complex fraction is a fraction in which the numerator or denominator or both contain one or more fractions or mixed numbers. **3.** $\dfrac{3}{28}$ **5.** $\dfrac{3}{2}$ $\left(\text{or } 1\dfrac{1}{2}\right)$ **7.** $\dfrac{5}{3}$ $\left(\text{or } 1\dfrac{2}{3}\right)$ **9.** $\dfrac{26}{135}$ **11.** -2 **13.** $\dfrac{21}{22}$ **15.** $\dfrac{235}{4}$ $\left(\text{or } 58\dfrac{3}{4}\right)$ **17.** $24y$

19. $\dfrac{1}{6}$ **21.** $\dfrac{7}{60}$ **23.** $\dfrac{31}{28}$ $\left(\text{or } 1\dfrac{3}{28}\right)$ **25.** $\dfrac{379}{60}$ $\left(\text{or } 6\dfrac{19}{60}\right)$ **27.** $\dfrac{341}{30}$ $\left(\text{or } 11\dfrac{11}{30}\right)$ **29.** $-\dfrac{4}{21}$ **31.** $\dfrac{15x-13}{15}$ **33.** $\dfrac{42y+13}{42}$

35. $\dfrac{14a+9}{35}$ **37.** $\dfrac{3-2x}{7x}$ **39.** $4\dfrac{7}{16}$ **41.** $\dfrac{128}{135}$ **43.** $\dfrac{11}{400}$ **45. a.** $\dfrac{\dfrac{4}{5}+\dfrac{2}{15}}{2\dfrac{1}{4}-\dfrac{7}{8}}$ **b.** $\left(\dfrac{4}{5}+\dfrac{2}{15}\right) \div \left(2\dfrac{1}{4}-\dfrac{7}{8}\right)$ **47.** The square of any number between 0 and 1 is always smaller than the original number. Answers will vary.

Section 2.12: Solving Equations with Fractions

Margin Examples 1. $x = -20$ **2.** $x = 42$ **3.** $x = -1$ **4.** $x = 6$ **5. a.** $y = -1\frac{3}{5}$ **b.** $x = -3\frac{1}{2}$ **6.** $x = \frac{9}{10}$ **7.** $x = \frac{15}{26}$ **8.** $x = 16$

Exercises 1. Write the equation. Simplify. Divide both sides by 2. Simplify. **3.** Write the equation. Multiply each term by 10. Simplify. Add -2 to both sides. Simplify. Divide both sides by 5. Simplify. **5.** $x = 12$ **7.** $y = 55$ **9.** $x = 5$ **11.** $n = 7$ **13.** $x = -11$ **15.** $x = -6$ **17.** $y = -4$ **19.** $n = -\frac{14}{5}$ **21.** $y = -\frac{1}{2}$ **23.** $x = \frac{1}{3}$ **25.** $x = -\frac{25}{2}$ **27.** $x = 5$ **29.** $x = -\frac{2}{3}$ **31.** $y = 39$ **33.** $x = -210$ **35.** $x = \frac{70}{3}$ **37.** $y = -\frac{10}{3}$ **39.** $y = \frac{14}{15}$ **41.** $x = -2$ **43.** $n = 2$ **45.** $x = \frac{55}{8}$ **47.** $x = 2$ **49.** $n = -\frac{3}{2}$ **51.** $x = -\frac{1}{10}$ **53.** $n = -\frac{8}{9}$ **55.** $y = -10$ **57.** $x = 64$ **59.** $y = -45$ **61.** $x = -\frac{3}{4}$ **63.** $n = -\frac{5}{6}$ **65.** $x = 81$

Section 2.13: Ratios and Proportions

Margin Examples 1. 4 apples : 5 oranges or 4 apples to 5 oranges or $\frac{4 \text{ apples}}{5 \text{ oranges}}$ **2.** $\frac{3 \text{ quarters}}{1 \text{ dollar}} = \frac{3 \text{ quarters}}{4 \text{ quarters}} = \frac{3}{4}$ **3.** $\frac{500 \text{ washers}}{4000 \text{ bolts}} = \frac{1 \text{ washer}}{8 \text{ bolts}}$ **4.** $\frac{36 \text{ inches}}{5 \text{ feet}} = \frac{3 \text{ feet}}{5 \text{ feet}} = \frac{3}{5}$ **5.** 3-liter bottle for $1.49 **6.** 16-ounce box for $3.00; 20¢/oz (12.5-ounce box); 18.8¢/ounce (16-ounce box) **7.** 1.3 and 2.1 are the extremes. 1.5 and 1.82 are the means. **8.** $5\frac{1}{2}$ and 3 are the extremes. 2 and $8\frac{1}{4}$ are the means. **9.** true **10.** true **11.** false **12.** true **13.** $x = 20$ **14.** $R = 30$ **15.** $x = 60$ **16.** $x = 0.8$ **17.** $z = 3\frac{3}{5}$ **18.** $z = \frac{3}{8}$

Exercises 1. $\frac{1}{2}$ **3.** $\frac{4}{1}$ or 4 **5.** $\frac{50 \text{ miles}}{1 \text{ hour}}$ **7.** $\frac{3}{4}$ **9.** $\frac{8}{7}$ **11.** $\frac{\$2 \text{ profit}}{\$5 \text{ invested}}$ **13.** $\frac{7}{25}$ **15.** $\frac{9}{41}$ **17.** 69.8¢/lb; 58.9¢/lb; 10 lb at $5.89 **19.** 113.7¢/oz; 66.6¢/oz; 12 oz at $7.99 **21.** 16.8¢/oz; 15.3¢/oz; 32 oz at $4.89 **23.** 29.1¢/oz; 33.3¢/oz; 27.5¢/oz; 40 oz at $10.99 **25.** 3.7¢/sq ft.; 4.9¢/sq ft.; 6.4¢/sq ft.; 6¢/sq ft.; 200 sq ft. at $7.39 **27. a.** 7 and 544 **b.** 8 and 476 **29.** true **31.** true **33.** false **35.** true **37.** true **39.** true **41.** true **43.** true **45.** true **47.** true **49.** $x = 12$ **51.** $x = 20$ **53.** $B = 40$ **55.** $x = 50$ **57.** $A = \frac{21}{2}$ **59.** $D = 100$ **61.** $x = 1$ **63.** $x = \frac{3}{5}$ **65.** $w = 24$ **67.** $y = 6$ **69.** $x = 60$ **71.** $B = 7.8$ **73.** $x = 1.56$ **75.** $A = 27.3$ **77.** b. **79.** c. **81.** c.

Chapter 2 Test

1. 2, 3, 6, 9 **2.** 2, 5, 10 **3.** none **4.** 11, 22, 33, 44, 55, 66, 77, 88, 99 **5.** 2 **6.** 2, 11, 13, 37 **7. a.** False; the number 5 is not a factor twice of $5 \cdot 4 \cdot 3 \cdot 2 \cdot 1$. **b.** True; $2 \cdot 5 \cdot 6 \cdot 7 \cdot 9 = 2 \cdot 5 \cdot 2 \cdot 3 \cdot 7 \cdot 9 = 27 \cdot 140$. **8.** $2^4 \cdot 5$ **9.** $2^2 \cdot 3^2 \cdot 5$ **10.** $3^2 \cdot 5^2$ **11. a.** $2 \cdot 3^2 \cdot 5$ **b.** 1, 2, 3, 5, 6, 9, 10, 15, 18, 30, 45, 90 **12.** 8 **13.** 15 **14.** 6 **15.** 840 **16.** $360x^3y^2$ **17.** $168a^4$ **18. a.** 378 seconds **b.** 7 laps and 6 laps, respectively **c.** No, they will not meet at another point because the faster runner does not pass the slower until the faster has completed 7 laps. **19.** integers, zero. **20.** any two examples of the form $\frac{a}{b} \cdot \frac{c}{d} = \frac{c}{d} \cdot \frac{a}{b}$ **21.** 0, undefined. **22.** $\frac{6}{5}$ **23.** $-\frac{2}{9x}$ **24.** $\frac{5a^2}{4b}$ **25.** $\frac{13}{16}$ **26.** $\frac{x^2}{2y}$ **27.** $405x^2$ **28.** $-\frac{12a}{35b}$ **29.** $\frac{4}{5a}$ **30.** $-\frac{13}{36}$ **31.** $\frac{5}{12}$ **32.** $\frac{25}{24}$ **33.** 75 gigabytes **34.** $\frac{3}{7}$ **35.** $6\frac{3}{4}$ **36.** $3\frac{21}{50}$ **37.** $-\frac{501}{100}$ **38.** $\frac{38}{13}$ **39.** $2\frac{1}{6}$ **40.** $4\frac{47}{180}$ **41.** $7\frac{11}{42}$ **42.** $14\frac{19}{30}$ **43.** -16 **44.** $11\frac{3}{5}$ **45.** $\frac{x-1}{9}$ **46.** $-\frac{1}{3}$ **47.** $x = 3$ **48.** $n = \frac{2}{15}$ **49.** $y = \frac{42}{19}$ **50.** $x = -\frac{9}{2}$ **51. a.** $27\frac{1}{2}$ in. **b.** $46\frac{7}{8}$ in.² **52. a.** larger **b.** $19\frac{1}{2}$ **53.** $1633\frac{73}{200}$ cm³ **54.** $\frac{7}{6}$ **55.** extremes. **56.** false; $4 \cdot 14 \neq 6 \cdot 9$ **57.** $\frac{3}{5}$ **58.** $\frac{2}{5}$ **59.** $\frac{55 \text{ miles}}{1 \text{ hour}}$ **60.** $2.50 / pair; $2.75 / pair; 5 pairs at $12.50 **61.** $x = 27$ **62.** $y = \frac{50}{3}$ **63.** $x = 75$ **64.** $y = 19.5$ **65.** $x = \frac{5}{6}$ **66.** $x = \frac{1}{90}$

Chapters 1-2 Cumulative Review

1. a. 5 **b.** 3 **c.** 125 **2. a.** B **b.** A **c.** D **d.** C **3.** 630,000 **4.** 180,000 **5.** -125 **6. a.** 2, 3, 5, 7, 11, 13, 17, 19, 23 **b.** 4, 9, 25, 49, 121, 169, 289, 361, 529 **7.** 1575 **8.** 30 **9.** 0, undefined. **10.** 44 **11.** 14,275 **12.** 588 **13.** 28,538 **14.** 204 **15.** $-\frac{1}{40}$ **16.** $\frac{1}{63}$ **17.** $\frac{1}{24}$ **18.** $-\frac{1}{4}$ **19.** $68\frac{23}{60}$ **20.** $-6\frac{21}{40}$ **21.** $-\frac{121}{81}$ or $-1\frac{40}{81}$ **22.** $\frac{7}{4}$ or $1\frac{3}{4}$ **23.** $\frac{123}{38}$ or $3\frac{9}{38}$

24. $-\dfrac{47}{6}$ or $-7\dfrac{5}{6}$ **25.** $\dfrac{x-2}{15}$ **26.** $x = -27$ **27.** $y = 7$ **28.** $x = -\dfrac{1}{2}$ **29.** $n = 13$ **30.** $x = 8$ **31.** $y = -\dfrac{31}{3}$ **32.** $x = -\dfrac{9}{10}$
33. $n = 14$ **34.** $x = 1$ **35.** $y = -\dfrac{1}{8}$ **36.** $x = 1$ **37.** $x = -\dfrac{3}{22}$ **38. a.** more **b.** $815 **39. a.** 15 gallons **b.** yes
40. $16\dfrac{3}{40}$ m **41. a.** $116\dfrac{1}{2}$ ft **b.** $807\dfrac{5}{8}$ ft^2 **42.** $x = \dfrac{5}{2}$ **43.** $A = 0.32$

Chapter 3: Decimal Numbers, Percents, and Square Roots

Section 3.1: Reading, Writing, and Rounding Decimal Numbers

Margin Examples **1.** 52.42; fifty-two and forty-two hundredths **2.** 12.003; twelve and three thousandths **3.** 10.004
4. 600.5 **5.** 700.007 **6.** 8.6 **7.** 5.04 **8.** 0.018 **9.** 240

Exercises **1.** 6.5 **3.** 19.075 **5.** 62.547 **7.** $13\dfrac{2}{100}$ **9.** $200\dfrac{6}{10}$ **11.** 0.4 **13.** 0.23 **15.** 5.028 **17.** 600.66 **19.** 3495.342
21. nine tenths **23.** six and five hundredths **25.** fifty and seven thousandths **27.** eight hundred and nine thousandths
29. five thousand and five thousandths **31. a.** 7 **b.** 8 **c.** 8; 7; 8; 8 **d.** 34.8 **33.** 89.0 **35.** 18.1 **37.** 14.3 **39.** 0.39 **41.** 8.00
43. 0.08 **45.** 0.057 **47.** 0.002 **49.** 32.458 **51.** 479 **53.** 164 **55.** 300 **57.** 5200 **59.** 76,500 **61.** 500 **63.** 62,000
65. 103,000 **67.** 7,305,000 **69.** 0.00076 **71.** nine hundred fourteen thousandths; one and nine hundredths; thirty-
nine and thirty-seven hundredths; three and thirty-seven hundredths **73.** two and eight hundred twenty-five ten-
thousandths **75.** three and fourteen thousand, one hundred fifty-nine hundred-thousandths **77.** thirty-five and eight
tenths; twenty-six and nine tenths; eighteen and nine tenths; twelve and three tenths; seven and two tenths **79.** one
hundred fourteen and eight tenths

Section 3.2: Addition and Subtraction with Decimal Numbers

Margin Examples **1.** 59.804 **2.** 50.78x **3.** 157.9 **4.** $8.68 **5.** −35.18 **6.** $x = -40.05$ **7.** −14; −13.04 **8.** 960; 1202.59
9. $6.09; $3.91 **10. a.** 4564.479 **b.** 188.0059
Exercises **1.** 3.3 **3.** 5.65 **5.** 4.7925 **7.** 117.385 **9.** 718.15 **11.** 1.55 **13.** 15.89 **15.** 54.946 **17.** 4.888 **19.** 30.464 **21.** −13.4x
23. 45.5y **25.** −0.7x **27.** 6.3t **29.** 10.3x − 4.1y **31.** $x = -107.3$ **33.** $y = -1.7$ **35.** $z = 18.39$ **37.** $w = 3.84$ **39.** $6950
41. 12 in. **43. a.** 152.274 million **b.** 32.085 million **45.** 15.85 in.; 64.9 in. **47.** 74.96 **49.** 9.9209 **51.** 89.9484 **53.** 0.660932
55. 1406.8 **57.** 510.989 **59.** −34.35

Section 3.3: Multiplication and Division with Decimal Numbers

Margin Examples **1.** 297.28916 **2.** −0.046 **3. a.** 83.6 **b.** −97.35 **c.** 14,820 **4.** 1872 **5.** 1.14 **6.** 4.95 **7.** 14.9 **8. a.** 1.6
b. 0.08346 **c.** 0.732 **9. a.** 2.8 **b.** 3.045 **10.** estimate: 4.0; quotient: 4.3 **11.** You can save $200. **12.** about 3 hits per game
13. The boat travles 27.54 miles. **14.** The hot dog costs $2.39. **15. a.** 22.7176 **b.** 5.1391
Exercises **1. a.** 0.72 **b.** 3.0 **c.** 6.0 **d.** 0.06 **e.** 5.0 **3. a.** 2 **b.** 2000 **c.** 75 **d.** 5 **e.** 0.02 **5.** 0.35 **7.** 10.8 **9.** 0.0004 **11.** −0.112 **13.** −1
15. 0.00429 **17.** 2.036 **19.** 0.002028 **21.** 1.632204 **23.** 3.24 **25.** −0.79 **27.** −0.006 **29.** 0.7 **31.** 21.3 **33.** 5.04 **35.** 2.290
37. 209.167 **39.** 376 **41.** 950 **43.** 730,000 **45.** 0.026483 **47.** 0.00178 **49. a.** 53.6 in. **b.** 179.56 in.^{2} **51.** $310.10
53. a. about 15 mpg **b.** 22 mpg **55.** $825 **57.** about 6 hrs **59.** 4.4 yards per carry **61.** −7711.7724 **63.** 11,052.2782
65. −28.8083 **67.** −233.1976 **69. a.** 25.994 mph **b.** 25.971 mph

Section 3.4: Measures of Center

Margin Examples **1.** 102 minutes **2.** 87 **3.** mode: 77; range: 31 **4.** 97
Exercises **1. a.** 80 **b.** 82 **c.** 85 **d.** 36 **3. a.** $48,625 **b.** $46,500 **c.** $63,000 **d.** $43,000 **5. a.** $16,480.60 **b.** $15,618.50 **c.** none
d. $11,502 **7. a.** 1135 miles **b.** 980 miles **c.** none **d.** 2020 miles **9. a.** 8,116,780 m^3 **b.** 296,200 m^3 **c.** none **d.** 39,090,500
m^3 **11. a.** 15.425 **b.** 14.5 **c.** 17 and 3 **d.** 44 **13.** 79 **15. a.** 3.48 farms **b.** 2.23 farms **17. a.** 806,814.3 **b.** 410,635.5 **19. a.** 3.3
b. 2.4 **c.** Answers will vary.

Section 3.5: Decimal Numbers, Fractions, and Scientific Notation

Margin Examples **1. a.** $\frac{7}{20}$ **b.** $\frac{1}{8}$ **c.** $\frac{12}{5}$ or $2\frac{2}{5}$ **2.** 0.2 **3.** –0.5625 **4.** 0.2222... **5.** 0.14285714285714...

6. decimal form: 12.96; fraction form: $\frac{324}{25}$ **7.** –0.375 **8.** fraction form: $\frac{3}{40}$; decimal form: 0.075 **9.** –18.75 **10. a.** $\frac{8}{9}$

b. 0.007 **11. a.** 6.39×10^7 **b.** 2.45×10^{-6}

Exercises **1.** $\frac{7}{10}$ **3.** $\frac{5}{10}$ **5.** $\frac{16}{1000}$ **7.** $\frac{835}{100}$ **9.** $\frac{1}{8}$ **11.** $1\frac{4}{5}$ **13.** $-2\frac{3}{4}$ **15.** $\frac{33}{100}$ **17.** $0.\overline{6}$ **19.** $-0.\overline{142857}$ **21.** 0.6875

23. $0.\overline{63}$ **25.** 0.208 **27.** 0.917 **29.** 5.714 **31.** –2.273 **33.** 1.165 **35.** 1.796 **37.** 71.22 **39.** –2.875 **41.** –0.625 **43.** 0.0567

45. –10.395 **47.** 120.31 **49.** –17.55 **51.** –1.7 **53.** $2\frac{3}{4}$ **55.** 0 **57.** $-\frac{5}{9}$ **59.** $15\frac{5}{8}$ **61.** 1.8×10^5 **63.** 1.24×10^{-4}

65. 8.9×10^2 **67.** 5.88×10^{12} **69.** 40,000,000,000,000,000,000,000 **71.** 57,638,370 gallons; 5.763837×10^7 gallons

73. 11,552,200,000,000,000; 154,400,000,000,000; 99,382,100,000,000,000 **75. a.** Examples will vary. **b.** Examples will vary.

c. $|x + y| < |x| + |y|$ when x and y are opposite in sign. **d.** $|x + y| = |x| + |y|$ when x and y have the same sign or one or both are zero. **77.** Explanations will vary. Digits are compared, place by place, until one digit is larger (or smaller).

Section 3.6: Basics of Percent

Margin Examples **1. a.** 89% **b.** 4.2% **c.** 35.75% **2.** Investment **b.** is better than investment **a.** because 40% was earned in investment **b.**, whereas only 35% was earned on investment **a.** **3.** Investment **b.** is better than investment **a.** because 5.5% was earned in investment **b.**, whereas only 5% was earned on investment **a.** **4. a.** 35% **b.** 217% **c.** 87.1% **d.** 10%

5. a. 0.213 **b.** 0.006 **c.** 1.75 **d.** 0.85 **6.** 60% **7.** 54% **8.** 280% **9.** 44.4% or $44\frac{4}{9}$% **10. a.** 18.52% in 1900; 57.69% in 2003

b. 4714.81% **11. a.** $\frac{8}{25}$ **b.** $\frac{67}{20}$ or $3\frac{7}{20}$ **c.** $\frac{181}{200}$ **12.** $\frac{317}{500}$

Exercises **1.** 66% **3.** 20% **5.** 99% **7.** 30% **9.** 48% **11.** 16.3% **13.** $24\frac{1}{2}$% **15. a.** $\frac{19}{100} = 19\%$ **b.** $\frac{17}{100} = 17\%$

c. Investment **a.** is better. **17. a.** $\frac{5}{100} = 5\%$ **b.** $\frac{4}{100} = 4\%$ **c.** Investment **a.** is better. **19.** 3% **21.** 300% **23.** 5.5%

25. 175% **27.** 108% **29.** 36% **31.** 0.02 **33.** 0.22 **35.** 0.25 **37.** 0.101 **39.** 0.065 **41.** 0.8 **43.** 5% **45.** 96%

47. $14\frac{2}{7}$% or 14.3% **49.** $137\frac{1}{2}$% or 137.5% **51.** $\frac{1}{20}$ **53.** $\frac{1}{8}$ **55.** $1\frac{1}{5}$ **57.** $\frac{1}{500}$ **59. b.** 0.875 **c.** 87.5% **61. a.** $\frac{3}{50}$

c. 6% **63. a.** $\frac{6}{25}$ **b.** 0.24 **65.** 0.15 **67.** 32% **69.** $\frac{7}{20}$ **71. a.** 10% **b.** 14% **c.** 11% **d.** 12% **73.** 0.2071 **75.** Answers will depend on the student's calculator.

Section 3.7: Applications of Percent

Margin Examples **1. a.** $264 **b.** $616 **2.** $643.72 **3.** $61 **4.** $1830 **5. a.** $9 **b.** 150% **c.** 60%

Exercises **1.** $7500 **3.** $770 **5.** 1.5% **7.** 136 **9.** 40 **11. a.** $16.88 **b.** $5.63 **13.** $37,500 **15. a.** $180 **b.** 40%

c. $28\frac{4}{7}$% (or 28.6%) **17.** 30%; 32%; 12% **19. a.** 86% **b.** 300 **c.** 42 **21. a.** $2550 **b.** $2703 **23. a.** $875 **b.** $700 **c.** $750.75

25. a. $156 **b.** $5460 **27. a.** He loses 10% of his original weight. **b.** He gains back $11\frac{1}{9}$% of his weight, because the gain of 20 pounds is based on 180 pounds, not the original weight of 200 pounds.

Section 3.8: Simple and Compound Interest

Margin Examples **1.** $525 **2.** $150 **3.** $6800 **4.** 16% **5.** $244.83; $4244.83 **6.** $3590.44 **7.** $590.44 **8. a.** $15,529.69

b. $5529.69 **9.** $38,900.38 **10.** $3240.71

Exercises **1.** $37.50 **3.** $275 **5.** $48 **7.** 2.4 months **9.** 10% **11. a.** $470.25 **b.** $29.75 **c.** to collect interest at 18% **13. a.** 16

b. 100 **c.** 1 month **d.** 0.94% **15. a.** $7803 **b.** $8118.24 **17. a.** $1200 **b.** $2346.64 **19. a.** $450.77 **b.** $459.23 **21. a.** $26,620

b. $26,801.91; no **c.** Explanations will vary. By compounding semiannually, interest is posted twice as frequently, so more interest is earned on interest accrued. **23.** $24,492.29 **25. a.** $73,424.76 **b.** $3313.07 **c.** $170.13 **d.** $2.24

27. $1,078,673.79 **29. a.** $1638.62 **b.** $638.62 **31. a.** $1628.89 **b.** $628.89 **33. a.** $120,025.52 **b.** $95,025.52 **35.** $1379.17

37. $46,729.50 **39.** $12,997.50 **41.** Answers will vary. The time for doubling depends on the rate of interest and the period of compounding and is independent of the principal. **43.** Responses will vary. $P = \$12,049.37$

Section 3.9: Square Roots

Margin Examples **1. a.** 6 **b.** 18 **c.** 14 **2. a.** 64 **b.** 10 **3.** 2.236 **4.** 2.739

Exercises **1.** yes, $121 = 11^2$ **3.** not a perfect square **5.** yes, $400 = 20^2$ **7.** yes, $225 = 15^2$ **9.** not a perfect square **11.** 7

13. 36 **15.** 16 **17.** 21 **19.** 206 **21. a.** Answers will vary. $\sqrt{39} > \sqrt{36} = 6$, so $\sqrt{39} > 6$ **b.** Answers will vary. $\sqrt{39} < \sqrt{49} = 7$, so $\sqrt{39} < 7$ **23.** $(1.4142)^2 = 1.99996164 < 2$ and $(1.4143)^2 = 2.00024449 > 2$. So, $1.4142 < \sqrt{2} < 1.4143$. **25. a.** 4 and 5 **b.** 4.7958 **27. a.** 3 and 4 **b.** 3.6056 **29. a.** 8 and 9 **b.** 8.4853 **31. a.** 7 and 8 **b.** 7.7460 **33. a.** 9 and 10 **b.** 9.7468 **35.** −0.4142 **37.** 6.4641 **39.** 5.8990 **41.** −1 **43.** 1, 8, 27, 64, 125, 216, 343, 512, 729, 1000

Chapter 3 Test

1. a. thirty-two and sixty-four thousandths **b.** $32\frac{8}{125}$ **2.** 2.032 **3. a.** 216.7 **b.** 216.70 **c.** 216.705 **d.** 220 **4. a.** 73.0 **b.** 73.01 **c.** 73.015 **d.** 70 **5.** 122.863 **6.** −126.49 **7.** 205.27 **8.** 50.716 **9.** −5.3075 **10.** 0.00006 **11.** 0.08217 **12.** 0.2182 **13.** 82,170 **14.** 10.80667 **15.** 25.81 **16.** 16.49 **17. a.** 25.4 **b.** 25 **c.** 25 **d.** 14 **18. a.** 6.7×10^7 **b.** 0.000083 **19. a.** 14.12 **b.** −1.9875 or $-1\frac{79}{80}$ **20. a.** 121 **b.** 5 **21. a.** 17.3205 **b.** 4.4641 **22. a.** 359 mi **b.** 24 gal **23.** 102% **24.** 7.25% **25.** 0.5% **26.** 15% **27.** 37.5% **28.** 0.24 **29.** 0.068 **30.** 0.055 **31.** $1\frac{3}{5}$ **32.** $\frac{12}{125}$ **33.** $\frac{5}{8}$ **34.** $5400 **35. a.** more **b.** 97% **c.** $2500 **d.** $75 **36. a.** 30% **b.** $23\frac{1}{13}\%$ (or 23.1%) **37.** 9% **38. a.** $13,828.17 **b.** $3828.17 **39. a.** $72.76 **b.** $103.95 **c.** $207,892.82

Chapters 1-3 Cumulative Review

1. identity; a **2.** In general, $a - b \neq b - a$. Any counterexample will also serve as an explanation. **3. a.** 9.9802; 10.0 **b.** 10.08; 10.1 **c.** Reducing precision before completing an arithmetic operation can lead to rounding errors. **4. a.** 90 (estimate) **b.** 90.93 **5.** −16 **6.** 58 **7.** 24 **8.** −27 **9.** 57 **10.** 27 **11. a.** $5\cdot 13$ **b.** $2^4 \cdot 3^2$ **c.** $2 \cdot 3^3 \cdot 5$ **12.** 6, 12, 18, 24, 30, 36, 42, 48, 54, 60, 66, 72, 78, 84, 90, 96 **13. a.** 560 **b.** $120a^2b^3$ **14. a.** yes **b.** 63 **15.** $A = 1.5625$ **16. a.** $\frac{3}{200}$ **b.** $6\frac{8}{25}$ **c.** $-1\frac{1}{20}$ **17.** $\frac{1}{5}$ **18.** $\frac{2x+1}{4}$ **19.** −6.55 **20.** $\frac{13}{15}$ **21.** $a = \frac{11}{5}$ or $a = 2.2$ **22.** $x = -\frac{7}{5}$ **23. a.** 8.65×10^5 **b.** 0.00341 **24. a.** 330 yd **b.** 6500 yd^2 **25. a.** More, because $5^2 = 25 < 29$ **b.** 5.385 **26. a.** $\frac{11}{16}$ **b.** $8\frac{1}{4}$ ft **27. a.** More than $1500 **b.** $2250 **28.** $143.75 **29.** $1716.59 **30. a.** 30 **b.** 6 months **c.** 20% **31. a.** $28\frac{4}{7}\%$ (or 28.6%) **b.** 25% **c.** 20% **32.** 27 min **33.** $38.92 **34. a.** 12.86 **b.** 16.25 **c.** 9.96 **d.** 11.74 **e.** 10.70 **f.** 16.79 **35.** He should pay cash since the installment plan will cost him $5760 extra. **36. a.** 30,294 gallons **b.** 30,668 gallons **c.** none **d.** 31,416 gallons **37. a.** 420 cm^3 **b.** 394 cm^2

Chapter 4: Geometry and Graphs

Section 4.1: Metric System: Weight and Volume

Margin Examples 1. a. 3500 **b.** 64 **c.** 1.23 **d.** 6000 **2. a.** 0.0059 **b.** 0.320 **c.** 65 **d.** 7 **3.** 0.65 dam **4.** 4300 centimeters **5.** 100 **6. a.** 2300 **b.** 8600 **c.** 6 **7. a.** 52 **b.** 3.60 **c.** 0.42 **8.** 4 950 000 hectares **9. a.** 121,000 g **b.** 3.5 t **c.** 4.576 t **d.** 6.7 g **10. a.** 43,000 g **b.** 0.250 t **c.** 23,000 mg **11. a.** 9000 **b.** 7500 **c.** 250 000 **d.** 35 000 **12. a.** 0.000 54 **b.** 0.005 982 **13. a.** 0.750 cubic centimeters **b.** 19 000 cubic centimeters **c.** 1 600 000 cubic centimeters **d.** 63 700 cubic decimeters **14. a.** 0.000 006 3 cubic meters **b.** 192 000 cubic centimeters **15. a.** 0.0937 hectoliters **b.** 0.353 liters **c.** 12 000 liters **16. a.** 1.952 liters **b.** 124 cubic centimeters **c.** 19.75 cubic meters **17. a.** 4,300,000,000 hertz **b.** 2.125 megatons

Exercises 1. 300 **3.** 1500 **5.** 450 **7.** 13 600 **9.** 182.5 **11.** 0.48 **13.** 0.065 **15.** 1.3 **17.** 0.0525 **19.** 5.5 **21.** 0.245 m **23.** 10 km **25.** 200 m **27.** 0.32 cm **29.** 17 350 mm **31.** 1300 **33.** 960 **35.** 5 **37.** 1400 **39.** 5000 **41.** 1300 cm^2 = 130 000 mm^2 **43.** 1150 dm^2 = 115 000 cm^2 = 11 500 000 mm^2 **45.** 4 dm^2 = 400 cm^2 = 40 000 mm^2 **47.** 670 **49.** 20 000 **51.** 0.02 **53.** 1.5 **55.** 75 000 **57.** 27 a = 2700 m^2 **59.** 350 000 a = 3500 ha **61.** 2000 **63.** 3.7 **65.** 5.6 **67.** 0.091 **69.** 700 **71.** 5000 kg **73.** 2000 kg **75.** 896 000 mg **77.** 75 kg **79.** 7 000 000 g **81.** 0.000 34 kg **83.** 0.016 g **85.** 0.0923 **87.** 7580 **89.** 2.963 **91.** 1000; 1000; 1000; 1 000 000 000 **93.** 900 **95.** 400 000 000 **97.** 0.063 **99.** 3100 mm^3 **101.** 5000 mm^3 **103.** 0.0764 L **105.** 30 mL **107.** 0.0053 L **109.** 72 **111.** 0.569 **113.** 7300; 7300 **115.** 150.3 **117.** 5,000,000 t **119.** 2,000,000 more pixels

Section 4.2: US Customary and Metric Systems

Margin Examples 1. a. 32° F **b.** 10° C **2.** 25° C **3.** 113° F **4.** 2.134 m **5.** 91.4 m **6.** 29.53 in. **7.** 1640 ft **8.** 341.85 cm^2 **9.** 6.48 ha **10.** 7.41 acres **11.** 409.032 ft^2 **12.** 15.14 L **13.** 16.91 qt **14.** 26.488 L **15.** 428.57 ft^3 **16.** 85.05 g **17.** 1.47 oz

Exercises 1. 77° F **3.** 122° F **5.** 10° C **7.** 0° C **9.** 2.74 m **11.** 96.6 km **13.** 124 mi **15.** 19.7 in. **17.** 19.35 cm^2 **19.** 55.8 m^2 **21.** 83.6 m^2 **23.** 405 ha **25.** 741 acres **27.** 53.82 ft^2 **29.** 4.65 in.^{2} **31.** 9.46 L **33.** 10.57 qt **35.** 11.09 gal **37.** 4.54 kg **39.** 453.6 g **41.** 3038.31 m^2 **43. a.** 11,145.6 cm^2 **b.** 1.116 m^2

Section 4.3: Introduction to Geometry

Margin Examples 1. 180° **2. a.** right **b.** obtuse **3.** ∠2 and ∠3 are complementary, ∠AOB and ∠1 are supplementary, ∠AOC and ∠COD are also supplementary, and ∠AOD is a straight angle. **4. a.** 15° **b.** 90° **c.** Yes, ∠TOU and ∠UOV.
5. a. ∠COD, ∠COA, ∠BOA, ∠BOE, ∠COE or ∠BOD. **b.** 120°, 120° **6.** scalene **7.** yes **8. a.** 22° **b.** obtuse **c.** $\overline{BR}$
d. $\overline{OB}$ and $\overline{OR}$ **e.** No, none of the angles are right angles.

Exercises 1. a. m∠2 = 75° **b.** m∠2 = 87° **c.** m∠2 = 45° **d.** m∠2 = 15° **3. a.** a right angle **b.** an acute angle
c. an acute angle **d.** 45° **5.** 25° **7.** 40° **9.** 55° **11.** 110° **13.** m∠A = 60°; m∠B = 54°; m∠C = 66°
15. m∠L = 117°; m∠M = 104°; m∠N = 111°; m∠O = 100°; m∠P = 108° **17. a.** straight **b.** right **c.** acute **d.** obtuse
19. a. m∠2 = 150° **b.** yes; ∠2 and ∠3 are supplementary **c.** ∠1 and ∠2; ∠1 and ∠4; ∠2 and ∠3; ∠3 and ∠4
21. m∠2=150°; m∠3 = 30°; m∠4=150° **23.** m∠1 = m∠4 = 140°; m∠4 = m∠6 = 140°; m∠1 = m∠6 = 140°;
m∠3 = m∠5 = 40°; m∠2 = m∠5 = 40° **25.** equilateral **27.** obtuse **29.** isosceles **31.** acute **33.** Yes, since 25 < 12 + 15.
35. a. m∠Z = 80° **b.** acute **c.** $\overline{YZ}$ **d.** $\overline{XZ}$ and $\overline{XY}$ **e.** no; no ∠ = 90°

Section 4.4: Length and Perimeter

Margin Exercises 1. 16 in.; 40.64 cm **2.** 36 mm; 3.6 cm **3.** 14.444 m; 47.376 ft **4.** 64.25 cm; 0.6425 m
Exercises 1. a. B **b.** C **c.** D **d.** A **e.** F **f.** E **3.** P = 104 mm, P = 4.098 in. or 4.095 in. **5.** C = 18.84 yd, C = 17.220 m
7. P = 107.1 mm, P = 10.71 cm, P = 4.21974 in. **9.** P = 27.42 mm, P = 2.742 cm, P = 1.080348 in. **11.** P = 254 mm, P = 25.4 cm,
P = 10 in. **13.** P = 600 mm, P = 60 cm, P = 23.62 in. **15. a.** 4 **b.** Answers will vary somewhat, but they should be about 3.3.
c. Answers will vary. The numbers will become closer to pi as the number of sides increases.

Section 4.5: Area

Margin Exercises 1. 6900 cm²; 1069.5 in.² **2.** 938.86 mm²; 9.3886 cm² **3.** 794.8125 ft²; 73.9176 m²
Exercises 1. a. C **b.** B **c.** D **d.** A **e.** F **f.** E **3.** A = 78.5 yd², A = 65.626 m² **5.** A = 38.5 mm², A = 0.385 cm² **7.** A = 706.5 mm²,
A = 7.065 cm², A = 1.095 in.² **9.** A = 36 842.4 mm², A = 368.424 cm², A = 57.12 in.² **11.** A = 10 600 mm², A = 106 cm²,
A = 16.43 in.² **13.** A = 50.13 mm², A = 0.5013 cm², A = 0.077 701 5 in.² **15.** A = 304 440 mm², A = 3044.4 cm², A = 472 in.²

Section 4.6: Volume

Margin Exercises 1. 220 mL; 220 cm³ **2.** 535.893 cm³; 0.536 L
Exercises 1. a. C **b.** E **c.** D **d.** B **e.** A **3.** V = 1.1775 ft³, V = 0.032 97 m³ **5.** V = 12.56 dm³, V = 0.012 56 m³ **7.** 96 mL
9. V = 224 000 mm³, V = 224 cm³, V = 13.44 in.³ **11.** V = 169 560 mm³, V = 169.56 cm³, V = 10.1736 in.³ **13.** V = 9106 mm³,
V = 9.106 cm³, V = 0.546 36 in.³

Section 4.7: Similarity

Margin Exercises 1. x = 5, y = 3 **2.** 3 meters **3.** x = 40°, y = 88° **4. a.** a = 115°, b = 3.94 cm **b.** a = 9 ft, b = 12 ft
Exercises 1. $x = 7\frac{1}{2}$, y = 15 **3.** x = 7, y = 5 **5.** x = 50°, y = 60° **7.** x = 50°, y = 50° **9.** $x = \frac{24}{5}, y = \frac{5}{2}$ **11.** x = 8 , y = 2
13. x = 100° , y = 90° **15.** $x = \frac{5}{3}$ km; y = 2 km **17.** c = 6 m , d = 12 m

Section 4.8: The Pythagorean Theorem

Margin Exercises 1. $5^2 + 12^2 = 25 + 144 = 169 = 13^2$ **2.** $2\sqrt{13}$ **3.** 14.14 inches **4.** 17.89 feet
Exercises 1. yes, $5^2 + 12^2 = 13^2$. **3.** yes, $6^2 + 8^2 = 10^2$. **5.** c = 2.24 ft **7.** x = 14.14 inches **9.** c = 20.62 ft
11. x = 6.00 cm **13.** x = 13.42 ft **15.** 10 ft; 18.47 ft **17. a.** C = 94.2 ft **b.** A = 706.5 ft² **c.** P = 84.85 ft **d.** A = 450 ft²
e. A = 256.5 ft² **19. a.** no **b.** home plate **c.** no **21.** 2.24 miles **23.** 855.9 feet long

Section 4.9: Reading Graphs

Margin Exercises 1. a. $2294 million **b.** $1281 million **c.** 55.8% **2.** $2000 **3. a.** 80° **b.** 60° **c.** 12° **4. a.** first class **b.** $\frac{18}{50} = 36\%$
c. third and fourth classes
Exercises 1. a. Social Science **b.** Chemistry & Physics, and Humanities **c.** about 3300 **d.** about 21.2% **3. a.** Sue
b. Bob and Sue **c.** 85.7% **d.** Bob and Sue; Bob and Sue; Yes in most cases. **e.** No, the vertical scales represent two
different types of quantities. **5. a.** news: 300 min; movies: 120 min; sitcoms: 156 min; soaps: 180 min; drama: 144 min;
childrens' shows: 120 min; commercials: 180 min **b.** news **c.** 480 minutes **7. a.** February and May **b.** 6 inches **c.** March
d. 3.58 inches **9. a.** August **b.** 6 home runs **c.** April and June **d.** 5 home runs **e.** about 22% **f.** about 17% **11. a.** West: 5%;

Northeast: 28%; Midwest: 35%; South: 32% **b.** West: 22%; Northeast: 19%; Midwest: 23%; South: 36% **c.** South **d.** 5%
e. West **f.** 5%, 1900 **g.** 22%, 2000 **h.** Midwest **13. a.** 8 **b.** 3 **c.** eighth class **d.** 2 **e.** 27, 29 **f.** 50 **g.** 10 **h.** 16% **15. a.** $530
b. repairs: 15.1%; gas: 15.1%; insurance: 13.2%; loan payment: 56.6% **c.** 7 : 8

Chapter 4 Test

1. a. $m\angle 2 = 55°$ **b.** $m\angle 4 = 165°$ **2. a.** obtuse **b.** straight **c.** $\angle AOD$ and $\angle DOB$; $\angle AOC$ and $\angle COB$ **3. a.** $m\angle x = 30°$
b. obtuse **c.** $\overline{RT}$ **4. a.** isosceles **b.** obtuse **c.** right **5.** 10 cm is longer; 90 mm longer **6.** The volume is the same:
$10\text{ mL} = 10\text{ cm}^3$. **7.** 0.56 m **8.** 3300 cm **9.** 11 m **10.** 0.0835 L **11.** 9.6 cm^2 **12.** 150 000 cm^2 **13.** 0.75 ha **14.** 7500 a
15. 140 000 mm^3 **16.** 0.032 m^3 **17.** 4 g **18.** 0.022 kg **19.** 19.7 inches **20.** 6.6 gal **21.** 124 mi **22.** 50.32 cm^2 **23.** 28.38 L
24. 1.092 m^3 **25. a.** 94.2 cm **b.** 706.5 cm^2 **26. a.** 40.2 in. **b.** 98.6 in.^{2} **27.** 904.32 in.^{3} = 14 819.092 cm^3 **28. a.** 11.5 mm
b. 6 mm^2 **29. a.** 13.71 m **b.** 11.2325 m^2 **30. a.** 30 ft **b.** 30 ft^2 **31.** 486 m^2 **32.** 90 cm^3 = 5.4 in.^{3}
33. $x = 3.75$; $y = 2.5$ **34. a.** The Pythagorean Theorem **b.** Yes, $10^2 + 24^2 = 26^2$
35. 1 cm^2 **36.** 35.5 ft **37. a.** 157 yd **b.** 188.4 yd **c.** 863.5 yd^2 **38.** $\dfrac{13}{25}$ **39.** $750,000
40. $468,000 **41.** $15.5 million **42.** 10% **43.** 100%

Chapters 1-4 Cumulative Review

1. a. commutative property of addition; $x = 15$ **b.** associative property of multiplication; $y = 3$ **c.** associative property
of addition; $a = 6$ **2. a.** -10 **b.** -0.81 **c.** $-\dfrac{5}{7}$ **3.** 2,400,000 **4.** 520.065 **5.** 5.1 **6.** 7.937 **7.** 735 **8.** -25 **9.** 1.15 **10.** 141
11. $12\dfrac{19}{24}$ **12.** 61.65 **13.** $6\dfrac{3}{32}$ **14.** $x = -8$ **15.** $x = 9$ **16.** $x = 6$ **17.** $y = -1$ **18.** $8950.45 **19.** $2000 **20.** food: $9000;
housing: $11,250; transportation: $6750; miscellaneous: $6750; savings: $2250; education: $4500; taxes: $4500
21. a. $0.10 per sq in. **b.** The larger pizza is the better buy because it is $0.05 cheaper per sq. in. **22. a.** 7.85 ft **b.** 4.90625 ft^2
23. Pay cash because he can save $5760. **24. a.** $18 **b.** $45 **25. a.** $x = 1.8$ **b.** $y = \dfrac{5}{3}\left(\text{or } y = 1\dfrac{2}{3}\right)$ **26. a.** mean = 45
b. median = 46 **c.** mode = 41 **d.** range = 35 **27. a.** 1.93×10^8 **b.** 3.86×10^{-4} **28. a.** $\sqrt{20}$ in. = $2\sqrt{5}$ in. **b.** 4.47 in.
29. $x = 3$ in.; $y = 25°$ **30.** $7x^3 + 8x - 6$; trinomial; third-degree **31.** $14y^2 + 2y - 1$; trinomial; second-degree
32. $37x^2 + 10$; binomial; second-degree

Chapter 5: Algebraic Expressions, Linear Equations, and Applications

Section 5.1: Simplifying and Evaluating Algebraic Expressions

Margin Exercises 1. $-3.9, 5, 0$; $4x, -2x, x$; $3xy^2, -6xy^2$ **2. a** $11y$ **b.** $2.1x$ **c.** $x^2 + 10z$ **d.** $11m - 10$ **e.** $6x$ **3. a.** 4; 9 **b.** -4; -9
4. a. $2x + 4$; -6 **b.** $4ab + 3a$; -15 **c.** $6x + 9$; 27

Exercises 1. -5, $\dfrac{1}{6}$, and 8 are like terms; $7x$ and $9x$ are like terms. **3.** $5xy$ and $-6xy$ are like terms; $-x^2$ and $2x^2$
are like terms; $3x^2y$ and $5x^2y$ are like terms. **5.** 24, 8.3, and -6 are like terms; $1.5xyz$, $-1.4xyz$, and xyz are like
terms. **7.** 64 **9.** -121 **11.** $15x$ **13.** $3x$ **15.** $-2n$ **17.** $5y^2$ **19.** $34x^2$ **21.** $7x + 2$ **23.** $x - 4y$ **25.** $13x^2 - 2y$ **27.** $2n + 3$
29. $7a - 8b$ **31.** $8x + y$ **33.** $2x^2 - x$ **35.** $-2n^2 + 2n - 2$ **37.** $3x^2 - xy + y^2$ **39.** $2x$ **41.** $-y$ **43. a.** $3x + 4$ **b.** 16
45. a. $-7x + 30$ **b.** 2 **47. a.** $10a + 13$ **b.** -7 **49. a.** $9y + 2$ **b.** 29 **51. a.** $-5a + b$ **b.** 9 **53. a.** $-3.1a^2 + 4a$ **b.** -20.4
55. a. $8a - ab^2$ **b.** -14 **57. a.** $3.9y - 18.2$ **b.** -6.5 **59. a.** $9a$ **b.** -18 **61. a.** $3b$ **b.** -3 **63.** -5^2 is the square of 5 multiplied by
-1 by the order of operations; while $\left(-5\right)^2$ is the square of -5. ($-5^2 = -25$ and $\left(-5\right)^2 = 25$)

Section 5.2: Translating English Phrases and Algebraic Expressions

Margin Exercises 1. a. $5y$ **b.** $x + 1$ **c.** $3(z + 2)$ **d.** $3 + 8x$ **e.** $6x - 4$ **f.** x^3 **g.** $5000 + 10r$ **2. a.** the product of 3 and a number
b. 7 times a number increased by 6 **c.** twice the difference between n and 7

Exercises 1. 4 times a number **3.** 5 more than a number **5.** 7 times the sum of a number and 1.1 **7.** -2 times the
difference between a number and 8 **9.** 6 divided by the difference between a number and 1 **11.** 5 times the sum of
twice a number and 3 **13.** 3 times a number plus 7; 3 times the sum of a number and 7 **15.** The product of 7 and a

number minus 3; 7 times the difference between a number and 3 **17.** $x + 6$ **19.** $x - 4$ **21.** $\dfrac{2x}{10}$ **23.** $3x - 5$ **25.** $8 - 2x$
27. $20 - 4.8x$ **29.** $9(x + 2)$ **31.** $4(x + 1) - 13$ **33.** $3(x + 6) + 8$ **35.** $3(7 - x) - 4$ **37. a.** $x - 6$ **b.** $6 - x$ **39. a.** $3x - 5$
b. $5 - 3x$ **41.** $24d$ **43.** $3.15x$ **45.** $365y$ **47.** $7t + 3$ **49.** $7t + 3$ **51.** $20 + 0.15m$ **53.** $2w + 2(2w - 3) = 6w - 6$
55. A phrase whose meaning is not clear or for which there may be two or more interpretations.

Section 5.3: Solving Linear Equations: $x + b = c$ and $ax = c$

Margin Exercises **1. a.** -7 is a solution **b.** -1.7 is not a solution **c.** 3.7 is a solution **d.** -12 is not a solution **2. a.** $x = 17$
b. $x = -12$ **c.** $x = \dfrac{9}{8}$ **3.** $z = 1.7$ **4. a.** $x = 11$ **b.** $x = 4$ **c.** $x = \dfrac{6}{5}$ **d.** $x = 15$ **5.** The original price of the wool coat was \$100.59.
Exercises **1.** -2 is a solution **3.** 4 is not a solution **5.** -4 is a solution **7.** -18 is a solution **9.** -28 is a solution
11. $x = 7$ **13.** $y = -4$ **15.** $x = -19$ **17.** $n = 37$ **19.** $z = -6$ **21.** $x = 5$ **23.** $y = -5.9$ **25.** $x = -1.2$ **27.** $x = \dfrac{11}{20}$ **29.** $x = 9$
31. $y = 8$ **33.** $x = 20$ **35.** $y = 10$ **37.** $x = -2$ **39.** $x = 12$ **41.** $n = 8$ **43.** $y = 2.1$ **45.** $x = \dfrac{20}{9}$ **47.** $x = -13.3$ **49.** $y = -12$
51. $x = -\dfrac{2}{5}$ **53.** $x = -4$ **55.** $x = \dfrac{5}{8}$ **57.** $n = 9.7$ **59.** $x = -4$ **61.** 1945 kanji **63.** 11,500 words **65.** $y = -50.753$
67. $x = -17.214$ **69.** $x = 246$ **71.** $x = -153.17$ **73. a.** Yes. It is stating that $6 + 3$ is equal to 9. **b.** No. If we substitute 4
for x, we get the statement $9 = 10$, which is not true.

Section 5.4: Solving Linear Equations: $ax + b = c$

Margin Exercises **1. a.** $x = -13$ **b.** $y = 2$ **2. a.** $z = 0.2$ **b.** $x = 8$ **3. a.** $x = 2$ **b.** $x = -108$
Exercises **1.** $x = -3$ **3.** $x = 2$ **5.** $x = 2$ **7.** $x = 2$ **9.** $y = -1$ **11.** $t = -1$ **13.** $x = -0.12$ **15.** $x = 4$ **17.** $x = 0$ **19.** $y = 0$
21. $x = -2$ **23.** $y = -6$ **25.** $n = 6$ **27.** $n = 8$ **29.** $x = 0$ **31.** $x = -7$ **33.** $x = -\dfrac{1}{8}$ **35.** $x = -\dfrac{13}{2}$ **37.** $x = -\dfrac{21}{5}$
39. $x = -\dfrac{8}{15}$ **41.** $y = \dfrac{28}{5}$ **43.** $x = 2$ **45.** $y = \dfrac{7}{5}$ **47.** $x = -4.5$ **49.** $x = -44$ **51.** $x = 2$ **53.** $x = -4$ **55.** $y = 0.5$ **57.** $x = 1.5$
59. $x = 0.2$ **61.** 14,000 tickets per hour **63.** $36.5°C$ **65.** $x = 6.1$ **67.** $x = 1.12$

Section 5.5: Solving Linear Equations: $ax + b = cx + d$

Margin Exercises **1. a.** $x = -6$ **b.** $x = -7$ **2.** $y = 2.1$ **3.** $x = 4$ **4. a.** $y = -4$ **b.** $x = -6$ **5. a.** conditional **b.** contradiction
c. identity
Exercises **1.** $x = -5$ **3.** $n = 3$ **5.** $y = 6$ **7.** $x = 3$ **9.** $n = 0$ **11.** $y = 0$ **13.** $z = -1$ **15.** $y = \dfrac{1}{5}$ **17.** $x = -3$ **19.** $x = -4$
21. $x = -21$ **23.** $y = 0$ **25.** $y = 1$ **27.** $x = -\dfrac{3}{2}$ **29.** $x = \dfrac{1}{4}$ **31.** $x = \dfrac{3}{17}$ **33.** $x = -\dfrac{1}{4}$ **35.** $x = \dfrac{8}{5}$ **37.** $x = \dfrac{2}{3}$ **39.** $x = 6$
41. $x = -11$ **43.** $x = \dfrac{1}{2}$ **45.** $x = -5$ **47.** $n = -1.5$ **49.** $x = 0$ **51.** conditional **53.** conditional **55.** contradiction
57. identity **59.** conditional **61.** 1800 square feet **63.** 240 sundaes **65.** $x = -50.21$ **67.** $x = 1.067$

Section 5.6: Applications: Number Problems and Consecutive Integers

Margin Exercises **1. a.** The number is 64. **b.** The number is -5. **c.** The two integers are 5 and 23. **2. a.** $-5, -3,$ and -1
3. a. \$2800 **b.** bagpipes: \$288.85; soccer net: \$218.45.
Exercises **1.** $x - 5 = 13 - x; 9$ **3.** $36 = 2x + 4; 16$ **5.** $7x = 2x + 35; 7$ **7.** $3x + 14 = 6 - x; -2$ **9.** $\dfrac{2x}{5} = x + 6; -10$ **11.** $4(x - 5) = x + 4; 8$
13. $\dfrac{2x + 5}{11} = 4 - x; 3$ **15.** $2x + 3x = 4(x + 3); 12$ **17.** $(2n + 1) + (2n + 3) = 60; 29, 31$ **19.** $n + (n + 1) + (n + 2) = 69; 22, 23, 24$
21. $n + (n + 1) + (n + 2) + (n + 3) = 74; 17, 18, 19, 20$ **23.** $171 - n = (n + 1) + (n + 2); 56, 57, 58$
25. $208 - 3n = (n + 1) + (n + 2) + (n + 3) - 50; 42, 43, 44, 45$ **27.** $2n + 2(2n + 2) = 4(2n + 4) - 54; 42, 44, 46$
29. $(2n + 2) + (2n + 4) - 2n = 66; 60, 62, 64$ **31.** $2(2n + 1) + 3(2n + 3) = 2(2n + 5) + 7; 3, 5, 7$ **33.** $c + (c + 49.50) = 125.74;$
calculator: \$38.12, textbook: \$87.62 **35.** $2x + 90,000 = 310,000; \$110,000$ **37.** $3x + 1500 = 12,000; \$3500$
39. $(x + 56) + x = 542; 243$ boys **41.** $(x + 68) + x = 158;$ \$45 million on electric guitars **43.** $2x + 0.28(250) = 140; \$35$
45. $x + x + 2 + x + 6 = 29; 7$ feet, 9 feet, 13 feet **47.** $x + x + 25,000 = 275,000;$ lot: \$125,000, house: \$150,000
49. The difference between five times a number and the number is equal to $8; 2$ **51.** Find two consecutive integers whose
sum is $33; 16, 17$ **53.** Find two consecutive integers such that 3 times the second is 53 more than the first; 25, 26
55. a. $n, n + 2, n + 4, n + 6$ **b.** $n, n + 2, n + 4, n + 6$ **c.** Yes. Answers will vary.

Section 5.7: Percent Problems

Margin Exercises **1. a.** 198 **b.** 336.8 **c.** 20% **2.** $624 **3.** $648.96 **4.** $10,300 **5.** $8700 **6. a.** has a percent of profit of 33.3% and **b.** has a 36.4% percent of profit. Therefore, **b.** is a better investment.

Exercises **1.** 91% **3.** 137% **5.** 62.5% **7.** 75% **9.** 0.69 **11.** 1.62 **13.** 0.075 **15.** 0.005 **17.** $\frac{7}{20}$ **19.** $\frac{13}{10}$ **21. a.** 32% **b.** 28% **c.** 12% **23. a.** 38.9% **b.** 19.4% **c.** 66.7% **25.** 61.56 **27.** 40 **29.** 2180 **31.** 125% **33.** 80 **35. a.** 8% **b.** 10% **c.** b **37.** $128.81 **39.** $20,000 **41.** 129 free throws **43.** $11.60; $171.60 **45. a.** $22.50 **b.** $9 **47. a.** 10% **b.** 11.11% **c.** The first percent is a percent of his original weight, while the second is a percent of his weight after he lost weight. **49.** 62,500 pounds **51.** 2000 baskets **53. a.** $8800 **b.** Sam **c.** Maria

Section 5.8: Working with Formulas

Margin Exercises **1. a.** $2020 **b.** 122° **c.** 12,500 lb **2. a.** $k = \dfrac{F}{Av^2}$ **b.** $y = \dfrac{400 - 25z}{16} = 25 - \dfrac{25z}{16}$ **c.** $x = 4 - 2y - 3z$

$$z = \dfrac{400 - 16y}{25} = 16 - \dfrac{16y}{25}$$

Exercises **1.** $120 **3.** $10,000 **5. a.** $183.75 **b.** $3683.75 **7.** 2 seconds **9.** 4 milliliters **11.** $1030 **13.** 14 rafters **15.** 336 in. or 28 ft **17.** $1030 **19.** 230 calculators **21.** $5 million **23.** $2400 **25.** 7 hours **27.** 1.625 miles per hour **29.** $b = P - a - c$ **31.** $m = \dfrac{F}{a}$ **33.** $w = \dfrac{A}{l}$ **35.** $n = \dfrac{R}{p}$ **37.** $P = A - I$ **39.** $m = 2A - n$ **41.** $t = \dfrac{I}{Pr}$ **43.** $b = \dfrac{P - a}{2}$ **45.** $\beta = 180° - \alpha - \gamma$ **47.** $h = \dfrac{V}{lw}$ **49.** $b = \dfrac{2A}{h}$ **51.** $\pi = \dfrac{A}{r^2}$ **53.** $g = \dfrac{mv^2}{2K}$ **55.** $y = \dfrac{6 - 2x}{3}$ **57.** $x = \dfrac{11 - 2y}{5}$ **59.** $b = \dfrac{2A - hc}{h}$ or $b = \dfrac{2A}{h} - c$ **61.** $x = \dfrac{8R + 36}{3}$ or $x = \dfrac{8R}{3} + 12$ **63.** $y = -x - 12$ **65.** $C = nt + 9$ **67.** $C = 325n + 5400$ **69. a.** 0; No, because the numerator will be zero and thus the whole fraction will be equal to zero for all values of s. **b.** $x < 70$ **c.** Answers will vary.

Section 5.9: Applications: Distance-Rate-Time, Interest, Average

Margin Exercises **1.** first part: $\frac{4}{3}$ or 1.333 hours; second part: $\frac{8}{3}$ or 2.667 hours. **2.** low-risk stock: $2500; high risk stock: $12,500 **3.** at least a 53 **4.** 40,000 students **5.** $2300

Exercises **1.** 1.68 mph **3.** 8.75 hours **5.** 3 hours **7.** 60 mph; 300 miles **9.** 7 hours **11.** 36 mph; 60 mph **13.** day: 56 mph; night: 69 mph **15.** 4.5 miles **17.** $14,000 at 5%; $11,000 at 6% **19.** 4.5% on $10,000; 5.5% on $6000 **21.** 6.5% on $4000; 6% on $3000 **23.** $24,000 at 4.5%; $18,000 at 6% **25.** $6720 at 5.5%; $5280 at 7% **27.** $600 at 2.5%; $800 at 4% **29.** $11,000 at 4%; $9500 at 5% **31.** $40.50 **33.** $113.75 **35.** 71 **37.** 62 min **39.** 6 hours **41.** 67 **43. a.** 53.8°F **b.** 44°F **c.** −17°F **45. a.** $24.2 billion **b.** $26 billion **c.** $24 billion

Chapter 5 Test

1. a. $5x^2 + 7x$ **b.** 6 **2. a.** $6y - 6$ **b.** 12 **3. a.** $5x + 2$ **b.** −8 **4. a.** $2y^2 + 7y$ **b.** 39 **5. a.** the product of 5 and a number increased by 18 **b.** 3 multiplied by the sum of a number and 6 **c.** 42 less the product of 7 and a number **d.** the product of 6 and a number decreased by 11 **6. a.** $6x - 3$ **b.** $2(x + 5)$ **c.** $2(x + 15) - 4$ **d.** $\dfrac{x}{10} + 2x$ **7.** $x = -4$ **8.** $y = -29$ **9.** $x = -3$ **10.** $x = -6$ **11.** $x = 5$ **12.** $y = 0$ **13.** $a = -\dfrac{9}{8}$ **14.** $x = \dfrac{3}{2}$ **15.** $x = 20$ **16.** $x = -2$ **17.** identity **18.** conditional **19.** $(2y + 5) + y = -22$; −9, −13 **20.** $2n + 3(n + 1) = 83$; 16, 17 **21.** $3(n + 2) = n + (n + 4) + 27$; 25, 27, 29 **22.** $2(1) + 1(3) + 2x = 21$; 8 2-point shots **23.** 111.6 **24.** 32% **25.** The $6000 investment is the better investment since it has a higher percent of profit, 6%, than the $10,000 investment, 5%. **26.** $1875 **27.** 3 meters, 4 meters, and 5 meters **28.** $730 **29.** $m = \dfrac{N - p}{rt}$ **30.** $y = \dfrac{7 - 5x}{3}$ **31.** $29.00 **32.** 4 hours **33.** $12,000 at 5%, $13,000 at 3.5% **34.** $332.50 **35.** 88 **36. a.** 18 months **b.** $490 **37. a.** 175.05 million **b.** 184.5 million **c.** 128.5 million

1. $2 \cdot 2 \cdot 5 \cdot 19$ **2.** $\dfrac{1}{6}$ **3.** 0 **4.** $\dfrac{49}{40}$ **5.** $-\dfrac{31}{5}$ **6. a.** $-2y - 12$ **b.** -18 **7. a.** $4x + 7$ **b.** -1 **8. a.** $x^2 + 13x$ **b.** -22 **9. a.** $x - 7$
b. -9 **10.** $13x - 5$ **11.** $\dfrac{a}{6}$ **12. a.** $2x - 3$ **b.** $\dfrac{x}{6} + 3x$ **c.** $2(x + 10) - 13$ **13. a.** the sum of 6 and 4 times a number
b. 18 times the difference of a number and 5 **14.** $x = -9.8$ **15.** $x = 18$ **16.** $y = -3$ **17.** $y = 6$ **18.** $x = -1$
19. $x = 3$ **20.** $x = -11$ **21.** $x = -4$ **22.** $x = \dfrac{31}{8}$ **23.** $y = \dfrac{24}{5}$ **24.** $y = \dfrac{15}{2}$ **25.** $y = 0$ **26.** conditional **27.** contradiction
28. contradiction **29.** identity **30.** identity **31.** $x = 2.4$; $y = 1.2$ **32.** $-\dfrac{5}{51}$ **33.** -5 **34.** $\dfrac{10}{3}$ cups or $3\dfrac{1}{3}$ cups
35. 12% **36.** 11 **37.** $506.94 **38.** 18, 20, 22 **39.** $555.80 **40.** $-2, -1, 0, 1$ **41.** These are equally good investments
since the percent of profit, 8%, is the same for each investment. **42.** hard drive: $84.60; battery: $46.15
43. $v = \dfrac{h + 16t^2}{t}$ **44.** $r = \dfrac{A - P}{Pt}$ **45.** $y = -\dfrac{14 - 5x}{3}$ **46.** $h = \dfrac{3V}{\pi r^2}$ **47.** $d = \dfrac{C}{\pi}$ **48.** $y = \dfrac{10 - 3x}{5}$ **49.** 1.8 hours (or 1 hour
48 minutes) **50.** $16,500 at 3%; $5500 at 4.5%

Chapter 6: Graphing Linear Equations

Section 6.1: The Cartesian Coordinate System

Margin Exercises 1.a. 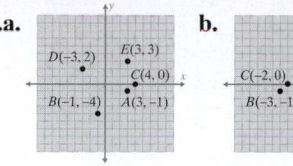 **b.** **2. a.** $(0, -3), (2, 5)$ satisfy the equation $y = 4x - 3$

b. $(0, 3), (5, 2)$
$(15, 0), (30, -3)$

c.

x	y	(x, y)
0	2	$(0, 2)$
$\dfrac{1}{3}$	1	$\left(\dfrac{1}{3}, 1\right)$
-2	8	$(-2, 8)$
$\dfrac{2}{3}$	0	$\left(\dfrac{2}{3}, 0\right)$

3. a. $(-5, -2), (0, 1), (5, 4)$ **b.** $(-5, 5), (0, 1), (5, -3)$

Exercises 1. $\{A(-5, 1), B(-3, 3), C(-1, 1), D(1, 2), E(2, -2)\}$

3. $\{A(-3, -2), B(-1, -3), C(-1, 3), D(0, 0), E(2, 1)\}$

5. $\{A(-4, 4), B(-3, -4), C(0, -4), D(0, 3), E(4, 1)\}$ **7.** $\{A(-3, -5), B(-1, 4), C(0, -1), D(3, 1), E(6, 0)\}$

9. $\{A(-5, 0), B(-2, 2), C(-1, -4), D(0, 6), E(2, 0)\}$

11. 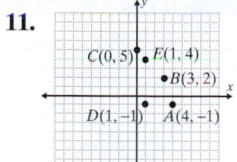 **13.** 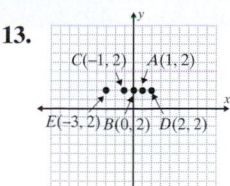 **15.** 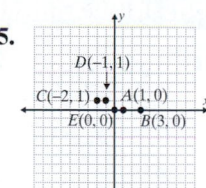 **17.**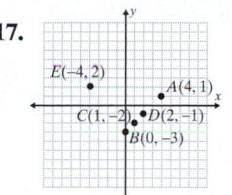

19. 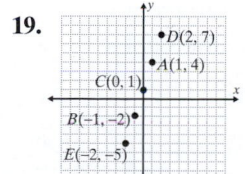 **21.** 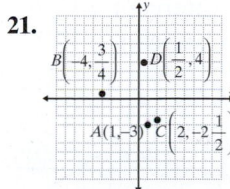 **23.** 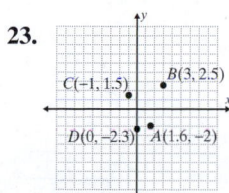 **25. b., c., d.** **27. a., c.** **29. a., c., d.**
31. a. $(0, -4)$ **b.** $(2, -2)$ **c.** $(4, 0)$ **d.** $(1, -3)$
33. a. $(0, 3)$ **b.** $(2, 2)$ **c.** $(6, 0)$ **d.** $(-2, 4)$
35. a. $(0, -8)$ **b.** $(1, -4)$ **c.** $(2, 0)$ **d.** $(3, 4)$

37. a. $(0, 2)$ **b.** $\left(-1, \dfrac{8}{3}\right)$ **c.** $(3, 0)$ **d.** $(6, -2)$ **39. a.** $\left(0, -\dfrac{7}{4}\right)$ **b.** $(1, -1)$ **c.** $\left(\dfrac{7}{3}, 0\right)$ **d.** $\left(3, \dfrac{1}{2}\right)$

41. $(0, 0), (-1, -3),$
$(-2, -6), (2, 6)$

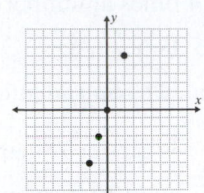

43. $(0, -3), (1, -1),$
$(-2, -7), (3, 3)$

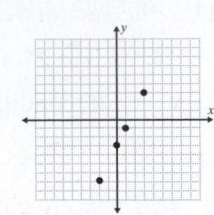

45. $(0, 9), (3, 0),$
$(1, 6), (4, -3)$

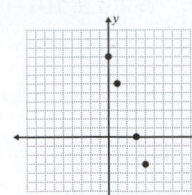

47. $(0, 2), (4, 5),$
$(-4, -1), \left(-1, \dfrac{5}{4}\right)$

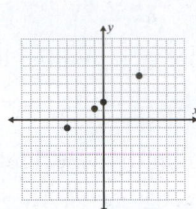

49. $\left(0, -\dfrac{9}{5}\right), (3, 0),$
$(-2, -3), \left(\dfrac{4}{3}, -1\right)$

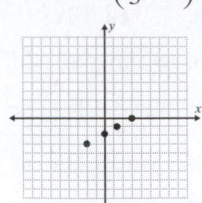

51. $(0, -5), (2, 0),$
$\left(-1, -\dfrac{15}{2}\right), (4, 5)$

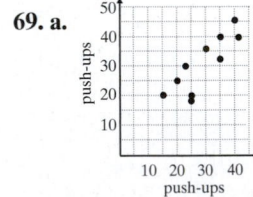

53. $(0, 2), (3.2, 0),$
$(1.92, 0.8), (3.52, -0.2)$

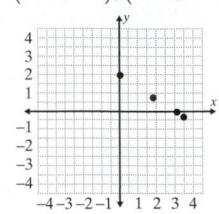

55. For example, $(-3, -3), (0, 1),$ and $(3, 5)$

57. For example, $(-3, -1), (0, 0),$ and $(3, 1)$

59. For example, $(-2, 3), (0, 3),$ and $(4, 3)$

61. For example, $(-4, -4), (0, -3),$ and $(4, -2)$

63. For example, $(-3, -3), (-2, 0),$ and $(-1, 3)$

65. a. $(100, 78.02), (200, 156.04), (300, 234.06),$
$(400, 312.08), (500, 390.1)$

b.

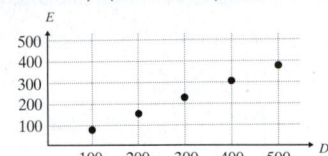

67. a. $(1, 16), (2, 64),$
$(3.5, 196), (4, 256),$
$(4.5, 324), (5, 400)$

b.

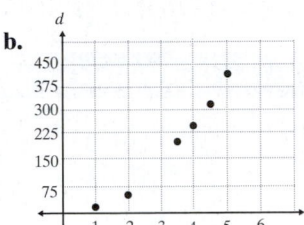

c. The time t is squared in the equation.

69. a.

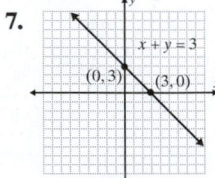

b. Yes, the more sit-ups a person can do it appears the more push-ups he/she can do.
c. 24 sit-ups, 33 sit-ups, 36 sit-ups, 45 sit-ups; Answers will vary.

71. Answers will vary. Not all scatterplots can be used to predict information related to the two variables graphed because not all variables are related.

Section 6.2: Graphing Linear Equations in Two Variables: $ax + by = c$

Margin Exercises

1. a.

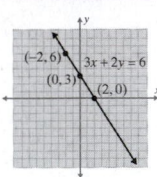

b.

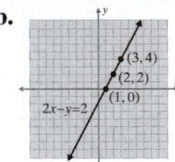

c.

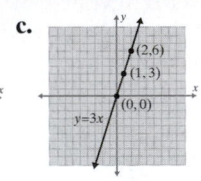

2. a.

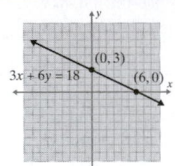

x-intercept $= (6, 0)$
y-intercept $= (0, 3)$

b.

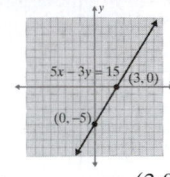

x-intercept $= (3, 0)$
y-intercept $= (0, -5)$

Exercises 1. a **3.** d **5.** f

7.

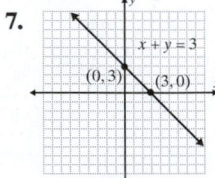

9.

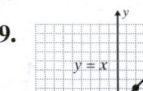

11.

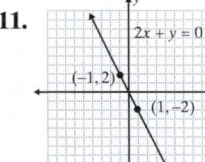

13.

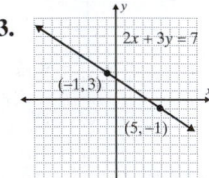

15.

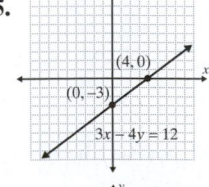

17.

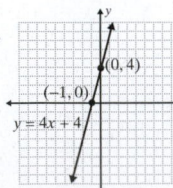

19.

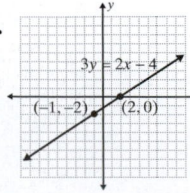

21.

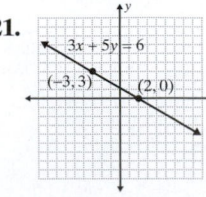

23.

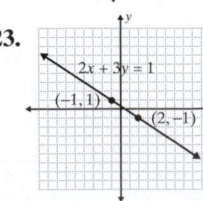

25.

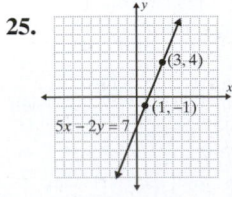

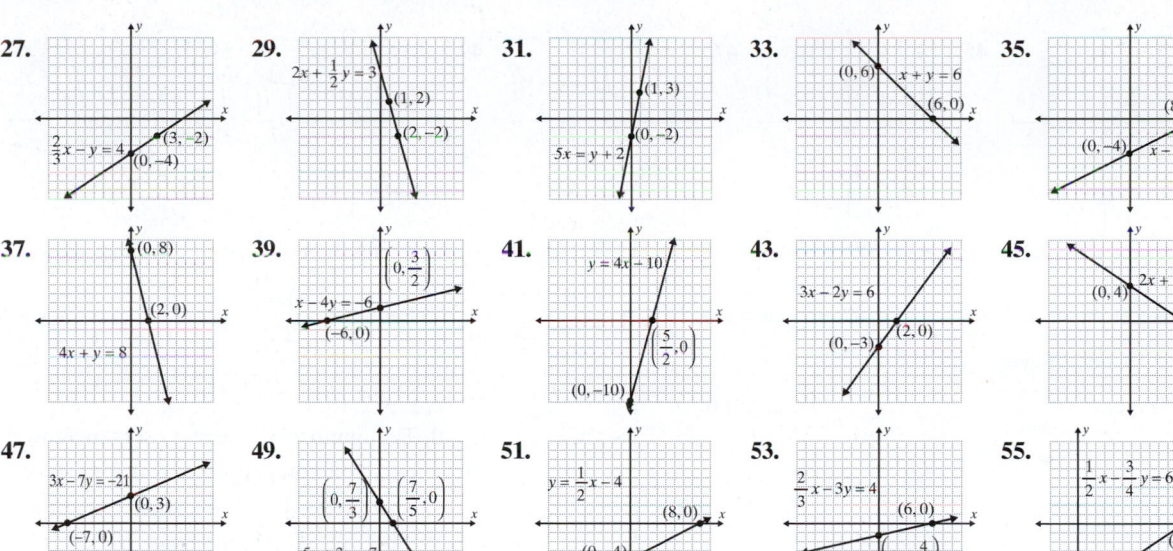

27. $\frac{2}{3}x - y = 4$, $(3,-2)$, $(0,-4)$

29. $2x + \frac{1}{2}y = 3$, $(1,2)$, $(2,-2)$

31. $5x = y + 2$, $(1,3)$, $(0,-2)$

33. $(0,6)$, $x + y = 6$, $(6,0)$

35. $(8,0)$, $(0,-4)$, $x - 2y = 8$

37. $(0,8)$, $(2,0)$, $4x + y = 8$

39. $\left(0,\frac{3}{2}\right)$, $x - 4y = -6$, $(-6,0)$

41. $y = 4x - 10$, $\left(\frac{5}{2},0\right)$, $(0,-10)$

43. $3x - 2y = 6$, $(2,0)$, $(0,-3)$

45. $2x + 3y = 12$, $(0,4)$, $(6,0)$

47. $3x - 7y = -21$, $(0,3)$, $(-7,0)$

49. $\left(0,\frac{7}{3}\right)$, $\left(\frac{7}{5},0\right)$, $5x + 3y = 7$

51. $y = \frac{1}{2}x - 4$, $(8,0)$, $(0,-4)$

53. $\frac{2}{3}x - 3y = 4$, $(6,0)$, $\left(0,-\frac{4}{3}\right)$

55. $\frac{1}{2}x - \frac{3}{4}y = 6$, $(12,0)$, $(0,-8)$

57. Two (unique) points determine a line.

Section 6.3: The Slope Intercept Form: $y=mx+b$

Margin Exercises

1. 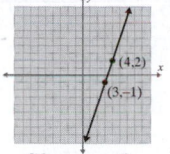 $(4,2)$, $(3,-1)$ Slope = 3

2. 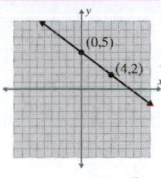 $(0,5)$, $(4,2)$ Slope = $\frac{-3}{4}$

3. a. The equation is $y = -2$ and the slope is 0.
b. The equation is $x = 2$ and the slope is undefined.

4. a. 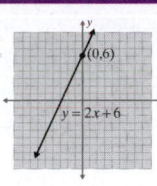 $(0,6)$, $y = 2x + 6$ The slope is 2, and the y-intercept is $(0,6)$

b. 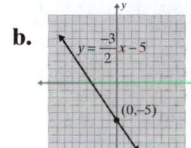 $y = \frac{-3}{2}x - 5$, $(0,-5)$ The slope is $\frac{-3}{2}$ and the y-intercept is $(0,-5)$

c. $y = \frac{2}{3}x - 3$ 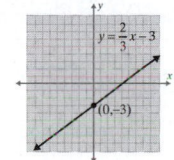 $y = \frac{2}{3}x - 3$, $(0,-3)$

Exercises **1.** $m = 5$ **3.** $m = -\frac{1}{7}$ **5.** $m = 0$ **7.** $m = \frac{1}{2}$ **9.** $m = 2$ **11.** $m = \frac{1}{5}$

13. horizontal line; $m = 0$

15. vertical line; m is undefined

17. horizontal line; $m = 0$

19. vertical line; m is undefined

21. 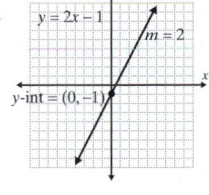 $y = 2x - 1$, $m = 2$, y-int = $(0,-1)$

23. 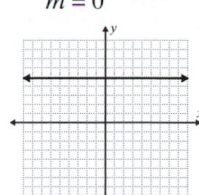 y-int = $(0,5)$, $y = -4x + 5$, $m = -4$

25. 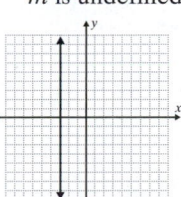 $y = \frac{2}{3}x - 3$, y-int = $(0,-3)$, $m = \frac{2}{3}$

27. 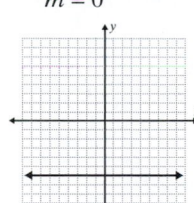 y-int = $(0,5)$, $m = -1$, $y = -x + 5$

29. 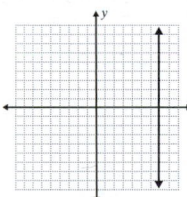 $m = -\frac{1}{5}$, y-int = $(0,2)$, $y = -\frac{1}{5}x + 2$

31. $m = -4$, $y = -4x - 3$, y-int = $(0,-3)$

33. 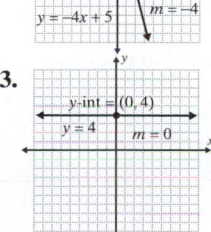 y-int = $(0,4)$, $y = 4$, $m = 0$

35. 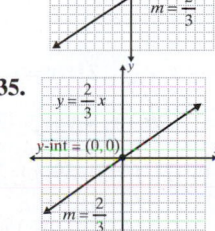 $y = \frac{2}{3}x$, y-int = $(0,0)$, $m = \frac{2}{3}$

37. 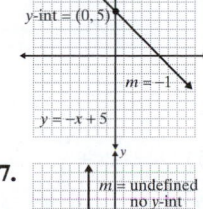 m = undefined no y-int, $x = -3$

39. 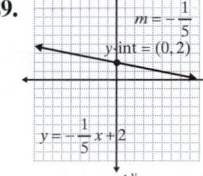 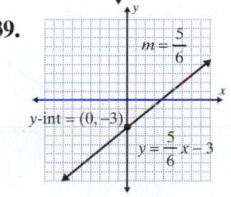 $m = \frac{5}{6}$, y-int = $(0,-3)$, $y = \frac{5}{6}x - 3$

41. 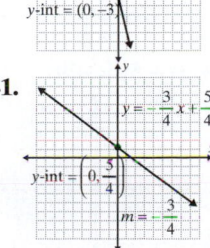 $y = -\frac{3}{4}x + \frac{5}{4}$, y-int = $\left(0,\frac{5}{4}\right)$, $m = -\frac{3}{4}$

43.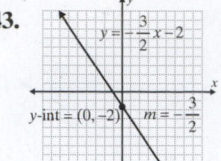

45.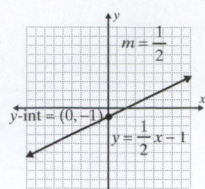

47.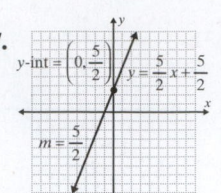

49.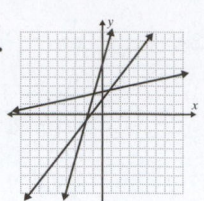

51.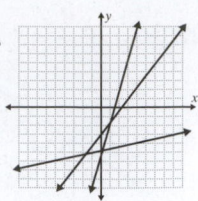

53. $y = -\dfrac{1}{2}x + 3$ **55.** $y = \dfrac{2}{5}x - 3$ **57.** $y = 4x - 5$ **59.** $y = x - 4$ **61.** $y = -\dfrac{5}{6}x - 3$ **63. a.** $m = \dfrac{3}{2}$ **b.** $(0, 7)$ **c.** $y = \dfrac{3}{2}x + 7$

65. a. $m = 0$ **b.** $(0, -6)$ **c.** $y = -6$ **67. a.** $m = \dfrac{1}{2}$ **b.** $(0, -3)$ **c.** $y = \dfrac{1}{2}x - 3$ **69. a.** $m = -\dfrac{1}{3}$ **b.** $(0, 2)$ **c.** $y = -\dfrac{1}{3}x + 2$

71. yes **73.** no **75.** yes

77. $4000/year 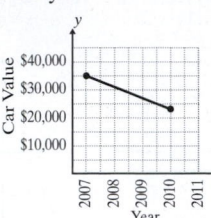 **79. a. and b.** 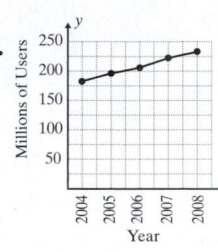 **c.** 13, 12, 10, 11

d. The number of internet users increased by 13 million people/year from '04–'05, 12 mppy from '05–'06, 10 mppy from '06–'07, and 11 mppy from '07–'08.

81. a. and b. 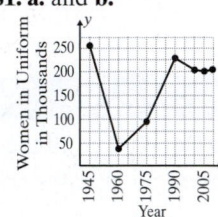 **c.** −15,647.07; 4354.53; 8676.67; −2441.7; −205.2; 450

d. Number of female active duty military personnel decreased by 15,647.07 women/year from 1945–1960; increased by 4354.53 wpy from '60–'75; increased by 8676.67 wpy from '75–'90; decreased by 2441.7 wpy from '90–'00; decreased by 205.2 wpy from '00–'05; and increased by 450 wpy from '05–'09.

83. Answers will vary. **85. a.** the x-axis **b.** the y-axis **87.** A grade of 12% means the slope of the road is 0.12. For every 100 feet of horizontal distance (run) there is 12 feet of vertical distance (rise).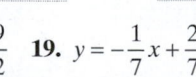

Section 6.4: The Point-Slope Form: $y - y_1 = m(x - x_1)$

Margin Exercises 1. $y = \dfrac{-5}{2}x - 1$ or $5x + 2y = -2$ **2.** 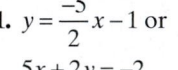 **3.** 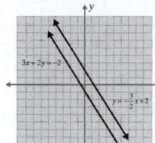 **4.** 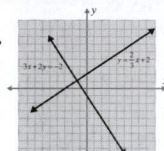 **5.**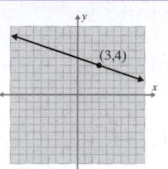

6. $2x + 5y = 6$

$y = -\dfrac{3}{2}x + 2$ or $3x + 2y = 4$ $y = \dfrac{2}{3}x + 2$ or $2x - 3y = -6$ $y = -\dfrac{2}{3}x + 3$ or $2x + 3y = 9$

Exercises 1. a. $m = 2$ **b.** $(3, 1)$ **c.** 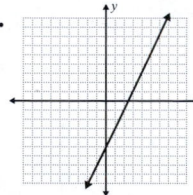 **3. a.** $m = -5$ **b.** $(0, -2)$ **c.**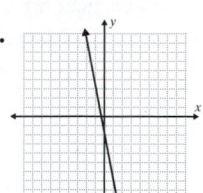

5. a. $m = -\dfrac{1}{4}$ **b.** $(-2, 3)$ **7.** $2x + y = -3$ **9.** $y = -2$ **11.** $x - 2y = -15$

c.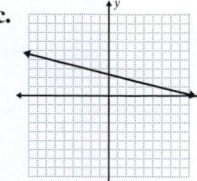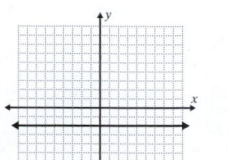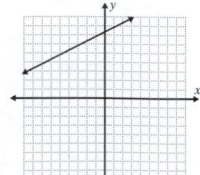

13. $3x - 5y = -29$ **15.** $2x - 3y = -5$ 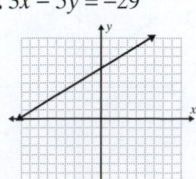 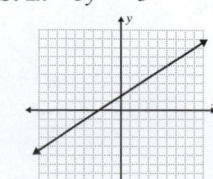 **17.** $y = \dfrac{1}{2}x + \dfrac{9}{2}$ **19.** $y = -\dfrac{1}{7}x + \dfrac{2}{7}$ **21.** $y = -\dfrac{5}{4}x + 2$ **23.** $y = -5$

25. $y = -x + 4$ **27.** $x - y = 1$ **29.** $3x + y = 2$ **31.** $-3x + 4y = 14$

33. $7x + 3y = -5$ **35.** $y = 6$ **37.** $y = 7$ **39.** $y = 7$

41. $y = 2x$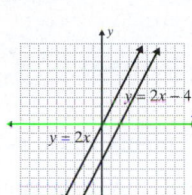

43. $y = 5x + 2$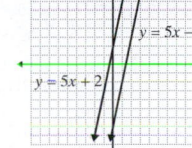

45. $y = -\dfrac{3}{5}x + \dfrac{7}{5}$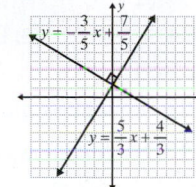

47. $y = -\dfrac{1}{3}x$

49. $y = -\dfrac{1}{2}x - 2$

51.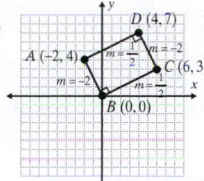

53. parallel

55. perpendicular

57. neither

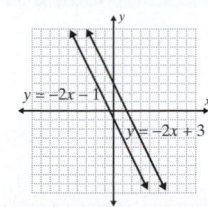
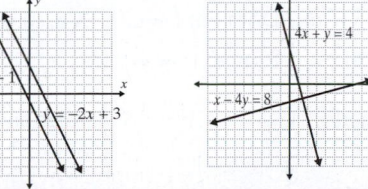
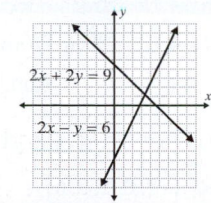

59. The Americans with Disabilities Act recommends a slope of $\dfrac{1}{12}$ for business use and a slope of $\dfrac{1}{6}$ for residential use. Answers will vary.

Section 6.5: Linear Inequalities in One or Two Variables

Margin Exercises 1. a. 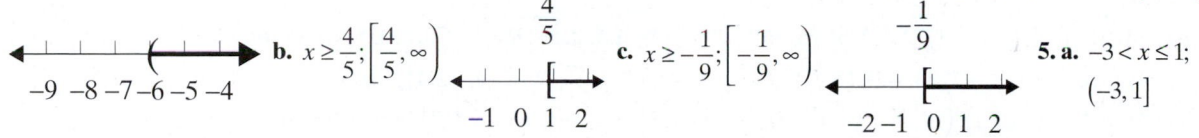 **b.** **c.** $-2 \le x < 1$ is a half-open interval.

d. $[8, \infty)$ is a half-open interval. **2.**  **3.** **4. a.** $x > -6$; $(-6, \infty)$

b. $x \ge \dfrac{4}{5}$; $\left[\dfrac{4}{5}, \infty\right)$ **c.** $x \ge -\dfrac{1}{9}$; $\left[-\dfrac{1}{9}, \infty\right)$ **5. a.** $-3 < x \le 1$; $(-3, 1]$

b. $-3 < x < 2$; $(-3, 2)$ **c.** $3 \le x \le \dfrac{27}{5}$; $\left[3, \dfrac{27}{5}\right]$

6. 400 milligrams or less **7.** at most 8 rose centerpieces

8. a. 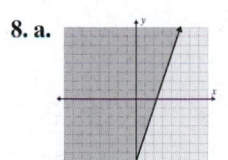 **b.** 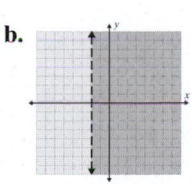 **c.** 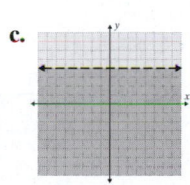 **9.** 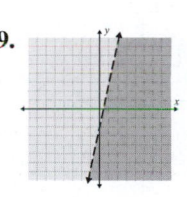 **b.**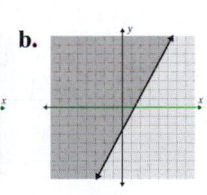

Exercises 1. **3.** **5.** **7.** no solution, $\varnothing$

9. $\{x \,|\, 3 \le x < 5\}$ **11.** $\{x \,|\, x \ge -2.5\}$ **13. a.** $\{x \,|\, -3 \le x \le 1\}$ **b.** $[-3, 1]$ **c.** closed interval **15. a.** $\{x \,|\, -8 < x \le -2\}$ **b.** $(-8, -2]$

c. half-open interval **17. a.** $\{x \,|\, x \le 1\}$ **b.** $(-\infty, 1]$ **c.** half-open interval **19. a.** $\{x \,|\, -4 < x < 4\}$ **b.** $(-4, 4)$ **c.** open interval

21. half-open interval **23.** open interval **25.**

half-open interval **27.** closed interval **29.** half-open interval

31. $(-\infty, 1)$ **33.** $(2, \infty)$ **35.** $\left(\dfrac{8}{3}, \infty\right)$

37. $[-0.4, \infty)$ **39.** $(-\infty, 2)$ **41.** $[3, \infty)$

43. $(-\infty, 2]$ **45.** $\left(-\dfrac{1}{3}, \infty\right)$ **47.** $\left(-\infty, -\dfrac{11}{2}\right)$

49. $[1, \infty)$ **51.** $\left(\dfrac{9}{2}, \infty\right)$ **53.** $(-\infty, -0.6)$

55. $(6.5, \infty)$ **57.** $(-\infty, -13)$ **59.** $\left(-\infty, -\dfrac{10}{9}\right]$

61. $[-4, \infty)$ **63.** $(-\infty, 1]$ **65.** $(-17, \infty)$

67. 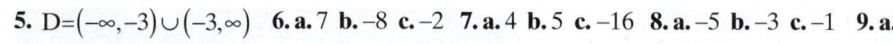 $(-\infty, -30)$ **69.** 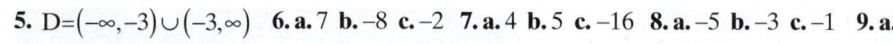 $\left(-\infty, -\dfrac{29}{2}\right]$ **71.** 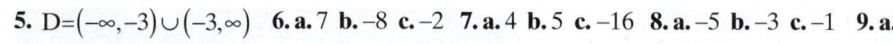 $(-9, 1)$

73. 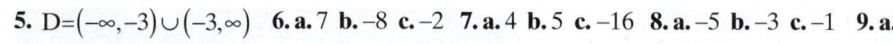 $\left[\dfrac{1}{2}, \dfrac{3}{2}\right]$ **75.** 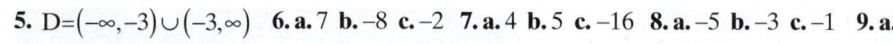 $[3, 15]$ **77.** 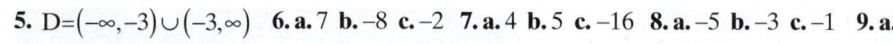 $(-10, -5)$

79. 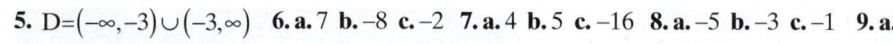 $(-2.8, -0.3)$ **81. a.** The student would need a score higher than 102 points, which is not possible. Thus he cannot earn an A in the course. **b.** The student must score at least 192 points to earn an A in the course. **83.** He must sell at least 10 cars.

Section 6.6: Introduction to Functions and Function Notation

Margin Exercises 1. a. $D = \{4, 7, 3\}$; **b.** $D = \{-2, -4, 0\}$;   **2. a.** Domain: $[-5, 6]$ **b.** Domain: $(-\infty, 3]$ **3. a.** s is not

$\qquad R = \{5, 3, 6\} \qquad R = \{3, -3, 0\} \qquad$ Range: $[-5, 5]$ Range: $(-\infty, \infty)$

a function **b.** t is a function   **4. a.** not a function $D = \{-7, -3, 0, 2, 4, 5\}$ **b.** function $D = (-\infty, \infty)$ **c.** not a function $D = [-5, 7]$

$\qquad\qquad\qquad\qquad\qquad\qquad R = \{-2, 0, 2, 3, 6\} \qquad\qquad\qquad R = [-2, \infty) \qquad\qquad\qquad R = [-1, 5]$

5. $D = (-\infty, -3) \cup (-3, \infty)$ **6. a.** 7 **b.** -8 **c.** -2 **7. a.** 4 **b.** 5 **c.** -16 **8. a.** -5 **b.** -3 **c.** -1 **9. a.** 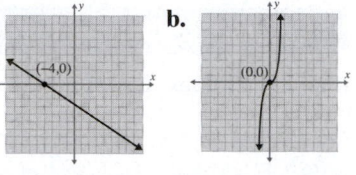   **b.**

or $x \neq -3$

c.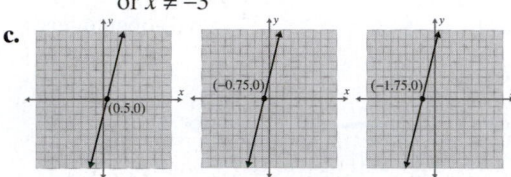

Exercises 1. $\{(-4, 0), (-1, 4), (1, 2), (2, 5), (6, -3)\}$; $D = \{-4, -1, 1, 2, 6\}$; $R = \{-3, 0, 2, 4, 5\}$; function

3. $\{(-5, -4), (-4, -2), (-2, -2), (1, -2), (2, 1)\}$; $D = \{-5, -4, -2, 1, 2\}$; $R = \{-4, -2, 1\}$; function

5. $\{(-4, -3), (-4, 1), (-1, -1), (-1, 3), (3, -4)\}$; $D = \{-4, -1, 3\}$; $R = \{-4, -3, -1, 1, 3\}$; not a function

7. $\{(-5, -5), (-5, 3), (0, 5), (1, -2), (1, 2)\}$; $D = \{-5, 0, 1\}$; $R = \{-5, -2, 2, 3, 5\}$; not a function

9. $D = \{-3, 0, 1, 2, 4\}$;   **11.** $D = \{-4, -3, 1, 2, 3\}$;   **13.** $D = \{-3, -1, 0, 2, 3\}$;   **15.** $D = \{-1\}$; $R = \{-2, 0, 2, 4, 6\}$;

$\quad R = \{-2, -1, 0, 5, 6\}$; function $R = \{4\}$; function $R = \{1, 2, 4, 5\}$; function   not a function

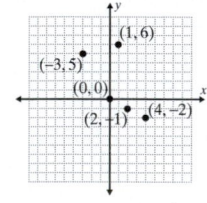   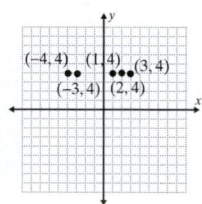   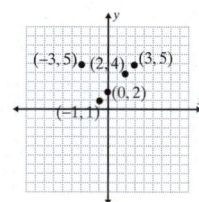 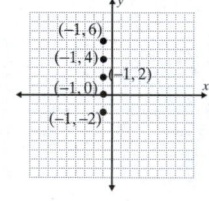

17. function; $D = (-\infty, \infty)$; $R = [0, \infty)$ **19.** function; $D = (-\infty, \infty)$; $R = (-\infty, \infty)$ **21.** not a function; $D = (-\infty, \infty)$; $R = (-\infty, \infty)$

23. not a function; $D = (-\infty, \infty)$; $R = (-\infty, \infty)$ **25.** function; $D = [-5, 5]$; $R = [-2, 2]$ **27.** not a function; $D = \{-3\}$; $R = (-\infty, \infty)$

29. $\left\{(-9, -26), \left(-\dfrac{1}{3}, 0\right), (0, 1), \left(\dfrac{4}{3}, 5\right), (2, 7)\right\}$ **31.** $\{(-2, -11), (-1, -2), (0, 1), (1, -2), (2, -11)\}$ **33.** $D = (-\infty, \infty)$

35. $D = (-\infty, 0) \cup (0, \infty)$ or $x \neq 0$ **37.** $D = (-\infty, 3) \cup (3, \infty)$ or $x \neq 3$ **39. a.** -4 **b.** -16 **c.** -10 **41. a.** 0 **b.** 12 **c.** 56

43. a. -3 **b.** 0 **c.** 3   **49.** 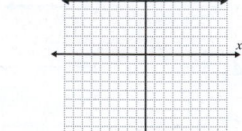   **51.** 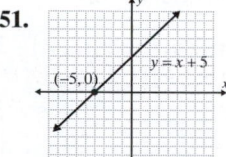   **53.** 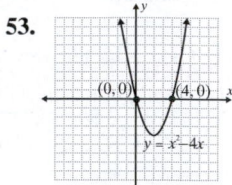   **55.**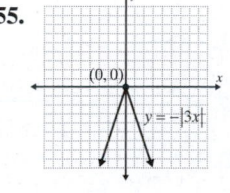

45. a. 3 **b.** 2 **c.** 0

47. a. -3 **b.** -2 **c.** 5

57. 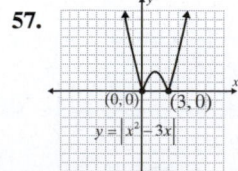   **59.** 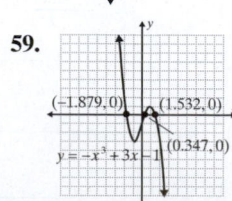   **61.** 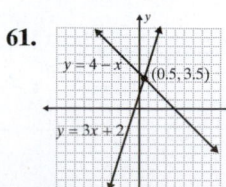   **63.**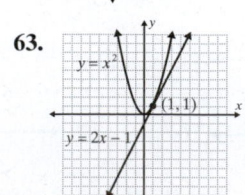

65. y-intercept $= (0, -5)$ (should be $(0, 5)$) **67.** y-intercept $= (0, -8)$ (should be $(0, -2)$) **69.** slope $= -3$ (should be the reciprocal, $-\dfrac{1}{3}$) and y-intercept $= (0, 2)$ (should be $(0, 0)$)

Chapter 6 Test

1. $\left\{ A(-3, 2), B(-2, -1), C(0, -3), D(1, 2), E(3, 4), F(5, 0), G(4, -2) \right\}$

2.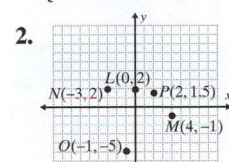

3. a. $(0, 2)$ **b.** $\left(\dfrac{2}{3}, 0\right)$ **c.** $(-2, 8)$ **d.** $(3, -7)$ **5.**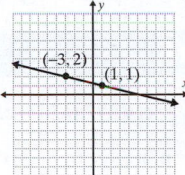

4. a. $\left(0, -\dfrac{6}{5}\right)$ **b.** $(6, 0)$ **c.** $(11, 1)$ **d.** $(-4, -2)$

6.

7. $5x - 3y = 9$ **8.** **9.** $m = \dfrac{9}{8}$ **10.** $m = -\dfrac{1}{5}$

11. $y = \dfrac{1}{3}x - \dfrac{4}{3}$; $m = \dfrac{1}{3}$; y-int $= \left(0, -\dfrac{4}{3}\right)$

12. $y = -\dfrac{4}{3}x + 1$; y-int $= (0, 1)$; $m = -\dfrac{4}{3}$

13. a. and b.

c. $11, -14$
d. Sam's speed increased 11 mph from the 1st to the 2nd hour and decreased 14 mph from the 2nd to the 3rd hour.

14. $5x + 3y = 36$ **15.** $8x + 7y = 10$ **16.** $y = 6$ **17.** $y = -\dfrac{3}{2}x + 7$ **18.** $y = -2$ **19.** See page 793.

20. $f = \left\{ (-3, 2), (-2, -1), (0, -3), (1, 2), (3, 4), (5, 0) \right\}$; $D = \{-3, -2, 0, 1, 3, 5\}$; $R = \{-3, -1, 0, 2, 4\}$; function

21. function; $D = (-\infty, \infty)$; $R = [0, \infty)$ **22. a.** 13 **b.** 5 **c.** 4

23. a. $y = 5 + 3x - 2x^2$ **b.** $(-1, 0), (2.5, 0)$ **24.** **25.**

26. **27. a.** $\{x \mid x < 5\}$ **b.** $(-\infty, 5)$ **c.** open interval **28. a.** $\{x \mid -1 \le x \le 4\}$ **b.** $[-1, 4]$ **c.** closed interval

29. $(-5, \infty)$ **30.** $\left(-\infty, -\dfrac{7}{2}\right)$

31. $(-1, 5)$

Chapters 1–6 Cumulative Review

1. 180 **2.** 2100 **3.** $120a^2b^2$ **4.** 13.125 **5.** 83.3 **6.** 11 **7.** 22 **8.** -0.8 **9.** $-\dfrac{1}{12}$ **10.** -40 **11.** $\dfrac{2}{3}$ **12.** undefined

13. 0 **14.** $-2x - 12$ **15.** 0 **16.** $3x^3 - x^2 + 4x - 1$ **17.** $-10x + 4$ **18.** $x = 2$ **19.** $x = -4$ **20.** $x = -4$ **21.** $x = -8$

22. conditional **23.** identity **24. a.** $n = 2A - m$ **b.** $f = \dfrac{\omega}{2\pi}$ **25.** $(4, \infty)$

26. $[-12.8, \infty)$ **27.** $(-\infty, 2]$ **28.** $(-\infty, -5]$

29. 3.2×10^7 **30.** $x = 0.375$

Chapter 7: Exponents, Polynomials, and Factoring Techniques

Section 7.1: Exponents

Margin Exercises **1. a.** x^5 **b.** y^9 **c.** 125 **d.** 243 **e.** -2187 **2. a.** $20y^9$ **b.** $-10x^7$ **c.** $-35a^2b^7$ **3. a.** 1 **b.** x^2 **c.** 1 **4. a.** x^2 **b.** y^3 **c.** 1 **5. a.** $3x^8$ **b.** $7x^6y^4$ **6. a.** $\dfrac{1}{7}$ **b.** $\dfrac{1}{x^7}$ **c.** $\dfrac{1}{x^5}$ **7. a.** 27 **b.** x^6 **c.** $\dfrac{1}{8^2}$ or $\dfrac{1}{64}$ **d.** $\dfrac{x^3}{y^3}$ **e.** $\dfrac{4}{x^3}$ **8. a.** 0.037 **b.** 1 **c.** 38.17152 **9. a.** x^{15} **b.** $\dfrac{1}{x^{12}}$ **c.** $\dfrac{1}{y^{15}}$ **d.** $\dfrac{1}{9^3}$ or $\dfrac{1}{729}$ **10. a.** $16x^2$ **b.** x^7y^7 **c.** $81a^2b^2$ **d.** $\dfrac{1}{a^3b^3}$ **e.** $\dfrac{x^9}{y^{12}}$ **11. a.** $\dfrac{x^7}{y^7}$ **b.** $\dfrac{25}{36}$ **c.** $\dfrac{27}{a^3}$ **d.** $\dfrac{x^3}{216}$ **12. a.** $\dfrac{-27x^3}{y^9}$ **b.** $\dfrac{16b^2}{a^2}$ **13.** $\dfrac{y^{15}}{x^{30}}$ **14.** $\dfrac{64}{225x^{18}y^2}$

Exercises **1.** 27 **3.** 512 **5.** $\dfrac{1}{3}$ **7.** $\dfrac{1}{25}$ **9.** 16 **11.** 24 **13.** -500 **15.** $\dfrac{3}{8}$ **17.** $-\dfrac{3}{25}$ **19.** x^5 **21.** y^2 **23.** $\dfrac{1}{x^3}$ **25.** $\dfrac{2}{x}$ **27.** $-\dfrac{8}{y^2}$ **29.** $\dfrac{5x^6}{y^4}$ **31.** 4 **33.** 49 **35.** $\dfrac{1}{10}$ **37.** $\dfrac{1}{8}$ **39.** x^2 **41.** x^2 **43.** x^4 **45.** $\dfrac{1}{x^4}$ **47.** x^6 **49.** x^2 **51.** y^2 **53.** $3x^3$ **55.** x^4 **57.** $36x^3$ **59.** $-14x^5$ **61.** $-12x^6$ **63.** $4y$ **65.** $3y^2$ **67.** $-2y^2$ **69.** $\dfrac{1}{x^2}$ **71.** 1000 **73.** 1 **75.** $-18x^5y^7$ **77.** $-\dfrac{2y^2}{x}$ **79.** $12a^3b^9c$ **81.** $-4a^{10}b^3c$ **83.** $\dfrac{5}{x}$ **85.** 1 **87.** 0.390625 **89.** 8875.147264 **91.** -81 **93.** -16 **95.** 1,000,000 **97.** $36x^6$ **99.** $-108x^6$ **101.** $\dfrac{5x^2}{y}$ **103.** $-\dfrac{2y^6}{27x^{15}}$ **105.** $\dfrac{27x^3}{y^3}$ **107.** 1 **109.** $4x^4y^4$ **111.** $\dfrac{y^2}{x^2}$ **113.** $\dfrac{1}{3xy^2}$ **115.** $-\dfrac{x^3y^6}{27}$ **117.** m^2n^4 **119.** $-\dfrac{y^2}{49x^2}$ **121.** $\dfrac{25x^6}{y^2}$ **123.** $\dfrac{x^6}{y^6}$ **125.** $\dfrac{36y^{14}}{x^4}$ **127.** $\dfrac{49y^4}{x^6}$ **129.** $\dfrac{24y^5}{x^7}$ **131.** $\dfrac{4x^3}{3}$ **133.** $\dfrac{16x^{13}}{243y^6}$ **135.** $96x^2y^4z^4$

Section 7.2: Introduction to Polynomials

Margin Exercises **1. a.** $9x^2$ second-degree monomial **b.** $9x^2 - 5x$ second-degree binomial **c.** $4y^3 - \dfrac{8}{3}y^2 + 8$ third-degree trinomial **d.** $4x^2 - 8$ second-degree binomial **e.** not a polynomial **2. a.** 23 **b.** -6 **c.** $24a - 35$

Exercises **1.** monomial **3.** binomial **5.** trinomial **7.** not a polynomial **9.** binomial **11.** $4y$; first-degree monomial; leading coefficient 4 **13.** $x^3 + 3x^2 - 2x$; third-degree trinomial; leading coefficient 1 **15.** $-2x^2$; second-degree monomial; leading coefficient -2 **17.** 0; monomial of no degree; leading coefficient 0 **19.** $6a^5 - 7a^3 - a^2$; fifth-degree trinomial; leading coefficient 6 **21.** $2y^3 + 4y$; third-degree binomial; leading coefficient 2 **23.** 4; monomial of degree 0; leading coefficient 4 **25.** $2x^3 + 3x^2 - x + 1$; third-degree polynomial; leading coefficient 2 **27.** $4x^4 - x^2 + 4x - 10$; fourth-degree polynomial; leading coefficient 4 **29.** $9x^3 + 4x^2 + 8x$; third-degree trinomial; leading coefficient 9 **31.** -16 **33.** -41 **35.** 4379 **37.** 13 **39.** -28 **41.** $3a^4 + 5a^3 - 8a^2 - 9a$ **43.** $3a + 11$ **45.** $10a + 25$

47. a. 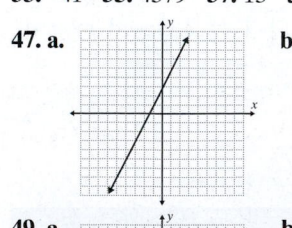 **b.** 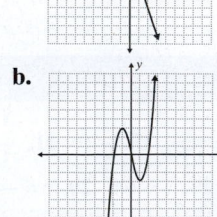 **c.**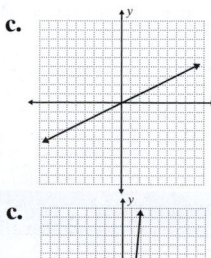

49. a. 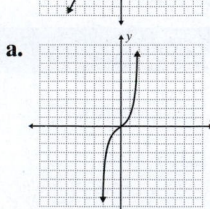 **b.** 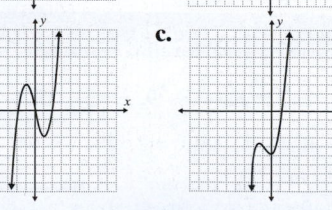 **c.**

Section 7.3: Addition and Subtraction with Polynomials

Margin Exercises **1. a.** $8x^3 + 4x^2 - 6x - 11$ **b.** $4x^3 - 3x^2 + 7x + 2$ **2. a.** $15x^4 - 3x^3 + 5x^2 + 1$ **b.** $6x^3 - 9x^2 - 8x + 1$ **3. a.** $8x - 21$ **b.** $4x - 3$

Exercises **1.** $3x^2 + 7x + 2$ **3.** $2x^2 + 11x - 7$ **5.** $3x^2$ **7.** $x^2 - 5x + 17$ **9.** $-x^2 + 2x - 7$ **11.** $2x^3 + 4x^2 - 3x - 7$
13. $-x^2 + 7x - 3$ **15.** $-x^3 + 2x^2 + 3x - 4$ **17.** $4x^3 + x^2 + 7$ **19.** $2x^3 + 11x^2 - 4x - 3$ **21.** $x^2 + x + 6$ **23.** $-3x^2 - 6x - 2$
25. $3x^2 - 4x - 8$ **27.** $-x^4 - 2x^3 - 16$ **29.** $-4x^4 - 7x^3 - 11x^2 + 5x + 13$ **31.** $-4x^2 + 6x + 1$ **33.** $3x^4 - 7x^3 + 3x^2 - 5x + 9$
35. $6x^2 - 7x + 18$ **37.** $8x^4 + 6x^2 + 15$ **39.** $2x^3 + 4x^2 + 3x - 10$ **41.** $4x - 13$ **43.** $-7x + 9$ **45.** $2x + 17$ **47.** $3x^3 + 13x^2 + 9$
49. $8x^2 - x - 2$ **51.** $3x - 19$ **53.** $x^2 - 2x + 13$ **55.** $7x - 3$ **57.** $4x^2 + 3x + 5$ **59.** $7x - 18$ **61.** $x^2 + 11x - 8$
63. $7x^3 - x^2 - 7$ **65.** Any monomial or algebraic sum of monomials. **67.** The largest of the degrees of its terms after like terms have been combined.

Section 7.4: Mutiplication with Polynomials

Margin Exercises **1. a.** $-18x^3 + 6x^2 + 18x$ **b.** $12x^2 + 3x - 9$ **c.** $12x^3 + 28x^2 + 2x - 2$ **d.** $x^4 - 16x^2 + 16x - 4$
e. $x^3 - 5x^2 - 12x + 36$
Exercises **1.** $-6x^5 - 15x^3$ **3.** $4x^7 - 12x^6 + 4x^5$ **5.** $-y^5 + 8y - 2$ **7.** $-4x^8 + 8x^7 - 12x^4$ **9.** $25x^5 - 5x^4 + 10x^3$
11. $a^7 + 2a^6 - 5a^3 + a^2$ **13.** $6x^2 - x - 2$ **15.** $9a^2 - 25$ **17.** $-10x^2 + 39x - 14$ **19.** $x^3 + 5x^2 + 8x + 4$ **21.** $x^2 + x - 12$
23. $a^2 - 2a - 48$ **25.** $x^2 - 3x + 2$ **27.** $3t^2 - 3t - 60$ **29.** $x^3 + 11x^2 + 24x$ **31.** $2x^2 - 7x - 4$ **33.** $6x^2 + 17x - 3$
35. $4x^2 - 9$ **37.** $16x^2 + 8x + 1$ **39.** $y^3 + 2y^2 + y + 12$ **41.** $3x^2 - 8x - 35$ **43.** $5x^3 + 6x^2 - 22x - 9$
45. $2x^4 + 7x^3 + 5x^2 + x - 15$ **47.** $3x^2 + 2x - 8$ **49.** $2x^2 + 3x - 5$ **51.** $7x^2 - 13x - 2$ **53.** $6x^2 - 13x - 8$
55. $4x^2 + 12x + 9$ **57.** $x^3 + 3x^2 - 4x - 12$ **59.** $4x^2 - 49$ **61.** $x^3 + 1$ **63.** $49a^2 - 28a + 4$ **65.** $2x^3 + x^2 - 5x - 3$
67. $x^3 + 6x^2 + 11x + 6$ **69.** $a^4 - a^2 + 2a - 1$ **71.** $t^4 + 6t^3 + 13t^2 + 12t + 4$ **73.** $2y^2 - 61$ **75.** $3a^2 - 17a + 11$
77. $-3x - 21$ **79.** $a^2 + 9a - 1$ **81.** Answers will vary.

Section 7.5: Special Products of Polynomials

Margin Exercises **1. a.** $3x^2 + 25x + 28$ **b.** $8x^2 - 18x + 10$ **c.** $x^2 - 16$ **2. a.** $x^2 - 36$ **b.** $16y^2 - 9$ **c.** $x^8 - 9$ **3. a.** $9x^2 + 30x + 25$
b. $16x^2 - 16x + 4$ **c.** $4x^2 - 32x + 64$ **d.** $4y^6 - 8y^3 + 4$
Exercises **1.** $x^2 - 14x + 49$; perfect square trinomial **3.** $x^2 + 8x + 16$; perfect square trinomial **5.** $x^2 - 9$; difference
of two squares **7.** $x^2 - 81$; difference of two squares **9.** $2x^2 + x - 3$ **11.** $9x^2 - 24x + 16$; perfect square trinomial
13. $25x^2 - 4$; difference of two squares **15.** $9x^2 - 12x + 4$; perfect square trinomial **17.** $x^2 - 16x + 64$; perfect square
trinomial **19.** $16x^2 - 25$; difference of two squares **21.** $25x^2 - 81$; difference of two squares **23.** $x^2 - 8x + 16$;
perfect square trinomial **25.** $4x^2 - 49$; difference of two squares **27.** $10x^4 - 11x^2 - 6$ **29.** $49x^2 + 14x + 1$; perfect square
trinomial **31.** $5x^2 + 11x + 2$ **33.** $4x^2 + 13x - 12$ **35.** $3x^2 - 25x + 42$ **37.** $x^2 + 10x + 25$ **39.** $x^4 - 1$ **41.** $x^4 + 6x^2 + 9$
43. $x^6 - 4x^3 + 4$ **45.** $x^4 + 3x^2 - 54$ **47.** $x^2 - \dfrac{4}{9}$ **49.** $x^2 - \dfrac{9}{16}$ **51.** $x^2 + \dfrac{6}{5}x + \dfrac{9}{25}$ **53.** $x^2 - \dfrac{5}{3}x + \dfrac{25}{36}$ **55.** $x^2 - \dfrac{1}{4}x - \dfrac{1}{8}$
57. $x^2 + \dfrac{5}{6}x + \dfrac{1}{6}$ **59.** $x^2 - 1.96$ **61.** $x^2 - 5x + 6.25$ **63.** $x^2 - 4.6225$ **65.** $x^2 + 2.48x + 1.5376$ **67.** $2.0164x^2 + 27.264x + 92.16$
69. $129.96x^2 - 12.25$ **71.** $93.24x^2 + 142.46x - 104.04$ **73. a.** $A(x) = 400 - 4x^2$ **b.** $P(x) = 4(20 - 2x) + 8x = 80$
75. $A(x) = 8x + 15$ **77. a.** $A(x) = 150 - 4x^2$ **b.** $P(x) = 2(10 - 2x) + 2(15 - 2x) + 8x = 50$ **c.** $V(x) = (10 - 2x)(15 - 2x)x$
$= 4x^3 - 50x^2 + 150x$ **79.** As indicated in the diagram, $(x + 5)^2 = x^2 + 2(5x) + 5^2$. Answers will vary.

Section 7.6: Greatest Common Factor (GCF) and Factoring by Grouping

Margin Exercises **1. a.** $\dfrac{x^2}{2} - 2x + 4$ **b.** $3y^2 - 2y + 1$ **2. a.** The GCF is 5. **b.** The GCF is $50xy$. **3. a.** $10y(1 + 3y)$
b. $-9b^2(b^2 - b - 1)$ **4.** $-11xz(3x + 5 + 7z^2)$ **5. a.** $(x^2 + 2)(2x - y)$ **b.** $(6y + 3v)(x - u)$ **6. a.** $(y - 1)(3x + 2)$ **b.** not factorable

Exercises **1.** 5 **3.** 8 **5.** 1 **7.** $10x^3$ **9.** $4a^2$ **11.** $13ab$ **13.** $15xy^2z^2$ **15.** x^4 **17.** $-4y$ **19.** $3x^3$ **21.** $2x^2y$ **23.** $m + 9$
25. $x - 6$ **27.** $b + 1$ **29.** $3y + 4x + 1$ **31.** $11(x - 11)$ **33.** $4y(4y^2 + 3)$ **35.** $-3a(2x - 3y)$ **37.** $5xy(2x - 5)$ **39.** $-2yz(9yz - 1)$
41. $8(y^2 - 4y + 1)$ **43.** $x(2y^2 - 3y - 1)$ **45.** $4m^2(2x^3 - 3y + z)$ **47.** $-7x^2z^3(8x^2 + 14xz + 5z^2)$ **49.** $x^4y^2(15 + 24x^2y^4 - 32x^3y)$
51. $(y + 3)(7y^2 + 2)$ **53.** $(x - 4)(3x + 1)$ **55.** $(x - 2)(4x^3 - 1)$ **57.** $(2y + 3)(10y - 7)$ **59.** $(x - 2)(a - b)$ **61.** $(b + c)(x + 1)$
63. $(x^2 + 6)(x + 3)$ **65.** not factorable **67.** $(3 - b)(x + y)$ **69.** $(y - 4)(5x + z)$ **71.** $(z^2 + 3)(a + 1)$ **73.** $(6x + 1)(a + 2)$
75. $(x + 1)(y + 1)$ **77.** $(2y - 7z)(5x - y)$ **79.** $(3x - 4u)(y - 2v)$ **81.** not factorable **83.** $(2c - 3d)(3a + b)$
85. Although both can be factored out of $-3x^2 + 3$, 3 is greater than -3 making it the greatest common factor.

Section 7.7: Factoring Trinomials: $x^2 + bx + c$

Margin Exercises **1. a.** $(x + 3)(x + 7)$ **b.** $(x - 5)(x + 4)$ **2. a.** $7y(y^2 - 7 + 2y)$ **b.** $11xy(x + 3)(x - 1)$
Exercises **1.** $\{1, 15\}, \{-1, -15\}, \{3, 5\}, \{-3, -5\}$ **3.** $\{1, 20\}, \{-1, -20\}, \{4, 5\}, \{-4, -5\}, \{2, 10\}, \{-2, -10\}$
5. $\{1, -6\}, \{6, -1\}, \{2, -3\}, \{3, -2\}$ **7.** $\{1, 16\}, \{-1, -16\}, \{4, 4\}, \{-4, -4\}, \{8, 2\}, \{-8, -2\}$

9. $\{1,-10\},\{10,-1\},\{5,-2\},\{2,-5\}$ **11.** $4,3$ **13.** $-7,2$ **15.** $8,-1$ **17.** $-6,-6$ **19.** $-5,-4$ **21.** $x+1$ **23.** $p-10$ **25.** $a+6$
27. $(x-4)(x+3)$ **29.** $(y+6)(y-5)$ **31.** not factorable **33.** $(x-4)(x-4)$ **35.** $(x+4)(x+3)$ **37.** $(y-1)(y-2)$
39. not factorable **41.** $(x+8)(x-9)$ **43.** $(z-6)(z-9)$ **45.** $x(x+7)(x+3)$ **47.** $5(x-4)(x+3)$ **49.** $10y(y-3)(y+2)$
51. $4p^2(p+1)(p+8)$ **53.** $2x^2(x-9)(x+2)$ **55.** $2(x^2-x-36)$ **57.** $2a^2(a-10)(a+6)$ **59.** $3y^3(y-8)(y+1)$
61. $x(x-2)(x-8)$ **63.** $5(a^2+2a-6)$ **65.** $20a^2(a+1)(a+1)$ **67.** base $=x+48$; height $=x$ **69.** $x+5$ **71.** This is not
an error, but the trinomial is not completely factored. The completely factored form of this trinomial is $2(x+2)(x+3)$.

Section 7.8: More on Factoring Trinomials: ax^2+bx+c

Margin Exercises 1. a. $(x+6)(x+2)$ **b.** $(4u-7)(2u+3)$ **2. a.** $4x(2x-1)(x-1)$ **b.** $(3x^2+7x-1)(7x)$ **3. a.** $(5a+9)(a+3)$
b. $3(b-2)(b+4)$
Exercises 1. $(x+2)(x+3)$ **3.** $(2x-5)(x+1)$ **5.** $(6x+5)(x+1)$ **7.** $-(x-2)(x-1)$ **9.** $(x-5)(x+2)$ **11.** $-(x-14)(x+1)$
13. not factorable **15.** $-x(2x+1)(x-1)$ **17.** $(t-1)(4t+1)$ **19.** $(5a-6)(a+1)$ **21.** $(7x-2)(x+1)$ **23.** $(2x-3)(4x+1)$
25. $(3x+4)(3x-5)$ **27.** $2(2x-5)(3x-2)$ **29.** $(3x-1)(x-2)$ **31.** $(3x-1)(3x-1)$ **33.** $(3y+2)(2y+1)$
35. $(x-1)(x-45)$ **37.** not factorable **39.** $2b(4a-3)(a-2)$ **41.** not factorable **43.** $(4x-1)(4x-1)$ **45.** $(8x-3)(8x-3)$
47. $2(3x-5)(x+2)$ **49.** $5(2x+3)(x+2)$ **51.** $-2(9x^2-36x+4)$ **53.** $-15(3y+4)(y-2)$ **55.** $3(2x-5)(2x-5)$
57. $3x(2x-1)(x+2)$ **59.** $3x(2x-9)(2x-9)$ **61.** $9xy^3(x^2+x+1)$ **63.** $4xy(3y-4)(4y-3)$ **65.** $7y^2(y-4)(3y-2)$
67. If the sign of the constant term is positive, the signs in the factors will both be positive or both be negative. If the sign
of the constant term is negative, the sign in one factor will be positive and the sign in the other factor will be negative.

Section 7.9: Special Factoring Techniques

Margin Exercises 1. a. $7y(u-7)(u+7)$ **b.** $(v^3-10)(v^3+10)$ **2. a.** not factorable **b.** $5(9x^2+4)$ **3. a.** $(z+20)^2$
b. $3xy(u-3v)^2$ **c.** $(v+4-z)(v+4+z)$ **4. a.** $(v-3)(v^2+3y+9)$ **b.** $6(2v^4-5)(4v^8+10v^4+25)$
Exercises 1. $(x-5)(x+5)$ **3.** $(9-y)(9+y)$ **5.** $2(x-8)(x+8)$ **7.** $4(x-2)(x+2)(x^2+4)$ **9.** not factorable
11. $(y-8)^2$ **13.** $-4(x-5)(x+5)$ **15.** $(3x-5)(3x+5)$ **17.** $(y-5)^2$ **19.** $(2x-1)^2$ **21.** $(5x+3)^2$ **23.** $(4x-5)^2$
25. $4x(x-4)(x+4)$ **27.** $2xy(x+8)^2$ **29.** $(y+3)^2$ **31.** $(x-10)^2$ **33.** $(x^2+5y)^2$ **35.** $(x-5)(x^2+5x+25)$
37. $(y+6)(y^2-6y+36)$ **39.** $(x+3y)(x^2-3xy+9y^2)$ **41.** not factorable **43.** $4(x-2)(x^2+2x+4)$
45. $2(3x-y)(9x^2+3xy+y^2)$ **47.** $y(x+y)(x^2-xy+y^2)$ **49.** $x^2y^2(1-y)(1+y+y^2)$
51. $3xy(2x+3y)(4x^2-6xy+9y^2)$ **53.** $(x^2-y^3)(x^4+x^2y^3+y^6)$ **55.** $(3x+y^2)(9x^2-3xy^2+y^4)$
57. $(2x+y)(4x^2-2xy+y^2)$ **59.** $8(y-1)(y^2+y+1)$ **61.** $(3x-y)(3x+y)$ **63.** $(x-2y)(x+2y)(x^2+4y^2)$
65. $(x-y-9)(x-y+9)$ **67.** $(x-y-6)(x-y+6)$ **69.** $(4x+1-y)(4x+1+y)$ **71. a.** x^2-16 **b.**

$x-4$
$x+4$

73. a. $xy+xy+x^2+y^2=x^2+2xy+y^2=(x+y)^2$ **b.**

$(x+y)(x+y)=(x+y)^2$

75. For a 3-digit integer: $abc=100a+10b+c=(99+1)a+(9+1)b+c=9(11a+b)+a+b+c$ So, if the sum $(a+b+c)$ is
divisible by 3 (or 9), then the number abc will be divisible by 3 (or 9).
For a 4-digit integer: $abcd=1000a+100b+10c+d=(999+1)a+(99+1)b+(9+1)c+d=9(111a+11b+c)+a+b+c+d$
So, if the sum $(a+b+c+d)$ is divisible by 3 (or 9), then the number $abcd$ will be divisible by 3 (or 9).

Section 7.10: Solving Quadratic Equations by Factoring

Margin Exercises 1. $y=7,\dfrac{5}{3}$ **2. a.** $v=0$ or $v=2$ **b.** $x=3$ (double root) **c.** $x=4$ or $x=-1$ **d.** $x=-\dfrac{7}{2}$ or $x=\dfrac{4}{3}$ **e.** $x=3$ or
$x=-1$ **3.** $x=0$ or $x=-2$ or $x=5$ **4.** $2x^2-11x+12=0$
Exercises 1. $x=2,3$ **3.** $x=-2,\dfrac{9}{2}$ **5.** $x=-3$ **7.** $x=-5$ **9.** $x=0,2$ **11.** $x=-6$ **13.** $x=-1,4$ **15.** $x=-3,4$ **17.** $x=-3,0$
19. $x=2,4$ **21.** $x=-4,3$ **23.** $x=-\dfrac{1}{2},3$ **25.** $x=-\dfrac{2}{3},2$ **27.** $x=-\dfrac{1}{2},4$ **29.** $x=-2,\dfrac{4}{3}$ **31.** $x=\dfrac{3}{2}$ **33.** $x=0,\dfrac{8}{5}$
35. $x=-2,2$ **37.** $x=1$ **39.** $x=2$ **41.** $x=-3,3$ **43.** $x=-5,10$ **45.** $x=-6,-2$ **47.** $x=\dfrac{1}{2}$ **49.** $x=0,2,4$ **51.** $x=-\dfrac{2}{3},-\dfrac{1}{2},0$
53. $x=-10,10$ **55.** $x=-5,5$ **57.** $x=-4$ **59.** $x=3$ **61.** $x=-1,3$ **63.** $x=-8,-2$ **65.** $x=-5,2$ **67.** $x=-5,7$ **69.** $x=-6,2$

71. $x = -1, \dfrac{2}{3}$ **73.** $x = -\dfrac{3}{2}, 4$ **75.** $y^2 - y - 6 = 0$ **77.** $2x^2 + 11x + 5 = 0$ **79.** $8x^2 - 10x + 3 = 0$ **81.** $x^3 - x^2 - 6x = 0$
83. $y^3 - 4y^2 - 3y + 18 = 0$ **85.** This allows us to use the zero-factor property which says that for the product to equal zero one of the factors must equal zero. Answers will vary. **87. a.** 640 ft; 384 ft **b.** 144 ft; 400 ft **c.** 7 seconds; $0 = -16(t + 7)(t - 7)$

Section 7.11: Applications of Quadratic Equations

Margin Exercises 1. a. –7 and –4 or 4 and 7 **b.** 12 houses **2.** –9, –11 **3.** 57.87 feet
Exercises 1. $x(x + 8) = -16$; $x = -4$, so the numbers are –4 and 4 **3.** $x^2 = 7x$; $x = 0, 7$ **5.** $x^2 + 3x = 28$; $x = 4$
7. $x(x + 7) = 78$; $x = -13, 6$; so the numbers are –13 and –6 or 13 and 6 **9.** $(x + 6)^2 + x^2 = 260$; $x = 8$; so the numbers
are 8 and 14 **11.** $x + (x + 8)^2 = 124$; $x = 3$; so the integers are 3 and 11 **13.** $x(2x - 5) = x + 56$; $x = -4$
15. $x(x + 1) = 72$; $x = 8$; so the integers are 8 and 9 **17.** $x^2 + (x + 1)^2 = 85$; $x = 6$; so the integers are 6 and 7
19. $x(x + 2) = 63$; $x = -9, 7$; so the integers are –9 and –7 or 7 and 9 **21.** $4x + (x + 1)^2 = 41$; $x = 4$; so the integers are
4 and 5 **23.** $2x(x + 1) = (x + 1)(x + 2) + 88$; $x = 10$; so the integers are 10, 11, and 12 **25.** $6x(x + 2) = (x + 1) + (x + 3)^2$;
$x = -2, 1$; so the integers are –2, –1, 0, and 1 or 1, 2, 3, and 4 **27.** $w(2w) = 72$; $w = 6$; so width is 6 in. and length is 12 in.
29. $w(4w) = 64$; $w = 4$; so width is 4 ft and length is 16 ft **31.** $l(l - 4) = 117$; $l = 13$; so width is 9 ft and length is 13 ft
33. $\dfrac{1}{2}b(b - 4) = 16$; $b = 8$; so base is 8 ft and height is 4 ft **35.** $\dfrac{1}{2}(h + 15)h = 63$; $h = 6$; so base is 21 in.
37. $w(16 - w) = 48$; $w = 4, 12$; so the rectangle is 4 in. by 12 in. **39.** $r(r + 13) = 140$; $r = 7$; so there are 7 trees in each row
41. $r(r + 7) = 144$; $r = 9$; so there are 9 rows **43.** $n(n + 1675) = 8400$; $n = 5$, so there are 5 floors **45.** $(w + 11)(w + 4) = 98$;
$w = 3$; so the rectangle is 3 cm by 10 cm **47.** $w(50 - 2w) = 300$; $w = 10, 15$; so width is 10 ft and length is 30 ft or width
is 15 ft and length is 20 ft **49.** $h^2 + (h - 34)^2 = (h + 2)^2$; $h = 48$; so the height of the pole is 48 ft **51.** $x^2 + (x - 49)^2 = (x + 1)^2$;
$x = 60$; so height is 60 ft **53.** $l^2 + (l - 28)^2 = (l + 8)^2$; $l = 60$, so the length of the mat is 60 inches **55.** \$1.50 per pound
57. \$16 or \$20 per reel **59.** $12^2 + 5^2 = 13^2; 20^2 + 21^2 = 29^2; 24^2 + 7^2 = 25^2; 14^2 + 48^2 = 50^2; 60^2 + 11^2 = 61^2$

Chapter 7 Test

1. $-10a^5$ **2.** 1 **3.** $\dfrac{4x^3}{y^7}$ **4.** $\dfrac{x}{3y^2}$ **5.** $\dfrac{x^2}{4y^2}$ **6.** $4x^2y^4$ **7.** $8x^2 + 3x$; second-degree binomial; leading coefficient 8
8. $-x^3 + 3x^2 + 3x - 1$; third-degree polynomial; leading coefficient –1 **9.** $5x^5 + 2x^4 - 11x + 3$; fifth-degree polynomial;
leading coefficient 5 **10. a.** 20 **b.** –110 **11.** $-3x + 2$ **12.** $20x - 8$ **13.** $5x^3 - x^2 + 6x + 5$ **14.** $7x^4 + 14x^2 + 4$
15. $15x^7 - 20x^6 + 15x^5 - 40x^4 - 10x^2$ **16.** $49x^2 - 9$; difference of two squares **17.** $16x^2 + 8x + 1$; perfect square trinomial
18. $36x^2 - 60x + 25$; perfect square trinomial **19.** $12x^2 + 24x - 15$ **20.** $6x^3 - 69x^2 + 189x$ **21.** $7x^2 + 7x + 14$
22. $10x^4 + 4x^3 - 15x^2 - 41x - 14$ **23.** $2x + \dfrac{3}{2} - \dfrac{3}{x}$ **24.** $\dfrac{5a}{3} + 2a^2 + \dfrac{1}{a}$ **25.** $2x^2 - 3x - 13$
26. a. $A(x) = 240 - (x^2 + 3x) = -(x^2 + 3x - 240)$ **b.** $P(x) = 64$ **27.** $7ab^2(4x - 3y)$ **28.** $6yz^2(3z - yz + 2)$ **29.** $(x - 5)(x - 4)$
30. $-(x + 7)^2$ **31.** $(x - 5)(y - 7)$ **32.** $6(x + 1)(x - 1)$ **33.** $2(6x - 5)(x + 1)$ **34.** $(x + 3)(3x - 8)$ **35.** $(4x - 5y)(4x + 5y)$
36. $x(x + 1)(2x - 3)$ **37.** $(2x - 3)(3x - 2)$ **38.** $(y + 7)(2x - 3)$ **39.** not factorable **40.** $-3x(x^2 - 2x + 2)$ **41.** $x = -1, 8$
42. $x = -6, 0$ **43.** $x = -\dfrac{3}{4}, 5$ **44.** $x = \dfrac{3}{2}, 4$ **45.** $x^2 + 5x - 24 = 0$ **46.** $4x^2 - 4x + 1 = 0$ **47.** 6, 20 or –30, –4 **48.** 3, 12
49. 18, 19 **50.** length = 15 centimeters; width = 11 centimeters **51.** 18 cm **52.** $P(x) = 4(3x + 5)$

Chapters 1-7 Cumulative Review

1. 120 **2.** $168x^2y$ **3.** 6 **4.** –138 **5.** $\dfrac{55}{48}$ **6.** $\dfrac{19}{60a}$ **7.** $\dfrac{3}{10}$ **8.** $\dfrac{75x}{23}$ **9.** $x = \dfrac{9}{10}$ **10.** $x = 21$ **11.** $x = 21$

12. conditional **13.** identity **14.** contradiction **15.** $x = \dfrac{y - b}{m}$ **16.** $y = \dfrac{10 - 3x}{5}$ **17.** 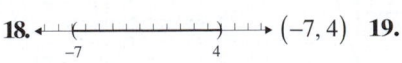 $[-6, \infty)$

18. 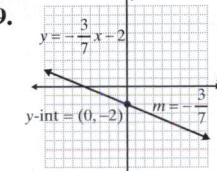 $(-7, 4)$ **19.** 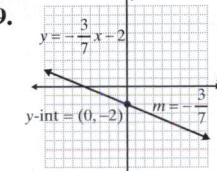 **20.** $y = 7$ 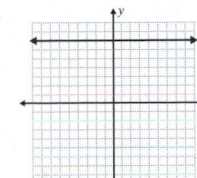 **21.** $x + y = -2$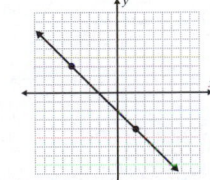

© Hawkes Learning Systems. All rights reserved.

22. $y = 3x + 4$

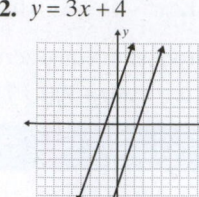

23. $y = -\dfrac{1}{3}x + \dfrac{2}{3}$

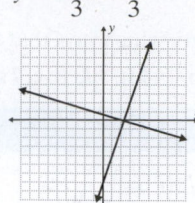

24. a.

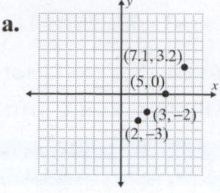

b. $D = \{2, 3, 5, 7.1\}$ **c.** $R = \{-3, -2, 0, 3.2\}$

d. It is a function because each first coordinate (domain) has only one corresponding second coordinate (range).

25. function; $D = (-\infty, \infty)$; $R = (-\infty, \infty)$

26. not a function; $D = (-\infty, 4]$; $R = (-\infty, \infty)$ **27. a.** 58 **b.** 4 **c.** $\dfrac{11}{4}$

28.

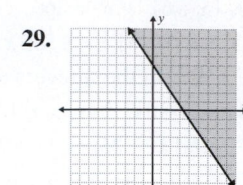

29.

30. $64x^6 y^3$ **31.** $-\dfrac{1}{8x^9 y^6}$ **32.** $7x^2 y$ **33.** $\dfrac{36x^4}{y^{10}}$ **34.** $\dfrac{y^2}{x^3}$ **35.** $\dfrac{9y^{14}}{x^4}$

36. a. $\left(5.6 \times 10^{-7}\right)\left(3 \times 10^{-4}\right)$; 1.68×10^{-10}

b. $\dfrac{\left(8.1 \times 10^4\right)\left(6.2 \times 10^3\right)}{\left(3 \times 10^{-3}\right)\left(2 \times 10^{-1}\right)}$; 8.37×10^{11}

37. $13x + 1$ **38.** $x^2 + 12x + 3$ **39.** $4x^2 + 5x - 8$ **40.** $-4x^2 - 2x + 3$ **41.** $2x^2 + x - 28$ **42.** $-3x^2 - 17x + 6$

43. $x^2 + 12x + 36$ **44.** $4x^2 - 28x + 49$ **45.** $2x - \dfrac{5}{4} + \dfrac{1}{y}$ **46.** $4(2x - 5)$ **47.** $6(x - 16)$ **48.** $(x + 2)(y + 3)$

49. $(x - 2)(a + b)$ **50.** $(x - 6)(x - 3)$ **51.** $(2x - 3)(3x + 4)$ **52.** $-4x(3x + 4)$ **53.** $8xy(2x - 3)$ **54.** $5x^2(x - 5)(x + 8)$

55. $(2x + 1)(2x - 1)$ **56.** $3(x + 4y)(x - 4y)$ **57.** not factorable **58.** $(x + 1)(3x + 2)$ **59.** $2x(x - 5)(x - 5)$ **60.** $4x(x^2 + 25)$

61. $5(x - 3y)(x^2 + 3xy + 9y^2)$ **62.** $x = -1, 7$ **63.** $x = 0, 7$ **64.** $x = -6, -2$ **65.** $x = -4, 7$ **66.** $x = -7, 0$ **67.** $x = -2, 0$

68. $x = -\dfrac{5}{4}, 1$ **69.** $x = -5, 0, 2$ **70.** $x^2 + 15x + 50 = 0$ **71.** $x^3 - 17x^2 + 52x = 0$ **72. a.** 13% **b.** 11% **c.** a **73.** 65 mph

74. $7500 at 6%; $2500 at 8% **75.** 13 inches by 17 inches **76.** 4, 11 **77.** 8 and 9 or -8 and -9 **78.** 5 m

Chapter 8: Systems of Linear Equations

Section 8.1: Systems of Equations: Solutions by Graphing

Margin Exercises 1. $6 - 2(4) = -2$ true statement; $18 + 8 = 26$ true statement **2.** $2 = -6 + 8$ true statement; $2 = 9 - 8$ false statement **3.** $(-3, 0)$ **4.** no solution **5.** infinitely many solutions **6.** $\left(\dfrac{3}{2}, 3\right)$ **7.** $(-4, -1)$

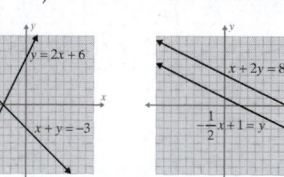
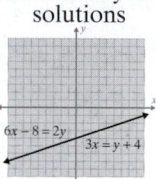
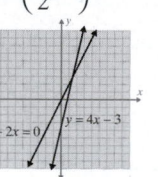
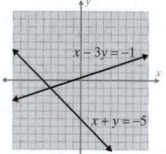

Exercises 1. c **3.** a, c **5.** $(0, 2)$ **7.** $(4, 2)$ **9.** $m_1 = -2, b_1 = 3; m_2 = -2, b_2 = \dfrac{5}{2}$ **11.** $m_1 = \dfrac{1}{2}, b_1 = 3; m_2 = \dfrac{1}{2}, b_2 = -\dfrac{1}{2}$

13. $(-3, -2)$
consistent

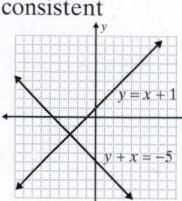

15. $(2, 0)$
consistent

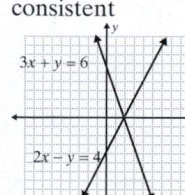

17. infinitely many solutions dependent

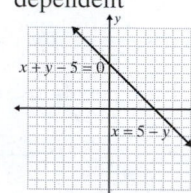

19. no solution inconsistent

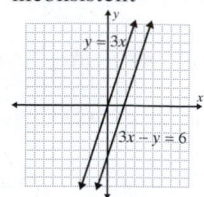

21. $(-3, -8)$
consistent

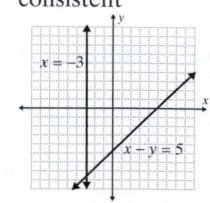

23. $(2, 3)$
consistent

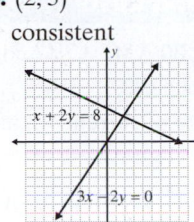

25. infinitely many solutions dependent

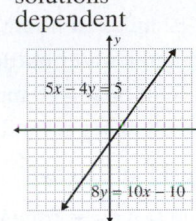

27. $(-1, -1)$
consistent

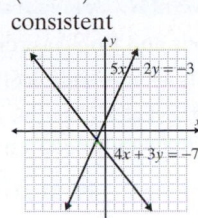

29. $(5, -1)$
consistent

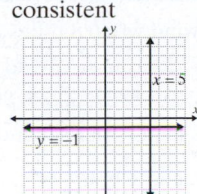

31. no solution inconsistent

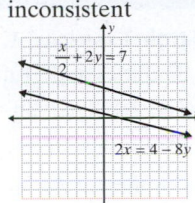

33. $(1, 3)$
consistent

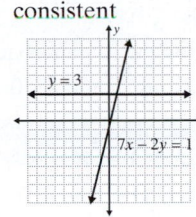

35. infinitely many solutions dependent

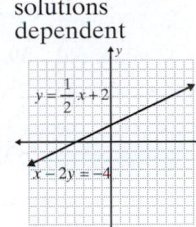

37. $(1, 1)$
consistent

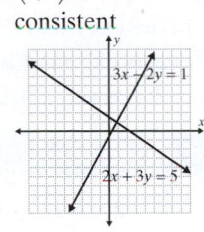

39. no solution inconsistent

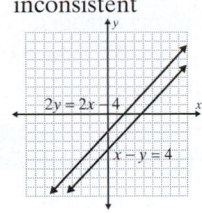

41. $(1, -1)$
consistent

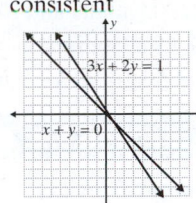

43. $(20, 5)$

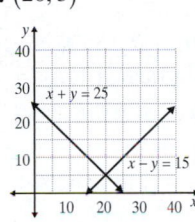

45. 5 gallons of 12%; 10 gallons of 3%

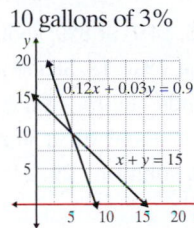

47. $(1, 4)$

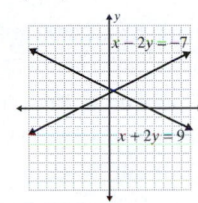

49. $(1.5, 2)$

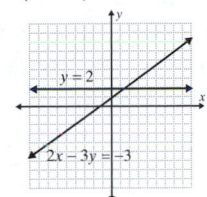

51. $(1.5, -3)$

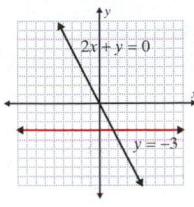

53. $(1.4, 2.1)$

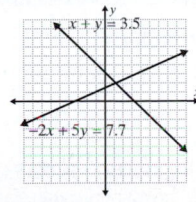

55. The solution to a consistent system of linear equations is a single point, which is easily written as an ordered pair.

Section 8.2: Systems of Equations: Solutions by Substitution

Margin Exercises **1.** $\left(7, \dfrac{29}{5}\right)$ **2.** no solution **3.** infinitely many solutions **4.** $x = -1.5$; $y = 7.5$

Exercises **1.** $(2, 4)$ **3.** dependent system **5.** $(-6, -2)$ **7.** $(4, 1)$ **9.** no solution **11.** $(3, 2)$ **13.** $(4, -5)$ **15.** $(3, -2)$

17. $\left(2, \dfrac{5}{2}\right)$ **19.** $\left(\dfrac{1}{2}, -4\right)$ **21.** $\left(\dfrac{7}{2}, -\dfrac{1}{2}\right)$ **23.** dependent system **25.** $(-2, 1)$ **27.** $\left(-\dfrac{4}{5}, -\dfrac{7}{5}\right)$ **29.** $\left(\dfrac{11}{7}, \dfrac{8}{7}\right)$ **31.** dependent

system **33.** $(3, 3)$ **35.** $(10, 20)$ **37.** no solution **39.** dependent system **41.** $\left(2, \dfrac{5}{3}\right)$ **43.** $(20, 5)$ **45.** 5 gallons of 12%;

10 gallons of 3% **47.** Answers will vary.

Section 8.3: Systems of Equations: Solutions by Addition

Margin Exercises **1.** $(2, -3)$ **2.** The system has infinitely many solutions. **3.** $(2.4, -4.6)$ **4.** $y = \dfrac{5}{2}x - 12$

Exercises **1.** consistent; $(4, 3)$ **3.** consistent; $\left(1, -\dfrac{3}{2}\right)$ **5.** inconsistent; no solution **7.** dependent; infinitely many

solutions **9.** consistent; $(-2, -3)$ **11.** consistent; $(7, 5)$ **13.** $(1, -5)$ **15.** infinitely many solutions **17.** $\left(\dfrac{22}{7}, -\dfrac{2}{7}\right)$

19. $(2, -2)$ **21.** infinitely many solutions **23.** $(2, -1)$ **25.** $(5, -6)$ **27.** $(3, -1)$ **29.** $(2, 4)$ **31.** $(-6, 2)$

33. no solution **35.** $(20, 10)$ **37.** $\left(2, \dfrac{10}{9}\right)$ **39.** no solution **41.** $\left(-\dfrac{45}{7}, \dfrac{92}{7}\right)$ **43.** $y = 5x - 7$; $m = 5$, $b = -7$

45. $y = -3$; $m = 0$, $b = -3$ **47.** $y = \dfrac{1}{2}x + \dfrac{3}{2}$; $m = \dfrac{1}{2}$, $b = \dfrac{3}{2}$ **49.** $4000 at 10%; $6000 at 6% **51.** 40 liters of 30% solution (x),

60 liters of 40% solution (y)

Section 8.4: Applications: Distance-Rate-Time, Number Problems, Amounts and Costs

Margin Exercises **1.** The wind was 0.8 miles an hour and Bob was running 9.2 miles an hour. **2.** 11:30 AM **3.** 93 and 57 **4.** 33 dimes and 11 nickels **5.** Tom is 13 and Jane is 4 **6.** A soda costs \$1.25 and a water bottle costs \$0.80.

Exercises **1.** 23, 33 **3.** 15, 21 **5.** 37, 50 **7.** 80°, 100° **9.** 55°, 55°, 70° **11.** rate of boat = 10 mph; rate of current = 2 mph **13.** Bolt's speed was 10.32 meters per second and the wind speed was 0.12 meters per second. **15.** He traveled $1\frac{1}{2}$ hours at the first rate and 2 hours at the second rate. **17.** Marcos traveled at 40 mph and Cana traveled at 51 mph. **19.** Steve traveled at 28 mph and Tim traveled at 7 mph. **21.** The westbound train was traveling 45 mph and the eastbound train was traveling 40 mph. **23.** 12 miles **25.** 20 nickels and 10 dimes **27.** 52 nickels and 130 pennies **29.** 800 adults and 2700 students attended **31.** $l = 14$ meters; $w = 8$ meters **33.** 100 yards × 45 yards **35.** $l = 33$ meters; $w = 17$ meters **37.** Priscilla was 21 years old, and Elvis was 32 years old. **39.** She bought 10 paperbacks and 5 hardbacks. **41.** 5000 general admission and 7500 reserved tickets were sold. **43.** 22 at \$625 and 25 and \$550

Section 8.5: Applications: Interest and Mixture

Margin Exercises **1.** Fergus has \$5000 at 9% and \$5800 invested at 12%. **2.** He should invest \$2500 at 9% and \$6500 at 5%. **3.** 50 ounces of the 18% solution and 100 ounces of the 12% solution **4.** 96 gallons of the 22% solution and 24 gallons of the 17% solution

Exercises **1.** \$5500 at 6%; \$3500 at 10% **3.** \$7400 at 5.5%; \$2600 at 6% **5.** \$450 at 8%; \$650 at 10% **7.** \$3500 in each or \$7000 total **9.** \$20,000 at 24%; \$11,000 at 18% **11.** \$800 at 5%; \$2100 at 7% **13.** \$8500 at 9%; \$3500 at 11% **15.** \$87,000 in bonds; \$37,000 in certificates **17.** 20 pounds of 20%; 30 pounds of 70% **19.** 20 ounces of 30%; 30 ounces of 20% **21.** 450 pounds of 35%; 1350 pounds of 15% **23.** 20 pounds of 40%; 30 pounds of 15% **25.** 10 g of acid; 20 g of the 40% solution **27.** 10 oz of salt; 50 oz of the 4% solution **29.** 2 lb of 72%; 4 lb of 42% **31.** 7.11 oz of 0.5% solution; 0.89 oz of 5% solution **33.** Answers will vary.

Section 8.6: Systems of Linear Equations: Three Variables

Margin Exercises **1.** $(1, -2, 4)$ **2.** no solution **3.** infinitely many solutions **4.** 18 dollar coins, 12 quarters, and 7 dimes.
Exercises **1.** $(1, 0, 1)$ **3.** $(1, 2, -1)$ **5.** infinitely many solutions **7.** $(4, 1, 1)$ **9.** $(1, 2, -1)$ **11.** $(-2, 3, 1)$ **13.** no solution **15.** $(3, -1, 2)$ **17.** $(2, 1, -3)$ **19.** $\left(\frac{1}{2}, \frac{1}{3}, -1\right)$ **21.** 34, 6, 27 **23.** 18 ones, 16 fives, 12 tens **25.** 19 cm, 24 cm, 30 cm **27.** 300 main floor, 200 mezzanine, and 80 balcony **29.** 3 lillies, 5 roses, and 8 daisies **31.** savings: \$30,000; bonds: \$55,000; stocks: \$15,000 **33.** 3 liters of 10%, 4.5 liters of 30%, 1.5 liters of 40% **35.** No. Graphically, the three planes intersect in one point (one solution) or in a line (infinitely many solutions) or they do not have a common intersection (no solution). **37.** $A = 4, B = 2, C = -1$

Section 8.7: Matrices and Gaussian Elimination

Margin Exercises **1. a.** $\begin{bmatrix} 2 & 1 & 1 \\ 1 & 2 & 1 \\ 3 & 1 & -1 \end{bmatrix}, \begin{bmatrix} 2 & 1 & 1 & | & 4 \\ 1 & 2 & 1 & | & 1 \\ 3 & 1 & -1 & | & -3 \end{bmatrix}$ **b.** $\begin{bmatrix} 3 & 1 & -1 & | & -3 \\ 3 & 6 & 3 & | & 3 \\ 2 & 1 & 1 & | & 4 \end{bmatrix}$ **c.** $\begin{bmatrix} 4 & 5 & 3 & | & 6 \\ 1 & 2 & 1 & | & 1 \\ 3 & 1 & -1 & | & -3 \end{bmatrix}$ **2.** $(3, -2)$ **3.** $(4, -3, -1)$ **4.** $(1, -1, 3)$

Exercises **1.** $\begin{bmatrix} 2 & 2 \\ 5 & -1 \end{bmatrix}, \begin{bmatrix} 2 & 2 & | & 13 \\ 5 & -1 & | & 10 \end{bmatrix}$ **3.** $\begin{bmatrix} 7 & -2 & 7 \\ -5 & 3 & 0 \\ 0 & 4 & 11 \end{bmatrix}, \begin{bmatrix} 7 & -2 & 7 & | & 2 \\ -5 & 3 & 0 & | & 2 \\ 0 & 4 & 11 & | & 8 \end{bmatrix}$ **5.** $\begin{bmatrix} 3 & 1 & -1 & 2 \\ 1 & -1 & 2 & -1 \\ 0 & 2 & 5 & 1 \\ 1 & 3 & 0 & 3 \end{bmatrix}, \begin{bmatrix} 3 & 1 & -1 & 2 & | & 6 \\ 1 & -1 & 2 & -1 & | & -8 \\ 0 & 2 & 5 & 1 & | & 2 \\ 1 & 3 & 0 & 3 & | & 14 \end{bmatrix}$

7. $\begin{cases} -3x + 5y = 1 \\ -x + 3y = 2 \end{cases}$ **9.** $\begin{cases} x + 3y + 4z = 1 \\ 2x - 3y - 2z = 0 \\ x + y = -4 \end{cases}$ **11. a.** $\begin{bmatrix} -1 & 7 \\ 1 & 4 \end{bmatrix}$ **b.** $\begin{bmatrix} 1 & 4 \\ -2 & 14 \end{bmatrix}$ **13. a.** $\begin{bmatrix} 1 & 3 & 7 \\ 4 & -1 & 6 \\ -8 & -2 & 5 \end{bmatrix}$ **b.** $\begin{bmatrix} 1 & 3 & 7 \\ -16 & 0 & -7 \\ 4 & -1 & 6 \end{bmatrix}$

15. $(-1, 2)$ **17.** $(-1, -1)$ **19.** $(-1, -2, 3)$ **21.** $(1, 0, 1)$ **23.** $(2, 1, -1)$ **25.** $(-2, 9, 1)$ **27.** no solution **29.** infinitely many solutions **31.** $(1, -3, 2)$ **33.** 52, 40, 77 **35.** bacon: \$3.09/lb; eggs: \$4.03/doz; bread: \$1.40/loaf **37.** $(0, -4)$

39. $(2, 1, 7)$ **41.** $(2, -3, 4)$ **43.** $\left(\dfrac{13}{12}, \dfrac{5}{4}, \dfrac{8}{3}\right)$ **45.** Solving the second equation for z, we can back substitute into the first equation, eliminating z. The result is the equation $x + 5y = 6$ which means the system has an infinite number of solutions.

Section 8.8: Systems of Linear Inequalities

Margin Exercises 1. a. 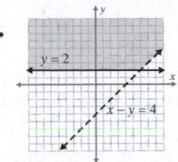 **b.** 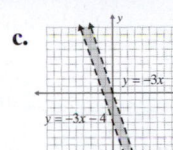 **c.** **2.**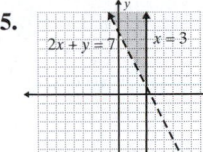

Exercises 1. 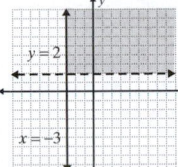 **3.** **5.** 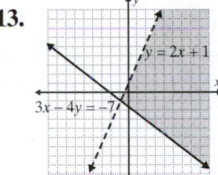 **7.**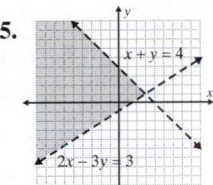

9. 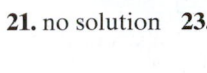 **11.** **13.** **15.** 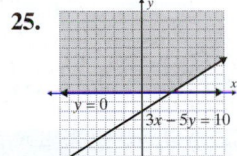 **17.**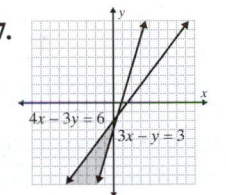

19. 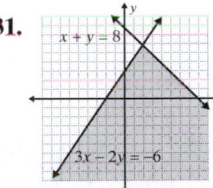 **21.** no solution **23.** **25.** 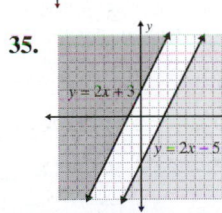 **27.**

29. 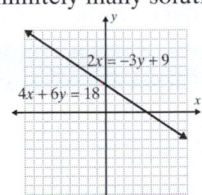 **31.** 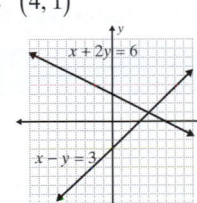 **33.** **35.** The solutions of the two inequalities do not overlap.

Chapter 8 Test

1. c **2.** infinitely many solutions **3.** $(4, 1)$ **4.** $(-8, -20)$ **5.** $(-2, 6)$ **6.** no solution **7.** $\left(4, -\dfrac{1}{2}\right)$

8. $(-3, 5)$ **9.** no solution **10.** $(3, -6)$ **11.** infinitely many solutions **12.** $(-1, 1, -2)$ **13.** no solution

14. a. $\begin{bmatrix} 1 & 2 & -3 \\ 1 & -1 & -1 \\ 1 & 3 & 2 \end{bmatrix} \begin{bmatrix} 1 & 2 & -3 & | & -11 \\ 1 & -1 & -1 & | & 2 \\ 1 & 3 & 2 & | & -4 \end{bmatrix}$ **b.** $(1, -3, 2)$ **15.** 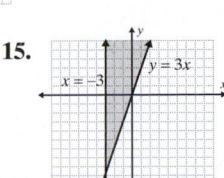 **16.**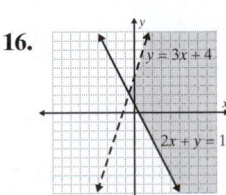

17. soy latte: $60.20; iced coffee: $18.90 **18.** speed of the boat: 14 mph; speed of the current: 2 mph **19.** pen price: $0.79; pencil price: $0.08 **20.** $1600 at 8% and $960 at 6% **21.** 17 inches by 13 inches **22.** 1600 lb of 83% and 400 lb of 68% **23.** nickels = 45; quarters = 60 **24.** 41¢ stamps: 60; 58¢ stamps: 20; 75¢ stamps: 10

1. $200,016.04$ **2.** 17.99 **3.** 0.4 **4.** 180% **5.** 0.015 **6.** $\dfrac{4}{3}$ **7.** $3\dfrac{1}{11}$ or $\dfrac{34}{11}$ **8.** $-\dfrac{4}{21}$ **9.** $10\dfrac{2}{15}$ **10.** $46\dfrac{5}{12}$ **11.** 12 **12.** $1\dfrac{4}{5}$

13. $-560,000$ **14.** -26.9 **15.** 75.74 **16.** 0.00049923 **17.** -80.6 **18.** 7 **19.** -11.11 **20.** -2.75 **21.** $2^2 \cdot 3^2 \cdot 11$

22. a. $C = 62.8$ ft **b.** $A = 314$ ft^2 **23.** $5\sqrt{5}$ in. or 11.18 in. **24.** $x = 3$ **25.** $x = 4$ **26.** $x = -\dfrac{9}{8}$ **27.** $x = \dfrac{29}{11}$ **28.** $\left(-\infty, -\dfrac{1}{4}\right]$

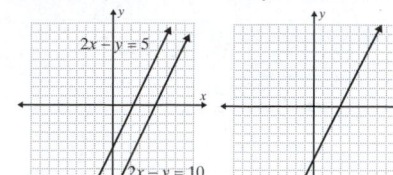 **29.** $\left(-\infty, -\dfrac{50}{3}\right)$ 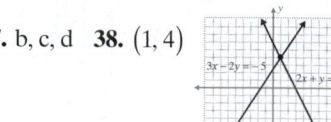 **30.** m is undefined; no y-intercept; vertical line

31. $m = \dfrac{3}{5}$; y-intercept $= \left(0, -\dfrac{2}{5}\right)$ **32.** $m = 4$; y-intercept $= (0, -1)$ **33.** $m = 0$; y-intercept $= \left(0, \dfrac{3}{2}\right)$; horizontal line

34. $2x - y = 10$ **35.** $y = 2x - 6$ **36. a.** not a function; $D = [-5, 5]$; $R = [-3, 3]$ **b.** function;

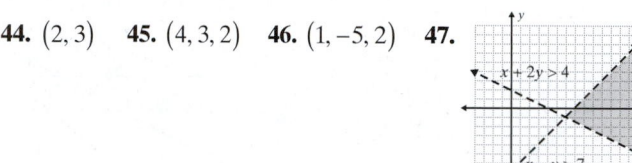 $D = \{-7, -3, 0, 5, 6\}$; $R = \{-2, 1, 4\}$ **c.** function; $D = [-3, 3]$; $R = [0, 6]$

37. b, c, d **38.** $(1, 4)$ 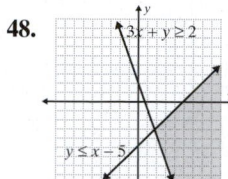 **39.** $(1, 3)$ **40.** $(-6, 2)$ **41.** $(-1, 4)$

42. no solution

43. infinitely many solutions

44. $(2, 3)$ **45.** $(4, 3, 2)$ **46.** $(1, -5, 2)$ **47.** **48.** **49.** 3 and 11 **50.** $4200 at 7%; $2800 at 8%

Chapter 9: Roots and Radicals

Section 9.1: Radical Expressions

Margin Exercises 1. a. $-9, 9$ **b.** $-14, 14$ **c.** $-7, 7$ **d.** not a real number **2. a.** $\dfrac{3}{8}$ **b.** 0.04 **3.** Because $64 < 67 < 81$, we have $\sqrt{64} < \sqrt{67} < \sqrt{81}$ and $8 < \sqrt{67} < 9$. The approximation 8.1854 is between 8 and 9 and is reasonable. **4. a.** 4 **b.** -5 **c.** $\dfrac{1}{10}$ **5. a.** 2 **b.** 4 **6. a.** 6.708203932 **b.** 27.386127875 **c.** 13.291502622

Exercises 1. 3 **3.** 9 **5.** 17 **7.** 13 **9.** 1 **11.** 5 **13.** 6 **15.** $\dfrac{1}{2}$ **17.** $\dfrac{3}{4}$ **19.** 0.2 **21.** -10 **23.** -0.04 **25.** -3 **27.** -5 **29.** $\dfrac{3}{5}$

31. Because $64 < 74 < 81$, we have $\sqrt{64} < \sqrt{74} < \sqrt{81}$ and $8 < \sqrt{74} < 9$. $(8.6023)^2 = 73.99956529$ **33.** Because $25 < 32 < 36$,

we have $\sqrt{25} < \sqrt{32} < \sqrt{36}$ and $5 < \sqrt{32} < 6$. $(5.6569)^2 = 32.00051761$ **35.** rational **37.** rational **39.** irrational

41. nonreal **43.** rational **45.** irrational **47.** 4 **49.** -2 **51.** 4 **53.** 13 **55.** 6.2450 **57.** 2.4960 **59.** 0.4472 **61.** 8.9443

63. -8.2462 **65.** 7.8990 **67.** -5.4641 **69.** There is no real number that results in a negative number when squared.

Section 9.2: Simplifying Radicals

Margin Exercises 1. a. $7\sqrt{2}$ **b.** $3\sqrt{5}$ **c.** $\dfrac{2\sqrt{3}}{5}$ **2. a.** $6z$ **b.** $5b\sqrt{3}$ **c.** $3cd\sqrt{5}$ **3. a.** $4x^4$ **b.** $5b^2\sqrt{2b}$ **c.** $2x^4 y^6\sqrt{3}$ **d.** $\dfrac{5z^9}{y^4}$

4. a. $2z\sqrt[3]{6}$ **b.** $-3a^2 b^4\sqrt[3]{3a^2}$ **c.** $7x^2 y^3\sqrt[3]{2y^2}$

Exercises 1. $2\sqrt{3}$ **3.** $12\sqrt{2}$ **5.** $-6\sqrt{2}$ **7.** $-2\sqrt{14}$ **9.** $-5\sqrt{5}$ **11.** $\dfrac{1}{2}$ **13.** $-\dfrac{\sqrt{11}}{8}$ **15.** $\dfrac{2\sqrt{7}}{5}$ **17.** $6x$ **19.** $2x\sqrt{2x}$

21. $2x^5 y\sqrt{6x}$ **23.** $5xy^3\sqrt{5x}$ **25.** $-3xy\sqrt{2}$ **27.** $2bc\sqrt{3ac}$ **29.** $5x^2 y^3 z^4\sqrt{3}$ **31.** $\dfrac{x^2\sqrt{5}}{3}$ **33.** $\dfrac{4a^2\sqrt{2a}}{9b^8}$ **35.** $\dfrac{10x^4\sqrt{2}}{17}$

37. 6 **39.** $2\sqrt[3]{7}$ **41.** -1 **43.** $-4\sqrt[3]{2}$ **45.** $5x\sqrt[3]{x}$ **47.** $-2x^2\sqrt[3]{x^2}$ **49.** $2a^2 b\sqrt[3]{9b}$ **51.** $6x^2 y\sqrt[3]{y^2}$ **53.** $2xy^2 z^3\sqrt[3]{3x^2 y}$ **55.** $\dfrac{\sqrt[3]{3}}{2}$

57. $\dfrac{5\sqrt[3]{3}}{2}$ **59.** $\dfrac{5y^4}{3x^2}$ **61.** 7 inches **63.** $\sqrt{6} \approx 2.45$ amperes **65.** 120 volts **67.** When $a < 0$.

Section 9.3: Rational Exponents

Margin Exercises 1. a. 6 **b.** 2 **c.** $\dfrac{1}{3}$ **d.** 0.2 **e.** no solution **2. a.** $\sqrt[5]{x^2}$ **b.** $8\sqrt[7]{z^6}$ **c.** $-b\sqrt[4]{b}$ **d.** $x^{\frac{5}{7}}$ **e.** $3s^{\frac{1}{2}}$ **f.** $-5^{\frac{1}{4}}$

3. a. $x^{\frac{7}{12}}$ **b.** $\dfrac{1}{a^{\frac{2}{9}}}$ **c.** $81b^{\frac{4}{5}}$ **d.** $\dfrac{z^{\frac{2}{9}}}{8}$ **e.** not a real number **f.** 4 **4. a.** $\sqrt[10]{x}$ **b.** $\sqrt[4]{x^3}$ **c.** $x\sqrt[3]{x}$ **5. a.** 256 **b.** 22.527227346

Exercises 1. $\sqrt[3]{8}$ **3.** $-\sqrt[6]{x}$ **5.** $\sqrt[5]{(2z)^2}=\sqrt[5]{4z^2}$ **7.** $13^{\frac{1}{7}}$ **9.** $(-9)^{\frac{1}{3}}$ **11.** 3 **13.** $\dfrac{1}{10}$ **15.** -512 **17.** -4 **19.** nonreal **21.** $\dfrac{3}{7}$

23. 16 **25.** $-\dfrac{1}{6}$ **27.** $\dfrac{5}{2}$ **29.** $\dfrac{1}{4}$ **31.** $\dfrac{3}{8}$ **33.** $-\dfrac{1}{1000}$ **35.** 64 **37.** 8.5499 **39.** 10,000,000 **41.** 99.6055 **43.** 0.0922

45. 1.6083 **47.** 0.2236 **49.** 7.7460 **51.** 2.0408 **53.** $8x$ **55.** $\dfrac{1}{3a^2}$ **57.** $8x^{\frac{5}{2}}$ **59.** $5a^{\frac{13}{6}}$ **61.** $x^{\frac{7}{12}}$ **63.** $x^{\frac{1}{2}}$ **65.** $\dfrac{1}{a^{\frac{9}{8}}}$

67. $a^{\frac{1}{4}}$ **69.** $\dfrac{a^2}{b^{\frac{6}{5}}}$ **71.** $8x^{\frac{3}{2}}y$ **73.** $\dfrac{x^{\frac{3}{2}}}{16y^{\frac{2}{5}}}$ **75.** $\dfrac{x^2y^4}{z^4}$ **77.** $\dfrac{y^{\frac{3}{2}}z^2}{x}$ **79.** $\dfrac{8b^{\frac{9}{4}}}{a^3c^3}$ **81.** $x^{\frac{1}{4}}y^{\frac{5}{4}}$ **83.** $\dfrac{y^{\frac{2}{3}}}{50x^{\frac{4}{3}}}$ **85.** $\dfrac{b^{\frac{5}{12}}}{a^{\frac{11}{12}}}$ **87.** $\dfrac{3x^{\frac{1}{2}}}{20y^{\frac{2}{3}}}$

89. $\sqrt[6]{a^5}$ **91.** $\sqrt[12]{y^7}$ **93.** $\sqrt[30]{x^{11}}$ **95.** $\sqrt[6]{y}$ **97.** $\sqrt[9]{x}$ **99.** $\sqrt[3]{7a}$ **101.** $\sqrt[24]{x}$ **103.** $a^{20}b^5c^{10}$ **105.** no: $\sqrt[5]{a}\cdot\sqrt{a}=a^{\frac{7}{10}};\sqrt[5]{a^2}=a^{\frac{2}{5}};a^{\frac{7}{10}}\neq a^{\frac{2}{5}}$

Section 9.4: Operations with Radicals

Margin Exercises 1. a. $9\sqrt{3a}$ **b.** $5\sqrt{5}+3\sqrt{3}$ **c.** $-\sqrt[3]{9x}$ **2. a.** 15 **b.** $13\sqrt{2}-13$ **c.** $8+3\sqrt{5}$ **d.** $12-2\sqrt{35}$ **e.** $3z-5$

3. a. $\dfrac{\sqrt{21}}{3}$ **b.** $\dfrac{\sqrt{14z}}{6z}$ **c.** $\dfrac{11\sqrt[3]{3x^2}}{3x}$ **4. a.** $-\dfrac{5\sqrt{3}+25}{22}$ **b.** $\dfrac{105-15\sqrt{6}}{43}$ **c.** $\dfrac{\sqrt{8}-\sqrt{3}}{5}$ **d.** $\dfrac{8+8\sqrt{z}}{1-z}$ **e.** $\dfrac{x\sqrt{x}-x\sqrt{y}+y\sqrt{x}-y\sqrt{y}}{x-y}$

5. a. 11.0711 **b.** 7 **c.** -8.6374

Exercises 1. $-6\sqrt{2}$ **3.** $7\sqrt{11}$ **5.** $-3\sqrt[3]{7x^2}$ **7.** $10\sqrt{3}$ **9.** $-7\sqrt{2}$ **11.** $6\sqrt{2}-\sqrt{3}$ **13.** $8\sqrt{3x}$ **15.** $20x\sqrt{5x}$ **17.** $14\sqrt{5}-3\sqrt{7}$

19. $-7\sqrt[3]{3x^2}-10\sqrt[3]{6x^2}$ **21.** $2x\sqrt{y}-y\sqrt{xy}$ **23.** $x^2y\sqrt{2x}$ **25.** $13+2\sqrt{2}$ **27.** $3x-9\sqrt{3x}+8$ **29.** $2-2\sqrt{7}$ **31.** -3

33. $13+4\sqrt{10}$ **35.** $\sqrt{10}+\sqrt{15}-\sqrt{6}-3$ **37.** $x-2\sqrt{6x}-18$ **39.** 58 **41.** $x+10\sqrt{xy}+25y$ **43.** $4x-9\sqrt{xy}-9y$ **45.** $\dfrac{\sqrt{21}}{3}$

47. $\dfrac{\sqrt[3]{70}}{2}$ **49.** $-\dfrac{\sqrt{6y}}{3y}$ **51.** $\dfrac{2\sqrt{10x}}{5y}$ **53.** $\dfrac{2\sqrt{2y}}{y}$ **55.** $\dfrac{y\sqrt[3]{2x}}{3x}$ **57.** $\dfrac{\sqrt[3]{30a^2b^2}}{5b^2}$ **59.** $\sqrt{2}-1$ **61.** $-\dfrac{\sqrt{15}+4\sqrt{3}}{11}$ **63.** $2\sqrt{3}-2\sqrt{2}$

65. $\dfrac{\sqrt{35}+\sqrt{15}}{4}$ **67.** $\dfrac{\sqrt{6}}{3}$ **69.** $-\dfrac{5+\sqrt{21}}{2}$ **71.** $\dfrac{2x+y\sqrt{x}-y^2}{x-y^2}$ **73.** 4.3397 **75.** 31.6 **77.** -57 **79.** -37.3569 **81.** 0.2831

83. 0.3820 **85.** $5\sqrt{6}+\sqrt{170}\approx 25.29$ ft **87.** Multiply both the numerator and the denominator by the conjugate of the denominator. This works because multiplying the denominator by its conjugate results in an expression with no square roots. Answers will vary.

Section 9.5: Solving Equations with Radicals

Margin Exercises 1. a. $x=-8$ or $x=8$ **b.** $y=-2$ **c.** $x=\dfrac{1}{2}$ or $x=-3$ **d.** no solution **2. a.** $x=2$ **b.** $x=9$, $x=25$ **3.** $x=8$

Exercises 1. $x=3$ **3.** no solution **5.** $x=-3$ **7.** $x=14$ **9.** $x=9$ **11.** no solution **13.** $x=6$ **15.** $x=-4,1$ **17.** $x=-5,\dfrac{5}{2}$

19. $x=-2$ **21.** $x=2,3$ **23.** $x=2,5$ **25.** $x=-5,5$ **27.** $x=4$ **29.** $x=4$ **31.** $x=3$ **33.** $x=2$ **35.** $x=2$ **37.** $x=7$ **39.** $x=4$

41. $x=0$ **43.** $x=5$ **45.** no solution **47.** $x=4$ **49.** $x=-1,3$ **51.** $x=5$ **53.** $x=2$ **55.** $x=1$ **57.** $x=-4$ **59.** $x=12$

61. $x=40$ **63.** $(a+b)^2=(a+b)(a+b)=a^2+2ab+b^2\neq a^2+b^2$

Section 9.6: Functions with Radicals

Margin Exercises 1. a. $[-2,\infty)$ **b.** $(-\infty,\infty)$ **2. a.** $f(0)=0, f(2)=10\ f(8)=20$ **b.** $f(1)=-1, f(2)=1\ f(15)=3$

3. 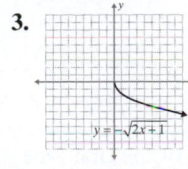 **4. a.**

x	$y1$
-7	3.1072
-1	1.8171
0	1.2599
2	-1.817
7	-2.962

b.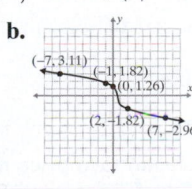

Exercises 1. a. $\sqrt{5}\approx 2.2361$ **b.** 3 **c.** $5\sqrt{2}\approx 7.0711$ **d.** 2 **3. a.** 3 **b.** -1 **c.** -2 **d.** $2\sqrt[3]{3}\approx 2.8845$ **5.** $[-8,\infty)$ **7.** $\left(-\infty,\dfrac{1}{2}\right]$

9. $(-\infty, \infty)$ **11.** $[0, \infty)$ **13.** $(-\infty, \infty)$ **15.** E **17.** B **19.** A

21. 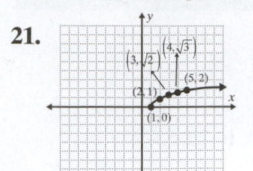 **23.** 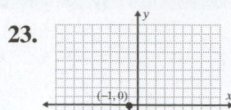 **25.** 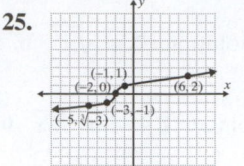 **27.** 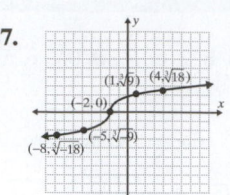 **29.**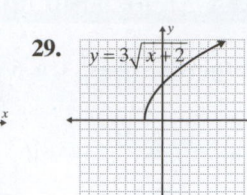

31. 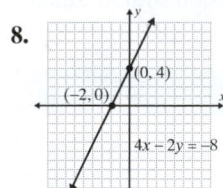 **33.** 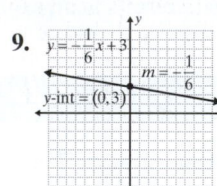 **35.** 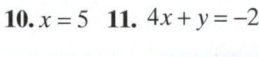 **37.** 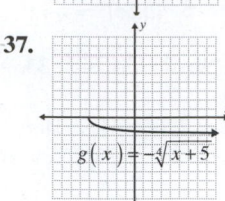 **39.**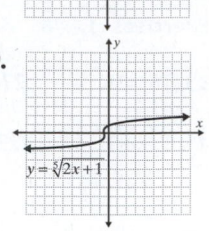

41. a. $\dfrac{1}{\sqrt{3+h}+\sqrt{3}}$ **b.** Slope of the line connecting $\left(3+h, f(3+h)\right)$ and $\left(3, f(3)\right)$

c. A line just touching the curve at one point. **d.** $\dfrac{1}{2\sqrt{3}}$; represents the slope of the line tangent to $f(x)$ at $x = 3$.

Chapter 9 Test

1. $4\sqrt{7}$ **2.** $2y^2\sqrt{30x}$ **3.** $2y\sqrt[3]{6x^2y^2}$ **4.** $\dfrac{7x^6 y\sqrt[3]{y}}{2z^2}$ **5.** $\sqrt[3]{4x^2}$ **6.** $2^{\frac{1}{2}}x^{\frac{1}{3}}y^{\frac{2}{3}}$ **7.** 4 **8.** $4x^{\frac{7}{6}}$ **9.** $\dfrac{8y^{\frac{3}{2}}}{x^3}$ **10.** $\sqrt[12]{x^{11}}$

11. $17\sqrt{3}$ **12.** $-xy\sqrt{y}$ **13.** $5-2\sqrt{6}$ **14.** $30-7\sqrt{3x}-6x$ **15. a.** $\dfrac{y\sqrt{10x}}{4x^2}$ **b.** $1+\sqrt{x}$ **16.** $x=14$ **17.** $x=-4$ **18.** $x=1$

19. $x=7$ **20. a.** $\left[-\dfrac{4}{3}, \infty\right)$ **b.** $(-\infty, \infty)$ **21. a.** 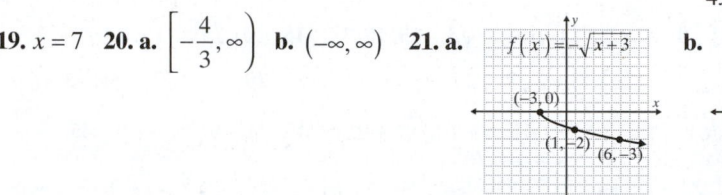 **b.** 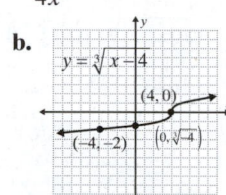 **22.** 2.6008 **23.** 0.125

24. 3.0711 **25.** 0.9758

Chapters 1-9 Cumulative Review

1. $1\dfrac{13}{18}$ **2.** $30\dfrac{13}{40}$ **3.** 23.69 **4.** 9.068 **5.** $x=\dfrac{1}{21}$ **6.** no solution **7.** 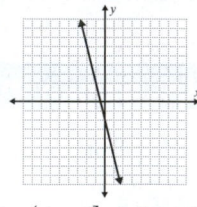 $(-\infty, 6)$

8. 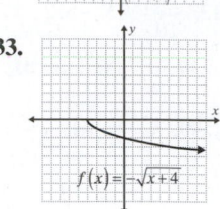 **9.** 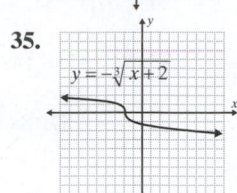 **10.** $x=5$ **11.** $4x+y=-2$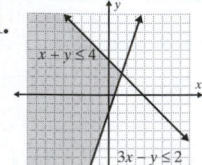

12. a. function; $D=(-\infty, \infty)$; $R=(-\infty, 0]$ **b.** not a function; $D=(-\infty, \infty)$; $R=(-\infty, 5]$ **c.** function; $D=[-2\pi, 2\pi]$; $R=[-2, 2]$

13. 1×10^3 **14. a.** -12 **b.** 0 **c.** -168 **15.** $2x^3-4x^2+8x-4$ **16.** $6x^2-23x+7$ **17.** $-5x^2+18x+8$ **18.** $49x^2-28x+4$

19. $(7+2x)(4-x)$ **20.** $4y^2(4y+1)^2$ **21.** $5(x-4)(x^2+4x+16)$ **22.** $(x+4)(x+1)(x-1)$ **23.** $(x-16)(x+3)=0$;

$x=-3, 16$ **24.** $(2x+3)(x-2)=0$; $x=-\dfrac{3}{2}, 2$ **25.** $(5x-2)(3x-1)=0$; $x=\dfrac{1}{3}, \dfrac{2}{5}$ **26. a.** $x^2-11x+28=0$

b. $x^3-11x^2-14x+24=0$ **27.** $(1, -2)$ **28.** $(14, 6)$ **29.** no solution **30.** $(1, 1, 1)$ **31.** 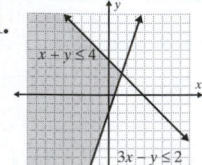 **32.** $\dfrac{1}{16}$ **33.** $x^{\frac{7}{3}}$

34. $12\sqrt{2}$ **35.** $2x^2y^3\sqrt[3]{2y}$ **36.** $\dfrac{\sqrt{10y}}{2y}$ **37.** $2\sqrt{30}-11$ **38.** $x=-1$ **39.** $x=-1$

40. $(-\infty, 3]$ **41. a.** 16π ft (or 50.24 ft) **b.** 64π ft^2 (or 200.96 ft^2)

42. $10\sqrt{5}$ m; 22.361 m **43. a.** \$1969 **b.** \$419 **44.** $-15, -14$ **45.** 15, 17

46. \$35,000 in 6% and \$15,000 in 10% **47.** burrito price is \$2.60 and taco price is \$1.35 **48.** 60 pounds of the first type (\$1.25 candy) and 40 pounds of the second type (\$2.50 candy) **49. a.** 8.4¢, 7.9¢, 7.1¢ **b.** 14 in. diameter for \$10.95 costs less per square inch. **50.** 26.5 meters

Chapter 10: Exponential and Logarithmic Functions

Section 10.1: Algebra of Functions

Margin Exercises 1. a. $2x^2 + 4x - 6$ **b.** $2x^2 + 2x - 12$ **c.** $2x^3 + 9x^2 - 27$ **2. a.** $-2x^2 + 2x - 1$ **b.** $\dfrac{x-1}{2x^2-x}, \left(x \neq \dfrac{1}{2}, 0\right)$

3. a. $x - 1 + \sqrt{x+3}$, $[-3, \infty)$ **b.** $\dfrac{\sqrt{x+3}}{x-1}$, $[-3,1) \cup (1, \infty)$ **4. a.** $6x^2 - 20x + 30$

Exercises 1. a. $2x - 3$ **b.** 7 **c.** $x^2 - 3x - 10$ **d.** $\dfrac{x+2}{x-5}, x \neq 5$ **3. a.** $x^2 + 3x - 4$ **b.** $x^2 - 3x + 4$ **c.** $3x^3 - 4x^2$ **d.** $\dfrac{x^2}{3x-4}, x \neq \dfrac{4}{3}$

5. a. $x^2 + x - 12$ **b.** $x^2 - x - 6$ **c.** $x^3 - 3x^2 - 9x + 27$ **d.** $x + 3, x \neq 3$ **7. a.** $3x^2 + x + 2$ **b.** $x^2 + x - 2$ **c.** $2x^4 + x^3 + 4x^2 + 2x$

d. $\dfrac{2x^2 + x}{x^2 + 2}$ **9. a.** $2x^2 + 2$ **b.** $8x$ **c.** $x^4 - 14x^2 + 1$ **d.** $\dfrac{x^2 + 4x + 1}{x^2 - 4x + 1}, x \neq 2 \pm \sqrt{3}$ **11.** 9 **13.** $-a^2 - a - 1$ **15.** 27 **17.** $\dfrac{8}{5}$

19. -31 **21.** $\sqrt{2x-6} + x + 4, D = [3, \infty)$ **23.** $3x^2 - 19x - 14, D = (-\infty, \infty)$ **25.** $\dfrac{x-5}{\sqrt{x+3}}, D = (-3, \infty)$ **27.** $-3x\sqrt{x-3}, D = [3, \infty)$

29. $\sqrt[3]{x+3} + \sqrt{5+x}$, $D = [-5, \infty)$

31. a. $f(x) + g(x)$; points $(-2,3)$, $(0,-4)$, $(2,-2)$ **b.** $f(x) - g(x)$; points $(-2,7)$, $(2,4)$, $(0,-2)$

33. a. $f(x) + g(x)$; points $(-1,5)$, $(-3,-4)$, $(2,-3)$ **b.** $f(x) - g(x)$; points $(-3,2)$, $(-1,-5)$, $(2,-5)$

35. a. $f(x) + g(x)$; points $(-1,4)$, $(1,6)$, $(-2,-6)$, $(0,-5)$ **b.** $f(x) - g(x)$; points $(-1,4)$, $(1,4)$, $(0,3)$, $(-2,-4)$

37. a. $f(x) + g(x)$; points $(3,6)$, $(1,4)$, $(-3,-2)$, $(-1,-2)$ **b.** $f(x) - g(x)$; points $(-1,6)$, $(-3,4)$, $(3,-2)$, $(1,-2)$

39. a. $f(x) + g(x)$; points $(-5,3)$, $(2,3)$, $(-1,3)$, $(3,1)$ **b.** $f(x) - g(x)$; points $(2,3)$, $(-1,1)$, $(3,3)$, $(-5,-1)$

41. graph of $(f+g)(x)$, $f(x)$, $g(x)$

43. graph of $f(x)$, $(f+g)(x)$, $g(x)$

45. graph of $(f+g)(x)$, $f(x)$, $g(x)$

47. graph of $(f+g)(x)$, $g(x)$, $f(x)$

49. graph of $g(x)$, $(f+g)(x)$, $f(x)$

51. 4 **53.** -12 **55.** $-\dfrac{3}{7}$

57. graph

59. graph

61. graph

63. graph of $(f+h)(x)$, $f(x)$, $h(x)$

65. graph of $f(x)$, $(f+g)(x)$, $g(x)$

67. graph of $(f+h)(x)$, $h(x)$, $f(x)$

69. graph of $(g+h)(x)$, $h(x)$, $g(x)$

71. In general, subtraction is not commutative. Answers will vary.

73. a. 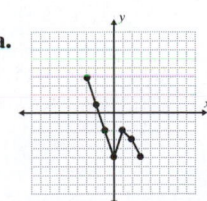 **b.** graph **c.** $\dfrac{f}{g}$ is undefined at $x = -2$ as $g(-2) = 0$.

Margin Exercises **1. a.** $-8r^2 - 15r - 7$ **b.** $b^2 - 8b^4$ **2. a.** $(f \circ g)(x) = (x^2 + 2)^2$ or $x^4 + 4x^2 + 4$ and $(g \circ f)(x) = x^4 + 2$

b. $f(g(x)) = 3\sqrt{x+3} + 3$ and $g(f(x)) = \sqrt{3x+6}$ **3. a.** one-to-one **b.** not one-to-one **4.** $f(g(x)) = (\sqrt{x+3})^2 - 3 = x + 3 - 3 = x$

for $x \geq -3$; $g(f(x)) = \sqrt{(x^2 - 3) + 3} = \sqrt{x^2} = x$ for $x \geq 0$ **5. a.** 4 **b.** -1 **6. a.** $f^{-1}(x) = \dfrac{x-2}{7}$ **b.** $f(x) = \dfrac{x^2 - 2}{2}$ for $x \geq 0$

Exercises **1. a.** $8r - 5$ **b.** $24a - 13$ **3. a.** $5x^2 - 20x + 24$ **b.** $45n^4 + 4$ **5. a.** $3n^2 - 6n - 9$ **b.** $48y^6 + 24y^3 - 9$

7. a. $f(g(2)) = 14$ **b.** $g(f(2)) = \dfrac{15}{2}$ **9. a.** $(f \circ g)(-5) = 49$ **b.** $(g \circ f)(-1) = 5$ **11.** $f(g(x)) = \sqrt{x^2} = x$, $g(f(x)) = (\sqrt{x})^2 = x$

13. $f(g(x)) = \sqrt{x-2}$, $g(f(x)) = \sqrt{x} - 2$ **15.** $f(g(x)) = \dfrac{1}{x^2} - 1$, $g(f(x)) = \dfrac{1}{(x-1)^2}$

17. $f(g(x)) = x^3 + 3x^2 + 4x + 3$, $g(f(x)) = x^3 + x + 2$ **19.** $f(g(x)) = \dfrac{1}{|x|}$, $g(f(x)) = \dfrac{1}{x}$

21. $f(g(x)) = \dfrac{1}{x^2 + 7x - 8}$, $g(f(x)) = \left(\dfrac{1}{x}\right)^2 + 7\left(\dfrac{1}{x}\right) - 8$ **23.** $f(g(x)) = (2x - 6)^{3n}$, $g(f(x)) = 2x^{3n} - 6$

25. $f(g(x)) = (x-8)^{\frac{3}{2}}$, $g(f(x)) = \sqrt{x^3 - 8}$ **27. a.** 21 **b.** 2 **c.** no; $f(g(x)) \neq x$ and $g(f(x)) \neq x$ **29. a.** -9 **b.** does not exist

c. $f(4) = \dfrac{1}{9}$ and $\dfrac{1}{9}$ is in the domain of g, so $g(f(4))$ is defined. However, $g(2) = -\dfrac{1}{2}$ and $-\dfrac{1}{2}$ is not in the domain

of f, so $f(g(2))$ is not defined.

31. 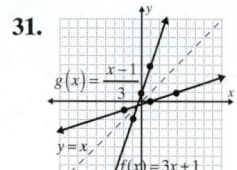   **33.** 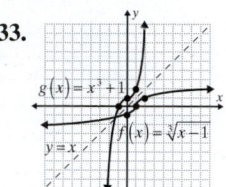   **35.** 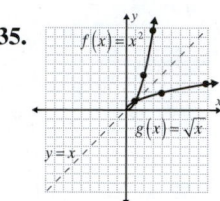   **37.** 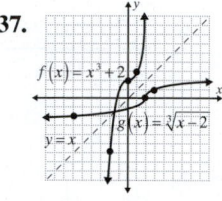   **39.**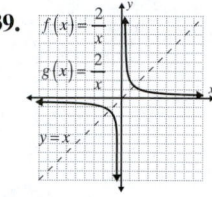

41. $f^{-1}(x) = \dfrac{x+3}{2}$ **43.** $g^{-1}(x) = x$ **45.** $f^{-1}(x) = \dfrac{x-1}{5}$ **47.** $g^{-1}(x) = \dfrac{3(x-2)}{2}$ **49.** $f^{-1}(x) = -x - 2$

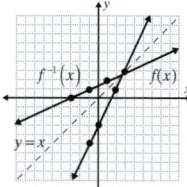   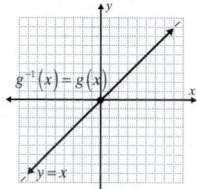   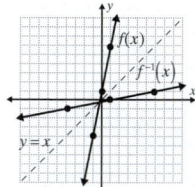   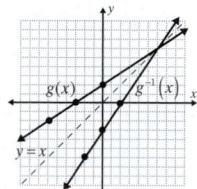   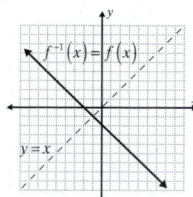

51. $f^{-1}(x) = \sqrt{x-1}$ **53.** $f^{-1}(x) = x^2, x \leq 0$ **55.** one-to-one **57.** not one-to-one **59.** not one-to-one

61. one-to-one **63.** one-to-one

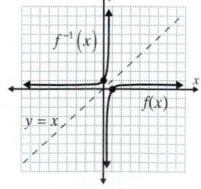   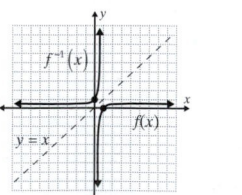   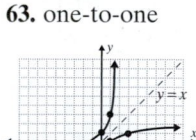

65. one-to-one **67.** not one-to-one **69.** one-to-one **71.** one-to-one **73.** one-to-one

75. not one-to-one **77.** $f^{-1}(x) = \sqrt[3]{x}$ **79.** $f^{-1}(x) = \dfrac{1}{x} + 3$ **81.** $f^{-1}(x) = \sqrt{x}$ **83.** $g^{-1}(x) = \sqrt[3]{x-2}$

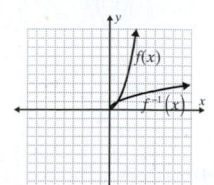   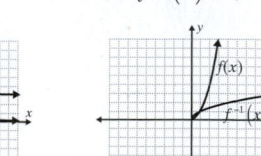   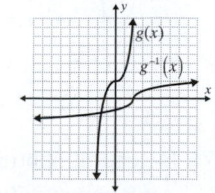

85. $f^{-1}(x) = x^2 - 5$, $x \geq 0$ **87.** $f^{-1}(x) = \sqrt{1-x}$

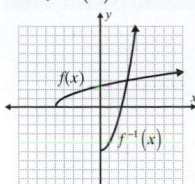

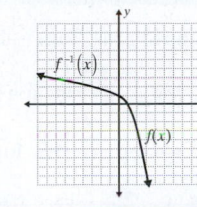

89. The domain of $f(g(x))$ can only include values of x in the domain of g, while $g(f(x))$ can only include values of x in the domain of f. Ex: $f(x) = \sqrt{x}$, $g(x) = -x$ Answers will vary.

Section 10.3: Exponential Functions

Margin Exercises 1. $y = 2{,}621{,}440{,}000$ or 2.62144×10^9 **2. a.** $A = 2000(1 + .05)^2 = \$2205$

b. $A = 2000\left(1 + \dfrac{.05}{12}\right)^{12(2)} = \$2209.88$ **3.** $\$1869.12$

Exercises 1.
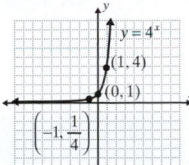
3.
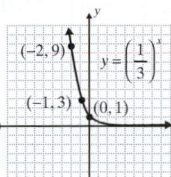
5.
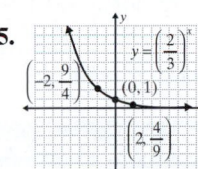
7.
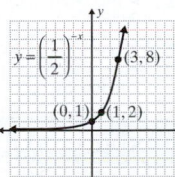

9.

11.
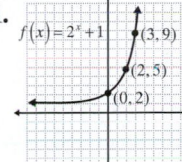
13.
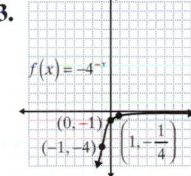
15.
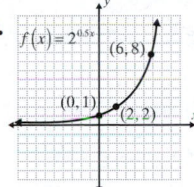
17.
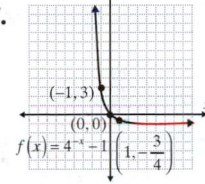

19.
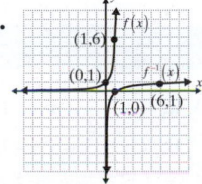

21. 48 **23.** 12.27 **25.** 4108.87 **27.** 25,000 bacteria **29. a.** $5877.31 **b.** $5920.98 **c.** $5943.79

d. $5967.30 **31.** $2154.99 **33.** $46.88 **35.** 250,000 people **37.** $\dfrac{1}{64}$ **39.** $491.52

41. $y = 50(2)^{-t}$ **43. a.**
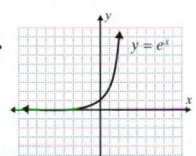
b.
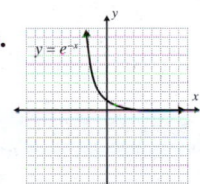
c.
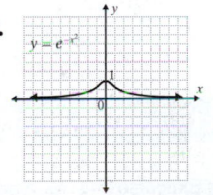

Section 10.4: Logarithmic Functions

Margin Exercises 1. $\log_2 8 = 3$ **2.** $2^5 = 32$ **3.** 3 **4.** 0 **5.** $\dfrac{1}{2}$ **6.** 2 **7, 8.**

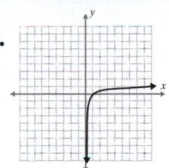

Exercises 1. $\log_7 49 = 2$ **3.** $\log_5 \dfrac{1}{25} = -2$ **5.** $\log_\pi 1 = 0$ **7.** $\log_{10} 100 = 2$ **9.** $\log_{10} 23 = k$ **11.** $\log_{2/3} \dfrac{4}{9} = 2$ **13.** $3^2 = 9$

15. $9^{\frac{1}{2}} = 3$ **17.** $7^{-1} = \dfrac{1}{7}$ **19.** $10^{1.74} = N$ **21.** $b^4 = 18$ **23.** $n^x = y^2$ **25.** $x = 16$ **27.** $x = 2$ **29.** $x = -3$ **31.** $x = \dfrac{1}{6}$

33. $x = 2$ **35.** $x = 32$ **37.** $x = 3.7$ **39.** $x = 2$

41.
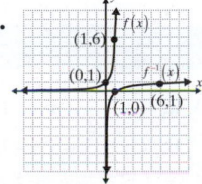
43.
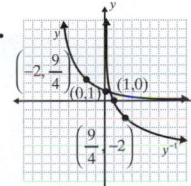
45.
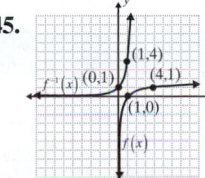
47.
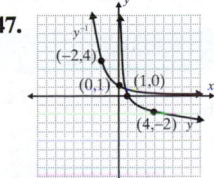
49.
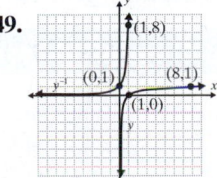

51. The two functions are symmetric about the line $y = x$.

Section 10.5: Properties of Logarithms

Margin Exercises 1. a. 2 **b.** 5 **c.** 2 **2. a.** $\log_8 15 + \log_8 k$ **b.** $\log_9 5 + \log_9 8$ **3. a.** $\log_3 4 - \log_3 t$ **b.** $\log_{10} 5 - \log_{10} 2$

4. a. 0.2386 **b.** 1.2041 **c.** 0.9542 **5. a.** $\log_b 4 + 4\log_b x$ **b.** $\dfrac{1}{2}\left(\log_b 4 - \log_b x\right)$ **6. a.** $\log_b\left(\dfrac{x^4}{z^2}\right)$ **b.** $\log_z\left(x^2 + x - 2\right)$ **c.** $\dfrac{5}{2}\log_b x$

Exercises 1. a. 2 **b.** 3 **c.** 5 **d.** 6 **3.** 5 **5.** -2 **7.** $\dfrac{1}{2}$ **9.** 10 **11.** $\sqrt{3}$ **13.** $\log_b 5 + 4\log_b x$ **15.** $\log_b 2 - 3\log_b x + \log_b y$

17. $\log_6 2 + \log_6 x - 3\log_6 y$ **19.** $2\log_b x - \log_b y - \log_b z$ **21.** $-2\log_5 x - 2\log_5 y$ **23.** $\dfrac{1}{3}\log_6 x + \dfrac{2}{3}\log_6 y$

25. $\dfrac{1}{2}\log_3 x + \dfrac{1}{2}\log_3 y - \dfrac{1}{2}\log_3 z$ **27.** $\log_5 21 + 2\log_5 x + \dfrac{2}{3}\log_5 y$ **29.** $-\dfrac{1}{2}\log_6 x - \dfrac{5}{2}\log_6 y$

31. $-9\log_b x - 6\log_b y + 3\log_b z$ **33.** $\log_b\left(\dfrac{9x}{5}\right)$ **35.** $\log_2 63x^2$ **37.** $\log_b x^2 y$ **39.** $\log_5\left(\dfrac{y^3}{\sqrt{x}}\right)$ or $\log_5\left(\dfrac{y^3\sqrt{x}}{x}\right)$

41. $\log_5\sqrt{\dfrac{x}{y}}$ or $\log_5\dfrac{\sqrt{xy}}{y}$ **43.** $\log_2\left(\dfrac{xz}{y}\right)$ **45.** $\log_b\left(\dfrac{xy^2}{\sqrt{z}}\right)$ or $\log_b\left(\dfrac{xy^2\sqrt{z}}{z}\right)$ **47.** $\log_5\left(2x^3 + x^2\right)$ **49.** $\log_2\left(x^2 + 2x - 3\right)$

51. $\log_b\left(x + 1\right)$ **53.** $\log_{10}\left(\dfrac{1}{2x - 3}\right)$ **55.** Answers will vary.

Section 10.6: Common Logarithms and Natural Logarithms

Margin Exercises 1. $\log 100 = 2$ **2. a.** 2.6021 **b.** 3.6990 **3. a.** 316.2278 **b.** 2152.7817 **c.** 3.1623 **4. a.** $\ln 2 = x$ **b.** $\ln c = 4$

5. a. 0.6931 **b.** -0.4463 **6. a.** $x = 2.3539 \times 10^{17}$ **b.** 0.9900

Exercises 1. $\log x = 1.5$ **3.** $\log\dfrac{1}{1000} = -3$ **5.** $\ln 27 = x$ **7.** $\ln 1 = 0$ **9.** $\log 3.2 = x$ **11.** $10^0 = 1$ **13.** $10^y = 5.4$

15. $e^{1.54} = x$ **17.** $e^1 = e$ **19.** $e^a = x$ **21.** 2.2380 **23.** 1.9465 **25.** -1.2418 **27.** 3.6243 **29.** error (undefined)

31. -5.3795 **33.** 204.1738 **35.** 0.0120 **37.** 0.9572 **39.** 175.9148 **41.** 2.4852×10^{-4} **43.** 1.0403 **45.** $\log x$ is a base 10

logarithm. $\ln x$ is a base e logarithm. **47.** $D = \left(-\infty, 0\right) \cup \left(0, \infty\right)$

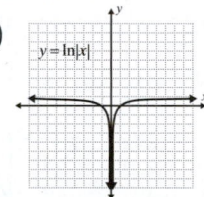

Section 10.7: Logarithmic and Exponential Equations and Change-of-Base

Margin Exercises 1. a. $x = 1$ or $x = 3$ **b.** $x = -1$ or $x = 2$ **2. a.** $x = \dfrac{\log 3}{\log 4} \approx 0.7925$ **b.** $x = \dfrac{\log 20}{\log 5} \approx 1.8614$ **3. a.** $x = 50$

b. $x = -1$ **4. a.** $\dfrac{\ln 4}{\ln 2} = 2$ **b.** $\dfrac{\ln 25}{\ln 8} \approx 1.5480$

Exercises 1. $x = 11$ **3.** $x = 9$ **5.** $x = \dfrac{7}{2}$ **7.** $x = \dfrac{6}{5}$ **9.** $x = \dfrac{7}{2}$ **11.** $x = -5$ **13.** $x = -3$ **15.** $x = -1, 4$ **17.** $x = -1, \dfrac{3}{2}$

19. $x = -2, 3$ **21.** $x = -2$ **23.** $x = -\dfrac{3}{2}, 0$ **25.** $x = -1, 3$ **27.** $x \approx 0.7154$ **29.** $x \approx 7.5098$ **31.** $x \approx -1.5947$ **33.** $x \approx 0.2473$

35. $x \approx -7.7003$ **37.** $t \approx -1.5193$ **39.** $x \approx 3.3219$ **41.** $x \approx -1.4307$ **43.** $x = 1$ **45.** $x \approx -4.1610$ **47.** $x \approx 1.2058$

49. $x \approx 1.1292$ **51.** $x \approx 0.9514$ **53.** $x \approx 1.2520$ **55.** $x \approx 25.1189$ **57.** $x \approx 31.6228$ **59.** $x = 0.0001$ **61.** $x \approx 4.9530$

63. $x \approx \pm 0.3329$ **65.** $x = 3$ **67.** $x = 100$ **69.** $x = 6$ **71.** $x = 20$ **73.** $x \approx 3.1893$ **75.** no solution **77.** no solution

79. $x = 105$ **81.** $x \approx 22.0855$ **83.** $x = -10, 8$ **85.** 2.2619 **87.** 0.3223 **89.** -0.6279 **91.** 2.4391 **93.** -1.2222

95. $x \approx 2.3219$ **97.** $x \approx 1.5480$ **99.** $x \approx 0.8390$ **101.** Answers will vary. $x = \dfrac{1}{2}$ **103.** $7 \cdot 7^x = 7^{1+x}$ and $49^x = 7^{2x}$. Since

$1 + x \neq 2x$, in general, $7 \cdot 7^x \neq 49^x$. Answers will vary.

Chapter 10 Test

1. a. 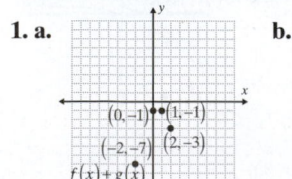 **b.** 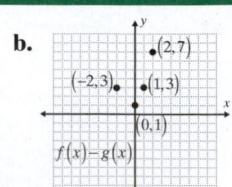 **2. a.** 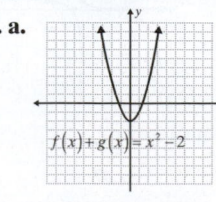 **b.** 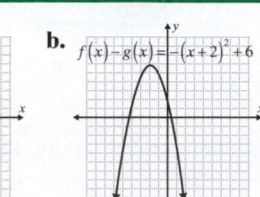 **3. a.** $\sqrt{x - 3} + x^2 + 1$

b. $\sqrt{x - 3} - x^2 - 1$

c. $\left(\sqrt{x - 3}\right)\left(x^2 + 1\right)$

d. $\dfrac{\sqrt{x - 3}}{x^2 + 1}$

4. a. $-4x^2 + 1$ **b.** $-8x^2 + 40x - 47$

5. one-to-one

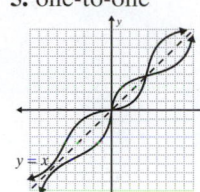

6. a. not inverses **b.** inverses

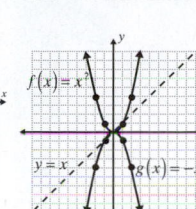

7. $f^{-1}(x) = \dfrac{1}{x+3} - 1$ **8. a.**

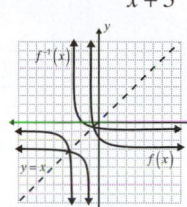

b.

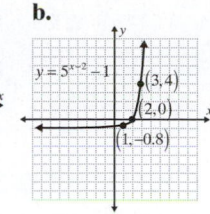

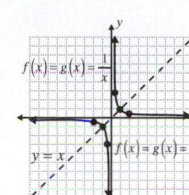

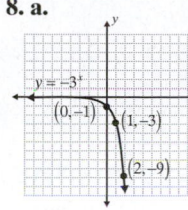

9. a. $\log 100{,}000 = 5$ **b.** $\log_{1/2} 8 = -3$ **c.** $x = \ln 15$ **10. a.** $10^4 = x$ **b.** $3^{-2} = \dfrac{1}{9}$ **c.** $4^x = 256$ **11.** $y^{-1} = \log_{1/2} x$

12. a. $x = 343$ **b.** $x = \dfrac{3}{2}$ **13. a.** $\ln(x+5) + \ln(x-5)$ **b.** $\dfrac{2}{3}\log x - \dfrac{1}{3}\log y$

14. a. $\ln(x^2 + x - 20)$ **b.** $\log \dfrac{x\sqrt{x}}{5}$ **15. a.** 2.7627 **b.** 6.4883 **16.** $x = -4$ **17.** $x = -3$

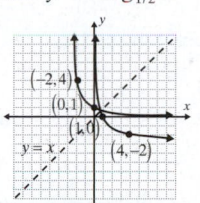

18. $x \approx 0.4518$ **19.** $x \approx 10.6873$ **20.** $x \approx 1.7925$ **21.** no solution **22.** $x \approx 21.0855$

23. 1.7965

Chapters 1–10 Cumulative Review

1. $\dfrac{37}{40}$ **2.** $-\dfrac{5}{6}$ **3.** $-\dfrac{7}{36}$ **4.** $-\dfrac{525}{512}$ **5.** 32 **6.** 7 **7.** $4xy\sqrt{5y}$ **8.** $\dfrac{2x\sqrt{10y}}{5y}$ **9.** $\dfrac{x^{\frac{1}{6}}}{y^{\frac{1}{3}}}$ **10.** $\dfrac{x^{\frac{1}{2}}}{2y^2}$ **11.** $x^2 y^{\frac{3}{2}}$ **12.** $x^{\frac{2}{3}}y$

13. $7x^2 - 26x - 8$ **14.** $-6y^2 - 18y + 60$ **15.** -6 **16.** 12 **17.** $3\log_b x + \dfrac{1}{2}\log_b y$ **18.** $\log_b\left(\dfrac{7x^2}{y}\right)$ **19.** $x = \dfrac{7}{2}$ **20.** $x = 6$

21. $x = -0.9212$ **22.** $x \approx 17.1825$ **23.** $x = -3$ **24.** $(1, -1, 2)$

25. $y = -\dfrac{2}{3}x + 2$ **26. a.** parallel **b.** no solution **27. a.** perpendicular **b.** $(2, 1)$ **28.** **29.**

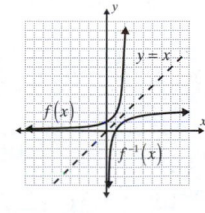

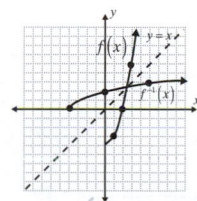

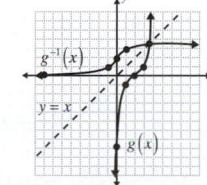

c. **c.**

30. a. function; $D = (-\infty, -1] \cup [1, \infty)$; $R = (-\infty, -2] \cup [2, \infty)$ **b.** not a function; $D = (-\infty, \infty)$; $R = (-\infty, \infty)$

c. function; $D = (-\infty, \infty)$; $R = [0, \infty)$ **31.** $\{(-2, 13), (-1, 8), (0, 5), (1, 4), (2, 5)\}$

32. $\{(-2, -28), (-1, -6), (0, 0), (1, -4), (2, -12)\}$ **33. a.** $x^2 + x + \sqrt{x+2}$ **b.** $x^2 + x - \sqrt{x+2}$ **c.** $(x^2 + x)\sqrt{x+2}$

d. $\dfrac{x^2 + x}{\sqrt{x+2}}$, $x > -2$ **34. a.** $4x^2 + 4x + 4$ **b.** $2x^2 + 7$

35. not a one-to-one function **36.** one-to-one function **37.** $f^{-1}(x) = \sqrt{x+4}$ **38.** $g^{-1}(x) = \sqrt[3]{x} + 2$ **39.** $-5, -3, -1$

40. -7 **41.** $\$750$

42. $\$11{,}618.22$

43. width: 16 inches; height: 12 inches **44.** 13 hardback; 30 paperback **45. a.** mean $= 320$ **b.** median $= 350$

c. mode $= 350$ **d.** range $= 200$ **46. a.** 125.6 ft; 94.2 ft **b.** 549.5 ft^2 **47.** -7.2288 **48.** 1.9266 **49.** 1.7372 **50.** 9.2103

Chapter 11: Rational Expressions

Section 11.1: Rational Expressions

Margin Exercises **1. a.** $x \neq \dfrac{1}{5}$ **b.** $x \neq 3, 4$ **c.** no restrictions. **2. a.** $\dfrac{2}{5}, (x \neq 3)$ **b.** $\dfrac{x^2 + 5x + 25}{x + 5}, (x \neq -5, 5)$ **c.** $-1, (x \neq 5)$

3. a. $\dfrac{x^4}{6y^6}, (x \neq 0, y \neq 0)$ **b.** $\dfrac{x + 3}{x^2}, (x \neq 0, 3)$ **c.** $\dfrac{x - 1}{x(x + 1)}$ or $\dfrac{x - 1}{x^2 + x}, (x \neq -1, 0, 1)$ **d.** $\dfrac{x^2 - x - 30}{3x + 9}, (x \neq -3, 2, 5)$

4. a. $\dfrac{1}{4x^2 y}$ **b.** $\dfrac{-x(x^2 - xy + y^2)}{y^2}$ or $\dfrac{-x^3 + x^2 y - xy^2}{y^2}$ **c.** $\dfrac{(x - 3)(x - 5)}{(3x + 1)(3x + 1)}$ or $\dfrac{x^2 - 8x + 15}{9x^2 + 6x + 1}$

Exercises **1.** $\dfrac{3x}{4y}; x \neq 0, y \neq 0$ **3.** $\dfrac{2x^3}{3y^3}; x \neq 0, y \neq 0$ **5.** $\dfrac{1}{x - 3}; x \neq 0, 3$ **7.** $7; x \neq 2$ **9.** $-\dfrac{3}{4}; x \neq 3$ **11.** $\dfrac{2x}{y}; x \neq -\dfrac{2}{3}, y \neq 0$

13. $\dfrac{x}{x - 1}; x \neq -6, 1$ **15.** $\dfrac{x^2 - 3x + 9}{x - 3}; x \neq -3, 3$ **17.** $\dfrac{x - 3}{y - 2}; y \neq -2, 2$ **19.** $\dfrac{x^2 + 2x + 4}{y + 5}; x \neq 2, y \neq -5$ **21.** $\dfrac{ab}{6y}$

23. $\dfrac{8x^2 y^3}{15}$ **25.** $\dfrac{x + 3}{x}$ **27.** $\dfrac{x - 1}{x + 1}$ **29.** $-\dfrac{1}{x - 8}$ **31.** $\dfrac{x - 2}{x}$ **33.** $\dfrac{4x + 20}{x(x + 1)}$ **35.** $\dfrac{x}{(x + 3)(x - 1)}$ **37.** $-\dfrac{x + 4}{x(x + 1)}$

39. $\dfrac{x + 2y}{(x - 3y)(x - 2y)}$ **41.** $\dfrac{x - 1}{x(2x - 1)}$ **43.** $\dfrac{1}{x + 1}$ **45.** $\dfrac{x + 2}{x - 2}$ **47.** $\dfrac{1}{3xy^6}$ **49.** $\dfrac{6y^7}{x^4}$ **51.** $\dfrac{x}{12}$ **53.** $\dfrac{6x + 18}{x^2}$ **55.** $\dfrac{6}{5x}$

57. $\dfrac{3x + 1}{x + 1}$ **59.** $\dfrac{x - 2}{2x - 1}$ **61.** $-\dfrac{x + 4}{x(2x - 1)}$ **63.** $\dfrac{x + 1}{x - 1}$ **65.** $\dfrac{6x^3 - x^2 + 1}{x^2(4x - 3)(x - 1)}$ **67.** $\dfrac{x^2 + 4x + 4}{x^2(2x - 5)}$ **69.** $\dfrac{x^2 - 6x + 5}{(x - 7)(x - 2)(x + 7)}$

71. $\dfrac{x^2 - 3x}{(x - 1)^2}$ **73.** $\dfrac{x^2 + 5x}{2x + 1}$ **75.** 1 **77.** $2x - 5$ feet **79. a.** A rational expression is an algebraic expression that can be

written in the form $\dfrac{P}{Q}$ where P and Q are polynomials and $Q \neq 0$. **b.** $\dfrac{x - 1}{(x + 2)(x - 3)}$ Answers will vary. **c.** $\dfrac{1}{x + 5}$

Answers will vary. **81. a.** $x = 4$ **b.** $x = -10, 10$

Section 11.2: Adding and Subtracting Rational Expressions

Margin Exercises **1. a.** $\dfrac{1}{x - 5}, (x \neq -5, 5)$ **b.** $\dfrac{3}{x + 5}, (x \neq -5, -3)$ **2. a.** $\dfrac{x^2 + 6x + 6}{x^2 + 5x + 6}$ **b.** $\dfrac{x^2 + 4x - 25}{2(x + 5)^2(x - 5)}$ **3. a.** $\dfrac{x + 2y}{3x - y}$

b. $\dfrac{x - 5}{x + 3}$ **c.** $\dfrac{x + 5}{x - 4}$ **4. a.** $\dfrac{x + 18}{x + 6}$ **b.** $\dfrac{15x^2 - 13xy - 4y^2}{3(x + y)^2(x - y)}$ **c.** $\dfrac{-x^2 - 2x + 12}{(x + 6)(x + 3)}$ **d.** $\dfrac{11y + 17}{(x - 2)(y + 1)(y + 3)}$

Exercises **1.** 3 **3.** 2 **5.** 1 **7.** $\dfrac{2}{x - 1}$ **9.** $\dfrac{14}{7 - x}$ **11.** 4 **13.** $\dfrac{x^2 - x + 1}{(x + 4)(x - 3)}$ **15.** $\dfrac{x - 2}{x + 2}$ **17.** $\dfrac{4x + 5}{2(7x - 2)}$ **19.** $\dfrac{6x + 15}{(x + 3)(x - 3)}$

21. $\dfrac{x^2 - 2x + 4}{(x + 2)(x - 1)}$ **23.** $\dfrac{-x^2 - 3x - 6}{(x + 3)(3 - x)}$ **25.** $\dfrac{8x^2 + 13x - 21}{6(x + 3)(x - 3)}$ **27.** $\dfrac{3x^2 - 20x}{(x + 6)(x - 6)}$ **29.** $\dfrac{-4x}{x - 7}$ **31.** $\dfrac{4x^2 - x - 12}{(x + 7)(x - 4)(x - 1)}$

33. $\dfrac{x - 6}{(x - 10)(x - 8)}$ **35.** $\dfrac{6x}{(x - 1)(x - 7)}$ **37.** $\dfrac{4x - 19}{(7x + 4)(x - 1)(x + 2)}$ **39.** $\dfrac{-7x - 9}{(4x + 3)(x - 2)}$ **41.** $\dfrac{4x^2 - 41x + 3}{(x + 4)(x - 4)}$

43. $\dfrac{x - 4}{2(x - 2)}$ **45.** $\dfrac{x^2 - 4x - 6}{(x + 2)(x - 2)(x - 1)}$ **47.** $\dfrac{3x^2 + 26x - 3}{(x + 7)(x - 3)(x + 1)}$ **49.** $\dfrac{6x + 2}{(x - 1)(x + 3)}$ **51.** $\dfrac{2x^2 + x - 4}{(x - 2)(y + 1)(x + 1)}$

53. $\dfrac{2x + 4xy - 15y}{(x + 3)(y + 2)(x - 5)}$ **55.** $\dfrac{-2x + 2}{x^2 + x + 1}$ **57.** $\dfrac{2x^2 - x - 5}{(x - 3)(x + 3)(x^2 + 1)}$ **59.** $\dfrac{5x^3 - x^2 + 6x - 4}{(2x + 1)(x - 1)(x + 2)(3x - 2)}$

61. Answers will vary.

Section 11.3: Simplifying Complex Fractions

Margin Exercises 1. $\dfrac{1}{3y}$ 2. a. $\dfrac{-6}{(x+6)^2}$ b. $-9xy$ 3. a. $\dfrac{-6}{(x+6)^2}$ b. $-9xy$ 4. $\dfrac{x+2}{x+4}$

Exercises 1. $\dfrac{4}{5xy}$ 3. $\dfrac{8}{7x^2y}$ 5. $\dfrac{2x^2+6x}{2x-1}$ 7. $\dfrac{2x-1}{2+3x}$ 9. $\dfrac{7}{2(x+2)}$ 11. $\dfrac{x}{x-1}$ 13. $\dfrac{4x}{3(x+6)}$ 15. $\dfrac{7x}{x+2}$ 17. $\dfrac{2x+6}{3(x-2)}$

19. $\dfrac{24y+9x}{2(9y-10x)}$ 21. $\dfrac{x}{x-1}$ 23. $\dfrac{xy}{x+y}$ 25. $\dfrac{1}{xy}$ 27. $\dfrac{y+x}{y-x}$ 29. $\dfrac{3-x}{x}$ 31. $\dfrac{x+1}{x+3}$ 33. $\dfrac{-1}{x(x+h)}$ 35. $\dfrac{-1}{x(x+h)}$

37. $-(x-2y)(x-y)$ 39. $\dfrac{2x}{x^2+1}$ 41. $\dfrac{(x-3)(x^2-2x+4)}{(x-4)(x-2)(x+1)}$ 43. $\dfrac{-5}{x+1}$ 45. $\dfrac{29}{4(4x+5)}$ 47. $\dfrac{x^2-3x-6}{x(x-1)}$

49. $\dfrac{x^2-4x-2}{(x-4)(x+4)}$ 51. a. $\dfrac{8}{5}$ b. 1 c. $\dfrac{x^4+x^3+3x^2+2x+1}{x^3+x^2+2x+1}$

Section 11.4: Solving Rational Equations

Margin Exercises 1. a. $x=11$ b. $x=16$ 2. 225 miles 3. a. $(x\neq-7,-2,0), x=-4,3$ b. $(x\neq0,2), x=\dfrac{2}{5}$

c. $(x\neq-5,0,5); \varnothing$, no solution 4. $l=\dfrac{SA-2wh}{2h+2w}$

Exercises 1. $x=7$ 3. $x\neq0,2; x=4$ 5. $x\neq-3,4; x=-10$ 7. $x\neq0; x=18$ 9. $x\neq6; x=-\dfrac{74}{9}$ 11. 360 defective computers

13. 28,800 students 15. 34 miles 17. 6.8 cups 19. width $=3$ inches; length $=7.5$ inches 21. $x=\dfrac{1}{4}$ 23. $x=6$

25. $x=4$ 27. $x\neq0; x=\dfrac{10}{3}$ 29. $x\neq0; x=-\dfrac{3}{4}$ 31. $x\neq0; x=-\dfrac{3}{16}$ 33. $x\neq-9,-\dfrac{1}{4},0; x=-2,1$

35. $x\neq\dfrac{3}{2},0,6; x=\dfrac{3}{5},9$ 37. $x\neq\dfrac{1}{2},4; x=-3$ 39. $x\neq-4,-1; x=2$ 41. $x\neq-1,\dfrac{1}{4}; x=\dfrac{2}{3}$ 43. $x\neq2,3; x=\dfrac{13}{10}$

45. $x\neq-\dfrac{2}{3},2$; no solution 47. $x\neq-1,\dfrac{1}{3},\dfrac{1}{2}; x=\dfrac{1}{5}$ 49. $r=\dfrac{S-a}{S}$ 51. $s=\dfrac{x-\bar{x}}{z}$ 53. $y=m(x-x_1)+y_1$

55. $R_{\text{total}}=\dfrac{R_1R_2}{R_1+R_2}$ 57. $P=\dfrac{A}{1+r}$ 59. a. $\dfrac{4x^2+41x-10}{x(x-1)}$ b. $x=\dfrac{1}{4},10$ and $x\neq0,1$ 61. a. $\dfrac{10x-6}{(x-2)(x+2)}$

b. $x=\dfrac{7}{3}$ and $x\neq2,-2$ 63. a. $\dfrac{x^2-153}{2(x-9)(x+9)}$ b. $x=3,-3$ and $x\neq9,-9$

Section 11.5: Applications: Rational Expressions

Margin Exercises 1. $\dfrac{5}{8}$ 2. a. $\dfrac{36}{5}$ hours or $7\dfrac{1}{5}$ hours b. mom: $\dfrac{9}{2}$ hours, or $4\dfrac{1}{2}$ hours; son: 9 hours

c. The pool will drain in $\dfrac{20}{3}$ hours, or $6\dfrac{2}{3}$ hours. 3. a. $c=2$ mph b. commercial airplane: 360 mph; private airplane 180 mph

Exercises 1. 72, 45 3. 9 5. $\dfrac{6}{13}$ 7. 36, 27 9. 7, 12 11. 45 shirts 13. 12 brushes 15. 1875 miles

17. person: 505 minutes; machine: 5.05 minutes 19. 45 minutes 21. Beth: 52 mph; Anna: 48 mph

23. commercial airliner: 300 mph; private plane: 120 mph 25. 6 hours 27. 50 mph 29. $\dfrac{9}{4}$ or $2\dfrac{1}{4}$ hours

31. $\dfrac{9}{2}$ or $4\dfrac{1}{2}$ days and 9 days 33. boat: 14 mph; current: 2 mph 35. 14 kilometers per hour

Section 11.6: Applications: Variation

Margin Exercises 1. $y=4$ 2. The spring will stretch 15 cm. 3. $y=2$ 4. 173 pounds 5. $z=648$ 6. a. 400 feet
b. 200 cubic inches c. 8000 pounds

Exercises 1. $\dfrac{7}{3}$ 3. 2 5. $-\dfrac{32}{9}$ 7. 36 9. 120 11. $\dfrac{56}{3}$ 13. 40 15. 54 17. $\dfrac{48}{5}$ 19. 27 21. 400 feet 23. $59.70

25. 4.71 feet 27. 6 m 29. 0.0073 cm 31. 16,000 lb 33. 9×10^{-11} N 35. 15,000 lb 37. 6400 lb 39. 1.8 ft^3

41. 1700 g per in.^{2} 43. 15 ohms 45. 2.56 ohms 47. 900 lb 49. 5 ft from the 300 lb weight (or 20 feet from the 75 lb weight) 51. a. When two variables vary directly, an increase in the value of one variable indicates an increase in the other, and the ratio of the two quantities is constant. b. When two variables vary inversely, an increase in the value of

one variable indicates a decrease in the other, and the product of the two quantities is constant. **c.** Joint variation is when a variable varies directly with more than one other variable. **d.** Combined variation is when a variable varies directly or inversely with more than one variable.

Section 11.7: Polynomial Division

Margin Exercises 1. $4x+1+\dfrac{5}{3x-2}$ **2.** $2x^2-5x+2$ **3.** $x^2+2x+2+\dfrac{10x+9}{x^2-2x+1}$ **4. a.** $3x^3+3x^2+1+\dfrac{8}{x-1}$

b. $8x^2-22x+80-\dfrac{329}{x+4}$ **5. a.** yes; x^2-2x+1 **b.** no **6. a.** $J(3)=-33$ **b.** $f(-2)=-57$ **c.**

$$\begin{array}{r|rrrr} 8 & 1 & -6 & -20 & 32 \\ & & 8 & 16 & -32 \\ \hline & 1 & 2 & -4 & 0 \end{array}$$

Exercises 1. $x-6+\dfrac{4}{x+4}$ **3.** $3x-4-\dfrac{7}{2x-1}$ **5.** $3x+4+\dfrac{1}{7x-1}$ **7.** $x-9$ **9.** $x^2-x-\dfrac{3}{x-8}$ **11.** $4x^2-6x+9-\dfrac{17}{x+2}$

13. $x^2+7x+55+\dfrac{388}{x-7}$ **15.** $2x^2-9x+18-\dfrac{30}{x+2}$ **17.** $7x^2+2x+1$ **19.** $x^2+2x+2-\dfrac{12}{2x+3}$ **21.** $x^2+3x+2-\dfrac{2}{x-4}$

23. $2x^2+x-3+\dfrac{18}{5x+3}$ **25.** $2x^2-8x+25-\dfrac{98}{x+4}$ **27.** $3x^2+2x-5-\dfrac{1}{3x-2}$ **29.** $3x^2-2x+5$ **31.** $x^3+2x+5+\dfrac{17}{x-3}$

33. $x^3+2x^2+6x+9+\dfrac{23}{x-2}$ **35.** $x^3+\dfrac{1}{2}x^2-\dfrac{3}{4}x-\dfrac{3}{8}+\dfrac{45}{16\left(x-\dfrac{1}{2}\right)}$ **37.** $3x+5+\dfrac{x-1}{x^2+2}$ **39.** $x^2+x-4+\dfrac{-8x+17}{x^2+4}$

41. $2x+3+\dfrac{6}{3x^2-2x-1}$ **43.** $3x^2-10x+12+\dfrac{-x-14}{x^2+x+1}$ **45.** $x^2-2x+7+\dfrac{-17x+14}{x^2+2x-3}$ **47.** x^2+3x+9

49. $x^5-x^4+x^3-x^2+x-1$ **51.** $x^4+x^3+x^2+x+1+\dfrac{2}{x-1}$ **53.** $x^4-\dfrac{1}{2}x^3-\dfrac{3}{4}x^2+\dfrac{3}{8}x+\dfrac{13}{16}-\dfrac{13}{32\left(x+\dfrac{1}{2}\right)}$

55. a. $x-9$ **b.** $c=3$; $P(3)=0$ **57. a.** $x^2-4x+33-\dfrac{265}{x+8}$ **b.** $c=-8$; $P(-8)=-265$ **59. a.** $4x^2-6x+9-\dfrac{17}{x+2}$

b. $c=-2$; $P(-2)=-17$ **61. a.** $x^2+7x+55+\dfrac{388}{x-7}$ **b.** $c=7$; $P(7)=388$ **63. a.** $2x^2-2x+6-\dfrac{27}{x+3}$ **b.** $c=-3$;

$P(-3)=-27$ **65. a.** $x^3+2x+5+\dfrac{17}{x-3}$ **b.** $c=3$; $P(3)=17$ **67. a.** $x^3+2x^2+6x+9+\dfrac{23}{x-2}$ **b.** $c=2$; $P(2)=23$

69. a. $x^3+\dfrac{1}{2}x^2-\dfrac{3}{4}x-\dfrac{3}{8}+\dfrac{45}{16\left(x-\dfrac{1}{2}\right)}$ **b.** $c=\dfrac{1}{2}$; $P\left(\dfrac{1}{2}\right)=\dfrac{45}{16}$ **71. a.** $x^4+x^3+x^2+x+1$ **b.** $c=1$; $P(1)=0$

73. a. $x^3-\dfrac{14}{5}x^2+\dfrac{56}{25}x-\dfrac{224}{125}+\dfrac{3396}{625\left(x+\dfrac{4}{5}\right)}$ **b.** $c=-\dfrac{4}{5}$; $P\left(-\dfrac{4}{5}\right)=\dfrac{3396}{625}$ **75.** $3x^3+4x^2+8x+12$; Answers will vary.

77. a. 19; $2x^2-4x+2+\dfrac{19}{x-2}$ **b.** -5; $2x^2-10x+20-\dfrac{5}{x+1}$ **c.** 55; $2x^2+10+\dfrac{55}{x-4}$ Answers will vary.

Chapter 11 Test

1. $\dfrac{x}{x+4}$; $x\neq -4,-3$ **2.** $\dfrac{1}{2x+5}$; $x\neq -\dfrac{5}{2}$ **3.** $\dfrac{x+3}{x+4}$ **4.** $\dfrac{3x-2}{3x+2}$ **5.** $\dfrac{-2x^2-13x}{(x+5)(x+2)(x-2)}$ **6.** $\dfrac{-2x^2-7x+18}{(3x+2)(x-4)(x+1)}$

7. $2x^2$ **8.** $\dfrac{x^2-7x+1}{(x+3)(x-3)}$ **9.** $\dfrac{3}{2xy^3}$ **10.** $\dfrac{-3x}{x-2}$ **11.** $-\dfrac{1}{xy}$ **12.** $\dfrac{12x+24}{(x-3)(x+3)^2}$ **13. a.** $\dfrac{7x+11}{2x(x+1)}$ **b.** $x=\dfrac{1}{3}$; $x\neq -1,0$

14. $x\neq -4$; $x=21$ **15.** $x\neq 0$; $x=-\dfrac{7}{2}$ **16.** $x\neq -4,1$; $x=-1$ **17.** $x\neq -2,-1$; $x=1$ **18.** $n=\dfrac{2S}{a_1+a_n}$ **19.** $x=\dfrac{y-b}{m}$

20. $x-6-\dfrac{2}{2x+3}$ **21.** $x-9+\dfrac{15x-12}{x^2+x-3}$ **22. a.** $x^2-4x+2-\dfrac{1}{x+5}$ **b.** $c=-5$; $P(-5)=-1$ **23. a.** $x^3+x^2+x+2+\dfrac{3}{x-1}$

b. $c=1$; $P(1)=3$ **24.** $\dfrac{2}{7}$ **25.** 4 hours   **26.** Carlos: 42 mph; Mario: 57 mph   **27.** 11 mph   **28.** $z=24$ **29.** $z=\dfrac{400}{9}$

30. $\dfrac{15}{2}$ cm   **31.** 5.13 in.^{3}

1.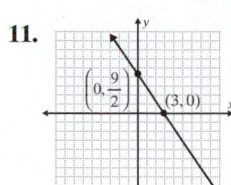

2. a. $\dfrac{19}{12}$ **b.** $-\dfrac{3}{14}$ **c.** $\dfrac{1}{4}$ **d.** $\dfrac{1}{12}$ **3.** $2(n+6)-5$ **4.** $3x^2-5x-28$ **5.** $4x^2-20x+25$

6. $6x-23$ **7.** $10x+4$ **8.** 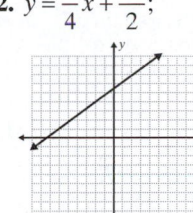 $(6,\infty)$ **9. a.**

b. $D=\{-1,0,5,6\}$ **c.** $R=\{0,5,2\}$

d. It is not a function because the x-coordinate -1 has more than one corresponding y-coordinate.

10. a. 8 **b.** -1 **c.** $-\dfrac{59}{27}$

11. **12.** $y=\dfrac{3}{4}x+\dfrac{11}{2};$ **13.** $y=\dfrac{1}{3}x+\dfrac{2}{3};$ **14.** $x=-3;$ **15.** $2x+y=-8;$

16. function; $D=\{-4,-2,0,2,4\}; R=\{-2\}$ **17.** function; $D=(-\infty,\infty); R=(-\infty,\infty)$ **18.**

19. $8x^2$ **20.** $5x^3y^2$ **21.** $(3x+y)(y+1)$ **22.** $(2x-5)(2x+3)$ **23.** $(3x-2)(2x-1)$

24. $(2x+5)(2x+3)$ **25.** $2(x+5)(x-2)$ **26.** $2x(3x+1)(x-4)$

27. $(3x^3+2y)(3x^3-2y)$ **28.** $(2x+5)(4x^2-10x+25)$ **29.** $x^2+3x-18=0$

30. $x^3+4x^2+3x=0$ **31.** $\dfrac{3}{5}x+2y+\dfrac{y^2}{x}$ **32.** $1-3y+\dfrac{8}{7}y^2$ **33.** $x-15-\dfrac{1}{x+1}$ **34.** $2x^2-x+3-\dfrac{2}{x+3}$

35. a. $x^2+2x+5+\dfrac{12}{x-2}$ **b.** $c=2; P(2)=12$ **36. a.** x^3+2x^2+x+3 **b.** $c=-1; P(-1)=0$

37. $\dfrac{4x^2}{x-3}; x\neq-\dfrac{1}{2},3$ **38.** $x-3; x\neq-3$ **39.** $\dfrac{1}{x+1}; x\neq-1,0$ **40.** $\dfrac{x+5}{2(x-3)}; x\neq3$ **41.** $x-y$ **42.** $\dfrac{x}{3(x+1)}$ **43.** $\dfrac{x+4}{3x}$

44. $\dfrac{3x^3+12x^2+12x}{x+3}$ **45.** $\dfrac{2x+1}{x-1}$ **46.** $\dfrac{2x^2+14x-8}{(x+3)(x-1)(x-2)}$ **47.** $\dfrac{2x+4}{(x-1)(x+4)}$ **48.** $\dfrac{x-4}{(x+2)(x-2)}$ **49.** $\dfrac{19}{14}$

50. $\dfrac{x+4}{6x}$ **51.** $x=-\dfrac{3}{5}$ **52.** $x=-5,6$ **53.** $x=-\dfrac{5}{2}$ **54.** $x=-4,0,5$ **55.** $x=2,5$ **56.** $x=-45; x\neq-6,\dfrac{1}{2}$ **57. a.** $\dfrac{9-2x}{x(x+3)}$

b. $x=\dfrac{9}{2}; x\neq-3,0$ **58. a.** $\dfrac{6x^2-16x-20}{(x-4)(x+2)}$ **b.** $x=7; x\neq-2,4$ **59.** $t=\dfrac{A-P}{Pr}$ **60.** $\dfrac{1}{2}$ year **61.** $-6,-5,-4$ or $7,8,9$

62. train: 60 miles per hour; airplane: 230 mph **63.** father: $2\dfrac{2}{3}$ hours; daughter: 8 hours **64.** 3 mph **65.** $7\dfrac{9}{13}$ cm

Chapter 12: Complex Numbers and Quadratic Equations

Section 12.1: Complex Numbers

Margin Exercises **1. a.** $10i$ **b.** $7i$ **c.** $3i\sqrt{2}$ **d.** $6i\sqrt{2}$ **2. a.** real: 0, imaginary: 5 **b.** real: 14, imaginary: $\sqrt{7}$ **c.** real: $\dfrac{6}{5}$, imaginary: $-\dfrac{11}{5}$ **d.** real: -13, imaginary: 0 **3. a.** $x=10$ and $y=-2$ **b.** $y=1$ and $x=-5$ **4. a.** $7+4i$ **b.** $-4\sqrt{2}i$

c. $\left(4+\sqrt{2}\right)+\left(8-\sqrt{5}i\right)$ **5. a.** $28+4i$ **b.** $2+26i$ **c.** 3 **6. a.** $\dfrac{8}{5}+\dfrac{16}{5}i$ **b.** $\dfrac{1}{9}-\dfrac{4\sqrt{5}}{9}i$ **c.** $-\dfrac{1}{4}-i$ **d.** $\dfrac{13}{19}-\dfrac{8\sqrt{3}i}{19}$ **7. a.** i **b.** -1 **c.** 1

Exercises **1.** real part: 4, imaginary part: -3 **3.** real part: -11, imaginary part: $\sqrt{2}$ **5.** real part: $\dfrac{3}{8}$, imaginary part: 0

7. real part: $\dfrac{4}{5}$, imaginary part: $\dfrac{7}{5}$ **9.** real part: $\dfrac{2}{3}$, imaginary part: $\sqrt{17}$ **11.** $7i$ **13.** $-8i$ **15.** $8\sqrt{2}$ **17.** $-12i\sqrt{3}$

19. $10i\sqrt{10}$ **21.** $x=6, y=-3$ **23.** $x=-2, y=\sqrt{5}$ **25.** $x=2, y=-8$ **27.** $x=2, y=-6$ **29.** $x=-\dfrac{4}{3}, y=-3$ **31.** -9

33. $14i$ **35.** $5+\left(\sqrt{6}+1\right)i$ **37.** $\sqrt{3}-5$ **39.** $-2-5i$ **41.** 2 **43.** $3+4i$ **45.** $16+24i$ **47.** $3+2i\sqrt{3}$ **49.** $-4+2i\sqrt{5}$

51. $5\sqrt{2}+10i$ **53.** $-7-11i$ **55.** $41+0i$ **57.** $5-i\sqrt{3}$ **59.** $4+4i\sqrt{7}$ **61.** $\left(8\sqrt{3}-3\right)+\left(4+6\sqrt{3}\right)i$ **63.** $\left(6-\sqrt{6}\right)-\left(3\sqrt{3}+2\sqrt{2}\right)i$

65. $0 - \dfrac{5}{4}i$ **67.** $-\dfrac{4}{5} + \dfrac{8}{5}i$ **69.** $-\dfrac{6}{5} - \dfrac{2}{5}i$ **71.** $\left(-\dfrac{1}{3} + \dfrac{3\sqrt{5}}{2}\right) - \left(\dfrac{3}{2} + \dfrac{\sqrt{5}}{3}\right)i$ **73.** i **75.** -1 **77.** i **79.** 1 **81.** 0 **83.** $x^2 + 9$

85. $x^2 + 2$ **87.** $5y^2 + 4$ **89.** $x^2 + 2x + 9$ **91.** $x^2 - 2x + 26$ **93. a.** yes **b.** no **95.** Given a complex number $(a + bi)$: $(a + bi)(a - bi) = a^2 - abi + abi - b^2 i^2 = a^2 + b^2$, which is the sum of squares of real numbers. Thus the product must be a positive real number.

Section 12.2: Completing the Square and the Square Root Property

Margin Exercises **1. a.** $x = 2, 8$ **b.** $x = 3$ **2.** $x = -5i, 5i$ **3. a.** $x = -6 \pm 2\sqrt{3}$ **b.** $\pm 4i$ **4. a.** $y^2 - 14y + \underline{49} = (y - 7)^2$

b. $x^2 + 9x + \dfrac{81}{4} = \left(x + \dfrac{9}{2}\right)^2$ **5. a.** $5 \pm 2\sqrt{14}$ **b.** $-1 \pm \sqrt{7}$ **c.** $x = \dfrac{-1 \pm \sqrt{19}}{2}$ **d.** $x = -2 \pm i\sqrt{7}$ **6. a.** $y^2 - 8y + 25$

b. $x^2 - 4x - 14$ **c.** $x^2 - 2x + 6$

Exercises **1.** $x = 3, 15$ **3.** $x = 5, 12$ **5.** $y = -11, 11$ **7.** $z = -6, 0, 2$ **9.** $x = -5$ **11.** $x = \pm 12$ **13.** $x = \pm 5i$ **15.** $x = \pm 2i\sqrt{6}$

17. $x = -1, 5$ **19.** $x = \pm\sqrt{5}$ **21.** $x = -3 \pm \sqrt{3}$ **23.** $x = 3 \pm 2i$ **25.** $x = -2 \pm i\sqrt{7}$ **27.** $x^2 - 12x + \underline{36} = (x - 6)^2$

29. $x^2 + 6x + \underline{9} = (x + 3)^2$ **31.** $x^2 - 5x + \dfrac{25}{4} = \left(x - \dfrac{5}{2}\right)^2$ **33.** $y^2 + y + \dfrac{1}{4} = \left(y + \dfrac{1}{2}\right)^2$ **35.** $x^2 + \dfrac{1}{3}x + \dfrac{1}{36} = \left(x + \dfrac{1}{6}\right)^2$

37. $x = -5, 1$ **39.** $y = -1 \pm \sqrt{6}$ **41.** $x = 5 \pm \sqrt{22}$ **43.** $x = 3 \pm i$ **45.** $x = 1, 11$ **47.** $y = 5 \pm \sqrt{21}$ **49.** $z = \dfrac{-3 \pm \sqrt{29}}{2}$

51. $x = \dfrac{-1 \pm i\sqrt{7}}{2}$ **53.** $x = \dfrac{-5 \pm \sqrt{17}}{2}$ **55.** $x = \dfrac{5 \pm \sqrt{10}}{3}$ **57.** $x = -1 \pm i\sqrt{5}$ **59.** $x = \dfrac{1 \pm i\sqrt{11}}{4}$ **61.** $y = \dfrac{-3 \pm i\sqrt{11}}{2}$

63. $y = -\dfrac{4}{3}, 1$ **65.** $x = 2 \pm \sqrt{2}$ **67.** $x^2 - 7 = 0$ **69.** $x^2 - 2x - 2 = 0$ **71.** $y^2 + 4y - 16 = 0$ **73.** $x^2 + 16 = 0$

75. $y^2 + 6 = 0$ **77.** $x^2 - 4x + 5 = 0$ **79.** $x^2 - 2x + 3 = 0$ **81.** $x^2 + 10x + 49 = 0$ **83.** See Example 5.

Section 12.3: The Quadratic Equation

Margin Exercises **1. a.** $\dfrac{7 \pm \sqrt{33}}{2}$ **b.** $\dfrac{3 \pm i\sqrt{31}}{10}$ **c.** $\pm 3i$ **2.** $x = 0, \dfrac{5 \pm \sqrt{17}}{2}$ **3. a.** 253, there are two real solutions.

b. 0, there is one real solution, a double root. **c.** -36, there are two non-real solutions **4. a.** $c = 9$ **b.** $a > 16$

Exercises **1.** 68; two real solutions **3.** 0; one real solution **5.** -44; two nonreal solutions **7.** 4; two real solutions

9. 19,600; two real solutions **11.** -11; two nonreal solutions **13.** $x = -2 \pm 2\sqrt{2}$ **15.** $x = -\dfrac{2}{3}$ **17.** $x = 1 \pm i\sqrt{6}$

19. $x = -3, \dfrac{1}{2}$ **21.** $x = \dfrac{-3 \pm \sqrt{5}}{4}$ **23.** $x = \dfrac{-3 \pm i\sqrt{3}}{4}$ **25.** $x = \dfrac{-3 \pm \sqrt{29}}{2}$ **27.** $x = -3, -1$ **29.** $x = \pm 2i\sqrt{2}$

31. $x = \dfrac{5 \pm \sqrt{17}}{2}$ **33.** $x = -\dfrac{1}{4}$ **35.** $x = \pm \dfrac{2\sqrt{3}}{3}$ **37.** $x = \dfrac{2}{3}$ **39.** $x = \dfrac{-4 \pm i\sqrt{2}}{2}$ **41.** $x = \dfrac{7 \pm i\sqrt{51}}{10}$ **43.** $x = -2, \dfrac{5}{3}$

45. $x = 3 \pm i\sqrt{2}$ **47.** $x = -1, 0$ **49.** $x = \dfrac{9 \pm \sqrt{65}}{2}, 0$ **51.** $x = \dfrac{-3 \pm \sqrt{5}}{2}, 0$ **53.** $x = \dfrac{2 \pm \sqrt{3}}{3}$ **55.** $x = \dfrac{1 \pm i\sqrt{3}}{6}$

57. $x = \dfrac{2 \pm i\sqrt{2}}{2}$ **59.** $x = \dfrac{-7 \pm \sqrt{17}}{4}$ **61.** $c < 16$ **63.** $c = \dfrac{81}{4}$ **65.** $a > 3$ **67.** $a > -\dfrac{1}{36}$ **69.** $a = \dfrac{49}{48}$ **71.** $c > \dfrac{4}{3}$

73. $x \approx 2.5993, 60.4007$ **75.** $x \approx -0.7862, 2.0110$ **77.** $x \approx -4.1334, -0.5806$ **79.** $x \approx -2.6933, 2.6933$

81. $x^4 - 13x^2 + 36 = 0$; , multiplied $(x - 2)(x + 2)(x - 3)(x + 3)$

Section 12.4: Applications

Margin Exercises **1.** length = 24 meters, width = 10 meters **2. a.** in 30 seconds **b.** at 5 seconds and at 25 seconds
3. length = 26 inches; width = 18 inches. **4.** 20 members attended the championship.

Exercises **1.** 6, 13 **3.** 4, 14 **5.** $-11, -6$ **7.** $-5 - 4\sqrt{3}$ **9.** $\dfrac{1 + \sqrt{33}}{4}$ **11.** $7\sqrt{2}$ cm, $7\sqrt{2}$ cm **13.** Mel: 30 mph; John: 40 mph

15. length = 10 feet; width = 4 feet **17.** 17 inches × 6 inches **19.** 70 trees **21.** 9 floors **23.** 3 in. **25.** 5 amperes
27. 12 signs **29. a.** \$307.20 **b.** 45 cents or 35 cents **31.** 8 workers **33.** 6 mph **35.** 4 hours and 12 hours
37. 10 lb of grade A, 20 lb of grade B **39. a.** 6.75 s **b.** 3 s, 3.75 s **41. a.** 7 s **b.** 2 s, 12 s **43.** 14.14 cm
45. a. $C = 94.20$ ft, $A = 706.50$ ft^2 **b.** $P = 84.85$ ft, $A = 450.00$ ft^2 **47. a.** no **b.** home plate **c.** no **49.** 855.86 feet
51. Answers will vary.

Margin Exercises **1. a.** Line of symmetry: $x = 0$ **b.** Line of symmetry: $x = 2$ **2. a.** Zeros: $x = 3, 7$
Vertex = $(0,5)$ Vertex = $(2,0)$ Line of symmetry: $x = 5$
Domain = $(-\infty, \infty)$ or   Domain = $(-\infty, \infty)$ or   Vertex = $(5, -4)$
$\{x | x \text{ is any real number}\}$ $\{x | x \text{ is any real number}\}$ Domain = $(-\infty, \infty)$ or
Range = $[5, \infty)$ or   Range = $[0, \infty)$ or $\{x | x \text{ is any real number}\}$
$\{y | y \geq 5\}$ $\{y | y \geq 0\}$ Range = $[-4, \infty)$ or
    $\{y | y \geq -4\}$

b. Zeros: $x = -3 \pm 2\sqrt{3}$ **c.** Zeros: no real zeros **3. a.** The hockey team will make its maximum revenue
Line of symmetry: $x = -3$ Line of symmetry: $x = 1$ of \$39,200 by charging a ticket price of \$14 per
Vertex = $(-3, 12)$ Vertex = $(1, 3)$ ticket.
Domain = $(-\infty, \infty)$ or Domain = $(-\infty, \infty)$ or
$\{x | x \text{ is any real number}\}$ $\{x | x \text{ is any real number}\}$ **b.** Two sides of the lot are 90 yards and the third side
Range = $(-\infty, 12]$ or Range = $[3, \infty)$ or is 180 yards for a maximum area of 16,200 yards.
$\{y | y \leq 12\}$ $\{y | y \geq 3\}$

Exercises **1.** $x = 0; (0, -4); D = (-\infty, \infty); R = [-4, \infty)$ **3.** $x = 0; (0, 9); D = (-\infty, \infty); R = [9, \infty)$

5. $x = 0; (0, 1); D = (-\infty, \infty); R = (-\infty, 1]$ **7.** $x = 0; \left(0, \dfrac{1}{5}\right); D = (-\infty, \infty); R = \left(-\infty, \dfrac{1}{5}\right]$

9. $x = -1; (-1, 0); D = (-\infty, \infty); R = [0, \infty)$ **11.** $x = 4; (4, 0); D = (-\infty, \infty); R = (-\infty, 0]$

13. $x = -3; (-3, -2); D = (-\infty, \infty); R = [-2, \infty)$ **15.** $x = -2; (-2, -6); D = (-\infty, \infty); R = [-6, \infty)$

17. $x = \dfrac{3}{2}; \left(\dfrac{3}{2}, \dfrac{7}{2}\right); D = (-\infty, \infty); R = \left(-\infty, \dfrac{7}{2}\right]$ **19.** $x = \dfrac{4}{5}; \left(\dfrac{4}{5}, -\dfrac{11}{5}\right); D = (-\infty, \infty); R = \left[-\dfrac{11}{5}, \infty\right)$

21. a. 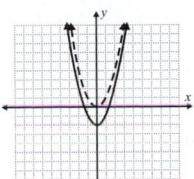 **b.** **c.** 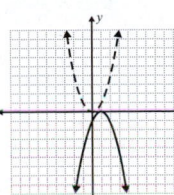 **d.**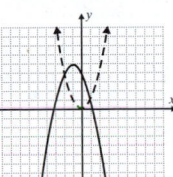

23. a. 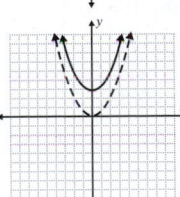 **b.** 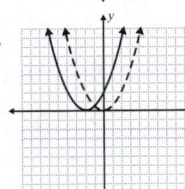 **c.** 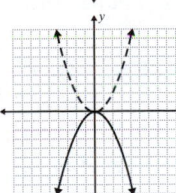 **d.**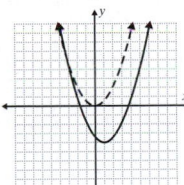

25. $y = 2(x - 1)^2$; **27.** $y = (x - 1)^2 - 4$; **29.** $y = (x + 3)^2 - 4$; **31.** $y = 2(x - 2)^2 - 3$; **33.** $y = -3(x + 2)^2 + 3$;
$(1, 0)$; $(1, -4)$; $(-3, -4)$; $(2, -3)$; $(-2, 3)$;
$R = [0, \infty)$; $R = [-4, \infty)$; $R = [-4, \infty)$; $R = [-3, \infty)$; $R = (-\infty, 3]$;
zeros: $x = 1$ zeros: $x = -1, 3$ zeros: $x = -5, -1$ zeros: $x = \dfrac{4 \pm \sqrt{6}}{2}$ zeros: $x = -1, -3$

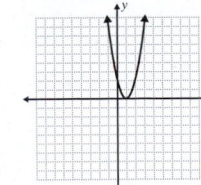 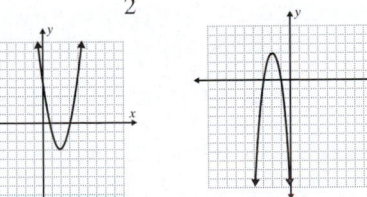

35. $y = 5(x-1)^2 + 3$;
$(1, 3)$;
$R = [3, \infty)$;
zeros: none

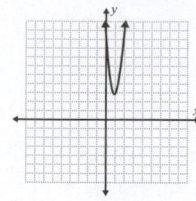

37. $y = -\left(x + \frac{5}{2}\right)^2 + \frac{17}{4}$;
$\left(-\frac{5}{2}, \frac{17}{4}\right)$;
$R = \left(-\infty, \frac{17}{4}\right]$;
zeros: $x = \dfrac{-5 \pm \sqrt{17}}{2}$

39. $y = 2\left(x + \frac{7}{4}\right)^2 - \frac{9}{8}$;
$\left(-\frac{7}{4}, -\frac{9}{8}\right)$;
$R = \left[-\frac{9}{8}, \infty\right)$;
zeros: $x = -\dfrac{5}{2}, -1$

41.

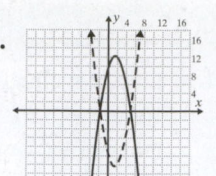

a. no **b.** yes
c. One function is a reflection of the other across the x-axis.

43.

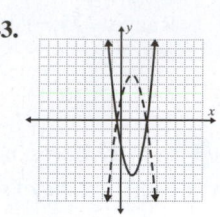

a. no **b.** yes
c. One function is a reflection of the other across the x-axis.

45. a. 3.5 s **b.** 196 ft **47. a.** 5 s **b.** 420 ft **49. a.** \$50 **b.** \$2500 **51.** \$20 **53.** zeros: $x \approx -0.7321, 2.7321$
55. zeros: $x \approx -1.1583, 2.1583$ **57.** no real zeros

59.

61.

63.

65.

67. a. a parabola **b.** $x = -\dfrac{b}{2a}$ **c.** $x = -\dfrac{b}{2a}$ **d.** No. A graph can be entirely above or below the x-axis.

69. Domain: $(-\infty, \infty)$; For $a > 0$, Range: $[k, \infty)$; For $a < 0$, Range: $(-\infty, k]$.

Chapter 12 Test

1. real part: 6, imaginary part: $-\dfrac{1}{3}$ **2.** real part: $-\dfrac{2}{5}$, imaginary part: $\dfrac{3}{5}$ **3.** $x = \dfrac{11}{2}, y = 3$ **4.** $16 + 4i$ **5.** $-5 + 16i$ **6.** $23 - 14i$ **7.** $x^2 + 4$ **8.** $\dfrac{8}{13} - \dfrac{1}{13}i$ **9.** $-i$ **10.** $x = -4, 2$ **11. a.** $x^2 - 30x + \underline{225} = (x - 15)^2$ **12.** $x = -2 \pm \sqrt{3}$

13. $x = -\dfrac{1}{2}, 0$ **14.** $x^2 - 2x - 4 = 0$ **15.** 73, two real solutions **16.** $b = \pm 2\sqrt{6}$ **17.** $x = \dfrac{-1 \pm i\sqrt{7}}{4}$ **18.** $x = \dfrac{3 \pm \sqrt{41}}{4}$

19. $x = \dfrac{2 \pm i\sqrt{2}}{2}$ **20.** $x = 3; (3, -1); D = (-\infty, \infty); R = [-1, \infty)$ **21.** $x = 3; (3, -9); D = (-\infty, \infty); R = [-9, \infty)$

22. $y = (x + 2)^2 - 10; (-2, -10)$;
$R = [-10, \infty)$;
zeros: $x = -2 \pm \sqrt{10}$

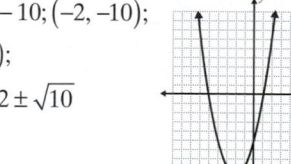

23. $y = (x - 3)^2 - 1; (3, -1)$;
$R = [-1, \infty)$;
zeros: $x = 2, 4$

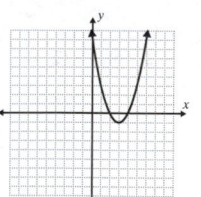

24. 8 m, 15 m **25. a.** $t = 7$ seconds **b.** $t = 6.46$ seconds **26.** width = 12 inches; length = 16 inches **27.** speed returning: 40 mph; speed going: 50 mph **28.** $y = x(x - 10); -5, 5$; minimum product $= -25$ **29.** $\dfrac{11}{2}$ in. $\times \dfrac{11}{2}$ in.

Chapters 1-12 Cumulative Review

1. $\dfrac{1}{x^5}$ **2.** $x^3 y^6$ **3.** $\dfrac{9\sqrt{y}}{x^2}$ **4.** $\dfrac{27x^2}{8y}$ **5.** $-3x^2 y^2 \sqrt[3]{y^2}$ **6.** $2x^2 y^3 \sqrt[4]{2xy^3}$ **7.** $\dfrac{9\sqrt{2}}{2}$ **8.** $\dfrac{4\sqrt{3}}{3}$ **9.** $5\left(\sqrt{3} - \sqrt{2}\right)$
10. $(x + 9)\left(\sqrt{x} - 3\right)$ **11.** $4 + 2i$ **12.** $19 + 4i$ **13.** $\dfrac{-11}{13} + \dfrac{16}{13}i$ **14.** $(x - 5)(y + 2)$ **15.** $(8x + 9)(8x - 9)$
16. $2\left(x^2 - 6y\right)\left(x^4 + 6x^2 y + 36y^2\right)$ **17. a.** -2 **b.** 37 **18.** $x - 9 + \dfrac{21x - 24}{x^2 + 2x - 1}$ **19.** $\dfrac{9x^2 - 2x - 2}{(x + 4)(2x - 1)(3x - 2)}$
20. $\dfrac{21x + 8}{(x - 8)(x + 1)(3x - 2)}$ **21.** $\dfrac{x + 3}{3(x - 5)}$ **22.** $\dfrac{2x^2 + 3x + 1}{2(2x - 3)(x - 1)}$ **23.** $x = 6$ **24.** $x = \dfrac{7}{8}$ **25.** $x = -\dfrac{3}{2}, \dfrac{2}{5}$

26. $x = \dfrac{-7 \pm \sqrt{33}}{8}$ **27.** $x = -1$ **28. a.** -47; no real solution **b.** 0; one real solution **29.** $4x^2 + 17x - 15 = 0$

30. $x^2 - 2x - 19 = 0$ **31.** $\left[-\dfrac{19}{2}, \infty \right)$ 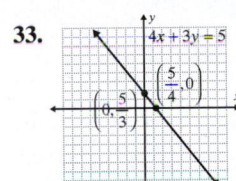 **32.** $\left(-\infty, \dfrac{7}{5} \right]$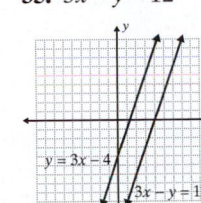

33. 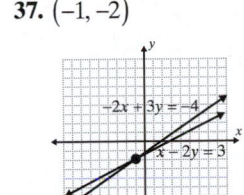 **34.** $y = x - 7$ 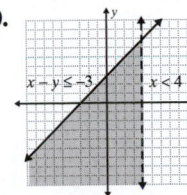 **35.** $3x - y = 12$ 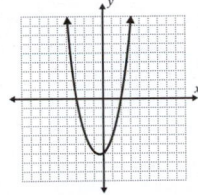 **36.** If any vertical line intersects the graph of a relation at more than one point, then the relation graphed is not a function. This works because it checks whether any first coordinate appears more than once.

37. $(-1, -2)$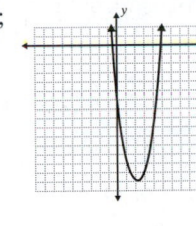

38. $(5, 4)$

39. $(1, -1, 3)$

40.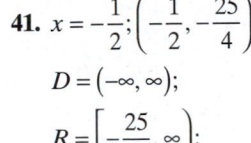

41. $x = -\dfrac{1}{2}; \left(-\dfrac{1}{2}, -\dfrac{25}{4} \right)$;

$D = (-\infty, \infty)$;

$R = \left[-\dfrac{25}{4}, \infty \right)$;

zeros: $x = -3, 2$

42. $x = \dfrac{1}{2}; \left(\dfrac{1}{2}, \dfrac{49}{4} \right)$;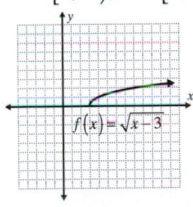

$D = (-\infty, \infty)$;

$R = \left(-\infty, \dfrac{49}{4} \right]$;

zeros: $x = -3, 4$

43. $x = \dfrac{9}{4}; \left(\dfrac{9}{4}, -\dfrac{121}{8} \right)$;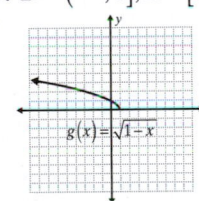

$D = (-\infty, \infty)$;

$R = \left[-\dfrac{121}{8}, \infty \right)$;

zeros: $x = -\dfrac{1}{2}, 5$

44. $y = -\dfrac{3}{2}x + 3$ **45.** $y = -\dfrac{3}{2}x + 10$

46. $\dfrac{50}{3}$ in.^{3} or $16.\overline{6}$ in.^{3} **47.** $10\,\text{mph}$

48. $20\,\text{m} \times 26\,\text{m}$ **49.** $12, 14, 16$

50. 2.14 hours **51.** 100

52. $\$2500$ at 9% and $\$7500$ at 11% **53.** 15 mg of drug A, 5 mg of drug B **54. a.** $10{,}000$ dozen **b.** $\$995{,}000$ **c.** $120{,}000$ golf balls

55. 5.9136 **56.** 14.8306 **57.** -5.8284 **58.** $D = [3, \infty); R = [0, \infty)$ **59.** $D = (-\infty, 1]; R = [0, \infty)$ **60.** $D = [-1, \infty); R = (-\infty, 0]$

Glossary

A

Absolute value   The distance a number is from 0 on a number line

Acute angle   An angle whose measure is between 0° and 90°

Acute triangle   A triangle in which all three angles are acute

Addends   The numbers being added in an addition problem

Addition principle of equality   If the same algebraic expression is added to both sides of an equation, the new equation has the same solutions as the original equation.

Additive identity   The number 0 is called the additive identity.

Additive identity property   The sum of any number and 0 is equal to the number itself.

Additive inverse   The opposite of an integer; two integers are additive inverses (or opposites) if their sum is equal to 0.

Adjacent angles   Two angles are adjacent if they have a common side.

Algebraic expression   A combination of variables and numbers using any of the operations of addition, subtraction, multiplication, or division, as well as exponents

Altitude of a triangle   The height of a triangle

Angle   Two rays with a common endpoint (called a vertex)

Area   The measure of the interior, or enclosed region, of a plane surface

Ascending order   The exponents on the terms of a polynomial increase in order from left to right.

Associative property of addition   The grouping of the numbers in addition can be changed.

Associative property of multiplication   The grouping of the numbers in multiplication can be changed.

Augmented matrix   The matrix derived from a system of equations that includes the coefficients and the constant terms

B

Bar graph   A graph used to emphasize comparative amounts

Base   In the expression a^n, the number a is called the base.

Binomial   A polynomial with two terms

C

Change in value   To calculate the change in value, take the end value and subtract the beginning value.

Circle   The set of all points in a plane that are some fixed distance from a fixed point called the center of the circle

Class   In a histogram, an interval (or range) of numbers that contain data items

Circle graph   A graph used to help in understanding percents or parts of a whole

Circumference   The perimeter of a circle

Class boundaries   In a histogram, numbers that are halfway between the upper limit of one class and the lower limit of the next class

Class width   In a histogram, the difference between the class boundaries of a class (the width of each bar)

Closed figure   A closed figure begins and ends at the same point.

Closed half-plane   If the boundary line is included in the solution set, then the half-plane is said to be closed.

Closed interval   Both endpoints of the interval are included

Coefficient   The number written next to a variable

Coefficient matrix   The matrix derived from a system of equations that includes the coefficients of the variables, but not constant terms

Column   Formed by the vertical entries of a matrix

Combined variation   If a variable varies either directly or inversely with more than one other variable, the variation is said to be combined variation.

Commission   A fee paid to an agent or salesperson for a service

Commutative property of addition   The order of the numbers in addition can be reversed.

Commutative property of multiplication   The order of the numbers in multiplication can be reversed.

Complementary angles   Two angles are complementary if the sum of their measures is 90°.

Complex algebraic expression   An expression that involves rational expressions and more than one operation

Complex conjugates   The two complex numbers $a + bi$ and $a - bi$ are called complex conjugates, or simply conjugates, of each other.

Complex fraction   Fraction in which the numerator or denominator or both contain one or more fractions or mixed numbers

Glossary   1374

Complex numbers   The set of numbers that includes all the real numbers and the even roots of negative numbers

Composite number   A counting number with more than two different factors (or divisors)

Compound interest   Interest paid on interest earned

Compound inequality   A mathematical expression that uses inequality symbols to compare the order of three expressions or values

Conditional equation   An equation that has a finite number (a countable number) of solutions

Congruent angles   Two angles with the same measure

Congruent triangles   Two triangles are congruent if the corresponding angles have the same measure and the lengths of the corresponding sides are equal.

Conjugates   The two expressions $(a - b)$ and $(a + b)$ are called conjugates; the product of conjugates results in the difference of two squares.

Consecutive even integers   Even integers are consecutive if each is 2 more than the previous even integer.

Consecutive integers   Integers are consecutive if each is 1 more than the previous integer.

Consecutive odd integers   Odd integers are consecutive if each is 2 more than the previous odd integer.

Consistent system   A system of equations that has exactly one solution

Constant (or constant term)   A term that consists of only a number

Constant of variation   The constant multiplier in a relationship of direct or inverse variation

Contradiction   An equation that simplifies to a statement that is never true (such as $0 = 2$) and has no solution

Coordinate   Either of the numbers in an ordered pair; may also refer to the number that corresponds to a point on a number line

Cube   A rectangular solid in which the length, width, and height are all equal

Cube root   The cube root of a number equals another number that when cubed results in the original number.

Cube of a number   In expressions with exponent 3, the base is said to be cubed.

D

Data   Value(s) measuring some characteristic of interest such as income, height, weight, grade point averages, scores on tests, and so on

Decimal notation   Decimal notation uses a decimal point, with whole numbers written to the left of the decimal point and fractions written to the right of the decimal point.

Decimal numbers   Numbers written in decimal notation are said to be decimal numbers (or simply decimals).

Decimal point   A period inserted between the whole number and fractional parts of a decimal number

Degree of a polynomial   The largest of the degrees of the polynomial's terms

Degree of a term   The sum of the exponents on the variables in a term

Denominator   The bottom number in a fraction

Dependent system   A system of equations that has an infinite number of solutions

Dependent variable   The second coordinate y in an ordered pair

Depreciation   The decrease in value of an item

Descending order   The exponents on the terms of a polynomial decrease in order from left to right.

Diameter   The distance from one point on a circle through the center to the point directly opposite it

Difference   The result of subtraction

Difference of cubes   A binomial that can be written in the form $x^3 - a^3$

Difference of squares   A binomial that can be written in the form $x^2 - a^2$

Digit   A symbol used in our number system; namely 0, 1, 2, 3, 4, 5, 6, 7, 8, and 9

Dimension of a matrix   The number of rows by the number of columns in a matrix

Direct variation   A variable quantity y varies directly as a variable x if there is a constant k such that $\frac{y}{x} = k$ or $y = kx$. When two variables vary directly, an increase in one indicates an increase in the other.

Discount   A reduction in the original selling price of an item; the difference between the original price and the sale price

Discriminant   In the quadratic formula, the expression $b^2 - 4ac$ is called the discriminant.

Distributive property   The product of a number and a sum is equal to the sum of the products of the number and each of the addends.

Dividend   The number being divided in a division problem

Divisible   If a number can be divided by another number so that the remainder is 0, then the dividend is divisible by the divisor.

Division algorithm (or long division)   The process (or series of steps) that we follow when dividing two numbers or two polynomials

Divisor   The number doing the dividing in a divsion problem

Domain of a function   The set of all first coordinates in a relation

Domain axis   In the graph of a relation, the horizontal axis (the x-axis)

Double solution   The special cases in which the two factors of a quadratic equation are the same and there is only one solution

E

Elementary row operations   An operation that can be done to a matrix that will not affect the solution set of the system

Elements   The items in the set

Empty set (null set)   A set with absolutely no elements

Equation   A statement that two algebraic expressions are equal

Equilateral triangle   A triangle in which all three sides have equal lengths

Equivalent equations   Equations with the same solutions

Even number   If an integer is divided by 2 and the remainder is 0, then the integer is even.

Exponent   A number placed above the base to show the number of times the base is multiplied by itself

Exponential notation   Notation of the form a^n, where a is the base, and n is the exponent

Extraneous roots   In rational expressions, solutions that are not solutions to the original equation; may be introduced by multiplying by the LCD

Extraneous solution   A number that is found when solving an equation but that does not satisfy the original equation

F

Factor   A number that is being multiplied; may also refer to a number that divides a given number

Factor theorem   If $x = c$ is a root of a polynomial equation in the form $P(x) = 0$, then $x - c$ is a factor of the polynomial $P(x)$.

Factoring   Given a product, the process used to find the factors

FOIL method   Procedure for multiplying two binomials; multiply the first terms, the outside terms, the inside terms, and the last terms

Formula   General statement (usually an equation) that relates two or more variables

Fraction   A number that can represent parts of a whole, the ratio of two numbers, or division; also called a rational number

Frequency   In a histogram, the number of data items in a class

Function   A relation in which each domain element has exactly one corresponding range element

Function notation   Notation of the form $f(x)$, where f is the name of the function, and x is the input variable

G

Gaussian elimination   Method that uses augmented matrices and elementary row operations to solve a system of linear equations

Graph   Visual representation of numerical information

Greatest common factor (GCF)   The largest integer or algebraic term that is a factor (or divisor) of all of the numbers or terms

H

Half-open interval   Only one endpoint of the interval is included

Half-plane   A straight line separates a plane into two half-planes.

Hemisphere   Half of a sphere

Histograms   Used to indicate data in classes (a range or interval of numbers)

Horizontal line   A line with a slope of 0

Hypotenuse   The longest side of a right triangle; the side opposite the right angle

I

Identity   An equation that leads to a statement that is always true (such as $0 = 0$) and has an infinite number of solutions

Imaginary part   The real number b in a complex number $a + bi$

Improper fraction   A fraction in which the numerator is greater than or equal to the denominator

Inconsistent system   A system of equations that has no solution

Independent variable   The first coordinate x in an ordered pair

Index of a radical   The index of the radical $\sqrt[n]{a}$ is the number n.

Inequality   A mathematical expression that includes the symbols $<, >, \leq, \geq,$ or $\neq$

Inflation   A measure of relative purchasing power

Integers   The set of numbers consisting of the whole numbers and their opposites

Interest   Money paid for the use of money

Intersect   Two lines intersect if there is one point on both lines

Intersection   The intersection of two (or more) sets is the set of all elements that belong to both sets.

Interval   The set of all real numbers between two endpoints is called an interval of real numbers.

Interval notation   Notation to represent intervals of real numbers where brackets indicate that an endpoint is included and parentheses indicate that an endpoint is not included

Inverse variation   A variable quantity varies inversely as a variable x if there is a constant k such that $x \cdot y = k$ or $y = \dfrac{k}{x}$. When two variables vary inversely, an increase in one indicates a decrease in the other.

Irrational numbers   Numbers that can be written as infinite nonrepeating decimals

Isosceles triangle   A triangle in which two or more sides have equal lengths.

J

Joint variation   If the combined variation is all direct variation (the variables are multiplied), then the variation is called joint variation.

L

Leading coefficient   The coefficient of the term with the largest degree of a polynomial

Least common denominator (LCD)   The least common multiple of the denominators of two or more fractions

Least common multiple (LCM)   The smallest number that is a multiple of each of the given numbers

Leg   Either of the two sides of a right triangle that are not the hypotenuse

Like radicals   Radicals that have the same index and radicand or can be simplified so that they have the same index and radicand

Like terms (similar terms)   Terms that are constants or terms that contain the same variables raised to the same powers

Line   A line has no beginning or end. Lines are labeled with small letters or by two points on the line.

Line graph   A graph used to indicate trends over a period of time

Line of symmetry   The line through the vertex of a parabola that divides the graph into two symmetrical parts

Linear equation in x   An equation that can be written in the form $ax + b = c$, where a, b, and c are constants and $a \neq 0$

Linear equation in three variables   An equation of the form $ax + by + cz = d$ where a, b, and c are not all 0

Linear function   A function represented by an equation of the form $y = mx + b$

Linear inequality   An inequality that contains only constant or linear terms

Lower class limit   In a histogram, the smallest whole number that belongs to a class.

M

Mass   The amount of material in an object

Matrix   A rectangular array of numbers

Mean   The sum of all the data divided by the number of data items; also referred to as the average or arithmetic average

Measure of an angle   The size of the angle; measured in degrees

Median   The middle data item

Metric system   The system of measurement used by about 90% of the world, but not often used in the United States

Minuend   The number, or quantity, from which another number (the subtrahend) is to be subtracted

Mixed number   The sum of a whole number and a proper fraction

Mode   The single data item that appears the most number of times

Monomial   A polynomial with one term

Multiples   To find the multiples of a number, multiply each of the counting numbers by that number.

Multiplication (or division) principle of equality   If both sides of an equation are multiplied by (or divided by) the same nonzero constant, the new equation has the same solutions as the original equation.

Multiplicative identity   The number 1 is called the multiplicative identity.

Multiplicative identity property   The product of any number and 1 is the number itself.

Multiplicative inverse   The reciprocal of a number; two numbers are multiplicative inverses if their product is equal to 1.

N

Natural (counting) numbers   The numbers 1, 2, 3, 4, ...

Negative integers   The opposites of the natural numbers; they lie to the left of 0 on a number line.

Nonreal complex numbers   Complex numbers of the form $a + bi$ where $b \neq 0$

Nonterminating decimal number   If the remainder of division is never 0, the decimal quotient is nonterminating.

Note   A loan for a period of 1 year or less

Numerator   The top number in a fraction

O

Obtuse angle   An angle whose measure is between 90° and 180°

Obtuse triangle   A triangle in which one angle is obtuse

Octant   One of the eight regions that are formed by the intersection of the xy-plane, the xz-plane, and the yz-plane.

Odd numbers   If an integer is divided by 2 and the remainder is 1, then the integer is odd.

Open half-plane   If the boundary line is not included in the solution set, then the half-plane is said to be open.

Open interval   Neither endpoint of the interval is included

Opposite   Two integers are opposites (or additive inverses) if their sum is equal to 0.

Ordered pair   A pair of numbers in the form (x, y) where the order of the numbers is critical

Ordered triple   Three numbers in the form (x, y, z) where the order of the numbers is critical

Origin   The point of intersection of the x-axis and the y-axis

P

Parabola   The graph of a quadratic function

Parallel lines   Lines that never intersect (cross each other) and whose slopes are equal

Parallelogram   A four-sided polygon with both pairs of opposite sides parallel

Pentagon   A 5-sided polygon

Percent   The ratio of a number to 100

Perfect cube   The cube of an integer

Perfect square   The square of an integer

Perfect square trinomial   The result of squaring a binomial

Perimeter   The distance around a figure; found by adding the lengths of the sides of the figure

Period   Group of three digits separated with commas

Perpendicular lines   Lines that intersect at 90° (right) angles and whose slopes are negative reciprocals of each other

Pi (π) The ratio of a circle's circumference to its diameter; approximated by 3.14

Plane   Flat surfaces, such as a table top or wall, represent planes.

Plane geometry   The study of the properties of figures in a plane

Point   A dot represents a point. Points are labeled with capital letters.

Point-slope form   The point-slope form for the equation of a line is $y - y_1 = m(x - x_1)$, where m is the slope of the line and (x_1, y_1) is any point on the line.

Polygon   A closed plane figure, with three or more sides, in which each side is a line segment

Polynomial   A polynomial is a monomial or the indicated sum or difference of monomials.

Positive integers   The natural numbers; they lie to the right of 0 on a number line

Prime factorization   The unique factorization of a composite number that contains only prime factors

Prime number   A counting number greater than 1 that has exactly two different factors (or divisors) — itself and 1

Principal   The initial amount of money that is invested or borrowed

Principal square root   Every positive real number has two square roots, one positive and one negative. The positive square root is called the principal square root.

Product   The result of multiplication

Profit   The difference between selling price and cost

Proper fraction   A fraction in which the numerator is less than the denominator

Proportion   A statement that two ratios are equal

Pythagorean theorem   In a right triangle, the legs, a and b, and the hypotenuse, c, have the following relationship: $a^2 + b^2 = c^2$.

Q

Quadrant   The x-axis and y-axis separate the Cartesian plane into four quadrants.

Quadratic equation   Equations that can be written in the form $ax^2 + bx + c = 0$ where a, b, and c are real numbers and $a \neq 0$

Quadratic formula   A formula that is used to find the solutions of the general quadratic equation $ax^2 + bx + c = 0$; the quadratic formula is $\dfrac{-b \pm \sqrt{b^2 - 4ac}}{2a}$

Quadratic function   A function of the form $y = ax^2 + bx + c$ where a, b, and c are real numbers and $a \neq 0$

Quotient The result of division

R

Radical The complete expression involving both the radical sign and the radicand

Radical sign The symbol $\sqrt{\ }$

Radical function Function of the form $y = \sqrt[n]{g(x)}$ in which the radicand contains a variable expression

Radicand The number, or expression, under the radical sign.

Radius The distance from the center of a circle to any point on the circle

Range The difference between the largest and smallest data items

Range of a function The set of all second coordinates in the relation

Range axis In the graph of a relation, the vertical axis (the y-axis)

Ratio A comparison of two quantities by division

Rational expressions Fractions in which the numerator and denominator are polynomials

Rational number A number that can be written in the form $\frac{a}{b}$ where a and b are integers and $b \neq 0$

Rationalizing a denominator The process used to remove radicals from the denominator of a rational expression

Ray Consists of a point (called the endpoint) and all the points on a line on one side of that point

Real numbers The set of numbers that consist of all rational and irrational numbers

Real part The real number a in a complex number $a + bi$

Reciprocals If the product of two nonzero fractions is 1, then the fractions are called reciprocals of each other.

Rectangle A polygon with four sides in which adjacent sides are perpendicular (meet at a 90° angle)

Regular hexagon A six-sided polygon where all sides have equal length and all angles have equal measure

Regular octagon An eight-sided polygon where all sides have equal length and all angles have equal measure

Relation A set of ordered pairs of real numbers

Relatively prime If the GCD of two numbers is 1, then the two numbers are said to be relatively prime.

Remainder The number left after division

Repeating decimal number Decimal number that does not terminate but has a repeating pattern to its digits

Restrictions on a variable Values of the variable that make a rational expression undefined

Right angle An angle whose measure is equal to 90°

Right triangle A triangle containing one right angle

Roster form The elements of a set are listed within braces

Rounding To find another number close to the given number

Row Formed by the horizontal entries of a matrix

Row echelon form The upper triangular form of a matrix with all 1's in the main diagonal

S

Sale price The new, reduced price of an item after a discount has been applied

Sales tax A tax charged on the actual selling price of goods sold by retailers

Scalene triangle A triangle in which no two sides are equal in length

Scientific notation Decimal numbers written as the product of a number greater than or equal to one and less than 10, and an integer power of 10

Semicircle Half of a circle

Set A collection of objects or numbers

Set-builder notation The elements of a set described by giving a condition (or restriction) for the variable

Similar triangles Two triangles are similar if the measures of the corresponding angles are equal and the lengths of the corresponding sides are proportional.

Simple interest Interest that involves only one payment at the end of the term of a loan.

Simplest form for cube roots A cube root is considered to be in simplest form when the radicand has no perfect cube as a factor.

Simplest form for square roots A square root is considered to be in simplest form when the radicand has no perfect square as a factor.

Slope The ratio of rise to run of a line

Slope-intercept form The slope-intercept form for the equation of a line is $y = mx + b$, where m is the slope of the line and $(0, b)$ is the y-intercept.

Solution A solution to an equation is a number that gives a true statement when substituted for the variable in the equation

Solution set The solutions to an equation

Sphere All points in three dimensions that are the same distance from a fixed point; a ball is an example of a sphere

Square A rectangle in which all four sides are the same length

Square root The square root of a number equals another

number that when squared results in the original number.

Square of a number   In expressions with exponent 2, the base is said to be squared.

Standard form of a linear equation   Equation of the form $Ax + By = C$, where A, B, and C are real numbers and where A and B are not both 0

Standard form of a quadratic equation   Equation of the form $ax^2 + bx + c = 0$ where a, b, and c are real numbers and $a \neq 0$.

Standard form of a complex number   The standard form of a complex number is $a + bi$ where a and b are real numbers

Statistics   The study of how to gather, organize, analyze, and interpret numerical information; in statistics, a particular measure or characteristic of a part, or sample, of a larger collection of items is called a *statistic*.

Straight angle   An angle whose measure is equal to 180°

Subtrahend   The number or quantity to be subtracted

Sum   The result of addition

Sum of cubes   A product that can be written in the form $x^3 + a^3$

Sum of squares   An expression of the form $x^2 + a^2$ that is not factorable

Supplementary angles   Two angles are supplementary if the sum of their measures is 180°.

Synthetic division   A simplified version of long division of polynomials in which the variables are omitted entirely and only coefficients are written

System of linear equations   A set of two or more equations

System of linear inequalities   A set of two or more inequalities

T

Term   Any constant or variable, or the indicated product and/or quotient of constants and variables

Terminating decimal number   If the remainder of division is eventually 0, the decimal quotient is said to be terminating.

Transversal   A line in a plane that intersects two or more lines in that plane in different points

Trapezoid   A four-sided polygon with one pair of opposite sides that are parallel

Triangle   A polygon with three sides

Trinomial   A polynomial with three terms

U

Union   The union of two (or more) sets is the set of all elements that belong to either one set or the other set or to both sets.

Unit fraction   A fraction equivalent to 1

Upper class limit   In a histogram, the largest whole number that belongs to a class

Upper triangular form   A matrix is in upper triangular form if its entries in the lower left triangular region are all 0's.

V

Variable   A symbol (generally a letter of the alphabet) that is used to represent an unknown number

Vertex of a parabola   The "turning point" of the curve that represents a quadratic function

Vertex of a polygon   Each point where two sides of a polygon meet is called a vertex.

Vertex of an angle   The common endpoint of the rays that form the angle

Vertical angles   The angles opposite each other created by two intersecting lines; vertical angles are congruent

Vertical lines   A line whose slope is undefined

Vertical line test   If any vertical line intersects the graph of a relation at more than one point, then the relation is not a function.

Volume   The measure of the space enclosed by a three-dimensional figure

W

Weight   Force of the Earth's gravitational pull on an object

Whole numbers   The number 0 and the natural numbers

X

x-axis   The horizontal number line

x-intercept   The point on the graph where a line crosses the x-axis

Y

y-axis   The vertical number line

y-intercept   The point on the graph where a line crosses the y-axis

Z

Zero-factor law   The product of any number and 0 is equal to 0.

Zero-factor property   If the product of two (or more) factors is 0, then at least one of the factors must be 0.

Index

Symbols

1
- as an exponent 64
- characteristics of 170

A

Absolute value 86–88, 374
Abstract numbers 285
ac-method of factoring 849
Acre 44
Acute angles 482
Acute triangles 489
Addends 11
- missing (in subtraction) 14
Addition
- additive inverse 104
- algebraic functions 1058
- associative property of 12, 94
- commutative property of 12, 94
- complex numbers 1255
- with decimal numbers 331
- estimating sums 16
- with fractions 223–227, 1157
- functions 1058
- identity 13
- with integers 93
- in vertical format 97
- key words for 46, 573, 648
- with mixed numbers 256
- polynomials 801–802
- properties of 12, 227
- radical expressions 1014
- rational expressions with common denominators 1157, 1170
- rational expressions with different denominators 1171
- terms used in 11
- using a calculator 19, 336
- with whole numbers 11–14
Addition method
- solving systems of linear equations 920, 946
Addition principle of equality 57, 140, 275, 580
Additive identity 13
Additive inverse 82, 96, 104
Adjacent angles 485
Algebraic expressions 566
- combining like terms 566
- evaluating 133–134
- simplifying 565, 803–804, 1186
- terms 565
- translated into English phrases 573
Algebraic fractions 1155
- addition with 1170
- subtraction with 1173

Algebraic notation 55, 717–718
Algebraic operations with functions 1058
Algorithm, division 32, 1222
Altitude, in geometry 44, 506
Ambiguous phrase 575
And
- in decimal numbers 4, 322, 431
- to indicate union 716
Angle(s)
- acute 482
- adjacent 485
- bisector 493
- complementary 483
- congruent 483
- measures of 480
- obtuse 482
- opposite 485
- right 482
- straight 482
- supplementary 483
- vertical 485
Antilog 1125
Applications
- amounts and costs 929, 931
- bacterial growth 1096–1097
- basic strategy for solving word problems 46, 397
- change in value 123
- circumference 1211
- combined variation 1213–1215
- commission 401, 619
- compound interest 411, 1097
- consecutive integers , 608–609
- continuously compounded interest 1099–1101
- cost 619–620, 642, 1292
- decimal numbers 349
- depreciation 417
- direct variation 1210
- discount 398, 618
- distance-rate-time 626, 637, 928, 1204–1206
- electricity 1220
- fractions 1201–1202
- geometry 877, 1291
- gravitational force 1213
- handicapped access 714
- Hooke's Law 1211
- inequalities 725
- inflation 416
- interest 639, 936
- inverse variation 1212–1213
- joint variation 1213–1215
- least common multiple 190
- levers 1220
- linear inequalities 725
- mean 124, 270, 640
- minimum and maximum values 1308
- mixture 938, 950
- number problems 606, 929, 1200
- percent 409
- percent of profit 382, 403, 620
- perimeter 629

pitch of a roof 685
Polya's four-step process for problem solving 397
pressure 1219
price per unit 287
profit 403, 619–620
projectiles 1289
proportions 1190–1191
Pythagorean Theorem 1288
rational expressions 1200
resistance 1220
revenue 1309–1310
sale price 618
sales tax 398, 618
simple interest 409, 626
strategy for solving word problems 46, 397, 875, 1200, 1288
systems of three linear equations 950–951
systems of two linear equations 927, 936
temperature 626
unit price 287
variation 1210
work 1201
Approximate answers; *See* Estimating
Are 44, 453
Area 506
- changing units of 450, 453, 472, 506
- concept of 30, 247, 450
- in metric system 450, 472, 506
- maximizing 1310
Arithmetic
- fundamental theorem of 178
Arithmetic average 124, 358–359; *See* Mean
Arithmetic rules for rational numbers (or fractions) 1157
Ascending order of a polynomial 75, 795
Associative property
- of addition 12, 94, 228
- of multiplication 26, 114, 202
Asymptotes 1093
Attack plan for word problems 875
Augmented matrix 954
Average 124, 270, 358–359; *See* Mean
Axis
- domain 737
- horizontal 662
- of symmetry 1298
- range 737
- vertical 662
- *x*- 662, 944
- *y*- 662, 944
- *z*- 944

B

Back substitution 913, 921, 946
Bacterial growth 1096–1097

Bar graph 532
Base 771
  in geometry 44
  of exponents 62
Base-10 logarithms 1123–1126
Base-*e* logarithms 1126–1128
Base
  of a logarithm 1105
  of an exponent 1092
Basic strategy for solving word
      problems 46, 397–398
Binomial(s) 75, 794
  expansion of 808, 815
  multiplication of 808, 815
  multiplying using FOIL 815
  multiplying $(x + a)(x - a)$ 816
Bisector
  angle 493
Borrowing (in subtraction) 15–16,
      259–260
Boundary line 726, 968
Boyle's Law 1219
Braces 67, 118, 269, 309, 715, 803,
      149
Brackets 67, 118, 269, 309, 803, 149
Building fractions to higher terms
      203; *See* Raising fractions to
      higher terms
Building rational expressions to
      higher terms 1158, 1171, 1175
Bytes 462

C

Calculators; *See also* Graphing
      calculators
  addition and subtraction with whole
      numbers 19–21
  exponents 65
  multiplicaton and division with whole
      numbers 37–39
  products and quotients with decimal
      numbers 352
  rules for order of operations 70
  sums and differences with decimal
      numbers 336
  used to calculate depreciation 418
  used to calculate square roots 427
  used to find compound interest
      414–416
Caret key 747, 1008
Carrying digits in addition 11
Cartesian coordinate system 661
Celsius 468
  conversion to Fahrenheit 468, 626
Centi-
  gram 454
  liter 460
  meter 447–448, 456
Change in value 123
Change-of-base 1137

Changing
  decimal numbers to fractions 368
  decimal numbers to percent 382–383
  fractions to decimal numbers 369
  fractions to higher terms 203
  fractions to mixed numbers 240–242
  fractions to percent 386
  mixed numbers to improper fractions
      237–239
  percents to decimal numbers 385
  percents to fractions 390
  the form of a function 1304–1305
Checking solutions 720
Circle(s)
  area of 506, 509, 1211
  key terms of 499
  perimeter (circumference) of 500,
      1211
Circle graph 532
Circular cylinder
  surface area 1196, 1287
  volume 1213–1214
Circumference of a circle 499, 1211
Class boundaries, in histograms 536
Class, in histograms 535
Class width, in histograms 536
Closed half-plane 726
Closed interval 717
Coefficient(s) 55, 74, 130, 295, 565,
      793, 954
  leading 794
  monomial 793
  numerical 189
Coefficient matrix 954
Collinear points 698
Column of a matrix 954
Combined variation 1213
Combining like radicals 1014
Combining like terms 566, 801, 887
Commission 401–402, 619
Common
  denominator 1170
  divisor 192
  errors 63, 227, 232, 391, 1160, 1175,
      1257, 1281
  factors 1158
  logarithms 1123
    inverse of 1125–1126
    using a calculator 1124
Commutative property
  of multiplication 201
  of addition 12, 94, 227
  of multiplication 26, 114
Complementary angles 483, 932
Completing the square 1268–1270,
      1278
  to solve quadratic equations 1270
Complex conjugates 1257, 1266
Complex fractions 1182
  simplifying 267–268
Complex numbers 1251–1255
  addition with 1255

  conjugates of 1257
  division with 1257
  equality of 1254
  imaginary part of 1252
  multiplication with 1256
  powers of *i* 1260–1261
  real part of 1252
  solving equations involving 1254
  standard form of 1252, 1257
  subtraction with 1255
Complex roots 1272
Components of an ordered pair 661
Composite functions 1073
Composite numbers 170–171
  factors of 181–182
  prime factorization of 179
Composition of two functions;
      See Composite functions
Compounded continuously 1098,
      1099
Compound inequalities 719,
      723–725
  and 716
  graphing of 723–725, 725–727
  or 716
Compound interest 411–415, 864,
      1097
  continuous 1099–1101
  formula for 413–415
Condensing logarithmic expressions
      1118–1119
Conditional equation 600
Condition on *x* 715
Congruent angles 483
Conjugates 1018, 1257
  complex 1266
  division of complex numbers using
      1257
Consecutive integers 607–609,
      608–609
  even 608
  odd 608
Consistent system of equations
      903–908, 945
Constant 55, 74, 130, 189
Constant of variation 1210–1215
Constant (or constant term) 565,
      579, 793, 954
  degree of zero 793
Constant term; *See* Constant
Continuously compounded interest
      1099–1101
Contradiction 600
Coordinate
  first 661
  second 661
Coordinate system, Cartesian 661
Correlation coefficient 357
Corresponding angles, of similar
      triangles 520

Corresponding sides, of similar triangles 520
Cost-of-living index 416–417
Cost problems 400, 619–620, 642, 1292
Counting numbers 3, 170
Cross products 291
Cubed 63, 148
Cube roots
  evaluating 988
  simplest form 998
  simplifying 998–999
  symbol for 988
Cubes 860, 988
  perfect 860, 988
  sums and differences of 859
Cubic
  equations 1281
  polynomials 800
  units 248, 308
Current value 418
Cylinder
  surface area 1196, 1287
  volume 1213–1214

D

Data 358
Deca-
  gram 454
  meter 447
Decay
  exponential 1093
Deci-
  gram 454
  liter 460
  meter 447
Decimal notation 321
Decimal numbers 321
  addition with 331
  applications 349
  changing fractions to 369
    using a calculator 369
  changing percents to 385
  changing to fractions 368
  changing to percents 383
  dividing by powers of 10 347
  division with 343
  finite 321
  infinite 321, 426
  multiplication with 340
  multiplying by powers of 10 342
  nonrepeating 321, 426
  nonterminating 321, 345, 369, 426
  notation 321–322
  place value in 321–322
  positive and negative 333
  reading and writing 321–324
  repeating 321, 345,
  rounding 325
  scientific notation 375
  subtraction with 332
  terminating 321, 345, 426

Decimal point 4, 321
Decimal system 3, 321; *See* Decimal numbers
Defined values 743
Degree
  measure of an angle 480, 519–520
  monomial 793
  of a monomial 74
  of a polynomial 74, 794
  of a term 565, 793
  zero 793
Deka-
  gram 454
  liter 460
  meter 447
Denominator
  cannot be zero (0) 117, 198
  defined 198
  least common 224–232, 1171
  rationalizing 1017–1018
    square root or cube root 1017
    with a sum or difference in the denominator 1019
Dependent 903
  systems of equations 945
  variable 662
Depreciation 416–418
  formula for 418, 634
Depreciation formula 634
Descending order, of a polynomial 75, 795
Diameter of a circle 499
Difference 575
  estimating 16, 334
  of integers 104
  of polynomials 802–803
  of two cubes 859
  of two squares 856, 858, 859, 1018
  of whole numbers 14
  using a calculator 336
Dimension of a matrix 954
Directly proportional to 1210
Direct variation 1210
Discount 398
Discriminant 1281–1282
  finding the 1282
  for determining number and type of solutions 1282
Distance-rate-time
  formula ($d = rt$) 626, 637, 1204
  problems 1204–1206
Distributive property 28, 131
  for use in multiplying polynomials 808–811
  while combining like terms 566
Divide exactly 33
Dividend 1222
  of whole numbers 32
Divides 163–166
Divisibility of products 164
Divisibility, tests for 161–164, 179

Division
  algorithm 32, 1222
  by a monomial 824
  complex fractions (dividing rational expressions) 1182
  complex numbers 1257
  with decimal numbers 343
  estimating 35–37
  fractions 1157
  with fractions 214
  functions 1058
  key terms 573, 648
  key words for 46
  long division 32, 1222
  meaning of 31–32
  polynomial 1222
  powers of 10 347
  principle of equality 57, 140, 275, 583
  radical expressions; *See* Rationalizing denominators
  rational expressions 1162
  rules for integers 116
  synthetic 1226–1228
  terms used in 31–32
  using a calculator 37, 352
  with whole numbers 31
  with integers 115–117
  with mixed numbers 249
  with zero 34, 117
Divisor 31, 163
  common 192
  greatest common 192
Domain 1034, 1058
  algebra of functions 1057–1058
  axis 737
  of exponential functions 1108
  of an inverse function 1079
  of linear functions 742
  of logarithmic functions 1108
  of non-linear functions 743, 744
  of quadratic functions 1300
  of radical functions 1035
  of a relation 737
Double root 866, 1282
Double subscript 957

E

e, base of natural logarithms 1100
  deriving 1099–1100
  evaluated using a calculator 1101
Electricity 1220
Elementary row operations 955
Elements of a set 715
Elimination method for solving systems 920–921
Ellipsis 3, 370
Empty set 715
  symbol for 715
Endpoint 480
English phrases translated into algebraic expressions 574–575
Entry in a matrix 954

Equal angles; *See* Congruent angles
Equality
  addition principle of 580
  multiplication principle of 583
Equality of complex numbers 1254
Equation(s) 55, 139, 579
  addition principle of equality 580
  complex numbers 1254
  conditional 600
  contradiction 600
  direct variation 1210
  distance 637
  equivalent 57, 140, 279, 581
  exponential 1132–1134
  extraneous solutions to 1026, 1194
  Fahrenheit to Celsius 626
  first-degree 55–59, 139, 579
  found using slope and/or points 691,
    702
  graph of 677–678
  horizontal lines 689–690
  identity 600
  inverse 1132
  inverse variation 1212
  linear 579, 678
  linear, in one variable 55–59, 139
  logarithmic 1105
  multiplication principle of equality
    583
  of the form
    $ax + b = c$ 590
    $ax + b = dx + c$ 597
    $ax = c$ 583
    $x + b = c$ 581
  parabolas 1299
  parallel 703
  perpendicular 703
  point-slope form 707
  quadratic 865, 1265
  with rational expressions 1193
  slope-intercept form 692
  solution of 661
  solutions to 56, 99, 107, 139
  solving 55–59, 67, 140, 274–277
  solving exponential 1132
  solving for specified variables 581,
    590, 597
  solving linear
    $ax + b = c$ 590
    $ax + b = dx + c$ 597
    $ax = c$ 583
    $x + b = c$ 581
  solving logarithmic 1135
  solving quadratic 865
  standard form 678
  variation 1210
  vertical lines 690–691
  with fractions 278
  with integers 99
  with radical expressions 1026
  with whole numbers 55–59
Equilateral triangle 488
Equivalent equations 57, 140, 279,
  581

Equivalent fractions; *See* Raising
  fractions to higher terms
Equivalent measures
  of area 506–507
  of liquid volume 460–461
  of volume 457–460
Equivalents (percent-decimal-frac-
  tion) 392
Eratosthenes, sieve of 171
Estimating
  products and quotients 35–37, 347
  sums and differences 16–17, 334
Evaluating
  equations for given values 627
  expressions 67, 76–77, 133, 568
  formulas 627
  functions 744
  logarithms 1106
  polynomials 76–77, 133
  radical functions 1036
  radicals 986, 988, 1004
    using a calculator 990, 1008, 1021
Even consecutive integers 608
Even integers 162, 172, 304
Exact divisibility 33, 161
Expanding logarithmic expressions
  1117
Exponent(s) 62, 771
  base of 771
  cubed 63, 148
  fractional 1003
  integer 778
  logarithms are 1105
  one 64
  power of an 771
  power rule for fractions 784
  properties of 1132
  quotient rule 775
  simplifying expressions with rational
    1006
  solving equations with 1133
  squared 63, 148
  summary of rules for 1005
  terms used with 62
  using a calculator 65
  zero 64, 774
Exponential decay 1093
Exponential equations 1132
  solving 1132–1133
Exponential
  expressions 62; *See also* Exponents
  functions 1091
    domain 1108
    general concepts of 1095
    range 1108
  growth 1092
Exponents 62
  cubed 63, 148
  one 64
  squared 63, 148
  terms used with 62
  using a calculator 65
  zero 64

Expressions; *See* Algebraic expressions
  algebraic 566
    translating 573
  combining like terms 131
  evaluating 67, 133, 568
  exponential 62
  radical 998
  rational 1155
  simplifying 566
    algebraic 803–804
    complex algebraic 1186
Extraneous
  roots 1194
  solutions 1026, 1194
Extremes of a proportion 291

F

Factor(s) 161, 163, 170, 825, 1158,
  1225
  greatest common 192
  of composite numbers 181
  prime 178–180, 191–193
Factoring 825
  completely 853, 890
  difference of two squares 856, 858,
    859
  not factorable 829, 841, 850, 853, 857,
    890
  polynomials
    by finding the GCF 828
    by grouping 850
    by the *ac*-method 849
    by the trial-and-error method 839,
      845, 849
  solving quadratic equations by
    865–869, 1265–1266
  special techniques 856
  sum of two squares 857
  sums and differences of cubes 859
  trinomials by factoring out a mono-
    mial first 840–841
Factorizations
  prime 178, 185, 192
Factors 25, 31–33
Factor theorem 871
Fahrenheit 468
  converting to Celsius 469
Finding equations given two points
  on a line 702–703
Finding equivalent fractions;
  See Raising fractions to
  higher terms
Finite decimal number 321
First-degree
  equations 55–58, 139, 274, 579;
    See Linear equation(s)
  solving 581
  inequality 719
  solving 720
FOIL method 815, 1016

x-axis 662
y-axis 662
Gravitational force 1213
Greater than 84
Greatest common divisor (GCD) 192–194
Greatest common factor (GCF) 192, 825, 889
Growth
  exponential 1092

H

Half-open interval 717
Half-plane(s) 726, 968
  intersection 968
  open and closed 726, 968
Hectare 44, 453
Hecto-
  gram 454
  liter 461
  meter 447
Height, in geometry 44, 506
Hertz 462
Histogram 535
Hooke's Law 1211
Horizontal
  asymptote 1093, 1108
  axis 662
    domain axis 737
  lines 689
  line test 1076
  parabolas 1299
  translation 1302
How to Solve Word Problems 397
Hypotenuse 44, 490, 526, 880, 1288, 1319
  finding length of 526–528

I

i
  powers of 1260
Identity 600
  for addition 13
  for multiplication 26
  multiplicative 202
Imaginary numbers 1252
  pure 1252
Imaginary part of a complex number 1252
Improper fractions 199
  changing mixed numbers to 238
  changing to mixed numbers 240–242
Inconsistent system of equations 903–905
Independent variable 662
Index of a radical 988, 1003
Inequalities; *See* linear inequalities
  checking solutions to 720
  compound 719, 723

first-degree 719
  graphing 89
  linear 719
    graphing 717, 727
    solving 720, 727
  reading 716
  reversing the "sense" 720
  solutions of 89
  symbols 84–85
Infinite
  nonrepeating decimals 321, 369–370, 426
  repeating decimals 321, 345, 369–370
  solutions
    systems of equations with 903
Infinity 717
Inflation 416–418
  formula for 416
Integers 82
  addition with 93
  and absolute value 86–88
  consecutive 608
    even 608
    odd 608
  division with 115–117
  even and odd 162, 172, 304
  multiplication with 113
  as rational numbers 198
  solving equations with 139
  subtraction with 104
Intercept
  x- 680
  y- 680, 692
Interest 409, 639–640
  applications 936
  compound 411–415, 1097
  continuously compounded 1099
  formulas for 409–416
  simple 409–411, 626, 936
Intersect command on TI-84 Plus 907
Intersection 716
  of two half-planes 968
Intervals 716
  closed 717
  graphing on a number line 717
  half-open 717
  notation 717
  open 717–718
Inverse
  additive 82, 96, 104
Inverse functions 1078, 1125, 1128
  domain of 1079
  how to find 1083
  range of 1079
Inverse ln of N 1128
Inverse log of N 1125
Inversely proportional to 1212
Inverse variation 1212
Irrational numbers 198–199, 321–322, 370, 426, 987
Isosceles triangle(s) 488, 932

J

Joint variation 1213

K

Key words
  list of 573
  that indicate operations 46
Kilo-
  gram 454
  liter 461
  meter 447

L

Land area, measures of 453
Lateral surface area 626
Law
  Hooke's 1211
  zero-factor 26
LCM; *See* Least common multiple
Leading coefficient 794
Least common denominator 224, 1171
Least common multiple 185, 224, 592, 1171, 1184
  for a set of polynomials 1171
  of a set of algebraic terms 189
Legs of a right triangle 490, 526, 880, 1288, 1319
Length
  changing units of 447
  in metric system 447, 470, 499
  units of 447
Less than 84
Levers 1220
Like radicals 1014
Like terms 130, 565, 1014
  combining 566
Limit 1099
Limiting value; *See* Limit
Line(s) 678
  boundary 726
  finding the equation of 702–703
  in geometry 479
  graph 534
  horizontal 689
  parallel 703
  perpendicular 703
  slope of 687–688
  straight 678
  summary of formulas and properties for 710
  of symmetry 1298
  vertical 690
Linear equation(s) 579, 678; *See* Equation(s)
  $ax + b = c$ 590
  $ax + b = dx + c$ 597
  $ax = c$ 583
  defined 579

finding points that satisfy 664, 677
graphing 679–680
  in two variables 678
  using intercepts 680
  using slope and a point 705
graphs of 678–679
in three variables 944
in two variables 678, 901
point-slope form 707–708
slope-intercept form 692–693
solution set 677
solving for specified variables 590, 597
standard form 678
systems of two 901
$x + b = c$ 581
Linear equation
  in x 55–59, 139
Linear equation(s) 274
Linear functions 742
  domain of 742
  range of 742
Linear inequalities 719; *See* In-equalities
  compound 719, 723
  graphing 717, 727
    boundary lines 726
    half-plane (open and closed) 726
    test point 727
    using a graphing calculator 730
  reversing the "sense" 720–721
  solving 720, 727
Linear polynomials 800
Liquid volume 460, 460–461
Liters 460
Ln function on a TI-84 Plus 1127
ln x; *See* Natural logarithms
Logarithm(s)
  base-10 1123
  base-e 1126
  change-of-base 1137
  common 1123
    finding the inverse 1125
  evaluating 1106
  natural 1126
    finding the inverse 1128
  power rule 1115
  product rule 1114
  properties of 1113, 1116
  properties of equations with 1132
  quotient rule 1115
  solving equations with 1135
  undefined values 1125
Logarithmic equations 1135
  graphing 1108
  solved using exponential form 1107
  solving 1135
Logarithmic functions 1105
  domain 1108
  range 1108
Log function on a TI-84 Plus 1124
log x; *See* Logarithms, Common logarithms

Long division 32, 1222–1223
Lower and upper class limits, in histograms 536
Lowest terms 203, 1158

M

Main diagonal of a matrix 958
Mass 453
  changing metric units of 455, 475
  metric units of 454, 475
  U.S. customary units of 475
Mathematicians
  Gauss, Carl Friedrich 958
  Pòlya, George 397
  Pythagoras 526, 880
Matrices 954
  augmented matrix 954
  coefficient matrix 954
  column 954
  dimension 954
  elementary row operations 955
  entry 954
  Gaussian elimination 958–959
  main diagonal 958
  row 954
  row echelon form 958, 962
  row equivalent 956
  square 954
  triangular form 958
  using a graphing calculator 962
Maximum value 1308
Mean 124, 270, 358, 640
Means of a proportion 291
Measurements
  changing metric units
    of area 451
    of length 448
    of mass 455
    of volume 458, 459
  converting between metric and U.S. customary system units
    of area 472, 506
    of length 469, 470, 499
    of mass 475
    of volume 474, 514
  of area 450, 472, 506
  of land area 453
  of length, metric 447, 470, 499
  of liquid volume 460
  of mass (weight), metric 453–455
  of temperature 468
  of volume 456, 460, 474
Measure of an angle 480
Measures of center 357
Median 358
  finding the 359
Mega-
  byte 462, 548
Mersenne, Marin 177
Mersenne primes 177
Meter 447

Method of addition for solving systems 920
Method of substitution for solving systems 913
Metric system 248, 447
  area 450–453, 472, 506
  bytes 462
  changing units
    of area 451
    of length 448
    of mass 455
    of volume 458, 459
  conversions to U.S. customary units
    of area 472, 506
    of length 470, 499
    of mass 475
    of temperature 468
    of volume 474, 514
  degrees Celsius 468
  hertz 462
  land area 453
  length 447, 470, 499
  liquid volume 460
  liter 460
  mass 453–455, 475
  meter 447
  notation 450
  pixels 462
  temperature 468
  volume 456, 460, 474
Metric ton 454
Milli-
  gram 454
  liter 461
  meter 447–450
Minimum value 1308
Minuend 14
Mixed numbers 237, 1222
  addition with 256
  changing fractions to 240–242
  changing percents to 390
  changing to fraction form 237–239
  division with 249
  multiplication with 244–248
  positive and negative 261–262
  subtraction with 258
Mixture problems 938, 950
Mnemonic devices
  FOIL 815
  PEMDAS 67
Mode 358
Monomial(s) 74, 793, 794
  degree of 74, 793
  division of a polynomial by a 824
  multiplication of a polynomial by a 808
Multiple(s) 171, 185
  least common 185, 189
Multiplication
  associative property of 202
  commutative property of 201
  complex numbers 1256
  estimating 35

Radicand 424, 985, 1003, 1045
Radius of a circle 499
Raising fractions to higher terms 203
Range 358, 1034
  for exponential functions 1108
  for logarithmic functions 1108
  of an inverse function 1079
  of a function 737–738
  of a linear function 742
  of quadratic functions 1300
Range axis 737
Rate
  in ratio 285
  of discount 398
  of sales tax 398
Rate of change 686
Ratio 285, 1190
Rational
  equations
    solving 1193
  exponents
    simplifying expressions with 1006
  expression(s) 824
    addition with 1170
    applications 1200
    building to higher terms 1158, 1171, 1175
    complex fractions 1182–1185
    division with 1162
    finding the least common denominator 1171
    multiplication with 1160
    opposite 1159, 1174
    proportions 1190
    reducing to lowest terms 1158
    solving equations with 1193
    subtraction with 1173
  numbers 198–200, 370, 987, 1155
    arithmetic rules for 1157
Rationalizing denominators 1017
  square root or cube root 1017
  with a sum or difference in the denominator 1019
Ratio of rise to run 685
Ray 480
Reading
  decimal numbers 322
  graphs 532
Real number(s) 1252
  intervals of 716
Real numbers 426
Real part of a complex number 1252
Reciprocal 1157, 1162
Reciprocals of fractions 213
Rectangle
  area of 506
  perimeter of 500
Rectangular pyramid
  volume of 514

Rectangular solid
  volume of 514
Reducing fractions to lowest terms 204
Reducing rational expressions 1158
ref; See row echelon form
Reflecting lines across the line
  $y = x$ 1079
Relation 1034
  as a set of ordered pairs 739
  definition of 737
  domain of 737
  range of 737
Relatively prime 193
Remainder 32, 345, 1222
  theorem 1229
Replacement set 12, 56, 294
Resistance 1220
Restrictions on a variable 715, 1156, 1158
Revenue 1309–1310
Right angle 482
Right circular cone
  volume of 514
Right circular cylinder
  volume of 514
Right triangles 44, 489, 526, 880, 1288
  hypotenuse 490, 526, 880, 1288
  legs 490, 526, 880, 1288
Roots; See Radical(s)
  cube 988
  double 1282
  extraneous 1026, 1194
  factor theorem 871
  finding equations using 870
  nth 1003
  square 985
Roster form 715
Rounding
  decimal numbers 325
  estimating answers using 16, 35
  whole numbers 6
Row echelon form 958, 962
Row equivalent 956
Row of a matrix 954
Row operations 955
Rules for
  division with integers 116
  multiplication with integers 114
  order of operations 66, 118, 269
    using a calculator 70
  rounding 6
  rounding decimal numbers 325
  solving equations 57–58, 140

S

Sale price 398, 618
Sales tax 398–401, 618
Sample 357

Scalene triangle 488
Scatter diagram 676
Scatter plot 676
Scientific notation 375
Second component of ordered pairs 661
Sets 715
  element 715
  empty 715
  finite 715
  infinite 715
  null 715
  roster form 715
  solution 677
Set-builder notation 715
Sets
  solution 579
Sides
  of an angle 480
  of a triangle 486
Sieve of Eratosthenes 171
Signed numbers 86
Similarity 520
Similar terms 565; See Like terms
Similar triangles 520
  symbols for 520
Simple interest 409–411, 626
  formula for 409
Simplest form of radical expressions 994, 998
Simplifying
  algebraic expressions 565, 803
  complex fractions 267–268, 1182
  expressions 130–133
  expressions with rational exponents 1006
  fractions; See Reducing fractions to lowest terms
  polynomials 130–133
  radicals 994
  rational expressions 1158
  square roots with variables 996
Simultaneous equations; See Systems of linear equations
Slope 685
  calculating the 687
  of horizontal lines 689
  negative 689
  of parallel lines 703
  of perpendicular lines 703
  positive 689
  rate-of-change 686
  of vertical lines 690
Slope-intercept form 692
Solution(s) 579
  complex 1266
  to compound inequalities 723
  consistent 903
  dependent 903
  to equations 56, 99, 107, 139, 274
  extraneous 1026, 1194
  to first-degree equations 590

inconsistent 903
to inequalities 89, 719
nonreal complex 1305
set 56, 139, 579, 677, 968
to systems of equations 901, 946
  by addition 920
  by graphing 904
  by substitution 913
to systems of linear inequalities 968
in three variables 944
in two variables 901
  ordered pairs 661
  points 677

Solving
addition principle of equality 581
compound inequalities 723–725
definition of 579
equations 55–59, 67, 107, 275
  with integers 140
  with whole numbers 57
equations with fractions 278
equations with radicals 1026
equations with rational expressions
  1193
formulas for specified variables
  629–631, 1195
linear equations
  $ax + b = c$ 590
  $ax + b = dx + c$ 597
  $ax = c$ 583
  $x + b = c$ 581
linear inequalities 720–722
multiplication principle of equality
  583
quadratic equations
  by completing the square 1270
  by factoring 865, 1265–1266
  by the quadratic formula 1279–1280
  by the square root property 1267
rational equations 1193
systems of equations
  by addition 920
  by graphing 904
  by substitution 913
  using Gaussian elimination 958–961
  with a graphing calculator 962–965
systems of linear inequalities 968
word problems 1288
Solving exponential equations
with different bases 1133
with the same base 1132
Solving logarithmic equations 1135
Special factoring techniques 856
Special products of polynomials
  815
Sphere
volume of 514
Square(s)
of a binomial difference 858
of a binomial sum 823, 858
completing the 1268
area of 506
the difference of 856, 858, 859
perimeter of 500

matrix 954
perfect 64, 424, 856
root(s) 424, 426, 985; *See* Radical(s)
  evaluating 986
  evaluating with a calculator 990
  negative 986
  of negative numbers 1251
  of x^2 996
  principal 425, 986
  properties 994, 1268
  simplest form 994
  simplifying 994
  symbol 424, 985
  table of 425
  using a calculator 427
  with variables 996–997
sum of two 857
units 30, 247, 450
Squared 63, 424, 426, 148
Standard deviation 357
Standard form
of a complex number 1252, 1257
linear equations 678
of a quadratic equation 1265, 1318
Standard notation 3
Statistic(s) 357–362
terms used in 358
Straight angle 482
Straight lines 678; *See* Linear equations
Strategy for solving word problems
  46, 397, 1200, 1288
Subscript 633, 687, 955
double 957
Substitution method
solving systems of linear equations
  913
Subtraction
borrowing in 14–15
complex numbers 1255
with decimal numbers 332
estimating differences 16
with fractions 228
functions 1058
with integers 104
in vertical format 106
key words for 46, 573, 648
with mixed numbers 258
of algebraic functions 1058
with radical expressions 1014–1015
relation to addition 14
terms used in 14
using a calculator 19, 336
with fractions 1157
with polynomials 802–803
with rational expressions 1173
with whole numbers 14
Subtraction principle, for solving
  equations 57, 140
Subtrahend 14
"such that" 715

Sum 801, 887; *See* Addition
of cubes 859
estimating 16, 334
of two squares 857, 1257, 1266
using a calculator 336
of whole numbers 11
Supplementary angles 483
Surface area
right circular cylinder 1196, 1287
Symbol(s)
for composite functions 1073
cube roots 988
for element of 715
for empty set 715
for inequality 84
for infinity 717
for intersection 716
inverse of a function 1079
for multiplication 25
for order 84–85
for percent 381
radical 985, 1003
for similar triangles 520
for slope 687
sqrt(−1) 1251, 1252
square root 985
for "such that" 715
for union 716
Symmetric about the line $y = x$ 1079
Synthetic division 1226–1228
Systems of linear equations
consistent 903
dependent 903
inconsistent 903
solved by addition 946–949
solved by graphing 904
solved by substitution 913
solved using a graphing calculator 907
solved using Gaussian elimination
  958–961
in three variables 946
in two variables 901
Systems of linear inequalities 968

T

Table function on a TI-84 Plus 1039
Tangent 1300
Temperature 626
Temperature conversions 468
Tera-
byte 462, 548
Term(s) 793
algebraic 74, 130
coefficient of 565, 793
constant 189, 565
degree of 793
like (similar) 565
in a proportion 291
unlike 566
Terminating decimal numbers 321,
  345, 369
Test-point 727